A GUIDE TO SPELEOLOGICAL LITERATURE

OF THE ENGLISH LANGUAGE

1794—1996

A GUIDE TO SPELEOLOGICAL LITERATURE OF THE ENGLISH LANGUAGE

1794—1996

Compiled and Edited by

Diana E. Northup
Emily Davis Mobley
Kenneth L. Ingham III
William W. Mixon

Production Staff

Anne D. Schultz
Trevor R. Shaw
Kathleen R. Hardy
Ken P. Kloppenborg, Jr.
Christina R. Lopez
Caroline J. Sanchez

CAVE BOOKS   SAINT LOUIS   1998

To Edward J. Zawlocki for so many years of friendship.
E.D.M.

Copyright ©1998 Cave Books

All rights reserved. This book or parts thereof may not be reproduced in any form without permission.

Cave Books, 756 Harvard Avenue, Saint Louis, MO, 63130, USA

Cave Books is the publications affiliate of **The Cave Research Foundation**

Cover Photograph: Anne D. Schutz, by Kenneth L. Ingham III

Cover Design: by Roger E. McClure

Adding to the Bibliography
To add references of speleological books missing from this Guide, visit our web site at **http://www.i-pi.com/speleobiblio**.
If you do not have access to the World Wide Web, send complete references to:
Diana E. Northup
1601 Rita Drive, NE.
Albuquerque, NM 87106-1127

Library of Congress Cataloging-in-Publication Data

   A guide to speleological literature of the English Language :
      1794-1996 / compiled and edited by Diana E. Northup ... [et al.].
        p. cm.
      Includes bibliographical references and index.
      ISBN 0-939748-51-7. – ISBN 0-939748-52-5 (pb)
      1. Speleology–Bibliography. I. Northup, Diana E., 1948-   .
Z6033.C3G85 1998
[GB601]
016.55144'7–dc21
                                                                               98-10758
                                                                              CIP

About the Editors

DIANA NORTHUP is Associate Professor in the General Library at the University of New Mexico, with over twenty-five years of reference experience in biological and medical libraries. She has M.S. degrees in both Biology and Library Science. She is a Fellow of the Cave Research Foundation, and of the National Speleological Society, from which she received a Ralph Stone Award for Research in 1987.

EMILY DAVIS MOBLEY is a bookseller specializing in current and rare books on caves, karst, and bats for twenty-two years. She has a large professional collection of her own, most of which she has read. She has an M.S. in Communication Arts, gives talks on caves and bats, and does research on them for writers and film makers. She is a Fellow of the National Speleological Society and has served on its Board of Governors as well as being chair of several NSS committees.

KENNETH INGHAM has an M.S. in Computer Science. He manages his own company as a consulting computer expert, and has written UNIX TOOL BUILDING (Academic Press, 1990). He is a member of the National Speleological Society, and has been a member of numerous scientific caving teams. He managed the database, typesetting, and general organization of this volume.

WILLIAM MIXON is a computer programmer with a B.S. in Physics and an M.A. in Mathematics. He has published over 350 reviews of speleological works for the NSS NEWS and other publications. He edited the WINDY CITY SPELEONEWS for sixteen years, four volumes of the SPELEO DIGEST, and four volumes for the Association for Mexican Cave Studies. He is a Fellow of the National Speleological Society and received its William J. Stevenson Award for Outstanding Service in 1987. He served on the NSS Board of Governors for twenty-five years and has chaired or served on a number of its committees.

Contributors

Kenneth C. Carstens
Anthropology Department
Murray State University
Murray, KY 42071

David C. Culver
Department of Biology
American University
4400 Massachusetts
Washington, DC 20016-8002

R. Laurence Davis
Department of Biology and Environmental Sciences
University of New Haven
300 Orange St.
West Haven, CT 06516

Richard A. DesJardins
P.O. Box 723
Sandia Park, NM 87047

Thomas D. Engel
16 Equinox Ct.
Apt 2A
Delmar, NY 12054-1726

Frederick V. Grady
1201 South Scott Street, Apt. 123
Arlington, VA 22204-4655

Chris Howes
51 Timbers Square
Roath, Cardiff CF2 3SH, Great Britain

George N. Huppert
Department of Geography
University of Wisconsin La Crosse
1725 State Street
La Crosse, WI 54601

Lara Lamb
Department of Anthropology
Northern Territory University
Darwin, NT 0909, Australia

Kathleen H. Lavoie
College of Arts and Sciences
101 Hudson Hall
State University of New York-Plattsburgh
Plattsburgh, NY 12901

William W. Mixon
14045 North Green Hills Loop
Austin, TX 78737

Arthur N. Palmer
Department of Earth Sciences
State University of New York
Oneonta, NY 13820-4015

J. Michael Queen
P.O. Box 510
Richmondville, NY 12149

Ira D. Sasowsky
Department of Geology
University of Akron
Akron, OH 44325-4101

Gary K. Soule
224 South 7th Avenue
Sturgeon Bay, WI 54235-2216

Rickard S. Toomey
Quaternary Studies Program
Illinois State Museum
RCC, 1011 E. Ash St
Springfield, IL 62703

Alan Warild
41 Northwood Street
Newtown, 2042 NSW Australia

Patty Jo Watson
Department of Anthropology
Washington University
Campus Box 1114
One Brookings Drive
St. Louis, MO 63130-4899

William B. White
Department of Geosciences and
Environmental Resources Research Institute
The Pennsylvania State University
University Park, PA 16802

Elizabeth L. White
Department of Geosciences and
Environmental Resources Research Institute
The Pennsylvania State University
University Park, PA 16802

Foreword

The Llano Estacado of southeastern New Mexico is an arid almost featureless plain that stretches east and north into the Texas panhandle, and is mostly noted for oil production and ranching. I grew up on the Llano far from caves, canyons, and oceans. The Llano is the birthplace of such diverse people as rock and roller Buddy Holly, country music outlaw Waylon Jennings, and the great Comanche leader Quanah Parker. One of its few cities is a place of which it has been said that "Happiness is Lubbock, Texas in my rear view mirror." The Llano, to put a point on it, is not exactly the vacation or cultural center of the Universe. But, it was, for my formative years, home, and where I first learned of caves, canyons, and the ocean.

On that seemly endless plain of grass and mesquite that was, for most people a very out of the way place, I discovered books and the wonderful portals of their pages. Books out of which strode such people as the adventurer Richard Halliburton, the diver Jacques Cousteau, the mountaineer Sir Edmund Hillery, the anthropologist Thor Hyerdahl, and the cave explorer William Halliday. I began to want to dive, climb, go into caves, and have adventures, not because of its geography, but because of its libraries and book stands. Where there are books, there is the world.

I was most likely the only person in Lovington, New Mexico, in the 1950s and 1960s who actually asked for inter-library loans and who ordered books from the local drug store/news stand with titles such as *The Silent World* and *Exploring American Caves*. It was, at best, difficult to find out about new or old titles of publications in which I was most interested in those years in Lovington. There was no *Titles in Speleology* or *Annotated Titles in American Diving* or anything even remotely similar in the local library. I had to rely on the assistance and good graces of the local librarian and conventional books in print lists (none of which were available via Internet in that Dawn of the Rock and Roll era).

My best time for reading the latest books (and reading them first) was during my high school years as a librarian's aid, with the responsibility of suggesting titles related to diving, caving, and geographical exploration, and of cataloging the new books as they arrived. I developed the habit of taking new books home to read before anyone else could get to them. I spirited the new books away (always to return them) by hiding them under my clothing and thereby was always able to read the latest caving and diving books before the librarian even saw them. This illegal practice came to a sudden end one December day when I tried to return three new books to the book storage room only to find the room locked! I quickly decided to go to the librarian and claim to have "found these books in the hall" and return them. That proved to be my undoing. I had fallen into the trap set by Ms. Aussie Ramsey, who looked me square in the eye and said in her best southern/prison matron voice: "You the one been commin' in my library stealin' my books!" Well, Ms. Ramsey, now you know, although I never confessed at the time, I am the one, but I have an explanation: I knew I was going to become a cave resource specialist and eventually meet Diana Northup and Emily Davis Mobley as they were preparing their *Guide to Speleological Literature* and that I would be asked to write the foreword to the *Guide*!

This volume, which could never have come to be when I was growing up on the Llano, was long in arriving but will certainly be around for decades to come. It is this, forever lasting, about books that makes them so wonderful and important. Diana Northup and Emily Davis Mobley have done a great service for all of us involved in this great symphony (as Donald Davis once called it) of speleology. They have included all the component parts of the piece: the long deep back ground-beat of archeology, the lively tone of biology, the driving brass of exploration, the important notes of geology, geomorphology, and hydrology without which there would be no music, and all the other sections of the composition of speleology: history, mineralogy, paleontology, and everything you need for management, conservation and guides for your special interest in the piece. And they have picked the very best in the field to play in the orchestra: Patty Jo Watson, Alan Warild, Chris Howes, Gary K. Soule, and Arthur and Margaret Palmer to name a few of the annotators and section introduction authors of this most valuable volume.

You should be able to find almost any speleological title you need or desire in this guide, but you will also find the guide itself to be one of the most interesting tools you will ever own or check out of a library. It is a reference that you can dip into time after time and that is, at the same time, a very exciting and interesting read. I am going to see to it that at least one copy ends up in the Lovington, New Mexico high school library and another in the Lovington public library, just in case there is someone else there who understands that geography sets no limits to your imagination. Thank you Diana, Emily, Kenneth, and Bill for this labor of love.

Ronal C. Kerbo NSS 11539FHM
Denver, Colorado January, 1997

Preface

In 1981, Ed Zawlocki left for the South Pole to winter over as the science officer in McMurdo station. He took with him a large pile of Tony Oldham's and Speleobooks' catalogs hoping that he would be able to use his free time to produce the beginnings of a bibliography of speleology. When he returned, he had about 500 titles on a disk. Over the next ten years he attempted to get collectors all over the US to fill out forms and send them to him, but most found the task far too daunting and asked if they could help fill in gaps at a later date.

Ed looked at the project as long term and never really pushed. Anyway, it was hard to push from his home in Japan and so the project languished, and was brought to the surface only at National Speleological Society conventions.

In the summer of 1991, Ed spent an exceptionally long time in the US. He met with George Huppert and Diana Northup at the convention that year hoping to move the project forward at a faster pace. He visited friends and made plans to move to New York that October. The day before he was to fly back to Japan he suddenly died. He left behind many friends who would miss him and a disk with about 1,000 bibliographical entries.

I, foolishly, decided the bibliography had to be finished and talked Diana Northup into heading the project as she was the only professional librarian caver I knew. Neither of us knew what we were getting into. At first, our personalities clashed (she hid under the table and I wondered what I was doing wrong!), but the goal made us find common ground. Next we found that Ed's work, although a good beginning, was full of holes, fiction, and bat books. We had literally to travel all over the world to get the information we wanted. Five years later, we have traveled to Australia, Great Britain, as well as to many states and have nearly 4,000 entries. We have mooched off of dozens of cavers, including each other, and talked others in to writing annotations and introductions. We even became good friends. We realized that the project would never really be "done" but that we were.

Ed had no realization of what he had started. I imagine him now looking down at us with a Tom Sawyer-like grin knowing that he had found a way to get someone else to finish a very tough job. We love you anyway, Ed.

Emily Davis Mobley NSS 12154LFHM
Albuquerque, November 1996

Acknowledgments

This work is the result of the help of and consultation with a great number of people. The authors are extremely grateful for this help. We especially thank the American Library Association for the financial support provided by The Whitney-Carnegie Award. Without the many speleo-bibliophiles who opened their homes and collections to us, this work would not have been possible: Online Computer Library Center (OCLC) for access to its World-Cat database and for permission to supplement the bibliography from their database, Thomas Cromer and Chris Sharpe, Ross A. Ellis, William R. Halliday, Francis G. Howarth, Chris Howes and Judith Calford, Anthony Jarret, Raymond Mansfield, Charles R. Pease, James Quinlan, Bruce W. Rogers, Geary M. and Susan Schindel, Trevor R. Shaw, Elery Hamilton Smith, Gordon L. Smith, Jr., William W. Torode and Gregory and Lynette Tunnock. We thank the National Speleological Society for bringing us all together to work on this project and the Cave Research Foundation for supporting the publication.

We thank George Huppert and Edward Zawlocki for their original inspiration for this project.

The manuscript was finely tuned by the firm and competent hand of Richard A. Watson.

Cataloging consultation was provided by Clare-Lise Benaud, Mary Ellen Hanson, and Todd Hollister, of the Cataloging Department of University of New Mexico General Library.

In addition to the production staff, many individuals helped with the editing of the final manuscript, input annotations, or revised text. We are very appreciative of the work of Kathylyn Beck, Donna Cromer, Richard A. and Jill DesJardins, Thomas D. Engel, James M. and Fredericka H. R. Hardy, Deborah Harrison, J. Michael Queen, and Dena Thomas.

Thanks everyone! We couldn't have done it without you!

Cave Conservation

Caves are fragile in many ways. Their features take hundreds of thousands of years to form. Cave animals such as blindfish are rare, and always live in precarious ecological balance in their underground environments. Cave features and cave life can be destroyed unknowingly by people who enter caves without informing themselves about cave conservation. Great and irreparable damage has been done by people who take stalactites and other flowstone features from caves and by those who disturb cave life such as bats, particularly in winter when they are hibernating. Caves are wonderful places for scientific research and recreational adventure, but before you enter a cave, please first learn about careful caving by contacting The National Speleological Society, 2813 Cave Avenue, Huntsville, AL 35810.

Contents

Foreword		ix
Preface		xi

1 Introduction — 1
  1.1 Explanation of data fields — 1

2 Archeology — 5
  2.1 General — 10
  2.2 Anthropology — 17
    2.2.1 Art — 17
    2.2.2 Rock Shelters — 19

3 Biology — 21
  3.1 General — 23
  3.2 Adaptation — 23
  3.3 Bacteria — 24
  3.4 Botany — 24
  3.5 Cave Ecology — 24
  3.6 Cave Fauna — 24
  3.7 Cave Flora — 29
  3.8 Ecology — 29
  3.9 Evolution — 30
  3.10 Fungi — 30
  3.11 Geomicrobiology — 30
  3.12 Guano — 30
  3.13 Physiology — 30
  3.14 Rabies — 31
  3.15 Systematics — 31

4 Exploration — 32
  4.1 General — 33

5 Geology — 40
  5.1 General — 41
  5.2 Carbonates — 46
  5.3 Cave Classifications — 46
    5.3.1 Cenotes — 46
    5.3.2 Dolomite Caves — 46
    5.3.3 Ice Caves — 46
    5.3.4 Lava Tubes — 47
    5.3.5 Sandstone Caves — 48
    5.3.6 Sea Caves — 48
    5.3.7 Sinkholes — 48
    5.3.8 Springs — 50
    5.3.9 Talus Caves — 50
    5.3.10 Underwater Caves — 50
  5.4 Evaporites — 51
    5.4.1 Gypsum — 51
  5.5 Geochemistry — 51
  5.6 Geomorphology — 51
    5.6.1 General — 53
  5.7 Geophysics — 53
  5.8 Hydrology — 53
    5.8.1 General — 58
  5.9 Hydrogeology — 63
  5.10 Karst — 63
  5.11 Mineralogy — 67
    5.11.1 Saltpeter — 68
    5.11.2 Speleothems — 68
  5.12 Paleokarst — 68
  5.13 Speleogenesis — 68
  5.14 Vulcanospeleology — 69

6 General Caving — 70
  6.1 Caving — 71
  6.2 Climbing — 71
  6.3 Diving — 72
  6.4 Equipment — 73
  6.5 General — 73
  6.6 Photography — 75
  6.7 Rescue — 76
  6.8 Safety — 77
  6.9 Surveying — 78
  6.10 Techniques — 78
  6.11 Training — 80

7 History — 81
  7.1 General — 82

8 Management/Conservation — 87
  8.1 Conservation — 89
  8.2 Development — 91
  8.3 Evaluation — 91
  8.4 Legal — 91
  8.5 Legislation — 92
  8.6 Management — 92
  8.7 Pollution — 97
  8.8 Restoration — 97

9 Paleontology — 98
  9.1 General — 99
  9.2 Paleoclimatology — 102

10 Regional Guides — 103
  10.1 General — 105
  10.2 Catalogs — 105
  10.3 Guidebooks — 106
  10.4 Listings — 118

11 Show Caves — 119
  11.1 General — 120

12 Science	**136**
12.1 Chemistry	136
12.2 Chronology	136
12.3 Communication	136
12.4 Computers	136
12.5 Dating	136
12.6 Engineering	136
12.7 Environmental Studies	136
12.8 Geography	137
12.9 Medicine	137
12.10 Meteorology	137
12.10.1 Climatology	137
12.11 Military	137
12.12 Mining	137
12.13 Physics	137
12.13.1 Cavity Detection	137
12.13.2 Radioactivity	138
12.13.3 Radon	138
12.14 Psychology	138
12.15 Speleology	138
13 Miscellaneous	**141**
13.1 Anthologies	141
13.2 Atlases	141
13.3 Audiovisuals	141
13.4 Bibliographies	143
13.5 Directories	145
13.6 Dictionaries	145
13.7 Education	145
13.8 Folklore	145
13.9 Glossaries	145
13.10 Handbooks	145
13.11 Historical	145
13.12 Humor	150
13.13 Indexes	150
13.14 Inventories	151
13.15 Juvenile	151
13.16 Manuals	152
13.17 Meetings	152
13.18 National Speleological Society	153
13.19 Periodicals	153
13.20 People	153
13.20.1 Autobiographies	153
13.20.2 Biographies	154
13.21 Philately	154
13.22 Postcards	154
13.23 Religion	154
13.24 Songs	154

14 Geographic Index	**156**
14.1 World	156
14.2 Americas	156
14.3 Antarctica	156
14.4 Arctic	156
14.5 Africa	156
14.5.1 Algeria	156
14.5.2 Egypt	156
14.5.3 Ethiopia	156
14.5.4 Kenya	156
14.5.5 Libya	156
14.5.6 Madagascar	156
14.5.7 Morocco	157
14.5.8 Nigeria	157
14.5.9 Nyasaland	157
14.5.10 Sierra Leone	157
14.5.11 South Africa	157
14.5.12 Tanzania	157
14.5.13 Tunisia	157
14.5.14 Zimbabwe	157
14.6 Europe	158
14.6.1 Albania	158
14.6.2 Alps	158
14.6.3 Austria	158
14.6.4 Belgium	158
14.6.5 Bohemia	158
14.6.6 British Isles	159
14.6.6.1 Great Britain	159
14.6.6.1.1 Derbyshire	163
14.6.6.1.2 England	163
14.6.6.1.3 Hebrides	165
14.6.6.1.4 Mendips	165
14.6.6.1.5 Sark	166
14.6.6.1.6 Scotland	166
14.6.6.1.7 Somerset	166
14.6.6.1.8 Wales	166
14.6.6.2 Ireland	167
14.6.6.3 Yorkshire	168
14.6.7 Bulgaria	168
14.6.8 Channel Islands	168
14.6.9 Crete	168
14.6.10 Cyprus	169
14.6.11 Czechoslovakia	169
14.6.12 Czech Republic	169
14.6.13 France	169
14.6.14 Germany	170
14.6.15 Greece	170
14.6.16 Hungary	171
14.6.17 Iceland	171
14.6.18 Italy	171
14.6.19 Malta	171
14.6.20 Moravia	171
14.6.21 Netherlands	172
14.6.22 Norway	172

14.6.23	Poland	172
14.6.24	Portugal	172
14.6.25	Pyrenees	172
14.6.26	Serbia	172
14.6.27	Slovakia	172
14.6.28	Slovenia	172
14.6.29	Spain	173
	14.6.29.1 Canary Islands	174
	14.6.29.2 Gibraltar	174
	14.6.29.3 Majorca	174
	14.6.29.4 Mallorca	174
14.6.30	Sweden	174
14.6.31	Switzerland	174
14.6.32	Turkey	174
14.6.33	Yugoslavia	174
14.7 Asia		174
14.7.1	China	175
14.7.2	Georgia SSR	175
14.7.3	Himalayas	175
14.7.4	Hong Kong	175
14.7.5	India	176
14.7.6	Indonesia	176
	14.7.6.1 Borneo	176
	14.7.6.2 Java	176
	14.7.6.3 Malaysia	176
	14.7.6.4 Sulawesi	177
14.7.7	Japan	177
	14.7.7.1 Okinawa	177
14.7.8	Kurdistan	177
14.7.9	Nepal	177
14.7.10	Thailand	177
14.7.11	Uzbekistan	177
14.7.12	USSR	177
14.7.13	Vietnam	177
14.8 Central America		177
14.8.1	Belize	178
14.8.2	Costa Rica	178
14.8.3	Guatemala	178
14.8.4	Honduras	178
14.9 North America		178
14.9.1	Canada	179
	14.9.1.1 British Columbia	179
	14.9.1.2 Northwest Territories	179
	14.9.1.3 Nova Scotia	179
	14.9.1.4 Ontario	179
	14.9.1.5 New Brunswick	179
14.9.2	Caribbean	179
	14.9.2.1 Anguilla	179
	14.9.2.2 Barbados	180
	14.9.2.3 Cuba	180
	14.9.2.4 Haiti	180
	14.9.2.5 Netherland Antillies	180
	14.9.2.6 Puerto Rico	180
	14.9.2.7 West Indies	180
	14.9.2.7.1 Bahamas	180
	14.9.2.7.2 Jamaica	181
	14.9.2.7.3 Tobago	181
	14.9.2.7.4 Trinidad	181
14.9.3	Mexico	181
	14.9.3.1 Chihuahua	182
	14.9.3.2 Oaxaca	182
	14.9.3.3 Yucatan	182
14.9.4	United States	183
	14.9.4.1 Regions	184
	14.9.4.1.1 Appalachians	184
	14.9.4.1.2 Black Hills	184
	14.9.4.1.3 New England	184
	14.9.4.1.4 Guadalupe Mountains	184
	14.9.4.1.5 Ozarks	184
	14.9.4.2 States	184
	14.9.4.2.1 Alabama	184
	14.9.4.2.2 Alaska	185
	14.9.4.2.3 Arizona	186
	14.9.4.2.4 Arkansas	186
	14.9.4.2.5 California	187
	14.9.4.2.6 Colorado	188
	14.9.4.2.7 Connecticut	189
	14.9.4.2.8 Delaware	189
	14.9.4.2.9 Florida	189
	14.9.4.2.10 Georgia	189
	14.9.4.2.11 Hawaii	190
	14.9.4.2.12 Idaho	190
	14.9.4.2.13 Illinois	191
	14.9.4.2.14 Indiana	191
	14.9.4.2.15 Iowa	193
	14.9.4.2.16 Kansas	193
	14.9.4.2.17 Kentucky	193
	14.9.4.2.18 Louisiana	198
	14.9.4.2.19 Maryland	198
	14.9.4.2.20 Massachusetts	199
	14.9.4.2.21 Michigan	199
	14.9.4.2.22 Minnesota	199
	14.9.4.2.23 Mississippi	199
	14.9.4.2.24 Missouri	199
	14.9.4.2.25 Montana	202
	14.9.4.2.26 Nebraska	202
	14.9.4.2.27 Nevada	202
	14.9.4.2.28 New Hampshire	203
	14.9.4.2.29 New Jersey	203
	14.9.4.2.30 New Mexico	203
	14.9.4.2.31 New York	206
	14.9.4.2.32 North Carolina	207
	14.9.4.2.33 Ohio	207
	14.9.4.2.34 Oklahoma	207
	14.9.4.2.35 Oregon	207
	14.9.4.2.36 Pennsylvania	208

- 14.9.4.2.37 South Carolina 209
- 14.9.4.2.38 South Dakota 209
- 14.9.4.2.39 Tennessee 210
- 14.9.4.2.40 Texas 211
- 14.9.4.2.41 Utah 214
- 14.9.4.2.42 Vermont 214
- 14.9.4.2.43 Virginia 214
- 14.9.4.2.44 Washington 217
- 14.9.4.2.45 West Virginia 218
- 14.9.4.2.46 Wisconsin 219
- 14.9.4.2.47 Wyoming 219

14.10 South America 219
- 14.10.1 Brazil 219
- 14.10.2 Chile 219
- 14.10.3 Colombia 219
- 14.10.4 Ecuador 219
- 14.10.5 Peru 220
- 14.10.6 Venezuela 220

14.11 Australasia 220
- 14.11.1 Australia 220
  - 14.11.1.1 New South Wales 223
  - 14.11.1.2 Queensland 224
  - 14.11.1.3 South Australia 224
  - 14.11.1.4 Tasmania 225
  - 14.11.1.5 Victoria 225
  - 14.11.1.6 Western Australia 225

14.12 Pacific 226
- 14.12.1 Guam 226
- 14.12.2 Mariana Islands 226
- 14.12.3 Marquesas Islands 226
- 14.12.4 Micronesia 226
- 14.12.5 New Caledonia 226
- 14.12.6 New Guinea 226
- 14.12.7 New Zealand 226
- 14.12.8 Palau 227
- 14.12.9 Papua New Guinea 227
- 14.12.10 Philippines 227
- 14.12.11 Polynesia 228

14.13 Other Geographic Terms 228
- 14.13.1 Mediterranean 228
- 14.13.2 Middle East 228
  - 14.13.2.1 Israel 228
  - 14.13.2.2 Iran 228
  - 14.13.2.3 Iraq 228
  - 14.13.2.4 Oman 228
  - 14.13.2.5 Syria 228

15 Bibliography 229

16 Index of Secondary Authors, Editors, Illustrators, Etc. 461

17 Title Index 480

ISO Country Codes and USPS State Codes 537

1 Introduction

The main body of this Guide to Speleological Literature is a selectively annotated bibliography of English language works of ten or more pages. Access is given by author, title, subject, and geography (where applicable). The original core bibliography, compiled by Edward Zawlocki, was supplemented by a search of the OCLC WorldCat database and by visiting the libraries of several prominent speleobibliophiles. The criteria for inclusion and exclusion is as follows:

- Works that contain more than fifty percent foreign language material are excluded (i.e. are less than fifty percent in English).

- Fiction. Some songbooks, poetry, folklore, etc. have been included if they include some history or other non-fiction information. Books that contain drawings and other cave art are included.

- Juvenile non-fiction literature is included.

- The work must be at least fifty percent about caves. Works with one chapter about caves, even if it's important, are not included.

- The closely related field of Bats has been *excluded* due to space constraints. This subject deserves its own bibliography.

- Articles in periodicals have been *excluded*. Series volumes (e.g., v.22 of the University of Wyoming Publications, *The Human Skeletal Remains from Pictograph and Ghost Caves, Montana*) and volumes of periodicals that entirely devoted to one subject (e.g., *Caves of Oregon*, no.49, *W.S.S. Bulletin* are included. If a record exists in OCLC's WorldCat database for a reprint of an article that appears separately, the item is included.

These criteria are completely arbitrary and were established to limit the size and time involved in the production of this guide, as well as to provide consistency in what was included. We realize that many pertinent items are missing from the bibliography and hope that a fresh recruit can be found to continue the work that we've started. A very serious attempt was made to proofread the entries by Trevor Shaw and William Mixon. Despite their best efforts, we realize that many errors remain that you will undoubtedly discover as your eye falls upon a given page. We were incredibly dismayed to find that every time we examined a page we had supposedly already proofed, we found errors even though the page had been proofed several times before. We imagined that gremlins were getting into the computer at night and having fun at our expense! We hope that you will forgive this, report them to us, and realize that a work of this magnitude would take many additional years to perfect.

Integrated into the subject listings are introductions to the literature of major subdisciplines of speleology and caving. These provide the basic history of books and monographs in a subdiscipline, analyses of the milestones, and discussions of areas in which little research has been done. The introductions cover what countries are or have been in the forefront and when any subdisciplines developed. The introductions give the reader a good sense of what the important works in a given field are.

Entries in the author chapter (Chapter 15) of the bibliography are the main entries and contain the fullest information. The format is shown and discussed below. In actual entries, punctuation will appear as shown in the example and the field names are replaced by the actual information. Optional fields are shown enclosed here in square brackets; when square brackets appear in an actual entry, they indicate that the enclosed information in inferred, as stated in the explanations of the individual fields. Items shown in the sample in bold or italic type appear in those styles in the actual entries.

Accession number Author *Title*. [Other people] [Edition] Place of publication: Publisher, **Date**. [Series title] [Illus types] [; Folded stuff] [Lining pages] [front: type] [;] Pagination [; End matter] [; Size].

[Notes.] [Annotation.]

[Other edition info.]

1.1 Explanation of data fields

Accession number Entries in the author listing are given consecutive numbers, which appear here between brackets.

Author This is the first author or editor only. Additional authors, editors, illustrators, and so on are listed in the *Other people* field. Publications that are part of a series of conference proceedings, guidebooks, or programs will often be listed under a series "author" such as *National Speleological Society Annual Convention Guidebook*, in order to make it easy to locate all of them. In such cases, the individual author or editor will appear in *Other people*. Other publications of agencies, especially government ones, or organizations may appear with the name of the organization as "author" especially if they are otherwise anonymous.

Honorifics such as PhD or Dr. are not included in names, but modifiers such as Jr. or III are. Generally, names are given in as full a form as is known to avoid confusion among authors with similar names; they may appear in a shorter form in the publication itself. Here and in the *Other people* field, names of individuals may be followed, after a comma, by birth and/or death years in normal type, respectively before and after a dash. If the role of the person, here or in the *Other people* field, is not that of author or coauthor, his role will be given by one of the following abbreviations in parentheses (in cases with insufficient information, a likely error is for a "compiler" or "editor" to not be identified as such). A known author of an apparently anonymous work will appear in square brackets. If no author information is available, the work will be listed under **Anon**.

abst	abstractor
advs	advisor
app	appendix writer
asst	assistant
auth	author (used only if we had to put two functions after a name)
cap trans	captions translator
chair	chairman
co-pub	co-publisher
co-surv	co-surveyor
collab	collaborator
comp	compiler
compos	compositor
consul	consultant
cont	contributor
conv	convenor
coord	coordinator
des	designer
ed	editor
eng adap	English adaptor
forw	forward
illus	illustrator (used for artist also)
intro	introduction writer
lect	lecturer
map des	map designer
narr	narrator
photo	photographer
prep	preparer
prod	producer
proj coord	project coordinator
proj offr	project officer
rev	reviser
scrn wrt	screen writer
spon	sponsor
surv	surveyor
tm ldr	team leader
tm mem	team member
trans	translator
trans rev	translation reviser
illus	illustrator

Title The title, in italics, of the work as shown on the title page, if any. The title includes any subtitles and appears with the original spelling, although additional punctuation of colons or commas may have been added for clarity. Only the first word and proper nouns in the title are capitalized. In the absence of a title page and the title was taken from the cover, "title from cover" appears in the *Notes* field. If the cover title differs from that on the title page, the cover title will appear in the *Notes* field.

Other people (optional) This field will include other people, such a coauthors, translators, or illustrators, who are named in the work. Individuals other than authors or coauthors will have their contributions defined in parentheses according to the same code used in the *Author* field, and their birth and death dates may be given in the same way as in that field. Multiple entries in this field are separated by commas. An "organization as author" may appear in this field.

Edition (optional) If the details in this main-entry paragraph are known to apply to some edition of the work other than the first or only, then this information will appear here. For example, the field might contain something like "Second edition". Where an edition is not specified, but it is clear that changes have been made from the preceding version, the year of publication is used in the edition field.

Place of publication As much as is known of the city, state, and country of publication. If the information has been inferred, it is enclosed in square brackets. If the place of publication is unknown, the abbreviation npop is used. States in the United States are specified with the two-letter postal abbreviations listed in Appendix . States in Australia are abbreviated ACT, VIC, SA, NT, NWS, WA, QLD, or TAS. States or analogous subdivisions in other countries are spelled out. Countries are specified by the ISO two-letter country codes in Appendix . The *Place of publication* field is separated from the *Publisher* field by a colon.

Publisher The company, organization, or person responsible for the publication of the work, if known. If not known, the abbreviation np appears. Inferred data will be enclosed in []. When it is clear from other data in the entry what is meant, something like "the museum", "the society", or "the author" may appear here. The *Publisher* field is separated from the *Date* field by a comma.

Date The year of publication, if known or inferred, in bold type. The month of publication may precede the year. Inferred dates, which are known but not printed in the publication, are given in square brackets. Approximate dates are shown, e.g., as c1865. If the publication date is unknown, the abbreviation nd appears.

Series title (optional) This field may contain either the title of a series in which the publisher has placed a book, such as Zephrus Press's Speleologia series, or the title, volume, and number information of the periodical or serial of which this entry is a monograph issue, such as "Smithsonian Contributions to Zoology 244" (vol and no are the abbreviations for volume and number, respectively, with an occasional v or no). There may be more than one such series designation if there are co-publishers; if so, they will be separated by a semicolon.

Format fields The following fields in the main paragraph describe the appearance of the work. This information is most likely to be incomplete. For works not seen by the compilers, little or none of it may be available. If the only information about illustration came from some external source, and not from viewing the actual work, that information appears, perhaps with the number of illustrations, in the Notes field. Even if a work has been examined, as most have, omissions from the lists of possible illustrations are likely, because it would be easy to miss, say, a single table when leafing through a thick book. Also, various compilers may have placed the line between a chart and a drawing or between a table and a chart somewhat differently, or considered photographs of people merely photos, rather than portraits. The format fields that are present are separated from each other by semicolons.

Illus types (optional) The kinds of illustrations that appear in the work, not including the covers. Illustrations are listed in one or more of the following categories, preceded by the label shown in parentheses and separated from one another by semicolons: black and white (b&w), really meaning illustrations printed in the same ink as the text on whatever color of paper was used; partly colored (part col) for illustrations in five or fewer colors besides black and white; and colored (col) for illustrations in full color, such as photographs. For each color category present, one or more of the following notations will appear in the following order, according to what is present in the work. All illustrations in the work, including those that may be separately mentioned again in the Frontispiece, Folded stuff: or Lining pages fields, are included here.

Ads Advertisements for goods or services, sometimes unrelated to the subject of the book.

Cartoons Caricatures or humorous drawings; does not include other sketches included in the broad definition of cartoon.

Charts Diagrams including elements of both type and drawing, such a flow-charts or stratigraphic section diagrams.

Drawings Non-photographic representational illustrations such as sketches, paintings, or engravings.

Graphs Diagrams displaying data in graph format, with coordinate axes.

Maps Geographical, topographical, or cave maps.

Overlays Illustrations, usually line drawings or maps, on transparent leaves that overlie some other graphic.

Photos Photographs, other than photomicrographs or portraits however reproduced.

Photomicrographs (photomicros) Photographs of material viewed through a microscope.

Portraits (ports) Illustrations, whether drawings or photographs, that primarily show a close-up of a person or persons.

Reproductions (repros). Accurate, except perhaps for color, reproductions of earlier works such as drawings, paintings, photographs, newspaper articles, or fliers. Illustrations made for the work itself are excluded; repros are generally something of historical interest. A simple reprint of a scientific illustration from a somewhat earlier paper is not a repro.

Tables Text arranged in a tabular format for easy use. Some tables, such as lists of map symbols, may contain graphic elements.

Folded stuff (optional) Maps, etc., that are larger than the leaves of the book and are folded and either loose, bound in the text, or in a pocket or separate envelope; the number and nature of the items is detailed.

Lining pages (optional) The lining pages are the pages directly inside the front or back covers. For example, on a softbound book, they are usually the reverse of the paper on which the cover is printed. This field notes what text or types of illustrations appear on the front and/or rear lining pages.

Frontispiece (optional) A frontispiece is an illustration that appears, typically alone on a page, preceding the title page. This field, if present, tells the nature of the illustration on the frontispiece.

Pagination This field tells the number of pages in the work. The exact details are tricky (and error-prone). Information appears in the following order, separated by commas, with all but the body-page count optional, according to what is present in the work.

Prefatory pages. These are the pages that precede page 1 of the main body of the work. They are often numbered with Roman numerals, and the total number of them is usually given in Roman numerals in the data. The count of prefatory pages does not include the front cover unless forced too by, for example, the first non-cover page being explicitly labeled page iii. If none of the prefatory pages are numbered in the work, then the total number of unnumbered prefatory pages is given in square brackets. If there are unnumbered prefatory pages preceding the numbered page i, then both a number of unnumbered prefatory pages (in [ ]) followed by a number of numbered prefatory pages will appear.

Main body pages. This is the number of pages in the main body of the work. Normally this will be pages 1 thought the end of the book. Sometimes it is clear from the pagination that the covers are meant to be included in the page count; this is particularly likely to happen in issues of periodicals intended to be bound in volumes. If so, then the covers are included in the count given here. If, as in a issue of a periodical, the page numbering does not begin with 1 (e.g., pages 153-228), the actual number of pages is calculated as though the numbering did start with 1 (76p in this example). If the body pages are unnumbered but have been counted exactly, the number will be given in square brackets. Other special cases include a work with separately numbered chapters, which may be noted "various pagination", perhaps together with an estimated or actual total, or an unpaged work with an estimated page count, e.g. "about 200p". Besides the fact that the main body page count will usually be the largest number in this field, it is the one followed by p, e.g. 246p.

In a few cases, there are additional numbered (starting with 1) or unnumbered pages following body, perhaps for appendixes. The number of such pages will follow the number of main-body pages, perhaps in square brackets if the following pages are not numbered in the work.

Plates are leaves not included in the numbering of the work. Typically they will contain photographs, folded maps, or similar material not printed on the same press at the same time as the main text. The number of such pages is given, followed by the notation pp pls (for pages of plates). If the numbering of the pages in the work is such that it includes a plate, whether or not the page number is actually printed on the plate, then that plate is included in the body-page count, not in the count of plates.

End matter (optional). This field, if present, specifies the sorts of end matter that follow the main text, such as appendixes (app), glossaries (gloss), addenda (add), bibliographies (bibl), or index. A bibliography listed here may be distributed throughout the work in bibliographies at the end of individual chapters.

Size (optional). The size, height by width in centimeters, of the leaves of the work (not including things like foldouts). If only one figure appears, it is the height. Example: 23 x 18 cm.

Notes (optional). This field, if present, will begin a second paragraph of the entry. It may contain additional information on the format of the work. If there are known to be no illustrations, it will so state. Illustration information too sketchy to appear in the main paragraph may appear here, such as "23 illus". Any other information helpful in identifying the work may appear in this field.

Annotation (optional). This field, if present, will follow the Notes field, if any, in a separate paragraph. These brief notes further describing the work have been, usually, provided by the author of the introductory chapter concerning works in this subject category or an abstract writer (see List of Contributors), although a number of them are from other sources.

Other edition info Subsequent editions with exactly the same author and title are listed below the first edition with only the information that has changed from the **original** edition. There may be several such paragraphs. Only information about the edition that differs from the information in the main entry data paragraph is given, and there may be additional notes such as "revised" to further describe the editions. Whether separate editions have received separate main entries is sometimes a bit arbitrary.

2 Archeology

Patty Jo Watson

Department of Anthropology
Washington University
Campus Box 1114
One Brookings Drive
St. Louis, MO, 63130-4899

In the most fundamental sense, archeology in caves is simply archeology, with all the characteristics of field archeology done anywhere. There is the compulsive documentation of every investigative move by means of notes, recording forms, sketches, drawings, maps, photos in black-and-white and color, videos; the obsession with proveniences (find spots) and with accurate comprehensive information about all *in situ* (undisturbed by the excavator) situations. And there is the same panoply of equipment and techniques that are required by archeology anywhere, from trowels, line-levels, and 2-meter tapes to computerized laser transits and state-of-the-science camera equipment. But, of course, archeology underground is different in one significant detail from archeology done in any other terrestrial locale: archeology done inside a cave interior means archeology done in the dark. Adequate lighting is a problem for every single individual at every moment, and often necessitates alterations in basic techniques that are perfectly straightforward above ground but are much less so in a cave, even those as simple as reading a tape accurately. There is another attribute of cave archeology that is also important to note: so far as is known to date, ancient people never actually *lived* in the dark zones of deep caves. They often entered, explored, mined, or quarried them, or used them as water sources or as storage places. They placed their dead in dark cave interiors, and they performed ceremonies there that sometimes involved altering the cave significantly by reshaping it (enlarging it, for instance), or decorating it (painting, sculpting, incising, or engraving cave walls, ceilings, or flowstone). Many of these activities would have necessitated spending long hours in the dark zone and/or making repeated visits to it, but so far as we know, no ancient human group ever took up residence in the dark zone of a cave. Therefore, deep cave archeology always means investigation of a special-purpose site with little or no stratification of the sort found in residential sites resulting form one or more episodes of continuous occupation. Deep cave sites are, for most practical purposes, subterranean examples of "surface sites" in the archeological sense, meaning that the archeological materials are scattered across a surface (in this case, cave floors and ledges) rather than piled up on it with items from the bottom of the pile being markedly older than items at the top. Because of the special preservation conditions in dry caves, it is often very difficult to date non-chronologically diagnostic cultural remains found in them by any of the ordinary techniques based on superposition and physical association. Rather, archeometric dates (most commonly radiocarbon determinations) are necessary.

Obviously the discussion above concerns cave archeology in true caves, i.e., subterranean spaces with more or less extensive dark zones. But excavation of cave entrances and of rock shelters (spaces beneath overhangs in bluffs or cliffs) is also often referred to as cave archeology, just as the rock overhangs themselves are frequently called caves. In both the account that follows and in the bibliography and reference list, "cave archeology" includes archeological research in both cave interiors and rock shelters, but in this introduction I attempt to maintain the distinction because the kinds of materials found, and hence the archeological methods used and the information obtained as a result of rock shelter/cave entrance excavation often differ from techniques and information characterizing archeology in the dark zones of caves. There is a further slight complication to the distinction between rock shelter archeology and archeology in the dark zones of caves because many cave entrances were occupied intermittently or continuously over long time periods. Sites in cave entrances are like rock shelter sites, but may have unusual and complicated deposits if the cave interior beyond them was also used at various times for various purposes.

Readers interested in archeology of dark zones in caves will find that literature less abundant and more specialized than that on rock shelter and cave mouth excavation, but book and monograph titles seldom make the distinction. Deep caves with extensive dark zones are not referred to as rock shelters, but rock shelters and small caves with no true dark zones are frequently called caves.

Books on Cave and Rock Shelter Archeology in the Old World

By "Old World" is meant Africa, Asia, and Europe, but some comments on Australia and Oceania are also included in this section.

For our purposes here, we may begin with the lively interest in relics of the human past that burst forth during the mid to late 19th century in western Europe. This interest was stimulated in large part by the consensus then reached among geologists and paleontologists that the earth was far older than the 6,000 years previously calculated from Old Testament data by Archbishop Ussher. Educated people in England and Europe were also influenced by Charles Darwin's publications and those of other evolutionists. Although there was no formalized discipline of archeology until the end of the 1800s, there was widespread interest in relics of the past, such as bones and stone tools, many of which were actively sought in quarries, rock shelters, and caves by collectors of all sorts. One such individual, Don Sautaola, discovered ancient artifacts and also glorious polychrome paintings of Ice Age animals in the dark zone of a cave on his property at Altamira in the Spanish Pyrenees (Anon. 1935, Carballo 1963). Altamira was the first painted cave known to modern scholars, but by now there are some 200, whose distribution includes southwestern France, northern Spain, and the Ural Mountains (most are concentrated in southern France and northern Spain, however). The literature on the decorated caves in Spain and France is vast and still expanding. Two new decorated caves were recently found in southern France, one in 1991 on the Mediterranean coast near Cassis (the Grotte Cosquer) and one in 1994 near Vallon Pont D'Arc in the Ariege region of southern central France (the Grotte Chauvet). These recent finds are described in Chauvet (1996) and Clottes (1996). But, in general, books in English are few. Fortunately, a massive synthesis written some forty years ago by the French prehistorian Andre Leroi-Gourhan has been translated into English (Leroi-Gourhan 1967); and two other, more recent books in English concerning Upper Paleolithic decorated caves and other art are available: Marshack (1991), White (1986). For earlier general summaries in English see Ann Sieveking (1962) and Peter Ucko (1967).

Although the decorated caves of Europe are not paralleled elsewhere, pictographs and petroglyphs (often referred to as "rock art") are widespread throughout the world, often being found in or near rock shelters, cave mouths, and cave interiors. The most abundant and best known rock art sites are in the Saharan region of northern Africa, and in several locales in South Africa and Australia. Works that describe these include, e.g., Breuil 1955, Goodall 1959, Black 1943, McCarthy 1967, Wartenstein 1964).

There is another body of early cave archeological/paleontological literature in England focused upon bone caves (e.g., Dawkins 1873). In general, 19th-20th century European publications on cave archeology resulted from attempts to extract "relics" (archeological or paleontological) and to date them by relative means (superposition, typology, and stratigraphy). Some caves (as well as many gravel pits and river terraces) yielded artifacts as well as human skeletal remains and bones of non-human, now-extinct animals (cave bears, saber-toothed tigers, woolly mammoths and rhinos, forest horses, bison, etc.), so that information about faunal assemblages and changes in them through time accumulated, together with evidence for simultaneous presence of various ancient types of humans (for a recent example and discussion of this situation, see Gargett 1996). Most of this evidence was published in journals produced by various antiquarian societies and associations of geologists or paleontologists (e.g., in England, the Society of Antiquaries and the Royal Geological Society; in France the Societe Prehistorique Francaise). There are some standard histories of archeology and summaries of the human paleontological record that discuss—in English—these early developments (e.g., Daniel 1975, Grayson 1983), but rarely is specific attention paid to cave archeology as such. In the period between World Wars I and II, the basic Pleistocene (Ice Age) sequence of human biological and cultural evolution established for western Europe began to be exported to other world areas, notably Africa and Asia. Cave and rock shelter archeology played an important part in these efforts, perhaps the most sensational being finds of artifacts and ancient hominid remains in China at the Choukoutien caves (find spot of *Homo pekinensis*, later renamed *Homo erectus*; see Chia 1975, Jia and Huang 1990), and discoveries of even earlier hominid types in South Africa (*Australopithecus africanus*). The South African skeletal remains are fragments of small hominids, now known to be several million years old, whose bodies together with those of many other animals had come to rest in small caves that were later partially or entirely filled in, the cave contents being brecciated in the process. The first of these places to be discovered, Taung, was the site of a modern limestone quarry where dynamite-blasting broke into the ancient caves (see Brain 1981).

In West Asia, excavations during the 1930s at three small caves—barely more than rock shelters—by British archeologists delineated another piece of the Pleistocene and early post-Pleistocene human story as discussed in two major volumes: Garrod and Bate (1937) and McCown and Keith (1939) (see also Coon 1957, Garrod 1968).

Although cave deposits have provided important information on the Paleolithic epoch (roughly from 12,000/14,000 to three million years ago), caves and rock shelters also yield evidence about post-Pleistocene developments. One example is Franchthi Cave in Greece, where sediments accumulated from the Upper Paleolithic to the late prehistoric/early historic periods (Jacobsen 1987). Another example is a series of very unusual finds in a little desert cave in Israel called Nahal Hamar, where Early Neolithic people hid or stored what seem to be shamanistic paraphernalia (Bar-Yosef 1988). And of course it was in small dry caves in the Judean Desert, used for storage and secret repositories during the Iron Age, that the Dead Sea scrolls were found (Allegro 1975, Baigent and Leigh 1991, Wilson 1977; see also Bar-Adon 1980).

In Hawaii, lava tube caves were used by pre- and proto-historic islanders, and some were sacred places (Cleghorn 1980, Davis 1980, Bonk 1969). In Australia and New Zealand, cave and rock shelter archeology has also provided valuable information about prehistoric developments (e.g., Bowdler 1989, Gould 1977, Richards 1961).

Books on Cave and Rock Shelter Archeology in the New World

While European geologists, paleontologists, and antiquarians were searching for souvenirs of remote eras in the earth's past, similar investigations were launched in the Americas. Henry Mercer, e.g., excavated inside North American and Central American caves hoping to locate evidence for the presence of ancient humans with extinct animals (Crothers 1987; Mercer 1896, reprinted 1975). There was also concern with possible remains of great antiquity in South America (e.g., Bird 1938, and see Jennings 1983b). The big question at issue was that of human entry into the Americas, and this is still a lively research issue with no complete consensus to date. Two South American sites have attracted considerable recent attention, Monte Verde in Chile and Pedra Furada in Brazil (Dillehay 1989, Guidon and Delibrias 1986). Monte Verde is an open site, but Pedra Furada is a large rock shelter. The controversial results from Pedra Furada include radiocarbon dates of 30,000 B.P. (Before Present) and older, but there is no detailed, book-length account for Pedra Furada as yet.

Recent summaries of the oldest evidence for peopling of the Americas, some of which comes from caves and rock shelters, can be found in Brian Fagan's *The Great Journey* (Fagan 1987), as well as in more technical accounts such as the volumes edited by Dillehay and Meltzer (1991) and Jennings (1983b). For the Andean region of South America, see, e.g., Lynch (1980), MacNeish (1980-1983), and Rick (1980).

In North, Central, and South America, as in the Old World, rock shelter and cave mouth excavations have contributed important stratigraphic and cultural information about human lifeways during the past several thousand years. A few examples for South America have just been noted above. For Central America see Flannery (1986), Hardy (1996), MacNeish (1964, 1971), Pendergast (1971), Rouse (1990). For America north of Mexico, there is a long list, but here are a few examples: Southwest- Haury (1950), Martin (1952). Great Basin- Aikens (1970), Cressman (1942, 1951), Heizer (1970), Jennings (1980). Canada/Alaska- Dixon (1981), Hrdlicka (1941). Eastern North America- Adovasio *et al.* (1982), Broyles (1958), DeJarnette *et al.* (1962), Funk (1994), Funkhouser and Webb (1929), Goldman-Finn and Driskell (1994), Griffin (1974), Harrington (1960).

In addition, for the Southwest, there is a large body of literature on the so-called cliff dwellings built within big rock shelters in dramatic locales such as the canyons cut into the Colorado Plateau. The most famous site is probably Cliff Palace on Mesa Verde, but there are many other such ancient rock shelter communities in what is now Mesa Verde National Park (Nordenskiold 1893, Rohn 1971 and 1977). For a clear and very important technical archeological account of cliff dwellings elsewhere, see Dean (1969).

Archeological investigation of deep cave interiors, although pioneered in North and Central America by Henry Mercer and taken up by a few avocational archeologists (e.g., Colonel Bennett Young, 1910), was not systematically pursued by professionals until the post World War II period. Following Nels Nelson's excavations in the entry chamber of Mammoth Cave in 1916 (Nelson 1917), and Alonzo Pond's recovery of a prehistoric body from the cave interior (Pond 1937), a brief summary of Mammoth Cave National Park prehistory with some discussion of archeological material inside Mammoth Cave was prepared for the National Park Service by Douglas Schwartz (1965). Further archeological work in the world's longest cave (the Mammoth Cave System) sponsored by the Cave Research Foundation and in collaboration with the National Park Service, began in the 1960s and is continuing (Carstens and Watson eds. 1996, Haskins 1988, Kennedy 1990, Watson 1969, Watson ed. 1974). Archeological study of other cave interiors in the midcontinental karst of the United States began in the 1980s and is continuing (e.g., Faulkner 1986, Jackson 1975, Munson and Munson 1990).

Uncharred botanical remains preserved in abundance in dry caves and rock shelters are the focus of specialized studies on prehistoric plant use and early agriculture in the Eastern United States, a temperate environment where such materials do not survive in open sites (Fritz 1994, Gremillion 1994). Several desiccated human bodies ("mummies") have also been preserved in dry caves or rock shelters in both Western and Eastern North America (George 1985, 1990; Guernsey 1921; Meloy 1971; Morris 1969).

The only extensive and intensive investigation of deep cave interiors in Latin America to date is in Central America where several archeologists are carrying out detailed studies of mortuary and other ceremonial cave sites (Brady 1989, Steele 1987, Veni ed. 1996).

As in the Old World, cave mouths and interiors as well as rock shelters (and bluff faces) were painted, incised, or engraved by ancient peoples, presumably primarily for ritual purposes or during the performance of religious ceremonies, although some pictographs and petroglyphs have been interpreted as historical records or other more mundane communications (see, e.g., Duncan 1993, Schaafsma 1980).

Conclusions

Because of certain characteristics, cave mouths and rock shelters have collected and preserved physical remains of ancient peoples (their bones) and of their activities (artifacts). The stable microenvironments of dry rock shelters and caves conserve for hundreds or thousands of years delicate materials that would not last more than a few days or weeks in the open (Crothers and Watson 1994; see, e.g., Gremillion and Sobolik 1996; Orchard 1920, Shafer and Bryant 1977). Because living space under a shelter overhang or in a cave entrance is limited, long sequences of occupational debris are often found there, and provide stratigraphic guides to local and regional culture history. Special resources in cave interiors—permanent water supplies, minerals, speleothems, access to the supernatural world—have drawn people at various times and places to explore, work, meditate, or worship in the dark zones of caves. Tangible evidence of such activities, if undisturbed by later intrusions, persist for millennia. Hence, caves and rock shelters are valuable repositories of archeological as well as geological and paleontological information, whose full potential is only beginning to be realized.

References not in Main Bibliography

Adovasio, James M., J. D. Gunn, J. Donahue, and R. Stuckenrath. 1982. Meadowcroft Rockshelter, 1973-1977: a Synopsis. pp 97-131: In J.E. Ericson, R.E. Taylor, and R. berger, eds. *Peopling of the New World.* Los Altos, California: Ballena Press.

Allegro, J. M. 1975. *The Dead Sea Scrolls: a Reappraisal.* London: Harmondsworth.

Baigent, Michael, and Richard Leigh. 1991. *The Dead Sea Scrolls Deception.* London: Jonathan Cape.

Bird, Junius B. 1938. Antiquity and Migrations of the Early Inhabitants of Patagonia. *The Geographical Review.* 28: 250-275.

Brady, James E. 1989. *An Investigation of Maya Ritual Use with Special Reference to Naj Tunich, Peten, Guatemala.* Ph.D. Dissertation, University of California-Los Angeles, Archaeology Program.

Cressman, Luther S. 1942. *Archaeological Researches in the Northern Great Basin.* Publication 538. Washington D.C.: Carnegie Institute of Washington.

Cressman, Luther S. 1951. Western Prehistory in the Light of Carbon 14 Dating. *Southwestern Journal of Anthropology.* 7: 289-313.

Crothers, George M. 1987. *Archaeological Survey of Big Bone Cave, Tennessee, and Diachronic Patterns of Cave Utilization in the Eastern Woodlands.* M.A. Thesis, University of Tennessee, Knoxville, Department of Anthropology.

Crothers, George M., and Patty Jo Watson. 1993. Archaeological Contexts in Deep Cave Sites: Examples from the Eastern Woodlands of North America. pp 53-63: In P. Goldberg, D. Nash, and M. Petraglia, eds. *Formation Processes in Archaeological Context.* Monographs in World Archaeology No. 17. Madison, Wisconsin: Prehistory Press.

Daniel, Glyn. 1975. *150 Years of Archaeology.* London: Duckworth.

Dean, Jeffrey. 1969. *Chronological Analysis of Tsegi Phase Sites in Northeastern Arizona.* Papers of the Laboratory of Tree-Ring Research 3. Tucson: University of Arizona Press.

DeJarnette, David L., E. Kurjack, J. Cambron, and others. 1962. Stanfield-Worley Bluff Shelter Excavations. *Journal of Alabama Archaeology.* 8, numbers 1 and 2.

Dillehay, Tom D. 1989. *Monte Verde: A Late Pleistocene Site in Chile.*, volume 1. Washington D.C.: Smithsonian Institution Press.

Dillehay, Tom D., and David J. Meltzer, editors. 1991. *The First Americans.* Boca Raton, Louisiana: CRC Press.

Duncan, Carol Diaz-Granados. 1993. *The Petroglyphs and Pictographs of Missouri: A Distributional, Stylistic, Contextual, Functional, and Temporal Analysis of the State's Rock Graphics.* Ph.D. Dissertation, Washington University, St. Louis; Department of Anthropology.

Fagan, Brian M. 1987. *The Great Journey: the Peopling of Ancient America.* London: Thames and Hudson.

Flannery, Kent, editor. 1986. *Guila Naquitz: Archaic Foraging and Early Agriculture in Oaxaca.* New York: Academic Press.

Fritz, Gayle J. 1994. In Color and in Time: Prehistoric Ozark Agriculture. In W. Green, ed. *Agricultural Origins and Development in the Midcontinent, Report 10.* Iowa City, Iowa: Office of the State Archaeologist. Pp. 105-126.

Garrod, Dorothy A.E., and Dorothea M.A. Bate. 1937. *The Stone Age of Mount Carmel, Vol. I.* Oxford: Oxford University Press.

Goldman-Finn, N.S., and Boyce N. Driskell. 1994. Preliminary Archaeological Papers on Dust Cave, Northwestern Alabama. *Journal of Alabama Archaeology.* 40, numbers 1 and 2. Moundville: Alabama Archaeological Society.

Gould, Richard A. 1977. Puntutjarpa Rockshelter and the Australian Desert Culture. *Anthropological Papers of the American Museum of Natural History.* 54.

Grayson, Donald K. 1983. *The Establishment of Human Antiquity.* New York: Academic Press.

Gremillion, Kristen J. 1994. Evidence of Plant Domestication from Kentucky Caves and Rockshelters. In: W. Green ed. *Agricultural Origins and Development in the Midcontinent.*, Report 10. Iowa City, Iowa: Office of the State Archaeologist. Pp. 87-103.

Gremillion, Kristen J., and Kristin D. Sobolik. 1996. Dietary Variability among Prehistoric Forager-Farmers of Eastern North America. *Current Anthropology.* 37: 529-539.

Guidon, N. and G. Delibrias. 1986. Carbon-14 Dates Point to Man in the Americas 32,000 Years Ago. *Nature.* 321: 769-771.

Hardy, Karen V. 1996. The Preceramic Sequence from the Tehuacan Valley: A Reevaluation. *Current Anthropology.* 37: 700-716.

Harrington, Mark. 1960. *The Ozark Bluff-Dwellers.* Indian Notes and Monographs, volume XII. New York: Museum of the American Indian, Heye Foundation.

Haskins, Valerie A. 1988. *The Prehistory of Prewitts Knob, Kentucky.* M.A. Thesis, Washington University, St. Louis, Department of Anthropology.

Haury, Emil W. 1950. *The Stratigraphy and Archaeology of Ventana Cave, Arizona.* Tucson: University of Arizona Press.

Jennings, Jesse, editor. 1983a. *Ancient North Americans.* San Francisco: W.H. Freeman.

Jennings, Jesse, editor. 1983b. *Ancient South Americans.* San Francisco: W.H. Freeman.

Jia, Lanpo, and Huang, Weiwen. 1990. *The Story of Peking Man: From Archaeology to Mystery.* Translated by Zhiqi Yin. Beijing and Hong Kong: Foreign Languages Press and Oxford University Press.

Kennedy, Mary C. 1990. *An Analysis of the Radiocarbon Dates from Salts and Mammoth Caves, Mammoth Cave National park, Kentucky.* M.A. Thesis, Washington University, St. Louis, Department of Anthropology.

MacNeish, Richard S. 1964. Ancient Mesoamerican Civilization. *Science.* 143: 531-537.

MacNeish, Richard S. 1971. Speculation about How and Why Food Production and Village Life Developed in the Tehuacan Valley, Mexico. *Archaeology.* 24: 307-315.

MacNeish, Richard S. 1980-1983. *The Prehistory of the Ayacucho Basin, Peru.* Three volumes. Ann Arbor, Michigan: University of Michigan Press and R.S. Peabody Foundation for Archaeology.

Marshack, Alexander. 1991. *The Roots of Civilization.* Second edition. Mount Kisco, N.Y.: Moyer Bell Limited.

McCown, Theodore D., and Sir Arthur Keith. 1939. *The Stone Age of Mount Carmel, Vol. II. The Fossil Human Remains from the Levalloiso-Mousterian.* Oxford: Oxford University Press.

Morris, Elizabeth Ann. 1969. *Basketmaker Caves in the Prayer Rock District, Northeastern Arizona.* Anthropological Papers of the University of Arizona No. 35. Tucson: University of Arizona Press.

Nordenskiold, Gustaf. 1893. *The Cliff Dwellers of the Mesa Verde, Southwestern Colorado.* Stockholm: P.A. Norstedt and Soner.

Pond, Alonzo. 1937. Lost John of Mummy Ledge. *Natural History.* 39:176-184.

Rick, John. 1980. *Prehistoric Hunters of the High Andes.* New York: Academic Press.

Rohn, Arthur. 1971. *Mug House.* Archeological Research Series No. 7-D. Washington D.C.: National Park Service, U.S. Department of the Interior.

Rohn, Arthur. 1977. *Cultural Change and Continuity on Chapin Mesa.* Lawrence, Kansas: The Regents Press of Kansas.

Schaafsma, Polly. 1980. *Indian Rock Art of the Southwest.* Santa Fe and Albuquerque, New Mexico: School of American Research and University of New Mexico Press.

Steele, Janet. 1987. *Blade Cave: an Archaeological Preservation Study in the Mazatec Region, Oaxaca, Mexico.* M.A. Thesis, University of Texas, San Antonio, Department of Anthropology.

Veni, George, guest editor. 1996. Speleology in Belize: An Introduction. *Journal of Cave and Karst Studies.* 58(2).

White, Randall K. 1986. *Dark Caves, Bright Visions; Life in Ice Age Europe.* New York: American Museum of Natural History in association with W.W. Norton.

Williams-Dean, Glenna. 1978. *Ethnobotany and Cultural Ecology of Prehistoric Man in Southwest Texas.* College Station, Texas: Texas A&M University, Anthropology Research Laboratory.

Young, Colonel Bennett. 1910. *Prehistoric Men of Kentucky.* Louisville, Kentucky: Filson Club Publications No. 25, J. P. Morton, Printers.

2.1 General

[14]Adams, William Henry Davenport, 1828–1891, 1886, Famous caves and catacombs: described and illustrated

[16]Adovasio, J. M., 1980, Archaeological testing at two rockshelters in the Tombigbee River Multi-Resource District, Alabama and Mississippi: an interim report

[18]Aikens, C. Melvin, 1970, Hogup Cave

[24]Alex, Lynn Marie, 1948–, 1991, The archaeology of the Beaver Creek shelter (39CU779), Wind Cave National Park, South Dakota: a preliminary statement

[29]Alexander, Hubert Griggs, 1935, Report on the excavation of Jemez Cave, New Mexico, Santa Fe, N.M.

[31]Alexander, Robert K., 1970, Archeological investigations at Parida Cave, Val Verde County, Texas

[53]Allison, Vernon Charles, 1891–, 1926, The antiquity of the deposits in Jacob's Cavern

[74]Andrews, Edward Wyllys, 1916–1971, 1970, Balankanche, throne of the tiger priest

[76]Andrews, Rhonda Lynette, 1980, Perishable industries from Hinds Cave, Val Verde County, Texas

[89]Anon, Feb 1903, The cave at Vari: excavations by American School of Classical Studies at Athens in February, 1901

[131]Anon, 1934, A guide to the Museum of the Torquay Natural History Society

[226]Armstrong, A. Leslie, 1931, Rhodesian archaeological expedition (1929): excavations in Bambata Cave and researches on prehistoric sites in Southern Rhodesia

[242]Athens, J. Stephen, Aug 1989, Prehistoric upland bird hunters: archaeological inventory survey and testing for the MPRC project area and the Bobcat Trail Road, Pohakuloa Training Area, Island of Hawaii

[247]Aubarbier, Jean-Luc, Jun 1985, Prehistoric sites in Perigord

[268]Ayer, Mary Youngman, 1936, The archaeological and faunal material from Williams Cave, Guadalupe Mountains, Texas

[274]Bahn, Paul G., 1988, Images of the ice age

[277]Bailey, John H., 1940, A stratified rock shelter in Vermont

[294]Balch, Herbert Ernest, [1927], Excavations at Chelm's Combe: Cheddar: conducted under the Excavations Committee of the Somerset Archaeological and Natural History Society 1925-26

[298]Balch, Herbert Ernest, 1929, Mendip: the great cave of Wookey Hole

[299]Balch, Herbert Ernest, 1914, Wookey Hole: its caves and cave dwellers

[300]Balch, Herbert Ernest , c1971, Fourteen years at the Badger Hole

[316]Bar-Adon, Pesah, 1907–, 1980, The cave of the treasure: the finds from the caves in Nabal Mishmar

[317]Bar-Yosef, Ofer, 1985, A cave in the desert, Nahal Hemar: 9,000-year-old finds: [exhibition, the Israel Museum, Jerusalem, spring 1985]

[318]Bar-Yosef, Ofer, 1988, Nahal Hemar Cave

[319]Bard, James C., 1979, Ezra's Retreat: a rockshelter/cave occupation site in the north central Great Basin

[320]Bard, James C., 1980, Test excavations at Painted Cave, Pershing County, Nevada: for Bureau of Land Management, Winnemucca District Office

[340]Bassie-Sweet, Karen, 1991, From the mouth of the dark cave: commemorative sculpture of the late classic Maya

[341]Basso, Dave, 1992, 2,000-year-old duck decoys from Lovelock Cave, Nevada

[349]Baumgartel, Elise J., 1892–, 1971, The cave of Manaccora, Monte Gargano

[374]Beldam, Joseph, 1795–1866, 1858, The origin and use of the Royston Cave, being the substance of a report some time since presented to the Royal Society of Antiquaries by the late Joseph Beldam ...

[380]Ben-Tor, Amnon, 1975, Two burial caves of the Proto-Urban Period at Azor, 1971: the first season of excavations at Tell-Yarmuth, 1970

[390]Benthall, Joseph L., , Daugherty's Cave: a stratified site in Russell County, Virginia

[392]Berenguer, Magin, 1973, Prehistoric man and his art: the caves of Ribadesella

[406]Black, Davidson, 1884–1934, 1925, The human skeletal remains from the Sha Kuo T'un cave deposit in comparison with those from Yang Shao Tsun and with recent North China skeletal material

[419]Blore, J. D., c1977, Archaeological excavation at North Face Cave, Little Ormes Head, Gwynedd 1962-1976

[432]Bonk, William J., 1969, Lua Nunu o Kamakalepo: a cave of refuge in Kau, Hawaii

[439]Booy, Theodoor Hendrik Nikolaas de, 1882–1919, 1915, Pottery from certain caves in eastern Santo Domingo, West Indies

[440]Bordes, Francois, 1972, A tale of two caves

[447]Bousman, C. Britt, 1973, An archaeological assessment of Carlsbad Caverns National Park

[448]Bowdler, Sandra, 1989, A Test excavation at Mulka's Cave (Bate's Cave) near Hyden, Western Australia: a report to the Department of Aboriginal Sites, Western Australian Museum

[452]Boyle, Mary Elizabeth, 1881–, 1925, Barma Grande; the great cave and its inhabitants

[460]Brain, Charles Kimberlin, 1958, The Transvaal apeman-bearing deposits

[467]Breternitz, David A., 1969, Archaeological investigations in Turkey Cave (NA2520) Navajo National Monument, 1963

[470]Breuil, Abbé Henri, 1877–1961, 1949, Beyond the bounds of history: scenes from the old stone age

2.1. GENERAL

[512]Brooks, Ian, 1985, Excavation techniques in Pin Hole Cave, Creswell Crags S.S.S.I., Derbyshire
[519]Brown, Edward, nd, The pictured cave of La Crosse Valley, near West Salem, Wisconsin
[533]Broyles, Bettye J., 1958, Russell Cave in northern Alabama
[534]Bruce, Murray D., 1980, Preliminary survey of Palawan, Philippines, by Traditional Explorations and the Sydney Speleological Society, January/February 1980
[542]Bryan, Kirk, 1941, Correlation of the deposits of Sandia Cave, New Mexico, with the glacial chronology
[554]Bunch, James H., , The archaeological excavation of two rock shelters near Cave Lake, Ely, Nevada
[566]Burrows, Russell, , Mystery cave of many faces: first in a series on the saga of Burrows' Cave
[589]Campbell, John B., 1944–, 1971, A new analysis of Kent's Cavern, Devonshire, England
[602]Carstens, Kenneth Charles, 1996, Of caves and shell mounds
[627]Cervino, Donald Jay, 1990, Water features of the European Upper Paleolithic decorated cave environment
[633]Champe, John L., Oct 1946, Ash Hollow Cave: a study of stratigraphic sequence in the central great plains
[642]Chauvet, Jean-Marie, 1996, Chauvet Cave: the discovery of the world's oldest cave paintings
[654]Chin, Lucas, 1980, Archaeological work in Sarawak: with special reference to Niah Caves
[655]Chin, Lucas, 1977, Summary of archaeological work in Sarawak: with special reference to Niah Caves
[658]Church, Flora, 1954–, Jun 1990, Mitigation of the Brady Run Rockshelter 3: a multi-component site in Washington Township, Lawrence County, Ohio
[663]Clark, John Desmond, 1916–, 1942, Further excavations (1939) at the Mumbwa caves, Northern Rhodesia
[665]Cleghorn, Paul L., Feb 1980, The Hilina Pali petroglyph cave, Hawai'i island: a report on preliminary archaeological investigations
[678]Coggins, Clemency, 1984, Cenote of sacrifice: Maya treasures from the sacred well at Chichen Itza
[687]Combes, John D., Jun 1969, The Excavation of Squirt Cave, 45WW25
[702]Cooke, Ian, 1937-, 1993, Mother and sun: the Cornish Fogou
[705]Coon, Carleton Stevens, 1904–, 1968, Cave explorations in Iran, 1949
[706]Coon, Carleton Stevens, 1904–, 1957, The seven caves: archaeological explorations in the Middle East
[707]Coon, Carleton Stevens, 1904–, 1968, Yengema Cave Report
[720]Cosgrove, Cornelius Burton, 1875–, 1947, Caves of the Upper Gila and Hueco areas in New Mexico and Texas
[727]Cowles, John, 1959, Cougar Mountain Cave in south central Oregon
[756]Crofton, Denis, 1901, All about Poole's Cavern, Buxton. An official guide to this unique natural curiosity, together with an inventory of the museum
[764]Crothers, George Martin, Apr 1986, Final report on the survey and assessment of the prehistoric and historic archaeological remains in Big Bear Cave, Van Buron County, Tennessee
[809]Daugherty, Richard D., 1922–, 1987, A data recovery study of Layser Cave (45-LE-223) in Lewis County, Washington
[810]Davaras, Costis, 1989, The Cave of Psychro
[811]Davenport, J. Walker, 1933, Archaeological exploration of Eagle Cave, Langtry, Texas
[832]Davis, Bertell D., 1980, Use and abandonment of habitation caves in the prehistoric settlement of southeastern Oahu: a proposed research design for the 1980 University of Hawaii Archaeological Field Program
[837]Davis, Wilbur A., 1964, Archaeological surveys of Crater Lake National Park and Oregon Caves National Monument, Oregon
[842]Dawkins, William Boyd, 1838–1929, 1903, On the discovery of an ossiferous cavern of Pliocene Age at Doveholes, Buxton (Derbyshire)
[849]Dehejia, Vidya, 1969, The Namakkal caves
[851]Delderfield, Eric Raymond, c1959, The story of Brixham Cavern
[865]Dibble, David S., Jan 1968, Bonfire shelter: a stratified bison kill site, Val Verde County, Texas
[866]Dick, Herbert W., 1965, Bat Cave
[870]Dickson, Don R., 1991, The Albertson site: a deeply and clearly stratified Ozark bluff shelter
[876]Dixon, E. James, Apr 1981, 1980 progress report of archeological reconnaissance and testing of Pleistocene cave and alluvial deposits, Porcupine River, Alaska
[883]Dopp, Katherine Elizabeth, 1863–, 1904, The early cave-men
[884]Dopp, Katherine Elizabeth, 1863–, 1906, The later cave men
[886]Dortch, C. E., 1976, Devils Lair: a search for ancient man in Western Australia
[887]Dortch, C. E., 1984, Devil's Lair, a study in prehistory
[897]Draper, John A., 1989, Archaeology of Chokecherry Cave (35GR500) in Grant County, southeastern Oregon
[899]Drew, David Phillip, 1943–, 1980, Dunmore Cave, County Kilkenny: a reassessment
[900]Drew, David Phillip, 1943–, 1978, Dunmore Cave: a short guide
[910]Dulluc, Brigitte, 1990, Discovering Lascaux
[943]Ediger, Donald, 1971, The well of sacrifice
[965]Ellis, Arthur Charles, 1930, An historical survey of Torquay from the earliest times, as illustrated by finds in Kent's Cavern, down to the present time
[977]Ellul, Joseph S., 1988, Malta's prediluvian culture at the stone age temples with special reference to Hagar Qim,

Ghar Dalam cart ruts, Il-Misqa, Il-Maqluba and Creation
[978]Elsasser, Albert B., 1963, The archaeology of Bowers Cave, Los Angeles County, California
[992]Enger, Walter D., 1942, Archaeology of Black Rock 3 Cave, Utah
[998]Epstein, Jeremiah F., 1960, Centipede and Damp caves: excavations in Val Verde County, Texas, 1958
[1008]Euler, Robert C., 1978, Archaeological and paleobiological studies at Stanton's Cave, Grand Canyon National Park, Arizona; a report of progress
[1009]Euler, Robert C. (ed), 1984, The Archaeology, geology, and paleobiology of Stanton's Cave: Grand Canyon National Park, Arizona
[1033]Farwell, Robin E., 1985, The Pictured Cliffs project: petroglyphs and talus shelters in San Juan County, New Mexico
[1047]Ferdon, Edwin N., 1946, An excavation of Hermit's cave, New Mexico
[1070]Flood, Josephine, 1983, Archaeology of the dreamtime
[1071]Flood, Josephine, 1990, The riches of ancient Australia: a journey into prehistory
[1093]Ford, Trevor David, 1925–(comp, ed), 1977, Limestone and caves of the Peak District
[1109]Fowke, Gerard, 1855–1933, 1922, Archeological investigations
[1110]Fowler, Don D., 1936–, Mar 1968, The archeology of Newark Cave, White Pine County, Nevada
[1111]Fox, Robert B., 1918–, 1964, The Tabon Caves: archaeological excavations on Palawan Island, Philippines (1962—64)
[1112]Fox, Robert B., 1918–, 1970, The Tabon Caves; archaeological explorations and excavations on Palawan Island, Philippines
[1113]Fox, W. Storrs, 1908, Notes on the excavation of Harborough Cave, near Brassington, Derbyshire
[1129]Fulton, William Shirley, 1941, A ceremonial cave in the Winchester Mountains, Arizona
[1130]Funk, Robert E., 1994, Archaeological and paleoenvironmental investigations in the Dutchess Quarry Caves, Orange County, New York
[1131]Funkhouser, William Delbert, 1881–, 1929, The so-called "ash caves" in Lee County, Kentucky
[1132]Furlong, Eustace Leopold, 1874–, 1906, The exploration of Samwel Cave
[1146]García Guinea, Miguel Angel, Mar 1969, Altamira, the beginning of art
[1157]Gargett, Robert H., 1996, Cave bears and modern human origins: the spatial taphonomy of Pod Hradem Cave, Czech Republic
[1162]Garrod, Dorothy Anne Elizabeth, 1892–, 1968, The paleolithic of southern Kurdistan: excavations in the caves of Zarzi and Hazar Merd
[1163]Garrod, Dorothy Anne Elizabeth, 1892–, 1982, The transition from Lower to Middle Palaeolithic and the origin of modern man: international symposium to commemorate the 50th anniversary of excavations in the Mount Carmel Caves by D.A.E. Garrod, University of Haifa, 6-14 October 1980
[1176]George, Angelo Isham, 1944–, 1985, Mummies of Short Cave, Kentucky, and the Great Catacomb Mystery
[1177]George, Angelo Isham, 1944–, 1994, Mummies, catacombs, and Mammoth Cave
[1179]George, Angelo Isham, 1944–, 1990, Prehistoric mummies from the Mammoth Cave area: foundations and concepts
[1184]Gerrard, John, 1960, Kent's Cavern
[1188]Gibb, Hugh, 1981, The Tabon caves
[1195]Gifford Pinchot National Forest, 1991, Layser Cave: silent voices, vital clues
[1196]Gifford, James C., 1973–, 1980, Archaeological explorations in caves of the Point of Pines region, Arizona
[1198]Gilbertson, D. D., 1984, In the shadow of extinction: a Quaternary archaeology and paleoecology of the lake, fissures, and smaller caves at Creswell Crags SSSI
[1218]Gleave, Joseph James, c1900, Yorkshire Caves: Victoria cave, Settle; Ingleborough Cave, Clapham; Yordas Cave, Ingleton
[1221]Goldberg, Paul, 1993, Formation processes in archaeological context
[1224]Goode, Glenn T., 1985, Archaeological testing of the cave at site 41BX22, Bexar County, Texas
[1225]Goodwin, Astley John Hilary, 1900–1959, 1935, Archaeology of the Cape St. Blaize cave and raised beach, Mossel Bay
[1255]Greer, John W., 1966, Report on preliminary archeological explorations at Carlsbad Caverns National Park, New Mexico
[1272]Grosscup, Gordon L., 1960, The culture history of Lovelock Cave, Nevada
[1274]Gruhn, Ruth1907–, 1961, The archaeology of Wilson Butte Cave south-central Idaho
[1277]Guernsey, Samuel James, 1868–1936, 1921, Basket-Maker caves of northeastern Arizona; report on the explorations, 1916-17
[1278]Guernsey, Samuel James, 1868–1936, 1931, Explorations in northeastern Arizona; report on the archaeological fieldwork of 1920-1923
[1280]Guilday, John E., 1977, The Clark's Cave bone deposit and the late Pleistocene paleoecology of the central Appalachian Mountains of Virginia
[1300]Haast, J., 1874, Researches and excavations carried on in and near Moa Bone Point Cave, Summer Road, in the year 1872
[1353]Hammatt, Hallett H., Apr 1990, Archaeological assessment and sensitivity map for the Pohakuloa training area (PTA), Hawaii Island, State of Hawaii
[1359]Hargrave, Lyndon Lane, 1896–, 1967, Feathers

2.1. GENERAL

from Sand Dune Cave: a Basketmaker cave near Navajo Mountain, Utah
[1361]Harpers Ferry Center. Division of Museum Services, 1976, Mammoth Cave National Park: collection management plan
[1362]Harrington, Mark Raymond, 1882–1971, Apr 1933, Gypsum cave, Nevada, report of the second sessions expedition
[1386]Hatt, Robert Torrens, 1902–, 1953, Faunal and archeological researches in Yucatan caves
[1392]Hay, Deborah, Oct 1986, Kahaluu data recovery project: excavations at site 50-10-37-7702, Kahaluu Habitation Cave, land of Kahaluu, North Kona, Island of Hawaii
[1398]Haynes, Caleb Vance, 1928–, 1986, Geochronology of Sandia Cave
[1420]Heizer, Robert Fleming, 1915–, 1970, Archaeological investigations in Lovelock Cave, Nevada
[1421]Heizer, Robert Fleming, 1915–, 1956, The archaeology of Humboldt Cave, Churchill County, Nevada
[1422]Heizer, Robert Fleming, 1915–, 1961, The archaeology of two sites at Eastgate, Churchill County, Nevada
[1423]Heizer, Robert Fleming, 1915–, 1942, Massacre Lake Cave, Tule Lake Cave and shore sites
[1451]Hevrah la-hakirat Erets-Yisrael ve-atikoteha, 1960, 1962, The Expedition to the Judean Desert
[1452]Hibben, Frank Cummings, 1910–, 1941, Evidences of early occupation in Sandia Cave, New Mexico, and other sites in the Sandia-Manzano Region
[1474]Hoebel, Edward Adamson, 1906–, 1946, The archaeology of Bone cave, Miller county, Missouri
[1513]Home, Everard, 1756–1832, 1794, Account of some remarkable caves in the principality of Bayreuth: likewise observations on the fossil bones by the late John Hunter
[1542]Howard, John Eliot, 1879, The caves of South Devon and their teachings
[1546]Howe, Bruce, 1912–, 1974, A stone age cave site in Tangier: preliminary report on the excavations at the Mugharet el 'Aliya, or High Cave, in Tangier
[1559]Hrdlicka, Ales, 1869–1943, 1941, Exploration of mummy caves in the Aleutian Islands ...
[1564]Hudgens, Bruce R., 1975, The archeology of Exhausted Cave: a study of prehistoric cultural ecology on the Coconino National Forest, Arizona
[1567]Hudson, Travis, 1982, Guide to Painted Cave
[1572]Humphreys, A. J. B., 1978, The re-excavation of Powerhouse Cave and an assessment of Dr. Frank Peabody's work on Holocene deposits in the Taung area
[1611]Inskeep, R. R., 1987, Nelson Bay Cave, Cape Province, South Africa: the Holocene levels
[1679]J., J. M., 1941, How the grottoes of an ancient church were discovered in the convent of the "Dames de Nazareth" at Nazareth in Galilee
[1688]Jacobsen, T. W., 1987, Excavations at Franchthi Cave, Greece
[1689]Jacobsen, T. W., 1987, Franchthi Cave and Paralia: maps, plans, and sections
[1705]Janetski, Joel C., 1983, An archaeological and geological assessment of Antelope Cave (NA 5507), Mohave County, Northwestern Arizona
[1718]Jennings, Jesse David, 1909–, 1980, Cowboy Cave
[1719]Jennings, Jesse David, 1909–, Oct 1957, Danger Cave
[1720]Jennings, Jesse David, 1909–, Sep 1975, Excavation of Cowboy Caves, June 3-July 26, 1975: preliminary report
[1721]Jennings, Jesse David, 1909–, 1951, Mogollone material, Tularosa Cave, New Mexico
[1722]Jennings, Jesse David, 1909–, Oct 1957, Shaw Cave, Wyoming
[1732]Jewish Museum (New York, N.Y.), 1967, Masada and the finds from the Bar-Kokhba caves: struggle for freedom
[1738]Johnson, Elden, 1956, The Lee Mill Cave
[1751]Jones, Thomas Rupert, 1819–1911, 1877, Lecture on the antiquity of man: illustrated by the contents of caves and relics of cave-folk
[1785]Keith, Arthur, 1866–1955, 1936, History from caves: a new theory of the origin of modern races of mankind, by Arthur Keith, being the presidential address given at Buxton, Derbyshire to the first speleological conference
[1821]King, Mary Elizabeth, 1974, The Salts Cave textiles: a preliminary account
[1823]Kirk, Ruth, 1970, The oldest man in America: an adventure in archeology
[1844]Kohary, Jana, 1988, Archaeology and ecology of Rose Cottage Cave, Orange Free State
[1856]Kozlowski, Janusz Krzysztof, 1982, Excavation in the Bacho Kiro Cave (Bulgaria): final report
[1857]Kozlowski, Janusz Krzysztof, 1994, Meso and Neolithic sequence from the Odmut Cave (Montenegro)
[1858]Kozlowski, Janusz Krzysztof, 1992, Temnata cave: excavations in Karlukovo karst area Bulgaria
[1868]Kühn, Herbert, 1895–, 1932, Symposium on cave sites and explorations in Upper Rhine regions
[1883]Lambert, Marjorie F., 1961, A survey and excavation of caves in Hidalgo County, New Mexico
[1892]Lan-po Chia, 1975, The cave home of Peking Man
[1893]Lan-po Chia, 1990, The story of Peking Man, from archaeology to mystery
[1901]Larsen, Helge, 1968, Trail Creek: final report on the excavation of two caves on Seward Peninsula, Alaska
[1916]Lawson, Andrew J., , Cave art
[1923]Lee, G., 1988, The rock art of Petroglyph Point and Fern Cave, Lava Beds National Monument: final report
[1925]Lee, Thomas A., 1988, San Pablo Cave and El Cayo

on the Usumacinta River, Chiapas, Mexico

[1954]Lockett, H. Claiborne, 1953, Woodchuck Cave: A basketmaker II site in Tsegi Canyon, Arizona

[1955]Logan, Wilfred David, 1923–, 1952, Graham Cave, an archaic site in Montgomery County, Missouri

[1965]Lothson, Gordon A., 1939–, 1980, The archaeology of the Fallen Arches Cave Site, 45SA41, Borigo timber sale, Gifford Pinchot National Forest: phase I survey and reconnaissance

[1967]Loud, Llewellyn Lemont, 1879–1946, 1965, Lovelock Cave

[1968]Loud, Llewellyn Lemont, 1879–1946, Feb 1991, Lovelock Cave: The republication of the rare 1929 Nevada classic

[1999]Lynch, Thomas F., 1938–, 1980, Guitarrero Cave: early man in the Andes

[2000]Lynott, Mark J., 1986, Archeological investigations at Limekiln Cave, 23SH109, Ozark National Scenic Riverways, southeast Missouri

[2009]MacEnery, John M., 1796–1841, 1859, Cavern researches, or, discoveries of organic remains, and of British and Roman reliques, in the caves of Kent's Hole, Anstis Cove, Chudleigh, and Berry Head

[2012]Mackie, Evan W., 1981, The caves at East Wemyss, Fife: an interim report on new investigations in 1980

[2013]MacLean, John Patterson, 1848–1939, 1890, An historical, archaeological and geological examination of Fingal's Cave in the island of Staffa

[2016]Madacheen, L. Michael, 1977, New Melones cave inventory and evaluation study: preliminary report: archeological caves

[2050]Marsh, Dorothy, 1980, Life at Russell Cave

[2061]Martin, George Castor, 1885–, 1933, Archaeological exploration of the Shumla caves; report of the George C. Martin expedition... June, July and August, 1933

[2064]Martin, Paul Sidney, 1899–, 1954, Caves of the Reserve area

[2065]Martin, Paul Sidney, 1899–, 1952, Mogollon cultural continuity and change; the stratigraphic analysis of Tularosa and Cordova caves

[2069]Martinez del Rio, Pablo, 1953, A preliminary report on the mortuary cave of Candelaria, Coahuila, Mexico

[2071]Mason, Edmund John, 1911–1993, 1979, Bone Cave (Ogof-yr-Esgyrn) Dan-yr-Ogof: the history of the Bone Cave (Ogof yr Esgyrn): a guide and background to the exhibits in Bone Cave

[2074]Mason, Gregory, 1889–, 1928, Pottery and other artifacts from caves in British Honduras and Guatemala

[2083]Massey, William C., 1976, A burial cave in Baja California, the Palmer Collection 1887

[2095]Maturango Museum of Indian Wells Valley, 1974, Excavation of two sites in the Coso Mountains of Inyo County, California

[2097]Mazonowicz, Douglas, 1973, In search of cave art

[2100]Mazonowicz, Douglas, 1970, Prehistoric paintings of France and Spain: a description of 34 actual size copies hand screenprinted by Douglas Mazonowicz

[2114]McCracken, Harold, 1894–, 1978, The Mummy Cave Project in northwestern Wyoming

[2135]McLoyd & Graham, 1894, Catalogue and description of a very large collection of prehistoric relics: obtained in the cliff houses and caves of Southeastern Utah

[2143]Meleen, Elmer E., Apr 1941, A preliminary report on rock shelters in Fall River County, South Dakota

[2145]Meloy, Harold Raymond, 1913–1985, , Mummies of Mammoth Cave: an account of the Indian mummies discovered in Short Cave, Salts Cave, and Mammoth Cave, Kentucky

[2146]Meloy, Harold Raymond, 1913–1985, 1968, Mummies of Mammoth Cave: Fawn Hoof, Little Alice, Lost John and others. An account of the Indian mummies discovered in Short Cave, Salts Cave, and Mammoth Cave, Kentucky

[2147]Mera, H. P., 1938, Reconnaissance and excavation in southeastern New Mexico

[2153]Mercer, Henry Chapman, 1856–1930, 1896, The hill-caves of Yucatan, a search for evidence of man's antiquity in the caverns of Central America. Being an account of the Corwith expedition of the Department of Archaeology and Palaeontology of the University of Pennsylvania

[2154]Mercer, Henry Chapman, 1856–1930, 1975, The hill-caves of Yucatan, a search for evidence of man's antiquity in the caverns of Central America. Being an account of the Corwith expedition of the Department of Archaeology and Paleontology of the University of Pennsylvania

[2159]Merk, Conrad, 1876, Excavations at the Kesserloch near Kesserloch near Thayngen Switzerland: a cave of the Reindeer Period

[2161]Meshram, Pradip Shaligram, 1954–, 1991, Early caves of Maharashtra: a cultural study

[2194]Miss, Christian J., nd, Archaeological evaluations of the Riparia (45WT1) and Ash Cave (45WW61) sites on the Lower Snake River

[2198]Missouri Speleological Survey, Dec 1977, Archaeological investigations: cave explorations in the Ozark region of central Missouri

[2270]Morgan, Llewellyn E. (auth, photo), nd, Guide to Dan-yr-Ogof Caves: Swansea Valley Caves

[2292]Movius, Hallam Leonard, 1907–, 1953, The Mousterian cave of Teshik-Tash, Southeastern Uzbekistan, Central Asia

[2294]Mturi, A. A., 1975, A guide to Tongoni ruins: with notes on other antiquities in Tanga and Pangani districts including the Amboni Caves: pamoja na maelezo kwa Kiswahili

[2300]Munson, Patrick J., 1990, The prehistoric and early historic archaeology of Wyandotte Cave and other caves in southern Indiana

2.1. GENERAL

[2489]Nelson, Nels Christian, 1875–, 1917, Contributions to the archaeology of Mammoth Cave and vicinity, Kentucky

[2535]Nusbaum, Jesse Logan, 1981, The 1926 re-excavation of Step House Cave, Mesa Verde National Park

[2536]Nusbaum, Jesse Logan, 1985, A basket-maker cave in Kane County, Utah

[2539]Oakley, Carey B., 1990, Archaeological testing at site 1Ms357, Cathedral Caverns, Marshall County, Alabama

[2541]Obermaier, Hugo, 1877–1946, 1930, Bushman art; rock paintings of south-west Africa, based on the photographic material collected by Reinhard Maack

[2573]Orchard, William C., 1920, Sandals and other fabrics from Kentucky caves

[2576]Orr, Phil C., 1952, Excavations in Moaning Cave

[2578]Orr, Phil C., 1952, Preliminary excavations of Pershing County caves

[2595]Palacio, Joseph O., 1977, Excavation at Hokeb Ha, Belize

[2620]Payen, Louis A., 1963, Preliminary report on the archeological investigation of Pinnacle Point Cave, Toulumne [sic] County, California

[2622]Peabody, Charles, 1867–1939, 1904, The exploration of Jacobs Cavern, McDonald County, Missouri

[2623]Peabody, Charles, 1867–1939, 1908, Pt. I: the exploration of Bushey Cavern near Cavetown, Maryland

[2636]Pemberton, Clive, 1972, Kents Caverns, home of prehistoric man and animals: the origin, story, and descriptive tour of the caves

[2637]Pemberton, Clive, 1947, The origin and story of Kents Cavern

[2638]Pemberton, Clive, 1964, The origin, story and descriptive tour of the caves

[2640]Pendergast, David M., 1971, Excavations at Eduardo Quiroz Cave, British Honduras (Belize)

[2641]Pendergast, David M., 1970, A. H. Anderson's excavations at Rio Frio Cave, British Honduras (Belize)

[2642]Pendergast, David M., Feb 1974, Excavations of Actum Polbilche, Belize

[2643]Pengelly, Hester (ed), 1897, A memoir of William Pengelly, of Torquay, F.R.S., Geologist, with a selection from his correspondence

[2649]Pennington, Rooke, 1877, Notes on the barrows and bone-caves of Derbyshire with an account of a descent into Elden Hole

[2663]Peyrony, E., 1959, Les Eyzies and the Vezere Valley; an illustrated guide for scholars and tourists

[2683]Plew, Mark G., 1986, The archaeology of Nahas Cave: material culture and chronology

[2688]Pollard, Anthony James, 1941–, 1992, Smoo Cave

[2698]Poulianos, Aris N., 1982, The cave of the Petralonian archanthropinae: a guide to the science behind the excavations

[2707]Prentice, Guy, Jan 1988, Mammoth Cave archeological inventory project interim report - 1987 investigations

[2708]Prentice, Guy, 1989, Mammoth Cave National Park archeological inventory project interim report - 1988 investigations

[2728]Prufer, Olaf H., 1989, Krill Cave: a stratified rockshelter in Summit County, Ohio

[2734]Putnam, Frederic Ward, 1839–1915, 1875, Archaeological researches in Kentucky and Indiana, 1874

[2761]Raphael, Max, 1889–1952, 1945, Prehistoric cave paintings

[2791]Reddell, James Russell, 1938–, 1977, A preliminary survey of the caves of the Yucatan Peninsula

[2798]Reed, Alan D., 1988, A stabilization assessment of Mantel's Cave, site 5MF1, Dinosaur National Monument, Colorado

[2817]Riddell, Francis A., 1951, The archaeology of two Kern County sites

[2821]Ringer, Árpád, 1989, Prehistoric remains in Hungary

[2836]Rodda, Jan, 1989, Mulka's Cave site management project: emphasising visitor survey, April-June 1988, with management evaluation and further recommendations for management; with conclusions of an archaeological test excavation

[2853]Ronquillo, Wilfredo P., Dec 1981, The technological and functional analyses of lithic flake tools from Rabel Cave, northern Luzon, Philippines

[2856]Ross, Richard E., Oct 1965, The archeology of Eagle Cave

[2861]Rouse, Irving, 1913–, 1990, Excavations at Maria de la Cruz Cave and Hacienda Grande Village Site, Loiza, Puerto Rico

[2864]Rozarie, Charles, 1964, The Archaeology at Lehman Caves National Monument: Nevada State Museum report

[2867]Ruspoli, Mario, 1987, The cave of Lascaux: the final photographs

[2885]Sarawak Museum, 1973, Summary of archaeological work in Sarawak: with special reference to Niah Caves

[2894]Sawtell, Ruth Otis, 1927, Primitive hearths in the Pyrenees; the story of a summer's exploration in the haunts of prehistoric man

[2911]Schroeder, Albert H., 1978, Pratt Cave studies: Guadalupe National Park, Texas

[2912]Schuetz, Mardith K., 1956, An analysis of Val Verde County cave material

[2918]Schwartz, Douglas W., 1965, Prehistoric man in Mammoth Cave

[2927]Seger, Joe D., 1988, Gezer V: the field I caves

[2932]Sellers, O. R., 1953, A Roman-Byzantine burial cave in Northern Palestine (the joint excavation of the American School of Oriental Research in Jerusalem and

McCormick Theological Seminary at Silet edhDhahr

[2941]Shafer, Harry J., 1977, Archeological and botanical studies at Hinds Cave, Val Verde County, Texas

[2942]Shafer, Harry J., 1975, A preliminary report of Hinds Cave, Val Verde County, Texas

[2970]Shippee, J. M., Dec 1966, The archaeology of Arnold Research Cave, Callaway County, Missouri

[2974]Shutler, Mary Elizabeth (ed), Oct 1963, Deer Creek cave, Elko County, Nevada

[2977]Sieveking, Ann, 1962, The caves of France and northern Spain

[2978]Sieveking, Ann, 1966, The caves of France and northern Spain, a guide

[2989]Sinclair, William John, 1877–1935, 1904, Euceratherium: a new ungulate from the Quaternary caves of California

[2997]Skjolsvold, Arne, Feb 1972, Excavations of Habitation Cave

[3002]Sloan, Tacoma G., 1960, Archaeological survey of Mammoth Cave National Park

[3015]Smith, Christopher, 1989, Mid Argyll cave and rock shelter survey

[3017]Smith, Elmer Richard, 1909–, 1952, The archaeology of Deadman Cave, Utah: a revision

[3018]Smith, Gerald Arthur, 1915–, 1957, The archaeology of Newberry Cave, San Bernardino County, Newberry, California

[3019]Smith, Gerald Arthur, 1915–, 1955, Preliminary report of the Schuiling Cave, Newberry, California

[3041]Snodgrasse, Richard Montgomery, 1958, The human skeletal remains from Pictograph and Ghost caves, Montana

[3046]Solecki, Ralph S., 1917–, 1972, Shanidar: the first Flower people

[3047]Sollas, William Johnson, 1849–1936, 1913, Paviland Cave: an Aurignacian station in Wales (The Huxley memorial lecture for 1913)

[3112]Steen, Charlie R., 1942, Ruins stabilization records for Canyon de Chelly National Monument 1942: Antelope House and Mummy Cave Ruins

[3114]Stein, Aurel, 1862–1943, 1980, Serindia: detailed report of explorations in Central Asia and westernmost China

[3115]Stekelis, Moshe, 1898–1967, 1952, The Abu Usba Cave (Mount Carmel)

[3125]Steward, Julian Haynes, 1902–1972, 1937, Ancient caves of the Great Salt Lake region

[3127]Stewart, Thomas Dale, 1901–, Apr 1977, The Neanderthal skeletal remains from Shanidar Cave, Iraq: a summary of findings to date

[3133]Stone, Andrea Joyce, 1995, Images from the underworld: Naj Tunich and the tradition of Maya cave painting

[3155]Studley, Cordelia A., 1884, Notes upon the human remains from the caves of Coahuila, Mexico

[3160]Surin, Phukhachon, 1991, Preliminary report of excavations at Moh-Khiew Cave, Krabi Province, Sakai Cave, Trang Province, and ethnoarchaeological research of hunter-gatherer group, socall[ed] Sakai or Semang at Trang Province: the Hoabinnian Research Project in Thailand

[3161]Sutcliffe, Antony John, c1942, Joint Mitnor Caves: Buckfastleigh

[3197]Tennessee Academy of Science, Jul 1930, Journal of the Tennessee Academy of Science. Cave number

[3200]Thomas, David Hurst, Jun 1985, The archaeology of Hidden Cave, Nevada

[3202]Thompson, Edward Herbert, 1860–1935, 1897, Cave of Loltun, Yucatan: report of explorations by the Museum, 1888-89 and 1890-91

[3220]Toll, Henry Wolcott, 1995, An analysis of variability and condition of cavate structures in Bandelier National Monument

[3228]Treganza, Adan Eduardo, 1916–1968, 1964, An ethno-archaeological examination of Samwel Cave

[3239]Trinkaus, Erik, 1985, The Shanidar Neandertals

[3240]Trinkaus, Erik, 1983, The Shanidar Neandertals

[3256]Turpin, Sloveig A. (comp), 1985, Seminole Sink (41VV620): excavation of a vertical shaft tomb, Val Verde County, Texas

[3303]U.S. National Park Service, 1978, Mammoth Cave National Park: collection management plan

[3331]University of California Archaeological Research Facility, Department of Anthropology, 1968, Papers on Great Basin prehistory

[3332]University of California Archaeological Survey, 1952, Papers in California archaeology: 17-18

[3333]University of California Archaeological Survey, 1955, Papers on California archaeology: 30-31

[3334]University of California Archaeological Survey, 1955, Papers on California archaeology: 32-33

[3358]Verneau, Rene, 1852–1938, 1908, The men of the Barma-grande (Baousse-Rousse) An account of the objects collected in the Museum praehistoricum, founded by Commendatore Th. Hanbury near Mentone

[3362]Vietzen, Raymond Charles, 1907–, , The saga of Glover's Cave

[3369]Vishoek (South Africa), 1949, The Peers' cave, tunnel cave and rock shelters at Skildergat, Fish Hoek, the home of pre-historic man

[3392]Waltham, Anthony Clive, 1942–[Tony] (ed), 1974, Limestone and caves of Northwest England

[3396]Ward, John A., 1990, The people of Burrows Cave: who they were, where they came from and when

[3407]Watson, Patty Jo, 1932–, 1974, Archeology of the Mammoth Cave Area

[3408]Watson, Patty Jo, 1932–, 1969, The prehistory of Salts Cave: Kentucky

[3417]Watters, David Robert, Dec 1987, Final report on

the archaeology of Fountain Cavern, Anguilla, West Indies: a report
[3438]Wendy, Herbert, 1956, In search of Adam: the story of mans quest for the truth about his earliest ancestors
[3447]West Texas Historical and Scientific Society, 1932, Publication no. 4
[3507]Wilson, George Herbert, 1874?–1958, nd, Cave hunting holidays in Peakland
[3535]Wright, R. V. S., 1971, Archaeology of the Gallus Site, Koonalda Cave
[3553]Zeoderborg, Harry, nd, The phantoms of stork fontein [sic]
[3557]Zingg, Robert Mowry, 1900–1957, 1940, Report on archaeology of southern Chihuahua

2.2 Anthropology

[322]Baring-Gould, Sabine, 1834–1924, 1968, Cliff castles and cave dwellings of Europe
[384]Benn, David W., 1948–, 1980, Hadfields Cave: a perspective on Late Woodland culture in northeastern Iowa
[463]Branigan, Keith, 1992, Romano-British cavemen: cave use in Roman Britain
[539]Bruun, Daniel, 1985, The cave dwellers of Southern Tunisia: recollections of a sojourn with the Khalifa of Matmata
[583]Callard, Thomas Karr, [1880], Contemporaneity of man with the extinct Mammalia, as taught by recent cavern exploration, and its bearing upon the question of man's antiquity
[999]Epstein, Sam, 1909–, 1959, All about prehistoric cave men
[1049]Ferguson, Samuel, 1810–1886, 1865, Account of Ogham inscriptions in the cave at Rathcroghan, Co. Roscommon
[1050]Fernandez, Carlos A., 1972, The Tasaday: cave-dwelling food gatherers of South Cotabato, Mindanao: preliminary report submitted to the Panamin Foundation, Inc. on June 1, 1972
[1054]Fewkes, Jesse Walter, 1850-1930, 1911, The cave dwellings of the Old and New Worlds
[1161]Garrigan, George A., 1994, Skyline Caverns and its geologic relationship to the Shenandoah Valley and paleoindian cultures
[1171]Geikie, James, 1839-1915, 1914, The antiquity of man in Europe, being the Munro lectures, 1913
[1358]Hardingham, B. G., 1943, Living in caves
[1542]Howard, John Eliot, 1879, The caves of South Devon and their teachings
[1571]Humphrey, Richard V., 1989, Mystery Hill: myth and mythology in the land of academe
[1790]Kempe, David Ronald Charles, 1927–, 1988, Living underground: a history of cave and cliff dwelling
[1874]Kunsky, Josef, 1903–1977, 1954, Homes of primeval man: wandering in the caves of Czechoslovakia
[2676]Pinkham, Mary R., 1954, From the cradle to the cave: the life story of "Dad" Truitt, "Cave Man of the Ozarks"
[2704]Powers, Richard M., 1963, The cave dwellers in the old stone age
[2899]Scheele, William E., 1959, The cave hunters
[3556]Zim, Herbert Spencer, 1909–, 1978, Caves and life

2.2.1 Art

[63]Anati, Emmanuel, 1993, Rock art: the state of research in rock art - 1993: archetypes, constants and universal paradigms
[90]Anon, 1988, Cave paintings of the Chumash Indians
[128]Anon, 1935, A guide to the Cave of Altamira and the town of Santillana del Mar (Province of Santander, Spain)
[176]Anon, 1984, Prehistoric painting, part 2
[247]Aubarbier, Jean-Luc, Jun 1985, Prehistoric sites in Perigord
[249]Aujoulat, Norbert, 1989, Lascaux revisited
[274]Bahn, Paul G., 1988, Images of the ice age
[286]Baker, Janet, 1994, A brief history of the Dunhuang Caves: a millennium of Chinese Buddhist art
[312]Bandi, Hans-Georg (ed), 1961, The art of the stone age: forty thousand years of rock art
[342]Bataille, Georges, 1897–1962, 1955, Prehistoric painting: Lascaux, or, the birth of art
[347]Baumann, Hans, 1914–, 1962, The caves of the great hunters
[389]Bent, James Theodore, 1852-1897, 1892, The ruined cities of Mashonaland; being a record of excavation and exploration in 1891
[391]Berenguer, Magin, 1994, Prehistoric cave art in northern Spain, Asturias
[392]Berenguer, Magin, 1973, Prehistoric man and his art: the caves of Ribadesella
[410]Black, Lindsay, 1943, Aboriginal art galleries of western New South Wales
[411]Black, Nancy Gail, 1965, A comparative study of stylistic and extrapictorial elements in the murals of Cave I, Ajanta and of the Brancacci Chapel
[471]Breuil, Abbé Henri, 1877–1961, 1954, Cave drawings: an exhibition of drawings by the Abbé Breuil of palaeolithic paintings and engravings
[472]Breuil, Abbé Henri, 1877–1961, 1955, The rock paintings of southern Africa
[494]Broderick, Alan Houghton, 1963, Abbé Breuil: prehistorian: a biography
[495]Broderick, Alan Houghton, 1963, Father of prehistory: the Abbé Henri Breuil: his life and times
[496]Broderick, Alan Houghton, 1949, Lascaux: a com-

mentary

[497]Broderick, Alan Houghton, nd, Prehistoric painting: with 56 plates in colour and monochrome and 7 line illustrations in the text

[523]Brown, Gerard Baldwin, 1849–1932, 1928, Art of the cave dweller—a study of the earliest artistic activities of man

[524]Brown, Gerard Baldwin, 1849–1932, 1932, The art of the cave dweller: a study of the earliest artistic activities of man

[598]Carballo, Jesus Maria, 1965, The cave of Altamira and other caves with paintings in the province of Santander

[611]Casteret, Norbert, 1897–1987, 1940, Ten years under the earth

[642]Chauvet, Jean-Marie, 1996, Chauvet Cave: the discovery of the world's oldest cave paintings

[643]Chauvet, Jean-Marie, 1996, Dawn of art: the Chauvet Cave: the oldest known paintings in the world

[667]Clottes, Jean, 1996, The cave beneath the sea: paleolithic images at Cosquer

[734]Crain, Sally Lucille 1938–(auth, prod), 1979, Prehistoric magic

[758]Crosby, Harry, 1926–, 1976, Cave paintings of Baja

[759]Crosby, Harry, 1926–, 1984, The cave paintings of Baja California

[760]Crosby, Harry, 1926–, 1975, The cave paintings of Baja California: the great murals of an unknown people

[788]Cuvay, Roxane, 1963, Cave painting

[789]Cuvay, Roxane, 1963, Prehistoric cave painting

[791]D'Arcy Galleries (New York, N.Y.), 1968, The Caves of Karawari

[800]Daniel, Glyn Edmund, 1955, Lascaux and Carnac

[910]Dulluc, Brigitte, 1990, Discovering Lascaux

[965]Ellis, Arthur Charles, 1930, An historical survey of Torquay from the earliest times, as illustrated by finds in Kent's Cavern, down to the present time

[1000]Epton, Nina Consuelo, 1955, The Valley of Pyrene

[1034]Faulkner, Charles H. (ed), 1986, The prehistoric native American art of Mud Glyph Cave

[1070]Flood, Josephine, 1983, Archaeology of the dreamtime

[1145]García Guinea, Miguel Angel, 1975, Altamira and prehistoric art in the caves of Santander

[1146]García Guinea, Miguel Angel, Mar 1969, Altamira, the beginning of art

[1147]García Guinea, Miguel Angel, 1971, Altamira: the origin of art

[1148]García Guinea, Miguel Angel, 1984, Santillana and Altamira

[1150]Gardner, Erle Stanley, 1962, The hidden heart of Baja [Mexico]

[1158]Garlake, Peter S., 1987, The painted caves: an introduction to the prehistoric art of Zimbabwe

[1171]Geikie, James, 1839-1915, 1914, The antiquity of man in Europe, being the Munro lectures, 1913

[1186]Ghosh, Leila, 1986, Ajanta and Ellora

[1213]Gimenez Reyna, Simeon, 1965, The cave of "La Pileta": Benaojan, Malaga, Spain

[1220]Godden, Elaine, 1988, Rock paintings of Aboriginal Australia

[1222]Goodall, Elizabeth, 1959, Prehistoric rock art of the Federation of Rhodesia & Nyasaland

[1243]Grattan-Smith, Thomas Edward, 1938, The cave of a thousand columns

[1265]Grigson, Geoffrey, 1905–, 1957, Painted caves

[1273]Grove, David C., 1970, The Olmec paintings of Oxtotitlan Cave Guerrero, Mexico

[1312]Hadingham, Evan, 1979, Secrets of the Ice Age: the world of the cave artists

[1514]Hooks, William H., 1977, Maria's Cave

[1636]International Film Bureau, 1976, Lascaux, cradle of man's art

[1662][International Union of Speleology], 1989, Caves in fine art exhibition

[1867]Kühn, Herbert, 1895–, 1956, The rock pictures of Europe

[1884]Laming-Emperaire, Annette, 1917–1977, 1950, The cave of Lascaux

[1885]Laming-Emperaire, Annette, 1917–1977, 1959, Lascaux: paintings and engravings

[1916]Lawson, Andrew J., , Cave art

[1919]Leakey, Richard E., 1981, A new era

[1934]Leroi-Gourhan, André, 1911–1986, 1967, Treasures of prehistoric art

[1966]Loubser, J. H. N., Oct 1993, A guide to the rock paintings of Tandjesberg

[2045]Marcus, Rebecca Brian, 1968, Prehistoric cave paintings

[2084]Mathpal, Yashodhar, 1984, Prehistoric rock paintings of Bhimbetka, Central India

[2096]Mazonowicz, Douglas, 1984, Cave art of France and Spain

[2097]Mazonowicz, Douglas, 1973, In search of cave art

[2098]Mazonowicz, Douglas, 1989, On the rocks: prehistoric art of France and Spain

[2099]Mazonowicz, Douglas, 1979, The painted caves of France & Spain

[2100]Mazonowicz, Douglas, 1970, Prehistoric paintings of France and Spain: a description of 34 actual size copies hand screenprinted by Douglas Mazonowicz

[2101]Mazonowicz, Douglas, 1966, Prehistoric paintings: a catalog of actual size copies in silkscreen

[2102]Mazonowicz, Douglas, 1970, The prehistoric rock paintings of Tassili n'Ajjer: a description of 15 actual size copies, hand screenpainted

[2103]Mazonowicz, Douglas, 1974, Voices from the stone age: a search for cave and canyon art

[2106] McCarthy, Frederick D., 1905–, 1967, Australian aboriginal rock art
[2107] McCarthy, Frederick D., 1905–, 1958, Australian aboriginal rock art
[2177] Miles, Sibella Elizabeth, 1864, The Grotto of Neptune ("Antro Di Nettuno"), Sardinia; a poem illustrative of three views of this interesting cavern, taken in July 1824 by the late Commander Alfred Miles and dedicated to his memory by his widow, Sibella Elizabeth Miles
[2285] Moulin, Raoul-Jean, 1966, Prehistoric painting
[2333] Myron, Robert, 1964, Prehistoric art
[2529] Nougier, Louis René, 1912–, 1958, The Cave of Rouffignac
[2530] Nougier, Louis-René, 1912–, 1961, Art treasures of prehistoric man in the caves of France and Spain
[2541] Obermaier, Hugo, 1877–1946, 1930, Bushman art; rock paintings of south-west Africa, based on the photographic material collected by Reinhard Maack
[2542] Obermaier, Hugo, 1877–1946, 1928, The caves of Altamira
[2653] Pericot Garcia, Luis, 1899–(ed), 1964, Prehistoric art of the western Mediterranean and the Sahara
[2761] Raphael, Max, 1889–1952, 1945, Prehistoric cave paintings
[2818] Riddell, William Hatton, 1938, Altamira: a note upon the Palaeolithic paintings in the Cave of Altamira near Santillane del Mar in the Spanish province of Santander
[2822] Ripoll Perello, Eduardo, 1980, The Cave of Las Monedas in Puente Viesgo (Santander)
[2823] Ripoll Perello, Eduardo, 1968, The painted shelters of La Gasulla (Castellon)
[2837] Rodriguez, Ortega Eduardo (auth, photo), 1970, The cave of Nerja
[2867] Ruspoli, Mario, 1987, The cave of Lascaux: the final photographs
[2883] Sandak, Inc., 1973, Cave paintings—Lascaux and Altamira
[2907] Schoon, Theo, 1985, Maori rock drawing: the Theo Schoon interpretations
[2952] Shaw, Trevor Royle, 1928–(comp), 1967, Cave illustrations before 1900: a catalogue of non-photographic illustrations of caves
[2976] Sieveking, Ann, 1979, The cave artists
[2977] Sieveking, Ann, 1962, The caves of France and northern Spain
[2978] Sieveking, Ann, 1966, The caves of France and northern Spain, a guide
[2985] Simpson, James Young, 1811–1870, 1867, Account of some ancient sculptures on the walls of caves in Fife
[3095] Spink, Walter M., 1990, Ajanta: a brief history and guide
[3121] Stern, Philip Van Doren, 1900–, 1973, The beginnings of art
[3133] Stone, Andrea Joyce, 1995, Images from the underworld: Naj Tunich and the tradition of Maya cave painting
[3177] Taketaro Shinkai, 1868–1927, 1921, Rock-carvings from the Yun-kang caves
[3181] Taralon, Jean, 1962, The grotto of Lascaux
[3230] Trezise, P. J., 1971, Rock art of South-east Cape York
[3326] Ucko, Peter J., 1967, Palaeolithic cave art
[3335] University of California Santa Barbara, Art Galleries, 1965, Prehistoric rock art of the Santa Barbara region: [exhibition] Art Gallery, University of California, Santa Barbara, October 12-Nov. 7, 1965
[3448] Westall, William, 1818, Views of the caves near Ingleton, Gordale Scar and Malham Cove in Yorkshire
[3479] Whitfield, Roderick, 1995, Dunhuang: caves of the singing sands: Buddhist art from the Silk Road
[3514] Windels, Fernand, 1949, The Lascaux Cave paintings

2.2.2 Rock Shelters

[16] Adovasio, J. M., 1980, Archaeological testing at two rockshelters in the Tombigbee River Multi-Resource District, Alabama and Mississippi: an interim report
[24] Alex, Lynn Marie, 1948–, 1991, The archaeology of the Beaver Creek shelter (39CU779), Wind Cave National Park, South Dakota: a preliminary statement
[277] Bailey, John H., 1940, A stratified rock shelter in Vermont
[294] Balch, Herbert Ernest, [1927], Excavations at Chelm's Combe: Cheddar: conducted under the Excavations Committee of the Somerset Archaeological and Natural History Society 1925-26
[296] Balch, Herbert Ernest, 1937, Mendip, its swallet caves and rock shelters
[319] Bard, James C., 1979, Ezra's Retreat: a rockshelter/cave occupation site in the north central Great Basin
[870] Dickson, Don R., 1991, The Albertson site: a deeply and clearly stratified Ozark bluff shelter
[1025] Fabun, Don, 1970, Shelter: the cave re-examined
[1053] Fewkes, Jesse Walter, 1850–1930, 1896, Preliminary account of an expedition to the cliff villages of the red rock country, and the Tusayan ruins of Sityatki and Awatobi, Arizona, in 1895
[1055] Field, Henry, 1902–, 1955, Caves and rockshelters in northern Iraq.
[1169] Gebauer, Herbert Daniel, 1985, Kurnool 1984: report of the speleological expedition to the district of Kurnool, Andhra Pradesh, India
[1220] Godden, Elaine, 1988, Rock paintings of Aboriginal Australia
[1246] Grayson, Donald K., May 1988, Danger Cave, Last Supper Cave, and Hanging Rock Shelter: the faunas
[1257] Gregory, John Walter, 1864–1932, 1930, Some

caves and a rock shelter at Loch Ryan and Portpatrick, Galloway

[1278]Guernsey, Samuel James, 1868–1936, 1931, Explorations in northeastern Arizona; report on the archaeological fieldwork of 1920-1923

[1790]Kempe, David Ronald Charles, 1927–, 1988, Living underground: a history of cave and cliff dwelling

[1913]Laville, Henri, 1980, Rock shelters of the Perigord: geological stratigraphy and archaeological succession

[2143]Meleen, Elmer E., Apr 1941, A preliminary report on rock shelters in Fall River County, South Dakota

[2530]Nougier, Louis-René, 1912–, 1961 , Art treasures of prehistoric man in the caves of France and Spain

[2728]Prufer, Olaf H., 1989, Krill Cave: a stratified rock-shelter in Summit County, Ohio

[3015]Smith, Christopher, 1989, Mid Argyll cave and rock shelter survey

[3220]Toll, Henry Wolcott, 1995, An analysis of variability and condition of cavate structures in Bandelier National Monument

[3230]Trezise, P. J., 1971, Rock art of South-east Cape York

[3332]University of California Archaeological Survey, 1952, Papers in California archaeology: 17-18

[3369]Vishoek (South Africa), 1949, The Peers' cave, tunnel cave and rock shelters at Skildergat, Fish Hoek, the home of pre-historic man

3 Biology

Kathleen H. Lavoie

College of Arts and Sciences
101 Hudson Hall
State University of New York-Plattsburgh
Plattsburgh, NY 12901

Biospeleology is the study of life found in caves. The earliest writings we have about visits to caves express wonder and surprise at the often bizarre and unique animals found there. The science of studying animals in caves is, in many ways, still in the early stages despite a long history. This can be attributed to the difficulty of working in caves, problems with getting to the often remote areas where animals are found, the very low numbers and often extreme rarity of cave animals, and the difficult of rearing many species outside the cave environment. Most cave species have been studied only anecdotally or descriptively. Others, such as bats, cave fish, and cave crickets, have received considerably more study.

There were isolated reports of cave fauna in the 1550's. In 1665 Kircher reported on the presence of cave fish in an intermittent lake. Valvasor, in 1689, reported on the Ohm, or *Proteus* from Lintvern karst springs in Slovenia. Invertebrates were first collected and described beginning in 1801, with various proposals to classify them according to where in the cave they were located and later by what we now describe as troglomorphic characteristics. Bedel and Simon cataloged all known species in 1875. The first laboratory for the study of the effects of light and other variables on the evolution of cave animals was established by Viré in the catacombs of Paris in 1897. British fauna are relatively scarce and were little known until the 19th century. Work in the United States was intermittent beginning in 1840 with a report on the Mammoth Cave blindfish by Davidson (1840). Pollution in karst regions was recognized as being important as early as 1892 by Martel (Roberts, 1947).

The establishment of modern biospeleology started in 1907 with the publication of an important essay by Racovitza. Racovitza reviewed the status of knowledge at the time and discussed the main unsolved problems in biospeleology, which Culver, Kane and Wong (1995) describe as still reading "like a research program for the study of subterranean life".

Most of the work indexed here is concerned with the distribution of organisms in particular caves or regions ranging from very small to very large. Of particular note are reports from various caving groups and from expeditions, as well as individuals. A classic is Banta's 1907 *The fauna of Mayfield's Cave. Irish hypogean fauna and Irish biological records, 1956-1971*, published by the British Cave Research Association. Results of 1969 through 1975 expeditions to Mexico are published by the National Academy of Lincei in Italy (Anon 1972, 1973, 1977). Barr (1961) published the *Caves of Tennessee*, which is both a guidebook to recreational caves in Tennessee, and contains distributional information in cave fauna in the area. Holsinger and Culver (1988) review the zoogeography and ecology of the invertebrate cave fauna of Virginia and parts of Eastern Tennessee. Botosaneanu (1986) edited the ambitious *Stygofauna mundi*, which purports to be a faunistic, distributional and ecological synthesis of the world fauna inhabiting subterranean waters (including the marine interstitial). Reddell (1965-1981) has published many checklists and annotated bibliographies of the cave fauna of Texas and Mexico, as well as his *Studies on the Cave and Endogean Fauna of North America* (Volumes 1 and 2). Gagne (1974) has a series on the cavernicolous fauna of Hawaiian lava tubes. Blatchley (1896) published Indiana caves and their fauna, which have expanded with the recent publication (1990) of Blatchley's *Gleanings From Nature: Ten Indiana Caves and the Animals That Inhabit Them*, Abeille de Perrin (1877) published *On the Collection of Cavern Insects*.

Related to a basic understanding of the natural history of organisms are works done to classify and characterize particular organisms or a group of organisms. H.H. Hobbs the II and III have widely published (1972-1977) on troglobitic decapods. Holsinger (1970-1986) has published many works cited here in the systematics, speciation and distribution of troglobitic amphipods, and on the Mammoth Cave shrimp. Various spiders, and scorpions are reported on by Briggs (1972, 1974). Mitchell (Mitchell and Reddell 1971, 1973; Mitchell, Russell, and Elliott 1977) has the definitive series on eyeless Mexican Characin fishes. Poulson (1975) edited the very significant *Symposium on Life Histories of Cave Beetles*.

There are also many works with a focus on conservation. Cooper (1982) has published on a recovery plan for the Alabama cave fish. Elliott (1978) reported on the results of a transplant experiment with the New Melones Cave harvestman. The Oklahoma Department of Wildlife Conservation (1990, 1991) has evaluated the hydrogeology of cave fish caves in the Ozarks. Gardner (1984) published a useful introduction to the inventory and evaluation of biological cave resources through the Missouri Department of Conservation. The literature in this work includes several significant books published with the goal of educating and informing the public, and introducing and exciting new scientists to the study of speleology.

Of particular note is *Speleology: The Study of Caves*, by Moore and Sullivan, first published in 1964, updated in 1978, and revised as *Speleology: Caves and the Cave*

Environment (1997). This book is widely available to the general public, and is probably the publication cited here with the widest distribution and the greatest impact. Now hard to come by, the 1966 McGraw-Hill book *The Life of the Cave*, by Mohr and Poulson, is beautifully illustrated with outstanding photographs and sketches. Many works for children also fall in this category, such as Bronins (1972) *The Cave: What Lives There*.

Many general works have been published both to serve as a basis for scientific review, and generally to inform. Some focus on biospeleology, others review the entire field of speleology. The most influential of these up to the 1980's is *Biospeleology: the Biology of Cavernicolous Animals* by Vandel (1965). Vandel reviewed biospeleology world-wide, with an emphasis on Europe, where most studies were done until modern cave science began in the United States in the 1950's. Many of Vandel's views on evolution have since been refuted, but the basic information in the text remains sound. Other works with a similar focus are Ford and Cullingford (1976), *The Science of Speleology*; Mohr and Gurnee (1965), *Cave Life*; and Chapman (1993), *Caves and Cave Life*. Juberthie is editing the ambitious *Encyclopaedia Biospeologica*, and published Tome 1 in 1994.

A recent book of monographs which is useful to both beginners and experts is *The Natural History of Biospeleology*, edited by A.I. Camacho (1992). Camacho recounts that the goal of the project was to recover the past and project it into the future. There is an introductory chapter which reviews the history of biospeleology, then the book is divided into six parts with anywhere from one to eight chapters per part. Part one deals with the subterranean environment from both a physical and biological perspective. Part two covers sampling methods, with an emphasis on aquatic environments. Part three covers troglomorphy and fossils. Part four is the most extensive, dealing with biological processes. This part includes adaptations, behavior, natural selection and evolution. Conservation issues are the focus of Part five. The final Part six is a more in-depth review of specific study cases from the Canary Islands, Mammoth Cave, anchialine environments and the chemoautotrophically-based Movile Cave in Romania.

A major index term used in this guide are studies relating to evolution and adaptation. Early researchers were fascinated by the "bizarre" life forms frequently encountered in caves, and spent a lot of effort looking for confirmation of their ideas about them. This approach is very evident in early writing, such as Packard's (1888) *The Cave Fauna of North America*. Packard frequently expresses disappointment with his results; e.g., he was not happy to find that cave crickets collected from deep inside a cave show the same eye morphology as those collected near an entrance. He invoked a complicated explanation of acceleration and retardation to explain differences in ovipositor length instead of attributing differences to a range of sizes and ages in crickets. Eigenmann (1909) published his *Cave Vertebrates of America: A Study in Degenerative Evolution*. Modern evolution studies are brought together in Barr's (1960) *Symposium: Speciation and Raciation in Cavernicoles*.

In his very important book, *Cave Life: Evolution and Ecology* (1982), Culver maintains that caves are good places to study population ecology and evolution not because they are unique, but because they are simple. They can serve as models of complicated interactions that take place in epigean environments, but without the extreme environmental conditions found in most other simplified natural ecosystems. Culver also points out that since caves are quite common, at least in some areas, there is the possibility of doing replicated or comparative studies. Some cave systems are also so extensive that multiple sites within one cave can be studied. The newest and very significant contribution to the literature is the book by Culver, Kane, and Wong (1995), *Adaptation and Natural Selection in Caves: The Evolution of Gammarus minus*. The authors of this important book argues that caves can serve as important models for the study of evolution, especially adaptation, by focusing on evolutionary trade-offs. They review early evolution theorists beginning with Lamarck and Darwin, Banta (1907), through the Neo-Larmarckists Hyatt, Cope, and Packard, and the Neo-Darwinists, Christiansen and Poulson. The modern synthesis presented in this book is that there has been an over-emphasis on loss of features, and that troglomorphy involves the evolution of both elaborated and reduced characteristics. The authors then make their point with the study of the isopod, *Gammarus minus*, as a model organism for the study of evolution. Experimental work related to the physiology of organisms or groups of organisms included here are mostly in the form of graduate student theses and dissertations.

There are several gems that are not easy to characterize or find. Hubbell (1978) has published the definitive monograph on systematics of the Hadenoecini cave crickets, which includes a wonderful paper by Hubbell and Norton, who review the reproductive behavior and ecology of *Hadenocus subterraneus*, and demonstrate that longer ovipositors are found in populations of cave crickets with the highest level of predation from cave beetles. Other studies are in human biology such as those by Siffre in the 1970's on human circadian rhythms that he conducted by isolating himself and others in caves.

Other valuable sources of information are various meeting and conference proceedings, such as those of the International Congress of Speleology (e.g. Beck, 1981), the Australian Speleological Federation, Cave Research Group of Great Britain, and annual reports, such as those of the Cave Research Foundation. The Texas Memorial Museum has published an important series of speleological monographs, including work by Elliott (1976) and Reddell (1981, 1982, 1986, 1992, 1995).

While not as extensive as the literature of cave archeology or geology, there are has several interesting and important books and monographs in biospeleology. The real abundance of literature is, however, in the journal literature, which is a sign of the relatively recent beginnings of study of life in caves.

References not in Main Bibliography

Bedel L. and E. Simon. 1875. Liste Générale des Articulés Cavernicoles de l'Europe. *Journal de Zoologie.* 4.

Briggs, Thomas S. 1972. *A cavernicolous whip-scorpion from the northern Mojave Desert, California (Schizomida: Schizomidae)..* Occasional Papers of the California Academy of Sciences no 98. San Francisco: California Academy of Sciences.

Kircher A. 1665. *Mundus Subterraneus, in xii Libros Digestus ....* Amsterdam: J. Janssonium & E. Weyerstraten, 1st ed., 2 vols in 1.

Racovitza E.G. 1907. Essai sur les Problemes Biospeleogiques.. Biospeleogica, 1 *Archives de Zoologie Expérimentale et Générale.*, series 4, 6: [371]-448.

Roberts E.E. 1947. Edouard Alfred Martel (1859-1938). *Yorks, Ramblers' Club Journal.* 7(24): [105]-116.

Valvasor J.W. and Reisp, Branko (ed). 1689. *Die Ehre dess Hertzogthums Crain und Historisch-topgraphische Beschreibung.* Laybach: W. M. Endter buchhandlern in Nurnberg, 4 vols.

Viré A. 1897. Le laboratorie des catacombs. *Bulletin du Muséum d'Histoire Naturelle. Paris.* 3.

3.1 General

[140]Anon, 1981, International Symposium on groundwater biology: symposium held in 1978 in Blacksburg, Virginia, U.S.A.
[619][Cave Research Group], Dec 1955, Fauna collected from caves as recorded in the C.R.G. fauna records: part I (1938-39)
[620][Cave Research Group], Oct 1956, Fauna collected from caves as recorded in the C.R.G. fauna records: part II (war years 1940-46 and 1945-46)
[622][Cave Research Group of Great Britain], Nov 1972, Hypogean fauna and biological records 1970-71
[753]Crawford, Rod, Jul 1977, An annotated bibliography of Pacific Northwest speleobiology
[862]Devil's Hole Pupfish Recovery Team (prep), Jul 1980, Devil's Hole Pupfish Recovery Plan

[1267]Grimes, Ken, February 1996, Field Guide to karst features in southeast South Australia and Western Victoria: for the Karst Studies Seminar, Naracoorte, February 1996
[1342]Hamilton-Smith, Elery (ed), February 1996, Abstracts of papers: karst studies seminar Naracoorte
[1408][Hazelton, Mary (ed)], Mar 1970, Hypogean fauna, biological records 1968
[1792]Kennard, Don (ed), 1975, Devil's Sinkhole area: headwaters of the Nueces River
[1876]Kurtén, Björn, 1976, The cave bear story: life and death of a vanished animal
[2085]Mattheij, Johannes Adrianus Maria, 1940–, 1970, The functional cell types in the pars distalis analogue of the pituitary gland in the blind mexican cave fish, *Anoptichthys jordani*
[2200]Missouri Speleological Survey, 1976, A biological study of Cathedral Cave, Crawford County, Missouri
[2896]Sbordoni, Valerio (ed), 1987, Symposium on speciation and adaptation to cave life: gradual vs. rectangular evolution: a symposium organized by the Unione Zoologica Italiana and the Société de Biospéologie, Roma, October 6-11, 1986: part I
[2897]Sbordoni, Valerio (ed), 1988, Symposium on speciation and adaptation to cave life: gradual vs. rectangular evolution: a symposium organized by the Unione Zoologica Italiana and the Société de Biospéologie, Roma, October 6-11, 1986: part II
[3197]Tennessee Academy of Science, Jul 1930, Journal of the Tennessee Academy of Science. Cave number

3.2 Adaptation

[2896]Sbordoni, Valerio (ed), 1987, Symposium on speciation and adaptation to cave life: gradual vs. rectangular evolution: a symposium organized by the Unione Zoologica Italiana and the Société de Biospéologie, Roma, October 6-11, 1986: part I
[2897]Sbordoni, Valerio (ed), 1988, Symposium on speciation and adaptation to cave life: gradual vs. rectangular evolution: a symposium organized by the Unione Zoologica Italiana and the Société de Biospéologie, Roma, October 6-11, 1986: part II
[2979]Siffre, Michel, 1939–, 1965, Beyond time
[2980]Siffre, Michel, 1939–, Jun 1976, Preliminary results of an experiment on human chronobiology and neurobiology in a subterranean environment. 1. Life on a bicircadian rhythm (P. Englender). 2. Life in continuous light (J. Chabert). Longitudinal analysis and computer correlation of neurologic, psychological and physiologic data collected in beyond-time cave experiments from 1968-1969

3.3 Bacteria

[2080]Mason-Williams, Ann, Sep 1958, A preliminary investigation into the bacterial & botanical flora of caves in South Wales
[2887]Sasowsky, Ira Daniel, 1959–(ed), 1994, Breakthroughs in karst geomicrobiology and redox geochemistry: abstracts and field trip guide for the symposium held February 16 - 19, 1994, Colorado Springs, Colorado

3.4 Botany

[771]Cullen, P., 1990, A survey of the vegetation
[2941]Shafer, Harry J., 1977, Archeological and botanical studies at Hinds Cave, Val Verde County, Texas

3.5 Cave Ecology

[327]Barr, Thomas Calhoun, 1931–, Dec 1976, Ecological effects of water pollutants in Mammoth Cave: final technical report to the National Park Service
[980]Encyclopaedia Britannica Educational Corporation, 1976, Caves: the dark wilderness

3.6 Cave Fauna

[3]Abeille de Perrin, Elzear, 1877, On the collection of cavern insects
[21]Akira Yoshida, Jan 1952, A list of the Arthropoda in the limestone caves in Kantô-Mountainland, with the descriptions of a new genus and three species
[35]Aley, Thomas John, 1938–, 1988, Control of exotic plants in Oregon Caves, Oregon Caves National Monument
[47]Aljančič, Marko (auth, ed), 1993, *Proteus*; the mysterious ruler of karst darkness
[95]Anon, 1964, Caves
[114]Anon, c1967, Exploration '66; University of Nottingham; biospeleological research expedition to Ireland; Riverview work project in Portugal; British speleological expeditions to Turkey
[196]Anon, 1972, Subterranean fauna of Mexico: Part I: some results of the first Italian zoological mission to Mexico sponsored by the National Academy of Lincei (October 10 - December 9, 1969)
[197]Anon, 1973, Subterranean fauna of Mexico: Part II: further results of the first Italian zoological mission to Mexico sponsored by the National Academy of Lincei (1969 and 1971)
[198]Anon, 1977, Subterranean fauna of Mexico: Part III: further results of the first Italian zoological mission to Mexico sponsored by the National Academy of Lincei (1973 and 1975)
[209]Anon, nd, Waitomo Day 1982: summary of papers: Waitomo Caves research programme
[243]Atkinson, Anne, 1995, Undara Volcano and its lava tubes: a geological wonder of Australia in Undara Volcanic National Park, North Queensland
[262]Australian Speleological Federation, Feb 1972, Proceedings of the eighth biennial conference of the Australian Speleological Federation
[279]Bailey, Vernon Orlando, 1864–1942, 1928, Animal life of the Carlsbad Cavern
[280]Bailey, Vernon Orlando, 1864–1942, 1933, Cave life of Kentucky: mainly in the Mammoth cave region
[314]Banta, Arthur Mangun, 1877–, Sep 1907, The fauna of Mayfield's Cave
[315]Banta, Arthur Mangun, 1877–, Apr 1906, The life history of the cave salamander, *Spelerpes maculicaudus* (Cope)
[326]Barr, Thomas Calhoun, 1931–, 1972, Caves of Tennessee
[329]Barr, Thomas Calhoun, 1931–, 1960, Symposium: speciation and raciation in cavernicoles
[399]Binkerd, Adam D., 1869, The Mammoth Cave and its denizens: a complete descriptive guide
[404]Bishop, Michael J., 1982, The mammal fauna of the early middle Pleistocene cavern infill site of Westbury-sub-Mendip, Somerset
[409]Black, Jeffrey H., 1971, The cave life of Oklahoma: a preliminary study (excluding chiroptera)
[413]Blatchley, William Stanley, 1859–1940, 1990, Gleanings from nature: ten Indiana caves and the animals that inhabit them
[414]Blatchley, William Stanley, 1859–1940, 1896, Indiana caves and their fauna
[442]Bosnak, Art D., 1981, Acute toxicity of cadmium, zinc, and total residual chlorine to epigean and hypogean isopods (Asellidae)
[443]Botosaneanu, Lazare (ed), 1986, Stygofauna mundi: a faunistic, distributional, and ecological synthesis of the world fauna inhabiting subterranean waters (including the marine interstitial)
[465]Breder, Charles Marcus, 1897–, Dec 1947, Comparative studies in the light sensitivity of blind characins from a series of Mexican caves
[478]Briggs, Thomas S., 1974, Phalangodidae from caves in the Sierra Nevada (California) with a redescription of the type genus (Opiliones, Phalangodidae)
[498]Bronin, Andrew, 1972, The cave: what lives there
[516]Brown, Arthur V., 1984, Cavefish (*Amblyopsis rosae*) in Arkansas: populations, incidence, habitat requirements, and mortality factors: a final report
[538]Brunton, Daniel Francis, 1948–, 1990, A biological inventory of the Warsaw Caves area of natural and scien-

3.6. CAVE FAUNA

tific interest, Peterborough County, Ontario
[550]Buikema, Arthur L., 1980, Studies on Gammaridea II: proceedings of the 4th International Colloquium on Gammarus and Niphargus, Blacksburg, Virginia, U.S.A., 10-16 September 1978
[575]Cahn, Phyllis H., 1958, Comparative optic development in *Astyanax mexicanus* and in two of its blind cave derivatives
[585]Camacho, Ana Isabel (ed), 1992, The natural history of biospeleology
[619][Cave Research Group], Dec 1955, Fauna collected from caves as recorded in the C.R.G. fauna records: part I (1938-39)
[620][Cave Research Group], Oct 1956, Fauna collected from caves as recorded in the C.R.G. fauna records: part II (war years 1940-46 and 1945-46)
[621]Cave Research Group of Great Britain, 1971, Hypogean fauna and biological records 1970, [and] fauna of Gibraltar caves
[622][Cave Research Group of Great Britain], Nov 1972, Hypogean fauna and biological records 1970-71
[624]Caving Club SC 33, c1994, 1993 Jamaica expedition
[629]Chamberlin, Joseph Conrad, 1898–, 1962, New and little-known false scorpions, principally from caves, belonging to the families Chthoniidae and Neobisiidae (Arachnida, Chelonethida)
[630]Chamberlin, Joseph Conrad, 1898–, 1947, The Vachoniidae—a new family of false scorpions: two new species from the caves of Yucatan
[631]Chamberlin, Ralph Vary, 1879–1967, 1930, A new geophiloid chilopod from Potter Creek Cave, California
[632]Chamberlin, Ralph Vary, 1879-1967, 1942, On centipeds and millipeds from Mexican caves
[638]Chapman, Philip, 1993, Caves and cave life
[653]Chilton, Charles, B. 1860–, 1894, The subterranean Crustacea of New Zealand: with some general remarks on the fauna of caves and wells
[656]Christiansen, Kenneth Allen, 1924-, 1982, Notes on Mexican cave Pseudosinella (Collembola: Entomobryidae) with the description of six new species
[682]Collier, Don (host), 1990, Life underground: caves, mines, minerals
[691]Conn, David Bruce, 1981, Cave life of Carter Caves State Park
[708]Cooper, John Edward, 1929–, May 1982, Recovery plan for the Alabama cavefish, *Speoplatyrhinus poulsoni* Cooper and Kuehne 1974
[732]Cox, Ulysses O., 1905, A revision of the cave fishes of North America
[777]Culver, David C., 1944–, 1995, Adaptation and natural selection in caves: the evolution of *Gammarus minus*
[778]Culver, David C., 1944–, 1994, Biogeography of subterranean crustaceans: the effects of different scales: a symposium held at the summer meeting of the Crustacean Society in Charleston, South Carolina, USA, in June 1992
[779]Culver, David C., 1944–, 1982, Cave life: evolution and ecology
[780]Curcic, Bozidar P. M., 1988, Cave-dwelling pseudoscorpions of the Dinaric Karst = Jamski pascipalci dinarskega krasa
[808]Dasque, Jean (prod), 1972, Garden of shadows
[848]Deeleman-Reinhold, Christa L., 1978, Revision of the cave-dwelling and related spiders of the genus *Troglohyphantes* Joseph (Linyphiidae) with special reference to the Yugoslav species
[864]Dewey, Jennifer, 1994, The creatures underneath
[939]Eberhard, Stefan, nd, The cave fauna at Ida Bay and the effect of quarry operation
[940]Eberhard, Stefan, 1991, The invertebrate cave fauna of Tasmania
[950]Eigenmann, Carl H., 1863–1927, 1909, Cave vertebrates of America: a study in degenerative evolution
[961]Elliott, William Rawleigh, 1946–, Jun 1993, Draft recovery plan for endangered karst invertebrates in Travis and Williamson Counties, Texas
[963]Elliott, William Rawleigh, 1946–, 1976, New cavernicolous Rhagidiidae from Idaho, Washington, and Utah (Prostigmata, Acari, Arachnida)
[964]Elliott, William Rawleigh, 1946–, 1978, The new Melones Cave harvestman transplant
[980]Encyclopaedia Britannica Educational Corporation, 1976, Caves: the dark wilderness
[1056]Fieseler, Ronald G. (ed), Apr 1975, The caves of Brewster and western Pecos Counties
[1093]Ford, Trevor David, 1925–(comp, ed), 1977, Limestone and caves of the Peak District
[1101]Forster, Raymond R., 1922–, 1965, Harvestmen of the sub-order Laniatores from New Zealand caves
[1115]Francis, Charles M. (comp), 1989, The management of edible bird's nest caves in Sabah
[1125]Frederikson, Rosalie, 1983, The New Zealand glowworm
[1138]Gagné, Wayne C., 1974, The cavernicolous fauna of Hawaiian lava tubes, part VI: Mesoveliidae or water treaders (Heteroptera)
[1139]Gagné, Wayne C., Jul 1974, The cavernicolous fauna of Hawaiian lava tubes, part VII. Emesinae or thread-legged bugs (Heteroptera: Redvuiidae)
[1153]Gardner, James Eugene, 1953–, Oct 1984, An introduction to the inventory and evaluation of biological cave resources
[1155]Gardner, James Eugene, 1953–, 1986, Invertebrate fauna from Missouri caves and springs
[1165]Garton, Emmel Ray, 1950–, 1993, The vertebrate fauna of West Virginia caves
[1185]Gertsch, Willis J., 1984, The spider family Nesticidae (Araneae) in North America, Central America, and the West Indies

[1192]Gidley, James William, 1866–1931, 1914, Preliminary report on a recently discovered Pleistocene cave deposit near Cumberland, Maryland
[1197]Gilbert Group, 1978, Caves, a deeper look at our Earth
[1201]Giles, Cathy, 1983, Cave fauna of Waitomo
[1219]Glennie, E. A., 1945, Cave Fauna
[1228]Gordon, Isabella, 1957, On *Spelaeogriphus*, a new cavernicolous crustacean from South Africa
[1266]Grimes, K. G., Oct 1995, South East Karst Province of South Australia: Australian Caves & Karst Management Association October 1995
[1289]Gunzi, Christiane, 1993, Cave life
[1320]Halliday, William R., 1926–, 1983, Ape Cave and the Mount St. Helens apes
[1334]Halliday, William R., 1926–, May 1981, Outline of major biological conclusions from 1980-81 Mount St. Helens eruptions
[1343]Hamilton-Smith, Elery, Oct 1962, Australian cave fauna: notes on collecting
[1393]Hay, William Perry, 1872–, 1903, Observations on the crustacean fauna of Nickajack cave, Tennessee, and vicinity
[1394]Hay, William Perry, 1872–, 1903, Observations on the crustacean fauna of the region about Mammoth Cave, Kentucky
[1395]Hayami, Itaru, 1933–, 1993, Submarine cave bivalvia from the Ryukyu Islands: systematics and evolutionary significance
[1399]Hazelton, Mary (ed), 1965, British hypogean fauna and biological records of the Cave Research Group part IX (1963)
[1400]Hazelton, Mary, Oct 1958, Fauna collected from caves as recorded in the C.R.G. fauna records: part III (1947)
[1401]Hazelton, Mary, Feb 1959, Fauna collected from caves, mines and wells as recorded in the C.R.G. fauna records: part IV (1948-1949)
[1402]Hazelton, Mary, Feb 1960, Fauna collected from caves, mines and wells as recorded in the C.R.G. fauna records: part V (1950-1953)
[1403]Hazelton, Mary, Sep 1960, Fauna collected from caves, mines and wells as recorded in the C.R.G. fauna records: part VI (1954-1955-1956)
[1404]Hazelton, Mary, Nov 1961, Fauna collected from caves, mines and wells as recorded in the C.R.G. fauna records: part VII (1957-1959)
[1405]Hazelton, Mary, Sep 1963, Fauna collected from caves, mines and wells as recorded in the C.R.G. fauna records: part VIII (1960-1962)
[1406][Hazelton, Mary (ed)], Sep 1967, Hypogean fauna and biological records 1964-1966
[1407][Hazelton, Mary (ed)], Dec 1968, Hypogean fauna, biological records 1967
[1408][Hazelton, Mary (ed)], Mar 1970, Hypogean fauna, biological records 1968
[1409]Hazelton, Mary (ed), 1974, Irish hypogean fauna and Irish biological records, 1856-1971
[1465]Hobbs, Horton Holcombe, 1914–, 1967, A new crayfish from Alabama caves with notes of the origin of the genera *Orcnectes* and *Cambaras* (Decapoda: Astracidae)
[1466]Hobbs, Horton Holcombe, 1914–, 1972, Origins and affinities of the troglobitic crayfishes of North America (Decapoda: Astacidae), II, genus em Orconectes
[1467]Hobbs, Horton Holcombe, 1914–, 1977, A review of the troglobitic decapod crustaceans of the Americas
[1468]Hobbs, Horton Holcombe, 1944–, Jan 1994, Assessment of the ecological resources of the caves of Russell Cave National Monument, Jackson County, Alabama and of selected caves at the Lookout Mountain unit of Chickamanga-Chattanooga National Military Park, Dade County, Georgia and Hamilton County, Tennessee
[1469]Hobbs, Horton Holcombe, 1944–, 1976, On the troglobitic shrimps of the Yucatan peninsula, Mexico (Decapoda: Atyidae and Palaemonidae)
[1476]Hoffman, Richard L., 1956, New genera and species of cavernicolous diplopods from Alabama
[1496]Holsinger, John Robert, 1934–, Jan 1985, Ecological analysis of the Kentucky cave shrimp, *Palaemonias ganteri* Hay, at Mammoth Cave National Park (phase V)
[1497]Holsinger, John Robert, 1934–, 1986, Ecological analysis of the Kentucky cave shrimp, *Palaemonias ganteri* Hay, at Mammoth Cave National Park (phase VI): preliminary observations on stream interstitial meiofauna communities and related abiotic factors
[1498]Holsinger, John Robert, 1934–, Oct 1983, Ecological analysis of the Kentucky cave shrimp, *Palaemonias ganteri* Hay, Mammoth Cave National Park
[1499]Holsinger, John Robert, 1934–, 1982, Ecological analysis of the Kentucky cave shrimp, *Palaemonias ganteri* Hay, Mammoth Cave National Park (phase I)
[1500]Holsinger, John Robert, 1934–, Oct 1983, Ecological analysis of the Kentucky cave shrimp, *Palaemonias ganteri* Hay, Mammoth Cave National Park (phase II)
[1501]Holsinger, John Robert, 1934–, Apr 1983, Ecological analysis of the Kentucky cave shrimp, *Palaemonias ganteri* Hay, Mammoth Cave National Park (phase III)
[1502]Holsinger, John Robert, 1934–, Apr 1972, The freshwater amphipod crustaceans Gammaridae of North America
[1503]Holsinger, John Robert, 1934–, Jun 1988, The invertebrate cave fauna of Virginia and a part of eastern Tennessee
[1504]Holsinger, John Robert, 1934–, 1976, The invertebrate cave fauna of West Virginia
[1505]Holsinger, John Robert, 1934–, 1970, Morphological variation in *Gammarus minus* Say (Amphipoda, Gam-

3.6. CAVE FAUNA

maridae) with emphasis on subterranean forms
[1506]Holsinger, John Robert, 1934–, Nov 1985, The origin and geographic distribution of troglobites
[1507]Holsinger, John Robert, 1934–, Feb 1985, Speciation in cave fauna
[1508]Holsinger, John Robert, 1934–, 1980, The subterranean amphipod crustacean fauna of an artesian well in Texas
[1509]Holsinger, John Robert, 1934–, 1974, Systematics of the subterranean amphipod genus *Stygobromus* (Gammaridae): part I. species of the western United States
[1510]Holsinger, John Robert, 1934–, 1978, Systematics of the subterranean Amphipod genus *Stygobromus* (Rangonyctidae), part II: species of the eastern United States
[1511]Holsinger, John Robert, 1934–, 1967, Systematics, speciation, and distribution of the subterranean amphipod genus *Stygonectes* (Gammaridae)
[1561]Hubbell, Theodore Huntington, 1897–1989, Aug 1978, The systematics and biology of the cave-crickets of the North American tribe Hademoecini (Orthoptera Saltatoria, Ensifera, Rhaphidophoridae, Dolichopodinae)
[1562]Hubbs, Carl Leavitt, 1894–1979, 1938, Fishes from the caves of Yucatan
[1573]Humphreys, William F. (ed), Dec 1993, The biogeography of Cape Range Western Australia: being the proceedings of a symposium held under the auspices of the Western Australian Museum in Perth on 21 November 1992 at the Art Gallery of Western Australia
[1574]Humphreys, William F., Apr 1991, Survey of caves in Cape Range North West Cape Peninsula Western Australia
[1584]Husmann, Siegfried, 1976, 1st international symposium on groundwater ecology, Schlitz,1975
[1585]Hutchison, Victor Hobbs, 1931–, Nov 1956, Notes on the plethodontid salamanders, *Eurycea lucifuga* (Rafinesque) and *Eurycea longicauda longicauda* (Green)
[1590]Hyman, Libbie Henrietta, 1888–1969, 1956, North American triclad Turbellaria, XIII: three new cave planarians
[1646]International Symposium on Cave Biology and Cave Paleontology (1975: Oudtshoorn, South Africa), 1975, Proceedings, U.I.S. international symposium on cave biology and cave paleontology, held in Oudtshoorn, South Africa from the 3rd- 6th August, 1975
[1685]Jackson, Harold , c1969, Sudwala
[1707]Jantschke, Herbert, 1994, Tunel de la Atlantida, Haria, Lanzarote, Canary Islands: the hydrodynamic, the chemistry and the minerals of the lava tube, the population density of *Munidopsis polymorpha*
[1761]Juberthie, Christian (ed), 1994, Encyclopaedia biospeologica tome 1
[1791]Kenk, Roman, 1977, Freshwater Triclads (Turbellaria) of North America, IX, the genus *Sphalloplana*
[1847]Komatsu, Toshihiro, 1961, Cave spiders of Japan; their taxonomy, chorology, and ecology
[1848]Kornicker, Louis S., 1919–, 1989, New Ostracoda (Halocyprida: Thaumatocyprididae and Halocyprididae) from anchialine caves in the Bahamas, Palau, and Mexico
[1849]Kornicker, Louis S., 1919–, 1990, Ostracoda (Halocyprididae) from anchialine caves in the Bahamas
[1850]Kornicker, Louis S., 1919–, 1992, Ostracoda (Halocypridina, Cladocopina) from anchialine caves in Jamaica, West Indies
[1851]Kornicker, Louis S., 1919–, 1996, The troglobitic halocyprid Ostracoda of anchialine caves in Cuba
[1852]Kornicker, Louis S., 1919–, 1989, Troglobitic Ostracoda (Myodocopa: Cyprididae, Thaumatocyprididae) from anchialine pools on Santa Cruz Island, Galapagos Islands
[1959]Longley, Glenn, Nov 1977, Preliminary report of biological investigation Valdina Farms sink hole - Medina Co., Texas
[1960]Longley, Glenn, Jun 1978, Status of *Trogloglanis pattersoni* Eigenmann, the toothless blindcat, and status of *Satan eurystomus* Hubbs and Bailey, the widemouth blindcat
[1961]Longley, Glenn, 1942–, Jan 1952, A list of the arthropoda in the limestone caves in Kantô-Mountainland, with the description of a new genus and three species
[1962]Longley, Glenn, 1942–, Feb 1977, Status of *Typhlomolge* (=Eurycea) *rathbuni*
[1979]Lübke, Anton, 1890–, 1958, The world of caves
[2085]Mattheij, Johannes Adrianus Maria, 1940–, 1970, The functional cell types in the pars distalis analogue of the pituitary gland in the blind mexican cave fish, *Anoptichthys jordani*
[2164]Meyer-Rochow, Victor Benno, 1990, The New Zealand glowworm
[2206]Missouri Speleological Survey, Apr 1975, A checklist of invertebrate species recorded from Missouri subterranean habitats
[2212]Missouri Speleological Survey, Oct 1974, The invertebrate fauna of Mystery Cave, Perry County, Missouri
[2239]Mitchell, Robert Wetsel, 1933–, Feb 1977, Mexican eyeless Characin fishes, genus *Astyanax*: environment, distribution, and evolution
[2240]Mitchell, Robert Wetsel, 1933–, 1972, A new family, genus, and species of cave-adapted planarian from Mexico (Turbellaria, Tricladida, Maricola)
[2241]Mitchell, Robert Wetsel, 1933–(ed), 1971, Studies on the cavernicole fauna of Mexico
[2242]Mitchell, Robert Wetsel, 1933–(ed), Jul 1973, Studies on the cavernicole fauna of Mexico and adjacent regions
[2248]Mohr, Charles E., 1907–, , Cave life
[2250]Mohr, Charles E., 1907–, 1966, The life of the cave
[2258]Montz, Gary R., Jun 1993, The aquatic invertebrates of Mystery Cave, Forestville State Park, Minnesota

[2277]Morris, Linda, 1985, The Hayes Cave site, South Maitland, Nova Scotia

[2283]Motaş, C., 1968, Emil Racoviţă 1868-1947

[2295]Muchmore, William B., 1965, North American cave pseudoscorpions of the genus *Kleptochthonius*, subgenus *Chamberlinochthonius* (Chelonethida, Chthoniidae)

[2521]Northup, Diana Eleanor, 1948–, 1992, Lechuguilla Cave: biological inventory

[2545]O'Donnell, Lisa, Aug 1994, Recovery plan for endangered karst invertebrates in Travis and Williamson Counties, Texas

[2548]Oklahoma Department of Wildlife Conservation, 1990, Survey and species determination of cave crayfish in Oklahoma

[2582]Owen, Daniel (comp), 1987, Proyek Kelelawar: final report of the Oxford University expedition to the Togian Islands, Sulawesi, Indonesia summer 1987

[2589]P. Tarrant Ltd, nd, Waitomo Caves New Zealand: a souvenir booklet of your Waitomo visit with story of life cycle of New Zealand glow-worm

[2590]Packard, Alpheus Spring, 1839–1905, 1977, The cave fauna of North America, with remarks on the anatomy of the brain and origin of the blind species

[2591]Packard, Alpheus Spring, 1839–1905, 1872, The Mammoth Cave and its inhabitants, or, descriptions of the fishes, insects and crustaceans found in the cave with figures of the various species and an account of allied forms, comprising notes upon their structure, development and habits, with remarks upon subterranean life in general

[2592]Packard, Alpheus Spring, 1839–1905, 1877, On a new cave fauna in Utah: and on new phyllopod Crustacea from the West

[2594]Page, Lawrence M., 1977, Status of the cypres darter, *Etheostoma proeliare*, and comments on the spring cavefish, *Chologaster agassizi*, in Max Creek, Johnson County, Illinois

[2618]Patton, Thomas Hudson, 1934–, Sep 1963, Fossil vertebrates from Miller's Cave, Llano County, Texas

[2631]Pearse, Arthur Sperry, 1877–1956, Jun 1938, Fauna of the caves of Yucatan

[2634]Peck, Steward Blaine, 1942–, May 1973, A systematic revision and evolutionary biology of the *Ptomaphagus* (Adelops) beetles of North America (Coleoptera; Leiodidae; Catopinae), with emphasis on cave-inhabiting species

[2693]Pope, Joyce, 1991, Life in the dark

[2699]Poulson, Thomas Layman, 1934–(ed), 1975, Symposium on life histories of cave beetles: symposium held at the 1973 annual convention of the National Speleological Society, Bloomington, Indiana

[2762]Rasquin, Priscilla, 1949, The influence of light and darkness on thyroid and pituitary activity of the characin *Astyanax mexicanus* and its cave derivatives

[2766]Ray, Michael Allen, nd, Underground worlds: tour guide training manual

[2771]Reddell, James Russell, 1938–, 1995, Catalogue, bibliography, and generic revision of the order Schizomida (Arachnida)

[2774]Reddell, James Russell, 1938–, Jun 1964, The caves of Comal county

[2775]Reddell, James Russell, 1938–, Nov 1965, The caves of Edwards County

[2778]Reddell, James Russell, 1938–(ed), Apr 1967, The caves of Medina county

[2780]Reddell, James Russell, 1938–, Feb 1973, The caves of San Saba County

[2784]Reddell, James Russell, 1938–(ed), Oct 1963, The caves of Willamson County

[2785]Reddell, James Russell, 1938–, Aug 1969, A checklist and annotated bibliography of the subterranean aquatic fauna of Texas

[2786]Reddell, James Russell, 1938–, 1965, A checklist of the cave fauna of Texas

[2789]Reddell, James Russell, 1938–(ed), Mar 1982, Further studies on the cavernicole fauna of Mexico and adjacent regions

[2790]Reddell, James Russell, 1938–, 1971, A preliminary bibliography of Mexican cave biology with a checklist of published records

[2791]Reddell, James Russell, 1938–, 1977, A preliminary survey of the caves of the Yucatan Peninsula

[2792]Reddell, James Russell, 1938–(ed), Jul 1981, A review of the cavernicole fauna of Mexico, Guatemala, and Belize

[2794]Reddell, James Russell, 1938–(ed), Dec 1986, Studies on the cave and endogean fauna of North America

[2795]Reddell, James Russell, 1938–(ed), Dec 1992, Studies on the cave and endogean fauna of North America II

[2796]Reddell, James Russell, 1938–, Sep 1971, Studies on the cavernicole fauna of Mexico

[2797]Reddell, James Russell, 1938–, 1977, Studies on the caves and cave fauna of the Yucatan Peninsula

[2806]Rentz, David C., 1972, A new genus and species of camel cricket from the Farallon Islands of California (Orthoptera: Gryllacrididae)

[2810]Rheams, Karen F., 1994, Hydrogeologic and biologic factors related to the occurrence of the Alabama cave shrimp (*Palaemonias alabamae*), Madison County, Alabama

[2826]Roberts, Allan, 1983, Underground life

[2855]Rosen, Donn Eric, 1929–, Oct 1962, Comments on the relationships of the North American cave fishes of the family Amblyopsidae

[2913]Schultz, Charles Bertrand, 1908–, 1935, The fauna of Burnet Cave, Guadalupe Mountains, New Mexico

[2944]Shaler, Nathaniel Southgate, 1841–1906, 1876, On

the antiquity of the caverns and cavern life of the Ohio valley
[2957]Shear, William A., 1974, North American cave millipeds II, an unusual new species (Dorypetalidae) from Southern California, and new records of *Speodesmus tuganbius* (Trichopolydesmidae) from New Mexico
[2958]Shear, William A., 1969, A synopsis of the cave millipeds of the United States: with an illustrated key to genera
[3011]Smith, Allyn Goodwin, 1957, Snails from California caves
[3033]Smith, Philip Wayne, 1921–, 1978, A summary of the life history and distribution of the spring cavefish, *Chologaster agassizi* Putnam, with population estimates for the species in southern Illinois
[3097]Sprent, J. K., (ed), 1970, Mount Etna caves: a collection of papers covering several aspects of the Mt. Etna and Limestone Ridge caves area of central Queensland
[3157]Sullivan, Gerardus Nicholas, 1927–, Jul 1960, Appendix: checklist of macroscopic troglobitic organisms of the United States
[3158]Sullivan, Gerardus Nicholas, 1927–, 1962, Checklist of troglobitic organisms of Middle America
[3217]Thurgate, Mia E., 1995, Sinkholes, caves and spring lakes: an introduction to the unusual aquatic ecosystems of the lower south east of South Australia: South Australian Underwater Speleological Society occasional paper number 1
[3263]U.S. Bureau of Land Management, Barstow Resource Area, 1982, Shoshone Cave (Whip-scorpion Habitat) wildlife habitat management plan
[3336]University of Nevada System, Laboratory of Desert Biology, Dec 1968, Final reports on the Lehman Caves studies
[3343]Valentine, Joseph Manson, 1902–, Nov 1952, New genera of Anophthalmid beetles from Cumberland Caves
[3344]Valli, Eric, 1990, The nest gatherers of Tiger Cave
[3350]Vandel, Albert, 1894–1980, 1965, Biospeleology: the biology of cavernicolous animals
[3386]Walters, R. (comp), 1986, The Crocodile Caves of Ankarana: 1986: an expedition to study and explore the limestone massif of Ankarana in northern Madagascar
[3392]Waltham, Anthony Clive, 1942–[Tony] (ed), 1974, Limestone and caves of Northwest England
[3425]Webb, Donald W., May 1994, The biological resources of Illinois caves and other subterranean environments: determination of the diversity, distribution, and status of the subterranean faunas of Illinois caves and how these faunas are related to groundwater quality
[3426]Webb, Donald W., 1995, Status report on the cave Amphipod *Gammarus acherondytes* Hubricht and Mackin (Crustacea: Amphipoda) in Southern Illinois
[3435]Wells, Patrick H., Apr 1959, Responses to light by cave crayfishes
[3449]Westcott, Richard L., Jun 1968, A new subfamily of blind beetle from Idaho ice caves, with notes on its bionomics and evolution (Coleoptera: Leiodidae)
[3473]White, William Blaine, 1934–(ed), 1976, Geology and biology of Pennsylvania caves
[3504]Willis, L. D., Dec 1986, A recovery plan for the Ozark cavefish (*Amblyopsis rosae*)
[3509]Wilson, Jane, 1990, Lemurs of the lost world: exploring the forests and Crocodile Caves of Madagascar
[3547]Youngsteadt, Norman W., Oct 1978, A survey of some cave invertebrates from northern Arkansas
[3556]Zim, Herbert Spencer, 1909–, 1978, Caves and life

3.7 Cave Flora

[755]Cribb, A. B., 1965, An ecological and taxonomic account of the algae of a semi-marine cavern, Paradise Cave, Queensland
[1200]Giles, Cathy, 1984, Cave entrance plants including Lampenflora
[1266]Grimes, K. G., Oct 1995, South East Karst Province of South Australia: Australian Caves & Karst Management Association October 1995
[1574]Humphreys, William F., Apr 1991, Survey of caves in Cape Range North West Cape Peninsula Western Australia
[2080]Mason-Williams, Ann, Sep 1958, A preliminary investigation into the bacterial & botanical flora of caves in South Wales

3.8 Ecology

[208]Anon, 1982, Waitomo Caves management plan 1982
[382]Benedict, Ellen Maring, 1974, Cave ecology: a course book for the cave ecology class at Malheur Environmental Field Station
[443]Botosaneanu, Lazare (ed), 1986, Stygofauna mundi: a faunistic, distributional, and ecological synthesis of the world fauna inhabiting subterranean waters (including the marine interstitial)
[585]Camacho, Ana Isabel (ed), 1992, The natural history of biospeleology
[638]Chapman, Philip, 1993, Caves and cave life
[677]Coggins, Allen R., 1981, The caves of the Tennessee State Natural Areas system
[717]Coronet Instructional Films, 1970, Cave ecology
[778]Culver, David C., 1944–, 1994, Biogeography of subterranean crustaceans: the effects of different scales: a symposium held at the summer meeting of the Crustacean Society in Charleston, South Carolina, USA, in June 1992
[1543]Howarth, Francis Gard, 1940–, 1972, Ecological studies on Hawaiian lava tubes

[1761]Juberthie, Christian (ed), 1994, Encyclopaedia biospeologica tome 1
[1844]Kohary, Jana, 1988, Archaeology and ecology of Rose Cottage Cave, Orange Free State
[1877]Kurtén, Björn, 1924–1988, 1958, Life and death of the Pleistocene cave bear: a study in paleoecology
[1960]Longley, Glenn, Jun 1978, Status of Trogloglanis pattersoni Eigenmann, the toothless blindcat, and status of Satan eurystomus Hubbs and Bailey, the widemouth blindcat
[2010]MacGregor, John R., 1978, Ecology of a limestone cave
[2164]Meyer-Rochow, Victor Benno, 1990, The New Zealand glowworm
[2239]Mitchell, Robert Wetsel, 1933–, Feb 1977, Mexican eyeless Characin fishes, genus *Astyanax*: environment, distribution, and evolution
[2250]Mohr, Charles E., 1907–, 1966, The life of the cave
[2588]Ozark National Forest (AR), 1980, Blanchard Springs Caverns: the amazing world below
[2699]Poulson, Thomas Layman, 1934–(ed), 1975, Symposium on life histories of cave beetles: symposium held at the 1973 annual convention of the National Speleological Society, Bloomington, Indiana
[2804]Reid-Cowan Productions, 1976, Caves: the dark wilderness
[2905]Schindel, Geary Michael, 1957–, Dec 1991, Guidebook: environmental hydrogeology of karst terranes in the vicinity of Nashville Tennessee
[3336]University of Nevada System, Laboratory of Desert Biology, Dec 1968, Final reports on the Lehman Caves studies
[3350]Vandel, Albert, 1894–1980, 1965, Biospeleology: the biology of cavernicolous animals
[3425]Webb, Donald W., May 1994, The biological resources of Illinois caves and other subterranean environments: determination of the diversity, distribution, and status of the subterranean faunas of Illinois caves and how these faunas are related to groundwater quality

3.9 Evolution

[575]Cahn, Phyllis H., 1958, Comparative optic development in *Astyanax mexicanus* and in two of its blind cave derivatives
[585]Camacho, Ana Isabel (ed), 1992, The natural history of biospeleology
[777]Culver, David C., 1944–, 1995, Adaptation and natural selection in caves: the evolution of *Gammarus minus*
[950]Eigenmann, Carl H., 1863–1927, 1909, Cave vertebrates of America: a study in degenerative evolution
[1233]Government of Australia, 1993, Nomination of Australian fossil sites (a serial nomination of sites at Murgon, Riversleigh and Naracoorte): The origin and evolution of Australia's mammals
[1506]Holsinger, John Robert, 1934–, Nov 1985, The origin and geographic distribution of troglobites
[1507]Holsinger, John Robert, 1934–, Feb 1985, Speciation in cave fauna
[1511]Holsinger, John Robert, 1934–, 1967, Systematics, speciation, and distribution of the subterranean amphipod genus *Stygonectes* (Gammaridae)
[2239]Mitchell, Robert Wetsel, 1933–, Feb 1977, Mexican eyeless Characin fishes, genus *Astyanax*: environment, distribution, and evolution
[2590]Packard, Alpheus Spring, 1839–1905, 1977, The cave fauna of North America, with remarks on the anatomy of the brain and origin of the blind species
[2634]Peck, Steward Blaine, 1942–, May 1973, A systematic revision and evolutionary biology of the *Ptomaphagus* (Adelops) beetles of North America (Coleoptera; Leiodidae; Catopinae), with emphasis on cave-inhabiting species
[3449]Westcott, Richard L., Jun 1968, A new subfamily of blind beetle from Idaho ice caves, with notes on its bionomics and evolution (Coleoptera: Leiodidae)

3.10 Fungi

[2521]Northup, Diana Eleanor, 1948–, 1992, Lechuguilla Cave: biological inventory

3.11 Geomicrobiology

[2887]Sasowsky, Ira Daniel, 1959–(ed), 1994, Breakthroughs in karst geomicrobiology and redox geochemistry: abstracts and field trip guide for the symposium held February 16 - 19, 1994, Colorado Springs, Colorado

3.12 Guano

[3378]Wallis, G. R., 1965, Glass sand occurrences, Kurnell Peninsula: preliminary investigations ; Guano deposits in Willi Willi Caves, Kempsey ; Geological report on Wallent's Somersby Clay Pit

3.13 Physiology

[2485]Natural Resources Seminar (3rd: 1976: Santa Fe), 1976, Radon radiation situation in NPS caves: January 19, 1976
[2762]Rasquin, Priscilla, 1949, The influence of light and darkness on thyroid and pituitary activity of the characin *Astyanax mexicanus* and its cave derivatives
[3350]Vandel, Albert, 1894–1980, 1965, Biospeleology: the biology of cavernicolous animals

[3435] Wells, Patrick H., Apr 1959, Responses to light by cave crayfishes

3.14 Rabies

[695] Constantine, Denny G., 1925–, Jun 1967, Rabies transmission by air in bat caves

3.15 Systematics

[2634] Peck, Steward Blaine, 1942–, May 1973, A systematic revision and evolutionary biology of the *Ptomaphagus* (Adelops) beetles of North America (Coleoptera; Leiodidae; Catopinae), with emphasis on cave-inhabiting species

4 Exploration

Thomas D. Engel

7D West Street
Voorheesville, NY 12186-9718

Tales of exploration fire our imagination. The word evokes images of Stanley meeting Livingstone, Balboa topping a mountain peak and seeing the Pacific, and Marco Polo at the Forbidden City. For cavers, who are themselves explorers at heart, there is Martel descending Gaping Gill and John Wilcox viewing the railing in Mammoth Cave after an arduous journey from Flint Ridge. Since few of us have stood in the limelight of such sublime discovery, we rely on books to transport us to our chosen place and time.

Many books of exploration, such as *The Voyage of the Beagle* by Charles Darwin, mention caves in passing. Books devoted primarily to caves started showing up at the end of the 18th Century, such as Everard Home's *Account of Some Remarkable Caves in the Principality of Bayreuth* published in 1794. This book is like the 19th Century books that follow it. While they seem to concentrate on cave exploration, they are really books about cave science. *Cave Hunting in Yucatan* by Henry Mercer (1897) is as much about archeology as it is about cave exploration. The same can be said of many books including *Cavern Researches, or, Discoveries of Organic Remains, and of British and Roman Reliques* by J. MacEnery (1859), *In the Land of Cave and Cliff Dwellers* by Frederick Schwatka (1893), and *Cave of Loltun, Yucatan* by Edward H. Thompson (1897). If these are really about archeology, there are other "exploration" books that are really about geology and so forth.

Around the turn of the 20th Century, books primarily about exploration of caves started to be published. Some science is thrown in–it was impossible to leave out–but the science is minimized. Exploration is paramount. In this regard, books on cave exploration come later than other exploration books common in the mid to late 1800's and which were typically of the "through darkest Africa with gun and camera" variety. Early examples are *Recent Cave Explorations in California* by John Campbell Merriam (1906), *Cave Explorers in Co. Fermanagh 1907* by Ernest Albert Baker (1907), *Reports on the Exploration of "Dog Holes" Cave, Warton Crag* by J. Wilfrid Jackson (1910), and *A Visit to Carlsbad Cavern: Recent Explorations of a Limestone Cave in the Guadalupe Mountains* by Willis T. Lee (1924).

The golden age of cave exploration books started in the 1930's with the publication of *The Exploration of Carlsbad Cavern* by Frank E. Nicholson (1930) and *Caving, Episodes of Underground Exploration* by Ernest Albert Baker (1932). In the next twenty-five years, a wide variety of cave exploration books were published that are still read today. Such books include: *Underground New England* by Clay Perry (1939); *Ten Years Under the Earth* (1938), *My Caves* (1947), *Cave Men New and Old* (1951), and *The Darkness Under the Earth* (1954) by Norbert Casteret; *Subterranean climbers* by Pierre Chevalier (1951); *Caves of Adventure* by Haroun Tazieff (1953); *The Caves Beyond* by Joe Lawrence and Roger Brucker (1955); and *One Thousand Metres Down* by Jean Cadoux (1957).

Many older cavers today can point to one or more of these books as the inspiration for their involvement in caving. It is also interesting to note that these books were available in the United States during the formative years of the National Speleological Society (NSS). In 1939, the NSS was an idea of a few cavers. *The Caves Beyond* recounts the Society's expedition to Floyd Collins' Crystal Cave.

There were significant changes and growth during next twenty years. Many of the books mentioned above were reprinted to inspire a new generation of cavers. Manuscripts that had gathered dust for years were published. Many of these were of local interest only; an example is *The Lesser Caves of Schoharie County* by Arthur Van Voris, written in the 1930's but published in 1972.

Due to decreasing air fares and such, there was, during this period, an increase in mobility of society as a whole–including cavers. Young cavers went forth and upon their return wrote up expedition reports. These were self-published and were written for cavers, not for a general audience. Such reports include: *CEGSA 1965-66 Nullarbor Expedition* by Joe Jennings (1961), *Deep Cave, Texas. A Preliminary Report of the 1965 Project "Deep"* by James Estes (1967), *Mullamullang Cave Expeditions 1966* edited by A. L. Hill (1966), *67 Expedition to Crete: BUSS* by the Birmingham [England] University Speleological Society([1968]), *Expedition 67 to the Gou* by P. Watkinson (1968), *The Discovery and Exploration of St. Cuthbert's Swallet* by David Irwin (1968), and *British Karst Research Expedition to the Himalaya 1970* by Tony [Anthony Clive] Waltham (1971). The preponderance of British titles is because others got into the expedition report habit quite a bit later. In fact, with few exceptions, Americans have just started producing these recently with reports on Gunong Buda and Cueva Cheve. The British have continued publishing. New expedition reports come out annually.

During the same period, mass market books on cave exploration were being published. Notable among these are: *More Years Under the Earth* by Norbert Casteret (1962), *Climbing Blind* by Richard Colette (1966), *Ghar Parau* by David Judson (1973), *Discovery at the Rio Ca-*

muy by Russell Gurnee (1974), and *The Longest Cave* by Brucker and Watson (1976).

Also at this time, cavers started looking back. Several histories of exploration were published, including: *The Story of Wookey Hole* by Edmund Mason (1963), *The History of Mendip Caving* by Cuthbert Johnson (1967), and *Gaping Gill: 150 Years of Exploration* (1984) by Howard Beck (1984).

Today, cavers continue to travel to the four corners of the globe – as well as below their own backyards – in search of new caves. As long as this continues, the literature of cave exploration will continue to grow in richness and diversity.

4.1 General

[4] Absolon, Karel B., 1980, The conquest (the riddle of an abyss and its sinking river)

[5] Absolon, Karel B., 1987, The conquest of the caves and underground rivers of Czechoslovakia's Macocha Abyss: a historical and technical study of their exploration

[51] Allen, Tim (tm ldr), c1992, Hang Vietnam; report of the British Speleological Expedition to the Bac-Sun Massiflang

[70] Anderson, Robert Cleve (ed), 1982, Capital area cavers bulletin number 1

[77] Anglo-Canadian Rocky Mountains Speleological Expedition (1983) (comp), 1986, The Anglo-Canadian Rocky Mountains Speleological Expeditions 1983 and 1984: a report on recent discoveries made by two caving expeditions to the Rocky Mountains of Canada by combined British and Canadian teams

[114] Anon, c1967, Exploration '66; University of Nottingham; biospeleological research expedition to Ireland; Riverview work project in Portugal; British speleological expeditions to Turkey

[115] Anon, 1882, Exploration of the caves and rivers of New South Wales (minutes, reports, correspondence, accounts)

[227] [Army Caving Association], nd, The 1984 Army Caving Association expedition to the Gouffre Berger

[228] [Army Caving Association], c1988, Army Caving Association: Peru 1987

[236] Aspey, Steve, [1985], Sheffield University Speleological Society Central Crete Expedition: Greece 1984

[237] Aspin, J., Mar 1952, The caverns of Upper Ease Gill

[243] Atkinson, Anne, 1995, Undara Volcano and its lava tubes: a geological wonder of Australia in Undara Volcanic National Park, North Queensland

[264] Australian Speleological Federation Digest Commission, 1976, Australian speleology 1972

[281] Baker, Ernest Albert, 1869–1941, 1907, Cave Explorers in Co. Fermanagh 1907

[283] Baker, Ernest Albert, 1869–1941, 1907, The netherworld of Mendip: explorations in the great caverns of Somerset, Yorkshire, Derbyshire, and elsewhere

[284] Baker, Ernest Albert, 1869–1941, 1970, Caving, episodes of underground exploration

[287] Baker, Percy Frederick, 1921–, 1979, The Baker extension to the Banwell Bone Cave

[302] Balcombe, F. Graham, 1936, Ireland 1936 the record of the party of S.J. Pick (Leicester) in County Clare, Easter 1936

[303] Balcombe, F. Graham, 1937, Ireland 1937 the record of S.J. Pick (Leicester) in County Clare May 1937

[304] Balcombe, F. Graham, [1936], The Log of the Wookey Hole: exploration expedition 1935

[366] Beck, Howard M., 1984, Gaping Gill: 150 years of exploration

[378] Belski, David Stanley, 1937–, 1992, GYPKAP 1987 annual report

[385] Bennett, R., Apr 1969, Bristol Exploration Club caving report no.13 [St. Cuthbert's Report]: part F: Gour Hall area

[386] Bennett, R. (Cerberus series), Oct 1982, Cerberus series: Maypole series

[387] Bennett, R. H., 1973, Balague '70

[393] Bernabei, Tullio, 1992, Caves and stories of central Asia

[402] [Birmingham University Speleological Society], [1968], 67 expedition to Crete: BUSS

[405] Bitterli, Thomas, 1990, Proyecto Cerro Rabon

[407] Black, Don F., 1993, I don't play golf: recollections of a rescue volunteer

[435] Boon, J. M., 1977, Down to the sunless sea

[438] Boothroyd, Colin, May 1993, Caves of Thunder expedition report: an expedition to the world's largest underground river, Irian Jaya, Indonesia 1992

[445] Boulton, John, 1871, Particulars of a first exploration of the extensive and newly discovered cavern, at Stainton, Low Furness

[449] Boy Scouts of America, 1991, Caving

[458] Bradshaw, D. R. (comp), nd, Report on the British speleological expedition to Vietnam; March April 1990

[474] Bridgemon, Rondal Rex, 1944–(ed), [1991], South China caves: information on the cave and karst of South China and a report on the 1988 joint expedition between the Institute of Karst Geology, the Speleological Society of South China Normal University, and the Cave Research Foundation

[475] Bridges, Thomas Charles, 1937, Adventures under ground [sic]

[481] Bristol Exploration Club, nd, Bristol exploration club: caving report no.19: 1975 expedition to the Pierre Saint-Martin

[483] British Broadcasting Corporation Television Service, 1980, Atea: in search of the world's deepest cave

[485][British Cave Research Association], Jun 1981, Matienzo, Spain

[486]British Cave Research Association, 1991, Peak and Speedwell Caverns, exploration and science: including the talks presented at the Peak-Speedwell symposium, Sheffield, November 1989

[489]British Speleological Association, c1968, 1967 expedition to the Gouffre Berger

[490][British Speleological Association], Nov 1973, Imperial College karst research expedition to the Peruvian Andes, 1972

[492]British Speleological Expedition: Members of the Expedition, 1966, British Speleological Expedition to the Cantabrian Mountains, northwest Spain 1965

[502]Brook, David B., 1944–(comp), Dec 1976, The British New Guinea speleological expedition

[503]Brook, David B., 1944–(ed), 1978, Caves of Mulu: the limestone caves of the Gunong Mulu National Park, Sarawak

[504]Brook, David B., 1944–(ed, prod), Oct 1969, The explorations journal of The University of Leeds Speleological Association

[513]Brooks, S. J. (ed), Mar 1995, Caving in the abode of the clouds: the caves and karst of Meghalaya, north east India

[515]Brown, A. L., Nov 1976, Lelet: report of the 1975 New Ireland Speleological Expedition

[531]Brown, Robert F., Oct 1964, Exploration in Wind Cave

[536]Brucker, Roger Warren, 1929–, 1987, The longest cave

[563]Burgess, Robert Forrest, 1976, The cave divers: illustrated with photographs and drawings

[574]Cadoux, Jean, 1930–, 1957, One thousand metres down

[586]Camberlin, G. Will, 1993, The exploration and survey of McBridge Cave, Jackson County, Alabama

[591]Campbell, Newell Paul, 1938–, Oct 1975, Glacier Park cave study

[592]Canada. National and Historic Parks Branch, 1972, Nahanni

[594]Canada. National Parks Service, 1974, Castle Guard Cave, challenge under the glacier

[606]Casteret, Norbert, 1897–1987, 1951, Cave men new and old

[607]Casteret, Norbert, 1897–1987, 1954, The darkness under the earth

[609]Casteret, Norbert, 1897–1987, 1962, More years under the earth

[610]Casteret, Norbert, 1897–1987, 1947, My caves

[611]Casteret, Norbert, 1897–1987, 1940, Ten years under the earth

[624]Caving Club SC 33, c1994, 1993 Jamaica expedition

[641]Chatterton, John B., Sep 1971, Report of the Queen Mary College Society spelaeological expedition to Yugoslavia

[645]Chekley, D., 1985, La Sima 56 (Picos de Europa-España)

[647]Chevalier, Pierre, 1951, Subterranean climbers; twelve years in the world's deepest chasm

[648]Chevalier, Pierre, 1975, Subterranean climbers; twelve years in the world's deepest chasm

[660]C'ilek, V'acav, c1985, Czechoslovak speleological expedition to Nepal Himalaya 85

[672]Coase, Alan C., Mar 1977, Dan yr Ogof and its associated caves

[685]Collins, S. J., Jan 1956, Surveying in Red Cliffe Caves: Bristol 1953-1954

[692]Conn, Herbert Dunn, 1981, The Jewel Cave adventure: fifty miles of discovery under South Dakota

[705]Coon, Carleton Stevens, 1904–, 1968, Cave explorations in Iran, 1949

[774]Cullingford, Cecil Howard Dunstan, 1904–1990, 1976, Caving

[775]Cullingford, Cecil Howard Dunstan, 1904–1990, 1951, Exploring caves

[841]Dawkins, William Boyd, 1838–1929, Nov 1879, Further discoveries in the Cresswell caves

[850]de Joly, Robert, 1975, Memoirs of a speleologist: the adventurous life of a famous French cave explorer

[869]Dickinson, Leo, 1989, Anything is possible

[877]Dixon, E. James, Jan 1980, Report of 1979 archeological and geological reconnaissance and testing of cave deposits, Porcupine River, Alaska

[891]Douglas, John Scott, 1956, Caves of mystery: the story of cave exploration

[918]Dunkley, John R., 1971, The exploration and speleogeography of Mammoth Cave, Jenolan

[919]Dunkley, John R., 1971, The exploration and speleography of Mammoth Cave, Jenolan

[923]Dunkley, John R., 1986, Caves of north-west Thailand: report of the Australian speleological expeditions, 1983-1986

[927]Dusar, Michiel, 1991, Geological and speleological reconnaissance of the East Yunnan Karst (P.R. China): preliminary results of a field trip between 17.12.1990 and 3.01.1991

[928]Duxbury, J., Jun 1991, The joint Anglo-Soviet speleological expedition to central Asia 1990 "Kugitang 90"

[934]Eaton, Joli (ed), [1988], GYPKAP 1987 annual report

[936]Eavis, A. J. (comp), 1981, Caves of Mulu '80: the limestone caves of the Gunong Mulu National Park, Sarawak

[937]Eavis, A. J. (comp), 1985, Caves of Mulu '84: the limestone caves of the Gunong Mulu National Park, Sarawak

[941]Eberhard, Stefan, 1986, Report of the Tasmanian

4.1. GENERAL

Caverneering Club 1986 speleological reconnaissance expedition to Precipitous Bluff
[986]Endless Caverns, Inc, 1926, Exploring the Endless Caverns of New Market, Virginia in the heart of the Shenandoah Valley: two authentic and thrilling accounts of the adventures of members of the American Museum of Natural History and of the Explorers Club, who vainly sought the end to the Endless Caverns
[987]Endless Caverns, Inc, 1947, Exploring the Endless Caverns of New Market, Virginia. Authentic and thrilling accounts of the adventures of members of the Explorers Club and of the American Museum of Natural History
[993]England, A. W., 1973, Caves, an anthology
[1006]Estes, James H. , Mar 1967, Deep Cave, Texas. A preliminary report of the 1965 Project "Deep"
[1013]Evans, M. J. , [1967], 67 expedition to Crete: BUSS Report
[1017]Exley, Irby Sheck, 1944–1994, 1994, Caverns measureless to man
[1020]Eyre, Jim, 1981, The cave explorers
[1021]Eyre, Jim, 1986, The Ease Gill system: forty years of exploration
[1022]Eyre, Jim, Mar 1967, Lancaster Hole and the Ease Gill Caverns, Casterton Fell, Westmorland
[1029]Farr, Martyn, 1950–, 1984, The great caving adventure
[1031]Farrer, James William, 1970, Further explorations in the Dowkerbottom Caves, in Craven
[1035]Faulkner, Trevor, 1942–, 1977, Report of the SWETC caving club expedition to Norway 1974
[1061]Fincham, Alan Goldworth, 1933–(comp), Jan 1967, The University of Leeds hydrological survey expedition to Jamaica 1963
[1074]Fogg, P. (ed), nd, China Caves Project 1987—1988: the Anglo—Chinese project in caves of South China
[1078]Ford, Derek Clifford, 1935–(scien adv), Jul 1966, The 1965/66 karst hydrology expedition to Jamaica: full report
[1116]Francis, Timothy, 1995, Mendip Caving Group: Belize '94
[1120]Franke, Herbert Werner, 1927–, 1958, Wilderness under the earth
[1169]Gebauer, Herbert Daniel, 1985, Kurnool 1984: report of the speleological expedition to the district of Kurnool, Andhra Pradesh, India
[1173]Gemmell, Arthur, 1915–, 1952, Underground adventure
[1202]Gill, David W., 1988, The untamed river expedition: Nakanai Mountains, East New Britain, Papua New Guinea
[1207]Gillett, J. E. (ed), c1983, Crewe Climbing and Potholing Club: Gouffre Berger Expedition
[1208]Gillieson, D. S., Dec 1977, Lelet: report of the 1976 New Ireland Speleological Expedition
[1237]Graham, Richard, 1970, Exploring the underground
[1291]Gurnee, Jeanne Marie, 1926–, Aug 1968, National Speleological Society field trip to Aguas Buenas Caves, Puerto Rico: February 1968
[1295]Gurnee, Russell Hampton, 1922–1995, 1974, Discovery at the Rio Camuy
[1296]Gurnee, Russell Hampton, 1922–1995, 1968, Discovery at the Rio Camuy, Puerto Rico
[1311]Hacker, Bradley (ed), [1996], Caves of Gunung Buda: report of the joint Sarawak Forest Department and USA caving expedition to Sarawak, Malaysia 1995
[1318]Halliday, William R., 1926–, 1959, Adventure is underground: the story of the great caves of the West and the men who explore them
[1337]Halliwell, Ric (ed), c1995, Gouffre Berger: August 1994
[1363]Harris, Shane R., c1992, UEA speleological expedition northern Turkey: August / September 1992
[1364]Harris, Stephen, 1976, Caves of New Caledonia: report of the 1975 Australian Expedition
[1382]Hassemer, Jerry Herman, 1934–(ed, assemb), Feb 1975, Colorado caves and karst 1974
[1383]Hassemer, Jerry Herman, 1934–(ed, assemb), Apr 1977, Colorado caves and karst 1975-1976
[1414]Heap, David, 1964, Potholing: beneath the northern Pennines
[1415]Heap, David (comp), 1969-1970, Report of the British speleological expedition to arctic Norway, 1969, incorporating the work of the 1968 Hulme Schools expedition
[1454]Hill, A. L. (ed), 1966, Mullamullang Cave expeditions 1966
[1480]Hogg, Garry, 1962, Deep down: great achievements in cave exploration
[1513]Home, Everard, 1756–1832, 1794, Account of some remarkable caves in the principality of Bayreuth: likewise observations on the fossil bones by the late John Hunter
[1541]Hovey, Horace Carter, 1833–1914, 1982, One hundred miles in Mammoth Cave - in 1880: an early exploration of America's most famous cavern
[1559]Hrdlicka, Ales, 1869–1943, 1941, Exploration of mummy caves in the Aleutian Islands ...
[1610]Inside Earth Publications, 1977, 1973, The Wilderness below
[1669]Irwin, David J., Aug 1970, Bristol Exploration Club caving report no 13 [St. Cuthbert's report]: Part H: Rabbit Warren extension
[1670]Irwin, David J., Jun 1970, Bristol Exploration Club caving report no 13: St. Cuthbert's report: Part E: Rabbit Warren
[1674]Irwin, David J. (ed), Oct 1974, Cave notes 74
[1676]Irwin, David J., Oct 1968, The discovery and explo-

ration of St. Cuthbert's Swallet

[1681] Jackson, Donald Dale, 1935–, 1982, Underground worlds

[1686] Jackson, John Wilfrid, 1880–1978, 1910-1913, [Fist [sic] to third] reports on the exploration of "Dog Holes" Cave, Warton Crag, near Carnforth, Lancashire

[1696] James, Ian (ed), [1986], Army caving expedition to South East Java: July and August 1986: exercise Phreatic Diamond

[1697] James, Julia M., 1980, Caves and karst of the Muller Range: report of the 1978 speleological expedition to the Atea Kananda, Southern Highlands, Papua New Guinea

[1699] James, Julia M., 1974, Papua New Guinea Speleological Expedition NSRE 1973: the report of the 1973 Niugini Speleological Research Expedition to the Muller Range

[1724] Jennings, Joseph Newell, 1916–1984, Aug 1961, CEGSA 1965-6 Nullarbor expedition

[1737] Johnson, Cuthbert, 1946–, 1967, The history of Mendip caving

[1746] Jones, H. (auth, ed), c1994, Honduras recce 1994: a short report

[1765] Judson, David M., 1973, Ghar Parau

[1793] Kenney, C. Howard, 1927–1980, 1985, Caving log 1942-1950

[1861] Kranjc, Andrej, 1943–(ed), 1993, The contribution to the history of the speleological explorations of the West Indies at 500—anniversary of the discovery of America

[1914] Lawrence, Joseph, 1924–, 1955, The caves beyond: the story of the Floyd Collins' Crystal Cave exploration

[1915] Lawrence, Joseph, 1924–, 1975, The caves beyond: the story of the Floyd Collins' Crystal Cave exploration

[1926] Lee, Willis Thomas, 1924, A visit to Carlsbad Cavern: recent explorations of a limestone cave in the Guadalupe Mountains of New Mexico reveal a natural wonder of the first magnitude

[1935] LeRoux, J. P., Jan 1993, The report of the 1992 caving expedition to the Chimanimani Mountains in the eastern highlands of Zimbabwe

[1936] Letcher, Montgomery E., 1839, Wonderful discovery: being an account of a recent exploration of the celebrated Mammoth Cave, in Edmonson County, Kentucky, by Dr. Rowan, Professor Simmons and others, of Louisville, to its termination in an inhabited region, in the interior of the earth

[1943] Limbert, Howard, [1986], Mexico 85/86

[1944] Limbert, Howard, [1995], The 1994 British/Vietnamese speleological expedition report March/April 1994

[1976] Lowe, David John, 1988, Tonga '86, Tonga '87 expedition report

[1982] Lumley, Mark, [1989], Mexico: the black holes expedition: the 1988 British expedition to explore the caves of the Sierra de Zongolica

[2001] Lyon, Ben, 1983, Venturing underground: the new speleo's guide

[2009] MacEnery, John M., 1796–1841, 1859, Cavern researches, or, discoveries of organic remains, and of British and Roman reliques, in the caves of Kent's Hole, Anstis Cove, Chudleigh, and Berry Head

[2015] MacMaster University Caving and Climbing Club, [1968], A description of the Sotano del Rio Iglesia

[2021] Maire, Richard (coord, surv, art), Jul 1981, Report of the French Speleological expeditions to PNG in collaboration with the Committee of French Speleological Expedition and the Scientific Committee of the French Federation of Speleology

[2035] [Manchester University Speleological Society], [1975], Report of the 1974 British speleological expedition to Matienzo area of Santander Province of northern Spain

[2038] Mann, Paul (ed), 1994, Oxford University Cave Club Cabeza Julagua expedition final report 28 June - 20 August 1993

[2048] Marochov, Nick (ed), 1992, Below Belize: Queen Mary College speleological expedition to Belize, 1988 and the British speleological expedition to Belize, 1989

[2049] Marochov, Nick (ed), [1992], Below Belize; Queen Mary College Speleological Expedition to Belize 1988 and the British Speleological Expedition to Belize 1989

[2054] Marshall, Des, 1993, Vercors caves

[2057] Martel, Edouard Alfred, 1859–1938, 1914, The caverns of Derbyshire: being an extract from Irlande at Cavernes Anglaises

[2073] Mason, Edmund John, 1911–1993, Jul 1963, The story of Wookey Hole

[2082] Masschelein, Jan (ed), nd, Teng Long Dong, the longest cave in China: report of the first Belgian-Chinese speleological expedition in 1988

[2086] Matthews, Geoff (ed), May 1971, Report of the Nottingham University Students Union Spelaeological Expedition 1970, Picos de Europa North-West Spain

[2108] McClurg, David Robert, 1929–, 1986, Adventure of caving: a practical guide for advanced and beginning cavers

[2109] McClurg, David Robert, 1929–(auth, photo), 1996, Adventure of caving: new updated edition

[2119] McDonald, Mike, c1993, The journal of the Joint Bristol Exploration Club (United Kingdom) and the National Mountaineering Federation of the Philippines caving expedition to the Philippines, January to April 1992

[2150] Mercer, Henry Chapman, 1856–1930, 1897, Cave hunting in Yucatan: a lecture delivered before the Society of Arts of the Massachusetts Institute of Technology, on December 10, 1896

[2151] Mercer, Henry Chapman, 1856–1930, 1897, An exploration of Durham Cave in 1893

[2157] Meredith, Michael Edward, 1943–, [1985], Java

4.1. GENERAL

caves 1983: report of a visit to Indonesia by Australian and British cavers
[2160]Merriam, John Campbell, 1869–1945, 1906, Recent cave explorations in California
[2184]Miller, Tom, [1984], The karst development and associated archeology of the Chiquibul, Belize: (preliminary report of the results of grant #2742-83)
[2254]Monico, Paul (auth, comp), 1969, ULSA explorations: journal II
[2302]Murie, Adolph, 1899–, 1959, Field investigation report: Lehman Caves - Wheeler Peak: portion of southern section of Snake Range, White Pine County, Nevada
[2303]Murie, Adolph, 1899–, Feb 1959, Second field investigation report, Lehman Caves - Wheeler Peak: October 13 to 17, 1958, October 29 to November 13, 1958: portion of southern section of Snake Range, White Pine County, Nevada
[2332]Myrick, Donal Richard, 1938–, 1972, Fern cave: the history of the discovery, exploration and mapping of the Fern Cave system
[2359]National Speleological Society, Mar 1977, A scrap book of articles about the C3 expedition into Floyd Collins Crystal Cave, Kentucky; from Sunday February 14 to February 20, 1954
[2501]Newbould, Ronald L., Mar 1975, The discovery of Barralong Cave, Jenolan Caves, N.S.W.: based on the diary of Ronald L. Newbould, guide, Jenolan Caves, co-discoverer with John P. Culley, senior guide, Jenolan Ceves [sic] of the "Barralong Cave" on the 7th June, 1964
[2509]Nicholson, Frank Ernest, Oct 1930, The exploration of Carlsbad Cavern
[2528]Nott, David, 1975, Into the lost world - a descent into prehistoric time
[2529]Nougier, Louis René, 1912–, 1958, The Cave of Rouffignac
[2538]Nymeyer, Robert, 1910–1983, 1978, Carlsbad, caves, and a camera
[2579]Orrock, Clive, c1985, Imperial College Caving Club Peru '84 Expedition (19th July - 19th Sept 1984)
[2585]Oxford University Cave Club, 1993, Oxford University Cave Club: Huerta del Rey expedition final report
[2586]Oxford University Cave Club, c1994, Oxford University Cave Club: La Verdelluenga 1994 final report
[2587]Oxford University Expedition to Northern Spain (1961), Nov 1965, Oxford University expedition to northern Spain, 1961
[2603]Palmer, Robert John, 1951–1997, 1985, The blue holes of the Bahamas
[2604]Palmer, Robert John, 1951–1997, 1989, Deep into blue holes: the story of the Andros Project
[2605]Palmer, Robert John, 1951–1997 (comp), Mar 1984, The report of the 1981 and 1982 British cave diving expeditions to Andros Island, Bahamas
[2606]Palmer, Robert John, 1951–1997 (ed), 1988, Report of the 1987 international blueholes research project
[2624]Peacock, Norma Dee, 1994, Studies in the Rio Corredor Basin: 1988 to 1991
[2655]Perry, Clair Willard [Clay], 1887–1961, 1946, New England's buried treasure
[2656]Perry, Clair Willard [Clay], 1887–1961, 1948, Underground empire: wonders and tales of New York caves
[2657]Perry, Clair Willard [Clay], 1887–1961, 1939, Underground New England
[2680]Pistole, Nancy (ed), 1995, Proyecto Cheve 1986-1993
[2686]Pohl, Erwin Robert, 1904–, Apr 1955, Vertical shafts in limestone caves
[2687]Pollack, John C., 1949–, 1977, National Speleological Society (USA) 1973 field trip to Greece
[2705]Pratchett, Nick, 1956, International expedition to Gouffre Berger, 1956: the exploits of the British members, Nick Pratchett & Bob Powell
[2709]Prestwich, Joseph, 1872, Report on the exploration of Brixham Cave, conducted by a committee of the geological society, and under the superintendence of Wm. Pengelly, etc
[2718]Price, Graham, Sep 1981, The Holloch: a general account and trip report
[2729]Publications Team (ed), May 1977, Cave Notes 75
[2752]Raines, Terry W., Sep 1988, First reports: 1984-1988
[2757]Ralston, Basil, 1989, Jenolan: the golden ages of caving
[2813]Richard, Colette, , Climbing blind
[2854]Rose, David, 1959–, , Beneath the mountains: exploring the deep caves of Asturias
[2875]St. Pierre, David, 1966, The caves of Graatadalen, northern Norway: report of the Southwest Essex Technical College Caving Club expeditions 1963-5
[2876]St. Pierre, David, Mar 1969, The Caves of Rana, Nordland, Norway
[2892]Savidge, Bob, [1995], Cwmbran Caving Club New Mexico expedition report; caving in the Guadalupe Mountains and Carlsbad National Park April 1994
[2893]Savory, James Henry, 1889–1962, 1989, A man deep in Mendip: the caving diaries of Harry Savory, 1910—1921
[2900]Scheltens, John (auth, comp), [1973], Wind Cave 1972
[2901][Scheltens, John(ed)], [1971], Wind Cave expedition 1971
[2902][Scheltens, John (ed)], [1970], Wind Cave expedition August 1970: a report to the National Park Service by the Windy City Grotto of the National Speleological Society on the results of the August 1970 Wind Cave expedition
[2909]Schreiber, Richard, Sep 1969, Ellison's Cave: Georgia's finest

[2920]Schwatka, Frederick, 1893, In the land of cave and cliff dwellers
[2923]Scott, Fred, 1942–, 1979, Preliminary investigations at Hayes Cave, Hants County, Nova Scotia, in 1978
[2928]Selby, Paul (ed), Feb 1975, Sieben Hengste 74: the karst and caves of the southwestern zone of the Sieben Hengste Ridge: a report by the members of the Croydon Caving Club on an expedition to the Seefeld area, north of Interlaken, Switzerland, summer 1974
[2934]Senior, Kevin J. (ed), Oct 1995, The Yangtze Gorges expedition: China caves project
[2954]Shaw, Trevor Royle, 1928–, 1992, History of cave science: the exploration and study of limestone caves, to 1990
[3000]Slaven, John F., nd, The world within
[3003]Sloane, Bruce, 1935–, 1977, Cavers, caves, and caving
[3006]Smart, Peter L. (comp), Jun 1982, Cathay Pacific Airways Mulu '80 expedition: British Malaysian speleological expedition to Sarawak
[3012]Smith, Anthony, 1926–, , Blind white fish in Persia
[3013]Smith, Anthony, 1926–, 1953, Blind white fish in Persia
[3027]Smith, Marion Otis, 1942–, 1977, The exploration & survey of Ellison's Cave, Georgia
[3061]South Wales Caving Club, c1968, South Wales Caving Club twenty first anniversary publication
[3075]Southhampton University Exploration Society, 1982, Southhampton University Exploration Society: Peru expedition
[3111]Steele, C. William, 1985, Yochib, the river cave: an account of the exploration of the Sumidero Yochib of Mexico, a dangerous and difficult cave
[3122]Stevens, John (ed), Jan 1992, An exploration journal of Llangattwg Mountain
[3126]Stewart, John, 1920–, 1972, Secret of the bats; the exploration of Carlsbad Caverns
[3139]Stone, William Curtis, 1952–, 1992, Inner space: the last terrestrial frontier
[3140]Stone, William Curtis, 1952–, Aug 1994, A report to the government of Mexico on the 1994 San Augustin expedition to Huautla de Jimenez, Oaxaca
[3150]Stratford, Tim (comp), 1992, Toros 89-92; the Camlik Project: a report on the speleological investigations carried out by the Swindon Speleological Society in the Camlik Region of Toros Mountain, Konya Turkey from 1989-1992
[3159]Sumia Konda, Jul 1993, The report of lava flow and lava caves on Mauna Loa
[3186]Tásler, Radko, 1991, Owen 90: New Zealand
[3191]Taylor, Michael Ray, 1959–, 1996, Cave passages: roaming the underground wilderness
[3193]Tazieff, Haroun, 1914–, 1953, Caves of adventure
[3194]Technical Aids in Caving Symposium (1972: Buxton High Peak College), 1972, Technical aids in caving symposium held on Sunday 5th March, 1972 at Buxton High Peak College, Harpur Hill, Buxton, Derbyshire
[3199]Thomas, Alan, 1931– (ed), 1989, The last adventure
[3205]Thompson, Peter, 1943–, 1976, Cave exploration in Canada: a special issue of the Canadian Caver magazine
[3219]Tintilozov, Zurab, 1978, Akhali Atoni's cave
[3245]Truluck, T. F., Oct 1992, SASA (Cape) expedition to Chinhoyi, Zimbabwe August 1992
[3259]Tyler, Ronald M., Dec 1971, Caving handbook: a guide to the underworld
[3337][University of Nottingham], [1965], University of Nottingham Spelaeological Expedition Exploration '64
[3338][University of Nottingham Union], 1969, University of Nottingham Union report of the expedition's co-ordinating committee for 1968
[3342]Vale, W. P., [1990], Aspex '90: Anglo -Soviet Pamirs Expedition 1990: Bajsuntai Khrebetuzbekistan SR, USSR
[3360]Vessely, Carol, [1989], Proyecto Papalo expedition report, 1986-1989: the exploration of Sistema Cuicateca - second deepest cave in the western hemisphere
[3380]Walsh, Frank K., 1982, Oregon Caves discovery & exploration: Oregon Caves National Monument
[3382]Walsh, John Michael, 1947–, 1972, Mexican caving, 1966-1971
[3386]Walters, R. (comp), 1986, The Crocodile Caves of Ankarana: 1986: an expedition to study and explore the limestone massif of Ankarana in northern Madagascar
[3387]Waltham, Anthony Clive, 1942–[Tony], Jun 1971, British karst research expedition to the Himalaya 1970: full report
[3390]Waltham, Anthony Clive, 1942–[Tony], 1986, China caves '85: the first Anglo-Chinese project in the caves of south China
[3402]Warton, Mike, 1993, A preliminary report of findings for the karst terrains feature known as Rugh Cavern, located along northern Seco Creek, northwestern Medina County, Texas
[3405]Watkinson, P., 1968, Expedition 67 to the Gou
[3422]Weaver, Herman Dwight, 1938–, Apr 1980, Missouri: the cave state
[3462]White, Jim, 1919–, 1951, Carlsbad Caverns National Park, New Mexico: its early explorations as told by Jim White
[3476]White, William Blaine, 1934–, Jan 1974, Reconnaissance geology of Timpanogos Cave: Wasatch County, Utah
[3487]Wilkins, Bob, 1989, Cuba Contact '88: an investigative visit by members of the Westminster Speleological Group to the karst of the Sierra de los Organos 18/9/88 - 7/10/88
[3496]Williams, Nick (auth, ed), 1992, Below Belize

4.1. GENERAL

1991: the report of the 1991 expedition to the Little Quartz Ridge Southern Belize February to April 1991

[3499] Williams, Nick (ed), 1990, Queen Mary College: below Belize 1988; the logistics report of the 1988 Speleological Expedition to Belize

[3503] Willis, Dick, 1986, Caving expeditions

[3507] Wilson, George Herbert, 1874?–1958, nd, Cave hunting holidays in Peakland

[3509] Wilson, Jane, 1990, Lemurs of the lost world: exploring the forests and Crocodile Caves of Madagascar

[3534] Wright, David P. (ed), [1988], 1st Claygate "Selachii" Venture Scout Unit expedition to Iceland: 21st July - 9th August 1987

[3543] Yorkshire Subterranean Society (Great Britain), 1976, Caves and caving, 1

[3544] Yorkshire Subterranean Society (Great Britain), 1979, Caves and caving, 2

5 Geology

William B. White
Elizabeth L. White

Department of Geosciences and
Environmental Resources Research Institute
The Pennsylvania State University
University Park, PA 16802

The geological aspects of caves and their associated karst landscapes have been approached from three different points of view. These might be called "top down", "inside out", and "the ways of the waters". "Top down" is geomorphology. Geomorphology of karst regions has traditionally been concerned with surface landforms rather than caves and the practitioners have traditionally been geographers. "Inside out" is what is here called cave geology. Cave geology is primarily concerned with caves rather than surface landforms or drainage basins and the practitioners have often been geologists who were cavers first and geologists second. Cave geology is a descendent of what was earlier called "Speleology", a term that is not used much at the present time. "The ways of the waters" refers to karst hydrology, a relatively late arrival in the English language literature. The concern is with drainage basins, flow paths, cave streams, wells, springs, and the quantitative measurement of water budgets as well as water quality issues. Some of the practitioners are cavers and some are not.

Each of these approaches is simply a different point of view on the same reality. Certainly, the subjects are interconnected and overlapping. The division of book-length works into "Geomorphology", "Geology", and "Hydrology" is to a great extent arbitrary. With these caveats in mind, we can now examine some of the central works on cave geology.

To place the subject in perspective, it is necessary to begin with works not written in English, specifically the writings of Edouard A. Martel. Martel not only was the founder of systematic cave exploration, he also was the founder of scientific cave exploration. In his earliest work, *Les Cevennes* (1890), Martel uses "la grottologie" for the scientific study of caves. By 1900, his term is "la speleologie" and speleology it has remained. The notion of a science of caves caught on quickly. In 1923, Georg Kyrle published *Grundriss der Theoretischen Speläologie*, a comprehensive treatise in which he discusses speleogenesis, the solutional sculpturing of caves, speleothems, cave sediments, and cave hydrology. However, Kyrle, like Martel before him, regards speleology as an all-encompassing science of caves and there are chapters on cave biology, paleontology, and anthropology as well. Félix Trombe (1952) continues this tradition in his *Traite de Speleologie* that furthermore includes chapters on the techniques of exploration. Thus, cave geology is embedded in "speleology" as only one of its components.

One of the first and most influential books in English is *British Caving* (Cullingford, 1953). Here one finds speleogenesis, cave deposits, cave meteorology, water tracing, archaeology and paleontology, cave fauna and flora, as well as chapters on cave exploration, cave diving, cave photography, cave surveying, and even safety and rescue. A replacement book, *The Science of Speleology*, by Trevor Ford and C. H. D. Cullingford appeared in 1976. The new book deals only with the science; exploration and technique have been deleted. Chapters on water chemistry and on volcanic caves make an appearance. Biological topics remain.

A most helpful book, especially for beginners, is George W. Moore and G. Nicholas Sullivan's *Speleology: The Study of Caves*. It first appeared as a slim paperback in 1964, then as a somewhat expanded second edition in 1978, and in 1997 as *Speleology: Caves and the Cave Environment*. There is discussion of cave origins, cave minerals, flora and fauna including microorganisms but little surface karst and hydrology.

So, how do we deal with cave geology? It's everywhere. It's nowhere. Although there are books that deal exclusively with karst hydrology and at least some books that deal exclusively with karst geomorphology, there are few books that deal exclusively with the geology of caves. Maybe that's as it should be. The idea that there is a "science" of speleology that includes everything that one might like to know about caves has proved infeasible. Instead of an inwardly focused study of caves, the current generation of cave scientists are finding that they need to look outward rather than inward. To understand caves, one must also understand the landscape, drainage basins, and rock units in which they occur. One must draw on geochemistry, fluid mechanics, crystallography, and many other disciplines that provide essential understanding of the processes that occur in caves.

This is indeed the approach of most authors of the decades of the 1980's and 1990's. Comprehensive books on the geology of caves in the broad sense but not pretending to span all of the many subjects that comprised "speleology" are Alfred Bögli's (1980) *Karst Hydrology and Physical Speleology*, Wolfgang Dreybrodt's (1988) *Processes in Karst Systems*, William B. White's (1988) *Geomorphology and Hydrology of Karst Terrains*, and Derek C. Ford and Paul W. Williams' (1989) *Karst Geomorphology and Hydrology*. Three of these are textbooks accessible to anyone with either some first hand experience with caves or with a minimum of geological training. Dreybrodt's book is a theoretical treatise that illustrates the input from science and engineering needed to really understand what is going on in caves.

Cave geology has come of age. The geological study of caves is now an integrated part of the geological sciences rather than a portion of an exotic borderland science called speleology. However, cave geology has not lost its roots. Many of the problems discussed in the most recent books also engaged Martel one hundred years ago.

References not in Main Bibliography

Kyrle, G. 1923. *Grundriss der Theoretischen Speläologie.* Vienna: Druck der Österreichischen Staatsdruckerei, 353pp.

Martel, Edouard A. 1890. *Les Cévennes et la region des Causses: (Lozere, Aveyron, Herault, Gard, Ardeche).* Paris: Librairie Ch. Delagrave, 408pp.

Trombe, F. 1952. *Traité de Spéléologie.* Paris: Payot, 376pp.

5.1 General

[25] Alexander, E. Calvin, 1987, Hydrogeologic study of Jewel Cave/Wind Cave: final report -year 2, 4th quarter of year 2, February 1, 1987 through April 30, 1987
[28] Alexander, Emmit Calvin, 1943–, 1986, Hydrogeologic study of Jewel Cave/Wind Cave: final report - phase I, 3rd and 4th quarters of phase I, 1 November 1985 through 30 April 1986
[65] Anderson, Arthur Wilhelm, 1892–, 1938, The Carlsbad Cavern of New Mexico: its history and geology
[67] Anderson, Arthur Wilhelm, 1892–, 1930, The Carlsbad Cavern of New Mexico: its history and geology; formations of the cavern, discovery and exploration, bats of the cavern, cavern geology, administration; the land nobody knows
[118] Anon, 1988, The geology of caves
[119] Anon, 1992, Geology, climate, hydrology and karst formation: field symposium in Australia 4 to 18 December 1992: Buchan - Mt. Gambier / Naracoorte - Nullarbor Plain humid temperate impounded karst, sub-humid temperate syngenetic karst, arid temperate karst: programme and abstracts
[234] Asher-Bolinder, Sigrid, 1992, A review of the deposition and alteration of filled-sink deposits of east-central Missouri
[239] Association of Missouri Geologists. Annual Meeting (19th: 1972: Rolla, Mo), 1972, Guidebook to the karst features and stratigraphy of the Rolla area
[262] Australian Speleological Federation, Feb 1972, Proceedings of the eighth biennial conference of the Australian Speleological Federation
[267] Avid Corporation Creative Learning, 1970, Formation of caves
[270] Back, William (ed), 1992, Hydrogeology of selected karst regions
[288] Bakó, Tamás, 1989, Paleokarst in Hungary
[309] Balogh, K., 1989, Karst hydrological and speleological features
[328] Barr, Thomas Calhoun, 1931–, 1956, The geology of Cumberland Caverns, Warren County, Tennessee
[354] Beck, Barry Frederic, 1944–(ed), 1993, Applied karst geology: proceedings of the fourth Multidisciplinary Conference on Sinkholes and Engineering and Environmental Impacts of Karst, Panama City, Florida 25-27 January 1993
[359] Beck, Barry Frederic, 1944–(ed), Oct 1985, Karst hydrogeology of central and northern Florida: a field trip guidebook produced in conjunction with the 1985 G.S.A. annual meeting and exposition, Orlando, FL., October, 1985
[363] Beck, Barry Frederic, 1944–(ed), 1984, Sinkholes: their geology, engineering and environmental impact, proceedings of the first Multidisciplinary Conference on Sinkholes, Orlando, Florida, 15-17 October 1984
[381] Bender, Lionel, 1989, Cave
[388] Benson, Richard C., 1979, Application of radar and seismic reflection techniques to cavity detection, Medford Cave site, Florida
[394] Bernasconi, R., 1976, The physico-chemical evolution of moonmilk
[395] Berthelsen, O. J., Mar 1992, Guide to cavern engineering
[427] Bolner, Katalin Tak'acs, 1992, ALCADI '92 international conference on speleo history field trip guide
[428] Bolton, David W., nd, The geology and speleogenesis of Red Run Cave Tucker County, West Virginia
[430] Bondesan, A., 1992, Morphometric analysis of dolines
[431] Bonin, Dan (ed), 1980, Hodag hunt 1980: guidebook September 12-14 1980
[456] Bradley, R. W., 1964, Geology of the Capitan Reef Complex of the Guadalupe Mountains, Culberson County, Texas and Eddy County, New Mexico. Field trip guidebook May 6, 7, 8, 9, 1964
[473] Bridgemon, Rondal Rex, 1944–, 1976, Wupatki National Monument earth cracks
[490] [British Speleological Association], Nov 1973, Imperial College karst research expedition to the Peruvian Andes, 1972
[491] [British Speleological Association], 1969, Limestone geomorphology: a study in Jamaica: University of Bristol karst expedition to Jamaica 1967

[525]Brown, Henry S., 1961, Linville Caverns through the ages: the geological story

[528]Brown, Michael Charles, 1972, Karst hydrology of the lower Maligne Basin, Jasper, Alberta

[542]Bryan, Kirk, 1941, Correlation of the deposits of Sandia Cave, New Mexico, with the glacial chronology

[543]Bryan, Kirk, nd, The geology and fossil vertebrates of Ventana Cave

[562]Burger, Andre (ed), 1975, Hydrogeology of karstic terrains: with a multilingual glossary of specific terms

[624]Caving Club SC 33, c1994, 1993 Jamaica expedition

[677]Coggins, Allen R., 1981, The caves of the Tennessee State Natural Areas system

[690]Conference on Karst Geology and Hydrology (1974), 1974, Abstracts of the fourth conference on karst geology and hydrology: West Virginia University, Mountainlair Theater of Mountainlair Building: May 3, 4, and 5, 1974

[709]Cooper, K. M., [1992], Morphology of extinct lava tubes and the implications for tube evolution, Chain of Craters Road, Hawaii Volcanoes National Park, Hawaii

[712]Cope, F. Wolverson, 1976, Geology explained in the Peak District

[718]Coronet Instructional Films, 1943, Limestone caverns

[740]Crawford, Nicholas Charles, 1942–, 1988, Hydrogeology of the Snail Shell Cave: overall creek drainage basin and the ecology of the Snail Shell Cave System

[741]Crawford, Nicholas Charles, 1942–, 1981, Karst hydrogeology and environmental problems in the Bowling Green area

[744]Crawford, Nicholas Charles, 1942–, 1979, The karst hydrogeology of the Cumberland Plateau escarpment of Tennessee: Part I: subterranean stream invasion, conduit cavern development, and slope retreat in the Lost Creek Cove area, White County, Tennessee

[745]Crawford, Nicholas Charles, 1942–, 1979, The karst hydrogeology of the Cumberland Plateau escarpment of Tennessee: Part II: karst valley development and the headward advance of the Sequatchie Valley in the Grassy Cove area, Cumberland County, Tennessee

[746]Crawford, Nicholas Charles, 1942–, 1980, The karst hydrogeology of the Cumberland Plateau escarpment of Tennessee: Part III: karst valley development in the Lost Cove area, Franklin County, Tennessee

[747]Crawford, Nicholas Charles, 1942–, 1987, The karst hydrogeology of the Cumberland Plateau escarpment of Tennessee: subterranean stream invasion, conduit cavern development, and slope retreat in the Lost Creek Cove area, White County, Tennessee

[781]Currens, James C., 1952–, 1979, Bibliography of karst geology in Kentucky

[795]Dalton, Richard F., 1976, Caves of New Jersey

[804]Darton, Nelson Horatio, 1865–1948, 1932, Western Texas and Carlsbad Caverns

[808]Dasque, Jean (prod), 1972, Garden of shadows

[830]Davies, William Edward, 1917–, 1977, Geology of caves

[831]Davies, William Edward, 1917–, Aug 1959, Report on sediments in Mammoth Cave, Kentucky

[853]de Longchène, M., 1993, The underground world of geological marvels

[880]Doe, Michael F., Aug 1985, Geology in the vicinity of Montezuma's Cave, southern Huachuca Mountains, Cochise County, Arizona: final [report]

[888]Dougherty, Percy H. (ed), 1985, Caves and karst of Kentucky

[911]Dumont, Kevin Allen, 1995, Karst hydrology and geomorphology of the Barrack Zourie Cave System, Schoharie County, New York

[925]Durden, C. C., 1928, The Carlsbad Cavern of New Mexico: it's [sic] history and geology

[927]Dusar, Michiel, 1991, Geological and speleological reconnaissance of the East Yunnan Karst (P.R. China): preliminary results of a field trip between 17.12.1990 and 3.01.1991

[929]Dyer, C. F., 1961, Geology and occurrence of ground water at Jewel Cave National Monument, South Dakota

[946]Edwards, Ira, 1925, Underground Geology at the Endless Caverns, New Market, VA

[980]Encyclopaedia Britannica Educational Corporation, 1976, Caves: the dark wilderness

[1002]Eraso, Adolfo (ed), Mar 1988, The Stor-Glomfjord project: karst research - the possibilities of leakage and the selection of damsite alternative

[1005]Erickson, Jon, 1948–, 1993, Craters, caverns, and canyons: delving beneath the earth's surface

[1039]Fellows, L. D., 1970, Guidebook to Ozark carbonate terrane, Rolla - Devils Elbow area, Missouri

[1079]Ford, Derek Clifford, 1935–, 1983, Castleguard Cave and karst, Columbia Icefield Area, Rocky Mountains of Canada: a symposium

[1085]Ford, Trevor David, 1925–(ed), Nov 1972, International seminar on karst denudation

[1093]Ford, Trevor David, 1925–(comp, ed), 1977, Limestone and caves of the Peak District

[1099]Ford, Trevor David, 1925–(chair), Jun 1971, Symposium of the origin and development of caves

[1134]Gabriel, Diaconu, 1990, Closani Cave: mineralogical and genetic study of carbonates and clays

[1140]Gams, Ivan (ed, trans), 1987, Man's impact in Dinaric Karst: guide-book

[1161]Garrigan, George A., 1994, Skyline Caverns and its geologic relationship to the Shenandoah Valley and paleoindian cultures

[1209]Gillieson, David S. (ed), 1992, Geology, climate, hydrology and karst formation: field symposium in Australia 4 to 18 December 1992: Buchan - Mt. Gambier /

5.1. GENERAL

Naracoorte - Nullarbor Plain humid temperate impounded karst, sub-humid temperate syngenetic karst, arid temperate karst: guidebook

[1211]Gillieson, David S., 1996, Caves: processes, development, and management

[1216]Giusti, Ennio V., 1976, Water resources of the North Coast Limestone area, Puerto Rico

[1229]Gospodarič, R. (ed), 1976, Underground water tracing: investigation in Slovenia 1972-1975

[1248]Greeley, Ronald, , Geology and morphology of selected lava tubes in the vicinity of Bend, Oregon

[1249]Greeley, Ronald, 1972, Geology of selected lava tubes in the Bend Area, Oregon

[1257]Gregory, John Walter, 1864–1932, 1930, Some caves and a rock shelter at Loch Ryan and Portpatrick, Galloway

[1266]Grimes, K. G., Oct 1995, South East Karst Province of South Australia: Australian Caves & Karst Management Association October 1995

[1267]Grimes, Ken, February 1996, Field Guide to karst features in southeast South Australia and Western Victoria: for the Karst Studies Seminar, Naracoorte, February 1996

[1310]Hack, John Tilton, 1913–, 1962, Geology of Luray Caverns Virginia

[1321]Halliday, William R., 1926–, Jan 1954, The basic geology of Neff Canyon Cave, Utah ; The speleogenesis of Neff Canyon Cave, Utah

[1322]Halliday, William R., 1926–, Dec 1992, Caves and associated features of open vertical volcanic conduits of the Kaupulehu lava flows xenolith nodule beds, Haulalai Volcano, Hawaii: basic speleological considerations

[1323]Halliday, William R., 1926–, Apr 1992, Caves and associated features of open vertical volcanic conduits of the Kaupulehu lava flows xenolith nodules beds, Hualalai Volcano, Hawaii: preliminary report

[1342]Hamilton-Smith, Elery (ed), February 1996, Abstracts of papers: karst studies seminar Naracoorte

[1417]Heath, Thomas, 1880, Creswell Caves vs Professor Boyd Dawkins

[1447]Herman, Janet S. (ed), 1990, Travertine-marl: stream deposits in Virginia

[1449]Hesler, Donald J., 1990, A hydrogeologic study of the Knox-Skull Cave System, Albany County, New York

[1458]Hill, Carol Ann, 1940–, 1988, The geology of Carlsbad Cavern

[1459]Hill, Carol Ann, 1940–, 1987, Geology of Carlsbad Cavern and other caves in the Guadalupe Mountains, New Mexico and Texas

[1460]Hill, Carol Ann, 1940–, 1996, Geology of the Delaware Basin, Guadalupe, Apache, and Glass Mountains, West Texas and New Mexico

[1462]Hill, John R., nd, The geologic story of Wyandotte Cave

[1464]Hindle, Brian Paul, 1960, Cave formation in northern England

[1560]Hubbard Scientific Company, 1972, Cavern formation

[1573]Humphreys, William F. (ed), Dec 1993, The biogeography of Cape Range Western Australia: being the proceedings of a symposium held under the auspices of the Western Australian Museum in Perth on 21 November 1992 at the Art Gallery of Western Australia

[1578]Huppert, George Nixon, 1944–, Jun 1988, Cave and karst-related theses in United States and Canadian Universities, 1899-1988

[1647]International Symposium on Changing Karst Environments (1994: Oxford, England and Huddersfield, England), Aug 1994, Conference abstracts: [of papers presented at Changing karst environments, hydrogeology, geomorphology and conservation, an international symposium held at the Universities of Oxford and Huddersfield, September 1994]

[1648]International Symposium on Changing Karst Environments (1994: Oxford, England and Huddersfield, England), Mar 1995, Papers presented at the international symposium on changing karst environments, Oxford and Huddersfield, September 1994

[1691]Jagnow, David Henry, Jan 1979, Cavern development in the Guadalupe Mountains

[1692]Jagnow, David Henry, 1992, Stories from stone: the geology of the Guadalupe Mountains

[1705]Janetski, Joel C., 1983, An archaeological and geological assessment of Antelope Cave (NA 5507), Mohave County, Northwestern Arizona

[1707]Jantschke, Herbert, 1994, Tunel de la Atlantida, Haria, Lanzarote, Canary Islands: the hydrodynamic, the chemistry and the minerals of the lava tube, the population density of *Munidopsis polymorpha*

[1725]Jennings, Joseph Newell, 1916–1984, 1971, Karst

[1727]Jennings, Joseph Newell, 1916–1984, 1963, The limestone ranges of the Fitzroy Basin, Western Australia; a tropical semi-arid karst

[1734]Jillson, Willard Rouse, 1890–, 1954, Geology of Crystal Cave in southern Pulaski County, Kentucky

[1735]JLM Visuals, 1979, Caves and other ground water features

[1742]Joiner, Thomas J., 1969, Hydrology of limestone terranes: geophysical investigations

[1757]Jones, William K., 1945–, Mar 1975, Karst hydrogeology: A summary of the principals of ground-water flow through soluble limestone aquifers prepared for the Smithsonian Institution short course on speleology

[1769]Kastning, Ernst H., 1944–, 1991, Appalachian Karst: proceedings of the Appalachian karst symposium: Radford, Virginia, March 23-26, 1991

[1770]Kastning, Ernst H., 1944–, 1975, Cavern development in the Helderberg Plateau, East-central New York

[1772] Kastning, Ernst H., 1944–, Apr 1989, Caves and karst of the New River Valley, Virginia: guidebook for a geologic fieldtrip, eighth annual New River symposium, Radford, Virginia

[1774] Kastning, Ernst H., 1944–, May 1977, Geochemistry of karst waters: a basic bibliography

[1788] Keller, David R. (ed), 1993, Paleokarst, karst-related diagenesis, reservoir development, and exploration concepts: examples from the Paleozoic section of the southern mid-continent: 1993 annual fieldtrip guidebook Permian Basin Section - SEPM Arbuckle Mountains, Oklahoma

[1808] Kerans, C., 1954–, 1990, Depositional systems and karst geology of the Ellenburger Group (Lower Ordovician), subsurface west Texas

[1809] Kerans, C., 1954–, 1989, Karst-controlled reservoir heterogeneity and an example from the Ellenburger Group (Lower Ordovician) of west Texas

[1829] Knapp, Brian J., 1992, Cave

[1862] Kranjc, Andrej, 1943–, 1995, Proceedings of international symposium "man on karst," Postojana, September 23-25, 1993

[1941] Libra, Bob, [1990], 1990 Friends of karst field trip: the Big Spring Basin, northeast Iowa

[1950] Livesay, Elizabeth Ann, 1962, Geology of the Mammoth Cave National Park area

[1953] Lobeck, Armin Kohl, 1886–1958, 1928, The geology and physiography of the Mammoth Cave national park

[1975] Lowe, David (comp), 1995, A dictionary of karst and caves

[1977] Lowry, D. C., 1970, Geology of the Western Australian part of the Eucla Basin

[1978] Lu, Yaoru, 1972, Engineering and geological conditions of karst

[1979] Lübke, Anton, 1890–, 1958, The world of caves

[2013] MacLean, John Patterson, 1848–1939, 1890, An historical, archaeological and geological examination of Fingal's Cave in the island of Staffa

[2014] Macleay, Kenneth, 1819–, 1811, Description of the spar cave, lately discovered in the Isle of Skye: with some geological remarks relative to that island

[2046] Marker, Margaret E., 1970, Echo cave: a tentative Quaternary chronology for the Eastern Transvaal

[2047] Marker, Margaret E., 1975, Lower southeast of South Australia: a karst province

[2094] Matthews, William Henry, 1919–, Feb 1963, The geologic story of Longhorn Cavern

[2125] McGill, William Mahone, 1897–1962, 1933, Caverns of Virginia

[2126] McGrain, Preston (comp), Apr 1954, Itinerary: geology of the Mammoth Cave Region, Barren, Edmonson, and Hart Counties, Kentucky

[2127] McGrain, Preston, 1917–, 1961, The geologic story of Diamond Caverns

[2128] McGrain, Preston, 1917–, 1954, Geology of the Carter and Cascade Caves area

[2176] Milanović, Petar T., 1981, Karst hydrogeology

[2197] Missouri Geological Survey and Water Resources, Dec 1960, Guidebook to the geology of the Rolla area emphasizing solution phenomena

[2209] Missouri Speleological Survey, Jan 1976, Hydrogeologic controls on solution of carbonate rocks in Christian County, Missouri

[2255] Monroe, Watson Hiner, 1907–, 1970, A glossary of karst terminology

[2256] Monroe, Watson Hiner, 1907–, 1976, The karst landforms of Puerto Rico: a discussion of a solution landscape formed in a tropical climate of moderately high rainfall

[2296] Muir, Robert Dalton, 1985, Castleguard

[2327] Mylroie, John Eglinton, 1949–(ed), 1988, Field guide to the karst geology of San Salvador Island, Bahamas: prepared for the 10th Friends of Karst meeting, February 11-15, 1988, College Center of the Finger Lakes, Bahamian Field Station, San Salvador Island, Bahamas

[2330] Mylroie, John Eglinton, 1949–, 1979, Western Kentucky Speleological Survey: annual report 1978

[2331] Mylroie, John Eglinton, 1949–(ed), 1979, Western Kentucky Speleological Survey: annual report 1979

[2353] National Research Council (U.S.). Transportation Research Board, 1976, Subsidence over mines and caverns, moisture and frost actions, and classification

[2422] National Speleological Society. Annual Convention. Guidebook. Geology Field Trip (1986: Tularosa, NM), 1986, Geology field trip guidebook, 45th National Speleological Society (1986) Tularosa, New Mexico, convention

[2423] National Speleological Society. Annual Convention. Guidebook. Geology Field Trip (1990: Yreka, CA), 1990, Geologic features of Mount Shasta and Medicine Lake Volcanoes

[2492] New Mexico Geological Society, 1954, Guidebook of southeastern New Mexico: fifth field conference, October 21-22-23 & 24, 1954

[2515] Nodine-Zeller, Doris E., Aug 1991, Karst-derived early Pennsylvanian conglomerate in Ness County, Kansas: subsurface Mississippian-Pennsylvanian boundary delineated in well core

[2534] Nunez Jimenez, Antonio, 1966, Karstological investigations in Cuba

[2583] Owen, Luella Agnes, Apr 1968, Cave regions of the Ozarks

[2587] Oxford University Expedition to Northern Spain (1961), Nov 1965, Oxford University expedition to northern Spain, 1961

[2596] Palmer, Arthur Nicholas, 1940–, 1981, A geological guide to Mammoth Cave National Park

5.1. GENERAL

[2597] Palmer, Arthur Nicholas, 1940–, 1993, Geology and origin of Mystery Cave, Forestville State Park, Minnesota

[2598] Palmer, Arthur Nicholas, 1940–, 1981, The geology of Wind Cave

[2600] Palmer, Arthur Nicholas, 1940–, 1984, Jewel Cave: a gift from the past

[2602] Palmer, Arthur Nicholas, 1940–, , Wind Cave: an ancient world beneath the hills

[2639] Pemble, Edna R., Nov 1972, Limestones caves and cavers

[2654] Perkal, Malissa Faye, 1978, Uranium decay series studies on archaeological materials: application to thermoluminescence dating and dating of cave deposits

[2668] Phillips, Richard, 1767–1840, 1810, A view of the earth, containing an account of its internal structure; its caves and subterranean passages; its mountains, its rivers and cataracts. Together with a brief view of the universe, to which are added, problems on the globes, directions for drawing maps and tables of latitudes and longitudes

[2670] Picknett, R.G., May 1968, Symposium on cave hydrology and water-tracing

[2685] Pohl, Erwin Robert, 1904–, 1964, Itinerary, geologic features of the Mississippian Plateaus in the Mammoth Cave and Elizabethtown areas

[2703] Powell, Richard Lewis, 1936–, Dec 1970, A guide to the selection of limestone caverns and springs in the United States as national landmarks: for the National Park Service

[2717] Price, Graham (ed), Aug 1977, Fairy Cave Quarry: a study of the caves

[2730] Pulina, Marian (ed), 1992, 2nd international symposium of glacier caves and karst in polar regions: proceedings

[2738] Quick, Peter Gunder, 1994, Vermont caves: a geologic and historical guide

[2763] Rauch, Henry W., nd, Lithologic controls on the development of solution porosity in carbonate aquifers

[2764] Rauch, Henry William (ed), 1974, Proceedings of the fourth conference on karst geology and hydrology

[2833] Robson, M., 1975, The decomposition of limestone breccia from the cave of Les Eyzies, Dordogne, France

[2858] Roswell Geological Society, 1957, Slaughter Canyon, New Cave and Capitan Reef exposures, Carlsbad Caverns National Park: field trip no. 10, April 13, 1957

[2873] Rutter, John, 1796–1851, 1829, Delineations of the north western division of the county of Somerset, and of its antediluvian bone caverns, with a geological sketch of the district

[2886] Sartor, James Doyne, 1962, Meteorological-geological investigations of the Wupatki blowhole system

[2959] Shelley, Maryann, Dec 1956, Karst and caves in the Caucasus

[3036] Smith, William Hovey, 1983, Geology of the Tennille Lime Sinks, Washington County, Georgia ; with an introduction to local geology

[3049] Sonderegger, John L., 1970, Hydrology of limestone terranes: geologic investigations

[3050] Sonderegger, John L., 1970, Hydrology of limestone terranes: photogeologic investigations

[3097] Sprent, J. K., (ed), 1970, Mount Etna caves: a collection of papers covering several aspects of the Mt. Etna and Limestone Ridge caves area of central Queensland

[3128] Stirling, James, 1884, On the caves perforating marble deposits, Limestone Creek (read 12th April 1883)

[3148] Stow, Marcellus Henry, 1902–, [1939], Description of points of scenic, historic, and geologic interest between Washington, D. C., and Luray, Virginia

[3180] Tankersley, Ken, 1975, The cavernous karst land forms of Jackson County, Kentucky

[3195] Tell, Leander, 1961, Erosionsforloppet, med sarskild hansyn till Lummelundagrottorna. The rate of erosion, with special reference to the caves of Lummelunda

[3197] Tennessee Academy of Science, Jul 1930, Journal of the Tennessee Academy of Science. Cave number

[3201] Thomas, Harold S., 1953, The geology of Kankee Caverns [sic]: a geological report on the underground storage of LP-gas on the Phillips pipe line terminal, Kankakee-Illinois

[3203] Thompson, James B., 1958, The geology of Jewel Cave

[3217] Thurgate, Mia E., 1995, Sinkholes, caves and spring lakes: an introduction to the unusual aquatic ecosystems of the lower south east of South Australia: South Australian Underwater Speleological Society occasional paper number 1

[3226] Tratman, Edgar Kingsley, 1899–1978 (ed), 1969, The Caves of North West Clare, Ireland

[3264] U.S. Bureau of Land Management, Roswell District Office, Feb 1993, Bureau of Land Management interim guide for oil & gas drilling operations in cave and karst areas

[3284] U.S. Geological Survey, 1977, Geology of caves

[3364] Vineyard, Jerry Daniel, 1935–, 1967, Guidebook to the geology between Springfield and Branson, Missouri, emphasizing stratigraphy and cavern development

[3371] Wagner, Georg, 1883-1912, 1966, The bears cave of Erpfingen

[3387] Waltham, Anthony Clive, 1942–[Tony], Jun 1971, British karst research expedition to the Himalaya 1970: full report

[3392] Waltham, Anthony Clive, 1942–[Tony] (ed), 1974, Limestone and caves of Northwest England

[3418] Wayland, John Walter, 1872–1962, 1930, The master sculptor; a brief treatise on erosion, and its wondrous effects in the Shenandoah Valley

[3433]Weller, James Marvin, 1899–, 1927, The geology of Edmonson County: a detailed presentation of the physical, stratigraphic, structural, and economic geology of this district

[3439]Wermund, E. G., 1978, Regional distribution of fractures in the southern Edwards Plateau and their relationship to tectonics and caves

[3441]Werner, Eberhard Wolfgang 1942–, 1972, Development of solution features, Cloverlick Valley, Pocahontas County

[3442]Werner, Eberhard Wolfgang, 1942–, Jul 1981, Guidebook to the karst of the Central Appalachians; prepared for the Eighth International Congress of Speleology, Bowling Green, Kentucky, U.S.A., July 18 to 24, 1981

[3444]West Texas Geological Society, 1969, Delaware Basin exploration: Guadalupe Mts., Hueco Mts., Franklin Mts., geology of the Carlsbad Caverns ; Nov. 6th, 7th and 8th, 1969

[3445]West Texas Geological Society, 1969, Delaware Basin exploration: Guadalupe Mts., Hueco Mts., Franklin Mts., geology of the Carlsbad Caverns, Nov. 6th, 7th and 8th, 1969

[3446]West Texas Geological Society, 1968, Guadalupe Mts., Hueco Mts., Franklin Mts., geology of the Carlsbad Caverns, Delaware Basin exploration, Oct. 31st, Nov. 1st and 2nd, 1968

[3469]White, William Blaine, 1934–, May 1992, The Appalachian valleys of central Pennsylvania

[3470]White, William Blaine, 1934–, 1984, Cave and karst-related papers in the mainstream scientific literature: a bibliography

[3472]White, William Blaine, 1934–(ed), 1976, Caves of Western Pennsylvania compiled by members of the Mid-Appalachian Region of the National Speleological Society

[3473]White, William Blaine, 1934–(ed), 1976, Geology and biology of Pennsylvania caves

[3476]White, William Blaine, 1934–, Jan 1974, Reconnaissance geology of Timpanogos Cave: Wasatch County, Utah

[3485]Wilford, Gerald Edward, 1964, The geology of Sarawak and Sabah caves

[3490]Wilkinson, Tony J., 1990, Franchthi Paralia—the sediments, stratigraphy, and offshore investigations

[3537]Wyllie, Diana, 1973, Caves—origins, development and formations

[3542]Yevjevich, Vujica, 1981, Karst water research needs

[3550]Zámbó, László (ed), 1993, Conference on the karst and cave research activities of education and research institutions in Hungary: papers: Jósvafő 17-19 May 1991

[3555]Zhu Xuewen, 1932-, 1988, Guilin karst

5.2 Carbonates

[345]Bathurst, R. C. G., 1972, Carbonate sediments and their diagenesis

[1887]LaMoreaux, Philip Elmer, 1920–(ed), 1984, Guide to the hydrology of carbonate rocks

[1888]LaMoreaux, Philip Elmer, 1920–, 1970, Hydrology of limestone terranes: annotated bibliography of carbonate rocks

[1890]LaMoreaux, Philip Elmer, 1920–(ed), 1986, Hydrology of limestone terranes: annotated bibliography of carbonate rocks, volume three

[1891]LaMoreaux, Philip Elmer, 1920–(ed), Jan 1975, Hydrology of limestone terranes: progress of knowledge about hydrology of carbonate terranes with an annotated bibliography of carbonate rocks

5.3 Cave Classifications

[3265]U.S. Bureau of Land Management, Roswell District Office, [1982], Cave inventory and classification systems

5.3.1 Cenotes

[1342]Hamilton-Smith, Elery (ed), February 1996, Abstracts of papers: karst studies seminar Naracoorte

[2629]Pearse, Arthur Sperry, 1877–1956, 1936, The cenotes of Yucatan

[2630]Pearse, Arthur Sperry, 1877–1956, Feb 1936, The cenotes of Yucatan: a zoological and hydrological study

5.3.2 Dolomite Caves

[2298]Müller, H. O. (comp, ed), 1985, Dolomite caves of the eastern Transvaal (South Africa): a free caver monograph 1985

5.3.3 Ice Caves

[13]Adams, William Henry Davenport, 1828–1891, 1886, Famous caverns and grottoes: described and illustrated

[291]Balch, Edwin Swift, 1856–1927, 1970, Glaciéres; or, freezing caverns

[292]Balch, Edwin Swift, 1856–1927, 1896, Ice caves and the causes of subterranean ice

[532]Browne, George Forrest, 1833-1930, 1865, Ice-caves of France and Switzerland. A narrative of subterranean exploration

[607]Casteret, Norbert, 1897–1987, 1954, The darkness under the earth

[1001]Eraso, Adolfo (ed), 1991, 1st international symposium of glacier caves and karst in polar regions. Proceedings

[1335] Halliday, William R., 1926–, 1972, The Paradise Ice Caves, Mount Rainier National Park, Washington
[1418] Hedges, James, 1975, The Ice Cave at Decorah, Iowa
[1430] Henderson, Junius, 1865–1937, Oct 1932, Caverns, ice caves, sinkholes, and natural bridges, part I & II
[1820] Kimball, Herbert Harvey, 1862–, 1901, Ice caves and frozen wells as meteorological phenomena ...
[1824] Kiver, Eugene P., 1973, Summit firn cave study, 1970-1973, Mount Rainier, Washington
[2299] Munich, Frederick J. (auth, prod), 1977, Fumaroles, ice caves and time: the variety and scope of research on Mt. Baker
[2611] Papadakis, Peggy, May 1969, The story of Crystal Ice Caves: within the Great Rift National Landmark, American Falls, Idaho
[2635] Pelech, Johann E., 1879, The valley of Stracena and the Dobschau ice-cavern (Hungary)
[2672] Pictet, Marc-Auguste, 1752–1825, 1823, On the ice-caves or natural ice-houses found in some of the caverns of the Jura and the Alps
[2730] Pulina, Marian (ed), 1992, 2nd international symposium of glacier caves and karst in polar regions: proceedings
[2830] Robinson, E. Russell, 1932–1981, 1978, The story of the Shoshone Indian ice caves
[2935] Seppala, Matti, 1973, Glacier cave observations on Llewellyn Glacier, British Columbia

5.3.4 Lava Tubes

[13] Adams, William Henry Davenport, 1828–1891, 1886, Famous caverns and grottoes: described and illustrated
[171] Anon, [1993], Oregon Caves: a pictorial souvenir guide
[229] Arnold, Charles L., 1986, Inside the caves: Lava Beds National Monument
[235] Aslett, James L., 1982, Lava beds underground
[242] Athens, J. Stephen, Aug 1989, Prehistoric upland bird hunters: archaeological inventory survey and testing for the MPRC project area and the Bobcat Trail Road, Pohakuloa Training Area, Island of Hawaii
[243] Atkinson, Anne, 1995, Undara Volcano and its lava tubes: a geological wonder of Australia in Undara Volcanic National Park, North Queensland
[265] Australian Speleological Federation Inc, 1988, Preprints of papers for the 17th Biennial Conference of the Australian Speleological Federation Tropicon Conference, Lake Tinaroo, Far North Queensland. 27th to 31st December 1988
[271] Baddeley, Glenn (ed, auth), , Vulcon Guidebook: lava features and limestone karst of Victoria and southeastern South Australia. Vulcon 1995 20th Biennial Conference. Australian Speleological Federation, Inc, Hamilton, Victoria 2-6 January 1995
[272] Baddeley, Glenn (ed), 1995, Vulcon precedings: papers submitted for presentation. Vulcon 20th Biennial Conference. Australian Speleological Federation, Inc, Hamilton, Victoria 2-6 January 1995
[709] Cooper, K. M., [1992], Morphology of extinct lava tubes and the implications for tube evolution, Chain of Craters Road, Hawaii Volcanoes National Park, Hawaii
[1138] Gagné, Wayne C., 1974, The cavernicolous fauna of Hawaiian lava tubes, part VI: Mesoveliidae or water treaders (Heteroptera)
[1139] Gagné, Wayne C., Jul 1974, The cavernicolous fauna of Hawaiian lava tubes, part VII. Emesinae or thread-legged bugs (Heteroptera: Redvuiidae)
[1248] Greeley, Ronald, , Geology and morphology of selected lava tubes in the vicinity of Bend, Oregon
[1249] Greeley, Ronald, 1972, Geology of selected lava tubes in the Bend Area, Oregon
[1250] Greeley, Ronald, 1971, Lava tubes of the Cave Basalt, Mount St. Helens, Washington
[1251] Greeley, Ronald, 1971, Observations of actively forming lava tubes and associated structures, Hawaii
[1331] Halliday, William R., 1926–, Oct 1995, Initial inventory of named caves and related features and cave-related place names in Hawaii
[1332] Halliday, William R., 1926–, Aug 1991, Introduction to Hawaiian caves: field guide for the 6th international symposium on volcano-speleology: Hilo, HI
[1372] Harter, J. W., May 1970, Classification of lava tubes
[1373] Harter, Russell George, 1947–, Jun 1971, Bibliography on lava tube caves
[1374] Harter, Russell George, 1947–, 1972, Bibliography on lava tube caves, supplement
[1543] Howarth, Francis Gard, 1940–, 1972, Ecological studies on Hawaiian lava tubes
[1656] International Symposium on Vulcanospeleology (3rd: 1982: Bend, Oregon, USA), 1993, Proceedings of the third international symposium on volcanospeleology: a special session of the 39th annual convention of the National Speleological Society Bend, Oregon, USA: July 30-August 1, 1982: with related biovulcanospeleological papers also presented at the 39th annual meeting of the National Speleological Society
[1657] International Symposium on Vulcanospeleology (5th: 1988: Izunagaoka Hot Springs, Japan), 1988, Excursion guide book: 5th international symposium on vulcanospeleology: [pre-activity: November 4-9, 1988: post-activity: November 13-20, 1988]
[1658] International Symposium on Vulcanospeleology (5th: 1988: Izunagaoka Hot Springs, Japan), 1988, 5th International symposium on vulcanospeleology: program Izunagaoka: November 9-11, 1988
[1659] International Symposium on Vulcanospeleology (6th: 1991: Hilo, Hawaii), 1992, 6th international sympo-

sium on vulcanospeleology, Hilo, Hawaii August 1991

[1660] International Symposium on Vulcanospeleology (7th: 1994: Canary Islands), 1994, VII International Symposium on Volcanospeleology, La Palma - El Hierro - Tenoride Canary Islands, 4-11 November, 1994, Program and Abstracts

[1661] International Symposium on Vulcanospeleology and Its Extraterrestrial Applications (1972: White Salmon, Washington), 1977, Proceedings of the international symposium on vulcanospeleology and its extraterrestrial applications: a special session of the 29th Annual Convention of the National Speleological Society, White Salmon, Washington, 16 August 1972

[1667] Ireton, Frank, Jul 1973, Tee-Maze cave system, Lincoln County, Idaho, U.S.A

[1740] Johnston, David Alexander, 1949-1980, 1981, Guide to some volcanic terranes in Washington, Idaho, Oregon and Northern California

[1882] Lamb, Susan, 1991, Lava Beds National Monument

[1905] Larson, Charles Victor, 1928–, 1987, Central Oregon caves

[1906] Larson, Charles Victor, 1928–, 1993, An illustrated glossary of lava tube features

[1907] Larson, Charles Victor, 1928–, 1989, Lava Beds caves

[1908] Larson, Charles Victor, 1928–, 1987, Lava River Cave

[1923] Lee, G., 1988, The rock art of Petroglyph Point and Fern Cave, Lava Beds National Monument: final report

[2121] McEldowney, Holly, Oct 1991, Survey of lava tubes in the former Puna Forest Reserve and on adjacent State of Hawaii lands

[2299] Munich, Frederick J. (auth, prod), 1977, Fumaroles, ice caves and time: the variety and scope of research on Mt. Baker

[2394] National Speleological Society. Annual Convention. Guidebook (1972: White Salmon, WA), 1972, Selected caves of the Pacific Northwest with particular reference to the vulcanospeleology of the state of Washington

[2414] National Speleological Society. Annual Convention. Guidebook (1993: Pendleton, OR), 1993, The 1993 NSS Convention guidebook

[2423] National Speleological Society. Annual Convention. Guidebook. Geology Field Trip (1990: Yreka, CA), 1990, Geologic features of Mount Shasta and Medicine Lake Volcanoes

[2523] Northwest Caving Association. National Speleological Society (1987), 1987, Northwest Caving Association regional meeting guidebook: caves of the Peterson Prairie area

[2732] Purcell, David, 1977, Guide to the lava tube caves of central Oregon

[2733] Purcell, David, 1977, Spelunking (the exploration of caves) guide to central Oregon

[2951] Sharpe, Grant William, 1925–, 1973, Evaluation of Mount St. Helens lava casts and caves, Cowlitz and Skamania Counties, Washington, for eligibility for registered natural landmark designation

[2981] Silver, Constance S., Mar 1982, The pictographs of Fern Cave, Lava Beds National Monument: agents of deterioration and prospects for conservation

[3159] Sumia Konda, Jul 1993, The report of lava flow and lava caves on Mauna Loa

[3298] U.S. National Park Service, 1990, General management plan, wilderness suitability study: El Malpais National Monument, New Mexico

[3404] Waters, Aaron Clement, 1905–, 1990, Selected caves and lava tube systems in and near Lava Beds National Monument, California

5.3.5 Sandstone Caves

[3175] Szentes, Georg, Jul 1989, Sandstone caves in Nigeria

5.3.6 Sea Caves

[180] Anon, 1994, Sea Lion Caves Oregon coast

[181] Anon, c1993, Sea Lion Caves: America's largest sea cave, on U.S. Highway 101...just North of Florence, Oregon, Oregon coast

[555] Bunnell, David Edward, 1952–, 1993, Sea caves of Anacapa Island

[556] Bunnell, David Edward, 1952–, 1988, Sea caves of Santa Cruz Island

[2005] MacCulloch, Donald B., 1951, The island of Staffa

[2006] MacCulloch, Donald B., 1927, The island of Staffa: with 12 photographs, two plans and one map, by the author

[2013] MacLean, John Patterson, 1848–1939, 1890, An historical, archaeological and geological examination of Fingal's Cave in the island of Staffa

[2251] Moller, Jacob, 1935–, 1982, Coastal caves, marine limits, and ice retreat in Lofotenvesteralen, North Norway

[2603] Palmer, Robert John, 1951–1997, 1985, The blue holes of the Bahamas

[2604] Palmer, Robert John, 1951–1997, 1989, Deep into blue holes: the story of the Andros Project

[2606] Palmer, Robert John, 1951–1997 (ed), 1988, Report of the 1987 international blueholes research project

[3142] Stoneman, John (narr), 1990, The cave divers

5.3.7 Sinkholes

[37] Aley, Thomas John, 1938–, 1972, Groundwater contamination and sinkhole collapse induced by leaky impoundments in soluble rock terrain

5.3. CAVE CLASSIFICATIONS

[38]Aley, Thomas John, 1938–, 1981, Hydrogeologic mapping of unincorporated Greene County, Missouri, to identify areas where sinkhole flooding and serious groundwater contamination could result from land development

[71]Anderson, Warren, 1975, Hydrology of three sinkhole basins in Southwestern Seminole County Florida

[112]Anon, 1994, Environmental Education Enterprises, Inc. short course showcase January 26-27-28, 1994, Grosvenor Resort Orlando, Fl: Winter Park sinkhole guidebook

[113]Anon, 1986, Environmental geology and water sciences: special issue on sinkholes

[234]Asher-Bolinder, Sigrid, 1992, A review of the deposition and alteration of filled-sink deposits of east-central Missouri

[266]Autin, Whitney J., 1984, Observations and significance of sinkhole development at Jefferson Island

[348]Baumgardner, Robert W., 1982, Formation of the Wink Sink: a salt dissolution and collapse feature, Winkler County, Texas

[354]Beck, Barry Frederic, 1944–(ed), 1993, Applied karst geology: proceedings of the fourth Multidisciplinary Conference on Sinkholes and Engineering and Environmental Impacts of Karst, Panama City, Florida 25-27 January 1993

[355]Beck, Barry Frederic, 1944–(ed), Jul 1984, A computer based inventory of recorded recent sinkholes in Florida

[356]Beck, Barry Frederic, 1944–(ed), 1989, Engineering and environmental impacts of sinkholes and karst: proceedings of the third Multidisciplinary Conference on Sinkholes and the Engineering and Environmental Impacts of Karst, St. Petersburg Beach, Florida, 2-4 October 1989

[358]Beck, Barry Frederic, 1944–(ed), 1995, Karst geohazards: engineering and environmental problems in karst terrane: proceedings of the fifth Multidisciplinary Conference on Sinkholes and the Engineering and Environmental Impact of Karst, Gatlinburg, Tennessee, 2-5 April 1995

[360]Beck, Barry Frederic, 1944–(ed), 1987, Karst hydrogeology: engineering and environmental applications: proceedings of the second Multidisciplinary Conference on Sinkholes and the Environmental Impacts of Karst, Orlando, Florida 9-11 February 1987

[361]Beck, Barry Frederic, 1944–(ed), 1991, The sinkhole hazard in Pinellas County: a geological summary for planning purposes

[362]Beck, Barry Frederic, 1944–(ed), 1986, Sinkholes in Florida: an introduction

[363]Beck, Barry Frederic, 1944–(ed), 1984, Sinkholes: their geology, engineering and environmental impact, proceedings of the first Multidisciplinary Conference on Sinkholes, Orlando, Florida, 15-17 October 1984

[364]Beck, Barry Frederic, 1944–(ed), 1987, Use of ground penetrating radar for detecting and evaluating the sinkhole hazard in Florida

[365]Beck, Barry Frederic, 1944–(ed), 1985, Water on and under the ground: an introduction to the urban hydrogeology of the Orlando area

[529]Brown, Paul Martin, 1973, An introduction to the limestone sinkholes of northeastern Michigan

[634]Chan, Y. C, 1994, Factors affecting sinkhole formation

[748]Crawford, Nicholas Charles, 1942–, 1988, Karst hydrologic problems of south central Kentucky: ground water contamination, sinkhole flooding, and sinkhole collapse: field trip guide

[874]Dinger, James S., 1991, Ordinance for the control of urban development in sinkhole areas in the blue grass karst region, Lexington, Kentucky

[906]Drnevich, Vincent P., Aug 1972, Location of solution channels and sinkholes at dam sites and backwater areas by seismic methods

[947]Ege, John R., 1984, Formation of solution-subsidence sinkholes above salt beds

[1003]Erchul, Ronald A., Mar 1988, Geotechnical applications of the self potential (SP) method. Report 1: the use of self potential in the detection of subsurface flowpatterns in and around sinkholes

[1260]Gries, John Paul, 1911-, Jun 1969, Continued investigation of water losses to sinkholes in the Pahasapa Limestone and their relation to resurgent springs, Black Hills, South Dakota

[1317]Hallberg, George R., May 1982, Sinkholes, hydrogeology, and ground-water quality in northeast Iowa

[1430]Henderson, Junius, 1865–1937, Oct 1932, Caverns, ice caves, sinkholes, and natural bridges, part I & II

[1471]Hockensmith, Brenda Louise, Feb 1987, Investigation of sinkhole occurrences at Goretown, near Loris, South Carolina

[1478]Hoffmann, Paul R., Mar 1975, Newsome sinks - our national landmark

[1520]Horne, Peter, 1988, Gouldens Hole - 5L8 - mapping project, 1987 - 1988

[1522]Horne, Peter, 1985, Piccaninnie Ponds mapping project, 1984/85

[1563]Huber, Gary, Oct 1989, Sinkholes: landowner perceptions of a unique source of groundwater contamination

[1613]International Association of Engineering Geology, 1973, Symposium: sink - holes and subsidence engineering - geological problems related to soluble rocks

[1666]Iowa. Department of Agriculture and Land Stewardship, Jul 1989, Agricultural drainage well research and demonstration project. Sinkhole/karst area demonstration project

[1704]Jammal & Associates, Inc, Mar 1982, The Winter Park sinkhole: a report to the city of Winter Park, Florida

[1789]Kemmerly, Philip R., 1980, Subhold collapse in

Montgomery County, Tennessee: an overview for the planning process

[2027]Malott, Clyde Arnett, 1887–1950, 1974, Significant features of the Indiana karst

[2353]National Research Council (U.S.). Transportation Research Board, 1976, Subsidence over mines and caverns, moisture and frost actions, and classification

[2483]National Water Well Association, 1987, Proceedings of environmental problems in karst terranes and their solutions conference, October 28-30, 1986, Bowling Green, Kentucky

[2505]Newton, John G., 1929–, 1987, Development of sinkholes resulting from man's activities in the eastern United States

[2506]Newton, John G., 1929–, 1973, Sinkhole problem along proposed route of interstate highway 459 near Greenwood, Alabama

[2507]Newton, John G., 1929–, 1971, Sinkhole problem in and near Roberts Industrial subdivision, Birmingham, Alabama: a reconnaissance

[2868]Russell, D. J., 1993, Role of the Sylvania Formation in sinkhole development, Essex County

[2904]Schiffer, Donna M., 1996, Hydrology of the Wolf Branch sinkhole basin, Lake County, east-central Florida

[2987]Sinclair, William C., 1985, Types, features, and occurrence of sinkholes in the karst of west-central Florida

[2988]Sinclair, William Campbell, 1928–, 1982, Sinkhole development resulting from ground-water withdrawal in the Tampa area, Florida

[3001]Slifer, Dennis William, 1947–, Apr 1989, Sinkhole dumps and the risk to groundwater in Virginia's karst areas: final report to the Virginia Environment Endowment

[3036]Smith, William Hovey, 1983, Geology of the Tennille Lime Sinks, Washington County, Georgia ; with an introduction to local geology

[3078]Sowers, George F., 1996, Building on sinkholes: design and construction of foundations in karst terrain

[3094]Spigner, B. C., Jul 1978, Review of sinkhole-collapse problems in a carbonate terrane

[3132]Stohr, C. J., 1974, Delineation of sinkholes using thermal infrared imagery

[3179]Talley, John H., Mar 1981, Sinkholes, Hockessein area, Delaware

[3217]Thurgate, Mia E., 1995, Sinkholes, caves and spring lakes: an introduction to the unusual aquatic ecosystems of the lower south east of South Australia: South Australian Underwater Speleological Society occasional paper number 1

[3241]Trommer, J. T., 1987, Potential for pollution of the Upper Floridan aquifer from five sinkholes and an internally drained basin in west-central Florida

[3260]United States Army Corps of Engineers, Dec 1987, Geophysical phase IV monitoring of potential sinkhole collapse at military ocean terminal. Sunny Point Access Railroad, Boiling Springs, N.C.

[3351]Vandike, James E., Oct 1988, Using digital watershed modeling to estimate sinkhole-flooding potential at Perryville, Missouri

[3401]Warren, William M., 1945–, 1976, Sinkhole occurrence in western Shelby County, Alabama

[3429]Wegrzyn, Mikolaj, Jun 1984, Rainstorm related terrain failures in Puerto Rico: final technical report

[3452]Wetterhall, W. S., 1965, Reconnaissance of springs and sinks in west-central Florida

[3512]Wilson, William L., Jul 1987, Investigation of karst-related subsidence near the southwest landfill, Alachua County, Florida, using ground penetrating radar: a report

[3513]Wilson, William L. (ed), 1988, Karst of the Orlando area: a guidebook prepared for the engineering and geology of karst terranes short course, Orlando, Florida, February 1-5, 1988

5.3.8 Springs

[537]Brune, Gunnar, 1981, Springs of Texas

[636]Chandler, Robert V. , 1987, Springs in Alabama

[792]Dake, Charles Laurence, 1883–, 1924, Subterranean stream piracy in the Ozarks

[1048]Ferguson, George E., 1947, Springs of Florida

[2679]Pipkin, Turk (ed), 1993, Barton Springs eternal: the soul of a city

[3217]Thurgate, Mia E., 1995, Sinkholes, caves and spring lakes: an introduction to the unusual aquatic ecosystems of the lower south east of South Australia: South Australian Underwater Speleological Society occasional paper number 1

[3346]Van Couvering, John A., 1962, Characteristics of large springs in Kentucky

[3348]Van Everdingen, R. O., 1972, Thermal and mineral springs in the Southern Rocky Mountains of Canada

[3365]Vineyard, Jerry Daniel, 1935–, 1974, Springs of Missouri

[3452]Wetterhall, W. S., 1965, Reconnaissance of springs and sinks in west-central Florida

5.3.9 Talus Caves

[3083]Speare, Eva A., nd, The story of Polar Caves

5.3.10 Underwater Caves

[595]Canadian Broadcasting Corporation, 1972, The Blue holes of Andros

5.4 Evaporites

[348]Baumgardner, Robert W., 1982, Formation of the Wink Sink: a salt dissolution and collapse feature, Winkler County, Texas

5.4.1 Gypsum

[378]Belski, David Stanley, 1937–, 1992, GYPKAP 1987 annual report
[934]Eaton, Joli (ed), [1988], GYPKAP 1987 annual report
[1924]Lee, Joli (Eaton) (ed), 1996, GYPKAP report volume 3: January 1992—April 1996

5.5 Geochemistry

[305]Ball, J. W., 1987, WATEQ4F: a personal computer FORTRAN transformation of the geochemical model WATEQ2 with revised data base
[345]Bathurst, R. C. G., 1972, Carbonate sediments and their diagenesis
[577]Caldwell, Douglas E., 1981, Limnological survey of cavern pools: proposal to National Park Service, Carlsbad Caverns National Park
[2193]Miotke, Franz-Dieter, 1934–, 1972, Genetic relationship between caves and landforms in the Mammoth Cave National Park area; a preliminary report
[2483]National Water Well Association, 1987, Proceedings of environmental problems in karst terranes and their solutions conference, October 28-30, 1986, Bowling Green, Kentucky
[2612]Parizek, R. R., Nov 1971, Hydrogeology and geochemistry of folded and faulted carbonate rocks of the central Appalachian type and related land use problems
[2887]Sasowsky, Ira Daniel, 1959–(ed), 1994, Breakthroughs in karst geomicrobiology and redox geochemistry: abstracts and field trip guide for the symposium held February 16 - 19, 1994, Colorado Springs, Colorado
[3214]Thrailkill, John Vernon, 1930–, 1970, Solution geochemistry of the water of limestone terrains
[3348]Van Everdingen, R. O., 1972, Thermal and mineral springs in the Southern Rocky Mountains of Canada

5.6 Geomorphology

William B. White
Elizabeth L. White

Geomorphology is the study of the shape of the Earth's surface, the various landforms—mountains, rivers, valleys—that make up the landscape, and of the geological processes that are responsible for the form and evolution of the landforms. As a distinct subject of scientific investigation, geomorphology dates back to the latter decades of the 19th century to the work of William Morris Davis in the United States and Albrecht Penck in Germany. In the United States, geomorphology is generally taught as part of geology; in Great Britain it is often part of physical geography.

Karst geomorphology has its roots in the study of Adriatic karst, particularly near Trieste, and also the regions that are now located in southwestern Slovenia, southern Bosnia, southern and southeastern Croatia, and parts of Serbia and Montenegro. Major writings by Cvijić, Glund, Penck, Katzer, and others in the late nineteenth and early 20th century who established the systematics and much of the nomenclature for karst landforms. At that time most of the karst lands were part of the Austro-Hungarian Empire. Thus it was that the foundations of karst geomorphology are mainly written in German.

The founding father of karst geomorphology is generally considered to be Jovan Cvijić. Cvijić wrote a Ph.D. thesis on karst under the supervision of Albrecht Penck and the resulting 111 page monograph *Das Karstphanomen* published in 1893, is the first systematic and organized discussion of the subject. Cvijić's later and more mature reflections on karst were prepared as a new monograph following World War I. The manuscript was delivered to the publisher, then lost, and finally resurfaced many years later. It was published, in French translation, as *La Géographie des Terrains Calcaires* by the Serbian Academy of Sciences in 1960.

Following World War II and the rise of Communist states in eastern Europe, karst researchers turned to practical matters of water supply, irrigation, flood control, and hydroelectric power generation. For post-World-War-II geomorphological research, there is the book by Laszlo Jakucs. Originally published in Hungarian as *A Karsztok Morfogenetikáia*, which appeared in English translation as *Morphogenetics of Karst Regions* in 1977. Jakucs' book draws predominantly from Hungarian, Romanian, Austrian, and Russian literature, thus providing a useful access to writings hidden behind a substantial language barrier.

In spite of the fundamental work on karst in eastern Europe in the early decades of the 1900s, there were few systematic studies of karst landscapes in western Europe until the publication of J. Corbel's *Les Karsts du Nord-Ouest de L'Europe et de Quelques Regions de Comparaisons* in 1957. Corbel is interested in climatic influences on karst development and this volume is an attempt to support his contention (later shown to be incorrect) that the rate of karst development is greatest in alpine and arctic climates. There is great emphasis on the arctic karst of Scandinavia.

The first English language book on karst geomorphology is *Karst* by J. N. Jennings in 1971. It is volume 7 of an introductory series on landforms intended for supplementary reading in undergraduate geomorphology or physical geography courses. As a result, the book is highly readable. Jennings prepared a second edition, actually a new book, *Karst Geomorphology* (1985), that was completed just before his death in 1984.

In Great Britain, the dominant figure in karst geomorphology is Marjorie Mary Sweeting. Like Jennings, Sweeting was educated at Cambridge as a physical geographer. She spent her entire career at Oxford. Many of the prominent British-trained karst geomorphologists were her graduate students. Her classic book *Karst Landforms*, which appeared in 1972, is longer and rather more advanced than Jenning's book. Sweeting's book was the definitive work on karst geomorphology for many years. She also edited *Karst Geomorphology* (1981), which reprinted many of the classic journal articles on karst including English translations of many papers that originally appeared in German.

In the United States, karst geomorphology attracted very little attention until the 1960s. There were publications on the southern Indiana karst, on the Mammoth Cave area, and on a few other areas but for most part there were no major works that addressed surface landforms rather than caves. Karst, if it was mentioned at all, received only superficial discussion in most geomorphology textbooks of the period.

The surface expression of karst is a regional affair. The landforms, their degree of development, and their distribution all depend on the geologic setting, the climatic setting, and the tectonic and geomorphic history of the region. In the ideal world, each major karst region would have a book-length work describing the geology, the landforms, and an interpretation of the geomorphic history. The world is considerably less than ideal but the closest thing to it is an important book, *Karst: Important Karst Regions of the Northern Hemisphere*, edited by M. Herak and V. T. Stringfield in 1972. Karst regions are subdivided on political rather than geologic boundaries but nevertheless this book provides systematic reviews and lengthy bibliographies concerning the karst of Yugoslavia, Italy, France, Germany, Austria, Hungary, Czechoslovakia, Poland, Romania, USSR, Great Britain, Jamaica, and the United States. The book does have its shortcomings. Conspicuous by their absence are China, Malaysia, and the rest of southeast Asia. The United States is covered in one short chapter, and Canada, Mexico, and Central America are omitted.

Until the opening of China to the West in the 1970s, there were few descriptions in English of the spectacular Chinese karst. Since then, Chinese karst has been displayed first in a series of picture books and more recently in books of more scientific depth. The first picture book appeared in 1976. In the best socialist tradition, *Karst in China* has no listed authors and is sprinkled with quotations from Mao Zedong printed in flaming red.

Other picture books include *Karst in China—Landscapes, Types, and Rules* (1986) by Lu Yaoru (in Chinese with supplemental volume of English translation) [listed under Institute of Hydrogeology and Engineering Geology, Chinese Academy of Geological Sciences] and *Guilin Karst* (1988) by Zhu Xuewen [see Chu, Hsueh-wen]. *Karst of China* (1991) by Yuan Daoxian and *Karst in China* (1995) by Marjorie M. Sweeting are works in scientific geomorphology.

Other than China, the best regional coverage of karst is Great Britain. The series of books *Limestones and Caves of Northwest England* (Waltham, 1974), *Limestones and Caves of the Mendip Hills* (Smith and Drew, 1975), *Limestones and Caves of the Peak District* (Ford, 1977), and *Limestones and Caves of Wales* (Ford, 1989) are in large part cave descriptions but each contains a lengthy and systematic description of the karst geomorphology of the region.

The karst seen today is only a snapshot. Karst landscapes have formed throughout geologic time. Some of these old landscapes were buried under later sediments and were not destroyed by continued erosion. Sometimes paleokarst landscapes are exhumed by later erosion or are intersected by mining or oil drilling activities. Because of their importance as repositories for ore deposits and as petroleum reservoirs there is a substantial technical literature that has been the subject of two review volumes, *Paleokarst* (1988) edited by N. P. James and P. W. Choquette and *Paleokarst: A Systematic and Regional Review* (1989) edited by P. Bosák, D. Ford, J. Glazek, and I. Horáček. These two volumes provide a comprehensive access to the world's paleokarst.

Since the 1970's the trend has been toward a more unified science of caves and karst geomorphology, except for regional studies, is rarely split off as a subject separate from cave geology and karst hydrology. This is reflected in the more recent textbooks, *Geomorphology and Hydrology of Karst Terrains* (White, 1988) and *Karst Geomorphology and Hydrology* (Ford and Williams, 1989). It seems likely that this trend will be maintained in the future.

References not in Main Bibliography

Corbel, Jean. 1957. *Les Karst du Nord-Ouest de L'Europe et de Quelques Regions de Comparaisons*. Institut des Études Rhodaniennes de L'Université de Lyon Mémoires et Documents 12, 541pp.

Cvijić, Jovan. 1960. *La Géographie des Terrains Cal-*

caires. Serbian Academy of Sciences and Arts Monograph 341, Belgrade, 212pp.

Cvijić, J. 1893. Das Karstphänom. *Geolgraphische Abhandlungen*, 5(3): 218-329.

5.6.1 General

[769]Cullen, James J., Jul 1979, Karst hydrology and geomorphology of eastern New York: a guidebook to the geology field trip, National Speleological Society Annual Convention, Pittsfield, Massachusetts, August 5-12, 1979
[901]Drew, David Phillip, 1943–, 1985, Karst processes and land forms
[1082]Ford, Derek Clifford, 1935–, Jul 1972, Guidebook to symposium SA6 "Karst geomorphology of the Canadian Rockies" 22nd international geographical congress Canada 1972
[1083]Ford, Derek Clifford, 1935–, 1989, Karst geomorphology and hydrology
[1271]Groom, Gillian Elisabeth, 1958, The geomorphology and speleogenesis of the Dachstein caves
[1342]Hamilton-Smith, Elery (ed), February 1996, Abstracts of papers: karst studies seminar Naracoorte
[1453]Higgins, C.G. (ed), 1990, Groundwater geomorphology
[1619]International Conference on Geomorphology (2nd: 1989 Frankfurt am Main, Germany), 1992, Karst
[1637]International Geographical Union, Sep 1995, International symposium on karren landforms: Mallorca 1995
[1694]Jakucs, László, 1977, Morphogenetics of karst regions: variants of karst evolution
[1725]Jennings, Joseph Newell, 1916–1984, 1971, Karst
[1726]Jennings, Joseph Newell, 1916–1984, 1985, Karst geomorphology
[1727]Jennings, Joseph Newell, 1916–1984, 1963, The limestone ranges of the Fitzroy Basin, Western Australia; a tropical semi-arid karst
[1775]Kastning, Ernst H., 1944–, Jul 1989, Karst geomorphology and hydrogeology
[2066]Martin, Ronald L., 1972, Cave development in the Bull Creek drainage basin of southwest Missouri
[2193]Miotke, Franz-Dieter, 1934–, 1972, Genetic relationship between caves and landforms in the Mammoth Cave National Park area; a preliminary report
[2329]Mylroie, John Eglinton, 1949–, 1977, Speleogenesis and karst geomorphology of the Helderberg Plateau, Schoharie County, New York
[2421]National Speleological Society. Annual Convention. Guidebook. Geology Field Trip (1979: Pittsfield, MA), Aug 1979, Karst hydrogeology and geomorphology of eastern New York: a guidebook to the geology field trip, National Speleological Annual Convention, Pittsfield, Massachusetts August 5-12, 1979
[2665]Pfeffer, Karl-Heinz (ed), 1989, Karst

[2741]Quinlan, James Francis, 1936–1995, 1980, Ground-water hydrology and geomorphology of the Mammoth Cave Region, Kentucky, and of the Mitchell Plain, Indiana
[2742]Quinlan, James Francis, 1936–1995, 1990, Hydrogeology and geomorphology of the Mammoth Cave area, Kentucky: Southeastern Friends of the Pleistocene 1990 field excursion, November 17, 1990
[3165]Sweeting, Marjorie Mary, 1920–1994 (ed), 1981, Karst geomorphology
[3166]Sweeting, Marjorie Mary, 1920–1994, 1995, Karst in China: its geomorphology and environment
[3167]Sweeting, Marjorie Mary, 1920–1994, 1972, Karst landforms
[3195]Tell, Leander, 1961, Erosionsforloppet, med sarskild hansyn till Lummelundagrottorna. The rate of erosion, with special reference to the caves of Lummelunda
[3226]Tratman, Edgar Kingsley, 1899–1978 (ed), 1969, The Caves of North West Clare, Ireland
[3244]Trudgill, Stephen Thomas, 1947–, 1985, Limestone geomorphology
[3474]White, William Blaine, 1934–, 1988, Geomorphology and hydrology of karst terrains
[3555]Zhu Xuewen, 1932-, 1988, Guilin karst

5.7 Geophysics

[570]Butler, Dwain K., Mar 1983, Cavity detection and delineation research: Report 1: Microgravimetric and magnetic surveys: Medford Cave Site, Florida
[571]Butler, Dwain K., 1983, Cavity detection and delineation research: Report 4: Microgravimetric survey: Manatee Springs site, Florida
[997]Environmental Consulting Engineers, , East Chestnut ridge hydroelectric characterization: a geophysical study of two karst features
[1108]Fountain, Lewis S., Jul 1980, Earth resistivity and hole-to-hole electromagnetic transmission tests at Medford Cave, Florida
[1822]Kirk, Keith G., Apr 1981, Handbook of geophysical cavity-locating techniques: with emphasis on electrical resistivity
[2510]Nicol, Allen Hankins, 1907–, 1951, Application of geophysical methods to location of cavities in residual soil and caverns in coralline limestone

5.8 Hydrology

Arthur N. Palmer

Department of Earth Sciences

State University of New York
Oneonta, NY 13820-4015

When one considers the enormous impact of karst hydrology on human welfare, its literature in English is surprisingly scant. Underground water is an essential ingredient in the origin of karst and caves, and its importance extends to studies of groundwater supply, water quality, structural and civil engineering, petroleum geology, and the genesis of certain ores. Yet today's popular hydrology books barely mention the subject. It is no wonder that the average groundwater specialist is woefully unprepared to solve groundwater problems in karst. Despite the fact that it is pursued by relatively few researchers, karst hydrology has a long and illustrious history, and its advances have more than kept pace with those in other fields of hydrology.

Books on the subject comprise at least two distinct groups with little overlap: those written by hydrologists who are strictly concerned with groundwater resources and problems, and those by authors who are primarily interested in the karst landscapes and caves that result from the action of groundwater. Bibliophiles interested in karst hydrology will have some difficulty acquiring the major works devoted to the subject, because many are symposium volumes of limited production.

No known books in the pre-Renaissance literature deal specifically with karst, although the ancient Greek, Roman, and Chinese civilizations had a practical knowledge of karst drainage and springs and even conducted underground water tracing. Testimony to the advanced practical nature of this early knowledge is shown by the large number of surviving Roman hydraulic structures, many of which are in karst areas.

The roots of modern scientific hydrology reach back to the 16th and 17th centuries, with the work of Palissy (1580), Perrault (1678), and Mariotté (1686), which validated the concept of the hydrologic cycle. This included some karst work, such as Palissey's inquiry into the origin of stalactites. The first exclusive studies of karst began to appear at about that same time. In the karst of southern China, Xu Xiake (pen name of Xu Hongsu) traveled widely during the years 1636-1639 and wrote extensive descriptions of karst, caves, cave deposits, and underground water flow (see Cai, Yang, and Maire, 1993). In the Balkans, Gaffarel (1654) and von Valvasor (1689) accurately described a great many karst features and their hydrologic function. However, the German physician Kircher (1665) interpreted underground water as rising from the ocean in turbulent eddies, a fanciful idea that had been held by some of the ancient Greeks. Despite a few wrong turns along the way, scientists by the end of the 18th century had a rather firm grasp on the basic ideas of how underground water moves through karst. e.g., in a book boldly titled *Theory of the earth, with proofs and illustrations*, Hutton (1795) clearly describes the origin of caves by the dissolution of limestone, and the deposition of dripstone by loss of carbon dioxide.

Until the 20th century, most karst studies were incorporated within regional geologic or geomorphic reports, and there was little attempt to unify all the aspects of karst or to explain them in a systematic way. Because karst studies were rather isolated from each other in time and location, most disagreements about karst boiled down to simple misunderstandings of fundamental geologic, hydrologic, or chemical concepts. Minor scientific papers on karst hydrology tended to repeat the same general ideas over and over, differing only in emphasis and in wording. But as karst studies became more numerous and rigorous in the early decades of the 20th century, researchers began to overlap in their field areas and topics of research. Fundamental disagreements arose, and their impact continues to make waves even today. The first widely known scientific studies of karst were those of Cvijić (1893), Grund (1903), Katzer (1909), and Martel (1921), whose works led to a serious conceptual rift. Grund viewed karst groundwater as being similar to any other kind of groundwater in a porous medium, in having a discrete water table and with cavities forming beneath it at random like the holes in Swiss cheese. Only the upper parts of the groundwater were considered to have significant water motion (the zone of "karst water") as opposed to the rather static "groundwater" below. Katzer and Martel disagreed sharply with this interpretation and instead viewed karst as an elaborate plumbing system with interconnected conduits free to respond somewhat independently to changes in flow such as floods. Martel was the most experienced speleologist of his day, with a good working knowledge of how water moves through caves, and he illustrated his claims with numerous cave maps and profiles. Today the two viewpoints have been reconciled and combined to some extent, and although the controversy has long been resolved in the eyes of most karst researchers, similar questions still lurk among non-karst hydrologists.

Another topic of heated debate is the relation of cave origin to the water table. This chiefly American discussion was aired in journal articles by some of the most celebrated geologists of the early 20th century. William Morris Davis (1930) and J. Harlen Bretz (1940) argued that most cave development takes place deep beneath the water table and is exposed at and near the surface only during a later erosion cycle. Allyn Swinnerton (1932), Clyde Malott (1938), and others championed a shallower origin for caves. Although such works represented a great deal of careful observation and thought, they were limited by the lack of a widespread data base of cave maps and by a weak understanding of the physical and chemical aspects of karst processes.

5.8. HYDROLOGY

In contrast, Lehmann (1932, in German) approached karst from a purely hydraulic standpoint, with little concern for its origin and evolution. This is perhaps the first major work to approach karst quantitatively. His book (in German) is a kind of plumber's manual for karst. In the application of quantitative methods to karst hydrology he was far ahead of his time, but despite this—or perhaps because of it—his work seems to have had little impact on later research in the field.

Extensive work during these same decades by Oscar Meinzer of the U.S. Geological Survey led to his being considered the father of American groundwater hydrology. It is gratifying to realize that he was quite sympathetic toward karst studies. These sympathies seem to have been lost by most traditional groundwater hydrologists of later years. In his last major work, a comprehensive volume on hydrology (Meinzer, 1942), he included a chapter on karst written by none other than Allyn Swinnerton. However, work in the field was still largely qualitative, anecdotal, and rather simplified.

Researchers eventually came to grips with the obvious truth that karst hydrology can be understood only by combining geology, hydraulics, and chemistry in roughly equal parts. Some of the first approaches of this kind were by geochemists with only a peripheral interest in karst. Experiments by Kaye (1957) and Weyl (1958) with strong acids appeared to demonstrate the importance of flow velocity on rates of cave origin. Although their methods and conclusions are largely inapplicable to real karst processes, their mode of thinking represents a great forward stride. Karst researchers were not far behind. White and Longyear (1962), Bögli (1964), Howard (1964), Curl (1968), Thrailkill (1968), Shuster and White (1971), Palmer (1972), Wigley (1975), White (1977) and others began to explain karst as hydrochemical systems with rather predictable, or at least quantifiable, behavior. In a National Speleological Society symposium that focused on karst hydrology (Moore, 1966), the conflict between traditional and progressive views of karst hydrology was clearly expressed. Nevertheless, books on karst during that general period by Trombe (1952, in French), Trimmel (1968, in German), Jennings (1971), Sweeting (1972), and Herak and Stringfield (1972) contain only passing reference to quantitative approaches, and are focused almost entirely on geomorphology. Jakucs (1977) followed a similar course, but his ideas of karst hydrology are of considerable interest, if not validity, because of their wide divergence from those of most other karst researchers. The book has been translated into English from the original Hungarian. Gams (1974) and Bleahu (1982) approach the subject in the same manner as Jennings and Sweeting, but have limited American readership because they are written in Slovene and Romanian respectively. In contrast to all of the above authors, Bögli (1980) relates karst geomorphology to both the chemical and hydraulic factors that produce them, the first person to do so in a major book.

Because of the difficulty of scaling down the large times and sizes involved in karst processes, only a few hardware models of karst hydrology are described in the literature. Bedinger (1966) constructed an electrical analog model, using a grid of resistors in an electrical circuit to represent local hydraulic conductivities, which is intended to simulate the evolution of permeability in karst. The model is based on flawed assumptions that dissolution rate is a function only of groundwater discharge, and that the aquifer is perfectly homogeneous, and as a result the conclusion that the shortest paths would enlarge fastest was pre-determined even before the model was constructed. Ewers (1982) ran solvent water through scale models of gypsum and salt blocks and described the evolution of the resulting dissolution conduits. The applicability of his experiments is limited by the fact that carbonate rocks dissolve with different kinetics from that of gypsum or salt. Furthermore, the hydraulic gradient in such a scale model must be identical to that in the field, but in the models gradients were unnaturally steep. Nevertheless, Ewers' models clearly show how the development of conduits distort the potential field within a karst aquifer, and how conduits compete with one another as the aquifer evolves.

Traditional groundwater specialists have added greatly to the literature on karst hydrology. Regional studies are focused mainly on water-supply issues (e.g., Piper, 1932; Brown, 1966; Stringfield and LeGrand, 1966; Giusti, 1978). Several symposium volumes bring together karst researchers with a variety of interests, including many authors whose concern with karst hydrology is directed toward the geomorphic aspects of karst and caves (e.g., volumes edited by Tolson and Doyle, 1977, and by Dilamarter and Csallany, 1977). The International Association of Hydrogeologists, with financial support from UNESCO, has done a great deal to promote the study of karst hydrology, mainly through its Commission on the Hydrology of Karst. The many publications issuing from the IAH are of highly varied quality. Volumes edited by Burger and Dubertret (1975, 1984), Yevjevich (1976), Mijatović (1984), LaMoreaux and others (1984), Castany and others (1984), and LaMoreaux (1986), have done a great deal to bring together karst researchers from many backgrounds and geographic locations. In addition, LaMoreaux and others (1975) summarize the field of carbonate hydrology and include a bibliography of works in the hydrology of carbonate terranes. In these works, the dichotomy in the subject becomes most apparent, for many of the researchers have little concern for the geomorphic aspects of karst, and particularly caves, which are essential to understanding karst groundwater. Some of these publications represent little more than a compilation of fragmentary and unrelated papers on repetitive topics, but the popularity of this approach has apparently waned in the last decade. Nevertheless, these volumes opened international channels of communication among researchers of diverse interests.

Two recent volumes are focused specifically on the geomorphic aspects of groundwater. Based loosely on symposia dealing with this subject, they include a great deal of work by traditional karst researchers (see volumes edited by LaFleur, 1984, and by Higgins and Coates, 1990). The National Ground Water Association (then called the National Water Well Association) hosted two symposia on groundwater problems in karst in the late 1980s, coordinated by James Quinlan of the National Park Service. These volumes (National Water Well Association, 1986, 1988) include a wealth of articles by many researchers, but unfortunately the production runs were very limited and they are long out of print. Similar symposium-based volumes on environmental and engineering aspects of karst have been produced by the former Florida Sinkhole Research Institute and P. E. LaMoreaux and Associates, Inc. including one specifically on karst hydrology (Beck and Wilson, 1987).

Several prominent karst researchers have recently produced books that draw together all aspects of the field: geology, hydrology, chemistry, and geomorphology: Jennings (1985), White (1988), Dreybrodt (1988), and Ford and Williams (1989). Smith, Atkinson, and Drew (1976) contribute a thorough chapter on karst hydrology in the book *Speleology, the Science of Caves*, by Ford and Cullingford (1976). Two other books approach karst strictly from the standpoint of traditional hydrology. Milanović (1981) deals primarily with groundwater supply and problems, and Bonacci (1987) mainly with surface water in karst. Neither author seems to have had more than a passing acquaintance with caves, although Milanović includes a short chapter that describes the practical role of speleology in hydrologic studies. Both are concerned mainly with karst problems in the Balkans. A more thorough understanding of the hydrologic function of caves has been provided by several extensive field studies. The most thoroughly documented studies are those of the National Scientific Research Center at Moulis, France, by Mangin (1975, in French) and in the Mammoth Cave region of Kentucky (e.g., Quinlan, 1981; Hess and White, 1993). The Mammoth Cave area is also the topic of a volume compiled by numerous researchers whose work spans several decades (White and White, 1989). Dye-tracing studies have led to several publications that range from local studies (e.g. Jones, 1973; Gospodarić and Habić, 1976) to a full-scale book on karst hydrology (Zötl, 1974, in German). In recent years there has been an attempt to bridge the gap between karst researchers and traditional groundwater hydrologists, and a fair amount of work is appearing in mainstream scientific journals. The recently formed Karst Waters Institute of Washington, D.C., has promoted karst research from an interdisciplinary standpoint (e.g., Sasowsky and Palmer, 1994).

Karst hydrology researchers still confront a number of major questions. The hydrochemistry of deep-seated karst processes is receiving considerable attention (e.g., Egemeier, 1981), but many details are still far from being resolved. Quantifying the various types of porosity in karst aquifers and their effect on groundwater flow patterns and contaminant transport remain elusive goals. There is considerable debate about how to monitor wastes in carbonate aquifers (Quinlan and Alexander, 1987), and whether these aquifers can be modeled digitally (e.g., Cullen and LaFleur, 1984; Sauter, 1992). Computer modeling of conduit evolution has been moderately successful (Palmer, 1984, 1991; Dreybrodt, 1990; Groves and Howard, 1994; Howard and Groves, 1995). Such fields contain great promise for future karst researchers, because only with an extensive knowledge of the geomorphic aspects of karst and caves can a full understanding of groundwater hydrology in carbonate rocks be attained. Literature on the subject deserves to grow considerably both in volume and sophistication in the next few decades.

Reference List of Items not in Main Bibliography

The following references include only those cited in the introduction that are not listed in the main part of the bibliography. They include journal articles, books in non-English languages, and books that are only marginally concerned with karst hydrology.

Bedinger, M. S. 1966. Electric-analog study of cave formation. *National Speleological Society Bulletin*, 28: 127-132.

Bleahu, M. 1982. *Relieful carstic*. Bucharest: Editura Albatros, 296pp.

Bögli, Alfred. 1964. Mischungskorrosion, ein Beitrag zur Verkarstungsproblem. *Erdkunde*, 18: 83-92.

Bretz, J. H. 1942. Vadose and phreatic features of limestone caverns. *Journal of Geology*, 50: 675-811.

Cai, Z., Yang, W., and R. Maire. 1993. Le geographe chinois Xu Xiake, un précurseur de la karstologie et de la spéléologie. *Karstologia* no. 21: 43-50.

Cullen, J. J., and R. G. LaFleurr. 1984. Theoretical considerations on simulation of karstic aquifers. pp248-280 IN: LaFleur, R. G. (ed.), *Groundwater as a geomorphic agent*. Boston: Allen and Unwin.

Curl, R. L. 1968. Solution kinetics of calcite. *Proceedings of the 4th International Congress of Speleology, Ljubljana, Slovenia* vol. 3: 61-66.

Cvijić, J. 1893. Das Karstphänomen: Vienna. *Geographische Abhandlungen Herausgegeben von A. Penck*, 5(3).

5.8. HYDROLOGY

Davis, W. M. 1930. Origin of limestone caverns. *Geological Society of America Bulletin*, 41: 475-628.

Dreybrodt, W. 1990. The role of dissolution kinetics in the development of karst aquifers in limestone: a model simulation of karst evolution. *Journal of Geology*, 98(5): 639-655.

Egemeier, S. J. 1981. Cavern development by thermal waters. *National Speleological Society Bulletin*, 43: 31-51.

Ewers, R. O. 1982. *Cavern Development in the Dimensions of Length and Breadth*. Ph.D. dissertation, McMaster University, Hamilton, Ontario, 398pp.

Gaffarel, J. 1654. *Le Monde Sousterrein (sic)*. Paris: Mesnil, 7pp.

Gams, I. 1974. *Kras*. Ljubljana: Izdala Slovenska Matica, 359pp.

Groves, C. G. and A. D. Howard. 1994. Early development of karst systems, 1. Preferential flow path enlargement under laminar flow. *Water Resources Research*, 30(10): 2837-2846.

Grund, A. 1903. Die Karsthydrographie (Karst hydrology). *Geographisches Abhandlung A. Penck*, 7(3), 200pp.

Hess, J. W., and W. B. White. 1993. Groundwater geochemistry of the carbonate aquifer, south-central Kentucky, U.S.A.. *Applied Geochemistry*, 8: 189-204.

Howard, A. D. 1964. Processes of limestone cave development. *International Journal of Speleology*, 1: 47-60.

Howard, A. D., and C. G. Groves. 1995. Early development of karst systems, 2. Turbulent flow. *Water Resources Research*, 31(1): 19-26.

Hutton, J. 1795. *Theory of the Earth, with Proofs and Illustrations*. vol. 2, Edinburgh: printed for Cadell, Davies, and W. Creech.

Katzer, F. 1909. *Karst und Karsthydrographie (Karst and karst hydrology)*. Sarajevo: D. A. Kajon, Zur Kunde der Balkanhalbinsel, no. 8, 94pp.

Kaye, C. A. 1957. The effect of solvent motion on limestone solution. *Journal of Geology*, 65: 35-46.

Kircher, A. 1665. *Mundus subterraneus*. Amsterdam: Janson, 2 vols.

Lehmann, O. 1932. *Die Hydrographie der Karstes (Karst hydrology)*. Leipzig: F. Deuticke, Enzyklopadie der Erdkunde, vol. 6, 212pp.

Malott, C.A. 1938. *The invasion theory of cavern development (abstract): Geological Society of America Proceedings, p. 323.*

Mangin, A. 1974-75. *Contribution a l'étude hydrodynamique des aquifres karstiques: Annales de Spéléologie, vol. 29, p. 283-601; vol. 30, p. 21-124.*

Mariotté E. 1686. Traités du mouvement des eaux et des autres corps fluides. *published in Leyden in 1717.*

Martel, E. A. 1921. Nouveau traité des eaux souterraines. *Paris, Librairie Octave Doin, 838 p.*

Meinzer, O. (ed.). 1942. Hydrology. *New York, Dover Publ., 712 p.*

Moore, G.W. (ed.). 1966. *Limestone hydrology*. National Speleological Society Bulletin., vol. 28, p. 109-166.

Palissy, B. 1580. Discours admirable de la nature des eaux et fontaines tant naturelles qu'artificielles. *Paris, Le Jeune, 361 p.*

Palmer, A.N. 1972. *Dynamics of a sinking stream system, Onesquethaw Cave, New York.* National Speleological Society Bulletin., vol. 34, p. 89-110.

Palmer, A.N. 1984. *Recent trends in karst geomorphology.* Journal of Geological Education., vol. 32, p. 247-253.

Palmer, A.N. 1991. *Origin and morphology of limestone caves.* Geological Society of America Bulletin., vol. 103, p. 1-21.

Perrault, P. 1678. *De l'origine des fontaines (On the origin of springs). Paris.*

Piper, A.M. 1932. Ground water in north-central Tennessee. *U.S. Geological Survey Water-Supply Paper 640, 238 p.*

Quinlan, J.F. 1981. *Hydrogeologic research techniques and instrumentation used in the Mammoth Cave region, Kentucky.* Geological Society of America field-trip guidebook for annual convention., vol. 3, p. 502-504.

Quinlan, J.F. and Alexander, E. C. 1987. *How often should samples be taken at relevant locations for reliable monitoring of pollutants from an agricultural, waste disposal, or spill site in a karst terrane?.* Proceedings of 2nd Multidisciplinary Conference on Sinkholes and Environmental Impacts of Karst., Florida Sinkhole Research Institute, Orlando, Florida, p. 277-286.

Sasowsky, I. D. and Palmer, M.V. (eds.). 1994. *Breakthroughs in karst geomicrobiology and redox geochemistry.* Karst Waters Institute Special Publication 1, symposium abstracts and field guide, 111 p.

Shuster, E.T., and White, W.B. 1971. Seasonal fluctuations in the chemistry of limestone springs: a possible means for characterizing carbonate aquifers. *Journal of Hydrology.*, vol. 14, p. 93-128.

Smith, D.I., Atkinson, T.C., and Drew, D.P. 1976, The hydrology of limestone terranes., In: Ford, T.D., and Cullingford, C.H.D. (eds.), *The Science of Speleology.* London, Academic Press, p. 179-212.

Swinnerton, A.C. 1932. Origin of limestone caverns. *Geological Society of America Bulletin.*, vol. 43, p. 663-694.

Thrailkill, J. 1968. Chemical and hydrologic factors in the excavation of limestone caves. *Geological Society of America Bulletin.*, vol. 79, p. 19-46.

Trimmel, H. 1968. *Höhlenkunde.* Braunschweig, Friederich Vieweg & Sohn, 300 p.

Trombe, F. 1952. *TraitÉ de spéléologie.* Paris, Payot.

Valvasor, J.W.F. von. 1689. *Die Ehre des Herzogthums Krains.* Ljubljana, Endter, 4 vols., 696, 835, 730, and 610 p.

Weyl, P.K. 1958. The solution kinetics of calcite. *Journal of Geology.*, vol. 66, p. 163-176.

White, W.B. 1977. Role of solution kinetics in the development of karst aquifers., In: Tolson, J.S., and Doyle, F.L. (eds.), *Karst hydrogeology.* Congress of the International Association of Hydrogeologists, 12th Memoirs, p. 503-517.

White, W.B., and Longyear, J. 1962. Some limitations on speleo-genetic speculation imposed by the hydraulics of groundwater flow in limestone. *Nittany Grotto Newsletter.*, no. 10, p. 155-167.

Wigley, T.M.L. 1975. Speleogenesis: a fundamental approach. *Proceedings of 6th International Congress of Speleology.*, Oloumec, Czechoslovakia, vol. 3, p. 317-324.

Zötl, J. 1974. *Karsthydrogeologie.* Vienna, Springer-Verlag.

5.8.1 General

[25] Alexander, E. Calvin, 1987, Hydrogeologic study of Jewel Cave/Wind Cave: final report -year 2, 4th quarter of year 2, February 1, 1987 through April 30, 1987

[26] Alexander, E. Calvin, 1989, Hydrologic study of Jewel Cave/Wind Cave: final report

[27] Alexander, E. Calvin, 1992, Practical tracing of groundwater, with emphasis on karst terranes: a short course manual presented on the occasion of the annual meeting of the Geological Society of America, October 24, 1992, Cincinnati, Ohio

[28] Alexander, Emmit Calvin, 1943–, 1986, Hydrogeologic study of Jewel Cave/Wind Cave: final report - phase I, 3rd and 4th quarters of phase I, 1 November 1985 through 30 April 1986

[32] Aley, Thomas John, 1938–, 1984, Cave and karst hydrology assessment project for Horsethief Cave, Wyoming. Report to the Worland District, BLM

[36] Aley, Thomas John, 1938–, Feb 1990, Delineation and hydrogeologic study of the key cave aquifer Lauder Dale County, Alabama

[37] Aley, Thomas John, 1938–, 1972, Groundwater contamination and sinkhole collapse induced by leaky impoundments in soluble rock terrain

[39] Aley, Thomas John, 1938–, Nov 1993, Karst and cave resource significance assessment: Ketchikan Area, Tongass National Forest, Alaska

[40] Aley, Thomas John, 1938–, 1964, Origin and hydrology of caves in the White Limestone of north central Jamaica: report of field work carried out under ONR Contract 3656 (03) NR 388 067

[41] Aley, Thomas John, 1938–, Aug 1975, A predictive hydrologic model for evaluating the effects of land use and management on the quantity and quality of water from Ozark Springs: final report

[44] Aley, Thomas John, 1938–, 1987, Water quality protection studies, Logan Cave, Arkansas: final report

[71] Anderson, Warren, 1975, Hydrology of three sinkhole basins in Southwestern Seminole County Florida

[142] Anon, 1964, Karst groundwater investigations: Greece

[208] Anon, 1982, Waitomo Caves management plan 1982

[209] Anon, nd, Waitomo Day 1982: summary of papers: Waitomo Caves research programme

[237] Aspin, J., Mar 1952, The caverns of Upper Ease Gill

[238] Association of Ground Water Scientists and Engineers (a division of NGWA), Dec 1991, Proceedings of the third conference on hydrogeology, ecology, monitoring, and management of ground water in karst terranes, December 4-6, 1991, Maxwell House/Clarion, Nashville, Tennessee

[245] Atkinson, Timothy C., Oct 1967, Mendip Karst Hydrology Project: phases one and two

[270] Back, William (ed), 1992, Hydrogeology of selected karst regions

[285] Baker, Eva G., , Water resources publications of the USGS for Tennessee, 1907-1987

[327] Barr, Thomas Calhoun, 1931–, Dec 1976, Ecological effects of water pollutants in Mammoth Cave: final technical report to the National Park Service

[350] Bayless, Edward Randall, 1961-, 1994, Directions of ground-water flow and locations of ground-water divides in the Lost River watershed near Orleans, Indiana

[359] Beck, Barry Frederic, 1944–(ed), Oct 1985, Karst hydrogeology of central and northern Florida: a field trip guidebook produced in conjunction with the 1985 G.S.A. annual meeting and exposition, Orlando, FL., October, 1985

[360] Beck, Barry Frederic, 1944–(ed), 1987, Karst hydrogeology: engineering and environmental applications: proceedings of the second Multidisciplinary Conference on Sinkholes and the Environmental Impacts of Karst, Orlando, Florida 9-11 February 1987

5.8. HYDROLOGY

[365]Beck, Barry Frederic, 1944–(ed), 1985, Water on and under the ground: an introduction to the urban hydrogeology of the Orlando area

[408]Black, Jackie (ed), 1973, Caves, ground water and karst features of the Cretaceous in South central Texas: fourteenth annual S.A.S.G.S. field conference, April 6-8, 1973

[423]Bögli, Alfred, 1912–, 1980, Karst hydrology and physical speleology

[429]Bonacci, Ognjen, 1987, Karst hydrology with special reference to the Dinaric karst

[491][British Speleological Association], 1969, Limestone geomorphology: a study in Jamaica: University of Bristol karst expedition to Jamaica 1967

[521]Brown, George Henry, 1906, The underground streams in the neighbourhood of Clapham and Malham

[528]Brown, Michael Charles, 1972, Karst hydrology of the lower Maligne Basin, Jasper, Alberta

[530]Brown, Richmond Flint, 1925–, 1966, Hydrology of the cavernous limestones of the Mammoth Cave area, Kentucky

[549]Buderer, Tina (comp), 1987, Proceedings of the first annual watershed conference, May 20-21, 1987, Springfield, Missouri

[558]Burdon, David Joseph, 1963, Handbook of karst hydrogeology with special reference to the carbonate aquifers of the Mediterranean Region

[559]Burdon, David Joseph, 1963, The karst groundwater resources of Parnassoc-Ghiona, Greece: a report

[561]Burger, Andre (ed), 1984, Hydrogeology of karstic terrains

[562]Burger, Andre (ed), 1975, Hydrogeology of karstic terrains: with a multilingual glossary of specific terms

[605]Castany, G. (ed), 1984, Hydrogeology of karstic terranes, case histories

[648]Chevalier, Pierre, 1975, Subterranean climbers; twelve years in the world's deepest chasm

[689]Conference on Hydrogeology, Ecology, Monitoring, and Management of Ground Water in Karst Terranes (3rd: 1991: Nashville, TN), 1991, Proceedings of the Third Conference on Hydrogeology, Ecology, Monitoring and Management of Ground Water in Karst Terranes: December 4 - 6, 1991, Maxwell House/Clarion, Nashville, Tennessee

[690]Conference on Karst Geology and Hydrology (1974), 1974, Abstracts of the fourth conference on karst geology and hydrology: West Virginia University, Mountainlair Theater of Mountainlair Building: May 3, 4, and 5, 1974

[739]Crawford, Nicholas Charles, 1942–, 1982, Hydrogeologic problems resulting from development upon karst terrain, Bowling Green, Kentucky. Guidebook prepared for Karst Hydrogeology Workshop August 31 - September 3, 1982 Nashville, Tennessee

[740]Crawford, Nicholas Charles, 1942–, 1988, Hydrogeology of the Snail Shell Cave: overall creek drainage basin and the ecology of the Snail Shell Cave System

[741]Crawford, Nicholas Charles, 1942–, 1981, Karst hydrogeology and environmental problems in the Bowling Green area

[742]Crawford, Nicholas Charles, 1942–, 1982, Karst hydrogeology of Tennessee. Guidebook prepared for Karst Hydrogeology Workshop August 31 - September 3, 1982 Nashville, Tennessee

[743]Crawford, Nicholas Charles, 1942–, 1980, The karst hydrogeology of the Cumberland Plateau Escarpment of Tennessee. Part IV. Erosional processes associated with subterranean stream invasion, conduit cavern development and slope retreat

[744]Crawford, Nicholas Charles, 1942–, 1979, The karst hydrogeology of the Cumberland Plateau escarpment of Tennessee: Part I: subterranean stream invasion, conduit cavern development, and slope retreat in the Lost Creek Cove area, White County, Tennessee

[745]Crawford, Nicholas Charles, 1942–, 1979, The karst hydrogeology of the Cumberland Plateau escarpment of Tennessee: Part II: karst valley development and the headward advance of the Sequatchie Valley in the Grassy Cove area, Cumberland County, Tennessee

[746]Crawford, Nicholas Charles, 1942–, 1980, The karst hydrogeology of the Cumberland Plateau escarpment of Tennessee: Part III: karst valley development in the Lost Cove area, Franklin County, Tennessee

[747]Crawford, Nicholas Charles, 1942–, 1987, The karst hydrogeology of the Cumberland Plateau escarpment of Tennessee: subterranean stream invasion, conduit cavern development, and slope retreat in the Lost Creek Cove area, White County, Tennessee

[748]Crawford, Nicholas Charles, 1942–, 1988, Karst hydrologic problems of south central Kentucky: ground water contamination, sinkhole flooding, and sinkhole collapse: field trip guide

[750]Crawford, Nicholas Charles, 1942–, 1979, Safford Centennial Society field trip, fall 1979: guidebook: karst hydrogeology of the Cumberland plateau escarpment in the Lost Creek Cave and Karst Cave areas of Tennessee

[752]Crawford, Nicholas Charles, 1942–, Sep 1984, Storm water drainage wells in the karst areas of Kentucky and Tennessee: extended inventory of drainage wells in Kentucky and Tennessee: underground water source protection program Grant No. G004358-83-0

[769]Cullen, James J., Jul 1979, Karst hydrology and geomorphology of eastern New York: a guidebook to the geology field trip, National Speleological Society Annual Convention, Pittsfield, Massachusetts, August 5-12, 1979

[786]Cushman, Robert Vittum, 1967, Hydrologic problems as related to wilderness management and water supplies for recreational areas at Mammoth Cave National

Park, Kentucky
[787]Cushman, Robert Vittum, 1965, Present and future water supply for Mammoth Cave National Park, Kentucky
[792]Dake, Charles Laurence, 1883–, 1924, Subterranean stream piracy in the Ozarks
[871]Dilamarter, Ronald R. (ed), 1971, Hydrologic problems in karst regions
[894]Downing, R.A. (ed), 1993, The hydrogeology of the chalk of north-west Europe
[902]Drew, David Phillip, 1943–, Dec 1968, Mendip karst hydrology research project: phase 3
[903]Drew, David Phillip, 1943–, 1969, Techniques for the tracing of subterranean drainage
[906]Drnevich, Vincent P., Aug 1972, Location of solution channels and sinkholes at dam sites and backwater areas by seismic methods
[929]Dyer, C. F., 1961, Geology and occurrence of ground water at Jewel Cave National Monument, South Dakota
[1003]Erchul, Ronald A., Mar 1988, Geotechnical applications of the self potential (SP) method. Report 1: the use of self potential in the detection of subsurface flowpatterns in and around sinkholes
[1004]Erchul, Ronald A., May 1989, Geotechnical applications of the self potential (SP) method. Report 2: the use of self potential to detect ground-water flow in karst
[1061]Fincham, Alan Goldworth, 1933–(comp), Jan 1967, The University of Leeds hydrological survey expedition to Jamaica 1963
[1078]Ford, Derek Clifford, 1935–(scien adv), Jul 1966, The 1965/66 karst hydrology expedition to Jamaica: full report
[1083]Ford, Derek Clifford, 1935–, 1989, Karst geomorphology and hydrology
[1093]Ford, Trevor David, 1925–(comp, ed), 1977, Limestone and caves of the Peak District
[1149]Garder, George D., c1978, Tracing of subsurface flow in karst regions using artificially colored spores
[1215]Giusti, Ennio V., 1978, Hydrogeology of the karst of Puerto Rico
[1216]Giusti, Ennio V., 1976, Water resources of the North Coast Limestone area, Puerto Rico
[1229]Gospodarič, R. (ed), 1976, Underground water tracing: investigation in Slovenia 1972-1975
[1266]Grimes, K. G., Oct 1995, South East Karst Province of South Australia: Australian Caves & Karst Management Association October 1995
[1267]Grimes, Ken, February 1996, Field Guide to karst features in southeast South Australia and Western Victoria: for the Karst Studies Seminar, Naracoorte, February 1996
[1287]Günay, Gültekin (ed), , Karst water resources: proceedings of a symposium held at Ankara, July 1985, and sponsored by the United Nations Development Program, the United Nations Educational, Scientific and Cultural Organization, the International Association of Hydrological Sciences, many government organizations of Turkey, and Hacettepe University of Ankara, Turkey
[1317]Hallberg, George R., May 1982, Sinkholes, hydrogeology, and ground-water quality in northeast Iowa
[1342]Hamilton-Smith, Elery (ed), February 1996, Abstracts of papers: karst studies seminar Naracoorte
[1377]Harvey, Edward Joseph, 1916–, 1977, Application of thermal imagery and aerial photography to hydrologic studies of karst terrane in Missouri
[1378]Harvey, Edward Joseph, 1916–, 1983, Hydrology of carbonate terrane, Niangua, Ossage Fork, and Grandglaize basins, Missouri
[1413]Headworth, H. G., Nov 1975, The hydrogeology of artesian boreholes at some watercress farms in Hampshire
[1438]Hendricks, Albert C., 1974, Quality of groundwater in a carbonate terrain in western Virginia
[1450]Hess, John Warren, 1947–, Jun 1974, Hydrograph analysis of carbonate aquifers
[1453]Higgins, C.G. (ed), 1990, Groundwater geomorphology
[1530]Hötzl, H. (ed), 1992, Tracer hydrology: proceedings of the 6th international symposium on water tracing Karlsruhe, Germany 21-26 September 1992
[1584]Husmann, Siegfried, 1976, 1st international symposium on groundwater ecology, Schlitz,1975
[1589]Hydrological and Engineering Geological Team of the Geological Bureau of the Kwangsi Chuan Autonomous Region, Aug 1976, On the underground river system of the Tisu karst area, Tu-an County, Kwangsi, China
[1614]International Association of Hydrogeologists, 1975, International Association of Hydrogeologists 12th international congress: karst hydrogeology: abstracts and program 21-27 September 1975, Huntsville, Alabama, 28 September-3 October 1975, Gulf Shores, Alabama
[1616]International Association of Hydrogeologists, 1988, Proceedings of the IAH 21st congress
[1617]International Association of Hydrogeologists, 1977, Proceedings of the twelfth international congress: karst hydrogeology: Huntsville, Alabama
[1645]International Symposium of Underground Water Tracing (3rd: 1976: Ljubljana, Yugoslavia), 1976, Undergound water tracing: investigations in Slovenia 1972—1975
[1647]International Symposium on Changing Karst Environments (1994: Oxford, England and Huddersfield, England), Aug 1994, Conference abstracts: [of papers presented at Changing karst environments, hydrogeology, geomorphology and conservation, an international symposium held at the Universities of Oxford and Huddersfield, September 1994]
[1648]International Symposium on Changing Karst Environments (1994: Oxford, England and Huddersfield,

5.8. HYDROLOGY

England), Mar 1995, Papers presented at the international symposium on changing karst environments, Oxford and Huddersfield, September 1994

[1650]International Symposium on Karst Hydrogeology (1st: 1979: Oymapinar), 1980, International seminar on karst hydrogeology: Oymapinar, October 1979: proceedings

[1652]International Symposium on Karsthydrology (1978: Budapest, Hungary), 1978, [Proceedings] nemzetkozi karszthidrologiai szimpozium = international symposium on karsthydrology = Mezhdunarodnyi simpozium po gidrologii karsta

[1653]International Symposium on Underground Water Tracing (5th: 1986: Athens), 1986, Abstracts

[1654]International Symposium on Underground Water Tracing (5th: 1986: Athens), 1986, Proceedings of the 5th International Symposium on Underground Water Tracing: Athens 1986

[1655]International Symposium on Underground Water Tracing (5th: 1986: Athens), 1986, Program

[1664]International Union of Speleology. Commission for Physical, Chemical and Hydrological Research of Karst, 1989, Communications = mitteilungen = soobshcheniia: international symposium on physical, chemical and hydrological research of karst: Kosice, Czechoslovakia, May 10-15, 1988

[1693]Jakeman, Anthony John, 1951–, 1983, Time series models for the prediction and characterisation of stream flow from rainfall in the Cave Creek system

[1703]Jameson, Roy Alan, 1951–, 1994—1995, The waters of Mystery Cave, Forestville State Park, Minnesota

[1735]JLM Visuals, 1979, Caves and other ground water features

[1756]Jones, William K., 1945–, 1973, Hydrology of limestone karst in Greenbriar County, West Virginia

[1757]Jones, William K., 1945–, Mar 1975, Karst hydrogeology: A summary of the principals of ground-water flow through soluble limestone aquifers prepared for the Smithsonian Institution short course on speleology

[1760]Jordon, Donald George, 1926–, 1970, Water and copper-mine tailings in karst terrane of Rio Tanama Basin, Puerto Rico

[1768]Karst Hydrogeology Symposium (1977: Oymapinar), 1978, Karst hydrogeology symposium: Oymapinar, October 1977: proceedings

[1775]Kastning, Ernst H., 1944–, Jul 1989, Karst geomorphology and hydrogeology

[1809]Kerans, C., 1954–, 1989, Karst-controlled reservoir heterogeneity and an example from the Ellenburger Group (Lower Ordovician) of west Texas

[1833]Knisel, Walter G., 1972, Response of karst aquifers to recharge

[1880]LaFleur, R.G. (ed), 1984, Groundwater as a geomorphic agent

[1887]LaMoreaux, Philip Elmer, 1920–(ed), 1984, Guide to the hydrology of carbonate rocks

[1888]LaMoreaux, Philip Elmer, 1920–, 1970, Hydrology of limestone terranes: annotated bibliography of carbonate rocks

[1889]LaMoreaux, Philip Elmer, 1920–(ed), 1989, Hydrology of limestone terranes: annotated bibliography of carbonate rocks, volume 4

[1891]LaMoreaux, Philip Elmer, 1920–(ed), Jan 1975, Hydrology of limestone terranes: progress of knowledge about hydrology of carbonate terranes with an annotated bibliography of carbonate rocks

[1894]Lane, Edward, 1935–, 1991, Environmental geology and hydrogeology of the Ocala area, Florida

[1896]Lange, Arthur L., Jul 1958, Stream piracy and cave development along Baker Creek, Nevada

[1922]Leclerc, Guy, 1972, Derivation of hydrologic frequency caves

[1933]Leontiadis, J., Jan 1972, The use of radioisotopes in tracing karst groundwater in Greece

[1941]Libra, Bob, [1990], 1990 Friends of karst field trip: the Big Spring Basin, northeast Iowa

[2025]Malott, Clyde Arnett, 1887–1950, 1977, A geologic profile of Sloans Valley, Pulaski County, Kentucky

[2174]Mijatovió, B.F., 1984, Hydrology of the Dinaric karst

[2175]Mikulec, Stjepan, 1976, Karst hydrology and water resources: proceedings of the U. S.—Yugoslavian symposium, Dubrovnik, June 2-7, 1975

[2176]Milanović, Petar T., 1981, Karst hydrogeology

[2209]Missouri Speleological Survey, Jan 1976, Hydrogeologic controls on solution of carbonate rocks in Christian County, Missouri

[2213]Missouri Speleological Survey, Apr 1973, An investigation into ground water pollution in caves

[2264]Moore, Gerald K., 1969, Limestone hydrology in the Upper Stones River Basin, central Tennessee

[2265]Moore, James D., 1975, Hydrology of limestone terranes: quantitative studies

[2268]Morfis, A. (ed), 1986, Karst hydrogeology of the central and eastern Peloponnesus (Greece)

[2269]Morfis, A. (ed), 1986, Proceedings of the 5th international symposium on underground water tracing, Athens 1986

[2297]Mull, D. S., Oct 1988, Application of dye-tracing techniques for determining solute-transport characteristics of ground water in karst terranes

[2329]Mylroie, John Eglinton, 1949–, 1977, Speleogenesis and karst geomorphology of the Helderberg Plateau, Schoharie County, New York

[2483]National Water Well Association, 1987, Proceedings of environmental problems in karst terranes and their solutions conference, October 28-30, 1986, Bowling Green, Kentucky

[2527] Nota, Dirk Johannes Gregorius, 1978, A hydrogeological study in the basin of the Gulp Creek: a reconnaissance in a small catchment area: 1, Groundwater flow characteristics

[2575] Orghidan, Traian, 1917–1985 (ed), 1984, Proceedings of the first symposium on theoretical and applied karstology held in Bucharest, Romania, 22-24 April 1983

[2607] Palmer, William O., 1970 1981, Springs, caves and their related features

[2633] Peck, Dallas L., Oct 1988, Karst hydrogeology in the United States of America: 21st congress of the International Association of Hydrogeologists: karst hydrogeology and karst environment protection October 1988 Guilin, China

[2651] The Pennsylvania State University, 1970, Field trip guidebook for the central Appalachian carbonate hydrology workshop, The Pennsylvania State University, University Park, Pennsylvania, May 21, 22, 1970

[2670] Picknett, R.G., May 1968, Symposium on cave hydrology and water-tracing

[2681] Pitty, Alistair F., 1966, An approach to the study of karst water: illustrated by results from Poole's Cavern, Buxton

[2712] Přibyl, Jan, 1973, Paleohydrography of the caves in the Moravsky Kras (Moravian Karst)

[2739] Quinlan, James Francis, 1936–1995, 1986, Ground water flow in the Mammoth Cave Area, Kentucky with emphasis on principles, contaminant dispersal, instrumentation for monitoring water quality, and other methods of study

[2740] Quinlan, James Francis, 1936–1995, Feb 1989, Ground water monitoring in karst terranes: recommended protocols and implicit assumptions

[2741] Quinlan, James Francis, 1936–1995, 1980, Ground-water hydrology and geomorphology of the Mammoth Cave Region, Kentucky, and of the Mitchell Plain, Indiana

[2743] Quinlan, James Francis, 1936–1995, 1977, Hydrology and water quality in the central Kentucky karst: phase 1

[2744] Quinlan, James Francis, 1936–1995, 1976, Hydrology of the Turnhole Spring Groundwater Basin and vicinity, Kentucky: an area that includes part of the Mammoth Cave National Park

[2745] Quinlan, James Francis, 1936–1995 (comp), 1994, Practical karst hydrogeology, with emphasis on groundwater monitoring, February 8-11, 1994, Holiday Inn West, Gainesville, Florida

[2763] Rauch, Henry W., nd, Lithologic controls on the development of solution porosity in carbonate aquifers

[2764] Rauch, Henry William (ed), 1974, Proceedings of the fourth conference on karst geology and hydrology

[2769] Rechtien, Richard D., 1972, The detection and mapping of subterranean water bearing channels: completion report

[2800] Reeds, Chester Albert, 1882–, 1928, Rivers that glow underground

[2891] Sauter, Martin, 1992, Quantification and forecasting of regional groundwater flow and transport in a karst aquifer (Gallusquelle, Maim, southwestern Germany)

[2904] Schiffer, Donna M., 1996, Hydrology of the Wolf Branch sinkhole basin, Lake County, east-central Florida

[2967] Shepard, Edward M., 1907, Underground waters of Missouri; their geology and utilization

[3049] Sonderegger, John L., 1970, Hydrology of limestone terranes: geologic investigations

[3050] Sonderegger, John L., 1970, Hydrology of limestone terranes: photogeologic investigations

[3113] Steidtmann, Edward, 1881–1948, 1936, Humidity and waters of a limestone cavern near Lexington, Virginia

[3151] Stringfield, Victor Timothy, 1902–, 1964, Hydrology of limestone terranes in the coastal plain of the southeastern states

[3152] Stringfield, Victor Timothy, 1902–, 1966, Hydrology of limestone terranes in the coastal plain of the southeastern United States

[3153] Stringfield, Victor Timothy, 1902–, 1977, Hydrology of limestone terranes: development of karst and is effects on the permeability and circulation of water in carbonate rocks, with special reference to the southeastern states

[3154] Stringfield, Victor Timothy, 1902–, 1974, Karst and paleohydrology of carbonate rock terranes in semiarid and arid regions with a comparison to humid karst of Alabama

[3167] Sweeting, Marjorie Mary, 1920–1994, 1972, Karst landforms

[3217] Thurgate, Mia E., 1995, Sinkholes, caves and spring lakes: an introduction to the unusual aquatic ecosystems of the lower south east of South Australia: South Australian Underwater Speleological Society occasional paper number 1

[3283] U.S. Geological Survey, 1967, Bibliography of the Ground Water Resources of New York through 1967, with subject and location cross-reference to the annotated bibliography

[3325] U.S. Office of Water Resources Research, 1975, The detection and mapping of subterranean water bearing channels: phase 2: final report

[3330] University of Bristol, Dec 1967, Karst hydrology expedition to Jamaica: preliminary report

[3347] Van Everdingen, R. O., 1981, Morphology, hydrology and hydrochemistry of Karst in permafrost terrain near Great Bear Lake, Northwest Territories

[3397] Ward, William Cruse, 1933–, , Geology and hydrology of the Yucatan and Quaternary geology of Northeast Yucatan Peninsula with a part on the History of Northern Quintana Roo

[3403] Waterhouse, J. D., nd, Pollution of underground

water
[3425]Webb, Donald W., May 1994, The biological resources of Illinois caves and other subterranean environments: determination of the diversity, distribution, and status of the subterranean faunas of Illinois caves and how these faunas are related to groundwater quality
[3429]Wegrzyn, Mikolaj, Jun 1984, Rainstorm related terrain failures in Puerto Rico: final technical report
[3474]White, William Blaine, 1934–, 1988, Geomorphology and hydrology of karst terrains
[3475]White, William Blaine, 1934–(ed), 1989, Karst hydrology: concepts from the Mammoth Cave area
[3500]Williams, Paul W., 1968, Contributions to the study of Karst
[3542]Yevjevich, Vujica, 1981, Karst water research needs

5.9 Hydrogeology

[1615]International Association of Hydrogeologists, 1975, Karst hydrogeology: 12th international congress: International Association of Hydrogeologists 21-27 September 1975, Huntsville, Alabama U.S.A.
[2420]National Speleological Society. Annual Convention. Guidebook. Geology Field Excursion (1978: New Braunfels, TX), 1978, Caves and karst hydrogeology of the southeastern Edwards Plateau, Texas: guidebook, geology field excursion, National Speleological Society annual convention, New Braunfels, Tex., June 18-23, 1978
[2421]National Speleological Society. Annual Convention. Guidebook. Geology Field Trip (1979: Pittsfield, MA), Aug 1979, Karst hydrogeology and geomorphology of eastern New York: a guidebook to the geology field trip, National Speleological Annual Convention, Pittsfield, Massachusetts August 5-12, 1979
[2547]Oklahoma Department of Wildlife Conservation, 1990, Hydrogeology of Ozark cavefish caves
[2612]Parizek, R. R., Nov 1971, Hydrogeology and geochemistry of folded and faulted carbonate rocks of the central Appalachian type and related land use problems
[2742]Quinlan, James Francis, 1936–1995, 1990, Hydrogeology and geomorphology of the Mammoth Cave area, Kentucky: Southeastern Friends of the Pleistocene 1990 field excursion, November 17, 1990
[2810]Rheams, Karen F., 1994, Hydrogeologic and biologic factors related to the occurrence of the Alabama cave shrimp (*Palaemonias alabamae*), Madison County, Alabama
[2865]Ruhe, R. V., Dec 1975, Geohydrology of karst terraine, Lost River watershed southern Indiana
[2905]Schindel, Geary Michael, 1957–, Dec 1991, Guidebook: environmental hydrogeology of karst terranes in the vicinity of Nashville Tennessee

[3212]Thrailkill, John Vernon, 1930–, [1984], Hydrogeology and environmental geology of the inner Bluegrass Karst Region Kentucky: field guide for the annual meeting of the southeastern and north-central sections Geological Society of America: Lexington, Kentucky April 4-6 1984
[3215]Thrailkill, John Vernon, 1930–, 1983, Studies in dye-tracing techniques and karst hydrogeology
[3223]Torres González, Arturo, Mar 1983, Geohydrology of the Rio Camuy Cave system Puerto Rico

5.10 Karst

[32]Aley, Thomas John, 1938–, 1984, Cave and karst hydrology assessment project for Horsethief Cave, Wyoming. Report to the Worland District, BLM
[39]Aley, Thomas John, 1938–, Nov 1993, Karst and cave resource significance assessment: Ketchikan Area, Tongass National Forest, Alaska
[55]American Cave Conservation Association, nd, Karst curriculum resource guide
[119]Anon, 1992, Geology, climate, hydrology and karst formation: field symposium in Australia 4 to 18 December 1992: Buchan - Mt. Gambier / Naracoorte - Nullarbor Plain humid temperate impounded karst, sub-humid temperate syngenetic karst, arid temperate karst: programme and abstracts
[133]Anon, 1981, Guilin: the crown of superb landscapes in China
[164]Anon, 1983, New directions in karst: Anglo-French symposium, 1983: field excursion notes
[178]Anon, 1979, Research of China karst
[179]Anon, 1979, Research of China karst
[248]Audyová, Jiřina, 1993, The Moravian karst: time and stone
[265]Australian Speleological Federation Inc, 1988, Preprints of papers for the 17th Biennial Conference of the Australian Speleological Federation Tropicon Conference, Lake Tinaroo, Far North Queensland. 27th to 31st December 1988
[269]Bachman, George Odell, Sep 1987, Karst in evaporites in southeastern New Mexico.
[321]B'ardossy, György, 1982, Karst bauxites: bauxite deposits on carbonate rocks
[354]Beck, Barry Frederic, 1944–(ed), 1993, Applied karst geology: proceedings of the fourth Multidisciplinary Conference on Sinkholes and Engineering and Environmental Impacts of Karst, Panama City, Florida 25-27 January 1993
[356]Beck, Barry Frederic, 1944–(ed), 1989, Engineering and environmental impacts of sinkholes and karst: proceedings of the third Multidisciplinary Conference on Sinkholes and the Engineering and Environmental Im-

pacts of Karst, St. Petersburg Beach, Florida, 2-4 October 1989

[358] Beck, Barry Frederic, 1944–(ed), 1995, Karst geohazards: engineering and environmental problems in karst terrane: proceedings of the fifth Multidisciplinary Conference on Sinkholes and the Engineering and Environmental Impact of Karst, Gatlinburg, Tennessee, 2-5 April 1995

[359] Beck, Barry Frederic, 1944–(ed), Oct 1985, Karst hydrogeology of central and northern Florida: a field trip guidebook produced in conjunction with the 1985 G.S.A. annual meeting and exposition, Orlando, FL., October, 1985

[360] Beck, Barry Frederic, 1944–(ed), 1987, Karst hydrogeology: engineering and environmental applications: proceedings of the second Multidisciplinary Conference on Sinkholes and the Environmental Impacts of Karst, Orlando, Florida 9-11 February 1987

[408] Black, Jackie (ed), 1973, Caves, ground water and karst features of the Cretaceous in South central Texas: fourteenth annual S.A.S.G.S. field conference, April 6-8, 1973

[429] Bonacci, Ognjen, 1987, Karst hydrology with special reference to the Dinaric karst

[441] Bosák, Pavel (ed), 1989, Paleokarst: a systematic and regional review

[446] Bounk, Michael J., Oct 1983, Karstification on the Silurian Escarpment in Fayette County, northeastern Iowa

[549] Buderer, Tina (comp), 1987, Proceedings of the first annual watershed conference, May 20-21, 1987, Springfield, Missouri

[561] Burger, Andre (ed), 1984, Hydrogeology of karstic terrains

[565] Burnsville Cove Symposium, 1982, Burnsville Cove Symposium

[749] Crawford, Nicholas Charles, 1942–, 1989, The karst landscape of Warren County

[751] Crawford, Nicholas Charles, 1942–, nd, Significant natural landscape features of Warren County, Kentucky

[782] Currens, James C., 1952–, 1993, Flooding of the Sinking Creek Karst area in Jessamine and Woodford Counties, Kentucky

[802] Daoxian Yuan, 1933–, 1981, A brief Introduction to China's research in Karst

[803] Daoxian Yuan, 1933–, 1991, Karst of China

[813] Davey, Adrian G., 1978, Nullarbor karst, a bibliography

[867] Dicken, Samuel Newton, 1901–, 1938, Soil erosion in the karst lands of Kentucky

[871] Dilamarter, Ronald R. (ed), 1971, Hydrologic problems in karst regions

[889] Dougherty, Percy H. (ed), 1993, Environmental karst

[901] Drew, David Phillip, 1943–, 1985, Karst processes and land forms

[904] Dreybrodt, Wolfgang, 1988, Processes in karst systems

[907] Drumm, E. C., Jun 1990, Subsidence of residual soils in a karst terrain

[938] Eberhard, Rolan, 1994, The Junee River karst system, Tasmania: a report to the Forestry Commission

[997] Environmental Consulting Engineers, , East Chestnut ridge hydroelectric characterization: a geophysical study of two karst features

[1004] Erchul, Ronald A., May 1989, Geotechnical applications of the self potential (SP) method. Report 2: the use of self potential to detect ground-water flow in karst

[1081] Ford, Derek Clifford, 1935–, Jun 1993, Environmental change in karst areas

[1084] Ford, Derek Clifford, 1935–, Mar 1973, Theme and resource inventory study of the karst regions of Canada: final report upon project A

[1085] Ford, Trevor David, 1925–(ed), Nov 1972, International seminar on karst denudation

[1102] Forti, Paolo (ed), 1985, International symposium on evaporite karst: preprints: Bologna October 22-25, 1985

[1106] Foster, David G. (ed), 1994, Learning to live with caves and karst: a cave and karst curriculum and resource guide

[1114] Fraile, D. Daniel Barettino, 1994, El Karst / Karst

[1121] Franklin, A.G., 1981, Foundation considerations in siting of nuclear facilities in karst terrains and other areas susceptible to ground collapse

[1141] Gams, Ivan, 1973, Slovenska Kraska Terminologija = Slovene karst terminology

[1149] Garder, George D., c1978, Tracing of subsurface flow in karst regions using artificially colored spores

[1174] Geographical Field Group, 1960, Coast and karst in Istria Rossa; a report of geographical field work carried out during August, 1960

[1209] Gillieson, David S. (ed), 1992, Geology, climate, hydrology and karst formation: field symposium in Australia 4 to 18 December 1992: Buchan - Mt. Gambier / Naracoorte - Nullarbor Plain humid temperate impounded karst, sub-humid temperate syngenetic karst, arid temperate karst: guidebook

[1210] Gillieson, David S. (ed), 1989, Resource management in limestone landscapes: international perspectives proceedings of the International Geographical Union study group man's impact on karst Sydney 15-21 August 1988

[1214] Gines, Joaquin, May 1987, Proceedings of the 1986 meeting of the International Geophysical Union study group on man's impact on karst

[1266] Grimes, K. G., Oct 1995, South East Karst Province of South Australia: Australian Caves & Karst Management Association October 1995

[1267] Grimes, Ken, February 1996, Field Guide to karst features in southeast South Australia and Western Victoria: for the Karst Studies Seminar, Naracoorte, February

5.10. KARST

1996

[1287] Günay, Gültekin (ed), , Karst water resources: proceedings of a symposium held at Ankara, July 1985, and sponsored by the United Nations Development Program, the United Nations Educational, Scientific and Cultural Organization, the International Association of Hydrological Sciences, many government organizations of Turkey, and Hacettepe University of Ankara, Turkey

[1288] Gunn, John (ed), 1994, An introduction to British limestone karst environments

[1377] Harvey, Edward Joseph, 1916–, 1977, Application of thermal imagery and aerial photography to hydrologic studies of karst terrane in Missouri

[1378] Harvey, Edward Joseph, 1916–, 1983, Hydrology of carbonate terrane, Niangua, Ossage Fork, and Grandglaize basins, Missouri

[1446] Herak, Milan (ed), May 1972, Karst: important karst regions of the northern hemisphere

[1531] Houshold, Ian (prep), 1995, Cave & karst guidebook and documentation: Australasian Cave and Karst Management Association eleventh Australasian Conference on cave and karst management Tasmania 29 April to 7 May 1995 hosted by The Parks & Wildlife Service - Tasmania

[1578] Huppert, George Nixon, 1944–, Jun 1988, Cave and karst-related theses in United States and Canadian Universities, 1899-1988

[1612] Institute of Hydrogeology and Engineering Geology. Chinese Academy of Geological Sciences, 1976, Karst in China

[1614] International Association of Hydrogeologists, 1975, International Association of Hydrogeologists 12th international congress: karst hydrogeology: abstracts and program 21-27 September 1975, Huntsville, Alabama, 28 September-3 October 1975, Gulf Shores, Alabama

[1615] International Association of Hydrogeologists, 1975, Karst hydrogeology: 12th international congress: International Association of Hydrogeologists 21-27 September 1975, Huntsville, Alabama U.S.A.

[1616] International Association of Hydrogeologists, 1988, Proceedings of the IAH 21st congress

[1617] International Association of Hydrogeologists, 1977, Proceedings of the twelfth international congress: karst hydrogeology: Huntsville, Alabama

[1618] International Conference (2nd: 1989: Blaubeuren, Germany), 1992, Proceedings of the karst-symposium-Blaubeuren

[1619] International Conference on Geomorphology (2nd: 1989 Frankfurt am Main, Germany), 1992, Karst

[1638] International Geographical Union, 1973, Symposium on karst - morphogenesis: papers: European regional conference

[1639] International Geographical Union, 1977, Symposium on karst - morphogenesis: papers: European regional conference (minus papers): a detailed exposition of the study tours

[1640] International Geographical Union, 1971, Symposium on karst-morphogenesis: European regional conference: Budapest - Aggtelek 5-9 August 1971

[1641] International Geographical Union Study Group on Man's Impact in Karst, 1987, Karst and man: proceedings of the international symposium on human influence in karst, 11-14th September 1987, Postojna, Yugoslavia

[1642] International Meeting on Karstic Sedimentology (1987: Han-sur- Lesse, Belgium), Sep 1988, Colloque international de sedimentologie karstique = international meeting on karstic sedimentology, Han-sur-Lesse, Belgique, 18-22 Mai 1987

[1647] International Symposium on Changing Karst Environments (1994: Oxford, England and Huddersfield, England), Aug 1994, Conference abstracts: [of papers presented at Changing karst environments, hydrogeology, geomorphology and conservation, an international symposium held at the Universities of Oxford and Huddersfield, September 1994]

[1648] International Symposium on Changing Karst Environments (1994: Oxford, England and Huddersfield, England), Mar 1995, Papers presented at the international symposium on changing karst environments, Oxford and Huddersfield, September 1994

[1649] International Symposium on Human Influence in Karst, 1987, Karst and man: proceedings of the international symposium on human influence in karst, 11 - 14 September, 1987, Postojna, Yugoslavia

[1650] International Symposium on Karst Hydrogeology (1st: 1979: Oymapinar), 1980, International seminar on karst hydrogeology: Oymapinar, October 1979: proceedings

[1651] International Symposium on Karst of Inner Plate Region with Monsoon Climate (1991: Guilin, China), 1992, World karst correlation: proceedings of the international symposium on karst of the Inner Plate Region with monsoon climate, July 1991, Guilin, China

[1652] International Symposium on Karsthydrology (1978: Budapest, Hungary), 1978, [Proceedings] nemzetkozi karszthidrologiai szimpozium = international symposium on karsthydrology = Mezhdunarodnyi simpozium po gidrologii karsta

[1664] International Union of Speleology. Commission for Physical, Chemical and Hydrological Research of Karst, 1989, Communications = mitteilungen = soobshcheniia: international symposium on physical, chemical and hydrological research of karst: Kosice, Czechoslovakia, May 10-15, 1988

[1694] Jakucs, László, 1977, Morphogenetics of karst regions: variants of karst evolution

[1702] James, Noel P., 1988, Paleokarst

[1725] Jennings, Joseph Newell, 1916–1984, 1971, Karst

[1726]Jennings, Joseph Newell, 1916–1984, 1985, Karst geomorphology

[1727]Jennings, Joseph Newell, 1916–1984, 1963, The limestone ranges of the Fitzroy Basin, Western Australia; a tropical semi-arid karst

[1728]Jennings, Joseph Newell, 1916–1984, Aug 1961, A preliminary report of the karst morphology of the Nullarbor Plains

[1760]Jordon, Donald George, 1926–, 1970, Water and copper-mine tailings in karst terrane of Rio Tanama Basin, Puerto Rico

[1768]Karst Hydrogeology Symposium (1977: Oymapinar), 1978, Karst hydrogeology symposium: Oymapinar, October 1977: proceedings

[1769]Kastning, Ernst H., 1944–, 1991, Appalachian Karst: proceedings of the Appalachian karst symposium: Radford, Virginia, March 23-26, 1991

[1808]Kerans, C., 1954–, 1990, Depositional systems and karst geology of the Ellenburger Group (Lower Ordovician), subsurface west Texas

[1816]Kiernan, Kevin, 1995, An atlas of Tasmanian karst

[1817]Kiernan, Kevin, Sep 1989, Bibliography of Tasmanian karst

[1819]Kiernan, Kevin, 1988, The management of soluble rock landscapes: an Australian perspective

[1833]Knisel, Walter G., 1972, Response of karst aquifers to recharge

[1835]Knott, David L., 1993, Current foundation engineering practice for structures in karst areas

[1836]Knott, David L., 1992, Guide for foundation engineering in Pennsylvania karst

[1853]Kosa, Attila (ed), 1981, Bir Al Ghanam karst study project: final report 1981

[1878]L'Association Francaise de Karstologie, Museum d'Histoire Naturelle, 1979, Actes du symposium international sur l'erosion karstique = proceedings of the international symposium on karstic erosion: Aix—en-Provence-Marseille-Nimes 10-14 Septembre 1979

[1886]LaMoreaux, Philip Elmer, 1920–(ed), 1993, Annotated bibliography of karst terranes: volume five: with three review articles

[1887]LaMoreaux, Philip Elmer, 1920–(ed), 1984, Guide to the hydrology of carbonate rocks

[1889]LaMoreaux, Philip Elmer, 1920–(ed), 1989, Hydrology of limestone terranes: annotated bibliography of carbonate rocks, volume 4

[1890]LaMoreaux, Philip Elmer, 1920–(ed), 1986, Hydrology of limestone terranes: annotated bibliography of carbonate rocks, volume three

[1895]Lane, Edward, 1935–, 1986, Karst in Florida

[1898]Langford, R.L. (ed), 1990, Karst geology in Hong Kong: proceedings of a conference on "Karst geology in Hong Kong" held at the University of Hong Kong on 6 January 1990

[1899]Langford, R. L. (ed), 1990, Karst geology in Hong Kong: programme and abstracts: a conference organized by the Geological Society of Hong Kong and Department of Geography and Geology, University of Hong Kong: Programme and Abstracts: University of Hong Kong: 4-6 January 1990

[1933]Leontiadis, J., Jan 1972, The use of radioisotopes in tracing karst groundwater in Greece

[1975]Lowe, David (comp), 1995, A dictionary of karst and caves

[1978]Lu, Yaoru, 1972, Engineering and geological conditions of karst

[2017]Maddox, Gary L., Nov 1993, Karst features of Florida Caverns State Park and Falling Waters State Recreation Area, Jackson and Washington Counties, Florida

[2026]Malott, Clyde Arnett, 1887–1950, 1970, Lost River at Wesley Chapel Gulf, Orange County, Indiana

[2027]Malott, Clyde Arnett, 1887–1950, 1974, Significant features of the Indiana karst

[2175]Mikulec, Stjepan, 1976, Karst hydrology and water resources: proceedings of the U. S.—Yugoslavian symposium, Dubrovnik, June 2-7, 1975

[2179]Millar, Ian R., 1989, General policy and guidelines for cave and karst management in areas managed by the Department of Conservation

[2180]Miller, Carter H., 1984, Preliminary seismic-velocity and magnetic studies of a carbonate rock-sinkhole area in Shelby County, Alabama

[2192]Miotke, F. M., 1974, Carbon dioxide and the soil atmosphere

[2266]Moore, Michael C., 1972, Southern Indiana Karst field trip guide, March 11 & 12, 1972

[2297]Mull, D. S., Oct 1988, Application of dye-tracing techniques for determining solute-transport characteristics of ground water in karst terranes

[2424]National Speleological National Speleological Society. Annual Convention. Guidebook. Geology Field Trip (1995: Blacksburg, VA), 1995, Origin of caves and karst in the Shenandoah Valley, Rochingham and Augusta Counties, Virginia: guidebook for a geologic fieldtrip, National Speleological Society Annual Convention, Blacksburg, VA, 16 July 1995

[2483]National Water Well Association, 1987, Proceedings of environmental problems in karst terranes and their solutions conference, October 28-30, 1986, Bowling Green, Kentucky

[2484]National Water Well Association, 1988, Proceedings of the second conference on environmental problems in karst terranes and their solutions conference, November 16-18, 1988, Maxwell House Hotel, Nashville, Tennessee

[2534]Nunez Jimenez, Antonio, 1966, Karstological investigations in Cuba

[2575]Orghidan, Traian, 1917–1985 (ed), 1984, Proceedings of the first symposium on theoretical and applied karstology held in Bucharest, Romania, 22-24 April 1983

[2617]Paterson, Keith, 1986, New directions in karst: proceedings of the Anglo-French Karst Symposium, September 1983

[2633]Peck, Dallas L., Oct 1988, Karst hydrogeology in the United States of America: 21st congress of the International Association of Hydrogeologists: karst hydrogeology and karst environment protection October 1988 Guilin, China

[2664]Pfeffer, Karl-Heinz (ed), 1990, International atlas of karst phenomena: sheets 8 - 12

[2665]Pfeffer, Karl-Heinz (ed), 1989, Karst

[2697]Potton, Craig, 1987, Images from a limestone landscape: a journey into the Punakaiki-Paparosa Region

[2710]Přibyl, Jan, 1986, Karst utilization in practice

[2713]Přibyl, Jan, 1986, Regularities of karst processes (karst landscape)

[2714]Přibyl, Jan, nd, Věkǎum budoucím

[2731]Pulina, Marian (ed), 1984, The dynamic of the contemporary karstic processes in the tropical area of Cuba: preliminary report of the field investigations performed by the expedition Guajaibon '84 in the winter season 1984

[2739]Quinlan, James Francis, 1936–1995, 1986, Ground water flow in the Mammoth Cave Area, Kentucky with emphasis on principles, contaminant dispersal, instrumentation for monitoring water quality, and other methods of study

[2740]Quinlan, James Francis, 1936–1995, Feb 1989, Ground water monitoring in karst terranes: recommended protocols and implicit assumptions

[2742]Quinlan, James Francis, 1936–1995, 1990, Hydrogeology and geomorphology of the Mammoth Cave area, Kentucky: Southeastern Friends of the Pleistocene 1990 field excursion, November 17, 1990

[2745]Quinlan, James Francis, 1936–1995 (comp), 1994, Practical karst hydrogeology, with emphasis on groundwater monitoring, February 8-11, 1994, Holiday Inn West, Gainesville, Florida

[2890]Sauro, U. (ed), 1991, Proceedings of the international conference on environmental changes in karst areas I.C.E.C.K.A. Italy, September 15-27th, 1991

[3043]Snyder, R. P., 1932–, 1982, Evaluation of breccia pipes in southeastern New Mexico and their relation to the Waste Isolation Pilot Plant (WIPP) site

[3078]Sowers, George F., 1996, Building on sinkholes: design and construction of foundations in karst terrain

[3153]Stringfield, Victor Timothy, 1902–, 1977, Hydrology of limestone terranes: development of karst and is effects on the permeability and circulation of water in carbonate rocks, with special reference to the southeastern states

[3154]Stringfield, Victor Timothy, 1902–, 1974, Karst and paleohydrology of carbonate rock terranes in semiarid and arid regions with a comparison to humid karst of Alabama

[3165]Sweeting, Marjorie Mary, 1920–1994 (ed), 1981, Karst geomorphology

[3166]Sweeting, Marjorie Mary, 1920–1994, 1995, Karst in China: its geomorphology and environment

[3167]Sweeting, Marjorie Mary, 1920–1994, 1972, Karst landforms

[3168]Sweeting, Marjorie Mary, 1920–1994, 1976, Karst processes

[3243]Trudgill, Stephen Thomas, 1947–, 1977, A bibliography of British karst, 1960-1977

[3249]Tulis, J. (comp), 1993, Slovakia: the karst and speleology

[3347]Van Everdingen, R. O., 1981, Morphology, hydrology and hydrochemistry of Karst in permafrost terrain near Great Bear Lake, Northwest Territories

[3441]Werner, Eberhard Wolfgang 1942–, 1972, Development of solution features, Cloverlick Valley, Pocahontas County

[3469]White, William Blaine, 1934–, May 1992, The Appalachian valleys of central Pennsylvania

[3482]Wilde, Kevan A., 1981, The cave and karst resource of New Zealand

[3483]Wilde, Kevan A., nd, Environmental monitoring of karst and caves

[3500]Williams, Paul W., 1968, Contributions to the study of Karst

[3512]Wilson, William L., Jul 1987, Investigation of karst-related subsidence near the southwest landfill, Alachua County, Florida, using ground penetrating radar: a report

[3513]Wilson, William L. (ed), 1988, Karst of the Orlando area: a guidebook prepared for the engineering and geology of karst terranes short course, Orlando, Florida, February 1-5, 1988

[3529]Worthy, Trevor, 1989, Inventory of New Zealand caves and karst of international, national and regional importance

[3540]Yauru Lu, 1931–, 1986, Karst in China - landscapes, types, rules

[3548]Yuan, Daoxian, 1933–, 1983, Problems of environmental protection of karst area

5.11 Mineralogy

[20]Ainsworth, William, 1834, An account of the caves of Ballybunian, county of Kerry: with some mineralogical details

[321]B'ardossy, György, 1982, Karst bauxites: bauxite deposits on carbonate rocks

[394]Bernasconi, R., 1976, The physico-chemical evolution of moonmilk

[1128]Fullerton, Iona, 1983, Cave minerals and

speleothems: a review of the literature with special reference to Waitomo
[1134]Gabriel, Diaconu, 1990, Closani Cave: mineralogical and genetic study of carbonates and clays
[1315]Haley, Boyd Raymond, 1922–, 1980, Mineral resources of the Belle Starr Caves wilderness study area, Sebastian and Scott Counties, Arkansas
[1456]Hill, Carol Ann, 1940–, 1976, Cave minerals
[1457]Hill, Carol Ann, 1940–, 1986, Cave minerals of the world
[2570]Ollerenshaw, Arthur Edward, c1963, The history of Blue John Stone: methods of mining and working ancient and modern
[2719]Prinz, W., 1980, The crystalline structures of the caves of Belgium

5.11.1 Saltpeter

[857]De Paepe, Duane, 1985, Gunpowder from Mammoth Cave: the saga of saltpetre mining before and during the War of 1812
[1037]Faust, Burton Sherwood, 1898–1967, 1964, Saltpetre caves and Virginia history
[1038]Faust, Burton Sherwood, 1898–1967, 1967, Saltpetre mining in Mammoth Cave, Ky
[1180]George, Angelo Isham, 1944–, Apr 1986, Saltpeter and gunpower manufacturing in Kentucky
[2755]Rains, Geo. W., 1861, Notes on making saltpetre from the earth of the caves
[3029]Smith, Marion Otis, 1942–, 1990, Saltpeter mining in East Tennessee

5.11.2 Speleothems

[205]Anon, 1983, Unusual cave and cavern formations
[718]Coronet Instructional Films, 1943, Limestone caverns
[1736]JLM Visuals, 1979, Caves and other ground water features II
[2004]Lyons, R. G., 1985, Rock to stalactite
[2260]Moore, George William, 1928–, 1954, The origin of helictites
[2719]Prinz, W., 1980, The crystalline structures of the caves of Belgium
[3024]Smith, John, 1894, Monograph of the stalactites and stalagmites of the Cleaves Cove, near Dalry, Ayrshire
[3198]Thayer, Charles W., 1966, Mud stalagmites and the conulite, a new speleothem

5.12 Paleokarst

[1788]Keller, David R. (ed), 1993, Paleokarst, karst-related diagenesis, reservoir development, and exploration concepts: examples from the Paleozoic section of the southern mid-continent: 1993 annual fieldtrip guidebook Permian Basin Section - SEPM Arbuckle Mountains, Oklahoma
[1808]Kerans, C., 1954–, 1990, Depositional systems and karst geology of the Ellenburger Group (Lower Ordovician), subsurface west Texas
[1809]Kerans, C., 1954–, 1989, Karst-controlled reservoir heterogeneity and an example from the Ellenburger Group (Lower Ordovician) of west Texas
[3043]Snyder, R. P., 1932–, 1982, Evaluation of breccia pipes in southeastern New Mexico and their relation to the Waste Isolation Pilot Plant (WIPP) site
[3536]Wright, V. Paul, 1953–(ed), 1991, Palaeokarsts and palaeokarstic reservoirs

5.13 Speleogenesis

[40]Aley, Thomas John, 1938–, 1964, Origin and hydrology of caves in the White Limestone of north central Jamaica: report of field work carried out under ONR Contract 3656 (03) NR 388 067
[110]Anon, 1960, Development of caves
[367]Beckway, Gregory W. (ed), 1971, Cavern formation
[428]Bolton, David W., nd, The geology and speleogenesis of Red Run Cave Tucker County, West Virginia
[808]Dasque, Jean (prod), 1972, Garden of shadows
[1057]Film Associates of California, 1965, Cavern formation
[1087]Ford, Trevor David, 1925–, 1973, Caves—origins, development and formations
[1099]Ford, Trevor David, 1925–(chair), Jun 1971, Symposium of the origin and development of caves
[1197]Gilbert Group, 1978, Caves, a deeper look at our Earth
[1271]Groom, Gillian Elisabeth, 1958, The geomorphology and speleogenesis of the Dachstein caves
[1321]Halliday, William R., 1926–, Jan 1954, The basic geology of Neff Canyon Cave, Utah ; The speleogenesis of Neff Canyon Cave, Utah
[1560]Hubbard Scientific Company, 1972, Cavern formation
[1565]Hudson, George Henry, 1855–1934, 1911, Joint caves of Valcour Island: their age and their origin
[1770]Kastning, Ernst H., 1944–, 1975, Cavern development in the Helderberg Plateau, East-central New York
[1863]Kranjc, Andrej, 1943–, 1989, Recent fluvial cave sediments, their origin and role in speleogenesis = Recentni fluvialni jamski sedimenti, njihovo nastajanje in vloga v speleogenezi
[1897]Lange, Arthur L., 1957, Studies on the origin of Montezuma Well and cave, Arizona
[2127]McGrain, Preston, 1917–, 1961, The geologic story

of Diamond Caverns
[2167]Middleton, Gregory J., 1965—1966, Speleology
[2173]Middleton, T. C., 1979, Cave development and formations
[2219]Missouri Speleological Survey, July-December 1986, Origin and development of Cave Spring, Shannon County, Missouri
[2260]Moore, George William, 1928–, 1954, The origin of helictites
[2261]Moore, George William, 1928–(ed), Jan 1960, Origin of limestone caves; a symposium with discussion
[2424]National Speleological National Speleological Society. Annual Convention. Guidebook. Geology Field Trip (1995: Blacksburg, VA), 1995, Origin of caves and karst in the Shenandoah Valley, Rochingham and Augusta Counties, Virginia: guidebook for a geologic fieldtrip, National Speleological Society Annual Convention, Blacksburg, VA, 16 July 1995
[2596]Palmer, Arthur Nicholas, 1940–, 1981, A geological guide to Mammoth Cave National Park
[2597]Palmer, Arthur Nicholas, 1940–, 1993, Geology and origin of Mystery Cave, Forestville State Park, Minnesota
[2804]Reid-Cowan Productions, 1976, Caves: the dark wilderness
[2931]Selfinger, Edward John, 1964, Experimental geology applied to speleogenesis
[3203]Thompson, James B., 1958, The geology of Jewel Cave
[3364]Vineyard, Jerry Daniel, 1935–, 1967, Guidebook to the geology between Springfield and Branson, Missouri, emphasizing stratigraphy and cavern development
[3525]Woods, Julian Edmund, 1862, Geological observations in South Australia: principally in the district southeast of Adelaide
[3537]Wyllie, Diana, 1973, Caves—origins, development and formations
[3556]Zim, Herbert Spencer, 1909–, 1978, Caves and life

5.14 Vulcanospeleology

[271]Baddeley, Glenn (ed, auth), , Vulcon Guidebook: lava features and limestone karst of Victoria and southeastern South Australia. Vulcon 1995 20th Biennial Conference. Australian Speleological Federation, Inc, Hamilton, Victoria 2-6 January 1995
[272]Baddeley, Glenn (ed), 1995, Vulcon precedings: papers submitted for presentation. Vulcon 20th Biennial Conference. Australian Speleological Federation, Inc, Hamilton, Victoria 2-6 January 1995
[1334]Halliday, William R., 1926–, May 1981, Outline of major biological conclusions from 1980-81 Mount St. Helens eruptions
[2882]Sameshima, Teruhiko, 1988, 5th international symposium on vulcanospeleology excursion guide book

6 General Caving

Alan Warild

41 Northwood Street
Newtown, 2042 NSW Australia

Books about caving are often technical descriptions of the equipment required and how best to use it. How-to-do-it caving books have been around for some time. They vary from a single chapter in anecdotal exploration books telling how one did it, usually when the authors feel their approach is novel enough to warrant it, to entire books dedicated to the technique of caving.

Single chapter classics are in the *Papua New Guinea Speleological Expedition*, NSRE 1973, with its section on single rope techniques as used on the expedition (taking rope instead of ladders was a new idea at the time), and in the *Ghar Parau* expedition report mainly about ladders. For years, cavers have been reading the French classics such as Casteret's *Ten Years Under the Earth*, partly as adventure stories, and partly to extract a range of technical tips on caving from the days when caves were bigger, harder, and darker than they are today.

Pure technical caving books come in a variety of forms, depending on whether their accent is on horizontal or vertical caving, cave diving, rescue, or mapping techniques. There has been a major increase in the number of technical caving books since the 1960s, mainly due to the advances in techniques with the advent of single rope techniques and scuba equipment, as well as the increase in the number of people interested in caving.

Probably the best known technical books in English language caving literature today are *Caving Practice and Equipment* and its precursor *British Caving* and *Caving Basics*, both written by groups of specialist authors. Unfortunately, the former appeared towards the end of the cable ladder era in Britain and was out-of-date for a long time until the second edition appeared.

THE techniques of caving book must be *Techniques de la Speleologie Alpine* by Dobrilla, but it exists only in French and is not included in this bibliography. One of the earliest attempts at providing vertical caving information is *Prusiking* by Thrun. Neil Montgomery's *Single Rope Techniques* is the widely distributed vertical caving book in English. It quickly became the ultimate reference for vertical cavers and it was ten years before another technical book come near it. When *On Rope*, and a short time later *Vertical* appeared, they became the standards, *On Rope* in US circles and *Vertical Caving* in many English speaking areas of the world. Of course periodicals have been able to keep up to date faster than books and thus *Nylon Highway*, published by the vertical section of the NSS, is worth looking at in this area.

Technical caving books (like cavers!) tend to be almost painfully parochial, so each caving community will put out its own instructional caving books to describe how they do it, and why they are, therefore, way ahead of anybody else. The popular techniques in a region are described in minute detail, while the alternatives are allowed only a short, often obviously misinformed paragraph, or neglected entirely. Just as *On Rope* could have been written only in the USA, *Vertical Caving* by Meredith and Martinez could have been written only in France, no matter which of the four translations you care to read.

A relatively recent addition to the range of technical books are rescue books. Safe-caving books and club pamphlets have been around for quite some time. Manuals on rescue are a newer phenomenon, sort of a cross between a first aid manual and a how-not-to-go-caving book. Like other technical books, they can become outdated as fast as they are written. One example of this is *A Manual of Cave Rescue Techniques* which, although a leader in its field, is rewritten far too infrequently. Most groups have produced rescue texts in loose leaf notebooks and replace outdated information quickly by replacing only the pages needed.

Another aspect of caving is mapping caves. Mapping techniques are usually covered, or at least touched on by the standard technical caving books but several books deal entirely with cave surveying. Like other technical books, advances in technology keep leaving surveying books behind, although in a different way. The data you need to collect, or the means of getting it, have not changed much over recent years. However, what you do with it afterwards has undergone the same revolution as every other thing can make use of a silicon chip. Until the end of the 70s, cave surveying books told you how to process the data when you got home. Any book written since then that doesn't take into account the very likely possibility of a caver reducing the survey data and, probably, plotting the map back at camp or field house is living in the past. Once again, the classics seem to come from Britain with Bryan Ellis' *Surveying Caves*, although *On Station* from the NSS gives a fair guide to cave surveying. John Ganter's *A Systematic Guide to Making Your First Cave Map* is a leader in this area. One of the simplest but clearest texts is *The Art of Cave Mapping* by the Missouri Speleological Survey.

Cave diving is such a different aspect of caving that it ranks as a sport in its own right. This is especially true of the type of cave diving where the divers set out to avoid any parts of the cave that are inconsiderate enough to have air in them. In this field, Martyn Farr's *The Darkness Beckons* is one of the all time great stories of cave and sump diving, and it also contains a vast amount of technical advice. On the how-to-do-it side, Sheck Exley's *Ba-*

sic *Cave Diving* has been reprinted over 10 times. After reading it, you no longer wonder how Sheck survived so long. In recent years, both the British and Americans have produced guides to cave diving. *Cave Diving Manual: An Overview* by the NSS and *Cave Diving* by the Cave Diving Group are classics in this field.

Photography has improved, but the cost of color printing has still kept the number of high quality color coffee table books to a minimum. One of the first color coffee table offerings is Bögli's *Luminous Darkness*, with truly excellent photographs of the time, featuring some great European caves. Waltham's *Caves* and *The World of Caves* are exceptional in their class, both in terms of photo quality, range of caves covered, and the spirit of adventure they carry. Time has moved on though, and both great new caves and new equipment have appeared. *Lechuguilla, Jewel of the Underground* from Speleo Projects embodies this like no other; a magnificent book about a magnificent cave.

Of course, one step better than looking through a great cave book is to take the photographs yourself. Little has been written about cave photograph techniques and the NSS periodical *Flash* is a very good reference. *Cave Photography : A Practical Guide* by Chris Howes takes one through the tedious task of getting the very best cave photograph but it is a small and incomplete guide. After reading a book like this, you are either inspired to use the ideas to get your own show-stopping photographs, or convinced to stick with your muddy "happy snaps" and leave the real photos to the experts. This field is in need of some strong authors.

Adventure caving books are those whose primary aim is to portray the excitement of caving, or a particular cave. They are story books, and the best story wins. For my money, it's Pierre Chevalier's *Subterranean Climbers*; how could you beat a story about bunch of ordinary guys who, during the Second World War, spent their weekends off cycling up a mountain to explore what was at the time the world's deepest cave. No special gear, backup, or hint of the "aren't we good" attitude and hype often prevalent in other adventure books. Dozens of expedition reports from all over the world express this lust for adventure. If Chevalier's or others adventures seem just too remote to be real, North American readers can find find something closer to home in William Halliday's *Adventure of Caving* and *Depths of the Earth*, Brucker and Watson's *The Longest Cave*, or Michael Ray Taylor's *Cave Passages*, while inhabitants of the waterlogged isles across the Atlantic revel in very British humor of any of Jim Eyre's books. In between the laughs, there are considerable insights to the British attitude to caves and caving. You'll have to read the books though, to find out what those attitudes are. In addition one can find more of a US flavor of humor in Thom Engel's *The Almost Complete Eclectic Caver*.

References not in Main Bibliography

Dobrilla, J. C. 1973. *Techniques de la speleologie alpin.* Levallois-Perret: G. Marbach.

6.1 Caving

[608]Casteret, Norbert, 1897–1987, 1955, The descent of Pierre Saint-Martin
[723]The Council of Southern Caving Clubs, 1974, Caving for beginners
[868]Dickey, F., 1974, Report of the Caver Proliferation Committee
[926]Durrant, Gillian A. , 1979, Himalaya underground - 1976 speleological expedition
[1247]Great American Film Factory, 1977, Underground wilderness
[1381]Hassemer, Jerry Herman, 1934–(ed), 1987, Caving basics: a comprehensive manual for beginning cavers
[1767]Kahrau, Wolfgang, 1972, Australian caves and caving
[2288]Mountain Rescue Committee (Great Britain), 1975, Mountain and cave rescue, with lists of official rescue teams and posts: the handbook of the Mountain Rescue Committee
[2289]Mountain Rescue Committee (Great Britain), 1964, Mountain rescue & cave rescue
[2767]Rea, G. Thomas (ed), 1992, Caving basics: a comprehensive guide for beginning cavers
[3213]Thrailkill, John Vernon, 1930–, 1954, Introduction to caving
[3225]Traister, Robert J., 1983, Cave exploring
[3399]Warild, Alan, 1988, Vertical: a technical manual for cavers
[3412]Watson, Richard Allan, 1931–, 1994, Caving

6.2 Climbing

[105]Anon, 1995, Crewe Climbing and Pot Holing Club; Peak rigging guide
[956]Elliot, Dave, 1946–, 1987, Single rope technique rigging guide
[957]Elliot, Dave, 1946–, 1986, Single rope technique: a training manual
[1956]Logan, William Steve, 1971, Self-belaying cable ladder - a rope climbing aid
[2158]Meredith, Michael Edward, 1943–, c1980, Vertical caving
[2257]Montgomery, Neil R., 1977, Single rope techniques: a guide for vertical cavers

[2593] Padgett, Allen, 1987, On rope: North American vertical rope techniques for caving, search and rescue, mountaineering
[3216] Thrun, Robert, 1940–, 1973, Prusiking
[3399] Warild, Alan, 1988, Vertical: a technical manual for cavers

6.3 Diving

[221] Anon, 1986, Wookey Hole; the caves and mill
[301] Balcombe, F. Graham (comp, ed), 1990, Cave diving: the Cave Diving Group manual
[304] Balcombe, F. Graham, [1936], The Log of the Wookey Hole: exploration expedition 1935
[310] Banbury, Jack, 1972, Proceedings of the fifth annual seminar on cave diving
[338] Bartrop, Richard N., 1987, Cave Diving Group: Derbyshire sump index 1987
[434] Boon, J. M., 1966, Cave diving on air
[435] Boon, J. M., 1977, Down to the sunless sea
[477] Briel, Larry I. (comp), 1968, Proceedings of the first annual seminar on cave diving
[560] Burge, John W., 1988, Basic underwater cave surveying
[563] Burgess, Robert Forrest, 1976, The cave divers: illustrated with photographs and drawings
[603] Carter, R. L., Aug 1994, Cave Diving Group: Peak District sump index 1994
[609] Casteret, Norbert, 1897–1987, 1962, More years under the earth
[613] Cave Diving Association of Australia, Inc, [c1977], Cave Diving Association of Australia, Inc: conference 1977 held at Mount Gambier, South Australia
[614] Cave Diving Section. National Speleological Society, 1986, NSS Cave Diving Section 1986 members' manual
[700] Cook, Thomas Howard, 1946–, 1978, The el cheapo book of home brew diving equipment
[822] Davies, Melvyn, Nov 1966, Cave sump index; South Wales
[825] Davies, Phillip, Apr 1957, An index of submerged cave passages; sumps
[852] DeLoach, Ned (ed), 1978, Ned DeLoach's diving guide to underwater Florida
[869] Dickinson, Leo, 1989, Anything is possible
[879] Dobson, Phil (comp), Aug 1981, The Cave Diving Group annotated ten year cumulative index to the cave dives for the years 1969-1978
[1015] Exeter University Spelaeological Society, [1970], Expedition to Morocco 1969
[1016] Exley, Irby Sheck, 1944–1994, 1979, Basic cave diving: a blueprint for survival
[1017] Exley, Irby Sheck, 1944–1994, 1994, Caverns measureless to man
[1018] Exley, Irby Sheck, 1944–1994, 1972, Mapping underwater caves
[1019] Exley, Irby Sheck, 1944–1994 (ed), 1982, N. S. S. cave diving manual
[1028] Farr, Martyn, 1950–, 1980, The darkness beckons: the history and development of cave diving
[1029] Farr, Martyn, 1950–, 1984, The great caving adventure
[1030] Farr, Martyn, 1950–, 1985, Wookey: the caves beyond
[1036] Faulkner, Trevor, 1942–, 1979, Sump index Norway
[1062] Finley, Claudette, 1980, Hand signals for diving
[1089] Ford, Trevor David, 1925–(rev), 1968, Derbyshire sump index
[1245] Gray, Eric (comp), 1985, Devonshire sump index
[1263] Griffiths, Julian (comp), 1981, Cave Diving Group: northern sump index
[1475] Hoffman, Ray J., 1988, Chronology of diving activities and underground surveys in Devils Hole and Devils Hole Cave, Nye County, Nevada, 1950-86
[1523] Horne, Peter, 1980, A report on South Australian diving fatalities
[1524] Horne, Peter, 1990, Research handbook for cave divers: an introduction to the realm and techniques of the underwater speleologist
[1525] Horne, Peter, 1987, South Australian diving fatalities 1950-1985
[1743] Jones, Gareth Llwyd, 1988, Cave Diving Group: Irish sump index
[1745] Jones, Gareth Llwyd, 1987, Irish sump index
[1843] Koehler, W. H., 1971, Lure of the labyrinth
[1846] Kollár, Attila K. (comp), 1989, Subaquatic caves in Hungary for divers
[1912] Lavaur, Guy de, 1908–1986, 1956, Caves and cave diving
[1938] Lewis, Ian D., 1982, Cave diving in Australia
[1951] Lloyd, Oliver Cromwell, 1911–1985, Jul 1975, A cave diver's training manual
[2041] Mansfield, Raymond Walter, 1978, Somerset sump index
[2044] Mansfield, Raymond Walter, Apr 1964, Sump index: section 1 (Somerset)
[2073] Mason, Edmund John, 1911–1993, Jul 1963, The story of Wookey Hole
[2118] McDonald, M. C. (prep), Dec 1993, The Somerset sump index
[2253] Monico, Paul (comp), 1995, CDG northern sump index 1995
[2275] Morris, David (comp), Dec 1985, Cave Diving Group: Welsh sump index
[2286] Mount, Tom, 1973, Cave diving manual
[2287] Mount, Tom, 1973, Safe cave diving

[2603] Palmer, Robert John, 1951–1997, 1985, The blue holes of the Bahamas
[2604] Palmer, Robert John, 1951–1997, 1989, Deep into blue holes: the story of the Andros Project
[2605] Palmer, Robert John, 1951–1997 (comp), Mar 1984, The report of the 1981 and 1982 British cave diving expeditions to Andros Island, Bahamas
[2606] Palmer, Robert John, 1951–1997 (ed), 1988, Report of the 1987 international blueholes research project
[2724] Prosser, J. Joseph, 1990, Cave diving communications
[2725] Prosser, J. Joseph, 1992, NSS cave diving manual: an overview
[2726] Prosser, J. Joseph (ed), 1986, The NSS instructor's training manual
[2727] Prosser, J. Joseph, Oct 1989, NSS student cave diver workbook: designed specifically for use by the student cave diver participating in the NSS full cave diver course
[2879] Saltsman, Dayton, 1995, The art of safe cave diving
[2903] Schenck, Bill, 1975, Proceedings of the 6th annual seminar, Lindenwood College, St. Charles, Missouri, June 16-17, 1973; Proceedings of the 7th annual seminar, Ramada Inn West, Jacksonville, Florida, June 15- 16, 1974 ; Research papers by N.A.C.D. instructor candidates
[3048] Somers, Lee H., 1971, Cave diving: equipment and procedures
[3139] Stone, William Curtis, 1952–, 1992, Inner space: the last terrestrial frontier
[3141] Stone, William Curtis, 1952–(ed), 1989, The Wakulla Springs project
[3142] Stoneman, John (narr), 1990, The cave divers
[3143] Storey, James Welborn, 1935-1992, Jun 1971, Advanced cave diving
[3145] Storey, James Welborn, 1935-1992 (ed), [1970], Cave diving notes: equipment - methods - danger
[3199] Thomas, Alan, 1931– (ed), 1989, The last adventure
[3377] Wallace, Malcolm (ed), [1995], Back in time for tea and medals
[3541] Yeadon, Geoff (auth, illus), 1981, The Cave Diving Group technical review no.3: line laying and following
[3558] Zumrick, John L., 1988, NSS cavern diving manual

6.4 Equipment

[434] Boon, J. M., 1966, Cave diving on air
[450] Boy Scouts of America, 1991, Venture caving
[457] Bradshaw, Chris, 1982, The manual of basic caving
[584] Callow, Philip, nd, Cave light
[674] Coase, D. A., Dec 1962, The B. E. C. method of caving ladder construction
[967] Ellis, Bryan M., 1934–, Oct 1962, The manufacture of lightweight caving equipment
[969] Ellis, Bryan M., 1934–, Oct 1958, A survey of headgear and lighting available for caving
[2108] McClurg, David Robert, 1929–, 1986, Adventure of caving: a practical guide for advanced and beginning cavers
[2109] McClurg, David Robert, 1929–(auth, photo), 1996, Adventure of caving: new updated edition
[2111] McClurg, David Robert, 1929–(ed), 1982, Caving short course: 1982 NSS Convention
[2211] Missouri Speleological Survey, Jan 1980, An introduction to caving: a guide for beginners
[3048] Somers, Lee H., 1971, Cave diving: equipment and procedures
[3184] Tarkington, Terry Warren, 1925–, May 1965, A manual for beginners
[3185] Tarkington, Terry Warren, 1925–, 1965, So you want to go caving! A manual for beginners
[3259] Tyler, Ronald M., Dec 1971, Caving handbook: a guide to the underworld

6.5 General

[1] A. I. D. International, 1977, Caves
[13] Adams, William Henry Davenport, 1828–1891, 1886, Famous caverns and grottoes: described and illustrated
[69] Anderson, Jennifer Ann, 1942–, 1974, Cave exploring
[95] Anon, 1964, Caves
[96] Anon, 1965, Caves
[97] Anon, c1972, Caves and Caving City Museum Queens Road Bristol 8
[101] Anon, nd, Caving in West Virginia
[118] Anon, 1988, The geology of caves
[190] Anon, c1967, Speleology in France: Ressources [sic] in Limousin Quercy Perigord
[191] Anon, 1946, Spelunking
[248] Audyová, Jiřina, 1993, The Moravian karst: time and stone
[259] Australian Speleological Federation, nd, Caving in Australia
[273] Baguley, Frank (ed), Aug 1993, Cambrian Caving Council handbook 1993
[307] Ballard, Jim, 1978, The spur book of caving
[325] Barr, Thomas Calhoun, 1931–, 1989, The cave community
[331] Barrington, Nicholas Robert, 1968, Adventure underground
[346] Bauer, Ernst W. , 1971, The mysterious world of caves
[357] Beck, Barry Frederic, 1944–(ed), 1980, An introduction to caves and cave exploring in Georgia
[368] Bedford, Bruce (ed), 1970, Descent handbook for

cavers

[369] Bedford, Bruce L., 1942–, 1975, Challenge underground

[370] Bedford, Bruce L., 1942–(ed), 1971, Descent handbook for cavers 1971/1972

[371] Bedford, Bruce L., 1942–(ed), 1987, The Descent magazine's caving yearbook

[381] Bender, Lionel, 1989, Cave

[424] Bögli, Alfred, 1912–, 1967, Luminous darkness: the wonderful world of caves

[437] Booth, Eugene, 1940–, 1977, Under the ground

[462] Brandt, Keith, 1949–, 1985, Caves

[475] Bridges, Thomas Charles, 1937, Adventures under ground [sic]

[514] Brooks, Steve (comp), 1995, The Australian cavers diary

[588] Campbell, Jim, 1978, The Grampian caving manual: an introduction to caving and potholing for the beginner

[657] Christopher, N. S. J. (ed), 1969, Cambrian Caving Council handbook 1969

[676] Coder, Kate, 1989, Pennsylvania's caves & caverns: an activity book

[696] Contor, Roger J., 1963, The underworld of Oregon Caves National Monument

[698] Cook, John Hawley, 1989, Of caves & cavemen

[721] [Council of Northern Caving Clubs], 1969, Northern cave handbook

[722] Council of Northern Caving Clubs, 1979, Northern caving; handbook of the Council of Northern Caving Clubs

[801] Daniel, Thomas W., 1973, Exploring Alabama caves

[836] Davis, R. V., 1978, Limestone caves: a concise explanation

[845] Deakin, P. R., 1975, British caves and potholes

[846] Dean, Anabel, 1984, Going underground: all about caves and caving

[860] [Descent Magazine], [1973], Descent handbook for cavers 1974

[885] Dorrell, Margaret, 1974, Caves through the ages

[893] Dowling, Richard, 1846–1898, 1900, Catmur's cave

[908] Duddington, C. L. , 1969, Caves

[959] Elliott, D. M. (ed), Jul 1953, Tasmanian Caverneering Club Handbook

[974] Ellis, Ross Andrew, 1940–, 1980, Australian caves and caving

[1032] Farrer, Reginald John, 1880–1920, 1926, On the caves of the world

[1076] Folsom, Franklin Brewster, 1907–, 1962, Exploring American caves

[1077] Folsom, Franklin Brewster, 1907–, 1956, Exploring American caves, their history, geology, lore, and location: a spelunker's guide

[1100] Forder, John (auth, photo) , 1992, Secrets of the moors and dales

[1120] Franke, Herbert Werner, 1927–, 1958, Wilderness under the earth

[1142] Gans, Roma, 1894–, 1976, Caves

[1190] Gibbons, Gail, 1993, Caves and Caverns

[1199] Gilbreath, Alice Marie , 1921–, 1978, Nature's underground palaces: caves and caverns

[1254] Greenberg, Judith E., 1990, Caves

[1262] Griffin, K. A. , 1968, Confidence in the cave. A guide for a safe speleological adventure

[1319] Halliday, William R., 1926–, 1974, American caves and caving: techniques, pleasures, and safeguards of modern cave exploration

[1328] Halliday, William R., 1926–, 1966, Depths of the earth: caves and cavers of the United States

[1329] Halliday, William R., 1926–, 1976, Depths of the earth: caves and cavers of the United States

[1340] Hamilton, Elizabeth, 1900–, 1956, The first book of caves

[1346] Hamilton-Smith, Elery (ed), 1958, Caving in Australia

[1366] Harrison, David Lakin, 1970, The world of American caves

[1376] Hartwig, Georg, 1813–1880, 1888, Marvels under our feet: from "the subterranean world"

[1412] Hazslinszky, Tamás, 1989, Colourful world of caves

[1430] Henderson, Junius, 1865–1937, Oct 1932, Caverns, ice caves, sinkholes, and natural bridges, part I & II

[1439] Hendrix, Charles E., 1950, The cave book

[1552] [Howes, Chris, 1951–(ed)], 1994, The Descent caver's handbook

[1568] Hughes, Thomas McKenny, 1887, On caves

[1608] Indiana. Division of State Parks, 1992, Exploring the cave world: cave packet for teachers, intermediate level

[1681] Jackson, Donald Dale, 1935–, 1982, Underground worlds

[1710] Jasinski, Marc, 1967, Caves and caving: a guide to the exploration, geology and biology of caves

[1711] Jean, David, 1976, Underground worlds

[1723] Jennings, Joseph Newell, 1916–1984, 1970, Caves

[1736] JLM Visuals, 1979, Caves and other ground water features II

[1762] Judson, David, 1961, Caving and potholing

[1811] Kerbo, Ronal Carrel, 1944–, 1981, Caves

[1815] Kidder, Daniel Parish, 1815–1891, 1848, The caves of the earth: their natural history, features, and incidents

[1859] Kramer, Stephen P., 1995, Caves

[1918] Laycock, George , 1976, Caves

[1937] Lewis, Charles C. (ed), 1993, Chronicles of the Reading Grotto, in which we go to the California convention, June 3rd to September 6th, 1966

[1963] Longsworth, Polly, 1959, Exploring caves

[1972] Lovelock, James, 1969, Caving

[1973] Lovelock, James, 1981, A caving manual
[1979] Lübke, Anton, 1890–, 1958, The world of caves
[1989] Lundelius, Ernest L., 1927–(ed), 1971, Natural history of Texas caves
[2001] Lyon, Ben, 1983, Venturing underground: the new speleo's guide
[2072] Mason, Edmund John, 1911–1993, 1977, Caves and caving in Britain
[2093] Matthews, Peter Gahan, 1938–(ed), Jan 1968, Speleo handbook
[2110] McClurg, David Robert, 1929–, 1973, The amateur's guide to caves & caving; skill-building ways to finding and exploring the underground wilderness
[2113] McClurg, James E., 1962, Caves and their mysteries
[2122] McEwan, Graham, 1994, Crypts, caves and catacombes: subterrenea of Derbyshire & Nottingham
[2249] Mohr, Charles E., 1907–(ed), 1955, Celebrated American caves
[2301] Murdoch, Judy, Apr 1994, Light on dark: caves of the south east of South Australia
[2354] National Speleological Society, 1960—1978, Caving information series
[2518] Northcott, T. C., 1934, How nature makes a cave
[2608] Palmer, William Thomas, 1877–, 1934, The complete hill walker, rock climber and cave explorer
[2609] Panayiotakis, Yioryos I., 1988, The Dictaean Cave
[2655] Perry, Clair Willard [Clay], 1887–1961, 1946, New England's buried treasure
[2677] Pinney, Roy, 1962, The complete book of cave exploration: an authoritative guide to the wonders, mysteries and excitement of caves and caving
[2689] Pond, Alonzo William, 1894–, 1969, Caverns of the world
[2692] Poole, Lynn, 1962, Deep in caves and caverns
[2766] Ray, Michael Allen, nd, Underground worlds: tour guide training manual
[2805] Religious Tract Society, 1800, The caves of the earth: their natural history, features, and incidents
[2820] Rigby, Susan, 1992, Caves
[2828] Robinson, Donald, 1964, Caving and potholing
[2829] Robinson, Donald, , Potholing and caving
[2832] Robison, Mabel Otis, 1959, Mystic wonderlands
[2914] Schultz, Ronald, 1951–, 1993, Looking inside caves and caverns
[2945] Shannon, Robert Terry, 1960, About caves
[2946] Shannon, Robert Terry, 1966, About caves
[2953] Shaw, Trevor Royle, 1928–, 1961, The deepest caves in the world and caves which have held the world depth record
[2968] Sherman, Geraldine, 1980, Caverns: a world of mystery and beauty
[2982] Silver, Donald M., 1947–, 1993, Cave
[2991] Single, Michael, 1992, Castles of the underworld
[2994] Skinner, A. D., Jul 1978, The Mole Creek Caves
[3077] Southwest Missouri State University Dept. of Geography and Geology, 1979, Speleology workshop
[3119] Stenuit, Robert, 1966, Caves and the marvelous world beneath us
[3120] Sterling, Dorothy, 1913–, , The story of caves
[3144] Storey, James Welborn, 1935–1992 (ed), 1965, American caving, illustrated; caving, climbing, camping
[3156] Styles, Frank Showell, 1908–, 1959, How underground Britain is explored
[3187] Tasmanian Caverneering Club, 1963, Caverneering handbook of the Tasmanian Caverneering Club
[3221] Toops, Bonnie, 1990, Let's explore: caves and caverns: a young explorer series
[3225] Traister, Robert J., 1983, Cave exploring
[3284] U.S. Geological Survey, 1977, Geology of caves
[3356] Vaughan, Jennifer, 1973, Caves
[3373] Wainwright, Alfred, 1991, Wainwright in the Limestone Dales
[3392] Waltham, Anthony Clive, 1942–[Tony] (ed), 1974, Limestone and caves of Northwest England
[3393] Waltham, Anthony Clive, 1942–[Tony], 1976, The world of caves
[3488] Wilkins, Frances, 1977, Caves
[3522] Wood, Jenny, 1990, Caves
[3523] Wood, Jenny, 1990, Caves
[3554] Zhaoyang Wang, 1983, Underground worlds: Guizhou

6.6 Photography

[92] Anon, 1882, The Caverns of Luray
[173] Anon, c1886, Photographic views of some of the important points of Mammoth Cave situated in Edmondson, Co. Kentucky, USA: the wonders of this cave and its magnitude, cannot be described, and must be seen to be appreciated. It is reached only by the Louisville & Nashville Railroads. These photographs were taken by magnesium light by W.H. Sesser, St. Joseph, Mich., with a Collins camera
[201] Anon, Dec 1969, Symposium on cave photography
[218] Anon, nd, Wookey Hole Caves: sixteen exclusive camera studies
[248] Audyová, Jiřina, 1993, The Moravian karst: time and stone
[262] Australian Speleological Federation, Feb 1972, Proceedings of the eighth biennial conference of the Australian Speleological Federation
[424] Bögli, Alfred, 1912–, 1967, Luminous darkness: the wonderful world of caves
[729] Cox, Edward, 1910, Souvenir of Cox's Stalactite Caves "visited by his majesty the late King Edward VII"
[790] d'Amboise, Valery (auth, illus, photo), 1986, Eternal

caves

[948]Eggleston, Peter, 1994, Underground Video Techniques

[1100]Forder, John (auth, photo), 1992, Secrets of the moors and dales

[1549]Howes, Chris, 1951–, 1987, Cave photography: a practical guide

[1556]Howes, Chris, 1951–, 1989, To photograph darkness: the history of underground and flash photography

[1995]Luray Caverns Corporation, 1882, Electric light views in the Caverns of Luray. The Caverns of Luray (at Luray, Page County, Virginia, a station on the Shenandoah Valley Railroad) as a resort for tourists ... are unexcelled, in their wonderful attractiveness, by any other creation of nature

[2177]Miles, Sibella Elizabeth, 1864, The Grotto of Neptune ("Antro Di Nettuno"), Sardinia; a poem illustrative of three views of this interesting cavern, taken in July 1824 by the late Commander Alfred Miles and dedicated to his memory by his widow, Sibella Elizabeth Miles

[2486]Nazarieff, de Serge, 1935–, 1981, Clair de Roche

[2893]Savory, James Henry, 1889–1962, 1989, A man deep in Mendip: the caving diaries of Harry Savory, 1910—1921

[3131]Stoddard, Sheena, 1994, An introduction to cave photography

[3224]Towler, George (photo), 1890, Views of Chapman Cave

[3489]Wilkinson, Charles Smith, 1887, Photographs of the Jenolan Caves (interior views photographed by means of the electric and magnisium [sic] lights

6.7 Rescue

[246]Attout, Jacques, 1956, Men of Pierre Saint-Martin

[313]Bannerman, Jackie (ed staff), 1993, Universal study guide for use with basic orientation and basic team member courses

[407]Black, Don F., 1993, I don't play golf: recollections of a rescue volunteer

[436]Boon, J. M., 1980, The Great San Agustín Rescue

[464]Brannan, Eldred Boyd, [1925], The entombment of Floyd Collins in Sand Cave, Kentucky

[482]British Association for Immediate Care, 1986, Rescue from remote places: cave rescue and sport diving emergencies

[597]Cansell, Anthony, c1969, The Upper Wharfedale Fell Rescue Association 1948-1968

[607]Casteret, Norbert, 1897–1987, 1954, The darkness under the earth

[1023]Eyre, Jim, 1988, Race against time: a history of the Cave Rescue Organisation

[1075]Fogle, A. R. (ed staff), 1990, Basic cave rescue orientation course study guide

[1122]Frantz, A. P., 1980, Cave rescue operations

[1375]Hartley, Howard W., 1925, Tragedy of Sand Cave

[1566]Hudson, Stephen Edward, 1950–, 1988, Manual of U.S. cave rescue techniques

[1583]Hurtt, Howard A., 1977, Rescue system outline: Lilburn Cave Project

[1974]Lovelock, James, 1963, Life and Death Underground

[2276]Morris, Deborah, 1956–, 1993, Trapped in a cave!: a true story

[2290]Mountain Rescue Committee (Great Britain), 1963, Mountain rescue, cave rescue

[2291]Mountain Rescue Council, 1994, Mountain & cave rescue: the handbook of the Mountain Rescue Council

[2305]Murray, Robert K., 1979, Trapped!: the story of the struggle to rescue Floyd Collins from a Kentucky cave in 1925, an ordeal which became one of the most sensational events of modern times

[2306]Murray, Robert K., 1982, Trapped!: the story of the struggle to rescue Floyd Collins from a Kentucky cave in 1925, an ordeal which became one of the worst sensational events of modern times

[2361]National Speleological Society. American Caving Accidents (1967), Aug 1969, American caving accidents 1967

[2362]National Speleological Society. American Caving Accidents (1967-1970), 1974, American caving accidents 1967-1970

[2363]National Speleological Society. American Caving Accidents (1968), Apr 1970, American caving accidents 1968

[2364]National Speleological Society. American Caving Accidents (1969), nd, American caving accidents 1969

[2365]National Speleological Society. American Caving Accidents (1970), Nov 1971, American caving accidents 1970

[2366]National Speleological Society. American Caving Accidents (1971), Aug 1974, American caving accidents 1971

[2367]National Speleological Society. American Caving Accidents (1972), Aug 1975, American caving accidents 1972: a report of the National Speleological Society

[2368]National Speleological Society. American Caving Accidents (1973), Aug 1975, American caving accidents 1973: a report of the National Speleological Society

[2369]National Speleological Society. American Caving Accidents (1974), Aug 1976, American caving accidents 1974: a report of the National Speleological Society

[2370]National Speleological Society. American Caving Accidents (1975), Aug 1977, American caving accidents 1975: a report of the National Speleological Society

[2371]National Speleological Society. American Caving Accidents (1976-1979), May 1981, American caving ac-

cidents 1976 through 1979
[2372]National Speleological Society. American Caving Accidents (1983), Nov 1984, 1983 American caving accidents
[2373]National Speleological Society. American Caving Accidents (1984), Nov 1985, 1984 American caving accidents
[2374]National Speleological Society. American Caving Accidents (1985), Nov 1986, 1985 American caving accidents
[2375]National Speleological Society. American Caving Accidents (1986), Nov 1987, American caving accidents [for 1986]
[2376]National Speleological Society. American Caving Accidents (1987), Dec 1988, American caving accidents [for 1987]
[2377]National Speleological Society. American Caving Accidents (1988), Dec 1989, American caving accidents [for 1988]
[2378]National Speleological Society. American Caving Accidents (1989), Dec 1990, American caving accidents 1989
[2379]National Speleological Society. American Caving Accidents (1990), Dec 1991, American caving accidents 1990
[2380]National Speleological Society. American Caving Accidents (1991), Dec 1992, American caving accidents 1991
[2381]National Speleological Society. American Caving Accidents (1992), Dec 1992, American caving accidents
[2382]National Speleological Society. American Caving Accidents (1993), Dec 1994, American caving accidents 1993
[2531]The NSW Cave Rescue Squad Inc, nd, The NSW Cave Rescue Squad Inc. Annual Report July 1988-June 1989
[2532]The NSW Cave Rescue Squad Inc., nd, The NSW Cave Rescue Squad Inc
[3008]Smith Daniel Irving, 1946–(ed), 1978, Handbook of cave rescue operations
[3501]Williams, Toni Lewis (ed), 1981, Manual of U.S. cave rescue techniques

6.8 Safety

[434]Boon, J. M., 1966, Cave diving on air
[544]Bryant, Thomas Charles, 1964, An introduction to caving and potholing for novices
[615]Cave Rescue Organisation, 1960, Caving safety code: first aid cave rescue procedure
[684]Collins, S. J., Aug 1958, The shoring of swallet cave entrances
[776]Cullingford, Cecil Howard Dunstan, 1904–1990 (ed), 1969, Manual of Caving Techniques
[1016]Exley, Irby Sheck, 1944–1994, 1979, Basic cave diving: a blueprint for survival
[1023]Eyre, Jim, 1988, Race against time: a history of the Cave Rescue Organisation
[1068]Fletcher, George, 1968, Notes on caving and potholing for beginners (and for those needing a boost in safety)
[1523]Horne, Peter, 1980, A report on South Australian diving fatalities
[1525]Horne, Peter, 1987, South Australian diving fatalities 1950-1985
[2288]Mountain Rescue Committee (Great Britain), 1975, Mountain and cave rescue, with lists of official rescue teams and posts: the handbook of the Mountain Rescue Committee
[2289]Mountain Rescue Committee (Great Britain), 1964, Mountain rescue & cave rescue
[2293]Mroczkowski, Donna Marie, 1950–, 1975, Safety and techniques
[2349]National Caving Association, [1988], Leadership and instructor qualifications in caving
[2361]National Speleological Society. American Caving Accidents (1967), Aug 1969, American caving accidents 1967
[2362]National Speleological Society. American Caving Accidents (1967-1970), 1974, American caving accidents 1967-1970
[2363]National Speleological Society. American Caving Accidents (1968), Apr 1970, American caving accidents 1968
[2364]National Speleological Society. American Caving Accidents (1969), nd, American caving accidents 1969
[2365]National Speleological Society. American Caving Accidents (1970), Nov 1971, American caving accidents 1970
[2366]National Speleological Society. American Caving Accidents (1971), Aug 1974, American caving accidents 1971
[2367]National Speleological Society. American Caving Accidents (1972), Aug 1975, American caving accidents 1972: a report of the National Speleological Society
[2368]National Speleological Society. American Caving Accidents (1973), Aug 1975, American caving accidents 1973: a report of the National Speleological Society
[2369]National Speleological Society. American Caving Accidents (1974), Aug 1976, American caving accidents 1974: a report of the National Speleological Society
[2370]National Speleological Society. American Caving Accidents (1975), Aug 1977, American caving accidents 1975: a report of the National Speleological Society
[2371]National Speleological Society. American Caving Accidents (1976-1979), May 1981, American caving accidents 1976 through 1979

[2372]National Speleological Society. American Caving Accidents (1983), Nov 1984, 1983 American caving accidents
[2373]National Speleological Society. American Caving Accidents (1984), Nov 1985, 1984 American caving accidents
[2374]National Speleological Society. American Caving Accidents (1985), Nov 1986, 1985 American caving accidents
[2375]National Speleological Society. American Caving Accidents (1986), Nov 1987, American caving accidents [for 1986]
[2376]National Speleological Society. American Caving Accidents (1987), Dec 1988, American caving accidents [for 1987]
[2377]National Speleological Society. American Caving Accidents (1988), Dec 1989, American caving accidents [for 1988]
[2378]National Speleological Society. American Caving Accidents (1989), Dec 1990, American caving accidents 1989
[2379]National Speleological Society. American Caving Accidents (1990), Dec 1991, American caving accidents 1990
[2380]National Speleological Society. American Caving Accidents (1991), Dec 1992, American caving accidents 1991
[2381]National Speleological Society. American Caving Accidents (1992), Dec 1992, American caving accidents
[2382]National Speleological Society. American Caving Accidents (1993), Dec 1994, American caving accidents 1993
[2724]Prosser, J. Joseph, 1990, Cave diving communications
[2725]Prosser, J. Joseph, 1992, NSS cave diving manual: an overview
[3048]Somers, Lee H., 1971, Cave diving: equipment and procedures
[3105]Standing, Peter, Aug 1975, Medical aspects of speleology
[3558]Zumrick, John L., 1988, NSS cavern diving manual

6.9 Surveying

[62]Amundson, Bob, Jan 1979, User's guide for flowchart and cavemap computer programs
[339]Bassham, Elbert F., 1971, Cave surveying techniques
[383]Benke, David, 1989, Using trigonometric functions in cave surveying
[483]British Broadcasting Corporation Television Service, 1980, Atea: in search of the world's deepest cave
[560]Burge, John W., 1988, Basic underwater cave surveying
[568]Butcher, A.L., Jul 1966, Cave surveying
[569]Butcher, Arthur Lepine, 1951, Cave survey
[683]Collins, S. J., Sep 1966, The presentation of cave survey data
[806]Dasher, George R., 1952–, 1994, On station: a complete handbook for surveying and mapping caves
[966]Ellis, Bryan M., 1934–, 1988, An introduction to cave surveying: a handbook of techniques for the preparation and interpretation of conventional cave surveys
[970]Ellis, Bryan M., 1934–, 1976, Surveying caves
[1018]Exley, Irby Sheck, 1944–1994, 1972, Mapping underwater caves
[1024]Fabre, Guilhem (comp), 1978, Speleological conventional signs
[1143]Ganter, John Hamilton, 1962–, May 1985, A systematic guide to making your first cave map
[1385]Hatherley, Paul (comp), 1987, Symposium on surveying caves
[1529]Hosley, Robert J., 1971, Cave surveying and mapping
[1764]Judson, David M., 1974, Cave surveying for expeditions
[2166]Middleton, Gregory J., 1991, Oliver Trickett: doyen of Australia's cave surveyors 1847-1934
[2199]Missouri Speleological Survey, January-December 1991, The art of cave mapping
[2210]Missouri Speleological Survey, 1981, An introduction to cave mapping
[2488]Nelson, Douglas R., 1974, Introductory cave surveying
[2671]Picknett, R.G. (fore), Jul 1970, Symposium on cave surveying
[2850]Rohrer, Thomas A., 1961, Caves and cave surveying
[2939]Sexton, Robert T., Dec 1958, Cave surveying in South Australia
[3398]Warden, D. E., nd, Cave surveying
[3519]Wolfe, James E., 1975, Map location and dimensional definition of subsurface caverns

6.10 Techniques

[27]Alexander, E. Calvin, 1992, Practical tracing of groundwater, with emphasis on karst terranes: a short course manual presented on the occasion of the annual meeting of the Geological Society of America, October 24, 1992, Cincinnati, Ohio
[69]Anderson, Jennifer Ann, 1942–, 1974, Cave exploring
[262]Australian Speleological Federation, Feb 1972, Proceedings of the eighth biennial conference of the Australian Speleological Federation
[301]Balcombe, F. Graham (comp, ed), 1990, Cave div-

6.10. TECHNIQUES

ing: the Cave Diving Group manual
[307]Ballard, Jim, 1978, The spur book of caving
[331]Barrington, Nicholas Robert, 1968, Adventure underground
[449]Boy Scouts of America, 1991, Caving
[450]Boy Scouts of America, 1991, Venture caving
[457]Bradshaw, Chris, 1982, The manual of basic caving
[544]Bryant, Thomas Charles, 1964, An introduction to caving and potholing for novices
[617]Cave Research Foundation, May 1981, CRF personnel manual
[671]Coase, Alan C., c1973, Caving and potholing techniques
[684]Collins, S. J., Aug 1958, The shoring of swallet cave entrances
[694]Cons, David, 1966, Cavecraft: an introduction to caving and potholing
[721][Council of Northern Caving Clubs], 1969, Northern cave handbook
[722]Council of Northern Caving Clubs, 1979, Northern caving; handbook of the Council of Northern Caving Clubs
[726]Coward, Julian, Jun 1983, Construction and testing of an underground radio: a report submitted to the Research and Educational Branch of Workers' Health, Safety and Compensation, Alberta Government
[773]Cullingford, Cecil Howard Dunstan, 1904–1990 (ed), 1953, British caving: an introduction to speleology
[774]Cullingford, Cecil Howard Dunstan, 1904–1990, 1976, Caving
[776]Cullingford, Cecil Howard Dunstan, 1904–1990 (ed), 1969, Manual of Caving Techniques
[819]Davidson, Joseph Killworth, 1938–(ed), May 1967, CRF personnel manual
[856]Dennewald, Fearless Fiona (comp), 1985, Outdoor recreation II: caving: journey to the centre of the earth: 19-20 October 1985
[903]Drew, David Phillip, 1943–, 1969, Techniques for the tracing of subterranean drainage
[956]Elliot, Dave, 1946–, 1987, Single rope technique rigging guide
[957]Elliot, Dave, 1946–, 1986, Single rope technique: a training manual
[959]Elliott, D. M. (ed), Jul 1953, Tasmanian Caverneering Club Handbook
[968]Ellis, Bryan M., 1934–, [1963], Mendip cave registry: handbook for members of the Executive Committee
[991]Engelbrecht, N. P., Aug 1972, Caving in the Transvaal
[1068]Fletcher, George, 1968, Notes on caving and potholing for beginners (and for those needing a boost in safety)
[1088]Ford, Trevor David, 1925–, 1973, Caving and potholing techniques
[1126]Freeman, John P. (ed), 1975, CRF Personnel Manual: Central Kentucky karst area
[1153]Gardner, James Eugene, 1953–, Oct 1984, An introduction to the inventory and evaluation of biological cave resources
[1319]Halliday, William R., 1926–, 1974, American caves and caving: techniques, pleasures, and safeguards of modern cave exploration
[1343]Hamilton-Smith, Elery, Oct 1962, Australian cave fauna: notes on collecting
[1380]Hassemer, Jerry Herman, 1934–(ed), 1982, Caving basics
[1381]Hassemer, Jerry Herman, 1934–(ed), 1987, Caving basics: a comprehensive manual for beginning cavers
[1566]Hudson, Stephen Edward, 1950–, 1988, Manual of U.S. cave rescue techniques
[1575]Hunt, Geoffrey, 1952–, 1975, Cave gating: a handbook
[1690]Jacobson, Donald, 1986, Caving: an introductory guide to spelunking
[1713]Jeffreys, Alan Lawrence, 1940–(abstr), Mar 1970, A bibliography of technical articles referring to practical caving subjects
[1762]Judson, David, 1961, Caving and potholing
[1763]Judson, David (ed), 1984, Caving practice and equipment
[1766]Kaczmarak, Michael B., Nov 1973, Basic guidelines for spelunking
[1909]Larson, Lane, 1953–, 1982, Caving: the Sierra Club guide to spelunking
[1956]Logan, William Steve, 1971, Self-belaying cable ladder - a rope climbing aid
[1963]Longsworth, Polly, 1959, Exploring caves
[1972]Lovelock, James, 1969, Caving
[1973]Lovelock, James, 1981, A caving manual
[2001]Lyon, Ben, 1983, Venturing underground: the new speleo's guide
[2108]McClurg, David Robert, 1929–, 1986, Adventure of caving: a practical guide for advanced and beginning cavers
[2109]McClurg, David Robert, 1929–(auth, photo), 1996, Adventure of caving: new updated edition
[2110]McClurg, David Robert, 1929–, 1973, The amateur's guide to caves & caving; skill-building ways to finding and exploring the underground wilderness
[2111]McClurg, David Robert, 1929–(ed), 1982, Caving short course: 1982 NSS Convention
[2112]McClurg, David Robert, 1929–, 1980, Exploring caves: a guide to the underground wilderness
[2158]Meredith, Michael Edward, 1943–, c1980, Vertical caving
[2167]Middleton, Gregory J., 1965—1966, Speleology
[2211]Missouri Speleological Survey, Jan 1980, An introduction to caving: a guide for beginners

[2236] Mitchell, Albert, [1937], Yorkshire caves and potholes: 1, North Ribblesdale
[2237] Mitchell, Albert, [1950], Yorkshire caves and potholes: 2, under Ingleborough
[2257] Montgomery, Neil R., 1977, Single rope techniques: a guide for vertical cavers
[2269] Morfis, A. (ed), 1986, Proceedings of the 5th international symposium on underground water tracing, Athens 1986
[2293] Mroczkowski, Donna Marie, 1950–, 1975, Safety and techniques
[2297] Mull, D. S., Oct 1988, Application of dye-tracing techniques for determining solute-transport characteristics of ground water in karst terranes
[2349] National Caving Association, [1988], Leadership and instructor qualifications in caving
[2354] National Speleological Society, 1960—1978, Caving information series
[2593] Padgett, Allen, 1987, On rope: North American vertical rope techniques for caving, search and rescue, mountaineering
[2677] Pinney, Roy, 1962, The complete book of cave exploration: an authoritative guide to the wonders, mysteries and excitement of caves and caving
[2740] Quinlan, James Francis, 1936–1995, Feb 1989, Ground water monitoring in karst terranes: recommended protocols and implicit assumptions
[2745] Quinlan, James Francis, 1936–1995 (comp), 1994, Practical karst hydrogeology, with emphasis on groundwater monitoring, February 8-11, 1994, Holiday Inn West, Gainesville, Florida
[2999] Slaven, John F., 1971, The speleoguide
[3060] South Wales Caving Club, Nov 1962, Some technical aids for cave exploration
[3080] Sparrow, Andy, 1991, A Mendip caver's ropework guide
[3131] Stoddard, Sheena, 1994, An introduction to cave photography
[3169] Sydney Speleological Society, nd, Prospective's handbook
[3172] Sykes, Les, [1994], C. N. C. C. eco-resin rigging system no 1
[3184] Tarkington, Terry Warren, 1925–, May 1965, A manual for beginners
[3185] Tarkington, Terry Warren, 1925–, 1965, So you want to go caving! A manual for beginners
[3187] Tasmanian Caverneering Club, 1963, Caverneering handbook of the Tasmanian Caverneering Club
[3194] Technical Aids in Caving Symposium (1972: Buxton High Peak College), 1972, Technical aids in caving symposium held on Sunday 5th March, 1972 at Buxton High Peak College, Harpur Hill, Buxton, Derbyshire
[3215] Thrailkill, John Vernon, 1930–, 1983, Studies in dye-tracing techniques and karst hydrogeology
[3216] Thrun, Robert, 1940–, 1973, Prusiking
[3259] Tyler, Ronald M., Dec 1971, Caving handbook: a guide to the underworld
[3399] Warild, Alan, 1988, Vertical: a technical manual for cavers
[3503] Willis, Dick, 1986, Caving expeditions
[3541] Yeadon, Geoff (auth, illus), 1981, The Cave Diving Group technical review no.3: line laying and following

6.11 Training

[2111] McClurg, David Robert, 1929–(ed), 1982, Caving short course: 1982 NSS Convention
[3259] Tyler, Ronald M., Dec 1971, Caving handbook: a guide to the underworld
[3558] Zumrick, John L., 1988, NSS cavern diving manual

7 History

Chris Howes

51 Timbers Square
Roath, Cardiff CF2 3SH, Great Britain

The subject of spelean history research is one which, perhaps, involves fewer individuals than other specialist areas and, at first glance, therefore produces fewer publications. The total number of entries under "history" in this bibliography makes this one of the smallest sections. However, this does not reflect the true extent of published research in spelean history.

As to specialist publications concerned with cave history, there are few. In the USA the American Spelean History Association was formed in 1968 and published the first issue of *The Journal of Spelean History* in the winter of that same year. Extremely active from its inauguration, the group has now reached volume 29 and the journal is an important source of documented research and references, particularly (although not exclusively) related to the US.

By way of comparison, a Special Interest Group of the British Cave Research Association, the Spelean History Group, is far less prolific. Other historical societies with an interest in a specific area or cave exist around the world (the Jenolan Caves Historical and Preservation Society, e.g.), with obvious restrictions on the scope of their publications. Excellent papers on all aspects of history appear with scattered regularity in the pages of major periodicals such as *Cave and Karst Science*, the *International Journal of Speleology* and the *NSS Bulletin*. Caving clubs and grottoes are also a rich source of information, and their journals frequently carry well-researched articles on local caves. The fact that their publications may have only a short print run and, perhaps, be of poor quality and elusive to locate does not diminish the quality of the contents. Occasionally, a complete journal is devoted to history, e.g. an anniversary publication; on other occasions the publication reaches a wider audience, such as Thom Engel's *A Chronicle of Selected Northeastern Caves*, itself a history guide for the 1979 NSS Convention.

Why then, given that there is an appreciable number of such papers and journals in existence, and a keen interest in cavers who wish to read of the historic background to their favourite cave or area, are there so few publications in book form? The answer, presumably, lies with the perceived level of interest and concurrent sales that a specialist publication will bring. Yet, such publications have been achieved and, when they have, the results have proved extremely successful.

Take, for example, Russ Gurnee's *Discovery of Luray Caverns, Virginia*. An eminently readable account, there is little doubt that this has proved a commercial success, perhaps in part due to its link with a major cave that therefore provides an obvious sales base. Another example of this is Robert Nymeyer and William Halliday's *Carlsbad Caverns : The Early Years*. Martyn Farr's *The Darkness Beckons* (1980) also presents a new facet of speleological research in providing a readable, well-documented account of the history and development of cave diving, so much so that a much enlarged second edition appeared in 1991 with a German translation the following year. Again, there is a link outside caving to the field of diving, and hence to wider sales. The argument continues to Chris Howes' *To Photograph Darkness*, in which he documents the history and development of underground flash photography, thereby crossing the market-place divide to reach photographic researchers who otherwise have no interest in caves. For research into cave diving or cave photography, these volumes represent good starting points.

To have a broad appeal is not, obviously, a criterion for a book to appear in this bibliography. However, the existence of a realistic market does aid publication in the first place and, in turn, helps explain the number of history books directly concerned with show caves. Equally, such sites have a longer history and are, therefore, of greater interest and offer more of a challenge to the spelean researcher, hence their greater numbers.

The books by Farr, Gurnee, and Howes are very definitely classified as spelean history: they present the results of research in the field to the reader, with the research itself based on original documentation and primary sources of information (contemporary accounts, newspaper clippings, and interviews). But what of the contemporary accounts, often themselves in book format? Here, it is necessary to make a clear distinction between history and historical publications.

To take two examples, Xavier Belle's monograph in Camacho's *The Natural History of Biospeleology* references Packard's *The Cave Fauna of North America*, published in 1888, and a researcher on the history of the Grotte de Han in Belgium might refer to *The Hades of the Ardenne* (1883), a fascinating description of a single visit to the cave. Both these examples are historical in nature; they are part of caving's history, but do not represent historical research and are not considered in this bibliography as history publications. There is a clear difference between a carefully researched history and a historical item.

Less clear is the distinction between history publications and volumes that are essentially autobiographies or accounts of exploration at a specific site, perhaps for the duration of a single expedition. Here, an element of judgment has to be applied which may not always match the requirements of a specific search for literature. Casteret's many books, written for the popular market, do not appear within this history classification (they are more properly suited to exploration), but Rose and Gregson's *Beneath the Mountains*, a historical account of exploration in northern Spain, does. Catcot's descriptive account of his 1775 descent into Pen Park Hole near Bristol, England, remains an important historical document, but is excluded under these criteria. Likewise, virtually all show cave guidebooks contain a limited account of the cave's discovery but as these publications appear under their own category, they also have been excluded. If your interest is in a specific site, the categories of regional guidebooks or exploration may yield the required information.

Other anomalies exist. Using the above distinction, excluded from history are such items as Hartwig's *The Subterranean World* and the original publication of Bullitt's *Rambles in the Mammoth Cave of 1844*. However, the 1973 reprint of Bullitt's work appears, due to the inclusion of a historical introduction by Harold Meloy; the same argument applies to such titles as Thompson's *The Sucker's Visit to the Mammoth Cave* (introduction by John F. Bridge) and Owen's *Cave Regions of the Ozarks and Black Hills* (Jerry D. Vineyard). Russ Gurnee produced *Cave Clippings of the Nineteenth Century* (1983) following the destruction by fire of much of his collection. Consisting of facsimile newspaper clippings on the subject of caves, it represents historical research by dint of the organisation of the collection itself and therefore appears here, although there is only a short historical perspective included in the volume. Theo Schuurman's catalogue of Belgian cave postcards is equally an item of historical research, linking cards to dates and changes in the caves over the years (visual documentation by drawings and photographs is incredibly useful to the spelean historian), although it is obviously also a catalogue of information, a tool to be used for further research. On the subject of guidebooks, editions such as Clymer's *Story of Howe Caverns* is so inextricably interwoven with history that it appears here.

With a smaller base of books to deal with, a higher percentage of volumes have been given brief annotations. In general, these have not been ascribed to books where the subject matter is evident from the title, unless these are of specific importance.

Finally, of the remaining selection of published research in book format, one volume remains outstanding and worthy of specific mention: Trevor Shaw's *History of Cave Science*. Originally a thesis, a limited publication of about 150 copies in 1979 was followed by an updated version in 1992. As a source of information on the scientific theories and other aspects of spelean history, with invaluable references and appendices, the second edition is an essential document to use as a starting point to other publications and hence to the fascination of further research using primary sources.

7.1 General

[2]Abadie, Bernard, [1972], Grottos of Bétharram

[5]Absolon, Karel B., 1987, The conquest of the caves and underground rivers of Czechoslovakia's Macocha Abyss: a historical and technical study of their exploration

[12]Adams, Laurie Branson, Aug 1984, History of Morrell Cave: Part I Tennessee

[14]Adams, William Henry Davenport, 1828–1891, 1886, Famous caves and catacombs: described and illustrated

[19]Ainsworth, Joseph, Apr 1960, Saint Mary's University Speleological Society: 1954-1959

[59]Ammen, Samuel Zenas, 1843–1929, 1884, The caverns of Luray: an illustrated guide-book to the caverns, explaining the manner of their formation, their peculiar growths, their geology, chemistry, etc

[60]Ammen, Samuel Zenas, 1843–1929, 1882, History and description of the Luray Cave (illustrated), including explanations of the manner of its formation, its peculiar growths, its geology, chemistry, &c.; also a map. The whole so arranged as to serve as a guide.

[61]Ammen, Samuel Zenas, 1843–1929, 1880, History and description of the Luray Cave (illustrated), including explanations of the manner of its formation, its peculiar growths, its geology, chemistry, etc.; also a map. The whole so arranged as to serve as a guide

[64]Anchors, William E., 1954–, Mar 1989, Ruskin and Jewel Caves: a brief history

[65]Anderson, Arthur Wilhelm, 1892–, 1938, The Carlsbad Cavern of New Mexico: its history and geology

[67]Anderson, Arthur Wilhelm, 1892–, 1930, The Carlsbad Cavern of New Mexico: its history and geology; formations of the cavern, discovery and exploration, bats of the cavern, cavern geology, administration; the land nobody knows

[143]Anon, nd, Kents Cavern Wellswood Torquay Devon, home of prehistoric man and animals

[177]Anon, 1990, Red hills

[193]Anon, 1934, The story of most interesting discoveries commencing August 1923; White Skar Cave; under Ingleborough, Yorkshire, 1 1/2 miles from Ingleton on the Hawes Road

[194]Anon, 1931, The story of newly discovered White Skar Caverns; under Ingleborough, Yorkshire, 1 1/2 miles from Ingleton on the Hawes Road

[208]Anon, 1982, Waitomo Caves management plan 1982

[212]Anon, 1992, Welcome to Waitomo Caves: a photographic insight into this spectacular region of New Zealand

[243]Atkinson, Anne, 1995, Undara Volcano and its lava tubes: a geological wonder of Australia in Undara Volcanic National Park, North Queensland

[246]Attout, Jacques, 1956, Men of Pierre Saint-Martin

7.1. GENERAL

[299]Balch, Herbert Ernest, 1914, Wookey Hole: its caves and cave dwellers
[375]Bell, Alan, 1918–, 1928, The witch of Wookey Hole
[425]Bohi, John W., 1963, History of Wind Cave
[426]Bohi, John W., 1962, Seventy-five years at Wind Cave: a history of the national park
[454]Bradford Pothole Club Members, c1974, A history of Gaping Gill and Ingleborough Cave
[463]Branigan, Keith, 1992, Romano-British cavemen: cave use in Roman Britain
[464]Brannan, Eldred Boyd, [1925], The entombment of Floyd Collins in Sand Cave, Kentucky
[493]Brizius, Janice J., 1900, History of the 7 caves
[525]Brown, Henry S., 1961, Linville Caverns through the ages: the geological story
[526]Brown, L. Carson, 1970, Manitouwadge: cave of the Great Spirit
[536]Brucker, Roger Warren, 1929–, 1987, The longest cave
[596]Cansell, Anthony, [after 1982], Stump Cross Caverns - the underground wonderland, their development and exploration
[597]Cansell, Anthony, c1969, The Upper Wharfedale Fell Rescue Association 1948-1968
[623]Cave Research Group of Great Britain Friends and Members, Jul 1969, A volume of essays presented to Brigadier E. A. Gelnnie on the occasion of his 80th birthday, July 18th 1969
[668]Clymer, Virgil H.,–1952 (comp, ed), , Story of Howe Caverns
[686]Colong Committee (Australia), 1985, The Colong story
[736]Cravens, Raymond L., 1992, The spirit of Lost River: fact and legend about the Lost River Cave and Valley
[765]Crowther, Patricia P. , 1984, The Grand Kentucky junction: memoirs
[768]Cudmore, Dana D., 1990, The remarkable Howe Caverns story
[797]Damon, Paul Herbert, 1934–, 1991, Caving in America: the story of the National Speleological Society, 1941-1991: commemorating 50 years of history and growth: including a special illustrated history of cave exploration in the society entitled—the last frontier for the pioneer
[798]Damon, Paul Herbert, 1934–, Jun 1976, The history of Laurel Caverns of Fayette County, Pennsylvania, known throughout the years as Delaney's Cave
[799]Damon, Paul Herbert, 1934–, Oct 1977, Thirty years with the Pittsburgh Grotto, National Speleological Society
[824]Davies, Paul, 1975, A pictorial history of Swildon's Hole
[847]Dean, Pauline, 1931–, 1989, Manitouwadge: cave of the Great Spirit
[854]Demek, Jaromír, 1989, Czech Speleological Society 1986-1989: published on occasion of 10th International Speleological Congress Hungary 1989
[855]den Hertog, Sonja (ed), 1994, Walking the valley: an oral record of caving and bushwalking in the Burragorang and beyond, during the 1930s
[863]De Watteville, Alastair, 1993, The island of Staffa: home of the world-renowned Fingal's cave
[892]Dowie, H. G., 1932, The history of Kent's Cavern; Torquay: with illustrations
[920]Dunkley, John R., 1986, Jenolan caves as they were in the nineteenth century
[925]Durden, C. C., 1928, The Carlsbad Cavern of New Mexico: it's [sic] history and geology
[958]Elliot, Ian, 1977, The discovery and exploration of the Yanchep Caves
[976]Ellis, Ross Andrew, 1940–(ed), nd, A history of the journal of the Sydney Speleological Society
[979]Ely, Glen Sample. (photo, ed), 1990, A history of the Guadalupe Mountains and Carlsbad Caverns, 1840—1940
[990]Engel, Thomas Daniel, 1954–, Jul 1979, A chronicle of selected northeastern caves: a history guide for the 1979 NSS Convention, Pittsfield, Massachusetts, August 5-12, 1979
[1022]Eyre, Jim, Mar 1967, Lancaster Hole and the Ease Gill Caverns, Casterton Fell, Westmorland
[1028]Farr, Martyn, 1950–, 1980, The darkness beckons: the history and development of cave diving
[1037]Faust, Burton Sherwood, 1898–1967, 1964, Saltpetre caves and Virginia history
[1038]Faust, Burton Sherwood, 1898–1967, 1967, Saltpetre mining in Mammoth Cave, Ky
[1056]Fieseler, Ronald G. (ed), Apr 1975, The caves of Brewster and western Pecos Counties
[1177]George, Angelo Isham, 1944–, 1994, Mummies, catacombs, and Mammoth Cave
[1178]George, Angelo Isham, 1944–, 1992, The New Madrid earthquake at Mammoth Cave (1811-1812)
[1180]George, Angelo Isham, 1944–, Apr 1986, Saltpeter and gunpower manufacturing in Kentucky
[1181]George, Angelo Isham, 1944–, 1991, Wyandotte Cave down through the centuries
[1232]Gough, William, nd, The cliffs & caves of Cheddar: a series of beautiful views in real photogravure published by William Gough, son of the discoverer of the famous cave
[1266]Grimes, K. G., Oct 1995, South East Karst Province of South Australia: Australian Caves & Karst Management Association October 1995
[1294]Gurnee, Russell Hampton, 1922–1995 (comp), 1983, Cave clippings of the nineteenth century
[1297]Gurnee, Russell Hampton, 1922–1995, , Discovery of Luray Caverns, Virginia
[1306]Habe, Francé, 1972, The Postojna Grottes with the Planina and Predjama caves

[1314]Halbert, Erik, Nov 1994, The history of the Sydney Speleological Society 1954-1994

[1316]Halladay, Orlynn J., 1972, The Lehman Caves story

[1330]Halliday, William R., 1926–, Feb 1978, History and publications of the Western Speleological Survey

[1333]Halliday, William R., 1926–, Sep 1954, Littoral caves of ancient Lake Bonneville

[1341]Hamilton, Holman, 1957, The Cave of the Winds' and the compromise of 1850

[1342]Hamilton-Smith, Elery (ed), February 1996, Abstracts of papers: karst studies seminar Naracoorte

[1375]Hartley, Howard W., 1925, Tragedy of Sand Cave

[1448]Herzlík, Bořivoj (trans), 1986, Czech Speleological Society 1982-1986: published on occasion of 9th International Speleological Congress Spain 1986

[1480]Hogg, Garry, 1962, Deep down: great achievements in cave exploration

[1512]Holt, Robert (comp), 1989, Howe Caverns; Howes Cave N.Y.; 60th Anniversary, 1929-1989

[1535]Hovey, Horace Carter, 1833–1914, 1970, Celebrated American caverns: especially Mammoth, Wyandot, and Luray: together with historical, scientific, and descriptive notices of caves and grottoes in other lands

[1536]Hovey, Horace Carter, 1833–1914, 1882, Celebrated American caverns: especially Mammoth, Wyandot, and Luray: together with historical, scientific, and descriptive notices of caves and grottoes in other lands

[1556]Howes, Chris, 1951–, 1989, To photograph darkness: the history of underground and flash photography

[1565]Hudson, George Henry, 1855–1934, 1911, Joint caves of Valcour Island: their age and their origin

[1665]Iorio, Ralph, 1968, The history of Timpanogos Cave National Monument: American Fork Canyon, Utah

[1671]Irwin, David J. (comp), Aug 1995, A catalogue of ephemera relating to Wookey Hole

[1673]Irwin, David J., 1981, A catalogue of the postcards of Gough's Cave, Cox's Cave & Wookey Hole, Somerset 1900-1980

[1676]Irwin, David J., Oct 1968, The discovery and exploration of St. Cuthbert's Swallet

[1679]J., J. M., 1941, How the grottoes of an ancient church were discovered in the convent of the "Dames de Nazareth" at Nazareth in Galilee

[1682]Jackson, George Frederick, 1906-1981, 1972, The history and exploration of Wyandotte Cave

[1683]Jackson, George Frederick, 1906-1981, 1975, The story of Wyandotte Cave

[1711]Jean, David, 1976, Underground worlds

[1716]Jenkins, D. W. (ed), 1957, South Wales Caving Club tenth anniversary publication

[1737]Johnson, Cuthbert, 1946–, 1967, The history of Mendip caving

[1739]Johnson, Rogor, [1979], Rogor Johnson's cave scrapbook

[1778]Keck, Ken, 1991, Abercrombie Caves: cave chronicles: a new look at an old wonder: including an account of the Bathurst convict rebellion of 1830

[1781]Kehret, Roger Alan, 1942–, 1974, Minnesota caves of history and legend: a collection of unique cave stories

[1785]Keith, Arthur, 1866–1955, 1936, History from caves: a new theory of the origin of modern races of mankind, by Arthur Keith, being the presidential address given at Buxton, Derbyshire to the first speleological conference

[1790]Kempe, David Ronald Charles, 1927–, 1988, Living underground: a history of cave and cliff dwelling

[1793]Kenney, C. Howard, 1927–1980, 1985, Caving log 1942-1950

[1830]Knauth, Otto, 1975, Cold Water Cave

[1861]Kranjc, Andrej, 1943–(ed), 1993, The contribution to the history of the speleological explorations of the West Indies at 500—anniversary of the discovery of America

[1872]Kunath, Carl Edwin, 1940–(ed), Apr 1968, The caves of the Stockton Plateau, Texas

[1873]Kunath, Carl Edwin, 1940–, Mar 1965, Ye olde history

[1900]Larimore, Betty, 1947, Exploring the Endless Caverns of New Market Virginia on US 11 in the heart of the Shenandoah Valley

[1979]Lübke, Anton, 1890–, 1958, The world of caves

[2018]Maempel, George Zammit, 1989, Ghar Dalam: cave and deposits

[2019]Maempel, George Zammit, 1989, Pioneers of Maltese geology

[2030]Mammoth Cave National Park Association, 1985, A national park in Kentucky

[2051]Marsh, John, 1942–, 1979, A history of Glacier House and Nakimu Caves, Glacier National Park, British Columbia

[2052]Marsh, John, 1942–, 1973, Nakimu Caves

[2056]Martel, Edouard Alfred, 1859–1938, c1950, Cavern of the dragon

[2062]Martin, George V., 1973, The Timpanogos Cave story: the romance of its exploration

[2067]Martin, Ronald L., 1990, Marvel Cave: Silver Dollar City, Branson, Missouri

[2068]Martin, Ronald L., 1974, Official guide to Marvel Cave, Silver Dollar City, Missouri

[2071]Mason, Edmund John, 1911–1993, 1979, Bone Cave (Ogof-yr-Esgyrn) Dan-yr-Ogof: the history of the Bone Cave (Ogof yr Esgyrn): a guide and background to the exhibits in Bone Cave

[2088]Matthews, Larry Edwin, 1946–, 1989, Cumberland Caverns

[2105]McCann, Gerald, [1992], In my torchlight: a guide's guide to the Cango Caves

[2116]McDonald, Alvin F., 1873–1893, nd, Private account of A. F. McDonald, permanent guide of Wind Cave

7.1. GENERAL

[2117]McDonald, Donald L., 1974, Official guide to Marvel Cave: the only complete authentic history of Marvel Cave ever published
[2130]McKenzie, David (ed), 1964, The caves of Bell and Coryell Counties
[2166]Middleton, Gregory J., 1991, Oliver Trickett: doyen of Australia's cave surveyors 1847-1934
[2208]Missouri Speleological Survey, 1970, A history of the caves of Camden County, Missouri
[2243]Mitchell, William Reginald, 1928–, 1961, The hollow mountains: the story of man's conquest of the caves and potholes of Northwest Yorkshire throughout 10,000 years
[2244]Mitchell, William Reginald, 1928–, 1989, Yorkshire's hollow mountains
[2271]Morgan, Robert, nd, Caves in world history
[2280]Moseley, John Judy, 1948, History, geology of Carlsbad Caverns: what' a hole
[2300]Munson, Patrick J., 1990, The prehistoric and early historic archaeology of Wyandotte Cave and other caves in southern Indiana
[2305]Murray, Robert K., 1979, Trapped!: the story of the struggle to rescue Floyd Collins from a Kentucky cave in 1925, an ordeal which became one of the most sensational events of modern times
[2306]Murray, Robert K., 1982, Trapped!: the story of the struggle to rescue Floyd Collins from a Kentucky cave in 1925, an ordeal which became one of the worst sensational events of modern times
[2325]Myers, Arthur John, 1918–, 1982, Guide to Alabaster Cavern and Woodward County, Oklahoma
[2326]Myers, Arthur John, 1918–, 1969, Guide to Alabaster Cavern and Woodward County, Oklahoma
[2501]Newbould, Ronald L., Mar 1975, The discovery of Barralong Cave, Jenolan Caves, N.S.W.: based on the diary of Ronald L. Newbould, guide, Jenolan Caves, co-discoverer with John P. Culley, senior guide, Jenolan Ceves [sic] of the "Barralong Cave" on the 7th June, 1964
[2509]Nicholson, Frank Ernest, Oct 1930, The exploration of Carlsbad Cavern
[2529]Nougier, Louis René, 1912–, 1958, The Cave of Rouffignac
[2537]Nymeyer, Robert, 1910–1983, 1991, Carlsbad Cavern, the early years: a photographic history of the cave and its people
[2570]Ollerenshaw, Arthur Edward, c1963, The history of Blue John Stone: methods of mining and working ancient and modern
[2604]Palmer, Robert John, 1951–1997, 1989, Deep into blue holes: the story of the Andros Project
[2615]Parris, Lloyd E., 1973, Caves of Colorado
[2639]Pemble, Edna R., Nov 1972, Limestones caves and cavers
[2647]Pengelly, William, [1870], The literature of Kent's Cavern prior to 1859
[2648]Pennick, Nigel, 1981, The subterranean kingdom: a survey of man-made structures beneath the earth
[2655]Perry, Clair Willard [Clay], 1887–1961, 1946, New England's buried treasure
[2656]Perry, Clair Willard [Clay], 1887–1961, 1948, Underground empire: wonders and tales of New York caves
[2657]Perry, Clair Willard [Clay], 1887–1961, 1939, Underground New England
[2700]Pound, Louise, 1872–1958, 1948, Nebraska cave lore
[2706]Preble, John Wesley, 1969, Sinks of Gandy Creek
[2715]Price, Elizabeth Ann, Sep 1981, Club history 1956—1981 25th anniversary publication: Cerberus Spelaeological Society
[2737]Quick, Dell (ed), 1981, SCG history: summer 1981
[2738]Quick, Peter Gunder, 1994, Vermont caves: a geologic and historical guide
[2749]Railton, Courtenaye Lewis, 1907–1971, 1953, The Ogof Ffynnon Ddu system: its discovery and exploration (1927-53): and a theory of its development
[2752]Raines, Terry W., Sep 1988, First reports: 1984-1988
[2757]Ralston, Basil, 1989, Jenolan: the golden ages of caving
[2774]Reddell, James Russell, 1938–, Jun 1964, The caves of Comal county
[2775]Reddell, James Russell, 1938–, Nov 1965, The caves of Edwards County
[2777]Reddell, James Russell, 1938–(ed), Mar 1970, The caves of Lubbock County
[2778]Reddell, James Russell, 1938–(ed), Apr 1967, The caves of Medina county
[2780]Reddell, James Russell, 1938–, Feb 1973, The caves of San Saba County
[2816]Rickels, Curtis Eddie, 1992, The three ring circus: how Rickwood Caverns became a state park
[2848]Rogers, Edmund B., 1958, Oregon Caves National Monument, Oregon: history of legislation through the 82nd Congress
[2859]Rother, Hubert, 1996, Lost caves of St. Louis
[2895]Say, L. W. (ed), 1976, Cheddar Caves Museum
[2916]Schuurmans, Theo J. C., 1987, Postcards of the Caves of Han printed by Nels of Brussels between 1905 and 1940
[2917]Schuurmans, Theo J. C., 1989, Postcards of the Caves of Han printed by Nels of Brussels between 1905-1940
[2944]Shaler, Nathaniel Southgate, 1841–1906, 1876, On the antiquity of the caverns and cavern life of the Ohio valley
[2952]Shaw, Trevor Royle, 1928–(comp), 1967, Cave illustrations before 1900: a catalogue of non-photographic illustrations of caves

[2954]Shaw, Trevor Royle, 1928–, 1992, History of cave science: the exploration and study of limestone caves, to 1990

[2955]Shaw, Trevor Royle, 1928–, 1979, History of cave science: the scientific investigation of limestone caves, to 1900

[2972]Shoemaker, Henry Wharton, 1880–1958, , Penn's grandest cavern: the history, legends and description of Penn's Cave in Centre County, Pennsylvania

[2973]Shoemaker, Henry Wharton, 1880–1958, , Penn's grandest cavern: the history, legends photographs and description of Penn's Cave in Centre County, Pennsylvania

[2993]Skelton, Harold E., 1974, Weyer's Cave's first century, 1874-1974

[3027]Smith, Marion Otis, 1942–, 1977, The exploration & survey of Ellison's Cave, Georgia

[3028]Smith, Marion Otis, 1942–(comp), 1992, Letters from TAG 1966-1969

[3029]Smith, Marion Otis, 1942–, 1990, Saltpeter mining in East Tennessee

[3038]Snider, George Washington, 1916, How I found the Cave of the winds ; being the true story as set forth by the discoverer, George W. Snider

[3061]South Wales Caving Club, c1968, South Wales Caving Club twenty first anniversary publication

[3062]South Wales Caving Club, 1957, South Wales Caving Club, 1946—1956

[3084]Speece, Jack H., 1977, The cave of Delaware

[3089]Speece, Jack Howard, 1947–(ed), 1979, Special convention issue: Symposium on the History of American Caving, Pittsfield, Mass., August 5-12, 1979

[3097]Sprent, J. K., (ed), 1970, Mount Etna caves: a collection of papers covering several aspects of the Mt. Etna and Limestone Ridge caves area of central Queensland

[3110]Steel, William Gladstone, 1931, Oregon, no 32, vol 1

[3123]Stevens, Martin V. B., 1876, The history, guide and description of Wyandotte Cave

[3126]Stewart, John, 1920–, 1972, Secret of the bats; the exploration of Carlsbad Caverns

[3190]Taylor, Maurice Clague, 1991, Three below Gower: the story of cave exploration in Gower by The Taylors'

[3197]Tennessee Academy of Science, Jul 1930, Journal of the Tennessee Academy of Science. Cave number

[3211]Thornycroft, L. B., 1948, The story of Wookey Hole in fact, fiction and photo

[3222]Torode, William Wallace, 1943–(comp), nd, Ellison's Cave: Georgia

[3226]Tratman, Edgar Kingsley, 1899–1978 (ed), 1969, The Caves of North West Clare, Ireland

[3229]Trexler, Keith A., 1975, Lehman Caves...its human story: from the beginning through 1965

[3231]Trickett, Oliver, 1906, The Abercrombie Caves

[3255]Turner, Percy (auth, comp), 1895, History of Manitou's caves, the Grand Caverns and Cave of the Winds: and points of interest in and about Manitou

[3258]Tyler, Larry H., 1980, History of Rockhouse Cave

[3341]Vachell, Eustace Tanfield, 1959, Kents Cavern: its origin and history

[3361]Vickery, Margaret Ray, 1960, Ozark stories of the Upper Current River

[3368]Virginia Region. National Speleological Society, , A history of the Virginia Region

[3379]Walsh, Frank K., 1971, Discovery and exploration of the Oregon caves: Oregon Caves National Monument

[3380]Walsh, Frank K., 1982, Oregon Caves discovery & exploration: Oregon Caves National Monument

[3381]Walsh, Frank K., 1976, Oregon Caves: discovery and exploration

[3400]Warnell, Norman, 1997, Mammoth Cave: forgotten stories of it's [sic] people

[3409]Watson, Richard Allan, 1931–, 1984, The Cave Research Foundation 1969-1973

[3410]Watson, Richard Allan, 1931–, 1984, The Cave Research Foundation 1974-1978

[3411]Watson, Richard Allan, 1931–, 1981, The Cave Research Foundation: origins and the first twelve years [1957-1968]

[3421]Weaver, Herman Dwight, 1938–, Aug 1977, Meramec Caverns: legendary hideout of Jesse James

[3422]Weaver, Herman Dwight, 1938–, Apr 1980, Missouri: the cave state

[3434]Wells, Oliver C. (ed), 1960, A history of the exploration of Swildon's Hole

[3456]Wheeler, Joy, 1969, Oberon-Jenolan district historical notebook

[3460]White, Jim, 1882–1946, 1932, The discovery and history of Carlsbad Caverns, New Mexico

[3462]White, Jim, 1919–, 1951, Carlsbad Caverns National Park, New Mexico: its early explorations as told by Jim White

[3480]Whittemore, Anne, 1979, A history of the Virginia Region

[3495]Williams, James Henry, 1843–, 1914, Legend of the Cave of the Winds

[3502]Williamson, George Charles, 1858–1942, 1930, The Guildford Caverns

[3518]Witcombe, Richard, 1992, Who Was Aveline Anyway? Mendip's cave names explained

[3524]Woodall, Brian, 1976, Peak Cavern: a guide to this famous show-cave; its formation, history & folklore

[3549]Zadnikar, Marjan, 1960, The castle of Predjama

8 Management/Conservation

George Huppert

Department of Geography
University of Wisconsin La Crosse
1725 State Street
La Crosse, WI 54601

Introduction

Cave conservation has been practiced since caves were first operated as commercial enterprises for paying tourists. Development of caves as tourist attractions has resulted in some minimal protection from complete destruction if only to protect the financial investment. Literature on cave management did not appear until hundreds of years after the first show caves opened. Research on cave conservation and management did not appear in English until well into the present century. Serious thought and research on cave conservation and management began in the early 1950s, generally as a reaction to a great increase in the use of caves for recreational purposes and the concurrent damage.

The vast majority of these research efforts are published in book chapter or article format, and thus are not included in this bibliography.

Literature From Speleological Organizations

Perhaps the earliest report on cave conservation in book form was produced by the Cave Research Foundation in the 1971 study of the Mammoth Cave region entitled, *Wilderness Resources in Mammoth Cave National Park: A Regional Approach*. This report, by Joseph Davidson and William Bishop presents a management plan to protect the cave and adjacent karst landforms in the park as a wilderness enclave within the greater regional tourist industry. The proposal for both surface and underground wilderness was eventually rejected. A year later, the National Caving Association of Great Britain produced *Caves and Conservation* edited by John Wilmut (1972). This is an inventory of British caves and the type and extent of damage that they suffer from many sources. Wilmut's paper provides a good model for similar reports on other cave regions of the world. It presents a format for baseline studies needed to assess the extent of future damage to caves.

A significant report by the National Speleological Society appeared in 1974, Louise Power's *A Handbook on Cave Conservation Legislation*. At the time, this was a useful reference work comprised of a fully annotated list, both of then current and and pending state cave protection laws in the United States. This handbook is woefully dated but the topic has been updated numerous times in article form by various authors.

By the mid and late 1970's, many grottos of the NSS began to take an active interest in various conservation projects. The grottos frequently had need of guidelines for their efforts. They often produced reports on the work done. The NSS Caver Proliferation Committee produced a statistically flawed report in 1974 covering the rapid growth of recreational caving. Hunt and Stitt (1975) produced a how-to manual, *Cave Gating*. It covers the many factors to be considered prior to gating a cave, how to build many types of gates, how to design locking mechanisms, and how to minimize environmental effects of a gate. A revised second edition was produced in 1981. This book is again in great need of revision in light of the tremendous advances in cave gating technology. A third report published by the NSS in 1975 is by Addis and Stitt, *A Management Plan for Knox Cave*. The Society was considering purchasing the cave and needed an overview of the property. Descriptions of the cave and of the local physical and ecological environment are presented and management problems and solutions are discussed. Finally, specific recommendations are given to the NSS regarding the cave. The authors encouraged the NSS to buy and manage the cave with certain restrictions. The report provided a model of management guidelines for the NSS's future purchases and consulting projects.

During the late 1970's, numerous reports by cave associations concerning cave and karst management plans and inventories. Some of these were contracted by government and other parties interested in the purchase of caves or their proper management. Most of these plans were done by government agencies and are covered in that section of this introduction. Two notable items produced by non-governmental agencies are two produced by the National Speleological Foundation, *The Ensueno Cave Study* (Gurnee, 1978), and *A Study of Harrison's Cave, Barbados, West Indies*. These books are in the same category as Addis and Stitt (1975) in providing a guide for the future. In 1990, the American Cave Conservation Association produced the *Hart County Solid Waste Management Plan* under contract from the county. This report went a long way toward educating local officials of the complexity and sensitivity of karst.

During the 1970's, perhaps two of the most significant ongoing series of published works on cave and karst protection were started. These are the proceedings of two continuing conferences devoted to management of all aspects of karst research. The oldest is the *Proceedings of the Australasian Conference on Cave and Karst Management* starting in 1976. However, the first conference occurred in 1973. Ten volumes of the conference proceedings have been produced to date. Not to be out done, cave associations and government agencies began to sponsor a similar series of meetings starting in 1975. The first *Proceedings of the National Cave Management Symposium* was published in 1976. Nine more volumes followed, with one a double volume containing the papers of the conferences of two years (1978 and 1980). These two series are a must for any serious researcher in all aspects of the management of karst resources. Combined, the two series contain some five hundred articles and abstracts on the topic of cave and karst protection. These are especially useful in that many of the articles are researched and written by actual cave managers in touch with the daily problems of caring for caves.

Another proceedings of interest is that produced for the 170th anniversary of Postojnska jama in 1988. The volume entitled *Cave Tourism* (Kranjc, 1989) contains a series of papers on the management of tourist caves.

After a somewhat slow start, the production of English language materials on the topic of cave and karst protection has greatly increased over the last few years. Most of these works are collections of un-refereed papers from conferences rather than peer reviewed books. They do have the advantage of presenting the views of a great diversity of authors.

Literature From Government Agencies

The Federal government controls perhaps a majority of the known caves in the United States but, until recently, it has published little concerning their protection and management. This has changed over the past two decades. Most of of what is published appears in the context of various master plans, environmental assessments, management plans, and wilderness proposals on specific projects or areas.

Perhaps the first government document devoted to cave conservation in the United States is Rothrock's (1937) *Caves and Their Conservation*, written for the National Park Service to use as a guide for instructing Conservation Corps supervisors in proper construction methods when constructing trails and other projects in show caves. Another early work was produced by the State of Tennessee (1959). It is a rather short guide for cave recreationists on state lands.

A little known report, Powell's *A Guide to the Selection of Limestone Caverns and Springs in the United States as National Landmarks*, (1970) is a massive work (nearly 300 pages) describing numerous caves across the country and their suitability for national landmark status. Springs are also included in the listing. Along with the cave descriptions is an excellent summary of the many theories of cave origin to date. Unfortunately, only twenty copies of this interesting report were produced.

Many states have produced various inventories of the status of caves within their borders. Some examples are *The Caves of the Tennessee State Natural Areas System* (Coggins, 1981); *An Inventory and Evaluation of Missouri State Parks Cave Resources* (Gardner and Gardner, 1982); and *Missouri Department of Conservation Cooperative Cave Inventory Project* (Gardner, 1984). Other English speaking nations have provided a wealth of information on cave and karst management. In 1992, John Gunn *et al.*, produced a report on reproducing natural-looking limestone landscapes in abandoned quarries. In 1988, Kevin Kiernan, a prolific writer, wrote, *The Management of Soluble Rock Landscapes: An Australian Perspective*. This book contains an extensive bibliography of cave management works. In 1986, Davey and White published the results of their study of *Management of Victorian Caves and Karst* for the Department of Conservation, Forests and Lands. There are other studies and inventories of individual caves far too numerous to go into any detail in this review.

Other Sources

This bibliography does not include entries of articles in the periodical literature; however, there are some significant items that need mention. First are periodicals that devote a single issue to conservation and management; second are periodicals devoted entirely to the topic.

The *National Speleological Society News* devotes an entire issue each year to conservation topics. This is usually the February issue. This has become a regular feature over the last few years and will likely continue. Other periodicals produce cave conservation issues less regularly. "Environmental Change in Karst Areas," is the topic of the June 1993 issue of *Environmental Geology*, edited by Derek Ford. Paul Williams edited papers for "Karst Terrains: Environmental Changes and Human Impact," published as *Catena* Supplement 25 in 1993.

There are some journals devoted to conservation and management. The oldest (1982) is likely *The Newsletter of the Conservation and Management Section*, retitled *The Cave Conservationist*. The American Cave Conservation Association produced a first issue of *Conservation Newsletter* in 1985. The second issue was produced later that year and the name changed to *American Caves*. The *Australian Cave and Karst Management Association Inc. Newsletter* was first published in 1988. Since then, the name was changed to reflect the wide range of interest in caves in southeast Asia by substituting Australasian for Australian. Various states are now producing conservation newsletters. *Michigan Karst* is a newsletter of the group's activities and it has a bit to go to catch up with the *MCKC Digest*, which is a journal with researched articles. Somewhere in between is the *IKC Update* covering cave conservation issues in the state of Indiana. I hope we will see more of these in the future.

Summation

There is an extensive body of literature on cave conservation and management. Like all such collections, some is difficult and some quite easy to find. Much is in periodical literature (nearly 1,000 articles) and deserves a bibliography of its own. The rest are mostly government and society reports. Fortunately, most of these works were done within the last 25 years and are still available.

8.1 Conservation

[144]Anon, 1988, The last horizon
[161]Anon, Apr 1988, Mount Etna action book: time is running out
[256]Australian National Parks and WIldlife Service, 1987, Australian Ranger Bulletin: feature: cave management
[262]Australian Speleological Federation, Feb 1972, Proceedings of the eighth biennial conference of the Australian Speleological Federation
[263]Australian Speleological Federation, Jul 1975, Proceedings of the tenth biennial conference of the Australian Speleological Federation
[488]British Columbia. Parks and Outdoor Recreation Division, Jan 1980, Cave resources in British Columbia: a discussion paper
[549]Buderer, Tina (comp), 1987, Proceedings of the first annual watershed conference, May 20-21, 1987, Springfield, Missouri
[585]Camacho, Ana Isabel (ed), 1992, The natural history of biospeleology
[600]Carlsbad Caverns National Park, Dec 1984, Land protection plan: Carlsbad Caverns National Park: prepared by Carlsbad Caverns National Park with assistance from the Southwest Regional Office, National Park Service.
[640]Chasen, Frederick Nutter, 1896–, 1931, Report on the "birds' nest" caves and industry of British North Borneo: with special reference to the Gomantong caves
[681]Coley, P. G., nd, The caves of Table Mountain: a report on their present condition, potential, and conservation requirements
[686]Colong Committee (Australia), 1985, The Colong story
[708]Cooper, John Edward, 1929–, May 1982, Recovery plan for the Alabama cavefish, *Speoplatyrhinus poulsoni* Cooper and Kuehne 1974
[735]Craven, Stephen Adrian, 1994, Cango Cave, Oudtshoorn district of the Cape Province, South Africa: an assessment of its development and management 1780-1992
[741]Crawford, Nicholas Charles, 1942–, 1981, Karst hydrogeology and environmental problems in the Bowling Green area
[812]Davey, Adrian G., 1986, Management of Victorian caves and karst: a report to the Caves Classification Committee, Department of Conservation, Forests and Lands
[816]Davey, Adrian G., 1992, World Heritage significance of karst and other landforms in the Nullarbor Region
[862]Devil's Hole Pupfish Recovery Team (prep), Jul 1980, Devil's Hole Pupfish Recovery Plan
[961]Elliott, William Rawleigh, 1946–, Jun 1993, Draft recovery plan for endangered karst invertebrates in Travis and Williamson Counties, Texas
[1014]Eversole, Jack (ed), 1973, The Mammoth Cave area: a planning proposal
[1081]Ford, Derek Clifford, 1935–, Jun 1993, Environmental change in karst areas
[1140]Gams, Ivan (ed, trans), 1987, Man's impact in Dinaric Karst: guide-book
[1223]Goode, Cecil E., 1986, World wonder saved: how Mammoth Cave became a national park
[1234]Graham, Andrew W. (ed), May 1972, Aspects of land use planning at Fanning River. With particular reference to the Fanning River caves
[1268]Grisafe, David A., Aug 1992, Stabilization of Dakota Sandstone surface of the Faris Cave petroglyphs, Kanopolis Lake Project, Kansas
[1350]Hamilton-Smith, Elery, 1976, Mt. Etna & the caves: a plan for action
[1371]Hart, William J., Sep 1967, A wilderness plan for Mammoth Cave National Park and the surrounding region
[1470]Hobbs, Horton Holcombe, 1944–(ed), , A study of environmental factors in Harrison's Cave, Barbados, West Indies
[1569]Humphrey, Patricia (ed), nd, A revision of the caver's code and a collection of articles on cave conservation
[1575]Hunt, Geoffrey, 1952–, 1975, Cave gating: a handbook
[1577]Huntsville Grotto of the National Speleological Society, May 1979, Cave preservation and residential development: a compatible approach
[1579]Huppert, George Nixon, 1944–, 1985, Selected bibliography of cave conservation and management
[1647]International Symposium on Changing Karst Environments (1994: Oxford, England and Huddersfield, England), Aug 1994, Conference abstracts: [of papers presented at Changing karst environments, hydrogeology, geomorphology and conservation, an international symposium held at the Universities of Oxford and Huddersfield, September 1994]
[1648]International Symposium on Changing Karst En-

vironments (1994: Oxford, England and Huddersfield, England), Mar 1995, Papers presented at the international symposium on changing karst environments, Oxford and Huddersfield, September 1994

[1818]Kiernan, Kevin, Nov 1989, Karst, caves and management at Mole Creek, Tasmania

[1862]Kranjc, Andrej, 1943–, 1995, Proceedings of international symposium "man on karst," Postojana, September 23-25, 1993

[1928]Lehman Caves National Monument (Nev.), 1980, Proposed natural resources management plan and environmental assessment, Lehman Caves National Monument, Nevada

[1931]Lénárt, László, 1993, A Bukki barlangok kutatasanak, vedelmenek es hasznositasanak legujabb eredmenyei = Newest results in research, protection and use of caves of Bukk Mountains: Miskolci Egyetem, 1993. November 11-13

[1962]Longley, Glenn, 1942–, Feb 1977, Status of *Typhlomolge* (=Eurycea) *rathbuni*

[1998]Luvless, Dungass F. (pseud), 1972, The speleobopper's guide to the caves of the Gruesome Chapel Valley

[2029]Mammoth Cave National Park Association, 1928, Mammoth Cave National Park Association campaign, 1927-1928

[2104]McAlpine, Donald F., 1983, Status and conservation of solution caves in New Brunswick

[2148]Meramec Valley Conservation Task Force. National Speleological Society, [1973], An endangered heritage: a story of the Meramec Valley, its caves, and their possible destruction by the U. S. Army Corps of Engineers

[2213]Missouri Speleological Survey, Apr 1973, An investigation into ground water pollution in caves

[2227]Missouri Speleological Survey, Jul 1973, Report of the Devil's Icebox - Rockbridge Park conservation task force of the National Speleological Society

[2328]Mylroie, John Eglinton, 1949–(ed), 1983, First international cave management symposium: proceedings [held at] College of Environmental Sciences, Murray State University, July 15-18, 1981

[2502]Newbould, Ronald L., 1976, Steam cleaning of Orient Cave

[2503]Newbould, Ronald L., May 1974, Steam cleaning of Orient Cave, Jenolan Caves, N.S.W

[2735]Queensland Conservation Council (prep), Mar 1973, The case against the Pike Creek Dam

[2736]Queensland Conservation Council (prep), Jun 1973, Pike Creek Dam: a preliminary criticism of the Queensland Irrigation and Water Supply's Commission's environmental impact study

[2869]Russell, Donald R., 1971, An essay on the caves of the Cherokee Nation: a conservation plea

[2870]Russell, William Hart, 1937–, Jul 1993, The Buttercup Creek karst: Travis and Williamson Counties, Texas: geology, biology, and land development

[2951]Sharpe, Grant William, 1925–, 1973, Evaluation of Mount St. Helens lava casts and caves, Cowlitz and Skamania Counties, Washington, for eligibility for registered natural landmark designation

[3082]Spate, Andy, 1991, Kubla Khan Cave State Reserve, Mole Creek, Tasmania: pilot management study

[3088]Speece, Jack Howard, 1947–, 1973, Alexander Caverns: the Carlsbad of Pennsylvania

[3097]Sprent, J. K., (ed), 1970, Mount Etna caves: a collection of papers covering several aspects of the Mt. Etna and Limestone Ridge caves area of central Queensland

[3098]Squire, Ralph E., Nov 1971, Report of study by National Speleological Society, Cave Conservation Task Force: New Melones Project

[3099]Squire, Ralph E., 1972, Stanislaus cave country: report of study

[3112]Steen, Charlie R., 1942, Ruins stabilization records for Canyon de Chelly National Monument 1942: Antelope House and Mummy Cave Ruins

[3182]Tardy, Janos, 1989, Geological nature conservation cave-protection

[3262]U.S. Bureau of Land Management, Dec 1993, Final Dark Canyon environmental impact statement

[3263]U.S. Bureau of Land Management, Barstow Resource Area, 1982, Shoshone Cave (Whip-scorpion Habitat) wildlife habitat management plan

[3267]U.S. Congress, House Committee on Interior and Insular Affairs, Mar 1988, Protecting cave resources on federal lands, and for other purposes: report (to accompany H.R. 1975) (including the cost estimate of the Congressional Budget Office)

[3268]U.S. Congress, House Committee on Natural Resources, May 1993, Lechuguilla Cave Protection Act of 1993: report (to accompany H.R. 698) (including cost estimate of the Congressional Budget Office)

[3269]U.S. Congress, House Committee on Public Lands, 1926, Shenandoah, Great Smoky Mountains, and Mammoth Cave National Parks: hearings before the Committee on the Public Lands, House of Representatives, Sixty-ninth Congress, first session, on H.R. 11287 [and] H.R. 12020, May 11, 1926

[3270]U.S. Congress, Senate Committee on Energy and Natural Resources, Sep 1988, Federal Cave Resources Protection Act of 1988: report (to accompany H.R. 1975)

[3271]U.S. Congress, Senate Committee on Energy and Natural Resources, Dec 1993, Lechuguilla Cave Protection Act of 1993: report (to accompany H.R. 698)

[3272]U.S. Congress, Senate Committee on Energy and Natural Resources, Subcommittee on Public Lands, National Parks and Forests, 1988, Federal Cave Resources Protection Act and restriction of dams in parks and monuments hearing before the Subcommittee on Public Lands, National Parks, and Forests of the Committee on Energy

and Natural Resources, United States Senate, One Hundredth Congress, second session, on S. 927/H.R. 1975 ... H.R. 1173 ... June 16, 1988
[3274] U.S. Environmental Protection Agency, Region 4, Apr 1981, Environmental impact statement; Mammoth Cave area, Kentucky, wastewater facilities
[3275] U.S. Environmental Protection Agency, Region 4, Apr 1981, Mammoth Cave area, Kentucky, 201 facilities plan environmental impact statement
[3280] U.S. Forest Service, Southern Region, 1975, Construction of phases II and III of the Blanchard Springs Caverns Project final environmental statement
[3281] U.S. Forest Service, Southern Region, 1974, Construction of phases II and III of the Blanchard Springs Caverns Project: draft environmental statement
[3286] U.S. National Park Service, Sep 1980, Carlsbad Caverns National Park, New Mexico: wilderness reevaluation study: preliminary - subject to change
[3289] U.S. National Park Service, Mar 1937, Caves and their conservation
[3292] U.S. National Park Service, Feb 1993, Draft, environmental impact statement, general management plan, development concept plan Timpanogos Cave National Monument
[3293] U.S. National Park Service, Jun 1993, Draft, general management plan, environmental impact statement Jewel Cave National Monument, South Dakota
[3295] U.S. National Park Service, 1973, Final environmental statement: proposed wilderness, Carlsbad Caverns National Park, New Mexico
[3299] U.S. National Park Service, Apr 1979, The Great Onyx Job Corps Civilian Conservation Center alternative relocation sites, Mammoth Cave National Park, Kentucky
[3314] U.S. National Park Service, 1973, Proposed wilderness, Carlsbad Caverns National Park, New Mexico
[3320] U.S. National Park Service, Aug 1972, Wilderness recommendation, Carlsbad Caverns National Park, New Mexico
[3367] Virginia Commission on the Conservation of Caves, 1979, Report of the Virginia Commission on the Conservation of Caves to the Governor and the General Assembly of Virginia
[3415] Watson, Richard Allan, 1931–, May 1967, The preservation of wilderness karst in central Kentucky, U.S.A
[3450] Western Australia Department of Conservation and Land Management, 1990, Leeuwin-Naturaliste National Park: cave permits draft issue plan
[3451] Western Land Surveys, Utah Division, 1977, Timpanogos Cave environmental impact report
[3463] White, John, 1973, Nature preserve potential of the Burton Cave area, Adams County, Illinois
[3464] White, John, 1973, Nature preserve potential of Twin Culvert Cave, Pike County, Illinois
[3465] White, John, 1973, Preservation of caves in Illinois
[3466] White, John, 1973, Preservation of Fogelpole Cave, Monroe County, Illinois
[3504] Willis, L. D., Dec 1986, A recovery plan for the Ozark cavefish (*Amblyopsis rosae*)
[3505] Wilmut, John (ed), 1972, Caves and conservation
[3548] Yuan, Daoxian, 1933–, 1983, Problems of environmental protection of karst area

8.2 Development

[351] Baynes, F. J., 1991, Ida Bay - Benders Quarry. An estimate of reserves within existing quarry limits
[1842] Koch, Donald L., Dec 1974, Report on Cold Water Cave: a summary of research results with inclusion of information on related to potential development of a new recreational facility by the state of Iowa
[2493] New Mexico Resource Management and Development Division, 1986, Bandera Crater/Ice Cave: a state park feasibility study
[3323] U.S. National Park Service, Sep 1974, Development concept plan, Slaughter Canyon, Carlsbad Caverns National Park, New Mexico: environmental assessment

8.3 Evaluation

[1290] Gurnee, Jeanne Marie, 1926–(ed), 1988, The Ensueño Cave Study: Ensueño Cave, Hatillo, Puerto Rico; expedition period: February 18-25, 1987
[1293] Gurnee, Jeanne Marie, 1926–(ed), Jul 1978, A study of Harrison's Cave, Barbados, West Indies
[1842] Koch, Donald L., Dec 1974, Report on Cold Water Cave: a summary of research results with inclusion of information on related to potential development of a new recreational facility by the state of Iowa
[2195] Missouri Civil Defense Agency, 1962, Missouri underground shelter space: a fallout shelter survey of mines and caves in Missouri
[2259] Moody, Larry D., Nov 1977, Warrens Cave Reserve stewardship plan: Gainesville, Alachua County, Florida
[2951] Sharpe, Grant William, 1925–, 1973, Evaluation of Mount St. Helens lava casts and caves, Cowlitz and Skamania Counties, Washington, for eligibility for registered natural landmark designation
[3079] Sowers, J. M., 1990, Cave management plan and environmental assessment, Lava Beds National Monument, California

8.4 Legal

[2144] Mellors, P. T., 1989, Legal aspects of access underground: a guide to the legal rights and obligations of people who explore cave, potholes and disused mine, and of those people who control access to them

8.5 Legislation

[1117]Frank, Cathleen, Jun 1986, Legislative history for Oregon Caves National Monument, 59th Congress through 96th Congress

[2847]Rogers, Edmund B., 1958, Mammoth Cave National Park, Kentucky: history of legislation through the 82nd Congress

[2848]Rogers, Edmund B., 1958, Oregon Caves National Monument, Oregon: history of legislation through the 82nd Congress

[2849]Rogers, Edmund B., 1958, Wind Cave National Park, South Dakota: history of legislation through the 82nd Congress

[3267]U.S. Congress, House Committee on Interior and Insular Affairs, Mar 1988, Protecting cave resources on federal lands, and for other purposes: report (to accompany H.R. 1975) (including the cost estimate of the Congressional Budget Office)

[3269]U.S. Congress, House Committee on Public Lands, 1926, Shenandoah, Great Smoky Mountains, and Mammoth Cave National Parks: hearings before the Committee on the Public Lands, House of Representatives, Sixty-ninth Congress, first session, on H.R. 11287 [and] H.R. 12020, May 11, 1926

[3270]U.S. Congress, Senate Committee on Energy and Natural Resources, Sep 1988, Federal Cave Resources Protection Act of 1988: report (to accompany H.R. 1975)

[3272]U.S. Congress, Senate Committee on Energy and Natural Resources, Subcommittee on Public Lands, National Parks and Forests, 1988, Federal Cave Resources Protection Act and restriction of dams in parks and monuments hearing before the Subcommittee on Public Lands, National Parks, and Forests of the Committee on Energy and Natural Resources, United States Senate, One Hundredth Congress, second session, on S. 927/H.R. 1975 ... H.R. 1173 ... June16,1988

[3305]U.S. National Park Service, Dec 1994, National Cave and Karst Research Institute Study: a draft report to Congress as required by Public Law 101-578 of November 15, 1990

[3367]Virginia Commission on the Conservation of Caves, 1979, Report of the Virginia Commission on the Conservation of Caves to the Governor and the General Assembly of Virginia

8.6 Management

[15]Addis, Robert Philip, 1945–, 1979, A study for the National Speleological Society: Knox Cave: Albany County, New York

[33]Aley, Thomas John, 1938–, 1981, Cave management investigations on the Ozark National Scenic Riverways, Missouri

[34]Aley, Thomas John, 1938–, 1980, Cave management investigations on the Ozark National Scenic Riverways, Missouri

[35]Aley, Thomas John, 1938–, 1988, Control of exotic plants in Oregon Caves, Oregon Caves National Monument

[41]Aley, Thomas John, 1938–, Aug 1975, A predictive hydrologic model for evaluating the effects of land use and management on the quantity and quality of water from Ozark Springs: final report

[42]Aley, Thomas John, 1938–, 1987, Restoration of natural cave features, Oregon Caves National Monument, Oregon: final report

[43]Aley, Thomas John, 1938–, 1988, Restoration of natural microclimate in Oregon Caves, Oregon Caves National Monument: final report

[56]American Cave Conservation Association, 1990, National cave management seminar: Albuquerque, New Mexico, March 12-16, 1990

[68]Anderson, Grant, 1991, Wellington Caves, resource and management study

[135]Anon, July 1994, Hastings Newdegate cave rehabilitation plan

[208]Anon, 1982, Waitomo Caves management plan 1982

[209]Anon, nd, Waitomo Day 1982: summary of papers: Waitomo Caves research programme

[250]Australasian Conference on Cave Tourism and Management , nd, Cave Management in Australasia VIII: proceedings of the eighth Australian conference on cave tourism and management, Paparoa National Part, Punakaiki, New Zealand, April 1989

[251]Australian Conference on Cave Tourism, Jun 1976, Cave Management in Australia: proceedings of the first Australian Conference on cave tourism, Jenolan Caves, N.S.W. 10th-13th July, 1973

[252]Australian Conference on Cave Tourism and Management, Aug 1977, Cave Management in Australia II: proceedings of the second Australian Conference on cave tourism and management, Hobart, Tasmania, 3rd-5th May, 1977

[253]Australian Conference on Cave Tourism and Management, Nov 1980, Cave Management in Australia III: proceedings of the third Australian Conference on cave tourism and management, Mount Gambier, South Australia, 30th April - 4th May, 1979

[254]Australian Conference on Cave Tourism and Management, Aug 1982, Cave Management in Australia IV: proceedings of the fourth Australian Conference on cave tourism and management, Yallingup, Western Australia Australia, September 1981

8.6. MANAGEMENT

[255]Australian Conference on Cave Tourism and Management, 1990, Cave Management in Australia V: proceedings of the Fifth Australian Conference on cave tourism and management, Lakes Entrance, Victoria, April 1983

[256]Australian National Parks and WIldlife Service, 1987, Australian Ranger Bulletin: feature: cave management

[260]Australian Speleological Federation, 1983, Draft management plan: Tantanoola Caves Conservation Park: Lower South East, South Australia

[261]Australian Speleological Federation, c1973, Master plan for the development of Jenolan Caves Reserve

[263]Australian Speleological Federation, Jul 1975, Proceedings of the tenth biennial conference of the Australian Speleological Federation

[377]Bell, Peter (ed), 1991, The 9th ACKMA conference proceedings; Margaret River, WA, Australia

[488]British Columbia. Parks and Outdoor Recreation Division, Jan 1980, Cave resources in British Columbia: a discussion paper

[567]Bush, Kent, 1995, Oregon Caves National Monument collections management plan

[587]Cameron McNamara Pty Ltd, 1989, Jenolan Caves Reserve: plan of management.

[600]Carlsbad Caverns National Park, Dec 1984, Land protection plan: Carlsbad Caverns National Park: prepared by Carlsbad Caverns National Park with assistance from the Southwest Regional Office, National Park Service.

[640]Chasen, Frederick Nutter, 1896–, 1931, Report on the "birds' nest" caves and industry of British North Borneo: with special reference to the Gomantong caves

[688]Commission on Cave Tourism and Management of the Australian Speleological Federation Study Team, Jun 1984, Jenolan Caves resort - some management issues

[689]Conference on Hydrogeology, Ecology, Monitoring, and Management of Ground Water in Karst Terranes (3rd: 1991: Nashville, TN), 1991, Proceedings of the Third Conference on Hydrogeology, Ecology, Monitoring and Management of Ground Water in Karst Terranes: December 4 - 6, 1991, Maxwell House/Clarion, Nashville, Tennessee

[708]Cooper, John Edward, 1929–, May 1982, Recovery plan for the Alabama cavefish, *Speoplatyrhinus poulsoni* Cooper and Kuehne 1974

[735]Craven, Stephen Adrian, 1994, Cango Cave, Oudtshoorn district of the Cape Province, South Africa: an assessment of its development and management 1780-1992

[812]Davey, Adrian G., 1986, Management of Victorian caves and karst: a report to the Caves Classification Committee, Department of Conservation, Forests and Lands

[814]Davey, Adrian G. (co-ord), 1978, Resource management of the Nullarbor Region, W.A.

[815]Davey, Adrian G., 1986, Victorian caves and karst: strategies for management and catalogue: a report to the Caves Classification Committee, Department of Conservation, Forests and Lands /

[817]Davey, Adrian G., 1978, Yallingup Cave Park - a management plan

[820]Davidson, Joseph Killworth, 1938–, 1971, Wilderness resources in Mammoth Cave National Park: a regional approach

[874]Dinger, James S., 1991, Ordinance for the control of urban development in sinkhole areas in the blue grass karst region, Lexington, Kentucky

[905]Driscoll, Ian, Apr 1977, The Binoomea Cut, Jenolan Caves: a paper read to the Jenolan Caves Historical and Preservation Society on Saturday 9th, February 1974

[938]Eberhard, Rolan, 1994, The Junee River karst system, Tasmania: a report to the Forestry Commission

[942]Edelbrock, Gay, 1993, Oregon Caves National Monument resource database: a user's guide

[961]Elliott, William Rawleigh, 1946–, Jun 1993, Draft recovery plan for endangered karst invertebrates in Travis and Williamson Counties, Texas

[964]Elliott, William Rawleigh, 1946–, 1978, The new Melones Cave harvestman transplant

[968]Ellis, Bryan M., 1934–, [1963], Mendip cave registry: handbook for members of the Executive Committee

[1014]Eversole, Jack (ed), 1973, The Mammoth Cave area: a planning proposal

[1026]Far West Cave Management Symposium (1979: Redding, CA), Jan 1980, Far West cave management symposium proceedings: Redding, California October 23-26, 1979

[1027]Far West Cave Management Symposium (1981: Portland, OR), Jul 1980, Far west cave management symposium proceedings: Portland Oregon April 14-16, 1981

[1115]Francis, Charles M. (comp), 1989, The management of edible bird's nest caves in Sabah

[1152]Gardner, James Eugene, 1953–, 1983, Cave resources of Ozark National Scenic Riverways: an inventory and evaluation phase II

[1156]Gardner, James Eugene, 1953–, Jan 1984, Missouri Department of Conservation Cooperative cave inventory project: a final report submitted to the Missouri Department of Conservation as part of the cooperative cave inventory project: a final report

[1210]Gillieson, David S. (ed), 1989, Resource management in limestone landscapes: international perspectives proceedings of the International Geographical Union study group man's impact on karst Sydney 15-21 August 1988

[1211]Gillieson, David S., 1996, Caves: processes, development, and management

[1223]Goode, Cecil E., 1986, World wonder saved: how Mammoth Cave became a national park

[1266] Grimes, K. G., Oct 1995, South East Karst Province of South Australia: Australian Caves & Karst Management Association October 1995

[1268] Grisafe, David A., Aug 1992, Stabilization of Dakota Sandstone surface of the Faris Cave petroglyphs, Kanopolis Lake Project, Kansas

[1292] Gurnee, Jeanne Marie, 1926–(ed), 1969, A study of Fountain National Park and Fountain Cavern: Anguilla, British West Indies

[1327] Halliday, William R., 1926–, Apr 1995, Considerations of Lower Uilani Cave for inclusion in the Hawaii Natural Area Reserve System

[1344] Hamilton-Smith, Elery, Jun 1989, Batu Caves, Kuala Lumpur - towards a plan of management

[1345] Hamilton-Smith, Elery, Sep 1974, Cave Reserves of the Katherine Area

[1347] Hamilton-Smith, Elery (prep), Jan 1989, Cutta Cutta Caves Nature Park: draft plan of management

[1348] [Hamilton-Smith, Elery], Sep 1994, Determining an environmental and social carrying capacity for Jenolan Caves Reserve: applying a visitor impact management system

[1349] [Hamilton-Smith, Elery], 1966, Draft management plan: Naracoorte Caves Conservation Park: South East South Australia

[1350] Hamilton-Smith, Elery, 1976, Mt. Etna & the caves: a plan for action

[1351] Hamilton-Smith, Elery (tm ldr), Nov 1988, Proposed developments at Waitomo Caves

[1352] Hamilton-Smith, Elery, Oct 1994, The Regional Caves Interpretive Centre: a planning review report for the Augusta Margaret River Tourist Bureau

[1361] Harpers Ferry Center. Division of Museum Services, 1976, Mammoth Cave National Park: collection management plan

[1396] Hayllar, Bruce (proj offr), Mar 1984, The development of leisure and educational resources at Jenolan Caves Resort

[1531] Houshold, Ian (prep), 1995, Cave & karst guidebook and documentation: Australasian Cave and Karst Management Association eleventh Australasian Conference on cave and karst management Tasmania 29 April to 7 May 1995 hosted by The Parks & Wildlife Service - Tasmania

[1532] Houshold, Ian, October 1987, Karst Landforms of the Lemonthyme and Southern Forests, Tasmania: report to the Commonwealth Commission of Inquiry into the Lemonthyme and Southern Forests

[1575] Hunt, Geoffrey, 1952–, 1975, Cave gating: a handbook

[1579] Huppert, George Nixon, 1944–, 1985, Selected bibliography of cave conservation and management

[1606] Indiana. Dept of Natural Resources. Engineering Division, 1968, Master plan: Wyandotte Caves, Harrison-Crawford, Blue River Recreation Complex

[1616] International Association of Hydrogeologists, 1988, Proceedings of the IAH 21st congress

[1701] James, Julia M., 1976, Timor Caves

[1786] Kell, Neil, [1993], The Winston Churchill Memorial Trust of Australia: project: to study recent developments in the design and installation of lighting to enhance the public education and enjoyment of show caves

[1810] Kerbo, Ronal Carrel, 1944–, Mar 1994, Cave resources management, planning, restoration, and redevelopment, Cathedral Caverns State Park, Grant, Alabama, 10-14 January 1994

[1818] Kiernan, Kevin, Nov 1989, Karst, caves and management at Mole Creek, Tasmania

[1819] Kiernan, Kevin, 1988, The management of soluble rock landscapes: an Australian perspective

[1860] Kranjc, Andrej (ed), 1989, Proceedings of international symposium at 170th anniversary of Postojnska jama, Postojna (Yugoslavia) Nov. 10-12 1988

[1928] Lehman Caves National Monument (Nev.), 1980, Proposed natural resources management plan and environmental assessment, Lehman Caves National Monument, Nevada

[1929] Lehman Caves National Monument, 1974, Proposed natural resources management plan, Lehman Caves National Monument, Nevada

[1931] Lénárt, László, 1993, A Bukki barlangok kutatasanak, vedelmenek es hasznositasanak legujabb eredmenyei = Newest results in research, protection and use of caves of Bukk Mountains: Miskolci Egyetem, 1993. November 11-13

[1945] Lindsley, R. P. (ed), 1977, Survey and assessment of cave resources at Buffalo National River, Arkansas

[2034] Mammoth Cave Operating Committee, Mammoth Cave, Ky, 1859, Mammoth cave

[2120] McEachern, J. Michael, 1978, An inventory and evaluation of the cave resources to be impacted by the New Melones Reservoir Project, Calaveras and Tuolumne Counties, California

[2168] Middleton, Gregory J., 1979, Wilderness caves: a preliminary survey of the caves of the Gordon-Franklin River System, South-West Tasmania

[2179] Millar, Ian R., 1989, General policy and guidelines for cave and karst management in areas managed by the Department of Conservation

[2227] Missouri Speleological Survey, Jul 1973, Report of the Devil's Icebox - Rockbridge Park conservation task force of the National Speleological Society

[2259] Moody, Larry D., Nov 1977, Warrens Cave Reserve stewardship plan: Gainesville, Alachua County, Florida

[2328] Mylroie, John Eglinton, 1949–(ed), 1983, First international cave management symposium: proceedings [held at] College of Environmental Sciences, Murray State University, July 15-18, 1981

8.6. MANAGEMENT

[2337]National Cave Management Symposium (1975), 1976, National cave management symposium proceedings: Albuquerque, New Mexico, October 6-10, 1975

[2338]National Cave Management Symposium (1976), 1977, Proceedings: national cave management symposium, Mountain View, Arkansas, October 26 - 29, 1976

[2339]National Cave Management Symposium (1977), 1978, National cave management symposium proceedings: Big Sky, Montana, October 3-7, 1977

[2340]National Cave Management Symposium (1978+80), 1982, National cave management symposia: proceedings

[2341]National Cave Management Symposium (1980), 1980, Fifth national cave management symposium: Mammoth Cave National Park, October 14-17, 1980

[2342]National Cave Management Symposium (1982), 1985, Cave management symposium: proceedings: Harrisonburg, Virginia November 5-7, 1982

[2343]National Cave Management Symposium (1984), Jan 1985, Proceedings of the 1984 national cave management symposium, Rolla, MO

[2344]National Cave Management Symposium (1987), 1989, 1987 cave management symposium: Rapid City, South Dakota, October 1987

[2345]National Cave Management Symposium (1989), 1993, Proceedings of the 1989 national cave management symposium: New Braunfels, Texas, U.S.A.

[2346]National Cave Management Symposium (1991), 1993, 1991 national cave management symposium proceedings: Bowling Green, Kentucky, October 23-26, 1991

[2347]National Cave Management Symposium (1993), 1995, Proceedings of the 1993 national cave management symposium: held in Carlsbad, New Mexico, October 27-30, 1993

[2348]National Cave Management Symposium (1995), 1995, Proceedings of the 1995 national cave management symposium: Spring Mill State Park: Mitchell, Indiana: October 25-28, 1995

[2352]National Parks and Public Land Division, Department of Conservation & Environment, Mar 1991, Draft strategy for the management of caves and karst in Victoria

[2491]Netherton, Shaaron, Aug 1985, Cave management plan for the Ely District

[2496]New South Wales National Parks and Wildlife Service, Jul 1987, Cooleman Plain karst area management plan: a supplementary plan to the Kosciusko National Park plan of management

[2514]Noble, Bruce J., 1991, Cultural resource management in Mammoth Cave National Park: a National Park Service—Kentucky Heritage Council cooperative project

[2522]Northwest Cave Research Institute, Jan 1987, A management plan for Bighorn Caverns, Montana

[2545]O'Donnell, Lisa, Aug 1994, Recovery plan for endangered karst invertebrates in Travis and Williamson Counties, Texas

[2703]Powell, Richard Lewis, 1936–, Dec 1970, A guide to the selection of limestone caverns and springs in the United States as national landmarks: for the National Park Service

[2765]Rautjoki, Harri, nd, North-West Nelson State Forest Park Honeycomb Hill Cave System: concept plan: West Coast Conservancy

[2836]Rodda, Jan, 1989, Mulka's Cave site management project: emphasising visitor survey, April-June 1988, with management evaluation and further recommendations for management; with conclusions of an archaeological test excavation

[2905]Schindel, Geary Michael, 1957–, Dec 1991, Guidebook: environmental hydrogeology of karst terranes in the vicinity of Nashville Tennessee

[3055]South Australia National Parks and Wildlife Service, Mar 1983, Draft management plan, Tantanoola Caves Conservation Park, lower South East, South Australia

[3056]South Australia National Parks and Wildlife Service, 1990, Naracoorte Caves Conservation Park management plan, South East, South Australia

[3057]South Australia National Parks and Wildlife Service, Feb 1986, Naracoorte Caves Conservation Park, South East, South Australia: draft management plan

[3058]South Australia National Parks and Wildlife Service, 1990, Tantanoola Caves Conservation Park management plan, South East, South Australia

[3076]Southwest Georgia Planning & Development Commission, 1968, Glory Hole Caverns

[3079]Sowers, J. M., 1990, Cave management plan and environmental assessment, Lava Beds National Monument, California

[3082]Spate, Andy, 1991, Kubla Khan Cave State Reserve, Mole Creek, Tasmania: pilot management study

[3090]Speleological Conference (1987), 1987, Proceedings - 16th biennial conference Australian Speleological Federation Inc

[3112]Steen, Charlie R., 1942, Ruins stabilization records for Canyon de Chelly National Monument 1942: Antelope House and Mummy Cave Ruins

[3170]Sydney Speleological Society, 1976, Timor Caves

[3182]Tardy, Janos, 1989, Geological nature conservation cave-protection

[3263]U.S. Bureau of Land Management, Barstow Resource Area, 1982, Shoshone Cave (Whip-scorpion Habitat) wildlife habitat management plan

[3266]U.S. Bureau of Land Management, Shoshone District, 1987, T-maze cave management plan

[3267]U.S. Congress, House Committee on Interior and Insular Affairs, Mar 1988, Protecting cave resources on federal lands, and for other purposes: report (to accompany H.R. 1975) (including the cost estimate of the Con-

[3268] U.S. Congress, House Committee on Natural Resources, May 1993, Lechuguilla Cave Protection Act of 1993: report (to accompany H.R. 698) (including cost estimate of the Congressional Budget Office)

[3269] U.S. Congress, House Committee on Public Lands, 1926, Shenandoah, Great Smoky Mountains, and Mammoth Cave National Parks: hearings before the Committee on the Public Lands, House of Representatives, Sixty-ninth Congress, first session, on H.R. 11287 [and] H.R. 12020, May 11, 1926

[3270] U.S. Congress, Senate Committee on Energy and Natural Resources, Sep 1988, Federal Cave Resources Protection Act of 1988: report (to accompany H.R. 1975)

[3271] U.S. Congress, Senate Committee on Energy and Natural Resources, Dec 1993, Lechuguilla Cave Protection Act of 1993: report (to accompany H.R. 698)

[3272] U.S. Congress, Senate Committee on Energy and Natural Resources, Subcommittee on Public Lands, National Parks and Forests, 1988, Federal Cave Resources Protection Act and restriction of dams in parks and monuments hearing before the Subcommittee on Public Lands, National Parks, and Forests of the Committee on Energy and Natural Resources, United States Senate, One Hundredth Congress, second session, on S. 927/H.R. 1975 ... H.R. 1173 ... June 16, 1988

[3273] U.S. Environmental Protection Agency, Region 4, Aug 1981, Environmental impact statement, Mammoth Cave area, Kentucky: wastewater facilities

[3274] U.S. Environmental Protection Agency, Region 4, Apr 1981, Environmental impact statement; Mammoth Cave area, Kentucky, wastewater facilities

[3275] U.S. Environmental Protection Agency, Region 4, Apr 1981, Mammoth Cave area, Kentucky, 201 facilities plan environmental impact statement

[3276] U.S. Forest Service, 1973, Operation of Blanchard Springs Caverns, Ozark-St. Francis National Forest, Arkansas: final environmental statement

[3279] U.S. Forest Service, 1964, A preliminary plan for the development of Blanchard Springs Cavern: Sylamore District, Ozark National Forest

[3280] U.S. Forest Service, Southern Region, 1975, Construction of phases II and III of the Blanchard Springs Caverns Project final environmental statement

[3281] U.S. Forest Service, Southern Region, 1974, Construction of phases II and III of the Blanchard Springs Caverns Project: draft environmental statement

[3290] U.S. National Park Service, Aug 1975, Development concept: Slaughter Canyon, Carlsbad Caverns National Park, New Mexico

[3291] U.S. National Park Service, Nov 1995, Draft general management plan, environmental impact statement: Carlsbad Caverns National Park, New Mexico

[3292] U.S. National Park Service, Feb 1993, Draft, environmental impact statement, general management plan, development concept plan Timpanogos Cave National Monument

[3293] U.S. National Park Service, Jun 1993, Draft, general management plan, environmental impact statement Jewel Cave National Monument, South Dakota

[3294] U.S. National Park Service, 1976, Final environmental statement: Mammoth Cave National Park master plan and wilderness suitability study, Kentucky

[3295] U.S. National Park Service, 1973, Final environmental statement: proposed wilderness, Carlsbad Caverns National Park, New Mexico

[3296] U.S. National Park Service, Apr 1976, Final master plan: Mammoth Cave National Park

[3297] U.S. National Park Service, Jun 1994, Final, general management plan, environmental impact statement Jewel Cave National Monument, South Dakota

[3298] U.S. National Park Service, 1990, General management plan, wilderness suitability study: El Malpais National Monument, New Mexico

[3303] U.S. National Park Service, 1978, Mammoth Cave National Park: collection management plan

[3304] U.S. National Park Service, [1973], Master plan: Carlsbad Caverns National Park, New Mexico

[3306] U.S. National Park Service, Feb 1980, Natural resources management program: an addendum to the natural resources management plan for Lehman Caves National Monument, Nevada

[3308] U.S. National Park Service, 1975, Oregon Caves National Monument, master plan

[3309] U.S. National Park Service, 1976, Oregon Caves National Monument, Oregon: final interpretive prospectus

[3312] U.S. National Park Service, 1973, Pollution abatement project, Carlsbad Caverns National Park, New Mexico

[3313] U.S. National Park Service, nd, Proposed Mammoth Cave National Park master plan and wilderness study, Kentucky: draft environmental impact statement

[3315] U.S. National Park Service, 1987, Statement for management: Jewel Cave National Monument

[3316] U.S. National Park Service, Jul 1986, Statement for management: Wind Cave National Park

[3318] U.S. National Park Service, Aug 1986, Timpanogos Cave National Monument: statement for management

[3323] U.S. National Park Service, Sep 1974, Development concept plan, Slaughter Canyon, Carlsbad Caverns National Park, New Mexico: environmental assessment

[3374] Waitomo District Council, March 1994, District plan discussion paper: tourism issues

[3383] Walsh, John Michael, 1947–(ed), 1991, A proposed cave access and cave management plan for the Devil's Sinkhole State Natural Area & Kickapoo Caverns State Park

[3450]Western Australia Department of Conservation and Land Management, 1990, Leeuwin-Naturaliste National Park: cave permits draft issue plan
[3451]Western Land Surveys, Utah Division, 1977, Timpanogos Cave environmental impact report
[3484]Wilde, Kevan A., 1992, West coast cave and karst management strategy and operational guidelines
[3511]Wilson, Paul (prep), 1977, Managing the limestone caves of Chillagoe and Mungana

8.7 Pollution

[37]Aley, Thomas John, 1938–, 1972, Groundwater contamination and sinkhole collapse induced by leaky impoundments in soluble rock terrain
[38]Aley, Thomas John, 1938–, 1981, Hydrogeologic mapping of unincorporated Greene County, Missouri, to identify areas where sinkhole flooding and serious groundwater contamination could result from land development
[238]Association of Ground Water Scientists and Engineers (a division of NGWA), Dec 1991, Proceedings of the third conference on hydrogeology, ecology, monitoring, and management of ground water in karst terranes, December 4-6, 1991, Maxwell House/Clarion, Nashville, Tennessee
[337]Barton, Nicholas James, 1935–, 1992, The lost rivers of London: a study of their effects upon London and Londoners, and the effects of London and Londoners upon them
[442]Bosnak, Art D., 1981, Acute toxicity of cadmium, zinc, and total residual chlorine to epigean and hypogean isopods (Asellidae)
[739]Crawford, Nicholas Charles, 1942–, 1982, Hydrogeologic problems resulting from development upon karst terrain, Bowling Green, Kentucky. Guidebook prepared for Karst Hydrogeology Workshop August 31 - September 3, 1982 Nashville, Tennessee
[741]Crawford, Nicholas Charles, 1942–, 1981, Karst hydrogeology and environmental problems in the Bowling Green area
[748]Crawford, Nicholas Charles, 1942–, 1988, Karst hydrologic problems of south central Kentucky: ground water contamination, sinkhole flooding, and sinkhole collapse: field trip guide
[889]Dougherty, Percy H. (ed), 1993, Environmental karst
[960]Elliott, Larry P., 1974, Study of potential gas accumulation in caves in Bowling Green, including relationship to water quality
[1563]Huber, Gary, Oct 1989, Sinkholes: landowner perceptions of a unique source of groundwater contamination
[1641]International Geographical Union Study Group on Man's Impact in Karst, 1987, Karst and man: proceedings of the international symposium on human influence in karst, 11-14th September 1987, Postojna, Yugoslavia
[2483]National Water Well Association, 1987, Proceedings of environmental problems in karst terranes and their solutions conference, October 28-30, 1986, Bowling Green, Kentucky
[2484]National Water Well Association, 1988, Proceedings of the second conference on environmental problems in karst terranes and their solutions conference, November 16-18, 1988, Maxwell House Hotel, Nashville, Tennessee
[2739]Quinlan, James Francis, 1936–1995, 1986, Ground water flow in the Mammoth Cave Area, Kentucky with emphasis on principles, contaminant dispersal, instrumentation for monitoring water quality, and other methods of study
[2742]Quinlan, James Francis, 1936–1995, 1990, Hydrogeology and geomorphology of the Mammoth Cave area, Kentucky: Southeastern Friends of the Pleistocene 1990 field excursion, November 17, 1990
[3001]Slifer, Dennis William, 1947–, Apr 1989, Sinkhole dumps and the risk to groundwater in Virginia's karst areas: final report to the Virginia Environment Endowment
[3241]Trommer, J. T., 1987, Potential for pollution of the Upper Floridan aquifer from five sinkholes and an internally drained basin in west-central Florida
[3273]U.S. Environmental Protection Agency, Region 4, Aug 1981, Environmental impact statement, Mammoth Cave area, Kentucky: wastewater facilities
[3312]U.S. National Park Service, 1973, Pollution abatement project, Carlsbad Caverns National Park, New Mexico
[3403]Waterhouse, J. D., nd, Pollution of underground water

8.8 Restoration

[42]Aley, Thomas John, 1938–, 1987, Restoration of natural cave features, Oregon Caves National Monument, Oregon: final report
[43]Aley, Thomas John, 1938–, 1988, Restoration of natural microclimate in Oregon Caves, Oregon Caves National Monument: final report
[2502]Newbould, Ronald L., 1976, Steam cleaning of Orient Cave
[2503]Newbould, Ronald L., May 1974, Steam cleaning of Orient Cave, Jenolan Caves, N.S.W

9 Paleontology

Frederick Grady

1201 South Scott Street, Apt. 123
Arlington, VA 22204-4655

The earliest literature on cave paleontology as a science dates from the late 18th century. Johann Freidrich Esper with his 1774 publication *Description des Zoolithes Nouvellement Decouvertes d'Animaux Quadrupeds Inconnus et des Cavernes qui les Renferment de Meme que de Plusiers Autres Grottes Remarkables qui se Trouvent dans le Margraviat de Bareinth au de la des Monts* provides what can be called the first scientific descriptions of fossil bones from caves. At the end of the century, Thomas Jefferson in 1799 produced *A Memoir on the Discovery of Certain Bones of a Quadruped of the Clawed Kind*. This is the first American description of fossil bones found in a cave.

During much of the 19th century, Europe was the center of cave paleontological studies. Georges Cuvier (1812) wrote *Researches sur les Ossements Fossiles de Quadrupèdes...*, which includes descriptions of bones found in caves of France and other areas of Europe. From the cave at Gailenreuth, Germany Couvier notes bones of hyaena, wolf, glutton, and reindeer. William Buckland in his 1823 *Reliquiae Diluvianae; or Observations on the Organic Remains Contained in Caves, Fissures, and Diluvial Gravel and Other Geological Phenomena, Attesting to the Action of the Universal Diluge* provides a comprehensive description of bone sites in England and continental Europe. Buckland's volume includes illustrations of several of the caves and many of the fossils found within them. The Reverend John MacEnery was a contemporary of Buckland's who did detailed studies in Kent's Cavern, England, which unfortunately were not published until after his death by William Pengelly in 1869 *The Literature on Kent's Cavern. part III Including the Whole of the Rev. J. MacEnery's Manuscript*.

The Middle to late 19th century was a particularly active time for British Cave Paleontological researches. Paleontological work was invariably associated with that of archeology and anthropology. Methods were developed to excavate by carefully noting positions of bones and artifacts both vertically and horizontally. William Pengelly is one of the first to excavate with such attention to detail as reported in his 1859 *On the Ossiferous Caverns and Fissures of Devonshire*. William Boyd Dawkins in his 1874 volume *Cave Hunting, Researches on the Evidence of Caves Respecting the Early Inhabitants of Europe*, includes a great deal of paleontological information as well as anthropological. In an appendix to the volume Dawkins gives detailed instructions on to how to map and excavate a cave, most of which are still applicable today.

Meanwhile in North America there was rather little in the way of publications on cave paleontology. Edward Drinker Cope, who in 1869 wrote *Remarks on Fossils from Limestone Caves of Virginia*, is one of the few researchers who did paleontological work on cave faunas. In the final decade of the 1800's, Henry C. Mercer worked with Cope and published several descriptions of the work one of the best known being his 1899 *The Bone Cave at Port Kennedy, Pennsylvania and Its Partial Excavation in 1894, 1895, and 1896*, in which he employed techniques similar to those recommended by Dawkins.

Cave paleontological studies continued into the 20th century by workers employing methods similar to those noted by Dawkins and Mercer. Barnum Brown, better known for his dinosaur work, excavated an important site in Arkansas (1908) *The Conard Fissure, a Pleistocene Bone Deposit in Northern Arkansas: with Descriptions of Two New Genera and Twenty New Species of Mammals*. J.W. Gidley of the Smithsonian Institution wrote of the first discoveries in Cumberland Cave, Maryland (1938) *Preliminary Report on a Recently Discovered Pleistocene Cave Deposit Near Cumberland, Maryland*.

Not until the middle of the present century were significant advances made in paleontological techniques with regard to cave studies. John Guilday, Paul Martin, and Allen McCrady (1964), *New Paris No 4: A Pleistocene Cave Deposit in Bedford County, Pennsylvania*, employed fine screening of cave sediments to recover small fossils such as mouse teeth. They also used carbon-14 dating to get a better idea of the absolute age of the deposit. The use of dilute acids to free bones from travertine coatings was also developed. C. K. Brain used this technique in his work on South African Caves, summarized in his 1981 volume *The Hunters and the Hunted: An Introduction to African Cave Taphonomy*. More recently, Peter Andrews (1990) *Owls, Caves, and Fossils*, employed a scanning electron microscope to study surface markings on bones from caves compared to those of animals killed by modern carnivores.

Two of the most significant volumes that cover a great deal of information on cave paleontology are Bjorn Kurten (1968) *Pleistocene Mammals of Europe* and Bjorn Kurten and Elaine Anderson (1980) *Pleistocene Mammals of North America*. Both of the volumes cover many cave sites but are now out-of-date and in need of revision. Kurten also published several volumes for the public including his 1976 *The Cave Bear Story: Life and Death of a Vanished Animal*.

References not in Main Bibliography

Cope, Edward D. 1869. Synopsis of the extinct Mammalia of the cave formations in the United States, with observations on some Myriapoda found in and near the same, and on some extinct mammals of the caves of Anguilla, W.I., and of other localities. *Proceedings of the American Philosophical Society [Philadelphia]*. 11: 171-192.

Cuvier, Georges, baron. 1812. *Recherches sur les Ossements Fossiles de Quadrupèdes, ou l'on Etablit les Caracteres de Plusieurs Especes d'Animaux que les Revolutions du Globe Paraissent Avoir Detruites.* Paris: Deterville, 4 vols.

Esper, Johann Friedrich. 1774. *Description des Zoolithes Nouvellement Decouvertes d'Animaux Quadrupedes Inconnus et des Cavernes qui les Renferment de Meme que de Plusieurs Autres Grottes Remarquables ....* Nuremburg: Chez les heritiers de fau G.W. Knorr, 121pp.

Jefferson, Thomas. 1799. A Memoir on the Discovery of Certain Bones of a Quadreuped of the Clawed Kind in the Western Parts of Virginia. *Transactions of the American Philosophical Society.* 4:246-260.

Kurten, Bjorn. 1968. *Pleistocene Mammals of Europe.* Chicago, Aldine Publishing Company, 317pp.

Kurten, Bjorn. 1980. *Pleistocene Mammals of North America.* New York: Columbia University Press, 442pp.

Kurten, Bjorn. 1976. *The Cave Bear Story: Life and Death of a Vanished Animal.* New York: Columbia University Press, 163pp.

Pengelly, William. 1859. On the Ossiferous Caverns and Fissures of Devonshire. *Proceedings of the Royal Institution of Great Britain.* 3(30):149-151.

Pengelly, William. 1869. The Literature of Kent's Cavern. Part III including the whole of the Rev. J. MacEnery's manuscript. *Report & transactions of the Devonshire Association for the Advancement of Science, Literature and Art.* 3(1): 191-482.

9.1 General

[6]Adam, David P., 1963, Pollen analysis of a stratigraphic section from Bat Cave, New Mexico

[48]Allen, E. E., 1946, Gower Caves: a survey of the Gower Caves with an account of recent excavations: part I

[49]Allen, E. E., 1948, Gower Caves: a survey of the Gower Caves with an account of recent excavations: part II

[50]Allen, Joel Asaph, 1838–1921, 1885, On an extinct type of dog from Ely Cave, Lee County, Virginia

[72]Andersson, Johan Gunnar, 1874–1960, nd, The cave-deposit at Sha Kuo Tun in Fengtien

[75]Andrews, Peter, 1940–, 1990, Owls, caves, and fossils: predation, preservation, and accumulation of small mammal bones in caves, with an analysis of the Pleistocene cave faunas from Westbury-sub-Mendip, Somerset, UK

[115]Anon, 1882, Exploration of the caves and rivers of New South Wales (minutes, reports, correspondence, accounts)

[268]Ayer, Mary Youngman, 1936, The archaeological and faunal material from Williams Cave, Guadalupe Mountains, Texas

[294]Balch, Herbert Ernest, [1927], Excavations at Chelm's Combe: Cheddar: conducted under the Excavations Committee of the Somerset Archaeological and Natural History Society 1925-26

[298]Balch, Herbert Ernest, 1929, Mendip: the great cave of Wookey Hole

[404]Bishop, Michael J., 1982, The mammal fauna of the early middle Pleistocene cavern infill site of Westbury-sub-Mendip, Somerset

[459]Brain, Charles Kimberlin, 1981, The hunters or the hunted? An introduction to African cave taphonomy

[466]Brenan, Edward, 1859, Notice of the occurrence of mammoth and other animal remains: discovered under limestone in a bone cave at Shandon, near Dungarvan, in the county of Waterford

[517]Brown, Barnum, 1873–1963, 1908, The Conard fissure, a Pleistocene bone deposit in northern Arkansas: with descriptions of two new genera and twenty new species of mammals

[543]Bryan, Kirk, nd, The geology and fossil vertebrates of Ventana Cave

[545]Buckland, William, 1784–1856, 1824, Reliquiae diluvianae, or observations on the organic remains contained in caves, fissures, and diluvial gravel, and on other geological phenomena attesting the action of an universal deluge /

[546]Buckland, William, 1784–1856, 1978, Reliquiae diluvianae: or, observations on the organic remains contained in caves, fissures, and diluvial gravel and on other geological phenomena attesting the action of a universal deluge

[547]Buckland, William, 1784–1856, 1823, Reliquiae diluvianae: or, observations on the organic remains contained in caves, fissures, and diluvial gravel and on other geological phenomena attesting the action of an universal deluge

[583]Callard, Thomas Karr, [1880], Contemporaneity of man with the extinct Mammalia, as taught by recent cavern exploration, and its bearing upon the question of man's antiquity

[611]Casteret, Norbert, 1897–1987, 1940, Ten years under the earth

[701]Cooke, Cranmer Kenrick, 1906–, 1978, The Redcliff

stone age site, Rhodesia

[703]Cooke, John H. , c1893, The Har Dalam Cavern, Malta and its fossiliferous contents by John H. Cooke with a report on the organic remains by Arthur Smith Woodward

[711]Cope, Edward Drinker, 1840–1897, 1883, On the contents of a bone cave in the island of Anguilla (West Indies)

[784]Curry, David A., May 1982, Joint Mitnor Cave

[794]Dalquest, Walter W., Jun 1969, The mammal fauna of Schulze Cave, Edwards County, Texas

[839]Dawkins, William Boyd, 1838–1929, 1868, The British Pleistocene Mammalia: part II

[840]Dawkins, William Boyd, 1838–1929, 1973, Cave hunting, researches on the evidence of caves respecting the early inhabitants of Europe

[873]Dimitrijevic, Vesna M., 1991, Quaternary mammals of the Smolucka Cave in southwest Serbia

[887]Dortch, C. E. , 1984, Devil's Lair, a study in prehistory

[895]Downs, Theodore, 1919–, 1959, Quaternary animals from Schuiling Cave in the Mojave Desert, California

[932]Eastmead, William, 1824, Historia rievallensis: containing the history of Kirkby Moorside ... to which is prefixed a dissertation on the animal remains, and other curious phenomena, in the recently discovered cave at Kirkdale

[1008]Euler, Robert C., 1978, Archaeological and paleobiological studies at Stanton's Cave, Grand Canyon National Park, Arizona; a report of progress

[1011]Evans, Glen Louis, 1911–, 1961, The Friesenhahn Cave

[1119]Frank, Ruben Milton, 1935–, Sep 1964, The vertebrate paleontology of Texas caves

[1130]Funk, Robert E., 1994, Archaeological and paleoenvironmental investigations in the Dutchess Quarry Caves, Orange County, New York

[1157]Gargett, Robert H., 1996, Cave bears and modern human origins: the spatial taphonomy of Pod Hradem Cave, Czech Republic

[1162]Garrod, Dorothy Anne Elizabeth, 1892–, 1968, The paleolithic of southern Kurdistan: excavations in the caves of Zarzi and Hazar Merd

[1184]Gerrard, John, 1960, Kent's Cavern

[1191]Gidley, James William, 1866–1931, 1938, The Pleistocene vertebrate fauna from Cumberland Cave, Maryland

[1192]Gidley, James William, 1866–1931, 1914, Preliminary report on a recently discovered Pleistocene cave deposit near Cumberland, Maryland

[1193]Gidley, James Williams, 1866–1931, 1918, A pleistocene cave deposit of Western Maryland

[1194]Gidley, James Williams, 1866–1931, 1921, Pleistocene peccaries from the Cumberland cave deposit

[1198]Gilbertson, D. D., 1984, In the shadow of extinction: a Quaternary archaeology and paleoecology of the lake, fissures, and smaller caves at Creswell Crags SSSI

[1233]Government of Australia, 1993, Nomination of Australian fossil sites (a serial nomination of sites at Murgon, Riversleigh and Naracoorte): The origin and evolution of Australia's mammals

[1246]Grayson, Donald K., May 1988, Danger Cave, Last Supper Cave, and Hanging Rock Shelter: the faunas

[1252]Green, H. Stephen, 1984, Pontnewydd Cave: a lower palaeolithic hominid site in Wales: the first report

[1261]Griffin, John Wallace, 1978, Investigations in Russell Cave: Russell Cave National Monument, Alabama

[1279]Guilday, John E., 1978, The Baker Bluff cave deposit, Tennessee, and the late Pleistocene faunal gradient

[1281]Guilday, John E., 1972, Jaguar (*Panthera onca*) remains from Big Bone Cave, Tennessee and east central North America

[1282]Guilday, John E., 1973, The late Pleistocene small mammals of Eagle Cave, Pendleton County, West Virginia

[1283]Guilday, John E. , 1971, The Welch Cave Peccaries (Platygonus) and associated fauna, Kentucky Pleistocene

[1284]Guilday, John E., 1925–1982, 1964, New Paris no. 4: a late Pleistocene cave deposit in Bedford County, Pennsylvania

[1285]Guilday, John E., 1925–1982, 1967, A new *Peromyscus* (Rodentia: Cricetidae) from the Pleistocene of Maryland

[1354]Hanihara, Kazuro, 1927- , 1978, Paleolithic site of Douara Cave and paleogeography of Palmyra Basin in Syria

[1360]Harlan, Richard, 1796–1843, 1831, Description of the fossil bones of the Megalonyx, discovered in White Cave, Kentucky

[1386]Hatt, Robert Torrens, 1902–, 1953, Faunal and archeological researches in Yucatan caves

[1416]Heath, Thomas, Aug 1879, An abstract description and history of the Bone Caves of Creswell Crags

[1491]Holman, J. Alan, 1931-, 1982, The pleistocene (Kansan) herpetofauna of Trout Cave, West Virginia

[1492]Holman, J. Alan, 1931–, Mar 1987, A Mid-Pleistocene (Irvingtonian) herpetofauna from a cave in southcentral Texas

[1493]Holman, J. Alan, 1931–, 1977, The Pleistocene (Kansan) herpetofauna of Cumberland Cave, Maryland

[1519]Horne, Peter, 1988, Fossil cave, 5L81 underwater palaeontological and surveying project, 1987-1988

[1570]Humphrey, Philip Strong, 1926–, Feb 1993, Avifauna of three Holocene cave deposits in southern Chile

[1646]International Symposium on Cave Biology and Cave Paleontology (1975: Oudtshoorn, South Africa), 1975, Proceedings, U.I.S. international symposium on cave biology and cave paleontology, held in Oudtshoorn, South Africa from the 3rd- 6th August, 1975

9.1. GENERAL

[1687]Jackson, John Wilfrid, 1880–1978, 1928, Report on the animal remains found in the Kilgreany Cave, Co. Waterford

[1826]Klippel, Walter E., nd, The paleontology of Cheek Bend Cave, Maury County, Tennessee: phase II: report to the Tennessee Valley Authority

[1827]Klippel, Walter E., Aug 1986, The unmodified vertebrate fauna from Granite Quarry Cave (23CT36), Carter County, Missouri

[1875]Kurtén, Björn, 1965, The Carnivora of the Palestine caves

[1876]Kurtén, Björn, 1976, The cave bear story: life and death of a vanished animal

[1877]Kurtén, Björn, 1924–1988, 1958, Life and death of the Pleistocene cave bear: a study in paleoecology

[1983]Lundelius, Ernest L., 1927–, 1982, The mammalian fauna of Madura Cave, Western Australia

[1984]Lundelius, Ernest L., 1927–, 1984, The mammalian fauna of Madura Cave, Western Australia, part VI, Macropodidae: Potoroinae

[1985]Lundelius, Ernest L., 1927–, Feb 1973, The mammalian fauna of Madura Cave, Western Australia: part I

[1986]Lundelius, Ernest L., 1927–, Aug 1975, The mammalian fauna of Madura Cave, Western Australia: part II

[1987]Lundelius, Ernest L., 1927–, Dec 1978, The mammalian fauna of Madura Cave, Western Australia: part III

[1988]Lundelius, Ernest L., 1927–, Feb 1981, The mammalian fauna of Madura Cave, Western Australia: Part IV

[2007]MacCurdy, George Grant, 1863–, 1914, La Combe, a paleolithic cave in the Dordogne

[2009]MacEnery, John M., 1796–1841, 1859, Cavern researches, or, discoveries of organic remains, and of British and Roman reliques, in the caves of Kent's Hole, Anstis Cove, Chudleigh, and Berry Head

[2023]Malatesta, Alberto, 1980, Dwarf deer and other late Pleistocene fauna of the Simonelli Cave in Crete

[2070]Maryland Governor's Advisory Committee on Promoting Paleontology, Oct 1993, Report

[2078]Mason, Revil J., nd, Cave of Hearths, Makapansgat, Transvaal

[2079]Mason, Revil J., Sep 1988, Kruger Cave, late Stone Age, Magaliesberg

[2115]McCrady, Edward, 1906–, 1954, New finds of Pleistocene jaguar skeletons from Tennessee caves

[2137]Mead, Jim I., Nov 1984, The late Wisconsinan vertebrate fauna from Deadman Cave, southern Arizona

[2141]Mehl, Maurice Goldsmith, 1887–1966, Dec 1962, Missouri's Ice Age animals

[2142]Mehl, Maurice Goldsmith, 1887–1966, 1969, Notes on Missouri Pleistocene peccaries

[2149]Mercer, Henry Chapman, 1856–1930, 1899, The bone cave at Port Kennedy, Pennsylvania, and its partial excavation in 1894, 1895 and 1896

[2152]Mercer, Henry Chapman, 1856–1930, nd, The finding of the remains of the fossil sloth at Big Bone Cave, Tennessee, in 1896

[2163]Meulen, A. J. van der, 1978, *Microtus* and *Pitymys* (Arvicolidae) from Cumberland Cave, Maryland, with a comparison of some new and old world species

[2182]Miller, Gerrit Smith, 1869–1923, Mar 1929, A second collection of mammals from caves near St. Michel, Haiti

[2183]Miller, Loye Holmes, 1874–1970, 1980, The Pleistocene birds of San Josecito cavern, Mexico

[2223]Missouri Speleological Survey, Jan 1975, A pleistocene fauna from Zoo Cave, Taney County, Missouri

[2226]Missouri Speleological Survey, January-June 1986, Remains of Quaternary vertebrates from Ozark caves and other miscellaneous sites

[2267]Morana, Martin, 1987, The prehistoric cave of Gher Dalam

[2304]Murray, Peter, 1977, Pleistocene vertebrate remains from a cave near Montagu, N.W. Tasmania

[2577]Orr, Phil C., 1956, Pleistocene man in Fishbone Cave, Pershing County, Nevada

[2613]Parmalee, Paul Woodburn, 1972, Pleistocene and recent faunas from the Brynjulfson Caves, Missouri

[2614]Parmalee, Paul Woodburn, 1969, Pleistocene and recent vertebrate faunas from Crankshaft Cave, Missouri

[2623]Peabody, Charles, 1867–1939, 1908, Pt. I: the exploration of Bushey Cavern near Cavetown, Maryland

[2636]Pemberton, Clive, 1972, Kents Caverns, home of prehistoric man and animals: the origin, story, and descriptive tour of the caves

[2637]Pemberton, Clive, 1947, The origin and story of Kents Cavern

[2638]Pemberton, Clive, 1964, The origin, story and descriptive tour of the caves

[2807]Repenning, Charles Albert, 1922–, 1989, The microtine rodents of the Cheetah Room fauna, Hamilton Cave, West Virginia, and the spontaneous origin of *Synaptomys*

[2808]Reynolds, Sidney Hugh, 1906, A monograph of the British Pleistocene Mammalia vol II part II: the bears

[2809]Reynolds, Sidney Hugh, 1902, A monograph of the Pleistocene mammalia vol II, part I: the cave hyaena

[2840]Roemer, Ferdinand Carl, 1818–1891, 1884, The bone caves of Ojcow in Poland

[2889]Sattler, Robert A., Apr 1989, Quantitative analysis of a late-Quaternary cave deposit, Porcupine River, Alaska

[2908]Schram, Frederick R., 1943–, 1970, Structural composition and dental variations in the murids of the Broom Cave fauna, late Pleistocene, Wombeyan Caves area, N.S.W., Australia

[2919]Schwartz, Jeffrey H., Jan 1994, A diverse hominoid fauna from the Late Middle Pleistocene breccia cave of Tham Khuyen, Socialist Republic of Vietnam

[2933]Semken, Holmes A., 1967, Mammalian remains from Rattlesnake Cave, Kinney County, Texas

[2984]Simpson, George Gaylord, 1902–, Feb 1949, A fossil deposit in a cave in St. Louis

[2989]Sinclair, William John, 1877–1935, 1904, Eucratherium: a new ungulate from the Quaternary caves of California

[2990]Sinclair, William John, 1877–1935, 1905, New mammalia from the Quaternary caves of California

[2995]Skinner, Morris F., 1942, The fauna of Papago Springs Cave, Arizona, and a study of Stockoceros; with three new Antilocaprines from Nebraska and Arizona

[3036]Smith, William Hovey, 1983, Geology of the Tennille Lime Sinks, Washington County, Georgia ; with an introduction to local geology

[3129]Stock, Chester, 1892–1950, 1930, 1932, Quaternary antelope remains from a second cave deposit in the Organ Mountains and further study of the quaternary antelopes of Shelter Cave, New Mexico

[3142]Stoneman, John (narr), 1990, The cave divers

[3163]Suzuki, Hisashi, 1912–(ed), 1973, The Palaeolithic site at Douara Cave in Syria: report of the fourth season of the Tokyo University Scientific Expedition to Western Asia

[3188]Taylor, Alisa Johanna, 1982, The mammalian fauna from the mid-Irvingtonian Fyllan Cave local fauna, Travis County, Texas

[3208]Thompson, Robert S., 1985, Paleoenvironmental investigations at Seed Cave (Windust Cave H- 45FR46), Franklin County, Washington

[3251]Turnbull, William D., 1973, Broom County *Cercartetus*, with observations on pygmy possum dental morphology, variation, and taxonomy

[3311]U.S. National Park Service, 1992, Paleontology

[3371]Wagner, Georg, 1883-1912, 1966, The bears cave of Erpfingen

[3427]Webb, William Snyder, 1882–, , The occurrence of the fossil remains of Pleistocene vertebrates in the caves of Barren County, Kentucky

[3436]Wells, R. T., 1975, World famous Fossil Cave Naracoorte

[3437]Wen-chung Pei, 1904–1982, 1940, The upper cave fauna of Choukoutien

[3458]Whidbey, Joseph, 1823, On some fossil bones discovered in caverns in the lime-stone quarries of Oreston

[3473]White, William Blaine, 1934–(ed), 1976, Geology and biology of Pennsylvania caves

[3528]Worthy, Trevor, 1993, Fossils of Honeycomb Hill

9.2 Paleoclimatology

[3162]Sutton, L. J., 1945, Meteorological conditions in caves and ancient tombs in Egypt

10   Regional Guides

William W. Mixon

14045 North Green Hills Loop
Austin, TX 78737

As few books about caves are truly world-wide in scope, virtually all cave books are regional studies in a sense. However, the primary topic covered here is books and other publications that are essentially catalogs of caves, that is, "Caves of ..." books, as these books do not fall readily into other categories such as geology or exploration. Books in this category are indexed under the term guidebooks, although many are not guidebooks in the strictest sense, as many do not contain all the information needed actually to visit the caves.

Catalogs of caves are of course a staple of cave-book publishing. Naturally, most catalogs in English cover caves in English-speaking countries. In Great Britain itself, caving guides have a long history, and it is there that they still flourish to the greatest extent. The earliest regional books in Great Britain are really mainly narrations of exploration, such as Baker and Balch's *The Netherworld of Mendip* (1907), but as more caves became known, these graded into more comprehensive regional surveys. Among those published before World War II are Balch's *Mendip-Cheddar, its Gorge and Caves* (1935, second edition 1947) and *Mendip, its Swallet Caves and Rock Shelters* (1937, second edition 1948). After the war, actual comprehensive guides to caves, especially in England, appeared, and some of these have gone through several commercially published editions. Notable is the Northern Caves series, by David Brook and a variety of coauthors. Five pocket-size volumes were published from 1972 through 1976, and each subsequently had updated editions. The five volumes become six when volume 4 was divided into 4A and 4B in the early 1980s. A new edition in three volumes (*Wharfedale and the Northeast*, 1988; *The Three Peaks*, 1991; *The Three Counties System and the North west*, 1991) totals about 850 pages. For each cave, the name, topographic-map coordinates, length and depth, and a short description are given. There are typically several caves per page, although the more significant systems have longer entries with fairly extensive descriptions of routes and so on. There are very few cave maps, and the fact that rigging instructions are still in terms of ladders and lifelines suggest that most of the information is reprinted from the earlier series. This format is fairly typical, and similar guides exist for Mendip (Irwin and Knibbs 1977, Irwin and Jarrett 1993, Barrington and Stanton 1970 and subsequent editions), Derbyshire or the Peak District (Ford 1974, Gill and Beck 1991 and various other editions), and so on. It is apparent that there is no stigma attached to publishing and distributing caving guides in the United Kingdom.

Scotland and Wales seem to be covered mainly by a variety of duplicated booklets written or published by Tony Oldham. Examples are *The Concise Caves of North Wales* (1989) and *The Caves of Scotland* (1975).

Ireland has two hardbound books, Tratman's *The Caves of North-West Clare, Ireland* (n.d. [1969]) and Self's *Caves of County Clare* (n.d. [1981]); the former is more than just a guide, having introductory material on the geology and geomorphology of the area and the caves. Coleman's *Caves of Ireland* (third edition, 1965) covers the entire island.

The British Cave Research Association's *Limestones and Caves of Britain* series deserves mention as examples of regional monographs of a scientific nature. So far, volumes on northwest England (Waltham, 1974), the Mendip Hills (Smith, 1975), the Peak District (Ford, 1977), and Wales (Ford, 1989) have appeared.

An unusual booklet is *Who Was Aveline, Anyway?*, by Richard Witcombe (1992), which is a glossary of cave and cave-feature names in Mendip, giving their origins. A neat idea.

In the United States, caves are usually cataloged by state. Unlike in the UK, most US cave guides have heavy emphasis on publishing as many cave maps as possible, sometimes consisting of little else. Many of the cave guides have been published by state geological surveys or similar agencies. The earliest state-wide example may be Stone's *Pennsylvania Caves* (1930, 1932). Other examples from major cave states include *Caves of Tennessee* (Barr, 1961), *Descriptions of Tennessee Caves* (Matthews, 1971), *Caves of Indiana* (Powell, 1961; that's me on the cover, although I didn't realize it for years), *Caverns of Virginia* (McGill, 1933), *Descriptions of Virginia Caves* (Holsinger, 1975), *Caverns of West Virginia* (Davies, 1958 and 1965), and *Caves of Missouri* (Bretz, 1956). This sort of publishing has pretty much come to an end, because beginning about 1960, cavers, especially in the western states where caves tend to be in unsupervised wilderness and less well known than in the east, began to oppose, for conservation and owner-relations reasons, publication of wild-cave information for uncontrolled distribution. Nevertheless, the most recent state-agency publications are from the west: Wyoming (Hill et al., 1976) and Montana (Campbell, 1978). These, however, give locations only to the square mile or even the topographic quadrangle.

In other cases, state cave catalogs have been published privately, either by individuals or caving organizations. Halliday's (1962) *Caves of California* was controversial when it was published, despite the small run of 500 copies and the fact that locations are given only to the square mile. California is the birthplace of the secrecy movement. That book has since become quite rare, even in libraries, from which most copies have been stolen, one suspects by cavers seeking to suppress the information. Another notable private publication is Douglas's (1964) *Caves of Virginia*, a 761-page hardbound book published by the author (as the "Virginia Cave Survey"). Parris's (1973) *Caves of Colorado* caused a general uproar because it was published by a commercial publisher meaning to give it the widest possible distribution.

Caving groups, mainly more or less formally organized state cave survey organizations, have produced cave catalogs for many states, either as books or as issues of a serial or periodical, the latter often being devoted to single counties within the state. The books are often informal publications, frequently, sometimes annually, updated and with very limited distribution. State catalogs are typically duplicated and often not really bound and often consist of little more than an up-to-date computer printout showing little besides name, location, length, and depth, plus a set of all available maps. An example is the *Alabama Caves* series. The 1965 edition (Tarkington, *et al.*) lists 617 caves and includes maps, one per page, of the majority of them. *Alabama Caves, 1980* (Varnedoe, 1981), published as an unbound photocopy, lists over 1,700 caves, and the stack of maps is four inches thick. Clearly, such publications can become very unwieldy, but of course they are still valuable to those working in the area. Notable serial publications include issues of *Missouri Speleology*, many of which are county surveys, the bulletins of the West Virginia Speleological Survey, and the bulletins of the Mid-Appalachian Region of the National Speleological Society. The Texas Speleological Survey published county surveys as issues of a serial with three volumes published in the 1960s and early 1970s; subsequently, occasional unnumbered county surveys have appeared. Publications such as these have not caught on in the far west, partly as a matter of philosophy and partly for lack of enough caves to warrant them. The extent to which locations are given is variable, even though the publications are not widely available outside the caving community. Only cavers in the east and southeast of the United States still seem willing to publish exact locations (usually usable only with a topographic map).

The National Speleological Society's Guidebook Series consists of guidebooks published in conjunction with each annual convention. In recent years, these have been good overall introductions to the caves of the area, with numerous cave descriptions and maps, but very seldom with location information sufficient for them really to serve as guidebooks. Nevertheless, many of them are the best available references on caves, including some of their scientific aspects, of a state or region. They are not meant to be comprehensive catalogs, however.

A catalog of Australian caves and karst features has been published by the Australian Speleological Federation. *Australian Karst Index 1985* (Matthews, 1985) is essentially a printout of a computer database. There is a list of known maps, but no maps are printed. Peter Crossley has compiled cave atlases for New Zealand (North Island, 1988; South Island 1990). These are pure map books. From the other side of the world, Alan Fincham's *Jamaica Underground* (1977) contains brief descriptions and available maps of all the caves on that island. And Gebauer's *Caves of India and Nepal* (n.d. [1983]) is a nice little book cataloging the few known caves in that large area.

Among non-English speaking countries with significant concentrations of caves, Mexico probably has the most extensive coverage in English, in the newsletters and bulletins of the Association for Mexican Cave Studies. Bulletin 1, *Caves of the Inter-American Highway* (Russell and Raines, 1967) is a path-breaking guidebook to the caves of northern Mexico. It is now rare and much sought after. Most AMCS publications have emphasized exploration or biology, but Bulletin 7 (Stone and Jameson, 1977) is a guide to caves of the San Juan Plateau, and a survey of the caves of the Yucatan by Reddell is in Bulletin 6 and was also printed separately (1977). Much older surveys of Yucatan caves were motivated by archaeology (Mercer, 1896, reprint 1975) or aquatic biology (Pearse, et al., 1938).

Naturally, most of the literature on wild caves around the world appears scattered among almost innumerable club publications. Use of *Current Titles in Speleology* or *Speleological Abstracts* is essential to track down most of this literature. Most local newsletters from the US are gleaned for valuable material for the annual *Speleo Digests*; a multi-year index of this publication, covering 1956 through 1980, has been produced (National Speleological Society. Speleo Digest (Sasowsky and Wheeland), 1991).

Any discussion of cave catalogs must mention, even though they are not regional because they cover the entire world, two highly useful references. Middleton and Waltham's *Underground Atlas* (1986) contains a summary of the karst in every country in the world, even if that summary is "of no speleological interest." Maps indicating areas of karst within many of the countries are included, and there are small-scale maps of fifty of the world's most significant caves. And the monumental *Atlas of the Great Caves of the World* (Courbon et al., 1989) contains maps and brief descriptions of the longest and deepest caves in all the cavernous countries in the world. It is a translation and minor updating of Courbon and Chabert's *Atlas des Grandes Cavités Mondiales*, (1986).

References not in Main Bibliography

Courbon, Paul and Claude Chabert. 1986. *Atlas des Grandes Cavités mondiales.* Publie avec le concours de l'Union Internationale de Speleologie et de la Federation Francaise de Speleologie, S.N.I.P./Offset, 255pp.

10.1 General

[1151]Gardner, James, 1982, An inventory and evaluation of Missouri state parks cave resources

10.2 Catalogs

[619][Cave Research Group], Dec 1955, Fauna collected from caves as recorded in the C.R.G. fauna records: part I (1938-39)
[620][Cave Research Group], Oct 1956, Fauna collected from caves as recorded in the C.R.G. fauna records: part II (war years 1940-46 and 1945-46)
[622][Cave Research Group of Great Britain], Nov 1972, Hypogean fauna and biological records 1970-71
[844]Dayton, Gordon O. (comp), Jul 1981, The caves of Mifflin County, Pennsylvania
[922]Dunkley, John R., 1994, Thailand caves catalogue
[972]Ellis, Martin, 1992, The caves of Tenerife
[1066]Fisher, John L. (ed), 1958, Caves of the Watertown Area, New York state
[1238]Grampian Speleological Group. Library, 1991, Grampian Speleological Group library catalogue
[1399]Hazelton, Mary (ed), 1965, British hypogean fauna and biological records of the Cave Research Group part IX (1963)
[1400]Hazelton, Mary, Oct 1958, Fauna collected from caves as recorded in the C.R.G. fauna records: part III (1947)
[1401]Hazelton, Mary, Feb 1959, Fauna collected from caves, mines and wells as recorded in the C.R.G. fauna records: part IV (1948-1949)
[1402]Hazelton, Mary, Feb 1960, Fauna collected from caves, mines and wells as recorded in the C.R.G. fauna records: part V (1950-1953)
[1403]Hazelton, Mary, Sep 1960, Fauna collected from caves, mines and wells as recorded in the C.R.G. fauna records: part VI (1954-1955-1956)
[1404]Hazelton, Mary, Nov 1961, Fauna collected from caves, mines and wells as recorded in the C.R.G. fauna records: part VII (1957-1959)
[1405]Hazelton, Mary, Sep 1963, Fauna collected from caves, mines and wells as recorded in the C.R.G. fauna records: part VIII (1960-1962)
[1406][Hazelton, Mary (ed)], Sep 1967, Hypogean fauna and biological records 1964-1966
[1407][Hazelton, Mary (ed)], Dec 1968, Hypogean fauna, biological records 1967
[1408][Hazelton, Mary (ed)], Mar 1970, Hypogean fauna, biological records 1968
[1428]Hempel, John Charles, 1949–(comp), 1969, Come crawl into the caves of western Pennsylvania
[1489]Holler, Cato Oliver, 1944–(ed), 1975, North Carolina cave survey vol. 1 no. 1
[1490]Holler, Cato Oliver, 1944–(ed), 1979, North Carolina cave survey vol. 2
[1592]Ibberson, Dale (ed), Jan 1983, The caves of Perry County, Pennsylvania
[1662][International Union of Speleology], 1989, Caves in fine art exhibition
[1671]Irwin, David J. (comp), Aug 1995, A catalogue of ephemera relating to Wookey Hole
[2039]Mansfield, Kay Patricia Russell (comp), Mar 1968, Caving periodicals in the Central Collection of Caving Publications
[2040]Mansfield, Raymond Walter, 1965, Mendip cave bibliography and survey catalogue, January, 1901—December, 1963
[2185]Mills, M. T., Jul 1990, The subterranean wonders of Sardinia
[2196]Missouri Division of Geological Survey and Water Resources, Mar 1952, Partial catalog of caves in Missouri
[2771]Reddell, James Russell, 1938–, 1995, Catalogue, bibliography, and generic revision of the order Schizomida (Arachnida)
[2772]Reddell, James Russell, 1938–(ed), nd, The caves of Langtry
[2773]Reddell, James Russell, 1938–(ed), Mar 1962, The caves of Bexar County
[2779]Reddell, James Russell, 1938–(ed), Jun 1963, The caves of northwest Texas
[2802]Reich, J. R. (comp), 1969, Caves of Pennsylvania's Piedmont region
[2870]Russell, William Hart, 1937–, Jul 1993, The Buttercup Creek karst: Travis and Williamson Counties, Texas: geology, biology, and land development
[2952]Shaw, Trevor Royle, 1928–(comp), 1967, Cave illustrations before 1900: a catalogue of non-photographic illustrations of caves
[3007]Smeltzer, Bernard L., Oct 1964, Caves of the southern Cumberland valley
[3085]Speece, Jack H. (comp, ed), Jul 1972, The caves of Blair County, Pennsylvania
[3086]Speece, Jack H. (ed), Feb 1969, Fulton County, Pennsylvania
[3363]Vineyard, Jerry Daniel, 1935–(comp), , Catalogue of the caves of Missouri
[3506]Wilson, B. E. (comp), Nov 1965, The Bradford Pothole Club library list

10.3 Guidebooks

[7]Adam, John, 1886, Aberbrothok illustrated
[9]Adamkó, Péter, 1989, The caves of Budapest
[79]Anon, nd, [Indiana cave list]
[94]Anon, 1986, The Caverns, rocks and ruins of America's Southwest
[122]Anon, [1994], The great Dan-yr-ogof day out
[133]Anon, 1981, Guilin: the crown of superb landscapes in China
[157]Anon, 1962, Mexico's caves and caverns
[175]Anon, c1950, Poole's Cavern: Buxton
[211]Anon, [1947], The weird wonders of Wookey Hole Caves
[219]Anon, [1938], Wookey Hole Caves: Wells, Somerset
[232]Ashbrook, Bert, 1995, Caves in the Richlands area of Greenbrier County, West Virginia
[233]Ashbrook, Bert, Dec 1990, The Caves of Northumberland County, Pennsylvania
[244]Atkinson, T. C. , 1977, Caves and karst of southern England and South Wales: guidebook for the International Congress of Speleology at Sheffield, 1977
[257]Australian Speleological Expedition, 1976, Caves of New Caledonia: report of the 1975 Australian Expedition
[278]Bailey, Thomas L., Apr 1918, Report on the caves of the eastern highland rim and Cumberland Mountains
[282]Baker, Ernest Albert, 1869–1941, [1903], Moors, crags & caves of the High Peak and the neighbourhood
[288]Bakó, Tamás, 1989, Paleokarst in Hungary
[296]Balch, Herbert Ernest, 1937, Mendip, its swallet caves and rock shelters
[306]Ballard, Jim, 1974, Guide to the sporting caves, potholes and mines of Derbyshire
[309]Balogh, K., 1989, Karst hydrological and speleological features
[326]Barr, Thomas Calhoun, 1931–, 1972, Caves of Tennessee
[330]Barrett, Charles Leslie, 1879–, 1944, Australian caves, cliffs, and waterfalls
[332]Barrington, Nicholas Robert (comp), 1962, The caves of Mendip
[333]Barrington, Nicholas Robert, 1970, The complete caves of Mendip
[334]Barrington, Nicholas Robert, 1977, Mendip: the complete caves and a view of the hills
[338]Bartrop, Richard N., 1987, Cave Diving Group: Derbyshire sump index 1987
[372]Bedford, Bruce L., 1942–, 1985, Underground Britain: a guide to the wild caves and show caves of England, Scotland and Wales
[373]Behrens, Georg Henning, 1662-1712, 1730, The natural history of Hartz-Forest, in His Majesty King George's German dominions. Being a succinct account of the caverns, lakes, springs, rivers, mountains, rocks, quarries, fossils, castles, gardens, the famous pagan idol Pustrich or spit-fire, dwarf-holes, etc....in the said forest: with several useful and entertaining physical observations
[431]Bonin, Dan (ed), 1980, Hodag hunt 1980: guidebook September 12-14 1980
[444]Botto, Larry (comp), [1993], A partial guidebook to the caves of the Berkshires: 1993 spring N.R.O. hosted by the Berkshire Hills Grotto (May 14th, 15th, 16th)
[451]Boyd, Susan (comp), 1966, A guide to major Wisconsin caves
[468]Bretz, J Harlen, 1882–1981 , 1961, Caves of Illinois
[469]Bretz, J Harlen, 1882–1981 , 1956, Caves of Missouri
[499]Brook, Alan, 1945–, 1991, Northern caves volume 2: the three peaks
[500]Brook, Alan, 1945–, 1975, Northern caves volume three: Ingleborough
[501]Brook, Alan, 1945–, 1976, Northern caves volume two: Penyghent and Malham
[505]Brook, David B., 1944–, 1983, Northern caves volume 4a: Scales Moor and King Scale
[506]Brook, David B., 1944–, 1983, Northern caves volume 4b: Leck and Casterton Fells
[507]Brook, David B., 1944–, 1974, Northern caves volume five: the Northern Dales
[508]Brook, David B., 1944–, 1975, Northern caves volume four: Whernside and Gragareth
[509]Brook, David B., 1944–, 1972, Northern caves volume one: Wharfedale and Nidderdale
[510]Brook, David B., 1944–, 1991, Northern Caves: volume 3: the three counties system and the north-west
[511]Brook, David B., 1944–, 1988, Northern caves: volume1. Wharfedale and the North-East
[513]Brooks, S. J. (ed), Mar 1995, Caving in the abode of the clouds: the caves and karst of Meghalaya, north east India
[555]Bunnell, David Edward, 1952–, 1993, Sea caves of Anacapa Island
[556]Bunnell, David Edward, 1952–, 1988, Sea caves of Santa Cruz Island
[557]Bunton, Stephen, 1984, Vertical caves of Tasmania: a caver's guidebook
[564]Burkhardt, Rudolf, 1973, The Speleological Club in Brno 1945-1973
[590]Campbell, Newell Paul, 1938–, 1978, Caves of Montana
[603]Carter, R. L., Aug 1994, Cave Diving Group: Peak District sump index 1994
[604]Cartwright, Roger, 1946, A guide to the Craven Dales
[625][Central Indiana Grotto of the National Speleological Society], Jun 1973, Selected caves of Orange county
[626]Čerňanská, Miroslava (ed), 1973, Welkome (sic) to our Slovak caves

10.3. GUIDEBOOKS

[637]Chapman, J. Roy, 1963, The caves of Georgia

[639]Charnock, S. J., Feb 1994, Underneath the arches: caving in the RibbleHead District

[649]Chillagoe Caving Club, nd, Broken River karst: a speleological field guide North Queensland Australia

[651]Chillagoe Caving Club (comp), 1982, A speleological field guide of the towers and caves of Chillagoe - Munanga - Rookwood areas in Far North Queensland, Australia

[652]Chillagoe Caving Club (comp), Jul 1988, A speleological field guide of the towers and caves of Chillagoe - Munanga - Rookwood areas in Far North Queensland, Australia. Supplement for caves CH301-CH400

[659]Cigna, Arrigo A. (ed), 1966, Caving in Italy

[661]Clapham, Christopher S., 1967, The Caves of Sof Omar

[664]Clarke, Owen (comp), 1990, Holiday caving in Mallorca

[669]Clyne, Patricia Edwards, 1980, Caves for kids: in historic New York: illustrated with maps and photographs

[679]Coleman, John Christopher, May 1965, The caves of Ireland

[697]Cook, Holly (ed), , 5th annual Hog-Fest: Lickford Valley Campgrounds, Corydon, Indiana, May 1-5, 1992

[720]Cosgrove, Cornelius Burton, 1875–, 1947, Caves of the Upper Gila and Hueco areas in New Mexico and Texas

[725]Courbon, Paul, 1936–, 1989, Atlas of the great caves of the world

[731]Cox, Guy, Apr 1973, Caves of the Catabrian Mountains, north-west Spain

[737]Crawford, Harriet E. W., 1979, Subterranean Britain: aspects of underground archeology

[738]Crawford, Nicholas Charles, 1942–, 1982, Guidebook to karst and caves of Tennessee: emphasis on the Cumberland plateau escarpment region and guidebook to karst and caves of the Ozark Region of Missouri and Arkansas: prepared for the Eighth International Congress of Speleology, Bowling Green, Kentucky, U. S. A. July 18 to 24 1981

[772]Cullinan, Mike, Jun 1975, The caves of Huntingdon County, Pennsylvania

[795]Dalton, Richard F., 1976, Caves of New Jersey

[796]Dalton, Richard F., 1970, New Jersey caves in brief

[805]Dasher, George R., 1952–(ed), Aug 1994, The caves and karst of the Buckeye Creek Basin, Greenbrier County, West Virginia

[815]Davey, Adrian G., 1986, Victorian caves and karst: strategies for management and catalogue: a report to the Caves Classification Committee, Department of Conservation, Forests and Lands /

[822]Davies, Melvyn, Nov 1966, Cave sump index; South Wales

[823]Davies, Melvyn, 1991, The caves of the South Gower coast: an archaeological assessment

[825]Davies, Phillip, Apr 1957, An index of submerged cave passages; sumps

[827]Davies, William Edward, 1917–, Jul 1958, Caverns of West Virginia

[828]Davies, William Edward, 1917–, 1965, Caverns of West Virginia (supplement)

[829]Davies, William Edward, 1917–, 1961, The caves of Maryland

[833]Davis, Harold (ed), 1966, Caves of Schoharie County, New York

[843]Dayton, Gordon O. (comp), Feb 1979, The caves of Centre County, PA

[858][Derbyshire Caving Association], 1987, Derbyshire caving 1987: the handbook of Derbyshire caving

[879]Dobson, Phil (comp), Aug 1981, The Cave Diving Group annotated ten year cumulative index to the cave dives for the years 1969-1978

[888]Dougherty, Percy H. (ed), 1985, Caves and karst of Kentucky

[890]Douglas, Henry Hulbert, 1906–1989, 1964, Caves of Virginia

[896]Dowswell, Peter N. F., Nov 1980, Limestone caves of Scotland, vol IV: the caves of Schichallion

[898]Drew, David Phillip, 1943–, 1984, Aillwee Cave and the caves of the Burren

[916]Dunkley, John R., Jan 1995, The caves of Thailand

[917]Dunkley, John R., 1978, Caves of the Nullarbor; a review of speleological investigations in the Nullarbor Plain, Southern Australia

[923]Dunkley, John R. , 1986, Caves of north-west Thailand: report of the Australian speleological expeditions, 1983-1986

[924]Dunlop, Bruce Thomas, , Jenolan Caves

[931]Dyson, H. Jane (ed), 1982, Wombeyan Caves

[949]Ehr, Bob (prod), 1974, Guide to the caves of Wisconsin

[951]Elliot, Dave, 1946–, 1975, Caves of Northern Derbyshire: part 1: the Eldon Hill area

[952]Elliot, Dave, 1946–, Mar 1975, Caves of Northern Derbyshire: part 2: Giants-Oxlow System

[953]Elliot, Dave, 1946–, May 1975, Caves of Northern Derbyshire: part 3: Perryfoot/Coal Pit Hole

[954]Elliot, Dave, 1946–, Feb 1976, Caves of Northern Derbyshire: part 4: Rushup Edge swallets

[955]Elliot, Dave, 1946–, nd, Caves of Northern Derbyshire: part 5: Treak Cliff Hill

[956]Elliot, Dave, 1946–, 1987, Single rope technique rigging guide

[962]Elliott, William Rawleigh, 1946–(ed), Dec 1985, A field guide to the caves of Kendall County

[975]Ellis, Ross Andrew, 1940–(ed), 1972, Bungonia Caves

[991]Engelbrecht, N. P., Aug 1972, Caving in the Transvaal

[995]Ensminger, Scott A., 1983, Canadian caves of the Niagara Gorge

[996]Ensminger, Scott A., 1987, The caves of Niagara County, New York

[1007]Eszterhás, István, 1989, The caves on the Tés Plateau and in the Bakony Mountains

[1040]Fells, Richard, 1989, A visitor's guide to underground Britain: caves, caverns, mines, tunnels, grottoes

[1041]Felter, Brian, 1979, The Virginia cave locator series: book 1: Frederick Co. and Clark Co

[1042]Felter, Brian, 1979, The Virginia cave locator series: book 2: Warren Co

[1043]Felter, Brian, 1980, The Virginia cave locator series: book 3: Shenandoah Co

[1044]Felter, Brian, 1980, The Virginia cave locator series: book 4: Page Co

[1045]Felter, Brian, 1982, The Virginia cave locator series: book 5: Rockingham Co

[1046]Felter, Brian, 1982, The Virginia cave locator series: book 6: Augusta Co

[1051]Fetvedt, John E. (ed), Aug 1984, Mystery Cave area: MSS Corn Feed August 1984

[1052]Fetvedt, John E. (ed), Apr 1984, North Country Region: Spring 1984, LaCrosse, Wisconsin

[1055]Field, Henry, 1902–, 1955, Caves and rockshelters in northern Iraq.

[1056]Fieseler, Ronald G. (ed), Apr 1975, The caves of Brewster and western Pecos Counties

[1060]Fincham, Alan Goldworth, 1933–, 1977, Jamaica underground: a register of data regarding the caves, sinkholes and underground rivers of the island

[1069]Flindall, Roger, 1976, The caverns and mines of Matlock Bath: 1 Nestus Mines: Rutland and Masson caverns

[1072]Florida Cave Survey, 1962, Cave locations listing

[1086]Ford, Trevor David, 1925–(comp), 1964, Caves of Derbyshire

[1090]Ford, Trevor David, 1925–, 1971, Ingleborough Cavern

[1092]Ford, Trevor David, 1925–, 1967, Ingleborough Cavern including: notes on Gaping Gill and the geology of Ingleborough

[1094]Ford, Trevor David, 1925–(ed), 1989, Limestones and caves of Wales

[1105]Foster, David G., 1991, Guidebook: 1991 National Cave Management Symposium, Bowling Green, KY Oct. 23-26

[1123]Franz, Richard, 1971, Caves of Maryland

[1135]Gádoros, Miklós, 1989, Speleotherapic and speleclimatological centres

[1159]Garrard, Ian M., 1983, Caving

[1164]Garton, Emmel Ray, 1950–, Jun 1976, Caves of North Central West Virginia

[1166]Gascone, William (ed), 1982, Caves of the North Outcrop

[1168]Gebauer, Herbert Daniel, 1983, Caves of India & Nepal: including the results of Speläologische Südasien Expedition 1981/82

[1172]Gem State Grotto of the National Speleological Society, 1968, Caves of the gem state

[1182]Georgia Speleological Survey, Oct 1991, Georgia Speleological Survey mapbook

[1183]Gerboth, David W., Aug 1989, A guidebook to the caves and sinks of the Spring Valley Caverns area, Fillmore County, Minnesota: 1989 Cornfeed Minnesota Speleological Survey

[1203]Gill, David W. (comp), 1991, Caves of the Peak District

[1227]Gordon, George Byron, 1870–1927, 1898, Caverns of Copan, Honduras: report on explorations by the museum, 1896-97

[1245]Gray, Eric (comp), 1985, Devonshire sump index

[1253]Green, John, nd, The Story of Cave Hill: Grottos, VA

[1259]Grenfell, Harold, 1971, The caves of Gower

[1263]Griffiths, Julian (comp), 1981, Cave Diving Group: northern sump index

[1264]Griffiths, Paul, 1975, Horne Lake Wonder Caves: a guide to Horne Lake Caves Park

[1275]Guernsey Press, 1931, Illustrated "Press" guide to the caves and cliffs of Sark: with map and plans of roads.

[1286]Gulden, Robert, Dec 1984, Caves of the eastern panhandle of West Virginia

[1298]Gurnee, Russell Hampton, 1922–1995, 1980, Gurnee guide to American caves: a comprehensive guide to the caves in the United States open to the public

[1324]Halliday, William R., 1926–, 1962, Caves of California: a special report of the Western Speleological Survey in cooperation with the National Speleological Society, June 1962

[1325]Halliday, William R., 1926–, 1963, Caves of Washington

[1332]Halliday, William R., 1926–, Aug 1991, Introduction to Hawaiian caves: field guide for the 6th international symposium on volcano-speleology: Hilo, HI

[1336]Halliday, William R., 1926–, 1961, A preliminary report on the caves of eight northern California counties

[1338]Hamer, John Lewis, 1946, The falls and caves of Ingleton

[1339]Hamer, John Lewis, 1951, The moors and fells of Ingleton

[1355]Hannan, Thomas, 1926, The beautiful Isle of Mull with Iona and the Isle of Saints

[1357]Hardcastle, Raymond A., Apr 1977, Caves and shelters in Bonanza King Canyon

[1365]Harris, Yvonne Dorothy, 1935–, 1990, The caves of Chlingitani

[1387]Hauer, Peter Marshall, 1945–1975, 1969, Caves of

10.3. GUIDEBOOKS

Massachusetts
[1410]Hazslinszky, Tamás (ed), 1975, International conference Baradla 150: field-trip guide: Budapest-Aggtelek, 26-29.08.1975
[1411]Hazslinszky, Tamás (auth, ed, trans), 1989, Show caves of Hungary
[1427]Hempel, John Charles, 1949–(comp), 1975, Caves of Monroe County, W. Va
[1435]Henderson, Kent, 1954–, 1987, Princess Margaret Rose Caves - Western Victoria
[1443]Hennessey, David, 1847–, 1900, The Caves of Shend
[1444]Hennings, Kevin (ed), c1971, 1971 guidebook: Upper Mississippi conference UMAC Sept. 17, 18, 19 at Eagle Cave
[1461]Hill, Chris, 1947–, 1976, Caves of Wyoming
[1463]Hills Speleology Club Ltd (prep), 1992, A guide to Mount Fairy Caves: a speleological field guide to the limestone and caves at Mount Fairy
[1477]Hoffmann, Paul, 1974, The caves of north central Morgan County, Alabama: a guide for the 1974 SERA Winter Business Meeting of the National Speleological Society at Decatur, Alabama, February 1974
[1479]Hogberg, Rudolph K., 1967, Guide to the caves of Minnesota
[1485]Holland, Eric George, 1955, A guide to Batu caves
[1486]Holland, Eric George (comp), 1967, Underground in Furness: a guide to the geology, mines, potholes and caves
[1487]Holland, Eric George, 1960, Underground in Furness: a guide to the geology, mines, potholes and caves
[1495]Holsinger, John Robert, 1934–, 1975, Descriptions of Virginia caves
[1515]Hooper, John, Apr 1985, Caves in Buckfastleigh Quarries
[1521]Horne, Peter (comp), Aug 1993, Lower South East cave reference book: an illustrated catalogue of the registered caves, sinkholes and associated karst features of the Lower South East Region of South Australia
[1535]Hovey, Horace Carter, 1833–1914, 1970, Celebrated American caverns: especially Mammoth, Wyandot, and Luray: together with historical, scientific, and descriptive notices of caves and grottoes in other lands
[1536]Hovey, Horace Carter, 1833–1914, 1882, Celebrated American caverns: especially Mammoth, Wyandot, and Luray: together with historical, scientific, and descriptive notices of caves and grottoes in other lands
[1548]Howell, C., Jul 1973, A Burrington cave atlas
[1576]Huntsville Grotto of the National Speleological Society, 1967, 30 favorite Alabama caves
[1594]India. Public Works Department, nd, The caves of Elephanta
[1595]Indiana Cave Capers. Guidebook (1971), 1971, 1971—18th Cave Cavers: location Bedford
[1596]Indiana Cave Capers. Guidebook (1974), 1974, 1974 Cave Capers guidebook: for Harrison & Crawford counties
[1597]Indiana Cave Capers. Guidebook (1975), 1975, Guidebook: Cave Capers 1975
[1598]Indiana Cave Capers. Guidebook (1976), 1976, Guidebook: Cave Capers 1976
[1599]Indiana Cave Capers. Guidebook (1977), 1977, 24th annual CIG Cave Capers
[1600]Indiana Cave Capers. Guidebook (1978), 1978, Cave Capers 1978
[1601]Indiana Cave Capers. Guidebook (1979), 1979, Cave Capers 1979 guidebook: Milltown, Indiana
[1602]Indiana Cave Capers. Guidebook (1980), 1980, 27th annual Cave Capers: 1980 guidebook
[1603]Indiana Cave Capers. Guidebook (1983), 1983, Cave Capers, 1983
[1604]Indiana Cave Capers. Guidebook (1987), 1987, Wander Indiana underground: 1987 Cave Capers, Spring Valley
[1605]Indiana Cave Capers. Guidebook (1989), 1989, Indiana Cave Capers guidebook: Hickory Hill Campground, Owen County, Indiana July 14-16, 1989
[1607]Indiana. Division of Geology, 1939, Guide[s] to the Indiana caverns
[1609]Innes-Smith, Robert, 1976, Castleton and its caves
[1631]International Congress of Speleology (9th:1977: Sheffield, England), 1977, Caves and karst of Ireland: guidebook for the International Congress of Speleology at Sheffield, 1977
[1632]International Congress of Speleology (9th:1977: Sheffield, England), 1977, Caves and karst of the Yorkshire Dales: guidebook for the International Congress of Speleology at Sheffield 1977
[1644]International Speleologist, Jun 1962, Caves of Okinawa
[1657]International Symposium on Vulcanospeleology (5th: 1988: Izunagaoka Hot Springs, Japan), 1988, Excursion guide book: 5th international symposium on vulcanospeleology: [pre-activity: November 4-9, 1988: post-activity: November 13-20, 1988]
[1668]Irving, David, nd, A guide to the caves in the Oak Ridge-Knoxville area
[1678]Irwin, David J., 1977, Mendip underground: a caver's guide
[1698]James, Julia M., 1980, Caves and karst of the Muller Range: report of the 1978 Speleological Expedition to the Atea Kananda, Southern Highlands, Papua New Guinea
[1712]Jeffery, R. G., nd, Caves of Plymouth and District
[1714]Jeffreys, Alan Lawrence, 1940–, Apr 1972, The Caves of Assynt
[1715]Jeffreys, Alan Lawrence, 1940–, Jan 1984, Scotland underground: a caver's guide to the limestone caves

of Scotland

[1717]Jenkins, David William, 1963, Caves in Wales and the Marches

[1719]Jennings, Jesse David, 1909–, Oct 1957, Danger Cave

[1729]Jennings, Joseph Newell, 1916–1984, 1987, Wee Jasper Caves

[1731]Jessey, Gary D., 1973, Letcher County's Pine Mountain caves: an in-depth pictorial essay of Letcher County's Pine Mountain caves and caverns

[1743]Jones, Gareth Llwyd, 1988, Cave Diving Group: Irish sump index

[1744]Jones, Gareth Llwyd, 1974, The caves of Fermanagh and Cavan

[1745]Jones, Gareth Llwyd, 1987, Irish sump index

[1749]Jones, Keith, 1992, Caves and mines of the Sychrhyd Gorge

[1750]Jones, R., 1974, Wilderness caves of the Gordon - Franklin River System

[1752]Jones, W. F., 1962, Some caves of Cyprus

[1753]Jones, Walter Bryan, 1895–1977, 1968, Caves of Madison County, Alabama

[1754]Jones, Walter Bryan, 1895–1977, 1960, Caves of Morgan County, Alabama

[1767]Kahrau, Wolfgang, 1972, Australian caves and caving

[1771]Kastning, Ernst H., 1944–, Jun 1988, Caverns of the Shawangunk (Shon-gum) and its environs, Southeastern New York

[1772]Kastning, Ernst H., 1944–, Apr 1989, Caves and karst of the New River Valley, Virginia: guidebook for a geologic fieldtrip, eighth annual New River symposium, Radford, Virginia

[1773]Kastning, Ernst H., 1944–, Mar 1991, Caves, karst, and environmental impact in the new river drainage basin of Virginia and West Virginia: guidebook for a geologic fieldtrip, Appalachian karst symposium, Radford, Virginia, 23 March 1991

[1782]Kehret, Roger Alan, 1942–, 1988, Minnesota karst country tours

[1796]Kentucky Speleofest. Guidebook (1972), 1972, Guidebook to the Kentucky Speleofest, Meade County, Kentucky

[1797]Kentucky Speleofest. Guidebook (1973), 1973, Guidebook to the Kentucky Speleofest, Warren County, Kentucky

[1798]Kentucky Speleofest. Guidebook (1974), 1974, Kentucky Speleofest guidebook: selected caves for the annual summer field conference, the Louisville Grotto, NSS: Squire Boon Caverns, Indiana

[1799]Kentucky Speleofest. Guidebook (1975), 1975, Kentucky Speleofest guidebook: annual summer field conference, the Louisville Grotto NSS

[1800]Kentucky Speleofest. Guidebook (1976), 1976, Guidebook to the 5th annual Kentucky Speleofest. Annual summer field conference of the Louisville Grotto NSS: Camp Carlson, Meade County, Kentucky

[1801]Kentucky Speleofest. Guidebook (1977), 1977, Guidebook to the 1977 Kentucky Speleofest, Cedar Hill Camp, Edmonson Co., Ky

[1802]Kentucky Speleofest. Guidebook (1978), 1978, Guidebook to the 1978 Kentucky Speleofest: Camp Carlson, Meade County, Kentucky

[1803]Kentucky Speleofest. Guidebook (1979), 1979, Guidebook: Kentucky Speleofest '79

[1804]Kentucky Speleofest. Guidebook (1980), 1980, Guidebook to the 1980 Kentucky Speleo-fest: annual summer field conference, Pulaski Co. Park, Ky

[1805]Kentucky Speleofest. Guidebook (1987), 1987, Guidebook for the 1987 Kentucky Speleofest: Renfro Valley KOA, Kentucky, May 22-25, 1987

[1806]Kentucky Speleofest. Guidebook (1989), 1989, 1989 Speleofest guidebook. May 26-29, 1989, Sandhill Campground, Whitley City, Kentucky

[1807]Kentucky Speleofest. Guidebook (1997), 1997, 1997 Speleofest: Cadiz, Ky, Trigg County Recreational Complex, Cadiz, Kentucky

[1825]Klamath Mountains Conservation Task Force, 1977, A preliminary report on the caves of the Marble Mountain Wilderness

[1831]Knight, E. Leslie, 1974, Caves of Mississippi

[1839]Knutson, Richard Stephen, nd, The caves of Deschutes County, Oregon

[1846]Kollár, Attila K. (comp), 1989, Subaquatic caves in Hungary for divers

[1864]Kranzel, Rich, 1983, Caves of Bucks County, Pennsylvania

[1865]Kranzel, Rich, Apr 1985, Caves of Lehigh County, Pennsylvania

[1870]Kunath, Carl Edwin, 1940–(ed), Oct 1995, The caves of Carta Valley, Edwards and Val Verde Counties, Texas

[1871]Kunath, Carl Edwin, 1940–(ed), 1978, The caves of McKittrick Hill, Eddy County, New Mexico: history and results of field work 1965-1976

[1872]Kunath, Carl Edwin, 1940–(ed), Apr 1968, The caves of the Stockton Plateau, Texas

[1879]La Rock, Edward J., Dec 1976, The caves of Snyder County, Pennsylvania

[1904]Larson, Charles Victor, 1928–, 1975, Caves of Oregon

[1910]Latrobe, G., 1964, Guide to the Coast, Caves, and Bays of Sark

[1911]Latrobe, G., 1968, A revised and enlarged guide to Sark containing a detailed description of the coast, caves and bays

[1917]Lawson, T. J., 1988, Caves of Assynt

[1932]Lénárt, László, 1989, Caves in the Bükk Mountains

10.3. GUIDEBOOKS

[1940]Lewis, Ian D., 1976, South Australian cave reference book

[1941]Libra, Bob, [1990], 1990 Friends of karst field trip: the Big Spring Basin, northeast Iowa

[1942]Lillis, Brian, nd, Poul Na Gollor Cave at Inch Ennis, Co Clare

[1948]Little, William Henry, 1919–1992, 1970, Coming to South Wales caves?

[1969]Love, Douglas L., 1972, The spelunker's guide to the caves of the Garrison Chapel Valley

[1970]Love, Douglas P., 1977, The spelunker's guide to the caves north of Campbellsburg, Indiana

[1971]Love, Douglas P., 1973, The spelunker's guide to the caves North of Campbellsburg, Washington County, Indiana

[2020]Maih, Lindsay (ed), 1993, The New Zealand Cave atlas: volume 2: South Island

[2041]Mansfield, Raymond Walter, 1978, Somerset sump index

[2044]Mansfield, Raymond Walter, Apr 1964, Sump index: section 1 (Somerset)

[2053]Marshall, Des, 1994, Mallorca Caves; an interim guide

[2055]Marshall, Wayne (ed), 1989, Sixth annual Florida caver's re-union (Cave Cavort): April 7-9, 1989, Hog Island Campground, Withlacoochee State Forest

[2064]Martin, Paul Sidney, 1899–, 1954, Caves of the Reserve area

[2066]Martin, Ronald L., 1972, Cave development in the Bull Creek drainage basin of southwest Missouri

[2089]Matthews, Larry Edwin, 1946–, 1971, Descriptions of Tennessee caves

[2093]Matthews, Peter Gahan, 1938–(ed), Jan 1968, Speleo handbook

[2118]McDonald, M. C. (prep), Dec 1993, The Somerset sump index

[2120]McEachern, J. Michael, 1978, An inventory and evaluation of the cave resources to be impacted by the New Melones Reservoir Project, Calaveras and Tuolumne Counties, California

[2123]McFarlane, Donald A., Oct 1989, A preliminary catalog of the caves on Anguilla, British West Indies

[2124]McGill, Greg, 1987, Alabama caves

[2125]McGill, William Mahone, 1897–1962, 1933, Caverns of Virginia

[2128]McGrain, Preston, 1917–, 1954, Geology of the Carter and Cascade Caves area

[2129]McGregor, Duncan J., 1962, Some features of karst topography in Indiana

[2130]McKenzie, David (ed), 1964, The caves of Bell and Coryell Counties

[2138]Medville, Douglas Michael, 1941–, Mar 1976, Caves and karst hydrology in northern Pocahontas County

[2139]Medville, Douglas Michael, 1941–, 1995, Caves and karst of Randolph County

[2140]Medville, Douglas Michael, 1941–, 1971, Caves of Randolph County

[2156]Meredith, Michael Edward, 1943–, 1992, Giant caves of Borneo

[2171]Middleton, John, 1942–, , The underground atlas: a gazetteer of the world's cave regions

[2172]Middleton, John, 1942–, nd, The underground atlas: a gazetteer of the world's cave regions

[2178]Mill, Lloyd (ed), 1980, Victorian caves and karst: a guidebook to the 13th A.S.F. conference: Melbourne 1980

[2186]Mills, Martin Taylor, 1974, The subterranean wonders of Hawaii

[2187]Mills, Martin Taylor, Dec 1981, The subterranean wonders of Iceland

[2188]Mills, Martin Taylor, 1977, The subterranean wonders of Kenya

[2189]Milner, A. J., 1972, Caves of the Alum Pot area

[2190]Mindick, Robert R., 1977, Commercial caves in Southern Indiana

[2191]Mindick, Robert R., 1977, Four non-commercial caves in Southern Indiana

[2201]Missouri Speleological Survey, Jul 1975, Caves of Christian County, Missouri: part 1 - eastern half

[2202]Missouri Speleological Survey, Oct 1975, Caves of Christian County, Missouri: part 2- western half

[2203]Missouri Speleological Survey, Jul 1981, Caves of Crawford County

[2204]Missouri Speleological Survey, July-December 1982, Caves of the Current River Valley

[2214]Missouri Speleological Survey, October-December 1979, J Harlen Bretz: a geologist's encounters with Missouri caves

[2215]Missouri Speleological Survey, Jan 1971, Jefferson County caves

[2216]Missouri Speleological Survey, Jan 1973, Laclede County caves

[2217]Missouri Speleological Survey, Apr 1961, Miller County caves

[2218]Missouri Speleological Survey, 1992, Mystery Cave

[2220]Missouri Speleological Survey, Jan 1963, Phelps County caves

[2221]Missouri Speleological Survey, 1963, Pike County caves

[2222]Missouri Speleological Survey, Apr 1971, Pit caves of Jefferson County

[2224]Missouri Speleological Survey, Apr 1964, Pulaski County caves

[2225]Missouri Speleological Survey, Jan 1965, Ralls County caves

[2228]Missouri Speleological Survey, 1992, Rimstone River Cave

[2229]Missouri Speleological Survey, Jul 1971, Sandstone caves of Jefferson County

[2230]Missouri Speleological Survey, [Dec 1966], Shannon County caves

[2231]Missouri Speleological Survey, Apr 1965, St. Charles County caves

[2232]Missouri Speleological Survey, Jul 1965, St. Genevieve County caves

[2233]Missouri Speleological Survey, Jan 1966, St. Louis County caves

[2234]Missouri Speleological Survey, Oct 1967, Washington County caves

[2235]Missouri Speleological Survey, January-December 1983, The wild caves of Camden County, Missouri

[2236]Mitchell, Albert, [1937], Yorkshire caves and potholes: 1, North Ribblesdale

[2237]Mitchell, Albert, [1950], Yorkshire caves and potholes: 2, under Ingleborough

[2245]Mockford, D. P., Oct 1974, Caves of the Bristol region

[2246]Mohapatra, R. P., 1939–, 1981, Udayagiri and Shandagiri caves

[2252]Molyneux, Arthur John Charles, 1935, The Sinoia caves, Lomagundi district

[2275]Morris, David (comp), Dec 1985, Cave Diving Group: Welsh sump index

[2282]Mostardi, Michael (comp), Dec 1991, Caves of Berks County, Pennsylvania

[2284]Mott, Kevin (ed), 1982, Nullarbor caving atlas

[2298]Müller, H. O. (comp, ed), 1985, Dolomite caves of the eastern Transvaal (South Africa): a free caver monograph 1985

[2309]MVOR (Mississippi Valley Ozark Regional). Guidebook (1968), 1968, Fall 1968 MVOR guidebook: Stone County, Arkansas

[2310]MVOR (Mississippi Valley Ozark Regional). Guidebook (1969), 1969, MVOR Guidebook Spring 1969

[2311]MVOR (Mississippi Valley Ozark Regional). Guidebook (1970), nd, 1970 guidebook UMAC-MVOR: Upper Mississippi area conference-Mississippi Valley Ozarks Regional Sept. 25, 26, & 27 at Eagle Cave

[2312]MVOR (Mississippi Valley Ozark Regional). Guidebook (1970), 1970, MVOR Spring "70" Guidebook

[2313]MVOR (Mississippi Valley Ozark Regional). Guidebook (1971), 1971, Fall 1971 MVOR: Glover's Cave

[2314]MVOR (Mississippi Valley Ozark Regional). Guidebook (1971), 1971, MVOR Spring 1971, April 30, May 1 & 2: Pulaski County

[2315]MVOR (Mississippi Valley Ozark Regional). Guidebook (1972), 1972, Guidebook: Fall 1972 MVOR, Perry County, Missouri

[2316]MVOR (Mississippi Valley Ozark Regional). Guidebook (1973), 1973, Guidebook: MVOR Spring '73

[2317]MVOR (Mississippi Valley Ozark Regional). Guidebook (1974), Sep 1974, Lake of the Ozarks MVOR, Fall 1974: Camden County, Miller County, Morgan County

[2318]MVOR (Mississippi Valley Ozark Regional). Guidebook (1975), 1975, MVOR: Spring 1975 in the Shawnee Hills

[2319]MVOR (Mississippi Valley Ozark Regional). Guidebook (1976), Sep 1976, Fall 1976 MVOR guidebook, Perry Co., Mo

[2320]MVOR (Mississippi Valley Ozark Regional). Guidebook (1977), 1977, Spring 1977 MVOR Guidebook: Middle Mississippi Valley Grotto, Shannon Co., Missouri

[2321]MVOR (Mississippi Valley Ozark Regional). Guidebook (1979), 1979, Perry County: spring M.V.O.R. '79

[2322]MVOR (Mississippi Valley Ozark Regional). Guidebook (1980), 1980, Guidebook to spring 1980 MVOR, Shannon County, Missouri

[2323]MVOR (Mississippi Valley–Ozark Region). Guidebook (1984), 1984, MVOR Guidebook: 1984 Fall

[2324]MVOR (Mississippi Valley–Ozark Region). Guidebook (1989), 1989, MVOR Guidebook: 1989 Spring

[2357]National Speleological Society, Apr 1948, The caves of Texas

[2383]National Speleological Society. Annual Convention. Guidebook (1960: NM), 1960, A guide book to Carlsbad Caverns National Park

[2384]National Speleological Society. Annual Convention. Guidebook (1962: [Black Hills], SD) , 1962, A guide to the caves of the Black Hills, S. D

[2385]National Speleological Society. Annual Convention. Guidebook (1963: Mountain Lake, VA), 1963, Guide book to the major caves in the vicinity of Mountain Lake, Virginia

[2386]National Speleological Society. Annual Convention. Guidebook (1964: New Braunfels, TX), May 1964, A guide to the caves of Texas and 1964 N.S.S. convention (New Braunfels, TX) field trips

[2387]National Speleological Society. Annual Convention. Guidebook (1965: Bloomington, IN), [1965], National Speleological Society convention 1965 guidebook, Indiana University Bloomington, Indiana June 12-20, 1965

[2388]National Speleological Society. Annual Convention. Guidebook (1966), 1966, Caves of the Sequoia Region California, issued in conjunction with the 25th anniversary convention of the National Speleological Society, David R McClurg, convention chairman

[2389]National Speleological Society. Annual Convention. Guidebook (1967: Birmingham and Huntsville, AL), 1967, The caves of Alabama: a guide for the 1967 convention of the National Speleological Society at Birmingham

10.3. GUIDEBOOKS

and Huntsville, Alabama, June 1967

[2390] National Speleological Society. Annual Convention. Guidebook (1968: Southwestern MO), Jul 1968, Guidebook to selected caves in southwestern Missouri: prepared for the 25th annual convention of the National Speleological Society

[2392] National Speleological Society. Annual Convention. Guidebook (1970: Pittsburgh, PA), 1970, NSS 1970 guidebook: Mid-Appalachian Region

[2393] National Speleological Society. Annual Convention. Guidebook (1971), 1971, NSS 71 guidebook

[2394] National Speleological Society. Annual Convention. Guidebook (1972: White Salmon, WA), 1972, Selected caves of the Pacific Northwest with particular reference to the vulcanospeleology of the state of Washington

[2395] National Speleological Society. Annual Convention. Guidebook (1973: Bloomington, IN), 1973, NSS 73 convention guidebook; National Speleological Society convention, Bloomington, Indiana, June 16-24, 1973

[2396] National Speleological Society. Annual Convention. Guidebook (1974: Decorah, IO), 1974, National Speleological Society 1974 convention Luther College, Decorah, Iowa 10-18 August, 1974

[2397] National Speleological Society. Annual Convention. Guidebook (1975: Frogtown, CA), 1975, National Speleological Society 1975 Convention, Frogtown, Ca

[2398] National Speleological Society. Annual Convention. Guidebook (1976: Morgantown, WV), 1976, 1976 annual convention guidebook, Morgantown, W. Va

[2399] National Speleological Society. Annual Convention. Guidebook (1977: Alpena, MI), 1977, Official 1977 guidebook, Alpena, Michigan: '77, the Lakeshore convention: Alpena, Michigan, NSS

[2400] National Speleological Society. Annual Convention. Guidebook (1978: New Braunfels, TX), 1978, An introduction to the caves of Texas: prepared for the 1978 National Speleological Society Convention New Braunfels, Texas

[2401] National Speleological Society. Annual Convention. Guidebook (1979: Pittsfield, MA), 1979, An introduction to caves of the northeast: guidebook for the 1979 National Speleological Convention, Pittsfield, Massachusetts

[2402] National Speleological Society. Annual Convention. Guidebook (1980: White Bear Lake, MN), Jul 1980, An introduction to caves of Minnesota, Iowa, and Wisconsin: guidebook for the 1980 National Speleological Society Convention, Lakewood Community College, White Bear Lake, Minnesota

[2403] National Speleological Society. Annual Convention. Guidebook (1982: Bend, OR), 1982, An introduction to caves of the Bend area: guidebook of the 1982 NSS Convention

[2404] National Speleological Society. Annual Convention. Guidebook (1983: Elkins, WV), 1983, An introduction to the caves of east-central West Virginia: guidebook for the 1983 National Speleological Society Convention, Elkins, West Virginia

[2405] National Speleological Society. Annual Convention. Guidebook (1984: Sheridan, WY), 1984, Guidebook of the 1984 NSS convention

[2406] National Speleological Society. Annual Convention. Guidebook (1985: Frankfort, KY), 1985, The caves of south eastern Kentucky: guidebook for the 1985 National Speleological Society Convention, Frankfort, Kentucky

[2407] National Speleological Society. Annual Convention. Guidebook (1986: Tularosa, NM), 1986, 1986 NSS convention guidebook, Tularosa, New Mexico

[2408] National Speleological Society. Annual Convention. Guidebook (1987: Sault Sainte Marie, MI), 1987, National Speleological Society official 1987 guidebook, August 3-7, 1987, Sault Sainte Marie, Michigan

[2409] National Speleological Society. Annual Convention. Guidebook (1988: Hot Springs, SD), 1988, 1988 NSS convention guidebook

[2410] National Speleological Society. Annual Convention. Guidebook (1989: Sewanee, TN), 1989, Caves and caving in TAG: a guidebook for the 1989 Convention of the National Speleological Society; Sewanee, Tennessee

[2411] National Speleological Society. Annual Convention. Guidebook (1990: Yreka, CA), 1990, NSS 1990 convention guidebook: Yreka, California: "where the lava meets the limestone"

[2412] National Speleological Society. Annual Convention. Guidebook (1991: Cobleskill, NY), 1991, Guide to the caves and karst of the Northeast

[2413] National Speleological Society. Annual Convention. Guidebook (1992: Salem, IN), 1992, Caving in the heartland: a guidebook for the 1992 Convention of the National Speleological Society, Salem, Indiana

[2414] National Speleological Society. Annual Convention. Guidebook (1993: Pendleton, OR), 1993, The 1993 NSS Convention guidebook

[2415] National Speleological Society. Annual Convention. Guidebook (1994: Brackettville, TX), 1994, The caves and karst of Texas: a guidebook for the 1994 Convention of the National Speleological Society with emphasis on the Southwestern Edwards Plateau: Brackettville, Texas, June 19-24, 1994

[2416] National Speleological Society. Annual Convention. Guidebook (1995: Blacksburg, VA), 1995, Underground in the Appalachians: a guidebook for the 1995 Convention of the National Speleological Society, Blacksburg, Virginia, July 17-21, 1995

[2417] National Speleological Society. Annual Convention. Guidebook (1996: Salida, CO), 1996, The caves and karst of Colorado: a guidebook for the 1996 convention of the National Speleological Society

[2418]National Speleological Society. Annual Convention. Guidebook (1997: Sullivan, MO), 1997, Exploring Missouri caves. A guidebook for the 1997 convention of the National Speleological Society Sullivan, Missouri June 23-27, 1997

[2419]National Speleological Society. Annual Convention. Guidebook. Geology and Biology Field Trip (1982: Bend, OR), 1982, Caves and other volcanic landforms of Central Oregon

[2420]National Speleological Society. Annual Convention. Guidebook. Geology Field Excursion (1978: New Braunfels, TX), 1978, Caves and karst hydrogeology of the southeastern Edwards Plateau, Texas: guidebook, geology field excursion, National Speleological Society annual convention, New Braunfels, Tex., June 18-23, 1978

[2421]National Speleological Society. Annual Convention. Guidebook. Geology Field Trip (1979: Pittsfield, MA), Aug 1979, Karst hydrogeology and geomorphology of eastern New York: a guidebook to the geology field trip, National Speleological Annual Convention, Pittsfield, Massachusetts August 5-12, 1979

[2422]National Speleological Society. Annual Convention. Guidebook. Geology Field Trip (1986: Tularosa, NM), 1986, Geology field trip guidebook, 45th National Speleological Society (1986) Tularosa, New Mexico, convention

[2424]National Speleological National Speleological Society. Annual Convention. Guidebook. Geology Field Trip (1995: Blacksburg, VA), 1995, Origin of caves and karst in the Shenandoah Valley, Rochingham and Augusta Counties, Virginia: guidebook for a geologic fieldtrip, National Speleological Society Annual Convention, Blacksburg, VA, 16 July 1995

[2425]National Speleological Society. Annual Convention. Guidebook. Preconvention Field Trip (1973: Bloomington, IN), 1973, C. H. U. G. Crawford Harrison Underground Gang presents twenty-two favorite Harrison Crawford caves: 1973 pre-convention activities

[2426]National Speleological Society. Annual Convention. Postconvention Guidebook (1992: Salem, IN), 1992, 1992 post convention guidebook: enter the karst of Indiana, you may never want to come out

[2447]National Speleological Society. Annual Convention. Supplemental Guidebook (1970: Pittsburgh, PA), 1970, Supplemental guidebook: 1970 NSS convention

[2448]National Speleological Society. Annual Convention. Supplemental Guidebook (1989: Sewanee, TN), 1989, A guide to East Tennessee Caves: prepared for the 1989 National Speleological Society Convention, Sewanee, Tennessee

[2487]Neighbor, Frank, Nov 1954, Major limestone caverns of the Black Hills, South Dakota

[2494]New South Wales Department of Tourism, 1973, Jenolan Caves, New South Wales

[2498]New South Wales. Department of Tourism, 1982, Limestone caves of New South Wales

[2511]Nicoll, Robert S., Dec 1976, Guidebook to the caves of southeastern New South Wales and eastern Victoria and caves around Canberra

[2516]North Country Region, 1979, North Country Region MECCA Summer 1979

[2519]Northeast Regional Organization [NRO]. National Speleological Society, Jul 1979, New Hampshire Caves

[2523]Northwest Caving Association. National Speleological Society (1987), 1987, Northwest Caving Association regional meeting guidebook: caves of the Peterson Prairie area

[2524]Northwest Caving Association. National Speleological Society (1992), 1992, 1992 Northwest Caving Association regional meeting cave guide [22-24 may 1992]

[2533]Nucolorvue Productions, 1974, Mt. Gambier and district: Port MacDonnell and Tantanoola caves

[2543]Ockenden, Allen C., 1991, Caves of the Mellte Valley

[2549]Oldham, Anthony Clive [Tony], 1939–, 1975, The caves of Carmarthen

[2550]Oldham, Anthony Clive [Tony], 1939–, 1981, The caves of Clydach

[2551]Oldham, Anthony Clive [Tony], 1939–, 1981, The caves of Co. Cork

[2552]Oldham, Anthony Clive [Tony], 1939–, 1978, The caves of Gower

[2553]Oldham, Anthony Clive [Tony], 1939–, 1977, The caves of north Wales

[2554]Oldham, Anthony Clive [Tony], 1939–, 1975, The caves of Scotland (except Assynt)

[2555]Oldham, Anthony Clive [Tony], 1939–, 1986, Caves of the central northern outcrop

[2556]Oldham, Anthony Clive [Tony], 1939–, 1993, The caves of the Little Neath Valley

[2557]Oldham, Anthony Clive [Tony], 1939–, 1991, The caves of the south eastern outcrop

[2558]Oldham, Anthony Clive [Tony], 1939–, 1992, The caves of the south eastern outcrop and caves and mines in the Forest of Dean

[2559]Oldham, Anthony Clive [Tony], 1939–, Jan 1985, The caves of the southern outcrop

[2560]Oldham, Anthony Clive [Tony], 1939–, 1979, The caves of west Wales

[2561]Oldham, Anthony Clive [Tony], 1939–, 1989, The complete caves of Devon

[2562]Oldham, Anthony Clive [Tony], 1939–, 1990, The concise caves of north Wales

[2563]Oldham, Anthony Clive [Tony], 1939–, 1972, Discovering caves

[2564]Oldham, Joyce Elizabeth Anne, Nov 1972, The caves of Devon

10.3. GUIDEBOOKS

[2565]Oldham, Joyce Elizabeth Anne, Apr 1985, The concise caves of Devon

[2566]Oldham, Joyce Elizabeth Anne, 1978, The limestones and caves of Devon

[2584]Owen, Luella Agnes, 1970, Cave regions of the Ozarks and Black Hills

[2615]Parris, Lloyd E., 1973, Caves of Colorado

[2625]Pearman, Harry, 1968, Caves and tunnels in Kent

[2626]Pearman, Harry, 1976, Caves and tunnels in southeast England

[2627]Pearman, Harry, 1963, Secret tunnels in Surrey

[2628]Pearman, Harry, 1983, Underground gazetteer of south-east England

[2632]Pearson, Les, 1983, Chillagoe souvenir guide

[2652]Pereda de la Reguera, Manuel, 1966, Guide Santillane on the sea and Altamira

[2656]Perry, Clair Willard [Clay], 1887–1961, 1948, Underground empire: wonders and tales of New York caves

[2657]Perry, Clair Willard [Clay], 1887–1961, 1939, Underground New England

[2661]Petrocheilou, Anna, 1984, The Greek caves

[2663]Peyrony, E., 1959, Les Eyzies and the Vezere Valley; an illustrated guide for scholars and tourists

[2669]Photochrom Co., Tunbridge Wells, 1900, Cheddar Gorge and caves

[2674]Pilkington, Graham (ed), 1986, Nullarbor caving atlas

[2678]Pinnock, Rick (ed), 1987, A guide to Glenrock caves: a speleological field guide to the limestone caves on Glenrock Station

[2682]Plante, Alan R. (comp), 1990, Berkshire Cave Guide

[2691]Poole, Gary A. (ed), 1978, Bexar County speleology vol 1

[2695]Porteous, Crichton, 1950, Caves and caverns of Peakland

[2702]Powell, Richard Lewis, 1936–, Oct 1961, Caves of Indiana

[2720]Proctor, Christopher J. (comp), Jul 1987, Atlas of the Berry Head Caves

[2721]Proctor, Christopher J., 1987, The caves of Chudleigh and Kingsteignton

[2722]Proctor, Christopher J., 1987, Caves of East Devon

[2723]Proctor, Christopher J., 1987, Caves of North Tor Bay

[2733]Purcell, David, 1977, Spelunking (the exploration of caves) guide to central Oregon

[2751]Raines, Terry W. (comp), 1989, Caves of Mexico

[2756]Raistrick, Arthur, 1897?–1991, 1948, Grassington and Upper Wharfedale

[2760]Rankin, Frank, 1989, Guide to the Wemyss caves

[2768]Rebmann, James R., 1972, Caves of Rockcastle County, Kentucky

[2772]Reddell, James Russell, 1938–(ed), nd, The caves of Langtry

[2773]Reddell, James Russell, 1938–(ed), Mar 1962, The caves of Bexar County

[2774]Reddell, James Russell, 1938–, Jun 1964, The caves of Comal county

[2775]Reddell, James Russell, 1938–, Nov 1965, The caves of Edwards County

[2776]Reddell, James Russell, 1938–(ed), Apr 1977, The caves of far west Texas

[2777]Reddell, James Russell, 1938–(ed), Mar 1970, The caves of Lubbock County

[2778]Reddell, James Russell, 1938–(ed), Apr 1967, The caves of Medina county

[2779]Reddell, James Russell, 1938–(ed), Jun 1963, The caves of northwest Texas

[2780]Reddell, James Russell, 1938–, Feb 1973, The caves of San Saba County

[2781]Reddell, James Russell, 1938–(ed), Jul 1962, The caves of San Saba County part I

[2782]Reddell, James Russell, 1938–(ed), nd, The caves of Travis County

[2783]Reddell, James Russell, 1938–(ed), Feb 1963, The caves of Val Verde County

[2784]Reddell, James Russell, 1938–(ed), Oct 1963, The caves of Willamson County

[2788]Reddell, James Russell, 1938–(ed), Apr 1989, A field guide to caves of Blanco, Gillespie, and Llano counties, Texas

[2791]Reddell, James Russell, 1938–, 1977, A preliminary survey of the caves of the Yucatan Peninsula

[2803]Reich, J. R. (comp), 1974, Caves of southeastern Pennsylvania

[2812]Rhodes, Andrew Jackson, 1829–1907, 1905, The wonders of Lost River: 40 miles of underground river with its caves and hidden waterfalls, blind fish and other wonders

[2834]Rocky Mountain Region. Guidebook (1991), 1991, Top secret, for cavers' eyes only: 1991 Rocky Mountain Regional guidebook, Black Hills, South Dakota

[2835]Rocky Mountain Region. Guidebook (1994), nd, Caves of the eastern White River Plateau: guidebook to the 1994 Rocky Mountain regional, Willow Peak, Colorado

[2838]Rodway School Speleological Group, 1974, Caves of the Avon Gorge Bristol, part 2: eastern or Clifton side

[2839]Rodway School Speleological Group, 1973, Caves of the Avon Gorge Bristol: part 1: western or Leigh Woods side

[2857]Ross, Sylvia H., 1969, Introduction to Idaho caves and caving

[2871]Russell, William Hart, 1937–, 1967, Caves of the Inter-American Highway: Nuevo Laredo, Tamulipas to Tamazunchale, San Luis Potosi

[2874]Ryder, Peter F., 1995, Caves of Skye

[2877] Salamon, Gábor, 1989, Aslóhegy shafts
[2878] Salamon, Gábor, 1989, The Stalactite Caves of Jósvafő and Aggtelek
[2880] Salvona, J., 1988, Scottish cave guides; the southern highlands
[2881] Salvona, J., 1989, The Southern highlands
[2882] Sameshima, Teruhiko, 1988, 5th international symposium on vulcanospeleology excursion guide book
[2884] Santamarta Cuenca, Pedro, 1978, The grottos of Majorca
[2905] Schindel, Geary Michael, 1957–, Dec 1991, Guidebook: environmental hydrogeology of karst terranes in the vicinity of Nashville Tennessee
[2915] Schuurmans, Theo J. C., 1987, A caver's view of the Clydach River: an introduction to the underground waters of Pwll Ddu, Llangynid and Llangattwg mountains
[2921] Schweiker, Roioli, , Caves of Albany County, New York
[2922] Schweiker, Roioli, 1958, Field guide to New York caves: Schoharie County
[2924] Scott, John, 1929–, 1959, Caves in Vermont, a spelunker's guide to their location and lore
[2929] Self, Charles A., , Caves of County Clare
[2937] Šerko, Alfred, 1910–1948, 1967, The Postojna grottoes and the other marvels of the karst
[2940] Sexton, Robert T., 1965, Caves of the coastal areas of South Australia
[2959] Shelley, Maryann, Dec 1956, Karst and caves in the Caucasus
[2969] Shimek, S. (ed), 1971, Bower Cave and Cave of Orpheus file
[2977] Sieveking, Ann, 1962, The caves of France and northern Spain
[2978] Sieveking, Ann, 1966, The caves of France and northern Spain, a guide
[2992] Sinise, Jerry, 1963, Texas show caves
[2998] Skřivánek, František, 1973, Caves in Czechoslovakia
[3004] Sloane, Howard N. (d. 1972), 1966, Visiting American caves
[3009] Smith, A. Richard (ed), 1971, The caves of Kimble county
[3010] Smith, A. Richard (ed), 1965, The caves of Kinney county
[3016] Smith, David Ingle (comp, ed), 1975, Limestones and caves of the Mendip Hills
[3023] Smith, Jim, [1984], From Katoomba to Jenolan Caves: the Six Foot Track, 1884-1984
[3026] Smith, Lee, nd, Down under Texas
[3030] Smith, Marion Otis, 1942–, c.1973, TAG pits
[3031] Smith, Peter B., 1973, The P8 Cave: Castleton
[3042] Snyder, Dean H. (comp, ed), Feb 1989, The caves of Northampton County, Pennsylvania
[3045] Soepadmo, E., 1971, A guide to Batu Caves
[3059] South Wales Caving Club, Feb 1993, Members handbook: South Wales Caving Club
[3063] Southeastern Regional Association (S.E.R.A.) Cave Carnival (1968: TN), [1968], 1968 SERA cave carnival
[3064] Southeastern Regional Association (S.E.R.A.) Cave Carnival (1975: Monterey, TN), 1975, Guidebook to the caves of the Monterey, Tennessee Area
[3065] Southeastern Regional Association (S.E.R.A.) Cave Carnival (1976: Guntersville, AL), 1976, SERA 1976 guidebook
[3066] Southeastern Regional Association (S.E.R.A.) Cave Carnival (1978: TN), 1978, Guidebook to the caves of the 1978 summer SERA cave carnival
[3067] Southeastern Regional Association (S.E.R.A.) Cave Carnival (1979: Ketner's Mill, TN), 1979, 1979 Southeastern Regional Association (S.E.R.A.) cave carnival at Ketner's Mill, TN
[3068] Southeastern Regional Association (S.E.R.A.) Cave Carnival (1983: AL), Aug 1983, 1983 SERA cave carnival: 5-7 August 1983
[3069] Southeastern Regional Association (S.E.R.A.) Cave Carnival (1989: [TN]), [1989], SERA 89
[3070] Southeastern Regional Association (S.E.R.A.) Cave Carnival (1990: Goose Pond Campground, Scottsboro, AL), 1990, SERA cave carnival, 1990
[3071] Southeastern Regional Association (S.E.R.A.) Cave Carnival (1992: AL), [1992], SERA cave carnival 1992
[3072] Southeastern Regional Association (S.E.R.A.) Cave Carnival (1994: Short Mountain, TN), 1994, Guidebook to the 1994 SERA cave carnival: June 10-12, 1994 CCWHA Campground at Short Mountain Cannon County, Tennessee
[3073] Southeastern Regional Association (S.E.R.A.) Cave Carnival (1995: Lafayette, GA), [1995], 1995 SERA summer cave carnival: the party on the farm: Smokey and Tina Cadwell's farm: Lafayette, Georgia May 19-21, 1995
[3074] Southeastern Regional Association (S.E.R.A.) Cave Carnival (1996: Camp Jackson, Jackson Co. AL), 1996, 45th Annual SERA Summer cave carnival: July 19-21, 1996, Camp Jackson, Jackson County, Alabama
[3084] Speece, Jack H., 1977, The cave of Delaware
[3093] Spencer, Edgar Winston, 1985, Guidebook to the Natural Bridge and Natural Bridge Caverns
[3096] Sprague, Stuart, c1960, Caves of Herkimer County
[3097] Sprent, J. K., (ed), 1970, Mount Etna caves: a collection of papers covering several aspects of the Mt. Etna and Limestone Ridge caves area of central Queensland
[3100] Standing, Ian James, Sep 1966, Council of Southern Caving Clubs handbook
[3118] Stenner, Roger D., Oct 1961, Some smaller Mendip caves: vol 1
[3124] Stevens, Paul James, 1944–, 1988, Caves of the Or-

10.3. GUIDEBOOKS

gan Cave Plateau: Greenbrier County, West Virginia
[3134]Stone, John M., 1914, The underground passages, caverns, etc., of Greenwich and Blackheath
[3135]Stone, Ralph Walter, 1876–1964, 1953, Caves of Pennsylvania
[3136]Stone, Ralph Walter, 1876–1964, 1953, Descriptions of Pennsylvania's undeveloped caves
[3137]Stone, Ralph Walter, 1876–1964, 1930, Pennsylvania caves
[3138]Stone, William Curtis, 1952–, 1977, Caves of the San Juan Plateau
[3146]Storrick, Gary D., May 1992, The caves and karst hydrology of southern Pocahontas County and the upper Spring Creek Valley
[3149]Stratford, Tim, 1978, Caves of South Wales
[3164]Sweatman, Cheyenne (ed), Oct 1985, 8th Annual TAG fall cave-in
[3171]Sydney University Speleological Society, Nov 1975, Nibicon log book reports December 1972 - February 1973
[3175]Szentes, Georg, Jul 1989, Sandstone caves in Nigeria
[3176]Szentes, Georg, 1992, Tropical karst and caves in the central Cordillera, Colombia
[3180]Tankersley, Ken, 1975, The cavernous karst land forms of Jackson County, Kentucky
[3183]Tarkington, Terry Warren, 1925–, Jun 1965, Alabama caves
[3189]Taylor, Kirk R. (comp), Jan 1993, Caves of Bedford County, Pennsylvania
[3196]Tell, Leander, 1895–, 1976, Fifty typical Swedish caves
[3209]Thornber, Norman, 1953, Britain underground
[3210]Thornber, Norman, 1959, Pennine underground
[3226]Tratman, Edgar Kingsley, 1899–1978 (ed), 1969, The Caves of North West Clare, Ireland
[3248]Tucker, J. H., Aug 1962, Some smaller Mendip caves: vol 2
[3249]Tulis, J. (comp), 1993, Slovakia: the karst and speleology
[3250]Tullis, Edward L., 1940, Black Hills caves
[3257][Tyers, John], 1962, A guide to Wind Cave and Wind Cave National Park. 19th annual convention of the National Speleological Society self guiding tour of the commercial areas of Wind Cave
[3261]U.S. Army, Engineer Intelligence Division, Aug 1962, Natural caves of West Germany Baden - Wuerttemberg
[3324]U.S. Office of Strategic Services, Research and Analysis Branch, Apr 1945, The caves of greater Germany
[3349]Van Voris, Arthur H., 1970, The lesser caverns of Schoharie County
[3352]Varnedoe, William Whitfield, 1923–, 1973, Alabama caves and caverns
[3353]Varnedoe, William Whitfield, 1923–, Mar 1981, Alabama caves, 1980
[3354]Varnedoe, William Whitfield, 1923–, Jan 1972, Interim Alabama cave survey: report number 1
[3355]Varnedoe, William Whitfield, 1923–, 1975, Interim report number 2: Alabama cave survey
[3357]Veni, George, 1957–, 1988, The caves of Bexar County
[3361]Vickery, Margaret Ray, 1960, Ozark stories of the Upper Current River
[3366]Vineyard, Jerry Daniel, 1935–(comp), 1960, Supplement to the catalogue of the caves of Missouri
[3370]Wagenaar Hummelinck, P. , 1979, De Grotten Van de Nederlande Antillen - Caves of the Netherland Antilles
[3375]Walker, Kevin, 1983, Caving in the Crickhowell area
[3376]Walker, Kevin, 1983, The Llangattock escarpment
[3385]Walter, Erin, Jul 1996, A guide to Austin's most visited caves, with maps and cross-sections of area caves
[3388]Waltham, Anthony Clive, 1942–[Tony], 1987, Caves and karst of the Yorkshire Dales: an excursion guidebook to the karst landforms and some accessible caves within Yorkshire Dales
[3389]Waltham, Anthony Clive, 1942–[Tony], 1984, Caves, crags, and gorges: a guide to the limestone country of England and Wales
[3391]Waltham, Anthony Clive, 1942–[Tony], 1987, Karst and caves in the Yorkshire Dales National Park
[3395]Waltham, Anthony Clive, 1942–[Tony], 1987, Yorkshire Dales: limestone country
[3404]Waters, Aaron Clement, 1905–, 1990, Selected caves and lava tube systems in and near Lava Beds National Monument, California
[3420]Weaver, Herman Dwight, 1938–(ed), 1982, Great American show caves
[3431]Weinel, John, 1979, Pittsburgh Grotto Picnic Fall M.A.R. '79: Guidebook, Sept. 14-15, 1979
[3432]Welch, Bruce R. (ed), 1976, The northern limestone
[3459]White, George W., 1926, The limestone caves and caverns of Ohio
[3471]White, William Blaine, 1934–(ed), 1964, Caves of Pennsylvania
[3472]White, William Blaine, 1934–(ed), 1976, Caves of Western Pennsylvania compiled by members of the Mid-Appalachian Region of the National Speleological Society
[3481]Wickersham, David L., 1988, A survey of the caves of Roanoke County, Virginia
[3508]Wilson, George Herbert, 1874?–1958, c1926, Some caves and crags of Peakland
[3515]Winder, Francis A., 1938, Unconventional guide to the Caverns of Castleton and Disaster
[3517]Wisconsin Speleological Society, 1978, 1978 hodag hunt guidebook
[3539]Yale Speleological Society, 1963, The caves of

Connecticut
[3545] Young, Ivan (ed), 1978, Appin cave guide
[3546] Young, James Jay, 1948–, 1993, Caves in Kansas
[3552] Zawislak, Ronald L., 1967, The Tennessee pit survey

10.4 Listings

[355] Beck, Barry Frederic, 1944–(ed), Jul 1984, A computer based inventory of recorded recent sinkholes in Florida
[378] Belski, David Stanley, 1937–, 1992, GYPKAP 1987 annual report
[433] Bookout, John , nd, Named caves of Oregon
[650] Chillagoe Caving Club (comp), nd, Numerical index - area maps
[922] Dunkley, John R., 1994, Thailand caves catalogue
[934] Eaton, Joli (ed), [1988], GYPKAP 1987 annual report
[1036] Faulkner, Trevor, 1942–, 1979, Sump index Norway
[1060] Fincham, Alan Goldworth, 1933–, 1977, Jamaica underground: a register of data regarding the caves, sinkholes and underground rivers of the island
[1331] Halliday, William R., 1926–, Oct 1995, Initial inventory of named caves and related features and cave-related place names in Hawaii
[1391] Hawkins, Thomas, 1993, The caves of Honduras: a first list
[2090] Matthews, Peter Gahan, 1938–(ed), 1985, Australian karst index 1985
[2092] Matthews, Peter Gahan, 1938–(ed), 1979, Checklist of Australian caves and karst, 1979
[2512] Nieland, James Raymond, 1949–, 1969, Little known caves of Oregon
[2703] Powell, Richard Lewis, 1936–, Dec 1970, A guide to the selection of limestone caverns and springs in the United States as national landmarks: for the National Park Service
[2787] Reddell, James Russell, 1938–(ed), Apr 1962, A checklist of the caves of Texas
[2793] Reddell, James Russell, 1938–, Jul 1966, A revised checklist of Texas caves
[2841] Rogers, Bruce W., Jan 1992, Karst features of Pohn Pei, Chuuk, and Waqab: a survey of caves and karst features in the States of Pohn Pei, Chuuk, and Waqab, Federated States of Micronesia
[2842] Rogers, Bruce W., May 1992, Karst features of Rota (Luta) Island, CNMI, A survey of caves and karst features on Rota (Luta) Island, Commonwealth of the Northern Mariana Islands
[2843] Rogers, Bruce W., May 1992, Karst features of Saipan, CNMI: a survey of caves and karst features on Saipan Island, Commonwealth of the Northern Mariana Islands
[2844] Rogers, Bruce W., May 1992, Karst features of the Palau Islands, a survey of caves and karst features in the Palau Archipelago
[2845] Rogers, Bruce W., May 1992, Karst features of the Territory of Guam: a survey of caves and karst features on Guam Island, Territory
[3005] Smart, James, 1994, A Philippine cave index and bibliography
[3485] Wilford, Gerald Edward, 1964, The geology of Sarawak and Sabah caves

11 Show Caves

Gary K. Soule

224 South 7th Avenue
Sturgeon Bay, WI, 54235-2216 US

Show caves are scattered throughout the world in at least forty countries. Europe has a large number, while the second largest continent in the world, Africa, has very few; only four show caves in South Africa, and one in Kenya plus more in northern Africa, are known. Asia has many show caves, Australia has a fair number, with still fewer in South America. European show caves predate those of North America due to that continent's longer history of settlement, and it was not until 1806 that Grand Caverns in Virginia became America's first show cave.

In 1972 Tony and Anne Oldham produced a nice regional guide to the show caves of Britain: *Discovering Caves*. Good guidebooks also exist for South Africa, particularly on Cango Caves, and for Australia on the Jenolan caves, which even include details on steam cleaning formations.

Many non-English speaking countries possess at least one English language guidebook. Examples include Hungary, Malaysia, Slovenia, Spain, Switzerland, and even the former Soviet Union. There are some fine guides to show caves for France, but nearly all are in French. The best work on European show caves is the 1975 *Guide des Grottes d'Europe* by Villy Aellen and Pierre Strinati, though it is available only in French or German.

In the United States, early tourist brochures and guidebooks appeared without delay following a cave's opening. e.g., the third edition of the Luray Caverns guidebook was published in 1882, only four years after its discovery. Today, more detailed guides are available, such as *Discovery of Luray Caverns, Virginia* by Russell H. Gurnee (1978).

One of the most important general show caves guides is Horace Hovey's *Celebrated American Caverns*. His book gives information on all of the great caves of the 19th century, including Mammoth, Wyandotte, Luray, and Howe.

Other important early works helped popularize Mammoth: *Rambles in Mammoth Cave* by Croghan and *An Excursion to the Mammoth Cave* by R. Davidson. Wyandotte had several works which helped to spread the word about its beauty: H. W. Rothrock's *Wyandotte Cave: where it is and what it is* and James P. Stelle's *The Wyandotte Cave of Crawford County Ind*.

Alonzo Pond, the first manager of Cave of the Mounds near Blue Mounds, Wisconsin, established trends for detail and quality in show cave guides. He released an extremely fine guidebook shortly after becoming manager in 1940. For the first time, geology, a map of the cave, and history were all skillfully combined into an interesting text that educated the reader. Other show cave operators emulated his work. Another pioneer in this field, George P. Jackson, worked on Wyandotte Cave in Southern Indiana. Today his numerous works are still highly respected.

Dwight Weaver, a show cave historian from Missouri, produces highly detailed, accurate guides that delve deeply into history. His Dark Pathway series, with fine illustrations by Paul Johnson, covers a number of show caves, including Mark Twain Cave, Meramec Caverns, and Onondaga Cave. Perhaps Dwight's skill comes from not only being a professional writer, but also having managed or been associated with Bridal Cave, Mark Twain Cave, Ozark Caverns, and what today is named Fantasy World Caverns in Missouri.

In contrast, Roger Kehret of Minnesota has produced a series of simple, straight-forward guides that present information to tourists. These include small booklets on Crystal Lake Cave and Spook Cave in Iowa, and Mystery Cave in Minnesota. Bill Anchors of Tennessee has written a nice book on Jewel and Ruskin Caves complete with their history and, not to be outdone, Cato and Susan Holler have more than covered Linville Caverns in North Carolina with *Hollow Hills of Sunnalee* (1989). The 1980 Ruby Falls guidebook by Ed Brinkley is extremely detailed, but it is better to have more than enough information, as long as it is accurate and relates to the cave.

Speleo-historians writing guides as a "labour of love" is a more common trend today. They write to help preserve history, not for any royalties. The time spent in research is enormous, and publishing costs are high and rising. As show cave tourism rises, so will the requirement for better and more detailed guides. Most show caves have increased the history content of their guides, as well as including a cave map. Perhaps show cave owners realize that tourists want something more than "This is the Fairies Grotto" or "Note the cave formation resembling an Angel". Science and conservation are the new themes for guidebooks, which are completed with chapters on geology, speleothems, and cave management.

Looking to the future, an English language show cave directory to the developed caves of the world is sorely needed. About 1980, the Union International de Spéléologie suggested the production of a detailed listing of all show caves. This would have presumably incorporated the *Gurnee Guide to American Caves* as well as an extensive listing of current and ex-commercial show caves in Canada and the United States by Gary Soule. However, perhaps due to the logistics or financing difficulties of such a massive venture, the project was not begun. The requirement, including full details of each cave, remains. An excellent beginning to this work would be a translation of the

Guide des Grottes d'Europe into English.

It is important to define what constitutes a true "show cave". A show cave can be classified as a cave where payment is made for entry, but as organized, paid instruction during "wild" trips increases this criterion becomes blurred. At least one site in the US is free, but has walkways, a shop, lighting, and guides. Clearly, this definition is not completely acceptable. In 1987 during a discussion with Russ Gurnee, we decided that while many factors must be taken into account and that there are many borderline cases, the main criterion is whether the cave has more or less fixed hours of operation that are advertised.

An increasing number of show cave owners may form groups similar to those in Arkansas, Missouri, Ohio and Pennsylvania, and even national associations. Primary examples are the National Caves Association in the United States and the Association Nationale des Exploitants de Cavernes Amenagees pour le Tourisme (ANECAT) in France. With the translation of many of the members' guides into other languages, the future of the English language guidebook looks bright.

Show caves have existed for hundreds of years and will continue to attract the public for years to come. As secondary attractions are added, guides will reflect these additional sources of income while the show cave remains the commercial anchor. The higher quality guides of the future will probably be produced by cavers, staff, and owner/managers who consider the cave as an important natural resource rather than merely as a source of income. To maintain standards, a good English language book or guide requires many hours to produce over many months or years; public and cavers alike will benefit from this trend.

Reference List of Item not in Main Bibliography

Aellen, Villy. 1975. *Guide des Grottes d'Europe Occidentale.* Neuchatel; Paris: Delachaux & Niestle, 316pp.

11.1 General

[2]Abadie, Bernard, [1972], Grottos of Bétharram
[6]Adam, David P., 1963, Pollen analysis of a stratigraphic section from Bat Cave, New Mexico
[8]Adamkó, Péter, 1992, The caves of Buda
[10]Adams, Aislinn (des, illus), 1984, Dunmore Cave
[15]Addis, Robert Philip, 1945–, 1979, A study for the National Speleological Society: Knox Cave: Albany County, New York
[17]Agee, James K., 1990, Oregon Caves forest and fire history
[22]Albi, S.A. Apa Poux (ed), 1990, Padirac Chasm and underground river
[23]Alderson, Laura White, 1984, Mammoth Cave: Kentucky's buried treasure
[25]Alexander, E. Calvin, 1987, Hydrogeologic study of Jewel Cave/Wind Cave: final report -year 2, 4th quarter of year 2, February 1, 1987 through April 30, 1987
[26]Alexander, E. Calvin, 1989, Hydrologic study of Jewel Cave/Wind Cave: final report
[28]Alexander, Emmit Calvin, 1943–, 1986, Hydrogeologic study of Jewel Cave/Wind Cave: final report - phase I, 3rd and 4th quarters of phase I, 1 November 1985 through 30 April 1986
[35]Aley, Thomas John, 1938–, 1988, Control of exotic plants in Oregon Caves, Oregon Caves National Monument
[42]Aley, Thomas John, 1938–, 1987, Restoration of natural cave features, Oregon Caves National Monument, Oregon: final report
[43]Aley, Thomas John, 1938–, 1988, Restoration of natural microclimate in Oregon Caves, Oregon Caves National Monument: final report
[57]Ammen, Samuel Zenas, 1843–1929, 1890, The caverns of Luray, the property of the Valley Land and Improvement Co., Luray, Va: the manner of their formation, their peculiar growths, their geology, chemistry, etc: an illustrated guide book
[58]Ammen, Samuel Zenas, 1843–1929, 1893, The caverns of Luray. The property of the Valley Land and Improvement Co., Luray, Va. The manner of their formation, their peculiar growths, their geology, chemistry, etc. An illustrated guide book
[59]Ammen, Samuel Zenas, 1843–1929, 1884, The caverns of Luray: an illustrated guide-book to the caverns, explaining the manner of their formation, their peculiar growths, their geology, chemistry, etc
[60]Ammen, Samuel Zenas, 1843–1929, 1882, History and description of the Luray Cave (illustrated), including explanations of the manner of its formation, its peculiar growths, its geology, chemistry, &c.; also a map. The whole so arranged as to serve as a guide.
[61]Ammen, Samuel Zenas, 1843–1929, 1880, History and description of the Luray Cave (illustrated), including explanations of the manner of its formation, its peculiar growths, its geology, chemistry, etc.; also a map. The whole so arranged as to serve as a guide
[65]Anderson, Arthur Wilhelm, 1892–, 1938, The Carlsbad Cavern of New Mexico: its history and geology
[66]Anderson, Arthur Wilhelm, 1892–, nd, The Carlsbad Cavern of New Mexico: its history and geology, formations of the cavern, discovery and exploration, bats of the cavern, cavern geology, administration: the land nobody knows
[67]Anderson, Arthur Wilhelm, 1892–, 1930, The Carlsbad Cavern of New Mexico: its history and geology; formations of the cavern, discovery and exploration, bats of

11.1. GENERAL

the cavern, cavern geology, administration; the land nobody knows

[73]Andre, Daniel, 1994, Bramabiau's underground river

[78]Anon, nd, 37 views of Naracoorte caves and Naracoorte South Australia

[80]Anon, c1990, Aillwee Cave: Ireland's premier show cave

[82]Anon, 1978, Beautiful Kweilin

[84]Anon, 1993, Carlsbad Cavern National Park. Guadalupe Mountains National Park

[85]Anon, 1985, Carlsbad Caverns National Park

[86]Anon, c1988, Carlsbad Caverns National Park, New Mexico

[87]Anon, 1982, Carlsbad Caverns, New Mexico

[91]Anon, 1953, Cavern: a pictorial souvenir of Carlsbad Caverns

[92]Anon, 1882, The Caverns of Luray

[93]Anon, c1880, The Caverns of Luray located on line of the Shenandoah Valley R.R.

[98]Anon, 1959, The caves of Arta situated on the sea coast of the municipal confines of Capdepera

[99]Anon, 1980, The caves of Kentucky: Mammoth Cave, Colossal Cavern, White's Cave

[100]Anon, c1905, The caves of Western Australia

[102]Anon, c1914, Cheddar

[104]Anon, pre-1920, Cox's Stalactite Cavern, Cheddar

[106]Anon, [1992], Dan-yr-ogof show caves

[107]Anon, 1865, A description of Howe's Cave: with a popular treatise on the formation of caves in lime rock, from the size of a quill to a mammoth

[108]Anon, 1850, A description of the Mammoth Cave of Kentucky, the Niagara River and Falls, and the Falls in summer and winter; the prairies, or life in the West; the Fairmont Water Works and scenes on the Schuylkill, xc. xc.: to illustrate Brewer's Panorama

[109]Anon, 1849, A description of the Mammoth Cave of Kentucky, the Niagara river and falls, and the falls in summer and winter; the prairies, or life in the West; the Fairmount Water Works and scenes on the Schuykill, etc. etc.: to illustrate Brewer's panorama.

[111]Anon, 1920, Dossey Domes: most beautiful of all the caverns at Green River Landing (One-half mile from Mammoth Cave, Ky.) ; Cliff Walk along Green River: finest scenery in the state

[116]Anon, nd, Fairy Cave

[117]Anon, nd, Focus on the Cango Caves

[120]Anon, nd, Gough's Caves, Cheddar, Somerset

[121]Anon, 1926, Grand Caverns: wonders of the subterranean world: Grottoes, Virginia

[124]Anon, nd, The Grottos of Han and of Rochefort (Belgium)

[125]Anon, 1870, Guide book for the Diamond Cave, Barren County, Kentucky. Located near "Old Bell Tavern," now the Mammoth Cave and Glosgow Junctions, on the L & N railroad, two miles from the R.R. immediately on the Mammoth Cave Road

[126]Anon, 1876, A guide manual to the Mammoth Cave of Kentucky

[127]Anon, 1876, Guide manual to the Mammoth Cave of Kentucky

[130]Anon, nd, Guide to the Kalk Bay and Muizenberg Mountains; (walks, caves, camp sites)

[131]Anon, 1934, A guide to the Museum of the Torquay Natural History Society

[132]Anon, 1987, Guilin tourist album

[134]Anon, Nov 1967, Handbook for caves and guiding staff

[136]Anon, nd, The historic Blue Grottoes of Virginia; Harrisonburg VA

[137]Anon, 1967, How nature made the beautiful caverns

[138]Anon, 1930, Illustrated souvenir grid to the caverns and glens at the foot of Ingleborough (with map)

[139]Anon, nd, Ingleborough Cavern: the finest show cave in Yorkshire

[141]Anon, nd, Jenolan Caves: Australia's underground fairyland

[143]Anon, nd, Kents Cavern Wellswood Torquay Devon, home of prehistoric man and animals

[145]Anon, , Luray Caverns

[146]Anon, 1900, Luray Caverns

[147]Anon, c1993, The magic of Sarawak's Mulu Caves; discover the latest wonder of the world

[148]Anon, 1935, Mammoth Cave in third dimension

[149]Anon, 1930, Mammoth Cave of Kentucky

[150]Anon, 1920, Mammoth Cave of Kentucky: the world's greatest subterranean wonder

[151]Anon, nd, Mammoth Cave, America's great natural wonder

[152]Anon, c1906, Mansion Inn, Luray Caverns

[153]Anon, Jul 1995, Marble Arch Caves

[154]Anon, 1990, Maribel Caves: an ideal health and summer resort.

[155]Anon, 1958, Marvel Cave: the story of America's largest privately owned cave

[156]Anon, nd, Marvelous Howe Caverns near Cobleskill, New York

[158]Anon, 1903, Monteagle Wonder Cave, or, The Cave of the Cumberland

[160]Anon, nd, The most beautiful caves in the world

[162]Anon, nd, The Naracoorte Caves: "how to reach them"

[163]Anon, nd, Natural Bridge Caverns: Texas' largest caverns: One of the great show caves of the world

[165]Anon, c1969, North Alabama's caves & caverns

[166]Anon, nd, The official guide to Cheddar Somerset with map and eleven illustrations

[167]Anon, 1976, Ohio Caverns

[168]Anon, 1930, The Ohio Caverns: where nature carved

a fairyland

[169] Anon, c1900, Oligio-Nunk, the place of caves. In the heart of Honeycomb Mountain ... Carter County, Kentucky

[171] Anon, [1993], Oregon Caves: a pictorial souvenir guide

[172] Anon, 1923, Oudtshoorn and the Cango Caves

[173] Anon, c1886, Photographic views of some of the important points of Mammoth Cave situated in Edmondson, Co. Kentucky, USA: the wonders of this cave and its magnitude, cannot be described, and must be seen to be appreciated. It is reached only by the Louisville & Nashville Railroads. These photographs were taken by magnesium light by W.H. Sesser, St. Joseph, Mich., with a Collins camera

[174] Anon, 1930, Pictures of Cheddar and Cox's cave by pen and camera

[175] Anon, c1950, Poole's Cavern: Buxton

[181] Anon, c1993, Sea Lion Caves: America's largest sea cave, on U.S. Highway 101...just North of Florence, Oregon, Oregon coast

[182] Anon, , Selected views of the beautiful caverns of Luray

[183] Anon, 1920, Shenandoah Caverns, the grotto of the gods, in the heart of the Valley of Virginia

[184] Anon, c1995, Shenandoah Caverns: a brief history

[185] Anon, nd, A short history of Kents Cavern

[186] Anon, c1987, Shri Amar Nath Ji Guide

[187] Anon, 1995, Skyline Caverns

[188] Anon, nd, Souvenir book of romantic Bridal Cave

[189] Anon, 1969, Souvenir book of the Oregon Caves National Monument: Marble Halls of Oregon

[192] Anon, c1995, The story of Howe Caverns

[193] Anon, 1934, The story of most interesting discoveries commencing August 1923; White Skar Cave; under Ingleborough, Yorkshire, 1 1/2 miles from Ingleton on the Hawes Road

[194] Anon, 1931, The story of newly discovered White Skar Caverns; under Ingleborough, Yorkshire, 1 1/2 miles from Ingleton on the Hawes Road

[195] Anon, nd, Stump Cross Caverns: the underground wonderland

[199] Anon, 1930, Subterranean wonders: Mammoth Cave and Colossal Cavern, Kentucky

[200] Anon, 1962, A symphony in limestone: the beautiful Olsen's Caves: 18 miles North of Rockhampton, Queensland: A description of Olsen's Caves

[202] Anon, 1970, There is a cavern in the town

[203] Anon, 1987, Timpanogos Cave public use and impact

[204] Anon, c1934, Twelve pictures of Cheddar and Cox's Cave

[206] Anon, 1907, Views of the Buchan Caves and pyramids

[207] Anon, 1984, Virginia Luray Caverns

[210] Anon, [c1910], Waitomo Ruakuri & Aranui New Zealand

[211] Anon, [1947], The weird wonders of Wookey Hole Caves

[212] Anon, 1992, Welcome to Waitomo Caves: a photographic insight into this spectacular region of New Zealand

[213] Anon, c1992, White Scar Caves: Britain's biggest tourist cave

[214] Anon, c1995, The Wind Cave and its territory; Garfagnana, the Apuan Alps, the Serchio Valley (Lucca, Tuscany, Italy)

[215] Anon, 1920, Wonderful, famous, spectacular Crystal Cave, Kutztown, Penna [sic]

[216] Anon, 1900, The wonders of the Grotto of Han

[218] Anon, nd, Wookey Hole Caves: sixteen exclusive camera studies

[219] Anon, [1938], Wookey Hole Caves: Wells, Somerset

[220] Anon, [1937], Wookey Hole Caves: Wells, Somerset

[221] Anon, 1986, Wookey Hole; the caves and mill

[225] Argus [pseud], 1898, Jenolan Caves and the Blue Mountains

[230] Arrell, Robert, 1984, Waitomo Caves: a century of tourism

[240] Atchison, Topeka, and Santa Fe Railway Company, 1941, Carlsbad Caverns National Park

[241] Atchison, Topeka, and Santa Fe Railway Company, 1927, Carlsbad Caverns, New Mexico

[250] Australasian Conference on Cave Tourism and Management , nd, Cave Management in Australasia VIII: proceedings of the eighth Australian conference on cave tourism and management, Paparoa National Part, Punakaiki, New Zealand, April 1989

[251] Australian Conference on Cave Tourism, Jun 1976, Cave Management in Australia: proceedings of the first Australian Conference on cave tourism, Jenolan Caves, N.S.W. 10th-13th July, 1973

[252] Australian Conference on Cave Tourism and Management, Aug 1977, Cave Management in Australia II: proceedings of the second Australian Conference on cave tourism and management, Hobart, Tasmania, 3rd-5th May, 1977

[253] Australian Conference on Cave Tourism and Management, Nov 1980, Cave Management in Australia III: proceedings of the third Australian Conference on cave tourism and management, Mount Gambier, South Australia, 30th April - 4th May, 1979

[254] Australian Conference on Cave Tourism and Management, Aug 1982, Cave Management in Australia IV: proceedings of the fourth Australian Conference on cave tourism and management, Yallingup, Western Australia Australia, September 1981

[255] Australian Conference on Cave Tourism and Man-

11.1. GENERAL

agement, 1990, Cave Management in Australia V: proceedings of the Fifth Australian Conference on cave tourism and management, Lakes Entrance, Victoria, April 1983

[261]Australian Speleological Federation, c1973, Master plan for the development of Jenolan Caves Reserve

[275]Bailey, Gilbert Stephen, 1822–1891, 1863, The great caverns of Kentucky: Diamond Cave, Mammoth Cave, Hundred Dome Cave

[276]Bailey, Gilbert Stephen, 1822–1891, Nov 1963, Hundred Dome Cave ("partial reproduction of the last known copy of the book, 'the great caverns of Kentucky'")

[279]Bailey, Vernon Orlando, 1864–1942, 1928, Animal life of the Carlsbad Cavern

[295]Balch, Herbert Ernest, 1948, The Mendip Caves

[297]Balch, Herbert Ernest, 1935, Mendip-Cheddar, its gorge and caves

[298]Balch, Herbert Ernest, 1929, Mendip: the great cave of Wookey Hole

[304]Balcombe, F. Graham, [1936], The Log of the Wookey Hole: exploration expedition 1935

[311]Bancroft, Hubert Howe, 1832–1918, 1886, Manitou Grand Caverns

[323]Barnett, John, 1927–, 1969, Carlsbad Caverns National Park

[324]Barnett, John, 1927—, 1981, Carlsbad Caverns National Park, Carlsbad, New Mexico: silent chambers, timeless beauty

[327]Barr, Thomas Calhoun, 1931–, Dec 1976, Ecological effects of water pollutants in Mammoth Cave: final technical report to the National Park Service

[328]Barr, Thomas Calhoun, 1931–, 1956, The geology of Cumberland Caverns, Warren County, Tennessee

[335]Bartel Photography Pty Ltd, 1985, Jenolan Caves New South Wales Australia

[336]Bartel Postcards, 1992, Jenolan Caves Australia

[344]Bates, Geoff, 1982, The Abercrombie Caves, New South Wales: a guide to these remarkable caves near Bathurst, N.S.W. and a description of the surrounding gold bearing country of Tuena and Trunkey Creek

[372]Bedford, Bruce L., 1942–, 1985, Underground Britain: a guide to the wild caves and show caves of England, Scotland and Wales

[398]Bigwood, K. (photo), c1980, Waitomo Caves Wonderland

[399]Binkerd, Adam D., 1869, The Mammoth Cave and its denizens: a complete descriptive guide

[400]Binkerd, Adam D., , Pictorial guide to the Mammoth Cave, Kentucky: a complete historic, descriptive and scientific account of the greatest subterranean wonder of the western world

[401]Bircham, Deric N., 1975, Waitomo tourist caves

[412]Blackhawk Hawks, 1970, Carlsbad Caverns

[422]Bogart, Robert C., 1960, Mark Twain Cave: an adventure

[425]Bohi, John W., 1963, History of Wind Cave

[426]Bohi, John W., 1962, Seventy-five years at Wind Cave: a history of the national park

[455]Bradley, J. F., Nov 1926, Glorious Kangaroo Island: its caves and beauty spots

[461]Branch, Pat, 1951, Illustrated guide book of your tour through Carlsbad Caverns

[476]Bridwell, Margaret Morris, 1905–, 1971, The story of Mammoth Cave National Park, Kentucky: a brief history

[479]Brinkley, Edward Booth, 1922–, 1964, The history of Ruby Falls

[518]Brown, Deana, nd, Caverns of Sonora: an awe-inspiring adventure you will remember forever

[520]Brown, George Henry, 1905, The Clapham Cave with illustrations from photographs by the author

[522]Brown, George Henry, 1908, Victoria Cave: its history and exploration

[527]Brown, Mary Mitchell, 1921, Nature underground: the Endless Caverns in the heart of the historic Shenandoah Valley

[530]Brown, Richmond Flint, 1925–, 1966, Hydrology of the cavernous limestones of the Mammoth Cave area, Kentucky

[531]Brown, Robert F., Oct 1964, Exploration in Wind Cave

[535]Brucker, Roger Warren, 1929–, 1966, The Flint Ridge Cave System: Mammoth Cave National Park, Kentucky; a Cave Research Foundation cartographic project in cooperation with the National Park Service

[536]Brucker, Roger Warren, 1929–, 1987, The longest cave

[540]Bryan, Benjamin, 1839, A description of the fluor spar or Blue John Mine at Castleton

[551]Bullington, Neal R., May 1968, Who Discovered Carlsbad Cavern?

[552][Bullitt, Alexander Clark, 1807–1868], 1973, Rambles in the Mammoth Cave, during the year 1844

[553]Bulpin, T. V., 1985, Kango: the story of the caves

[567]Bush, Kent, 1995, Oregon Caves National Monument collections management plan

[573]Byfield, Colin Charles, 1977, Let's look at Jenolan Caves: an introduction to the features of Australia's famous limestone caverns

[576]Caiar, Ruth, 1957, One man's dream; the story of Jim White, discoverer and explorer of the Carlsbad Caverns: a biography

[577]Caldwell, Douglas E., 1981, Limnological survey of cavern pools: proposal to National Park Service, Carlsbad Caverns National Park

[578]California. Department of Parks and Recreation, 1971, Mitchell Caverns State Reserve

[579] Call, Richard Ellsworth, 1856–1917, 1902, Grand excursion to Mammoth Cave

[580] Call, Richard Ellsworth, 1856–1917, 1897, Mammoth Cave, Kentucky

[581] Call, Richard Ellsworth, 1856–1917, c1890, The Mammoth Cave, Kentucky; a sketch

[587] Cameron McNamara Pty Ltd, 1989, Jenolan Caves Reserve: plan of management.

[596] Cansell, Anthony, [after 1982], Stump Cross Caverns - the underground wonderland, their development and exploration

[599] Carlsbad Cave National Monument, 1930, Carlsbad Caverns, New Mexico

[600] Carlsbad Caverns National Park, Dec 1984, Land protection plan: Carlsbad Caverns National Park: prepared by Carlsbad Caverns National Park with assistance from the Southwest Regional Office, National Park Service.

[602] Carstens, Kenneth Charles, 1996, Of caves and shell mounds

[618] Cave Research Foundation, 1992, The quadrangle maps of Carlsbad Cavern, New Mexico

[635] Chandler, Peter, 1987, A trip through time: a drive-yourself guide to the landforms and geology of the Waitomo Caves area

[646] Chenger, John D., 1995, Visitor's guide to Laurel Caverns Park

[662] Clark, E. John, nd, Jenolan Caves, New South Wales, Australia

[666] Close, Danny, 1981, Mammoth Cave... kids love it!

[670] Coase, Alan C., nd, Cathedral Cave situated between Brecon and Swansea on the A. 4067 road (750 yards North of Craig-y-Nos Castle) Dan-yr-Ogof Swansea Valley Caves

[673] Coase, Alan C., 1978, Official guide to Dan-yr-Ogof showcave

[675] Coates, Dorothy, 1960, Cheddar caves in the Cheddar Gorge: an illustrated survey & guide

[688] Commission on Cave Tourism and Management of the Australian Speleological Federation Study Team, Jun 1984, Jenolan Caves resort - some management issues

[692] Conn, Herbert Dunn, 1981, The Jewel Cave adventure: fifty miles of discovery under South Dakota

[693] Conneely, Jim, c1989, Aillwee Cave: Ballyvaughan, Burren Co., Clare

[699] Cook, Samuel, 1889, The Jenolan caves: an excursion in Australian wonderland

[719] Corrie, A. P., 1899, The Wombeyan Caves and the Bowral, Mittagong & Moss Vale tourist districts

[724] By "A Country Doctor", c1929, The wonders of Castleton

[728] Cox, Edward, c1914, Souvenir of Cox's Stalactite Caves

[729] Cox, Edward, 1910, Souvenir of Cox's Stalactite Caves "visited by his majesty the late King Edward VII"

[730] Cox, Edward Travers, 1821-1907, 1903, Wyandote Cave, pronounced by eminent scientists and noted writers of wide experience and extensive travel to be the greatest natural attraction of its kind in the world

[735] Craven, Stephen Adrian, 1994, Cango Cave, Oudtshoorn district of the Cape Province, South Africa: an assessment of its development and management 1780-1992

[754] Crestwood House, 1988, National parks: v. 2: Carlsbad Caverns

[756] Crofton, Denis, 1901, All about Poole's Cavern, Buxton. An official guide to this unique natural curiosity, together with an inventory of the museum

[757] Croghan, John, 1790–1849, 1845, Rambles in Mammoth Cave, during the year 1844, by a visitor

[765] Crowther, Patricia P., 1984, The Grand Kentucky junction: memoirs

[767] Crysler, Mildred G., 1955, Adventures in underground fairylands: a fantasy tour of the Carlsbad Caverns

[768] Cudmore, Dana D., 1990, The remarkable Howe Caverns story

[785] Curt Teich & Co, Inc, 1963, Carlsbad Caverns National Park, New Mexico

[786] Cushman, Robert Vittum, 1967, Hydrologic problems as related to wilderness management and water supplies for recreational areas at Mammoth Cave National Park, Kentucky

[787] Cushman, Robert Vittum, 1965, Present and future water supply for Mammoth Cave National Park, Kentucky

[790] d'Amboise, Valery (auth, illus, photo), 1986, Eternal caves

[798] Damon, Paul Herbert, 1934–, Jun 1976, The history of Laurel Caverns of Fayette County, Pennsylvania, known throughout the years as Delaney's Cave

[804] Darton, Nelson Horatio, 1865–1948, 1932, Western Texas and Carlsbad Caverns

[817] Davey, Adrian G., 1978, Yallingup Cave Park - a management plan

[818] Davidge, G. W., nd, Narracoorte [sic]: caves and town 20: book of views

[820] Davidson, Joseph Killworth, 1938–, 1971, Wilderness resources in Mammoth Cave National Park: a regional approach

[821] Davidson, Robert, 1808–1876, 1840, An excursion to the Mammoth Cave, and the barrens of Kentucky: with some notices of the early settlement of the state

[831] Davies, William Edward, 1917–, Aug 1959, Report on sediments in Mammoth Cave, Kentucky

[835] Davis, M. J., 1954, Nobody here but us bats!: through Carlsbad Caverns with cartoons and comments

[851] Delderfield, Eric Raymond, c1959, The story of Brixham Cavern

[857] De Paepe, Duane, 1985, Gunpowder from Mammoth Cave: the saga of saltpetre mining before and during the

11.1. GENERAL

War of 1812

[861]Despain, Joel, 1995, Crystal cave: a guidebook to the underground world of Sequoia National Park

[872]Dillaye, Stephen D., 1848, A glance at Mammoth Cave, Kentucky

[892]Dowie, H. G., 1932, The history of Kent's Cavern; Torquay: with illustrations

[905]Driscoll, Ian, Apr 1977, The Binoomea Cut, Jenolan Caves: a paper read to the Jenolan Caves Historical and Preservation Society on Saturday 9th, February 1974

[912]Dunham, Wayland A., 1898–, 1939, Enchanted corridors

[913]Dunkley, John R. (comp), nd, A bibliography of the Jenolan Caves: part 1: speleological literature

[914]Dunkley, John R. (comp), 1988, A bibliography of the Jenolan Caves: part 2: literature

[915]Dunkley, John R., 1978, The caves of Jenolan: 1, the exploration and speleogeography of Mammoth Cave, Jenolan

[918]Dunkley, John R., 1971, The exploration and speleogeography of Mammoth Cave, Jenolan

[920]Dunkley, John R., 1986, Jenolan caves as they were in the nineteenth century

[921]Dunkley, John R., Nov 1985, Preliminary checklist of references to Jenolan Caves

[924]Dunlop, Bruce Thomas, , Jenolan Caves

[925]Durden, C. C., 1928, The Carlsbad Cavern of New Mexico: it's [sic] history and geology

[929]Dyer, C. F., 1961, Geology and occurrence of ground water at Jewel Cave National Monument, South Dakota

[930]Dyer, William Henry, 1979, A brief story of St Clement's caves

[946]Edwards, Ira, 1925, Underground Geology at the Endless Caverns, New Market, VA

[958]Elliot, Ian, 1977, The discovery and exploration of the Yanchep Caves

[973]Ellis, Robert D. (scrn wrt), 1993, Carlsbad Caverns National Park

[981]Endless Caverns, Inc, 1930, Endless caverns, New Market, Virginia, in the heart of the Shenandoah Valley...3 minute drive off U.S. 11 (Lee highway) over new boulevard

[982]Endless Caverns, Inc, 1939, Endless Caverns, Virginia. "Considered the most beautiful of the Shenandoah Valley caverns" ...

[983]Endless Caverns, Inc, 1921, Endless Caverns, wonderful and spectacular

[984]Endless Caverns, Inc, nd, Endless Caverns, wonderful and spectacular, New Market, Virginia, in the heart of the Shenandoah Valley

[985]Endless Caverns, Inc, 1987, Endless Caverns: on U.S. 11, 3 miles south of New Market, Virginia: illuminated by indirect electric flood lighting

[986]Endless Caverns, Inc, 1926, Exploring the Endless Caverns of New Market, Virginia in the heart of the Shenandoah Valley: two authentic and thrilling accounts of the adventures of members of the American Museum of Natural History and of the Explorers Club, who vainly sought the end to the Endless Caverns

[987]Endless Caverns, Inc, 1947, Exploring the Endless Caverns of New Market, Virginia. Authentic and thrilling accounts of the adventures of members of the Explorers Club and of the American Museum of Natural History

[988]Endless Caverns, Inc, 1946, Nature underground; the Endless Caverns in the heart of the historic Shenandoah Valley

[994]English, Jackie (comp, photo), c1980, Mitchelstown Cave: its discovery and history

[1012]Evans, Gregory T. , 1976, Oregon Caves National Monument: a manual for cave guides

[1014]Eversole, Jack (ed), 1973, The Mammoth Cave area: a planning proposal

[1030]Farr, Martyn, 1950–, 1985, Wookey: the caves beyond

[1063]Finley-Holiday Film Corporation, 1980, Carlsbad Caverns National Park & Guadalupe Mountains National Park

[1064]Finley-Holiday Film Corporation, 1988, Luray Caverns in Virginia's Shenandoah Valley

[1065]Finley-Holiday Film Corporation, 1990, Mammoth Caves

[1067]Fleming, John, 1973, Blanchard Springs Caverns story

[1069]Flindall, Roger, 1976, The caverns and mines of Matlock Bath: 1 Nestus Mines: Rutland and Masson caverns

[1073]Floyd, E. F., 1887, The New Calaveras Cave, Murphys, Calaveras County, California: W.J. Mercer, proprietor: description and guide

[1091]Ford, Trevor David, 1925–, 1975, Ingleborough Cavern and Gaping Gill

[1096]Ford, Trevor David, 1925–, Jul 1977, The story of Poole's Cavern

[1097]Ford, Trevor David, 1925–, 1962, The story of the Speedwell Cavern

[1103]Forwood, William Stump, 1830–1892, 1870, An historical and descriptive narrative of the Mammoth Cave of Kentucky: including explanations of the causes concerned in its formation, its atmospheric conditions, its chemistry, geology, zoology, etc., with full scientific details of the eyeless fishes

[1107]Foster, J. J., 1890, The Jenolan Caves

[1124]Fraser, L.A., Jul 1892, Weyer Cave. Cave of the Fountains. Madison Cave: a descriptive sketch of the grottoes of the Shenandoah

[1133]G. P. Slide Co, 1970, Luray Caverns

[1136]GAF Corporation , 1950, Carlsbad Caverns National Park

[1137]GAF Corporation, 1970, Luray Caverns, Va
[1160]Garrett, Willie Ann, 1949, The Carlsbad Cavern National Park
[1161]Garrigan, George A., 1994, Skyline Caverns and its geologic relationship to the Shenandoah Valley and paleoindian cultures
[1167]Gatacre, E. V., Mar 1975, Wookey Hole [Including articles on] (paper making, Lady Bangor's fairground collection [and] Madame Tussaud's store)
[1176]George, Angelo Isham, 1944–, 1985, Mummies of Short Cave, Kentucky, and the Great Catacomb Mystery
[1177]George, Angelo Isham, 1944–, 1994, Mummies, catacombs, and Mammoth Cave
[1178]George, Angelo Isham, 1944–, 1992, The New Madrid earthquake at Mammoth Cave (1811-1812)
[1179]George, Angelo Isham, 1944–, 1990, Prehistoric mummies from the Mammoth Cave area: foundations and concepts
[1181]George, Angelo Isham, 1944–, 1991, Wyandotte Cave down through the centuries
[1184]Gerrard, John, 1960, Kent's Cavern
[1204]Gill, George, nd, Stump Cross Caverns
[1205]Gill, Thomas B., [1936], Guide to the home of the rainbow: Cox's Cave, Cheddar, Somerset
[1206]Gill, Thomas B., [1936], Guide to the world famous Gough's Caves, Cheddar, Somerset
[1223]Goode, Cecil E., 1986, World wonder saved: how Mammoth Cave became a national park
[1226]Goodwin, J. West, , Mammoth Cave: America's great natural wonders
[1230]Gough, A. G. H., 1912, Guide to Gough's Caves, Cheddar; how they were found, the finest in the world (electrically illuminated) eighteen beautiful views of the cave prehistoric man: - how he was found
[1231]Gough, A. G. H., nd, Pictorial guide to the caves Cheddar
[1236]Graham, Joseph J., 1939, The Florida Caverns at Marianna, Florida
[1239]Grange Cavern Military Museum, 1980, Souvenir guide to the world's largest underground military museum: the Grange Cavern
[1240]Grant, Blanche Chloe, 1874–1948, 1927, Carlsbad Caverns
[1241]Grant, Blanche Chloe, 1874–1948, 1928, Cavern guide book, Carlsbad Caverns, Carlsbad, N. M
[1255]Greer, John W., 1966, Report on preliminary archeological explorations at Carlsbad Caverns National Park, New Mexico
[1256]The Gregory Publishing Co Pty Ltd, nd, Jenolan Caves (N. S. W.) 'Nature's Masterpiece': a complete description of the geology, discovery, development, and features of the Jenolan limestone system beautifully illustrated, with high-class maps
[1269]Grobet, André-H., May 1966, The underground lake of St. Léonard: Valais
[1276]Guernsey, Daniel Riley, 1905, Lost in the Mammoth Cave
[1294]Gurnee, Russell Hampton, 1922–1995 (comp), 1983, Cave clippings of the nineteenth century
[1297]Gurnee, Russell Hampton, 1922–1995, , Discovery of Luray Caverns, Virginia
[1298]Gurnee, Russell Hampton, 1922–1995, 1980, Gurnee guide to American caves: a comprehensive guide to the caves in the United States open to the public
[1299]Gwin, J. N., 1875, A visit to the Mammoth Cave
[1301]Habe, France, 1988, Jame Škocjan = Grotte Skocjan
[1302]Habe, France, 1969, Postojna
[1303]Habe, France, 1976, Postojna Cave
[1304]Habe, France, 1973, Postojna Caves
[1305]Habe, France, 1977, The Postojna Caves and other tourist caves in Slovenia
[1306]Habe, France, 1972, The Postojna Grottes with the Planina and Predjama caves
[1307]Habe, France, 1973, The Postojna Grottoes
[1308]Habe, France, 1980, Predjama: the castle and the cave
[1309]Habe, France, 1978, Show caves in Slovenia - Guide
[1313]Halaburda, Sue, 1984, Mammoth Cave
[1316]Halladay, Orlynn J., 1972, The Lehman Caves story
[1329]Halliday, William R., 1926–, 1976, Depths of the earth: caves and cavers of the United States
[1344]Hamilton-Smith, Elery, Jun 1989, Batu Caves, Kuala Lumpur - towards a plan of management
[1347]Hamilton-Smith, Elery (prep), Jan 1989, Cutta Cutta Caves Nature Park: draft plan of management
[1348][Hamilton-Smith, Elery], Sep 1994, Determining an environmental and social carrying capacity for Jenolan Caves Reserve: applying a visitor impact management system
[1351]Hamilton-Smith, Elery (tm ldr), Nov 1988, Proposed developments at Waitomo Caves
[1361]Harpers Ferry Center. Division of Museum Services, 1976, Mammoth Cave National Park: collection management plan
[1367]Harrisonburg-Massanutten Corp (Harrisonburg, Va), 1928, About Massanutten caverns; the jeweled chamber that hides the secret of Nakwisi-Gatusi
[1368]Harrisonburg-Massanutten Corp (Harrisonburg, Va), 1925, Massanutten Caverns, Harrisonburg, Va. The most beautiful and unusual caverns yet discovered...
[1369]Harrisonburg-Massanutten Corp (Harrisonburg, Va), 1940, Massanutten Caverns, Harrisonburg, Va.: the most beautiful and unusual caverns yet discovered, in the heart of the Shenandoah Valley of Virginia
[1370]Harrisonburg-Massanutten Corp (Harrisonburg, Va), nd, The wonderful Massanutten Caverns in the

11.1. GENERAL

Shenandoah Valley, Harrisonburg, Va

[1371]Hart, William J., Sep 1967, A wilderness plan for Mammoth Cave National Park and the surrounding region

[1388]Havard, Ward L., nd, The Orient Cave, Jenolan Caves

[1389]Havard, Ward L., nd, The Romance of Jenolan Caves

[1394]Hay, William Perry, 1872–, 1903, Observations on the crustacean fauna of the region about Mammoth Cave, Kentucky

[1396]Hayllar, Bruce (proj offr), Mar 1984, The development of leisure and educational resources at Jenolan Caves Resort

[1397]Hayllar, Bruce (proj offr), Jan 1985, Jenolan Caves Resort teachers kit

[1411]Hazslinszky, Tamás (auth, ed, trans), 1989, Show caves of Hungary

[1419]Hedinger, J. M., 1820, A short description of Castleton in Derbyshire: its natural curiosities and mineral productions

[1424]Hellenic Speleological Society, 1983, International meeting on the show caves and their problems

[1425]Helm, Tex, 1969, Carlsbad Caverns; National Park, New Mexico

[1431]Henderson, Kent, 1992, The Wombeyan and Abercrombie caves

[1432]Henderson, Kent, 1954–, 1985, The Buchan experience: a guide to the Buchan and Murrindal Caves - East Gippsland, Victoria

[1433]Henderson, Kent, 1954–, 1990, Jenolan: a guide to Australia's famous caves

[1434]Henderson, Kent, 1954–, 1986, The Naracoorte and Tantanoola Caves: a guide to the famous caves of Southeast South Australia

[1436]Henderson, Kent, 1954–, 1988, The Wellington and Abercrombie Caves

[1437]Henderson, Kent, 1954–, 1985, The Wombeyan experience: a guide to the Wombeyan Caves New South Wales

[1458]Hill, Carol Ann, 1940–, 1988, The geology of Carlsbad Cavern

[1462]Hill, John R., nd, The geologic story of Wyandotte Cave

[1470]Hobbs, Horton Holcombe, 1944–(ed), , A study of environmental factors in Harrison's Cave, Barbados, West Indies

[1472]Hodgkinson, Gerald William, c1964, Wookey Hole: the cave of mystery and history

[1473]Hodgkinson, Olive, 1970, The story of Wookey Hole

[1481]Holiday Film Corporation, 1970 1980, Carlsbad Caverns, New Mexico

[1482]Holiday Film Corporation, 1984, Luray Caverns

[1483]Holiday Film Corporation, 1970, Mammoth Cave National Park

[1484]Holiday Film Corporation, 1991, Meramec Caverns, Stanton, Missouri

[1488]Holler, Cato Oliver, 1944–, 1989, Hollow Hills of Sunnalee: The Linville Caverns story

[1496]Holsinger, John Robert, 1934–, Jan 1985, Ecological analysis of the Kentucky cave shrimp, *Palaemonias ganteri* Hay, at Mammoth Cave National Park (phase V)

[1497]Holsinger, John Robert, 1934–, 1986, Ecological analysis of the Kentucky cave shrimp, *Palaemonias ganteri* Hay, at Mammoth Cave National Park (phase VI): preliminary observations on stream interstitial meiofauna communities and related abiotic factors

[1498]Holsinger, John Robert, 1934–, Oct 1983, Ecological analysis of the Kentucky cave shrimp, *Palaemonias ganteri* Hay, Mammoth Cave National Park

[1499]Holsinger, John Robert, 1934–, 1982, Ecological analysis of the Kentucky cave shrimp, *Palaemonias ganteri* Hay, Mammoth Cave National Park (phase I)

[1500]Holsinger, John Robert, 1934–, Oct 1983, Ecological analysis of the Kentucky cave shrimp, *Palaemonias ganteri* Hay, Mammoth Cave National Park (phase II)

[1501]Holsinger, John Robert, 1934–, Apr 1983, Ecological analysis of the Kentucky cave shrimp, *Palaemonias ganteri* Hay, Mammoth Cave National Park (phase III)

[1512]Holt, Robert (comp), 1989, Howe Caverns; Howes Cave N.Y.; 60th Anniversary, 1929-1989

[1516]Horn, Edward Courtright, 1871-, 1901, Mazes and marvels of Wind cave

[1517]Horne, Julia, 1994, Jenolan Caves: when the tourists came

[1527]Hosford, Jessie, 1953, Listen to the rain; a story of the Carlsbad Caverns

[1528]Hoskins, R. Taylor, 1940, Caverns of enchantment

[1534]Hovey, Edmund Otis, 1801–1877, 1902, Great American caverns. The underground beauties and wonders of Mammoth, Wyandotte, Luray, and other famous caves ...

[1535]Hovey, Horace Carter, 1833–1914, 1970, Celebrated American caverns: especially Mammoth, Wyandot, and Luray: together with historical, scientific, and descriptive notices of caves and grottoes in other lands

[1536]Hovey, Horace Carter, 1833–1914, 1882, Celebrated American caverns: especially Mammoth, Wyandot, and Luray: together with historical, scientific, and descriptive notices of caves and grottoes in other lands

[1537]Hovey, Horace Carter, 1833–1914, 1891, Guide book to the Mammoth Cave of Kentucky: historical, scientific, and descriptive

[1538]Hovey, Horace Carter, 1833–1914, 1909, Hovey's hand-book of the Mammoth Cave of Kentucky: a practical guide to the regulation routes with maps and illustrations

[1539]Hovey, Horace Carter, 1833–1914, 1912, Mammoth Cave of Kentucky: with an account of Colossal Cavern

[1540] Hovey, Horace Carter, 1833–1914, 1897, Mammoth Cave of Kentucky; an illustrated manual

[1541] Hovey, Horace Carter, 1833–1914, 1982, One hundred miles in Mammoth Cave - in 1880: an early exploration of America's most famous cavern

[1544] Howe Caverns, Inc, nd, Marvelous Howe Caverns

[1545] Howe Caverns, Inc, 1930, The story of Howe caverns...

[1547] Howe, Warren, 1988, The Howe Family From Massachusetts to New York: A story of Lester Howe, his ancestors, and the Howe Caverns

[1581] Hurley, Frank, 1890–1962, 1952, The Blue Mountains and Jenolan Caves: a camera study

[1582] Hurley, Frank, 1890–1962 (photo), nd, Gems of Jenolan

[1591] Hyslop, R. C., 1967, Augusta Jewel Caves and other points of interest

[1593] Imperial Film Company, 1965, Mammoth Cave, Kentucky

[1606] Indiana. Dept of Natural Resources. Engineering Division, 1968, Master plan: Wyandotte Caves, Harrison-Crawford, Blue River Recreation Complex

[1609] Innes-Smith, Robert, 1976, Castleton and its caves

[1663] International Union of Speleology. Commission for Cave Protection and Cave Tourism, 1989, Cave tourism: proceedings of international symposium at 170-anniversary of Postojnska jama, Postojna, Yugoslavia, Nov. 10-12, 1988

[1665] Iorio, Ralph, 1968, The history of Timpanogos Cave National Monument: American Fork Canyon, Utah

[1671] Irwin, David J. (comp), Aug 1995, A catalogue of ephemera relating to Wookey Hole

[1675] Irwin, David J., Apr 1986, Cheddar Caves bibliography

[1682] Jackson, George Frederick, 1906-1981, 1972, The history and exploration of Wyandotte Cave

[1683] Jackson, George Frederick, 1906-1981, 1975, The story of Wyandotte Cave

[1684] Jackson, George Frederick, 1906-1981, 1953, Wyandotte Cave

[1685] Jackson, Harold, c1969, Sudwala

[1695] James, C. H., (photo), 1882, Electric light views in the Caverns of Luray at Luray Station (Page Co., Virginia), Shenandoah Valley Railroad

[1708] Jarman, R. A., 1991, Ingleborough Cavern and Gaping Gill

[1730] Jensen, Douglas N., 1988, Discover Timpanogos Cave National Monument

[1733] Jillson, Willard Rouse, 1890–, 1953, A bibliography of Mammoth Cave, 1798-1949

[1741] Johnston, Frances Benjamin, 1864–1952, 1893, Mammoth Cave by flash-light

[1747] Jones, Ivor Wynne, 1986, Llechwedd

[1748] Jones, Ivor Wynne, 1976, Llechwedd Slate Caverns

[1755] Jones, William Basil, 1822–1897, 1844, Wonderful curiosity, or, a correct narrative of the celebrated Mammoth Cave of Kentucky with incidents and anecdotes

[1759] Jordan, J. Murray, 1897, The Caverns of Luray and vicinity

[1776] Kauffman, D. L., 1913, The beautiful Luray Caverns, Luray, Virginia

[1779] Keck, Ken E., Nov 1985, An introduction to Abercrombie Caves

[1780] Kehret, Roger Alan, 1942–, 1976, Crystal Lake Cave Iowa

[1783] Kehret, Roger Alan, 1942–, 1974, Your guide to Mystery Cave

[1784] Kehret, Roger Alan, 1942–, 1976, Your guide to Spook Cave

[1786] Kell, Neil, [1993], The Winston Churchill Memorial Trust of Australia: project: to study recent developments in the design and installation of lighting to enhance the public education and enjoyment of show caves

[1794] Kent, Paul, 1983, Tour caves of the Ozarks: Missouri Arkansas: the secret beneath the surface

[1812] Kerry, Charles (photo), nd, Right Imperial: Carlotta Arch: Devil's Coach House

[1813] Kessler, Hubert, 1907–1994, 1971, Aggtelek

[1814] Keyes, Elizabeth, 1930, Titania's cave; a fairy story of the Endless Caverns of Virginia

[1828] Kmecl, Matjaž, 1990, Postojna caves: enter traveller into this immensity!

[1834] Knoepfel, William H., 1853, An account of Knoepfel's Schoharie Cave, Schoharie County, New York: with the history of its discovery, subterranean lake, minerals and natural curiosities

[1837] Knox, Orion, 1976, Natural Bridge Caverns

[1841] Kobert, Bill, nd, Caverns of the Ozarks

[1860] Kranjc, Andrej (ed), 1989, Proceedings of international symposium at 170th anniversary of Postojnska jama, Postojna (Yugoslavia) Nov. 10-12 1988

[1869] Kuhns, Roger J., 1982, Speleology of Lewis and Clark Caverns, Montana

[1882] Lamb, Susan, 1991, Lava Beds National Monument

[1900] Larimore, Betty, 1947, Exploring the Endless Caverns of New Market Virginia on US 11 in the heart of the Shenandoah Valley

[1907] Larson, Charles Victor, 1928–, 1989, Lava Beds caves

[1921] Leck Studio, 1920, The wonderful Carlsbad Caverns: near Carlsbad, New Mexico

[1926] Lee, Willis Thomas, 1924, A visit to Carlsbad Cavern: recent explorations of a limestone cave in the Guadalupe Mountains of New Mexico reveal a natural wonder of the first magnitude

[1929] Lehman Caves National Monument, 1974, Pro-

11.1. GENERAL

posed natural resources management plan, Lehman Caves National Monument, Nevada

[1936]Letcher, Montgomery E., 1839, Wonderful discovery: being an account of a recent exploration of the celebrated Mammoth Cave, in Edmonson County, Kentucky, by Dr. Rowan, Professor Simmons and others, of Louisville, to its termination in an inhabited region, in the interior of the earth

[1939]Lewis, Ian D., 1977, Discover Naracoorte caves

[1946]Link, Louis W., , Lewis and Clark Cavern, Montana

[1950]Livesay, Elizabeth Ann, 1962, Geology of the Mammoth Cave National Park area

[1953]Lobeck, Armin Kohl, 1886–1958, 1928, The geology and physiography of the Mammoth Cave national park

[1957]Long, Abijah, 1956, The big cave: early history and authentic facts concerning the history and discovery of the world famous Carlsbad Caverns of New Mexico

[1958]Long, Gary F. (auth, des, photo), nd, Tuckaleechee Caverns: 31 photos in natural color: Townsend, Tennessee: greatest sight under the Smokies

[1981]Luce, Pat (comp), 1940, Let's see Marvel Cave: your personal guide to America's largest privately owned cave

[1990]Luray Caverns Corporation, 1953, The beautiful caverns of Luray, Luray, Virginia, in the Shenandoah Valley

[1991]Luray Caverns Corporation, 1906, The beautiful caverns of Luray, Luray, Virginia. In the Shenandoah Valley, three miles of subterranean splendor, brilliantly lighted by electricity

[1992]Luray Caverns Corporation, 1949, The beautiful caverns of Luray: Luray, Virginia

[1993]Luray Caverns Corporation, 1930, The beautiful caverns of Luray: Luray, Virginia, in the Shenandoah Valley: miles of subterranean splendor, brilliantly lighted by electricity

[1994]Luray Caverns Corporation, nd, Caverns of Luray

[1996]Luray Caverns Corporation, 1900, Luray Caverns

[1997]Luray Caverns Corporation, 1900, 1909, Luray Caverns of the beautiful Shenandoah Valley on the Norfolk and Western Railway, Luray, Virginia. Luray, Virginia Mansion Inn

[2003]Lyons, Joy Medley, 1991, Mammoth cave

[2008]MacDougall, J. H., [1922], Western Australia's wonderland

[2017]Maddox, Gary L., Nov 1993, Karst features of Florida Caverns State Park and Falling Waters State Recreation Area, Jackson and Washington Counties, Florida

[2022]Mais, Stuart Peter Brodie, c1934, 12 natural colour photographs of Cheddar and Cox's Cave "The home of the rainbow"

[2029]Mammoth Cave National Park Association, 1928, Mammoth Cave National Park Association campaign, 1927-1928

[2030]Mammoth Cave National Park Association, 1985, A national park in Kentucky

[2031]Mammoth Cave National Park Association, 1927, Now how! A national park in Kentucky

[2032]Mammoth Cave National Park Association, nd, A privilege—a duty—an opportunity for Kentucky

[2033]Mammoth Cave Operating Committee, Mammoth Cave, Ky, 1940, Caverns of enchantment

[2034]Mammoth Cave Operating Committee, Mammoth Cave, Ky, 1859, Mammoth cave

[2056]Martel, Edouard Alfred, 1859–1938, c1950, Cavern of the dragon

[2058]Martel, Edouard Alfred, 1859–1938, 1923, The Grottoes of the Drach near Manacor (Majorca)

[2059]Martel, Edouard Alfred, 1859–1938, 1980, Padirac; its history and a short description

[2062]Martin, George V., 1973, The Timpanogos Cave story: the romance of its exploration

[2063]Martin, Horace, 1851, Pictorial guide to the Mammoth Cave, Kentucky

[2067]Martin, Ronald L., 1990, Marvel Cave: Silver Dollar City, Branson, Missouri

[2068]Martin, Ronald L., 1974, Official guide to Marvel Cave, Silver Dollar City, Missouri

[2073]Mason, Edmund John, 1911–1993, Jul 1963, The story of Wookey Hole

[2075]Mason, Otis Tufton, 1838–1908, , The caverns of Luray

[2076]Mason, Otis Tufton, 1838–1908, 1882, The caverns of Luray

[2077]Mason, Otis Tufton, 1838–1908, 1881, Report of a visit to the Luray Cavern, in Page County, Virginia, under the auspices of the Smithsonian institution, July 13 and 14, 1880

[2081]Mass, Nuri, 1946, The wizard of Jenolan

[2088]Matthews, Larry Edwin, 1946–, 1989, Cumberland Caverns

[2094]Matthews, William Henry, 1919–, Feb 1963, The geologic story of Longhorn Cavern

[2105]McCann, Gerald, [1992], In my torchlight: a guide's guide to the Cango Caves

[2116]McDonald, Alvin F., 1873–1893, nd, Private account of A. F. McDonald, permanent guide of Wind Cave

[2117]McDonald, Donald L., 1974, Official guide to Marvel Cave: the only complete authentic history of Marvel Cave ever published

[2133]McLean, John Scott, 1942–, Feb 1976, Factors altering the microclimate in Carlsbad Caverns, New Mexico

[2134]McLean, John Scott, 1942–, May 1971, The microclimate in Carlsbad Caverns, New Mexico

[2145]Meloy, Harold Raymond, 1913–1985, , Mummies of Mammoth Cave: an account of the Indian mummies

discovered in Short Cave, Salts Cave, and Mammoth Cave, Kentucky
[2146]Meloy, Harold Raymond, 1913–1985, 1968, Mummies of Mammoth Cave: Fawn Hoof, Little Alice, Lost John and others. An account of the Indian mummies discovered in Short Cave, Salts Cave, and Mammoth Cave, Kentucky
[2162]Messling, Gordon, 1978, Caverns of Sonora
[2165]Miccolis, Maria, 1980, The Castellana Grottoes
[2169]Middleton, Gregory J., 1970, Yarrangobilly Caves, Kosciusko National Park
[2190]Mindick, Robert R., 1977, Commercial caves in Southern Indiana
[2207]Missouri Speleological Survey, January-December 1987, The commercial caves of Camden County, Missouri
[2238]Mitchell, Jack, 1964, Jack Mitchell, caveman
[2249]Mohr, Charles E., 1907–(ed), 1955, Celebrated American caves
[2267]Morana, Martin, 1987, The prehistoric cave of Gher Dalam
[2270]Morgan, Llewellyn E. (auth, photo), nd, Guide to Dan-yr-Ogof Caves: Swansea Valley Caves
[2272]Morgan, Robert, nd, Through Mammoth Cave, giving a graphic description of all the routes, avenues, passes, passages, pits, domes, streams, rivers, seas, lakes, and other wonders to be found in that greatest of all natural curiosities in the whole realm of this great world
[2273]Morgan, Robert, 1881, Through Mammoth Cave, giving a graphic description of all the routes, avenues, passes, passages, pits, domes, seas, lakes, rivers, streams, and other wonders of that greatest natural curiosity in the world
[2279]Morrison, Geo. D., c1920, New entrance to Mammoth Cave: a history of the cave together with descriptive guide to routes and illustrations of points of interest
[2280]Moseley, John Judy, 1948, History, geology of Carlsbad Caverns: what' a hole
[2281]Moseley, John Judy, 1941, What' a hole: condensed Carlsbad Cavern [i.e. Caverns] information
[2307]Mussaeus, Thomas Allen, 1887–, 1939, The lure of cave lore
[2308]Mussaeus, Thomas Allen, 1887–, 1950, The lure of cave lore; being a random narrative upon caverns in general and in particular the properties of the Skyline Caverns, Front Royal, Virginia
[2351]National Conservation Commission of Barbados, [1986], Harrison's Cave: Barbados
[2355]National Speleological Society (comp), 1951, America's wonderland of caves: palaces under the earth: a directory of commercially operated caves
[2360]National Speleological Society, 1990, Wind Cave: the world below
[2383]National Speleological Society. Annual Convention. Guidebook (1960: NM), 1960, A guide book to Carlsbad Caverns National Park
[2489]Nelson, Nels Christian, 1875–, 1917, Contributions to the archaeology of Mammoth Cave and vicinity, Kentucky
[2490]Nelson, Raymond L., Nov 1959, Mammoth Cave tour leader manual
[2492]New Mexico Geological Society, 1954, Guidebook of southeastern New Mexico: fifth field conference, October 21-22-23 & 24, 1954
[2494]New South Wales Department of Tourism, 1973, Jenolan Caves, New South Wales
[2495]New South Wales Immigration and Tourist Bureau, 1915, The systems of limestone caves: Jenolan, Yarrangobilly, Wombeyan, Abercrombie and Wellington
[2497]New South Wales Teaching Resources Centre, 1971, A visit to the Jenolan Caves: reference notes for the teacher
[2501]Newbould, Ronald L., Mar 1975, The discovery of Barralong Cave, Jenolan Caves, N.S.W.: based on the diary of Ronald L. Newbould, guide, Jenolan Caves, co-discoverer with John P. Culley, senior guide, Jenolan Ceves [sic] of the "Barralong Cave" on the 7th June, 1964
[2502]Newbould, Ronald L., 1976, Steam cleaning of Orient Cave
[2503]Newbould, Ronald L., May 1974, Steam cleaning of Orient Cave, Jenolan Caves, N.S.W
[2504]Newell, Amy L., 1995, The caves of Put-in-Bay
[2508]Nguyen Tai Duong (comp) , [1994], Vietnam cavern tourism
[2509]Nicholson, Frank Ernest, Oct 1930, The exploration of Carlsbad Cavern
[2513]Noah, Samuel, 1835, A description of Weast's Cave
[2514]Noble, Bruce J., 1991, Cultural resource management in Mammoth Cave National Park: a National Park Service—Kentucky Heritage Council cooperative project
[2517]Northcott, T. C., c1995, How nature made Luray Caverns, Virginia
[2518]Northcott, T. C., 1934, How nature makes a cave
[2526]Norton, John S., 1990, The Cave of the Winds
[2537]Nymeyer, Robert, 1910–1983, 1991, Carlsbad Cavern, the early years: a photographic history of the cave and its people
[2538]Nymeyer, Robert, 1910–1983, 1978, Carlsbad, caves, and a camera
[2539]Oakley, Carey B., 1990, Archaeological testing at site 1Ms357, Cathedral Caverns, Marshall County, Alabama
[2540]Oberhansley, Frank R., May , Crystal Cave in Sequoia National Park
[2544]O'Connor, Donal, 1991, Crag Cave: Castle Island: Co. Kerry: Ireland's most exciting show cave
[2546]Ohio Caverns, Inc., 1971, Ohio Caverns: where nature carved a fairyland
[2563]Oldham, Anthony Clive [Tony], 1939–, 1972, Dis-

11.1. GENERAL

covering caves

[2567] Oliva, Deane, 1991, A visitor's guide to Mammoth Cave National Park
[2568] Oliver, Douglas L., 1930, A boy scout in the grand cavern
[2569] Ollerenshaw, Arthur Edward, 1964, Blue John Caverns and Blue John Mine, Castleton, Sheffield
[2580] Osterreichische Fremden Verkehrswerbung (ed), , Caves in Austria
[2581] Osterreichische Fremden Verkehrswerbung (ed), nd, Caves in Austria
[2584] Owen, Luella Agnes, 1970, Cave regions of the Ozarks and Black Hills
[2589] P. Tarrant Ltd, nd, Waitomo Caves New Zealand: a souvenir booklet of your Waitomo visit with story of life cycle of New Zealand glow-worm
[2591] Packard, Alpheus Spring, 1839–1905, 1872, The Mammoth Cave and its inhabitants, or, descriptions of the fishes, insects and crustaceans found in the cave with figures of the various species and an account of allied forms, comprising notes upon their structure, development and habits, with remarks upon subterranean life in general
[2596] Palmer, Arthur Nicholas, 1940–, 1981, A geological guide to Mammoth Cave National Park
[2598] Palmer, Arthur Nicholas, 1940–, 1981, The geology of Wind Cave
[2599] Palmer, Arthur Nicholas, 1940–, 1982, A guide to the historic section of Mammoth Cave
[2600] Palmer, Arthur Nicholas, 1940–, 1984, Jewel Cave: a gift from the past
[2602] Palmer, Arthur Nicholas, 1940–, , Wind Cave: an ancient world beneath the hills
[2610] Panton, James Hoyes, 1847–1898, 1890, The Mammoth Cave of Kentucky
[2611] Papadakis, Peggy, May 1969, The story of Crystal Ice Caves: within the Great Rift National Landmark, American Falls, Idaho
[2636] Pemberton, Clive, 1972, Kents Caverns, home of prehistoric man and animals: the origin, story, and descriptive tour of the caves
[2637] Pemberton, Clive, 1947, The origin and story of Kents Cavern
[2638] Pemberton, Clive, 1964, The origin, story and descriptive tour of the caves
[2650] Pennsylvania Railroad, 1885, A summer series of personally-conducted pleasure tours to the Luray Caverns or Gettysburg via Pennsylvania Railroad
[2658] Perry, Eugene S., 1946, Morrison Cave, Lewis and Clark Cavern State Park, Montana
[2659] Petersen, David, 1994, Carlsbad Caverns National Park
[2660] Petrocheilou, Anna, 1970, The Diros Caves of Mani: Alepotrypa and Glyphada
[2661] Petrocheilou, Anna, 1984, The Greek caves
[2662] Petrocheilou, Anna, Dec 1972, The Perama Caves of Loannina
[2666] Phillips, Harry, nd, Jenolan Caves, (the underground wonderland) New South Wales, Australia
[2667] Phillips, Harry, nd, Nature's masterpiece: Jenolan Caves N.S.W
[2673] Pierpont, E. de, 1947, The wonders of the Grotto of Han (Belgium): guidebook, and description with engravings and plan and a notice upon the very interesting Grotto of Rochefort
[2685] Pohl, Erwin Robert, 1904–, 1964, Itinerary, geologic features of the Mississippian Plateaus in the Mammoth Cave and Elizabethtown areas
[2690] Pond, Alonzo William, 1894–, 1941, Guide book of Cave of the Mounds ... Blue Mound, Wis
[2694] Popham, William Lee, 1885–1953, 1993, Mammoth Cave romance
[2707] Prentice, Guy, Jan 1988, Mammoth Cave archeological inventory project interim report - 1987 investigations
[2708] Prentice, Guy, 1989, Mammoth Cave National Park archeological inventory project interim report-1988 investigations
[2734] Putnam, Frederic Ward, 1839–1915, 1875, Archaeological researches in Kentucky and Indiana, 1874
[2741] Quinlan, James Francis, 1936–1995, 1980, Ground-water hydrology and geomorphology of the Mammoth Cave Region, Kentucky, and of the Mitchell Plain, Indiana
[2746] Radcliffe, F.G., 1925, The wonder caves of Maoriland: Waitomo, Ruakuri & Aranui
[2747] Radlauer, Ruth, 1926–, 1981, Carlsbad Caverns National Park
[2748] Radlauer, Ruth, 1926–, 1986, Mammoth Cave National Park
[2757] Ralston, Basil, 1989, Jenolan: the golden ages of caving
[2758] Randall, Robert L., 1970, Along the way to the caves: Timpanogos Cave National Monument, Utah
[2759] Randolph, Helen Fitz, 1924, Mammoth Cave and the cave region of Kentucky, illustrated: with bibliography of Mammoth Cave (Willard Rouse Jillson); first accurate underground survey (H. Bruce Hoffman); introduction (H. C. Nelson)
[2799] Reeds, Chester Albert, 1882–, 1925, The Endless Caverns of the Shenandoah Valley, an account of the wonderful work of water: with special reference as to how the caverns and the Shenandoah Valley were formed
[2812] Rhodes, Andrew Jackson, 1829–1907, 1905, The wonders of Lost River: 40 miles of underground river with its caves and hidden waterfalls, blind fish and other wonders
[2814] Richards, James Harray, Dec , Waitomo caves
[2818] Riddell, William Hatton, 1938, Altamira: a note

upon the Palaeolithic paintings in the Cave of Altamira near Santillane del Mar in the Spanish province of Santander

[2819] Riely, Samuel L., 1945, Story of Wyandotte Cave

[2825] Roberson, Gary T., c1989, Official souvenir book of Marengo Cave: U.S. National Landmark: Marengo, Indiana, the heart of cave country

[2831] Robison, Mabel Otis, 1966, The hole in the mountain

[2836] Rodda, Jan, 1989, Mulka's Cave site management project: emphasising visitor survey, April-June 1988, with management evaluation and further recommendations for management; with conclusions of an archaeological test excavation

[2837] Rodriguez, Ortega Eduardo (auth, photo), 1970, The cave of Nerja

[2847] Rogers, Edmund B., 1958, Mammoth Cave National Park, Kentucky: history of legislation through the 82nd Congress

[2849] Rogers, Edmund B., 1958, Wind Cave National Park, South Dakota: history of legislation through the 82nd Congress

[2851] Rolsh Photographies, nd, Cave wonderlands of Western Australia; Jewel Caves, Lake Cave, Mammoth Cave, Yallingup Cave

[2860] Rothrock, H. W., 1889, Wyandotte Cave: where it is and what it is

[2862] Royse, John, nd, Ancient Castleton Caves: their place in time and history, also what people want to know about Castleton

[2863] Royse, W., 1891, Descriptive account of the Blue John Mines and Caverns at Castleton, Derbyshire

[2864] Rozarie, Charles, 1964, The Archaeology at Lehman Caves National Monument: Nevada State Museum report

[2866] Rusling, James Fowler, 1864, A trip to the Mammoth Cave, Ky

[2872] Rust, Pauline Carrington, 1887, Legend of the Luray Caverns, founded upon the discovery of a skeleton in one of its chasms

[2878] Salamon, Gábor, 1989, The Stalactite Caves of Jósvafő and Aggtelek

[2884] Santamarta Cuenca, Pedro, 1978, The grottos of Majorca

[2895] Say, L. W. (ed), 1976, Cheddar Caves Museum

[2900] Scheltens, John (auth, comp), [1973], Wind Cave 1972

[2901] [Scheltens, John(ed)], [1971], Wind Cave expedition 1971

[2902] [Scheltens, John (ed)], [1970], Wind Cave expedition August 1970: a report to the National Park Service by the Windy City Grotto of the National Speleological Society on the results of the August 1970 Wind Cave expedition

[2906] Schmidt, Jeremy, 1987, Lehman Caves

[2910] Schröder, Karl-Heinz, 10 1980, Guide through the Kubacher Crystal Cave

[2916] Schuurmans, Theo J. C., 1987, Postcards of the Caves of Han printed by Nels of Brussels between 1905 and 1940

[2918] Schwartz, Douglas W., 1965, Prehistoric man in Mammoth Cave

[2926] Scott, W. Ray, , A pictorial study of Mammoth Cave and Mammoth Cave National Park

[2936] Šerko, Alfred, 1910–1948, 1953, The Cave of Postojna and other curiosities of the karst

[2937] Šerko, Alfred, 1910–1948, 1967, The Postojna grottoes and the other marvels of the karst

[2943] Shahan, Myrtle, 1949, Creation of Carlsbad Caverns

[2947] Shapcott, L. E., [1934], The story of Yanchep: the western wonderland

[2948] Sharp, Mabel, 1968, White Scar Cave: a famous underground system which you can explore in your best clothes

[2949] Sharp, Mabel, 1959, White Scar Cave: a picture guide

[2950] Sharp, Mabel, 1965, White Scar Cave: a picture guide to a famous underground system you can explore in your best clothes

[2961] Shenandoah Caverns Corporation, 1932, Shenandoah Caverns of Virginia

[2962] Shenandoah Caverns Corporation, 1928, Shenandoah Caverns: Valley of Virginia

[2963] Shenandoah Caverns Corporation, Shenandoah Caverns, Va, 1929, Shenandoah Caverns, Valley of Virginia. A symphony in stone

[2964] Shenandoah Caverns Corporation, Shenandoah Caverns, Va, 1926, Shenandoah Caverns, Valley of Virginia. The grotto of the gods

[2965] Shenandoah Valley Railroad, 1889, Through the Shenandoah Valley: Caverns of Luray, Natural Bridge, grottoes of the Shenandoah and the chronicle of a leisurely journey through the uplands of Virginia

[2971] Shoemaker, Henry Wharton, 1880–1958, 1929, The legends of the caverns of Centre County, Pennsylvania

[2972] Shoemaker, Henry Wharton, 1880–1958, , Penn's grandest cavern: the history, legends and description of Penn's Cave in Centre County, Pennsylvania

[2973] Shoemaker, Henry Wharton, 1880–1958, , Penn's grandest cavern: the history, legends photographs and description of Penn's Cave in Centre County, Pennsylvania

[2983] Silverman, Sharon Hernes, 1991, Going underground: your guide to caves in the Mid-Atlantic

[2992] Sinise, Jerry, 1963, Texas show caves

[2994] Skinner, A. D., Jul 1978, The Mole Creek Caves

[2996] Skinner, R. K., nd, Hastings Caves State Reserve

11.1. GENERAL

Tasmania: a visitors' guide
[3002] Sloan, Tacoma G., 1960, Archaeological survey of Mammoth Cave National Park
[3014] Smith, Burrell F., 1988, The cave and her men: Oregon Caves National Monument: its history and formation
[3020] Smith, H., 1882, Kent's Cavern
[3022] Smith, James, Oct 1964, The early days of Olsen's Caves as seen through a geologist's eyes
[3031] Smith, Peter B., 1973, The P8 Cave: Castleton
[3034] Smith, Rodney, 1981, Waitomo Caves
[3035] Smith, Rodney, 1980, Waitomo: Glow-worm Cave, Aranui Cave, Ruakuri Cave and surrounding district
[3037] Smithsonian Institution, 1880, The caverns of Luray, located on line of the Shenandoah Valley R.R
[3039] Snider, George Washington, 1888, Manitou Grand Caverns, Manitou Springs, Colorado
[3040] Snider, George Washington, 1885, The tourists' gem: the Manitou Grand Caverns, the largest and most wonderful subterranean in the Rocky Mountains, and other attractions for tourists
[3044] Society for Visual Education, 1945, Carlsbad Caverns National Park
[3051] Sotnak, Lewann, 1988, Carlsbad Caverns
[3052] South African Spelaelogical Association Members, 1968, Cango; the story of the Cango Caves of South Africa
[3053] South African Spelaeological Association, 1970, Cango; the story of the Cango Caves of South Africa
[3054] South African Spelaeological Association Members, 1958, Cango; the story of the Cango Caves of South Africa
[3076] Southwest Georgia Planning & Development Commission, 1968, Glory Hole Caverns
[3083] Speare, Eva A., nd, The story of Polar Caves
[3088] Speece, Jack Howard, 1947–, 1973, Alexander Caverns: the Carlsbad of Pennsylvania
[3106] Stanton, William Iredale, 1980, Cheddar Caves: illustrated official guide
[3108] Stanton, William Iredale, c1987, Timeless Cheddar Caves
[3109] Starr, Joan, 1927–, 1985, The Wellington caves: treasure trove of fossils
[3110] Steel, William Gladstone, 1931, Oregon, no 32, vol 1
[3116] Stelle, James Parish, 1884, The Wyandotte Cave of Crawford County, Ind
[3123] Stevens, Martin V. B., 1876, The history, guide and description of Wyandotte Cave
[3126] Stewart, John, 1920–, 1972, Secret of the bats; the exploration of Carlsbad Caverns
[3147] Story, Isabelle Florence, 1888–1970, 1935, Carlsbad Caverns National Park
[3178] Talent, J. A., nd, The Buchan Caves
[3203] Thompson, James B., 1958, The geology of Jewel Cave
[3204] Thompson, John, 1874–, 1909, Mammoth Cave, Kentucky; an historical sketch containing a brief description of some of the principal places of interest in the Mammoth Cave. Also a short description of Colossal Cavern
[3206] Thompson, Ralph Seymour, 1970, The sucker's visit to Mammoth Cave
[3207] Thompson, Ralph Seymour, 1879, The sucker's visit to the Mammoth Cave including a history of the experience and adventures of a party who undertook to see the cave and have some fun going there. A full and accurate description of the cave and the science of its formation with an account of the living inhabitants
[3211] Thornycroft, L. B., 1948, The story of Wookey Hole in fact, fiction and photo
[3218] Thynne, Daphne Winifred Louise (Marchioness of Bath), 1956, Cheddar caves
[3219] Tintilozov, Zurab, 1978, Akhali Atoni's cave
[3229] Trexler, Keith A., 1975, Lehman Caves...its human story: from the beginning through 1965
[3231] Trickett, Oliver, 1906, The Abercrombie Caves
[3232] Trickett, Oliver, 1899, Guide to the Jenolan Caves New South Wales
[3233] Trickett, Oliver, 1906, Guide to the limestone caverns of New South Wales: Jenolan, Wombeyan, and Yarrangobilly
[3234] Trickett, Oliver, 1906, Guide to Wombeyan Caves NSW
[3235] Trickett, Oliver, 1917, Guide to Yarrangobilly Caves New South Wales
[3237] Trickett, Oliver, 1906, The Wellington Caves
[3238] Trimble, Stephen, 1950–, 1983, Window into the earth: Timpanogos Cave
[3242] Trower, Harold Edward, c1928, Capri: the story of the blue grotto
[3246] Tsen, Darrell Nyuk Choi, 1993, The show caves of Mulu Sarawak
[3247] Tsen, Darrell Nyuk Choi, 1991, The show caves of Mulu Sarawak
[3252] Turner, Brian B., 1973, Natural stone bridge and caves; fascinating geology, history & legends
[3254] Turner, James William, 1848–, 1912, Wonders of the great Mammoth Cave of Kentucky containing thorough and accurate historical and descriptive sketches of this marvelous underground world with a chapter on the geology of cave formation
[3255] Turner, Percy (auth, comp), 1895, History of Manitou's caves, the Grand Caverns and Cave of the Winds: and points of interest in and about Manitou
[3257] [Tyers, John], 1962, A guide to Wind Cave and Wind Cave National Park. 19th annual convention of the National Speleological Society self guiding tour of the commercial areas of Wind Cave
[3273] U.S. Environmental Protection Agency, Region 4, Aug 1981, Environmental impact statement, Mammoth

Cave area, Kentucky: wastewater facilities

[3274] U.S. Environmental Protection Agency, Region 4, Apr 1981, Environmental impact statement; Mammoth Cave area, Kentucky, wastewater facilities

[3275] U.S. Environmental Protection Agency, Region 4, Apr 1981, Mammoth Cave area, Kentucky, 201 facilities plan environmental impact statement

[3276] U.S. Forest Service, 1973, Operation of Blanchard Springs Caverns, Ozark-St. Francis National Forest, Arkansas: final environmental statement

[3277] U.S. Forest Service, 1924, The Oregon Caves, Siskiyou National Forest

[3278] U.S. Forest Service, 1900, The Oregon Caves: Siskiyou National Forest

[3279] U.S. Forest Service, 1964, A preliminary plan for the development of Blanchard Springs Cavern: Sylamore District, Ozark National Forest

[3281] U.S. Forest Service, Southern Region, 1974, Construction of phases II and III of the Blanchard Springs Caverns Project: draft environmental statement

[3282] U.S. Forest Service, Southern Region, 1973, Welcome to Blanchard Springs Caverns

[3286] U.S. National Park Service, Sep 1980, Carlsbad Caverns National Park, New Mexico: wilderness reevaluation study: preliminary - subject to change

[3287] U.S. National Park Service, 1976, Carlsbad Caverns National Park: final interpretive prospectus

[3288] U.S. National Park Service, 1986, Carlsbad Caverns: official map & guide

[3290] U.S. National Park Service, Aug 1975, Development concept: Slaughter Canyon, Carlsbad Caverns National Park, New Mexico

[3291] U.S. National Park Service, Nov 1995, Draft general management plan, environmental impact statement: Carlsbad Caverns National Park, New Mexico

[3292] U.S. National Park Service, Feb 1993, Draft, environmental impact statement, general management plan, development concept plan Timpanogos Cave National Monument

[3293] U.S. National Park Service, Jun 1993, Draft, general management plan, environmental impact statement Jewel Cave National Monument, South Dakota

[3294] U.S. National Park Service, 1976, Final environmental statement: Mammoth Cave National Park master plan and wilderness suitability study, Kentucky

[3295] U.S. National Park Service, 1973, Final environmental statement: proposed wilderness, Carlsbad Caverns National Park, New Mexico

[3296] U.S. National Park Service, Apr 1976, Final master plan: Mammoth Cave National Park

[3297] U.S. National Park Service, Jun 1994, Final, general management plan, environmental impact statement Jewel Cave National Monument, South Dakota

[3299] U.S. National Park Service, Apr 1979, The Great Onyx Job Corps Civilian Conservation Center alternative relocation sites, Mammoth Cave National Park, Kentucky

[3300] U.S. National Park Service, 1971, Lehman Caves National Monument

[3301] U.S. National Park Service, 1987, Mammoth Cave National Park

[3302] U.S. National Park Service, 1974, Mammoth Cave National Park, Kentucky

[3303] U.S. National Park Service, 1978, Mammoth Cave National Park: collection management plan

[3304] U.S. National Park Service, [1973], Master plan: Carlsbad Caverns National Park, New Mexico

[3306] U.S. National Park Service, Feb 1980, Natural resources management program: an addendum to the natural resources management plan for Lehman Caves National Monument, Nevada

[3308] U.S. National Park Service, 1975, Oregon Caves National Monument, master plan

[3309] U.S. National Park Service, 1976, Oregon Caves National Monument, Oregon: final interpretive prospectus

[3310] U.S. National Park Service, 1992, Oregon Caves National Monument, Oregon: official map and guide

[3312] U.S. National Park Service, 1973, Pollution abatement project, Carlsbad Caverns National Park, New Mexico

[3313] U.S. National Park Service, nd, Proposed Mammoth Cave National Park master plan and wilderness study, Kentucky: draft environmental impact statement

[3314] U.S. National Park Service, 1973, Proposed wilderness, Carlsbad Caverns National Park, New Mexico

[3316] U.S. National Park Service, Jul 1986, Statement for management: Wind Cave National Park

[3317] U.S. National Park Service, 1976, Timpanogos Cave National Monument, Utah

[3318] U.S. National Park Service, Aug 1986, Timpanogos Cave National Monument: statement for management

[3319] U.S. National Park Service, 1984, Voices of the cave

[3320] U.S. National Park Service, Aug 1972, Wilderness recommendation, Carlsbad Caverns National Park, New Mexico

[3321] U.S. National Park Service, 1937, Wind Cave National Park, South Dakota

[3322] U.S. National Park Service, 1979, Wind Cave: National Park Service, South Dakota

[3328] United Air Lines, Inc., 1985, Luray Caverns in Virginia's Shenandoah Valley

[3336] University of Nevada System, Laboratory of Desert Biology, Dec 1968, Final reports on the Lehman Caves studies

[3339] University of Wisconsin, Recreation Resources Center, 1977, Seen a good cave lately?: Cave of the Mounds, Blue Mounds, Wis.: "a visitor study", 1977

11.1. GENERAL

[3340] Upper Valley Regional Park Authority (VA), 1975, The story of Cave Hill
[3345] Valsero, Juan José Durán, nd, Cueva de Nerja: English edition
[3359] Vernon, Robert Orion, 1912–, 1950, Florida Caverns: a nature-made wonderland
[3371] Wagner, Georg, 1883-1912, 1966, The bears cave of Erpfingen
[3372] Wagoner, John J., 1985, Mammoth Cave
[3379] Walsh, Frank K., 1971, Discovery and exploration of the Oregon caves: Oregon Caves National Monument
[3381] Walsh, Frank K., 1976, Oregon Caves: discovery and exploration
[3400] Warnell, Norman, 1997, Mammoth Cave: forgotten stories of it's [sic] people
[3406] Watson, J. R. (ed), Feb 1979, Cave tourism in Western Australia: proceedings of a seminar held at Busselton November 10-11, 1978
[3413] Watson, Richard Allan, 1931–, 1963, The Mammoth Cave National Park research center
[3414] Watson, Richard Allan, 1931–, 1963, The Mammoth Cave National Park Research Center: a Cave Research Foundation study
[3415] Watson, Richard Allan, 1931–, May 1967, The preservation of wilderness karst in central Kentucky, U.S.A
[3418] Wayland, John Walter, 1872–1962, 1930, The master sculptor; a brief treatise on erosion, and its wondrous effects in the Shenandoah Valley
[3419] Weaver, Herman Dwight, 1938–, 1972, Adventures at Mark Twain Cave
[3420] Weaver, Herman Dwight, 1938–(ed), 1982, Great American show caves
[3421] Weaver, Herman Dwight, 1938–, Aug 1977, Meramec Caverns: legendary hideout of Jesse James
[3422] Weaver, Herman Dwight, 1938–, Apr 1980, Missouri: the cave state
[3423] Weaver, Herman Dwight, 1938–, , Onondaga, the mammoth cave of Missouri
[3432] Welch, Bruce R. (ed), 1976, The northern limestone
[3433] Weller, James Marvin, 1899–, 1927, The geology of Edmonson County: a detailed presentation of the physical, stratigraphic, structural, and economic geology of this district
[3436] Wells, R. T., 1975, World famous Fossil Cave Naracoorte
[3446] West Texas Geological Society, 1968, Guadalupe Mts., Hueco Mts., Franklin Mts., geology of the Carlsbad Caverns, Delaware Basin exploration, Oct. 31st, Nov. 1st and 2nd, 1968
[3451] Western Land Surveys, Utah Division, 1977, Timpanogos Cave environmental impact report
[3453] Weyer, Bernard, 1881, A description of Weyer's Cave in Augusta County, Va
[3456] Wheeler, Joy, 1969, Oberon-Jenolan district historical notebook
[3457] Whelchel, Sandy, 1985, A day at the cave
[3460] White, Jim, 1882–1946, 1932, The discovery and history of Carlsbad Caverns, New Mexico
[3462] White, Jim, 1919–, 1951, Carlsbad Caverns National Park, New Mexico: its early explorations as told by Jim White
[3475] White, William Blaine, 1934–(ed), 1989, Karst hydrology: concepts from the Mammoth Cave area
[3476] White, William Blaine, 1934–, Jan 1974, Reconnaissance geology of Timpanogos Cave: Wasatch County, Utah
[3486] Wilkes, Frank G., 1962, Bibliography of Mammoth Cave National Park, Mammoth Cave, Kentucky
[3493] Williams, Dave, 1987, Waitomo Caves: Glowworm Cave, Ruakuri Cave, Aranui Cave
[3510] Wilson, John, 1800–1849, 1849, A visit to the Mammoth Cave of Kentucky
[3515] Winder, Francis A., 1938, Unconventional guide to the Caverns of Castleton and Disaster
[3516] Wintner, Radar H., 1977, Lake Shasta Caverns
[3520] Wood, Andrew, 1957, The mammoth cavern
[3521] Wood, Frances Elizabeth, 1963, Rocky Mountain, Mesa Verde, Carlsbad Caverns
[3524] Woodall, Brian, 1976, Peak Cavern: a guide to this famous show-cave; its formation, history & folklore
[3526] Workman, C. Lindsay (narr), 1991, Carlsbad Caverns National Park
[3527] World Travelogues Corporation, 1984, Lewis and Clark Caverns
[3531] Wright, Charles W., , A guide manual to the Mammoth Cave of Kentucky
[3532] Wright, Charles W., 1860, A guide manual to the Mammoth Cave of Kentucky
[3533] Wright, Charles W., 1858, The Mammoth Cave of Kentucky
[3538] Wynne, Annette, 1924, The trip thru fairyland; Great Onyx cave
[3549] Zadnikar, Marjan, 1960, The castle of Predjama
[3551] Zannes, Tom (prod, dir), 1993, Spirit of exploration: discovering the wonders of Carlsbad Caverns National Park
[3553] Zeoderborg, Harry, nd, The phantoms of stork fontein [sic]

12 Science

12.1 Chemistry

[209]Anon, nd, Waitomo Day 1982: summary of papers: Waitomo Caves research programme
[2192]Miotke, F. M., 1974, Carbon dioxide and the soil atmosphere

12.2 Chronology

[2571]Ollier, Clifford David, Oct 1966, Speleochronology

12.3 Communication

[3498]Williams, Nick (comp), Mar 1992, British Cave Research Association Cave Radio and Electronics Group bibliography of underground communications

12.4 Computers

[62]Amundson, Bob, Jan 1979, User's guide for flowchart and cavemap computer programs

12.5 Dating

[53]Allison, Vernon Charles, 1891–, 1926, The antiquity of the deposits in Jacob's Cavern
[2654]Perkal, Malissa Faye, 1978, Uranium decay series studies on archaeological materials: application to thermoluminescence dating and dating of cave deposits

12.6 Engineering

[54]American Cave Conservation Association, 1990, Hart County solid waste management plan
[1002]Eraso, Adolfo (ed), Mar 1988, The Stor-Glomfjord project: karst research - the possibilities of leakage and the selection of damsite alternative
[3078]Sowers, George F., 1996, Building on sinkholes: design and construction of foundations in karst terrain
[3264]U.S. Bureau of Land Management, Roswell District Office, Feb 1993, Bureau of Land Management interim guide for oil & gas drilling operations in cave and karst areas

12.7 Environmental Studies

[43]Aley, Thomas John, 1938–, 1988, Restoration of natural microclimate in Oregon Caves, Oregon Caves National Monument: final report
[238]Association of Ground Water Scientists and Engineers (a division of NGWA), Dec 1991, Proceedings of the third conference on hydrogeology, ecology, monitoring, and management of ground water in karst terranes, December 4-6, 1991, Maxwell House/Clarion, Nashville, Tennessee
[356]Beck, Barry Frederic, 1944–(ed), 1989, Engineering and environmental impacts of sinkholes and karst: proceedings of the third Multidisciplinary Conference on Sinkholes and the Engineering and Environmental Impacts of Karst, St. Petersburg Beach, Florida, 2-4 October 1989
[358]Beck, Barry Frederic, 1944–(ed), 1995, Karst geohazards: engineering and environmental problems in karst terrane: proceedings of the fifth Multidisciplinary Conference on Sinkholes and the Engineering and Environmental Impact of Karst, Gatlinburg, Tennessee, 2-5 April 1995
[360]Beck, Barry Frederic, 1944–(ed), 1987, Karst hydrogeology: engineering and environmental applications: proceedings of the second Multidisciplinary Conference on Sinkholes and the Environmental Impacts of Karst, Orlando, Florida 9-11 February 1987
[889]Dougherty, Percy H. (ed), 1993, Environmental karst
[1356]Hanwell, James, 1935–, Oct 1970, The great storms and floods of July 1968 in Mendip
[1468]Hobbs, Horton Holcombe, 1944–, Jan 1994, Assessment of the ecological resources of the caves of Russell Cave National Monument, Jackson County, Alabama and of selected caves at the Lookout Mountain unit of Chickamanga-Chattanooga National Military Park, Dade County, Georgia and Hamilton County, Tennessee
[1518]Horne, Peter, 1988, Barnoolut Estate sinkholes environmental assessment project, 27 December 1988
[1836]Knott, David L., 1992, Guide for foundation engineering in Pennsylvania karst
[1894]Lane, Edward, 1935–, 1991, Environmental geology and hydrogeology of the Ocala area, Florida
[1949]Lively, Richard, 1993, Radon concentrations, radon decay product activity, meteorological conditions and ventilation in Mystery Cave
[2484]National Water Well Association, 1988, Proceedings of the second conference on environmental problems in karst terranes and their solutions conference, November 16-18, 1988, Maxwell House Hotel, Nashville, Tennessee
[2890]Sauro, U. (ed), 1991, Proceedings of the international conference on environmental changes in karst areas I.C.E.C.K.A. Italy, September 15-27th, 1991
[3317]U.S. National Park Service, 1976, Timpanogos Cave National Monument, Utah
[3483]Wilde, Kevan A., nd, Environmental monitoring of karst and caves

12.8 Geography

[1174]Geographical Field Group, 1960, Coast and karst in Istria Rossa; a report of geographical field work carried out during August, 1960

12.9 Medicine

[1217]Glanvill, Peter, 1969, First aid for cavers
[1951]Lloyd, Oliver Cromwell, 1911–1985, Jul 1975, A cave diver's training manual
[3105]Standing, Peter, Aug 1975, Medical aspects of speleology

12.10 Meteorology

[1820]Kimball, Herbert Harvey, 1862–, 1901, Ice caves and frozen wells as meteorological phenomena ...
[3113]Steidtmann, Edward, 1881–1948, 1936, Humidity and waters of a limestone cavern near Lexington, Virginia
[3428]Wefer, Fred L., Jun 1991, An annotated bibliography of cave meteorology

12.10.1 Climatology

[713]Corbel, Jean, 1975, The major caves of France and their relationships with climatic factors
[859]de Saussure, Raymond, Mar 1953, Report of the California-Nevada Speleological Survey June 13-September 7, 1952
[1266]Grimes, K. G., Oct 1995, South East Karst Province of South Australia: Australian Caves & Karst Management Association October 1995
[1326]Halliday, William R., 1926–, Oct 1958, A consideration of the effect of surface hot-cold cycles on caves
[1342]Hamilton-Smith, Elery (ed), February 1996, Abstracts of papers: karst studies seminar Naracoorte
[2133]McLean, John Scott, 1942–, Feb 1976, Factors altering the microclimate in Carlsbad Caverns, New Mexico
[2134]McLean, John Scott, 1942–, May 1971, The microclimate in Carlsbad Caverns, New Mexico
[3428]Wefer, Fred L., Jun 1991, An annotated bibliography of cave meteorology

12.11 Military

[1239]Grange Cavern Military Museum, 1980, Souvenir guide to the world's largest underground military museum: the Grange Cavern

12.12 Mining

[2036]Maness, Lindsey Vance, 1994, North Carolina coastal plain caves and their impact on mining and quarrying

12.13 Physics

12.13.1 Cavity Detection

[308]Ballard, Robert F., Jul 1983, Cavity detection and delineation research: report 5: electromagnetic (radar) techniques applied to cavity detection
[343]Bates, Edward R., Jun 1973, Detection of subsurface cavities
[364]Beck, Barry Frederic, 1944–(ed), 1987, Use of ground penetrating radar for detecting and evaluating the sinkhole hazard in Florida
[570]Butler, Dwain K., Mar 1983, Cavity detection and delineation research: Report 1: Microgravimetric and magnetic surveys: Medford Cave Site, Florida
[571]Butler, Dwain K., 1983, Cavity detection and delineation research: Report 4: Microgravimetric survey: Manatee Springs site, Florida
[710]Cooper, Stafford S., 1983, Cavity detection and delineation research: Report 3: Acoustic resonance and self-potential applications: Medford Cave and Manatee Springs sites, Florida
[783]Curro, Joseph R., Jun 1963, Cavity detection and delineation research; Prepared for Office, Chief of Engineers. Report 2: Seismic methodology: Medford Cave Site, Florida
[909]Duff, B. M., 1980, Ground penetrating electromagnetic tests at Medford Cave, Florida and Waterways Experiment Station, Vicksburg, Mississippi
[1822]Kirk, Keith G., Apr 1981, Handbook of geophysical cavity-locating techniques: with emphasis on electrical resistivity
[1881]Laine, Edwin F., 1980, Detection of water-filled and air-filled underground cavities
[2510]Nicol, Allen Hankins, 1907–, 1951, Application of geophysical methods to location of cavities in residual soil and caverns in coralline limestone
[2769]Rechtien, Richard D., 1972, The detection and mapping of subterranean water bearing channels: completion report
[2852]Romberg, Frederick, 1961, The detection of subsurface voids by gravimetry
[3174]Symposium on Detection of Subsurface Cavities (1977: Vicksburg, MS), Oct 1977, Symposium on Detection of Subsurface Cavities, 12-15 July, 1977
[3260]United States Army Corps of Engineers, Dec 1987, Geophysical phase IV monitoring of potential sinkhole collapse at military ocean terminal. Sunny Point Access

Railroad, Boiling Springs, N.C.
[3325]U.S. Office of Water Resources Research, 1975, The detection and mapping of subterranean water bearing channels: phase 2: final report
[3519]Wolfe, James E., 1975, Map location and dimensional definition of subsurface caverns

12.13.2 Radioactivity

[1949]Lively, Richard, 1993, Radon concentrations, radon decay product activity, meteorological conditions and ventilation in Mystery Cave

12.13.3 Radon

[2485]Natural Resources Seminar (3rd: 1976: Santa Fe), 1976, Radon radiation situation in NPS caves: January 19, 1976

12.14 Psychology

[2350]National Caving Association, Jul 1990, What do cavers think and do?

12.15 Speleology

[263]Australian Speleological Federation, Jul 1975, Proceedings of the tenth biennial conference of the Australian Speleological Federation
[264]Australian Speleological Federation Digest Commission, 1976, Australian speleology 1972
[265]Australian Speleological Federation Inc, 1988, Preprints of papers for the 17th Biennial Conference of the Australian Speleological Federation Tropicon Conference, Lake Tinaroo, Far North Queensland. 27th to 31st December 1988
[283]Baker, Ernest Albert, 1869–1941, 1907, The netherworld of Mendip: explorations in the great caverns of Somerset, Yorkshire, Derbyshire, and elsewhere
[291]Balch, Edwin Swift, 1856–1927, 1970, Glaciéres; or, freezing caverns
[292]Balch, Edwin Swift, 1856–1927, 1896, Ice caves and the causes of subterranean ice
[415]Block, Guy de (ed), c1974, Bibliographie Speleologique Belge
[416]Block, Guy de (ed), 1981, Bibliographie Speleologique Belge: editions speleologiques belges 1975-1979
[417]Block, Guy de (ed), 1965, Bibliographie Speleologique Belge. Editions speleologiques belges. 1907-1964. Avec en appendice la liste des ouvrages a caractere speleologique edites en Belgique
[486]British Cave Research Association, 1991, Peak and Speedwell Caverns, exploration and science: including the talks presented at the Peak-Speedwell symposium, Sheffield, November 1989
[610]Casteret, Norbert, 1897–1987, 1947, My caves
[616]Cave Research Associates, 1959, Cave Studies No. 1-11
[733]Crabtree, P. W. (ed), 1962–, Speleological Abstracts: key to Britain's speleological literature
[773]Cullingford, Cecil Howard Dunstan, 1904–1990 (ed), 1953, British caving: an introduction to speleology
[816]Davey, Adrian G., 1992, World Heritage significance of karst and other landforms in the Nullarbor Region
[859]de Saussure, Raymond, Mar 1953, Report of the California-Nevada Speleological Survey June 13-September 7, 1952
[878]Dixon, Richard L., 1965, Speleological investigation of Nakimu Caves, British Columbia
[888]Dougherty, Percy H. (ed), 1985, Caves and karst of Kentucky
[917]Dunkley, John R., 1978, Caves of the Nullarbor; a review of speleological investigations in the Nullarbor Plain, Southern Australia
[931]Dyson, H. Jane (ed), 1982, Wombeyan Caves
[936]Eavis, A. J. (comp), 1981, Caves of Mulu '80: the limestone caves of the Gunong Mulu National Park, Sarawak
[975]Ellis, Ross Andrew, 1940–(ed), 1972, Bungonia Caves
[1001]Eraso, Adolfo (ed), 1991, 1st international symposium of glacier caves and karst in polar regions. Proceedings
[1010]European Regional Conference of Speleology (1980: Sofia, Bulgaria), 1980, Abstracts of papers
[1094]Ford, Trevor David, 1925–(ed), 1989, Limestones and caves of Wales
[1095]Ford, Trevor David, 1925–(ed), 1976, The science of speleology
[1244]Graves, John (ed), , The River Styx-Salt Spring Cave system
[1292]Gurnee, Jeanne Marie, 1926–(ed), 1969, A study of Fountain National Park and Fountain Cavern: Anguilla, British West Indies
[1429]Hempel, John Charles, 1949–(comp), 1989, Report of the 1982 expedition to Barra Honda National Park
[1580]Huppert, George Nixon, 1944–, 1981, Speleography of Papoose Cave, Idaho County, Idaho
[1620]International Congress of Speleology (2nd: 1958: Bari - Lecce - Salerno, Italy), 1963, Actes du deuxième congrès international de spèlèologie Bari - Lecce - Alerno 5-12 Octobre 1958
[1621]International Congress of Speleology (4th: 1965: Postojina, Yugoslavia), 1968, Proceedings of the 4th International Congress of Speleology in Yugoslavia, Postojina

12.15. SPELEOLOGY

- Ljubljana - Dubrovnik 12-26 ix 1965
[1622]International Congress of Speleology (5th: 1969: Stuttgart, Germany), 1969, 5 Internationaler Kongress für Speläologie Stuttgart 1969
[1624]International Congress of Speleology (6th: 1973: Olomouc, Czechoslovakia), 1975-1977, Proceedings of the 6th International Congress of speleology = Actes du 6e Congrès international de spéléologie
[1625]International Congress of Speleology (7th: 1977: Sheffield, England), 1977, Proceedings of the 7th International Speleological Congress, Sheffield, 1977
[1626]International Congress of Speleology (8th: 1981: Bowling Green, Kentucky, USA), 1981, Excursion E11 supplemental materials: karst of the central Appalachians: field excursion on the occasion of the eighth International Congress of Speleology
[1627]International Congress of Speleology (8th: 1981: Bowling Green, Kentucky, USA), 1981, Proceedings of the eighth international congress of speleology: a meeting of the International Union of Speleology
[1628]International Congress of Speleology (8th: 1981: Bowling Green, Kentucky, USA), 1981, Program
[1629]International Congress of Speleology (8th: 1981: Bowling Green, Kentucky, USA), 1981, Second circular
[1630]International Congress of Speleology (8th: 1981: Bowling Green, Kentucky, USA), 1981, Third circular
[1633]International Congress of Speleology (9th: 1986: Barcelona, Spain), 1986, Communications 1-7 August 1986
[1634]International Congress of Speleology (10th: 1989: Budapest, Hungary), 1989, Proceedings: 13-20 August 1989
[1635]International Congress of Speleology (11th: 1993: Beijing, China), 1993, Proceedings of the XI international congress of speleology, August 2 to 8, 1993, Beijing China
[1643]International Speleological Conference (1964-: Brno), 1965, Problems of the speleological research: proceedings of the international speleological conference held in Brno June 29-July 4, 1964
[1697]James, Julia M., 1980, Caves and karst of the Muller Range: report of the 1978 speleological expedition to the Atea Kananda, Southern Highlands, Papua New Guinea
[1710]Jasinski, Marc, 1967, Caves and caving: a guide to the exploration, geology and biology of caves
[1840]Knutson, Richard Stephen, nd, The speleological potential of the Big Bar Limestone deposit, Hell's Canyon, Oregon-Idaho
[1871]Kunath, Carl Edwin, 1940–(ed), 1978, The caves of McKittrick Hill, Eddy County, New Mexico: history and results of field work 1965-1976
[1872]Kunath, Carl Edwin, 1940–(ed), Apr 1968, The caves of the Stockton Plateau, Texas
[1902]Larson, Charles Victor, 1928–, 1977, Bibliography of Oregon speleology, 1977
[2024]Malayan Nature Society, May 1965, Malaysian caves
[2087]Matthews, Larry Edwin, 1946–(ed), 1974, Bibliography of Tennessee speleology
[2249]Mohr, Charles E., 1907–(ed), 1955, Celebrated American caves
[2262]Moore, George William, 1928–, 1997, Speleology: caves and the cave environment
[2263]Moore, George William, 1928–, 1964, Speleology: the study of caves
[2274]Morland, Timothy Edward, 1959, A glossary of French speleological terms
[2357]National Speleological Society, Apr 1948, The caves of Texas
[2401]National Speleological Society. Annual Convention. Guidebook (1979: Pittsfield, MA), 1979, An introduction to caves of the northeast: guidebook for the 1979 National Speleological Convention, Pittsfield, Massachusetts
[2404]National Speleological Society. Annual Convention. Guidebook (1983: Elkins, WV), 1983, An introduction to the caves of east-central West Virginia: guidebook for the 1983 National Speleological Society Convention, Elkins, West Virginia
[2441]National Speleological Society. Annual Convention. Program with Abstracts (1990: Yreka, CA), 1990, Program of the National Speleological Society convention July 8 through July 14, 1990: Yreka, California
[2525]Northwest Regional Association Symposium on Cave Science and Technology, 1980—1982, Northwest Regional Association symposium on cave science and technology: 1980: Feb 16-18, 1980, Seattle, WA, 1981: Feb 14-16, 1981, Seattle, WA. 1982: Feb 13-15, 1982, Boise, Idaho
[2528]Nott, David, 1975, Into the lost world - a descent into prehistoric time
[2607]Palmer, William O., 1970 1981, Springs, caves and their related features
[2677]Pinney, Roy, 1962, The complete book of cave exploration: an authoritative guide to the wonders, mysteries and excitement of caves and caving
[2770]Reddell, James Russell, 1938–, Oct 1968, A bibliographic guide to Texas speleology
[2954]Shaw, Trevor Royle, 1928–, 1992, History of cave science: the exploration and study of limestone caves, to 1990
[2955]Shaw, Trevor Royle, 1928–, 1979, History of cave science: the scientific investigation of limestone caves, to 1900
[3032]Smith, Philip M., 1960, Speleological research in the Mammoth Cave Region, Kentucky: elements of an integrated program
[3077]Southwest Missouri State University Dept. of Ge-

ography and Geology, 1979, Speleology workshop

[3081] Spate, Andy, 1988, A brief introduction to speleology

[3090] Speleological Conference (1987), 1987, Proceedings - 16th biennial conference Australian Speleological Federation Inc

[3091] Speleological Conference (1991: Margaret River, WA, AU), 1991, Proceedings of the 18th biennial speleological conference: 30 December 1990 to 5 January 1991 at Margaret River, W.A

[3092] Speleology Workshop (1979: 2nd: Southwest Missouri State University), 1979, 1979 Speleology workshop manual

[3097] Sprent, J. K., (ed), 1970, Mount Etna caves: a collection of papers covering several aspects of the Mt. Etna and Limestone Ridge caves area of central Queensland

[3205] Thompson, Peter, 1943–, 1976, Cave exploration in Canada: a special issue of the Canadian Caver magazine

[3227] Tratman, Edgar Kingsley, 1899–1978 (ed), Feb 1963, Reports on the investigations of Pen Park Hole, Bristol

[3305] U.S. National Park Service, Dec 1994, National Cave and Karst Research Institute Study: a draft report to Congress as required by Public Law 101-578 of November 15, 1990

[3409] Watson, Richard Allan, 1931–, 1984, The Cave Research Foundation 1969-1973

[3410] Watson, Richard Allan, 1931–, 1984, The Cave Research Foundation 1974-1978

[3411] Watson, Richard Allan, 1931–, 1981, The Cave Research Foundation: origins and the first twelve years [1957-1968]

[3413] Watson, Richard Allan, 1931–, 1963, The Mammoth Cave National Park research center

[3414] Watson, Richard Allan, 1931–, 1963, The Mammoth Cave National Park Research Center: a Cave Research Foundation study

[3467] White, M., 1978, Western Region speleo-education seminar program with abstracts

[3470] White, William Blaine, 1934–, 1984, Cave and karst-related papers in the mainstream scientific literature: a bibliography

13 Miscellaneous

13.1 Anthologies

[993]England, A. W., 1973, Caves, an anthology
[2450]National Speleological Society. Speleo Digest (1956), Mar 1957, Speleo digest, 1956
[2451]National Speleological Society. Speleo Digest (1957), Apr 1958, Speleo digest, 1957
[2452]National Speleological Society. Speleo Digest (1958), May 1959, Speleo digest, 1958
[2453]National Speleological Society. Speleo Digest (1959), Apr 1961, Speleo digest, 1959
[2454]National Speleological Society. Speleo Digest (1960), Apr 1962, Speleo digest, 1960
[2455]National Speleological Society. Speleo Digest (1961), Apr 1963, Speleo digest, 1961
[2456]National Speleological Society. Speleo Digest (1962), Apr 1964, Speleo digest, 1962
[2457]National Speleological Society. Speleo Digest (1963), Apr 1965, Speleo digest, 1963
[2458]National Speleological Society. Speleo Digest (1964), Apr 1966, Speleo digest, 1964
[2459]National Speleological Society. Speleo Digest (1965), Aug 1967, Speleo digest, 1965
[2460]National Speleological Society. Speleo Digest (1966), Apr 1969, Speleo digest, 1966
[2461]National Speleological Society. Speleo Digest (1967), Jun 1974, Speleo digest, 1967
[2462]National Speleological Society. Speleo Digest (1968), Jul 1974, Speleo digest, 1968
[2463]National Speleological Society. Speleo Digest (1969), 1976, Speleo digest, 1969
[2464]National Speleological Society. Speleo Digest (1970), 1970, Speleo digest, 1970
[2465]National Speleological Society. Speleo Digest (1971), 1978, Speleo digest, 1971
[2466]National Speleological Society. Speleo Digest (1972), 1980, Speleo digest, 1972
[2467]National Speleological Society. Speleo Digest (1973), 1980, Speleo digest, 1973
[2468]National Speleological Society. Speleo Digest (1974), 1981, Speleo digest, 1974
[2469]National Speleological Society. Speleo Digest (1975), 1982, Speleo digest, 1975
[2470]National Speleological Society. Speleo Digest (1976), 1983, Speleo digest, 1976
[2471]National Speleological Society. Speleo Digest (1977), 1983, Speleo digest, 1977
[2472]National Speleological Society. Speleo Digest (1978), 1980, Speleo digest, 1978
[2473]National Speleological Society. Speleo Digest (1979), 1981, Speleo digest, 1979
[2474]National Speleological Society. Speleo Digest (1980), 1985, Speleo digest 1980
[2475]National Speleological Society. Speleo Digest (1981), 1984, Speleo digest, 1981
[2476]National Speleological Society. Speleo Digest (1982), 1986, Speleo digest, 1982
[2477]National Speleological Society. Speleo Digest (1983), 1991, Speleo digest, 1983
[2478]National Speleological Society. Speleo Digest (1985), 1987, Speleo digest, 1985
[2479]National Speleological Society. Speleo Digest (1986), 1990, Speleo digest, 1986
[2480]National Speleological Society. Speleo Digest (1987), 1991, Speleo digest, 1987
[2481]National Speleological Society. Speleo Digest (1989), 1990, Speleo digest, 1989
[2482]National Speleological Society. Speleo Digest (1993), 1994, Speleo digest, 1993
[2986]Simpson, Lou, 1995, Sex, lies, and survey tape

13.2 Atlases

[761]Crossley, P. C. (comp), 1988, The New Zealand cave atlas: North Island
[762]Crossley, P. C. (comp), 1990, New Zealand cave atlas: South Island
[763]Crossley, P. C. (comp), 1981, The New Zealand cave atlas
[2664]Pfeffer, Karl-Heinz (ed), 1990, International atlas of karst phenomena: sheets 8 - 12
[3236]Trickett, Oliver, 1898, Notes on the limestone caves of New South Wales with plans

13.3 Audiovisuals

[1]A. I. D. International, 1977, Caves
[84]Anon, 1993, Carlsbad Cavern National Park. Guadalupe Mountains National Park
[85]Anon, 1985, Carlsbad Caverns National Park
[87]Anon, 1982, Carlsbad Caverns, New Mexico
[90]Anon, 1988, Cave paintings of the Chumash Indians
[94]Anon, 1986, The Caverns, rocks and ruins of America's Southwest
[95]Anon, 1964, Caves
[96]Anon, 1965, Caves
[101]Anon, nd, Caving in West Virginia
[110]Anon, 1960, Development of caves
[118]Anon, 1988, The geology of caves
[144]Anon, 1988, The last horizon
[177]Anon, 1990, Red hills
[180]Anon, 1994, Sea Lion Caves Oregon coast
[191]Anon, 1946, Spelunking

[203]Anon, 1987, Timpanogos Cave public use and impact
[205]Anon, 1983, Unusual cave and cavern formations
[207]Anon, 1984, Virginia Luray Caverns
[223]Appalshop, Inc, 1971, Line Fork falls and caves
[249]Aujoulat, Norbert, 1989, Lascaux revisited
[267]Avid Corporation Creative Learning, 1970, Formation of caves
[325]Barr, Thomas Calhoun, 1931–, 1989, The cave community
[352]Baysinger, David (narr, videographer), 1986, Silent Splendor
[353]Baysinger, David. , 1988, Lechuguilla cave: the hidden giant
[367]Beckway, Gregory W. (ed), 1971, Cavern formation
[383]Benke, David, 1989, Using trigonometric functions in cave surveying
[396]BFA Educational Media, 1961, Caverns and geysers
[412]Blackhawk Hawks, 1970, Carlsbad Caverns
[483]British Broadcasting Corporation Television Service, 1980, Atea: in search of the world's deepest cave
[592]Canada. National and Historic Parks Branch, 1972, Nahanni
[594]Canada. National Parks Service , 1974, Castle Guard Cave, challenge under the glacier
[595]Canadian Broadcasting Corporation, 1972, The Blue holes of Andros
[682]Collier, Don (host), 1990, Life underground: caves, mines, minerals
[698]Cook, John Hawley, 1989, Of caves & cavemen
[717]Coronet Instructional Films, 1970, Cave ecology
[718]Coronet Instructional Films, 1943, Limestone caverns
[734]Crain, Sally Lucille 1938–(auth, prod), 1979, Prehistoric magic
[758]Crosby, Harry, 1926–, 1976, Cave paintings of Baja
[808]Dasque, Jean (prod), 1972, Garden of shadows
[945]Educational Enrichment Materials, Inc, 1980, Cave science
[973]Ellis, Robert D. (scrn wrt), 1993, Carlsbad Caverns National Park
[979]Ely, Glen Sample. (photo, ed), 1990, A history of the Guadalupe Mountains and Carlsbad Caverns, 1840—1940
[980]Encyclopaedia Britannica Educational Corporation, 1976, Caves: the dark wilderness
[1057]Film Associates of California, 1965, Cavern formation
[1058]Film Images, 1957, Caves of Pierre Saint-Martain
[1063]Finley-Holiday Film Corporation, 1980, Carlsbad Caverns National Park & Guadalupe Mountains National Park
[1064]Finley-Holiday Film Corporation, 1988, Luray Caverns in Virginia's Shenandoah Valley
[1065]Finley-Holiday Film Corporation, 1990, Mammoth Caves
[1080]Ford, Derek Clifford, 1935–(prod), 1974, Castleguard Cave: challenge under the glacier
[1087]Ford, Trevor David, 1925–, 1973, Caves—origins, development and formations
[1088]Ford, Trevor David, 1925–, 1973, Caving and potholing techniques
[1127]Friedberg, Lionel, 1992, Mysteries underground
[1133]G. P. Slide Co, 1970, Luray Caverns
[1136]GAF Corporation , 1950, Carlsbad Caverns National Park
[1137]GAF Corporation , 1970, Luray Caverns, Va
[1188]Gibb, Hugh, 1981, The Tabon caves
[1195]Gifford Pinchot National Forest, 1991, Layser Cave: silent voices, vital clues
[1197]Gilbert Group, 1978, Caves, a deeper look at our Earth
[1212]Gilmore, Elizabeth (prod, ed, auth), 1986, Silent Splendor
[1237]Graham, Richard, 1970, Exploring the underground
[1247]Great American Film Factory, 1977, Underground wilderness
[1458]Hill, Carol Ann, 1940–, 1988, The geology of Carlsbad Cavern
[1481]Holiday Film Corporation, 1970 1980, Carlsbad Caverns, New Mexico
[1482]Holiday Film Corporation, 1984, Luray Caverns
[1483]Holiday Film Corporation, 1970, Mammoth Cave National Park
[1484]Holiday Film Corporation, 1991, Meramec Caverns, Stanton, Missouri
[1560]Hubbard Scientific Company, 1972, Cavern formation
[1593]Imperial Film Company, 1965, Mammoth Cave, Kentucky
[1610]Inside Earth Publications, 1977, 1973, The Wilderness below
[1636]International Film Bureau, 1976, Lascaux, cradle of man's art
[1706]Jankowski, Laurence (exe prod), 1985, Let's explore a cave
[1721]Jennings, Jesse David, 1909–, 1951, Mogollone material, Tularosa Cave, New Mexico
[1730]Jensen, Douglas N. , 1988, Discover Timpanogos Cave National Monument
[1735]JLM Visuals, 1979, Caves and other ground water features
[1736]JLM Visuals, 1979, Caves and other ground water features II
[1830]Knauth, Otto, 1975, Cold Water Cave
[1919]Leakey, Richard E., 1981, A new era
[1964]Longul, Wally, 1987, The last horizon
[2010]MacGregor, John R., 1978, Ecology of a limestone

cave
[2098] Mazonowicz, Douglas, 1989, On the rocks: prehistoric art of France and Spain
[2099] Mazonowicz, Douglas, 1979, The painted caves of France & Spain
[2173] Middleton, T. C., 1979, Cave development and formations
[2299] Munich, Frederick J. (auth, prod), 1977, Fumaroles, ice caves and time: the variety and scope of research on Mt. Baker
[2334] Naden, C. J., 1979, I can read about caves
[2360] National Speleological Society, 1990, Wind Cave: the world below
[2588] Ozark National Forest (AR), 1980, Blanchard Springs Caverns: the amazing world below
[2607] Palmer, William O., 1970 1981, Springs, caves and their related features
[2684] Plimpton, George (host), 1987, The Los Tayos challenge
[2696] PostCard Videos, Inc, nd, Howe Caverns and New York Central Leatherstocking Region
[2804] Reid-Cowan Productions, 1976, Caves: the dark wilderness
[2991] Single, Michael, 1992, Castles of the underworld
[3000] Slaven, John F., nd, The world within
[3026] Smith, Lee, nd, Down under Texas
[3044] Society for Visual Education, 1945, Carlsbad Caverns National Park
[3139] Stone, William Curtis, 1952–, 1992, Inner space: the last terrestrial frontier
[3142] Stoneman, John (narr), 1990, The cave divers
[3239] Trinkaus, Erik, 1985, The Shanidar Neandertals
[3300] U.S. National Park Service, 1971, Lehman Caves National Monument
[3301] U.S. National Park Service, 1987, Mammoth Cave National Park
[3319] U.S. National Park Service, 1984, Voices of the cave
[3328] United Air Lines, Inc., 1985, Luray Caverns in Virginia's Shenandoah Valley
[3329] Universitah ha-petuhah, 1988, Israeli scenery Soreq Cavern
[3526] Workman, C. Lindsay (narr), 1991, Carlsbad Caverns National Park
[3527] World Travelogues Corporation, 1984, Lewis and Clark Caverns
[3537] Wyllie, Diana, 1973, Caves—origins, development and formations
[3551] Zannes, Tom (prod, dir), 1993, Spirit of exploration: discovering the wonders of Carlsbad Caverns National Park

13.4 Bibliographies

[11] Adams, Laurie Branson (comp), Oct 1986, Annotated bibliography of North Carolina speleology
[88] Anon, Jul 1993, Catalogue of literature on Chinese karst caves
[414] Blatchley, William Stanley, 1859–1940, 1896, Indiana caves and their fauna
[415] Block, Guy de (ed), c1974, Bibliographie Speleologique Belge
[416] Block, Guy de (ed), 1981, Bibliographie Speleologique Belge: editions speleologiques belges 1975-1979
[417] Block, Guy de (ed), 1965, Bibliographie Speleologique Belge. Editions speleologiques belges. 1907-1964. Avec en appendice la liste des ouvrages a caractere speleologique edites en Belgique
[487] [British Cave Research Association], 1989, British Cave Research Association, library catalogue
[628] Chabert, Claude, 1971, E.A. Martel 1859 - 1939 bibliographie
[733] Crabtree, P. W. (ed), 1962—, Speleological Abstracts: key to Britain's speleological literature
[753] Crawford, Rod, Jul 1977, An annotated bibliography of Pacific Northwest speleobiology
[781] Currens, James C., 1952–, 1979, Bibliography of karst geology in Kentucky
[813] Davey, Adrian G., 1978, Nullarbor karst, a bibliography
[826] Davies, William Edward, 1917–, nd, Bibliography of North American speleology
[913] Dunkley, John R. (comp), nd, A bibliography of the Jenolan Caves: part 1: speleological literature
[914] Dunkley, John R. (comp), 1988, A bibliography of the Jenolan Caves: part 2: literature
[921] Dunkley, John R., Nov 1985, Preliminary checklist of references to Jenolan Caves
[1144] Ganter, John Hamilton, 1962–, 1989, The Xilitla region: an annotated bibliography
[1170] Gebauer, Herbert Daniel, Nov 1995, Speleological bibliography of South Asia including the Himalayan regions
[1175] George, Angelo Isham, 1944–, 1991, Bibliography of Wyandotte Cave
[1330] Halliday, William R., 1926–, Feb 1978, History and publications of the Western Speleological Survey
[1373] Harter, Russell George, 1947–, Jun 1971, Bibliography on lava tube caves
[1374] Harter, Russell George, 1947–, 1972, Bibliography on lava tube caves, supplement
[1550] Howes, Chris, 1951–, 1987, Cave references in Scientific American
[1578] Huppert, George Nixon, 1944–, Jun 1988, Cave and karst-related theses in United States and Canadian

Universities, 1899-1988

[1579] Huppert, George Nixon, 1944–, 1985, Selected bibliography of cave conservation and management
[1675] Irwin, David J., Apr 1986, Cheddar Caves bibliography
[1713] Jeffreys, Alan Lawrence, 1940–(abstr), Mar 1970, A bibliography of technical articles referring to practical caving subjects
[1733] Jillson, Willard Rouse, 1890–, 1953, A bibliography of Mammoth Cave, 1798-1949
[1774] Kastning, Ernst H., 1944–, May 1977, Geochemistry of karst waters: a basic bibliography
[1775] Kastning, Ernst H., 1944–, Jul 1989, Karst geomorphology and hydrogeology
[1817] Kiernan, Kevin, Sep 1989, Bibliography of Tasmanian karst
[1838] Knutson, Richard Stephen (comp), 1966, Bibliography of Oregon speleology: initial compilation
[1886] LaMoreaux, Philip Elmer, 1920–(ed), 1993, Annotated bibliography of karst terranes: volume five: with three review articles
[1888] LaMoreaux, Philip Elmer, 1920–, 1970, Hydrology of limestone terranes: annotated bibliography of carbonate rocks
[1889] LaMoreaux, Philip Elmer, 1920–(ed), 1989, Hydrology of limestone terranes: annotated bibliography of carbonate rocks, volume 4
[1890] LaMoreaux, Philip Elmer, 1920–(ed), 1986, Hydrology of limestone terranes: annotated bibliography of carbonate rocks, volume three
[1891] LaMoreaux, Philip Elmer, 1920–(ed), Jan 1975, Hydrology of limestone terranes: progress of knowledge about hydrology of carbonate terranes with an annotated bibliography of carbonate rocks
[1902] Larson, Charles Victor, 1928–, 1977, Bibliography of Oregon speleology, 1977
[1903] Larson, Charles Victor, 1928–, nd, A bibliography of Sea Lion Cave
[1927] Legault, Myrna Diaz (comp), Oct 1990, Bibliography of Puerto Rican caves, karst and limestone geology; Rio Camuy Cave System
[2040] Mansfield, Raymond Walter, 1965, Mendip cave bibliography and survey catalogue, January, 1901—December, 1963
[2042] Mansfield, Raymond Walter (ed), Jul 1970, Speleological abstracts: key to Britain's speleological literature
[2043] Mansfield, Raymond Walter (ed), Dec 1971, Speleological abstracts: the literature of 1968
[2087] Matthews, Larry Edwin, 1946–(ed), 1974, Bibliography of Tennessee speleology
[2132] McLane, Alvin Ray, 1974, A bibliography of Nevada caves
[2336] Nash, D. A., Sep 1989, A Peak Cavern bibliography
[2770] Reddell, James Russell, 1938–, Oct 1968, A bibliographic guide to Texas speleology
[2771] Reddell, James Russell, 1938–, 1995, Catalogue, bibliography, and generic revision of the order Schizomida (Arachnida)
[2785] Reddell, James Russell, 1938–, Aug 1969, A checklist and annotated bibliography of the subterranean aquatic fauna of Texas
[2790] Reddell, James Russell, 1938–, 1971, A preliminary bibliography of Mexican cave biology with a checklist of published records
[2815] Richardson, Douglas Turnbull (comp), Nov 1962, White Rose Pothole Club library catalog November 1962
[2827] Roberts, Jon, Nov 1982, Mendip Caving Group: 1982 library list
[2956] Shaw, Trevor Royle, 1928–, Jul 1972, Mendip Cave bibliography: part II: books pamphlets, manuscripts and maps; 3rd Century to December 1968
[3005] Smart, James, 1994, A Philippine cave index and bibliography
[3101] Standing, Ian James (ed), 1963, Speleological abstracts: the publications of 1963
[3102] Standing, Ian James (ed), 1964, Speleological abstracts: the publications of 1964
[3103] Standing, Ian James (ed), 1969, Speleological abstracts: the publications of 1965
[3104] Standing, Ian James (ed), Sep 1969, Speleological abstracts: the publications of 1966
[3236] Trickett, Oliver, 1898, Notes on the limestone caves of New South Wales with plans
[3243] Trudgill, Stephen Thomas, 1947–, 1977, A bibliography of British karst, 1960-1977
[3283] U.S. Geological Survey, 1967, Bibliography of the Ground Water Resources of New York through 1967, with subject and location cross-reference to the annotated bibliography
[3428] Wefer, Fred L., Jun 1991, An annotated bibliography of cave meteorology
[3440] Werner, Eberhard Wolfgand, 1942–, 1974, Index of the literature pertaining to West Virginia caves and karst
[3468] White, Susan, Aug 1986, A bibliography of Victorian caves & karst
[3470] White, William Blaine, 1934–, 1984, Cave and karst-related papers in the mainstream scientific literature: a bibliography
[3486] Wilkes, Frank G., 1962, Bibliography of Mammoth Cave National Park, Mammoth Cave, Kentucky
[3497] Williams, Nick, Jan 1994, Bibliography of Belizean caving
[3498] Williams, Nick (comp), Mar 1992, British Cave Research Association Cave Radio and Electronics Group bibliography of underground communications
[3503] Willis, Dick, 1986, Caving expeditions

13.5 Directories

[614]Cave Diving Section. National Speleological Society, 1986, NSS Cave Diving Section 1986 members' manual

[875]District of Columbia Grotto of the National Speleological Society, May 1991, Members manual

13.6 Dictionaries

[1141]Gams, Ivan, 1973, Slovenska Kraska Terminologija = Slovene karst terminology

[1854]Kósa, Attila, 1995, The caver's living dictionary

[1975]Lowe, David (comp), 1995, A dictionary of karst and caves

13.7 Education

[55]American Cave Conservation Association, nd, Karst curriculum resource guide

[1106]Foster, David G. (ed), 1994, Learning to live with caves and karst: a cave and karst curriculum and resource guide

13.8 Folklore

[23]Alderson, Laura White, 1984, Mammoth Cave: Kentucky's buried treasure

[3400]Warnell, Norman, 1997, Mammoth Cave: forgotten stories of it's [sic] people

[3455]Whealdon, Everett, 1983, The legend of Parker's Cave

13.9 Glossaries

[562]Burger, Andre (ed), 1975, Hydrogeology of karstic terrains: with a multilingual glossary of specific terms

[1455]Hill, Arthur H., 1952, A brief glossary of Welsh topographic names for walkers and cavers

[1906]Larson, Charles Victor, 1928–, 1993, An illustrated glossary of lava tube features

[2255]Monroe, Watson Hiner, 1907–, 1970, A glossary of karst terminology

[2274]Morland, Timothy Edward, 1959, A glossary of French speleological terms

13.10 Handbooks

[81]Anon, 1968, Australian Speleological Handbook

[273]Baguley, Frank (ed), Aug 1993, Cambrian Caving Council handbook 1993

13.11 Historical

[3]Abeille de Perrin, Elzear, 1877, On the collection of cavern insects

[7]Adam, John, 1886, Aberbrothok illustrated

[13]Adams, William Henry Davenport, 1828–1891, 1886, Famous caverns and grottoes: described and illustrated

[14]Adams, William Henry Davenport, 1828–1891, 1886, Famous caves and catacombs: described and illustrated

[20]Ainsworth, William, 1834, An account of the caves of Ballybunian, county of Kerry: with some mineralogical details

[50]Allen, Joel Asaph, 1838–1921, 1885, On an extinct type of dog from Ely Cave, Lee County, Virginia

[57]Ammen, Samuel Zenas, 1843–1929, 1890, The caverns of Luray, the property of the Valley Land and Improvement Co., Luray, Va: the manner of their formation, their peculiar growths, their geology, chemistry, etc: an illustrated guide book

[58]Ammen, Samuel Zenas, 1843–1929, 1893, The caverns of Luray. The property of the Valley Land and Improvement Co., Luray, Va. The manner of their formation, their peculiar growths, their geology, chemistry, etc. An illustrated guide book

[59]Ammen, Samuel Zenas, 1843–1929, 1884, The caverns of Luray: an illustrated guide-book to the caverns, explaining the manner of their formation, their peculiar growths, their geology, chemistry, etc

[60]Ammen, Samuel Zenas, 1843–1929, 1882, History and description of the Luray Cave (illustrated), including explanations of the manner of its formation, its peculiar growths, its geology, chemistry, &c.; also a map. The whole so arranged as to serve as a guide.

[61]Ammen, Samuel Zenas, 1843–1929, 1880, History and description of the Luray Cave (illustrated), including explanations of the manner of its formation, its peculiar growths, its geology, chemistry, etc.; also a map. The whole so arranged as to serve as a guide

[92]Anon, 1882, The Caverns of Luray

[107]Anon, 1865, A description of Howe's Cave: with a popular treatise on the formation of caves in lime rock, from the size of a quill to a mammoth

[108]Anon, 1850, A description of the Mammoth Cave of Kentucky, the Niagara River and Falls, and the Falls in summer and winter; the prairies, or life in the West; the Fairmont Water Works and scenes on the Schuylkill, xc. xc.: to illustrate Brewer's Panorama

[109]Anon, 1849, A description of the Mammoth Cave of Kentucky, the Niagara river and falls, and the falls in summer and winter; the prairies, or life in the West; the Fairmount Water Works and scenes on the Schuykill, etc. etc.: to illustrate Brewer's panorama.

[115]Anon, 1882, Exploration of the caves and rivers of New South Wales (minutes, reports, correspondence, accounts)

[123]Anon, 1800, The grottoes of Adelsberg and the *Proteus* of Anguinus: a few pages from the journal of a continental tour

[125]Anon, 1870, Guide book for the Diamond Cave, Barren County, Kentucky. Located near "Old Bell Tavern," now the Mammoth Cave and Glosgow Junctions, on the L & N railroad, two miles from the R.R. immediately on the Mammoth Cave Road

[126]Anon, 1876, A guide manual to the Mammoth Cave of Kentucky

[127]Anon, 1876, Guide manual to the Mammoth Cave of Kentucky

[129]Anon, 1818, Guide to the Grand Cavern within the mountain of Abraham's Heights, Matlock Bath

[154]Anon, 1990, Maribel Caves: an ideal health and summer resort.

[170]Anon, 1828, On the cold caves of the Monte Testaccio at Rome

[173]Anon, c1886, Photographic views of some of the important points of Mammoth Cave situated in Edmondson, Co. Kentucky, USA: the wonders of this cave and its magnitude, cannot be described, and must be seen to be appreciated. It is reached only by the Louisville & Nashville Railroads. These photographs were taken by magnesium light by W.H. Sesser, St. Joseph, Mich., with a Collins camera

[199]Anon, 1930, Subterranean wonders: Mammoth Cave and Colossal Cavern, Kentucky

[207]Anon, 1984, Virginia Luray Caverns

[225]Argus [pseud], 1898, Jenolan Caves and the Blue Mountains

[275]Bailey, Gilbert Stephen, 1822–1891, 1863, The great caverns of Kentucky: Diamond Cave, Mammoth Cave, Hundred Dome Cave

[291]Balch, Edwin Swift, 1856–1927, 1970, Glaciéres; or, freezing caverns

[292]Balch, Edwin Swift, 1856–1927, 1896, Ice caves and the causes of subterranean ice

[311]Bancroft, Hubert Howe, 1832–1918, 1886, Manitou Grand Caverns

[373]Behrens, Georg Henning, 1662–1712, 1730, The natural history of Hartz-Forest, in His Majesty King George's German dominions. Being a succinct account of the caverns, lakes, springs, rivers, mountains, rocks, quarries, fossils, castles, gardens, the famous pagan idol Pustrich or spit-fire, dwarf-holes, etc....in the said forest: with several useful and entertaining physical observations

[374]Beldam, Joseph, 1795–1866, 1858, The origin and use of the Royston Cave, being the substance of a report some time since presented to the Royal Society of Antiquaries by the late Joseph Beldam ...

[389]Bent, James Theodore, 1852-1897, 1892, The ruined cities of Mashonaland; being a record of excavation and exploration in 1891

[400]Binkerd, Adam D., , Pictorial guide to the Mammoth Cave, Kentucky: a complete historic, descriptive and scientific account of the greatest subterranean wonder of the western world

[445]Boulton, John, 1871, Particulars of a first exploration of the extensive and newly discovered cavern, at Stainton, Low Furness

[466]Brenan, Edward, 1859, Notice of the occurrence of mammoth and other animal remains: discovered under limestone in a bone cave at Shandon, near Dungarvan, in the county of Waterford

[532]Browne, George Forrest, 1833-1930, 1865, Ice-caves of France and Switzerland. A narrative of subterranean exploration

[539]Bruun, Daniel, 1985, The cave dwellers of Southern Tunisia: recollections of a sojourn with the Khalifa of Matmata

[540]Bryan, Benjamin, 1839, A description of the fluor spar or Blue John Mine at Castleton

[541]Bryan, Benjamin, 1840, A description of the grand Devonshire cavern at Matlock

[545]Buckland, William, 1784–1856, 1824, Reliquiae diluvianae, or observations on the organic remains contained in caves, fissures, and diluvial gravel, and on other geological phenomena attesting the action of an universal deluge /

[546]Buckland, William, 1784–1856, 1978, Reliquiae diluvianae: or, observations on the organic remains contained in caves, fissures, and diluvial gravel and on other geological phenomena attesting the action of a universal deluge

[547]Buckland, William, 1784–1856, 1823, Reliquiae diluvianae: or, observations on the organic remains contained in caves, fissures, and diluvial gravel, and on other geological phenomena attesting the action of an universal deluge

[552][Bullitt, Alexander Clark, 1807–1868], 1973, Rambles in the Mammoth Cave, during the year 1844

[580]Call, Richard Ellsworth, 1856–1917, 1897, Mammoth Cave, Kentucky

[581]Call, Richard Ellsworth, 1856–1917, c1890, The Mammoth Cave, Kentucky; a sketch

[583]Callard, Thomas Karr, [1880], Contemporaneity of man with the extinct Mammalia, as taught by recent cavern exploration, and its bearing upon the question of man's antiquity

[601]Carr, Joseph, 1865, Rambles about Ingleton, in caves by rivers and on mountains in the spring of 1865.

[612]Catcott, George Symes, 1792, A descriptive account of a descent made into Penpark-hole, in the parish of Westbury-upon-Trim, in the county of Gloucester, in the

13.11. HISTORICAL

year 1775

[699] Cook, Samuel, 1889, The Jenolan caves: an excursion in Australian wonderland

[704] Cooke, Robert L., 1834, A description of Weyer's Cave

[711] Cope, Edward Drinker, 1840–1897, 1883, On the contents of a bone cave in the island of Anguilla (West Indies)

[719] Corrie, A. P., 1899, The Wombeyan Caves and the Bowral, Mittagong & Moss Vale tourist districts

[757] Croghan, John, 1790–1849, 1845, Rambles in Mammoth Cave, during the year 1844, by a visitor

[793] Daldy, Isbister & Co., 1878, Half hours underground: volcanoes, mines, and caves

[807] Dashwood, Francis, 1975, West Wycombe caves: [a brief history and description]

[821] Davidson, Robert, 1808–1876, 1840, An excursion to the Mammoth Cave, and the barrens of Kentucky: with some notices of the early settlement of the state

[834] Davis, Henry Harrison, 1851, An excursion from Lancaster, up the vale of Lune, and from Kirkby Lonsdale, to the caves of Yorkshire

[839] Dawkins, William Boyd, 1838–1929, 1868, The British Pleistocene Mammalia: part II

[840] Dawkins, William Boyd, 1838–1929, 1973, Cave hunting, researches on the evidence of caves respecting the early inhabitants of Europe

[872] Dillaye, Stephen D., 1848, A glance at Mammoth Cave, Kentucky

[932] Eastmead, William, 1824, Historia rievallensis: containing the history of Kirkby Moorside ... to which is prefixed a dissertation on the animal remains, and other curious phenomena, in the recently discovered cave at Kirkdale

[971] Ellis, Edward Sylvester, 1840–1916, 1900, Bear cavern

[1031] Farrer, James William, 1970, Further explorations in the Dowkerbottom Caves, in Craven

[1049] Ferguson, Samuel, 1810–1886, 1865, Account of Ogham inscriptions in the cave at Rathcroghan, Co. Roscommon

[1053] Fewkes, Jesse Walter, 1850–1930, 1896, Preliminary account of an expedition to the cliff villages of the red rock country, and the Tusayan ruins of Sityatki and Awatobi, Arizona, in 1895

[1073] Floyd, E. F., 1887, The New Calaveras Cave, Murphys, Calaveras County, California: W.J. Mercer, proprietor: description and guide

[1103] Forwood, William Stump, 1830–1892, 1870, An historical and descriptive narrative of the Mammoth Cave of Kentucky: including explanations of the causes concerned in its formation, its atmospheric conditions, its chemistry, geology, zoology, etc., with full scientific details of the eyeless fishes

[1107] Foster, J. J., 1890, The Jenolan Caves

[1124] Fraser, L.A., Jul 1892, Weyer Cave. Cave of the Fountains. Madison Cave: a descriptive sketch of the grottoes of the Shenandoah

[1226] Goodwin, J. West, , Mammoth Cave: America's great natural wonders

[1227] Gordon, George Byron, 1870–1927, 1898, Caverns of Copan, Honduras: report on explorations by the museum, 1896-97

[1242] Grant, James, 1831–1920, 1868, Superficial geology of the valley of the Ottawa and the Wakefield cave

[1299] Gwin, J. N., 1875, A visit to the Mammoth Cave

[1300] Haast, J., 1874, Researches and excavations carried on in and near Moa Bone Point Cave, Summer Road, in the year 1872

[1360] Harlan, Richard, 1796–1843, 1831, Description of the fossil bones of the Megalonyx, discovered in White Cave, Kentucky

[1376] Hartwig, Georg, 1813–1880, 1888, Marvels under our feet: from "the subterranean world"

[1417] Heath, Thomas, 1880, Creswell Caves vs Professor Boyd Dawkins

[1419] Hedinger, J. M., 1820, A short description of Castleton in Derbyshire: its natural curiosities and mineral productions

[1441] Henkel, David S., 1919, A description of the Endless Caverns of New Market

[1442] Henkel, David S., 1880, A description of the New Market Endless Caverns

[1513] Home, Everard, 1756–1832, 1794, Account of some remarkable caves in the principality of Bayreuth: likewise observations on the fossil bones by the late John Hunter

[1533] Housman, John, 1800, A descriptive tour, and guide to the lakes, caves, mountains, and other natural curiosities, in Cumberland, Westermoreland, Lancashire, and a part of the West Riding of Yorkshire

[1536] Hovey, Horace Carter, 1833–1914, 1882, Celebrated American caverns: especially Mammoth, Wyandot, and Luray: together with historical, scientific, and descriptive notices of caves and grottoes in other lands

[1537] Hovey, Horace Carter, 1833–1914, 1891, Guide book to the Mammoth Cave of Kentucky: historical, scientific, and descriptive

[1540] Hovey, Horace Carter, 1833–1914, 1897, Mammoth Cave of Kentucky; an illustrated manual

[1542] Howard, John Eliot, 1879, The caves of South Devon and their teachings

[1558] Howison, Allan Moore, 1889, The Grottoes of the Shenandoah, consisting of the Weyer and the Fountain Caves, the two most wonderful caverns in the world, at Grottoes Station of Shenandoah Valley Railway, Augusta County, Virginia. A. M. Howison, Secretary.

[1568] Hughes, Thomas McKenny, 1887, On caves

[1587]Hutton, John, 1740–1806, 1780, A tour to the caves in the environs of Ingleborough and Settle in the West Riding of Yorkshire: with some philosophical conjectures on the deluge, remarks on the origin of fountains, and observations on the ascent and descent of vapours, occasioned by facts peculiar to the places visited. Also a large glossary of old and original words made use of in common conversation in the north of England. In a letter to a friend

[1588]Hutton, John, 1740–1806, 1781, A tour to the caves in the environs of Ingleborough and Settle in the West—Riding of Yorkshire: with some philosophical conjectures on the deluge, remarks on the origin of fountains, and observations on the ascent and descent of vapours, occasioned by facts peculiar to the places visited. Also a large glossary of old and original words made use of in common conversation in the north of England. In a letter to a friend

[1695]James, C. H., (photo), 1882, Electric light views in the Caverns of Luray at Luray Station (Page Co., Virginia), Shenandoah Valley Railroad

[1741]Johnston, Frances Benjamin, 1864–1952, 1893, Mammoth Cave by flash-light

[1751]Jones, Thomas Rupert, 1819–1911, 1877, Lecture on the antiquity of man: illustrated by the contents of caves and relics of cave-folk

[1755]Jones, William Basil, 1822–1897, 1844, Wonderful curiosity, or, a correct narrative of the celebrated Mammoth Cave of Kentucky with incidents and anecdotes

[1758]Jordan, George, 1972, The great cave of Dry Fork of Cheat River, Virginia: one of the greatest wonders of the world examined and explored Giving a complete description of this subterraneous cavern, of its rooms and passages, which are of enormous size, together with a great stream of water, a lake, and springs, minerals, fossils, and many curiosities, that were found in singular abundance, &c.

[1759]Jordan, J. Murray, 1897, The Caverns of Luray and vicinity

[1795]Kentucky Geological Survey, 1876, Memoirs of the Geological Survey of Kentucky

[1815]Kidder, Daniel Parish, 1815–1891, 1848, The caves of the earth: their natural history, features, and incidents

[1834]Knoepfel, William H., 1853, An account of Knoepfel's Schoharie Cave, Schoharie County, New York: with the history of its discovery, subterranean lake, minerals and natural curiosities

[1845]Kohler, S. D. F., 1971, The Crystal Cave at Virginsville

[1936]Letcher, Montgomery E., 1839, Wonderful discovery: being an account of a recent exploration of the celebrated Mammoth Cave, in Edmonson County, Kentucky, by Dr. Rowan, Professor Simmons and others, of Louisville, to its termination in an inhabited region, in the interior of the earth

[1994]Luray Caverns Corporation, nd, Caverns of Luray

[1995]Luray Caverns Corporation, 1882, Electric light views in the Caverns of Luray. The Caverns of Luray (at Luray, Page County, Virginia, a station on the Shenandoah Valley Railroad) as a resort for tourists ... are unexcelled, in their wonderful attractiveness, by any other creation of nature

[2009]MacEnery, John M., 1796–1841, 1859, Cavern researches, or, discoveries of organic remains, and of British and Roman reliques, in the caves of Kent's Hole, Anstis Cove, Chudleigh, and Berry Head

[2011]Mackie, Charles, 1864–, 1830, Historical description of the chapel and castle of Roslin, and the caverns of Hawthornden

[2014]Macleay, Kenneth, 1819–, 1811, Description of the spar cave, lately discovered in the Isle of Skye: with some geological remarks relative to that island

[2034]Mammoth Cave Operating Committee, Mammoth Cave, Ky, 1859, Mammoth cave

[2063]Martin, Horace, 1851, Pictorial guide to the Mammoth Cave, Kentucky

[2075]Mason, Otis Tufton, 1838–1908, , The caverns of Luray

[2076]Mason, Otis Tufton, 1838–1908, 1882, The caverns of Luray

[2077]Mason, Otis Tufton, 1838–1908, 1881, Report of a visit to the Luray Cavern, in Page County, Virginia, under the auspices of the Smithsonian institution, July 13 and 14, 1880

[2116]McDonald, Alvin F., 1873–1893, nd, Private account of A. F. McDonald, permanent guide of Wind Cave

[2135]McLoyd & Graham, 1894, Catalogue and description of a very large collection of prehistoric relics: obtained in the cliff houses and caves of Southeastern Utah

[2136]McMurtrie, James, Aug 1882, The Mendip Caverns

[2149]Mercer, Henry Chapman, 1856–1930, 1899, The bone cave at Port Kennedy, Pennsylvania, and its partial excavation in 1894, 1895 and 1896

[2150]Mercer, Henry Chapman, 1856–1930, 1897, Cave hunting in Yucatan: a lecture delivered before the Society of Arts of the Massachusetts Institute of Technology, on December 10,1896

[2151]Mercer, Henry Chapman, 1856–1930, 1897, An exploration of Durham Cave in 1893

[2153]Mercer, Henry Chapman, 1856–1930, 1896, The hill-caves of Yucatan, a search for evidence of man's antiquity in the caverns of Central America. Being an account of the Corwith expedition of the Department of Archaeology and Palaeontology of the University of Pennsylvania

[2154]Mercer, Henry Chapman, 1856–1930, 1975, The hill-caves of Yucatan, a search for evidence of man's antiquity in the caverns of Central America. Being an account of the Corwith expedition of the Department of Archaeology and Paleontology of the University of Pennsylvania

[2155]Mercer, W. J., 1887, The New Calaveras Cave,

13.11. HISTORICAL

Murphys, Calaveras County, California: W.J. Mercer, proprietor: description and guide
[2159]Merk, Conrad, 1876, Excavations at the Kesserloch near Kesserloch near Thayngen Switzerland: a cave of the Reindeer Period
[2181]Miller, Edward Sherman, 1866–, 1893, Joint thesis upon Maidstone Cave
[2247]Mohler, John Leonard, 1840–1937, 1852, A description of Weyer's Cave in Augusta County, Virginia
[2272]Morgan, Robert, nd, Through Mammoth Cave, giving a graphic description of all the routes, avenues, passes, passages, pits, domes, streams, rivers, seas, lakes, and other wonders to be found in that greatest of all natural curiosities in the whole realm of this great world
[2273]Morgan, Robert, 1881, Through Mammoth Cave, giving a graphic description of all the routes, avenues, passes, passages, pits, domes, seas, lakes, rivers, streams, and other wonders of that greatest natural curiosity in the world
[2513]Noah, Samuel, 1835, A description of Weast's Cave
[2583]Owen, Luella Agnes, Apr 1968, Cave regions of the Ozarks
[2584]Owen, Luella Agnes, 1970, Cave regions of the Ozarks and Black Hills
[2590]Packard, Alpheus Spring, 1839–1905, 1977, The cave fauna of North America, with remarks on the anatomy of the brain and origin of the blind species
[2592]Packard, Alpheus Spring, 1839–1905, 1877, On a new cave fauna in Utah: and on new phyllopod Crustacea from the West
[2610]Panton, James Hoyes, 1847–1898, 1890, The Mammoth Cave of Kentucky
[2635]Pelech, Johann E., 1879, The valley of Stracena and the Dobschau ice-cavern (Hungary)
[2643]Pengelly, Hester (ed), 1897, A memoir of William Pengelly, of Torquay, F.R.S., Geologist, with a selection from his correspondence
[2644]Pengelly, William, [1872], Kent's Cavern: a lecture
[2645]Pengelly, William, 1874, Kent's Cavern and its wonders: a lecture delivered in the public hall Warrington on December 22nd 1873
[2646]Pengelly, William, 1876, Kent's Cavern: its testimony to the antiquity of man: a lecture delivered in the City Hall, Glasgow on Wednesday, 22nd December, 1875
[2649]Pennington, Rooke, 1877, Notes on the barrows and bone-caves of Derbyshire with an account of a descent into Elden Hole
[2650]Pennsylvania Railroad, 1885, A summer series of personally-conducted pleasure tours to the Luray Caverns or Gettysburg via Pennsylvania Railroad
[2668]Phillips, Richard, 1767–1840, 1810, A view of the earth, containing an account of its internal structure; its caves and subterranean passages; its mountains, its rivers and cataracts. Together with a brief view of the universe, to which are added, problems on the globes, directions for drawing maps and tables of latitudes and longitudes
[2672]Pictet, Marc-Auguste, 1752–1825, 1823, On the ice-caves or natural ice-houses found in some of the caverns of the Jura and the Alps
[2709]Prestwich, Joseph, 1872, Report on the exploration of Brixham Cave, conducted by a committee of the geological society, and under the superintendence of Wm. Pengelly, etc
[2755]Rains, Geo. W., 1861, Notes on making saltpetre from the earth of the caves
[2805]Religious Tract Society, 1800, The caves of the earth: their natural history, features, and incidents
[2840]Roemer, Ferdinand Carl, 1818–1891, 1884, The bone caves of Ojcow in Poland
[2860]Rothrock, H. W., 1889, Wyandotte Cave: where it is and what it is
[2863]Royse, W., 1891, Descriptive account of the Blue John Mines and Caverns at Castleton, Derbyshire
[2866]Rusling, James Fowler, 1864, A trip to the Mammoth Cave, Ky
[2873]Rutter, John, 1796–1851, 1829, Delineations of the north western division of the county of Somerset, and of its antediluvian bone caverns, with a geological sketch of the district
[2920]Schwatka, Frederick, 1893, In the land of cave and cliff dwellers
[2938]Seward, William, 1801, A tour to Yordes Cave
[2944]Shaler, Nathaniel Southgate, 1841–1906, 1876, On the antiquity of the caverns and cavern life of the Ohio valley
[2965]Shenandoah Valley Railroad, 1889, Through the Shenandoah Valley: Caverns of Luray, Natural Bridge, grottoes of the Shenandoah and the chronicle of a leisurely journey through the uplands of Virginia
[2985]Simpson, James Young, 1811–1870, 1867, Account of some ancient sculptures on the walls of caves in Fife
[3020]Smith, H., 1882, Kent's Cavern
[3021]Smith, J. P., 1890, Paper on Dunald Mill Hole: read by J. P. Smith before the Barrow Naturalists Field Club
[3024]Smith, John, 1894, Monograph of the stalactites and stalagmites of the Cleaves Cove, near Dalry, Ayrshire
[3025]Smith, John Moyr, 1883, The Hades of Ardenne: a visit to the Caves of Han described and illustrated
[3037]Smithsonian Institution, 1880, The caverns of Luray, located on line of the Shenandoah Valley R.R
[3039]Snider, George Washington, 1888, Manitou Grand Caverns, Manitou Springs, Colorado
[3040]Snider, George Washington, 1885, The tourists' gem: the Manitou Grand Caverns, the largest and most wonderful subterranean in the Rocky Mountains, and other attractions for tourists
[3116]Stelle, James Parish, 1884, The Wyandotte Cave of Crawford County, Ind

[3123] Stevens, Martin V. B., 1876, The history, guide and description of Wyandotte Cave
[3128] Stirling, James, 1884, On the caves perforating marble deposits, Limestone Creek (read 12th April 1883)
[3130] Stoddard, Seneca Ray, 1844–1917, 1889, Howe's Cave
[3155] Studley, Cordelia A., 1884, Notes upon the human remains from the caves of Coahuila, Mexico
[3173] Symonds, William Samuel, 1818–1887, 1883, The Seven Straits; or, notes on glacial drifts, bone caverns, and old glaciers: some within reach of the Malvern Hills
[3202] Thompson, Edward Herbert, 1860–1935, 1897, Cave of Loltun, Yucatan: report of explorations by the Museum, 1888-89 and 1890-91
[3207] Thompson, Ralph Seymour, 1879, The sucker's visit to the Mammoth Cave including a history of the experience and adventures of a party who undertook to see the cave and have some fun going there. A full and accurate description of the cave and the science of its formation with an account of the living inhabitants
[3224] Towler, George (photo), 1890, Views of Chapman Cave
[3232] Trickett, Oliver, 1899, Guide to the Jenolan Caves New South Wales
[3236] Trickett, Oliver, 1898, Notes on the limestone caves of New South Wales with plans
[3255] Turner, Percy (auth, comp), 1895, History of Manitou's caves, the Grand Caverns and Cave of the Winds: and points of interest in and about Manitou
[3448] Westall, William, 1818, Views of the caves near Ingleton, Gordale Scar and Malham Cove in Yorkshire
[3453] Weyer, Bernard, 1881, A description of Weyer's Cave in Augusta County, Va
[3454] Weyer, Bernard, 1849, A description of Weyer's Cave, in Augusta County, Virginia
[3461] Chisholm, T.O., 1892, Grand Avenue Cave. A description in detail of one of America's greatest natural wonders
[3477] Whitehouse, Frederic Cope, 1882, Is Fingal's Cave artificial?
[3489] Wilkinson, Charles Smith, 1887, Photographs of the Jenolan Caves (interior views photographed by means of the electric and magnisium [sic] lights
[3494] Williams, David, 1829, Some account of the fissures and caverns hitherto discovered in the western district of the Mendip range of hills: comprised in a letter from Rev. D. Williams to the Rev. Patterson
[3510] Wilson, John, 1800–1849, 1849, A visit to the Mammoth Cave of Kentucky
[3525] Woods, Julian Edmund, 1862, Geological observations in South Australia: principally in the district southeast of Adelaide
[3531] Wright, Charles W., , A guide manual to the Mammoth Cave of Kentucky
[3532] Wright, Charles W., 1860, A guide manual to the Mammoth Cave of Kentucky
[3533] Wright, Charles W., 1858, The Mammoth Cave of Kentucky

13.12 Humor

[45] Alfie [pseud for S. J. Collins], 1971, Reflections: a look at the 'spelaeodes' and other caving sagas
[46] Alfie [pseud for S. J. Collins], 1969, The spelaeodes
[397] Biddle, Roger (ed), Apr 1976, Collected caving songs: volume 1: the songs of Mendip
[639] Charnock, S. J., Feb 1994, Underneath the arches: caving in the RibbleHead District
[989] Engel, Thomas Daniel, 1954–, 1992, The almost complete eclectic caver
[1426] Helmer, William, 1961, "There we was!...."
[1440] Hendy, Phil, 1978, Muddy oxbows: a tourist trip into the seldom-trodden inner recesses of the human brain
[1866] Kristofferson, N. R., 1972, Kaver komix
[1998] Luvless, Dungass F. (pseud), 1972, The speleobopper's guide to the caves of the Gruesome Chapel Valley
[3384] Walsh, John Michael, 1947–(ed), 1982, Texas cave humor, the first twenty-five years

13.13 Indexes

[376] Bell, Anne (comp), [1986], South Wales Caving Club Newsletter: index to numbers 1 - 100: 1946 - 1985
[480] Bristol (England). Public Libraries, 1968, Caving periodicals in the central collection of caving publications: a catalogue of the caving periodicals held by Bristol Central Reference Library and a selected short title catalogue of books
[484] British Cave Research Association, 1985, BCRA library catalogue
[1390] Hawkins, R. J. (comp), 1972, Australasian speleo map index no. 1
[1523] Horne, Peter, 1980, A report on South Australian diving fatalities
[1525] Horne, Peter, 1987, South Australian diving fatalities 1950-1985
[1551] Howes, Chris, 1951–, 1995, CMW Bran CC Journal index vol. 1-20; 1967-1994 including newsletters, newssheets & the Silver Jubilee Journal
[1553] Howes, Chris, 1951–, 1991, The Descent index: issues 1-100.
[1554] Howes, Chris, 1951–, 1995, ISCA CC journal index volumes 1-17; 1984-1994
[1555] Howes, Chris, 1951–, 1994, The Red Dragon Journal of the Cambrian Caving Council index issues (1)-(20) 1974-1993

[1557]Howes, Chris, 1951–, 1994, Wide world magazine Index volumes 1-50 1898-1923
[1677]Irwin, David J., Dec 1987, An index to the publications of The Bristol Exploration Club 1947-1987
[1940]Lewis, Ian D., 1976, South Australian cave reference book
[2170]Middleton, Gregory. J. (comp), Sep 1988, Index to references to caves in three New South Wales government publications 1870-1919
[2358]National Speleological Society, 1986, Cumulative index for the National Speleological Society Bulletin, volumes 1 through 45, and Occasional papers of the N.S.S., numbers 1 through 4
[2449]National Speleological Society. Speleo Digest, 1991, Speleo Digest 25 - year cumulative index, 1956 through 1980
[2619]Pavey, Andrew (comp), Nov 1972, An index to cave maps in N.S.W
[2708]Prentice, Guy, 1989, Mammoth Cave National Park archeological inventory project interim report-1988 investigations
[2888]Sasowsky, Ira Daniel, 1959–, 1991, Index to volumes 46 through 50 of the National Speleological Society bulletin
[3005]Smart, James, 1994, A Philippine cave index and bibliography
[3024]Smith, John, 1894, Monograph of the stalactites and stalagmites of the Cleaves Cove, near Dalry, Ayrshire
[3440]Werner, Eberhard Wolfgang, 1942–, 1974, Index of the literature pertaining to West Virginia caves and karst
[3529]Worthy, Trevor, 1989, Inventory of New Zealand caves and karst of international, national and regional importance

13.14 Inventories

[766]Crush, Kenneth, Sep 1977, An inventory of caves, Harry S. Truman dam and reservoir
[1084]Ford, Derek Clifford, 1935–, Mar 1973, Theme and resource inventory study of the karst regions of Canada: final report upon project A
[1151]Gardner, James, 1982, An inventory and evaluation of Missouri state parks cave resources
[1152]Gardner, James Eugene, 1953–, 1983, Cave resources of Ozark National Scenic Riverways: an inventory and evaluation phase II
[1154]Gardner, James Eugene, 1953–, Sep 1982, An inventory and evaluation of cave resources of Mark Twain National Forest: a final report submitted to the Mark Twain National Forest, United States Department of Agriculture in compliance with a cooperative cave inventory agreement
[1156]Gardner, James Eugene, 1953–, Jan 1984, Missouri Department of Conservation Cooperative cave inventory project: a final report submitted to the Missouri Department of Conservation as part of the cooperative cave inventory project: a final report
[1468]Hobbs, Horton Holcombe, 1944–, Jan 1994, Assessment of the ecological resources of the caves of Russell Cave National Monument, Jackson County, Alabama and of selected caves at the Lookout Mountain unit of Chickamanga-Chattanooga National Military Park, Dade County, Georgia and Hamilton County, Tennessee
[1818]Kiernan, Kevin, Nov 1989, Karst, caves and management at Mole Creek, Tasmania
[2016]Madacheen, L. Michael, 1977, New Melones cave inventory and evaluation study: preliminary report: archeological caves
[2521]Northup, Diana Eleanor, 1948–, 1992, Lechuguilla Cave: biological inventory
[3082]Spate, Andy, 1991, Kubla Khan Cave State Reserve, Mole Creek, Tasmania: pilot management study

13.15 Juvenile

[1]A. I. D. International, 1977, Caves
[96]Anon, 1965, Caves
[381]Bender, Lionel, 1989, Cave
[437]Booth, Eugene, 1940–, 1977, Under the ground
[462]Brandt, Keith, 1949–, 1985, Caves
[470]Breuil, Abbé Henri, 1877–1961, 1949, Beyond the bounds of history: scenes from the old stone age
[498]Bronin, Andrew, 1972, The cave: what lives there
[666]Close, Danny, 1981, Mammoth Cave... kids love it!
[676]Coder, Kate, 1989, Pennsylvania's caves & caverns: an activity book
[734]Crain, Sally Lucille 1938–(auth, prod), 1979, Prehistoric magic
[793]Daldy, Isbister & Co., 1878, Half hours underground: volcanoes, mines, and caves
[846]Dean, Anabel, 1984, Going underground: all about caves and caving
[864]Dewey, Jennifer, 1994, The creatures underneath
[883]Dopp, Katherine Elizabeth, 1863–, 1904, The early cave-men
[884]Dopp, Katherine Elizabeth, 1863–, 1906, The later cave men
[908]Duddington, C. L. , 1969, Caves
[993]England, A. W., 1973, Caves, an anthology
[999]Epstein, Sam, 1909–, 1959, All about prehistoric cave men
[1104]Fossen, Joyce, 1985, Caves of Paha Sapa, and other findings of Dr. Knows-It
[1106]Foster, David G. (ed), 1994, Learning to live with caves and karst: a cave and karst curriculum and resource guide

[1142]Gans, Roma, 1894–, 1976, Caves
[1190]Gibbons, Gail, 1993, Caves and Caverns
[1199]Gilbreath, Alice Marie , 1921–, 1978, Nature's underground palaces: caves and caverns
[1247]Great American Film Factory, 1977, Underground wilderness
[1254]Greenberg, Judith E., 1990, Caves
[1289]Gunzi, Christiane, 1993, Cave life
[1313]Halaburda, Sue, 1984, Mammoth Cave
[1340]Hamilton, Elizabeth, 1900–, 1956, The first book of caves
[1384]Hastings, Valerie, 1959, Jo and Coney's Cavern
[1494]Holmes, Frank, 1951, Monkey Merry and St. Crida Cave
[1514]Hooks, William H., 1977, Maria's Cave
[1706]Jankowski, Laurence (exe prod), 1985, Let's explore a cave
[1811]Kerbo, Ronal Carrel, 1944–, 1981, Caves
[1815]Kidder, Daniel Parish, 1815–1891, 1848, The caves of the earth: their natural history, features, and incidents
[1829]Knapp, Brian J., 1992, Cave
[1859]Kramer, Stephen P., 1995, Caves
[1918]Laycock, George , 1976, Caves
[1963]Longsworth, Polly, 1959, Exploring caves
[1964]Longul, Wally, 1987, The last horizon
[2045]Marcus, Rebecca Brian, 1968, Prehistoric cave paintings
[2081]Mass, Nuri, 1946, The wizard of Jenolan
[2113]McClurg, James E., 1962, Caves and their mysteries
[2131]McKinnon, Judith, 1991, Under the ground
[2276]Morris, Deborah, 1956–, 1993, Trapped in a cave!: a true story
[2278]Morris, Neil, 1996, Caves
[2334]Naden, C. J., 1979, I can read about caves
[2335]Naegele, Thomas A., 1987, Caving: a child's adventure
[2659]Petersen, David, 1994, Carlsbad Caverns National Park
[2693]Pope, Joyce, 1991, Life in the dark
[2701]Powell, Hazel Rowena, 1953, Adventures underground in the caves of Missouri
[2704]Powers, Richard M., 1963, The cave dwellers in the old stone age
[2747]Radlauer, Ruth, 1926–, 1981, Carlsbad Caverns National Park
[2748]Radlauer, Ruth, 1926–, 1986, Mammoth Cave National Park
[2820]Rigby, Susan, 1992, Caves
[2826]Roberts, Allan, 1983, Underground life
[2831]Robison, Mabel Otis, 1966, The hole in the mountain
[2899]Scheele, William E., 1959, The cave hunters
[2914]Schultz, Ronald, 1951–, 1993, Looking inside caves and caverns
[2918]Schwartz, Douglas W., 1965, Prehistoric man in Mammoth Cave
[2945]Shannon, Robert Terry, 1960, About caves
[2946]Shannon, Robert Terry, 1966, About caves
[2968]Sherman, Geraldine, 1980, Caverns: a world of mystery and beauty
[2982]Silver, Donald M., 1947–, 1993, Cave
[3120]Sterling, Dorothy, 1913–, , The story of caves
[3121]Stern, Philip Van Doren, 1900–, 1973, The beginnings of art
[3221]Toops, Bonnie, 1990, Let's explore: caves and caverns: a young explorer series
[3356]Vaughan, Jennifer, 1973, Caves
[3488]Wilkins, Frances, 1977, Caves
[3522]Wood, Jenny, 1990, Caves
[3523]Wood, Jenny, 1990, Caves
[3556]Zim, Herbert Spencer, 1909–, 1978, Caves and life

13.16 Manuals

[258]Australian Speleological Federation, [1972], Administrative handbook
[420]Blue Mountains Speleological Club, 1977, Policy organisation and rules
[1270]Groenhout, Ron (ed), Sep 1968, Guide for prospective cavers
[2091]Matthews, Peter Gahan, 1938–, Jul 1983, Australian karst index data preparation manual
[2490]Nelson, Raymond L., Nov 1959, Mammoth Cave tour leader manual
[2499]New Zealand Speleological Society, Feb 1989, The little red cavers book
[2500]New Zealand Speleological Society, 1985, The little red cavers book: NZ Speleological Society information manual
[2724]Prosser, J. Joseph, 1990, Cave diving communications
[2725]Prosser, J. Joseph, 1992, NSS cave diving manual: an overview
[2727]Prosser, J. Joseph, Oct 1989, NSS student cave diver workbook: designed specifically for use by the student cave diver participating in the NSS full cave diver course
[2766]Ray, Michael Allen, nd, Underground worlds: tour guide training manual

13.17 Meetings

[2427]National Speleological Society. Annual Convention. Proceedings (1976: Morgantown, WV), 1977, Proceedings of the 1976 NSS annual convention at Morgantown, West Virginia 24 June - 3 July 1976

[2428]National Speleological Society. Annual Convention. Program (1976: Morgantown, WV), 1976, 1976 National Speleological Society 35th Annual Convention Guidebook *[program]* with Abstracts

[2429]National Speleological Society. Annual Convention. Program (1977: Alpena, MI), 1977, Program 1977 Annual Meeting of the National Speleological Society, August 1-5, Alpena, Michigan]

[2430]National Speleological Society. Annual Convention. Program (1978: New Braunfels, TX), 1978, Deep in the karst of Texas: general information program: abstracts NSS-New Braunfels 1978

[2431]National Speleological Society. Annual Convention. Program (1979: Pittsfield, MA), 1979, The program of the 1979 N.S.S. Convention, August 5-12, 1979, Pittsfield, Mass

[2432]National Speleological Society. Annual Convention. Program (1980: White Bear Lake, MN), 1980, 1980 NSS Convention Program, Caving in the City, July 27 - August 1, Lakewood College, White Bear Lake, Minnesota

[2433]National Speleological Society. Annual Convention. Program (1982: Bend, OR), 1982, The program of the annual convention of the National Speleological Society, June 27-July 3, Bend, Oregon

[2434]National Speleological Society. Annual Convention. Program (1983: Elkins, WV), 1983, Program, 1983 NSS Annual Convention

[2435]National Speleological Society. Annual Convention. Program (1984: Sheridan, Wyoming), 1984, The program of the 1984 NSS Convention, June 25-29, 1984, Sheridan, Wyoming

[2436]National Speleological Society. Annual Convention. Program (1985: Frankfort, KY), 1985, Program for the 1985 NSS Convention, Frankfort, Kentucky, June 23-29

[2437]National Speleological Society. Annual Convention. Program (1986: Tularosa, NM), 1986, Program, 1986 NSS Convention, June 22-28, 1986, Tularosa, New Mexico

[2438]National Speleological Society. Annual Convention. Program (1987: Sault Sainte Marie, MI), 1987, Program 1987 NSS Convention, August 3-7, 1987, Sault Sainte Marie, Michigan

[2439]National Speleological Society. Annual Convention. Program (1988: Hot Springs, SD), 1988, The program of the 1988 NSS convention: June 27-July 1, 1988: Hot Springs, South Dakota

[2440]National Speleological Society. Annual Convention. Program (1989: Swanee, TN), 1992, 1989 NSS convention program: July 31 - August 4, 1989

[2442]National Speleological Society. Annual Convention. Program (1991: Cobleskill, NY), 1991, Program: 1991 National Speleological Society convention: July 1-5: Cobleskill, NY

[2443]National Speleological Society. Annual Convention. Program (1992: Salem, IN), 1992, Program: 1992 NSS convention: the heartland of the karstland: Salem, Indiana

[2444]National Speleological Society. Annual Convention. Program (1993: Pendleton, OR), 1993, [The program of the annual convention of the National Speleological Society, 1-6 August, Pendleton, Oregon]

[2445]National Speleological Society. Annual Convention. Program (1994: Brackettville, TX), 1994, National Speleological Society, program: 1994 NSS Convention: Brackettville, Texas: June 18-25

[2446]National Speleological Society. Annual Convention. Program with Abstracts (1986: Tularosa, NM), 1986, National Speleological Society convention program with abstracts

13.18 National Speleological Society

[868]Dickey, F., 1974, Report of the Caver Proliferation Committee

[3117]Stellmack, John Arnold, 1926–, Jul 1970, Sort of an annual report from the standing committees and the secretary-treasurers of the NSS, including a summary annual report of the National Speleological Foundation, 1969-70

13.19 Periodicals

[480]Bristol (England). Public Libraries, 1968, Caving periodicals in the central collection of caving publications: a catalogue of the caving periodicals held by Bristol Central Reference Library and a selected short title catalogue of books

[1555]Howes, Chris, 1951–, 1994, The Red Dragon Journal of the Cambrian Caving Council index issues (1)-(20) 1974-1993

[1557]Howes, Chris, 1951–, 1994, Wide world magazine Index volumes 1-50 1898-1923

[2358]National Speleological Society, 1986, Cumulative index for the National Speleological Society Bulletin, volumes 1 through 45, and Occasional papers of the N.S.S., numbers 1 through 4

13.20 People

13.20.1 Autobiographies

[407]Black, Don F., 1993, I don't play golf: recollections of a rescue volunteer

[435]Boon, J. M., 1977, Down to the sunless sea

[850]de Joly, Robert, 1975, Memoirs of a speleologist: the adventurous life of a famous French cave explorer
[1017]Exley, Irby Sheck, 1944–1994, 1994, Caverns measureless to man
[1793]Kenney, C. Howard, 1927–1980, 1985, Caving log 1942-1950
[1937]Lewis, Charles C. (ed), 1993, Chronicles of the Reading Grotto, in which we go to the California convention, June 3rd to September 6th, 1966
[2238]Mitchell, Jack, 1964, Jack Mitchell, caveman
[2893]Savory, James Henry, 1889–1962, 1989, A man deep in Mendip: the caving diaries of Harry Savory, 1910—1921
[2986]Simpson, Lou, 1995, Sex, lies, and survey tape

13.20.2 Biographies

[403]Bishop, Michael J. (ed), 1982, The cave hunters: biographical sketches of the lives of Sir William Boyd Dawkins (1837-1929) and Dr. J. Wilfrid Jackson (1880-1978)
[494]Broderick, Alan Houghton, 1963, Abbé Breuil: prehistorian: a biography
[495]Broderick, Alan Houghton, 1963, Father of prehistory: the Abbé Henri Breuil: his life and times
[576]Caiar, Ruth, 1957, One man's dream; the story of Jim White, discoverer and explorer of the Carlsbad Caverns: a biography
[608]Casteret, Norbert, 1897–1987, 1955, The descent of Pierre Saint-Martin
[1375]Hartley, Howard W., 1925, Tragedy of Sand Cave
[1623]International Congress of Speleology (6th: 1973: Olomouc, Czechoslovakia), 1973, Pantheon of Czech speleologists
[2002]Lyons, David, 1981, Floyd Collins, greatest cave explorer ever known
[2166]Middleton, Gregory J., 1991, Oliver Trickett: doyen of Australia's cave surveyors 1847-1934
[2283]Motaş, C., 1968, Emil Racoviţă 1868-1947
[2305]Murray, Robert K., 1979, Trapped!: the story of the struggle to rescue Floyd Collins from a Kentucky cave in 1925, an ordeal which became one of the most sensational events of modern times
[2306]Murray, Robert K., 1982, Trapped!: the story of the struggle to rescue Floyd Collins from a Kentucky cave in 1925, an ordeal which became one of the worst sensational events of modern times
[2643]Pengelly, Hester (ed), 1897, A memoir of William Pengelly, of Torquay, F.R.S., Geologist, with a selection from his correspondence
[3038]Snider, George Washington, 1916, How I found the Cave of the winds ; being the true story as set forth by the discoverer, George W. Snider
[3107]Stanton, William Iredale, Oct 1969, Pioneer under the Mendips: Herbert Ernest Balch of Wells, a short biography
[3126]Stewart, John, 1920–, 1972, Secret of the bats; the exploration of Carlsbad Caverns
[3253]Turner, Cleon, 1968, Discovery of the Onyx Cave and my biography
[3377]Wallace, Malcolm (ed), [1995], Back in time for tea and medals

13.21 Philately

[770]Cullen, James J., 1979, Spelean stamps: a special topical collection of postage stamps of interest to cavers

13.22 Postcards

[1672]Irwin, David J. (comp), Mar 1983, A catalogue of postcards of caves in Yorkshire & Derbyshire (including some misc sites in the Midlands)
[1673]Irwin, David J., 1981, A catalogue of the postcards of Gough's Cave, Cox's Cave & Wookey Hole, Somerset 1900-1980
[2916]Schuurmans, Theo J. C., 1987, Postcards of the Caves of Han printed by Nels of Brussels between 1905 and 1940
[2917]Schuurmans, Theo J. C., 1989, Postcards of the Caves of Han printed by Nels of Brussels between 1905-1940

13.23 Religion

[1855]Koul, Samsar Chand, c1980, The mysterious cave of Amar Nath, Kashmir India
[3479]Whitfield, Roderick, 1995, Dunhuang: caves of the singing sands: Buddhist art from the Silk Road

13.24 Songs

[83]Anon, [1955], A book of songs and poems for the Hunters Lodge and similar places
[397]Biddle, Roger (ed), Apr 1976, Collected caving songs: volume 1: the songs of Mendip
[715][Cornwall-Smith, Nick], [1976], HWYL [songbook]
[716]Cornwell-Smith, Nick (collect, ed, comp), 1993, They words, they words they 'orrible words: an anthology of caving songs
[944]Edmonds, Francis Samuel, 1928, The dream of Kent's Cavern and ballads of the Bay
[1920]Leakey, Robert Dove, [1947], Potholer's songs
[2060]Martin, Dave (ed), Feb 1968, Orpheus Caving Club songbook

13.24. SONGS

[2930]Self, Charles A., Dec 1979, Not the UBSS song book

[2966]Shepard, Dave, Mar 1971, Songs poems and curses

14 Geographic Index

14.1 World

[13]Adams, William Henry Davenport, 1828–1891, 1886, Famous caverns and grottoes: described and illustrated
[14]Adams, William Henry Davenport, 1828–1891, 1886, Famous caves and catacombs: described and illustrated
[217]Anon, c1933, Wonders of the world
[312]Bandi, Hans-Georg (ed), 1961, The art of the stone age: forty thousand years of rock art
[443]Botosaneanu, Lazare (ed), 1986, Stygofauna mundi: a faunistic, distributional, and ecological synthesis of the world fauna inhabiting subterranean waters (including the marine interstitial)
[725]Courbon, Paul, 1936–, 1989, Atlas of the great caves of the world
[1028]Farr, Martyn, 1950–, 1980, The darkness beckons: the history and development of cave diving
[1446]Herak, Milan (ed), May 1972, Karst: important karst regions of the northern hemisphere
[1761]Juberthie, Christian (ed), 1994, Encyclopaedia biospeologica tome 1
[2171]Middleton, John, 1942–, , The underground atlas: a gazetteer of the world's cave regions
[2172]Middleton, John, 1942–, nd, The underground atlas: a gazetteer of the world's cave regions
[2271]Morgan, Robert, nd, Caves in world history
[2689]Pond, Alonzo William, 1894–, 1969, Caverns of the world
[2805]Religious Tract Society, 1800, The caves of the earth: their natural history, features, and incidents
[2952]Shaw, Trevor Royle, 1928–(comp), 1967, Cave illustrations before 1900: a catalogue of non-photographic illustrations of caves
[2953]Shaw, Trevor Royle, 1928–, 1961, The deepest caves in the world and caves which have held the world depth record
[3393]Waltham, Anthony Clive, 1942–[Tony], 1976, The world of caves

14.2 Americas

[1467]Hobbs, Horton Holcombe, 1914–, 1977, A review of the troglobitic decapod crustaceans of the Americas

14.3 Antarctica

[1001]Eraso, Adolfo (ed), 1991, 1st international symposium of glacier caves and karst in polar regions. Proceedings

14.4 Arctic

[1001]Eraso, Adolfo (ed), 1991, 1st international symposium of glacier caves and karst in polar regions. Proceedings

14.5 Africa

[459]Brain, Charles Kimberlin, 1981, The hunters or the hunted? An introduction to African cave taphonomy
[2103]Mazonowicz, Douglas, 1974, Voices from the stone age: a search for cave and canyon art
[2653]Pericot Garcia, Luis, 1899–(ed), 1964, Prehistoric art of the western Mediterranean and the Sahara

14.5.1 Algeria

[2102]Mazonowicz, Douglas, 1970, The prehistoric rock paintings of Tassili n'Ajjer: a description of 15 actual size copies, hand screenpainted

14.5.2 Egypt

[3154]Stringfield, Victor Timothy, 1902–, 1974, Karst and paleohydrology of carbonate rock terranes in semiarid and arid regions with a comparison to humid karst of Alabama
[3162]Sutton, L. J., 1945, Meteorological conditions in caves and ancient tombs in Egypt

14.5.3 Ethiopia

[661]Clapham, Christopher S., 1967, The Caves of Sof Omar

14.5.4 Kenya

[2188]Mills, Martin Taylor, 1977, The subterranean wonders of Kenya

14.5.5 Libya

[1853]Kosa, Attila (ed), 1981, Bir Al Ghanam karst study project: final report 1981

14.5.6 Madagascar

[3386]Walters, R. (comp), 1986, The Crocodile Caves of Ankarana: 1986: an expedition to study and explore the limestone massif of Ankarana in northern Madagascar
[3509]Wilson, Jane, 1990, Lemurs of the lost world: exploring the forests and Crocodile Caves of Madagascar

14.5.7 Morocco

[1015]Exeter University Spelaeological Society, [1970], Expedition to Morocco 1969
[1546]Howe, Bruce, 1912–, 1974, A stone age cave site in Tangier: preliminary report on the excavations at the Mugharet el 'Aliya, or High Cave, in Tangier

14.5.8 Nigeria

[3175]Szentes, Georg, Jul 1989, Sandstone caves in Nigeria

14.5.9 Nyasaland

[1222]Goodall, Elizabeth, 1959, Prehistoric rock art of the Federation of Rhodesia & Nyasaland

14.5.10 Sierra Leone

[707]Coon, Carleton Stevens, 1904–, 1968, Yengema Cave Report

14.5.11 South Africa

[117]Anon, nd, Focus on the Cango Caves
[130]Anon, nd, Guide to the Kalk Bay and Muizenberg Mountains; (walks, caves, camp sites)
[172]Anon, 1923, Oudtshoorn and the Cango Caves
[460]Brain, Charles Kimberlin, 1958, The Transvaal ape-man-bearing deposits
[472]Breuil, Abbé Henri, 1877–1961, 1955, The rock paintings of southern Africa
[553]Bulpin, T. V., 1985, Kango: the story of the caves
[681]Coley, P. G., nd, The caves of Table Mountain: a report on their present condition, potential, and conservation requirements
[735]Craven, Stephen Adrian, 1994, Cango Cave, Oudtshoorn district of the Cape Province, South Africa: an assessment of its development and management 1780-1992
[991]Engelbrecht, N. P., Aug 1972, Caving in the Transvaal
[1225]Goodwin, Astley John Hilary, 1900–1959, 1935, Archaeology of the Cape St. Blaize cave and raised beach, Mossel Bay
[1228]Gordon, Isabella, 1957, On *Spelaeogriphus*, a new cavernicolous crustacean from South Africa
[1258]Gregory, W. L., 1980, The Walkberg Cave System
[1572]Humphreys, A. J. B., 1978, The re-excavation of Powerhouse Cave and an assessment of Dr. Frank Peabody's work on Holocene deposits in the Taung area
[1611]Inskeep, R. R., 1987, Nelson Bay Cave, Cape Province, South Africa: the Holocene levels
[1685]Jackson, Harold , c1969, Sudwala
[1787]Keller, Charles M., 1973, Montagu Cave in prehistory: a descriptive analysis
[1844]Kohary, Jana, 1988, Archaeology and ecology of Rose Cottage Cave, Orange Free State
[1966]Loubser, J. H. N., Oct 1993, A guide to the rock paintings of Tandjesberg
[2046]Marker, Margaret E., 1970, Echo cave: a tentative Quaternary chronology for the Eastern Transvaal
[2078]Mason, Revil J., nd, Cave of Hearths, Makapansgat, Transvaal
[2079]Mason, Revil J., Sep 1988, Kruger Cave, late Stone Age, Magaliesberg
[2105]McCann, Gerald, [1992], In my torchlight: a guide's guide to the Cango Caves
[2298]Müller, H. O. (comp, ed), 1985, Dolomite caves of the eastern Transvaal (South Africa): a free caver monograph 1985
[3052]South African Spelaelogical Association Members, 1968, Cango; the story of the Cango Caves of South Africa
[3053]South African Spelaeological Association, 1970, Cango; the story of the Cango Caves of South Africa
[3054]South African Spelaeological Association Members, 1958, Cango; the story of the Cango Caves of South Africa
[3369]Vishoek (South Africa), 1949, The Peers' cave, tunnel cave and rock shelters at Skildergat, Fish Hoek, the home of pre-historic man
[3553]Zeoderborg, Harry, nd, The phantoms of stork fontein [sic]

14.5.12 Tanzania

[2294]Mturi, A. A., 1975, A guide to Tongoni ruins: with notes on other antiquities in Tanga and Pangani districts including the Amboni Caves: pamoja na maelezo kwa Kiswahili

14.5.13 Tunisia

[539]Bruun, Daniel, 1985, The cave dwellers of Southern Tunisia: recollections of a sojourn with the Khalifa of Matmata

14.5.14 Zimbabwe

[226]Armstrong, A. Leslie, 1931, Rhodesian archaeological expedition (1929): excavations in Bambata Cave and researches on prehistoric sites in Southern Rhodesia
[231]Arsenis, Mylda L., 1970, The caves of Chinoyi: a Rhodesian story
[389]Bent, James Theodore, 1852-1897, 1892, The ruined cities of Mashonaland; being a record of excavation and exploration in 1891

[663]Clark, John Desmond, 1916–, 1942, Further excavations (1939) at the Mumbwa caves, Northern Rhodesia
[701]Cooke, Cranmer Kenrick, 1906–, 1978, The Redcliff stone age site, Rhodesia
[1158]Garlake, Peter S., 1987, The painted caves: an introduction to the prehistoric art of Zimbabwe
[1222]Goodall, Elizabeth, 1959, Prehistoric rock art of the Federation of Rhodesia & Nyasaland
[1935]LeRoux, J. P., Jan 1993, The report of the 1992 caving expedition to the Chimanimani Mountains in the eastern highlands of Zimbabwe
[2252]Molyneux, Arthur John Charles, 1935, The Sinoia caves, Lomagundi district
[3245]Truluck, T. F., Oct 1992, SASA (Cape) expedition to Chinhoyi, Zimbabwe August 1992

14.6 Europe

[322]Baring-Gould, Sabine, 1834–1924, 1968, Cliff castles and cave dwellings of Europe
[347]Baumann, Hans, 1914–, 1962, The caves of the great hunters
[545]Buckland, William, 1784–1856, 1824, Reliquiae diluvianae, or observations on the organic remains contained in caves, fissures, and diluvial gravel, and on other geological phenomena attesting the action of an universal deluge /
[546]Buckland, William, 1784–1856, 1978, Reliquiae diluvianae: or, observations on the organic remains contained in caves, fissures, and diluvial gravel and on other geological phenomena attesting the action of a universal deluge
[547]Buckland, William, 1784–1856, 1823, Reliquiae diluvianae: or, observations on the organic remains contained in caves, fissures, and diluvial gravel and on other geological phenomena attesting the action of an universal deluge
[627]Cervino, Donald Jay, 1990, Water features of the European Upper Paleolithic decorated cave environment
[840]Dawkins, William Boyd, 1838–1929, 1973, Cave hunting, researches on the evidence of caves respecting the early inhabitants of Europe
[894]Downing, R.A. (ed), 1993, The hydrogeology of the chalk of north-west Europe
[1020]Eyre, Jim, 1981, The cave explorers
[1120]Franke, Herbert Werner, 1927–, 1958, Wilderness under the earth
[1171]Geikie, James, 1839-1915, 1914, The antiquity of man in Europe, being the Munro lectures, 1913
[1312]Hadingham, Evan, 1979, Secrets of the Ice Age: the world of the cave artists
[1663]International Union of Speleology. Commission for Cave Protection and Cave Tourism , 1989, Cave tourism: proceedings of international symposium at 170-anniversary of Postojnska jama, Postojna, Yugoslavia, Nov. 10-12, 1988
[1867]Kühn, Herbert, 1895–, 1956, The rock pictures of Europe
[2813]Richard, Colette, , Climbing blind
[2976]Sieveking, Ann, 1979, The cave artists
[3121]Stern, Philip Van Doren, 1900–, 1973, The beginnings of art

14.6.1 Albania

[1118]Frank, Helmut, 1966, Dunkle Portale

14.6.2 Alps

[2672]Pictet, Marc-Auguste, 1752–1825, 1823, On the ice-caves or natural ice-houses found in some of the caverns of the Jura and the Alps

14.6.3 Austria

[421]Bock, Hermann, 1872–, nd, In Salzburg's Netherworld
[1271]Groom, Gillian Elisabeth, 1958, The geomorphology and speleogenesis of the Dachstein caves
[2580]Osterreichische Fremden Verkehrswerbung (ed), , Caves in Austria
[2581]Osterreichische Fremden Verkehrswerbung (ed), nd, Caves in Austria
[3324]U.S. Office of Strategic Services, Research and Analysis Branch, Apr 1945, The caves of greater Germany

14.6.4 Belgium

[124]Anon, nd, The Grottos of Han and of Rochefort (Belgium)
[216]Anon, 1900, The wonders of the Grotto of Han
[2673]Pierpont, E. de, 1947, The wonders of the Grotto of Han (Belgium): guidebook, and description with engravings and plan and a notice upon the very interesting Grotto of Rochefort
[2719]Prinz, W., 1980, The crystalline structures of the caves of Belgium
[2916]Schuurmans, Theo J. C., 1987, Postcards of the Caves of Han printed by Nels of Brussels between 1905 and 1940
[3025]Smith, John Moyr, 1883, The Hades of Ardenne: a visit to the Caves of Han described and illustrated

14.6.5 Bohemia

[1874]Kunsky, Josef, 1903–1977, 1954, Homes of primeval man: wandering in the caves of Czechoslovakia

14.6.6 British Isles

14.6.6.1 Great Britain

[75]Andrews, Peter, 1940–, 1990, Owls, caves, and fossils: predation, preservation, and accumulation of small mammal bones in caves, with an analysis of the Pleistocene cave faunas from Westbury-sub-Mendip, Somerset, UK

[116]Anon, nd, Fairy Cave

[120]Anon, nd, Gough's Caves, Cheddar, Somerset

[129]Anon, 1818, Guide to the Grand Cavern within the mountain of Abraham's Heights, Matlock Bath

[131]Anon, 1934, A guide to the Museum of the Torquay Natural History Society

[166]Anon, nd, The official guide to Cheddar Somerset with map and eleven illustrations

[175]Anon, c1950, Poole's Cavern: Buxton

[185]Anon, nd, A short history of Kents Cavern

[195]Anon, nd, Stump Cross Caverns: the underground wonderland

[202]Anon, 1970, There is a cavern in the town

[237]Aspin, J., Mar 1952, The caverns of Upper Ease Gill

[244]Atkinson, T. C., 1977, Caves and karst of southern England and South Wales: guidebook for the International Congress of Speleology at Sheffield, 1977

[282]Baker, Ernest Albert, 1869–1941, [1903], Moors, crags & caves of the High Peak and the neighbourhood

[284]Baker, Ernest Albert, 1869–1941, 1970, Caving, episodes of underground exploration

[306]Ballard, Jim, 1974, Guide to the sporting caves, potholes and mines of Derbyshire

[337]Barton, Nicholas James, 1935–, 1992, The lost rivers of London: a study of their effects upon London and Londoners, and the effects of London and Londoners upon them

[366]Beck, Howard M., 1984, Gaping Gill: 150 years of exploration

[368]Bedford, Bruce (ed), 1970, Descent handbook for cavers

[370]Bedford, Bruce L., 1942–(ed), 1971, Descent handbook for cavers 1971/1972

[371]Bedford, Bruce L., 1942–(ed), 1987, The Descent magazine's caving yearbook

[372]Bedford, Bruce L., 1942–, 1985, Underground Britain: a guide to the wild caves and show caves of England, Scotland and Wales

[379]Bemrose, Henry Howe, 1934, The caves in Derbyshire

[385]Bennett, R., Apr 1969, Bristol Exploration Club caving report no.13 [St. Cuthbert's Report]: part F: Gour Hall area

[397]Biddle, Roger (ed), Apr 1976, Collected caving songs: volume 1: the songs of Mendip

[445]Boulton, John, 1871, Particulars of a first exploration of the extensive and newly discovered cavern, at Stainton, Low Furness

[463]Branigan, Keith, 1992, Romano-British cavemen: cave use in Roman Britain

[482]British Association for Immediate Care, 1986, Rescue from remote places: cave rescue and sport diving emergencies

[484]British Cave Research Association, 1985, BCRA library catalogue

[486]British Cave Research Association, 1991, Peak and Speedwell Caverns, exploration and science: including the talks presented at the Peak-Speedwell symposium, Sheffield, November 1989

[487][British Cave Research Association], 1989, British Cave Research Association, library catalogue

[499]Brook, Alan, 1945–, 1991, Northern caves volume 2: the three peaks

[500]Brook, Alan, 1945–, 1975, Northern caves volume three: Ingleborough

[501]Brook, Alan, 1945–, 1976, Northern caves volume two: Penyghent and Malham

[505]Brook, David B., 1944–, 1983, Northern caves volume 4a: Scales Moor and King Scale

[506]Brook, David B., 1944–, 1983, Northern caves volume 4b: Leck and Casterton Fells

[507]Brook, David B., 1944–, 1974, Northern caves volume five: the Northern Dales

[508]Brook, David B., 1944–, 1975, Northern caves volume four: Whernside and Gragareth

[509]Brook, David B., 1944–, 1972, Northern caves volume one: Wharfedale and Nidderdale

[510]Brook, David B., 1944–, 1991, Northern Caves: volume 3: the three counties system and the north-west

[511]Brook, David B., 1944–, 1988, Northern caves: volume1. Wharfedale and the North-East

[512]Brooks, Ian, 1985, Excavation techniques in Pin Hole Cave, Creswell Crags S.S.S.I., Derbyshire

[548]Buckrose, J. E., 1868–1931, 1913, Rambles in the North Yorkshire Dales

[569]Butcher, Arthur Lepine, 1951, Cave survey

[583]Callard, Thomas Karr, [1880], Contemporaneity of man with the extinct Mammalia, as taught by recent cavern exploration, and its bearing upon the question of man's antiquity

[589]Campbell, John B., 1944–, 1971, A new analysis of Kent's Cavern, Devonshire, England

[596]Cansell, Anthony, [after 1982], Stump Cross Caverns - the underground wonderland, their development and exploration

[604]Cartwright, Roger, 1946, A guide to the Craven Dales

[609]Casteret, Norbert, 1897–1987, 1962, More years under the earth

[612]Catcott, George Symes, 1792, A descriptive account

of a descent made into Penpark-hole, in the parish of Westbury-upon-Trim, in the county of Gloucester, in the year 1775

[619][Cave Research Group], Dec 1955, Fauna collected from caves as recorded in the C.R.G. fauna records: part I (1938-39)

[620][Cave Research Group], Oct 1956, Fauna collected from caves as recorded in the C.R.G. fauna records: part II (war years 1940-46 and 1945-46)

[622][Cave Research Group of Great Britain], Nov 1972, Hypogean fauna and biological records 1970-71

[644]Checkley, Dave Lancaster, 1977, LUSS expeditions to Tresviso and the Picos de Europa in northern Spain, 1974-1977

[675]Coates, Dorothy, 1960, Cheddar caves in the Cheddar Gorge: an illustrated survey & guide

[702]Cooke, Ian, 1937-, 1993, Mother and sun: the Cornish Fogou

[712]Cope, F. Wolverson, 1976, Geology explained in the Peak District

[714]Cordingley, John, Sep 1986, The Peak Cavern system - a caver's guide

[722]Council of Northern Caving Clubs, 1979, Northern caving; handbook of the Council of Northern Caving Clubs

[733]Crabtree, P. W. (ed), 1962—, Speleological Abstracts: key to Britain's speleological literature

[737]Crawford, Harriet E. W., 1979, Subterranean Britain: aspects of underground archeology

[756]Crofton, Denis, 1901, All about Poole's Cavern, Buxton. An official guide to this unique natural curiosity, together with an inventory of the museum

[773]Cullingford, Cecil Howard Dunstan, 1904–1990 (ed), 1953, British caving: an introduction to speleology

[775]Cullingford, Cecil Howard Dunstan, 1904–1990, 1951, Exploring caves

[784]Curry, David A., May 1982, Joint Mitnor Cave

[807]Dashwood, Francis, 1975, West Wycombe caves: [a brief history and description]

[824]Davies, Paul, 1975, A pictorial history of Swildon's Hole

[836]Davis, R. V., 1978, Limestone caves: a concise explanation

[839]Dawkins, William Boyd, 1838–1929, 1868, The British Pleistocene Mammalia: part II

[842]Dawkins, William Boyd, 1838–1929, 1903, On the discovery of an ossiferous cavern of Pliocene Age at Doveholes, Buxton (Derbyshire)

[845]Deakin, P. R., 1975, British caves and potholes

[851]Delderfield, Eric Raymond, c1959, The story of Brixham Cavern

[860][Descent Magazine], [1973], Descent handbook for cavers 1974

[892]Dowie, H. G., 1932, The history of Kent's Cavern; Torquay: with illustrations

[930]Dyer, William Henry, 1979, A brief story of St Clement's caves

[932]Eastmead, William, 1824, Historia rievallensis: containing the history of Kirkby Moorside ... to which is prefixed a dissertation on the animal remains, and other curious phenomena, in the recently discovered cave at Kirkdale

[935]Eatough, J. A., [1972], Mendip's vanishing grottoes: a photographic record of Balch Cave by J. A. Eatough and Shatter Cave by A. E. Mc. R. Pearce

[944]Edmonds, Francis Samuel, 1928, The dream of Kent's Cavern and ballads of the Bay

[956]Elliot, Dave, 1946–, 1987, Single rope technique rigging guide

[1020]Eyre, Jim, 1981, The cave explorers

[1021]Eyre, Jim, 1986, The Ease Gill system: forty years of exploration

[1022]Eyre, Jim, Mar 1967, Lancaster Hole and the Ease Gill Caverns, Casterton Fell, Westmorland

[1023]Eyre, Jim, 1988, Race against time: a history of the Cave Rescue Organisation

[1028]Farr, Martyn, 1950–, 1980, The darkness beckons: the history and development of cave diving

[1030]Farr, Martyn, 1950–, 1985, Wookey: the caves beyond

[1031]Farrer, James William, 1970, Further explorations in the Dowkerbottom Caves, in Craven

[1040]Fells, Richard, 1989, A visitor's guide to underground Britain: caves, caverns, mines, tunnels, grottoes

[1069]Flindall, Roger, 1976, The caverns and mines of Matlock Bath: 1 Nestus Mines: Rutland and Masson caverns

[1086]Ford, Trevor David, 1925–(comp), 1964, Caves of Derbyshire

[1090]Ford, Trevor David, 1925–, 1971, Ingleborough Cavern

[1091]Ford, Trevor David, 1925–, 1975, Ingleborough Cavern and Gaping Gill

[1092]Ford, Trevor David, 1925–, 1967, Ingleborough Cavern including: notes on Gaping Gill and the geology of Ingleborough

[1093]Ford, Trevor David, 1925–(comp, ed), 1977, Limestone and caves of the Peak District

[1096]Ford, Trevor David, 1925–, Jul 1977, The story of Poole's Cavern

[1097]Ford, Trevor David, 1925–, 1962, The story of the Speedwell Cavern

[1098]Ford, Trevor David, 1925–, 1955, The story of Treak Cliff Cavern

[1184]Gerrard, John, 1960, Kent's Cavern

[1203]Gill, David W. (comp), 1991, Caves of the Peak District

[1204]Gill, George, nd, Stump Cross Caverns

14.6. EUROPE

[1218] Gleave, Joseph James, c1900, Yorkshire Caves: Victoria cave, Settle; Ingleborough Cave, Clapham; Yordas Cave, Ingleton

[1231] Gough, A. G. H., nd, Pictorial guide to the caves Cheddar

[1232] Gough, William, nd, The cliffs & caves of Cheddar: a series of beautiful views in real photogravure published by William Gough, son of the discoverer of the famous cave

[1239] Grange Cavern Military Museum, 1980, Souvenir guide to the world's largest underground military museum: the Grange Cavern

[1265] Grigson, Geoffrey, 1905–, 1957, Painted caves

[1288] Gunn, John (ed), 1994, An introduction to British limestone karst environments

[1338] Hamer, John Lewis, 1946, The falls and caves of Ingleton

[1339] Hamer, John Lewis, 1951, The moors and fells of Ingleton

[1356] Hanwell, James, 1935–, Oct 1970, The great storms and floods of July 1968 in Mendip

[1399] Hazelton, Mary (ed), 1965, British hypogean fauna and biological records of the Cave Research Group part IX (1963)

[1400] Hazelton, Mary, Oct 1958, Fauna collected from caves as recorded in the C.R.G. fauna records: part III (1947)

[1401] Hazelton, Mary, Feb 1959, Fauna collected from caves, mines and wells as recorded in the C.R.G. fauna records: part IV (1948-1949)

[1402] Hazelton, Mary, Feb 1960, Fauna collected from caves, mines and wells as recorded in the C.R.G. fauna records: part V (1950-1953)

[1403] Hazelton, Mary, Sep 1960, Fauna collected from caves, mines and wells as recorded in the C.R.G. fauna records: part VI (1954-1955-1956)

[1404] Hazelton, Mary, Nov 1961, Fauna collected from caves, mines and wells as recorded in the C.R.G. fauna records: part VII (1957-1959)

[1405] Hazelton, Mary, Sep 1963, Fauna collected from caves, mines and wells as recorded in the C.R.G. fauna records: part VIII (1960-1962)

[1406] [Hazelton, Mary (ed)], Sep 1967, Hypogean fauna and biological records 1964-1966

[1407] [Hazelton, Mary (ed)], Dec 1968, Hypogean fauna, biological records 1967

[1408] [Hazelton, Mary (ed)], Mar 1970, Hypogean fauna, biological records 1968

[1413] Headworth, H. G., Nov 1975, The hydrogeology of artesian boreholes at some watercress farms in Hampshire

[1414] Heap, David, 1964, Potholing: beneath the northern Pennines

[1416] Heath, Thomas, Aug 1879, An abstract description and history of the Bone Caves of Creswell Crags

[1417] Heath, Thomas, 1880, Creswell Caves vs Professor Boyd Dawkins

[1440] Hendy, Phil, 1978, Muddy oxbows: a tourist trip into the seldom-trodden inner recesses of the human brain

[1455] Hill, Arthur H., 1952, A brief glossary of Welsh topographic names for walkers and cavers

[1464] Hindle, Brian Paul, 1960, Cave formation in northern England

[1486] Holland, Eric George (comp), 1967, Underground in Furness: a guide to the geology, mines, potholes and caves

[1487] Holland, Eric George, 1960, Underground in Furness: a guide to the geology, mines, potholes and caves

[1515] Hooper, John, Apr 1985, Caves in Buckfastleigh Quarries

[1526] Horsley, Dave, 1989, Oxford University Cave Club Juracao expedition 28th June-17th August 1989

[1533] Housman, John, 1800, A descriptive tour, and guide to the lakes, caves, mountains, and other natural curiosities, in Cumberland, Westermoreland, Lancashire, and a part of the West Riding of Yorkshire

[1542] Howard, John Eliot, 1879, The caves of South Devon and their teachings

[1548] Howell, C., Jul 1973, A Burrington cave atlas

[1552] [Howes, Chris, 1951–(ed)], 1994, The Descent caver's handbook

[1587] Hutton, John, 1740–1806, 1780, A tour to the caves in the environs of Ingleborough and Settle in the West Riding of Yorkshire: with some philosophical conjectures on the deluge, remarks on the origin of fountains, and observations on the ascent and descent of vapours, occasioned by facts peculiar to the places visited. Also a large glossary of old and original words made use of in common conversation in the north of England. In a letter to a friend

[1632] International Congress of Speleology (9th:1977: Sheffield, England), 1977, Caves and karst of the Yorkshire Dales: guidebook for the International Congress of Speleology at Sheffield 1977

[1669] Irwin, David J., Aug 1970, Bristol Exploration Club caving report no 13 [St. Cuthbert's report]: Part H: Rabbit Warren extension

[1670] Irwin, David J., Jun 1970, Bristol Exploration Club caving report no 13: St. Cuthbert's report: Part E: Rabbit Warren

[1677] Irwin, David J., Dec 1987, An index to the publications of The Bristol Exploration Club 1947-1987

[1678] Irwin, David J., 1977, Mendip underground: a caver's guide

[1686] Jackson, John Wilfrid, 1880–1978, 1910-1913, [Fist [sic] to third] reports on the exploration of "Dog Holes" Cave, Warton Crag, near Carnforth, Lancashire

[1687] Jackson, John Wilfrid, 1880–1978, 1928, Report on the animal remains found in the Kilgreany Cave, Co. Waterford

[1708]Jarman, R. A., 1991, Ingleborough Cavern and Gaping Gill

[1710]Jasinski, Marc, 1967, Caves and caving: a guide to the exploration, geology and biology of caves

[1712]Jeffery, R. G., nd, Caves of Plymouth and District

[1716]Jenkins, D. W. (ed), 1957, South Wales Caving Club tenth anniversary publication

[1717]Jenkins, David William, 1963, Caves in Wales and the Marches

[1737]Johnson, Cuthbert, 1946–, 1967, The history of Mendip caving

[1751]Jones, Thomas Rupert, 1819–1911, 1877, Lecture on the antiquity of man: illustrated by the contents of caves and relics of cave-folk

[1762]Judson, David, 1961, Caving and potholing

[1972]Lovelock, James, 1969, Caving

[1974]Lovelock, James, 1963, Life and Death Underground

[2040]Mansfield, Raymond Walter, 1965, Mendip cave bibliography and survey catalogue, January, 1901—December, 1963

[2041]Mansfield, Raymond Walter, 1978, Somerset sump index

[2042]Mansfield, Raymond Walter (ed), Jul 1970, Speleological abstracts: key to Britain's speleological literature

[2043]Mansfield, Raymond Walter (ed), Dec 1971, Speleological abstracts: the literature of 1968

[2057]Martel, Edouard Alfred, 1859–1938, 1914, The caverns of Derbyshire: being an extract from Irlande at Cavernes Anglaises

[2072]Mason, Edmund John, 1911–1993, 1977, Caves and caving in Britain

[2136]McMurtrie, James, Aug 1882, The Mendip Caverns

[2253]Monico, Paul (comp), 1995, CDG northern sump index 1995

[2290]Mountain Rescue Committee (Great Britain), 1963, Mountain rescue, cave rescue

[2291]Mountain Rescue Council, 1994, Mountain & cave rescue: the handbook of the Mountain Rescue Council

[2349]National Caving Association, [1988], Leadership and instructor qualifications in caving

[2561]Oldham, Anthony Clive [Tony], 1939–, 1989, The complete caves of Devon

[2563]Oldham, Anthony Clive [Tony], 1939–, 1972, Discovering caves

[2564]Oldham, Joyce Elizabeth Anne, Nov 1972, The caves of Devon

[2565]Oldham, Joyce Elizabeth Anne, Apr 1985, The concise caves of Devon

[2566]Oldham, Joyce Elizabeth Anne, 1978, The limestones and caves of Devon

[2570]Ollerenshaw, Arthur Edward, c1963, The history of Blue John Stone: methods of mining and working ancient and modern

[2625]Pearman, Harry, 1968, Caves and tunnels in Kent

[2627]Pearman, Harry, 1963, Secret tunnels in Surrey

[2628]Pearman, Harry, 1983, Underground gazetteer of south-east England

[2636]Pemberton, Clive, 1972, Kents Caverns, home of prehistoric man and animals: the origin, story, and descriptive tour of the caves

[2637]Pemberton, Clive, 1947, The origin and story of Kents Cavern

[2638]Pemberton, Clive, 1964, The origin, story and descriptive tour of the caves

[2644]Pengelly, William, [1872], Kent's Cavern: a lecture

[2645]Pengelly, William, 1874, Kent's Cavern and its wonders: a lecture delivered in the public hall Warrington on December 22nd 1873

[2646]Pengelly, William, 1876, Kent's Cavern: its testimony to the antiquity of man: a lecture delivered in the City Hall, Glasgow on Wednesday, 22nd December, 1875

[2647]Pengelly, William, [1870], The literature of Kent's Cavern prior to 1859

[2648]Pennick, Nigel, 1981, The subterranean kingdom: a survey of man-made structures beneath the earth

[2649]Pennington, Rooke, 1877, Notes on the barrows and bone-caves of Derbyshire with an account of a descent into Elden Hole

[2669]Photochrom Co., Tunbridge Wells, 1900, Cheddar Gorge and caves

[2695]Porteous, Crichton, 1950, Caves and caverns of Peakland

[2715]Price, Elizabeth Ann, Sep 1981, Club history 1956—1981 25th anniversary publication: Cerberus Spelaeological Society

[2716]Price, Elizabeth Ann [Liz], 1984, Bath freestone workings

[2720]Proctor, Christopher J. (comp), Jul 1987, Atlas of the Berry Head Caves

[2721]Proctor, Christopher J., 1987, The caves of Chudleigh and Kingsteignton

[2722]Proctor, Christopher J., 1987, Caves of East Devon

[2723]Proctor, Christopher J., 1987, Caves of North Tor Bay

[2756]Raistrick, Arthur, 1897?–1991, 1948, Grassington and Upper Wharfedale

[2801]Reeve, Terry, 1979, Caves & swallets in chalk

[2808]Reynolds, Sidney Hugh, 1906, A monograph of the British Pleistocene Mammalia vol II part II: the bears

[2809]Reynolds, Sidney Hugh, 1902, A monograph of the Pleistocene mammalia vol II, part I: the cave hyaena

[2828]Robinson, Donald, 1964, Caving and potholing

[2829]Robinson, Donald, , Potholing and caving

[2873]Rutter, John, 1796–1851, 1829, Delineations of the north western division of the county of Somerset, and of its antediluvian bone caverns, with a geological sketch of the district

14.6. EUROPE

[2874]Ryder, Peter F., 1995, Caves of Skye
[2893]Savory, James Henry, 1889-1962, 1989, A man deep in Mendip: the caving diaries of Harry Savory, 1910—1921
[2915]Schuurmans, Theo J. C., 1987, A caver's view of the Clydach River: an introduction to the underground waters of Pwll Ddu, Llangynid and Llangattwg mountains
[2938]Seward, William, 1801, A tour to Yordes Cave
[2948]Sharp, Mabel, 1968, White Scar Cave: a famous underground system which you can explore in your best clothes
[2949]Sharp, Mabel, 1959, White Scar Cave: a picture guide
[2950]Sharp, Mabel, 1965, White Scar Cave: a picture guide to a famous underground system you can explore in your best clothes
[3021]Smith, J. P., 1890, Paper on Dunald Mill Hole: read by J. P. Smith before the Barrow Naturalists Field Club
[3100]Standing, Ian James, Sep 1966, Council of Southern Caving Clubs handbook
[3101]Standing, Ian James (ed), 1963, Speleological abstracts: the publications of 1963
[3102]Standing, Ian James (ed), 1964, Speleological abstracts: the publications of 1964
[3103]Standing, Ian James (ed), 1969, Speleological abstracts: the publications of 1965
[3104]Standing, Ian James (ed), Sep 1969, Speleological abstracts: the publications of 1966
[3106]Stanton, William Iredale, 1980, Cheddar Caves: illustrated official guide
[3107]Stanton, William Iredale, Oct 1969, Pioneer under the Mendips: Herbert Ernest Balch of Wells, a short biography
[3122]Stevens, John (ed), Jan 1992, An exploration journal of Llangattwg Mountain
[3134]Stone, John M., 1914, The underground passages, caverns, etc., of Greenwich and Blackheath
[3156]Styles, Frank Showell, 1908-, 1959, How underground Britain is explored
[3161]Sutcliffe, Antony John, c1942, Joint Mitnor Caves: Buckfastleigh
[3172]Sykes, Les, [1994], C. N. C. C. eco-resin rigging system no 1
[3173]Symonds, William Samuel, 1818-1887, 1883, The Seven Straits; or, notes on glacial drifts, bone caverns, and old glaciers: some within reach of the Malvern Hills
[3199]Thomas, Alan, 1931- (ed), 1989, The last adventure
[3209]Thornber, Norman, 1953, Britain underground
[3210]Thornber, Norman, 1959, Pennine underground
[3218]Thynne, Daphne Winifred Louise (Marchioness of Bath), 1956, Cheddar caves
[3224]Towler, George (photo), 1890, Views of Chapman Cave

[3243]Trudgill, Stephen Thomas, 1947-, 1977, A bibliography of British karst, 1960-1977
[3341]Vachell, Eustace Tanfield, 1959, Kents Cavern: its origin and history
[3373]Wainwright, Alfred, 1991, Wainwright in the Limestone Dales
[3389]Waltham, Anthony Clive, 1942-[Tony], 1984, Caves, crags, and gorges: a guide to the limestone country of England and Wales
[3392]Waltham, Anthony Clive, 1942-[Tony] (ed), 1974, Limestone and caves of Northwest England
[3395]Waltham, Anthony Clive, 1942-[Tony], 1987, Yorkshire Dales: limestone country
[3416]Watson, Sally, 1954-, 1991,, Secret underground Bristol
[3430]Weigh, Cliff, 1988, The South Street caves, Dorking
[3434]Wells, Oliver C. (ed), 1960, A history of the exploration of Swildon's Hole
[3448]Westall, William, 1818, Views of the caves near Ingleton, Gordale Scar and Malham Cove in Yorkshire
[3458]Whidbey, Joseph, 1823, On some fossil bones discovered in caverns in the lime-stone quarries of Oreston
[3494]Williams, David, 1829, Some account of the fissures and caverns hitherto discovered in the western district of the Mendip range of hills: comprised in a letter from Rev. D. Williams to the Rev. Patterson
[3502]Williamson, George Charles, 1858-1942, 1930, The Guildford Caverns
[3505]Wilmut, John (ed), 1972, Caves and conservation
[3506]Wilson, B. E. (comp), Nov 1965, The Bradford Pothole Club library list
[3507]Wilson, George Herbert, 1874?-1958, nd, Cave hunting holidays in Peakland
[3524]Woodall, Brian, 1976, Peak Cavern: a guide to this famous show-cave; its formation, history & folklore

14.6.6.1.1 Derbyshire

[541]Bryan, Benjamin, 1840, A description of the grand Devonshire cavern at Matlock
[858][Derbyshire Caving Association], 1987, Derbyshire caving 1987: the handbook of Derbyshire caving
[954]Elliot, Dave, 1946-, Feb 1976, Caves of Northern Derbyshire: part 4: Rushup Edge swallets
[1113]Fox, W. Storrs, 1908, Notes on the excavation of Harborough Cave, near Brassington, Derbyshire
[2336]Nash, D. A., Sep 1989, A Peak Cavern bibliography

14.6.6.1.2 England

[105]Anon, 1995, Crewe Climbing and Pot Holing Club; Peak rigging guide
[138]Anon, 1930, Illustrated souvenir grid to the caverns and glens at the foot of Ingleborough (with map)

[213]Anon, c1992, White Scar Caves: Britain's biggest tourist cave
[273]Baguley, Frank (ed), Aug 1993, Cambrian Caving Council handbook 1993
[287]Baker, Percy Frederick, 1921–, 1979, The Baker extension to the Banwell Bone Cave
[294]Balch, Herbert Ernest, [1927], Excavations at Chelm's Combe: Cheddar: conducted under the Excavations Committee of the Somerset Archaeological and Natural History Society 1925-26
[300]Balch, Herbert Ernest, c1971, Fourteen years at the Badger Hole
[338]Bartrop, Richard N., 1987, Cave Diving Group: Derbyshire sump index 1987
[374]Beldam, Joseph, 1795–1866, 1858, The origin and use of the Royston Cave, being the substance of a report some time since presented to the Royal Society of Antiquaries by the late Joseph Beldam ...
[386]Bennett, R. (Cerberus series), Oct 1982, Cerberus series: Maypole series
[404]Bishop, Michael J., 1982, The mammal fauna of the early middle Pleistocene cavern infill site of Westbury-sub-Mendip, Somerset
[454]Bradford Pothole Club Members, c1974, A history of Gaping Gill and Ingleborough Cave
[480]Bristol (England). Public Libraries, 1968, Caving periodicals in the central collection of caving publications: a catalogue of the caving periodicals held by Bristol Central Reference Library and a selected short title catalogue of books
[504]Brook, David B., 1944–(ed, prod), Oct 1969, The explorations journal of The University of Leeds Speleological Association
[520]Brown, George Henry, 1905, The Clapham Cave with illustrations from photographs by the author
[522]Brown, George Henry, 1908, Victoria Cave: its history and exploration
[540]Bryan, Benjamin, 1839, A description of the fluor spar or Blue John Mine at Castleton
[603]Carter, R. L., Aug 1994, Cave Diving Group: Peak District sump index 1994
[639]Charnock, S. J., Feb 1994, Underneath the arches: caving in the RibbleHead District
[685]Collins, S. J., Jan 1956, Surveying in Red Cliffe Caves: Bristol 1953-1954
[721][Council of Northern Caving Clubs], 1969, Northern cave handbook
[724]By "A Country Doctor", c1929, The wonders of Castleton
[825]Davies, Phillip, Apr 1957, An index of submerged cave passages; sumps
[951]Elliot, Dave, 1946–, 1975, Caves of Northern Derbyshire: part 1: the Eldon Hill area
[952]Elliot, Dave, 1946–, Mar 1975, Caves of Northern Derbyshire: part 2: Giants-Oxlow System
[953]Elliot, Dave, 1946–, May 1975, Caves of Northern Derbyshire: part 3: Perryfoot/Coal Pit Hole
[955]Elliot, Dave, 1946–, nd, Caves of Northern Derbyshire: part 5: Treak Cliff Hill
[1198]Gilbertson, D. D., 1984, In the shadow of extinction: a Quaternary archaeology and paleoecology of the lake, fissures, and smaller caves at Creswell Crags SSSI
[1245]Gray, Eric (comp), 1985, Devonshire sump index
[1263]Griffiths, Julian (comp), 1981, Cave Diving Group: northern sump index
[1419]Hedinger, J. M., 1820, A short description of Castleton in Derbyshire: its natural curiosities and mineral productions
[1513]Home, Everard, 1756–1832, 1794, Account of some remarkable caves in the principality of Bayreuth: likewise observations on the fossil bones by the late John Hunter
[1586]Hutton, John, 1740–1806, 1970, A tour to the caves in the environs of Ingleborough and settle in the West Riding of Yorkshire
[1588]Hutton, John, 1740–1806, 1781, A tour to the caves in the environs of Ingleborough and Settle in the West— Riding of Yorkshire: with some philosophical conjectures on the deluge, remarks on the origin of fountains, and observations on the ascent and descent of vapours, occasioned by facts peculiar to the places visited. Also a large glossary of old and original words made use of in common conversation in the north of England. In a letter to a friend
[1609]Innes-Smith, Robert, 1976, Castleton and its caves
[1672]Irwin, David J. (comp), Mar 1983, A catalogue of postcards of caves in Yorkshire & Derbyshire (including some misc sites in the Midlands)
[1674]Irwin, David J. (ed), Oct 1974, Cave notes 74
[1675]Irwin, David J., Apr 1986, Cheddar Caves bibliography
[1676]Irwin, David J., Oct 1968, The discovery and exploration of St. Cuthbert's Swallet
[2044]Mansfield, Raymond Walter, Apr 1964, Sump index: section 1 (Somerset)
[2118]McDonald, M. C. (prep), Dec 1993, The Somerset sump index
[2122]McEwan, Graham, 1994, Crypts, caves and catacombes: subterrenea of Derbyshire & Nottingham
[2144]Mellors, P. T., 1989, Legal aspects of access underground: a guide to the legal rights and obligations of people who explore cave, potholes and disused mine, and of those people who control access to them
[2254]Monico, Paul (auth, comp), 1969, ULSA explorations: journal II
[2569]Ollerenshaw, Arthur Edward, 1964, Blue John Caverns and Blue John Mine, Castleton, Sheffield
[2626]Pearman, Harry, 1976, Caves and tunnels in southeast England

14.6. EUROPE

[2681]Pitty, Alistair F., 1966, An approach to the study of karst water: illustrated by results from Poole's Cavern, Buxton
[2729]Publications Team (ed), May 1977, Cave Notes 75
[2838]Rodway School Speleological Group, 1974, Caves of the Avon Gorge Bristol, part 2: eastern or Clifton side
[2839]Rodway School Speleological Group, 1973, Caves of the Avon Gorge Bristol: part 1: western or Leigh Woods side
[2862]Royse, John, nd, Ancient Castleton Caves: their place in time and history, also what people want to know about Castleton
[2863]Royse, W., 1891, Descriptive account of the Blue John Mines and Caverns at Castleton, Derbyshire
[3031]Smith, Peter B., 1973, The P8 Cave: Castleton
[3227]Tratman, Edgar Kingsley, 1899–1978 (ed), Feb 1963, Reports on the investigations of Pen Park Hole, Bristol
[3388]Waltham, Anthony Clive, 1942–[Tony], 1987, Caves and karst of the Yorkshire Dales: an excursion guidebook to the karst landforms and some accessible caves within Yorkshire Dales
[3391]Waltham, Anthony Clive, 1942–[Tony], 1987, Karst and caves in the Yorkshire Dales National Park
[3491]William Pengelly Cave Studies Trust, Apr 1985, Caves in Buckfastleigh quarries
[3492]William Pengelly Cave Studies Trust, May 1982, Joint Mitnor Cave
[3508]Wilson, George Herbert, 1874?–1958, c1926, Some caves and crags of Peakland
[3515]Winder, Francis A., 1938, Unconventional guide to the Caverns of Castleton and Disaster

14.6.6.1.3 Hebrides

[1355]Hannan, Thomas, 1926, The beautiful Isle of Mull with Iona and the Isle of Saints

14.6.6.1.4 Mendips

[83]Anon, [1955], A book of songs and poems for the Hunters Lodge and similar places
[102]Anon, c1914, Cheddar
[104]Anon, pre-1920, Cox's Stalactite Cavern, Cheddar
[174]Anon, 1930, Pictures of Cheddar and Cox's cave by pen and camera
[204]Anon, c1934, Twelve pictures of Cheddar and Cox's Cave
[211]Anon, [1947], The weird wonders of Wookey Hole Caves
[218]Anon, nd, Wookey Hole Caves: sixteen exclusive camera studies
[219]Anon, [1938], Wookey Hole Caves: Wells, Somerset
[220]Anon, [1937], Wookey Hole Caves: Wells, Somerset
[221]Anon, 1986, Wookey Hole; the caves and mill
[245]Atkinson, Timothy C., Oct 1967, Mendip Karst Hydrology Project: phases one and two
[283]Baker, Ernest Albert, 1869–1941, 1907, The netherworld of Mendip: explorations in the great caverns of Somerset, Yorkshire, Derbyshire, and elsewhere
[293]Balch, Herbert Ernest, 1926, The caves of Mendip
[295]Balch, Herbert Ernest, 1948, The Mendip Caves
[296]Balch, Herbert Ernest, 1937, Mendip, its swallet caves and rock shelters
[297]Balch, Herbert Ernest, 1935, Mendip-Cheddar, its gorge and caves
[298]Balch, Herbert Ernest, 1929, Mendip: the great cave of Wookey Hole
[299]Balch, Herbert Ernest, 1914, Wookey Hole: its caves and cave dwellers
[304]Balcombe, F. Graham, [1936], The Log of the Wookey Hole: exploration expedition 1935
[332]Barrington, Nicholas Robert (comp), 1962, The caves of Mendip
[333]Barrington, Nicholas Robert, 1970, The complete caves of Mendip
[334]Barrington, Nicholas Robert, 1977, Mendip: the complete caves and a view of the hills
[375]Bell, Alan, 1918–, 1928, The witch of Wookey Hole
[728]Cox, Edward, c1914, Souvenir of Cox's Stalactite Caves
[729]Cox, Edward, 1910, Souvenir of Cox's Stalactite Caves "visited by his majesty the late King Edward VII"
[902]Drew, David Phillip, 1943–, Dec 1968, Mendip karst hydrology research project: phase 3
[968]Ellis, Bryan M., 1934–, [1963], Mendip cave registry: handbook for members of the Executive Committee
[1167]Gatacre, E. V., Mar 1975, Wookey Hole [Including articles on] (paper making, Lady Bangor's fairground collection [and] Madame Tussaud's store)
[1205]Gill, Thomas B., [1936], Guide to the home of the rainbow: Cox's Cave, Cheddar, Somerset
[1206]Gill, Thomas B., [1936], Guide to the world famous Gough's Caves, Cheddar, Somerset
[1230]Gough, A. G. H., 1912, Guide to Gough's Caves, Cheddar; how they were found, the finest in the world (electrically illuminated) eighteen beautiful views of the cave prehistoric man: - how he was found
[1472]Hodgkinson, Gerald William, c1964, Wookey Hole: the cave of mystery and history
[1473]Hodgkinson, Olive, 1970, The story of Wookey Hole
[1671]Irwin, David J. (comp), Aug 1995, A catalogue of ephemera relating to Wookey Hole
[1673]Irwin, David J., 1981, A catalogue of the postcards of Gough's Cave, Cox's Cave & Wookey Hole, Somerset 1900-1980
[2022]Mais, Stuart Peter Brodie, c1934, 12 natural colour

photographs of Cheddar and Cox's Cave "The home of the rainbow"
[2073]Mason, Edmund John, 1911–1993, Jul 1963, The story of Wookey Hole
[2245]Mockford, D. P., Oct 1974, Caves of the Bristol region
[2827]Roberts, Jon, Nov 1982, Mendip Caving Group: 1982 library list
[2895]Say, L. W. (ed), 1976, Cheddar Caves Museum
[2956]Shaw, Trevor Royle, 1928–, Jul 1972, Mendip Cave bibliography: part II: books pamphlets, manuscripts and maps; 3rd Century to December 1968
[3016]Smith, David Ingle (comp, ed), 1975, Limestones and caves of the Mendip Hills
[3080]Sparrow, Andy, 1991, A Mendip caver's ropework guide
[3108]Stanton, William Iredale, c1987, Timeless Cheddar Caves
[3118]Stenner, Roger D., Oct 1961, Some smaller Mendip caves: vol 1
[3211]Thornycroft, L. B., 1948, The story of Wookey Hole in fact, fiction and photo
[3248]Tucker, J. H., Aug 1962, Some smaller Mendip caves: vol 2
[3518]Witcombe, Richard, 1992, Who Was Aveline Anyway? Mendip's cave names explained

14.6.6.1.5 Sark

[1275]Guernsey Press, 1931, Illustrated "Press" guide to the caves and cliffs of Sark: with map and plans of roads.

14.6.6.1.6 Scotland

[7]Adam, John, 1886, Aberbrothok illustrated
[863]De Watteville, Alastair, 1993, The island of Staffa: home of the world-renowned Fingal's cave
[896]Dowswell, Peter N. F., Nov 1980, Limestone caves of Scotland, vol IV: the caves of Schichallion
[1257]Gregory, John Walter, 1864–1932, 1930, Some caves and a rock shelter at Loch Ryan and Portpatrick, Galloway
[1714]Jeffreys, Alan Lawrence, 1940–, Apr 1972, The Caves of Assynt
[1715]Jeffreys, Alan Lawrence, 1940–, Jan 1984, Scotland underground: a caver's guide to the limestone caves of Scotland
[1917]Lawson, T. J., 1988, Caves of Assynt
[2005]MacCulloch, Donald B., 1951, The island of Staffa
[2006]MacCulloch, Donald B., 1927, The island of Staffa: with 12 photographs, two plans and one map, by the author
[2011]Mackie, Charles, 1864–, 1830, Historical description of the chapel and castle of Roslin, and the caverns of Hawthornden
[2013]MacLean, John Patterson, 1848–1939, 1890, An historical, archaeological and geological examination of Fingal's Cave in the island of Staffa
[2014]Macleay, Kenneth, 1819–, 1811, Description of the spar cave, lately discovered in the Isle of Skye: with some geological remarks relative to that island
[2554]Oldham, Anthony Clive [Tony], 1939–, 1975, The caves of Scotland (except Assynt)
[2688]Pollard, Anthony James, 1941–, 1992, Smoo Cave
[2760]Rankin, Frank, 1989, Guide to the Wemyss caves
[2880]Salvona, J., 1988, Scottish cave guides; the southern highlands
[2881]Salvona, J., 1989, The Southern highlands
[2925]Scott, Thea, 1961, Fingal's cave
[2985]Simpson, James Young, 1811–1870, 1867, Account of some ancient sculptures on the walls of caves in Fife
[3015]Smith, Christopher, 1989, Mid Argyll cave and rock shelter survey
[3477]Whitehouse, Frederic Cope, 1882, Is Fingal's Cave artificial?
[3545]Young, Ivan (ed), 1978, Appin cave guide

14.6.6.1.7 Somerset

[2717]Price, Graham (ed), Aug 1977, Fairy Cave Quarry: a study of the caves

14.6.6.1.8 Wales

[30]Alexander, John Martin, 1934–, 1959, The survey of Pant Mawr Pot, South Wales
[48]Allen, E. E., 1946, Gower Caves: a survey of the Gower Caves with an account of recent excavations: part I
[49]Allen, E. E., 1948, Gower Caves: a survey of the Gower Caves with an account of recent excavations: part II
[106]Anon, [1992], Dan-yr-ogof show caves
[122]Anon, [1994], The great Dan-yr-ogof day out
[376]Bell, Anne (comp), [1986], South Wales Caving Club Newsletter: index to numbers 1 - 100: 1946 - 1985
[419]Blore, J. D., c1977, Archaeological excavation at North Face Cave, Little Ormes Head, Gwynedd 1962-1976
[657]Christopher, N. S. J. (ed), 1969, Cambrian Caving Council handbook 1969
[670]Coase, Alan C., nd, Cathedral Cave situated between Brecon and Swansea on the A. 4067 road (750 yards North of Craig-y-Nos Castle) Dan-yr-Ogof Swansea Valley Caves
[672]Coase, Alan C., Mar 1977, Dan yr Ogof and its associated caves
[673]Coase, Alan C., 1978, Official guide to Dan-yr-Ogof showcave

14.6. EUROPE

[822] Davies, Melvyn, Nov 1966, Cave sump index; South Wales
[823] Davies, Melvyn, 1991, The caves of the South Gower coast: an archaeological assessment
[1094] Ford, Trevor David, 1925–(ed), 1989, Limestones and caves of Wales
[1166] Gascone, William (ed), 1982, Caves of the North Outcrop
[1252] Green, H. Stephen, 1984, Pontnewydd Cave: a lower palaeolithic hominid site in Wales: the first report
[1259] Grenfell, Harold, 1971, The caves of Gower
[1551] Howes, Chris, 1951–, 1995, CMW Bran CC Journal index vol. 1-20; 1967-1994 including newsletters, newssheets & the Silver Jubilee Journal
[1554] Howes, Chris, 1951–, 1995, ISCA CC journal index volumes 1-17; 1984-1994
[1555] Howes, Chris, 1951–, 1994, The Red Dragon Journal of the Cambrian Caving Council index issues (1)-(20) 1974-1993
[1717] Jenkins, David William, 1963, Caves in Wales and the Marches
[1747] Jones, Ivor Wynne, 1986, Llechwedd
[1748] Jones, Ivor Wynne, 1976, Llechwedd Slate Caverns
[1749] Jones, Keith, 1992, Caves and mines of the Sychrhyd Gorge
[1930] Leitch, David Edwin, 1960, Ogof Agen Allwedd in relation to the Mynydd Llangattwg
[1948] Little, William Henry, 1919–1992, 1970, Coming to South Wales caves?
[2071] Mason, Edmund John, 1911–1993, 1979, Bone Cave (Ogof-yr-Esgyrn) Dan-yr-Ogof: the history of the Bone Cave (Ogof yr Esgyrn): a guide and background to the exhibits in Bone Cave
[2080] Mason-Williams, Ann, Sep 1958, A preliminary investigation into the bacterial & botanical flora of caves in South Wales
[2270] Morgan, Llewellyn E. (auth, photo), nd, Guide to Dan-yr-Ogof Caves: Swansea Valley Caves
[2275] Morris, David (comp), Dec 1985, Cave Diving Group: Welsh sump index
[2543] Ockenden, Allen C., 1991, Caves of the Mellte Valley
[2549] Oldham, Anthony Clive [Tony], 1939–, 1975, The caves of Carmarthen
[2550] Oldham, Anthony Clive [Tony], 1939–, 1981, The caves of Clydach
[2552] Oldham, Anthony Clive [Tony], 1939–, 1978, The caves of Gower
[2553] Oldham, Anthony Clive [Tony], 1939–, 1977, The caves of north Wales
[2555] Oldham, Anthony Clive [Tony], 1939–, 1986, Caves of the central northern outcrop
[2556] Oldham, Anthony Clive [Tony], 1939–, 1993, The caves of the Little Neath Valley
[2557] Oldham, Anthony Clive [Tony], 1939–, 1991, The caves of the south eastern outcrop
[2558] Oldham, Anthony Clive [Tony], 1939–, 1992, The caves of the south eastern outcrop and caves and mines in the Forest of Dean
[2559] Oldham, Anthony Clive [Tony], 1939–, Jan 1985, The caves of the southern outcrop
[2560] Oldham, Anthony Clive [Tony], 1939–, 1979, The caves of west Wales
[2562] Oldham, Anthony Clive [Tony], 1939–, 1990, The concise caves of north Wales
[2574] O'Reilly, P. M., 1969, Ogof Ffynnon Ddu: Penwyllt, Breconshire
[2749] Railton, Courtenaye Lewis, 1907–1971, 1953, The Ogof Ffynnon Ddu system: its discovery and exploration (1927-53): and a theory of its development
[2750] Railton, Courtenaye Lewis, 1907–1971, 1958, The survey of Tunnel Cave: S. Wales
[3047] Sollas, William Johnson, 1849–1936, 1913, Paviland Cave: an Aurignacian station in Wales (The Huxley memorial lecture for 1913)
[3059] South Wales Caving Club, Feb 1993, Members handbook: South Wales Caving Club
[3061] South Wales Caving Club, c1968, South Wales Caving Club twenty first anniversary publication
[3062] South Wales Caving Club, 1957, South Wales Caving Club, 1946—1956
[3149] Stratford, Tim, 1978, Caves of South Wales
[3190] Taylor, Maurice Clague, 1991, Three below Gower: the story of cave exploration in Gower by The Taylors'
[3375] Walker, Kevin, 1983, Caving in the Crickhowell area
[3376] Walker, Kevin, 1983, The Llangattock escarpment
[3378] Wallis, G. R., 1965, Glass sand occurrences, Kurnell Peninsula: preliminary investigations ; Guano deposits in Willi Willi Caves, Kempsey ; Geological report on Wallent's Somersby Clay Pit

14.6.6.2 Ireland

[10] Adams, Aislinn (des, illus), 1984, Dunmore Cave
[20] Ainsworth, William, 1834, An account of the caves of Ballybunian, county of Kerry: with some mineralogical details
[80] Anon, c1990, Aillwee Cave: Ireland's premier show cave
[114] Anon, c1967, Exploration '66; University of Nottingham; biospeleological research expedition to Ireland; Riverview work project in Portugal; British speleological expeditions to Turkey
[153] Anon, Jul 1995, Marble Arch Caves
[281] Baker, Ernest Albert, 1869–1941, 1907, Cave Explorers in Co. Fermanagh 1907
[302] Balcombe, F. Graham, 1936, Ireland 1936 the record

of the party of S.J. Pick (Leicester) in County Clare, Easter 1936
[303]Balcombe, F. Graham, 1937, Ireland 1937 the record of S.J. Pick (Leicester) in County Clare May 1937
[466]Brenan, Edward, 1859, Notice of the occurrence of mammoth and other animal remains: discovered under limestone in a bone cave at Shandon, near Dungarvan, in the county of Waterford
[679]Coleman, John Christopher, May 1965, The caves of Ireland
[680]Coleman, John Christopher, 1944, The Polnagollum cave, Co. Clare
[693]Conneely, Jim, c1989, Aillwee Cave: Ballyvaughan, Burren Co., Clare
[898]Drew, David Phillip, 1943–, 1984, Aillwee Cave and the caves of the Burren
[899]Drew, David Phillip, 1943–, 1980, Dunmore Cave, County Kilkenny: a reassessment
[900]Drew, David Phillip, 1943–, 1978, Dunmore Cave: a short guide
[994]English, Jackie (comp, photo), c1980, Mitchelstown Cave: its discovery and history
[1049]Ferguson, Samuel, 1810–1886, 1865, Account of Ogham inscriptions in the cave at Rathcroghan, Co. Roscommon
[1059]Fincham, A., 1962, Speleological surveys in Co. Sligo and Co. Leitrim, Eire
[1409]Hazelton, Mary (ed), 1974, Irish hypogean fauna and Irish biological records, 1856-1971
[1631]International Congress of Speleology (9th:1977: Sheffield, England), 1977, Caves and karst of Ireland: guidebook for the International Congress of Speleology at Sheffield, 1977
[1743]Jones, Gareth Llwyd, 1988, Cave Diving Group: Irish sump index
[1744]Jones, Gareth Llwyd, 1974, The caves of Fermanagh and Cavan
[1942]Lillis, Brian, nd, Poul Na Gollor Cave at Inch Ennis, Co Clare
[1952]Lloyd, Oliver Cromwell, 1911–1985, Nov 1964, Doolin-St. Catherine's Cave: Co. Clare, Eire
[1980]Luce, A. A., 1931, Berkeley's description of the cave of Dunmore
[2012]Mackie, Evan W., 1981, The caves at East Wemyss, Fife: an interim report on new investigations in 1980
[2544]O'Connor, Donal, 1991, Crag Cave: Castle Island: Co. Kerry: Ireland's most exciting show cave
[2551]Oldham, Anthony Clive [Tony], 1939–, 1981, The caves of Co. Cork
[2929]Self, Charles A., , Caves of County Clare
[3226]Tratman, Edgar Kingsley, 1899–1978 (ed), 1969, The Caves of North West Clare, Ireland

14.6.6.3 Yorkshire

[139]Anon, nd, Ingleborough Cavern: the finest show cave in Yorkshire
[193]Anon, 1934, The story of most interesting discoveries commencing August 1923; White Skar Cave; under Ingleborough, Yorkshire, 1 1/2 miles from Ingleton on the Hawes Road
[194]Anon, 1931, The story of newly discovered White Skar Caverns; under Ingleborough, Yorkshire, 1 1/2 miles from Ingleton on the Hawes Road
[521]Brown, George Henry, 1906, The underground streams in the neighbourhood of Clapham and Malham
[601]Carr, Joseph, 1865, Rambles about Ingleton, in caves by rivers and on mountains in the spring of 1865.
[834]Davis, Henry Harrison, 1851, An excursion from Lancaster, up the vale of Lune, and from Kirkby Lonsdale, to the caves of Yorkshire
[1100]Forder, John (auth, photo) , 1992, Secrets of the moors and dales
[1173]Gemmell, Arthur, 1915–, 1952, Underground adventure
[2236]Mitchell, Albert, [1937], Yorkshire caves and potholes: 1, North Ribblesdale
[2237]Mitchell, Albert, [1950], Yorkshire caves and potholes: 2, under Ingleborough
[2243]Mitchell, William Reginald, 1928–, 1961, The hollow mountains: the story of man's conquest of the caves and potholes of Northwest Yorkshire throughout 10,000 years
[2244]Mitchell, William Reginald, 1928–, 1989, Yorkshire's hollow mountains

14.6.7 Bulgaria

[1020]Eyre, Jim, 1981, The cave explorers
[1856]Kozlowski, Janusz Krzysztof, 1982, Excavation in the Bacho Kiro Cave (Bulgaria): final report
[1858]Kozlowski, Janusz Krzysztof, 1992, Temnata cave: excavations in Karlukovo karst area Bulgaria

14.6.8 Channel Islands

[1910]Latrobe, G., 1964, Guide to the Coast, Caves, and Bays of Sark
[1911]Latrobe, G., 1968, A revised and enlarged guide to Sark containing a detailed description of the coast, caves and bays

14.6.9 Crete

[402][Birmingham University Speleological Society], [1968], 67 expedition to Crete: BUSS
[2609]Panayiotakis, Yioryos I., 1988, The Dictaean Cave

14.6.10 Cyprus

[1752]Jones, W. F., 1962, Some caves of Cyprus

14.6.11 Czechoslovakia

[5]Absolon, Karel B., 1987, The conquest of the caves and underground rivers of Czechoslovakia's Macocha Abyss: a historical and technical study of their exploration
[564]Burkhardt, Rudolf, 1973, The Speleological Club in Brno 1945-1973
[854]Demek, Jaromír, 1989, Czech Speleological Society 1986-1989: published on occasion of 10th International Speleological Congress Hungary 1989
[1448]Herzlík, Bořivoj (trans), 1986, Czech Speleological Society 1982-1986: published on occasion of 9th International Speleological Congress Spain 1986
[1623]International Congress of Speleology (6th: 1973: Olomouc, Czechoslovakia), 1973, Pantheon of Czech speleologists
[2712]Přibyl, Jan, 1973, Paleohydrography of the caves in the Moravsky Kras (Moravian Karst)
[2714]Přibyl, Jan, nd, Věkǎum budoucím
[2998]Skřivánek, František, 1973, Caves in Czechoslovakia

14.6.12 Czech Republic

[1157]Gargett, Robert H., 1996, Cave bears and modern human origins: the spatial taphonomy of Pod Hradem Cave, Czech Republic
[3324]U.S. Office of Strategic Services, Research and Analysis Branch, Apr 1945, The caves of greater Germany

14.6.13 France

[2]Abadie, Bernard, [1972], Grottos of Bétharram
[22]Albi, S.A. Apa Poux (ed), 1990, Padirac Chasm and underground river
[73]Andre, Daniel, 1994, Bramabiau's underground river
[176]Anon, 1984, Prehistoric painting, part 2
[190]Anon, c1967, Speleology in France: Ressources [sic] in Limousin Quercy Perigord
[227][Army Caving Association], nd, The 1984 Army Caving Association expedition to the Gouffre Berger
[246]Attout, Jacques, 1956, Men of Pierre Saint-Martin
[247]Aubarbier, Jean-Luc, Jun 1985, Prehistoric sites in Perigord
[249]Aujoulat, Norbert, 1989, Lascaux revisited
[274]Bahn, Paul G., 1988, Images of the ice age
[342]Bataille, Georges, 1897–1962, 1955, Prehistoric painting: Lascaux, or, the birth of art
[387]Bennett, R. H., 1973, Balague '70
[440]Bordes, Francois, 1972, A tale of two caves
[481]Bristol Exploration Club, nd, Bristol exploration club: caving report no.19: 1975 expedition to the Pierre Saint-Martin
[489]British Speleological Association, c1968, 1967 expedition to the Gouffre Berger
[496]Broderick, Alan Houghton, 1949, Lascaux: a commentary
[532]Browne, George Forrest, 1833-1930, 1865, Ice-caves of France and Switzerland. A narrative of subterranean exploration
[574]Cadoux, Jean, 1930–, 1957, One thousand metres down
[608]Casteret, Norbert, 1897–1987, 1955, The descent of Pierre Saint-Martin
[610]Casteret, Norbert, 1897–1987, 1947, My caves
[611]Casteret, Norbert, 1897–1987, 1940, Ten years under the earth
[642]Chauvet, Jean-Marie, 1996, Chauvet Cave: the discovery of the world's oldest cave paintings
[643]Chauvet, Jean-Marie, 1996, Dawn of art: the Chauvet Cave: the oldest known paintings in the world
[647]Chevalier, Pierre, 1951, Subterranean climbers; twelve years in the world's deepest chasm
[648]Chevalier, Pierre, 1975, Subterranean climbers; twelve years in the world's deepest chasm
[667]Clottes, Jean, 1996, The cave beneath the sea: paleolithic images at Cosquer
[713]Corbel, Jean, 1975, The major caves of France and their relationships with climatic factors
[734]Crain, Sally Lucille 1938–(auth, prod), 1979, Prehistoric magic
[790]d'Amboise, Valery (auth, illus, photo), 1986, Eternal caves
[800]Daniel, Glyn Edmund, 1955, Lascaux and Carnac
[850]de Joly, Robert, 1975, Memoirs of a speleologist: the adventurous life of a famous French cave explorer
[910]Dulluc, Brigitte, 1990, Discovering Lascaux
[1000]Epton, Nina Consuelo, 1955, The Valley of Pyrene
[1020]Eyre, Jim, 1981, The cave explorers
[1058]Film Images, 1957, Caves of Pierre Saint-Martain
[1207]Gillett, J. E. (ed), c1983, Crewe Climbing and Potholing Club: Gouffre Berger Expedition
[1265]Grigson, Geoffrey, 1905–, 1957, Painted caves
[1269]Grobet, André-H., May 1966, The underground lake of St. Léonard: Valais
[1337]Halliwell, Ric (ed), c1995, Gouffre Berger: August 1994
[1636]International Film Bureau, 1976, Lascaux, cradle of man's art
[1832]Knight, J. C., 1962, The Lapiaz Superieure du Pla Segoune
[1884]Laming-Emperaire, Annette, 1917–1977, 1950, The cave of Lascaux
[1885]Laming-Emperaire, Annette, 1917–1977, 1959,

Lascaux: paintings and engravings
[1913]Laville, Henri, 1980, Rock shelters of the Perigord: geological stratigraphy and archaeological succession
[1916]Lawson, Andrew J., , Cave art
[1919]Leakey, Richard E., 1981, A new era
[2007]MacCurdy, George Grant, 1863–, 1914, La Combe, a paleolithic cave in the Dordogne
[2054]Marshall, Des, 1993, Vercors caves
[2059]Martel, Edouard Alfred, 1859–1938, 1980, Padirac; its history and a short description
[2098]Mazonowicz, Douglas, 1989, On the rocks: prehistoric art of France and Spain
[2099]Mazonowicz, Douglas, 1979, The painted caves of France & Spain
[2100]Mazonowicz, Douglas, 1970, Prehistoric paintings of France and Spain: a description of 34 actual size copies hand screenprinted by Douglas Mazonowicz
[2103]Mazonowicz, Douglas, 1974, Voices from the stone age: a search for cave and canyon art
[2274]Morland, Timothy Edward, 1959, A glossary of French speleological terms
[2529]Nougier, Louis René, 1912–, 1958, The Cave of Rouffignac
[2530]Nougier, Louis-René, 1912–, 1961, Art treasures of prehistoric man in the caves of France and Spain
[2663]Peyrony, E., 1959, Les Eyzies and the Vezere Valley; an illustrated guide for scholars and tourists
[2705]Pratchett, Nick, 1956, International expedition to Gouffre Berger, 1956: the exploits of the British members, Nick Pratchett & Bob Powell
[2761]Raphael, Max, 1889–1952, 1945, Prehistoric cave paintings
[2833]Robson, M., 1975, The decomposition of limestone breccia from the cave of Les Eyzies, Dordogne, France
[2867]Ruspoli, Mario, 1987, The cave of Lascaux: the final photographs
[2883]Sandak, Inc., 1973, Cave paintings—Lascaux and Altamira
[2894]Sawtell, Ruth Otis, 1927, Primitive hearths in the Pyrenees; the story of a summer's exploration in the haunts of prehistoric man
[2977]Sieveking, Ann, 1962, The caves of France and northern Spain
[2978]Sieveking, Ann, 1966, The caves of France and northern Spain, a guide
[2979]Siffre, Michel, 1939–, 1965, Beyond time
[3181]Taralon, Jean, 1962, The grotto of Lascaux
[3514]Windels, Fernand, 1949, The Lascaux Cave paintings

14.6.14 Germany

[373]Behrens, Georg Henning, 1662–1712, 1730, The natural history of Hartz-Forest, in His Majesty King George's German dominions. Being a succinct account of the caverns, lakes, springs, rivers, mountains, rocks, quarries, fossils, castles, gardens, the famous pagan idol Pustrich or spit-fire, dwarf-holes, etc....in the said forest: with several useful and entertaining physical observations
[1868]Kühn, Herbert, 1895–, 1932, Symposium on cave sites and explorations in Upper Rhine regions
[2891]Sauter, Martin, 1992, Quantification and forecasting of regional groundwater flow and transport in a karst aquifer (Gallusquelle, Maim, southwestern Germany)
[2910]Schröder, Karl-Heinz, 10 1980, Guide through the Kubacher Crystal Cave
[3261]U.S. Army, Engineer Intelligence Division, Aug 1962, Natural caves of West Germany Baden - Wuerttemberg
[3324]U.S. Office of Strategic Services, Research and Analysis Branch, Apr 1945, The caves of greater Germany

14.6.15 Greece

[142]Anon, 1964, Karst groundwater investigations: Greece
[236]Aspey, Steve, [1985], Sheffield University Speleological Society Central Crete Expedition: Greece 1984
[559]Burdon, David Joseph, 1963, The karst groundwater resources of Parnassoc-Ghiona, Greece: a report
[810]Davaras, Costis, 1989, The Cave of Psychro
[1013]Evans, M. J. , [1967], 67 expedition to Crete: BUSS Report
[1020]Eyre, Jim, 1981, The cave explorers
[1688]Jacobsen, T. W., 1987, Excavations at Franchthi Cave, Greece
[1689]Jacobsen, T. W., 1987, Franchthi Cave and Paralia: maps, plans, and sections
[1933]Leontiadis, J., Jan 1972, The use of radioisotopes in tracing karst groundwater in Greece
[2023]Malatesta, Alberto, 1980, Dwarf deer and other late Pleistocene fauna of the Simonelli Cave in Crete
[2268]Morfis, A. (ed), 1986, Karst hydrogeology of the central and eastern Peloponnesus (Greece)
[2660]Petrocheilou, Anna, 1970, The Diros Caves of Mani: Alepotrypa and Glyphada
[2661]Petrocheilou, Anna, 1984, The Greek caves
[2662]Petrocheilou, Anna, Dec 1972, The Perama Caves of Loannina
[2687]Pollack, John C., 1949–, 1977, National Speleological Society (USA) 1973 field trip to Greece
[2698]Poulianos, Aris N., 1982, The cave of the Petralonian archanthropinae: a guide to the science behind the excavations
[3490]Wilkinson, Tony J., 1990, Franchthi Paralia—the sediments, stratigraphy, and offshore investigations

14.6.16 Hungary

[8]Adamkó, Péter , 1992, The caves of Buda
[9]Adamkó, Péter, 1989, The caves of Budapest
[288]Bakó, Tamás, 1989, Paleokarst in Hungary
[289]Balázs, Dénes (ed), 1989, Special issue on the occasion of 10th international congress held in Hungary 1989
[290]Balázs, Dénes (ed), 1977, Special issue on the occasion of 7th international congress held in England 1977
[309]Balogh, K., 1989, Karst hydrological and speleological features
[427]Bolner, Katalin Tak'acs , 1992, ALCADI '92 international conference on speleo history field trip guide
[1007]Eszterhás, István, 1989, The caves on the Tés Plateau and in the Bakony Mountains
[1135]Gádoros, Miklós, 1989, Speleotherapic and speleclimatological centres
[1410]Hazslinszky, Tamás (ed), 1975, International conference Baradla 150: field-trip guide: Budapest-Aggtelek, 26-29.08.1975
[1411]Hazslinszky, Tamás (auth, ed, trans), 1989, Show caves of Hungary
[1412]Hazslinszky, Tamás, 1989, Colourful world of caves
[1813]Kessler, Hubert, 1907–1994, 1971, Aggtelek
[1846]Kollár, Attila K. (comp), 1989, Subaquatic caves in Hungary for divers
[1931]Lénárt, László, 1993, A Bukki barlangok kutatasanak, vedelmenek es hasznositasanak legujabb eredmenyei = Newest results in research, protection and use of caves of Bukk Mountains: Miskolci Egyetem, 1993. November 11-13
[1932]Lénárt, László, 1989, Caves in the Bükk Mountains
[2821]Ringer, Árpád, 1989, Prehistoric remains in Hungary
[2877]Salamon, Gábor, 1989, Aslóhegy shafts
[2878]Salamon, Gábor, 1989, The Stalactite Caves of Jósvafő and Aggtelek
[3182]Tardy, Janos, 1989, Geological nature conservation cave-protection
[3550]Zámbó, László (ed), 1993, Conference on the karst and cave research activities of education and research institutions in Hungary: papers: Jósvafő 17-19 May 1991

14.6.17 Iceland

[2187]Mills, Martin Taylor, Dec 1981, The subterranean wonders of Iceland
[3534]Wright, David P. (ed), [1988], 1st Claygate "Selachii" Venture Scout Unit expedition to Iceland: 21st July - 9th August 1987

14.6.18 Italy

[170]Anon, 1828, On the cold caves of the Monte Testaccio at Rome
[214]Anon, c1995, The Wind Cave and its territory; Garfagnana, the Apuan Alps, the Serchio Valley (Lucca, Tuscany, Italy)
[430]Bondesan, A., 1992, Morphometric analysis of dolines
[452]Boyle, Mary Elizabeth, 1881–, 1925, Barma Grande; the great cave and its inhabitants
[659]Cigna, Arrigo A. (ed), 1966, Caving in Italy
[1140]Gams, Ivan (ed, trans), 1987, Man's impact in Dinaric Karst: guide-book
[2165]Miccolis, Maria, 1980, The Castellana Grottoes
[2177]Miles, Sibella Elizabeth, 1864, The Grotto of Neptune ("Antro Di Nettuno"), Sardinia; a poem illustrative of three views of this interesting cavern, taken in July 1824 by the late Commander Alfred Miles and dedicated to his memory by his widow, Sibella Elizabeth Miles
[2185]Mills, M. T., Jul 1990, The subterranean wonders of Sardinia
[3242]Trower, Harold Edward, c1928, Capri: the story of the blue grotto
[3358]Verneau, Rene, 1852–1938, 1908, The men of the Barma-grande (Baousse-Rousse) An account of the objects collected in the Museum praehistoricum, founded by Commendatore Th. Hanbury near Mentone
[3530]Wreschner, E., 1967, The Geula Caves. Mount Carmel

14.6.19 Malta

[703]Cooke, John H. , c1893, The Har Dalam Cavern, Malta and its fossiliferous contents by John H. Cooke with a report on the organic remains by Arthur Smith Woodward
[977]Ellul, Joseph S., 1988, Malta's prediluvian culture at the stone age temples with special reference to Hagar Qim, Ghar Dalam cart ruts, Il-Misqa, Il-Maqluba and Creation
[2018]Maempel, George Zammit, 1989, Ghar Dalam: cave and deposits
[2019]Maempel, George Zammit, 1989, Pioneers of Maltese geology
[2267]Morana, Martin, 1987, The prehistoric cave of Gher Dalam

14.6.20 Moravia

[4]Absolon, Karel B., 1980, The conquest (the riddle of an abyss and its sinking river)
[1874]Kunsky, Josef, 1903–1977, 1954, Homes of primeval man: wandering in the caves of Czechoslovakia
[2711]Přibyl, Jan (ed), 1973, Largest cave system of the Czech Socialist Republic in the Moravský kras (Moravian

karst)
[2714]Přibyl, Jan, nd, Věkåum budoucím

14.6.21 Netherlands

[2527]Nota, Dirk Johannes Gregorius, 1978, A hydrogeological study in the basin of the Gulp Creek: a reconnaissance in a small catchment area: 1, Groundwater flow characteristics
[2898]Schaik, David Cornelis van, 1945, The old town of Maastricht and the caves of mount St. Peter

14.6.22 Norway

[1002]Eraso, Adolfo (ed), Mar 1988, The Stor-Glomfjord project: karst research - the possibilities of leakage and the selection of damsite alternative
[1035]Faulkner, Trevor, 1942–, 1977, Report of the SWETC caving club expedition to Norway 1974
[1036]Faulkner, Trevor, 1942–, 1979, Sump index Norway
[1415]Heap, David (comp), 1969-1970, Report of the British speleological expedition to arctic Norway, 1969, incorporating the work of the 1968 Hulme Schools expedition
[2251]Moller, Jacob, 1935–, 1982, Coastal caves, marine limits, and ice retreat in Lofotenvesteralen, North Norway
[2875]St. Pierre, David, 1966, The caves of Graatadalen, northern Norway: report of the Southwest Essex Technical College Caving Club expeditions 1963-5
[2876]St. Pierre, David, Mar 1969, The Caves of Rana, Nordland, Norway

14.6.23 Poland

[2840]Roemer, Ferdinand Carl, 1818–1891, 1884, The bone caves of Ojcow in Poland

14.6.24 Portugal

[114]Anon, c1967, Exploration '66; University of Nottingham; biospeleological research expedition to Ireland; Riverview work project in Portugal; British speleological expeditions to Turkey

14.6.25 Pyrenees

[606]Casteret, Norbert, 1897–1987, 1951, Cave men new and old
[608]Casteret, Norbert, 1897–1987, 1955, The descent of Pierre Saint-Martin
[609]Casteret, Norbert, 1897–1987, 1962, More years under the earth
[610]Casteret, Norbert, 1897–1987, 1947, My caves
[611]Casteret, Norbert, 1897–1987, 1940, Ten years under the earth
[2894]Sawtell, Ruth Otis, 1927, Primitive hearths in the Pyrenees; the story of a summer's exploration in the haunts of prehistoric man
[3193]Tazieff, Haroun, 1914–, 1953, Caves of adventure

14.6.26 Serbia

[873]Dimitrijevic, Vesna M., 1991, Quaternary mammals of the Smolucka Cave in southwest Serbia

14.6.27 Slovakia

[626]Čerňanská, Miroslava (ed), 1973, Welkome (sic) to our Slovak caves
[1874]Kunsky, Josef, 1903–1977, 1954, Homes of primeval man: wandering in the caves of Czechoslovakia
[2635]Pelech, Johann E., 1879, The valley of Stracena and the Dobschau ice-cavern (Hungary)
[3249]Tulis, J. (comp), 1993, Slovakia: the karst and speleology

14.6.28 Slovenia

[47]Aljančič, Marko (auth, ed), 1993, *Proteus*; the mysterious ruler of karst darkness
[123]Anon, 1800, The grottoes of Adelsberg and the *Proteus* of Anguinus: a few pages from the journal of a continental tour
[780]Curcic, Bozidar P. M., 1988, Cave-dwelling pseudoscorpions of the Dinaric Karst = Jamski pascipalci dinarskega krasa
[1140]Gams, Ivan (ed, trans), 1987, Man's impact in Dinaric Karst: guide-book
[1229]Gospodarič, R. (ed), 1976, Underground water tracing: investigation in Slovenia 1972-1975
[1301]Habe, Francé, 1988, Jame Škocjan = Grotte Skocjan
[1302]Habe, Francé, 1969, Postojna
[1303]Habe, Francé, 1976, Postojna Cave
[1304]Habe, Francé, 1973, Postojna Caves
[1305]Habe, Francé, 1977, The Postojna Caves and other tourist caves in Slovenia
[1306]Habe, Francé, 1972, The Postojna Grottes with the Planina and Predjama caves
[1307]Habe, Francé, 1973, The Postojna Grottoes
[1308]Habe, Francé, 1980, Predjama: the castle and the cave
[1309]Habe, Francé, 1978, Show caves in Slovenia - Guide
[1645]International Symposium of Underground Water Tracing (3rd: 1976: Ljubljana, Yugoslavia), 1976, Under-

gound water tracing: investigations in Slovenia 1972–1975
[1828]Kmecl, Matjaž, 1990, Postojna caves: enter traveller into this immensity!
[2936]Šerko, Alfred, 1910–1948, 1953, The Cave of Postojna and other curiosities of the karst
[2937]Šerko, Alfred, 1910–1948, 1967, The Postojna grottoes and the other marvels of the karst
[3549]Zadnikar, Marjan, 1960, The castle of Predjama

14.6.29 Spain

[98]Anon, 1959, The caves of Arta situated on the sea coast of the municipal confines of Capdepera
[128]Anon, 1935, A guide to the Cave of Altamira and the town of Santillana del Mar (Province of Santander, Spain)
[176]Anon, 1984, Prehistoric painting, part 2
[274]Bahn, Paul G., 1988, Images of the ice age
[391]Berenguer, Magin, 1994, Prehistoric cave art in northern Spain, Asturias
[392]Berenguer, Magin, 1973, Prehistoric man and his art: the caves of Ribadesella
[485][British Cave Research Association], Jun 1981, Matienzo, Spain
[492]British Speleological Expedition: Members of the Expedition, 1966, British Speleological Expedition to the Cantabrian Mountains, northwest Spain 1965
[598]Carballo, Jesus Maria, 1965, The cave of Altamira and other caves with paintings in the province of Santander
[608]Casteret, Norbert, 1897–1987, 1955, The descent of Pierre Saint-Martin
[610]Casteret, Norbert, 1897–1987, 1947, My caves
[611]Casteret, Norbert, 1897–1987, 1940, Ten years under the earth
[644]Checkley, Dave Lancaster, 1977, LUSS expeditions to Tresviso and the Picos de Europa in northern Spain, 1974-1977
[645]Chekley, D., 1985, La Sima 56 (Picos de Europa-España)
[731]Cox, Guy, Apr 1973, Caves of the Catabrian Mountains, north-west Spain
[972]Ellis, Martin, 1992, The caves of Tenerife
[1145]García Guinea, Miguel Angel, 1975, Altamira and prehistoric art in the caves of Santander
[1146]García Guinea, Miguel Angel, Mar 1969, Altamira, the beginning of art
[1147]García Guinea, Miguel Angel, 1971, Altamira: the origin of art
[1148]García Guinea, Miguel Angel, 1984, Santillana and Altamira
[1213]Gimenez Reyna, Simeon, 1965, The cave of "La Pileta": Benaojan, Malaga, Spain
[1214]Gines, Joaquin, May 1987, Proceedings of the 1986 meeting of the International Geophysical Union study group on man's impact on karst
[1265]Grigson, Geoffrey, 1905–, 1957, Painted caves
[1514]Hooks, William H., 1977, Maria's Cave
[1916]Lawson, Andrew J., , Cave art
[2035][Manchester University Speleological Society], [1975], Report of the 1974 British speleological expedition to Matienzo area of Santander Province of northern Spain
[2038]Mann, Paul (ed), 1994, Oxford University Cave Club Cabeza Julagua expedition final report 28 June - 20 August 1993
[2053]Marshall, Des, 1994, Mallorca Caves; an interim guide
[2056]Martel, Edouard Alfred, 1859–1938, c1950, Cavern of the dragon
[2058]Martel, Edouard Alfred, 1859–1938, 1923, The Grottoes of the Drach near Manacor (Majorca)
[2086]Matthews, Geoff (ed), May 1971, Report of the Nottingham University Students Union Spelaeological Expedition 1970, Picos de Europa North-West Spain
[2099]Mazonowicz, Douglas, 1979, The painted caves of France & Spain
[2100]Mazonowicz, Douglas, 1970, Prehistoric paintings of France and Spain: a description of 34 actual size copies hand screenprinted by Douglas Mazonowicz
[2103]Mazonowicz, Douglas, 1974, Voices from the stone age: a search for cave and canyon art
[2530]Nougier, Louis-René, 1912–, 1961 , Art treasures of prehistoric man in the caves of France and Spain
[2542]Obermaier, Hugo, 1877–1946, 1928, The caves of Altamira
[2585]Oxford University Cave Club, 1993, Oxford University Cave Club: Huerta del Rey expedition final report
[2586]Oxford University Cave Club, c1994, Oxford University Cave Club: La Verdelluenga 1994 final report
[2587]Oxford University Expedition to Northern Spain (1961), Nov 1965, Oxford University expedition to northern Spain, 1961
[2652]Pereda de la Reguera, Manuel, 1966, Guide Santillane on the sea and Altamira
[2761]Raphael, Max, 1889–1952, 1945, Prehistoric cave paintings
[2818]Riddell, William Hatton, 1938, Altamira: a note upon the Palaeolithic paintings in the Cave of Altamira near Santillane del Mar in the Spanish province of Santander
[2822]Ripoll Perello, Eduardo, 1980, The Cave of Las Monedas in Puente Viesgo (Santander)
[2823]Ripoll Perello, Eduardo, 1968, The painted shelters of La Gasulla (Castellon)
[2837]Rodriguez, Ortega Eduardo (auth, photo), 1970, The cave of Nerja
[2854]Rose, David, 1959–, , Beneath the mountains: ex-

ploring the deep caves of Asturias
[2883]Sandak, Inc., 1973, Cave paintings—Lascaux and Altamira
[2977]Sieveking, Ann, 1962, The caves of France and northern Spain
[2978]Sieveking, Ann, 1966, The caves of France and northern Spain, a guide
[3337][University of Nottingham], [1965], University of Nottingham Spelaeological Expedition Exploration '64
[3338][University of Nottingham Union], 1969, University of Nottingham Union report of the expedition's co-ordinating committee for 1968
[3345]Valsero, Juan José Durán, nd, Cueva de Nerja: English edition

14.6.29.1 Canary Islands

[972]Ellis, Martin, 1992, The caves of Tenerife
[1707]Jantschke, Herbert, 1994, Tunel de la Atlantida, Haria, Lanzarote, Canary Islands: the hydrodynamic, the chemistry and the minerals of the lava tube, the population density of *Munidopsis polymorpha*

14.6.29.2 Gibraltar

[621]Cave Research Group of Great Britain, 1971, Hypogean fauna and biological records 1970, [and] fauna of Gibraltar caves

14.6.29.3 Majorca

[2884]Santamarta Cuenca, Pedro, 1978, The grottos of Majorca

14.6.29.4 Mallorca

[664]Clarke, Owen (comp), 1990, Holiday caving in Mallorca
[1637]International Geographical Union, Sep 1995, International symposium on karren landforms: Mallorca 1995

14.6.30 Sweden

[3195]Tell, Leander, 1961, Erosionsforloppet, med sarskild hansyn till Lummelundagrottorna. The rate of erosion, with special reference to the caves of Lummelunda
[3196]Tell, Leander, 1895–, 1976, Fifty typical Swedish caves

14.6.31 Switzerland

[532]Browne, George Forrest, 1833-1930, 1865, Ice-caves of France and Switzerland. A narrative of subterranean exploration

[2159]Merk, Conrad, 1876, Excavations at the Kesserloch near Kesserloch near Thayngen Switzerland: a cave of the Reindeer Period
[2718]Price, Graham, Sep 1981, The Holloch: a general account and trip report
[2928]Selby, Paul (ed), Feb 1975, Sieben Hengste 74: the karst and caves of the southwestern zone of the Sieben Hengste Ridge: a report by the members of the Croydon Caving Club on an expedition to the Seefeld area, north of Interlaken, Switzerland, summer 1974

14.6.32 Turkey

[114]Anon, c1967, Exploration '66; University of Nottingham; biospeleological research expedition to Ireland; Riverview work project in Portugal; British speleological expeditions to Turkey
[1363]Harris, Shane R., c1992, UEA speleological expedition northern Turkey: August / September 1992
[3150]Stratford, Tim (comp), 1992, Toros 89-92; the Camlik Project: a report on the speleological investigations carried out by the Swindon Speleological Society in the Camlik Region of Toros Mountain, Konya Turkey from 1989-1992

14.6.33 Yugoslavia

[641]Chatterton, John B., Sep 1971, Report of the Queen Mary College Society spelaeological expedition to Yugoslavia
[848]Deeleman-Reinhold, Christa L., 1978, Revision of the cave-dwelling and related spiders of the genus *Troglohyphantes* Joseph (Linyphiidae) with special reference to the Yugoslav species
[1020]Eyre, Jim, 1981, The cave explorers
[1141]Gams, Ivan, 1973, Slovenska Kraska Terminologija = Slovene karst terminology
[1857]Kozlowski, Janusz Krzysztof, 1994, Meso and Neolithic sequence from the Odmut Cave (Montenegro)
[2175]Mikulec, Stjepan, 1976, Karst hydrology and water resources: proceedings of the U. S.—Yugoslavian symposium, Dubrovnik, June 2-7, 1975

14.7 Asia

[393]Bernabei, Tullio, 1992, Caves and stories of central Asia
[928]Duxbury, J., Jun 1991, The joint Anglo-Soviet speleological expedition to central Asia 1990 "Kugitang 90"
[1170]Gebauer, Herbert Daniel, Nov 1995, Speleological bibliography of South Asia including the Himalayan regions

14.7. ASIA

[1657]International Symposium on Vulcanospeleology (5th: 1988: Izunagaoka Hot Springs, Japan), 1988, Excursion guide book: 5th international symposium on vulcanospeleology: [pre-activity: November 4-9, 1988: post-activity: November 13-20, 1988]
[3114]Stein, Aurel, 1862–1943, 1980, Serindia: detailed report of explorations in Central Asia and westernmost China

14.7.1 China

[72]Andersson, Johan Gunnar, 1874–1960, nd, The cave-deposit at Sha Kuo Tun in Fengtien
[82]Anon, 1978, Beautiful Kweilin
[88]Anon, Jul 1993, Catalogue of literature on Chinese karst caves
[132]Anon, 1987, Guilin tourist album
[133]Anon, 1981, Guilin: the crown of superb landscapes in China
[178]Anon, 1979, Research of China karst
[179]Anon, 1979, Research of China karst
[286]Baker, Janet, 1994, A brief history of the Dunhuang Caves: a millennium of Chinese Buddhist art
[406]Black, Davidson, 1884–1934, 1925, The human skeletal remains from the Sha Kuo T'un cave deposit in comparison with those from Yang Shao Tsun and with recent North China skeletal material
[474]Bridgemon, Rondal Rex, 1944–(ed), [1991], South China caves: information on the cave and karst of South China and a report on the 1988 joint expedition between the Institute of Karst Geology, the Speleological Society of South China Normal University, and the Cave Research Foundation
[802]Daoxian Yuan, 1933–, 1981, A brief Introduction to China's research in Karst
[803]Daoxian Yuan, 1933–, 1991, Karst of China
[927]Dusar, Michiel, 1991, Geological and speleological reconnaissance of the East Yunnan Karst (P.R. China): preliminary results of a field trip between 17.12.1990 and 3.01.1991
[1074]Fogg, P. (ed), nd, China Caves Project 1987—1988: the Anglo—Chinese project in caves of South China
[1235]Graham, David Crockett, 1932, The ancient caves of Szechwan Province, China
[1589]Hydrological and Engineering Geological Team of the Geological Bureau of the Kwangsi Chuan Autonomous Region, Aug 1976, On the underground river system of the Tisu karst area, Tu-an County, Kwangsi, China
[1612]Institute of Hydrogeology and Engineering Geology. Chinese Academy of Geological Sciences, 1976, Karst in China
[1651]International Symposium on Karst of Inner Plate Region with Monsoon Climate (1991: Guilin, China), 1992, World karst correlation: proceedings of the international symposium on karst of the Inner Plate Region with monsoon climate, July 1991, Guilin, China
[1892]Lan-po Chia, 1975, The cave home of Peking Man
[1893]Lan-po Chia, 1990, The story of Peking Man, from archaeology to mystery
[2082]Masschelein, Jan (ed), nd, Teng Long Dong, the longest cave in China: report of the first Belgian-Chinese speleological expedition in 1988
[2934]Senior, Kevin J. (ed), Oct 1995, The Yangtze Gorges expedition: China caves project
[3114]Stein, Aurel, 1862–1943, 1980, Serindia: detailed report of explorations in Central Asia and westernmost China
[3166]Sweeting, Marjorie Mary, 1920–1994, 1995, Karst in China: its geomorphology and environment
[3177]Taketaro Shinkai, 1868–1927, 1921, Rock-carvings from the Yun-kang caves
[3390]Waltham, Anthony Clive, 1942–[Tony], 1986, China caves '85: the first Anglo-Chinese project in the caves of south China
[3394]Waltham, Anthony Clive, 1942–[Tony] (ed), 1993, Xingwen: China caves project 1989-1992
[3437]Wen-chung Pei, 1904–1982, 1940, The upper cave fauna of Choukoutien
[3479]Whitfield, Roderick, 1995, Dunhuang: caves of the singing sands: Buddhist art from the Silk Road
[3540]Yauru Lu, 1931–, 1986, Karst in China - landscapes, types, rules
[3554]Zhaoyang Wang, 1983, Underground worlds: Guizhou
[3555]Zhu Xuewen, 1932-, 1988, Guilin karst

14.7.2 Georgia SSR

[2959]Shelley, Maryann, Dec 1956, Karst and caves in the Caucasus
[3219]Tintilozov, Zurab, 1978, Akhali Atoni's cave

14.7.3 Himalayas

[926]Durrant, Gillian A., 1979, Himalaya underground - 1976 speleological expedition
[3387]Waltham, Anthony Clive, 1942–[Tony], Jun 1971, British karst research expedition to the Himalaya 1970: full report

14.7.4 Hong Kong

[395]Berthelsen, O. J., Mar 1992, Guide to cavern engineering
[1898]Langford, R.L. (ed), 1990, Karst geology in Hong Kong: proceedings of a conference on "Karst geology in

Hong Kong" held at the University of Hong Kong on 6 January 1990
[1899] Langford, R. L. (ed), 1990, Karst geology in Hong Kong: programme and abstracts: a conference organized by the Geological Society of Hong Kong and Department of Geography and Geology, University of Hong Kong: Programme and Abstracts: University of Hong Kong: 4-6 January 1990

14.7.5 India

[411] Black, Nancy Gail, 1965, A comparative study of stylistic and extrapictorial elements in the murals of Cave I, Ajanta and of the Brancacci Chapel
[513] Brooks, S. J. (ed), Mar 1995, Caving in the abode of the clouds: the caves and karst of Meghalaya, north east India
[849] Dehejia, Vidya, 1969, The Namakkal caves
[1168] Gebauer, Herbert Daniel, 1983, Caves of India & Nepal: including the results of Speläologische Südasien Expedition 1981/82
[1169] Gebauer, Herbert Daniel, 1985, Kurnool 1984: report of the speleological expedition to the district of Kurnool, Andhra Pradesh, India
[1186] Ghosh, Leila, 1986, Ajanta and Ellora
[1594] India. Public Works Department, nd, The caves of Elephanta
[1855] Koul, Samsar Chand, c1980, The mysterious cave of Amar Nath, Kashmir India
[2084] Mathpal, Yashodhar, 1984, Prehistoric rock paintings of Bhimbetka, Central India
[2161] Meshram, Pradip Shaligram, 1954–, 1991, Early caves of Maharashtra: a cultural study
[2246] Mohapatra, R. P., 1939–, 1981, Udayagiri and Shandagiri caves
[2975] Siddiqui, Safiuddin, 1971, A pictorial guide to Aurangabad, Daulatabad, Ellora, and Ajanta
[3095] Spink, Walter M., 1990, Ajanta: a brief history and guide

14.7.6 Indonesia

[438] Boothroyd, Colin, May 1993, Caves of Thunder expedition report: an expedition to the world's largest underground river, Irian Jaya, Indonesia 1992
[2582] Owen, Daniel (comp), 1987, Proyek Kelelawar: final report of the Oxford University expedition to the Togian Islands, Sulawesi, Indonesia summer 1987

14.7.6.1 Borneo

[503] Brook, David B., 1944–(ed), 1978, Caves of Mulu: the limestone caves of the Gunong Mulu National Park, Sarawak
[936] Eavis, A. J. (comp), 1981, Caves of Mulu '80: the limestone caves of the Gunong Mulu National Park, Sarawak
[937] Eavis, A. J. (comp), 1985, Caves of Mulu '84: the limestone caves of the Gunong Mulu National Park, Sarawak
[2156] Meredith, Michael Edward, 1943–, 1992, Giant caves of Borneo
[3485] Wilford, Gerald Edward, 1964, The geology of Sarawak and Sabah caves

14.7.6.2 Java

[1696] James, Ian (ed), [1986], Army caving expedition to South East Java: July and August 1986: exercise Phreatic Diamond
[2157] Meredith, Michael Edward, 1943–, [1985], Java caves 1983: report of a visit to Indonesia by Australian and British cavers

14.7.6.3 Malaysia

[147] Anon, c1993, The magic of Sarawak's Mulu Caves; discover the latest wonder of the world
[640] Chasen, Frederick Nutter, 1896–, 1931, Report on the "birds' nest" caves and industry of British North Borneo: with special reference to the Gomantong caves
[654] Chin, Lucas, 1980, Archaeological work in Sarawak: with special reference to Niah Caves
[655] Chin, Lucas, 1977, Summary of archaeological work in Sarawak: with special reference to Niah Caves
[1311] Hacker, Bradley (ed), [1996], Caves of Gunung Buda: report of the joint Sarawak Forest Department and USA caving expedition to Sarawak, Malaysia 1995
[1344] Hamilton-Smith, Elery, Jun 1989, Batu Caves, Kuala Lumpur - towards a plan of management
[1485] Holland, Eric George, 1955, A guide to Batu caves
[2024] Malayan Nature Society, May 1965, Malaysian caves
[2156] Meredith, Michael Edward, 1943–, 1992, Giant caves of Borneo
[2572] Ong, Johnney, 1994, Mysterious caves of Langkawi, Malaysia
[2885] Sarawak Museum, 1973, Summary of archaeological work in Sarawak: with special reference to Niah Caves
[3006] Smart, Peter L. (comp), Jun 1982, Cathay Pacific Airways Mulu '80 expedition: British Malaysian speleological expedition to Sarawak
[3045] Soepadmo, E., 1971, A guide to Batu Caves
[3246] Tsen, Darrell Nyuk Choi, 1993, The show caves of Mulu Sarawak
[3247] Tsen, Darrell Nyuk Choi, 1991, The show caves of Mulu Sarawak

14.7.6.4 Sulawesi

[2582]Owen, Daniel (comp), 1987, Proyek Kelelawar: final report of the Oxford University expedition to the Togian Islands, Sulawesi, Indonesia summer 1987

14.7.7 Japan

[21]Akira Yoshida, Jan 1952, A list of the Arthropoda in the limestone caves in Kantô-Mountainland, with the descriptions of a new genus and three species
[1395]Hayami, Itaru, 1933–, 1993, Submarine cave bivalvia from the Ryukyu Islands: systematics and evolutionary significance
[1847]Komatsu, Toshihiro, 1961, Cave spiders of Japan; their taxonomy, chorology, and ecology
[1961]Longley, Glenn, 1942–, Jan 1952, A list of the arthropoda in the limestone caves in Kantô-Mountainland, with the description of a new genus and three species
[2882]Sameshima, Teruhiko, 1988, 5th international symposium on vulcanospeleology excursion guide book

14.7.7.1 Okinawa

[1644]International Speleologist, Jun 1962, Caves of Okinawa
[2510]Nicol, Allen Hankins, 1907–, 1951, Application of geophysical methods to location of cavities in residual soil and caverns in coralline limestone

14.7.8 Kurdistan

[1162]Garrod, Dorothy Anne Elizabeth, 1892–, 1968, The paleolithic of southern Kurdistan: excavations in the caves of Zarzi and Hazar Merd

14.7.9 Nepal

[660]C'ilek, V'acav , c1985, Czechoslovak speleological expedition to Nepal Himalaya 85
[1168]Gebauer, Herbert Daniel, 1983, Caves of India & Nepal: including the results of Speläologische Südasien Expedition 1981/82

14.7.10 Thailand

[103]Anon, nd, Chiang Dao Cave
[916]Dunkley, John R., Jan 1995, The caves of Thailand
[922]Dunkley, John R., 1994, Thailand caves catalogue
[923]Dunkley, John R. , 1986, Caves of north-west Thailand: report of the Australian speleological expeditions, 1983-1986
[3160]Surin, Phukhachon, 1991, Preliminary report of excavations at Moh-Khiew Cave, Krabi Province, Sakai Cave, Trang Province, and ethnoarchaeological research of hunter-gatherer group, socall[ed] Sakai or Semang at Trang Province: the Hoabinnian Research Project in Thailand
[3344]Valli, Eric, 1990, The nest gatherers of Tiger Cave

14.7.11 Uzbekistan

[2292]Movius, Hallam Leonard, 1907–, 1953, The Mousterian cave of Teshik-Tash, Southeastern Uzbekistan, Central Asia

14.7.12 USSR

[3342]Vale, W. P., [1990], Aspex '90: Anglo -Soviet Pamirs Expedition 1990: Bajsuntai Khrebetuzbekistan SR, USSR

14.7.13 Vietnam

[51]Allen, Tim (tm ldr), c1992, Hang Vietnam; report of the British Speleological Expedition to the Bac-Sun Massiflang
[458]Bradshaw, D. R. (comp), nd, Report on the British speleological expedition to Vietnam; March April 1990
[1944]Limbert, Howard, [1995], The 1994 British/Vietnamese speleological expedition report March/April 1994
[2508]Nguyen Tai Duong (comp) , [1994], Vietnam cavern tourism
[2919]Schwartz, Jeffrey H., Jan 1994, A diverse hominoid fauna from the Late Middle Pleistocene breccia cave of Tham Khuyen, Socialist Republic of Vietnam

14.8 Central America

[340]Bassie-Sweet, Karen, 1991, From the mouth of the dark cave: commemorative sculpture of the late classic Maya
[1185]Gertsch, Willis J., 1984, The spider family Nesticidae (Araneae) in North America, Central America, and the West Indies
[2242]Mitchell, Robert Wetsel, 1933–(ed), Jul 1973, Studies on the cavernicole fauna of Mexico and adjacent regions
[2789]Reddell, James Russell, 1938–(ed), Mar 1982, Further studies on the cavernicole fauna of Mexico and adjacent regions
[3158]Sullivan, Gerardus Nicholas, 1927–, 1962, Checklist of troglobitic organisms of Middle America

14.8.1 Belize

[1116]Francis, Timothy, 1995, Mendip Caving Group: Belize '94
[2048]Marochov, Nick (ed), 1992, Below Belize: Queen Mary College speleological expedition to Belize, 1988 and the British speleological expedition to Belize, 1989
[2049]Marochov, Nick (ed), [1992], Below Belize; Queen Mary College Speleological Expedition to Belize 1988 and the British Speleological Expedition to Belize 1989
[2184]Miller, Tom, [1984], The karst development and associated archeology of the Chiquibul, Belize: (preliminary report of the results of grant #2742-83)
[2595]Palacio, Joseph O., 1977, Excavation at Hokeb Ha, Belize
[2640]Pendergast, David M., 1971, Excavations at Eduardo Quiroz Cave, British Honduras (Belize)
[2641]Pendergast, David M., 1970, A. H. Anderson's excavations at Rio Frio Cave, British Honduras (Belize)
[2642]Pendergast, David M., Feb 1974, Excavations of Actum Polbilche, Belize
[2792]Reddell, James Russell, 1938–(ed), Jul 1981, A review of the cavernicole fauna of Mexico, Guatemala, and Belize
[3496]Williams, Nick (auth, ed), 1992, Below Belize 1991: the report of the 1991 expedition to the Little Quartz Ridge Southern Belize February to April 1991
[3497]Williams, Nick, Jan 1994, Bibliography of Belizean caving
[3499]Williams, Nick (ed), 1990, Queen Mary College: below Belize 1988; the logistics report of the 1988 Speleological Expedition to Belize

14.8.2 Costa Rica

[407]Black, Don F., 1993, I don't play golf: recollections of a rescue volunteer
[1429]Hempel, John Charles, 1949–(comp), 1989, Report of the 1982 expedition to Barra Honda National Park
[2624]Peacock, Norma Dee, 1994, Studies in the Rio Corredor Basin: 1988 to 1991

14.8.3 Guatemala

[2074]Mason, Gregory, 1889–, 1928, Pottery and other artifacts from caves in British Honduras and Guatemala
[2792]Reddell, James Russell, 1938–(ed), Jul 1981, A review of the cavernicole fauna of Mexico, Guatemala, and Belize
[3133]Stone, Andrea Joyce, 1995, Images from the underworld: Naj Tunich and the tradition of Maya cave painting

14.8.4 Honduras

[1227]Gordon, George Byron, 1870–1927, 1898, Caverns of Copan, Honduras: report on explorations by the museum, 1896-97
[1391]Hawkins, Thomas, 1993, The caves of Honduras: a first list
[1746]Jones, H. (auth, ed), c1994, Honduras recce 1994: a short report
[2074]Mason, Gregory, 1889–, 1928, Pottery and other artifacts from caves in British Honduras and Guatemala

14.9 North America

[329]Barr, Thomas Calhoun, 1931–, 1960, Symposium: speciation and raciation in cavernicoles
[1185]Gertsch, Willis J., 1984, The spider family Nesticidae (Araneae) in North America, Central America, and the West Indies
[1319]Halliday, William R., 1926–, 1974, American caves and caving: techniques, pleasures, and safeguards of modern cave exploration
[1466]Hobbs, Horton Holcombe, 1914–, 1972, Origins and affinities of the troglobitic crayfishes of North America (Decapoda: Astacidae), II, genus em Orconectes
[1502]Holsinger, John Robert, 1934–, Apr 1972, The freshwater amphipod crustaceans Gammaridae of North America
[1561]Hubbell, Theodore Huntington, 1897–1989, Aug 1978, The systematics and biology of the cave-crickets of the North American tribe Hademoecini (Orthoptera Saltatoria, Ensifera, Rhaphidophoridae, Dolichopodinae)
[1578]Huppert, George Nixon, 1944–, Jun 1988, Cave and karst-related theses in United States and Canadian Universities, 1899-1988
[1590]Hyman, Libbie Henrietta, 1888–1969, 1956, North American triclad Turbellaria, XIII: three new cave planarians
[1791]Kenk, Roman, 1977, Freshwater Triclads (Turbellaria) of North America, IX, the genus *Sphalloplana*
[2295]Muchmore, William B., 1965, North American cave pseudoscorpions of the genus *Kleptochthonius*, subgenus *Chamberlinochthonius* (Chelonethida, Chthoniidae)
[2590]Packard, Alpheus Spring, 1839–1905, 1977, The cave fauna of North America, with remarks on the anatomy of the brain and origin of the blind species
[2794]Reddell, James Russell, 1938–(ed), Dec 1986, Studies on the cave and endogean fauna of North America
[2795]Reddell, James Russell, 1938–(ed), Dec 1992, Studies on the cave and endogean fauna of North America II
[2871]Russell, William Hart, 1937–, 1967, Caves of the

Inter-American Highway: Nuevo Laredo, Tamulipas to Tamazunchale, San Luis Potosi

14.9.1 Canada

[77]Anglo-Canadian Rocky Mountains Speleological Expedition (1983) (comp), 1986, The Anglo-Canadian Rocky Mountains Speleological Expeditions 1983 and 1984: a report on recent discoveries made by two caving expeditions to the Rocky Mountains of Canada by combined British and Canadian teams
[526]Brown, L. Carson, 1970, Manitouwadge: cave of the Great Spirit
[528]Brown, Michael Charles, 1972, Karst hydrology of the lower Maligne Basin, Jasper, Alberta
[538]Brunton, Daniel Francis, 1948–, 1990, A biological inventory of the Warsaw Caves area of natural and scientific interest, Peterborough County, Ontario
[592]Canada. National and Historic Parks Branch, 1972, Nahanni
[593]Canada. National Parks Branch, 1914, The Nakimu Caves, Glacier Dominion Park, B.C.
[594]Canada. National Parks Service, 1974, Castle Guard Cave, challenge under the glacier
[726]Coward, Julian, Jun 1983, Construction and testing of an underground radio: a report submitted to the Research and Educational Branch of Workers' Health, Safety and Compensation, Alberta Government
[847]Dean, Pauline, 1931–, 1989, Manitouwadge: cave of the Great Spirit
[995]Ensminger, Scott A., 1983, Canadian caves of the Niagara Gorge
[1079]Ford, Derek Clifford, 1935–, 1983, Castleguard Cave and karst, Columbia Icefield Area, Rocky Mountains of Canada: a symposium
[1080]Ford, Derek Clifford, 1935–(prod), 1974, Castleguard Cave: challenge under the glacier
[1082]Ford, Derek Clifford, 1935–, Jul 1972, Guidebook to symposium SA6 "Karst geomorphology of the Canadian Rockies" 22nd international geographical congress Canada 1972
[1084]Ford, Derek Clifford, 1935–, Mar 1973, Theme and resource inventory study of the karst regions of Canada: final report upon project A
[1187]Gibb, George Duncan, 1821–1876, 1861, On Canadian caverns (read before the British Association for the Advancement of Science, at Aberdeen, 16th Sept. 1859)
[1242]Grant, James, 1831–1920, 1868, Superficial geology of the valley of the Ottawa and the Wakefield cave
[1435]Henderson, Kent, 1954–, 1987, Princess Margaret Rose Caves - Western Victoria
[2051]Marsh, John, 1942–, 1979, A history of Glacier House and Nakimu Caves, Glacier National Park, British Columbia
[2052]Marsh, John, 1942–, 1973, Nakimu Caves
[2296]Muir, Robert Dalton, 1985, Castleguard
[2868]Russell, D. J., 1993, Role of the Sylvania Formation in sinkhole development, Essex County
[3205]Thompson, Peter, 1943–, 1976, Cave exploration in Canada: a special issue of the Canadian Caver magazine
[3347]Van Everdingen, R. O., 1981, Morphology, hydrology and hydrochemistry of Karst in permafrost terrain near Great Bear Lake, Northwest Territories
[3348]Van Everdingen, R. O., 1972, Thermal and mineral springs in the Southern Rocky Mountains of Canada

14.9.1.1 British Columbia

[488]British Columbia. Parks and Outdoor Recreation Division, Jan 1980, Cave resources in British Columbia: a discussion paper
[878]Dixon, Richard L., 1965, Speleological investigation of Nakimu Caves, British Columbia
[1264]Griffiths, Paul, 1975, Horne Lake Wonder Caves: a guide to Horne Lake Caves Park
[2935]Seppala, Matti, 1973, Glacier cave observations on Llewellyn Glacier, British Columbia

14.9.1.2 Northwest Territories

[592]Canada. National and Historic Parks Branch, 1972, Nahanni

14.9.1.3 Nova Scotia

[2277]Morris, Linda, 1985, The Hayes Cave site, South Maitland, Nova Scotia
[2923]Scott, Fred, 1942–, 1979, Preliminary investigations at Hayes Cave, Hants County, Nova Scotia, in 1978

14.9.1.4 Ontario

[2399]National Speleological Society. Annual Convention. Guidebook (1977: Alpena, MI), 1977, Official 1977 guidebook, Alpena, Michigan: '77, the Lakeshore convention: Alpena, Michigan, NSS

14.9.1.5 New Brunswick

[2104]McAlpine, Donald F., 1983, Status and conservation of solution caves in New Brunswick

14.9.2 Caribbean

14.9.2.1 Anguilla

[1292]Gurnee, Jeanne Marie, 1926–(ed), 1969, A study of Fountain National Park and Fountain Cavern: Anguilla, British West Indies

[2123]McFarlane, Donald A., Oct 1989, A preliminary catalog of the caves on Anguilla, British West Indies

14.9.2.2 Barbados

[1293]Gurnee, Jeanne Marie, 1926–(ed), Jul 1978, A study of Harrison's Cave, Barbados, West Indies
[1470]Hobbs, Horton Holcombe, 1944–(ed), , A study of environmental factors in Harrison's Cave, Barbados, West Indies
[2351]National Conservation Commission of Barbados, [1986], Harrison's Cave: Barbados

14.9.2.3 Cuba

[1851]Kornicker, Louis S., 1919–, 1996, The troglobitic halocyprid Ostracoda of anchialine caves in Cuba
[2534]Nunez Jimenez, Antonio, 1966, Karstological investigations in Cuba
[2731]Pulina, Marian (ed), 1984, The dynamic of the contemporary karstic processes in the tropical area of Cuba: preliminary report of the field investigations performed by the expedition Guajaibon '84 in the winter season 1984
[3487]Wilkins, Bob, 1989, Cuba Contact '88: an investigative visit by members of the Westminster Speleological Group to the karst of the Sierra de los Organos 18/9/88 - 7/10/88

14.9.2.4 Haiti

[2182]Miller, Gerrit Smith, 1869–1923, Mar 1929, A second collection of mammals from caves near St. Michel, Haiti

14.9.2.5 Netherland Antillies

[3370]Wagenaar Hummelinck, P. , 1979, De Grotten Van de Nederlande Antillen - Caves of the Netherland Antilles

14.9.2.6 Puerto Rico

[1215]Giusti, Ennio V., 1978, Hydrogeology of the karst of Puerto Rico
[1216]Giusti, Ennio V., 1976, Water resources of the North Coast Limestone area, Puerto Rico
[1290]Gurnee, Jeanne Marie, 1926–(ed), 1988, The Ensueño Cave Study: Ensueño Cave, Hatillo, Puerto Rico; expedition period: February 18-25, 1987
[1291]Gurnee, Jeanne Marie, 1926–, Aug 1968, National Speleological Society field trip to Aguas Buenas Caves, Puerto Rico: February 1968
[1295]Gurnee, Russell Hampton, 1922–1995, 1974, Discovery at the Rio Camuy
[1296]Gurnee, Russell Hampton, 1922–1995, 1968, Discovery at the Rio Camuy, Puerto Rico
[1760]Jordon, Donald George, 1926–, 1970, Water and copper-mine tailings in karst terrane of Rio Tanama Basin, Puerto Rico
[1927]Legault, Myrna Diaz (comp), Oct 1990, Bibliography of Puerto Rican caves, karst and limestone geology; Rio Camuy Cave System
[2256]Monroe, Watson Hiner, 1907–, 1976, The karst landforms of Puerto Rico: a discussion of a solution landscape formed in a tropical climate of moderately high rainfall
[2861]Rouse, Irving, 1913–, 1990, Excavations at Maria de la Cruz Cave and Hacienda Grande Village Site, Loiza, Puerto Rico
[3223]Torres González, Arturo, Mar 1983, Geohydrology of the Rio Camuy Cave system Puerto Rico
[3429]Wegrzyn, Mikolaj, Jun 1984, Rainstorm related terrain failures in Puerto Rico: final technical report

14.9.2.7 West Indies

[439]Booy, Theodoor Hendrik Nikolaas de, 1882–1919, 1915, Pottery from certain caves in eastern Santo Domingo, West Indies
[711]Cope, Edward Drinker, 1840–1897, 1883, On the contents of a bone cave in the island of Anguilla (West Indies)
[1185]Gertsch, Willis J., 1984, The spider family Nesticidae (Araneae) in North America, Central America, and the West Indies
[1861]Kranjc, Andrej, 1943–(ed), 1993, The contribution to the history of the speleological explorations of the West Indies at 500—anniversary of the discovery of America
[3417]Watters, David Robert, Dec 1987, Final report on the archaeology of Fountain Cavern, Anguilla, West Indies: a report

14.9.2.7.1 Bahamas

[595]Canadian Broadcasting Corporation, 1972, The Blue holes of Andros
[1029]Farr, Martyn, 1950–, 1984, The great caving adventure
[1848]Kornicker, Louis S., 1919–, 1989, New Ostracoda (Halocyprida: Thaumatocyprididae and Halocyprididae) from anchialine caves in the Bahamas, Palau, and Mexico
[1849]Kornicker, Louis S., 1919–, 1990, Ostracoda (Halocyprididae) from anchialine caves in the Bahamas
[2327]Mylroie, John Eglinton, 1949–(ed), 1988, Field guide to the karst geology of San Salvador Island, Bahamas: prepared for the 10th Friends of Karst meeting, February 11-15, 1988, College Center of the Finger Lakes, Bahamian Field Station, San Salvador Island, Bahamas
[2603]Palmer, Robert John, 1951–1997, 1985, The blue holes of the Bahamas

14.9. NORTH AMERICA

[2604]Palmer, Robert John, 1951–1997, 1989, Deep into blue holes: the story of the Andros Project
[2605]Palmer, Robert John, 1951–1997 (comp), Mar 1984, The report of the 1981 and 1982 British cave diving expeditions to Andros Island, Bahamas
[2606]Palmer, Robert John, 1951–1997 (ed), 1988, Report of the 1987 international blueholes research project

14.9.2.7.2 Jamaica

[40]Aley, Thomas John, 1938–, 1964, Origin and hydrology of caves in the White Limestone of north central Jamaica: report of field work carried out under ONR Contract 3656 (03) NR 388 067
[491][British Speleological Association], 1969, Limestone geomorphology: a study in Jamaica: University of Bristol karst expedition to Jamaica 1967
[624]Caving Club SC 33, c1994, 1993 Jamaica expedition
[1060]Fincham, Alan Goldworth, 1933–, 1977, Jamaica underground: a register of data regarding the caves, sinkholes and underground rivers of the island
[1061]Fincham, Alan Goldworth, 1933–(comp), Jan 1967, The University of Leeds hydrological survey expedition to Jamaica 1963
[1078]Ford, Derek Clifford, 1935–(scien adv), Jul 1966, The 1965/66 karst hydrology expedition to Jamaica: full report
[1850]Kornicker, Louis S., 1919–, 1992, Ostracoda (Halocypridina, Cladocopina) from anchialine caves in Jamaica, West Indies
[3330]University of Bristol, Dec 1967, Karst hydrology expedition to Jamaica: preliminary report

14.9.2.7.3 Tobago

[224]Aquing, Felix, 1974, Survey of Lopinot caves

14.9.2.7.4 Trinidad

[224]Aquing, Felix, 1974, Survey of Lopinot caves

14.9.3 Mexico

[74]Andrews, Edward Wyllys, 1916–1971, 1970, Balankanche, throne of the tiger priest
[157]Anon, 1962, Mexico's caves and caverns
[196]Anon, 1972, Subterranean fauna of Mexico: Part I: some results of the first Italian zoological mission to Mexico sponsored by the National Academy of Lincei (October 10 - December 9, 1969)
[197]Anon, 1973, Subterranean fauna of Mexico: Part II: further results of the first Italian zoological mission to Mexico sponsored by the National Academy of Lincei (1969 and 1971)
[198]Anon, 1977, Subterranean fauna of Mexico: Part III: further results of the first Italian zoological mission to Mexico sponsored by the National Academy of Lincei (1973 and 1975)
[405]Bitterli, Thomas, 1990, Proyecto Cerro Rabon
[436]Boon, J. M., 1980, The Great San Agustín Rescue
[465]Breder, Charles Marcus, 1897–, Dec 1947, Comparative studies in the light sensitivity of blind characins from a series of Mexican caves
[632]Chamberlin, Ralph Vary, 1879-1967, 1942, On centipeds and millipeds from Mexican caves
[656]Christiansen, Kenneth Allen, 1924-, 1982, Notes on Mexican cave Pseudosinella (Collembola: Entomobryidae) with the description of six new species
[678]Coggins, Clemency, 1984, Cenote of sacrifice: Maya treasures from the sacred well at Chichen Itza
[758]Crosby, Harry, 1926–, 1976, Cave paintings of Baja
[759]Crosby, Harry, 1926–, 1984, The cave paintings of Baja California
[760]Crosby, Harry, 1926–, 1975, The cave paintings of Baja California: the great murals of an unknown people
[943]Ediger, Donald, 1971, The well of sacrifice
[1020]Eyre, Jim, 1981, The cave explorers
[1029]Farr, Martyn, 1950–, 1984, The great caving adventure
[1144]Ganter, John Hamilton, 1962–, 1989, The Xilitla region: an annotated bibliography
[1150]Gardner, Erle Stanley, 1962, The hidden heart of Baja [Mexico]
[1273]Grove, David C., 1970, The Olmec paintings of Oxtotitlan Cave Guerrero, Mexico
[1386]Hatt, Robert Torrens, 1902–, 1953, Faunal and archeological researches in Yucatan caves
[1469]Hobbs, Horton Holcombe, 1944–, 1976, On the troglobitic shrimps of the Yucatan peninsula, Mexico (Decapoda: Atyidae and Palaemonidae)
[1562]Hubbs, Carl Leavitt, 1894–1979, 1938, Fishes from the caves of Yucatan
[1848]Kornicker, Louis S., 1919–, 1989, New Ostracoda (Halocyprida: Thaumatocyprididae and Halocyprididae) from anchialine caves in the Bahamas, Palau, and Mexico
[1925]Lee, Thomas A., 1988, San Pablo Cave and El Cayo on the Usumacinta River, Chiapas, Mexico
[1943]Limbert, Howard, [1986], Mexico 85/86
[1982]Lumley, Mark, [1989], Mexico: the black holes expedition: the 1988 British expedition to explore the caves of the Sierra de Zongolica
[2015]MacMaster University Caving and Climbing Club, [1968], A description of the Sotano del Rio Iglesia
[2069]Martinez del Rio, Pablo, 1953, A preliminary report on the mortuary cave of Candelaria, Coahuila, Mexico
[2083]Massey, William C., 1976, A burial cave in Baja California, the Palmer Collection 1887
[2085]Mattheij, Johannes Adrianus Maria, 1940–, 1970,

The functional cell types in the pars distalis analogue of the pituitary gland in the blind mexican cave fish, *Anoptichthys jordani*

[2150]Mercer, Henry Chapman, 1856–1930, 1897, Cave hunting in Yucatan: a lecture delivered before the Society of Arts of the Massachusetts Institute of Technology, on December 10,1896

[2153]Mercer, Henry Chapman, 1856–1930, 1896, The hill-caves of Yucatan, a search for evidence of man's antiquity in the caverns of Central America. Being an account of the Corwith expedition of the Department of Archaeology and Palaeontology of the University of Pennsylvania

[2154]Mercer, Henry Chapman, 1856–1930, 1975, The hill-caves of Yucatan, a search for evidence of man's antiquity in the caverns of Central America. Being an account of the Corwith expedition of the Department of Archaeology and Paleontology of the University of Pennsylvania

[2183]Miller, Loye Holmes, 1874–1970, 1980, The Pleistocene birds of San Josecito cavern, Mexico

[2239]Mitchell, Robert Wetsel, 1933–, Feb 1977, Mexican eyeless Characin fishes, genus *Astyanax*: environment, distribution, and evolution

[2240]Mitchell, Robert Wetsel, 1933–, 1972, A new family, genus, and species of cave-adapted planarian from Mexico (Turbellaria, Tricladida, Maricola)

[2241]Mitchell, Robert Wetsel, 1933–(ed), 1971, Studies on the cavernicole fauna of Mexico

[2242]Mitchell, Robert Wetsel, 1933–(ed), Jul 1973, Studies on the cavernicole fauna of Mexico and adjacent regions

[2629]Pearse, Arthur Sperry, 1877–1956, 1936, The cenotes of Yucatan

[2630]Pearse, Arthur Sperry, 1877–1956, Feb 1936, The cenotes of Yucatan: a zoological and hydrological study

[2751]Raines, Terry W. (comp), 1989, Caves of Mexico

[2752]Raines, Terry W., Sep 1988, First reports: 1984-1988

[2753]Raines, Terry W., 1972, Sotanito de Ahuacatlán: Sierra Madre Oriental; Jalpan; Ahuacatlán

[2754]Raines, Terry W. (comp), Aug 1968, Sotano de las Golondrinas: description

[2789]Reddell, James Russell, 1938–(ed), Mar 1982, Further studies on the cavernicole fauna of Mexico and adjacent regions

[2790]Reddell, James Russell, 1938–, 1971, A preliminary bibliography of Mexican cave biology with a checklist of published records

[2792]Reddell, James Russell, 1938–(ed), Jul 1981, A review of the cavernicole fauna of Mexico, Guatemala, and Belize

[2796]Reddell, James Russell, 1938–, Sep 1971, Studies on the cavernicole fauna of Mexico

[2920]Schwatka, Frederick, 1893, In the land of cave and cliff dwellers

[3111]Steele, C. William, 1985, Yochib, the river cave: an account of the exploration of the Sumidero Yochib of Mexico, a dangerous and difficult cave

[3138]Stone, William Curtis, 1952–, 1977, Caves of the San Juan Plateau

[3140]Stone, William Curtis, 1952–, Aug 1994, A report to the government of Mexico on the 1994 San Augustin expedition to Huautla de Jimenez, Oaxaca

[3155]Studley, Cordelia A., 1884, Notes upon the human remains from the caves of Coahuila, Mexico

[3360]Vessely, Carol, [1989], Proyecto Papalo expedition report, 1986-1989: the exploration of Sistema Cuicateca - second deepest cave in the western hemisphere

[3382]Walsh, John Michael, 1947–, 1972, Mexican caving, 1966-1971

14.9.3.1 Chihuahua

[3557]Zingg, Robert Mowry, 1900–1957, 1940, Report on archaeology of southern Chihuahua

14.9.3.2 Oaxaca

[2680]Pistole, Nancy (ed), 1995, Proyecto Cheve 1986-1993

14.9.3.3 Yucatan

[418]Bloomgarden, Richard, 1984, The easy guide to Chichen Itza, Balankanchen and Izamal

[630]Chamberlin, Joseph Conrad, 1898–, 1947, The Vachoniidae—a new family of false scorpions: two new species from the caves of Yucatan

[1386]Hatt, Robert Torrens, 1902–, 1953, Faunal and archeological researches in Yucatan caves

[2629]Pearse, Arthur Sperry, 1877–1956, 1936, The cenotes of Yucatan

[2630]Pearse, Arthur Sperry, 1877–1956, Feb 1936, The cenotes of Yucatan: a zoological and hydrological study

[2631]Pearse, Arthur Sperry, 1877–1956, Jun 1938, Fauna of the caves of Yucatan

[2791]Reddell, James Russell, 1938–, 1977, A preliminary survey of the caves of the Yucatan Peninsula

[2797]Reddell, James Russell, 1938–, 1977, Studies on the caves and cave fauna of the Yucatan Peninsula

[3154]Stringfield, Victor Timothy, 1902–, 1974, Karst and paleohydrology of carbonate rock terranes in semiarid and arid regions with a comparison to humid karst of Alabama

[3202]Thompson, Edward Herbert, 1860–1935, 1897, Cave of Loltun, Yucatan: report of explorations by the Museum, 1888-89 and 1890-91

[3397]Ward, William Cruse, 1933–, , Geology and hydrology of the Yucatan and Quaternary geology of Northeast Yucatan Peninsula with a part on the History of Northern Quintana Roo

14.9.4 United States

[753]Crawford, Rod, Jul 1977, An annotated bibliography of Pacific Northwest speleobiology

[797]Damon, Paul Herbert, 1934–, 1991, Caving in America: the story of the National Speleological Society, 1941-1991: commemorating 50 years of history and growth: including a special illustrated history of cave exploration in the society entitled—the last frontier for the pioneer

[1029]Farr, Martyn, 1950–, 1984, The great caving adventure

[1076]Folsom, Franklin Brewster, 1907–, 1962, Exploring American caves

[1077]Folsom, Franklin Brewster, 1907–, 1956, Exploring American caves, their history, geology, lore, and location: a spelunker's guide

[1294]Gurnee, Russell Hampton, 1922–1995 (comp), 1983, Cave clippings of the nineteenth century

[1298]Gurnee, Russell Hampton, 1922–1995, 1980, Gurnee guide to American caves: a comprehensive guide to the caves in the United States open to the public

[1318]Halliday, William R., 1926–, 1959, Adventure is underground: the story of the great caves of the West and the men who explore them

[1328]Halliday, William R., 1926–, 1966, Depths of the earth: caves and cavers of the United States

[1329]Halliday, William R., 1926–, 1976, Depths of the earth: caves and cavers of the United States

[1366]Harrison, David Lakin, 1970, The world of American caves

[1509]Holsinger, John Robert, 1934–, 1974, Systematics of the subterranean amphipod genus *Stygobromus* (Gammaridae): part I. species of the western United States

[1510]Holsinger, John Robert, 1934–, 1978, Systematics of the subterranean Amphipod genus *Stygobromus* (Rangonyctidae), part II: species of the eastern United States

[1534]Hovey, Edmund Otis, 1801–1877, 1902, Great American caverns. The underground beauties and wonders of Mammoth, Wyandotte, Luray, and other famous caves ...

[1535]Hovey, Horace Carter, 1833–1914, 1970, Celebrated American caverns: especially Mammoth, Wyandot, and Luray: together with historical, scientific, and descriptive notices of caves and grottoes in other lands

[1536]Hovey, Horace Carter, 1833–1914, 1882, Celebrated American caverns: especially Mammoth, Wyandot, and Luray: together with historical, scientific, and descriptive notices of caves and grottoes in other lands

[1690]Jacobson, Donald, 1986, Caving: an introductory guide to spelunking

[1937]Lewis, Charles C. (ed), 1993, Chronicles of the Reading Grotto, in which we go to the California convention, June 3rd to September 6th, 1966

[2103]Mazonowicz, Douglas, 1974, Voices from the stone age: a search for cave and canyon art

[2110]McClurg, David Robert, 1929–, 1973, The amateur's guide to caves & caving; skill-building ways to finding and exploring the underground wilderness

[2249]Mohr, Charles E., 1907–(ed), 1955, Celebrated American caves

[2355]National Speleological Society (comp), 1951, America's wonderland of caves: palaces under the earth: a directory of commercially operated caves

[2412]National Speleological Society. Annual Convention. Guidebook (1991: Cobleskill, NY), 1991, Guide to the caves and karst of the Northeast

[2505]Newton, John G., 1929–, 1987, Development of sinkholes resulting from man's activities in the eastern United States

[2633]Peck, Dallas L., Oct 1988, Karst hydrogeology in the United States of America: 21st congress of the International Association of Hydrogeologists: karst hydrogeology and karst environment protection October 1988 Guilin, China

[2703]Powell, Richard Lewis, 1936–, Dec 1970, A guide to the selection of limestone caverns and springs in the United States as national landmarks: for the National Park Service

[2832]Robison, Mabel Otis, 1959, Mystic wonderlands

[2958]Shear, William A., 1969, A synopsis of the cave millipeds of the United States: with an illustrated key to genera

[2983]Silverman, Sharon Hernes, 1991, Going underground: your guide to caves in the Mid-Atlantic

[3003]Sloane, Bruce, 1935–, 1977, Cavers, caves, and caving

[3004]Sloane, Howard N. (d. 1972), 1966, Visiting American caves

[3089]Speece, Jack Howard, 1947–(ed), 1979, Special convention issue: Symposium on the History of American Caving, Pittsfield, Mass., August 5-12, 1979

[3144]Storey, James Welborn, 1935–1992 (ed), 1965, American caving, illustrated; caving, climbing, camping

[3151]Stringfield, Victor Timothy, 1902–, 1964, Hydrology of limestone terranes in the coastal plain of the southeastern states

[3157]Sullivan, Gerardus Nicholas, 1927–, Jul 1960, Appendix: checklist of macroscopic troglobitic organisms of the United States

[3225]Traister, Robert J., 1983, Cave exploring

[3305]U.S. National Park Service, Dec 1994, National Cave and Karst Research Institute Study: a draft report to Congress as required by Public Law 101-578 of November 15, 1990

[3420]Weaver, Herman Dwight, 1938–(ed), 1982, Great American show caves

[3467]White, M., 1978, Western Region speleo-education seminar program with abstracts

[3501]Williams, Toni Lewis (ed), 1981, Manual of U.S. cave rescue techniques

14.9.4.1 Regions

14.9.4.1.1 Appalachians

[1769]Kastning, Ernst H., 1944–, 1991, Appalachian Karst: proceedings of the Appalachian karst symposium: Radford, Virginia, March 23-26, 1991
[2416]National Speleological Society. Annual Convention. Guidebook (1995: Blacksburg, VA), 1995, Underground in the Appalachians: a guidebook for the 1995 Convention of the National Speleological Society, Blacksburg, Virginia, July 17-21, 1995
[2612]Parizek, R. R., Nov 1971, Hydrogeology and geochemistry of folded and faulted carbonate rocks of the central Appalachian type and related land use problems
[2651]The Pennsylvania State University, 1970, Field trip guidebook for the central Appalachian carbonate hydrology workshop, The Pennsylvania State University, University Park, Pennsylvania, May 21, 22, 1970

14.9.4.1.2 Black Hills

[1104]Fossen, Joyce, 1985, Caves of Paha Sapa, and other findings of Dr. Knows-It
[2384]National Speleological Society. Annual Convention. Guidebook (1962: [Black Hills], SD), 1962, A guide to the caves of the Black Hills, S. D
[2584]Owen, Luella Agnes, 1970, Cave regions of the Ozarks and Black Hills

14.9.4.1.3 New England

[2655]Perry, Clair Willard [Clay], 1887–1961, 1946, New England's buried treasure
[2657]Perry, Clair Willard [Clay], 1887–1961, 1939, Underground New England

14.9.4.1.4 Guadalupe Mountains

[979]Ely, Glen Sample. (photo, ed), 1990, A history of the Guadalupe Mountains and Carlsbad Caverns, 1840—1940
[1926]Lee, Willis Thomas, 1924, A visit to Carlsbad Cavern: recent explorations of a limestone cave in the Guadalupe Mountains of New Mexico reveal a natural wonder of the first magnitude

14.9.4.1.5 Ozarks

[792]Dake, Charles Laurence, 1883–, 1924, Subterranean stream piracy in the Ozarks
[1841]Kobert, Bill, nd, Caverns of the Ozarks
[2547]Oklahoma Department of Wildlife Conservation, 1990, Hydrogeology of Ozark cavefish caves
[2584]Owen, Luella Agnes, 1970, Cave regions of the Ozarks and Black Hills
[2676]Pinkham, Mary R., 1954, From the cradle to the cave: the life story of "Dad" Truitt, "Cave Man of the Ozarks"
[3077]Southwest Missouri State University Dept. of Geography and Geology, 1979, Speleology workshop
[3361]Vickery, Margaret Ray, 1960, Ozark stories of the Upper Current River
[3424]Weaver, Herman Dwight, 1938–, 1992, The wilderness underground: caves of the Ozark plateau
[3504]Willis, L. D., Dec 1986, A recovery plan for the Ozark cavefish (*Amblyopsis rosae*)

14.9.4.2 States

14.9.4.2.1 Alabama

[36]Aley, Thomas John, 1938–, Feb 1990, Delineation and hydrogeologic study of the key cave aquifer Lauder Dale County, Alabama
[165]Anon, c1969, North Alabama's caves & caverns
[533]Broyles, Bettye J., 1958, Russell Cave in northern Alabama
[586]Camberlin, G. Will, 1993, The exploration and survey of McBridge Cave, Jackson County, Alabama
[636]Chandler, Robert V., 1987, Springs in Alabama
[708]Cooper, John Edward, 1929–, May 1982, Recovery plan for the Alabama cavefish, *Speoplatyrhinus poulsoni* Cooper and Kuehne 1974
[801]Daniel, Thomas W., 1973, Exploring Alabama caves
[1261]Griffin, John Wallace, 1978, Investigations in Russell Cave: Russell Cave National Monument, Alabama
[1465]Hobbs, Horton Holcombe, 1914–, 1967, A new crayfish from Alabama caves with notes of the origin of the genera *Orcnectes* and *Cambaras* (Decapoda: Astracidae)
[1468]Hobbs, Horton Holcombe, 1944–, Jan 1994, Assessment of the ecological resources of the caves of Russell Cave National Monument, Jackson County, Alabama and of selected caves at the Lookout Mountain unit of Chickamanga-Chattanooga National Military Park, Dade County, Georgia and Hamilton County, Tennessee
[1476]Hoffman, Richard L., 1956, New genera and species of cavernicolous diplopods from Alabama
[1477]Hoffmann, Paul, 1974, The caves of north central Morgan County, Alabama: a guide for the 1974 SERA Winter Business Meeting of the National Speleological Society at Decatur, Alabama, February 1974
[1478]Hoffmann, Paul R., Mar 1975, Newsome sinks - our national landmark
[1576]Huntsville Grotto of the National Speleological Society, 1967, 30 favorite Alabama caves

14.9. NORTH AMERICA

[1577]Huntsville Grotto of the National Speleological Society, May 1979, Cave preservation and residential development: a compatible approach

[1742]Joiner, Thomas J., 1969, Hydrology of limestone terranes: geophysical investigations

[1753]Jones, Walter Bryan, 1895–1977, 1968, Caves of Madison County, Alabama

[1754]Jones, Walter Bryan, 1895–1977, 1960, Caves of Morgan County, Alabama

[1810]Kerbo, Ronal Carrel, 1944–, Mar 1994, Cave resources management, planning, restoration, and redevelopment, Cathedral Caverns State Park, Grant, Alabama, 10-14 January 1994

[2050]Marsh, Dorothy, 1980, Life at Russell Cave

[2124]McGill, Greg, 1987, Alabama caves

[2180]Miller, Carter H., 1984, Preliminary seismic-velocity and magnetic studies of a carbonate rock-sinkhole area in Shelby County, Alabama

[2265]Moore, James D., 1975, Hydrology of limestone terranes: quantitative studies

[2332]Myrick, Donal Richard, 1938–, 1972, Fern cave: the history of the discovery, exploration and mapping of the Fern Cave system

[2389]National Speleological Society. Annual Convention. Guidebook (1967: Birmingham and Huntsville, AL), 1967, The caves of Alabama: a guide for the 1967 convention of the National Speleological Society at Birmingham and Huntsville, Alabama, June 1967

[2410]National Speleological Society. Annual Convention. Guidebook (1989: Sewanee, TN), 1989, Caves and caving in TAG: a guidebook for the 1989 Convention of the National Speleological Society ; Sewanee, Tennessee

[2506]Newton, John G., 1929–, 1973, Sinkhole problem along proposed route of interstate highway 459 near Greenwood, Alabama

[2507]Newton, John G., 1929–, 1971, Sinkhole problem in and near Roberts Industrial subdivision, Birmingham, Alabama: a reconnaissance

[2539]Oakley, Carey B., 1990, Archaeological testing at site 1Ms357, Cathedral Caverns, Marshall County, Alabama

[2810]Rheams, Karen F., 1994, Hydrogeologic and biologic factors related to the occurrence of the Alabama cave shrimp (*Palaemonias alabamae*), Madison County, Alabama

[2816]Rickels, Curtis Eddie, 1992, The three ring circus: how Rickwood Caverns became a state park

[3028]Smith, Marion Otis, 1942–(comp), 1992, Letters from TAG 1966-1969

[3030]Smith, Marion Otis, 1942–, c.1973, TAG pits

[3049]Sonderegger, John L., 1970, Hydrology of limestone terranes: geologic investigations

[3050]Sonderegger, John L., 1970, Hydrology of limestone terranes: photogeologic investigations

[3065]Southeastern Regional Association (S.E.R.A.) Cave Carnival (1976: Guntersville, AL), 1976, SERA 1976 guidebook

[3068]Southeastern Regional Association (S.E.R.A.) Cave Carnival (1983: AL), Aug 1983, 1983 SERA cave carnival: 5-7 August 1983

[3069]Southeastern Regional Association (S.E.R.A.) Cave Carnival (1989: [TN]), [1989], SERA 89

[3070]Southeastern Regional Association (S.E.R.A.) Cave Carnival (1990: Goose Pond Campground, Scottsboro, AL), 1990, SERA cave carnival, 1990

[3071]Southeastern Regional Association (S.E.R.A.) Cave Carnival (1992: AL), [1992], SERA cave carnival 1992

[3073]Southeastern Regional Association (S.E.R.A.) Cave Carnival (1995: Lafayette, GA), [1995], 1995 SERA summer cave carnival: the party on the farm: Smokey and Tina Cadwell's farm: Lafayette, Georgia May 19-21, 1995

[3074]Southeastern Regional Association (S.E.R.A.) Cave Carnival (1996: Camp Jackson, Jackson Co. AL), 1996, 45th Annual SERA Summer cave carnival: July 19-21, 1996, Camp Jackson, Jackson County, Alabama

[3154]Stringfield, Victor Timothy, 1902–, 1974, Karst and paleohydrology of carbonate rock terranes in semiarid and arid regions with a comparison to humid karst of Alabama

[3164]Sweatman, Cheyenne (ed), Oct 1985, 8th Annual TAG fall cave-in

[3183]Tarkington, Terry Warren, 1925–, Jun 1965, Alabama caves

[3343]Valentine, Joseph Manson, 1902–, Nov 1952, New genera of Anophthalmid beetles from Cumberland Caves

[3352]Varnedoe, William Whitfield, 1923–, 1973, Alabama caves and caverns

[3353]Varnedoe, William Whitfield, 1923–, Mar 1981, Alabama caves, 1980

[3354]Varnedoe, William Whitfield, 1923–, Jan 1972, Interim Alabama cave survey: report number 1

[3355]Varnedoe, William Whitfield, 1923–, 1975, Interim report number 2: Alabama cave survey

[3401]Warren, William M., 1945–, 1976, Sinkhole occurrence in western Shelby County, Alabama

14.9.4.2.2 Alaska

[39]Aley, Thomas John, 1938–, Nov 1993, Karst and cave resource significance assessment: Ketchikan Area, Tongass National Forest, Alaska

[876]Dixon, E. James, Apr 1981, 1980 progress report of archeological reconnaissance and testing of Pleistocene cave and alluvial deposits, Porcupine River, Alaska

[877]Dixon, E. James, Jan 1980, Report of 1979 archeological and geological reconnaissance and testing of cave deposits, Porcupine River, Alaska

[1559]Hrdlicka, Ales, 1869–1943, 1941, Exploration of

mummy caves in the Aleutian Islands ...
[1901]Larsen, Helge, 1968, Trail Creek: final report on the excavation of two caves on Seward Peninsula, Alaska
[2889]Sattler, Robert A., Apr 1989, Quantitative analysis of a late-Quaternary cave deposit, Porcupine River, Alaska

14.9.4.2.3 Arizona

[467]Breternitz, David A., 1969, Archaeological investigations in Turkey Cave (NA2520) Navajo National Monument, 1963
[473]Bridgemon, Rondal Rex, 1944–, 1976, Wupatki National Monument earth cracks
[543]Bryan, Kirk, nd, The geology and fossil vertebrates of Ventana Cave
[682]Collier, Don (host), 1990, Life underground: caves, mines, minerals
[880]Doe, Michael F., Aug 1985, Geology in the vicinity of Montezuma's Cave, southern Huachuca Mountains, Cochise County, Arizona: final [report]
[1008]Euler, Robert C., 1978, Archaeological and paleobiological studies at Stanton's Cave, Grand Canyon National Park, Arizona; a report of progress
[1009]Euler, Robert C. (ed), 1984, The Archaeology, geology, and paleobiology of Stanton's Cave: Grand Canyon National Park, Arizona
[1053]Fewkes, Jesse Walter, 1850–1930, 1896, Preliminary account of an expedition to the cliff villages of the red rock country, and the Tusayan ruins of Sityatki and Awatobi, Arizona, in 1895
[1129]Fulton, William Shirley, 1941, A ceremonial cave in the Winchester Mountains, Arizona
[1196]Gifford, James C., 1973–, 1980, Archaeological explorations in caves of the Point of Pines region, Arizona
[1277]Guernsey, Samuel James, 1868–1936, 1921, Basket-Maker caves of northeastern Arizona; report on the explorations, 1916-17
[1278]Guernsey, Samuel James, 1868–1936, 1931, Explorations in northeastern Arizona; report on the archaeological fieldwork of 1920-1923
[1379]Hassemer, Jerry Herman, 1934–, Mar 1980, Cave development in Marble Canyon, Grand Canyon National Park, Arizona
[1564]Hudgens, Bruce R., 1975, The archeology of Exhausted Cave: a study of prehistoric cultural ecology on the Coconino National Forest, Arizona
[1705]Janetski, Joel C., 1983, An archaeological and geological assessment of Antelope Cave (NA 5507), Mohave County, Northwestern Arizona
[1897]Lange, Arthur L., 1957, Studies on the origin of Montezuma Well and cave, Arizona
[1954]Lockett, H. Claiborne, 1953, Woodchuck Cave: A basketmaker II site in Tsegi Canyon, Arizona
[2064]Martin, Paul Sidney, 1899–, 1954, Caves of the Reserve area
[2137]Mead, Jim I., Nov 1984, The late Wisconsinan vertebrate fauna from Deadman Cave, southern Arizona
[2886]Sartor, James Doyne, 1962, Meteorological-geological investigations of the Wupatki blowhole system
[2995]Skinner, Morris F., 1942, The fauna of Papago Springs Cave, Arizona, and a study of Stockoceros; with three new Antilocaprines from Nebraska and Arizona
[3112]Steen, Charlie R., 1942, Ruins stabilization records for Canyon de Chelly National Monument 1942: Antelope House and Mummy Cave Ruins
[3154]Stringfield, Victor Timothy, 1902–, 1974, Karst and paleohydrology of carbonate rock terranes in semiarid and arid regions with a comparison to humid karst of Alabama

14.9.4.2.4 Arkansas

[44]Aley, Thomas John, 1938–, 1987, Water quality protection studies, Logan Cave, Arkansas: final report
[516]Brown, Arthur V., 1984, Cavefish (*Amblyopsis rosae*) in Arkansas: populations, incidence, habitat requirements, and mortality factors: a final report
[517]Brown, Barnum, 1873–1863, 1908, The Conard fissure, a Pleistocene bone deposit in northern Arkansas: with descriptions of two new genera and twenty new species of mammals
[738]Crawford, Nicholas Charles, 1942–, 1982, Guidebook to karst and caves of Tennessee: emphasis on the Cumberland plateau escarpment region and guidebook to karst and caves of the Ozark Region of Missouri and Arkansas: prepared for the Eighth International Congress of Speleology, Bowling Green, Kentucky, U. S. A. July 18 to 24 1981
[870]Dickson, Don R., 1991, The Albertson site: a deeply and clearly stratified Ozark bluff shelter
[1067]Fleming, John, 1973, Blanchard Springs Caverns story
[1315]Haley, Boyd Raymond, 1922–, 1980, Mineral resources of the Belle Starr Caves wilderness study area, Sebastian and Scott Counties, Arkansas
[1794]Kent, Paul, 1983, Tour caves of the Ozarks: Missouri Arkansas: the secret beneath the surface
[1945]Lindsley, R. P. (ed), 1977, Survey and assessment of cave resources at Buffalo National River, Arkansas
[2204]Missouri Speleological Survey, July-December 1982, Caves of the Current River Valley
[2226]Missouri Speleological Survey, January-June 1986, Remains of Quaternary vertebrates from Ozark caves and other miscellaneous sites
[2309]MVOR (Mississippi Valley Ozark Regional). Guidebook (1968), 1968, Fall 1968 MVOR guidebook: Stone County, Arkansas
[2584]Owen, Luella Agnes, 1970, Cave regions of the

Ozarks and Black Hills

[2588]Ozark National Forest (AR), 1980, Blanchard Springs Caverns: the amazing world below
[3276]U.S. Forest Service, 1973, Operation of Blanchard Springs Caverns, Ozark-St. Francis National Forest, Arkansas: final environmental statement
[3279]U.S. Forest Service, 1964, A preliminary plan for the development of Blanchard Springs Cavern: Sylamore District, Ozark National Forest
[3280]U.S. Forest Service, Southern Region, 1975, Construction of phases II and III of the Blanchard Springs Caverns Project final environmental statement
[3281]U.S. Forest Service, Southern Region, 1974, Construction of phases II and III of the Blanchard Springs Caverns Project: draft environmental statement
[3282]U.S. Forest Service, Southern Region, 1973, Welcome to Blanchard Springs Caverns
[3547]Youngsteadt, Norman W., Oct 1978, A survey of some cave invertebrates from northern Arkansas

14.9.4.2.5 California

[90]Anon, 1988, Cave paintings of the Chumash Indians
[229]Arnold, Charles L., 1986, Inside the caves: Lava Beds National Monument
[235]Aslett, James L., 1982, Lava beds underground
[478]Briggs, Thomas S., 1974, Phalangodidae from caves in the Sierra Nevada (California) with a redescription of the type genus (Opiliones, Phalangodidae)
[555]Bunnell, David Edward, 1952–, 1993, Sea caves of Anacapa Island
[556]Bunnell, David Edward, 1952–, 1988, Sea caves of Santa Cruz Island
[572]Byer, Charles A., 1938, The caves of Capistrano
[578]California. Department of Parks and Recreation, 1971, Mitchell Caverns State Reserve
[631]Chamberlin, Ralph Vary, 1879–1967, 1930, A new geophiloid chilopod from Potter Creek Cave, California
[859]de Saussure, Raymond, Mar 1953, Report of the California-Nevada Speleological Survey June 13-September 7, 1952
[861]Despain, Joel, 1995, Crystal cave: a guidebook to the underground world of Sequoia National Park
[895]Downs, Theodore, 1919–, 1959, Quaternary animals from Schuiling Cave in the Mojave Desert, California
[964]Elliott, William Rawleigh, 1946–, 1978, The new Melones Cave harvestman transplant
[978]Elsasser, Albert B., 1963, The archaeology of Bowers Cave, Los Angeles County, California
[1073]Floyd, E. F., 1887, The New Calaveras Cave, Murphys, Calaveras County, California: W.J. Mercer, proprietor: description and guide
[1132]Furlong, Eustace Leopold, 1874–, 1906, The exploration of Samwel Cave
[1324]Halliday, William R., 1926–, 1962, Caves of California: a special report of the Western Speleological Survey in cooperation with the National Speleological Society, June 1962
[1336]Halliday, William R., 1926–, 1961, A preliminary report on the caves of eight northern California counties
[1357]Hardcastle, Raymond A., Apr 1977, Caves and shelters in Bonanza King Canyon
[1423]Heizer, Robert Fleming, 1915–, 1942, Massacre Lake Cave, Tule Lake Cave and shore sites
[1567]Hudson, Travis, 1982, Guide to Painted Cave
[1583]Hurtt, Howard A., 1977, Rescue system outline: Lilburn Cave Project
[1740]Johnston, David Alexander, 1949-1980, 1981, Guide to some volcanic terranes in Washington, Idaho, Oregon and Northern California
[1825]Klamath Mountains Conservation Task Force, 1977, A preliminary report on the caves of the Marble Mountain Wilderness
[1882]Lamb, Susan, 1991, Lava Beds National Monument
[1923]Lee, G., 1988, The rock art of Petroglyph Point and Fern Cave, Lava Beds National Monument: final report
[2016]Madacheen, L. Michael, 1977, New Melones cave inventory and evaluation study: preliminary report: archeological caves
[2095]Maturango Museum of Indian Wells Valley, 1974, Excavation of two sites in the Coso Mountains of Inyo County, California
[2120]McEachern, J. Michael, 1978, An inventory and evaluation of the cave resources to be impacted by the New Melones Reservoir Project, Calaveras and Tuolumne Counties, California
[2155]Mercer, W. J., 1887, The New Calaveras Cave, Murphys, Calaveras County, California: W.J. Mercer, proprietor: description and guide
[2160]Merriam, John Campbell, 1869–1945, 1906, Recent cave explorations in California
[2238]Mitchell, Jack, 1964, Jack Mitchell, caveman
[2388]National Speleological Society. Annual Convention. Guidebook (1966), 1966, Caves of the Sequoia Region California, issued in conjunction with the 25th anniversary convention of the National Speleological Society, David R McClurg, convention chairman
[2397]National Speleological Society. Annual Convention. Guidebook (1975: Frogtown, CA), 1975, National Speleological Society 1975 Convention, Frogtown, Ca
[2411]National Speleological Society. Annual Convention. Guidebook (1990: Yreka, CA), 1990, NSS 1990 convention guidebook: Yreka, California: "where the lava meets the limestone"
[2423]National Speleological Society. Annual Convention. Guidebook. Geology Field Trip (1990: Yreka, CA), 1990, Geologic features of Mount Shasta and Medicine

Lake Volcanoes

[2540]Oberhansley, Frank R., May , Crystal Cave in Sequoia National Park

[2576]Orr, Phil C., 1952, Excavations in Moaning Cave

[2620]Payen, Louis A., 1963, Preliminary report on the archeological investigation of Pinnacle Point Cave, Toulumne [sic] County, California

[2737]Quick, Dell (ed), 1981, SCG history: summer 1981

[2806]Rentz, David C., 1972, A new genus and species of camel cricket from the Farallon Islands of California (Orthoptera: Gryllacrididae)

[2817]Riddell, Francis A., 1951, The archaeology of two Kern County sites

[2846]Rogers, Bruce W., 1970, A preliminary report on Vanished River Cave, Santa Cruz

[2957]Shear, William A., 1974, North American cave millipeds II, an unusual new species (Dorypetalidae) from Southern California, and new records of *Speodesmus tuganbius* (Trichopolydesmidae) from New Mexico

[2981]Silver, Constance S., Mar 1982, The pictographs of Fern Cave, Lava Beds National Monument: agents of deterioration and prospects for conservation

[2989]Sinclair, William John, 1877–1935, 1904, Euceratherium: a new ungulate from the Quaternary caves of California

[2990]Sinclair, William John, 1877–1935, 1905, New mammalia from the Quaternary caves of California

[3011]Smith, Allyn Goodwin, 1957, Snails from California caves

[3018]Smith, Gerald Arthur, 1915–, 1957, The archaeology of Newberry Cave, San Bernardino County, Newberry, California

[3019]Smith, Gerald Arthur, 1915–, 1955, Preliminary report of the Schuiling Cave, Newberry, California

[3079]Sowers, J. M., 1990, Cave management plan and environmental assessment, Lava Beds National Monument, California

[3098]Squire, Ralph E., Nov 1971, Report of study by National Speleological Society, Cave Conservation Task Force: New Melones Project

[3099]Squire, Ralph E., 1972, Stanislaus cave country: report of study

[3228]Treganza, Adan Eduardo, 1916–1968, 1964, An ethno-archaeological examination of Samwel Cave

[3263]U.S. Bureau of Land Management, Barstow Resource Area, 1982, Shoshone Cave (Whip-scorpion Habitat) wildlife habitat management plan

[3332]University of California Archaeological Survey, 1952, Papers in California archaeology: 17-18

[3333]University of California Archaeological Survey, 1955, Papers on California archaeology: 30-31

[3334]University of California Archaeological Survey, 1955, Papers on California archaeology: 32-33

[3335]University of California Santa Barbara, Art Galleries, 1965, Prehistoric rock art of the Santa Barbara region: [exhibition] Art Gallery, University of California, Santa Barbara, October 12-Nov. 7, 1965

[3404]Waters, Aaron Clement, 1905–, 1990, Selected caves and lava tube systems in and near Lava Beds National Monument, California

[3455]Whealdon, Everett, 1983, The legend of Parker's Cave

[3478]Whitfield, Henry C., Jul 1963, King Caverns report

[3516]Wintner, Radar H., 1977, Lake Shasta Caverns

14.9.4.2.6 Colorado

[311]Bancroft, Hubert Howe, 1832–1918, 1886, Manitou Grand Caverns

[352]Baysinger, David (narr, videographer), 1986, Silent Splendor

[1212]Gilmore, Elizabeth (prod, ed, auth), 1986, Silent Splendor

[1382]Hassemer, Jerry Herman, 1934–(ed, assemb), Feb 1975, Colorado caves and karst 1974

[1383]Hassemer, Jerry Herman, 1934–(ed, assemb), Apr 1977, Colorado caves and karst 1975-1976

[2417]National Speleological Society. Annual Convention. Guidebook (1996: Salida, CO), 1996, The caves and karst of Colorado: a guidebook for the 1996 convention of the National Speleological Society

[2418]National Speleological Society. Annual Convention. Guidebook (1997: Sullivan, MO), 1997, Exploring Missouri caves. A guidebook for the 1997 convention of the National Speleological Society Sullivan, Missouri June 23-27, 1997

[2526]Norton, John S., 1990, The Cave of the Winds

[2535]Nusbaum, Jesse Logan, 1981, The 1926 reexcavation of Step House Cave, Mesa Verde National Park

[2615]Parris, Lloyd E., 1973, Caves of Colorado

[2798]Reed, Alan D., 1988, A stabilization assessment of Mantel's Cave, site 5MF1, Dinosaur National Monument, Colorado

[2835]Rocky Mountain Region. Guidebook (1994), nd, Caves of the eastern White River Plateau: guidebook to the 1994 Rocky Mountain regional, Willow Peak, Colorado

[3039]Snider, George Washington, 1888, Manitou Grand Caverns, Manitou Springs, Colorado

[3040]Snider, George Washington, 1885, The tourists' gem: the Manitou Grand Caverns, the largest and most wonderful subterranean in the Rocky Mountains, and other attractions for tourists

[3255]Turner, Percy (auth, comp), 1895, History of Manitou's caves, the Grand Caverns and Cave of the Winds: and points of interest in and about Manitou

[3457]Whelchel, Sandy, 1985, A day at the cave

[3495]Williams, James Henry, 1843–, 1914, Legend of the

Cave of the Winds

14.9.4.2.7 Connecticut

[3539]Yale Speleological Society, 1963, The caves of Connecticut

14.9.4.2.8 Delaware

[3084]Speece, Jack H., 1977, The cave of Delaware
[3179]Talley, John H., Mar 1981, Sinkholes, Hockessein area, Delaware

14.9.4.2.9 Florida

[71]Anderson, Warren, 1975, Hydrology of three sinkhole basins in Southwestern Seminole County Florida
[355]Beck, Barry Frederic, 1944–(ed), Jul 1984, A computer based inventory of recorded recent sinkholes in Florida
[359]Beck, Barry Frederic, 1944–(ed), Oct 1985, Karst hydrogeology of central and northern Florida: a field trip guidebook produced in conjunction with the 1985 G.S.A. annual meeting and exposition, Orlando, FL., October, 1985
[361]Beck, Barry Frederic, 1944–(ed), 1991, The sinkhole hazard in Pinellas County: a geological summary for planning purposes
[362]Beck, Barry Frederic, 1944–(ed), 1986, Sinkholes in Florida: an introduction
[364]Beck, Barry Frederic, 1944–(ed), 1987, Use of ground penetrating radar for detecting and evaluating the sinkhole hazard in Florida
[365]Beck, Barry Frederic, 1944–(ed), 1985, Water on and under the ground: an introduction to the urban hydrogeology of the Orlando area
[388]Benson, Richard C., 1979, Application of radar and seismic reflection techniques to cavity detection, Medford Cave site, Florida
[570]Butler, Dwain K., Mar 1983, Cavity detection and delineation research: Report 1: Microgravimetric and magnetic surveys: Medford Cave Site, Florida
[571]Butler, Dwain K., 1983, Cavity detection and delineation research: Report 4: Microgravimetric survey: Manatee Springs site, Florida
[710]Cooper, Stafford S., 1983, Cavity detection and delineation research: Report 3: Acoustic resonance and self-potential applications: Medford Cave and Manatee Springs sites, Florida
[783]Curro, Joseph R., Jun 1963, Cavity detection and delineation research; Prepared for Office, Chief of Engineers. Report 2: Seismic methodology: Medford Cave Site, Florida
[852]DeLoach, Ned (ed), 1978, Ned DeLoach's diving guide to underwater Florida
[909]Duff, B. M., 1980, Ground penetrating electromagnetic tests at Medford Cave, Florida and Waterways Experiment Station, Vicksburg, Mississippi
[1048]Ferguson, George E., 1947, Springs of Florida
[1072]Florida Cave Survey, 1962, Cave locations listing
[1108]Fountain, Lewis S., Jul 1980, Earth resistivity and hole-to-hole electromagnetic transmission tests at Medford Cave, Florida
[1236]Graham, Joseph J., 1939, The Florida Caverns at Marianna, Florida
[1704]Jammal & Associates, Inc, Mar 1982, The Winter Park sinkhole: a report to the city of Winter Park, Florida
[1894]Lane, Edward, 1935–, 1991, Environmental geology and hydrogeology of the Ocala area, Florida
[1895]Lane, Edward, 1935–, 1986, Karst in Florida
[2017]Maddox, Gary L., Nov 1993, Karst features of Florida Caverns State Park and Falling Waters State Recreation Area, Jackson and Washington Counties, Florida
[2055]Marshall, Wayne (ed), 1989, Sixth annual Florida caver's re-union (Cave Cavort): April 7-9, 1989, Hog Island Campground, Withlacoochee State Forest
[2259]Moody, Larry D., Nov 1977, Warrens Cave Reserve stewardship plan: Gainesville, Alachua County, Florida
[2904]Schiffer, Donna M., 1996, Hydrology of the Wolf Branch sinkhole basin, Lake County, east-central Florida
[2987]Sinclair, William C., 1985, Types, features, and occurrence of sinkholes in the karst of west-central Florida
[2988]Sinclair, William Campbell, 1928–, 1982, Sinkhole development resulting from ground-water withdrawal in the Tampa area, Florida
[3141]Stone, William Curtis, 1952–(ed), 1989, The Wakulla Springs project
[3142]Stoneman, John (narr), 1990, The cave divers
[3241]Trommer, J. T., 1987, Potential for pollution of the Upper Floridan aquifer from five sinkholes and an internally drained basin in west-central Florida
[3359]Vernon, Robert Orion, 1912–, 1950, Florida Caverns: a nature-made wonderland
[3452]Wetherhall, W. S., 1965, Reconnaissance of springs and sinks in west-central Florida
[3512]Wilson, William L., Jul 1987, Investigation of karst-related subsidence near the southwest landfill, Alachua County, Florida, using ground penetrating radar: a report

14.9.4.2.10 Georgia

[357]Beck, Barry Frederic, 1944–(ed), 1980, An introduction to caves and cave exploring in Georgia
[637]Chapman, J. Roy, 1963, The caves of Georgia
[881]Dogwood City Grotto of the National Speleological Society, Jul 1974, Pigeon Mountain and Ellison's Cave system: a report to the Georgia Department of Natural

Resources by the Dogwood City Grotto of the National Speleological Society, Post Office Box 12072, Atlanta, Georgia 30305

[1182]Georgia Speleological Survey, Oct 1991, Georgia Speleological Survey mapbook

[1468]Hobbs, Horton Holcombe, 1944–, Jan 1994, Assessment of the ecological resources of the caves of Russell Cave National Monument, Jackson County, Alabama and of selected caves at the Lookout Mountain unit of Chickamanga-Chattanooga National Military Park, Dade County, Georgia and Hamilton County, Tennessee

[2410]National Speleological Society. Annual Convention. Guidebook (1989: Sewanee, TN), 1989, Caves and caving in TAG: a guidebook for the 1989 Convention of the National Speleological Society ; Sewanee, Tennessee

[2909]Schreiber, Richard, Sep 1969, Ellison's Cave: Georgia's finest

[3027]Smith, Marion Otis, 1942–, 1977, The exploration & survey of Ellison's Cave, Georgia

[3028]Smith, Marion Otis, 1942–(comp), 1992, Letters from TAG 1966-1969

[3030]Smith, Marion Otis, 1942–, c.1973, TAG pits

[3036]Smith, William Hovey, 1983, Geology of the Tennille Lime Sinks, Washington County, Georgia ; with an introduction to local geology

[3069]Southeastern Regional Association (S.E.R.A.) Cave Carnival (1989: [TN]), [1989], SERA 89

[3071]Southeastern Regional Association (S.E.R.A.) Cave Carnival (1992: AL), [1992], SERA cave carnival 1992

[3073]Southeastern Regional Association (S.E.R.A.) Cave Carnival (1995: Lafayette, GA), [1995], 1995 SERA summer cave carnival: the party on the farm: Smokey and Tina Cadwell's farm: Lafayette, Georgia May 19-21, 1995

[3076]Southwest Georgia Planning & Development Commission, 1968, Glory Hole Caverns

[3164]Sweatman, Cheyenne (ed), Oct 1985, 8th Annual TAG fall cave-in

[3222]Torode, William Wallace, 1943–(comp), nd, Ellison's Cave: Georgia

14.9.4.2.11 Hawaii

[242]Athens, J. Stephen, Aug 1989, Prehistoric upland bird hunters: archaeological inventory survey and testing for the MPRC project area and the Bobcat Trail Road, Pohakuloa Training Area, Island of Hawaii

[432]Bonk, William J., 1969, Lua Nunu o Kamakalepo: a cave of refuge in Kau, Hawaii

[665]Cleghorn, Paul L., Feb 1980, The Hilina Pali petroglyph cave, Hawai'i island: a report on preliminary archaeological investigations

[709]Cooper, K. M., [1992], Morphology of extinct lava tubes and the implications for tube evolution, Chain of Craters Road, Hawaii Volcanoes National Park, Hawaii

[832]Davis, Bertell D., 1980, Use and abandonment of habitation caves in the prehistoric settlement of southeastern Oahu: a proposed research design for the 1980 University of Hawaii Archaeological Field Program

[1138]Gagné, Wayne C., 1974, The cavernicolous fauna of Hawaiian lava tubes, part VI: Mesoveliidae or water treaders (Heteroptera)

[1139]Gagné, Wayne C., Jul 1974, The cavernicolous fauna of Hawaiian lava tubes, part VII. Emesinae or thread-legged bugs (Heteroptera: Redvuiidae)

[1251]Greeley, Ronald, 1971, Observations of actively forming lava tubes and associated structures, Hawaii

[1322]Halliday, William R., 1926–, Dec 1992, Caves and associated features of open vertical volcanic conduits of the Kaupulehu lava flows xenolith nodule beds, Haualalai Volcano, Hawaii: basic speleological considerations

[1323]Halliday, William R., 1926–, Apr 1992, Caves and associated features of open vertical volcanic conduits of the Kaupulehu lava flows xenolith nodules beds, Hualalai Volcano, Hawaii: preliminary report

[1327]Halliday, William R., 1926–, Apr 1995, Considerations of Lower Uilani Cave for inclusion in the Hawaii Natural Area Reserve System

[1331]Halliday, William R., 1926–, Oct 1995, Initial inventory of named caves and related features and cave-related place names in Hawaii

[1332]Halliday, William R., 1926–, Aug 1991, Introduction to Hawaiian caves: field guide for the 6th international symposium on volcano-speleology: Hilo, HI

[1353]Hammatt, Hallett H., Apr 1990, Archaeological assessment and sensitivity map for the Pohakuloa training area (PTA), Hawaii Island, State of Hawaii

[1392]Hay, Deborah, Oct 1986, Kahaluu data recovery project: excavations at site 50-10-37-7702, Kahaluu Habitation Cave, land of Kahaluu, North Kona, Island of Hawaii

[1543]Howarth, Francis Gard, 1940–, 1972, Ecological studies on Hawaiian lava tubes

[2121]McEldowney, Holly, Oct 1991, Survey of lava tubes in the former Puna Forest Reserve and on adjacent State of Hawaii lands

[2186]Mills, Martin Taylor, 1974, The subterranean wonders of Hawaii

[3159]Sumia Konda, Jul 1993, The report of lava flow and lava caves on Mauna Loa

[3519]Wolfe, James E., 1975, Map location and dimensional definition of subsurface caverns

14.9.4.2.12 Idaho

[159]Anon, 1960, Montpelier and Bear Lake County

[963]Elliott, William Rawleigh, 1946–, 1976, New cavernicolous Rhagidiidae from Idaho, Washington, and Utah

(Prostigmata, Acari, Arachnida)
[1172]Gem State Grotto of the National Speleological Society, 1968, Caves of the gem state
[1274]Gruhn, Ruth1907–, 1961, The archaeology of Wilson Butte Cave south-central Idaho
[1580]Huppert, George Nixon, 1944–, 1981, Speleography of Papoose Cave, Idaho County, Idaho
[1667]Ireton, Frank, Jul 1973, Tee-Maze cave system, Lincoln County, Idaho, U.S.A
[1740]Johnston, David Alexander, 1949-1980, 1981, Guide to some volcanic terranes in Washington, Idaho, Oregon and Northern California
[2524]Northwest Caving Association. National Speleological Society (1992), 1992, 1992 Northwest Caving Association regional meeting cave guide [22-24 may 1992]
[2611]Papadakis, Peggy, May 1969, The story of Crystal Ice Caves: within the Great Rift National Landmark, American Falls, Idaho
[2683]Plew, Mark G., 1986, The archaeology of Nahas Cave: material culture and chronology
[2830]Robinson, E. Russell, 1932–1981, 1978, The story of the Shoshone Indian ice caves
[2857]Ross, Sylvia H., 1969, Introduction to Idaho caves and caving
[3449]Westcott, Richard L., Jun 1968, A new subfamily of blind beetle from Idaho ice caves, with notes on its bionomics and evolution (Coleoptera: Leiodidae)

14.9.4.2.13 Illinois

[468]Bretz, J Harlen, 1882–1981, 1961, Caves of Illinois
[566]Burrows, Russell, , Mystery cave of many faces: first in a series on the saga of Burrows' Cave
[2312]MVOR (Mississippi Valley Ozark Regional). Guidebook (1970), 1970, MVOR Spring "70" Guidebook
[2318]MVOR (Mississippi Valley Ozark Regional). Guidebook (1975), 1975, MVOR: Spring 1975 in the Shawnee Hills
[2324]MVOR (Mississippi Valley–Ozark Region). Guidebook (1989), 1989, MVOR Guidebook: 1989 Spring
[2516]North Country Region, 1979, North Country Region MECCA Summer 1979
[2594]Page, Lawrence M., 1977, Status of the cypres darter, *Etheostoma proeliare*, and comments on the spring cavefish, *Chologaster agassizi*, in Max Creek, Johnson County, Illinois
[3033]Smith, Philip Wayne, 1921–, 1978, A summary of the life history and distribution of the spring cavefish, *Chologaster agassizi* Putnam, with population estimates for the species in southern Illinois
[3201]Thomas, Harold S., 1953, The geology of Kankee Caverns [sic]: a geological report on the underground storage of LP-gas on the Phillips pipe line terminal, Kankakee-Illinois
[3396]Ward, John A., 1990, The people of Burrows Cave: who they were, where they came from and when
[3425]Webb, Donald W., May 1994, The biological resources of Illinois caves and other subterranean environments: determination of the diversity, distribution, and status of the subterranean faunas of Illinois caves and how these faunas are related to groundwater quality
[3426]Webb, Donald W., 1995, Status report on the cave Amphipod *Gammarus acherondytes* Hubricht and Mackin (Crustacea: Amphipoda) in Southern Illinois
[3463]White, John, 1973, Nature preserve potential of the Burton Cave area, Adams County, Illinois
[3464]White, John, 1973, Nature preserve potential of Twin Culvert Cave, Pike County, Illinois
[3465]White, John, 1973, Preservation of caves in Illinois
[3466]White, John, 1973, Preservation of Fogelpole Cave, Monroe County, Illinois

14.9.4.2.14 Indiana

[52]Allison, Harold, nd, Indiana caves and unique geological features
[79]Anon, nd, [Indiana cave list]
[314]Banta, Arthur Mangun, 1877–, Sep 1907, The fauna of Mayfield's Cave
[343]Bates, Edward R., Jun 1973, Detection of subsurface cavities
[350]Bayless, Edward Randall, 1961-, 1994, Directions of ground-water flow and locations of ground-water divides in the Lost River watershed near Orleans, Indiana
[413]Blatchley, William Stanley, 1859–1940, 1990, Gleanings from nature: ten Indiana caves and the animals that inhabit them
[414]Blatchley, William Stanley, 1859–1940, 1896, Indiana caves and their fauna
[625][Central Indiana Grotto of the National Speleological Society], Jun 1973, Selected caves of Orange county
[697]Cook, Holly (ed), , 5th annual Hog-Fest: Lickford Valley Campgrounds, Corydon, Indiana, May 1-5, 1992
[730]Cox, Edward Travers, 1821-1907, 1903, Wyandote Cave, pronounced by eminent scientists and noted writers of wide experience and extensive travel to be the greatest natural attraction of its kind in the world
[1175]George, Angelo Isham, 1944–, 1991, Bibliography of Wyandotte Cave
[1181]George, Angelo Isham, 1944–, 1991, Wyandotte Cave down through the centuries
[1462]Hill, John R., nd, The geologic story of Wyandotte Cave
[1595]Indiana Cave Capers. Guidebook (1971), 1971, 1971—18th Cave Cavers: location Bedford
[1596]Indiana Cave Capers. Guidebook (1974), 1974, 1974 Cave Capers guidebook: for Harrison & Crawford

counties

[1597] Indiana Cave Capers. Guidebook (1975), 1975, Guidebook: Cave Capers 1975
[1598] Indiana Cave Capers. Guidebook (1976), 1976, Guidebook: Cave Capers 1976
[1599] Indiana Cave Capers. Guidebook (1977), 1977, 24th annual CIG Cave Capers
[1600] Indiana Cave Capers. Guidebook (1978), 1978, Cave Capers 1978
[1601] Indiana Cave Capers. Guidebook (1979), 1979, Cave Capers 1979 guidebook: Milltown, Indiana
[1602] Indiana Cave Capers. Guidebook (1980), 1980, 27th annual Cave Capers: 1980 guidebook
[1603] Indiana Cave Capers. Guidebook (1983), 1983, Cave Capers, 1983
[1604] Indiana Cave Capers. Guidebook (1987), 1987, Wander Indiana underground: 1987 Cave Capers, Spring Valley
[1605] Indiana Cave Capers. Guidebook (1989), 1989, Indiana Cave Capers guidebook: Hickory Hill Campground, Owen County, Indiana July 14-16, 1989
[1606] Indiana. Dept of Natural Resources. Engineering Division, 1968, Master plan: Wyandotte Caves, Harrison-Crawford, Blue River Recreation Complex
[1607] Indiana. Division of Geology, 1939, Guide[s] to the Indiana caverns
[1682] Jackson, George Frederick, 1906-1981, 1972, The history and exploration of Wyandotte Cave
[1683] Jackson, George Frederick, 1906-1981, 1975, The story of Wyandotte Cave
[1684] Jackson, George Frederick, 1906-1981, 1953, Wyandotte Cave
[1969] Love, Douglas L., 1972, The spelunker's guide to the caves of the Garrison Chapel Valley
[1970] Love, Douglas P., 1977, The spelunker's guide to the caves north of Campbellsburg, Indiana
[1971] Love, Douglas P., 1973, The spelunker's guide to the caves North of Campbellsburg, Washington County, Indiana
[1998] Luvless, Dungass F. (pseud), 1972, The speleobopper's guide to the caves of the Gruesome Chapel Valley
[2026] Malott, Clyde Arnett, 1887–1950, 1970, Lost River at Wesley Chapel Gulf, Orange County, Indiana
[2027] Malott, Clyde Arnett, 1887–1950, 1974, Significant features of the Indiana karst
[2028] Malott, Clyde Arnett, 1887–1950, 1970, The swallow-holes of Lost River, Orange County, Indiana
[2129] McGregor, Duncan J., 1962, Some features of karst topography in Indiana
[2190] Mindick, Robert R., 1977, Commercial caves in Southern Indiana
[2191] Mindick, Robert R., 1977, Four non-commercial caves in Southern Indiana
[2266] Moore, Michael C., 1972, Southern Indiana Karst field trip guide, March 11 & 12, 1972
[2300] Munson, Patrick J., 1990, The prehistoric and early historic archaeology of Wyandotte Cave and other caves in southern Indiana
[2387] National Speleological Society. Annual Convention. Guidebook (1965: Bloomington, IN), [1965], National Speleological Society convention 1965 guidebook, Indiana University Bloomington, Indiana June 12-20, 1965
[2395] National Speleological Society. Annual Convention. Guidebook (1973: Bloomington, IN), 1973, NSS 73 convention guidebook; National Speleological Society convention, Bloomington, Indiana, June 16-24, 1973
[2413] National Speleological Society. Annual Convention. Guidebook (1992: Salem, IN), 1992, Caving in the heartland: a guidebook for the 1992 Convention of the National Speleological Society, Salem, Indiana
[2425] National Speleological Society. Annual Convention. Guidebook. Preconvention Field Trip (1973: Bloomington, IN), 1973, C. H. U. G. Crawford Harrison Underground Gang presents twenty-two favorite Harrison Crawford caves: 1973 pre-convention activities
[2426] National Speleological Society. Annual Convention. Postconvention Guidebook (1992: Salem, IN), 1992, 1992 post convention guidebook: enter the karst of Indiana, you may never want to come out
[2702] Powell, Richard Lewis, 1936–, Oct 1961, Caves of Indiana
[2734] Putnam, Frederic Ward, 1839–1915, 1875, Archaeological researches in Kentucky and Indiana, 1874
[2741] Quinlan, James Francis, 1936–1995, 1980, Ground-water hydrology and geomorphology of the Mammoth Cave Region, Kentucky, and of the Mitchell Plain, Indiana
[2811] Rhodes, Andrew Jackson, 1829–1907, 1973, Presenting the wonders of Lost River
[2812] Rhodes, Andrew Jackson, 1829–1907, 1905, The wonders of Lost River: 40 miles of underground river with its caves and hidden waterfalls, blind fish and other wonders
[2819] Riely, Samuel L., 1945, Story of Wyandotte Cave
[2824] Roberson, Gary, c1970, ISS report on Binkley's Cave system
[2825] Roberson, Gary T., c1989, Official souvenir book of Marengo Cave: U.S. National Landmark: Marengo, Indiana, the heart of cave country
[2860] Rothrock, H. W., 1889, Wyandotte Cave: where it is and what it is
[2865] Ruhe, R. V., Dec 1975, Geohydrology of karst terraine, Lost River watershed southern Indiana
[3116] Stelle, James Parish, 1884, The Wyandotte Cave of Crawford County, Ind
[3123] Stevens, Martin V. B., 1876, The history, guide and description of Wyandotte Cave

14.9. NORTH AMERICA

[3285]U.S. Geological Survey, 1980, Indiana cave areas: topographic quadrangle maps

14.9.4.2.15 Iowa

[384]Benn, David W., 1948–, 1980, Hadfields Cave: a perspective on Late Woodland culture in northeastern Iowa

[446]Bounk, Michael J., Oct 1983, Karstification on the Silurian Escarpment in Fayette County, northeastern Iowa

[1317]Hallberg, George R., May 1982, Sinkholes, hydrogeology, and ground-water quality in northeast Iowa

[1418]Hedges, James, 1975, The Ice Cave at Decorah, Iowa

[1445]Henry, Thomas, 1993, A guide to Maquoketa Caves State Park

[1666]Iowa. Department of Agriculture and Land Stewardship, Jul 1989, Agricultural drainage well research and demonstration project. Sinkhole/karst area demonstration project

[1780]Kehret, Roger Alan, 1942–, 1976, Crystal Lake Cave Iowa

[1784]Kehret, Roger Alan, 1942–, 1976, Your guide to Spook Cave

[1830]Knauth, Otto, 1975, Cold Water Cave

[1842]Koch, Donald L., Dec 1974, Report on Cold Water Cave: a summary of research results with inclusion of information on related to potential development of a new recreational facility by the state of Iowa

[1941]Libra, Bob, [1990], 1990 Friends of karst field trip: the Big Spring Basin, northeast Iowa

[2396]National Speleological Society. Annual Convention. Guidebook (1974: Decorah, IO), 1974, National Speleological Society 1974 convention Luther College, Decorah, Iowa 10-18 August, 1974

[2402]National Speleological Society. Annual Convention. Guidebook (1980: White Bear Lake, MN), Jul 1980, An introduction to caves of Minnesota, Iowa, and Wisconsin: guidebook for the 1980 National Speleological Society Convention, Lakewood Community College, White Bear Lake, Minnesota

14.9.4.2.16 Kansas

[177]Anon, 1990, Red hills

[1268]Grisafe, David A., Aug 1992, Stabilization of Dakota Sandstone surface of the Faris Cave petroglyphs, Kanopolis Lake Project, Kansas

[2515]Nodine-Zeller, Doris E., Aug 1991, Karst-derived early Pennsylvanian conglomerate in Ness County, Kansas: subsurface Mississippian-Pennsylvanian boundary delineated in well core

[3424]Weaver, Herman Dwight, 1938–, 1992, The wilderness underground: caves of the Ozark plateau

[3546]Young, James Jay, 1948–, 1993, Caves in Kansas

14.9.4.2.17 Kentucky

[23]Alderson, Laura White, 1984, Mammoth Cave: Kentucky's buried treasure

[99]Anon, 1980, The caves of Kentucky: Mammoth Cave, Colossal Cavern, White's Cave

[108]Anon, 1850, A description of the Mammoth Cave of Kentucky, the Niagara River and Falls, and the Falls in summer and winter; the prairies, or life in the West; the Fairmont Water Works and scenes on the Schuylkill, xc. xc.: to illustrate Brewer's Panorama

[109]Anon, 1849, A description of the Mammoth Cave of Kentucky, the Niagara river and falls, and the falls in summer and winter; the prairies, or life in the West; the Fairmount Water Works and scenes on the Schuykill, etc. etc.: to illustrate Brewer's panorama.

[111]Anon, 1920, Dossey Domes: most beautiful of all the caverns at Green River Landing (One-half mile from Mammoth Cave, Ky.) ; Cliff Walk along Green River: finest scenery in the state

[125]Anon, 1870, Guide book for the Diamond Cave, Barren County, Kentucky. Located near "Old Bell Tavern," now the Mammoth Cave and Glosgow Junctions, on the L & N railroad, two miles from the R.R. immediately on the Mammoth Cave Road

[126]Anon, 1876, A guide manual to the Mammoth Cave of Kentucky

[127]Anon, 1876, Guide manual to the Mammoth Cave of Kentucky

[148]Anon, 1935, Mammoth Cave in third dimension

[149]Anon, 1930, Mammoth Cave of Kentucky

[150]Anon, 1920, Mammoth Cave of Kentucky: the world's greatest subterranean wonder

[151]Anon, nd, Mammoth Cave, America's great natural wonder

[169]Anon, c1900, Oligio-Nunk, the place of caves. In the heart of Honeycomb Mountain ... Carter County, Kentucky

[173]Anon, c1886, Photographic views of some of the important points of Mammoth Cave situated in Edmondson, Co. Kentucky, USA: the wonders of this cave and its magnitude, cannot be described, and must be seen to be appreciated. It is reached only by the Louisville & Nashville Railroads. These photographs were taken by magnesium light by W.H. Sesser, St. Joseph, Mich., with a Collins camera

[199]Anon, 1930, Subterranean wonders: Mammoth Cave and Colossal Cavern, Kentucky

[223]Appalshop, Inc, 1971, Line Fork falls and caves

[275]Bailey, Gilbert Stephen, 1822–1891, 1863, The great caverns of Kentucky: Diamond Cave, Mammoth Cave, Hundred Dome Cave

[276]Bailey, Gilbert Stephen, 1822–1891, Nov 1963, Hundred Dome Cave ("partial reproduction of the last known copy of the book, 'the great caverns of Kentucky' ")

[280]Bailey, Vernon Orlando, 1864–1942, 1933, Cave life of Kentucky: mainly in the Mammoth cave region

[327]Barr, Thomas Calhoun, 1931–, Dec 1976, Ecological effects of water pollutants in Mammoth Cave: final technical report to the National Park Service

[399]Binkerd, Adam D., 1869, The Mammoth Cave and its denizens: a complete descriptive guide

[400]Binkerd, Adam D., , Pictorial guide to the Mammoth Cave, Kentucky: a complete historic, descriptive and scientific account of the greatest subterranean wonder of the western world

[464]Brannan, Eldred Boyd, [1925], The entombment of Floyd Collins in Sand Cave, Kentucky

[476]Bridwell, Margaret Morris, 1905–, 1971, The story of Mammoth Cave National Park, Kentucky: a brief history

[530]Brown, Richmond Flint, 1925–, 1966, Hydrology of the cavernous limestones of the Mammoth Cave area, Kentucky

[535]Brucker, Roger Warren, 1929–, 1966, The Flint Ridge Cave System: Mammoth Cave National Park, Kentucky; a Cave Research Foundation cartographic project in cooperation with the National Park Service

[536]Brucker, Roger Warren, 1929–, 1987, The longest cave

[552][Bullitt, Alexander Clark, 1807–1868], 1973, Rambles in the Mammoth Cave, during the year 1844

[579]Call, Richard Ellsworth, 1856–1917, 1902, Grand excursion to Mammoth Cave

[580]Call, Richard Ellsworth, 1856–1917, 1897, Mammoth Cave, Kentucky

[581]Call, Richard Ellsworth, 1856–1917, c1890, The Mammoth Cave, Kentucky; a sketch

[602]Carstens, Kenneth Charles, 1996, Of caves and shell mounds

[666]Close, Danny, 1981, Mammoth Cave... kids love it!

[691]Conn, David Bruce, 1981, Cave life of Carter Caves State Park

[736]Cravens, Raymond L., 1992, The spirit of Lost River: fact and legend about the Lost River Cave and Valley

[741]Crawford, Nicholas Charles, 1942–, 1981, Karst hydrogeology and environmental problems in the Bowling Green area

[748]Crawford, Nicholas Charles, 1942–, 1988, Karst hydrologic problems of south central Kentucky: ground water contamination, sinkhole flooding, and sinkhole collapse: field trip guide

[749]Crawford, Nicholas Charles, 1942–, 1989, The karst landscape of Warren County

[751]Crawford, Nicholas Charles, 1942–, nd, Significant natural landscape features of Warren County, Kentucky

[757]Croghan, John, 1790–1849, 1845, Rambles in Mammoth Cave, during the year 1844, by a visitor

[765]Crowther, Patricia P. , 1984, The Grand Kentucky junction: memoirs

[781]Currens, James C., 1952–, 1979, Bibliography of karst geology in Kentucky

[782]Currens, James C., 1952–, 1993, Flooding of the Sinking Creek Karst area in Jessamine and Woodford Counties, Kentucky

[787]Cushman, Robert Vittum, 1965, Present and future water supply for Mammoth Cave National Park, Kentucky

[820]Davidson, Joseph Killworth, 1938–, 1971, Wilderness resources in Mammoth Cave National Park: a regional approach

[821]Davidson, Robert, 1808–1876, 1840, An excursion to the Mammoth Cave, and the barrens of Kentucky: with some notices of the early settlement of the state

[831]Davies, William Edward, 1917–, Aug 1959, Report on sediments in Mammoth Cave, Kentucky

[857]De Paepe, Duane, 1985, Gunpowder from Mammoth Cave: the saga of saltpetre mining before and during the War of 1812

[867]Dicken, Samuel Newton, 1901–, 1938, Soil erosion in the karst lands of Kentucky

[871]Dilamarter, Ronald R. (ed), 1971, Hydrologic problems in karst regions

[872]Dillaye, Stephen D., 1848, A glance at Mammoth Cave, Kentucky

[888]Dougherty, Percy H. (ed), 1985, Caves and karst of Kentucky

[960]Elliott, Larry P., 1974, Study of potential gas accumulation in caves in Bowling Green, including relationship to water quality

[1014]Eversole, Jack (ed), 1973, The Mammoth Cave area: a planning proposal

[1038]Faust, Burton Sherwood, 1898–1967, 1967, Saltpetre mining in Mammoth Cave, Ky

[1065]Finley-Holiday Film Corporation, 1990, Mammoth Caves

[1103]Forwood, William Stump, 1830–1892, 1870, An historical and descriptive narrative of the Mammoth Cave of Kentucky: including explanations of the causes concerned in its formation, its atmospheric conditions, its chemistry, geology, zoology, etc., with full scientific details of the eyeless fishes

[1131]Funkhouser, William Delbert, 1881–, 1929, The so-called "ash caves" in Lee County, Kentucky

[1176]George, Angelo Isham, 1944–, 1985, Mummies of Short Cave, Kentucky, and the Great Catacomb Mystery

[1177]George, Angelo Isham, 1944–, 1994, Mummies, catacombs, and Mammoth Cave

[1178]George, Angelo Isham, 1944–, 1992, The New

Madrid earthquake at Mammoth Cave (1811-1812)

[1179]George, Angelo Isham, 1944–, 1990, Prehistoric mummies from the Mammoth Cave area: foundations and concepts

[1180]George, Angelo Isham, 1944–, Apr 1986, Saltpeter and gunpower manufacturing in Kentucky

[1223]Goode, Cecil E., 1986, World wonder saved: how Mammoth Cave became a national park

[1226]Goodwin, J. West, , Mammoth Cave: America's great natural wonders

[1276]Guernsey, Daniel Riley, 1905, Lost in the Mammoth Cave

[1283]Guilday, John E. , 1971, The Welch Cave Peccaries (Platygonus) and associated fauna, Kentucky Pleistocene

[1299]Gwin, J. N., 1875, A visit to the Mammoth Cave

[1313]Halaburda, Sue, 1984, Mammoth Cave

[1360]Harlan, Richard, 1796–1843, 1831, Description of the fossil bones of the Megalonyx, discovered in White Cave, Kentucky

[1361]Harpers Ferry Center. Division of Museum Services, 1976, Mammoth Cave National Park: collection management plan

[1371]Hart, William J., Sep 1967, A wilderness plan for Mammoth Cave National Park and the surrounding region

[1375]Hartley, Howard W., 1925, Tragedy of Sand Cave

[1394]Hay, William Perry, 1872–, 1903, Observations on the crustacean fauna of the region about Mammoth Cave, Kentucky

[1483]Holiday Film Corporation, 1970, Mammoth Cave National Park

[1496]Holsinger, John Robert, 1934–, Jan 1985, Ecological analysis of the Kentucky cave shrimp, *Palaemonias ganteri* Hay, at Mammoth Cave National Park (phase V)

[1497]Holsinger, John Robert, 1934–, 1986, Ecological analysis of the Kentucky cave shrimp, *Palaemonias ganteri* Hay, at Mammoth Cave National Park (phase VI): preliminary observations on stream interstitial meiofauna communities and related abiotic factors

[1498]Holsinger, John Robert, 1934–, Oct 1983, Ecological analysis of the Kentucky cave shrimp, *Palaemonias ganteri* Hay, Mammoth Cave National Park

[1499]Holsinger, John Robert, 1934–, 1982, Ecological analysis of the Kentucky cave shrimp, *Palaemonias ganteri* Hay, Mammoth Cave National Park (phase I)

[1500]Holsinger, John Robert, 1934–, Oct 1983, Ecological analysis of the Kentucky cave shrimp, *Palaemonias ganteri* Hay, Mammoth Cave National Park (phase II)

[1501]Holsinger, John Robert, 1934–, Apr 1983, Ecological analysis of the Kentucky cave shrimp, *Palaemonias ganteri* Hay, Mammoth Cave National Park (phase III)

[1528]Hoskins, R. Taylor, 1940, Caverns of enchantment

[1537]Hovey, Horace Carter, 1833–1914, 1891, Guide book to the Mammoth Cave of Kentucky: historical, scientific, and descriptive

[1538]Hovey, Horace Carter, 1833–1914, 1909, Hovey's hand-book of the Mammoth Cave of Kentucky: a practical guide to the regulation routes with maps and illustrations

[1539]Hovey, Horace Carter, 1833–1914, 1912, Mammoth Cave of Kentucky: with an account of Colossal Cavern

[1540]Hovey, Horace Carter, 1833–1914, 1897, Mammoth Cave of Kentucky; an illustrated manual

[1541]Hovey, Horace Carter, 1833–1914, 1982, One hundred miles in Mammoth Cave - in 1880: an early exploration of America's most famous cavern

[1593]Imperial Film Company, 1965, Mammoth Cave, Kentucky

[1731]Jessey, Gary D., 1973, Letcher County's Pine Mountain caves: an in-depth pictorial essay of Letcher County's Pine Mountain caves and caverns

[1733]Jillson, Willard Rouse, 1890–, 1953, A bibliography of Mammoth Cave, 1798-1949

[1734]Jillson, Willard Rouse, 1890–, 1954, Geology of Crystal Cave in southern Pulaski County, Kentucky

[1741]Johnston, Frances Benjamin, 1864–1952, 1893, Mammoth Cave by flash-light

[1755]Jones, William Basil, 1822–1897, 1844, Wonderful curiosity, or, a correct narrative of the celebrated Mammoth Cave of Kentucky with incidents and anecdotes

[1795]Kentucky Geological Survey, 1876, Memoirs of the Geological Survey of Kentucky

[1796]Kentucky Speleofest. Guidebook (1972), 1972, Guidebook to the Kentucky Speleofest, Meade County, Kentucky

[1797]Kentucky Speleofest. Guidebook (1973), 1973, Guidebook to the Kentucky Speleofest, Warren County, Kentucky

[1798]Kentucky Speleofest. Guidebook (1974), 1974, Kentucky Speleofest guidebook: selected caves for the annual summer field conference, the Louisville Grotto, NSS: Squire Boon Caverns, Indiana

[1799]Kentucky Speleofest. Guidebook (1975), 1975, Kentucky Speleofest guidebook: annual summer field conference, the Louisville Grotto NSS

[1800]Kentucky Speleofest. Guidebook (1976), 1976, Guidebook to the 5th annual Kentucky Speleofest. Annual summer field conference of the Louisville Grotto NSS: Camp Carlson, Meade County, Kentucky

[1801]Kentucky Speleofest. Guidebook (1977), 1977, Guidebook to the 1977 Kentucky Speleofest, Cedar Hill Camp, Edmonson Co., Ky

[1802]Kentucky Speleofest. Guidebook (1978), 1978, Guidebook to the 1978 Kentucky Speleofest: Camp Carlson, Meade County, Kentucky

[1803]Kentucky Speleofest. Guidebook (1979), 1979, Guidebook: Kentucky Speleofest '79

[1804]Kentucky Speleofest. Guidebook (1980), 1980, Guidebook to the 1980 Kentucky Speleo-fest: annual summer field conference, Pulaski Co. Park, Ky

[1805] Kentucky Speleofest. Guidebook (1987), 1987, Guidebook for the 1987 Kentucky Speleofest: Renfro Valley KOA, Kentucky, May 22-25, 1987

[1806] Kentucky Speleofest. Guidebook (1989), 1989, 1989 Speleofest guidebook. May 26-29, 1989, Sandhill Campground, Whitley City, Kentucky

[1807] Kentucky Speleofest. Guidebook (1997), 1997, 1997 Speleofest: Cadiz, Ky, Trigg County Recreational Complex, Cadiz, Kentucky

[1821] King, Mary Elizabeth, 1974, The Salts Cave textiles: a preliminary account

[1914] Lawrence, Joseph, 1924–, 1955, The caves beyond: the story of the Floyd Collins' Crystal Cave exploration

[1915] Lawrence, Joseph, 1924–, 1975, The caves beyond: the story of the Floyd Collins' Crystal Cave exploration

[1936] Letcher, Montgomery E., 1839, Wonderful discovery: being an account of a recent exploration of the celebrated Mammoth Cave, in Edmonson County, Kentucky, by Dr. Rowan, Professor Simmons and others, of Louisville, to its termination in an inhabited region, in the interior of the earth

[1950] Livesay, Elizabeth Ann, 1962, Geology of the Mammoth Cave National Park area

[1953] Lobeck, Armin Kohl, 1886-1958, 1928, The geology and physiography of the Mammoth Cave national park

[2002] Lyons, David, 1981, Floyd Collins, greatest cave explorer ever known

[2003] Lyons, Joy Medley, 1991, Mammoth cave

[2025] Malott, Clyde Arnett, 1887-1950, 1977, A geologic profile of Sloans Valley, Pulaski County, Kentucky

[2029] Mammoth Cave National Park Association, 1928, Mammoth Cave National Park Association campaign, 1927-1928

[2030] Mammoth Cave National Park Association, 1985, A national park in Kentucky

[2031] Mammoth Cave National Park Association, 1927, Now how! A national park in Kentucky

[2032] Mammoth Cave National Park Association, nd, A privilege—a duty—an opportunity for Kentucky

[2033] Mammoth Cave Operating Committee, Mammoth Cave, Ky, 1940, Caverns of enchantment

[2034] Mammoth Cave Operating Committee, Mammoth Cave, Ky, 1859, Mammoth cave

[2063] Martin, Horace, 1851, Pictorial guide to the Mammoth Cave, Kentucky

[2126] McGrain, Preston (comp), Apr 1954, Itinerary: geology of the Mammoth Cave Region, Barren, Edmonson, and Hart Counties, Kentucky

[2127] McGrain, Preston, 1917–, 1961, The geologic story of Diamond Caverns

[2128] McGrain, Preston, 1917–, 1954, Geology of the Carter and Cascade Caves area

[2145] Meloy, Harold Raymond, 1913-1985, , Mummies of Mammoth Cave: an account of the Indian mummies discovered in Short Cave, Salts Cave, and Mammoth Cave, Kentucky

[2146] Meloy, Harold Raymond, 1913-1985, 1968, Mummies of Mammoth Cave: Fawn Hoof, Little Alice, Lost John and others. An account of the Indian mummies discovered in Short Cave, Salts Cave, and Mammoth Cave, Kentucky

[2272] Morgan, Robert, nd, Through Mammoth Cave, giving a graphic description of all the routes, avenues, passes, passages, pits, domes, streams, rivers, seas, lakes, and other wonders to be found in that greatest of all natural curiosities in the whole realm of this great world

[2273] Morgan, Robert, 1881, Through Mammoth Cave, giving a graphic description of all the routes, avenues, passes, passages, pits, domes, seas, lakes, rivers, streams, and other wonders of that greatest natural curiosity in the world

[2279] Morrison, Geo. D., c1920, New entrance to Mammoth Cave: a history of the cave together with descriptive guide to routes and illustrations of points of interest

[2305] Murray, Robert K., 1979, Trapped!: the story of the struggle to rescue Floyd Collins from a Kentucky cave in 1925, an ordeal which became one of the most sensational events of modern times

[2306] Murray, Robert K., 1982, Trapped!: the story of the struggle to rescue Floyd Collins from a Kentucky cave in 1925, an ordeal which became one of the worst sensational events of modern times

[2313] MVOR (Mississippi Valley Ozark Regional). Guidebook (1971), 1971, Fall 1971 MVOR: Glover's Cave

[2323] MVOR (Mississippi Valley–Ozark Region). Guidebook (1984), 1984, MVOR Guidebook: 1984 Fall

[2330] Mylroie, John Eglinton, 1949–, 1979, Western Kentucky Speleological Survey: annual report 1978

[2331] Mylroie, John Eglinton, 1949–(ed), 1979, Western Kentucky Speleological Survey: annual report 1979

[2359] National Speleological Society, Mar 1977, A scrap book of articles about the C3 expedition into Floyd Collins Crystal Cave, Kentucky; from Sunday February 14 to February 20, 1954

[2406] National Speleological Society. Annual Convention. Guidebook (1985: Frankfort, KY), 1985, The caves of south eastern Kentucky: guidebook for the 1985 National Speleological Society Convention, Frankfort, Kentucky

[2489] Nelson, Nels Christian, 1875–, 1917, Contributions to the archaeology of Mammoth Cave and vicinity, Kentucky

[2490] Nelson, Raymond L., Nov 1959, Mammoth Cave tour leader manual

[2514] Noble, Bruce J., 1991, Cultural resource manage-

ment in Mammoth Cave National Park: a National Park Service—Kentucky Heritage Council cooperative project
[2567]Oliva, Deane, 1991, A visitor's guide to Mammoth Cave National Park
[2573]Orchard, William C., 1920, Sandals and other fabrics from Kentucky caves
[2591]Packard, Alpheus Spring, 1839–1905, 1872, The Mammoth Cave and its inhabitants, or, descriptions of the fishes, insects and crustaceans found in the cave with figures of the various species and an account of allied forms, comprising notes upon their structure, development and habits, with remarks upon subterranean life in general
[2596]Palmer, Arthur Nicholas, 1940–, 1981, A geological guide to Mammoth Cave National Park
[2599]Palmer, Arthur Nicholas, 1940–, 1982, A guide to the historic section of Mammoth Cave
[2610]Panton, James Hoyes, 1847–1898, 1890, The Mammoth Cave of Kentucky
[2685]Pohl, Erwin Robert, 1904–, 1964, Itinerary, geologic features of the Mississippian Plateaus in the Mammoth Cave and Elizabethtown areas
[2694]Popham, William Lee, 1885–1953, 1993, Mammoth Cave romance
[2707]Prentice, Guy, Jan 1988, Mammoth Cave archeological inventory project interim report - 1987 investigations
[2708]Prentice, Guy, 1989, Mammoth Cave National Park archeological inventory project interim report-1988 investigations
[2734]Putnam, Frederic Ward, 1839–1915, 1875, Archaeological researches in Kentucky and Indiana, 1874
[2739]Quinlan, James Francis, 1936–1995, 1986, Ground water flow in the Mammoth Cave Area, Kentucky with emphasis on principles, contaminant dispersal, instrumentation for monitoring water quality, and other methods of study
[2741]Quinlan, James Francis, 1936–1995, 1980, Ground-water hydrology and geomorphology of the Mammoth Cave Region, Kentucky, and of the Mitchell Plain, Indiana
[2742]Quinlan, James Francis, 1936–1995, 1990, Hydrogeology and geomorphology of the Mammoth Cave area, Kentucky: Southeastern Friends of the Pleistocene 1990 field excursion, November 17, 1990
[2743]Quinlan, James Francis, 1936–1995, 1977, Hydrology and water quality in the central Kentucky karst: phase 1
[2744]Quinlan, James Francis, 1936–1995, 1976, Hydrology of the Turnhole Spring Groundwater Basin and vicinity, Kentucky: an area that includes part of the Mammoth Cave National Park
[2748]Radlauer, Ruth, 1926–, 1986, Mammoth Cave National Park
[2759]Randolph, Helen Fitz, 1924, Mammoth Cave and the cave region of Kentucky, illustrated: with bibliography of Mammoth Cave (Willard Rouse Jillson); first accurate underground survey (H. Bruce Hoffman); introduction (H. C. Nelson)
[2768]Rebmann, James R., 1972, Caves of Rockcastle County, Kentucky
[2847]Rogers, Edmund B., 1958, Mammoth Cave National Park, Kentucky: history of legislation through the 82nd Congress
[2866]Rusling, James Fowler, 1864, A trip to the Mammoth Cave, Ky
[2918]Schwartz, Douglas W., 1965, Prehistoric man in Mammoth Cave
[2926]Scott, W. Ray, , A pictorial study of Mammoth Cave and Mammoth Cave National Park
[3002]Sloan, Tacoma G., 1960, Archaeological survey of Mammoth Cave National Park
[3032]Smith, Philip M., 1960, Speleological research in the Mammoth Cave Region, Kentucky: elements of an integrated program
[3180]Tankersley, Ken, 1975, The cavernous karst land forms of Jackson County, Kentucky
[3204]Thompson, John, l874–, 1909, Mammoth Cave, Kentucky; an historical sketch containing a brief description of some of the principal places of interest in the Mammoth Cave. Also a short description of Colossal Cavern
[3206]Thompson, Ralph Seymour, 1970, The sucker's visit to Mammoth Cave
[3207]Thompson, Ralph Seymour, 1879, The sucker's visit to the Mammoth Cave including a history of the experience and adventures of a party who undertook to see the cave and have some fun going there. A full and accurate description of the cave and the science of its formation with an account of the living inhabitants
[3212]Thrailkill, John Vernon, 1930–, [1984], Hydrogeology and environmental geology of the inner Bluegrass Karst Region Kentucky: field guide for the annual meeting of the southeastern and north-central sections Geological Society of America: Lexington, Kentucky April 4-6 1984
[3214]Thrailkill, John Vernon, 1930–, 1970, Solution geochemistry of the water of limestone terrains
[3253]Turner, Cleon, 1968, Discovery of the Onyx Cave and my biography
[3254]Turner, James William, 1848–, 1912, Wonders of the great Mammoth Cave of Kentucky containing thorough and accurate historical and descriptive sketches of this marvelous underground world with a chapter on the geology of cave formation
[3273]U.S. Environmental Protection Agency, Region 4, Aug 1981, Environmental impact statement, Mammoth Cave area, Kentucky: wastewater facilities
[3274]U.S. Environmental Protection Agency, Region 4, Apr 1981, Environmental impact statement; Mammoth

Cave area, Kentucky, wastewater facilities

[3275]U.S. Environmental Protection Agency, Region 4, Apr 1981, Mammoth Cave area, Kentucky, 201 facilities plan environmental impact statement

[3294]U.S. National Park Service, 1976, Final environmental statement: Mammoth Cave National Park master plan and wilderness suitability study, Kentucky

[3296]U.S. National Park Service, Apr 1976, Final master plan: Mammoth Cave National Park

[3299]U.S. National Park Service, Apr 1979, The Great Onyx Job Corps Civilian Conservation Center alternative relocation sites, Mammoth Cave National Park, Kentucky

[3301]U.S. National Park Service, 1987, Mammoth Cave National Park

[3302]U.S. National Park Service, 1974, Mammoth Cave National Park, Kentucky

[3303]U.S. National Park Service, 1978, Mammoth Cave National Park: collection management plan

[3313]U.S. National Park Service, nd, Proposed Mammoth Cave National Park master plan and wilderness study, Kentucky: draft environmental impact statement

[3319]U.S. National Park Service, 1984, Voices of the cave

[3346]Van Couvering, John A., 1962, Characteristics of large springs in Kentucky

[3362]Vietzen, Raymond Charles, 1907–, , The saga of Glover's Cave

[3372]Wagoner, John J., 1985, Mammoth Cave

[3400]Warnell, Norman, 1997, Mammoth Cave: forgotten stories of it's [sic] people

[3407]Watson, Patty Jo, 1932–, 1974, Archeology of the Mammoth Cave Area

[3408]Watson, Patty Jo, 1932–, 1969, The prehistory of Salts Cave: Kentucky

[3412]Watson, Richard Allan, 1931–, 1994, Caving

[3413]Watson, Richard Allan, 1931–, 1963, The Mammoth Cave National Park research center

[3414]Watson, Richard Allan, 1931–, 1963, The Mammoth Cave National Park Research Center: a Cave Research Foundation study

[3415]Watson, Richard Allan, 1931–, May 1967, The preservation of wilderness karst in central Kentucky, U.S.A

[3427]Webb, William Snyder, 1882–, , The occurrence of the fossil remains of Pleistocene vertebrates in the caves of Barren County, Kentucky

[3433]Weller, James Marvin, 1899–, 1927, The geology of Edmonson County: a detailed presentation of the physical, stratigraphic, structural, and economic geology of this district

[3442]Werner, Eberhard Wolfgang, 1942–, Jul 1981, Guidebook to the karst of the Central Appalachians; prepared for the Eighth International Congress of Speleology, Bowling Green, Kentucky, U.S.A., July 18 to 24, 1981

[3461]Chisholm, T.O., 1892, Grand Avenue Cave. A description in detail of one of America's greatest natural wonders

[3475]White, William Blaine, 1934–(ed), 1989, Karst hydrology: concepts from the Mammoth Cave area

[3486]Wilkes, Frank G., 1962, Bibliography of Mammoth Cave National Park, Mammoth Cave, Kentucky

[3510]Wilson, John, 1800–1849, 1849, A visit to the Mammoth Cave of Kentucky

[3520]Wood, Andrew, 1957, The mammoth cavern

[3531]Wright, Charles W., , A guide manual to the Mammoth Cave of Kentucky

[3532]Wright, Charles W., 1860, A guide manual to the Mammoth Cave of Kentucky

[3533]Wright, Charles W., 1858, The Mammoth Cave of Kentucky

[3538]Wynne, Annette, 1924, The trip thru fairyland; Great Onyx cave

14.9.4.2.18 Louisiana

[266]Autin, Whitney J., 1984, Observations and significance of sinkhole development at Jefferson Island

14.9.4.2.19 Maryland

[829]Davies, William Edward, 1917–, 1961, The caves of Maryland

[1123]Franz, Richard, 1971, Caves of Maryland

[1191]Gidley, James William, 1866–1931, 1938, The Pleistocene vertebrate fauna from Cumberland Cave, Maryland

[1192]Gidley, James William, 1866–1931, 1914, Preliminary report on a recently discovered Pleistocene cave deposit near Cumberland, Maryland

[1193]Gidley, James Williams, 1866–1931, 1918, A pleistocene cave deposit of Western Maryland

[1194]Gidley, James Williams, 1866–1931, 1921, Pleistocene peccaries from the Cumberland cave deposit

[1285]Guilday, John E., 1925–1982, 1967, A new *Peromyscus* (Rodentia: Cricetidae) from the Pleistocene of Maryland

[1493]Holman, J. Alan, 1931–, 1977, The Pleistocene (Kansan) herpetofauna of Cumberland Cave, Maryland

[2070]Maryland Governor's Advisory Committee on Promoting Paleontology, Oct 1993, Report

[2163]Meulen, A. J. van der, 1978, *Microtus* and *Pitymys* (Arvicolidae) from Cumberland Cave, Maryland, with a comparison of some new and old world species

[2398]National Speleological Society. Annual Convention. Guidebook (1976: Morgantown, WV), 1976, 1976 annual convention guidebook, Morgantown, W. Va

[2623]Peabody, Charles, 1867–1939, 1908, Pt. I: the exploration of Bushey Cavern near Cavetown, Maryland

14.9.4.2.20 Massachusetts

[444]Botto, Larry (comp), [1993], A partial guidebook to the caves of the Berkshires: 1993 spring N.R.O. hosted by the Berkshire Hills Grotto (May 14th, 15th, 16th)

[990]Engel, Thomas Daniel, 1954–, Jul 1979, A chronicle of selected northeastern caves: a history guide for the 1979 NSS Convention, Pittsfield, Massachusetts, August 5-12, 1979

[1387]Hauer, Peter Marshall, 1945–1975, 1969, Caves of Massachusetts

[2401]National Speleological Society. Annual Convention. Guidebook (1979: Pittsfield, MA), 1979, An introduction to caves of the northeast: guidebook for the 1979 National Speleological Convention, Pittsfield, Massachusetts

[2682]Plante, Alan R. (comp), 1990, Berkshire Cave Guide

14.9.4.2.21 Michigan

[529]Brown, Paul Martin, 1973, An introduction to the limestone sinkholes of northeastern Michigan

[2399]National Speleological Society. Annual Convention. Guidebook (1977: Alpena, MI), 1977, Official 1977 guidebook, Alpena, Michigan: '77, the Lakeshore convention: Alpena, Michigan, NSS

[2408]National Speleological Society. Annual Convention. Guidebook (1987: Sault Sainte Marie, MI), 1987, National Speleological Society official 1987 guidebook, August 3-7, 1987, Sault Sainte Marie, Michigan

14.9.4.2.22 Minnesota

[1051]Fetvedt, John E. (ed), Aug 1984, Mystery Cave area: MSS Corn Feed August 1984

[1052]Fetvedt, John E. (ed), Apr 1984, North Country Region: Spring 1984, LaCrosse, Wisconsin

[1183]Gerboth, David W., Aug 1989, A guidebook to the caves and sinks of the Spring Valley Caverns area, Fillmore County, Minnesota: 1989 Cornfeed Minnesota Speleological Survey

[1479]Hogberg, Rudolph K., 1967, Guide to the caves of Minnesota

[1703]Jameson, Roy Alan, 1951–, 1994—1995, The waters of Mystery Cave, Forestville State Park, Minnesota

[1738]Johnson, Elden, 1956, The Lee Mill Cave

[1781]Kehret, Roger Alan, 1942–, 1974, Minnesota caves of history and legend: a collection of unique cave stories

[1782]Kehret, Roger Alan, 1942–, 1988, Minnesota karst country tours

[1783]Kehret, Roger Alan, 1942–, 1974, Your guide to Mystery Cave

[1949]Lively, Richard, 1993, Radon concentrations, radon decay product activity, meteorological conditions and ventilation in Mystery Cave

[2258]Montz, Gary R., Jun 1993, The aquatic invertebrates of Mystery Cave, Forestville State Park, Minnesota

[2396]National Speleological Society. Annual Convention. Guidebook (1974: Decorah, IO), 1974, National Speleological Society 1974 convention Luther College, Decorah, Iowa 10-18 August, 1974

[2402]National Speleological Society. Annual Convention. Guidebook (1980: White Bear Lake, MN), Jul 1980, An introduction to caves of Minnesota, Iowa, and Wisconsin: guidebook for the 1980 National Speleological Society Convention, Lakewood Community College, White Bear Lake, Minnesota

[2597]Palmer, Arthur Nicholas, 1940–, 1993, Geology and origin of Mystery Cave, Forestville State Park, Minnesota

14.9.4.2.23 Mississippi

[16]Adovasio, J. M., 1980, Archaeological testing at two rockshelters in the Tombigbee River Multi-Resource District, Alabama and Mississippi: an interim report

[1831]Knight, E. Leslie, 1974, Caves of Mississippi

14.9.4.2.24 Missouri

[33]Aley, Thomas John, 1938–, 1981, Cave management investigations on the Ozark National Scenic Riverways, Missouri

[34]Aley, Thomas John, 1938–, 1980, Cave management investigations on the Ozark National Scenic Riverways, Missouri

[38]Aley, Thomas John, 1938–, 1981, Hydrogeologic mapping of unincorporated Greene County, Missouri, to identify areas where sinkhole flooding and serious groundwater contamination could result from land development

[41]Aley, Thomas John, 1938–, Aug 1975, A predictive hydrologic model for evaluating the effects of land use and management on the quantity and quality of water from Ozark Springs: final report

[53]Allison, Vernon Charles, 1891–, 1926, The antiquity of the deposits in Jacob's Cavern

[155]Anon, 1958, Marvel Cave: the story of America's largest privately owned cave

[188]Anon, nd, Souvenir book of romantic Bridal Cave

[234]Asher-Bolinder, Sigrid, 1992, A review of the deposition and alteration of filled-sink deposits of east-central Missouri

[239]Association of Missouri Geologists. Annual Meeting (19th: 1972: Rolla, Mo), 1972, Guidebook to the karst features and stratigraphy of the Rolla area

[422]Bogart, Robert C., 1960, Mark Twain Cave: an adventure

[469]Bretz, J Harlen, 1882–1981, 1956, Caves of Missouri

[738]Crawford, Nicholas Charles, 1942–, 1982, Guidebook to karst and caves of Tennessee: emphasis on the Cumberland plateau escarpment region and guidebook to karst and caves of the Ozark Region of Missouri and Arkansas: prepared for the Eighth International Congress of Speleology, Bowling Green, Kentucky, U. S. A. July 18 to 24 1981

[766]Crush, Kenneth, Sep 1977, An inventory of caves, Harry S. Truman dam and reservoir

[1039]Fellows, L. D., 1970, Guidebook to Ozark carbonate terrane, Rolla - Devils Elbow area, Missouri

[1109]Fowke, Gerard, 1855–1933, 1922, Archeological investigations

[1151]Gardner, James, 1982, An inventory and evaluation of Missouri state parks cave resources

[1154]Gardner, James Eugene, 1953–, Sep 1982, An inventory and evaluation of cave resources of Mark Twain National Forest: a final report submitted to the Mark Twain National Forest, United States Department of Agriculture in compliance with a cooperative cave inventory agreement

[1155]Gardner, James Eugene, 1953–, 1986, Invertebrate fauna from Missouri caves and springs

[1377]Harvey, Edward Joseph, 1916–, 1977, Application of thermal imagery and aerial photography to hydrologic studies of karst terrane in Missouri

[1378]Harvey, Edward Joseph, 1916–, 1983, Hydrology of carbonate terrane, Niangua, Ossage Fork, and Grandglaize basins, Missouri

[1474]Hoebel, Edward Adamson, 1906–, 1946, The archaeology of Bone cave, Miller county, Missouri

[1484]Holiday Film Corporation, 1991, Meramec Caverns, Stanton, Missouri

[1680]Jack, James L., nd, Moore's Cave: its discovery and significance

[1794]Kent, Paul, 1983, Tour caves of the Ozarks: Missouri Arkansas: the secret beneath the surface

[1827]Klippel, Walter E., Aug 1986, The unmodified vertebrate fauna from Granite Quarry Cave (23CT36), Carter County, Missouri

[1955]Logan, Wilfred David, 1923–, 1952, Graham Cave, an archaic site in Montgomery County, Missouri

[1981]Luce, Pat (comp), 1940, Let's see Marvel Cave: your personal guide to America's largest privately owned cave

[2000]Lynott, Mark J., 1986, Archeological investigations at Limekiln Cave, 23SH109, Ozark National Scenic Riverways, southeast Missouri

[2066]Martin, Ronald L., 1972, Cave development in the Bull Creek drainage basin of southwest Missouri

[2067]Martin, Ronald L., 1990, Marvel Cave: Silver Dollar City, Branson, Missouri

[2068]Martin, Ronald L., 1974, Official guide to Marvel Cave, Silver Dollar City, Missouri

[2117]McDonald, Donald L., 1974, Official guide to Marvel Cave: the only complete authentic history of Marvel Cave ever published

[2141]Mehl, Maurice Goldsmith, 1887–1966, Dec 1962, Missouri's Ice Age animals

[2142]Mehl, Maurice Goldsmith, 1887–1966, 1969, Notes on Missouri Pleistocene peccaries

[2148]Meramec Valley Conservation Task Force. National Speleological Society, [1973], An endangered heritage: a story of the Meramec Valley, its caves, and their possible destruction by the U. S. Army Corps of Engineers

[2195]Missouri Civil Defense Agency, 1962, Missouri underground shelter space: a fallout shelter survey of mines and caves in Missouri

[2196]Missouri Division of Geological Survey and Water Resources, Mar 1952, Partial catalog of caves in Missouri

[2197]Missouri Geological Survey and Water Resources, Dec 1960, Guidebook to the geology of the Rolla area emphasizing solution phenomena

[2198]Missouri Speleological Survey, Dec 1977, Archaeological investigations: cave explorations in the Ozark region of central Missouri

[2200]Missouri Speleological Survey, 1976, A biological study of Cathedral Cave, Crawford County, Missouri

[2201]Missouri Speleological Survey, Jul 1975, Caves of Christian County, Missouri: part 1 - eastern half

[2202]Missouri Speleological Survey, Oct 1975, Caves of Christian County, Missouri: part 2- western half

[2203]Missouri Speleological Survey, Jul 1981, Caves of Crawford County

[2204]Missouri Speleological Survey, July-December 1982, Caves of the Current River Valley

[2205]Missouri Speleological Survey, January-June 1982, Caves of the Jacks Fork Valley

[2206]Missouri Speleological Survey, Apr 1975, A checklist of invertebrate species recorded from Missouri subterranean habitats

[2207]Missouri Speleological Survey, January-December 1987, The commercial caves of Camden County, Missouri

[2208]Missouri Speleological Survey, 1970, A history of the caves of Camden County, Missouri

[2209]Missouri Speleological Survey, Jan 1976, Hydrogeologic controls on solution of carbonate rocks in Christian County, Missouri

[2212]Missouri Speleological Survey, Oct 1974, The invertebrate fauna of Mystery Cave, Perry County, Missouri

[2213]Missouri Speleological Survey, Apr 1973, An investigation into ground water pollution in caves

[2214]Missouri Speleological Survey, October-December 1979, J Harlen Bretz: a geologist's encounters with Missouri caves

[2215]Missouri Speleological Survey, Jan 1971, Jefferson County caves

[2216]Missouri Speleological Survey, Jan 1973, Laclede

14.9. NORTH AMERICA

County caves
[2217]Missouri Speleological Survey, Apr 1961, Miller County caves
[2218]Missouri Speleological Survey, 1992, Mystery Cave
[2219]Missouri Speleological Survey, July-December 1986, Origin and development of Cave Spring, Shannon County, Missouri
[2220]Missouri Speleological Survey, Jan 1963, Phelps County caves
[2221]Missouri Speleological Survey, 1963, Pike County caves
[2222]Missouri Speleological Survey, Apr 1971, Pit caves of Jefferson County
[2223]Missouri Speleological Survey, Jan 1975, A pleistocene fauna from Zoo Cave, Taney County, Missouri
[2224]Missouri Speleological Survey, Apr 1964, Pulaski County caves
[2225]Missouri Speleological Survey, Jan 1965, Ralls County caves
[2227]Missouri Speleological Survey, Jul 1973, Report of the Devil's Icebox - Rockbridge Park conservation task force of the National Speleological Society
[2228]Missouri Speleological Survey, 1992, Rimstone River Cave
[2229]Missouri Speleological Survey, Jul 1971, Sandstone caves of Jefferson County
[2230]Missouri Speleological Survey, [Dec 1966], Shannon County caves
[2231]Missouri Speleological Survey, Apr 1965, St. Charles County caves
[2232]Missouri Speleological Survey, Jul 1965, St. Genevieve County caves
[2233]Missouri Speleological Survey, Jan 1966, St. Louis County caves
[2234]Missouri Speleological Survey, Oct 1967, Washington County caves
[2235]Missouri Speleological Survey, January-December 1983, The wild caves of Camden County, Missouri
[2310]MVOR (Mississippi Valley Ozark Regional). Guidebook (1969), 1969, MVOR Guidebook Spring 1969
[2314]MVOR (Mississippi Valley Ozark Regional). Guidebook (1971), 1971, MVOR Spring 1971, April 30, May 1 & 2: Pulaski County
[2315]MVOR (Mississippi Valley Ozark Regional). Guidebook (1972), 1972, Guidebook: Fall 1972 MVOR, Perry County, Missouri
[2316]MVOR (Mississippi Valley Ozark Regional). Guidebook (1973), 1973, Guidebook: MVOR Spring '73
[2317]MVOR (Mississippi Valley Ozark Regional). Guidebook (1974), Sep 1974, Lake of the Ozarks MVOR, Fall 1974: Camden County, Miller County, Morgan County
[2319]MVOR (Mississippi Valley Ozark Regional). Guidebook (1976), Sep 1976, Fall 1976 MVOR guidebook, Perry Co., Mo
[2320]MVOR (Mississippi Valley Ozark Regional). Guidebook (1977), 1977, Spring 1977 MVOR Guidebook: Middle Mississippi Valley Grotto, Shannon Co., Missouri
[2321]MVOR (Mississippi Valley Ozark Regional). Guidebook (1979), 1979, Perry County: spring M.V.O.R. '79
[2322]MVOR (Mississippi Valley Ozark Regional). Guidebook (1980), 1980, Guidebook to spring 1980 MVOR, Shannon County, Missouri
[2390]National Speleological Society. Annual Convention. Guidebook (1968: Southwestern MO), Jul 1968, Guidebook to selected caves in southwestern Missouri: prepared for the 25th annual convention of the National Speleological Society
[2583]Owen, Luella Agnes, Apr 1968, Cave regions of the Ozarks
[2613]Parmalee, Paul Woodburn, 1972, Pleistocene and recent faunas from the Brynjulfson Caves, Missouri
[2614]Parmalee, Paul Woodburn, 1969, Pleistocene and recent vertebrate faunas from Crankshaft Cave, Missouri
[2622]Peabody, Charles, 1867–1939, 1904, The exploration of Jacobs Cavern, McDonald County, Missouri
[2701]Powell, Hazel Rowena, 1953, Adventures underground in the caves of Missouri
[2859]Rother, Hubert, 1996, Lost caves of St. Louis
[2967]Shepard, Edward M., 1907, Underground waters of Missouri; their geology and utilization
[2970]Shippee, J. M., Dec 1966, The archaeology of Arnold Research Cave, Callaway County, Missouri
[2984]Simpson, George Gaylord, 1902–, Feb 1949, A fossil deposit in a cave in St. Louis
[3092]Speleology Workshop (1979: 2nd: Southwest Missouri State University), 1979, 1979 Speleology workshop manual
[3258]Tyler, Larry H., 1980, History of Rockhouse Cave
[3325]U.S. Office of Water Resources Research, 1975, The detection and mapping of subterranean water bearing channels: phase 2: final report
[3351]Vandike, James E., Oct 1988, Using digital watershed modeling to estimate sinkhole-flooding potential at Perryville, Missouri
[3363]Vineyard, Jerry Daniel, 1935–(comp), , Catalogue of the caves of Missouri
[3364]Vineyard, Jerry Daniel, 1935–, 1967, Guidebook to the geology between Springfield and Branson, Missouri, emphasizing stratigraphy and cavern development
[3365]Vineyard, Jerry Daniel, 1935–, 1974, Springs of Missouri
[3366]Vineyard, Jerry Daniel, 1935–(comp), 1960, Supplement to the catalogue of the caves of Missouri
[3419]Weaver, Herman Dwight, 1938–, 1972, Adventures

at Mark Twain Cave

[3421]Weaver, Herman Dwight, 1938–, Aug 1977, Meramec Caverns: legendary hideout of Jesse James

[3422]Weaver, Herman Dwight, 1938–, Apr 1980, Missouri: the cave state

[3423]Weaver, Herman Dwight, 1938–, , Onondaga, the mammoth cave of Missouri

[3424]Weaver, Herman Dwight, 1938–, 1992, The wilderness underground: caves of the Ozark plateau

14.9.4.2.25 Montana

[590]Campbell, Newell Paul, 1938–, 1978, Caves of Montana

[591]Campbell, Newell Paul, 1938–, Oct 1975, Glacier Park cave study

[1869]Kuhns, Roger J., 1982, Speleology of Lewis and Clark Caverns, Montana

[1946]Link, Louis W., , Lewis and Clark Cavern, Montana

[2391]National Speleological Society. Annual Convention. Guidebook (1969: Lovell, WY), 1969, Caves of the Big Horn - Pryor Mountains of Montana and Wyoming. Issued in conjunction with the 28th anniversary convention of the National Speleological Society at Lovell, Wyoming, Harley A Leach, convention chairman

[2522]Northwest Cave Research Institute, Jan 1987, A management plan for Bighorn Caverns, Montana

[2658]Perry, Eugene S., 1946, Morrison Cave, Lewis and Clark Cavern State Park, Montana

[3041]Snodgrasse, Richard Montgomery, 1958, The human skeletal remains from Pictograph and Ghost caves, Montana

[3527]World Travelogues Corporation, 1984, Lewis and Clark Caverns

14.9.4.2.26 Nebraska

[633]Champe, John L., Oct 1946, Ash Hollow Cave: a study of stratigraphic sequence in the central great plains

[2700]Pound, Louise, 1872–1958, 1948, Nebraska cave lore

14.9.4.2.27 Nevada

[319]Bard, James C., 1979, Ezra's Retreat: a rockshelter/cave occupation site in the north central Great Basin

[320]Bard, James C., 1980, Test excavations at Painted Cave, Pershing County, Nevada: for Bureau of Land Management, Winnemucca District Office

[341]Basso, Dave, 1992, 2,000-year-old duck decoys from Lovelock Cave, Nevada

[554]Bunch, James H., , The archaeological excavation of two rock shelters near Cave Lake, Ely, Nevada

[859]de Saussure, Raymond, Mar 1953, Report of the California-Nevada Speleological Survey June 13-September 7, 1952

[862]Devil's Hole Pupfish Recovery Team (prep), Jul 1980, Devil's Hole Pupfish Recovery Plan

[1110]Fowler, Don D., 1936–, Mar 1968, The archeology of Newark Cave, White Pine County, Nevada

[1272]Grosscup, Gordon L., 1960, The culture history of Lovelock Cave, Nevada

[1316]Halladay, Orlynn J., 1972, The Lehman Caves story

[1362]Harrington, Mark Raymond, 1882–1971, Apr 1933, Gypsum cave, Nevada, report of the second sessions expedition

[1420]Heizer, Robert Fleming, 1915–, 1970, Archaeological investigations in Lovelock Cave, Nevada

[1421]Heizer, Robert Fleming, 1915–, 1956, The archaeology of Humboldt Cave, Churchill County, Nevada

[1422]Heizer, Robert Fleming, 1915–, 1961, The archaeology of two sites at Eastgate, Churchill County, Nevada

[1475]Hoffman, Ray J., 1988, Chronology of diving activities and underground surveys in Devils Hole and Devils Hole Cave, Nye County, Nevada, 1950-86

[1719]Jennings, Jesse David, 1909–, Oct 1957, Danger Cave

[1896]Lange, Arthur L., Jul 1958, Stream piracy and cave development along Baker Creek, Nevada

[1928]Lehman Caves National Monument (Nev.), 1980, Proposed natural resources management plan and environmental assessment, Lehman Caves National Monument, Nevada

[1929]Lehman Caves National Monument, 1974, Proposed natural resources management plan, Lehman Caves National Monument, Nevada

[1967]Loud, Llewellyn Lemont, 1879–1946, 1965, Lovelock Cave

[1968]Loud, Llewellyn Lemont, 1879–1946, Feb 1991, Lovelock Cave: The republication of the rare 1929 Nevada classic

[2132]McLane, Alvin Ray, 1974, A bibliography of Nevada caves

[2302]Murie, Adolph, 1899–, 1959, Field investigation report: Lehman Caves - Wheeler Peak: portion of southern section of Snake Range, White Pine County, Nevada

[2303]Murie, Adolph, 1899–, Feb 1959, Second field investigation report, Lehman Caves - Wheeler Peak: October 13 to 17, 1958, October 29 to November 13, 1958: portion of southern section of Snake Range, White Pine County, Nevada

[2491]Netherton, Shaaron, Aug 1985, Cave management plan for the Ely District

[2577]Orr, Phil C., 1956, Pleistocene man in Fishbone Cave, Pershing County, Nevada

[2578]Orr, Phil C., 1952, Preliminary excavations of Pershing County caves

[2864] Rozarie, Charles, 1964, The Archaeology at Lehman Caves National Monument: Nevada State Museum report
[2906] Schmidt, Jeremy, 1987, Lehman Caves
[2974] Shutler, Mary Elizabeth (ed), Oct 1963, Deer Creek cave, Elko County, Nevada
[2995] Skinner, Morris F., 1942, The fauna of Papago Springs Cave, Arizona, and a study of Stockoceros; with three new Antilocaprines from Nebraska and Arizona
[3200] Thomas, David Hurst, Jun 1985, The archaeology of Hidden Cave, Nevada
[3229] Trexler, Keith A., 1975, Lehman Caves...its human story: from the beginning through 1965
[3300] U.S. National Park Service, 1971, Lehman Caves National Monument
[3306] U.S. National Park Service, Feb 1980, Natural resources management program: an addendum to the natural resources management plan for Lehman Caves National Monument, Nevada
[3331] University of California Archaeological Research Facility, Department of Anthropology, 1968, Papers on Great Basin prehistory
[3336] University of Nevada System, Laboratory of Desert Biology, Dec 1968, Final reports on the Lehman Caves studies

14.9.4.2.28 New Hampshire

[1571] Humphrey, Richard V., 1989, Mystery Hill: myth and mythology in the land of academe
[2401] National Speleological Society. Annual Convention. Guidebook (1979: Pittsfield, MA), 1979, An introduction to caves of the northeast: guidebook for the 1979 National Speleological Convention, Pittsfield, Massachusetts
[2519] Northeast Regional Organization [NRO]. National Speleological Society, Jul 1979, New Hampshire Caves
[3083] Speare, Eva A., nd, The story of Polar Caves

14.9.4.2.29 New Jersey

[795] Dalton, Richard F., 1976, Caves of New Jersey
[796] Dalton, Richard F., 1970, New Jersey caves in brief

14.9.4.2.30 New Mexico

[6] Adam, David P., 1963, Pollen analysis of a stratigraphic section from Bat Cave, New Mexico
[29] Alexander, Hubert Griggs, 1935, Report on the excavation of Jemez Cave, New Mexico, Santa Fe, N.M.
[65] Anderson, Arthur Wilhelm, 1892–, 1938, The Carlsbad Cavern of New Mexico: its history and geology
[66] Anderson, Arthur Wilhelm, 1892–, nd, The Carlsbad Cavern of New Mexico: its history and geology, formations of the cavern, discovery and exploration, bats of the cavern, cavern geology, administration: the land nobody knows
[67] Anderson, Arthur Wilhelm, 1892–, 1930, The Carlsbad Cavern of New Mexico: its history and geology; formations of the cavern, discovery and exploration, bats of the cavern, cavern geology, administration; the land nobody knows
[84] Anon, 1993, Carlsbad Cavern National Park. Guadalupe Mountains National Park
[85] Anon, 1985, Carlsbad Caverns National Park
[86] Anon, c1988, Carlsbad Caverns National Park, New Mexico
[87] Anon, 1982, Carlsbad Caverns, New Mexico
[91] Anon, 1953, Cavern: a pictorial souvenir of Carlsbad Caverns
[240] Atchison, Topeka, and Santa Fe Railway Company, 1941, Carlsbad Caverns National Park
[241] Atchison, Topeka, and Santa Fe Railway Company, 1927, Carlsbad Caverns, New Mexico
[269] Bachman, George Odell, Sep 1987, Karst in evaporites in southeastern New Mexico.
[279] Bailey, Vernon Orlando, 1864–1942, 1928, Animal life of the Carlsbad Cavern
[323] Barnett, John, 1927–, 1969, Carlsbad Caverns National Park
[324] Barnett, John, 1927—, 1981, Carlsbad Caverns National Park, Carlsbad, New Mexico: silent chambers, timeless beauty
[353] Baysinger, David. , 1988, Lechuguilla cave: the hidden giant
[378] Belski, David Stanley, 1937–, 1992, GYPKAP 1987 annual report
[412] Blackhawk Hawks, 1970, Carlsbad Caverns
[447] Bousman, C. Britt, 1973, An archaeological assessment of Carlsbad Caverns National Park
[456] Bradley, R. W., 1964, Geology of the Capitan Reef Complex of the Guadalupe Mountains, Culberson County, Texas and Eddy County, New Mexico. Field trip guidebook May 6, 7, 8, 9, 1964
[461] Branch, Pat, 1951, Illustrated guide book of your tour through Carlsbad Caverns
[542] Bryan, Kirk, 1941, Correlation of the deposits of Sandia Cave, New Mexico, with the glacial chronology
[551] Bullington, Neal R., May 1968, Who Discovered Carlsbad Cavern?
[576] Caiar, Ruth, 1957, One man's dream; the story of Jim White, discoverer and explorer of the Carlsbad Caverns: a biography
[577] Caldwell, Douglas E., 1981, Limnological survey of cavern pools: proposal to National Park Service, Carlsbad Caverns National Park
[599] Carlsbad Cave National Monument, 1930, Carlsbad Caverns, New Mexico

[600]Carlsbad Caverns National Park, Dec 1984, Land protection plan: Carlsbad Caverns National Park: prepared by Carlsbad Caverns National Park with assistance from the Southwest Regional Office, National Park Service.
[618]Cave Research Foundation, 1992, The quadrangle maps of Carlsbad Cavern, New Mexico
[720]Cosgrove, Cornelius Burton, 1875–, 1947, Caves of the Upper Gila and Hueco areas in New Mexico and Texas
[754]Crestwood House, 1988, National parks: v. 2: Carlsbad Caverns
[767]Crysler, Mildred G., 1955, Adventures in underground fairylands: a fantasy tour of the Carlsbad Caverns
[785]Curt Teich & Co, Inc, 1963, Carlsbad Caverns National Park, New Mexico
[804]Darton, Nelson Horatio, 1865–1948, 1932, Western Texas and Carlsbad Caverns
[835]Davis, M. J., 1954, Nobody here but us bats!: through Carlsbad Caverns with cartoons and comments
[866]Dick, Herbert W., 1965, Bat Cave
[925]Durden, C. C., 1928, The Carlsbad Cavern of New Mexico: it's [sic] history and geology
[934]Eaton, Joli (ed), [1988], GYPKAP 1987 annual report
[973]Ellis, Robert D. (scrn wrt), 1993, Carlsbad Caverns National Park
[979]Ely, Glen Sample. (photo, ed), 1990, A history of the Guadalupe Mountains and Carlsbad Caverns, 1840—1940
[1033]Farwell, Robin E., 1985, The Pictured Cliffs project: petroglyphs and talus shelters in San Juan County, New Mexico
[1047]Ferdon, Edwin N., 1946, An excavation of Hermit's cave, New Mexico
[1063]Finley-Holiday Film Corporation, 1980, Carlsbad Caverns National Park & Guadalupe Mountains National Park
[1127]Friedberg, Lionel, 1992, Mysteries underground
[1136]GAF Corporation , 1950, Carlsbad Caverns National Park
[1160]Garrett, Willie Ann, 1949, The Carlsbad Cavern National Park
[1240]Grant, Blanche Chloe, 1874–1948, 1927, Carlsbad Caverns
[1241]Grant, Blanche Chloe, 1874–1948, 1928, Cavern guide book, Carlsbad Caverns, Carlsbad, N. M
[1255]Greer, John W., 1966, Report on preliminary archeological explorations at Carlsbad Caverns National Park, New Mexico
[1398]Haynes, Caleb Vance, 1928–, 1986, Geochronology of Sandia Cave
[1425]Helm, Tex, 1969, Carlsbad Caverns; National Park, New Mexico
[1452]Hibben, Frank Cummings, 1910–, 1941, Evidences of early occupation in Sandia Cave, New Mexico, and other sites in the Sandia-Manzano Region
[1458]Hill, Carol Ann, 1940–, 1988, The geology of Carlsbad Cavern
[1459]Hill, Carol Ann, 1940–, 1987, Geology of Carlsbad Cavern and other caves in the Guadalupe Mountains, New Mexico and Texas
[1460]Hill, Carol Ann, 1940–, 1996, Geology of the Delaware Basin, Guadalupe, Apache, and Glass Mountains, West Texas and New Mexico
[1481]Holiday Film Corporation, 1970 1980, Carlsbad Caverns, New Mexico
[1527]Hosford, Jessie, 1953, Listen to the rain; a story of the Carlsbad Caverns
[1691]Jagnow, David Henry, Jan 1979, Cavern development in the Guadalupe Mountains
[1692]Jagnow, David Henry, 1992, Stories from stone: the geology of the Guadalupe Mountains
[1721]Jennings, Jesse David, 1909–, 1951, Mogollone material, Tularosa Cave, New Mexico
[1871]Kunath, Carl Edwin, 1940–(ed), 1978, The caves of McKittrick Hill, Eddy County, New Mexico: history and results of field work 1965-1976
[1883]Lambert, Marjorie F., 1961, A survey and excavation of caves in Hidalgo County, New Mexico
[1921]Leck Studio, 1920, The wonderful Carlsbad Caverns: near Carlsbad, New Mexico
[1924]Lee, Joli (Eaton) (ed), 1996, GYPKAP report volume 3: January 1992—April 1996
[1957]Long, Abijah, 1956, The big cave: early history and authentic facts concerning the history and discovery of the world famous Carlsbad Caverns of New Mexico
[2064]Martin, Paul Sidney, 1899–, 1954, Caves of the Reserve area
[2065]Martin, Paul Sidney, 1899–, 1952, Mogollon cultural continuity and change; the stratigraphic analysis of Tularosa and Cordova caves
[2133]McLean, John Scott, 1942–, Feb 1976, Factors altering the microclimate in Carlsbad Caverns, New Mexico
[2134]McLean, John Scott, 1942–, May 1971, The microclimate in Carlsbad Caverns, New Mexico
[2147]Mera, H. P., 1938, Reconnaissance and excavation in southeastern New Mexico
[2280]Moseley, John Judy, 1948, History, geology of Carlsbad Caverns: what' a hole
[2281]Moseley, John Judy, 1941, What' a hole: condensed Carlsbad Cavern [i.e. Caverns] information
[2383]National Speleological Society. Annual Convention. Guidebook (1960: NM), 1960, A guide book to Carlsbad Caverns National Park
[2407]National Speleological Society. Annual Convention. Guidebook (1986: Tularosa, NM), 1986, 1986 NSS convention guidebook, Tularosa, New Mexico
[2422]National Speleological Society. Annual Convention. Guidebook. Geology Field Trip (1986: Tularosa,

NM), 1986, Geology field trip guidebook, 45th National Speleological Society (1986) Tularosa, New Mexico, convention

[2492]New Mexico Geological Society, 1954, Guidebook of southeastern New Mexico: fifth field conference, October 21-22-23 & 24, 1954

[2493]New Mexico Resource Management and Development Division, 1986, Bandera Crater/Ice Cave: a state park feasibility study

[2509]Nicholson, Frank Ernest, Oct 1930, The exploration of Carlsbad Cavern

[2521]Northup, Diana Eleanor, 1948–, 1992, Lechuguilla Cave: biological inventory

[2537]Nymeyer, Robert, 1910–1983, 1991, Carlsbad Cavern, the early years: a photographic history of the cave and its people

[2538]Nymeyer, Robert, 1910–1983, 1978, Carlsbad, caves, and a camera

[2568]Oliver, Douglas L., 1930, A boy scout in the grand cavern

[2659]Petersen, David, 1994, Carlsbad Caverns National Park

[2747]Radlauer, Ruth, 1926–, 1981, Carlsbad Caverns National Park

[2831]Robison, Mabel Otis, 1966, The hole in the mountain

[2858]Roswell Geological Society, 1957, Slaughter Canyon, New Cave and Capitan Reef exposures, Carlsbad Caverns National Park: field trip no. 10, April 13, 1957

[2892]Savidge, Bob, [1995], Cwmbran Caving Club New Mexico expedition report; caving in the Guadalupe Mountains and Carlsbad National Park April 1994

[2913]Schultz, Charles Bertrand, 1908–, 1935, The fauna of Burnet Cave, Guadalupe Mountains, New Mexico

[2943]Shahan, Myrtle, 1949, Creation of Carlsbad Caverns

[2957]Shear, William A., 1974, North American cave millipeds II, an unusual new species (Dorypetalidae) from Southern California, and new records of *Speodesmus tuganbius* (Trichopolydesmidae) from New Mexico

[3043]Snyder, R. P., 1932–, 1982, Evaluation of breccia pipes in southeastern New Mexico and their relation to the Waste Isolation Pilot Plant (WIPP) site

[3044]Society for Visual Education, 1945, Carlsbad Caverns National Park

[3051]Sotnak, Lewann, 1988, Carlsbad Caverns

[3126]Stewart, John, 1920–, 1972, Secret of the bats; the exploration of Carlsbad Caverns

[3129]Stock, Chester, 1892–1950, 1930, 1932, Quaternary antelope remains from a second cave deposit in the Organ Mountains and further study of the quaternary antelopes of Shelter Cave, New Mexico

[3147]Story, Isabelle Florence, 1888–1970, 1935, Carlsbad Caverns National Park

[3192]Taylor, Michael Ray, 1959–(ed), 1991, Lechuguilla: jewel of the underground

[3220]Toll, Henry Wolcott, 1995, An analysis of variability and condition of cavate structures in Bandelier National Monument

[3262]U.S. Bureau of Land Management, Dec 1993, Final Dark Canyon environmental impact statement

[3268]U.S. Congress, House Committee on Natural Resources, May 1993, Lechuguilla Cave Protection Act of 1993: report (to accompany H.R. 698) (including cost estimate of the Congressional Budget Office)

[3271]U.S. Congress, Senate Committee on Energy and Natural Resources, Dec 1993, Lechuguilla Cave Protection Act of 1993: report (to accompany H.R. 698)

[3286]U.S. National Park Service, Sep 1980, Carlsbad Caverns National Park, New Mexico: wilderness reevaluation study: preliminary - subject to change

[3287]U.S. National Park Service, 1976, Carlsbad Caverns National Park: final interpretive prospectus

[3288]U.S. National Park Service, 1986, Carlsbad Caverns: official map & guide

[3290]U.S. National Park Service, Aug 1975, Development concept: Slaughter Canyon, Carlsbad Caverns National Park, New Mexico

[3291]U.S. National Park Service, Nov 1995, Draft general management plan, environmental impact statement: Carlsbad Caverns National Park, New Mexico

[3295]U.S. National Park Service, 1973, Final environmental statement: proposed wilderness, Carlsbad Caverns National Park, New Mexico

[3298]U.S. National Park Service, 1990, General management plan, wilderness suitability study: El Malpais National Monument, New Mexico

[3304]U.S. National Park Service, [1973], Master plan: Carlsbad Caverns National Park, New Mexico

[3312]U.S. National Park Service, 1973, Pollution abatement project, Carlsbad Caverns National Park, New Mexico

[3314]U.S. National Park Service, 1973, Proposed wilderness, Carlsbad Caverns National Park, New Mexico

[3320]U.S. National Park Service, Aug 1972, Wilderness recommendation, Carlsbad Caverns National Park, New Mexico

[3323]U.S. National Park Service, Sep 1974, Development concept plan, Slaughter Canyon, Carlsbad Caverns National Park, New Mexico: environmental assessment

[3444]West Texas Geological Society, 1969, Delaware Basin exploration: Guadalupe Mts., Hueco Mts., Franklin Mts., geology of the Carlsbad Caverns ; Nov. 6th, 7th and 8th, 1969

[3445]West Texas Geological Society, 1969, Delaware Basin exploration: Guadalupe Mts., Hueco Mts., Franklin Mts., geology of the Carlsbad Caverns, Nov. 6th, 7th and

8th, 1969

[3446]West Texas Geological Society, 1968, Guadalupe Mts., Hueco Mts., Franklin Mts., geology of the Carlsbad Caverns, Delaware Basin exploration, Oct. 31st, Nov. 1st and 2nd, 1968

[3447]West Texas Historical and Scientific Society, 1932, Publication no. 4

[3460]White, Jim, 1882–1946, 1932, The discovery and history of Carlsbad Caverns, New Mexico

[3462]White, Jim, 1919–, 1951, Carlsbad Caverns National Park, New Mexico: its early explorations as told by Jim White

[3521]Wood, Frances Elizabeth, 1963, Rocky Mountain, Mesa Verde, Carlsbad Caverns

[3526]Workman, C. Lindsay (narr), 1991, Carlsbad Caverns National Park

[3551]Zannes, Tom (prod, dir), 1993, Spirit of exploration: discovering the wonders of Carlsbad Caverns National Park

14.9.4.2.31 New York

[15]Addis, Robert Philip, 1945–, 1979, A study for the National Speleological Society: Knox Cave: Albany County, New York

[107]Anon, 1865, A description of Howe's Cave: with a popular treatise on the formation of caves in lime rock, from the size of a quill to a mammoth

[156]Anon, nd, Marvelous Howe Caverns near Cobleskill, New York

[192]Anon, c1995, The story of Howe Caverns

[444]Botto, Larry (comp), [1993], A partial guidebook to the caves of the Berkshires: 1993 spring N.R.O. hosted by the Berkshire Hills Grotto (May 14th, 15th, 16th)

[582]Callahan, Thomas P., c1974, The history of the Lockport caves

[668]Clymer, Virgil H.,–1952 (comp, ed), , Story of Howe Caverns

[669]Clyne, Patricia Edwards, 1980, Caves for kids: in historic New York: illustrated with maps and photographs

[768]Cudmore, Dana D., 1990, The remarkable Howe Caverns story

[769]Cullen, James J., Jul 1979, Karst hydrology and geomorphology of eastern New York: a guidebook to the geology field trip, National Speleological Society Annual Convention, Pittsfield, Massachusetts, August 5-12, 1979

[833]Davis, Harold (ed), 1966, Caves of Schoharie County, New York

[911]Dumont, Kevin Allen, 1995, Karst hydrology and geomorphology of the Barrack Zourie Cave System, Schoharie County, New York

[990]Engel, Thomas Daniel, 1954–, Jul 1979, A chronicle of selected northeastern caves: a history guide for the 1979 NSS Convention, Pittsfield, Massachusetts, August 5-12, 1979

[996]Ensminger, Scott A., 1987, The caves of Niagara County, New York

[1066]Fisher, John L. (ed), 1958, Caves of the Watertown Area, New York state

[1130]Funk, Robert E., 1994, Archaeological and paleoenvironmental investigations in the Dutchess Quarry Caves, Orange County, New York

[1449]Hesler, Donald J., 1990, A hydrogeologic study of the Knox-Skull Cave System, Albany County, New York

[1512]Holt, Robert (comp), 1989, Howe Caverns; Howes Cave N.Y.; 60th Anniversary, 1929-1989

[1544]Howe Caverns, Inc, nd, Marvelous Howe Caverns

[1545]Howe Caverns, Inc, 1930, The story of Howe caverns...

[1547]Howe, Warren, 1988, The Howe Family From Massachusetts to New York: A story of Lester Howe, his ancestors, and the Howe Caverns

[1565]Hudson, George Henry, 1855–1934, 1911, Joint caves of Valcour Island: their age and their origin

[1770]Kastning, Ernst H., 1944–, 1975, Cavern development in the Helderberg Plateau, East-central New York

[1771]Kastning, Ernst H., 1944–, Jun 1988, Caverns of the Shawangunk (Shon-gum) and its environs, Southeastern New York

[1834]Knoepfel, William H., 1853, An account of Knoepfel's Schoharie Cave, Schoharie County, New York: with the history of its discovery, subterranean lake, minerals and natural curiosities

[2329]Mylroie, John Eglinton, 1949–, 1977, Speleogenesis and karst geomorphology of the Helderberg Plateau, Schoharie County, New York

[2401]National Speleological Society. Annual Convention. Guidebook (1979: Pittsfield, MA), 1979, An introduction to caves of the northeast: guidebook for the 1979 National Speleological Convention, Pittsfield, Massachusetts

[2421]National Speleological Society. Annual Convention. Guidebook. Geology Field Trip (1979: Pittsfield, MA), Aug 1979, Karst hydrogeology and geomorphology of eastern New York: a guidebook to the geology field trip, National Speleological Annual Convention, Pittsfield, Massachusetts August 5-12, 1979

[2520]Northeastern Regional Organization [NRO]. National Speleological Society (comp), nd, Gage Caverns

[2601]Palmer, Arthur Nicholas, 1940–, Jun 1963, Knox Cave: Albany County, NY

[2656]Perry, Clair Willard [Clay], 1887–1961, 1948, Underground empire: wonders and tales of New York caves

[2696]PostCard Videos, Inc, nd, Howe Caverns and New York Central Leatherstocking Region

[2921]Schweiker, Roioli, , Caves of Albany County, New York

[2922]Schweiker, Roioli, 1958, Field guide to New York

caves: Schoharie County

[3096] Sprague, Stuart, c1960, Caves of Herkimer County
[3130] Stoddard, Seneca Ray, 1844–1917, 1889, Howe's Cave
[3252] Turner, Brian B., 1973, Natural stone bridge and caves; fascinating geology, history & legends
[3283] U.S. Geological Survey, 1967, Bibliography of the Ground Water Resources of New York through 1967, with subject and location cross-reference to the annotated bibliography
[3349] Van Voris, Arthur H., 1970, The lesser caverns of Schoharie County

14.9.4.2.32 North Carolina

[11] Adams, Laurie Branson (comp), Oct 1986, Annotated bibliography of North Carolina speleology
[525] Brown, Henry S., 1961, Linville Caverns through the ages: the geological story
[1488] Holler, Cato Oliver, 1944–, 1989, Hollow Hills of Sunnalee: The Linville Caverns story
[1489] Holler, Cato Oliver, 1944–(ed), 1975, North Carolina cave survey vol. 1 no. 1
[1490] Holler, Cato Oliver, 1944–(ed), 1979, North Carolina cave survey vol. 2
[2036] Maness, Lindsey Vance, 1994, North Carolina coastal plain caves and their impact on mining and quarrying

14.9.4.2.33 Ohio

[167] Anon, 1976, Ohio Caverns
[168] Anon, 1930, The Ohio Caverns: where nature carved a fairyland
[493] Brizius, Janice J., 1900, History of the 7 caves
[658] Church, Flora, 1954–, Jun 1990, Mitigation of the Brady Run Rockshelter 3: a multi-component site in Washington Township, Lawrence County, Ohio
[2504] Newell, Amy L., 1995, The caves of Put-in-Bay
[2546] Ohio Caverns, Inc., 1971, Ohio Caverns: where nature carved a fairyland
[2728] Prufer, Olaf H., 1989, Krill Cave: a stratified rockshelter in Summit County, Ohio
[2944] Shaler, Nathaniel Southgate, 1841–1906, 1876, On the antiquity of the caverns and cavern life of the Ohio valley
[3459] White, George W., 1926, The limestone caves and caverns of Ohio

14.9.4.2.34 Oklahoma

[409] Black, Jeffrey H., 1971, The cave life of Oklahoma: a preliminary study (excluding chiroptera)
[453] Bozeman, Sue, 1987, The D. C. Jester cave system
[1788] Keller, David R. (ed), 1993, Paleokarst, karst-related diagenesis, reservoir development, and exploration concepts: examples from the Paleozoic section of the southern mid-continent: 1993 annual fieldtrip guidebook Permian Basin Section - SEPM Arbuckle Mountains, Oklahoma
[2325] Myers, Arthur John, 1918–, 1982, Guide to Alabaster Cavern and Woodward County, Oklahoma
[2326] Myers, Arthur John, 1918–, 1969, Guide to Alabaster Cavern and Woodward County, Oklahoma
[2548] Oklahoma Department of Wildlife Conservation, 1990, Survey and species determination of cave crayfish in Oklahoma
[2869] Russell, Donald R., 1971, An essay on the caves of the Cherokee Nation: a conservation plea
[3424] Weaver, Herman Dwight, 1938–, 1992, The wilderness underground: caves of the Ozark plateau

14.9.4.2.35 Oregon

[17] Agee, James K., 1990, Oregon Caves forest and fire history
[35] Aley, Thomas John, 1938–, 1988, Control of exotic plants in Oregon Caves, Oregon Caves National Monument
[42] Aley, Thomas John, 1938–, 1987, Restoration of natural cave features, Oregon Caves National Monument, Oregon: final report
[43] Aley, Thomas John, 1938–, 1988, Restoration of natural microclimate in Oregon Caves, Oregon Caves National Monument: final report
[171] Anon, [1993], Oregon Caves: a pictorial souvenir guide
[180] Anon, 1994, Sea Lion Caves Oregon coast
[181] Anon, c1993, Sea Lion Caves: America's largest sea cave, on U.S. Highway 101...just North of Florence, Oregon, Oregon coast
[189] Anon, 1969, Souvenir book of the Oregon Caves National Monument: Marble Halls of Oregon
[382] Benedict, Ellen Maring, 1974, Cave ecology: a course book for the cave ecology class at Malheur Environmental Field Station
[433] Bookout, John, nd, Named caves of Oregon
[567] Bush, Kent, 1995, Oregon Caves National Monument collections management plan
[696] Contor, Roger J., 1963, The underworld of Oregon Caves National Monument
[727] Cowles, John, 1959, Cougar Mountain Cave in south central Oregon
[837] Davis, Wilbur A., 1964, Archaeological surveys of Crater Lake National Park and Oregon Caves National Monument, Oregon
[897] Draper, John A., 1989, Archaeology of Chokecherry Cave (35GR500) in Grant County, southeastern Oregon

[912] Dunham, Wayland A., 1898–, 1939, Enchanted corridors
[933] Easton, T. S., 1934, The secret of the Wallowa Cave
[942] Edelbrock, Gay, 1993, Oregon Caves National Monument resource database: a user's guide
[1012] Evans, Gregory T. , 1976, Oregon Caves National Monument: a manual for cave guides
[1117] Frank, Cathleen, Jun 1986, Legislative history for Oregon Caves National Monument, 59th Congress through 96th Congress
[1248] Greeley, Ronald, , Geology and morphology of selected lava tubes in the vicinity of Bend, Oregon
[1249] Greeley, Ronald, 1972, Geology of selected lava tubes in the Bend Area, Oregon
[1740] Johnston, David Alexander, 1949-1980, 1981, Guide to some volcanic terranes in Washington, Idaho, Oregon and Northern California
[1838] Knutson, Richard Stephen (comp), 1966, Bibliography of Oregon speleology: initial compilation
[1839] Knutson, Richard Stephen, nd, The caves of Deschutes County, Oregon
[1840] Knutson, Richard Stephen, nd, The speleological potential of the Big Bar Limestone deposit, Hell's Canyon, Oregon-Idaho
[1902] Larson, Charles Victor, 1928–, 1977, Bibliography of Oregon speleology, 1977
[1903] Larson, Charles Victor, 1928–, nd, A bibliography of Sea Lion Cave
[1904] Larson, Charles Victor, 1928–, 1975, Caves of Oregon
[1905] Larson, Charles Victor, 1928–, 1987, Central Oregon caves
[1908] Larson, Charles Victor, 1928–, 1987, Lava River Cave
[2403] National Speleological Society. Annual Convention. Guidebook (1982: Bend, OR), 1982, An introduction to caves of the Bend area: guidebook of the 1982 NSS Convention
[2414] National Speleological Society. Annual Convention. Guidebook (1993: Pendleton, OR), 1993, The 1993 NSS Convention guidebook
[2419] National Speleological Society. Annual Convention. Guidebook. Geology and Biology Field Trip (1982: Bend, OR), 1982, Caves and other volcanic landforms of Central Oregon
[2512] Nieland, James Raymond, 1949–, 1969, Little known caves of Oregon
[2732] Purcell, David, 1977, Guide to the lava tube caves of central Oregon
[2733] Purcell, David, 1977, Spelunking (the exploration of caves) guide to central Oregon
[2848] Rogers, Edmund B., 1958, Oregon Caves National Monument, Oregon: history of legislation through the 82nd Congress
[3014] Smith, Burrell F., 1988, The cave and her men: Oregon Caves National Monument: its history and formation
[3110] Steel, William Gladstone, 1931, Oregon, no 32, vol 1
[3277] U.S. Forest Service, 1924, The Oregon Caves, Siskiyou National Forest
[3278] U.S. Forest Service, 1900, The Oregon Caves: Siskiyou National Forest
[3307] U.S. National Park Service, 1976, Oregon Caves
[3308] U.S. National Park Service, 1975, Oregon Caves National Monument, master plan
[3309] U.S. National Park Service, 1976, Oregon Caves National Monument, Oregon: final interpretive prospectus
[3310] U.S. National Park Service, 1992, Oregon Caves National Monument, Oregon: official map and guide
[3379] Walsh, Frank K., 1971, Discovery and exploration of the Oregon caves: Oregon Caves National Monument
[3380] Walsh, Frank K., 1982, Oregon Caves discovery & exploration: Oregon Caves National Monument
[3381] Walsh, Frank K., 1976, Oregon Caves: discovery and exploration

14.9.4.2.36 Pennsylvania

[215] Anon, 1920, Wonderful, famous, spectacular Crystal Cave, Kutztown, Penna [sic]
[233] Ashbrook, Bert, Dec 1990, The Caves of Northumberland County, Pennsylvania
[646] Chenger, John D., 1995, Visitor's guide to Laurel Caverns Park
[772] Cullinan, Mike, Jun 1975, The caves of Huntingdon County, Pennsylvania
[798] Damon, Paul Herbert, 1934–, Jun 1976, The history of Laurel Caverns of Fayette County, Pennsylvania, known throughout the years as Delaney's Cave
[799] Damon, Paul Herbert, 1934–, Oct 1977, Thirty years with the Pittsburgh Grotto, National Speleological Society
[843] Dayton, Gordon O. (comp), Feb 1979, The caves of Centre County, PA
[844] Dayton, Gordon O. (comp), Jul 1981, The caves of Mifflin County, Pennsylvania
[882] Donehoo, George Patterson, 1862–1934, 1930, Pennsylvania's Historic Indian Cave at Franklinville, PA, Huntington County...
[1284] Guilday, John E., 1925–1982, 1964, New Paris no. 4: a late Pleistocene cave deposit in Bedford County, Pennsylvania
[1428] Hempel, John Charles, 1949–(comp), 1969, Come crawl into the caves of western Pennsylvania
[1592] Ibberson, Dale (ed), Jan 1983, The caves of Perry County, Pennsylvania
[1845] Kohler, S. D. F., 1971, The Crystal Cave at Virginsville

[1864]Kranzel, Rich, 1983, Caves of Bucks County, Pennsylvania

[1865]Kranzel, Rich, Apr 1985, Caves of Lehigh County, Pennsylvania

[1879]La Rock, Edward J., Dec 1976, The caves of Snyder County, Pennsylvania

[2149]Mercer, Henry Chapman, 1856–1930, 1899, The bone cave at Port Kennedy, Pennsylvania, and its partial excavation in 1894, 1895 and 1896

[2151]Mercer, Henry Chapman, 1856–1930, 1897, An exploration of Durham Cave in 1893

[2282]Mostardi, Michael (comp), Dec 1991, Caves of Berks County, Pennsylvania

[2392]National Speleological Society. Annual Convention. Guidebook (1970: Pittsburgh, PA), 1970, NSS 1970 guidebook: Mid-Appalachian Region

[2398]National Speleological Society. Annual Convention. Guidebook (1976: Morgantown, WV), 1976, 1976 annual convention guidebook, Morgantown, W. Va

[2763]Rauch, Henry W., nd, Lithologic controls on the development of solution porosity in carbonate aquifers

[2802]Reich, J. R. (comp), 1969, Caves of Pennsylvania's Piedmont region

[2803]Reich, J. R. (comp), 1974, Caves of southeastern Pennsylvania

[2971]Shoemaker, Henry Wharton, 1880–1958, 1929, The legends of the caverns of Centre County, Pennsylvania

[2972]Shoemaker, Henry Wharton, 1880–1958, , Penn's grandest cavern: the history, legends and description of Penn's Cave in Centre County, Pennsylvania

[2973]Shoemaker, Henry Wharton, 1880–1958, , Penn's grandest cavern: the history, legends photographs and description of Penn's Cave in Centre County, Pennsylvania

[3007]Smeltzer, Bernard L., Oct 1964, Caves of the southern Cumberland valley

[3042]Snyder, Dean H. (comp, ed), Feb 1989, The caves of Northampton County, Pennsylvania

[3085]Speece, Jack H. (comp, ed), Jul 1972, The caves of Blair County, Pennsylvania

[3086]Speece, Jack H. (ed), Feb 1969, Fulton County, Pennsylvania

[3088]Speece, Jack Howard, 1947–, 1973, Alexander Caverns: the Carlsbad of Pennsylvania

[3135]Stone, Ralph Walter, 1876–1964, 1953, Caves of Pennsylvania

[3136]Stone, Ralph Walter, 1876–1964, 1953, Descriptions of Pennsylvania's undeveloped caves

[3137]Stone, Ralph Walter, 1876–1964, 1930, Pennsylvania caves

[3189]Taylor, Kirk R. (comp), Jan 1993, Caves of Bedford County, Pennsylvania

[3431]Weinel, John, 1979, Pittsburgh Grotto Picnic Fall M.A.R. '79: Guidebook, Sept. 14-15, 1979

[3469]White, William Blaine, 1934–, May 1992, The Appalachian valleys of central Pennsylvania

[3471]White, William Blaine, 1934–(ed), 1964, Caves of Pennsylvania

[3472]White, William Blaine, 1934–(ed), 1976, Caves of Western Pennsylvania compiled by members of the Mid-Appalachian Region of the National Speleological Society

[3473]White, William Blaine, 1934–(ed), 1976, Geology and biology of Pennsylvania caves

14.9.4.2.37 South Carolina

[1471]Hockensmith, Brenda Louise, Feb 1987, Investigation of sinkhole occurrences at Goretown, near Loris, South Carolina

[3094]Spigner, B. C., Jul 1978, Review of sinkhole-collapse problems in a carbonate terrane

14.9.4.2.38 South Dakota

[24]Alex, Lynn Marie, 1948–, 1991, The archaeology of the Beaver Creek shelter (39CU779), Wind Cave National Park, South Dakota: a preliminary statement

[25]Alexander, E. Calvin, 1987, Hydrogeologic study of Jewel Cave/Wind Cave: final report -year 2, 4th quarter of year 2, February 1, 1987 through April 30, 1987

[26]Alexander, E. Calvin, 1989, Hydrologic study of Jewel Cave/Wind Cave: final report

[28]Alexander, Emmit Calvin, 1943–, 1986, Hydrogeologic study of Jewel Cave/Wind Cave: final report - phase I, 3rd and 4th quarters of phase I, 1 November 1985 through 30 April 1986

[425]Bohi, John W., 1963, History of Wind Cave

[426]Bohi, John W., 1962, Seventy-five years at Wind Cave: a history of the national park

[531]Brown, Robert F., Oct 1964, Exploration in Wind Cave

[692]Conn, Herbert Dunn, 1981, The Jewel Cave adventure: fifty miles of discovery under South Dakota

[929]Dyer, C. F., 1961, Geology and occurrence of ground water at Jewel Cave National Monument, South Dakota

[1104]Fossen, Joyce, 1985, Caves of Paha Sapa, and other findings of Dr. Knows-It

[1260]Gries, John Paul, 1911-, Jun 1969, Continued investigation of water losses to sinkholes in the Pahasapa Limestone and their relation to resurgent springs, Black Hills, South Dakota

[1341]Hamilton, Holman, 1957, The Cave of the Winds' and the compromise of 1850

[1516]Horn, Edward Courtright, 1871-, 1901, Mazes and marvels of Wind cave

[2116]McDonald, Alvin F., 1873–1893, nd, Private account of A. F. McDonald, permanent guide of Wind Cave

[2143]Meleen, Elmer E., Apr 1941, A preliminary report on rock shelters in Fall River County, South Dakota

[2356]National Speleological Society, , Black Hills symposium
[2360]National Speleological Society , 1990, Wind Cave: the world below
[2384]National Speleological Society. Annual Convention. Guidebook (1962: [Black Hills], SD) , 1962, A guide to the caves of the Black Hills, S. D
[2409]National Speleological Society. Annual Convention. Guidebook (1988: Hot Springs, SD) , 1988, 1988 NSS convention guidebook
[2487]Neighbor, Frank, Nov 1954, Major limestone caverns of the Black Hills, South Dakota
[2584]Owen, Luella Agnes, 1970, Cave regions of the Ozarks and Black Hills
[2598]Palmer, Arthur Nicholas, 1940–, 1981, The geology of Wind Cave
[2600]Palmer, Arthur Nicholas, 1940–, 1984, Jewel Cave: a gift from the past
[2602]Palmer, Arthur Nicholas, 1940–, , Wind Cave: an ancient world beneath the hills
[2834]Rocky Mountain Region. Guidebook (1991), 1991, Top secret, for cavers' eyes only: 1991 Rocky Mountain Regional guidebook, Black Hills, South Dakota
[2849]Rogers, Edmund B., 1958, Wind Cave National Park, South Dakota: history of legislation through the 82nd Congress
[2900]Scheltens, John (auth, comp), [1973], Wind Cave 1972
[2901][Scheltens, John(ed)], [1971], Wind Cave expedition 1971
[2902][Scheltens, John (ed)], [1970], Wind Cave expedition August 1970: a report to the National Park Service by the Windy City Grotto of the National Speleological Society on the results of the August 1970 Wind Cave expedition
[3250]Tullis, Edward L., 1940, Black Hills caves
[3257][Tyers, John], 1962, A guide to Wind Cave and Wind Cave National Park. 19th annual convention of the National Speleological Society self guiding tour of the commercial areas of Wind Cave
[3293]U.S. National Park Service, Jun 1993, Draft, general management plan, environmental impact statement Jewel Cave National Monument, South Dakota
[3297]U.S. National Park Service, Jun 1994, Final, general management plan, environmental impact statement Jewel Cave National Monument, South Dakota
[3311]U.S. National Park Service, 1992, Paleontology
[3315]U.S. National Park Service, 1987, Statement for management: Jewel Cave National Monument
[3316]U.S. National Park Service, Jul 1986, Statement for management: Wind Cave National Park
[3321]U.S. National Park Service, 1937, Wind Cave National Park, South Dakota
[3322]U.S. National Park Service, 1979, Wind Cave: National Park Service, South Dakota

14.9.4.2.39 Tennessee

[12]Adams, Laurie Branson, Aug 1984, History of Morrell Cave: Part I Tennessee
[64]Anchors, William E., 1954–, Mar 1989, Ruskin and Jewel Caves: a brief history
[158]Anon, 1903, Monteagle Wonder Cave, or, The Cave of the Cumberland
[278]Bailey, Thomas L., Apr 1918, Report on the caves of the eastern highland rim and Cumberland Mountains
[285]Baker, Eva G., , Water resources publications of the USGS for Tennessee, 1907-1987
[326]Barr, Thomas Calhoun, 1931–, 1972, Caves of Tennessee
[328]Barr, Thomas Calhoun, 1931–, 1956, The geology of Cumberland Caverns, Warren County, Tennessee
[479]Brinkley, Edward Booth, 1922–, 1964, The history of Ruby Falls
[677]Coggins, Allen R., 1981, The caves of the Tennessee State Natural Areas system
[738]Crawford, Nicholas Charles, 1942–, 1982, Guidebook to karst and caves of Tennessee: emphasis on the Cumberland plateau escarpment region and guidebook to karst and caves of the Ozark Region of Missouri and Arkansas: prepared for the Eighth International Congress of Speleology, Bowling Green, Kentucky, U. S. A. July 18 to 24 1981
[740]Crawford, Nicholas Charles, 1942–, 1988, Hydrogeology of the Snail Shell Cave: overall creek drainage basin and the ecology of the Snail Shell Cave System
[742]Crawford, Nicholas Charles, 1942–, 1982, Karst hydrogeology of Tennessee. Guidebook prepared for Karst Hydrogeology Workshop August 31 - September 3, 1982 Nashville, Tennessee
[743]Crawford, Nicholas Charles, 1942–, 1980, The karst hydrogeology of the Cumberland Plateau Escarpment of Tennessee. Part IV. Erosional processes associated with subterranean stream invasion, conduit cavern development and slope retreat
[744]Crawford, Nicholas Charles, 1942–, 1979, The karst hydrogeology of the Cumberland Plateau escarpment of Tennessee: Part I: subterranean stream invasion, conduit cavern development, and slope retreat in the Lost Creek Cove area, White County, Tennessee
[745]Crawford, Nicholas Charles, 1942–, 1979, The karst hydrogeology of the Cumberland Plateau escarpment of Tennessee: Part II: karst valley development and the headward advance of the Sequatchie Valley in the Grassy Cove area, Cumberland County, Tennessee
[746]Crawford, Nicholas Charles, 1942–, 1980, The karst hydrogeology of the Cumberland Plateau escarpment of Tennessee: Part III: karst valley development in the Lost

14.9. NORTH AMERICA

Cove area, Franklin County, Tennessee
[747]Crawford, Nicholas Charles, 1942–, 1987, The karst hydrogeology of the Cumberland Plateau escarpment of Tennessee: subterranean stream invasion, conduit cavern development, and slope retreat in the Lost Creek Cove area, White County, Tennessee
[750]Crawford, Nicholas Charles, 1942–, 1979, Safford Centennial Society field trip, fall 1979: guidebook: karst hydrogeology of the Cumberland plateau escarpment in the Lost Creek Cave and Karst Cave areas of Tennessee
[764]Crothers, George Martin, Apr 1986, Final report on the survey and assessment of the prehistoric and historic archaeological remains in Big Bear Cave, Van Buron County, Tennessee
[1034]Faulkner, Charles H. (ed), 1986, The prehistoric native American art of Mud Glyph Cave
[1279]Guilday, John E., 1978, The Baker Bluff cave deposit, Tennessee, and the late Pleistocene faunal gradient
[1281]Guilday, John E., 1972, Jaguar (*Panthera onca*) remains from Big Bone Cave, Tennessee and east central North America
[1393]Hay, William Perry, 1872–, 1903, Observations on the crustacean fauna of Nickajack cave, Tennessee, and vicinity
[1503]Holsinger, John Robert, 1934–, Jun 1988, The invertebrate cave fauna of Virginia and a part of eastern Tennessee
[1668]Irving, David, nd, A guide to the caves in the Oak Ridge-Knoxville area
[1789]Kemmerly, Philip R., 1980, Subhold collapse in Montgomery County, Tennessee: an overview for the planning process
[1826]Klippel, Walter E., nd, The paleontology of Cheek Bend Cave, Maury County, Tennessee: phase II: report to the Tennessee Valley Authority
[1958]Long, Gary F. (auth, des, photo), nd, Tuckaleechee Caverns: 31 photos in natural color: Townsend, Tennessee: greatest sight under the Smokies
[2087]Matthews, Larry Edwin, 1946–(ed), 1974, Bibliography of Tennessee speleology
[2088]Matthews, Larry Edwin, 1946–, 1989, Cumberland Caverns
[2089]Matthews, Larry Edwin, 1946–, 1971, Descriptions of Tennessee caves
[2115]McCrady, Edward, 1906–, 1954, New finds of Pleistocene jaguar skeletons from Tennessee caves
[2152]Mercer, Henry Chapman, 1856–1930, nd, The finding of the remains of the fossil sloth at Big Bone Cave, Tennessee, in 1896
[2264]Moore, Gerald K., 1969, Limestone hydrology in the Upper Stones River Basin, central Tennessee
[2410]National Speleological Society. Annual Convention. Guidebook (1989: Sewanee, TN), 1989, Caves and caving in TAG: a guidebook for the 1989 Convention of the National Speleological Society ; Sewanee, Tennessee
[2448]National Speleological Society. Annual Convention. Supplemental Guidebook (1989: Sewanee, TN), 1989, A guide to East Tennessee Caves: prepared for the 1989 National Speleological Society Convention, Sewanee, Tennessee
[2621]Payne, R. M., nd, Wonder cave of Tennessee
[2905]Schindel, Geary Michael, 1957–, Dec 1991, Guidebook: environmental hydrogeology of karst terranes in the vicinity of Nashville Tennessee
[3028]Smith, Marion Otis, 1942–(comp), 1992, Letters from TAG 1966-1969
[3029]Smith, Marion Otis, 1942–, 1990, Saltpeter mining in East Tennessee
[3030]Smith, Marion Otis, 1942–, c.1973, TAG pits
[3063]Southeastern Regional Association (S.E.R.A.) Cave Carnival (1968: TN), [1968], 1968 SERA cave carnival
[3064]Southeastern Regional Association (S.E.R.A.) Cave Carnival (1975: Monterey, TN), 1975, Guidebook to the caves of the Monterey, Tennessee Area
[3066]Southeastern Regional Association (S.E.R.A.) Cave Carnival (1978: TN), 1978, Guidebook to the caves of the 1978 summer SERA cave carnival
[3067]Southeastern Regional Association (S.E.R.A.) Cave Carnival (1979: Ketner's Mill, TN), 1979, 1979 Southeastern Regional Association (S.E.R.A.) cave carnival at Ketner's Mill, TN
[3069]Southeastern Regional Association (S.E.R.A.) Cave Carnival (1989: [TN]), [1989], SERA 89
[3071]Southeastern Regional Association (S.E.R.A.) Cave Carnival (1992: AL), [1992], SERA cave carnival 1992
[3072]Southeastern Regional Association (S.E.R.A.) Cave Carnival (1994: Short Mountain, TN), 1994, Guidebook to the 1994 SERA cave carnival: June 10-12, 1994 CCWHA Campground at Short Mountain Cannon County, Tennessee
[3074]Southeastern Regional Association (S.E.R.A.) Cave Carnival (1996: Camp Jackson, Jackson Co. AL), 1996, 45th Annual SERA Summer cave carnival: July 19-21, 1996, Camp Jackson, Jackson County, Alabama
[3164]Sweatman, Cheyenne (ed), Oct 1985, 8th Annual TAG fall cave-in
[3197]Tennessee Academy of Science, Jul 1930, Journal of the Tennessee Academy of Science. Cave number
[3552]Zawislak, Ronald L., 1967, The Tennessee pit survey

14.9.4.2.40 Texas

[19]Ainsworth, Joseph, Apr 1960, Saint Mary's University Speleological Society: 1954-1959
[31]Alexander, Robert K., 1970, Archeological investiga-

tions at Parida Cave, Val Verde County, Texas
[76]Andrews, Rhonda Lynette, 1980, Perishable industries from Hinds Cave, Val Verde County, Texas
[163]Anon, nd, Natural Bridge Caverns: Texas' largest caverns: One of the great show caves of the world
[268]Ayer, Mary Youngman, 1936, The archaeological and faunal material from Williams Cave, Guadalupe Mountains, Texas
[348]Baumgardner, Robert W., 1982, Formation of the Wink Sink: a salt dissolution and collapse feature, Winkler County, Texas
[408]Black, Jackie (ed), 1973, Caves, ground water and karst features of the Cretaceous in South central Texas: fourteenth annual S.A.S.G.S. field conference, April 6-8, 1973
[518]Brown, Deana, nd, Caverns of Sonora: an awe-inspiring adventure you will remember forever
[537]Brune, Gunnar, 1981, Springs of Texas
[720]Cosgrove, Cornelius Burton, 1875–, 1947, Caves of the Upper Gila and Hueco areas in New Mexico and Texas
[794]Dalquest, Walter W., Jun 1969, The mammal fauna of Schulze Cave, Edwards County, Texas
[804]Darton, Nelson Horatio, 1865–1948, 1932, Western Texas and Carlsbad Caverns
[811]Davenport, J. Walker, 1933, Archaeological exploration of Eagle Cave, Langtry, Texas
[865]Dibble, David S., Jan 1968, Bonfire shelter: a stratified bison kill site, Val Verde County, Texas
[961]Elliott, William Rawleigh, 1946–, Jun 1993, Draft recovery plan for endangered karst invertebrates in Travis and Williamson Counties, Texas
[962]Elliott, William Rawleigh, 1946–(ed), Dec 1985, A field guide to the caves of Kendall County
[998]Epstein, Jeremiah F., 1960, Centipede and Damp caves: excavations in Val Verde County, Texas, 1958
[1006]Estes, James H. , Mar 1967, Deep Cave, Texas. A preliminary report of the 1965 Project "Deep"
[1011]Evans, Glen Louis, 1911–, 1961, The Friesenhahn Cave
[1056]Fieseler, Ronald G. (ed), Apr 1975, The caves of Brewster and western Pecos Counties
[1119]Frank, Ruben Milton, 1935–, Sep 1964, The vertebrate paleontology of Texas caves
[1224]Goode, Glenn T., 1985, Archaeological testing of the cave at site 41BX22, Bexar County, Texas
[1244]Graves, John (ed), , The River Styx-Salt Spring Cave system
[1459]Hill, Carol Ann, 1940–, 1987, Geology of Carlsbad Cavern and other caves in the Guadalupe Mountains, New Mexico and Texas
[1460]Hill, Carol Ann, 1940–, 1996, Geology of the Delaware Basin, Guadalupe, Apache, and Glass Mountains, West Texas and New Mexico
[1492]Holman, J. Alan, 1931–, Mar 1987, A Mid-Pleistocene (Irvingtonian) herpetofauna from a cave in southcentral Texas
[1508]Holsinger, John Robert, 1934–, 1980, The subterranean amphipod crustacean fauna of an artesian well in Texas
[1692]Jagnow, David Henry, 1992, Stories from stone: the geology of the Guadalupe Mountains
[1709]Jasek, James F. , 1982, Inner Space Cavern
[1792]Kennard, Don (ed), 1975, Devil's Sinkhole area: headwaters of the Nueces River
[1808]Kerans, C., 1954–, 1990, Depositional systems and karst geology of the Ellenburger Group (Lower Ordovician), subsurface west Texas
[1809]Kerans, C., 1954–, 1989, Karst-controlled reservoir heterogeneity and an example from the Ellenburger Group (Lower Ordovician) of west Texas
[1837]Knox, Orion, 1976, Natural Bridge Caverns
[1870]Kunath, Carl Edwin, 1940–(ed), Oct 1995, The caves of Carta Valley, Edwards and Val Verde Counties, Texas
[1872]Kunath, Carl Edwin, 1940–(ed), Apr 1968, The caves of the Stockton Plateau, Texas
[1873]Kunath, Carl Edwin, 1940–, Mar 1965, Ye olde history
[1959]Longley, Glenn, Nov 1977, Preliminary report of biological investigation Valdina Farms sink hole - Medina Co., Texas
[1960]Longley, Glenn, Jun 1978, Status of Trogloglanis pattersoni Eigenmann, the toothless blindcat, and status of Satan eurystomus Hubbs and Bailey, the widemouth blindcat
[1962]Longley, Glenn, 1942–, Feb 1977, Status of *Typhlomolge* (=*Eurycea*) *rathbuni*
[1989]Lundelius, Ernest L., 1927–(ed), 1971, Natural history of Texas caves
[2061]Martin, George Castor, 1885–, 1933, Archaeological exploration of the Shumla caves; report of the George C. Martin expedition... June, July and August, 1933
[2094]Matthews, William Henry, 1919–, Feb 1963, The geologic story of Longhorn Cavern
[2130]McKenzie, David (ed), 1964, The caves of Bell and Coryell Counties
[2162]Messling, Gordon, 1978, Caverns of Sonora
[2357]National Speleological Society, Apr 1948, The caves of Texas
[2386]National Speleological Society. Annual Convention. Guidebook (1964: New Braunfels, TX), May 1964, A guide to the caves of Texas and 1964 N.S.S. convention (New Braunfels, TX) field trips
[2400]National Speleological Society. Annual Convention. Guidebook (1978: New Braunfels, TX), 1978, An introduction to the caves of Texas: prepared for the 1978 National Speleological Society Convention New Braunfels, Texas

14.9. NORTH AMERICA

[2415] National Speleological Society. Annual Convention. Guidebook (1994: Brackettville, TX), 1994, The caves and karst of Texas: a guidebook for the 1994 Convention of the National Speleological Society with emphasis on the Southwestern Edwards Plateau: Brackettville, Texas, June 19-24, 1994

[2420] National Speleological Society. Annual Convention. Guidebook. Geology Field Excursion (1978: New Braunfels, TX), 1978, Caves and karst hydrogeology of the southeastern Edwards Plateau, Texas: guidebook, geology field excursion, National Speleological Society annual convention, New Braunfels, Tex., June 18-23, 1978

[2545] O'Donnell, Lisa, Aug 1994, Recovery plan for endangered karst invertebrates in Travis and Williamson Counties, Texas

[2618] Patton, Thomas Hudson, 1934–, Sep 1963, Fossil vertebrates from Miller's Cave, Llano County, Texas

[2679] Pipkin, Turk (ed), 1993, Barton Springs eternal: the soul of a city

[2691] Poole, Gary A. (ed), 1978, Bexar County speleology vol 1

[2770] Reddell, James Russell, 1938–, Oct 1968, A bibliographic guide to Texas speleology

[2772] Reddell, James Russell, 1938–(ed), nd, The caves of Langtry

[2773] Reddell, James Russell, 1938–(ed), Mar 1962, The caves of Bexar County

[2774] Reddell, James Russell, 1938–, Jun 1964, The caves of Comal county

[2775] Reddell, James Russell, 1938–, Nov 1965, The caves of Edwards County

[2776] Reddell, James Russell, 1938–(ed), Apr 1977, The caves of far west Texas

[2777] Reddell, James Russell, 1938–(ed), Mar 1970, The caves of Lubbock County

[2778] Reddell, James Russell, 1938–(ed), Apr 1967, The caves of Medina county

[2779] Reddell, James Russell, 1938–(ed), Jun 1963, The caves of northwest Texas

[2780] Reddell, James Russell, 1938–, Feb 1973, The caves of San Saba County

[2781] Reddell, James Russell, 1938–(ed), Jul 1962, The caves of San Saba County part I

[2782] Reddell, James Russell, 1938–(ed), nd, The caves of Travis County

[2783] Reddell, James Russell, 1938–(ed), Feb 1963, The caves of Val Verde County

[2784] Reddell, James Russell, 1938–(ed), Oct 1963, The caves of Willamson County

[2785] Reddell, James Russell, 1938–, Aug 1969, A checklist and annotated bibliography of the subterranean aquatic fauna of Texas

[2786] Reddell, James Russell, 1938–, 1965, A checklist of the cave fauna of Texas

[2787] Reddell, James Russell, 1938–(ed), Apr 1962, A checklist of the caves of Texas

[2788] Reddell, James Russell, 1938–(ed), Apr 1989, A field guide to caves of Blanco, Gillespie, and Llano counties, Texas

[2789] Reddell, James Russell, 1938–(ed), Mar 1982, Further studies on the cavernicole fauna of Mexico and adjacent regions

[2793] Reddell, James Russell, 1938–, Jul 1966, A revised checklist of Texas caves

[2856] Ross, Richard E., Oct 1965, The archeology of Eagle Cave

[2870] Russell, William Hart, 1937–, Jul 1993, The Buttercup Creek karst: Travis and Williamson Counties, Texas: geology, biology, and land development

[2911] Schroeder, Albert H., 1978, Pratt Cave studies: Guadalupe National Park, Texas

[2912] Schuetz, Mardith K., 1956, An analysis of Val Verde County cave material

[2933] Semken, Holmes A., 1967, Mammalian remains from Rattlesnake Cave, Kinney County, Texas

[2941] Shafer, Harry J., 1977, Archeological and botanical studies at Hinds Cave, Val Verde County, Texas

[2942] Shafer, Harry J., 1975, A preliminary report of Hinds Cave, Val Verde County, Texas

[2992] Sinise, Jerry, 1963, Texas show caves

[3009] Smith, A. Richard (ed), 1971, The caves of Kimble county

[3010] Smith, A. Richard (ed), 1965, The caves of Kinney county

[3026] Smith, Lee, nd, Down under Texas

[3188] Taylor, Alisa Johanna, 1982, The mammalian fauna from the mid-Irvingtonian Fyllan Cave local fauna, Travis County, Texas

[3256] Turpin, Sloveig A. (comp), 1985, Seminole Sink (41VV620): excavation of a vertical shaft tomb, Val Verde County, Texas

[3357] Veni, George, 1957–, 1988, The caves of Bexar County

[3383] Walsh, John Michael, 1947–(ed), 1991, A proposed cave access and cave management plan for the Devil's Sinkhole State Natural Area & Kickapoo Caverns State Park

[3384] Walsh, John Michael, 1947–(ed), 1982, Texas cave humor, the first twenty-five years

[3385] Walter, Erin, Jul 1996, A guide to Austin's most visited caves, with maps and cross-sections of area caves

[3402] Warton, Mike, 1993, A preliminary report of findings for the karst terrains feature known as Rugh Cavern, located along northern Seco Creek, northwestern Medina County, Texas

[3439] Wermund, E. G., 1978, Regional distribution of fractures in the southern Edwards Plateau and their relationship to tectonics and caves

[3444]West Texas Geological Society, 1969, Delaware Basin exploration: Guadalupe Mts., Hueco Mts., Franklin Mts., geology of the Carlsbad Caverns ; Nov. 6th, 7th and 8th, 1969

[3445]West Texas Geological Society, 1969, Delaware Basin exploration: Guadalupe Mts., Hueco Mts., Franklin Mts., geology of the Carlsbad Caverns, Nov. 6th, 7th and 8th, 1969

[3446]West Texas Geological Society, 1968, Guadalupe Mts., Hueco Mts., Franklin Mts., geology of the Carlsbad Caverns, Delaware Basin exploration, Oct. 31st, Nov. 1st and 2nd, 1968

14.9.4.2.41 Utah

[18]Aikens, C. Melvin, 1970, Hogup Cave

[203]Anon, 1987, Timpanogos Cave public use and impact

[963]Elliott, William Rawleigh, 1946–, 1976, New cavernicolous Rhagidiidae from Idaho, Washington, and Utah (Prostigmata, Acari, Arachnida)

[992]Enger, Walter D., 1942, Archaeology of Black Rock 3 Cave, Utah

[1246]Grayson, Donald K., May 1988, Danger Cave, Last Supper Cave, and Hanging Rock Shelter: the faunas

[1321]Halliday, William R., 1926–, Jan 1954, The basic geology of Neff Canyon Cave, Utah ; The speleogenesis of Neff Canyon Cave, Utah

[1333]Halliday, William R., 1926–, Sep 1954, Littoral caves of ancient Lake Bonneville

[1359]Hargrave, Lyndon Lane, 1896–, 1967, Feathers from Sand Dune Cave: a Basketmaker cave near Navajo Mountain, Utah

[1665]Iorio, Ralph, 1968, The history of Timpanogos Cave National Monument: American Fork Canyon, Utah

[1718]Jennings, Jesse David, 1909–, 1980, Cowboy Cave

[1719]Jennings, Jesse David, 1909–, Oct 1957, Danger Cave

[1720]Jennings, Jesse David, 1909–, Sep 1975, Excavation of Cowboy Caves, June 3-July 26, 1975: preliminary report

[1730]Jensen, Douglas N. , 1988, Discover Timpanogos Cave National Monument

[2062]Martin, George V., 1973, The Timpanogos Cave story: the romance of its exploration

[2135]McLoyd & Graham, 1894, Catalogue and description of a very large collection of prehistoric relics: obtained in the cliff houses and caves of Southeastern Utah

[2536]Nusbaum, Jesse Logan, 1985, A basket-maker cave in Kane County, Utah

[2592]Packard, Alpheus Spring, 1839–1905, 1877, On a new cave fauna in Utah: and on new phyllopod Crustacea from the West

[2758]Randall, Robert L., 1970, Along the way to the caves: Timpanogos Cave National Monument, Utah

[3017]Smith, Elmer Richard, 1909–, 1952, The archaeology of Deadman Cave, Utah: a revision

[3125]Steward, Julian Haynes, 1902–1972, 1937, Ancient caves of the Great Salt Lake region

[3238]Trimble, Stephen, 1950–, 1983, Window into the earth: Timpanogos Cave

[3292]U.S. National Park Service, Feb 1993, Draft, environmental impact statement, general management plan, development concept plan Timpanogos Cave National Monument

[3317]U.S. National Park Service, 1976, Timpanogos Cave National Monument, Utah

[3318]U.S. National Park Service, Aug 1986, Timpanogos Cave National Monument: statement for management

[3451]Western Land Surveys, Utah Division, 1977, Timpanogos Cave environmental impact report

[3476]White, William Blaine, 1934–, Jan 1974, Reconnaissance geology of Timpanogos Cave: Wasatch County, Utah

14.9.4.2.42 Vermont

[277]Bailey, John H., 1940, A stratified rock shelter in Vermont

[444]Botto, Larry (comp), [1993], A partial guidebook to the caves of the Berkshires: 1993 spring N.R.O. hosted by the Berkshire Hills Grotto (May 14th, 15th, 16th)

[2181]Miller, Edward Sherman, 1866–, 1893, Joint thesis upon Maidstone Cave

[2401]National Speleological Society. Annual Convention. Guidebook (1979: Pittsfield, MA), 1979, An introduction to caves of the northeast: guidebook for the 1979 National Speleological Convention, Pittsfield, Massachusetts

[2738]Quick, Peter Gunder, 1994, Vermont caves: a geologic and historical guide

[2924]Scott, John, 1929–, 1959, Caves in Vermont, a spelunker's guide to their location and lore

[3198]Thayer, Charles W., 1966, Mud stalagmites and the conulite, a new speleothem

14.9.4.2.43 Virginia

[50]Allen, Joel Asaph, 1838–1921, 1885, On an extinct type of dog from Ely Cave, Lee County, Virginia

[57]Ammen, Samuel Zenas, 1843–1929, 1890, The caverns of Luray, the property of the Valley Land and Improvement Co., Luray, Va: the manner of their formation, their peculiar growths, their geology, chemistry, etc: an illustrated guide book

[58]Ammen, Samuel Zenas, 1843–1929, 1893, The caverns of Luray. The property of the Valley Land and Improvement Co., Luray, Va. The manner of their formation,

14.9. NORTH AMERICA

their peculiar growths, their geology, chemistry, etc. An illustrated guide book

[59] Ammen, Samuel Zenas, 1843–1929, 1884, The caverns of Luray: an illustrated guide-book to the caverns, explaining the manner of their formation, their peculiar growths, their geology, chemistry, etc

[60] Ammen, Samuel Zenas, 1843–1929, 1882, History and description of the Luray Cave (illustrated), including explanations of the manner of its formation, its peculiar growths, its geology, chemistry, &c.; also a map. The whole so arranged as to serve as a guide.

[61] Ammen, Samuel Zenas, 1843–1929, 1880, History and description of the Luray Cave (illustrated), including explanations of the manner of its formation, its peculiar growths, its geology, chemistry, etc.; also a map. The whole so arranged as to serve as a guide

[92] Anon, 1882, The Caverns of Luray

[93] Anon, c1880, The Caverns of Luray located on line of the Shenandoah Valley R.R.

[121] Anon, 1926, Grand Caverns: wonders of the subterranean world: Grottoes, Virginia

[136] Anon, nd, The historic Blue Grottoes of Virginia; Harrisonburg VA

[137] Anon, 1967, How nature made the beautiful caverns

[145] Anon, , Luray Caverns

[146] Anon, 1900, Luray Caverns

[152] Anon, c1906, Mansion Inn, Luray Caverns

[182] Anon, , Selected views of the beautiful caverns of Luray

[183] Anon, 1920, Shenandoah Caverns, the grotto of the gods, in the heart of the Valley of Virginia

[184] Anon, c1995, Shenandoah Caverns: a brief history

[187] Anon, 1995, Skyline Caverns

[207] Anon, 1984, Virginia Luray Caverns

[390] Benthall, Joseph L., , Daugherty's Cave: a stratified site in Russell County, Virginia

[527] Brown, Mary Mitchell, 1921, Nature underground: the Endless Caverns in the heart of the historic Shenandoah Valley

[565] Burnsville Cove Symposium, 1982, Burnsville Cove Symposium

[704] Cooke, Robert L., 1834, A description of Weyer's Cave

[890] Douglas, Henry Hulbert, 1906–1989, 1964, Caves of Virginia

[946] Edwards, Ira, 1925, Underground Geology at the Endless Caverns, New Market, VA

[981] Endless Caverns, Inc, 1930, Endless caverns, New Market, Virginia, in the heart of the Shenandoah Valley...3 minute drive off U.S. 11 (Lee highway) over new boulevard

[982] Endless Caverns, Inc, 1939, Endless Caverns, Virginia. "Considered the most beautiful of the Shenandoah Valley caverns" ...

[983] Endless Caverns, Inc, 1921, Endless Caverns, wonderful and spectacular

[984] Endless Caverns, Inc, nd, Endless Caverns, wonderful and spectacular, New Market, Virginia, in the heart of the Shenandoah Valley

[985] Endless Caverns, Inc, 1987, Endless Caverns: on U.S. 11, 3 miles south of New Market, Virginia: illuminated by indirect electric flood lighting

[986] Endless Caverns, Inc, 1926, Exploring the Endless Caverns of New Market, Virginia in the heart of the Shenandoah Valley: two authentic and thrilling accounts of the adventures of members of the American Museum of Natural History and of the Explorers Club, who vainly sought the end to the Endless Caverns

[987] Endless Caverns, Inc, 1947, Exploring the Endless Caverns of New Market, Virginia. Authentic and thrilling accounts of the adventures of members of the Explorers Club and of the American Museum of Natural History

[988] Endless Caverns, Inc, 1946, Nature underground; the Endless Caverns in the heart of the historic Shenandoah Valley

[1037] Faust, Burton Sherwood, 1898–1967, 1964, Saltpetre caves and Virginia history

[1041] Felter, Brian, 1979, The Virginia cave locator series: book 1: Frederick Co. and Clark Co

[1042] Felter, Brian, 1979, The Virginia cave locator series: book 2: Warren Co

[1043] Felter, Brian, 1980, The Virginia cave locator series: book 3: Shenandoah Co

[1044] Felter, Brian, 1980, The Virginia cave locator series: book 4: Page Co

[1045] Felter, Brian, 1982, The Virginia cave locator series: book 5: Rockingham Co

[1046] Felter, Brian, 1982, The Virginia cave locator series: book 6: Augusta Co

[1064] Finley-Holiday Film Corporation, 1988, Luray Caverns in Virginia's Shenandoah Valley

[1124] Fraser, L.A., Jul 1892, Weyer Cave. Cave of the Fountains. Madison Cave: a descriptive sketch of the grottoes of the Shenandoah

[1133] G. P. Slide Co, 1970, Luray Caverns

[1137] GAF Corporation, 1970, Luray Caverns, Va

[1161] Garrigan, George A., 1994, Skyline Caverns and its geologic relationship to the Shenandoah Valley and paleoindian cultures

[1253] Green, John, nd, The Story of Cave Hill: Grottos, VA

[1280] Guilday, John E., 1977, The Clark's Cave bone deposit and the late Pleistocene paleoecology of the central Appalachian Mountains of Virginia

[1297] Gurnee, Russell Hampton, 1922–1995, , Discovery of Luray Caverns, Virginia

[1310] Hack, John Tilton, 1913–, 1962, Geology of Luray Caverns Virginia

[1367] Harrisonburg-Massanutten Corp (Harrisonburg, Va), 1928, About Massanutten caverns; the jeweled chamber that hides the secret of Nakwisi-Gatusi

[1368] Harrisonburg-Massanutten Corp (Harrisonburg, Va), 1925, Massanutten Caverns, Harrisonburg, Va. The most beautiful and unusual caverns yet discovered...

[1369] Harrisonburg-Massanutten Corp (Harrisonburg, Va), 1940, Massanutten Caverns, Harrisonburg, Va.: the most beautiful and unusual caverns yet discovered, in the heart of the Shenandoah Valley of Virginia

[1370] Harrisonburg-Massanutten Corp (Harrisonburg, Va), nd, The wonderful Massanutten Caverns in the Shenandoah Valley, Harrisonburg, Va

[1438] Hendricks, Albert C., 1974, Quality of groundwater in a carbonate terrain in western Virginia

[1441] Henkel, David S., 1919, A description of the Endless Caverns of New Market

[1442] Henkel, David S., 1880, A description of the New Market Endless Caverns

[1447] Herman, Janet S. (ed), 1990, Travertine-marl: stream deposits in Virginia

[1482] Holiday Film Corporation, 1984, Luray Caverns

[1495] Holsinger, John Robert, 1934–, 1975, Descriptions of Virginia caves

[1503] Holsinger, John Robert, 1934–, Jun 1988, The invertebrate cave fauna of Virginia and a part of eastern Tennessee

[1558] Howison, Allan Moore, 1889, The Grottoes of the Shenandoah, consisting of the Weyer and the Fountain Caves, the two most wonderful caverns in the world, at Grottoes Station of Shenandoah Valley Railway, Augusta County, Virginia. A. M. Howison, Secretary.

[1695] James, C. H., (photo), 1882, Electric light views in the Caverns of Luray at Luray Station (Page Co., Virginia), Shenandoah Valley Railroad

[1759] Jordan, J. Murray, 1897, The Caverns of Luray and vicinity

[1772] Kastning, Ernst H., 1944–, Apr 1989, Caves and karst of the New River Valley, Virginia: guidebook for a geologic fieldtrip, eighth annual New River symposium, Radford, Virginia

[1773] Kastning, Ernst H., 1944–, Mar 1991, Caves, karst, and environmental impact in the new river drainage basin of Virginia and West Virginia: guidebook for a geologic fieldtrip, Appalachian karst symposium, Radford, Virginia, 23 March 1991

[1776] Kauffman, D. L., 1913, The beautiful Luray Caverns, Luray, Virginia

[1814] Keyes, Elizabeth, 1930, Titania's cave; a fairy story of the Endless Caverns of Virginia

[1900] Larimore, Betty, 1947, Exploring the Endless Caverns of New Market Virginia on US 11 in the heart of the Shenandoah Valley

[1990] Luray Caverns Corporation, 1953, The beautiful caverns of Luray, Luray, Virginia, in the Shenandoah Valley

[1991] Luray Caverns Corporation, 1906, The beautiful caverns of Luray, Luray, Virginia. In the Shenandoah Valley, three miles of subterranean splendor, brilliantly lighted by electricity

[1992] Luray Caverns Corporation, 1949, The beautiful caverns of Luray: Luray, Virginia

[1993] Luray Caverns Corporation, 1930, The beautiful caverns of Luray: Luray, Virginia, in the Shenandoah Valley: miles of subterranean splendor, brilliantly lighted by electricity

[1994] Luray Caverns Corporation, nd, Caverns of Luray

[1995] Luray Caverns Corporation, 1882, Electric light views in the Caverns of Luray. The Caverns of Luray (at Luray, Page County, Virginia, a station on the Shenandoah Valley Railroad) as a resort for tourists ... are unexcelled, in their wonderful attractiveness, by any other creation of nature

[1996] Luray Caverns Corporation, 1900, Luray Caverns

[1997] Luray Caverns Corporation, 1900, 1909, Luray Caverns of the beautiful Shenandoah Valley on the Norfolk and Western Railway, Luray, Virginia. Luray, Virginia Mansion Inn

[2075] Mason, Otis Tufton, 1838–1908, , The caverns of Luray

[2076] Mason, Otis Tufton, 1838–1908, 1882, The caverns of Luray

[2077] Mason, Otis Tufton, 1838–1908, 1881, Report of a visit to the Luray Cavern, in Page County, Virginia, under the auspices of the Smithsonian institution, July 13 and 14, 1880

[2125] McGill, William Mahone, 1897–1962, 1933, Caverns of Virginia

[2247] Mohler, John Leonard, 1840–1937, 1852, A description of Weyer's Cave in Augusta County, Virginia

[2307] Mussaeus, Thomas Allen, 1887–, 1939, The lure of cave lore

[2308] Mussaeus, Thomas Allen, 1887–, 1950, The lure of cave lore; being a random narrative upon caverns in general and in particular the properties of the Skyline Caverns, Front Royal, Virginia

[2385] National Speleological Society. Annual Convention. Guidebook (1963: Mountain Lake, VA), 1963, Guide book to the major caves in the vicinity of Mountain Lake, Virginia

[2393] National Speleological Society. Annual Convention. Guidebook (1971), 1971, NSS 71 guidebook

[2416] National Speleological Society. Annual Convention. Guidebook (1995: Blacksburg, VA) , 1995, Underground in the Appalachians: a guidebook for the 1995 Convention of the National Speleological Society, Blacksburg, Virginia, July 17-21, 1995

[2424] National Speleological National Speleological Society. Annual Convention. Guidebook. Geology Field Trip (1995: Blacksburg, VA), 1995, Origin of caves and karst in the Shenandoah Valley, Rochingham and Augusta Counties, Virginia: guidebook for a geologic fieldtrip, National Speleological Society Annual Convention, Blacksburg, VA, 16 July 1995

[2513]Noah, Samuel, 1835, A description of Weast's Cave

[2517]Northcott, T. C., c1995, How nature made Luray Caverns, Virginia

[2518]Northcott, T. C., 1934, How nature makes a cave

[2650]Pennsylvania Railroad, 1885, A summer series of personally-conducted pleasure tours to the Luray Caverns or Gettysburg via Pennsylvania Railroad

[2799]Reeds, Chester Albert, 1882–, 1925, The Endless Caverns of the Shenandoah Valley, an account of the wonderful work of water: with special reference as to how the caverns and the Shenandoah Valley were formed

[2872]Rust, Pauline Carrington, 1887, Legend of the Luray Caverns, founded upon the discovery of a skeleton in one of its chasms

[2960]Shenandoah Caverns Corporation, 1922, Legends of Shenandoah Caverns: in the Valley of Virginia

[2961]Shenandoah Caverns Corporation, 1932, Shenandoah Caverns of Virginia

[2962]Shenandoah Caverns Corporation, 1928, Shenandoah Caverns: Valley of Virginia

[2963]Shenandoah Caverns Corporation, Shenandoah Caverns, Va, 1929, Shenandoah Caverns, Valley of Virginia. A symphony in stone

[2964]Shenandoah Caverns Corporation, Shenandoah Caverns, Va, 1926, Shenandoah Caverns, Valley of Virginia. The grotto of the gods

[2965]Shenandoah Valley Railroad, 1889, Through the Shenandoah Valley: Caverns of Luray, Natural Bridge, grottoes of the Shenandoah and the chronicle of a leisurely journey through the uplands of Virginia

[2993]Skelton, Harold E., 1974, Weyer's Cave's first century, 1874-1974

[3001]Slifer, Dennis William, 1947–, Apr 1989, Sinkhole dumps and the risk to groundwater in Virginia's karst areas: final report to the Virginia Environment Endowment

[3037]Smithsonian Institution, 1880, The caverns of Luray, located on line of the Shenandoah Valley R.R

[3093]Spencer, Edgar Winston, 1985, Guidebook to the Natural Bridge and Natural Bridge Caverns

[3113]Steidtmann, Edward, 1881–1948, 1936, Humidity and waters of a limestone cavern near Lexington, Virginia

[3148]Stow, Marcellus Henry, 1902–, [1939], Description of points of scenic, historic, and geologic interest between Washington, D. C., and Luray, Virginia

[3328]United Air Lines, Inc., 1985, Luray Caverns in Virginia's Shenandoah Valley

[3340]Upper Valley Regional Park Authority (VA), 1975, The story of Cave Hill

[3367]Virginia Commission on the Conservation of Caves, 1979, Report of the Virginia Commission on the Conservation of Caves to the Governor and the General Assembly of Virginia

[3368]Virginia Region. National Speleological Society, , A history of the Virginia Region

[3418]Wayland, John Walter, 1872–1962, 1930, The master sculptor; a brief treatise on erosion, and its wondrous effects in the Shenandoah Valley

[3442]Werner, Eberhard Wolfgang, 1942–, Jul 1981, Guidebook to the karst of the Central Appalachians; prepared for the Eighth International Congress of Speleology, Bowling Green, Kentucky, U.S.A., July 18 to 24, 1981

[3453]Weyer, Bernard, 1881, A description of Weyer's Cave in Augusta County, Va

[3454]Weyer, Bernard, 1849, A description of Weyer's Cave, in Augusta County, Virginia

[3480]Whittemore, Anne, 1979, A history of the Virginia Region

[3481]Wickersham, David L., 1988, A survey of the caves of Roanoke County, Virginia

14.9.4.2.44 Washington

[687]Combes, John D., Jun 1969, The Excavation of Squirt Cave, 45WW25

[809]Daugherty, Richard D., 1922–, 1987, A data recovery study of Layser Cave (45-LE-223) in Lewis County, Washington

[963]Elliott, William Rawleigh, 1946–, 1976, New cavernicolous Rhagidiidae from Idaho, Washington, and Utah (Prostigmata, Acari, Arachnida)

[1195]Gifford Pinchot National Forest, 1991, Layser Cave: silent voices, vital clues

[1250]Greeley, Ronald, 1971, Lava tubes of the Cave Basalt, Mount St. Helens, Washington

[1320]Halliday, William R., 1926–, 1983, Ape Cave and the Mount St. Helens apes

[1325]Halliday, William R., 1926–, 1963, Caves of Washington

[1335]Halliday, William R., 1926–, 1972, The Paradise Ice Caves, Mount Rainier National Park, Washington

[1740]Johnston, David Alexander, 1949-1980, 1981, Guide to some volcanic terranes in Washington, Idaho, Oregon and Northern California

[1823]Kirk, Ruth, 1970, The oldest man in America: an adventure in archeology

[1824]Kiver, Eugene P., 1973, Summit firn cave study, 1970-1973, Mount Rainier, Washington

[1907]Larson, Charles Victor, 1928–, 1989, Lava Beds caves

[1965]Lothson, Gordon A., 1939–, 1980, The archaeology of the Fallen Arches Cave Site, 45SA41, Borigo timber sale, Gifford Pinchot National Forest: phase I survey and reconnaissance

[2194]Miss, Christian J., nd, Archaeological evaluations of the Riparia (45WT1) and Ash Cave (45WW61) sites on the Lower Snake River

[2299]Munich, Frederick J. (auth, prod), 1977, Fumaroles, ice caves and time: the variety and scope of research on Mt. Baker

[2394]National Speleological Society. Annual Convention. Guidebook (1972: White Salmon, WA), 1972, Selected caves of the Pacific Northwest with particular reference to the vulcanospeleology of the state of Washington

[2523]Northwest Caving Association. National Speleological Society (1987), 1987, Northwest Caving Association regional meeting guidebook: caves of the Peterson Prairie area

[2951]Sharpe, Grant William, 1925–, 1973, Evaluation of Mount St. Helens lava casts and caves, Cowlitz and Skamania Counties, Washington, for eligibility for registered natural landmark designation

[3208]Thompson, Robert S., 1985, Paleoenvironmental investigations at Seed Cave (Windust Cave H- 45FR46), Franklin County, Washington

14.9.4.2.45 West Virginia

[70]Anderson, Robert Cleve (ed), 1982, Capital area cavers bulletin number 1

[101]Anon, nd, Caving in West Virginia

[232]Ashbrook, Bert, 1995, Caves in the Richlands area of Greenbrier County, West Virginia

[428]Bolton, David W., nd, The geology and speleogenesis of Red Run Cave Tucker County, West Virginia

[805]Dasher, George R., 1952–(ed), Aug 1994, The caves and karst of the Buckeye Creek Basin, Greenbrier County, West Virginia

[827]Davies, William Edward, 1917–, Jul 1958, Caverns of West Virginia

[828]Davies, William Edward, 1917–, 1965, Caverns of West Virginia (supplement)

[1164]Garton, Emmel Ray, 1950–, Jun 1976, Caves of North Central West Virginia

[1165]Garton, Emmel Ray, 1950–, 1993, The vertebrate fauna of West Virginia caves

[1282]Guilday, John E., 1973, The late Pleistocene small mammals of Eagle Cave, Pendleton County, West Virginia

[1286]Gulden, Robert, Dec 1984, Caves of the eastern panhandle of West Virginia

[1427]Hempel, John Charles, 1949–(comp), 1975, Caves of Monroe County, W. Va

[1491]Holman, J. Alan, 1931-, 1982, The pleistocene (Kansan) herpetofauna of Trout Cave, West Virginia

[1504]Holsinger, John Robert, 1934–, 1976, The invertebrate cave fauna of West Virginia

[1756]Jones, William K., 1945–, 1973, Hydrology of limestone karst in Greenbriar County, West Virginia

[1758]Jordan, George, 1972, The great cave of Dry Fork of Cheat River, Virginia: one of the greatest wonders of the world examined and explored Giving a complete description of this subterraneous cavern, of its rooms and passages, which are of enormous size, together with a great stream of water, a lake, and springs, minerals, fossils, and many curiosities, that were found in singular abundance, &c.

[1773]Kastning, Ernst H., 1944–, Mar 1991, Caves, karst, and environmental impact in the new river drainage basin of Virginia and West Virginia: guidebook for a geologic fieldtrip, Appalachian karst symposium, Radford, Virginia, 23 March 1991

[2138]Medville, Douglas Michael, 1941–, Mar 1976, Caves and karst hydrology in northern Pocahontas County

[2139]Medville, Douglas Michael, 1941–, 1995, Caves and karst of Randolph County

[2140]Medville, Douglas Michael, 1941–, 1971, Caves of Randolph County

[2276]Morris, Deborah, 1956–, 1993, Trapped in a cave!: a true story

[2392]National Speleological Society. Annual Convention. Guidebook (1970: Pittsburgh, PA), 1970, NSS 1970 guidebook: Mid-Appalachian Region

[2393]National Speleological Society. Annual Convention. Guidebook (1971), 1971, NSS 71 guidebook

[2398]National Speleological Society. Annual Convention. Guidebook (1976: Morgantown, WV), 1976, 1976 annual convention guidebook, Morgantown, W. Va

[2404]National Speleological Society. Annual Convention. Guidebook (1983: Elkins, WV), 1983, An introduction to the caves of east-central West Virginia: guidebook for the 1983 National Speleological Society Convention, Elkins, West Virginia

[2447]National Speleological Society. Annual Convention. Supplemental Guidebook (1970: Pittsburgh, PA), 1970, Supplemental guidebook: 1970 NSS convention

[2706]Preble, John Wesley, 1969, Sinks of Gandy Creek

[2807]Repenning, Charles Albert, 1922–, 1989, The microtine rodents of the Cheetah Room fauna, Hamilton Cave, West Virginia, and the spontaneous origin of *Synaptomys*

[2960]Shenandoah Caverns Corporation, 1922, Legends of Shenandoah Caverns: in the Valley of Virginia

[3087]Speece, Jack H., 1978, George Washington Cave

[3124]Stevens, Paul James, 1944–, 1988, Caves of the Organ Cave Plateau: Greenbrier County, West Virginia

[3146]Storrick, Gary D., May 1992, The caves and karst hydrology of southern Pocahontas County and the upper Spring Creek Valley

[3440]Werner, Eberhard Wolfgang, 1942–, 1974, Index of the literature pertaining to West Virginia caves and karst

[3441]Werner, Eberhard Wolfgang 1942–, 1972, Develop-

ment of solution features, Cloverlick Valley, Pocahontas County
[3442]Werner, Eberhard Wolfgang, 1942–, Jul 1981, Guidebook to the karst of the Central Appalachians; prepared for the Eighth International Congress of Speleology, Bowling Green, Kentucky, U.S.A., July 18 to 24, 1981

14.9.4.2.46 Wisconsin

[154]Anon, 1990, Maribel Caves: an ideal health and summer resort.
[431]Bonin, Dan (ed), 1980, Hodag hunt 1980: guidebook September 12-14 1980
[451]Boyd, Susan (comp), 1966, A guide to major Wisconsin caves
[519]Brown, Edward, nd, The pictured cave of La Crosse Valley, near West Salem, Wisconsin
[718]Coronet Instructional Films, 1943, Limestone caverns
[949]Ehr, Bob (prod), 1974, Guide to the caves of Wisconsin
[1444]Hennings, Kevin (ed), c1971, 1971 guidebook: Upper Mississippi conference UMAC Sept. 17, 18, 19 at Eagle Cave
[2311]MVOR (Mississippi Valley Ozark Regional). Guidebook (1970), nd, 1970 guidebook UMAC-MVOR: Upper Mississippi area conference-Mississippi Valley Ozarks Regional Sept. 25, 26, & 27 at Eagle Cave
[2396]National Speleological Society. Annual Convention. Guidebook (1974: Decorah, IO), 1974, National Speleological Society 1974 convention Luther College, Decorah, Iowa 10-18 August, 1974
[2402]National Speleological Society. Annual Convention. Guidebook (1980: White Bear Lake, MN), Jul 1980, An introduction to caves of Minnesota, Iowa, and Wisconsin: guidebook for the 1980 National Speleological Society Convention, Lakewood Community College, White Bear Lake, Minnesota
[2690]Pond, Alonzo William, 1894–, 1941, Guide book of Cave of the Mounds ... Blue Mound, Wis
[3327]Umbreit, Tom, 1976, Wisconsin Speleological Society present: Hodag Hunt 1976, Flying J Campgrounds, Gotham, Wisconsin
[3339]University of Wisconsin, Recreation Resources Center, 1977, Seen a good cave lately?: Cave of the Mounds, Blue Mounds, Wis.: "a visitor study", 1977
[3443]Wernerus, Mathias, 1929, The grottos at Dickeyville
[3517]Wisconsin Speleological Society, 1978, 1978 hodag hunt guidebook

14.9.4.2.47 Wyoming

[1104]Fossen, Joyce, 1985, Caves of Paha Sapa, and other findings of Dr. Knows-It
[1461]Hill, Chris, 1947–, 1976, Caves of Wyoming
[1722]Jennings, Jesse David, 1909–, Oct 1957, Shaw Cave, Wyoming
[2114]McCracken, Harold, 1894–, 1978, The Mummy Cave Project in northwestern Wyoming
[2356]National Speleological Society, , Black Hills symposium
[2391]National Speleological Society. Annual Convention. Guidebook (1969: Lovell, WY), 1969, Caves of the Big Horn - Pryor Mountains of Montana and Wyoming. Issued in conjunction with the 28th anniversary convention of the National Speleological Society at Lovell, Wyoming, Harley A Leach, convention chairman
[2405]National Speleological Society. Annual Convention. Guidebook (1984: Sheridan, WY), 1984, Guidebook of the 1984 NSS convention
[2409]National Speleological Society. Annual Convention. Guidebook (1988: Hot Springs, SD) , 1988, 1988 NSS convention guidebook
[2584]Owen, Luella Agnes, 1970, Cave regions of the Ozarks and Black Hills
[3250]Tullis, Edward L., 1940, Black Hills caves

14.10 South America

14.10.1 Brazil

[1947]Lino, Clayton F., 1989, Cavernas, o fascinante Brasil subterraneo = Caves, the fascination of underground Brazil

14.10.2 Chile

[1570]Humphrey, Philip Strong, 1926–, Feb 1993, Avifauna of three Holocene cave deposits in southern Chile

14.10.3 Colombia

[3176]Szentes, Georg, 1992, Tropical karst and caves in the central Cordillera, Colombia

14.10.4 Ecuador

[1852]Kornicker, Louis S., 1919–, 1989, Troglobitic Ostracoda (Myodocopa: Cypridinidae, Thaumatocyprididae) from anchialine pools on Santa Cruz Island, Galapagos Islands
[2684]Plimpton, George (host), 1987, The Los Tayos challenge

14.10.5 Peru

[228][Army Caving Association], c1988, Army Caving Association: Peru 1987
[490][British Speleological Association], Nov 1973, Imperial College karst research expedition to the Peruvian Andes, 1972
[1999]Lynch, Thomas F., 1938–, 1980, Guitarrero Cave: early man in the Andes
[2579]Orrock, Clive, c1985, Imperial College Caving Club Peru '84 Expedition (19th July - 19th Sept 1984)
[3075]Southhampton University Exploration Society, 1982, Southhampton University Exploration Society: Peru expedition

14.10.6 Venezuela

[2528]Nott, David, 1975, Into the lost world - a descent into prehistoric time

14.11 Australasia

[250]Australasian Conference on Cave Tourism and Management, nd, Cave Management in Australasia VIII: proceedings of the eighth Australian conference on cave tourism and management, Paparoa National Part, Punakaiki, New Zealand, April 1989
[1531]Houshold, Ian (prep), 1995, Cave & karst guidebook and documentation: Australasian Cave and Karst Management Association eleventh Australasian Conference on cave and karst management Tasmania 29 April to 7 May 1995 hosted by The Parks & Wildlife Service - Tasmania

14.11.1 Australia

[81]Anon, 1968, Australian Speleological Handbook
[134]Anon, Nov 1967, Handbook for caves and guiding staff
[144]Anon, 1988, The last horizon
[222]Anon, 1970, Yarrangobilly Caves—Kosciusko National Park
[251]Australian Conference on Cave Tourism, Jun 1976, Cave Management in Australia: proceedings of the first Australian Conference on cave tourism, Jenolan Caves, N.S.W. 10th-13th July, 1973
[252]Australian Conference on Cave Tourism and Management, Aug 1977, Cave Management in Australia II: proceedings of the second Australian Conference on cave tourism and management, Hobart, Tasmania, 3rd-5th May, 1977
[253]Australian Conference on Cave Tourism and Management, Nov 1980, Cave Management in Australia III: proceedings of the third Australian Conference on cave tourism and management, Mount Gambier, South Australia, 30th April - 4th May, 1979
[254]Australian Conference on Cave Tourism and Management, Aug 1982, Cave Management in Australia IV: proceedings of the fourth Australian Conference on cave tourism and management, Yallingup, Western Australia Australia, September 1981
[255]Australian Conference on Cave Tourism and Management, 1990, Cave Management in Australia V: proceedings of the Fifth Australian Conference on cave tourism and management, Lakes Entrance, Victoria, April 1983
[259]Australian Speleological Federation, nd, Caving in Australia
[260]Australian Speleological Federation, 1983, Draft management plan: Tantanoola Caves Conservation Park: Lower South East, South Australia
[262]Australian Speleological Federation, Feb 1972, Proceedings of the eighth biennial conference of the Australian Speleological Federation
[263]Australian Speleological Federation, Jul 1975, Proceedings of the tenth biennial conference of the Australian Speleological Federation
[264]Australian Speleological Federation Digest Commission, 1976, Australian speleology 1972
[265]Australian Speleological Federation Inc, 1988, Preprints of papers for the 17th Biennial Conference of the Australian Speleological Federation Tropicon Conference, Lake Tinaroo, Far North Queensland. 27th to 31st December 1988
[271]Baddeley, Glenn (ed, auth), , Vulcon Guidebook: lava features and limestone karst of Victoria and southeastern South Australia. Vulcon 1995 20th Biennial Conference. Australian Speleological Federation, Inc, Hamilton, Victoria 2-6 January 1995
[272]Baddeley, Glenn (ed), 1995, Vulcon precedings: papers submitted for presentation. Vulcon 20th Biennial Conference. Australian Speleological Federation, Inc, Hamilton, Victoria 2-6 January 1995
[330]Barrett, Charles Leslie, 1879–, 1944, Australian caves, cliffs, and waterfalls
[335]Bartel Photography Pty Ltd, 1985, Jenolan Caves New South Wales Australia
[344]Bates, Geoff, 1982, The Abercrombie Caves, New South Wales: a guide to these remarkable caves near Bathurst, N.S.W. and a description of the surrounding gold bearing country of Tuena and Trunkey Creek
[351]Baynes, F. J. , 1991, Ida Bay - Benders Quarry. An estimate of reserves within existing quarry limits
[377]Bell, Peter (ed), 1991, The 9th ACKMA conference proceedings; Margaret River, WA, Australia
[410]Black, Lindsay, 1943, Aboriginal art galleries of western New South Wales

14.11. AUSTRALASIA

[448] Bowdler, Sandra, 1989, A Test excavation at Mulka's Cave (Bate's Cave) near Hyden, Western Australia: a report to the Department of Aboriginal Sites, Western Australian Museum

[455] Bradley, J. F., Nov 1926, Glorious Kangaroo Island: its caves and beauty spots

[514] Brooks, Steve (comp), 1995, The Australian cavers diary

[573] Byfield, Colin Charles, 1977, Let's look at Jenolan Caves: an introduction to the features of Australia's famous limestone caverns

[613] Cave Diving Association of Australia, Inc, [c1977], Cave Diving Association of Australia, Inc: conference 1977 held at Mount Gambier, South Australia

[651] Chillagoe Caving Club (comp), 1982, A speleological field guide of the towers and caves of Chillagoe - Munanga - Rookwood areas in Far North Queensland, Australia

[652] Chillagoe Caving Club (comp), Jul 1988, A speleological field guide of the towers and caves of Chillagoe - Munanga - Rookwood areas in Far North Queensland, Australia. Supplement for caves CH301-CH400

[699] Cook, Samuel, 1889, The Jenolan caves: an excursion in Australian wonderland

[771] Cullen, P., 1990, A survey of the vegetation

[812] Davey, Adrian G., 1986, Management of Victorian caves and karst: a report to the Caves Classification Committee, Department of Conservation, Forests and Lands

[813] Davey, Adrian G., 1978, Nullarbor karst, a bibliography

[814] Davey, Adrian G. (co-ord), 1978, Resource management of the Nullarbor Region, W.A.

[816] Davey, Adrian G., 1992, World Heritage significance of karst and other landforms in the Nullarbor Region

[817] Davey, Adrian G., 1978, Yallingup Cave Park - a management plan

[887] Dortch, C. E., 1984, Devil's Lair, a study in prehistory

[915] Dunkley, John R., 1978, The caves of Jenolan: 1, the exploration and speleogeography of Mammoth Cave, Jenolan

[919] Dunkley, John R., 1971, The exploration and speleography of Mammoth Cave, Jenolan

[931] Dyson, H. Jane (ed), 1982, Wombeyan Caves

[938] Eberhard, Rolan, 1994, The Junee River karst system, Tasmania: a report to the Forestry Commission

[939] Eberhard, Stefan, nd, The cave fauna at Ida Bay and the effect of quarry operation

[974] Ellis, Ross Andrew, 1940–, 1980, Australian caves and caving

[975] Ellis, Ross Andrew, 1940–(ed), 1972, Bungonia Caves

[1070] Flood, Josephine, 1983, Archaeology of the dreamtime

[1071] Flood, Josephine, 1990, The riches of ancient Australia: a journey into prehistory

[1107] Foster, J. J., 1890, The Jenolan Caves

[1159] Garrard, Ian M., 1983, Caving

[1209] Gillieson, David S. (ed), 1992, Geology, climate, hydrology and karst formation: field symposium in Australia 4 to 18 December 1992: Buchan - Mt. Gambier / Naracoorte - Nullarbor Plain humid temperate impounded karst, sub-humid temperate syngenetic karst, arid temperate karst: guidebook

[1220] Godden, Elaine, 1988, Rock paintings of Aboriginal Australia

[1233] Government of Australia, 1993, Nomination of Australian fossil sites (a serial nomination of sites at Murgon, Riversleigh and Naracoorte): The origin and evolution of Australia's mammals

[1270] Groenhout, Ron (ed), Sep 1968, Guide for prospective cavers

[1343] Hamilton-Smith, Elery, Oct 1962, Australian cave fauna: notes on collecting

[1345] Hamilton-Smith, Elery, Sep 1974, Cave Reserves of the Katherine Area

[1346] Hamilton-Smith, Elery (ed), 1958, Caving in Australia

[1347] Hamilton-Smith, Elery (prep), Jan 1989, Cutta Cutta Caves Nature Park: draft plan of management

[1348] [Hamilton-Smith, Elery], Sep 1994, Determining an environmental and social carrying capacity for Jenolan Caves Reserve: applying a visitor impact management system

[1349] [Hamilton-Smith, Elery], 1966, Draft management plan: Naracoorte Caves Conservation Park: South East South Australia

[1350] Hamilton-Smith, Elery, 1976, Mt. Etna & the caves: a plan for action

[1352] Hamilton-Smith, Elery, Oct 1994, The Regional Caves Interpretive Centre: a planning review report for the Augusta Margaret River Tourist Bureau

[1389] Havard, Ward L., nd, The Romance of Jenolan Caves

[1390] Hawkins, R. J. (comp), 1972, Australasian speleo map index no. 1

[1396] Hayllar, Bruce (proj offr), Mar 1984, The development of leisure and educational resources at Jenolan Caves Resort

[1431] Henderson, Kent, 1992, The Wombeyan and Abercrombie caves

[1433] Henderson, Kent, 1954–, 1990, Jenolan: a guide to Australia's famous caves

[1436] Henderson, Kent, 1954–, 1988, The Wellington and Abercrombie Caves

[1518] Horne, Peter, 1988, Barnoolut Estate sinkholes environmental assessment project, 27 December 1988

[1519] Horne, Peter, 1988, Fossil cave, 5L81 underwater

palaeontological and surveying project, 1987-1988

[1523]Horne, Peter, 1980, A report on South Australian diving fatalities

[1524]Horne, Peter, 1990, Research handbook for cave divers: an introduction to the realm and techniques of the underwater speleologist

[1525]Horne, Peter, 1987, South Australian diving fatalities 1950-1985

[1573]Humphreys, William F. (ed), Dec 1993, The biogeography of Cape Range Western Australia: being the proceedings of a symposium held under the auspices of the Western Australian Museum in Perth on 21 November 1992 at the Art Gallery of Western Australia

[1693]Jakeman, Anthony John, 1951–, 1983, Time series models for the prediction and characterisation of stream flow from rainfall in the Cave Creek system

[1701]James, Julia M., 1976, Timor Caves

[1723]Jennings, Joseph Newell, 1916–1984, 1970, Caves

[1724]Jennings, Joseph Newell, 1916–1984, Aug 1961, CEGSA 1965-6 Nullarbor expedition

[1727]Jennings, Joseph Newell, 1916–1984, 1963, The limestone ranges of the Fitzroy Basin, Western Australia; a tropical semi-arid karst

[1728]Jennings, Joseph Newell, 1916–1984, Aug 1961, A preliminary report of the karst morphology of the Nullarbor Plains

[1767]Kahrau, Wolfgang, 1972, Australian caves and caving

[1778]Keck, Ken, 1991, Abercrombie Caves: cave chronicles: a new look at an old wonder: including an account of the Bathurst convict rebellion of 1830

[1819]Kiernan, Kevin, 1988, The management of soluble rock landscapes: an Australian perspective

[1938]Lewis, Ian D., 1982, Cave diving in Australia

[1939]Lewis, Ian D., 1977, Discover Naracoorte caves

[1964]Longul, Wally, 1987, The last horizon

[1977]Lowry, D. C., 1970, Geology of the Western Australian part of the Eucla Basin

[1983]Lundelius, Ernest L., 1927–, 1982, The mammalian fauna of Madura Cave, Western Australia

[1984]Lundelius, Ernest L., 1927–, 1984, The mammalian fauna of Madura Cave, Western Australia, part VI, Macropodidae: Potoroinae

[1985]Lundelius, Ernest L., 1927–, Feb 1973, The mammalian fauna of Madura Cave, Western Australia: part I

[1986]Lundelius, Ernest L., 1927–, Aug 1975, The mammalian fauna of Madura Cave, Western Australia: part II

[1987]Lundelius, Ernest L., 1927–, Dec 1978, The mammalian fauna of Madura Cave, Western Australia: part III

[1988]Lundelius, Ernest L., 1927–, Feb 1981, The mammalian fauna of Madura Cave, Western Australia: Part IV

[2008]MacDougall, J. H., [1922], Western Australia's wonderland

[2037]Mann, Ian (ed), [1987], Speleotec '87 guidebook: a guide to the 16th biennial conference of the Australian Federation Inc

[2047]Marker, Margaret E., 1975, Lower southeast of South Australia: a karst province

[2081]Mass, Nuri, 1946, The wizard of Jenolan

[2090]Matthews, Peter Gahan, 1938–(ed), 1985, Australian karst index 1985

[2092]Matthews, Peter Gahan, 1938–(ed), 1979, Checklist of Australian caves and karst, 1979

[2106]McCarthy, Frederick D., 1905–, 1967, Australian aboriginal rock art

[2107]McCarthy, Frederick D., 1905–, 1958, Australian aboriginal rock art

[2166]Middleton, Gregory J., 1991, Oliver Trickett: doyen of Australia's cave surveyors 1847-1934

[2170]Middleton, Gregory. J. (comp), Sep 1988, Index to references to caves in three New South Wales government publications 1870-1919

[2284]Mott, Kevin (ed), 1982, Nullarbor caving atlas

[2301]Murdoch, Judy, Apr 1994, Light on dark: caves of the south east of South Australia

[2494]New South Wales Department of Tourism, 1973, Jenolan Caves, New South Wales

[2497]New South Wales Teaching Resources Centre, 1971, A visit to the Jenolan Caves: reference notes for the teacher

[2498]New South Wales. Department of Tourism, 1982, Limestone caves of New South Wales

[2511]Nicoll, Robert S., Dec 1976, Guidebook to the caves of southeastern New South Wales and eastern Victoria and caves around Canberra

[2531]The NSW Cave Rescue Squad Inc, nd, The NSW Cave Rescue Squad Inc. Annual Report July 1988-June 1989

[2533]Nucolorvue Productions, 1974, Mt. Gambier and district: Port MacDonnell and Tantanoola caves

[2632]Pearson, Les, 1983, Chillagoe souvenir guide

[2639]Pemble, Edna R., Nov 1972, Limestones caves and cavers

[2674]Pilkington, Graham (ed), 1986, Nullarbor caving atlas

[2675]Pilkington, Graham, 1982, Speleovision field notes

[2757]Ralston, Basil, 1989, Jenolan: the golden ages of caving

[2836]Rodda, Jan, 1989, Mulka's Cave site management project: emphasising visitor survey, April-June 1988, with management evaluation and further recommendations for management; with conclusions of an archaeological test excavation

[2908]Schram, Frederick R., 1943–, 1970, Structural composition and dental variations in the murids of the Broom Cave fauna, late Pleistocene, Wombeyan Caves area, N.S.W., Australia

[2939]Sexton, Robert T., Dec 1958, Cave surveying in

14.11. AUSTRALASIA

South Australia
[2994]Skinner, A. D., Jul 1978, The Mole Creek Caves
[3023]Smith, Jim, [1984], From Katoomba to Jenolan Caves: the Six Foot Track, 1884-1984
[3081]Spate, Andy, 1988, A brief introduction to speleology
[3090]Speleological Conference (1987), 1987, Proceedings - 16th biennial conference Australian Speleological Federation Inc
[3091]Speleological Conference (1991: Margaret River, WA, AU), 1991, Proceedings of the 18th biennial speleological conference: 30 December 1990 to 5 January 1991 at Margaret River, W.A
[3109]Starr, Joan, 1927–, 1985, The Wellington caves: treasure trove of fossils
[3154]Stringfield, Victor Timothy, 1902–, 1974, Karst and paleohydrology of carbonate rock terranes in semiarid and arid regions with a comparison to humid karst of Alabama
[3169]Sydney Speleological Society, nd, Prospective's handbook
[3170]Sydney Speleological Society, 1976, Timor Caves
[3232]Trickett, Oliver, 1899, Guide to the Jenolan Caves New South Wales
[3251]Turnbull, William D., 1973, Broom County *Cercartetus*, with observations on pygmy possum dental morphology, variation, and taxonomy
[3450]Western Australia Department of Conservation and Land Management, 1990, Leeuwin-Naturaliste National Park: cave permits draft issue plan
[3456]Wheeler, Joy, 1969, Oberon-Jenolan district historical notebook
[3500]Williams, Paul W., 1968, Contributions to the study of Karst
[3511]Wilson, Paul (prep), 1977, Managing the limestone caves of Chillagoe and Mungana

14.11.1.1 New South Wales

[68]Anderson, Grant, 1991, Wellington Caves, resource and management study
[115]Anon, 1882, Exploration of the caves and rivers of New South Wales (minutes, reports, correspondence, accounts)
[141]Anon, nd, Jenolan Caves: Australia's underground fairyland
[222]Anon, 1970, Yarrangobilly Caves—Kosciusko National Park
[225]Argus [pseud], 1898, Jenolan Caves and the Blue Mountains
[261]Australian Speleological Federation, c1973, Master plan for the development of Jenolan Caves Reserve
[336]Bartel Postcards, 1992, Jenolan Caves Australia
[587]Cameron McNamara Pty Ltd, 1989, Jenolan Caves Reserve: plan of management.
[662]Clark, E. John, nd, Jenolan Caves, New South Wales, Australia
[686]Colong Committee (Australia), 1985, The Colong story
[688]Commission on Cave Tourism and Management of the Australian Speleological Federation Study Team, Jun 1984, Jenolan Caves resort - some management issues
[719]Corrie, A. P., 1899, The Wombeyan Caves and the Bowral, Mittagong & Moss Vale tourist districts
[855]den Hertog, Sonja (ed), 1994, Walking the valley: an oral record of caving and bushwalking in the Burragorang and beyond, during the 1930s
[905]Driscoll, Ian, Apr 1977, The Binoomea Cut, Jenolan Caves: a paper read to the Jenolan Caves Historical and Preservation Society on Saturday 9th, February 1974
[913]Dunkley, John R. (comp), nd, A bibliography of the Jenolan Caves: part 1: speleological literature
[914]Dunkley, John R. (comp), 1988, A bibliography of the Jenolan Caves: part 2: literature
[918]Dunkley, John R., 1971, The exploration and speleogeography of Mammoth Cave, Jenolan
[920]Dunkley, John R., 1986, Jenolan caves as they were in the nineteenth century
[921]Dunkley, John R., Nov 1985, Preliminary checklist of references to Jenolan Caves
[924]Dunlop, Bruce Thomas, , Jenolan Caves
[976]Ellis, Ross Andrew, 1940–(ed), nd, A history of the journal of the Sydney Speleological Society
[1189]Gibbens, Ashley, 1975, Church Creek Caves
[1256]The Gregory Publishing Co Pty Ltd, nd, Jenolan Caves (N. S. W.) 'Nature's Masterpiece': a complete description of the geology, discovery, development, and features of the Jenolan limestone system beautifully illustrated, with high-class maps
[1314]Halbert, Erik, Nov 1994, The history of the Sydney Speleological Society 1954-1994
[1388]Havard, Ward L., nd, The Orient Cave, Jenolan Caves
[1397]Hayllar, Bruce (proj offr), Jan 1985, Jenolan Caves Resort teachers kit
[1437]Henderson, Kent, 1954–, 1985, The Wombeyan experience: a guide to the Wombeyan Caves New South Wales
[1463]Hills Speleology Club Ltd (prep), 1992, A guide to Mount Fairy Caves: a speleological field guide to the limestone and caves at Mount Fairy
[1517]Horne, Julia, 1994, Jenolan Caves: when the tourists came
[1581]Hurley, Frank, 1890–1962, 1952, The Blue Mountains and Jenolan Caves: a camera study
[1582]Hurley, Frank, 1890–1962 (photo), nd, Gems of Jenolan
[1729]Jennings, Joseph Newell, 1916–1984, 1987, Wee Jasper Caves

[1779]Keck, Ken E., Nov 1985, An introduction to Abercrombie Caves
[1812]Kerry, Charles (photo), nd, Right Imperial: Carlotta Arch: Devil's Coach House
[2093]Matthews, Peter Gahan, 1938–(ed), Jan 1968, Speleo handbook
[2167]Middleton, Gregory J., 1965—1966, Speleology
[2169]Middleton, Gregory J., 1970, Yarrangobilly Caves, Kosciusko National Park
[2495]New South Wales Immigration and Tourist Bureau, 1915, The systems of limestone caves: Jenolan, Yarrangobilly, Wombeyan, Abercrombie and Wellington
[2496]New South Wales National Parks and Wildlife Service, Jul 1987, Cooleman Plain karst area management plan: a supplementary plan to the Kosciusko National Park plan of management
[2501]Newbould, Ronald L., Mar 1975, The discovery of Barralong Cave, Jenolan Caves, N.S.W.: based on the diary of Ronald L. Newbould, guide, Jenolan Caves, co-discoverer with John P. Culley, senior guide, Jenolan Ceves [sic] of the "Barralong Cave" on the 7th June, 1964
[2502]Newbould, Ronald L., 1976, Steam cleaning of Orient Cave
[2503]Newbould, Ronald L., May 1974, Steam cleaning of Orient Cave, Jenolan Caves, N.S.W
[2532]The NSW Cave Rescue Squad Inc., nd, The NSW Cave Rescue Squad Inc
[2619]Pavey, Andrew (comp), Nov 1972, An index to cave maps in N.S.W
[2666]Phillips, Harry, nd, Jenolan Caves, (the underground wonderland) New South Wales, Australia
[2667]Phillips, Harry, nd, Nature's masterpiece: Jenolan Caves N.S.W
[2678]Pinnock, Rick (ed), 1987, A guide to Glenrock caves: a speleological field guide to the limestone caves on Glenrock Station
[2908]Schram, Frederick R., 1943–, 1970, Structural composition and dental variations in the murids of the Broom Cave fauna, late Pleistocene, Wombeyan Caves area, N.S.W., Australia
[3171]Sydney University Speleological Society, Nov 1975, Nibicon log book reports December 1972 - February 1973
[3231]Trickett, Oliver, 1906, The Abercrombie Caves
[3233]Trickett, Oliver, 1906, Guide to the limestone caverns of New South Wales: Jenolan, Wombeyan, and Yarrangobilly
[3234]Trickett, Oliver, 1906, Guide to Wombeyan Caves NSW
[3235]Trickett, Oliver, 1917, Guide to Yarrangobilly Caves New South Wales
[3236]Trickett, Oliver, 1898, Notes on the limestone caves of New South Wales with plans
[3237]Trickett, Oliver, 1906, The Wellington Caves

[3432]Welch, Bruce R. (ed), 1976, The northern limestone
[3456]Wheeler, Joy, 1969, Oberon-Jenolan district historical notebook
[3489]Wilkinson, Charles Smith, 1887, Photographs of the Jenolan Caves (interior views photographed by means of the electric and magnisium [sic] lights

14.11.1.2 Queensland

[161]Anon, Apr 1988, Mount Etna action book: time is running out
[200]Anon, 1962, A symphony in limestone: the beautiful Olsen's Caves: 18 miles North of Rockhampton, Queensland: A description of Olsen's Caves
[649]Chillagoe Caving Club, nd, Broken River karst: a speleological field guide North Queensland Australia
[650]Chillagoe Caving Club (comp), nd, Numerical index - area maps
[755]Cribb, A. B., 1965, An ecological and taxonomic account of the algae of a semi-marine cavern, Paradise Cave, Queensland
[1234]Graham, Andrew W. (ed), May 1972, Aspects of land use planning at Fanning River. With particular reference to the Fanning River caves
[2735]Queensland Conservation Council (prep), Mar 1973, The case against the Pike Creek Dam
[2736]Queensland Conservation Council (prep), Jun 1973, Pike Creek Dam: a preliminary criticism of the Queensland Irrigation and Water Supply's Commission's environmental impact study
[3022]Smith, James, Oct 1964, The early days of Olsen's Caves as seen through a geologist's eyes
[3097]Sprent, J. K., (ed), 1970, Mount Etna caves: a collection of papers covering several aspects of the Mt. Etna and Limestone Ridge caves area of central Queensland
[3230]Trezise, P. J., 1971, Rock art of South-east Cape York

14.11.1.3 South Australia

[78]Anon, nd, 37 views of Naracoorte caves and Naracoorte South Australia
[162]Anon, nd, The Naracoorte Caves: "how to reach them"
[818]Davidge, G. W., nd, Narracoorte [sic]: caves and town 20: book of views
[917]Dunkley, John R., 1978, Caves of the Nullarbor; a review of speleological investigations in the Nullarbor Plain, Southern Australia
[1267]Grimes, Ken, February 1996, Field Guide to karst features in southeast South Australia and Western Victoria: for the Karst Studies Seminar, Naracoorte, February 1996
[1434]Henderson, Kent, 1954–, 1986, The Naracoorte and

14.11. AUSTRALASIA

Tantanoola Caves: a guide to the famous caves of Southeast South Australia

[1520]Horne, Peter, 1988, Gouldens Hole - 5L8 - mapping project, 1987 - 1988

[1521]Horne, Peter (comp), Aug 1993, Lower South East cave reference book: an illustrated catalogue of the registered caves, sinkholes and associated karst features of the Lower South East Region of South Australia

[1522]Horne, Peter, 1985, Piccaninnie Ponds mapping project, 1984/85

[1940]Lewis, Ian D., 1976, South Australian cave reference book

[2940]Sexton, Robert T., 1965, Caves of the coastal areas of South Australia

[3055]South Australia National Parks and Wildlife Service, Mar 1983, Draft management plan, Tantanoola Caves Conservation Park, lower South East, South Australia

[3056]South Australia National Parks and Wildlife Service, 1990, Naracoorte Caves Conservation Park management plan, South East, South Australia

[3057]South Australia National Parks and Wildlife Service, Feb 1986, Naracoorte Caves Conservation Park, South East, South Australia: draft management plan

[3058]South Australia National Parks and Wildlife Service, 1990, Tantanoola Caves Conservation Park management plan, South East, South Australia

[3217]Thurgate, Mia E., 1995, Sinkholes, caves and spring lakes: an introduction to the unusual aquatic ecosystems of the lower south east of South Australia: South Australian Underwater Speleological Society occasional paper number 1

[3436]Wells, R. T., 1975, World famous Fossil Cave Naracoorte

[3525]Woods, Julian Edmund, 1862, Geological observations in South Australia: principally in the district southeast of Adelaide

[3535]Wright, R. V. S., 1971, Archaeology of the Gallus Site, Koonalda Cave

14.11.1.4 Tasmania

[351]Baynes, F. J., 1991, Ida Bay - Benders Quarry. An estimate of reserves within existing quarry limits

[557]Bunton, Stephen, 1984, Vertical caves of Tasmania: a caver's guidebook

[771]Cullen, P., 1990, A survey of the vegetation

[938]Eberhard, Rolan, 1994, The Junee River karst system, Tasmania: a report to the Forestry Commission

[939]Eberhard, Stefan, nd, The cave fauna at Ida Bay and the effect of quarry operation

[940]Eberhard, Stefan, 1991, The invertebrate cave fauna of Tasmania

[941]Eberhard, Stefan, 1986, Report of the Tasmanian Caverneering Club 1986 speleological reconnaissance expedition to Precipitous Bluff

[1750]Jones, R., 1974, Wilderness caves of the Gordon - Franklin River System

[1816]Kiernan, Kevin, 1995, An atlas of Tasmanian karst

[1817]Kiernan, Kevin, Sep 1989, Bibliography of Tasmanian karst

[1818]Kiernan, Kevin, Nov 1989, Karst, caves and management at Mole Creek, Tasmania

[2168]Middleton, Gregory J., 1979, Wilderness caves: a preliminary survey of the caves of the Gordon-Franklin River System, South-West Tasmania

[2304]Murray, Peter, 1977, Pleistocene vertebrate remains from a cave near Montagu, N.W. Tasmania

[2996]Skinner, R. K., nd, Hastings Caves State Reserve Tasmania: a visitors' guide

[3082]Spate, Andy, 1991, Kubla Khan Cave State Reserve, Mole Creek, Tasmania: pilot management study

14.11.1.5 Victoria

[206]Anon, 1907, Views of the Buchan Caves and pyramids

[815]Davey, Adrian G., 1986, Victorian caves and karst: strategies for management and catalogue: a report to the Caves Classification Committee, Department of Conservation, Forests and Lands /

[1267]Grimes, Ken, February 1996, Field Guide to karst features in southeast South Australia and Western Victoria: for the Karst Studies Seminar, Naracoorte, February 1996

[1432]Henderson, Kent, 1954–, 1985, The Buchan experience: a guide to the Buchan and Murrindal Caves - East Gippsland, Victoria

[2178]Mill, Lloyd (ed), 1980, Victorian caves and karst: a guidebook to the 13th A.S.F. conference: Melbourne 1980

[2352]National Parks and Public Land Division, Department of Conservation & Environment, Mar 1991, Draft strategy for the management of caves and karst in Victoria

[3128]Stirling, James, 1884, On the caves perforating marble deposits, Limestone Creek (read 12th April 1883)

[3178]Talent, J. A., nd, The Buchan Caves

[3468]White, Susan, Aug 1986, A bibliography of Victorian caves & karst

14.11.1.6 Western Australia

[100]Anon, c1905, The caves of Western Australia

[886]Dortch, C. E., 1976, Devils Lair: a search for ancient man in Western Australia

[917]Dunkley, John R., 1978, Caves of the Nullarbor; a review of speleological investigations in the Nullarbor Plain, Southern Australia

[958]Elliot, Ian, 1977, The discovery and exploration of

the Yanchep Caves
[1454]Hill, A. L. (ed), 1966, Mullamullang Cave expeditions 1966
[1574]Humphreys, William F., Apr 1991, Survey of caves in Cape Range North West Cape Peninsula Western Australia
[1591]Hyslop, R. C. , 1967, Augusta Jewel Caves and other points of interest
[2851]Rolsh Photographies, nd, Cave wonderlands of Western Australia; Jewel Caves, Lake Cave, Mammoth Cave, Yallingup Cave
[2947]Shapcott, L. E., [1934], The story of Yanchep: the western wonderland
[3406]Watson, J. R. (ed), Feb 1979, Cave tourism in Western Australia: proceedings of a seminar held at Busselton November 10-11, 1978

14.12 Pacific

[2997]Skjolsvold, Arne, Feb 1972, Excavations of Habitation Cave

14.12.1 Guam

[2845]Rogers, Bruce W., May 1992, Karst features of the Territory of Guam: a survey of caves and karst features on Guam Island, Territory

14.12.2 Mariana Islands

[2842]Rogers, Bruce W., May 1992, Karst features of Rota (Luta) Island, CNMI, A survey of caves and karst features on Rota (Luta) Island, Commonwealth of the Northern Mariana Islands
[2843]Rogers, Bruce W., May 1992, Karst features of Saipan, CNMI: a survey of caves and karst features on Saipan Island, Commonwealth of the Northern Mariana Islands

14.12.3 Marquesas Islands

[2997]Skjolsvold, Arne, Feb 1972, Excavations of Habitation Cave

14.12.4 Micronesia

[1848]Kornicker, Louis S., 1919–, 1989, New Ostracoda (Halocyprida: Thaumatocyprididae and Halocyprididae) from anchialine caves in the Bahamas, Palau, and Mexico
[2841]Rogers, Bruce W., Jan 1992, Karst features of Pohn Pei, Chuuk, and Waqab: a survey of caves and karst features in the States of Pohn Pei, Chuuk, and Waqab, Federated States of Micronesia

14.12.5 New Caledonia

[257]Australian Speleological Expedition, 1976, Caves of New Caledonia: report of the 1975 Australian Expedition
[1364]Harris, Stephen, 1976, Caves of New Caledonia: report of the 1975 Australian Expedition

14.12.6 New Guinea

[502]Brook, David B., 1944–(comp), Dec 1976, The British New Guinea speleological expedition
[515]Brown, A. L., Nov 1976, Lelet: report of the 1975 New Ireland Speleological Expedition
[791]D'Arcy Galleries (New York, N.Y.), 1968, The Caves of Karawari
[1208]Gillieson, D. S., Dec 1977, Lelet: report of the 1976 New Ireland Speleological Expedition
[2093]Matthews, Peter Gahan, 1938–(ed), Jan 1968, Speleo handbook

14.12.7 New Zealand

[160]Anon, nd, The most beautiful caves in the world
[210]Anon, [c1910], Waitomo Ruakuri & Aranui New Zealand
[212]Anon, 1992, Welcome to Waitomo Caves: a photographic insight into this spectacular region of New Zealand
[230]Arrell, Robert, 1984, Waitomo Caves: a century of tourism
[398]Bigwood, K. (photo), c1980, Waitomo Caves Wonderland
[401]Bircham, Deric N., 1975, Waitomo tourist caves
[635]Chandler, Peter, 1987, A trip through time: a drive-yourself guide to the landforms and geology of the Waitomo Caves area
[653]Chilton, Charles, B. 1860–, 1894, The subterranean Crustacea of New Zealand: with some general remarks on the fauna of caves and wells
[761]Crossley, P. C. (comp), 1988, The New Zealand cave atlas: North Island
[762]Crossley, P. C. (comp), 1990, New Zealand cave atlas: South Island
[763]Crossley, P. C. (comp), 1981, The New Zealand cave atlas
[1101]Forster, Raymond R., 1922–, 1965, Harvestmen of the sub-order Laniatores from New Zealand caves
[1125]Frederikson, Rosalie, 1983, The New Zealand glowworm
[1128]Fullerton, Iona, 1983, Cave minerals and speleothems: a review of the literature with special reference to Waitomo
[1201]Giles, Cathy, 1983, Cave fauna of Waitomo

[1351] Hamilton-Smith, Elery (tm ldr), Nov 1988, Proposed developments at Waitomo Caves
[1777] Kawiti, Walter Brown, 1969, Waiomio's limestone caves; linked with Maori tribal legend, rich in natural history
[2020] Maih, Lindsay (ed), 1993, The New Zealand Cave atlas: volume 2: South Island
[2164] Meyer-Rochow, Victor Benno, 1990, The New Zealand glowworm
[2499] New Zealand Speleological Society, Feb 1989, The little red cavers book
[2500] New Zealand Speleological Society, 1985, The little red cavers book: NZ Speleological Society information manual
[2589] P. Tarrant Ltd, nd, Waitomo Caves New Zealand: a souvenir booklet of your Waitomo visit with story of life cycle of New Zealand glow-worm
[2616] Paterson, J., 1983, Teana-au caves information
[2697] Potton, Craig, 1987, Images from a limestone landscape: a journey into the Punakaiki-Paparosa Region
[2746] Radcliffe, F.G., 1925, The wonder caves of Maoriland: Waitomo, Ruakuri & Aranui
[2765] Rautjoki, Harri, nd, North-West Nelson State Forest Park Honeycomb Hill Cave System: concept plan: West Coast Conservancy
[2814] Richards, James Harray, Dec , Waitomo caves
[2907] Schoon, Theo, 1985, Maori rock drawing: the Theo Schoon interpretations
[3034] Smith, Rodney, 1981, Waitomo Caves
[3035] Smith, Rodney, 1980, Waitomo: Glow-worm Cave, Aranui Cave, Ruakuri Cave and surrounding district
[3186] Tásler, Radko, 1991, Owen 90: New Zealand
[3374] Waitomo District Council, March 1994, District plan discussion paper: tourism issues
[3482] Wilde, Kevan A., 1981, The cave and karst resource of New Zealand
[3484] Wilde, Kevan A., 1992, West coast cave and karst management strategy and operational guidelines
[3493] Williams, Dave, 1987, Waitomo Caves: Glow-worm Cave, Ruakuri Cave, Aranui Cave
[3528] Worthy, Trevor, 1993, Fossils of Honeycomb Hill
[3529] Worthy, Trevor, 1989, Inventory of New Zealand caves and karst of international, national and regional importance

14.12.8 Palau

[2844] Rogers, Bruce W., May 1992, Karst features of the Palau Islands, a survey of caves and karst features in the Palau Archipelago

14.12.9 Papua New Guinea

[483] British Broadcasting Corporation Television Service, 1980, Atea: in search of the world's deepest cave
[1202] Gill, David W., 1988, The untamed river expedition: Nakanai Mountains, East New Britain, Papua New Guinea
[1697] James, Julia M., 1980, Caves and karst of the Muller Range: report of the 1978 speleological expedition to the Atea Kananda, Southern Highlands, Papua New Guinea
[1698] James, Julia M., 1980, Caves and karst of the Muller Range: report of the 1978 Speleological Expedition to the Atea Kananda, Southern Highlands, Papua New Guinea
[1699] James, Julia M., 1974, Papua New Guinea Speleological Expedition NSRE 1973: the report of the 1973 Niugini Speleological Research Expedition to the Muller Range
[1700] James, Julia M. (ed), 1974, The report of the 1973 Niugini Speleological research expedition to the Muller Range
[2021] Maire, Richard (coord, surv, art), Jul 1981, Report of the French Speleological expeditions to PNG in collaboration with the Committee of French Speleological Expedition and the Scientific Committee of the French Federation of Speleology

14.12.10 Philippines

[534] Bruce, Murray D. , 1980, Preliminary survey of Palawan, Philippines, by Traditional Explorations and the Sydney Speleological Society, January/February 1980
[1050] Fernandez, Carlos A., 1972, The Tasaday: cave-dwelling food gatherers of South Cotabato, Mindanao: preliminary report submitted to the Panamin Foundation, Inc. on June 1, 1972
[1111] Fox, Robert B., 1918–, 1964, The Tabon Caves: archaeological excavations on Palawan Island, Philippines (1962—64)
[1112] Fox, Robert B., 1918–, 1970, The Tabon Caves; archaeological explorations and excavations on Palawan Island, Philippines
[1188] Gibb, Hugh, 1981, The Tabon caves
[2119] McDonald, Mike, c1993, The journal of the Joint Bristol Exploration Club (United Kingdom) and the National Mountaineering Federation of the Philippines caving expedition to the Philippines, January to April 1992
[2853] Ronquillo, Wilfredo P., Dec 1981, The technological and functional analyses of lithic flake tools from Rabel Cave, northern Luzon, Philippines
[3005] Smart, James, 1994, A Philippine cave index and bibliography

14.12.11 Polynesia

[1976]Lowe, David John, 1988, Tonga '86, Tonga '87 expedition report
[2997]Skjolsvold, Arne, Feb 1972, Excavations of Habitation Cave

14.13 Other Geographic Terms

14.13.1 Mediterranean

[558]Burdon, David Joseph, 1963, Handbook of karst hydrogeology with special reference to the carbonate aquifers of the Mediterranean Region
[2653]Pericot Garcia, Luis, 1899–(ed), 1964, Prehistoric art of the western Mediterranean and the Sahara

14.13.2 Middle East

[706]Coon, Carleton Stevens, 1904–, 1957, The seven caves: archaeological explorations in the Middle East
[1875]Kurtén, Björn, 1965, The Carnivora of the Palestine caves

14.13.2.1 Israel

[316]Bar-Adon, Pesah, 1907–, 1980, The cave of the treasure: the finds from the caves in Nabal Mishmar
[317]Bar-Yosef, Ofer, 1985, A cave in the desert, Nahal Hemar: 9,000-year-old finds: [exhibition, the Israel Museum, Jerusalem, spring 1985]
[318]Bar-Yosef, Ofer, 1988, Nahal Hemar Cave
[380]Ben-Tor, Amnon, 1975, Two burial caves of the Proto-Urban Period at Azor, 1971: the first season of excavations at Tell-Yarmuth, 1970
[1451]Hevrah la-hakirat Erets-Yisrael ve-atikoteha, 1960, 1962, The Expedition to the Judean Desert
[1679]J., J. M., 1941, How the grottoes of an ancient church were discovered in the convent of the "Dames de Nazareth" at Nazareth in Galilee
[1732]Jewish Museum (New York, N.Y.), 1967, Masada and the finds from the Bar-Kokhba caves: struggle for freedom
[2927]Seger, Joe D., 1988, Gezer V: the field I caves
[2932]Sellers, O. R., 1953, A Roman-Byzantine burial cave in Northern Palestine (the joint excavation of the American School of Oriental Research in Jerusalem and McCormick Theological Seminary at Silet edhDhahr
[3115]Stekelis, Moshe, 1898–1967, 1952, The Abu Usba Cave (Mount Carmel)
[3329]Universitah ha-petuhah, 1988, Israeli scenery Soreq Cavern

14.13.2.2 Iran

[705]Coon, Carleton Stevens, 1904–, 1968, Cave explorations in Iran, 1949
[1029]Farr, Martyn, 1950–, 1984, The great caving adventure
[1765]Judson, David M., 1973, Ghar Parau
[3012]Smith, Anthony, 1926–, , Blind white fish in Persia
[3013]Smith, Anthony, 1926–, 1953, Blind white fish in Persia

14.13.2.3 Iraq

[1055]Field, Henry, 1902–, 1955, Caves and rockshelters in northern Iraq.
[3046]Solecki, Ralph S., 1917–, 1972, Shanidar: the first Flower people
[3127]Stewart, Thomas Dale, 1901–, Apr 1977, The Neanderthal skeletal remains from Shanidar Cave, Iraq: a summary of findings to date
[3239]Trinkaus, Erik, 1985, The Shanidar Neandertals
[3240]Trinkaus, Erik, 1983, The Shanidar Neandertals

14.13.2.4 Oman

[838]Davison, W. Donald (auth, photo), Oct 1985, Majlis Al Jinn Cave, Sultanate of Oman

14.13.2.5 Syria

[1354]Hanihara, Kazuro, 1927- , 1978, Paleolithic site of Douara Cave and paleogeography of Palmyra Basin in Syria
[3163]Suzuki, Hisashi, 1912–(ed), 1973, The Palaeolithic site at Douara Cave in Syria: report of the fourth season of the Tokyo University Scientific Expedition to Western Asia

15 Bibliography

[1] **A. I. D. International** *Caves.* [London, GB]: A.I.D. International, **1977**. illus: b&w: tables; partly col: photomicros.

Title from cassette. For use in schools. 1 filmstrip, 1 sound cassette. Juvenile.

[2] **Abadie, Bernard** *Grottos of Bétharram.* Pau, FR: Marrimpoucy Jeane, [**1972**]. illus: b&w: maps, photos, ports; col: photos; 62p; 18 x 13 cm.

28 photos, 2 maps.
History and description of this show cave.

[3] **Abeille de Perrin, Elzear** *On the collection of cavern insects.* Cambridge, MS, US: University Press, **1877**. Kentucky Geological Survey 1873-1891 Memoirs ... v 2 pt 8. 14p; 30 cm.

[4] **Absolon, Karel B.** *The conquest (the riddle of an abyss and its sinking river).* Bethesda, MD, US: np, **1980**. illus: b&w: drawings, maps, photos; [x], x, 137p, 58p pls; bibl, index.

Unpublished? May be a posthumous translation. Copy examined was photocopy and difficult to determine exact nature of illustrations.

[5] **Absolon, Karel B.** *The conquest of the caves and underground rivers of Czechoslovakia's Macocha Abyss: a historical and technical study of their exploration.* Rockville, MD, US: Kabel Publishers, **1987**. illus: b&w: drawings, maps, photos; xiii, 103p, [6]; bibl, index.

"First limited number edition." May be a posthumous translation.

[6] **Adam, David P.** *Pollen analysis of a stratigraphic section from Bat Cave, New Mexico.* Parker, Marion L. npop: np, **1963**. [iv], 13p, 6p pls; bibl; 28 cm.

Cover title. "... work done under supervision of Dr. Paul S. Martin ..." Illus.

[7] **Adam, John** *Aberbrothok illustrated.* npop: np, **1886**. illus: b&w: drawings; 86p.

Engravings (124 etchings) with historical and topographical notes, on Scottish caves. Covers most of the caves in this area.

[8] **Adamkó, Péter** *The caves of Buda.* Dénes, György; Lél-Őssy, Szabolcs; Kandó, Michael (trans). Budapest, HU: Mayor of Budapest, **1992**. illus: b&w: maps; col: photos; map on rear lining page; 48p.

[9] **Adamkó, Péter** *The caves of Budapest.* Karpát, József; Kiss, Attila; Lél-Őssy, Sándor; Takacs-Solner, Katalin; Vér, Zsolt; Rádai, Ödön (trans); Lénárt, László (ed); Szenthe, István (ed); Varga, Csaba (ed). Budapest, HU: 10th International Congress of Speleology, **1989**. Field Trip Guide E1. illus: b&w: maps; N; 28p, 28p pls.

[10] **Adams, Aislinn (des, illus)** *Dunmore Cave.* IE: National Parks and Monuments Service, **1984**. illus: b&w: maps, photos; cave map on front lining page and bibl on rear lining page; 21p; bibl; 15 x 21 cm.

Other editions list David Phillip Drew and David Huddart as the authors.

[11] **Adams, Laurie Branson (comp)** *Annotated bibliography of North Carolina speleology.* Asheville, NC, US: Laurie Branson Adams, Oct, **1986**. [iv], 91p.

No illus.

[12] **Adams, Laurie Branson** *History of Morrell Cave: Part I Tennessee.* Old Fort, NC, US: Flitterhouse Grotto. National Speleological Society, Aug, **1984**. [i], 11p; bibl.

No illus.

[13] **Adams, William Henry Davenport**, 1828-1891 *Famous caverns and grottoes: described and illustrated.* 1886 ed. London, GB: Thomas Nelson and Sons, **1886**. illus: b&w: ads, drawings, maps, photos; b&w front; xii, 185p, [3]; 19 cm.

1890 edition is identical to the 1886.
1890 ed.

[14] **Adams, William Henry Davenport**, 1828-1891 *Famous caves and catacombs: described and illustrated.* 1886 ed. London, GB: Thomas Nelson and Sons, **1886**. illus: b&w: drawings, maps, photos; b&w front; xii, 204p; 19 cm.

Subtitle, "described and illustrated" appears on cover, but not on title page (except in reprint).

1888 ed. npop: np.

1972 reprint ed. Freeport, NY, US: Books for Libraries Press, **1972**. Essay Index Reprint Series. illus: b&w: drawings; b&w front; xii, 204p; 19 cm.

Microfilm edition: 1983 Library of Congress.

[15] **Addis, Robert Philip**, 1945– *A study for the National Speleological Society: Knox Cave: Albany County, New York.* Wilbraham, MA, US: Speleobooks, **1979**. New York Cave Survey Bulletin 3. illus: b&w: charts, drawings, maps, photos, ports; 1 map folded in back; front: b&w port; ix, 139p, [1]; app, bibl.

Detailed case study in cave management. 22 figs. This is the author's MBA thesis from SUNY Binghamton.

[16] **Adovasio, J. M.** *Archaeological testing at two rockshelters in the Tombigbee River Multi-Resource District, Alabama and Mississippi: an interim report.* United States Interagency Archeological Services [Atlanta]; United States Army Corps of Engineers Nashville District. [Pittsburgh, PA, US]: University of Pittsburgh. Dept of Anthropology; Cultural Resources Management Program, **1980**. 41p; 29 cm.

Illus.

[17] **Agee, James K.** *Oregon Caves forest and fire history.* Potash, Laura L. Seattle, WA, US: National Park Service, Cooperative Park Studies Unit, College of Forest Resources, University of Washington, **1990**. Report, Cooperative Park Studies Unit Seattle WA, no 90-1. illus: b&w: maps; ii, 66p, [11]; bibl; 28 cm.

Illus.

[18] **Aikens, C. Melvin** *Hogup Cave.* Salt Lake City, UT,

US: University of Utah Press, **1970**. Anthropological Papers (Salt Lake City) no 93 April 1970. illus: b&w: drawings, graphs, maps, photos, tables; xiii, 286p; app, bibl, errat; 28 cm.

129 figs, 18 tables.

One of a series of publications defining the Desert Archaic, a particular kind of cultural adaptation in the intermontane region of the United States. Unique preservation of the organic cultural artifacts.

[19] **Ainsworth, Joseph** *Saint Mary's University Speleological Society: 1954-1959.* San Antonio, TX, US: Saint Mary's University Speleological Society, Apr, **1960**. [iii], 19p.

[20] **Ainsworth, William** *An account of the caves of Ballybunian, county of Kerry: with some mineralogical details.* Jennings, Jesse D. (ed). Dublin, IE: William Curry Jun and Company, **1834**. illus: b&w: drawings, tables; table folded in text; front: b&w drawing; [v], 96p, 2p pls; 23 cm.

Accounts like this shaped our early expectations of caves and their mineral wealth.

[21] **Akira Yoshida** *A list of the Arthropoda in the limestone caves in Kantô-Mountainland, with the descriptions of a new genus and three species.* Sizumu Nomura. Tokyo, JP: Japanese Society for the Study of Insects, Jan, **1952**. The Chūhō no 6.

[22] **Albi, S.A. Apa Poux (ed)** *Padirac Chasm and underground river.* "As de Coeur Edition". npop: np, **1990**. illus: b&w: repros; partly col: charts; col: maps, photos; [24].

"Edited by S.A. Apa Poux Albi with the aid of Messrs. and Mrs. Maury."

[23] **Alderson, Laura White** *Mammoth Cave: Kentucky's buried treasure.* Clark, Chip (photo). [Washington, DC, US]: National Geographic Society, **1984**. illus: col: maps; fold-out col map; [13]; 27 cm.

Reprinted from the Autumn 1984 issue of *National Geographic Traveler*.

[24] **Alex, Lynn Marie**, 1948– *The archaeology of the Beaver Creek shelter (39CU779), Wind Cave National Park, South Dakota: a preliminary statement.* South Dakota School of Mines and Technology. Denver, CO, US: United States National Park Service, **1991**. Selections from the Division of Cultural Resources 3. x, 60p; bibl; 28 cm.

Illus.

[25] **Alexander, E. Calvin** *Hydrogeologic study of Jewel Cave/Wind Cave: final report -year 2, 4th quarter of year 2, February 1, 1987 through April 30, 1987.* Davis, Marsha. Minneapolis, MN, US: The Department of Geology and Geophysics, University of Minnesota, **1987**. illus: b&w: maps; iii, 71p; bibl; 28 cm.

Cover title. Performed for the U.S. National Park Service under contract CX- 1200-S-A047 (U of Minn. 0645-5647)."August, 1987."

[26] **Alexander, E. Calvin** *Hydrologic study of Jewel Cave/Wind Cave: final report.* Davis, Marsha; Alexander, Scott C. Minneapolis, MN, US: The Department of Geology and Geophysics, University of Minnesota, **1989**. illus: b&w: maps; 196p; bibl; 28 cm.

Cover title: Performed for the U.S. National Park Service under contract CX- 1200-5-A047 (Univ. of Minn. #0645-5647), contract period May 1, 1985 to April 30, 1988. "Submitted March, 1989." Illus.

[27] **Alexander, E. Calvin** *Practical tracing of groundwater, with emphasis on karst terranes: a short course manual presented on the occasion of the annual meeting of the Geological Society of America, October 24, 1992, Cincinnati, Ohio.* Quinlan, James Francis, 1936-1995. Boulder, CO, US: Geological Society of America, **1992**. illus: b&w: drawings, graphs, maps, tables; vol 2 has 1 folded map in pocket in text; bibl.

Contains reprints. Cover title: Geological Society of America short course manual on practical tracing of ground water, with emphasis on karst terranes. Pagination: vol I: ii, 38 + [3], i-ix, 1—79, [22], 1-13p; vol II: [88], 1-15, [25], 2-20, [5], 1-3, [14]p (some blank);1st edition of same title was by James F. Quinlan and E. Calvin Alexander, Jr. 1990.

[28] **Alexander, Emmit Calvin**, 1943– *Hydrogeologic study of Jewel Cave/Wind Cave: final report - phase I, 3rd and 4th quarters of phase I, 1 November 1985 through 30 April 1986.* Davis, Marsha. Minneapolis, MN, US: The Department of Geology and Geophysics, University of Minnesota, **1986**. 7p, [55]; 28 cm.

Cover title. Performed for the U.S. National Park Service under contract CX- 1200-5-A047 (U of Minn 0645-5647). "October, 1986." Illus.

[29] **Alexander, Hubert Griggs** *Report on the excavation of Jemez Cave, New Mexico, Santa Fe, N.M..* Reiter, Paul. Albuquerque, NM, US: University of New Mexico; School of American Research (Santa Fe, N.M.), **1935**. University of New Mexico Bulletin Monograph series v 1 no 3 whole no 278. illus: b&w: maps; iii, 67p, 11p pls; 28 cm.

Illus.

[30] **Alexander, John Martin**, 1934– *The survey of Pant Mawr Pot, South Wales.* Jones, John Clive, 1934–. npop: Cave Research Group of Great Britain, **1959**. Cave Research Group of Great Britain Publications no 9. illus: b&w: maps; fold out map in rear; iii, 11p; 25 x 20 cm.

[31] **Alexander, Robert K.** *Archeological investigations at Parida Cave, Val Verde County, Texas.* Rickind, David H. (app); Bacon, Mary R. (ed); Comstock, V. N. (ed). Austin, TX, US: np, **1970**. illus: b&w: drawings, maps, photos, tables; viii, 103p; app, bibl.

[32] **Aley, Thomas John**, 1938– *Cave and karst hydrology assessment project for Horsethief Cave, Wyoming.*

Report to the Worland District, BLM. Aley, Catherine. Protem, MO, US: Ozark Underground Lab, **1984**. 104p.

Results of impact studies on Horsethief Cave. Surface land use for mining and recreation is covered.

[33] Aley, Thomas John, 1938– *Cave management investigations on the Ozark National Scenic Riverways, Missouri.* United States National Park Service. Protem, MO, US: Ozark Underground Laboratory, **1981**. ii, 157p; bibl; 30 cm.

Col illus. Cover title: "Cave management investigations, Ozark National Scenic Riverways. Phase 2. Conducted for the Midwest Region, National Park Service, under purchase order PX 6000-9-0828. March, 1981."

[34] Aley, Thomas John, 1938– *Cave management investigations on the Ozark National Scenic Riverways, Missouri.* United States National Park Service. Protem, MO, US: Ozark Underground Laboratory, **1980**. iii, 111p, [13]; bibl; 30 cm.

Illus, some col. Phase 1. Conducted for the Midwest Region, National Park Service, under purchase order PX 6000-8-0918.

[35] Aley, Thomas John, 1938– *Control of exotic plants in Oregon Caves, Oregon Caves National Monument.* Aley, Catherine; Engineering and Design Associates; United States National Park Service. Protem, MO, US: Ozark Underground Laboratory, **1988**. one folded map in pocket; 118p; bibl; 28 cm.

Running title: "Oregon Caves plant control study. Contract report to Engineering and Design Associates, contractor to the National Park Service. November, 1985."

[36] Aley, Thomas John, 1938– *Delineation and hydrogeologic study of the key cave aquifer Lauder Dale County, Alabama.* Protem, MO, US: Ozark Underground Laboratory, Feb, **1990**. illus: b&w: maps, tables; four b&w maps folded in rear pocket; 154p; app, bibl.

An investigation and report prepared for the US Fish and Wildlife Service under contract number 14-16-0004-88-073. Pagination includes appendix numbering A1-A40.

[37] Aley, Thomas John, 1938– *Groundwater contamination and sinkhole collapse induced by leaky impoundments in soluble rock terrain.* Williams, James Hadley, 1929–; Massello, James W. Rolla, MO, US: Missouri Geological Survey and Water Resources, **1972**. Engineering Geology Series no 5. illus: b&w: drawings, maps, photos, tables; viii, 32p; bibl; 28 cm.

10 figs.

[38] Aley, Thomas John, 1938– *Hydrogeologic mapping of unincorporated Greene County, Missouri, to identify areas where sinkhole flooding and serious groundwater contamination could result from land development.* Thomson, Kenneth C. Protem, MO, US: Ozark Underground Laboratory, **1981**. illus: b&w: maps; 5 leaves of plates (folded); 11p; 35 cm.

[39] Aley, Thomas John, 1938– *Karst and cave resource significance assessment: Ketchikan Area, Tongass National Forest, Alaska.* Ketchikan, AK, US: U.S. Dept. of Agriculture, Forest Service, Alaska Region, Tongass National Forest, Ketchikan Area, Nov, **1993**. illus: b&w: maps; 123p; bibl; 28 cm.

Illus. "Report of the Karst Resources Panel: Thomas Aley, Catherine Aley, William R. Elliott, Peter W. Huntoon."

[40] Aley, Thomas John, 1938– *Origin and hydrology of caves in the White Limestone of north central Jamaica: report of field work carried out under ONR Contract 3656 (03) NR 388 067.* Parsons, James J. [Berkeley, CA, US]: [University of California, Dept of Geography], **1964**. illus: b&w: maps, photos; 291p; 28 cm.

Illus.

One of the few scientific works on the caves of Jamaica. It will be interesting to see how this holds up to modern theory.

[41] Aley, Thomas John, 1938– *A predictive hydrologic model for evaluating the effects of land use and management on the quantity and quality of water from Ozark Springs: final report.* Ozark Underground Laboratory. Springfield, MO, US: Ozark Underground Laboratory, Aug, **1975**. illus: b&w: drawings, graphs, maps, tables; [xv]; app, bibl.

Prepared for the National Forests in Missouri under contract 05-1277. Later issued by Missouri Speleological Survey as Missouri Speleology, v 18, 1978: [iv], 185p; illus (maps); bibl; 28 cm. Cover title: "Ozark hydrology: a predictive model."

An interesting springboard from which other studies might arise.

[42] Aley, Thomas John, 1938– *Restoration of natural cave features, Oregon Caves National Monument, Oregon: final report.* Aley, Catherine; Engineering and Design Associates; United States National Park Service. Protem, MO, US: Ozark Underground Laboratory, **1987**. 40p; bibl; 28 cm.

Illus. Running title: Cave restoration in Oregon Caves. Contract report to Engineering and Design Associates, prime contractor to the National Park Service. March, 1987.

[43] Aley, Thomas John, 1938– *Restoration of natural microclimate in Oregon Caves, Oregon Caves National Monument: final report.* Aley, Catherine; Engineering and Design Associates; United States National Park Service. Protem, MO, US: Ozark Underground Laboratory, **1988**. one folded map in pocket; 107p; bibl; 28 cm.

Illus. Running title: Oregon Caves microclimate study. Contract report to Engineering and Design Associates, contractor to the National Park Service. July, 1988. Report on methods to restore original climate patterns in a show cave.

[44] **Aley, Thomas John**, 1938– *Water quality protection studies, Logan Cave, Arkansas: final report.* Aley, Catherine; Arkansas Game and Fish Commission; United States Fish and Wildlife Service. Protem, MO, US: Ozark Underground Laboratory, **1987**. four folded maps in pocket; 61p, [29]; bibl; 28 cm.

Illus. Running title: Logan Cave protection study. March 6, 1987.

[45] **Alfie [pseud for S. J. Collins]** *Reflections: a look at the 'spelaeodes' and other caving sagas.* "JOK" [pseud for Jock Orr](illus). Cheddar, GB: Barton Productions / Cheddar Valley Press, **1971**. illus: b&w: ads, cartoons; 91p.

Includes 14 sagas with many sketches.

[46] **Alfie [pseud for S. J. Collins]** *The spelaeodes.* "JOK" [pseud for Jock Orr](illus). Bristol, GB: Bristol Exploration Club, **1969**. illus: b&w: cartoons; 32p; 20 x 16 cm.

[47] **Aljančič, Marko (auth, ed)** *Proteus; the mysterious ruler of karst darkness.* Et al. Ljubjana, YU: Vitrum, **1993**. illus: b&w: maps, photos, repros; partly col: maps, photos; 75p, [1]; 23.5 x 16 cm.

[48] **Allen, E. E.** *Gower Caves: a survey of the Gower Caves with an account of recent excavations: part I.* Rutter, J. G. (co-surv); Thompson, A. G. (photo). Swansea, GB: Vaughan Thomas, **1946**. illus: b&w: drawings, maps, photos; front: b&w photo; 28p, 12p pls; bibl.

49 figs.

[49] **Allen, E. E.** *Gower Caves: a survey of the Gower Caves with an account of recent excavations: part II.* Rutter, J. G. (co-surv); Webber, T. R. (photo). Swansea, GB: Welsh Guides, **1948**. illus: b&w: maps, photos; partly col: maps; 60p, 12p pls; app, bibl.

[50] **Allen, Joel Asaph**, 1838-1921 *On an extinct type of dog from Ely Cave, Lee County, Virginia.* Shaler, Nathaniel Southgate, 1841-1906. Cambridge, MA, US: Museum of Comparative Zoology, **1885**. Memoirs of the Museum of Comparative Zoology at Harvard College vol X no 2. 13p.

Each plate preceded by leaf with descriptive letterpress.

[51] **Allen, Tim (tm ldr)** *Hang Vietnam; report of the British Speleological Expedition to the Bac-Sun Massiflang.* Sheffield, GB: np, **c1992**. illus: b&w: maps; five fold-out maps; 36p; bibl; 30 x 21 cm.

"Published" in photocopy form.

[52] **Allison, Harold** *Indiana caves and unique geological features.* Washington, IN, US: Harold Allison, **nd**. 36p.

No illus.

[53] **Allison, Vernon Charles**, 1891– *The antiquity of the deposits in Jacob's Cavern.* New York, NY, US: American Museum Press, **1926**. Anthropological Papers of the American Museum of Natural History vol XIX pt VI, pp 293-335. illus: b&w: drawings, maps, photos, tables; 43p, [1]; 25 cm.

24 figs.

[54] **American Cave Conservation Association** *Hart County solid waste management plan.* Horse Cave, KY, US: np, **1990**. iii, 32p; app.

[55] **American Cave Conservation Association** *Karst curriculum resource guide.* Horse Cave, KY, US: American Cave Conservation Association, **nd**. Various pagination.

A collection of resource materials for teaching the concepts of caves and karst.

[56] **American Cave Conservation Association** *National cave management seminar: Albuquerque, New Mexico, March 12-16, 1990.* United States National Park Service. npop: np, **1990**. bibl; 28 cm.

Title from cover. Looseleaf. Illus. Contains papers, outlines and other handouts for the sessions at the Seminar.

[57] **Ammen, Samuel Zenas**, 1843-1929 *The caverns of Luray, the property of the Valley Land and Improvement Co., Luray, Va: the manner of their formation, their peculiar growths, their geology, chemistry, etc: an illustrated guide book.* Pennell, Joseph, 1857-1926 (illus); Lee, Alexander Y. (illus); Valley Land and Improvement Company [Luray, Va.]; Shenandoah Valley Railroad. Philadelphia, PA, US: Allen, Lane & Scott's Printing House, **1890**. illus: b&w: ads, drawings, maps, ports; map folded in front; 47p; 24 cm.

Issued by the Luray Cave and Hotel Company, affiliated with the Shenandoah Valley Railroad Company. On original paper cover: Shenandoah Valley Railroad. A revision of the author's History and description of Luray cave (Baltimore, J. W. Borst & Co., 1880 [i.e. 1881]) This is a slight revision of the edition issued by the Luray Cave and Hotel Company.

[58] **Ammen, Samuel Zenas**, 1843-1929 *The caverns of Luray. The property of the Valley Land and Improvement Co., Luray, Va. The manner of their formation, their peculiar growths, their geology, chemistry, etc. An illustrated guide book.* Pennell, Joseph, 1857-1926 (illus); Lee, Alexander Y. (illus); Valley Land and Improvement Company [Luray, Va.]; Shenandoah Valley Railroad. Baltimore, MD, US: Crowl & Lehner, Printers, **1893**. illus: b&w: ads, drawings, maps, ports; map folded in front; front; 49p, [1]; 23 cm.

On cover: Forfold and Western Railroad. Issued by the Valley Land and Improvement Company. A revision of the author's History and description of Luray cave (Baltimore, J. W. Borst & Co., 1880 [i.e. 1881]) A slight revision of the edition issued in 1890 by the Luray Cave and Hotel Company.

[59] **Ammen, Samuel Zenas**, 1843-1929 *The caverns of Luray: an illustrated guide-book to the caverns, explaining the manner of their formation, their peculiar growths,*

their geology, chemistry, etc. 4th ed. Philadelphia, PA, US: Allen, Lane & Scott's Printing House, **1884**. 46p; 24 cm.

Illus.

5th ed. Philadelphia, PA, US: Allen, Lane & Scott's Printing House, **1886**. illus: b&w: maps, ports; 47p; bibl; 24 cm.

[60] Ammen, Samuel Zenas, 1843-1929 *History and description of the Luray Cave (illustrated), including explanations of the manner of its formation, its peculiar growths, its geology, chemistry, &c.; also a map. The whole so arranged as to serve as a guide..* 3rd ed. Baltimore, MD, US: J. W. Borst & Co Publishers, **1882**. illus: b&w: ads, drawings, maps, ports; front: folded map; 48p; [4]; 22 cm.

Title from cover. Wrapper title: The Caverns of Luray.

[61] Ammen, Samuel Zenas, 1843-1929 *History and description of the Luray Cave (illustrated), including explanations of the manner of its formation, its peculiar growths, its geology, chemistry, etc.; also a map. The whole so arranged as to serve as a guide.* Baltimore, MD, US: J. W. Borst & Co Publishers, **1880**. illus: b&w: ads, maps, photos; front: folded map; 18p; 24 cm.

1993 microfilm edition done by SOLINET/ASERL Cooperative Microfilming Project. 1983 microfilm edition done by Library of Congress.

[62] Amundson, Bob *User's guide for flowchart and cavemap computer programs.* Douty, Bill. [WV, US]: West Virginia Association for Cave Studies, Jan, **1979**. illus: b&w: charts, drawings, maps, tables; [ii], 31p, [20]; app.

[63] Anati, Emmanuel *Rock art: the state of research in rock art - 1993: archetypes, constants and universal paradigms.* Colombo, LK: Central Cultural Fund, **1993**. Central Cultural Fund publication no 122. 150p; bibl; 30 cm.

Illus. Text in English, prefatory material in English and French.

[64] Anchors, William E., 1954– *Ruskin and Jewel Caves: a brief history.* [Dickson, TN, US]: W. E. Anchors, Mar, **1989**. illus: b&w: photos, ports; 70p, [1]; bibl; 22 cm.

[65] Anderson, Arthur Wilhelm, 1892– *The Carlsbad Cavern of New Mexico: its history and geology.* Bingham, Gaby L. (rev); Wroth, James Stewart, 1885–; Ryan, Joseph Edward (illus). 12th ed, rev. [Carlsbad, NM, US]: Myer Ptg Co, **1938**. Folded map pasted on front lining paper; [32], 3p pls; 26 cm.

Illus. Some records in OCLC list Wroth's birth year as 1865. Since there is little change in the Geology section written by Wroth, we have assumed the same Wroth is responsible for all editions and have arbitrarily chosen the 1885 date as his birth year.

14th ed, rev. Wroth, James Stewart, 1885–; Ryan, Joseph Edward (illus). [Carlsbad, NM, US]: Bradford-Robinson Ptg. Co, **1939**. illus: b&w: maps; folded map pasted on front lining paper; [32]; 25 cm.

Illus.

5th ed. Wroth, James Stewart, 1885–. [Carlsbad, NM, US]: Carlsbad Printing Co Press, **1931**. 1 folded map; [32], 3p pls; 25 cm.

Illus.

9th ed?. Wroth, James Stewart, 1885–; Ryan, Joseph Edward. npop: [G. L. Bingham], **1928**. illus: b&w: drawings, photos; 48p, [48]; 25 cm.

Cover title: The Carlsbad Cavern: Carlsbad New Mexico: Its history and geology.

This is of more interest for historical purposes than for scientific ones: we have come a long way.

[66] Anderson, Arthur Wilhelm, 1892– *The Carlsbad Cavern of New Mexico: its history and geology, formations of the cavern, discovery and exploration, bats of the cavern, cavern geology, administration: the land nobody knows.* Wroth, James Stewart, 1885–; Ryan, Joseph Edward (illus). 19th ed. [Carlsbad, NM, US]: Bradford-Robinson Ptg Co, **nd**. b&w front, folded; [36]; 24 cm.

Illus.

20th ed. Wroth, James Stewart, 1885–; Ryan, Joseph Edward (illus). [Carlsbad, NM, US]: Bradford-Robinson Ptg Co, **1939**. illus: b&w: photos, tables; partly col: drawings, maps; partly col, folded map; b&w front; [36]; 25 cm.

27th ed. np, **1958**. b&w front, folded; [36].

Copyright 1928, revised 1939, 1946, 1957 and 1958.

[67] Anderson, Arthur Wilhelm, 1892– *The Carlsbad Cavern of New Mexico: its history and geology; formations of the cavern, discovery and exploration, bats of the cavern, cavern geology, administration; the land nobody knows.* Wroth, James Stewart, 1885–; Ryan, Joseph Edward (illus). 1930. [Carlsbad, NM, US]: Carlsbad Printing Co Press, **1930**. [32]; 25 cm.

Illus.

[68] Anderson, Grant *Wellington Caves, resource and management study.* [Wellington, NSW, AU]: Wellington Shire Council, **1991**. illus: b&w: maps; bibl; 31 cm.

Title from cover. v. 1. Resource inventory - v. 2. Management issues - v. 3. Bibliography. Illus.

[69] Anderson, Jennifer Ann, 1942– *Cave exploring.* New York, NY, US: Association Press, **1974**. illus: b&w: drawings, photos; 126p; bibl, index; 22 cm.

Forward by John A. Stellmack, Past President, NSS.

A classic used for beginning cavers for over 20 years. Simple and inexpensive; it has a very wide distribution, and besides one of the authors of this bibliography has a photo in it.

[70] Anderson, Robert Cleve (ed) *Capital area cavers bulletin number 1.* Baker, Linda (ed). Washington, DC, US: District of Columbia Grotto of the NSS, Potomac

Speleological Club, **1982**. 28 cm.

[71] Anderson, Warren *Hydrology of three sinkhole basins in Southwestern Seminole County Florida.* Hughes, G. H.; Bureau of Geology, Division of Resource Management, Florida Department of Natural Resources; Board of County Commissioners of Sem. Tallahassee, FL, US: United States Geological Survey, **1975**. USGS Report of Investigations no 81. illus: b&w: graphs, maps; vi, 35p; bibl.

15 figs.

[72] Andersson, Johan Gunnar, 1874-1960 *The cave-deposit at Sha Kuo Tun in Fengtien.* Peking, CN: Geological Survey of China, **nd**. China Geological Survey Palaeontologia Sinica ser D v 1 fasc 1. 58p, [26]; 30 cm.

Last 26 pages in Chinese. Illus. XII pls (1 col).

[73] Andre, Daniel *Bramabiau's underground river.* Camprieu, FR: Editions: Abime de Bramabiau, **1994**. illus: b&w: cartoons, drawings, maps, repros; partly col: maps; col: drawings, photos; Text on both.; 32p.

[74] Andrews, Edward Wyllys, 1916-1971 *Balankanche, throne of the tiger priest.* New Orleans, LA, US: np, **1970**. Publication 32 Middle American Research Institute Tulane University. illus: b&w: drawings, maps, photos, repros, tables; col: drawings; front: repros; xi, 182p; add, app, bibl; 28 cm.

60 figs. 33 1/3rpm record included in back pocket "oraciones rezadas on la gruta de balankanche" LP recording of prayers during cave ceremony. OCLC also lists 1983 and 1985 dates. Appendices (p. 72-164): The ceremony of Tsikul tan ti yuntsiloob at Balankanche: 1. Description of the ceremony, by A. Barrera Vasquez.—2. Transcription and translation of the Maya text [into Spanish] by R. Arzapalo.

[75] Andrews, Peter, 1940– *Owls, caves, and fossils: predation, preservation, and accumulation of small mammal bones in caves, with an analysis of the Pleistocene cave faunas from Westbury-sub-Mendip, Somerset, UK.* Cook, Jill (SEMs). Chicago, IL, US: University of Chicago Press, **1990**. illus: b&w: drawings, graphs, maps, photos, photomicrographs, tables; vii, 231p; app, bibl, index; 29 cm.

This is the first comprehensive, fully illustrated account of small mammal taphonomy, emphasizing the importance of cave-roosting owls.

[76] Andrews, Rhonda Lynette *Perishable industries from Hinds Cave, Val Verde County, Texas.* Adovasio, J. M.; Dirkmaat, Dennis C. (cont). Pittsburgh, PA, US: University of Pittsburgh, Dept of Anthropology, **1980**. Ethnology Monographs number five. illus: b&w: graphs, maps, photos, tables; 375p; bibl; 26 cm.

21 tables, 152 figs.

[77] Anglo-Canadian Rocky Mountains Speleological Expedition (1983) (comp) *The Anglo-Canadian Rocky Mountains Speleological Expeditions 1983 and 1984: a report on recent discoveries made by two caving expeditions to the Rocky Mountains of Canada by combined British and Canadian teams.* Keyworth, Nottingham, GB: British Geological Survey, **1986**. illus: b&w: maps, photos; 107p; 30 x 21 cm.

Cover title: Canada; ACRMSE 83 & 84.

[78] Anon *37 views of Naracoorte caves and Naracoorte South Australia.* Naracoorte, SA, AU: Wm Vivian, **nd**. illus: b&w: photos; [24].

No text except photo captions. Cover title: Naracoorte: views of the caves and township.

[79] Anon *[Indiana cave list].* npop: np, **nd**. 56p; index. No illus.

[80] Anon *Aillwee Cave: Ireland's premier show cave.* Co Clare, IE: Aillwee Cave Company Ltd, **c1990**. illus: b&w: drawings, ports; partly col: maps; col: photos; "Notes from the diary" on rear lining page; 16p.

[81] Anon *Australian Speleological Handbook.* npop: np, **1968**.

[82] Anon *Beautiful Kweilin.* Peking, CN: Foreign Languages Press, **1978**. illus: b&w: photos; partly col: maps; col: photos; [iv], 109p; 26 cm.

Col illus. Title in Chinese: Mei li ti Kuei-lin. Picture book of caves and karst for tourists.

[83] Anon *A book of songs and poems for the Hunters Lodge and similar places.* [Mendip, GB]: np, **[1955]**. [38]; 21 x 17 cm.

Photocopy examined. No illus.

[84] Anon *Carlsbad Cavern National Park. Guadalupe Mountains National Park.* Whittier, CA, US: Finley-Holiday Film Corp, **1993**. National Park & Monument Series.

1 videocassette (53 min), sd, col; 1/2 in, VHS format.

A visual analysis and history are provided on an expedition in Carlsbad Caverns and Guadalupe Mountains National Parks.

[85] Anon *Carlsbad Caverns National Park.* Whittier, CA, US: Holiday Film Corp, **1985**.

Title on cassette: Carlsbad Caverns, New Mexico. Presents photographs with narration of Carlsbad Caverns, New Mexico. 40 slides: col. + 1 sound cassette (14 min).

[86] Anon *Carlsbad Caverns National Park, New Mexico.* Carlsbad, NM, US: Cavern Supply Co, **c1988**. illus: b&w: maps; col: photos; [24]; 28 x 21.5 cm.

[87] Anon *Carlsbad Caverns, New Mexico.* npop: Oriolo Educational Publications; Jarelco, **1982**. Natural Wonders of the U.S.

1 sound cassette (ca. 30 min.): 1 7/8 ips.

[88] Anon *Catalogue of literature on Chinese karst caves.* npop: The Literature Service Department of the Library of Chinese Academy of Science [and] the Scientific Committee of the XI International Congress of Speleology, Jul, **1993**. [i], 135p; bibl.

[89] Anon *The cave at Vari: excavations by American*

School of Classical Studies at Athens in February, 1901. npop: American School of Classical Studies at Athens, Feb, **1903**. illus: b&w: maps; fold-out map; 86p.

pp 263-349, 14 pls. Reprinted from the American Journal of Archaeology, Second Series. Journal of the Archaeological Institute of America, Vol VII (1903), No 3.

[90] Anon *Cave paintings of the Chumash Indians*. Santa Barbara, CA, US: Instructional development, University of California at Santa Barbara, CA, **1988**.

1 videocassette (30 min): sd, col; 1/2 in.

[91] Anon *Cavern: a pictorial souvenir of Carlsbad Caverns*. White's City, NM, US: The News, **1953**. xxix; 37 cm.

Special section of The News, White's City, New Mexico, vol. XIV, no. 30, Wednesday, July 22, 1953, second edition. Title from cover. Illus.

[92] Anon *The Caverns of Luray*. npop: np, **1882**. illus: b&w: photos; [17]; 30 x 25.5 cm.

Cover title: Photographic views in the Caverns of Luray by C. H. James. Taken by means of the Thomson - Houston electric light with which the caverns are brilliantly illuminated. No text.

[93] Anon *The Caverns of Luray located on line of the Shenandoah Valley R.R.*. npop: np, **c1880**. illus: b&w: ads, maps; 15p, [1]; 23 x 15.5 cm.

"From the Official Report of the Smithsonian Institution."

[94] Anon *The Caverns, rocks and ruins of America's Southwest*. Coeur d'Alene, ID, US: Celebration Productions, **1986**. Travel America Series.

1 videocassette (VHS) (47 min), sd col, 1/2 in.

Travel through New Mexico and Arizona exploring Carlsbad Caverns, Petrified Forest, Red Rocks, and others.

[95] Anon *Caves*. 1964 rev ed. [US]: American Educational Projections Corp [production company], **1964**. Animal Homes Series.

Distributed by Curriculum Materials Corp. 1 filmstrip (30 fr), col, 35 mm.

[96] Anon *Caves*. npop: Troll Associates, **1965**. Talking Cassette Encyclopedia.

1 cassette, mono, 2-track. OCLC lists both 1965 and 1971 as publication dates. Juvenile.

[97] Anon *Caves and Caving City Museum Queens Road Bristol 8*. Bristol, GB: City Museum, **c1972**. illus: b&w: charts, drawings, maps, photos; [i], 41p, [1]; add, bibl, errat; 21 x 14.5 cm.

Title from cover. Prepared for an exhibit at the Museum that included a 35 min, 16mm film with the same title, and a caving quiz and answer sheet.

[98] Anon *The caves of Arta situated on the sea coast of the municipal confines of Capdepera*. npop: Fernando Soler Queralt, **1959**. illus: b&w: repros; col: photos; 27p; 21 x 14 cm.

Colored reproduction.

[99] Anon *The caves of Kentucky: Mammoth Cave, Colossal Cavern, White's Cave*. Chicago, IL, US: Union Book & Pub. Co, **1980**. America Her Grandeur and Her Beauty pt 10. [16]; 35 cm.

Title from cover. Illus. OCLC lists both 1900 and 1980 as publication date.

[100] Anon *The caves of Western Australia*. Perth, WA, AU: A.E. Forgaith, **c1905**. illus: b&w: photos; [16]; 17.5 x 25 cm.

Title from cover. Drop head title: The limestone caves of Western Australia. Leaves printed one side only.

Photographs of the show caves of Western Australia with 1 p text.

[101] Anon *Caving in West Virginia*. Stanford, CT, US: np, **nd**. Trailside: Make Your Own Adventure.

1 videocassette (40 min), sd, col, 1/2 in. VHS. Video recordings for the hearing impaired.

"Crawl into one of the most extraordinary and one of the longest wild cave systems on the East Coast, the Organ Caves of West Virginia. You'll explore underground streams and formation rooms while learning caving techniques like rappelling and chimneying."

[102] Anon *Cheddar*. Reigate, GB: Francis Frith & Co, **c1914**. illus: b&w: photos; [16].

Photographs, with no tint, mostly of Cox's Cave. Photographs laid on thick, brown paper leaves. No text.

[103] Anon *Chiang Dao Cave*. npop: np, **nd**. illus: b&w: photos.

This is a small, very poorly printed book on the Chiang Dao Cave which is located near Chian Mia, Thailand. Book has three poor cave photos. Pagination unknown.

[104] Anon *Cox's Stalactite Cavern, Cheddar*. npop: np, **pre-1920**. illus: b&w: photos; 11p pls; 14 x 15 cm.

No text. Another edition exists with only a different cover title (Cox's Stalactite Cavern and Cheddar Cliffs).

[105] Anon *Crewe Climbing and Pot Holing Club; Peak rigging guide*. [GB]: [Crewe Climbing and Potholing Club], **1995**. illus: b&w: maps; 38p, [1]; 30 x 21 cm.

[106] Anon *Dan-yr-ogof show caves*. Swansea, GB: np, **[1992]**. illus: b&w: ads; col: photos; col photos on lining pages; 31p, [1]; 30 x 21 cm.

On cover: Souvenir colour guide. Printed by W. Walters, Son & Co.

[107] Anon *A description of Howe's Cave: with a popular treatise on the formation of caves in lime rock, from the size of a quill to a mammoth*. Albany, NY, US: Weed, Parsons, and Company, Printers, **1865**. illus: b&w: maps; 1 fold-out map in front; 28p, 1p pls; 24 cm.

Illustrated with numerous engravings. Engravings were supplied separately and are very rare.

[108] Anon *A description of the Mammoth Cave of Kentucky, the Niagara River and Falls, and the Falls in summer and winter; the prairies, or life in the West; the Fair-*

mont Water Works and scenes on the Schuylkill, xc. xc.: to illustrate Brewer's Panorama. Boston, MA, US: J.M. Hewes & Co, **1850**. "Recommendations" on back lining page; 44p, [2]; 23.5 x 15.5 cm.

Microfilm version 1993.

[109] **Anon** *A description of the Mammoth Cave of Kentucky, the Niagara river and falls, and the falls in summer and winter; the prairies, or life in the West; the Fairmount Water Works and scenes on the Schuykill, etc. etc.: to illustrate Brewer's panorama.*. Philadelphia, PA, US: United States Job Printing office, **1849**. 16p; 21 cm.

Cover title: The Mammoth cave of Kentucky—Niagara river and falls—The mansion and tomb of Washington—The natural bridge of Virginia—The prairies of Illinois—Opinions of distinguished artists, of the press, &c. Rare book ad lists author as Brewer, George St. P.

[110] **Anon** *Development of caves*. npop: C. S. Hammond, **1960**. Hammond Earth Science Transparency Series 8562. illus: partly col: maps; 26 x 31 cm.

4 transparencies on frame, 26 x 31 cm, col, and teacher's manual.

[111] **Anon** *Dossey Domes: most beautiful of all the caverns at Green River Landing (One-half mile from Mammoth Cave, Ky.) ; Cliff Walk along Green River: finest scenery in the state*. Dossey Domes Cavern and Dossey Cliff Walk. [Mammoth Cave, KY], US: np, **1920**. [14], 10p pls; 14 cm.

Cover title: Dossey Domes Cavern and Dossey Cliff Walk. Illus.

[112] **Anon** *Environmental Education Enterprises, Inc. short course showcase January 26-27-28, 1994, Grosvenor Resort Orlando, Fl: Winter Park sinkhole guidebook*. Columbus, OH, US: Environmental Education Enterprises, Inc, **1994**. illus: b&w: drawings, graphs, maps, photos; 21p.

Running title: Hydrogeology and history of the Winter Park sinkhole.

[113] **Anon** *Environmental geology and water sciences: special issue on sinkholes*. New York, NY, US: Springer-Verlag, **1986**. illus: b&w: ads, drawings, graphs, maps, photos, tables; text on lining pages; [ii], 104p, [2]; bibl.

Entire issue devoted to sinkholes.

[114] **Anon** *Exploration '66; University of Nottingham; biospeleological research expedition to Ireland; Riverview work project in Portugal; British speleological expeditions to Turkey*. [Nottingham], GB: University of Nottingham, **c1967**. illus: b&w: ads, drawings, maps, photos; contents on front lining page; ad on rear lining page; 47p; 22 x 14 cm.

[115] **Anon** *Exploration of the caves and rivers of New South Wales (minutes, reports, correspondence, accounts)*. npop: np, **1882**. Notes and Proceedings N.S.W. Parliament 1882 vol 5. illus: b&w: drawings, maps, photos, tables; two folded in text; 52p, [37].

"Presented to Parliament by command." Copy examined was a reproduction.

[116] **Anon** *Fairy Cave*. npop: np, **nd**. [16]; 14 x 22 cm.

A small softbound booklet. Has red cover. Eight full page photos and six pages of text.

[117] **Anon** *Focus on the Cango Caves*. npop: np, **nd**. illus: col: photos.

Contains 78 color photos of which 27 are cave scenes, or closely related to caves.

[118] **Anon** *The geology of caves*. Maumee, OH, US: Instructional Video, **1988**. Instructional Video Earth Sciences Series.

1 videocassette (20 min), sd, col, 1/2 in. Title on cassette label and container: Adventure in caves.

This video shows the spectacular size and beauty of caves, and explores how caves developed and how they have been decorated by nature.

[119] **Anon** *Geology, climate, hydrology and karst formation: field symposium in Australia 4 to 18 December 1992: Buchan - Mt. Gambier / Naracoorte - Nullarbor Plain humid temperate impounded karst, sub-humid temperate syngenetic karst, arid temperate karst: programme and abstracts*. Canberra, ACT, AU: Dept. of Geography and Oceanography, University College, University of New South Wales, Australian Defense Force Academy, **1992**. Project 299 of I.G.C.P. illus: b&w: maps; [18]; bibl.

[120] **Anon** *Gough's Caves, Cheddar, Somerset*. [GB]: np, **nd**. illus: b&w: photos; [18]; 23 x 15 cm.

2 pages of text, printed on one side only. See similar publications by Gough.

[121] **Anon** *Grand Caverns: wonders of the subterranean world: Grottoes, Virginia*. Chicago, IL, US: Rand McNally, **1926**. pamphlet folded to 24 x 11 cm; [12]; 24 x 21 cm.

[122] **Anon** *The great Dan-yr-ogof day out*. Swansea, GB: np, **[1994]**. illus: b&w: tables; col: maps, photos; lining pages included in pagination; 21p, [1]; 29.5 x 20.5 cm.

Cover title: The great Dan-yr-ogof day out: souvenir brochure. Printed by Walter Printers Ltd.

[123] **Anon** *The grottoes of Adelsberg and the Proteus of Anguinus: a few pages from the journal of a continental tour*. London, GB: Spottiswoode, **1800**. 16p; 19 cm.

At head of title: Appendix III. OCLC lists both 1800 and 1899 as date of publication. pp. 309-324

[124] **Anon** *The Grottos of Han and of Rochefort (Belgium)*. npop: American Express Co. Inc, **nd**. illus: b&w: drawings, maps, photos; 12p.

Back cover lists: Editions d'Art. Aguste GODENNE. 79. Rue de l'Ange Namur.

[125] **Anon** *Guide book for the Diamond Cave, Barren County, Kentucky. Located near "Old Bell Tavern," now the Mammoth Cave and Glosgow Junctions, on the L & N*

railroad, two miles from the R.R. immediately on the Mammoth Cave Road. Louisville, KY, US: Courier-Journal Steam Job Print, **1870**. 21p; 12 cm.

One OCLC record lists place of publication as Glasgow, KY and publisher as C.G. Smith, Printer, 1860 as date, and 16 p. Written in pencil in one copy: "by Richardson, J.T. Richardson, Discoverer."

[126] Anon *A guide manual to the Mammoth Cave of Kentucky*. Glasgow, KY, US: Printed at the Glasgow Times Office, **1876**. 38p; 18 cm.

[127] Anon *Guide manual to the Mammoth Cave of Kentucky*. Glasgow, KY, US: [Glasgow Times], **1876**. 38p; 18 cm.

Printed at the Glasgow Times Office. In pencil inside cover, "circa 1875".

[128] Anon *A guide to the Cave of Altamira and the town of Santillana del Mar (Province of Santander, Spain)*. 2nd. Madrid, ES: Junta Protectora de la Cueva de Altamira, **1935**. illus: b&w: maps, photos; 44p, [1], 16p pls; bibl, index; 19 cm.

OCLC lists a 1927 edition with 42p.

[129] Anon *Guide to the Grand Cavern within the mountain of Abraham's Heights, Matlock Bath*. Manchester, GB: np, **1818**. illus: b&w: repros; front: b&w engraving of Ossians Hall; 12p; 17 x 10 cm.

[130] Anon *Guide to the Kalk Bay and Muizenberg Mountains; (walks, caves, camp sites)*. Capetown, ZA: The Cape Peninsula Fire Protection Committee, **nd**. illus: partly col: maps; fold-out map on back; rules on front lining page; 16p; 21 x 13 cm.

Includes both wild and show caves.

[131] Anon *A guide to the Museum of the Torquay Natural History Society*. 10th ed. np, **1934**. illus: b&w: charts, photos; 16p.

12th ed. np, **1941**. illus: b&w: charts, photos; 16p.

16th ed. npop: np, **nd**. illus: b&w: charts, photos; 16p.

6th ed. Torquay, [GB]: [Torquay Natural History Society], **1926**. illus: b&w: charts, photos; fold-out map of cave; Photos on lining pages; vary by ed; 15p; 21.5 x 14 cm.

Cover title: The Torquay Natural History Society; A guide to the Society's museum with plan and notes on Kent's Cavern

9th ed. np, **1931**. illus: b&w: charts, photos; 16p.

[132] Anon *Guilin tourist album*. Hong Kong, HK: Tai Dao Publishing, **1987**. illus: col: photos; color photos on endpieces; front: col photo.

Bilingual Chinese/English tourist picture book.

[133] Anon *Guilin: the crown of superb landscapes in China*. Fakui Yao (cover des); Zi Meng (photo); Guiging Zhou (photo); Jiemin Xie (photo). Beijing, CN: China Travel and Tourism Press, **1981**. illus: b&w: tables; col: maps, photos, tables; [xviii], 41p, [1].

[134] Anon *Handbook for caves and guiding staff*. Sydney, NSW, AU: The New South Wales Department of Tourist Activities, Nov, **1967**. illus: b&w: graphs, maps; 65p; index.

[135] Anon *Hastings Newdegate cave rehabilitation plan*. TAS, AU: Parks and Wildlife Service, July, **1994**. illus: b&w: maps, photos, tables; [iii], 59p; 30 x 21 cm.

[136] Anon *The historic Blue Grottoes of Virginia; Harrisonburg VA*. Brooklyn, NY, US: Albertype Company, **nd**. illus: b&w: photos; [5], 12p pls; 20.5 x 25.5 cm.

Printed in black and blue.

[137] Anon *How nature made the beautiful caverns*. npop: Luray Caverns, **1967**. 16p.

Col illus.

[138] Anon *Illustrated souvenir grid to the caverns and glens at the foot of Ingleborough (with map)*. npop: np, **1930**. illus: b&w: ads, maps, photos; 23p; 18 x 12.5 cm.

9 unnumbered pages of ads.

[139] Anon *Ingleborough Cavern: the finest show cave in Yorkshire*. npop: np, **nd**. illus: b&w: drawings, photos; drawing on front text and photo on rear; [12]; 12.5 x 9.5 cm.

[140] Anon *International Symposium on groundwater biology: symposium held in 1978 in Blacksburg, Virginia, U.S.A.*. [IT]: Societá Speleologica Italiana, **1981**. International Journal of Speleology v11 no1-2. illus: b&w: charts, drawings, graphs, maps, photos, tables; 171p; bibl.

[141] Anon *Jenolan Caves: Australia's underground fairyland*. Sydney, NSW, AU: New South Wales Department of Tourism, **nd**. illus: partly col: maps; col: drawings, photos; text on lining pages and back cover; [11].

Cover title: Jenolan Caves: New South Wales. Included with a 45RPM record and slides of the cave titled: Donald Smith Australian Opera Star sings in the Cathedral Cave, Jenolan Caves. Different versions exist.

[142] Anon *Karst groundwater investigations: Greece*. Rome, IT: United Nations Special Fund, Food and Agriculture Organization of the United Nations, **1964**. illus: b&w: charts, drawings, graphs, maps, tables; 2 folded in rear pocket; chart folded in text; xi, 99p, 6p pls; app, bibl.

[143] Anon *Kents Cavern Wellswood Torquay Devon, home of prehistoric man and animals*. Torquay, GB: np, **nd**. illus: b&w: photos; 16p; 14 x 11 cm.

Several editions examined showed different arrangement and number of photos, and different arrangement of materials.

[144] Anon *The last horizon*. Chapel Hill, NC, US: TV Ontario, **1988**.

Title from data sheet. A Canadian videorecording VHS. 1 videocassette (30 min), sd, col, 1/2 in. + 1 teacher's guide.

Explores an underground freshwater lake under the parched Nullarbor Plain in Australia. Photographing it for the first time, it is unbelievably clean and transparent and is host to a colony of strange microorganisms.

[145] Anon *Luray Caverns*. 1904 ed. npop: np. 24p.

[146] Anon *Luray Caverns*. [Philadelphia, PA], US: J. Murray Jordan, **1900**. illus: b&w: photos; 15p; 20 x 25 cm.

Illus

[147] Anon *The magic of Sarawak's Mulu Caves; discover the latest wonder of the world*. Sarawak, MY: Ministry of Environment and Tourism, **c1993**. illus: col: drawings, maps, photos; col art and text on lining pages; [12]; 21 x 31 cm.

Slick ad brochure with some text.

[148] Anon *Mammoth Cave in third dimension*. Chicago, IL, US: Orthovis Co, **1935**. illus: b&w: photos; 31p.

A collection of 24 stereo pictures of Mammoth Cave. Book contains an ortho-scope for viewing the pictures.

[149] Anon *Mammoth Cave of Kentucky*. npop: np, **1930**. [11]; 23 x 21 cm.

Title from cover. Folded to 23 x 11 cm. Illus. OCLC indicates some ambiguity as whether the publication date is 1930 or 1939.

[150] Anon *Mammoth Cave of Kentucky: the world's greatest subterranean wonder*. [Mammoth Cave, KY], US: [Mammoth Cave Hotel], **1920**. illus: b&w: maps; [10]; 23 x 20 cm.

Illus. OCLC lists some ambiguity about whether the publication date is 1920 or 1929.

[151] Anon *Mammoth Cave, America's great natural wonder*. npop: np, **nd**. 16p.

Illus.

[152] Anon *Mansion Inn, Luray Caverns*. Natural Bridge, VA, US, **c1906**. 24p.

This small softbound booklet was published shortly after 1906.

[153] Anon *Marble Arch Caves*. Burns, Conor (photo). npop: Devensih Gallery Publications, Jul, **1995**. illus: b&w: drawings; partly col: maps; col: photos; 32p; 21 x 15 cm.

Cover title: Marble Arch Caves: Published to coincide with Martel Centenary 1895-1995.

[154] Anon *Maribel Caves: an ideal health and summer resort.*. npop: np, **1990**. 19p; 10 x 17 cm.

Reprint. Originally published: [189-?]. OCLC indicates some ambiguity about whether the reprint publication date was 1990 or 1993. Illus.

[155] Anon *Marvel Cave: the story of America's largest privately owned cave*. Neosho, MO, US: Marvel Cave Association, **1958**. illus: b&w: maps, photos, repros; 35p; [4].

[156] Anon *Marvelous Howe Caverns near Cobleskill, New York*. Cobleskill, NY, US: Howe Caverns, Inc, **nd**. illus: b&w: photos; [13].

[157] Anon *Mexico's caves and caverns*. MX: Pemex Travel Club, **1962**. illus: b&w: photos; 32p; 22 x 11 cm.

Foldout souvenir brochure. 1964 also seen as the date of publication.

[158] Anon *Monteagle Wonder Cave, or, The Cave of the Cumberland*. [Monteagle, TN, US]: [McQuiddy Printing Co], **1903**. illus: b&w: ads; [32]; 9 x 16 cm.

Includes advertising matter. Illus.

[159] Anon *Montpelier and Bear Lake County*. npop: np, **1960**. 21 cm.

Minnetonka Cave. Unpaged, illus. OCLC indicates some ambiguity about whether the publication date was 1960 or 1969.

[160] Anon *The most beautiful caves in the world*. Whangarei, NZ: F. G. Radcliffe, **nd**. illus: b&w: photos; [24].

Very limited text. Cover title: "The most beautiful caves in the world" Waitomo Ruakuri & Aranui New Zealand.

[161] Anon *Mount Etna action book: time is running out*. Rockhampton, QLD, AU: The Mount Etna Committee, Apr, **1988**. illus: b&w: drawings, maps, photos, repros; col: photos; col photos on lining pages; 24p.

[162] Anon *The Naracoorte Caves: "how to reach them"*. Adelaide, SA, AU: State Tourist Bureau, **nd**. illus: b&w: ads, maps, photos; map folded on back cover; 36p.

Cover title: South Australia. Naracoorte Caves.

[163] Anon *Natural Bridge Caverns: Texas' largest caverns: One of the great show caves of the world*. Berkeley, CA, US: np, **nd**. illus: b&w: drawings, photos; col: photos; text on lining pages; [24].

Cover title: Natural Bridge Caverns: Registered U.S. Natural Landmark: San Antonio, New Braunfels, Texas.

[164] Anon *New directions in karst: Anglo-French symposium, 1983: field excursion notes*. npop: np, **1983**. illus: b&w: drawings, graphs, maps; [i], 37p.

[165] Anon *North Alabama's caves & caverns*. [Montgomery, AL, US]: np, **c1969**. illus: col: maps, photos, repros; text on lining pages; [32]; gloss; 26 cm.

Title from cover. Unpaged, illus.

[166] Anon *The official guide to Cheddar Somerset with map and eleven illustrations*. 2nd. Cheddar, GB: Cheddar Parish Council, **nd**. illus: b&w: ads, maps, photos; 24p; 18 x 12 cm.

"Issued by authority of the Cheddar Parish Council; the Parish Hall Cheddar."

[167] Anon *Ohio Caverns*. npop: np, **1976**. illus: col: photos; 16p.

Description of cave in central Ohio, USA.

[168] Anon *The Ohio Caverns: where nature carved a fairyland*. [Urbana, OH, US]: Gaumer Publishing Company, **1930**. illus: b&w: maps; [11]; 24 cm.

Title from cover. Running title: The beautiful Ohio Caverns. Illus.

[169] Anon *Oligio-Nunk, the place of caves. In the heart* of Honeycomb Mountain ... Carter County, Kentucky. npop: np, **c1900**. illus: b&w: photos; ad on rear lining page; [32]; 13 cm.

[Brochure for the Fast Flying Virginian Train, C&O Route]

[170] Anon *On the cold caves of the Monte Testaccio at Rome*. Edinburgh; London, GB: William Blackwood ; T. Cadell, **1828**. 12p, 1p pls; bibl; 22 cm.

pp 205-216. Microopaque version: New York: Readex Microprint, 1977, 23 x 15 cm. (Landmarks II)

[171] Anon *Oregon Caves: a pictorial souvenir guide*. Portland, OR, US: J & H Sales Co Inc, **[1993]**. illus: partly col: maps; col: drawings, photos; glossary on rear lining page ; text on front lining page; 24p; gloss; 25 x 18 cm.

[172] Anon *Oudtshoorn and the Cango Caves*. Johannesburg, ZA: Oudtshoorn Municipality and South African Railways and Harbors, **1923**. illus: b&w: maps, photos; 63p.

[173] Anon *Photographic views of some of the important points of Mammoth Cave situated in Edmondson, Co. Kentucky, USA: the wonders of this cave and its magnitude, cannot be described, and must be seen to be appreciated. It is reached only by the Louisville & Nashville Railroads. These photographs were taken by magnesium light by W.H. Sesser, St. Joseph, Mich., with a Collins camera.* npop: np, **c1886**. illus: b&w: photos; [22]; 36.5 x 45 cm.

Large photographs in album; no text with the exception of photo mounts. Exceptionally rare.

[174] Anon *Pictures of Cheddar and Cox's cave by pen and camera*. npop: np, **1930**. [16]; 13 x 20.5 cm.

Photographs with 3p text. Title from cover.

[175] Anon *Poole's Cavern: Buxton*. Buxton, GB: np, **c1950**. illus: b&w: photos; 15p; 18.5 x 12 cm.

All about Poole's Cavern. 4 pages of text from lecture by William Boyd Dawkins.

[176] Anon *Prehistoric painting, part 2*. npop: np, **1984**. 5 x 5 cm.

21 slides, b&w with col, 5 x 5 cm. 81. Lascaux, Hall of Bulls—82. Lascaux, Hall of Bull—83. Altamira, Bison—84. Altamira, Bison—85. Altamira, Bison—86. Lascaux, stags—87. Lascaux, small horses—88. Lascaux.

[177] Anon *Red hills*. Topeka, KS, US: KTWU, **1990**. Sunflower Journeys.

1 videocassette (VHS) (28 min): sd, col, 1/2 in. Caves and caving in Kansas.

[178] Anon *Research of China karst*. Peking, CN: Academic Press, **1979**. 336p; 26 cm.

Illus.

[179] Anon *Research of China karst*. Ti chih yen chiu so; Yen jung yen chiu tsu. [Peking, CN]: Academia Sinica, Institute of Geology, **1979**. 17p; 26 cm.

Translation of table of contents, and English summary of each chapter of: Chung-kuo yen jung yen chiu. Translated title of Chinese original: Karst research in China.

[180] Anon *Sea Lion Caves Oregon coast*. Beverly Hills, CA, US: Panorama International Productions, **1994**.

1 videocassettes (30 min), sd, col, 1/2 in, VHS format.

[181] Anon *Sea Lion Caves: America's largest sea cave, on U.S. Highway 101...just North of Florence, Oregon, Oregon coast*. Florence, OR, US: np, **c1993**. illus: b&w: drawings, maps; partly col: drawings, maps, photos; text and photos on lining pages; [18]; 23.5 x 15 cm.

Title from cover. Different versions exist with [16-18]pages, some list J & H Sales, Portland, OR, as publisher.

[182] Anon *Selected views of the beautiful caverns of Luray*. Luray, VA, US: Luray Caverns Corporation. illus: b&w: photos; [4], 18p pls.

On cover: The Beautiful caverns of Luray in the Page Valley of the Shenandoah. Consists mainly of full-page B&W photos

[183] Anon *Shenandoah Caverns, the grotto of the gods, in the heart of the Valley of Virginia*. Frederick, MD, US: Marken & Bielfeld, Printers, **1920**. illus: partly col: maps; [16]; 23 cm.

Illus. OCLC indicates some ambiguity about whether the publication date is 1920 or 1929.

[184] Anon *Shenandoah Caverns: a brief history*. npop: np, **c1995**. illus: b&w: maps, photos; 16p; 21.5 x 14 cm.

Cover title: Shenandoah Caverns Virginia.

[185] Anon *A short history of Kents Cavern*. Wellswood-Torquay, GB: np, **nd**. illus: b&w: photos; text on rear lining page; 16p; 13.5 x 11 cm.

Similar to Kents Cavern Wellswood Torquay Devon home of prehistoric man and animals.

[186] Anon *Shri Amar Nath Ji Guide*. Delhi, IN: Amnt Book Depot, **c1987**. illus: b&w: photos; 32p; 17.5 x 12 cm.

[187] Anon *Skyline Caverns*. Front Royal, VA, US: Skyline Caverns, **1995**. illus: b&w: maps, photos; partly col: photos; col: photos; [24]; gloss; 23 x 15 cm.

[188] Anon *Souvenir book of romantic Bridal Cave*. Aurora, MO, US: NWN Color Press, **nd**. illus: b&w: drawings, photos; col: ads, drawings, photos; color photos and drawings on lining pages; 20p; 28 cm.

[189] Anon *Souvenir book of the Oregon Caves National Monument: Marble Halls of Oregon*. [Chicago, IL, US]: [C. Curt Teich & Co, Inc], **1969**. illus: b&w: drawings, maps; col: photos; [22]; gloss.

[190] Anon *Speleology in France: Ressources [sic] in Limousin Quercy Perigord*. npop: np, **c1967**. illus: b&w: maps; [12]; 16 x 20.5 cm.

How to go caving, who to call, etc. in Perigord region, France. Printed one side of leaves.

[191] Anon *Spelunking*. New York, NY, US: Radim, **1946**.

1 film reel (22 min), sd, b&w, 16 mm.

[192] Anon *The story of Howe Caverns*. NY, US: Howe

Caverns Corp, **c1995**. illus: b&w: charts, drawings, photos; partly col: maps; col: photos; contents on front lining page; area attractions on rear lining page; 31p, [1]; 23 x 15 cm.

Many earlier editions exist; most attributed to Clymer.

[193] Anon *The story of most interesting discoveries commencing August 1923; White Skar Cave; under Ingleborough, Yorkshire, 1 1/2 miles from Ingleton on the Hawes Road*. 1934. npop: np, **1934**. illus: b&w: ads, maps, photos; maps folded in front and rear; ad on rear lining page; 31p; 18.5 x 12 cm.

Cover subtitle: Illuminated by electric light. Poem on rear cover.

1937. npop: np, **1937**. illus: b&w: maps, photos; ad on rear lining page; 40p.

1938 edition has no obvious changes.

1948. Morecombe, Yorkshire, GB: Boro' Advertiser Ltd., **1948**. illus: b&w: ads, maps; map on front lining page; ad on rear lining page; 39p, [1].

Printed by the Boro' Advertiser Ltd Evston Road Morecombe.

[194] Anon *The story of newly discovered White Skar Caverns; under Ingleborough, Yorkshire, 1 1/2 miles from Ingleton on the Hawes Road*. 2nd edition. Yorkshire, GB: Yorkshire Herald Newspaper Company Ltd., **1931**. illus: b&w: ads, maps, photos; maps folded in front and rear; ad on rear lining page; 24p; 18.5 x 12 cm.

Cover subtitle: Illuminated by electric light. Poem on rear cover. Printed by Yorkshire Herald Newspaper Company Ltd.

[195] Anon *Stump Cross Caverns: the underground wonderland*. npop: np, **nd**. illus: b&w: photos; photo on front lining page; text on rear lining page; [12]; 12.5 x 9.5 cm.

[196] Anon *Subterranean fauna of Mexico: Part I: some results of the first Italian zoological mission to Mexico sponsored by the National Academy of Lincei (October 10 - December 9, 1969)*. Rome, IT: National Academy of Lincei, **1972**. Problemi Attuali de Scienza e di Cultura, Sezione: Missioni ed Esplorazioni - I, Quaderno N. 171. illus: b&w: ads, drawings, maps, photos; 172p, [5], 10p pls; bibl.

[197] Anon *Subterranean fauna of Mexico: Part II: further results of the first Italian zoological mission to Mexico sponsored by the National Academy of Lincei (1969 and 1971)*. Rome, IT: National Academy of Lincei, **1973**. Problemi Attuali de Scienza e di Cultura Sezione: Missioni ed Esplorazioni - I Quaderno N.171. illus: b&w: ads, drawings, maps, photos, photomicrographs; col: photos; 2 b&w maps folded in text; 372p, [8], 10p pls; bibl, index.

[198] Anon *Subterranean fauna of Mexico: Part III: further results of the first Italian zoological mission to Mexico sponsored by the National Academy of Lincei (1973 and 1975)*. Rome, IT: National Academy of Lincei, **1977**. Problemi Attuali de Scienza e di Cultura Sezione: Missioni ed Esplorazioni - I, Quaderno N. 171. illus: b&w: ads, drawings, maps, photos, photomicrographs, tables; col: photos; 2 b&w maps folded in text; 379p, [8], 11p pls; bibl, index.

[199] Anon *Subterranean wonders: Mammoth Cave and Colossal Cavern, Kentucky*. 10th ed, new series. npop: np, **1930**. illus: b&w: maps; 46p, 2p pls; 20 cm.

Publication date suspect.

11th ed, new series. npop: np, **nd**. illus: b&w: maps, photos; 1 map folded in rear, labeled 1897; 48p, 2p pls; 20 cm.

1900 ed. npop: np, **1900**. 1 folded map dated 1897; 31p, 2p pls; 23 cm.

Illus. Cover title. "For distribution through the newspapers represented by Frederic J. Haskin" (Haskin, Frederic Jennings, 1872-1944).

6th ed, new series. npop: np, **1930**. illus: b&w: maps; 46p, 2p pls; 20 cm.

Running title: Subterranean wonders of Kentucky.

[22nd ed]. [Louisville, KY, US]: Louisville & Nashville Railroad Passenger Department, **nd**. illus: b&w: maps; 1 folded map dated 1897; 46p, 2p pls; 20 cm.

13th ed, new series and 17th ed, new series had no publication date. Microfiche. Louisville: Lost Cause Press, 1969, 3 microfiche.

[200] Anon *A symphony in limestone: the beautiful Olsen's Caves: 18 miles North of Rockhampton, Queensland: A description of Olsen's Caves*. 3rd ed. [Olsen's Caves, QLD, AU]: np, **1962**. illus: b&w: ads, maps, photos, ports; 32p.

Title from cover.

[201] Anon *Symposium on cave photography*. Herefordshire, GB: Cave Research Group of Great Britain, Dec, **1969**. The Transactions of the Cave Research Group of Great Britain v 11 number 4 pp 215-276. illus: b&w: charts, drawings, maps, photos, tables; previous publications list on back lining page; 62p, 14p pls; bibl; 30 x 21 cm.

[202] Anon *There is a cavern in the town*. 11th ed. npop: np, **1970**. 20p.

Show cave guide to St. Clements Cave, Hastings. Illus.

[203] Anon *Timpanogos Cave public use and impact*. Salt Lake City, UT, US: Video graphics, **1987**.

1 videocassette (VHS)(9 min), sd, col, 1/2in.

Discusses public access to Timpanogos Cave, Utah, including available tours and information to use in planning a trip to the cave.

[204] Anon *Twelve pictures of Cheddar and Cox's Cave*. [Cheddar, GB]: [Cox's Cave], **c1934**. illus: b&w: photos; contents listed on front lining page; imprint on rear lining page; [16]; 14.5 x 20.5 cm.

Photographs mounted on thick brown leaves. 4 p of text.

[205] **Anon** *Unusual cave and cavern formations.* Crafton, WI, US: JLM Visuals, **1983**. Natural Science Slide Program 1475. illus: partly col: maps, photos; 20p.

Title from guide. Series statement from container. 20 slides, col, plus 1 script.

[206] **Anon** *Views of the Buchan Caves and pyramids.* Melbourne, VIC, AU: TC Lothian, **1907**. illus: b&w: maps; [15], 18p pls; 14 x 23 cm.

Copy examined was a photocopy. Tourist brochure. Printed one side only.

[207] **Anon** *Virginia Luray Caverns.* World Travelogues Corporation. Williamstown, MA, US: World Travelogues Corp, **1984**.

76 slides, col.

[208] **Anon** *Waitomo Caves management plan 1982.* npop: np, **1982**. illus: b&w: maps, tables; [ii], i, 52p; bibl; 29 x 21 cm.

[209] **Anon** *Waitomo Day 1982: summary of papers: Waitomo Caves research programme.* npop: np, **nd**. illus: b&w: charts, drawings, graphs; 39p, [3]; bibl; 30 x 21 cm.

[210] **Anon** *Waitomo Ruakuri & Aranui New Zealand.* Auckland, NZ: Whangares, **[c1910]**. illus: b&w: photos; [24]; 17 x 22 cm.

32 b&w photos of show caves with 1 p intro.

[211] **Anon** *The weird wonders of Wookey Hole Caves.* Wells, GB: np, **[1947]**. illus: b&w: ads, drawings, maps, photos; advertisements on rear lining page; front: b&w photo; [14]; 19.5 x 13 cm.

17 b&w photos

[212] **Anon** *Welcome to Waitomo Caves: a photographic insight into this spectacular region of New Zealand.* Aukland, NZ: First Class Publications Ltd, **1992**. illus: b&w: photos, ports; col: maps, photos; introduction (includes b&w photo), history on front lining page; col maps on rear lining page; 24p; 28 x 22 cm.

[213] **Anon** *White Scar Caves: Britain's biggest tourist cave.* Gravatt, T. (photo); Newman, Gaj (photo); Donaldson, D. R. (photo). Ingleton, GB: White Scar Caves Limited, **c1992**. illus: b&w: repros, tables; partly col: tables; col: maps, photos; text on lining pages; Various pagination, 15p, [1]; 24 x 17 cm.

[214] **Anon** *The Wind Cave and its territory; Garfagnana, the Apuan Alps, the Serchio Valley (Lucca, Tuscany, Italy).* npop: np, **c1995**. illus: b&w: ads; col: maps, photos; 47p; 23 x 15.5 cm.

[215] **Anon** *Wonderful, famous, spectacular Crystal Cave, Kutztown, Penna [sic].* [Pennsylvania, US]: np, **1920**. illus: b&w: maps; 16p; 12 x 16 cm.

On pictorial wrappers: Crystal Cave, a natural wonder, Kutztown, Pa.

[216] **Anon** *The wonders of the Grotto of Han.* Boyer, Paul (photo); Ruhir, Ed (photo). Brussels, BE: Scientific Printing House of Charles Bulens, **1900**. illus: b&w: ads, drawings, maps, photos; color fold-out map of cave in back; front lining page is business information; vi, 40p; 21.5 x 13 cm.

Cover title: Grotto of Han; Belgium. Tipped in back in a different quality paper is "useful information for travelers and tourists" which is mostly ads.

[217] **Anon** *Wonders of the world.* Gibbs, Philip (forw). London, GB: Odhams Press, **c1933**.

No date, no author, date is probably around 1933, Gibbs gives the foreword. Very well illustrated with literally hundreds of caves referenced.

[218] **Anon** *Wookey Hole Caves: sixteen exclusive camera studies.* Royal Tunbridge Wells, Kent, GB: Photochrom Co Ltd, **nd**. illus: b&w: photos; [2], 16p pls; 24 x 15 cm.

[219] **Anon** *Wookey Hole Caves: Wells, Somerset.* Taunton, GB: np, **[1938]**. illus: b&w: ads, drawings, maps, photos; photo and text on rear lining page; [12]; 18 x 12 cm.

Title from cover; 9 b&w photos. Earlier edition of The weird wonders of Wookey Hole Caves.

[220] **Anon** *Wookey Hole Caves: Wells, Somerset.* Hassall, John (photo). npop: np, **[1937]**. illus: b&w: photos; [18]; 23 x 15 cm.

Many editions; none dated; most have two pages text; all printed one side only; some have text at front, others at rear. Cover has different photo on different versions; one has cut-out window to view photo. Cover photo b&w in earliest version examined; all others had color photo.

[221] **Anon** *Wookey Hole; the caves and mill.* npop: np, **1986**. illus: col: drawings, maps, photos; col photos on lining pages; 38p, [2]; 27 x 21 cm.

Glossy modern booklet with info on papermaking mill and carousel as well as cave; more photos than text.

[222] **Anon** *Yarrangobilly Caves—Kosciusko National Park.* npop: National Parks and Wildlife Service of NSW, **1970**. illus: b&w: maps, photos; one map on front lining page; 16p, [2].

[223] **Appalshop, Inc** *Line Fork falls and caves.* Whitesburg, KY, US: Appalshop, Inc, **1971**.

1 film reel (10 min), sd, b&w, 16 mm.

Exploration of Pine Mountain caves in eastern Kentucky combined with a comic approach to finding a place in the mountains where one can still drink from a stream.

[224] **Aquing, Felix** *Survey of Lopinot caves.* [Port of Spain], TT: Government of Trinidad and Tobago, Ministry of Planning and Development, National Environment and Conservation Council, **1974**. Student Study Project Series 1974 Research paper- National Environment and Conservation Council no 1/74. illus: b&w: maps; 11p, [6]; 29 cm.

Illus. Standard short cave evaluation for possible use for tourism.

[225] **Argus [pseud]** *Jenolan Caves and the Blue Moun-*

tains. Parramatta, NSW, AU: Cumberland Argus Printing Works, **1898**. illus: b&w: ads, drawings, maps, photos; folded illus of caves in text; [xiv], 54p, [6].

Cover title: Jenolan. J. R. Dunkley [1988], *A bibliography of the Jenolan Caves part 2: literature*, p.8 lists "later editions 1899 (?), 1900, 1901, 1902, and 1903."

[226] Armstrong, A. Leslie *Rhodesian archaeological expedition (1929): excavations in Bambata Cave and researches on prehistoric sites in Southern Rhodesia.* London, GB: Royal Anthropological Institute of Great Britain and Ireland, **1931**. 28p.

Cover title. Reprint from the Journal of the Royal Anthropological Institute, vol. LXI, January-June, 1931, pp 239-276. Illus

[227] [Army Caving Association] *The 1984 Army Caving Association expedition to the Gouffre Berger.* [GB]: np, **nd**. illus: b&w: cartoons, drawings, maps, photos, tables; 1 foldout map at end.; iii, 55p, 1p pls.

[228] [Army Caving Association] *Army Caving Association: Peru 1987.* Woolridge, Gerry (photo). npop: Army Caving Association, **c1988**. illus: b&w: ads, maps, photos; col: photos; [ii], 95p; 30 x 21 cm.

[229] Arnold, Charles L. *Inside the caves: Lava Beds National Monument.* Albany, CA, US: C.L. Arnold, **1986**. illus: b&w: maps; col: drawings, photos; [iii], 35p; 23 cm.

[230] Arrell, Robert *Waitomo Caves: a century of tourism.* Waitomo, NZ: Waitomo Caves Museum Society, Inc, **1984**. illus: b&w: graphs, maps, photos, ports; b&w maps on lining pages; iv, 71p; app, bibl.

[231] Arsenis, Mylda L. *The caves of Chinoyi: a Rhodesian story.* Longshaw, Rose, (illus). [Salisbury, Rhodesia]: College Press, **1970**. 48p; 21 cm.

Illus.

[232] Ashbrook, Bert *Caves in the Richlands area of Greenbrier County, West Virginia.* [Barrackville], WV, US: West Virginia Speleological Survey for the Greater Allentown Grotto of the National Speleological Society and the West Virginia Association for Cave Studies, **1995**. Monograph series (West Virginia Speleological Survey) no 1. illus: b&w: charts, drawings, graphs, maps, photos, tables; four folded maps in pocket; vi, 37p; bibl; 28 cm.

Illus.

[233] Ashbrook, Bert *The Caves of Northumberland County, Pennsylvania.* Greater Allentown Grotto of the National Speleological Society. State College, PA, US: Mid-Appalachian Region of the National Speleological Society, Dec, **1990**. MAR Bulletin 17. illus: b&w: maps, photos; 4 fold-out maps; 44p; bibl; 27 cm.

[234] Asher-Bolinder, Sigrid *A review of the deposition and alteration of filled-sink deposits of east-central Missouri.* [Denver, CO, US]: Dept. of the Interior, U.S. Geological Survey, **1992**. U.S. Geological Survey open-file report 92-14. illus: b&w: maps; i, 15p; bibl; 28 cm.

Microfiche. [Denver, Colo.: U.S. Geological Survey, 1993] 2 microfiches.

[235] Aslett, James L. *Lava beds underground.* Carman, S. (illus); Donati, Anne (illus); Waters, Aaron C. CA, US: Lava Beds Natural History Association, **1982**. illus: b&w: drawings, maps, photos; [40]; gloss; 26 cm.

Title from cover. Chiefly maps.

[236] Aspey, Steve *Sheffield University Speleological Society Central Crete Expedition: Greece 1984.* Et al. [GB]: Sheffield University Speleological Society, **[1985]**. illus: b&w: charts, maps; 55p; bibl; 30 x 21 cm.

Photocopy examined.

[237] Aspin, J. *The caverns of Upper Ease Gill.* Gemmell, Arthur, 1915–; Jewett, A. Middleton, Leeds, GB: The Northern Pennine Club, Mar, **1952**. illus: b&w: maps; 1 large map in back flap; 23p, [5]; app.

[238] Association of Ground Water Scientists and Engineers (a division of NGWA) *Proceedings of the third conference on hydrogeology, ecology, monitoring, and management of ground water in karst terranes, December 4-6, 1991, Maxwell House/Clarion, Nashville, Tennessee.* Dublin, OH, US: U.S. EPA and the Association of Ground Water Scientists and Engineers (a division of NGWA), Dec, **1991**. illus: b&w: charts, drawings, graphs, maps, photos, tables; xiii, 793p; bibl.

This conference, along with the "Sinkhole Conference" are the two most important applied karst meetings. The proceedings are of mixed quality, but contain many important applied papers.

[239] Association of Missouri Geologists. Annual Meeting (19th: 1972: Rolla, Mo) *Guidebook to the karst features and stratigraphy of the Rolla area.* Rolla, MO, US: University of Missouri, **1972**. illus: b&w: maps; 109p; bibl; 28 cm.

A second, rev ed was issued in 1977 with same number of pages.

[240] Atchison, Topeka, and Santa Fe Railway Company *Carlsbad Caverns National Park.* npop: The Company, **1941**. illus: col: photos, ports; 23p; 23 cm.

Title from cover.

[241] Atchison, Topeka, and Santa Fe Railway Company *Carlsbad Caverns, New Mexico.* 1927 ed. [Chicago, IL, US]: Rand McNally, **1927**. illus: b&w: maps; 62p; 23 cm.

Illus. Pages listed as 62 columns.

1928 ed. The Railway, Rand McNally, **1928**.

1930 ed. np, **1930**.

[242] Athens, J. Stephen *Prehistoric upland bird hunters: archaeological inventory survey and testing for the MPRC project area and the Bobcat Trail Road, Pohakuloa Training Area, Island of Hawaii.* Kaschko, Michael W. Honolulu, HI, US: International Archaeological Research Institute, Inc, Aug, **1989**. illus: b&w: graphs, maps, photos, tables; xi, 295p; app, bibl.

Final report submitted to US Army Engineer District, Pacific Ocean Division Fort Shafter, Hawaii, 96858-5440. Contract No. DACA83-87-C-0017. 28 tables, 80 photos, 22 figs.

[243] Atkinson, Anne *Undara Volcano and its lava tubes: a geological wonder of Australia in Undara Volcanic National Park, North Queensland.* Atkinson, Vernon; Smith, Dick (forw). Brisbane, QLD, AU: Vernon and Anne Atkinson, **1995**. illus: b&w: charts, drawings, photos, tables; partly col: charts, maps; col: charts, drawings, graphs, maps, photos, photomicros, ports; overview on front lining page; about the authors, about Undara Volcanic National Park on rear lining page; x, 86p, [1]; app, bibl, errat, gloss; 30 x 21 cm.

Hardcover and softcover versions exist.

[244] Atkinson, T. C. *Caves and karst of southern England and South Wales: guidebook for the International Congress of Speleology at Sheffield, 1977.* Smart, P. L. Bridgwater, Somerset, GB: 7th International Speleological Congress Committee; [Distributed by] British Cave Research Association, **1977**. illus: b&w: charts, graphs, maps, photos, tables; 83p; bibl; 21 cm.

"Guidebook for the International Speleological Congress 1977" - cover.

[245] Atkinson, Timothy C. *Mendip Karst Hydrology Project: phases one and two.* Drew, David Phillip, 1943–; High, Colin. Pagbourne, Berks, GB: Wessex Cave Club, Oct, **1967**. Series 2 no 1 Occasional Publications (Wessex Cave Club). illus: b&w: graphs, photos, tables; [iii], 38p; bibl; 25.5 x 20 cm.

[246] Attout, Jacques *Men of Pierre Saint-Martin.* London, GB: T. Werner Laurie, Limited, **1956**. illus: b&w: maps, photos; vii, 158p, [11]; app, gloss.

Translated from the French *Les hommmes de la Pierre Saint-Martin*, 1954.

The story of the 1954 expedition to explore Pierre Saint-Martin, and of the recovery of the remains of Marcel Loubins, who was killed during the 1952 expedition. Told by the expedition chaplain. A gripping story which is well written and tells a great deal about cavers and caving at the time.

[247] Aubarbier, Jean-Luc *Prehistoric sites in Perigord.* Binet, Michel; Moyon, Angela (trans). Rennes, FR: Ovest France, Jun, **1985**. illus: b&w: maps; col: photos; 32p; 23 x 16.5 cm.

[248] Audyová, Jiřina *The Moravian karst: time and stone.* Simičková, Hedvika (trans); Audy, Igor (photo). 1st ed. Boskovice, CZ: Format, **1993**. illus: col: photos; [142]; 25 x 21 cm.

A color photo book with very little text.

[249] Aujoulat, Norbert *Lascaux revisited.* Ruspoli, Mario; Willemont, Jacques; Caisse National des Monuments Historiques et des Sites; Institut National de L'Audiovisuel. Aspen, CO, US: Crystal Productions (distributor), **1989**.

1 videocassette (35 min), sd, col, 1/2 in.

[250] Australasian Conference on Cave Tourism and Management *Cave Management in Australasia VIII: proceedings of the eighth Australian conference on cave tourism and management, Paparoa National Part, Punakaiki, New Zealand, April 1989.* Wilde, Kevan A. (ed); Department of Conservation, Buller District, West Coast Region (host). [Hokitika, NZ]: Australasian Cave and Karst Management Association, **nd**. illus: b&w: drawings, graphs, maps, photos, tables; [v], 183p; app, bibl.

Formerly called Cave Management in Australia. Pagination: [v], 183p, [v], 72p.

[251] Australian Conference on Cave Tourism *Cave Management in Australia: proceedings of the first Australian Conference on cave tourism, Jenolan Caves, N.S.W. 10th-13th July, 1973.* Hamilton-Smith, Elery (ed). Broadway, NSW, AU: Australian Speleological Federation, Jun, **1976**. illus: b&w: graphs, maps, tables; iii, 109p; bibl.

[252] Australian Conference on Cave Tourism and Management *Cave Management in Australia II: proceedings of the second Australian Conference on cave tourism and management, Hobart, Tasmania, 3rd-5th May, 1977.* Middleton, Gregory J. (ed). Sandy Bay, TAS, AU: National Parks and Wildlife Service, Tasmania and Australian Speleological Federation, Aug, **1977**. illus: b&w: charts, drawings, maps, tables; partly col: drawings, maps; iv, 119p; app, bibl.

[253] Australian Conference on Cave Tourism and Management *Cave Management in Australia III: proceedings of the third Australian Conference on cave tourism and management, Mount Gambier, South Australia, 30th April - 4th May, 1979.* Robinson, A. C. (ed). Adelaide, SA, AU: National Parks and Wildlife Service, South Australia and Australian Speleological Federation, Nov, **1980**. illus: b&w: drawings, graphs, maps, photos, repros, tables; [viii], 169p, [1].

[254] Australian Conference on Cave Tourism and Management *Cave Management in Australia IV: proceedings of the fourth Australian Conference on cave tourism and management, Yallingup, Western Australia Australia, September 1981.* Watson, R. J. (ed); Busseton Tourist Bureau (host). Busselton, WA, AU: National Parks Authority, Western Australia and Australian Speleological Federation, Aug, **1982**. illus: b&w: charts, maps, photos, tables; [iv], 82p; app, bibl.

[255] Australian Conference on Cave Tourism and Management *Cave Management in Australia V: proceedings of the Fifth Australian Conference on cave tourism and management, Lakes Entrance, Victoria, April 1983.* Hamilton-Smith, Elery (ed); Buchan Caves Advisory Committee (host); Australian Speleological Federation

(host). Carlton South, VIC, AU: Australasian Cave and Karst Management Association, **1990**. [iii], 67p; bibl.
No illus.

[256] Australian National Parks and WIldlife Service *Australian Ranger Bulletin: feature: cave management*. Canberra, ACT, AU: Australian National Parks and WIldlife Service, **1987**. Australian Ranger Bulletin vol 4(3). illus: b&w: drawings, maps, photos; 38p; bibl.

[257] Australian Speleological Expedition *Caves of New Caledonia: report of the 1975 Australian Expedition*. Harris, S. [Hobart, TAS, AU]: Harris, Gillieson, Gleeson, Landsberg, **1976**. illus: b&w: maps; 68p; 30 cm.
Illus.

[258] Australian Speleological Federation *Administrative handbook*. Melbourne, VIC, AU: Australian Speleological Federation, **[1972]**. Various pagination.
Limited edition. Published for internal purposes within the federation only. No illus.

[259] Australian Speleological Federation *Caving in Australia*. Broadway, NSW, AU: University of Technolgy Students' Union, **nd**. 23p; bibl.
No illus.

[260] Australian Speleological Federation *Draft management plan: Tantanoola Caves Conservation Park: Lower South East, South Australia*. Programmes Branch, National Parks and Wildlife Service (comp). Adelaide, SA, AU: Department of Environment and Planning, **1983**. illus: b&w: graphs, maps, photos, ports, tables; text on front lining page; ix, 71p.

[261] Australian Speleological Federation *Master plan for the development of Jenolan Caves Reserve*. Dunkley, J. R. (conv). Broadway, NSW, AU: Australian Speleological Federation, **c1973**. illus: b&w: maps; partly col: maps; vi, 74p; 30 cm.

[262] Australian Speleological Federation *Proceedings of the eighth biennial conference of the Australian Speleological Federation*. Goede, Albert (ed, prod); Cockerill, R. (ed, prod). Broadway, NSW, AU: Australian Speleological Federation, Feb, **1972**. illus: b&w: drawings, graphs, maps, photos, tables; 100p; bibl.
Conference held at the Hutchins School Board House, Sandy Bay, Tasmania, 27th to 31st December, 1970.

[263] Australian Speleological Federation *Proceedings of the tenth biennial conference of the Australian Speleological Federation*. Goede, Albert (ed, prod); Cockerill, R. (ed, prod). Broadway, NSW, AU: Australian Speleological Federation, Jul, **1975**. illus: b&w: drawings, maps, photos, tables; [iii], 110p, 8p pls; bibl.
Hosted by the University of Queensland Speleological Society, University of Queensland, St. Lucia, Brisbane, 27-29 December 1974. Cover title has wrong year for conference.

[264] Australian Speleological Federation Digest Commission *Australian speleology 1972*. Gillieson, Dave S. (ed). Brisbane, QLD, AU: Australian Speleological Federation Digest Commission, **1976**. illus: b&w: drawings, maps, tables; [v], 108p; app, bibl.
Reprints of articles.

[265] Australian Speleological Federation Inc *Preprints of papers for the 17th Biennial Conference of the Australian Speleological Federation Tropicon Conference, Lake Tinaroo, Far North Queensland. 27th to 31st December 1988*. Australian Speleological Foundation; Tropicon in Australia's bicentenary year. Caims, QLD, AU: Australian Speleological Federation Inc, **1988**. illus: b&w: charts, drawings, graphs, maps, photos, tables; [iv], 139p; bibl; 30 cm.
Cover title: Tropicon in Australia's bicentenary year: pre-prints of papers to be presented at Tinaroo Falls Far North Queensland: 27th to 31st December 1988. Available from Australian Speleological Federation, Tropicon Subcommittee.

[266] Autin, Whitney J. *Observations and significance of sinkhole development at Jefferson Island*. Baton Rouge, LA, US: Dept. of Natural Resources, Louisiana Geological Survey, **1984**. Louisiana Geological Survey Geological Pamphlet no 7. xiv, 75p; bibl; 28 cm.
Illus.

[267] Avid Corporation Creative Learning *Formation of caves*. npop: np, **1970**.
With captions. 32 fr, col, 35 mm.
Discusses types of caves and shows how caves are formed.

[268] Ayer, Mary Youngman *The archaeological and faunal material from Williams Cave, Guadalupe Mountains, Texas*. Philadelphia, PA, US: Academy of Natural Sciences of Philadelphia, **1936**. Proceedings of the Academy of Natural Sciences of Philadelphia v 88 pp 599-619. 21p; 28 cm.
Illus.

[269] Bachman, George Odell *Karst in evaporites in southeastern New Mexico.*. United States Department of Energy. Albuquerque, NM, US: Sandia National Laboratories, Sep, **1987**. Sandia National Laboratories SAND86-7078. illus: b&w: charts, drawings, maps, photos; 81p; bibl, index; 28 cm.
This is a detailed look at the present and paleokarst of the area, especially as it applies to the disposal of radioactive waste at the Waste Isolation Pilot Project (WIPP) site, southern New Mexico. This is a summary of this well respected author's work in this classic area.

[270] Back, William (ed) *Hydrogeology of selected karst regions*. Herman, Janet S.; Paloc, Henri (eds). Hannover, DE: International Association of Hydrogeologists, **1992**. International Contributions to Hydrogeology vol 13. illus: b&w: charts, drawings, graphs, maps, photos, tables; xx, 494p; bibl.

[271] Baddeley, Glenn (ed, auth) *Vulcon Guidebook:*

lava features and limestone karst of Victoria and southeastern South Australia. Vulcon 1995 20th Biennial Conference. Australian Speleological Federation, Inc, Hamilton, Victoria 2-6 January 1995. Grimes, Ken; White, Susan; Mott, Kevin; Watson, Tony. Melbourne, VIC, AU: Victorian Speleological Association, Inc. illus: b&w: drawings, maps, tables; vi, 116p; app, bibl, gloss.

Cover title: Vulcon Guidebook: 20th ASF Conference, Hamilton Victoria1995.

[272] Baddeley, Glenn (ed) *Vulcon precedings: papers submitted for presentation. Vulcon 20th Biennial Conference. Australian Speleological Federation, Inc, Hamilton, Victoria 2-6 January 1995.* Melbourne, VIC, AU: Victorian Speleological Association, Inc, **1995**. illus: b&w: charts, drawings, graphs, maps, tables; vi, 100p; bibl.

Cover title: Vulcon precedings: 20th ASF Conference, Hamilton Victoria1995.

[273] Baguley, Frank (ed) *Cambrian Caving Council handbook 1993.* Swansea, GB: Cambian Caving Council, Aug, **1993**. Various pagination, [70].

Very odd pagination; letters and numbers used. No illus.

[274] Bahn, Paul G. *Images of the ice age.* Vertut, Jean (photo). Leicester, GB: Windward, an imprint owned by W.H. Smith and Son, **1988**. illus: b&w: charts, drawings, maps, photos, tables; col: photos; front: col photo; 240p; add, bibl, index.

[275] Bailey, Gilbert Stephen, 1822-1891 *The great caverns of Kentucky: Diamond Cave, Mammoth Cave, Hundred Dome Cave.* Chicago, IL, US: Church & Goodman, **1863**. illus: b&w: maps; 63p; 15 cm.

Cover title: Caverns of Kentucky. Microfilm version: Microfiche. Louisville, Ky: Lost Cause Press.

[276] Bailey, Gilbert Stephen, 1822-1891 *Hundred Dome Cave ("partial reproduction of the last known copy of the book, 'the great caverns of Kentucky' ").* [KY, US]: Marco Development Company of Kentucky, Nov, **1963**. illus: b&w: maps; 63p.

Pagination: [5], 40-63 p.

[277] Bailey, John H. *A stratified rock shelter in Vermont.* Fort Ticonderoga, NY, US: The Champlain Valley Archaeological Society, **1940**. Bulletin of Champlain Valley Archaeological Society v l no 3. illus: b&w: maps; 30p; bibl; 23 cm.

[278] Bailey, Thomas L. *Report on the caves of the eastern highland rim and Cumberland Mountains.* Nashville, TN, US: Tennessee State Geological Survey, Apr, **1918**. The Resources of Tennessee v 8(2): pp85-138. illus: b&w: tables; 59p.

Locations and descriptions of 109 caves (mostly) and rock shelters in part of the state of Tennessee. No maps. A notable early survey.

[279] Bailey, Vernon Orlando, 1864-1942 *Animal life of the Carlsbad Cavern.* Baltimore, MD, US: The Williams & Wilkins Company, **1928**. Monographs of the American Society of Mammalogists no 3. illus: b&w: ads, maps, photos, repros; col: maps; front; xiii, 195p; index; 21 cm.

[4]p at end with publishers ads and section, Sans Tache. Describes the area in and around Carlsbad Caverns; includes bats. Some photographs. 8 chapters with 67 illus.

[280] Bailey, Vernon Orlando, 1864-1942 *Cave life of Kentucky: mainly in the Mammoth cave region.* Bailey, Florence Merriam, 1863–; Giovannoli, Leonard. Notre Dame, IN, US: The University Press, **1933**. illus: b&w: drawings, photos, repros, tables; 256p; bibl, index; 24 cm.

Prepared in the Bureau of Biological Survey, U. S. Department of Agriculture ... in cooperation with the Kentucky State Geological Survey ... "Reprinted from 'The American midland naturalist', vol. XIV, no. 5, pp. 385-635, 1933." 90 figs.

[281] Baker, Ernest Albert, 1869-1941 *Cave Explorers in Co. Fermanagh 1907.* Et al. Belfast, IE: W. & G. Baird Ltd., **1907**. 30p; 16 x 11.5 cm.

Printed by W. & G. Baird Limited. Photocopy examined. No illus.

[282] Baker, Ernest Albert, 1869-1941 *Moors, crags & caves of the High Peak and the neighbourhood.* London, GB: John Heywood Ltd, [**1903**]. illus: b&w: maps, photos; 2 maps folded in text; front; 207p, [2], 32p pls; 23 cm.

No date on copy examined. Only 30% caves.

[283] Baker, Ernest Albert, 1869-1941 *The netherworld of Mendip: explorations in the great caverns of Somerset, Yorkshire, Derbyshire, and elsewhere.* Balch, Herbert Ernest. London, GB: Clifton J. Baker & Son, **1907**. illus: b&w: drawings, maps, photos; xii, 172p, 56p pls; index; 23 cm.

The chapters dealing with the scientific results are by H. E. Balch ... the accounts of actual experiences, in which the sporting side is predominant, are by E. A. Baker.

[284] Baker, Ernest Albert, 1869-1941 *Caving, episodes of underground exploration.* Mellor, D. C. (fore). 1970 reprint ed. Wakefield, Yorkshire, GB: S. R. Publishers, Limited, **1970**. illus: b&w: drawings, maps, photos; b&w front; xv, 252p; bibl, index; 23 cm.

All plates (photos) placed at end of book; with a new foreword by D.C. Mellor. Five additional prefatory pages.

1975 reprint ed. Mellor, D. C. (fore). Teaneck, NJ, US: Zephryrus Press, **1975**. Series Speleologia. illus: b&w: drawings, maps, photos; b&w front; xv, 252p, [4]; bibl, index; 23 cm.

All plates (photos) placed at end of book). Blank pages for notes ([4]p at end).

1st ed. London, GB: Chapman & Hall Ltd, **1932**. illus: b&w: drawings, maps, photos; b&w front; xv, 252p, [4], 55p pls; bibl, index; 23 cm.

Reissued in a cheaper edition in 1935.

Speleology and caving treated as a sport. Cave stories from Britain, Ireland, France and Belgium from 1900 to 1935.

[285] Baker, Eva G. *Water resources publications of the USGS for Tennessee, 1907-1987.* Massingill, Renda C. Nashville, TN, US: US Geological Survey. Open-file Report U.S. Geological Survey 87-552. iii, 34p; bibl; 28 cm.

A handy bibliography touching a well studied karst area. The USGS has done quite a bit here, mostly with engineering and environmental problems.

[286] Baker, Janet *A brief history of the Dunhuang Caves: a millennium of Chinese Buddhist art.* npop: Western Conference of the Association for Asian Studies, **1994**. Selected papers in Asian studies, new series paper no 501994. 36p; bibl; 28 cm.

Illus.

[287] Baker, Percy Frederick, 1921– *The Baker extension to the Banwell Bone Cave.* Tucker, J. H. [Banwell, GB]: [P.F. Baker], **1979**. illus: b&w: maps, ports; map on rear lining page; "Percy Baker in the alcove leading to the Frozen Rivers Grotto" on front; 14p, [1]; bibl; 26 cm.

Illus.

[288] Bakó, Tamás *Paleokarst in Hungary.* Tádal, Ödön (trans); Hazslinszky, Tamás (ed). Budapest, HU: 10th International Congress of Speleology, **1989**. Field Trip Guide D4. illus: b&w: maps; 12p, 12p pls.

[289] Balázs, Dénes (ed) *Special issue on the occasion of 10th international congress held in Hungary 1989.* Székely, Kinga (ed); Lóczy, Dénes (trans); Pololy, Judit (trans); Ráday, Ödön (trans). Budapest, HU: Hungarian Speleological Society, **1989**. Karst and Cave: Bulletin of the Hungarian Speleological Society. illus: b&w: charts, drawings, graphs, maps, photos, repros, tables; col: photos; color photos on lining pages; 112p, 112p pls; bibl.

[290] Balázs, Dénes (ed) *Special issue on the occasion of 7th international congress held in England 1977.* Budapest, HU: Hungarian Speleological Society, **1977**. Karst and Cave: Bulletin of the Hungarian Speleological Society. illus: b&w: charts, drawings, graphs, maps, photos, repros, tables; text on front lining page; b&w photos rear lining page; 76p; bibl.

[291] Balch, Edwin Swift, 1856-1927 *Glaciéres; or, freezing caverns.* Halliday, William Ross, 1926–(intro). 1970 reprint ed. New York, NY, US: Johnson Reprint Corporation, **1970**. Classics in Speleology. [xiv], xxxiii, 237p; bibl, index; 22 cm.

1st ed. Philadelphia, PA, US: Allen, Lane & Scott, **1900**. illus: b&w: drawings, photos; front; [xiii], 237p, 19p pls; bibl, index; 25 cm.

An account of exploration, and an explanation, of ice caves, for a long time considered the definitive landmark of documentation on the subject in American caving history. With an excellent new introduction by William R. Halliday in the 1970 reprint, covering Balch's life, works and theories. This work is a credit to the authors careful observation and good common sense. It remains useful today.

[292] Balch, Edwin Swift, 1856-1927 *Ice caves and the causes of subterranean ice.* Philadelphia, PA, US: Press of Allen, Lane & Scott, **1896**. 26p; 25 cm.

A hint of more ambitious things to come

[293] Balch, Herbert Ernest *The caves of Mendip.* London, GB: Folk Press, Limited, **1926**. The Somerset Folk series no 26. illus: b&w: ads, drawings, photos, repros; folded pl and ads; front: b&w folded drawing; 82p, [4], 19p pls.

The caves of Mendip and Cheddar took up the greater part of the authors life and work: one can see why he found them so interesting.

[294] Balch, Herbert Ernest *Excavations at Chelm's Combe: Cheddar: conducted under the Excavations Committee of the Somerset Archaeological and Natural History Society 1925-26.* Cooper, N. C.; Clay, R. C. C.; Jackson, J. Wilfred. Sherborne, GB: np, **[1927]**. illus: b&w: charts, drawings, maps, photos; 32p; 22 x 14 cm.

Printed by J.C. and A.T. Savell Printers.

A good example of early systematic work

[295] Balch, Herbert Ernest *The Mendip Caves.* Bristol, GB: John Wright & Sons, **1948**. 366p.

One version retains original pagination and includes a third edition of the first book and second editions of the latter two books.

A single volume of the of the 3 Mendip books: Mendip, the great cave of Wookey Hole; Mendip-Cheddar, its gorge; and Mendip—its Swallet Caves and rock shelters. Each was reprinted a number of times, any of these being bound together in any combination and also being available as separate titles. The work concerns the history and exploration of the caves since c.1900, as well as local history; a valuable resource for Mendip cave studies. It is a measure of Balchs' work that, decades after its first publication, it is still being reprinted.

[296] Balch, Herbert Ernest *Mendip, its swallet caves and rock shelters.* 1st ed. Wells, GB: Clare, Son & Co, Ltd, **1937**. The Mendip Series. illus: b&w: drawings, maps, photos, tables; table folded in text; front: b&w drawing; 211p; app, index.

27 pls.

2nd ed. Bristol, GB: John Wright & Sons, Ltd, **1948**. The Mendip Series. illus: b&w: drawings, maps, photos, tables; 156p, 22p pls; index; 19 cm.

[297] Balch, Herbert Ernest *Mendip-Cheddar, its gorge and caves.* 1st ed. Wells, GB: Clare, Son, & Co, Ltd, **1935**. illus: b&w: drawings, maps, photos, tables; col front; 177p; index; 18 cm.

24 pls, 19 figs.

2nd ed. London, GB: Simpkin, Marshall, **1947**. vii, 102p; index; 18 cm.

Illus: 24 pls, 19 figs.

[298] Balch, Herbert Ernest *Mendip: the great cave of Wookey Hole*. 1st ed. Wells, GB: Clare, Son, & Co, Ltd, **1929**. illus: b&w: photos, repros; col: drawings; 126p, 30p pls; index; 19 cm.

Running title: The great cave of Wookey Hole.

More of historic than scientific interest, his observational skills are still apparent.

2nd ed. npop: np, **1932**. illus: b&w: drawings, photos; col: drawings; col front; 163p, 30p pls; index; 19 cm. 20 pls.

3rd ed. Bristol; London, GB: John Wright & Sons, Ltd; Simpkin, Marshall, Ltd, **1947**. illus: b&w: drawings, photos; col: drawings; col drawing front; vii, 108p, 24p pls; index; 19 cm.

[299] Balch, Herbert Ernest *Wookey Hole: its caves and cave dwellers*. Dawkins, William Boyd, 1838–(intro); Hassall, John (illus); Savory, J. H. (photo, illus). London, GB: Humphrey Milford, Oxford University Press, **1914**. illus: b&w: drawings, maps, photos, repros, tables; partly col: maps; xiv, 268p, 19p pls; bibl, index; 33 cm.

36 pls (including restorations), 55 figs.

This was and remains a classic, although more of historic interest than scientific.

[300] Balch, Herbert Ernest *Fourteen years at the Badger Hole*. Ashworth, H. W. W. (ed). Wells, GB: Wells Natural History and Archaeological Society, c**1971**. illus: b&w: maps, photos, ports, repros, tables; [ii], 24p, [4], 1p pls; 25.5 x 20 cm.

Account of archeological excavations in the cave 1938-1952.

[301] Balcombe, F. Graham (comp, ed) *Cave diving: the Cave Diving Group manual*. Cave Diving Group; Cordingley, John N.; Palmer, R. J.; Stevenson, R. A. (comp, ed); Bedford, Bruco L. (publishing, ed). Castle Cary, GB: Mendip Publishing, **1990**. illus: b&w: charts, drawings, graphs, maps, photos, tables; col: charts, photos; 268p, 8p pls; bibl, gloss, index; 22 cm.

An important manual for the advancement of cave diving techniques.

[302] Balcombe, F. Graham *Ireland 1936 the record of the party of S.J. Pick (Leicester) in County Clare, Easter 1936*. npop: np, **1936**. 13p; 21.5 x 17 cm.

No illus.

[303] Balcombe, F. Graham *Ireland 1937 the record of S.J. Pick (Leicester) in County Clare May 1937*. [Sunninghill Ascot], GB: np, **1937**. illus: b&w: ads; 13p, [2]; 21.5 x 17 cm.

[304] Balcombe, F. Graham *The Log of the Wookey Hole: exploration expedition 1935*. Powell, "Mossy"; Harris, "Digger"; Frost, Frank; Bufton, Bill; Tucknott, Bill. GB: F. G. Balcombe, **[1936]**. illus: b&w: maps, photos; v, 235p.

Forward by H. E. Balch.

A limited edition report of one of the first organized push dive expeditions in which the divers wore suits with copper helmets and had air pumped to them. Illustrated with original tipped in photographs. Mimeographed in Balcombe's basement.

[305] Ball, J. W. *WATEQ4F: a personal computer FORTRAN transformation of the geochemical model WATEQ2 with revised data base*. Nordstrom, Darrell Kirk; Zachman, Dieter W. Denver, CO, US: US Geological Survey, **1987**. Open File Reports no. 87-50. iii, 108p; bibl; 28 cm.

WATEQ is a handy and robust program for calculating mineral equilibria given physical and chemical input data: it is particularly well suited to cave-related studies. References cite other material useful in speleology.

[306] Ballard, Jim *Guide to the sporting caves, potholes and mines of Derbyshire*. Manchester, GB: Cicerone Press, **1974**.

[307] Ballard, Jim *The spur book of caving*. Bourne End, Bocking Hampshire, GB: Spurbooks, Ltd, **1978**. A Spurbook Venture Guide. illus: b&w: drawings, maps; 64p; bibl, gloss; 19 cm.

An extensively distributed public press booklet sold everywhere in Britain as part of a series. Helped popularize British caving.

[308] Ballard, Robert F. *Cavity detection and delineation research: report 5: electromagnetic (radar) techniques applied to cavity detection*. United States Army Corps of Engineers Geotechnical Laboratory. Vicksburg, MS: U.S. Army Engineer Waterways Experiment Station, Jul, **1983**. Technical report GL-83-1 Report 5 (U.S. Army Engineer Waterways Experiment Station). illus: b&w: charts, drawings, graphs, photos, tables; [ii], 90p, [16]; app, bibl; 27 cm.

Cover title: "July 1983. Prepared for Office, Chief of Engineers".

[309] Balogh, K. *Karst hydrological and speleological features*. Et al. (20 other authors); Tádal, Ödön (trans); Maucha, László (ed). Budapest, HU: 10th International Congress of Speleology, **1989**. Field Trip Guide D1. illus: b&w: charts, graphs, maps; 76p, 56p pls.

[310] Banbury, Jack *Proceedings of the fifth annual seminar on cave diving*. Miami, FL, US: National Association for Cave Diving, **1972**. v, 120p; bibl; 28 cm.

Held in Gainesville, Fl, December 1972. Illus.

[311] Bancroft, Hubert Howe, 1832-1918 *Manitou Grand Caverns*. npop: np, **1886**.

Discovered in 1881, opened to visitors in 1885. 1 portfolio. partial microfilm reel (6 exposures).

[312] Bandi, Hans-Georg (ed) *The art of the stone age: forty thousand years of rock art*. Koep, Ann E. (trans). New York, NY, US: Crown Publishers, Inc, **1961**. Out of the World. illus: b&w: charts, drawings, maps; col: photos; 249p; app, bibl, gloss, index.

Cover title: The art of the stone age. Col photos tipped

in.

[313] Bannerman, Jackie (ed staff) *Universal study guide for use with basic orientation and basic team member courses.* Fregeau, Annette; Hempel, John Charles, 1949–; Sullivan, Michele (ed staff). Dailey, WV, US: Eastern Region. National Cave Rescue Commission, **1993**. illus: b&w: charts, maps, photos, tables; Various pagination, 300p; app, gloss.

Cover title: Universal study guide for cave rescue training: a student guide for cave rescue courses.

[314] Banta, Arthur Mangun, 1877– *The fauna of Mayfield's Cave*. Washington, DC, US: Carnegie Institution of Washington, Sep, **1907**. Carnegie Institution of Washington Publication no 67. illus: b&w: ads, drawings, graphs, maps, photos, tables; front: b&w photo; 114p; bibl.

[315] Banta, Arthur Mangun, 1877– *The life history of the cave salamander, Spelerpes maculicaudus (Cope)*. McAtee, Waldo Lee, 1883-1962. npop: np, Apr, **1906**. Proceedings of the United States National Museum v 30. 17p; 24 cm.

pp 67-83.

[316] Bar-Adon, Pesah, 1907– *The cave of the treasure: the finds from the caves in Nabal Mishmar*. Pommerantz, Inna (trans). Jerusalem, IL: Israel Exploration Society, **1980**. Judean Desert Studies. illus: b&w: maps; 243p; bibl; 32 cm.

Illus. On 4th leaf preceding t.p.: The Israel Exploration Society. The Institute of Archaeology, the Hebrew University of Jerusalem. The Department of Antiquities and Museums. Ministry of Education and Culture. Translated from the Hebrew by Inna Pommerantz.

[317] Bar-Yosef, Ofer *A cave in the desert, Nahal Hemar: 9,000-year-old finds: [exhibition, the Israel Museum, Jerusalem, spring 1985]*. Muzeon Yisrael (Jerusalem). Jerusalem, IL: The Museum, **1985**. Cat / Israel Museum Jerusalem no 258. illus: b&w: maps; 31p, 29p pls; 18 x 21 cm.

English and Hebrew. Pagination listed as 16,15, [29]p. Illus (some col).

[318] Bar-Yosef, Ofer *Nahal Hemar Cave*. Alon, David; Schick, Tamar. Jerusalem, IL: Department of Antiquities and Museums, Ministry of Education and Culture, Israel Exploration Society, **1988**. Atiqot English Series v 18. 81p, 14p pls; bibl; 27 cm.

Illus (some col).

[319] Bard, James C. *Ezra's Retreat: a rockshelter/cave occupation site in the north central Great Basin*. Busby, Colin I.; Kobori, Larry S.; Spencer, Lee. Davis, CA, US: University of California, **1979**. Publication, Center for Archaeological Research at Davis no 6. illus: b&w: maps, photos; xi, 255p; bibl; 22 cm.

Illus.

Some interesting material from an archeological and paleontological/paleoenvironmental standpoint

[320] Bard, James C. *Test excavations at Painted Cave, Pershing County, Nevada: for Bureau of Land Management, Winnemucca District Office*. Busby, Colin I.; Kobori, Larry S.; Markoe, Glenn, 1951–; Basin Research Associates. Reno, NV, US: United States Bureau of Land Management Winnemucca District, **1980**. Technical Contributions to the Study of Cultural Resources technical report no 5. v, 153p; bibl; 28 cm.

Illus.

[321] B'ardossy, György *Karst bauxites: bauxite deposits on carbonate rocks*. Amsterdam, CA: Elsevier Scientific Publishing Co., **1982**. Developments in Economic Geology 14. illus: b&w: charts, drawings, graphs, maps, photos, photomicrographs, tables; col: photos; Pocket contains appendices—14 folded leaves plus 5 folded leaves of illustration (part colored); 441p, 8p pls; bibl, index.

180 figs, 52 tables.

This is an important text for those interested in these deposits, an integral yet remote part of karst phenomena. It presents a wealth of descriptive data, analyses and theory: quantitative analysis is stresses. Although the text is densely written, good diagrams, tables, black and white and color photos enhance its utility.

[322] Baring-Gould, Sabine, 1834-1924 *Cliff castles and cave dwellings of Europe*. 1968 reprint ed. Detroit, MI, US: Singing Tree Press, **1968**. 323p; 22 cm.

1st ed. London, GB: Seeley, & Co, Limited, **1911**. illus: b&w: maps, photos; 324p, 39p pls; bibl, index; 23 cm.

All about dwellers in rock. American ed: Philadelphia: J.B. Lippincott, 1911. 51 illus.

[323] Barnett, John, 1927– *Carlsbad Caverns National Park*. 1969 ed. Fresno, CA, US: Awani Press, **1969**. illus: b&w: maps, photos; col: photos; b&w maps on lining pages; 35p; 18 cm.

Cover title: America's scenic wonderland: Carlsbad Caverns National Park

1971 ed. Roswell, NM, US: John Barnett in cooperation with the Carlsbad Caverns Natural History Association, **1971**. illus: b&w: drawings, photos; col: photos; 32p; 28 cm.

Large softbound booklet with 41 color and 1 b&w photo. Some photos are the same as in the smaller book by the same author. 1973: 33p

[324] Barnett, John, 1927— *Carlsbad Caverns National Park, Carlsbad, New Mexico: silent chambers, timeless beauty*. [Carlsbad, NM, US]: Carlsbad Caverns Natural History Association, **1981**. illus: b&w: photos; partly col: drawings; col: photos; one folded plate; 29p, [4]; 28 cm.

Cover title: Carlsbad Caverns: silent chambers, timeless beauty. 1980 edition has 29 pages.

[325] Barr, Thomas Calhoun, 1931– *The cave community*. Krekeler, Carl H., 1920-. Chicago, IL, US: Encyclopaedia Britannica Educational Corporation, **1989**. EBE Biology Program.

1 videocassette (13 min), sd, col, 1/2 in + 1 teacher's guide. Serial no. X01851. VHS format. Originally released by Encyclopaedia Britannica Educational Corporation in 1960 as a 16 mm. motion picture.

[326] Barr, Thomas Calhoun, 1931– *Caves of Tennessee.* 1972 reprint ed. **1972**. Tennessee Division of Geology Bulletin 64. illus: b&w: maps, photos; one map folded in rear pocket; vii, 567p; app, bibl, index.

150 figs. Reprint Edition:

1st ed. Nashville, TN, US: Tennessee Department of Conservation and Commerce, **1961**. Tennessee Division of Geology Bulletin 64. illus: b&w: maps, photos; one map folded in pocket; vii, 567p; 23 cm.

The major work cataloging caves in the state. Descriptions and locations of about seven hundred caves. Few maps. Introductory text includes twenty pages on biology.

[327] Barr, Thomas Calhoun, 1931– *Ecological effects of water pollutants in Mammoth Cave: final technical report to the National Park Service.* United States National Park Service. rev. Lexington, KY, US: University of Kentucky, Dec, **1976**. illus: b&w: drawings, maps; 45p; bibl; 30 cm.

Contract no. CX50050204.

[328] Barr, Thomas Calhoun, 1931– *The geology of Cumberland Caverns, Warren County, Tennessee.* Nashville, TN, US: Tennessee Academy of Science, Geology and Geography Section, **1956**. Annual Spring Field Trip, Geology and Geography Section Tennessee Academy of Science 1956. illus: b&w: maps; ii, 10p; 28 cm.

[329] Barr, Thomas Calhoun, 1931– *Symposium: speciation and raciation in cavernicoles.* Causey, Nell B.; Chamberlin, Joseph C.; Christiansen, Kenneth; Goodnight, Clarence J.; Goodnight, Marie L; Hobbs, Horton Holcombe, 1914–; Hyman, Libbie H.; Malcolm, David R.; Sullivan, Gerardus Nicholas, 1927–[Brother Nicholas]; Park, Orlando. South Bend, IN, US: University of Notre Dame, **1960**. illus: b&w: maps, tables; 160p; app, bibl; 23 cm.

Presented at the 1959 Annual Meeting of the American Association for the Advancement of Science, December 28, sponsored by the National Speleological Society and the Society of Systematic Zoology. Arranged by Thomas C. Barr, Jr. Reprinted from the American Midland naturalist, Vol. 64, No. 1, pp. 1-160, July, 1960.

A symposium on the discussion on nearly all major groups of cave fauna in relation to modern ideas about speciation. This symposium heralds the emergence of cave biology after decades of neglect.

[330] Barrett, Charles Leslie, 1879– *Australian caves, cliffs, and waterfalls.* Melbourne, AU: Georgian House, **1944**. illus: b&w: photos; front: b&w photo; 126p, [1]; 24 x 18 cm.

A curious assortment of outcrops and incrops are presented, from which some geologic observations are possible.

[331] Barrington, Nicholas Robert *Adventure underground.* Leicester, GB: Robert Maxwell At Pergamon Press, **1968**. Take Home Books. illus: b&w: photos; front: b&w photo of G. B. Cavern Somerset; 15p.

No title page. Title from cover.

[332] Barrington, Nicholas Robert (comp) *The caves of Mendip.* 1962 rev ed. **1962**. illus: b&w: maps; 81p; bibl, gloss.

1964 rev ed. 1964. 91p.

1st ed. Clapham via Lancaster, Yorkshire, GB: The Dalesman Publishing Company Ltd, **1957**. illus: b&w: maps; 75p; bibl, gloss; 19 x 13 cm.

[333] Barrington, Nicholas Robert *The complete caves of Mendip.* Stanton, William. 1st ed. Cheddar, Somerset, GB: Barton Productions in conjunction with Cheddar Valley Press, **1970**. illus: b&w: ads, maps, photos; text on lining pages ; front: b&w photo; 131p; bibl, gloss; 19 cm.

2nd rev ed. Stanton, William. **1972**. illus: b&w: maps; 157p; gloss; 19 cm.

[334] Barrington, Nicholas Robert *Mendip: the complete caves and a view of the hills.* Stanton, William. 3rd rev ed. Cheddar, Somerset, GB: Barton Productions in conjunction with Cheddar Valley Press, **1977**. illus: b&w: ads, maps, photos; 236p; bibl, gloss; 19 cm.

1st and 2nd eds published under title: The complete caves of Mendip.

[335] Bartel Photography Pty Ltd *Jenolan Caves New South Wales Australia.* 3rd ed. Bondi Junction, NSW, AU: Bartel Photography Pty Ltd, **1985**. illus: partly col: maps; col: photos; 1 col photo with multiple foldings in text; text on front lining page; map on rear lining page; [20].

Minimal text.

[336] Bartel Postcards *Jenolan Caves Australia.* 6th ed. Waterloo, NSW, AU: Bartel Postcards, **1992**. illus: b&w: maps; col: photos; text on front lining page; map on rear lining page; [16].

Cover title: Jenolan Caves Australia: 28 page pictorial book with a 4 page centerfold. Limited text.

[337] Barton, Nicholas James, 1935– *The lost rivers of London: a study of their effects upon London and Londoners, and the effects of London and Londoners upon them.* 1992 rev ed. London, GB: Historical Publications, **1992**. illus: b&w: maps, tables; 167p; bibl; 26 cm.

Illus.

1st ed. London, GB: Phoenix House, **1962**. illus: b&w: drawings, maps, photos; 1 map folded in rear; front: b&w drawing; 148p, 14p pls; app, bibl, index; 23 cm.

Second and third printings, 1962; fourth printing 1965. Reissued? 1985: New Barnett, Herts [GB]: Historical Publications Ltd.

[338] Bartrop, Richard N. *Cave Diving Group: Derbyshire sump index 1987.* [Bristol, GB]: The Cave Group Diving, **1987**. illus: b&w: maps; ii, 48p; index; 30 x 21 cm.

[339] Bassham, Elbert F. *Cave surveying techniques.* Green, John W. (ed). El Paso, TX, US: El Paso Archaeological Society, **1971**. El Paso Archaeological Society Handbook Series no 1. illus: b&w: drawings, maps; iii, 12p, [4]; bibl; 29 cm.

[340] Bassie-Sweet, Karen *From the mouth of the dark cave: commemorative sculpture of the late classic Maya.* Norman and London, OK, US: University of Oklahoma Press, **1991**. illus: b&w: charts, drawings, maps, photos, tables; front: b&w photo; xv, 287p; app, bibl, index.

[341] Basso, Dave *2,000-year-old duck decoys from Lovelock Cave, Nevada.* Loud, Llewellyn Lemont, 1879-1946; Harrington, Mark Raymond, 1882–1971; Napton, Lewis K.; Tuohy, Donald R. Sparks, NV, US: Falcon Hill Press, **1992**. illus: b&w: ports; 15p; 27 cm.

Illus.

[342] Bataille, Georges, 1897-1962 *Prehistoric painting: Lascaux, or, the birth of art.* Wainhouse, Austryn (trans). Lausanee, FR: Skira, **1955**. The Great Centuries of Painting. illus: b&w: drawings, maps, photos, tables; col: drawings; 149p, [3]; 29 cm.

68 col pls. No date in front. 1955 listed in back. Distributed in the United States by the World Publishing Company, Cleveland, OH. Reprinted by Macmillan (London), 1980?

[343] Bates, Edward R. *Detection of subsurface cavities.* Vicksburg, MS, US: US Army Engineer waterways experiment station, Jun, **1973**. US Army Engineer Waterway Experiment Station Miscellaneous paper S 73 40. illus: b&w: drawings, graphs, maps, photos; xvii, 63p, [18]; bibl, gloss.

Study done on Indiana caves. 27 figs.

[344] Bates, Geoff *The Abercrombie Caves, New South Wales: a guide to these remarkable caves near Bathurst, N.S.W. and a description of the surrounding gold bearing country of Tuena and Trunkey Creek.* Waramanga, ACT, AU: Geoff Bates, **1982**. illus: b&w: drawings, maps, photos; 48p; 25 cm.

Extensive references are made to this Australian cave's history, including a useful summary and surveys. Some of the descriptions of the surrounding area are just as interesting as those of the caves.

[345] Bathurst, R. C. G. *Carbonate sediments and their diagenesis.* 2nd ed. New York, US: Elsevier, **1972**. illus: b&w: charts, drawings, graphs, maps, photos, photomicrographs, tables; app, bibl, index.

This is a must for anyone seriously interested in carbonate rocks, their origin, description, geochemistry, dissolution and repreciptation as speleothems. It remains a classic and is unlikely to be superseded in the near future.

[346] Bauer, Ernst W. *The mysterious world of caves.* London, GB: Collins Publishers Franklin Watts, Inc, **1971**. International Library. illus: b&w: drawings, graphs, photos; col: maps, photos; col photo and map on front and rear lining pages respectively; 129p; bibl, gloss, index; 26 cm.

A popular press book; good information on underground water and submarine springs.

[347] Baumann, Hans, 1914– *The caves of the great hunters.* 1962 ed. London, GB: Hutchinson & Co (Publishers) Ltd, **1962**. illus: b&w: drawings, maps; partly col: drawings; col: photos; 1 partly col drawing on lining pages; [v], 183p, [1], 20p pls; app; 22 cm.

Second impression has no date. 37 col photos.

1st ed. Renner, Hans Peter (illus); McHugh Pietsch, Isabel (trans); McHugh Pietsch, Florence (trans); Burges, L. (photo); Burges, H. (photo). London, GB: Hutchinson & Co (Publishers) Ltd, **1954**. illus: b&w: drawings, maps, photos; partly col: drawings; map with drawings on lining pages; 158p, [2]; 22 cm.

Translation of Höhlen der grössen Jäger. Issued in the US by Pantheon Books, [New York]. Cover subtitle: How four boys discover an ice age cave - the cradle of man's art.

Describes the discoveries and dramatic drawings of Europe's prehistoric caves—Lascaux, Altamira, Pindal, El Castillo, Niaux, Valltorta Gorge and the Gasulla Gorge. Caves such as these, and the papers which deal with them, clearly bridge the boundary between paleontology, archeology, anthropology and geology.

[348] Baumgardner, Robert W. *Formation of the Wink Sink: a salt dissolution and collapse feature, Winkler County, Texas.* Hoadley, Ann D.; Goldstein, Arthur G. Austin, TX, US: Bureau of Economic Geology, University of Texas at Austin, **1982**. Report of investigations Bureau of Economic Geology, University of Texas at Austin no 114. illus: b&w: drawings, maps, photos, ports; iv, 38p; bibl; 28 cm.

Illus.

An informative, well illustrated case study of deep subsurface evaporite dissolution and collapse - one of the few books dealing with this subject.

[349] Baumgartel, Elise J., 1892– *The cave of Manaccora, Monte Gargano.* New York, NY, US: Johnson, **1971**. Papers of the British School at Rome v 19 (new ser v 6) [no 2] v 21 (new ser v 8) [no 1]. bibl; 25 cm.

Illus. 2 vols: pt. 1. The site.—pt. 2. The contents of the three archaeological strata. OCLC lists both 1951 and 1971 as pub date.

[350] Bayless, Edward Randall, 1961- *Directions of ground-water flow and locations of ground-water divides in the Lost River watershed near Orleans, Indiana.* Taylor, Charles J.; Hopkins, Mark S. Indianapolis, IN, US: United States Geological Survey, **1994**. Water-resources investigations report 94-4195. illus: b&w: maps; Two large folded separate maps.; [iv], 25p; app, bibl; 28 cm.

Prepared in cooperation with the U.S. Army Corps of Engineers.

[351] Baynes, F. J. *Ida Bay - Benders Quarry. An estimate of reserves within existing quarry limits*. Hobart, TAS, AU: Department of Parks, Wildlife, and Heritage, **1991**. Ida Bay Karst Study Report to the Department of Parks, Wildlife, and Heritage. 30p.

A look at the quarry's limestone reserves and its effect on the local caves.

[352] Baysinger, David (narr, videographer) *Silent Splendor*. Denver Museum of Natural History Audio/Video Department. [Denver, CO, US]: Denver Museum of Natural History, **1986**.

1 videocassette (67 min), sd, col, 1/2 in. Title at end of program: Silent Splendor extension.

Presents the story of Cave of the Winds near Colorado Springs, Colorado, and the newly-discovered cave named Silent Splendor. The extension (40 min.) consists of interviews with geologists and caving experts.

[353] Baysinger, David. *Lechuguilla cave: the hidden giant*. Gilmore, Elizabeth; Denver Museum of Natural History. Denver, CO, US: Denver Museum of Natural History, **1988**.

1 videocassette (29 min), sd, col, 1/2 in. VHS format. Two version were produced; one was a "caver version" and is longer.

Join 20th century explorers as they enter the realm of a wild cave below the desert of southern New Mexico in Carlsbad Caverns National Park.

[354] Beck, Barry Frederic, 1944– *Applied karst geology: proceedings of the fourth Multidisciplinary Conference on Sinkholes and Engineering and Environmental Impacts of Karst, Panama City, Florida 25-27 January 1993*. Rotterdam, NL: A. A. Balkema, **1993**. illus: b&w: charts, drawings, graphs, maps, photos, tables; vii, 295p; bibl, index; 29 cm.

Sponsored by the Division of Sponsored Research at the University of Central Florida, Orlando, Florida and co-sponsored by P. E. LaMoreaux & Associates, Inc, Tuscaloosa, AL.

See also the proceedings of the other Sinkhole Conferences.

[355] Beck, Barry Frederic, 1944– *A computer based inventory of recorded recent sinkholes in Florida*. Orlando, FL, US: Florida Sinkhole Research Institute, College of Engineering, University of Central Florida, Jul, **1984**. STAR 83-019. 12p.

This marks the beginning of a long involvement with sinkholes

[356] Beck, Barry Frederic, 1944– *Engineering and environmental impacts of sinkholes and karst: proceedings of the third Multidisciplinary Conference on Sinkholes and the Engineering and Environmental Impacts of Karst, St. Petersburg Beach, Florida, 2-4 October 1989*. Hagen, Adrianne; Cavin, Scott; Barfus, Brian; Merkle, Virginia (asst); Florida sinkhole Research Institute. Rotterdam, NL: A. A. Balkema, **1989**. illus: b&w: charts, drawings, graphs, maps, photos, tables; viii, 384p; bibl; 29 cm.

See the other proceedings

[357] Beck, Barry Frederic, 1944– *An introduction to caves and cave exploring in Georgia*. Atlanta, GA, US: Georgia Geologic Survey, **1980**. Geologic Guide 5. illus: b&w: cartoons, drawings, maps, photos; b&w front; iv, 43p; bibl; 23 cm.

34 figs.

An attempt to fill the need for caving information in Georgia. Purposely written with no cave names or locations. An educational booklet, not a guidebook or catalog.

[358] Beck, Barry Frederic, 1944– *Karst geohazards: engineering and environmental problems in karst terrane: proceedings of the fifth Multidisciplinary Conference on Sinkholes and the Engineering and Environmental Impact of Karst, Gatlinburg, Tennessee, 2-5 April 1995*. Pearson, Felicity M.; P. E. LaMoreaux & Associates; National Groundwater Association [U. S.]. Rotterdam, NL: A.A. Balkema, **1995**. ix, 581p; bibl, index; 29 cm.

Illus. U.S. edition published in Brookfield, VT.

These proceedings and those of the earlier conferences (4th -1993; 3rd -1989; 2nd - 1987; 1st - 1984) present a wealth of geologic, environmental and engineering data and interpretations dealing with sinkholes. Although most papers deal with US examples, there is a large number of foreign studies and references.

[359] Beck, Barry Frederic, 1944– *Karst hydrogeology of central and northern Florida: a field trip guidebook produced in conjunction with the 1985 G.S.A. annual meeting and exposition, Orlando, FL., October, 1985*. Orlando, FL, US: Florida Sinkhole Research Institute, College of Engineering, University of Central Florida, Oct, **1985**. Report, Florida Sinkhole Research Institute 85-86-1. 46p; bibl; 28 cm.

Illus.

A good guide to this terrific area.

[360] Beck, Barry Frederic, 1944– *Karst hydrogeology: engineering and environmental applications: proceedings of the second Multidisciplinary Conference on Sinkholes and the Environmental Impacts of Karst, Orlando, Florida 9-11 February 1987*. Wilson, William L. (ed). Rotterdam, NL: A.A. Balkema, **1987**. illus: b&w: drawings, graphs, maps, photos, tables; viii, 467p; bibl; 29 cm.

A very important source for techniques of applied karst hydrology. Readers should review the other conference proceedings.

[361] Beck, Barry Frederic, 1944– *The sinkhole hazard in Pinellas County: a geological summary for planning purposes*. Sayed, Sayed. [Orlando, FL, US]: Florida Sinkhole Research Institute in cooperation with the Pinellas County Dept of Public Works, **1991**. Report, Florida sinkhole Research Institute, University of Central Florida 90-91-1. 139p; 28 cm.

Illus.

Environmental hazards are the focus of this one. Dr Beck clearly distinguishes himself as "Mr Sinkhole" with this and his other work.

[362] Beck, Barry Frederic, 1944– *Sinkholes in Florida: an introduction*. Sinclair, William Campbell, 1928–. Orlando, FL, US: Florida Sinkhole Research Institute, College of Engineering, University of Central Florida in cooperation with the United States Geological Survey, **1986**. Report, Florida Sinkhole Research Institute no 85-86-4. iv, 16p; 28 cm.

Illus.

This provides a good summary of sinkholes, their hydrology, geology and environmental aspects in this, one of the most sinkhole pocked areas in North America.

[363] Beck, Barry Frederic, 1944– *Sinkholes: their geology, engineering and environmental impact, proceedings of the first Multidisciplinary Conference on Sinkholes, Orlando, Florida, 15-17 October 1984*. Florida Sinkhole Research Institute, College of Engineering, Univ of Central Florida. Rotterdam, NL: A A Balkema, **1984**. illus: b&w: charts, drawings, graphs, maps, photos, tables; x, 429p; bibl; 29 cm.

See other conference proceedings.

[364] Beck, Barry Frederic, 1944– *Use of ground penetrating radar for detecting and evaluating the sinkhole hazard in Florida*. Orlando, FL, US: Florida Sinkhole Research Institute, College of Engineering, University of Central Florida, **1987**. Report, Florida Sinkhole Research Institute no 87-88-3. 94p; 28 cm.

Illus.

[365] Beck, Barry Frederic, 1944– *Water on and under the ground: an introduction to the urban hydrogeology of the Orlando area*. [Orlando, FL, US]: Florida Sinkhole Research Institute, University of Central Florida, **1985**. Report, Florida Sinkhole Research Institute 85-86-3. 23p; 28 cm.

1985 annual meeting and exposition of the Geological Society of America, October 28-31, 1985.

An interesting case study: one wonders whether it changed the long term urban-planning in the area.

[366] Beck, Howard M. *Gaping Gill: 150 years of exploration*. London, GB: Robert Hale, **1984**. illus: b&w: drawings, maps, photos; 190p, 24p pls; app, bibl, gloss, index; 23 cm.

A detailed account of Gaping Gill exploration from the 1800s to diving attempts in the early 1980s. Gaping Gill's first descent in 1895 is taken by some cavers to mark the beginning of British sport caving, although much work was done before this date.

[367] Beckway, Gregory W. (ed) *Cavern formation*. Northbrook, IL, US: Hubbard Scientific Co., **1971**.

A guide attached inside container. 1 filmloop (4 min), si, col, super 8 mm.

[368] Bedford, Bruce (ed) *Descent handbook for cavers*. Kent, GB: np, **1970**. illus: b&w: ads; 28p; 21 x 14.5 cm.

[369] Bedford, Bruce L., 1942– *Challenge underground*. London, GB: George Allen & Unwin, Ltd, **1975**. illus: b&w: maps; 192p, 6p pls; gloss, index; 23 cm.

Imprint covered by label which reads: Zephyrus Press, Teaneck [N.J.]. Illus. Whether a separate reprint edition by Zephyrus Press exists is unclear.

[370] Bedford, Bruce L., 1942– *Descent handbook for cavers 1971/1972*. Kent, GB: np, **1971**. illus: b&w: ads; 34p; 21 x 14.5 cm.

[371] Bedford, Bruce L., 1942– *The Descent magazine's caving yearbook*. Gloucester, GB: Ambit Publications Ltd, **1987**. illus: b&w: ads, photos; 106p, [5]; 20.5 x 14.5 cm.

Cover title: The Descent magazine's caving yearbook 1987/88

[372] Bedford, Bruce L., 1942– *Underground Britain: a guide to the wild caves and show caves of England, Scotland and Wales*. London, GB: Willow Books, William Collins Sons & Co, Ltd, **1985**. illus: b&w: drawings, maps, photos; col: photos; col front; bibl, gloss, index; 27 cm.

[373] Behrens, Georg Henning, 1662-1712 *The natural history of Hartz-Forest, in His Majesty King George's German dominions. Being a succinct account of the caverns, lakes, springs, rivers, mountains, rocks, quarries, fossils, castles, gardens, the famous pagan idol Pustrich or spit-fire, dwarf-holes, etc....in the said forest: with several useful and entertaining physical observations*. Andree, John, 1699?–1785 (trans). London, GB: T. Osborne, **1730**. illus: b&w: ads; [xvi], 164p, [12]; index; 19 x 12 cm.

Approximately one-third cave-related. Printed by W. Pearson. Translation by John Andree of: *Hercynia curiosa*, first published at Nordhausen, 1703. Advertisements at end. Microfilm edition: New Haven, Conn: Research Publications, [1981].

[374] Beldam, Joseph, 1795-1866 *The origin and use of the Royston Cave, being the substance of a report some time since presented to the Royal Society of Antiquaries by the late Joseph Beldam ....* 1st ed. Royston, GB: John Warren, **1858**. 56p; 21 cm.

2nd ed. Royston, GB: John Warren, **1877**. 49p; 21.4 cm.

3rd ed. Royston, GB: J. Brothers, **1884**. 52p, [2]; 18 cm.

Cave may be man-made.

4th ed. Royston, GB: Warren Brothers, **1898**. x, 51p; 18.5 cm.

5th ed. Warren Brothers, Jun, **1904**. illus: b&w: drawings; viii, 51p, 4p pls.

[375] **Bell, Alan**, 1918– *The witch of Wookey Hole.* Wookey Hole, Somerset, GB: G.W. Hodgkinson, **1928**. illus: b&w: drawings, maps; partly col: photos; front: b&w map; 45p, [1]; 19 cm.

Cover title: Wookey Hole: the cave & its history: A description and history of the three great caverns, their ancient occupation and the legend of the Witch of Wookey. [4]p booklet tipped in at end.

[376] **Bell, Anne (comp)** *South Wales Caving Club Newsletter: index to numbers 1 - 100: 1946 - 1985.* Peny-cae, Swansea, GB: South Wales Caving Club, [**1986**]. [ii], 45p; 30 x 21 cm.

No illus.

[377] **Bell, Peter (ed)** *The 9th ACKMA conference proceedings; Margaret River, WA, Australia.* Carlton, VI, AU: Australasian Cave and Karst Management Association, **1991**. 91p, [4], 2p pls.

23 articles on cave management and protection in Australasia.

[378] **Belski, David Stanley**, 1937– *GYPKAP 1987 annual report.* npop: Southwestern Region. National Speleological Society, **1992**. illus: b&w: charts, maps, photos, tables; 6 loose folded map pls; text on lining pages; 56p; bibl.

[379] **Bemrose, Henry Howe** *The caves in Derbyshire.* Nottingham, GB: University of Nottingham, **1934**. Nottingham Eng University Abbott Memorial Lecture 1934. [ii], 13p; 14 x 21.5 cm.

On the cover is printed "Delivered by Alderman Henry H. Bemrose." Title from cover.

[380] **Ben-Tor, Amnon** *Two burial caves of the Proto-Urban Period at Azor, 1971: the first season of excavations at Tell-Yarmuth, 1970.* [Jerusalem], IL: Institute of Archaeology, Hebrew University of Jerusalem, **1975**. Qedem 1. 86p, 17p pls; 29 cm.

Illus.

[381] **Bender, Lionel** *Cave.* New York, NY, US: Franklin Watts, **1989**. The Story of the Earth Series. illus: col: drawings, photos; 32p; gloss, index; 30 cm.

Juvenile.

"A geological explanation of the reasons and ways in which caves are formed with a discussion of their uses throughout history."

[382] **Benedict, Ellen Maring** *Cave ecology: a course book for the cave ecology class at Malheur Environmental Field Station.* npop: np, **1974**.

Limited edition. "...over 150 pages of reprints of cave maps, cave research reports, cave descriptions and technical articles, mainly about southeastern Oregon caves."

[383] **Benke, David** *Using trigonometric functions in cave surveying.* TI-IN Network (Television Station: San Antonio, Tex.). npop: np, **1989**.

Reproduction of an off-the-air recording. A teleference broadcast over the TI-IN Network, January 16, 1989.

1 videorecording (55 min), sd, col, 1/2 in.

[384] **Benn, David W.**, 1948– *Hadfields Cave: a perspective on Late Woodland culture in northeastern Iowa.* Iowa City, IA, US: Office of the State Archaeologist, University of Iowa, **1980**. Report Office of the State Archaeologist 13. xii, 238p; bibl.

Illus. Microfiche (2 sheets (196p, 11 x 15 cm) in pocket).

[385] **Bennett, R.** *Bristol Exploration Club caving report no.13 [St. Cuthbert's Report]: part F: Gour Hall area.* Irwin, David J.; Webster, M. (photo). Somerset, GB: Bristol Exploration Club, Apr, **1969**. illus: b&w: maps, photos, tables; 1 fold-out map; 14p, 3p pls; 25 x 20 cm.

[386] **Bennett, R. (Cerberus series)** *Cerberus series: Maypole series.* Irwin, David J. (Cerberus Series); Stenner, R. D. (Maypole Series); King, R. S. (Maypole Series). [GB]: Bristol Exploration Club, Oct, **1982**. Caving Report no 13 [St. Cuthbert's Report] part G. illus: b&w: photos; 2 maps folded in rear, loose; 16p, 5p pls; 26 x 20 cm.

[387] **Bennett, R. H.** *Balague '70.* Wells, Somerset, GB: Bristol Exploration Club, **1973**. Caving Report no 14. illus: b&w: drawings, maps; folded map loose in rear; [i], 11p, 4p pls; 25 x 21 cm.

[388] **Benson, Richard C.** *Application of radar and seismic reflection techniques to cavity detection, Medford Cave site, Florida.* Glaccum, Robert A.; Technos [Firm]; United States Army Engineer Waterways Experiment Station. [Miami, FL, US]: Technos, Inc, **1979**. Technos T 79 053. illus: col: photos; 3 folded maps; Various pagination; 28 cm.

[389] **Bent, James Theodore**, 1852-1897 *The ruined cities of Mashonaland; being a record of excavation and exploration in 1891.* Swan, Robert McNair Wilson, 1858-1904. London, GB: Longmans, Green and Co, **1892**. illus: b&w: maps, ports; maps partly folded; xi, 376p; 23 cm.

One source reports useful accounts of caves and rock paintings in Rhodesia (3rd ed). Not examined.

[390] **Benthall, Joseph L.** *Daugherty's Cave: a stratified site in Russell County, Virginia.* [Richmond, VA, US]: Archaeological Society of Virginia. Special Publication Archaeological Society of Virginia no 18. vi, 99p; bibl; 28 cm.

Illus.

[391] **Berenguer, Magin** *Prehistoric cave art in northern Spain, Asturias.* Hinds, Henry (trans); Arias de la Canal, Fredo (pref). Ciudad de Mexico, MX: Frente de Afirmacion Hispanista, **1994**. 286p; bibl, index; 30 cm.

153 illus (some col). Spainish title: Arte prehistorico en cuevas del norte de Espana Asturias. English and Spanish.

[392] **Berenguer, Magin** *Prehistoric man and his art: the caves of Ribadesella.* Heron, Michael (trans). 1973 ed. London, GB: Souvenir Press, **1973**. illus: b&w: drawings, maps, photos; 1 foldout b&w drawing; 168p, 1p pls; bibl.

1975 ed. Park Ridge, NJ, US: Noyes Press, **1975**. 1 folded leaf of pls; 168p, 1p pls; bibl; 23 cm.

Illus.

[393] Bernabei, Tullio *Caves and stories of central Asia.* De Vivo, Antonio (auth, trans). Padova, IT: Centro Editoriale Veneto, **1992**. illus: b&w: drawings, maps, photos, ports, tables; partly col: drawings, maps; col: drawings, maps, photos, repros; six folded maps (some b&w, some col) in text; one b&w, gray scale topo map folded in rear pocket; front: col photo; 307p, [2].

Complete English translation in back. Main body in Italian: Grotte e storie dell'Asia Centrale: le esplorazioni geografiche del Progetto Samarcanda

[394] Bernasconi, R. *The physico-chemical evolution of moonmilk.* Mansker, William L. (trans). npop: np, **1976**. Cave geology v 1 (3). 26p; bibl; 28 cm.

Illus, pp. 63-88.

Translated from the French: *Memoir V of Rassegna Speleologica Italiana*; Proceedings of the international symposium on speleology held at Varenna, Italy; p. 75-100, 1961

[395] Berthelsen, O. J. *Guide to cavern engineering.* Hong Kong. Geotechnical Control Office; Berdal Stromme Consulting Engineers. Hong Kong, HK, Mar, **1992**. Geoguide 4. 156p; bibl; 30 cm.

Illus.

[396] BFA Educational Media *Caverns and geysers.* Santa Monica, CA, US: BFA Educational Media, **1961**.

1 film reel (14 min), sd, b&w, 16 mm.

[397] Biddle, Roger (ed) *Collected caving songs: volume 1: the songs of Mendip.* Edinburgh, GB: Grampian Speleological Group, Apr, **1976**. The Grampian Speleological Group Occasional Publications no 3. illus: b&w: cartoons, drawings; vii, 35p; 30 x 21 cm.

Cover title: the caving songs of Mendip. Printed one side of paper only. Reprinted 1990 with 8 pref p and 42pp.

[398] Bigwood, K. (photo) *Waitomo Caves Wonderland.* Wellington, NZ: A. H. & A. W. Reed, c**1980**. Reed's Kowhai Colour Books no 9. illus: col: photos; printed text and captions on lining pages; [24]; 11.5 x 18 cm.

Other version lists Bascands Ltd, Christchurch, NZ as the publisher; otherwise they are the same. 27 col photos.

[399] Binkerd, Adam D. *The Mammoth Cave and its denizens: a complete descriptive guide.* Cincinnati, OH, US: R. Clarke, **1869**. vi, 95p; 24 cm.

Illus. Microfilm edition: Atlanta, Ga.: SOLINET, 1993. 1 microfilm reel ; 35 mm. One copy examined is dated 1868 on copyright page and has 36pp.

[400] Binkerd, Adam D. *Pictorial guide to the Mammoth Cave, Kentucky: a complete historic, descriptive and scientific account of the greatest subterranean wonder of the western world.* Cincinnati, OH, US: Press of Geo. P. Houston. 112p; 20 cm.

Illus. Microfilm version: Atlanta, Ga.: SOLINET, 1993. 1 microfilm reel ; 35 mm. Microopaque. Louisville, Ky.: Lost Cause Press, [1959?] 2 microcards. 8 x 13 cm. (Kentucky culture series, 112). OCLC also lists a 1958 version.

[401] Bircham, Deric N. *Waitomo tourist caves.* Wellington, NZ: A.H. & A.W. Reed Ltd, **1975**. illus: b&w: maps, photos, ports, repros; col: photos; text on lining pages; 30p.

[402] [Birmingham University Speleological Society] *67 expedition to Crete: BUSS.* npop: [Burmingham University Speleological Society], [**1968**]. illus: b&w: maps; 34p; 28 x 22 cm.

Pages bound with plastic holder; looks mimeographed.

[403] Bishop, Michael J. (ed) *The cave hunters: biographical sketches of the lives of Sir William Boyd Dawkins (1837-1929) and Dr. J. Wilfrid Jackson (1880-1978).* [Derbyshire, GB]: Derbyshire Museum Service, **1982**. illus: b&w: photos, ports, repros; 48p; bibl; 22 x 15 cm.

Microfiche edition: New York: New York Public Library, 1987. 1 microfiche: negative ; 11 x 15 cm.

[404] Bishop, Michael J. *The mammal fauna of the early middle Pleistocene cavern infill site of Westbury-sub-Mendip, Somerset.* London, GB: Palaeontological Association, **1982**. Special Papers in Paleontology no 28 0038 6804. 108p; bibl; 25 cm.

Illus.

[405] Bitterli, Thomas *Proyecto Cerro Rabon.* Jeannin, Pierre-Yves; Meyers, Karlin; Rouiller, Phillipe. Basel, CH: Speleo Projects, **1990**. illus: b&w: graphs, maps, photos, tables; Photo on front; map on rear.; 58p; bibl.

In English and German

[406] Black, Davidson, 1884-1934 *The human skeletal remains from the Sha Kuo T'un cave deposit in comparison with those from Yang Shao Tsun and with recent North China skeletal material.* Peking, CN: The Geological survey of China, **1925**. China Geological Survey Paleontologica Sinica ser D v 1 fasc 3. illus: b&w: tables; 148p, [18] ; 29 cm.

Last 18 pages in Chinese. 14 plates.

[407] Black, Don F. *I don't play golf: recollections of a rescue volunteer.* NY, US: Carlton Press, **1993**. 159p.

Gruesome tales of cave and surface rescue. Not for the faint of heart.

[408] Black, Jackie (ed) *Caves, ground water and karst features of the Cretaceous in South central Texas: fourteenth annual S.A.S.G.S. field conference, April 6-8, 1973.* Clark, Michael R. (ed). San Antonio, TX, US: St. Mary's University of Texas Student Geological Society, **1973**. illus: b&w: cartoons, charts, drawings, maps, photos, tables; Various pagination, x, 139p; app, bibl; 28 cm.

Cover title.

[409] Black, Jeffrey H. *The cave life of Oklahoma: a preliminary study (excluding chiroptera).* Oklahoma City,

OK, US: National Speleological Society, **1971**. Oklahoma underground: Central Oklahoma Grotto vol IV, nos 1 & 2. 56p; 28 cm.
Illus.

[410] Black, Lindsay *Aboriginal art galleries of western New South Wales*. Melbourne, VIC, AU: np, **1943**. illus: b&w: maps; 76p, [4]; 22 cm.
Printed by J. R. Stevens. Illus.

[411] Black, Nancy Gail *A comparative study of stylistic and extrapictorial elements in the murals of Cave I, Ajanta and of the Brancacci Chapel*. [Honolulu, HI, US], **1965**. University of Hawaii (Honolulu) Theses for the Degree of Master of Arts no 618. v, 312p; bibl.
Typescript. 43 mounted photos.

[412] Blackhawk Hawks *Carlsbad Caverns*. Davenport, IA, US: Eastin-Phelan Corp, **1970**.
30 slides, col, 2 x 2 in.
Photographs of New Mexico's famous Carlsbad Caverns.

[413] Blatchley, William Stanley, 1859-1940 *Gleanings from nature: ten Indiana caves and the animals that inhabit them*. npop: Indiana Karst Conservancy, **1990**. Indiana Karst Conservancy Special Publication #4. illus: b&w: drawings, maps, photos; [iii], 71p; 22 cm.
Reprint of chapter from the book Gleanings from nature by W. S. Blatchley. Indianapolis: Nature Publishing Co., 1899. Photos are photocopied. Publication date listed as both 1990 and 1992.

[414] Blatchley, William Stanley, 1859-1940 *Indiana caves and their fauna*. [Indianapolis, IN, US]: Department of Geology and Natural Resources, **1896**.
Detached from: Indiana Department of Geology and Natural Resources annual report, 1896.

[415] Block, Guy de (ed) *Bibliographie Speleologique Belge*. Fontaine, J. P. (ed). [Bruxelles, BE]: [Commission Belge de Bibliographie, Equipe Speleo de Bruxelles], **c1974**. illus: b&w: tables; 97p; 21 cm.
Softbound, continuation of bibliography. This part contains refs. nos. 663 thru 1160 plus 13 'x' items. No illus.

[416] Block, Guy de (ed) *Bibliographie Speleologique Belge: editions speleologiques belges 1975-1979*. Fontaine, J. P. (ed). Bruxelles, BE: Equipe Speleo de Bruxelles, **1981**. 147p; 21 cm.
This softbound book is the 3rd in the Belgium Biblio series. The 1st covered 1907-1964 and was published in 1968. The 2nd (1965-1969) was published in 1974. This issue contains references #1161 thru 1657 and a complete author index for all three biblios.

[417] Block, Guy de (ed) *Bibliographie Speleologique Belge. Editions speleologiques belges. 1907-1964. Avec en appendice la liste des ouvrages a caractere speleologique edites en Belgique*. Fontaine, J. P. (ed). Bruxelles, BE: Commission Belge de Bibliographie, Equipe Speleo de Bruxelles, **1965**. Bibliographia Belgica 99. illus: b&w: tables; 80p; 21 cm.

[418] Bloomgarden, Richard *The easy guide to Chichen Itza, Balankanchen and Izamal*. [16th ed, rev]. Shawnee Mission, KS, US: Editur, S.A.; Forsyth Travel Library, **1984**. illus: col: maps; 34p; bibl; 21 cm.
Title from cover. Illus (some col).

[419] Blore, J. D. *Archaeological excavation at North Face Cave, Little Ormes Head, Gwynedd 1962-1976*. Merseyside, GB: [J. D. Blore], **c1977**. illus: b&w: charts, drawings, graphs, maps; fold-out sediment chart; 24p; bibl; 22 x 17 cm.

[420] Blue Mountains Speleological Club *Policy organisation and rules*. Blacktown, NSW, AU: Outer Western Regional Council for Social Development for Blue Mountains Speleological Club, **1977**. 24p; 17 x 21 cm.
Title from cover. Illus.

[421] Bock, Hermann, 1872– *In Salzburg's Netherworld*. Montagu, Adelinae (trans). npop: np, **nd**. illus: b&w: ads, maps, photos; 1 b&w map folded in text; 19p, [26].
Translation of Der Lamprechtsofen bei Lofer which was published in Mitteilungen für Höhlenkunde vol 4, no 3, 1911. Includes [25]p original German.

[422] Bogart, Robert C. *Mark Twain Cave: an adventure*. Neace, Bob (des, illus). Hannibal, MO, US: Robert C. Bogart, **1960**. illus: b&w: drawings; partly col: maps; col: drawings, photos; text on lining pages; [i], 24p, [1].
Cover title: Mark Twain Cave.

[423] Bögli, Alfred, 1912– *Karst hydrology and physical speleology*. Schmid, June C. (trans). Berlin, DE: Springer-Verlag, **1980**. illus: b&w: ads, drawings, graphs, maps, photos, tables; xiii, 284p, [2]; app, bibl, index; 25 cm.
Translation of Karsthydrographie und physische Speläologie. 12 pls, 160 figs.
A useful and classic overview, long on technical/theoretical hydrology, hydrochemistry, kinetics, and karst geomorphology, short on general geology and post-formational cave features. The book suffers in part from some poorly labeled and/or captioned figures and a labored translation.

[424] Bögli, Alfred, 1912– *Luminous darkness: the wonderful world of caves*. Franke, Herbert W., 1927–; Charleston, B. M. (trans); Griffin, H. (trans). Chicago, IL, US: Rand McNally & Company, **1967**. illus: b&w: photos; col: photos; 83p, [1], 53p pls; 28 cm.
Translation of Leuchtende Finsternis, 1965, Berne, Kümmerly, & Frey. Same text and photos as Radiant darkness: the wonderful world of caves (UK ed, 1967). A second edition, 1968 may exist.
A major step in coffee-table, color picture books, this is an important, ground-breaking effort for legitimatizing cave photography, and, for that matter caving and cave science.

[425] Bohi, John W. *History of Wind Cave*. Pierre, SD,

US: State Publishing Company, **1963**. illus: b&w: maps, photos, repros; 91p; errat, index; 23 cm.

Reprint with loose errata from South Dakota Historical Collections. Bibliography scattered throughout as footnotes.

[426] Bohi, John W. *Seventy-five years at Wind Cave: a history of the national park.* South Dakota State Historical Society. Pierre, SD, US: South Dakota State Historical Society, **1962**. 103p; bibl; 24 cm.

Caption title. Illus.

[427] Bolner, Katalin Tak'acs ALCADI '92 *international conference on speleo history field trip guide.* Székely, Kinga. Budapest, HU: np, **1992**. Yellow Card Covers. illus: partly col: maps; [32]; 20.5 cm.

Describes geology and caves of areas: Buda Hills, Aggtelele Karst, Büklc Mountains.

[428] Bolton, David W. *The geology and speleogenesis of Red Run Cave Tucker County, West Virginia.* npop: np, **nd**. illus: b&w: drawings, maps, tables; partly col: maps; vi, 34p; app, bibl.

16 figs.

[429] Bonacci, Ognjen *Karst hydrology with special reference to the Dinaric karst.* Vidović-Čulić, Zjena (trans). Berlin-Heidelberg, DE: Springer-Verlag, **1987**. Springer Series in Physical Environment. illus: b&w: graphs, maps, photos, tables; x, 184p.

119 figs.

[430] Bondesan, A. *Morphometric analysis of dolines.* Meneghel, M.; Sauro, U. Trieste, IT: Societá Speleologica Italiana, **1992**. International Journal of Speleology Vol 21 no1-4. illus: b&w: charts, graphs, tables; text on lining pages; 55p; bibl.

[431] Bonin, Dan (ed) *Hodag hunt 1980: guidebook September 12-14 1980.* WI, US: [Wisconsin Speleological Society], **1980**. 15th Annual Field Conference of the Wisconsin Speleological Society. illus: b&w: charts, maps; partly col: maps; Various pagination.

Section two: geology field trip guide by Michael J. Barden with a contribution from Lou R. Goodman.

[432] Bonk, William J. *Lua Nunu o Kamakalepo: a cave of refuge in Kau, Hawaii.* npop: np, **1969**. vi, 26p; 28 cm.

Caption title. Originally published in: Archaeology on the island of Hawaii, 1969, pp. 75-92, 221-228. Illus.

[433] Bookout, John *Named caves of Oregon.* Bassett, Terry. [Portland, OR, US]: [Oregon Grotto], **nd**. illus: b&w: maps; 12p, [3].

[434] Boon, J. M. *Cave diving on air.* Bristol, GB: Cave Diving Group, **1966**. Cave Diving Group Technical Review no 1. 33p; 27 cm.

[435] Boon, J. M. *Down to the sunless sea.* Edmonton, Alberta, CA: Stalactite Press, **1977**. illus: b&w: maps; [4] folded leaves of maps; [v], 105p, 4p pls; 24 cm.

An autobiographical account of cave diving, in particular including Swildon's Hole (Mendip, England), Predjama (Slovenia), and caves in Yorkshire (England), Ireland and Jamaica. Surveys of major systems are included.

[436] Boon, J. M. *The Great San Agustín Rescue.* Edmonton, Alberta, CA: Stalactite Press, **1980**. 20p.

A quick and dirty personal account of a 1980 rescue of a Polish caver. No jokes about Polish rope - it really broke!

[437] Booth, Eugene, 1940– *Under the ground.* Collard, Derek. Milwaukee, WI, US: Raintree Childrens books, **1977**. A Raintree Spotlight Book. 21p; 24 cm.

Col illus. Juvenile.

"Pictures and questions stimulate discussion about animals, tunnels, pipes, and even men underground."

[438] Boothroyd, Colin *Caves of Thunder expedition report: an expedition to the world's largest underground river, Irian Jaya, Indonesia 1992.* Checkley, Dave; Gill, David W.; Gregory, Adrian; Jones, Steve; Newman, Gavin; Senior, Kev; Willis, Simon; Wyeth, John. Cumbria, GB: Lyon Equipment, May, **1993**. illus: b&w: maps, tables; 74p; app, bibl; 30 x 20 cm.

[439] Booy, Theodoor Hendrik Nikolaas de, 1882-1919 *Pottery from certain caves in eastern Santo Domingo, West Indies.* Lancaster, PA, US: Press of the New Era Printing Company, **1915**. 29p; 26 cm.

Cover title. Reprinted from the American anthropologist (n.S.) vol. XVII, no. 1, January-March 1915, pp 69-97. Illus, 6 pls.

[440] Bordes, Francois *A tale of two caves.* New York, NY, US: Harper & Row, **1972**. Harper's Case Studies in Archaeology. illus: b&w: drawings, graphs, maps, photos, tables; ix, 169p; bibl, gloss, index; 25 cm.

Archeological research and methods of excavations in Combe-Grenal and Pech L'Aze, two French Caves.

[441] Bosák, Pavel (ed) *Paleokarst: a systematic and regional review.* Ford, Derek Clifford, 1935–(ed); Glazek, Jerzy (ed); Horáček, Ivan (ed). Amsterdam, NL: Elsevier, **1989**. Developments in Earth Surface Processes 1. illus: b&w: charts, drawings, graphs, maps, photos, tables; 725p; bibl, index; 25 cm.

Published in co-edition with Academia Publishing House of the Czechoslovak Academy of Sciences, Prague.

More regional than systematic, and hardly comprehensive, there is still a lot presented on parts of the world from which English-language texts are rare.

[442] Bosnak, Art D. *Acute toxicity of cadmium, zinc, and total residual chlorine to epigean and hypogean isopods (Asellidae).* Morgan, Eric L. [Cookerville, TN, US]: np, **1981**. illus: b&w: drawings, graphs, maps, tables; iv, 37p.

14 tables, 5 figs.

[443] Botosaneanu, Lazare (ed) *Stygofauna mundi: a faunistic, distributional, and ecological synthesis of the world fauna inhabiting subterranean waters (including the marine interstitial).* Leiden, NL: E.J. Brill, **1986**. illus: b&w: drawings, maps, tables; folded maps of zones in rear; vi, 740p.

Partly in English, partly in French.

A bestiary of aquatic subsurface organisms, with taxonomic keys, distribution records, and in many cases line drawings of the species involved. All aquatic cave groups are covered.

[444] Botto, Larry (comp) *A partial guidebook to the caves of the Berkshires: 1993 spring N.R.O. hosted by the Berkshire Hills Grotto (May 14th, 15th, 16th)*. Plante, Alan (comp). npop: np, **[1993]**. illus: b&w: ads, maps, photos; 1 foldout map.; [v], 41p.

Title from cover.

[445] Boulton, John *Particulars of a first exploration of the extensive and newly discovered cavern, at Stainton, Low Furness*. Ulverston, GB: William Kitchin, **1871**. 16p; 17.5 x 11 cm.

[446] Bounk, Michael J. *Karstification on the Silurian Escarpment in Fayette County, northeastern Iowa*. Geological Society of Iowa. npop: np, Oct, **1983**. Guidebook: Geological Society of Iowa no 40; Field trip (Geological Society of Iowa) no 40. illus: b&w: photos; 26p; bibl; 28 cm.

October 22, 1983.

[447] Bousman, C. Britt *An archaeological assessment of Carlsbad Caverns National Park*. [Dallas, TX, US]: Archaeology Research Program, Southern Methodist University, **1973**. illus: b&w: maps, photos, tables; map folded in text; iii, 46p; bibl; 28 cm.

Date stamped on tp. Sponsored by the National Park Service.

[448] Bowdler, Sandra *A Test excavation at Mulka's Cave (Bate's Cave) near Hyden, Western Australia: a report to the Department of Aboriginal Sites, Western Australian Museum*. [Perth, WA, AU]: Centre for Prehistory, Department of Aboriginal Sites, University of Western Australia, **1989**. illus: b&w: maps; 37p, 25p pls; bibl; 31 cm.

Title from cover. Illus.

[449] Boy Scouts of America *Caving*. Irving, TX, US, **1991**. Venture (Boy Scouts of America). illus: b&w: photos; col: photos; 61p; bibl; 21 cm.

Cover title: Venture Caving. "No. 3458"---tp verso.

The first time Scout groups were given clear directions for caving trips. Prepared by caver/Scout leaders.

[450] Boy Scouts of America *Venture caving*. Irving, TX, US: Boy Scouts of America, **1991**. illus: b&w: charts, drawings, maps, tables; col: photos; text on front lining page; 61p; bibl.

[451] Boyd, Susan (comp) *A guide to major Wisconsin caves*. Kuhlen, Barbara (comp). Madison, WI, US: The Wisconsin Speleological Society, **1966**. illus: b&w: maps; [xv].

[452] Boyle, Mary Elizabeth, 1881– *Barma Grande; the great cave and its inhabitants*. [Grimaldi di Ventimiglia], IT: Abbo Freres & Cie, **1925**. illus: b&w: photos; 3 foldout pp; 32p, 5p pls; 18 cm.

1928 ed examined with no differences noted.

[453] Bozeman, Sue *The D. C. Jester cave system*. Johnson, Kenneth S.; Jagnow, Becky; Baker, Bruce; Kowalski, Gary; Riley, Patty. Oklahoma City, OK, US: Central Oklahoma Grotto, National Speleological Society, **1987**. Oklahoma Underground v 14 1987. illus: b&w: cartoons, charts, maps, photos, repros, tables; Folded map in pocket; 56p; bibl.

[454] Bradford Pothole Club Members *A history of Gaping Gill and Ingleborough Cave*. 1st ed. Kendal, GB: Titus Wilson, **c1974**. illus: b&w: maps, photos, repros; 13p; bibl; 21 x 15 cm.

2nd ed. **1976**. bibl.

[455] Bradley, J. F. *Glorious Kangaroo Island: its caves and beauty spots*. Bell, Harold. SA, AU, Nov, **1926**. illus: b&w: ads, photos; ads on lining pages; [i], 80p.

[456] Bradley, R. W. *Geology of the Capitan Reef Complex of the Guadalupe Mountains, Culberson County, Texas and Eddy County, New Mexico. Field trip guidebook May 6, 7, 8, 9, 1964*. Roswell, NM, US: Roswell Geological Society, **1964**. Roswell Geological Society Guidebook no 14. 2 folded col maps in pocket; 124p; bibl; 29 cm.

Field trip guidebook with a short chapter about Carlsbad Caverns. Includes, as separates, two geologic quadrangle maps.

[457] Bradshaw, Chris *The manual of basic caving*. Bath, Avon, GB: The Resurgence Press, **1982**. illus: b&w: ads, cartoons, charts, drawings, maps, photos, tables; col: photos; ads on lining pages; front: b&w photo; iv, 98p; bibl.

[458] Bradshaw, D. R. (comp) *Report on the British speleological expedition to Vietnam; March April 1990*. npop: British Cave Research Association, **nd**. illus: b&w: maps, tables; col: photos; 2 folded maps in rear; [i], 23p, 2p pls; 30 x 21 cm.

[459] Brain, Charles Kimberlin *The hunters or the hunted? An introduction to African cave taphonomy*. Chicago, IL, US: University of Chicago Press, **1981**. illus: b&w: charts, drawings, graphs, maps, photos; front: b&w port; x, 365p; app, bibl, index; 31 cm.

121 tables, 226 figs.

This serves a an example of, and introduction to, taphonometric studies applicable both to African and non-African cave sites. This combines paleontologic, stratigraphic and paleoecologic techniques in analyzing these death assemblages.

[460] Brain, Charles Kimberlin *The Transvaal ape-man-bearing deposits*. Ewer, R. F. (app). Pretoria, ZA: Transvaal Museum, **1958**. Transvaal Museum Memoir no 11. illus: b&w: drawings, maps, tables; 131p; app, bibl, index; 29 cm.

This is of use to anthropologists and paleontologists as well, especially those looking at taphonomy and cave stratigraphy.

[461] Branch, Pat *Illustrated guide book of your tour through Carlsbad Caverns.* npop: Grace & Tech Helm Travel Production, **1951**. illus: b&w: photos.

51 duotone photos.

[462] Brandt, Keith, 1949– *Caves.* Schneider, Rex (illus). Mahwah, NJ, US: Troll Associates, **1985**. illus: col: drawings; 30p; 24 cm.

Audio cassette version also published in 1984 to accompany book. Juvenile.

[463] Branigan, Keith *Romano-British cavemen: cave use in Roman Britain.* Dearne, M. J. Oxford, GB: Oxbow Books, **1992**. Oxbow Monograph 19. illus: b&w: drawings, maps, tables; [i], iii, 118p; bibl; 24 x 17 cm.

Distributed in the US by David Brown Book Co.

[464] Brannan, Eldred Boyd *The entombment of Floyd Collins in Sand Cave, Kentucky.* Alexander, Cecil Hume. Denver, CO, US: The Bradford-Robinson Ptg Co, **[1925]**. illus: b&w: photos; 93p, 4p pls.

1925? publication date obtained from NSS Library.

Rare and unusual booklet.

[465] Breder, Charles Marcus, 1897– *Comparative studies in the light sensitivity of blind characins from a series of Mexican caves.* Rasquin, Priscilla. New York, NY, US: [American Museum of Natural History], Dec, **1947**. Bulletin of the American Museum of Natural History v 89 article 5. illus: b&w: drawings, graphs, maps, tables; 29p; bibl; 27 cm.

323-351 p. Illus.

[466] Brenan, Edward *Notice of the occurrence of mammoth and other animal remains: discovered under limestone in a bone cave at Shandon, near Dungarvan, in the county of Waterford.* Dublin, IE: Printed at the University Press, by M.H. Gill, **1859**. 10p, 4p pls; 21 cm.

Illus.

[467] Breternitz, David A. *Archaeological investigations in Turkey Cave (NA2520) Navajo National Monument, 1963.* Flagstaff, AZ, US: Northern Arizona Society of Science and Art, Inc, **1969**. Technical Series no 8. illus: b&w: charts, maps, photos, tables; iii, 26p; app, bibl.

[468] Bretz, J Harlen, 1882-1981 *Caves of Illinois.* Harris, Seymour Edwin, 1897–. Urbana, IL, US: Illinois State Geological Survey, **1961**. Illinois State Geological Survey Report of Investigations 215. illus: b&w: charts, maps, photos; b&w front; 87p; bibl; 25 cm.

Reprint edition was issued in 1966. 19 figs, 9 pls.

Following geological introduction, descriptions of all known caves in state of Illinois, with maps (about a dozen) where available. Bretz's general descriptions include considerable geology.

[469] Bretz, J Harlen, 1882-1981 *Caves of Missouri.* Rolla, MO, US: State of Missouri Geological Survey and Water Resources, **1956**. State of Missouri Division of Geological Survey and Water Resources v 39 Second Series [Reports]. illus: b&w: drawings, graphs, maps, photos, repros; xxi, 490p, [1]; bibl, errat, gloss, index; 26 cm.

168 figs.

This is principally a county-by-county description of the wealth of caves (show caves, and wild caves, totaling 437) in this state, by an outstanding geologist and speleologist. Each description is rich in geological description and interpretations, making it a good example of what regional cave-inventories can and should be like. Few maps, mostly of the show caves are included.

[470] Breuil, Abbé Henri, 1877-1961 *Beyond the bounds of history: scenes from the old stone age.* Boyle, Mary Elizabeth (trans). London, GB: P. R. Gawthorn, **1949**. illus: b&w: photos; col: drawings; 100p.

9 b&w photos in front section of book and there are 31 color drawings in the second half. These are called "scenes from the old stone age." Book appears to be intended for children. Reprint of the 1949 ed was done in 1979 by AMS Press, New York. Juvenile.

[471] Breuil, Abbé Henri, 1877-1961 *Cave drawings: an exhibition of drawings by the Abbé Breuil of palaeolithic paintings and engravings.* Daniel, Glyn Edmund. [London, GB]: The Arts Council, **1954**. illus: b&w: drawings; 45p; 22 cm.

2nd ed 1975.

An exhibition of drawings by the author of paleolithic painting and engravings.

[472] Breuil, Abbé Henri, 1877-1961 *The rock paintings of southern Africa.* Paris, FR: Trianon Press, **1955**. Various pagination.

Illus. Title from half-title. Imprint varies: v. 2, London, Abbé Breuil Publ; v. 4, Clairvaux, France, Calouste Gulbenkian Foundation; v. 5, Clairvaux, France, Singer-Polignac Foundation. Volume titles: v 1: The white lady of the Brandberg. v 2: Philipp Cave. v 3: Tsisab Ravine. v 4: Anibib & Omandumba and other Erongo sites. v 5: Southern Rhodesia. v 6: The Sphinx and White Ghost shelters.

[473] Bridgemon, Rondal Rex, 1944– *Wupatki National Monument earth cracks.* Yellow Springs, OH, US: Cave Research Foundation, **1976**. illus: col: tables; 59p; bibl; 22 x 28 cm.

Illus.

[474] Bridgemon, Rondal Rex, 1944– *South China caves: information on the cave and karst of South China and a report on the 1988 joint expedition between the Institute of Karst Geology, the Speleological Society of South China Normal University, and the Cave Research Foundation.* Lindsley, Karen Bradley, 1948–(ed). Tucson, AZ, US: Cave Research Foundation, **[1991]**. illus: b&w: charts, maps, photos; 62p; app, bibl.

[475] Bridges, Thomas Charles *Adventures under*

ground [sic]. London, GB: Thomas Nelson & Sons Ltd, **1937**. illus: b&w: ads, photos; col: drawings; front: col drawing; xii, 212p, 8p pls.

[476] Bridwell, Margaret Morris, 1905– *The story of Mammoth Cave National Park, Kentucky: a brief history.* 11th ed. **1971**.

1952 ed. [Mammoth Cave, KY, US]: [Mammoth Cave National Park], **1952**. 63p; 22 cm.

Illus.

1959 rev ed. npop: np, **1959**.

Illus.

3rd ed. npop: np, **1963**. illus: b&w: drawings, photos; b&w drawings on lining pages; 64p.

[477] Briel, Larry I. (comp) *Proceedings of the first annual seminar on cave diving.* Desautels, David A. Gainesville, FL, US: National Association of Cave Divers, **1968**. ii, 43p; bibl; 28 cm.

Illus. "Compiled from notes by Frances E. Calvin."

[478] Briggs, Thomas S. *Phalangodidae from caves in the Sierra Nevada (California) with a redescription of the type genus (Opiliones, Phalangodidae).* San Francisco, CA, US: The California Academy of Sciences, **1974**. Occasional Papers of the California Academy of Sciences no 108. 15p; 26 cm.

Title from cover. Illus.

[479] Brinkley, Edward Booth, 1922– *The history of Ruby Falls.* Chattanooga, TN, US: Hudson Printing and Lithography Co, **1964**. illus: b&w: photos, ports; [i], 77p.

dob from pencil note in book, 2nd printing 1969.

[480] Bristol (England). Public Libraries *Caving periodicals in the central collection of caving publications: a catalogue of the caving periodicals held by Bristol Central Reference Library and a selected short title catalogue of books.* Mansfield, Kay; Mansfield, Raymond Walter. Sherbourne, Dorset, GB: Mendip Cave Registry, **1968**. 14p; 26 cm.

[481] Bristol Exploration Club *Bristol exploration club: caving report no.19: 1975 expedition to the Pierre Saint-Martin.* [GB]: np, **nd**. illus: b&w: maps, photos, tables; [i], 37p; bibl.

[482] British Association for Immediate Care *Rescue from remote places: cave rescue and sport diving emergencies.* Ipswich, GB: British Association for Immediate Care, **1986**. BASICS Monographs on Immediate Care no 3. 16p; bibl; 21 cm.

Title from cover. Illus.

[483] British Broadcasting Corporation Television Service *Atea: in search of the world's deepest cave.* British Broadcasting Corporation. Television Service. Wilmette, IL, US: Films Incorporated, **1980**.

1 videocassette (VHS) (30 min) sd, col.

A 1978 expedition explores, maps, and measures the massive Atea river cave to determine if it is the world's deepest cave. The team combats swift water, scales rock walls and plows through mud.

[484] British Cave Research Association *BCRA library catalogue.* [GB]: British Cave Research Association, **1985**. [iv], 147p; 30 cm.

No illus.

[485] [British Cave Research Association] *Matienzo, Spain.* Bridgwater, Somerset, GB: British Cave Research Association, Jun, **1981**. Transactions of the British Cave Research Association v 8 (2). illus: b&w: charts, maps, photos, tables; 1 fold-out map; text on lining pages; 60p; bibl, errat.

British expeditions 1969-1980.

[486] British Cave Research Association *Peak and Speedwell Caverns, exploration and science: including the talks presented at the Peak-Speedwell symposium, Sheffield, November 1989.* London, GB: British Cave Research Association, **1991**. Cave Science v 18 no 1. illus: b&w: charts, graphs, maps, photos, repros, tables; 1 fold-out map; text; 60p; bibl; 30 cm.

Title from contents page. Illus.

[487] [British Cave Research Association] *British Cave Research Association, library catalogue.* [GB]: British Cave Research Association, **1989**. [vi], 183p; 30 x 21 cm.

No illus.

[488] British Columbia. Parks and Outdoor Recreation Division *Cave resources in British Columbia: a discussion paper.* [Victoria, British Columbia, CA]: The Division, Jan, **1980**. illus: b&w: charts, maps, repros, tables; [iii], 20p; app; 29 cm.

A discourse in the philosophical and practical benefits and problems of cave conservation.

[489] British Speleological Association *1967 expedition to the Gouffre Berger.* GB: British Speleological Association, c1968. illus: b&w: photos; [16]; 25 x 20 cm.

List of donating firms and personnel on last page. Text on covers included in pagination.

[490] [British Speleological Association] *Imperial College karst research expedition to the Peruvian Andes, 1972.* GB: British Speleological Association, Nov, **1973**. Journal of the British Speleological Association: Cave Science no 52. illus: b&w: charts, graphs, maps, photos, tables; 4 fold-out maps; [i], 34p, 12p pls.

Cover title: Imperial college expedition to the karst of Peru.

[491] [British Speleological Association] *Limestone geomorphology: a study in Jamaica: University of Bristol karst expedition to Jamaica 1967.* GB: British Speleological Association, **1969**. Journal of the British Speleological Association: Cave Science no 43-44. illus: b&w: charts, graphs, maps, photos, tables; 2 fold-out maps; front: b&w photos; 83p.

[492] British Speleological Expedition: Members of

the Expedition *British Speleological Expedition to the Cantabrian Mountains, northwest Spain 1965*. Eyre, Jim (illus). Berwick upon Tweed, GB: A. C. Huntington, **1966**. illus: b&w: cartoons, maps; 7 loose maps; 37p, [1]; 25.5 x 20.5 cm.

"In conjunction with the Espeleologos de Penalba Leon."

[493] Brizius, Janice J. *History of the 7 caves*. Caniff, Milton Arthur, 1907–(cover illus). npop: np, **1900**. illus: b&w: maps; 22 cm.

Map on inside covers. Illus, unpaged.

[494] Broderick, Alan Houghton *Abbé Breuil: prehistorian: a biography*. London, GB: Hutchinson & Co (Publishers) Ltd, **1963**. illus: b&w: drawings, maps, photos; b&w front; 256p, 16p pls; bibl, index.

This is the biography of the father of the study of prehistoric cave art. Well illustrated, covering prehistoric caverns from Altamira to Lascaux.

[495] Broderick, Alan Houghton *Father of prehistory: the Abbé Henri Breuil: his life and times*. New York, NY, US: William Morrow & Company, **1963**. illus: b&w: drawings, photos, ports; vii, 306p, 16p pls.

[496] Broderick, Alan Houghton *Lascaux: a commentary*. Thaon, Maurice (photo). London, GB: Lindsay Drummond Ltd, **1949**. illus: b&w: maps, photos, tables; col front; 142p; 25 cm.

Contains 46 pls (b&w photos) and 9 chaps.

[497] Broderick, Alan Houghton *Prehistoric painting: with 56 plates in colour and monochrome and 7 line illustrations in the text*. Central Institute of Art and Design. London, GB: Avalon Press, Ltd, **nd**. illus: b&w: drawings, photos, repros; col: repros; 37p, 43p pls; bibl.

[498] Bronin, Andrew *The cave: what lives there*. Stahl, Ben F. (illus). New York, NY, US: Coward, McCann & Geog Hogan, **1972**. illus: b&w: drawings; partly col: drawings; [32].

Juvenile.

[499] Brook, Alan, 1945– *Northern caves volume 2: the three peaks*. Brook, David, 1944–; Griffiths, J.; Long, M. H. Clapham, GB: Dalesman, **1991**. illus: b&w: ads, maps, tables; col: photos; Front and rear endpapers have maps; 286p; index.

[500] Brook, Alan, 1945– *Northern caves volume three: Ingleborough*. Brook, David, 1944–; Davies, G. M.; Long, M. H. Clapham, North Yorkshire, GB: The Dalesman Publishing Company Ltd, **1975**. illus: b&w: ads, maps; 144p; 18 x 12 cm.

Describes an area with more deep potholes than the rest of England. 8 maps, 21 surveys, over 200 caves.

[501] Brook, Alan, 1945– *Northern caves volume two: Penyghent and Malham*. Brook, David, 1944–; Davies, G. M.; Long, M. H. Clapham, North Yorkshire, GB: The Dalesman Publishing Company, Ltd, **1976**. illus: b&w: ads, maps; 120p; index.

Popular book, covering classic areas; 8 maps, 12 surveys, caving code.

[502] Brook, David B., 1944– *The British New Guinea speleological expedition*. Bridgwater, Somerset, GB: British Cave Research Association, Dec, **1976**. Transactions of the British Cave Research Association vol 3 no 3-4. illus: b&w: charts, drawings, graphs, maps, photos, tables; 2 fold-out maps, 1 large folded map in pocket; text on lining pages; 127p.

"Glossary" is actually short dictionary of Telefol to English.

[503] Brook, David B., 1944– *Caves of Mulu: the limestone caves of the Gunong Mulu National Park, Sarawak*. Waltham, Anthony Clive, 1942–[Tony] (ed); Royal Geographical Society - Sarawak Government Expedition 1977-78. London, GB: The Royal Geographical Society, **1978**. illus: b&w: maps, photos; [4]p maps folded in center; 44p; 30 cm.

Reprint edition in 1979.

[504] Brook, David B., 1944– *The explorations journal of The University of Leeds Speleological Association*. Crabtree, H. (ed). Leeds, GB: [University of Leeds Speleological Association], Oct, **1969**. illus: b&w: maps, photos; folded map loose in rear; [iv], v, 73p; 30 x 23 cm.

[505] Brook, David B., 1944– *Northern caves volume 4a: Scales Moor and King Scale*. Davies, G. M.; Long, M. H. 2nd ed. Chapman, GB: Dalesman Books, **1983**. Northern Caves. illus: b&w: ads, maps, photos; 82p, 3p pls; 18.5 x 12 cm.

First published 1975 as one volume and reprinted in 1976.

[506] Brook, David B., 1944– *Northern caves volume 4b: Leck and Casterton Fells*. Davies, G. M.; Long, M. H.; Sutcliffe, J. R. 2nd ed. Chapman, GB: Dalesman Books, **1983**. Northern Caves. illus: b&w: ads, maps, photos; 83p, 3p pls; 18.5 x 12 cm.

First published 1975 as one volume and reprinted in 1976.

[507] Brook, David B., 1944– *Northern caves volume five: the Northern Dales*. Davies, G. M.; Long, M. H.; Ryder, P. F. 1st ed. Clapham, North Yorkshire, GB: The Dalesman Publishing Company, Ltd, **1974**. illus: b&w: maps; 144p; index.

West coast of Cumbria to Northumberland border; also Flamborough Head to Nottinghams.

2nd ed. 1978. index.

[508] Brook, David B., 1944– *Northern caves volume four: Whernside and Gragareth*. Davies, G. M.; Long, M. H.; Sutcliffe, J. R. Clapham, North Yorkshire, GB: The Dalesman Publishing Company, Ltd, **1975**. illus: b&w: ads, maps; 141p, [3]; index.

[509] Brook, David B., 1944– *Northern caves volume one: Wharfedale and Nidderdale*. Coe, R. G.; Davies, C. M.; Long, M. H. 1st ed. npop: np, **1972**. 142p.

One of three volumes in the new series cataloging the known caves of northern England. The others are A. Brook, et al. 1991 (The Three Peaks) and D. Brook, et al. 1994 (The Three-Counties System and the Northwest). They are pocket-size cavers' guides, with plastic covers for field use. A total of nearly nine hundred pages, with over a thousand caves described and located. Some of the major systems have small-scale maps. Emphasis is on route-finding and rigging information.

2nd ed (rev & enl). Clapham, GB: Dalesman, **1975**.
3rd ed. 1979. illus: b&w: maps; 144p; index.

[510] Brook, David B., 1944– *Northern Caves: volume 3: the three counties system and the north-west.* Griffiths, J.; Long, M. H.; Ryder, P. F. Skipton, GB: Dalesman, **1991**. illus: b&w: ads, maps, tables; col: photos; front endpaper has map; 287p; index; 17 cm.

[511] Brook, David B., 1944– *Northern caves: volume1. Wharfedale and the North-East.* Davies, G. M.; Long, M. H.; Ryder, P. F. Clapham (via Lancaster), GB: Dalesman, **1988**. illus: b&w: ads, maps, photos; map on rear lining page; 282p; 17 cm.

[512] Brooks, Ian *Excavation techniques in Pin Hole Cave, Creswell Crags S.S.S.I., Derbyshire.* Derbyshire (England) County Council; Nottinghamshire (England) County Council; Creswell Crags Visitor Centre. [Derbyshire, GB]: Prehistoric Archaeological Survey Team, **1985**. 11p.

[513] Brooks, S. J. (ed) *Caving in the abode of the clouds: the caves and karst of Meghalaya, north east India.* Smart, C. M. (ed); Bristol Exploration Club (comp); Orpheus Caving Club (comp). Wiltshire, Near Bradford on Avon, GB: Meghalaya 94, Mar, **1995**. illus: b&w: maps, photos, tables; [ii], 67p, [3]; app.

Last 17 pages numbered with Roman numerals. Cover title: Meghalaya: caving in the abode of the clouds.

[514] Brooks, Steve (comp) *The Australian cavers diary.* Nedlands, WA, AU: Western Australian Speleological Group, **1995**. illus: b&w: ads, drawings, photos; [144].

Title from cover. A calendar for 1995 with quotes and photos.

[515] Brown, A. L. *Lelet: report of the 1975 New Ireland Speleological Expedition.* Bourke, R. M.; Shannon, C. H. C. Keravat, PG: Papua New Guinea Cave Exploration Group, Nov, **1976**. Niugini Caver vol 4 no 3 pp 85-136. illus: b&w: maps, photos, tables; text on front lining page; 52p; app.

[516] Brown, Arthur V. *Cavefish (Amblyopsis rosae) in Arkansas: populations, incidence, habitat requirements, and mortality factors: a final report.* Willis, Lawrence D.; Arkansas Game and Fish Commission, Federal Aid in Fish and Wildlife Restoration (AK). Fayetteville, AR, US: Dept of Zoology, University of Arkansas, **1984**. ii, 61p; bibl; 28 cm.

Illus. Federal aid project E-1-6, utilizing funds available under terms of P-L93-205, The Endangered Species Act of 1973, as amended. November 1984.

[517] Brown, Barnum, 1873-1863 *The Conard fissure, a Pleistocene bone deposit in northern Arkansas: with descriptions of two new genera and twenty new species of mammals.* [Cambridge, MA, US]: [E. W. Wheeler, printer], **1908**. Memoirs of the American Museum of Natural History vol IX pt IV. illus: partly col: charts; 54p; 36 x 29 cm.

Title from cover. pp 155-208, illus, pls XIV-[XXV].

[518] Brown, Deana *Caverns of Sonora: an awe-inspiring adventure you will remember forever.* npop: np, **nd**. illus: col: photos; [38].

[519] Brown, Edward *The pictured cave of La Crosse Valley, near West Salem, Wisconsin.* Rice, John A. npop: np, **nd**. 15p; 23 cm.

Illus. Additional notes on the pictured cave by Hon. John A. Rice: pp 183-187.

[520] Brown, George Henry *The Clapham Cave with illustrations from photographs by the author.* Settle, GB: np, **1905**. illus: b&w: maps, photos; 28p; 17 x 11 cm.

[521] Brown, George Henry *The underground streams in the neighbourhood of Clapham and Malham.* Settle, GB: np, **1906**. 23p; 18 x 11.5 cm.

[522] Brown, George Henry *Victoria Cave: its history and exploration.* 2nd ed rev, enl. npop: J. W. Lambert, **1908**. illus: b&w: ads; 30p.

3rd ed. Settle, GB: J. W. Lambert & Sons, **1925**. 31p; 17.5 x 12 cm.

Another version that is 18.5 x 12 cm may exist.

[523] Brown, Gerard Baldwin, 1849-1932 *Art of the cave dweller—a study of the earliest artistic activities of man.* London, GB: John Murray, **1928**. illus: b&w: ads, drawings, maps, photos, tables; 2 b&w maps folded in text; col front; xix, 280p, [4]; index.

169 illus (95 half-tones, 4 b&w maps). One copy examined noted R.V. Coleman, National Book Buyers Service, NY. Another copy examined had 1929 penciled in.

[524] Brown, Gerard Baldwin, 1849-1932 *The art of the cave dweller: a study of the earliest artistic activities of man.* Cheap ed [sic]. London, GB: John Murray, T. and A. Constable, **1932**. illus: b&w: maps; [2] folded lvs of pls; xix, 280p, 2p pls; index; 23 cm.

Illus (some col).

[525] Brown, Henry S. *Linville Caverns through the ages: the geological story.* npop: H. Brown, **1961**. illus: b&w: photos; [20]; 23 cm.

Small booklet telling the geological story of Linville Caverns, N.C. Has 13 figures including 11 lower quality b&w photos.

[526] Brown, L. Carson *Manitouwadge: cave of the Great Spirit.* 4th ed. [Toronto, CA]: Ontario Department of Mines and Northern Affairs, **1970**. 15p.

Reprinted from Canadian Geographical Journal.

5th ed. Royal Canadian Geographical Society; Ontario Ministry of Natural Resources. Ottawa, CA: Royal Canadian Geographical Society, **1972**. illus: b&w: maps, ports; 28 cm.

Illus.

[527] Brown, Mary Mitchell *Nature underground: the Endless Caverns in the heart of the historic Shenandoah Valley*. Flanagan, J. R. (illus). 1921 ed. New Market, VA, US: np, **1921**. [35]; 23 x 30 cm.

Partly col illus.

1924 ed. [31].

Two columns to the page.

1946 ed. New Market, VA, US: Endless Caverns, Inc. illus: col: drawings; 30p; 23 x 31 cm.

Description of the cave. Color paintings of scenes in the cave, with text.

[528] Brown, Michael Charles *Karst hydrology of the lower Maligne Basin, Jasper, Alberta*. Castro Valley, CA, US: Cave Research Associates, **1972**. Cave Studies no 13. illus: b&w: charts, graphs, maps, photos, tables; col: photos; col photo front; viii, 84p; app, bibl; 24 cm.

[529] Brown, Paul Martin *An introduction to the limestone sinkholes of northeastern Michigan*. [Alpena, MI, US]: Jesse Besser Museum, **1973**. 28p; bibl; 23 cm.

Illus.

[530] Brown, Richmond Flint, 1925– *Hydrology of the cavernous limestones of the Mammoth Cave area, Kentucky*. Washington, DC, US: United States Government Printing Office, **1966**. Geological Survey (U.S.) Water Supply Paper 1837; Geological Survey of Alabama Special Report 22. illus: b&w: maps; v, 64p; bibl; 24 cm.

Illus.

One of the earliest scientific texts on Mammoth Cave. This serves as a standard to which subsequent work on the area may be compared.

[531] Brown, Robert F. *Exploration in Wind Cave*. npop: National Speleological Society, Oct, **1964**. illus: b&w: maps; [4]p fold-out map; 50p, 4p pls.

This is the complete report of the 1959 NSS Wind Cave Expedition.

[532] Browne, George Forrest, 1833-1930 *Ice-caves of France and Switzerland. A narrative of subterranean exploration*. London, GB: Longmans, Green & Co, **1865**. illus: b&w: drawings; viii, 315p; app; 21 cm.

At least two different versions exist; one with colored edges & inside covers and a ribbed binding; the other ed is slightly larger and not as fancy. Contains 13 illus.

Mainly exploration narratives, but does survey known glaciers. Contains supplementary information on glaciers including formation and structure of cave ice. This classic describes 19 ice cave explored by the author and also has four chapters of scientific observations.

[533] Broyles, Bettye J. *Russell Cave in northern Alabama*. [Knoxville, TN, US]: Tennessee Archaeological Society, **1958**. Miscellaneous Papers Tennessee Archaeological Society no 4. 29p, [6]; bibl; 28 cm.

Illus.

A definitive publication describing the archaic culture (8000-1000 BC) in the deep Southeastern United States.

[534] Bruce, Murray D. *Preliminary survey of Palawan, Philippines, by Traditional Explorations and the Sydney Speleological Society, January/February 1980*. Sydney Speleological Society. Palawan Expedition, stage I. Sydney, NSW, AU: Traditional Explorations, **1980**. illus: b&w: maps; viii, 47p; 30 cm.

At head of title: Palawan Expedition, stage I. Some paper reprints.

[535] Brucker, Roger Warren, 1929– *The Flint Ridge Cave System: Mammoth Cave National Park, Kentucky; a Cave Research Foundation cartographic project in cooperation with the National Park Service*. Burns, Denver P. Washington, DC, US: Cave Research Foundation, **1966**. illus: b&w: maps; 3p, 31p pls; bibl; 37 cm.

Scale of maps ca. 1:3,000. Portfolio of 35 plates.

This is the first edition of a folio of 35 plates showing the cave passages surveyed in the Flint Ridge Cave System in Mammoth Cave National Park, Kentucky.

[536] Brucker, Roger Warren, 1929– *The longest cave*. 1987 reprint ed. Carbondale, IL, US: Southern Illinois University Press, **1987**. illus: b&w: drawings; app, bibl, gloss, index.

Reprint ed contains afterword.

1st ed. Watson, Richard Allan, 1931–. New York, NY, US: Alfred A. Knopf, **1976**. illus: b&w: drawings, maps, photos; xx, 316p, [11]; app, bibl, gloss, index; 25 cm.

32 b&w photos, drawings, 7 maps. Fourth printing: 1980.

An account of 20 years of expeditions of the Flint Ridge/Mammoth Cave systems in Kentucky, leading to a connection of the two caves, and including a chronological listing of exploration.

[537] Brune, Gunnar *Springs of Texas*. Fort Worth, TX, US: Branch-Smith, Inc, **1981**. illus: b&w: drawings, graphs, maps, photos, tables; 9 maps folded in envelope loose in back; b&w maps on lining pages; xviii, 566p; bibl, gloss, index; 29 cm.

[538] Brunton, Daniel Francis, 1948– *A biological inventory of the Warsaw Caves area of natural and scientific interest, Peterborough County, Ontario*. Aurora, ONT, CA: Parks and Recreational Areas Section, Ontario Ministry of Natural Resources, Central Region, **1990**. vii, 76p; 28 cm.

[539] Bruun, Daniel *The cave dwellers of Southern Tunisia: recollections of a sojourn with the Khalifa of Matmata*. 1985 fasc reprint. London, GB: Darf, **1985**. note.

1st ed. London, GB: W. Thacker & Co, **1898**. illus:

b&w: ads, drawings, maps, photos, repros; front: b&w port; xii, 333p, [5]; note; 23 cm.

Translated from the Danish by L.A.E.B. of: *Huleboerne i syd Tunis; erindringer fra et ophold hos kalifaen i Matmata*. Microfiche ed: New York, N.Y., New York Public Library.

[540] Bryan, Benjamin *A description of the fluor spar or Blue John Mine at Castleton*. Derbyshire, GB: Henry Mozley and Sons, **1839**. 16p; 16 x 10 cm.

Photocopy examined. No illus.

[541] Bryan, Benjamin *A description of the grand Devonshire cavern at Matlock*. Matlock, Derbyshire, GB: Benjamin Bryan, **1840**. 16p; 16.5 x 9 cm.

Photocopy examined.

[542] Bryan, Kirk *Correlation of the deposits of Sandia Cave, New Mexico, with the glacial chronology*. Hibben, Frank C. Washington, DC, US: Smithsonian Institution, **1941**. 20p, 1p pls; bibl; 25 cm.

Illus. Title from cover. "Appendix to 'Evidences of early occupation in Sandia Cave, New Mexico, and other sites in the Sandia-Manzano region,' by Frank C. Hibben. Reprinted from Smithsonian Miscellaneous Collections, vol. 99, no. 23."

[543] Bryan, Kirk *The geology and fossil vertebrates of Ventana Cave*. npop: np, **nd**. 52p; bibl; 23 cm.

Part III: pp 75-126.

[544] Bryant, Thomas Charles *An introduction to caving and potholing for novices*. Kenney, C. Howard. Bristol, GB: Spelaeo Service, **1964**. [iv], 18p, [2]; index; 26 x 20 cm.

[545] Buckland, William, 1784-1856 *Reliquiae diluvianae, or observations on the organic remains contained in caves, fissures, and diluvial gravel, and on other geological phenomena attesting the action of an universal deluge /*. 2nd ed. **1824**. index.

[546] Buckland, William, 1784-1856 *Reliquiae diluvianae: or, observations on the organic remains contained in caves, fissures, and diluvial gravel and on other geological phenomena attesting the action of a universal deluge*. 1978 reprint ed. New York, NY, US: Arno Press, **1978**. History of Geology. 2 folded lvs pls; vii, 303p, 17p pls; index; 26 cm.

Reprint of the 1823 ed.

Some of these interpretations are still presented by defenders of the Great Flood, so it is useful to see the original arguments, and contrast them to modern theory.

[547] Buckland, William, 1784-1856 *Reliquiae diluvianae: or, observations on the organic remains contained in caves, fissures, and diluvial gravel and on other geological phenomena attesting the action of an universal deluge*. 1st ed. London, GB: J. Murray, **1823**. illus: b&w: charts, maps; partly col: tables; vii, 303p, 27p pls; index; 28 x 23 cm.

Micro-opaque ed: New York: Readex Microprint, 1968, 5 cards; 23 x 15 cm. (Landmarks of science)

[548] Buckrose, J. E., 1868-1931 *Rambles in the North Yorkshire Dales*. London, GB: Mills & Boon, **1913**. x, 191p, 26p pls; 20 cm.

Illus, some col. 50% cave related.

[549] Buderer, Tina (comp) *Proceedings of the first annual watershed conference, May 20-21, 1987, Springfield, Missouri*. Watershed Management Coordinating Committee. [MO, US]: Missouri Dept of Conservation, **1987**. illus: b&w: maps; vi, 209p; bibl, index; 28 cm.

Illus.

[550] Buikema, Arthur L. *Studies on Gammaridea II: proceedings of the 4th International Colloquium on Gammarus and Niphargus, Blacksburg, Virginia, U.S.A., 10-16 September 1978*. Holsinger, John Robert, 1934–. Leiden, NL: Brill, **1980**. Crustaceana supplement 6. illus: b&w: maps; partly col: photos; vi, 252p, 6p pls; bibl; 25 cm.

English and French, with summaries in English French, and German.

[551] Bullington, Neal R. *Who Discovered Carlsbad Cavern?* npop: np, May, **1968**. 25p.

Contains many pages of excerpts from documentation about early visits and knowledge of the Caverns. Park Naturalist Neal Bullington has put a lot of effort into this report.

[552] [Bullitt, Alexander Clark, 1807-1868 *Rambles in the Mammoth Cave, during the year 1844*. Meloy, Harold (intro). 1973 reprint ed. New York, NY, US: Johnson Reprint Corp, **1973**. Classics in Speleology. illus: b&w: maps, repros; map folded in front; xliv, 101p, 6p pls; bibl; 19 cm.

1985 reprint ed. Meloy, Harold (intro). St Louis, MO, US: Cave Books, **1985**. illus: b&w: maps, repros; map folded in front; xliv, 101p, 6p pls; bibl; 19 cm.

The reprints includes a new 33p introduction (numbered separately) by Harold Meloy.

1st ed. Bishop, Stephen; Croghan, John, 1790-1849. Louisville, KY, US: Morton & Griswold, **1845**. Selected Americana from Sabin's Dictionary of Books relating to America, from its Discovery to the Present Time. illus: b&w: maps, repros; map folded in front of title page; xii, 101p, 6p pls; errat; 20 cm.

This small book is collector's item. The author's name does not appear in it. It is stated just as "by a Visiter (sic)." Formerly attributed to John Croghan. Half title: Mammoth cave. "Map of the explored parts of the Mammoth Cave of Ky., by Stephen Bishop, one of the guides." Microfiche edition:Woodbridge, CT: Research Publications, 1992. 2 microfiches: negative.

[553] Bulpin, T. V. *Kango: the story of the caves*. Oosthuizen, Hans. Cape Town, ZA: Treasury of Travel, **1985**. Treasury of Travel Series 2. illus: b&w: drawings, maps, photos; col: photos; text on lining pages; [34].

[554] Bunch, James H. *The archaeological excavation*

of two rock shelters near Cave Lake, Ely, Nevada. Turner, T. Hal. Carson City, NV, US: Nevada Dept of Transportation. NDOT Archaeological Technical Report Series no 1. illus: b&w: maps; vi, 80p; bibl; 28 cm.

Illus.

[555] Bunnell, David Edward, 1952– *Sea caves of Anacapa Island*. Santa Barbara, CA, US: McNally & Loftin Publishers, **1993**. illus: b&w: drawings, maps, photos; [viii], 207p; 27 cm.

Descriptions, locations, and maps of 135 caves. Some useful geologic data is included. This and his volume on Santa Cruz are among the few specifically on sea-caves.

[556] Bunnell, David Edward, 1952– *Sea caves of Santa Cruz Island*. Santa Barbara, CA, US: McNally & Loftin Publishers, **1988**. illus: b&w: drawings, maps, photos; [viii], 123p; bibl, index; 26 cm.

First of series on sea caves of California. Brief geological and biological introductions, then descriptions, locations, and maps of 112 caves.

[557] Bunton, Stephen *Vertical caves of Tasmania: a caver's guidebook*. Eberhard, Rolan. Miranda, NSW, AU: Adventure Presentations, **1984**. illus: b&w: maps, photos; front: b&w photo; iv, 81p; app, index; 30 cm.

A cavers' guide to about three dozen caves, with descriptions, rigging details, and maps.

[558] Burdon, David Joseph *Handbook of karst hydrogeology with special reference to the carbonate aquifers of the Mediterranean Region*. Papakis, Nicholas. Athens, GR: United Nations Special Fund, Karst Groundwater Investigations; Institute for Geology and Subsurface Research, **1963**. illus: b&w: charts, drawings, graphs, tables; 1 graph and 1 drawing folded in text; viii, 276p; bibl; 28 cm.

"This Handbook has been prepared for use at the Training course of Karst hydrology to be given by the United Nations Special Fund Karst Groundwater Investigation Project in Greece, executed by the Food and Agriculture Organization of the United Nations in co-operation with the Institute for Geology and Sub-Surface Research."

[559] Burdon, David Joseph *The karst groundwater resources of Parnassoc-Ghiona, Greece: a report*. Papakis, Nicholas. Athens, GR: Institute for Geology and Subsurface Research; United Nations Special Fund, Karst Groundwater Investigations, **1963**. illus: b&w: drawings, graphs, maps, tables; 18 maps folded in text; xxiv, 301p; app, bibl.

Apps with separate numbering.

[560] Burge, John W. *Basic underwater cave surveying*. Branford, FL, US: Cave Diving Section of the National Speleological Society, **1988**. illus: b&w: charts, drawings, photos, tables; [vi], 134p; app, gloss, index; 23 cm.

Copyright 1987. pp 129-134 blank for notes.

[561] Burger, Andre (ed) *Hydrogeology of karstic terrains*. Dubertret, L. (ed). Paris, FR: International Association of Hydrogeologists, **1984**. illus: b&w: photos; col: maps; 264p; gloss.

Illus.

[562] Burger, Andre (ed) *Hydrogeology of karstic terrains: with a multilingual glossary of specific terms*. Dubertret, L. (ed); IAH Commission for Hydrogeology of Karst. Paris, FR: International Association of Hydrogeologists, **1975**. International of Geological Sciences Series B number 3. illus: b&w: charts, drawings, graphs, maps, tables; partly col: maps; 190p; bibl, gloss.

Cover title: Hydrogeology of karstic terrains. Partly English, partly French.

[563] Burgess, Robert Forrest *The cave divers: illustrated with photographs and drawings*. New York, NY, US: Dodd, Mead, & Company, **1976**. illus: b&w: drawings, maps, photos; xiii, 239p; app, bibl, index; 24 cm.

The stories of dozens of explorations in underwater caves and sinkholes. An account of cave diving from 1913 through the 1970s. An important advance in the distribution of cave diving information.

[564] Burkhardt, Rudolf *The Speleological Club in Brno 1945-1973*. Havel, Hugo; Mayer, Stanislaw; Musil, František; Panoš, Vladimír; Ryšav, Přemysl. Olomouc, CZ: Speleological Club in Brno and the Department of Geography, Faculty of Natural Sciences of the Palack University in Olomouc, **1973**. illus: b&w: maps, photos, tables; [iv], 28p, [5].

[565] Burnsville Cove Symposium *Burnsville Cove Symposium*. Davis, Nevin W.; White, William Blaine, 1934–. Huntsville, AL, US: National Speleological Society, **1982**. The NSS bulletin v. 44 no 3. illus: b&w: maps; 1 folded leaf of pls; 55p; bibl; 28 cm.

Title from cover. Illus, pp 47-101

[566] Burrows, Russell *Mystery cave of many faces: first in a series on the saga of Burrows' Cave*. 1992 ed. ix, 252p; 25 cm.

1st ed. Rydholm, Fred. Marquette, MI, US: Superior Heartland, **1991**. xxiv, 252p; bibl, index; 24 cm.

Illus.

[567] Bush, Kent *Oregon Caves National Monument collections management plan*. Seattle, WA, US: United States Dept of the Interior, National Park Service, Pacific Northwest Region, **1995**. Collection management plan. [xii], 187p; 28 cm.

Cover title: Collections management plan, Oregon Caves National Monument.

[568] Butcher, A.L. *Cave surveying*. Railton, C. L. GB: Cave Research Group of Great Britain, Jul, **1966**. Transactions of the Cave Research Group of Great Britain vol 8 no 2. illus: b&w: charts, drawings, maps, tables; [ii], 37p; bibl.

[569] Butcher, Arthur Lepine *Cave survey*. Leamington Spa, Warwickshire, GB: Cave Research Group of Great Britain, **1951**. Cave Research Group of Great Britain Publications series no 3. illus: b&w: charts, maps; x, 40p; bibl; 25 cm.

[570] Butler, Dwain K. *Cavity detection and delineation research: Report 1: Microgravimetric and magnetic surveys: Medford Cave Site, Florida.* United States Army Corps of Engineers Geotechnical Laboratory. Vicksburg; Springfield, MS; VA, US: The Station; available from NTIS, Mar, **1983**. Technical report (U.S. Army Engineer Waterways Experiment Station) no GL 83 1 Report 1. illus: b&w: charts, graphs, maps, photos, tables; 1 chart folded in text; Various pagination, 136p; bibl; 27 cm.

Cover title: "March 1983. Prepared for Office, Chief of Engineers, U.S. Army under CWIS Work Unit No. 31150."

This is the first of five reports (J.R. Curro, Seismic Methodology; S.S. Cooper, Acoustic resonance and self-potential; D.K. Butler, et al, Microgravimetric survey; R.F. Ballard, Electromagnetic (radar) techniques). Describes specialized techniques used in the detection of shallow voids and tunnels.

[571] Butler, Dwain K. *Cavity detection and delineation research: Report 4: Microgravimetric survey: Manatee Springs site, Florida.* Whitten, Charlie B.; Smith, Fred L.; United States Army Corps of Engineers Geotechnical Laboratory. Vicksburg; Springfield, MS;VA, US: The Station ; available from NTIS, **1983**. Technical report (U.S. Army Engineer Waterways Experiment Station) ; no. GL-83-1 Report 4. illus: b&w: charts, graphs, maps, tables; Various pagination, 59p; bibl; 27 cm.

Cover title. "March 1983. Prepared for Office, Chief of Engineers, U.S. Army under CWIS Work Unit No.31150."

[572] Byer, Charles A. *The caves of Capistrano.* npop: McKay, **1938**. 240p; 21 cm.

[573] Byfield, Colin Charles *Let's look at Jenolan Caves: an introduction to the features of Australia's famous limestone caverns.* Hughes, ACT, AU: Marcol Publications for the N.S.W. Dept of Tourism, **1977**. A *See and Know Australia* resource book. illus: b&w: maps; 28p; bibl; 24 cm.

Title from cover.

1979 rev ed. 1979.
1984 rev ed. 1984.

[574] Cadoux, Jean, 1930– *One thousand metres down.* 1957 American ed. New York, NY, US: A. S. Barnes and Company, **1957**. b&w front photo; 22 cm.

1957 British ed. Irving, Robert Lock Graham, 1877– (trans). London, GB: George Allen and Unwin Ltd, **1957**. illus: b&w: maps, photos; xii, 178p, [1], 22p pls; 23 cm.

Cover title: One Thousand metres down: a journey to the starless river. Besides being a translation of Opération-1000, this edition contains 5 additional pages describing events since the French version was prepared. 22 pls.

The full story of explorations in the Gouffre Berger in southeast France to a depth of over 3,750 feet from its discovery in 1953 to the summer of 1956. From 1956 to 1967 this was the deepest known cave in the world.

[575] Cahn, Phyllis H. *Comparative optic development in* Astyanax mexicanus *and in two of its blind cave derivatives.* New York, NY, US: American Museum of Natural History, **1958**. Bulletin of the American Museum of Natural History vol 115 article 2 pp 69-112. illus: b&w: photomicrographs, tables; 44p; bibl.

[576] Caiar, Ruth *One man's dream; the story of Jim White, discoverer and explorer of the Carlsbad Caverns: a biography.* White, Jim, 1919–. New York, NY, US: Pageant Press Inc, **1957**. illus: b&w: photos, ports; b&w front port; [v], 111p, 8p pls; app; 21 cm.

Appendix: Facts about Carlsbad Caverns.

Jim White's discovery and exploration of Carlsbad Caverns, NM.

[577] Caldwell, Douglas E. *Limnological survey of cavern pools: proposal to National Park Service, Carlsbad Caverns National Park.* University of New Mexico Biology Department. [Albuquerque, NM, US]: np, **1981**. illus: b&w: maps; 32p; bibl; 28 cm.

Title from cover.

[578] California. Department of Parks and Recreation *Mitchell Caverns State Reserve.* [Sacramento, CA, US]: np, **1971**. illus: partly col: repros; col: cartoons.

[579] Call, Richard Ellsworth, 1856-1917 *Grand excursion to Mammoth Cave.* Boston, MA, US: np, **1902**. [12].

A reprint from the pages of the National Magazine, 9 illus. This is a very rare item. Not listed in OCLC.

[580] Call, Richard Ellsworth, 1856-1917 *Mammoth Cave, Kentucky.* [Louisville, KY, US]: Passenger Department, Louisville & Nashville R.R., **1897**. illus: b&w: maps; [12]; 24 cm.

Illus. Caption title: In Kentucky's Mammoth Cave by R. Ellsworth Call, Ph.D. A reprint from the pages of the national magazine published in Boston, Mass. Master microform held by: LCP. Microopaque. Louisville, Ky., Lost Cause Press, 19—.

[581] Call, Richard Ellsworth, 1856-1917 *The Mammoth Cave, Kentucky; a sketch.* 2nd? ed. [Louisville, KY, US]: Passenger Dept, Louisville & Nashville Railroad Co, **c1890**. illus: b&w: drawings, photos; partly col: maps; front: b&w drawing; 29p; 19 cm.

Different cataloging records indicate different pagination (28p, 32p) and dates (1890, 1899, 1902). Different versions have same text, but different pictures. Master microform held by: LCP. Microfiche. Louisville, Ky., Lost Cause Press, 19—. On cover: Two hundred miles underground.

[582] Callahan, Thomas P. *The history of the Lockport caves.* Lockport, NY, US: Lockport Hydraulic Raceway Co, LTD, **c1974**. illus: b&w: cartoons, maps, repros; [v], 29p, [13].

Limited edition; copy examined appeared to be no 48; copy incomplete.

[583] Callard, Thomas Karr *Contemporaneity of man with the extinct Mammalia, as taught by recent cavern exploration, and its bearing upon the question of man's antiquity.* London, GB: D. Bogue, **[1880]**. [i], 34p; 20.5 x 13 cm.

Photocopy examined. No illus.

[584] Callow, Philip *Cave light.* Bradford, GB: np, **nd**. 43p; 22 cm.

This edition is limited to 500 copies of which 26 have been lettered A to Z and signed by the author—-Colophon.

[585] Camacho, Ana Isabel (ed) *The natural history of biospeleology.* [Madrid, ES]: Museo Nacional de Ciencias Naturales: Consejo Superior de Investigaciones Cientificas, **1992**. Monografias del Museo Nacional de Ciencias Naturales. illus: b&w: drawings, graphs, photos, tables; xxi, 680p, [1]; bibl, index; 24 cm.

A multi-authored volume with a thorough discussion of both the cave environment and methods for collecting fauna and their interactions. Numerous case studies are presented as well. An excellent coverage of the history of biospeleology; fully referenced.

[586] Camberlin, G. Will *The exploration and survey of McBridge Cave, Jackson County, Alabama.* Smith, Marian O. npop: np, **1993**. illus: b&w: maps, photos; title page on front lining page; b&w photos on rear lining page; 10p.

In two parts.

[587] Cameron McNamara Pty Ltd *Jenolan Caves Reserve: plan of management..* Tourism Commission of New South Wales. New South Wales. Dept of Mines. North Sydney, NSW, AU: np, **1989**. illus: b&w: maps; some folded lvs; 115p; bibl; 30 cm.

2 vols (69, 46 lvs), illus. Prepared for Tourism Commission of New South Wales, New South Wales Department of Lands by Cameron McNamara Consultants.

[588] Campbell, Jim *The Grampian caving manual: an introduction to caving and potholing for the beginner.* Edinburgh: Grampian Speleological Group, **1978**. Occasional publications, Grampian Speleological Group no 4. vi, 44p; bibl, index; 30 cm.

Illus.

[589] Campbell, John B., 1944– *A new analysis of Kent's Cavern, Devonshire, England.* Sampson, Clavil Garth, 1941–. [Eugene, OR, US]: [Dept. of Anthropology, University of Oregon], **1971**. University of Oregon Anthropological Papers no 3. vii, 40p; bibl; 28 cm.

[590] Campbell, Newell Paul, 1938– *Caves of Montana.* Montana Bureau of Mines and Geology. Butte, MT, US: Montana College of Mineral Science and Technology, **1978**. Bulletin 105 State of Montana Bureau of Mines and Geology. illus: b&w: charts, drawings, graphs, maps, tables; maps folded in pocket; vi, 169p, 2p pls; bibl, index. 84 figures.

Catalog of 302 caves in the state; some are shelters. Descriptions; locations, given only to section (square mile); maps of 65 caves.

[591] Campbell, Newell Paul, 1938– *Glacier Park cave study.* npop: np, Oct, **1975**. 21p.

6 illus. Unpublished report of the explorations of Poia Lake Cave and West Tunnel Cave, both located in Glacier National park. Caves were explored by author and Jens & Kathy Munthe 3-7 Sept 1975.

[592] Canada. National and Historic Parks Branch *Nahanni.* Floquet, Francois (direct); Bertolino, Daniel (direct). npop: np, **1972**.

28 min sd col, 16 mm. French version also issued.

Covers an expedition by speleologist Jean Poirel and his crew who parachute near the headwaters of the South Nahanni River in western Canada, travel downstream by inflatable boats, scale the canyon walls above the river, and explore Nahanni's caves.

[593] Canada. National Parks Branch *The Nakimu Caves, Glacier Dominion Park, B.C..* Ottawa, CA: Dominion Parks Branch, Dept of the Interior, **1914**. illus: b&w: maps, photos; folded map; 29p; 24 x 13 cm.

Title from cover. 10 b/w photos.

[594] Canada. National Parks Service *Castle Guard Cave, challenge under the glacier.* Ford, Derek Clifford, 1935–(prod); Perou, Syd (direct, photo); Hurson, Tim (script); Strange, Cy (narr); Gradidge, Havelock (film ed); Canada. Dept of Indian Affairs and Northern Development. [Montreal, CA]: National Film Board of Canada, **1974**.

2 reels, 51 min, sd, col, 16 mm.

Records the recent exploration of the perilous, but fascinating Castle Guard Cave, largest cavern known in Canada, located under the Columbia Icefield in Banff National Park, Alberta.

[595] Canadian Broadcasting Corporation *The Blue holes of Andros.* Bitman, Roman (prod, direct); Whitehead, William (script); McCallum, Tom (photo); Wallace, Bob (photo). npop: np, **1972**. The Nature of Things [Motion picture Canadian Broadcasting Corp].

30 min, sd, color, 16 mm.

Visits the vast network of underwater caves in the Bahamas.

[596] Cansell, Anthony *Stump Cross Caverns - the underground wonderland, their development and exploration.* Paleley, Bridge, GB: B. Gill, **[after 1982]**. illus: b&w: drawings, maps; back has color photos; front has title, contents, and photo; 16p; 21 x 15 cm.

Cover title: Stump Cross caverns; official guide to the show caves with a short account of the discovery and exploration.

[597] Cansell, Anthony *The Upper Wharfedale Fell Rescue Association 1948-1968.* Skipton, GB: np, **c1969**. illus: b&w: ads; ads on lining pages; 35p; 21 x 14 cm.

[598] Carballo, Jesus Maria *The cave of Altamira and other caves with paintings in the province of Santander.* 1965 ed. [Santander], ES: Patronato de las Cuevas Prehistoricas de la Provincia de Santander, **1965**. [i], 111p, 6p pls; 17 cm.

Col illus. Title on cover: Prehistoric caves Altamira, Santander. Spanish editions have titles: La cueva de Altamira y otras cuevas con pinturas en la provincia de Santander; Cuevas prehistoricas en Santander: Altamira.

3rd ed. [Santander, ES]: Patronato de las cuevas prehistoricas de la provincia de Santander, **1963**. illus: b&w: drawings, maps, photos; col: photos; one fold-out map tipped in; xvi, 96p; 16.5 x 12 cm.

Translated from Spanish. Cover title: Prehistoric caves: Altamira Santander. 1963 and 1965 editions examined.

[599] Carlsbad Cave National Monument *Carlsbad Caverns, New Mexico.* [Chicago, IL, US]: Published by The Atchison, Topeka and Santa Fe Railway Co, **1930**. 62p.

Title from cover. Illus.

[600] Carlsbad Caverns National Park *Land protection plan: Carlsbad Caverns National Park: prepared by Carlsbad Caverns National Park with assistance from the Southwest Regional Office, National Park Service..* United States National Park Service Southwest Region. Santa Fe, NM, US: np, Dec, **1984**. illus: b&w: maps; 20p; 28 cm.

Cover title: Carlsbad Caverns National Park: land protection plan.

[601] Carr, Joseph *Rambles about Ingleton, in caves by rivers and on mountains in the spring of 1865..* Lancaster, GB: np, **1865**. iv, 32p; 16 x 9.5 cm.

Cover title: Rambles about Ingleton

[602] Carstens, Kenneth Charles *Of caves and shell mounds.* Watson, Patty Jo, 1932–. Tuscaloosa, AL, US: University of Alabama Press, **1996**. illus: b&w: drawings, maps, tables; xvii, 209p; bibl, index; 24 cm.

Illus. Based on a symposium of the Southeastern Archeological Conference held in Tampa, Florida, 1989.

Collection of papers about the archaeology of the Big Bend and Mammoth Cave areas along Kentucky's Green River. Fourteen papers written in honor of Patty Jo Watson.

[603] Carter, R. L. *Cave Diving Group: Peak District sump index 1994.* Cordingley, J. N. (ed). [Bristol, GB]: The Cave Diving Group, Aug, **1994**. illus: b&w: maps; p 79 on rear lining page; 79p; 30 x 21 cm.

[604] Cartwright, Roger *A guide to the Craven Dales.* Skipton, GB: J.J. Waterfall, **1946**. illus: b&w: photos; v, 84p.

Illus. Contains many cave descriptions and photos.

[605] Castany, G. (ed) *Hydrogeology of karstic terranes, case histories.* Groba, E.; Romejn, E. (eds). Paris, FR: International Association of Hydrogeologists, **1984**. International Contributions to Hydrologeology vol 1. 264p.

[606] Casteret, Norbert, 1897-1987 *Cave men new and old.* Irving, Robert Lock Graham, 1877–(trans). London, GB: J.M. Dent & Sons, Ltd, **1951**. illus: b&w: maps, photos, ports; xiii, 178p, 16p pls; index.

Among the first to popularize caving, Casteret's series of accounts of his personal experiences in cave exploration set the pattern for this genre of writing.

[607] Casteret, Norbert, 1897-1987 *The darkness under the earth.* 1954 ed. London, GB: J. M. Dent & Sons, Ltd, **1954**. illus: b&w: maps, photos; xiv, 168p, 15p pls; index; 22 cm.

Translation of Tenebres. Divided into two parts: *The Joys of Speleology* and *The Dangers of Speleology.*

1955 ed. New York, NY, US: Henry Holt and Company, Inc, **1955**. illus: b&w: maps, photos; xii, 174p, 8p pls; index; 22 cm.

[608] Casteret, Norbert, 1897-1987 *The descent of Pierre Saint-Martin.* Warrington, John (trans). 1955 ed. London, GB: J.M. Dent & Sons Ltd, **1955**. illus: b&w: drawings, maps, photos; front: b&w photo; xi, 160p, 16p pls; index; 22 cm.

"Translated and re-arranged from ... *Trente ans sous terre.*"

1956 ed. Warrington, John (trans). New York, NY, US: Philosophical Library, **1956**. front: b&w photo; xi, 160p, 15p pls; app.

[609] Casteret, Norbert, 1897-1987 *More years under the earth.* Dinnage, Rosemary (trans). London, GB: Neville Spearman, **1962**. illus: b&w: photos; 164p, 24p pls; 23 cm.

28 b&w photos. Other British printings: 1962, London: Scientific Book Club. Translation of *Aux pays des eaux folles.*

Popular non-scientific book - an interesting read.

[610] Casteret, Norbert, 1897-1987 *My caves.* Irving, Robert Lock Graham, 1877–(trans). London, GB: J. M. Dent & Sons, Ltd, **1947**. illus: b&w: photos, ports; xi, 172p; index; 22 cm.

26 illus. Sequel to the author's Ten years under the earth.

[611] Casteret, Norbert, 1897-1987 *Ten years under the earth.* 1940 ed. London, GB: Readers Union, by arrangement with J. M. Dent, **1940**. xv, 240p, 24p pls; index; 22 cm.

1952 ed. Harmondsworth, GB: Penguin Books, **1952**. Penguin Books. Travel and adventure ; 84. 220p, 32p pls; index; 18 cm.

46 illus.

1963 ed. London, GB: J.M. Dent & Sons, Ltd, **1963**. xiv, 240p, 24p pls; 22 cm.

1975 reprint ed. Watson, Red (pref). Teaneck, NJ, US: Zephyrus Press, **1975**. Series Speleologia. illus:

b&w: drawings, maps, photos; xv, 239p, 40p pls; index; 22 cm.

Reprint of the 1939 ed published by J. M. Dent & Sons, London.

1st ed. Mussey, June Barrows, 1910–(ed, trans); Martel, E. A. (pref). New York, NY, US: Greystone Press, **1938**. illus: b&w: drawings, maps, ports; front: b&w drawing; xx, 283p, 24p pls; 22 cm.

Translator's note: "This volume contains the essential parts of two successive books from Monsieur Casteret's pen: *Dix ans sous terre...* and *Au fond des gouffres*."

First half tells of exploration of Cigalere, a large cave in the Pyrenees. Second part is about cave diving and exploring underground streams. This widely known caving book has been published in at least 12 languages.

[612] Catcott, George Symes *A descriptive account of a descent made into Penpark-hole, in the parish of Westbury-upon-Trim, in the county of Gloucester, in the year 1775*. Bristol, GB: Printed by J. Rudhall, in Smallstreet, and sold by the booksellers in Bristol; and Meyler and Bull, Bath, **1792**. folded front map; x, 45p, [1]; 21 cm.

Half-title p.[i].

[613] Cave Diving Association of Australia, Inc *Cave Diving Association of Australia, Inc: conference 1977 held at Mount Gambier, South Australia*. North Adelaide; Melbourne, SA;VIC, AU: Cave Diving Association of Australia Inc, **[c1977]**. Occasional Paper no 1. illus: b&w: charts; [iv], 32p.

Proceedings published after the conference.

[614] Cave Diving Section. National Speleological Society *NSS Cave Diving Section 1986 members' manual*. Branford, FL, US: National Speleological Society, Inc, Cave Diving Section, **1986**. [24]; 28 cm.

Title from cover. Illus.

[615] Cave Rescue Organisation *Caving safety code: first aid cave rescue procedure*. Manchester, [GB]: The Cave Rescue Organisation, **1960**. illus: b&w: drawings; 15p, [1].

[616] Cave Research Associates *Cave Studies No. 1-11*. Graham, R.; Lange, A. R.; de Saussure, R.; Western Speleological Institute. Castro Valley, CA, US: Cave Research Associates, **1959**. 23 cm.

First issue is a 1959 reprint in 1 v. of the separately published no 1-11, Jul, 1953-May, 1959.

[617] Cave Research Foundation *CRF personnel manual*. 3rd ed. Washington, DC, US: Cave Research Foundation, May, **1981**. illus: b&w: cartoons, charts, drawings, maps, photos, tables; Various pagination, xxi, 155p; bibl, index; 28 cm.

[618] Cave Research Foundation *The quadrangle maps of Carlsbad Cavern, New Mexico*. Dell, David; United States National Park Service. 1992 prelim ed. Golden, CO, US: The Cave Research Foundation, **1992**. illus: b&w: maps; [16]; 28 x 43 cm.

Title from P. [2] of cover. Cover title: Carlsbad Cavern, Carlsbad Caverns National Park. Caption title: 1992 preliminary edition of the quadrangle maps of Carlsbad Cavern. Scales differ. "November 7, 1992."

Atlas of quadrangle maps.

[619] [Cave Research Group] *Fauna collected from caves as recorded in the C.R.G. fauna records: part I (1938-39)*. [GB]: Cave Research Group, Dec, **1955**. Cave Research of Great Britain, Biological Supplement. illus: b&w: tables; 12p.

[620] [Cave Research Group] *Fauna collected from caves as recorded in the C.R.G. fauna records: part II (war years 1940-46 and 1945-46)*. [GB]: Cave Research Group, Oct, **1956**. Cave Research of Great Britain, Biological Supplement. illus: b&w: drawings, tables; 15p.

[621] Cave Research Group of Great Britain *Hypogean fauna and biological records 1970, [and] fauna of Gibraltar caves*. [Ledbury, GB]: Cave Research Group of Great Britain, **1971**. The Transactions of the Cave Research Group of Great Britain v 13(3). illus: b&w: maps, photos, tables; text on rear lining page; front: b&w photo; 93p, 1p pls; bibl; 30 cm.

Title from cover.

[622] [Cave Research Group of Great Britain] *Hypogean fauna and biological records 1970-71*. GB: Cave Research Group of Great Britain, Nov, **1972**. Transactions of the Cave Research Group of Great Britain vol 14 no 4. illus: b&w: drawings, maps, photos, photomicrographs, tables; text on lining pages; front: photomicrograph; 89p.

Cover has titles of several articles but inside title is monographic.

[623] Cave Research Group of Great Britain Friends and Members *A volume of essays presented to Brigadier E. A. Gelnnie on the occasion of his 80th birthday, July 18th 1969*. GB: Cave Research Group of Great Britain, Jul, **1969**. Transactions of the Cave Research Group of Great Britain vol 11(2). illus: b&w: charts, maps, photos, tables; text on rear lining page; front: b&w photo; 65p, 10p pls.

[624] Caving Club SC 33 *1993 Jamaica expedition*. Heule, BE: Caving Cub SC 33 v.z.w., **c1994**. illus: b&w: ads, charts, drawings, maps, photos, repros, tables; col: photos; back cover is a fold-out map of Dromilly Cave; photocopies of newspaper articles on the club on lining pages; 65p; bibl; 29 x 20 cm.

Photocopy examined.

[625] [Central Indiana Grotto of the National Speleological Society] *Selected caves of Orange county*. Indianapolis, IN, US: Speleological Studies Group and Central Indiana Grotto of the National Speleological Society, Jun, **1973**. illus: b&w: maps; [i], 15p, [13]; bibl.

Cover title: Selected caves of Orange county Indiana.

[626] Čerňanská, Miroslava (ed) *Welkome (sic) to our*

Slovak caves. [SK]: Czechoslovak Press Agency, **1973**. illus: b&w: maps; partly col: ads; col: photos; map of area on front lining page; 15p; 31 x 24 cm.

Title from cover. Lining pages included in pagination.

[627] Cervino, Donald Jay *Water features of the European Upper Paleolithic decorated cave environment*. npop: np, **1990**. illus: b&w: maps; vi, 60p; bibl; 29 cm.

Honors essay: Dept. of Anthropology, University of North Carolina at Chapel Hill, 1990. Illus.

[628] Chabert, Claude *E.A. Martel 1859 - 1939 bibliographie*. de Courval, M. [Paris, FR]: Imprimerie Margelin & Cie; Club Alpin Francais, **1971**. Travaux scientifiques du Speleo-club de Paris. illus: b&w: drawings; 104p; 24 cm.

103p. At head of title: C. Chabert. M. de Courval. French language bibliography containing Martel's English publications.

[629] Chamberlin, Joseph Conrad, 1898– *New and little-known false scorpions, principally from caves, belonging to the families Chthoniidae and Neobisiidae (Arachnida, Chelonethida)*. New York, NY, US: [American Museum of Natural History], **1962**. Bulletin of the American Museum of Natural History v 123 article 6 pp 303-352. 50p; 21 cm.

[630] Chamberlin, Joseph Conrad, 1898– *The Vachoniidae—a new family of false scorpions: two new species from the caves of Yucatan*. Salt Lake City, US: University of Utah, **1947**. Biological series v 10 no 4; Bulletin of the University of Utah v 38 no 7. illus: b&w: drawings; 15p; 26 cm.

[631] Chamberlin, Ralph Vary, 1879-1967 *A new geophiloid chilopod from Potter Creek Cave, California*. Berkeley, CA, US: University of California Press, **1930**. University of California Publications in Zoology v 33 no 14. illus: b&w: drawings, tables; 27 cm.

18 figs, 4 tables.

[632] Chamberlin, Ralph Vary, 1879-1967 *On centipeds and millipeds from Mexican caves*. Salt Lake City, UT, US: University of Utah, **1942**. Biological series vol 7 no 42; Bulletin of the University of Utah [new ser] v 33 no 4. illus: b&w: drawings; 19p.

[633] Champe, John L. *Ash Hollow Cave: a study of stratigraphic sequence in the central great plains*. Lincoln, NE, US, Oct, **1946**. University of Nebraska Studies New Series no 1. illus: b&w: charts, graphs, maps, photos; ix, 130p, 22p pls; app, bibl, index.

[634] Chan, Y. C *Factors affecting sinkhole formation*. Hong Kong, HK: Geotechnical Engineering Office, **1994**. GEO report no 28. 46p; bibl; 30 cm.

Illus. Originally produced in July 1988 as GCO Technical Note no. TN 4/88.

[635] Chandler, Peter *A trip through time: a drive-yourself guide to the landforms and geology of the Waitomo Caves area*. 3rd ed. Waitomo Caves, NZ: Waitomo Caves Museum Society, Inc, **1987**. illus: b&w: charts, drawings, maps, photos; bw photo of Marokopa Falls on front lining page; [iii], 50p; app, bibl, gloss.

Cover title: A trip through time; A guide to landforms: Waitomo Caves - Marokopa Coast. First edition 1984, second edition 1985.

[636] Chandler, Robert V. *Springs in Alabama*. Moore, James D. Tuscaloosa, AL, US: Geological Survey of Alabama, Water Resources Division, **1987**. Geological Survey of Alabama Circular 134. illus: b&w: graphs, maps, tables; 1 map folded in rear pocket; [v], 95p, [1]; app, bibl.

[637] Chapman, J. Roy *The caves of Georgia*. Geological Survey of Georgia. Atlanta, GA, US: Geological Department of Mines, Mining and Geology, **1963**. illus: b&w: cartoons; 20p; 29 cm.

Reprint. Illus. This is a 5" by 8" stapled together. 2nd version on 8 by 11 pages; reprint of article originally prepared for publishing in the Georgia Mineral Newsletter, 1948.

[638] Chapman, Philip *Caves and cave life*. London, GB: Harper Collins Publishers, **1993**. New Naturalist 79. illus: b&w: drawings, maps, photos; col: ads; 219p, [5]; bibl, gloss, index; 23 cm.

97 illustrations.

Highly readable general introduction to the cave environment and inhabitants of caves. It deals not only with species limited to caves (troglobites) but also with many part-time inhabitants including birds and bats.

[639] Charnock, S. J. *Underneath the arches: caving in the RibbleHead District*. [Yorkshire], GB: np, Feb, **1994**. illus: b&w: cartoons, drawings; 24p; gloss; 21 x 15 cm.

[640] Chasen, Frederick Nutter, 1896– *Report on the "birds' nest" caves and industry of British North Borneo: with special reference to the Gomantong caves*. npop: printed at the Govt Print Office, **1931**. illus: b&w: tables; 24p; 24 x 18 cm.

Title from cover.

[641] Chatterton, John B. *Report of the Queen Mary College Society spelaeological expedition to Yugoslavia*. McCulloch, Stuart J. London, GB: Queen Mary College Caving Society, Sep, **1971**. illus: b&w: graphs, maps; 1 map slipped in back; 45p, 5p pls; 30 x 21.5 cm.

[642] Chauvet, Jean-Marie *Chauvet Cave: the discovery of the world's oldest cave paintings*. Deschamps, Eliette Brunel; Hillaire, Christian; Clottes, Jean (epilog); Bahn, Paul G. (trans, forw). London, GB: Thames and Hudson, **1996**. illus: b&w: maps, tables; col: photos; 135p; bibl, index.

Published in US by Harry N. Abrams as Dawn of art; the Chauvet cave; the oldest known paintings in the world.

[643] Chauvet, Jean-Marie *Dawn of art: the Chauvet Cave: the oldest known paintings in the world*. Deschamps, Eliette Brunel; Hillaire, Christian; Clottes, Jean (epilog); Bahn, Paul G. (forw); Hillaire, Christian. New York, NY, US: H.N. Abrams Inc Publishers, **1996**. illus: b&w: maps; partly col: maps; col: photos; 135p; bibl, index, note; 32 cm.

Col illus.

[644] Checkley, Dave Lancaster *LUSS expeditions to Tresviso and the Picos de Europa in northern Spain, 1974-1977*. [Lancaster, GB]: [Lancaster University Speleological Society], **1977**. illus: b&w: maps; 1 map folded in pocket; 68p; 30 cm.

[645] Chekley, D. *La Sima 56 (Picos de Europa-España)*. Jones, H.; Foster, S.; Lancaster University Speleological Society. Universidad Politecnica de Madrid. Seccion Espeleologia Ingenieros Industriales. Madrid, ES: University of Lancaster, Universidad Politeéchrca de Madrid, **1985**. illus: b&w: charts, drawings, maps, photos, tables; 85p, [2]; index; 24 cm.

Cover title: Sima 56. English and Spanish in parallel columns. Captions also in English. On cover: Lancaster University Speleological Society. Seccion espeleologia, ingenieros, industriales.

[646] Chenger, John D. *Visitor's guide to Laurel Caverns Park*. Damon, Paul Herbert, 1934–; Kennedy, Jim "Crash". Uniontown, PA, US: John Chenger, **1995**. illus: b&w: charts, maps, photos; col: photos; Folded map glued inside back cover.; 58p.

[647] Chevalier, Pierre *Subterranean climbers; twelve years in the world's deepest chasm*. Hatt, E. M. (trans). London, GB: Faber and Faber Limited, **1951**. illus: b&w: drawings, maps, photos; b&w map on front; 223p, 16p pls; app, index; 23 cm.

Personal account of Dent deColles, France; Translation of Escalades souterraines. 23 photos, 9 drawings, 10 maps (incl front).

The story of the 1936-1947 exploration of the Dent de Crolles cave network in SE France linking the Glaz and Guiers-Mort caves during the Nazi occupation of France. Included are 23 b&w photos and 10 maps; appendices include the hydrogeology of the Dent de Crolles, cave depths chart, and other exploration-related charts.

[648] Chevalier, Pierre *Subterranean climbers; twelve years in the world's deepest chasm*. Teaneck, NJ, US: Zephyrus Press Reprint, **1975**. 223p, 9p pls; 22 cm.

Softbound reprint of the 1951 ed, 23 photos, 10 sketch maps, 9 sketches, 7 appendixes. Includes short 6 p introduction to new ed and this has 2 page list of deepest caves.

[649] Chillagoe Caving Club *Broken River karst: a speleological field guide North Queensland Australia*. Queensland National Parks and Wildlife Service. npop: Chillagoe Caving Club and Queensland National Parks and Wildlife Service, **nd**.

Un-numbered colored pages; photos are photocopied.

[650] Chillagoe Caving Club (comp) *Numerical index - area maps*. [QLD, AU]: Chillagoe Caving Club, **nd**. illus: b&w: maps, tables; 41p.

Title page notes that this is confidential information, not to be released to the public.

[651] Chillagoe Caving Club (comp) *A speleological field guide of the towers and caves of Chillagoe - Munanga - Rookwood areas in Far North Queensland, Australia*. Cairns, QLD, AU: Chillagoe Caving Club, **1982**. illus: b&w: maps, photos; [i], 187p; app.

Cover title: Chillagoe karst: A speleological field guide of the towers and caves of Chillagoe - Munanga - Rookwood areas in Far North Queensland, Australia.

[652] Chillagoe Caving Club (comp) *A speleological field guide of the towers and caves of Chillagoe - Munanga - Rookwood areas in Far North Queensland, Australia. Supplement for caves CH301-CH400*. Cairns, QLD, AU: Chillagoe Caving Club, Jul, **1988**. front: b&w photo; [ii], 40p.

Cover title: Chillagoe karst: A speleological field guide of the towers and caves of Chillagoe - Munanga - Rookwood areas in Far North Queensland, Australia. Supplement for caves CH301-CH400

[653] Chilton, Charles, B. 1860– *The subterranean Crustacea of New Zealand: with some general remarks on the fauna of caves and wells*. London, GB: Printed for the Linnean Society by Taylor and Francis, Red Lion Court, Fleet Street [etc.], **1894**. The Linnean Society of London Transactions 2nd ser Zoology vol 6 pt 2. 122p, 8p pls; bibl; 28 cm.

pp 163-284. Illus.

[654] Chin, Lucas *Archaeological work in Sarawak: with special reference to Niah Caves*. rev ed. [Kuching, MY]: Sarawak Museum, **1980**. Sarawak Museum Occasional Paper no 1. illus: b&w: photos; Text on front lining page; bibliography on rear lining page; [iv], 36p; bibl; 16 cm.

Illus. Another OCLC record lists Vanguard Press as the publisher.

[655] Chin, Lucas *Summary of archaeological work in Sarawak: with special reference to Niah Caves*. [Kuching, MY]: Borneo Literature Bureau for Sarawak Museum, **1977**. Sarawak Museum Occasional Paper no 1 Kertas risalah semasa Muzium Sarawak no 1. iv, 28p; bibl; 16 cm.

Illus.

[656] Christiansen, Kenneth Allen, 1924- *Notes on Mexican cave Pseudosinella (Collembola: Entomobryidae) with the description of six new species*. [Mexico City, MX], **1982**. Folia entomologica mexicana no 53, Latin American documents reel 1033, item 18. 22p; bibl; 23 cm.

Abstract in English and Spanish. Illus.

[657] Christopher, N. S. J. (ed) *Cambrian Caving Council handbook 1969*. Somerset, GB: Cambrian Caving Council, **1969**. [28]; 25 x 20 cm.

[658] Church, Flora, 1954– *Mitigation of the Brady*

Run Rockshelter 3: a multi-component site in Washington Township, Lawrence County, Ohio. Nass, John P.; Ericksen, Annette G.; Skinner, Shaune M.; P & L Systems, Inc; Ohio Dept of Natural Resources. Columbus, OH, US: Archaeological Services Consultants Inc, Jun, **1990**. illus: b&w: maps, photos; viii, 143p; bibl; 28 cm.

Submitted to: P&L Systems ... Columbus, Ohio. Lead agency: O.D.N.R.

[659] Cigna, Arrigo A. (ed) *Caving in Italy.* Bologna, IT: Sociéta Speleologica Italiana, **1966**. Supplement 15 Speleologia. 2 folded maps; text on front lining; map on rear lining page; 32p; bibl.

Distributed to participants of the 9th International Congress of Speleology. Text on rear cover.

[660] C'ilek, V'acav *Czechoslovak speleological expedition to Nepal Himalaya 85.* Kácha, Stenislav; Hašek, Zdeněk. npop: np, **c1985**. illus: b&w: drawings, maps; 11p; bibl; 30 x 22.5 cm.

Cover title: Czech Speleological Expedition (Report) Himalaya '85.

[661] Clapham, Christopher S. *The Caves of Sof Omar.* Robson, Eric. Addis Ababa, ET: Ethiopian Tourist Organization, **1967**. illus: b&w: drawings, maps, photos; partly col: maps; col: photos; 1 map folded in rear pocket; 11p, [1]; 22 cm.

[662] Clark, E. John *Jenolan Caves, New South Wales, Australia.* Bondi, NSW, AU: Bartel Photography Pty Ltd, **nd**. illus: col: photos; text on front lining page, map on back lining page; [16].

Limited text.

[663] Clark, John Desmond, 1916– *Further excavations (1939) at the Mumbwa caves, Northern Rhodesia.* Cape Town, ZA: Royal society of South Africa, Cape Town, **1942**. 69p; bibl; 25 cm.

pp [133]—201, pls xv-xvii. At head of title: Reprint from the Transactions of the Royal Society of South Africa, vol. XXIX, part III. "Read October 16, 1940."

[664] Clarke, Owen (comp) *Holiday caving in Mallorca.* [GB]: Cwmbran Caving Club and [Clarke, Owen], **1990**. illus: b&w: photos; 24p; 30 x 42 cm.

Ghastly production, but interesting information; photocopy printing.

[665] Cleghorn, Paul L. *The Hilina Pali petroglyph cave, Hawai'i island: a report on preliminary archaeological investigations.* Honolulu, HI, US: Dept of Anthropology, Bernice P. Bishop Museum, Feb, **1980**. Report - Department of Anthropology Bernice Pauahi Bishop Museum 80 1. illus: b&w: ads, drawings, maps, photos, tables; v, 32p, [2]; bibl; 28 cm.

[666] Close, Danny *Mammoth Cave... kids love it!.* npop: [Danny Close], **1981**. illus: b&w: drawings; [i], 23p.

Juvenile.

[667] Clottes, Jean *The cave beneath the sea: paleolithic images at Cosquer.* Courtin, Jean; Garner, Marilyn (trans). New York, NY, US: H.N. Abrams, Inc, Publishers, **1996**. illus: b&w: charts, drawings, maps, photos; partly col: charts, maps, tables; col: charts, drawings, maps, photos; 200p; bibl, gloss, note; 31 cm.

Translation of: La grotte Cosquer. Col illus.

[668] Clymer, Virgil H.,–1952 (comp, ed) *Story of Howe Caverns.* 1954 ed.

20,000 copies.

1st ed. Howes Cave, NY, US: Howe Caverns, Inc, **1937**. illus: b&w: maps, photos, ports, repros; 72p; 19 cm.

10,000 copies. Also published in paper. Many later editions published with few changes.

A guidebook to Howe Caverns, NY, with extensive notes on its history, including its commercial development and installation of the elevator.

2nd ed. Cobleskill, NY, US, **1941**. illus: col: photos; map folded in back; front: photo.

3rd ed. **1946**. illus: b&w: maps, photos, repros; col: photos; Folded map glued to inside right cover.; B&W photo; 72p, 4p pls.

4th ed. 1949.

20,000 copies.

[669] Clyne, Patricia Edwards *Caves for kids: in historic New York: illustrated with maps and photographs.* Monroe, NY, US: Library Research Associates, **1980**. illus: b&w: maps, photos; xi, 162p; bibl; 21 cm.

Cover title: Caves for kids: illustrated with maps and photographs. 28 pls, 18 maps; tables on the uses of caves.

"A guide to exploring 19 caves in New York, including historical information about the caverns and notes on caving."

[670] Coase, Alan C. *Cathedral Cave situated between Brecon and Swansea on the A. 4067 road (750 yards North of Craig-y-Nos Castle) Dan-yr-Ogof Swansea Valley Caves.* Swansea, GB: Dan-yr-Ogof Caves Limited, **nd**. illus: col: photos; map on rear lining page; 31p.

Cover title: Cathedral Caves: Official Guide.

[671] Coase, Alan C. *Caving and potholing techniques.* Diana Wyllie, Ltd (prod). London, GB: Diana Wyllie Limited, **c1973**. [22].

Notes to accompany Filmstrip DW-154 of the same title by Ford, Trevor David. No illus.

[672] Coase, Alan C. *Dan yr Ogof and its associated caves.* Judson, David; Bray, L. G. (cont); Edington, A. (cont); Farr, M. (cont); Mason, E.J. (cont). Bridgwater, Somerset, GB: British Cave Research Association, Mar, **1977**. Transactions of the British Cave Research Association vol 4 no1-2. illus: b&w: charts, graphs, maps, photos, tables; 4 loose folded maps; text on front lining page; 102p; bibl.

[673] Coase, Alan C. *Official guide to Dan-yr-Ogof showcave.* Abercrave, [GB]: Dan-yr-Ogof Showcave,

1978. 27p, 8p pls; 19 cm.

Illus, some col.

[674] Coase, D. A. *The B. E. C. method of caving ladder construction.* [GB]: Bristol Exploration Club, Dec, **1962**. Caving Report no10. illus: b&w: drawings; 29p.

[675] Coates, Dorothy *Cheddar caves in the Cheddar Gorge: an illustrated survey & guide.* Hanley, GB: English Life Publications, Ltd, **1960**. illus: b&w: maps, photos; col: photos; maps on lining pages; 32p; 20 cm.

OCLC listed 34p. Subtitle added from a copy with no date and which listed place of pub as Derby.

[676] Coder, Kate *Pennsylvania's caves & caverns: an activity book.* Taylor, Audrey; Molosky, Ann. Huntingdon, PA, US: Lincoln Caverns, Inc, **1989**. illus: b&w: drawings; 32p; gloss; 28 cm.

Later published as Caves and caverns: an activity book. Juvenile.

[677] Coggins, Allen R. *The caves of the Tennessee State Natural Areas system.* Nashville, TN, US: Tennessee Department of Conservation, **1981**. 104p, [9].

The occurrence geology and ecology of caves in the state's natural areas. Visitor use, science, and exploration are covered.

[678] Coggins, Clemency *Cenote of sacrifice: Maya treasures from the sacred well at Chichen Itza.* Shane, Orrin C. (ed). Austin, TX, US: University of Texas Press, **1984**. illus: b&w: charts, maps, photos; col: photos; Color photo; 176p; bibl, index; 29 cm.

Catalogue of an exhibition organized by the Science Museum of Minnesota in cooperation with the Peabody Museum of Archaeology and Ethnology, Harvard University.

[679] Coleman, John Christopher *The caves of Ireland.* Tralee, IE: Anvil Books, May, **1965**. illus: b&w: maps, photos; front; 88p, 8p pls; app, bibl, gloss, index; 26 cm.

35 plates (b&w photos) bet pp 40-41, 21 numbered figures, front, maps.

Catalog of and guide to 300 caves of the entire island. Locations, descriptions, references, and a few maps.

[680] Coleman, John Christopher *The Polnagollum cave, Co. Clare.* Dublin, IE: Hodges, **1944**. Proceedings of the Royal Irish Academy v 50 sec B no 5. 28p.

[681] Coley, P. G. *The caves of Table Mountain: a report on their present condition, potential, and conservation requirements.* African Spelaeological Association. Cape Section. npop: np, **nd**. 13p; bibl; 30 cm.

Compiled in collaboration with the committee of the Cape Section of the South African Spelaeological Association.

[682] Collier, Don (host) *Life underground: caves, mines, minerals.* Yule, Lauray (host); Kleespie, Tom; Duncan, Dan; Asarco Inc. Tucson, AZ, US: produced by KUAT-TV in cooperation with the Arizona-Sonora Desert Museum, **1990**. The Desert Peaks.

1 videocassette (30 min), sd, col, 1/2 in, VHS format. "Major underwriting by ASARCO."

This episode features a visit to the recently discovered Kartchner Caverns and a discussion of mining in Arizona at the turn of the century.

[683] Collins, S. J. *The presentation of cave survey data.* npop, [GB]: Bristol Exploration Club, Sep, **1966**. Bristol Exploration Club Caving Report No 12. illus: b&w: drawings; partly col: drawings; 41p, 11p pls; bibl.

[684] Collins, S. J. *The shoring of swallet cave entrances.* npop, [GB]: Bristol Exploration Club, Aug, **1958**. Caving Report no 4. illus: b&w: drawings, tables; 27p; 26 x 20 cm.

[685] Collins, S. J. *Surveying in Red Cliffe Caves: Bristol 1953-1954.* npop, [GB]: Bristol Exploration Club, Jan, **1956**. Caving Report no 1. illus: b&w: maps; 8p, 2p pls; 26 x 20 cm.

[686] Colong Committee (Australia) *The Colong story.* Sydney, AU: The Colong Committee, **1985**. illus: b&w: maps; 36p, [10], 2p pls; 30 cm.

Illus.

[687] Combes, John D. *The Excavation of Squirt Cave, 45WW25.* [Pullman, WA, US]: Laboratory of Anthropology, Washington State University, Jun, **1969**. 49p; 28 cm.

Illus.

[688] Commission on Cave Tourism and Management of the Australian Speleological Federation Study Team *Jenolan Caves resort - some management issues.* Broadway, NSW, AU: Australian Speleological Federation, Jun, **1984**. illus: b&w: drawings, maps, photos, repros, tables; some maps folded in text; xi, 165p; app, bibl.

Copy examined was a photocopy.

[689] Conference on Hydrogeology, Ecology, Monitoring, and Management of Ground Water in Karst Terranes (3rd: 1991: Nashville, TN) *Proceedings of the Third Conference on Hydrogeology, Ecology, Monitoring and Management of Ground Water in Karst Terranes: December 4 - 6, 1991, Maxwell House/Clarion, Nashville, Tennessee.* Association of Ground Water Scientists and Engineers (U.S.). Dublin, OH, US: National Ground Water Association, **1991**. Ground Water Management no10. illus: b&w: charts, drawings, graphs, maps, photos; xiii, 793p; bibl, index; 28 cm.

Presented by the U.S. EPA and the Association of Ground Water Scientists and Engineers, a division of NGWA.

Much neat material on a breadth of subjects. It is surprising what shows up where, as in this unexpected find.

[690] Conference on Karst Geology and Hydrology (1974) *Abstracts of the fourth conference on karst geology and hydrology: West Virginia University, Mountainlair Theater of Mountainlair Building: May 3, 4, and 5, 1974.* [WV, US]: Department of Geology and Geography, West Virginia University and West Virginia Geological and Economic Survey, **1974**. illus: b&w: tables; v, 30p.

[691] Conn, David Bruce *Cave life of Carter Caves State Park*. Marshall, Bette Yvonne (ed). Morehead, KY, US: Appalachian Development Center, Morehead State University, **1981**. illus: b&w: charts, drawings, maps, photos; [iv], 50p; gloss; 22 cm.

Illus.

[692] Conn, Herbert Dunn *The Jewel Cave adventure: fifty miles of discovery under South Dakota*. 1981 reprint ed. St Louis, MO, US: Cave Books, **1981**. Speleologia. One folded map laid in; bibl, index.

1st ed. Conn, Jan E. Teaneck, NJ, US: Zephyrus Press, **1977**. Speleologia. illus: b&w: maps, photos; 1 map folded in back pocket.; 238p, [1]; app, bibl, index; 26 cm.

89 b&w photos, 15 maps, 3 songs. Title page: Jewel Cave National Monument is part of the National Park System. The exploration described here was undertaken with the cooperation of the National Park Service and the book is published under the auspices of the National Speleological Society.

This book unassumingly describes the rigors and joys of two lives dedicated to the Jewel Cave system.

[693] Conneely, Jim *Aillwee Cave: Ballyvaughan, Burren Co., Clare*. Galway, IE: np, **c1989**. illus: b&w: maps; col: photos; text on lining pages; 10p; 20.5 x 14.5 cm.

[694] Cons, David *Cavecraft: an introduction to caving and potholing*. London, GB: George G. Harrap & Co, Ltd, **1966**. illus: b&w: drawings, photos; front: b&w photo; 184p; gloss; 22.5 cm.

[695] Constantine, Denny G., 1925– *Rabies transmission by air in bat caves*. Atlanta, GA, US: National Communicable Disease Center, Jun, **1967**. Health Service Publication #1617. ix, 51p.

Illus.

[696] Contor, Roger J. *The underworld of Oregon Caves National Monument*. npop: Crater Lake Natural History Association, **1963**. illus: b&w: drawings; b&w; [vi], 37p, [1]; bibl, gloss; 22 cm.

Different versions exist with minor differences, but all dated 1963 and all have 37 pages.

[697] Cook, Holly (ed) *5th annual Hog-Fest: Lickford Valley Campgrounds, Corydon, Indiana, May 1-5, 1992*. Black, David. Georgetown, IN, US: [Harrison Crawford Grotto of the National Speleological Society]. illus: b&w: maps; fold-out map following text; drawing on front cover; [i], 40p, [1].

Title from cover.

[698] Cook, John Hawley *Of caves & cavemen*. Cook, Emory, 1915–; Cook Laboratories. Stamford, CT, US: Cook, **1989**. American Storytellers vol 2.

1 sound disc, analog, 33 1/3 rpm, 12 in. "Road Recordings."

An old geologist unravels the legend of the caveman and describes the perils and pleasures of the underground.

[699] Cook, Samuel *The Jenolan caves: an excursion in Australian wonderland*. London, GB: Eyre & Spottiswoode, **1889**. illus: b&w: maps, photos; front; xi, 190p, [1]; 26 cm.

24 full-page plates included in text.

A classic book on Jenolan, Australia. Much of the text concerns cave descriptions, but each includes a date and brief details of their discovery. It is also notable for the inclusion of Charles Kerry's superb photographs, the first time cave photographs were printed in ink and appeared in book form.

[700] Cook, Thomas Howard, 1946– *The el cheapo book of home brew diving equipment*. npop: [Tom Cook], **1978**. illus: b&w: drawings; [40].

Mimeographed.

[701] Cooke, Cranmer Kenrick, 1906– *The Redcliff stone age site, Rhodesia*. Klein, Richard G. [Salisbury]: [National Museums and Monuments of Rhodesia], **1978**. Occasional Papers of the National Museums and Monuments of Rhodesia Series A Human Sciences v 4 pt 2 pp 45-80. 36p; bibl; 30 cm.

Title from cover. Illus. Preliminary analysis of the mammalian fauna from the Redcliff stone age cave site, Rhodesia 1978.

[702] Cooke, Ian, 1937- *Mother and sun: the Cornish Fogou*. Penzance, GB: Men-An-Tol Studio, **1993**. illus: b&w: maps; 351p; 26 cm.

Illus. Archeological excavations of cave dwellings in Cornwall (County), England.

[703] Cooke, John H. *The Har Dalam Cavern, Malta and its fossiliferous contents by John H. Cooke with a report on the organic remains by Arthur Smith Woodward*. Woodward, Arthur Smith, 1864-1944. MT: G. Muscat, **c1893**. illus: b&w: drawings, repros; 36p; 21.3 x 14 cm.

A shortened version of this paper was published in Proceedings of the Royal Society v.54, pp. 274-283.

[704] Cooke, Robert L. *A description of Weyer's Cave*. 1st ed. Staunton, VA, US: Seminary Press, **1834**. illus: b&w: maps; folded map; 33p, 1p pls; 19 cm.

2nd ed. np, **1836**. 36p; 17 cm.

3rd ed, rev. Raleigh, NC, US: Office of Carolina Cultivator, **1856**. illus: b&w: maps; folded map; 47p; 14 x 20 cm.

[705] Coon, Carleton Stevens, 1904– *Cave explorations in Iran, 1949*. 1968 facsimile ed. **1968**. 124p; 21 cm.

Photocopy of 1951 edition. Ann Arbor, Mich.: University Microfilms, 1968.

1st ed. Philadelphia, PA, US: The University Museum, University of Pennsylvania, **1951**. Museum Monographs University of Pennsylvania. illus: b&w: drawings, maps, photos, tables; [iii], 124p, [1]; add, app, errat; 27.5 x 21 cm.

Eight chapters and 15 plates. "Errata et corrigenda" leaf

inserted.

[706] Coon, Carleton Stevens, 1904– *The seven caves: archaeological explorations in the Middle East*. 1st ed. New York, NY, US: Alfred A. Knopf, **1957**. illus: b&w: charts, drawings, maps, photos; xx, 338p, [18], 31p pls; index; 22 cm.

Published simultaneously in Canada by McClelland and Stewart Limited.

1981 reprint ed. Westport, CT, US: Greenwood Press, **1981**. xx, 338p, 16p pls; 22 cm.

Reprint of the ed. published by Knopf, New York. Includes index.

[707] Coon, Carleton Stevens, 1904– *Yengema Cave Report*. Bricker, Harvey M. (collab); Johnson, Frederick (collab); Lamberg-Karlovsky, C. C. (collab). Philadelphia, PA, US: The University Museum University of Pennsylvania, **1968**. University of Pennsylvania Museum Monographs. illus: b&w: charts, drawings, graphs, maps, photos, tables; 1 col pl, folded; [vii], 77p, 35p pls; bibl; 28 cm.

Report on Yengema Cave in Sierra Leone. 13 tables, 2 charts, 6 figs in text (graphs, map, drawing), 35 plates (b&w drawings, photos).

[708] Cooper, John Edward, 1929– *Recovery plan for the Alabama cavefish, Speoplatyrhinus poulsoni Cooper and Kuehne 1974*. Atlanta, GA, US: United States Fish and Wildlife Service. Region IV, May, **1982**. ii, 72p; bibl; 28 cm.

Cover title: Alabama cavefish recovery plan. Illus.

[709] Cooper, K. M. *Morphology of extinct lava tubes and the implications for tube evolution, Chain of Craters Road, Hawaii Volcanoes National Park, Hawaii*. Hauakukaua, J. P. npop: United States Geological Survey, [**1992**]. Open-file Report 92-352. illus: b&w: charts, graphs, maps, tables; 14p; bibl.

[710] Cooper, Stafford S. *Cavity detection and delineation research: Report 3: Acoustic resonance and self-potential applications: Medford Cave and Manatee Springs sites, Florida*. United States Army Corps of Engineers Geotechnical Laboratory. Vicksburg, MS, US: The Station, **1983**. Technical report U.S. Army Engineer Waterways Experiment Station no GL 83 1 Report 3. illus: b&w: graphs, maps, tables; 21p, [40]; bibl; 27 cm.

Cover title: May 1983. Prepared for Office, Chief of Engineers, U.S. Army.

[711] Cope, Edward Drinker, 1840-1897 *On the contents of a bone cave in the island of Anguilla (West Indies)*. Washington, DC, US: Smithsonian Institution, **1883**. Smithsonian Contributions to Knowledge vol xxv art 3; Smithsonian Institution Publication 489. iii, 30p, [9], 5p pls; 33 cm.

Master microform held by: University Microfilms International.

[712] Cope, F. Wolverson *Geology explained in the Peak District*. Newton Abbot, GB: North Pomfret, VT: David & Charles, **1976**. illus: partly col: charts; 192p; bibl, index; 23 cm.

Excellent and well illustrated book on the caves of Derbyshire. Illus.

[713] Corbel, Jean *The major caves of France and their relationships with climatic factors*. Landis, Carolyn (trans); Landis, Charles (trans). npop: np, **1975**. Cave geology. State College, PA vol 1 no 3. illus: b&w: maps; 15p; bibl; 28 cm.

pp 41-55

[714] Cordingley, John *The Peak Cavern system - a caver's guide*. Manchester, GB: Vitagraph Books, Sep, **1986**. illus: b&w: cartoons, maps, photos, repros; front: b&w repro; 64p; bibl, gloss.

[715] [Cornwall-Smith, Nick] *HWYL [songbook]*. Shropshire, GB: South Wales Caving Club, [**1976**]. 35p, 1p pls; 20 x 13 cm.

No illus.

[716] Cornwell-Smith, Nick (collect, ed, comp) *They words, they words they 'orrible words: an anthology of caving songs*. [GB]: [N. Cornwall-Smith], **1993**. 248p; index; 12 x 15 cm.

[717] Coronet Instructional Films *Cave ecology*. Lawrence, KS, US: Centron Educational Films, **1970**.

1 film reel (13 min), sd, col, 16 mm.

Illustrates basic ecological principles through the study of a cave. Focuses on energy sources, discussing how the limited amount of energy in a cave causes cave organisms to differ from corresponding forest organisms. Develops the food web concept.

[718] Coronet Instructional Films *Limestone caverns*. Pond, Alonzo W. (ed collab). npop: np, **1943**.

11 min, sd, color, 16 mm., Kodachrome, with teacher's guide. Also issued in b&w.

Shows the physical and chemical action of water and atmosphere on strata and deposits, using Wisconsin's Cave of the Mounds as the locale. Pictures formations of stalactites, stalagmites, spattercones, helictites and oolites. Includes descriptions of other caves.

[719] Corrie, A. P. *The Wombeyan Caves and the Bowral, Mittagong & Moss Vale tourist districts*. Paramatta, NSW, AU: The Argus Printing Works, **1899**. illus: b&w: photos; 31p.

Copy examined was a photocopy. Rare.

[720] Cosgrove, Cornelius Burton, 1875– *Caves of the Upper Gila and Hueco areas in New Mexico and Texas*. 1st ed. Cambridge, MA, US: Museum, **1947**. Papers of the Peabody Museum of American Archaeology and Ethnology, Harvard University. vol 24 no 2. illus: b&w: maps; xv, 181p; bibl; 27 cm.

Master microform held by: University Microfilms International, 19—. 1 reel. 35 mm.

1975 reprint ed. Millwood, NY, US: Kraus Reprint

Co, **1975**.

[721] [Council of Northern Caving Clubs] *Northern cave handbook.* [GB]: Council of Northern Caving Clubs, **1969**. illus: b&w: maps, tables; 48p; add, bibl; 25.5 x 20.5 cm.

One copy examined with 1970 addendum.

[722] Council of Northern Caving Clubs *Northern caving; handbook of the Council of Northern Caving Clubs.* Jones, Ceris (illus). Bradford, GB: Council of Northern Caving Clubs, **1979**. illus: b&w: drawings; text on lining pages; 36p; 21 x 15 cm.

Frequent updates; 1985, 1993, 1994, 1995 updates examined. Later publications were produced in binders that allowed old pages to be removed and new pages to be added. 1990's editions illustrated by Ceris Jones. Different publishers were used.

[723] The Council of Southern Caving Clubs *Caving for beginners.* Wells, Somerset, GB: The Council of Southern Caving Clubs, **1974**. 19p; app, bibl.

[724] By "A Country Doctor" *The wonders of Castleton.* New Mills, GB: Stephen Evans, The Kinder Press, **c1929**. illus: b&w: ads, drawings; Illus of the rope walk; [22]; 22 x 13 cm.

Pagination includes covers. Ads on back cover.

[725] Courbon, Paul, 1936– *Atlas of the great caves of the world.* Chabert, Claud, 1939–; Bosted, Peter, 1954–; Lindsley, Karen Bradley, 1948–. St Louis, MO, US: Cave Books, **1989**. illus: b&w: graphs, maps, photos; [iv], 368p; bibl, index; 28 cm.

Enlarged and translated from: Atlas des grands gouffres du monde.

English translation and minor update of Atlas des grandes cavites mondiales by the first two authors (1986). Lists of longest and deepest caves in each cavernous country, with brief descriptions and references for most significant. About two hundred cave maps. Summary lists of deepest and longest caves, deepest pits, etc. in the world. Invaluable reference.

[726] Coward, Julian *Construction and testing of an underground radio: a report submitted to the Research and Educational Branch of Workers' Health, Safety and Compensation, Alberta Government.* Drummond, Ian M. Edmonton, Alberta, CA: [Alberta Speleological Society], Jun, **1983**. illus: b&w: drawings, graphs, maps, photos, tables; iv, 60p, [3]; bibl; 28 cm.

A project funded by the Occupational Health and Safety Heritage Grant Program.

[727] Cowles, John *Cougar Mountain Cave in south central Oregon.* npop: Daily New Press, **1959**. 50p; 24 cm.

Title from cover. Record of artifacts recovered from the cave. Illus.

[728] Cox, Edward *Souvenir of Cox's Stalactite Caves.* Cheddar, GB: E. Cox, **c1914**. illus: b&w: photos; 24p; 15 x 21 cm.

Guide book, quoting remarks from earlier visitors. Leaves printed on one side only.

[729] Cox, Edward *Souvenir of Cox's Stalactite Caves "visited by his majesty the late King Edward VII".* Cheddar, GB: E. Cox, **1910**. illus: b&w: photos; [20]; 18 x 26 cm.

Photos tipped onto pages.

[730] Cox, Edward Travers, 1821-1907 *Wyandote Cave, pronounced by eminent scientists and noted writers of wide experience and extensive travel to be the greatest natural attraction of its kind in the world.* Collett, John, 1828-1899; Levette, Gilbert M.; Rothrock, H. W. New Albany, IN, US: Public Press, **1903**. illus: b&w: ads; 1 folded leaf; 37p, [7]; 19 cm.

On cover: Wyandotte Cave, greatest natural wonder in the world: what it is! where it is! how to reach it!: something concerning this weird, but attractive realm of mystery, with its 23 miles of under-ground avenues and palaces.

[731] Cox, Guy *Caves of the Catabrian Mountains, north-west Spain.* GB: British Speleological Association, Apr, **1973**. Journal of the British Speleological Association: Cave Science no 51. illus: b&w: maps, photos; foldout geological map at rear; ii, 66p, 3p pls.

[732] Cox, Ulysses O. *A revision of the cave fishes of North America.* npop: United States Department of Commerce and Labor, Bureau of Fisheries, **1905**. illus: b&w: charts, drawings, photos, tables; 17p, 5p pls.

Appendix to the report of the commissioner of fisheries to the Secretary of Commerce and Labor for the year ending June 30, 1904.

[733] Crabtree, P. W. (ed) *Speleological Abstracts: key to Britain's speleological literature.* Settle, Yorkshire, GB: British Speleological Association, **1962**—. Vol 1 no 1. iv, 54p; 26 x 20 cm.

No illus.

[734] Crain, Sally Lucille 1938–(auth, prod) *Prehistoric magic.* WSJK (Television station: Knoxville, Tenn.); Tennessee Technological University. Agency for Instructional Television. Bloomington, IN, US: Agency for Instructional Television, **1979**. The Young at Art Videorecording no 8. illus: partly col: photos.

1 cassette, 15 min, sd, col, 3/4 in & teacher's guide. Juvenile.

Discusses the mystic quality of the ancient cave paintings of Lascaux, France and the techniques used by the prehistoric artists. Children then learn to create their own cave paintings.

[735] Craven, Stephen Adrian *Cango Cave, Oudtshoorn district of the Cape Province, South Africa: an assessment of its development and management 1780-1992.* Cape Town, ZA: South African Speleological Association, **1994**. South African Spelaeological Association Bulletin vol 34. illus: b&w: charts, graphs, maps, photos, repros, tables; partly col: charts; 3 fold-out maps; xii, 116p, 3p pls; app, bibl.

[736] Cravens, Raymond L. *The spirit of Lost River: fact and legend about the Lost River Cave and Valley.* Bowling Green, KY, US: The Friends of Lost River, **1992**. 79p; bibl; 15 cm.

Illus.

[737] Crawford, Harriet E. W. *Subterranean Britain: aspects of underground archeology.* Barnes, John Wykeham, 1921–. London: J. Baker, **1979**. illus: b&w: maps, photos; xv, 201p; bibl, index; 24 cm.

Comprehensive guide to Britain's underground, not only caves.

[738] Crawford, Nicholas Charles, 1942– *Guidebook to karst and caves of Tennessee: emphasis on the Cumberland plateau escarpment region and guidebook to karst and caves of the Ozark Region of Missouri and Arkansas: prepared for the Eighth International Congress of Speleology, Bowling Green, Kentucky, U. S. A. July 18 to 24 1981.* Vineyard, Jerry Daniel, 1935–. Huntsville, AL, US: National Speleological Society, **1982**. illus: b&w: drawings, maps, photos; col: maps; col maps folded in back pocket; b&w photo; [ii], 176p; bibl; 22 cm.

In English, French and German. Cover title: Guidebook to karst and caves of Tennessee and Missouri.

[739] Crawford, Nicholas Charles, 1942– *Hydrogeologic problems resulting from development upon karst terrain, Bowling Green, Kentucky. Guidebook prepared for Karst Hydrogeology Workshop August 31 - September 3, 1982 Nashville, Tennessee.* npop: np, **1982**. illus: b&w: graphs, maps, photos; 1 map folded in text; 1 graph folded in text; v, 34p; bibl.

Sponsored by Groundwater Section, Region IV, United States Environmental Protection Agency. Coordinated by Center for Cave and Karst Studies, Department of Geography and Geology, Western Kentucky University

[740] Crawford, Nicholas Charles, 1942– *Hydrogeology of the Snail Shell Cave: overall creek drainage basin and the ecology of the Snail Shell Cave System.* Barr, Thomas Calhoun, 1931–. Nashville, TN, US: Tennessee Dept. of Conservation, **1988**. illus: b&w: maps, photos; map folded in front pocket; 9 maps on 7 folded leaves in rear pocket; 171p, [11]; add, bibl; 30 cm.

Title from cover. At head of title: Tennessee white paper. Loose-leaf. 171 lvs in various foliations.

[741] Crawford, Nicholas Charles, 1942– *Karst hydrogeology and environmental problems in the Bowling Green area.* Bowling Green, KY, US: Center for Cave and Karst Studies, Dept of Geography and Geology, Western Kentucky University, **1981**. Report of investigations no 3. illus: b&w: drawings, graphs, maps, photos; folded map and folded graph in pocket; [ii], 21p; bibl; 28 cm.

Illus.

[742] Crawford, Nicholas Charles, 1942– *Karst hydrogeology of Tennessee. Guidebook prepared for Karst Hydrogeology Workshop August 31 - September 3, 1982 Nashville, Tennessee.* npop: np, **1982**. illus: b&w: charts, drawings, graphs, maps, photos, tables; 4 plates folded in pocket in back; ix, 102p, 3p pls; bibl.

Sponsored by Groundwater Section, Region IV, United States Environmental Protection Agency. Coordinated by Center for Cave and Karst Studies Department of Geography and Geology Western Kentucky University

[743] Crawford, Nicholas Charles, 1942– *The karst hydrogeology of the Cumberland Plateau Escarpment of Tennessee. Part IV. Erosional processes associated with subterranean stream invasion, conduit cavern development and slope retreat.* Bowling Green, KY, US: Center for Cave and Karst Studies, Department of Geography and Geology, Western Kentucky University, **1980**. Cave and Karst Studies Series no 4. illus: b&w: drawings, graphs, photos, tables; partly col: maps; 6 tables folded in apps; [xi], 152p; app, bibl.

[744] Crawford, Nicholas Charles, 1942– *The karst hydrogeology of the Cumberland Plateau escarpment of Tennessee: Part I: subterranean stream invasion, conduit cavern development, and slope retreat in the Lost Creek Cove area, White County, Tennessee.* Bowling Green, KY, US: Center for Cave and Karst Studies, Department of Geography and Geology, Western Kentucky State University, **1979**. Cave and Karst Study Series no 1. illus: b&w: charts, drawings, graphs, maps, photos, tables; partly col: maps; col: maps; two partly colored maps (plates) folded in rear pocket; 1 geologic cross section folded loose in text; two folded charts in text; [v], 75p; bibl.

Seminal publication on this karst region. Relates structure to observed forms.

[745] Crawford, Nicholas Charles, 1942– *The karst hydrogeology of the Cumberland Plateau escarpment of Tennessee: Part II: karst valley development and the headward advance of the Sequatchie Valley in the Grassy Cove area, Cumberland County, Tennessee.* Bowling Green, KY, US: Center for Cave and Karst Studies, Department of Geography and Geology, Western Kentucky State University, **1979**. Cave and Karst Study Series no 2. illus: b&w: drawings, graphs, maps, photos; partly col: maps; col: maps; two partly colored maps (plates 3 and 4) folded in rear pocket; one map folded in text; [iv], 50p; bibl.

[746] Crawford, Nicholas Charles, 1942– *The karst hydrogeology of the Cumberland Plateau escarpment of Tennessee: Part III: karst valley development in the Lost Cove area, Franklin County, Tennessee.* Bowling Green, KY, US: Center for Cave and Karst Studies, Department of Geography and Geology, Western Kentucky State University, **1980**. Cave and Karst Study Series no 3. illus: b&w: drawings, graphs, maps, photos; partly col: maps; col: maps; two partly colored maps (plates 5 and 6) folded in rear pocket; one drawing folded in text; [iii], 20p, [1];

[747] Crawford, Nicholas Charles, 1942– *The karst hydrogeology of the Cumberland Plateau escarpment of Tennessee: subterranean stream invasion, conduit cavern development, and slope retreat in the Lost Creek Cove area, White County, Tennessee.* Bowling Green, KY, US: Center for Cave and Karst Studies, Department of Geography and Geology, Western Kentucky State University, **1987**. State of Tennessee Department of Conservation Division of Geology, Report of Investigations no 44 part 1. illus: b&w: charts, drawings, graphs, maps, tables; partly col: maps; 1 map folded in rear pocket, partly col; folded drawings; ix, 43p; bibl.

[748] Crawford, Nicholas Charles, 1942– *Karst hydrologic problems of south central Kentucky: ground water contamination, sinkhole flooding, and sinkhole collapse: field trip guide.* [Dublin, OH, US]: National Water Well Association, **1988**. illus: b&w: maps; x, 157p; 27 cm.

Illus. "Second Conference on Environmental Problems in Karst Terranes and Their Solutions, November 16-18, 1988, Nashville, Tennessee. The field guides for field trips #2 and #3"—Cover.

[749] Crawford, Nicholas Charles, 1942– *The karst landscape of Warren County.* [Bowling Green, KY, US]: City-County Planning Commission of Warren County, **1989**. Warren County comprehensive plan technical report [no.23]. illus: b&w: maps; Various pagination; 28 cm.

Illus. Prepared in conjunction with the Center for Local Government, Western Kentucky University—Added t.p.

[750] Crawford, Nicholas Charles, 1942– *Safford Centennial Society field trip, fall 1979: guidebook: karst hydrogeology of the Cumberland plateau escarpment in the Lost Creek Cave and Karst Cave areas of Tennessee.* Bowling Green, KY, US: Department of Geography and Geology, Western Kentucky University, **1979**. illus: b&w: charts, drawings; partly col: maps; map folded in pocket; 4 maps, partly col, folded in rear pocket; 1 chart folded in text; 49p.

[751] Crawford, Nicholas Charles, 1942– *Significant natural landscape features of Warren County, Kentucky.* Smith, James H.; Western Kentucky University. Center for Local Government. Bowling Green, KY, US: City-County Planning Commission of Warren County, **nd**. Warren County Comprehensive Plan Technical Report no 8. illus: b&w: maps; partly col: photomicros; Various pagination; bibl; 28 cm.

Prepared in conjunction with the Center for Local Government, Western Kentucky University—Added tp.

[752] Crawford, Nicholas Charles, 1942– *Storm water drainage wells in the karst areas of Kentucky and Tennessee: extended inventory of drainage wells in Kentucky and Tennessee: underground water source protection program Grant No. G004358-83-0.* Groves, Christopher G. Bowling Green, KY, US: Center for Cave and Karst Studies, Department of Geography and Geology, Western Kentucky University, Sep, **1984**. illus: b&w: drawings, maps, photos; [xix], 52p; bibl.

Prepared for United States Environmental Protection Agency, Region V. 34 figures

[753] Crawford, Rod *An annotated bibliography of Pacific Northwest speleobiology.* US: Washington Speleological Survey, Jul, **1977**. Western Speleological Survey serial no 56; Washington Speleological Survey Bulletin no 11. 12p; bibl; 28 cm.

Title from cover.

[754] Crestwood House *National parks: v. 2: Carlsbad Caverns.* Mankato, MN, US: Crestwood House, **1988**. index; 28 cm.

Col illus.

[755] Cribb, A. B. *An ecological and taxonomic account of the algae of a semi-marine cavern, Paradise Cave, Queensland.* St. Lucia, AU: University of Queensland Press, **1965**. Queensland University Dept of Botany, Papers vol 4 no 16. 24p; 25 cm.

pp 259-282, illus.

[756] Crofton, Denis *All about Poole's Cavern, Buxton. An official guide to this unique natural curiosity, together with an inventory of the museum.* Dawkins, William Boyd, 1838–. Liverpool, GB: W. J. Blevin, **1901**. illus: b&w: ads, repros; front: untitled print (wood engraving) of cave interior; 30p; 18 x 12 cm.

Cover title: A description of Poole's Cavern, Buxton with a paper on the vestiges on ancient human habitation in the caves....and a later description to which is also added a catalogue of Red Fern's Museum, adjoining the cavern.

[757] Croghan, John, 1790-1849 *Rambles in Mammoth Cave, during the year 1844, by a visitor.* Louisville, KY, US: Morton & Griswold, **1845**. Kentucky Culture Series no 17. illus: b&w: maps, repros; xii, 101p, 12p pls; errat; 18 cm.

Actually written by Bullitt, Alexander Clark, 1807-1868 (See entries under Bullitt). Included here because of the numerous listings in OCLC under Croghan. Microfilm versions: Microopaque. 3 microcards. (Kentucky culture series, no. 17); Microfiche. [Louisville, Ky.: Lost Cause Press, 196-]; 4 microfiches ; 11 x 15 cm.

[758] Crosby, Harry, 1926– *Cave paintings of Baja.* Venice, CA, US: Environmental Communications, **1976**.

30 slides, col, 2x2 in, & script. Title from data sheet.

A survey of the paleolithic rock paintings of Baja California, including an example of what several astronomers consider pictures of the supernova event of A.D. 1054. Shows details of paintings and the spectacular and difficult terrain which has hidden and protected the art.

[759] Crosby, Harry, 1926– *The cave paintings of Baja California.* rev ed. La Jolla, CA, US: Copley Books, **1984**. illus: b&w: maps; partly col: maps, photos; map on lining paper; ix, 189p; bibl, index; 29 cm.

Illus (some col).

[760] Crosby, Harry, 1926– *The cave paintings of Baja California: the great murals of an unknown people*. [La Jolla, CA, US]: Copley Books, **1975**. illus: partly col: maps, photos; ix, 174p; bibl, index; 26 cm.

Illus. Sequel to The King's Highway in Baja California.

[761] Crossley, P. C. (comp) *The New Zealand cave atlas: North Island*. Waitomo Caves, NZ: New Zealand Speleological Society Inc, **1988**. illus: b&w: maps; 311p, 9p pls; index.

Aside from a table of lengths and depth, purely cave maps, about 150 of them.

[762] Crossley, P. C. (comp) *New Zealand cave atlas: South Island*. NZ: New Zealand Speleological Society, **1990**. illus: b&w: charts, maps, tables; col: photos; 3 foldout maps; [250]; bibl, index.

Almost entirely maps.

[763] Crossley, P. C. (comp) *The New Zealand cave atlas*. Hurst, B. P. (comp); West, R. G. (comp). Auckland, NZ: The Department of Geography, Auckland University, **1981**. illus: b&w: maps; [257]; index; 30 cm.

Copy examined had no date. Includes price list of separately available cave maps. Produced for the New Zealand Speleological Society Inc. Map scales differ.

[764] Crothers, George Martin *Final report on the survey and assessment of the prehistoric and historic archaeological remains in Big Bear Cave, Van Buron County, Tennessee*. Knoxville, TN, US: Department of Anthropology, Univ of Tennessee, Apr, **1986**. illus: b&w: drawings, maps, photos, tables; 2 maps folded in text; iv, 66p; bibl.

Project conducted and reported in accordance with Tennessee Department of Conservation contract 102087.

[765] Crowther, Patricia P. *The Grand Kentucky junction: memoirs*. Wells, Stephen G.; Wilcox John P.; Watson, Richard Allan, 1931–; Brucker, Roger Warren, 1929–; Eller, P. Gary. St. Louis, MO, US: Cave Books; Cave Research Foundation, **1984**. viii, 96p; 28 cm.

Limited edition of 1,000, of which 50 are signed by authors and specially bound.

The story of the connection of the Mammoth Cave and Flint Ridge Systems told first-hand by the individuals involved.

[766] Crush, Kenneth *An inventory of caves, Harry S. Truman dam and reservoir*. Zollweg, James. npop: np, Sep, **1977**. illus: b&w: tables; [iii], 68p.

[767] Crysler, Mildred G. *Adventures in underground fairylands: a fantasy tour of the Carlsbad Caverns*. New York, NY, US: Exposition Press, **1955**. 199p; 21 cm.

Illus.

[768] Cudmore, Dana D. *The remarkable Howe Caverns story*. Woodstock, NY, US: Overlook Press, **1990**. illus: b&w: maps, photos, repros; [x], 166p; 21 cm.

Illus.

[769] Cullen, James J. *Karst hydrology and geomorphology of eastern New York: a guidebook to the geology field trip, National Speleological Society Annual Convention, Pittsfield, Massachusetts, August 5-12, 1979*. Mylroie, John Eglinton, 1949–; Palmer, Arthur Nicholas, 1940–. Pittsfield, MA, US: Private Printing, Jul, **1979**. 83p; bibl; 28 cm.

[770] Cullen, James J. *Spelean stamps: a special topical collection of postage stamps of interest to cavers*. Cullen, Vera. npop: [Jim & Vera Cullen], **1979**. illus: b&w: tables; vi, 46p; bibl.

Text is one large table.

[771] Cullen, P. *A survey of the vegetation*. Hobart, TAS, AU: Department of Parks, Wildlife, and Heritage, **1990**. Ida Bay Karst Study Report to the Department of Parks, Wildlife, and Heritage. 11p.

[772] Cullinan, Mike *The caves of Huntingdon County, Pennsylvania*. Speece, Jack Howard, 1947–. State College, PA, US: National Speleological Society, Jun, **1975**. Bulletin #9 Mid-Appalachian Region of the National Speleological Society. illus: b&w: maps, photos; 113p; add, bibl.

[773] Cullingford, Cecil Howard Dunstan, 1904-1990 *British caving: an introduction to speleology*. 1st ed. [London, GB]: Routledge and Kegan Paul, Limited, **1953**. illus: b&w: drawings, maps, photos, tables; xvi, 468p, 48p pls; app, bibl, gloss, index; 26 cm.

48 plates, 87 figs, 12 tables.

An important compilation of many talented cavers. A first major step in distribution of techniques information.

2nd ed, rev. Members of the Cave Research Group. **1962**. xvi, 592p; app, bibl, gloss, index; 26 cm.

48 pls. Sponsored by the Cave Research Group Great Britain, this book covers aspects of caving, and is often used as a textbook. First pub in 1953, this edition has been revised and expanded.

[774] Cullingford, Cecil Howard Dunstan, 1904-1990 *Caving*. Gloucester, GB: Thornhill Press, **1976**. A Thornhill Guide 7. illus: b&w: drawings, photos; 32p; bibl; 19 x 11 cm.

[775] Cullingford, Cecil Howard Dunstan, 1904-1990 *Exploring caves*. Cobb, David (illus). London, GB: Geoffrey Cumberlege, Oxford University Press, **1951**. Compass books no 10. illus: b&w: drawings, photos; b&w photo; 148p; app, bibl, index; 19 cm.

Touches on the folklore related to caving in Britain; describes some episodes of cave exploration in Britain, including some descriptions of well-known caves.

[776] Cullingford, Cecil Howard Dunstan, 1904-1990 *Manual of Caving Techniques*. The Cave Research Group of Great Britain. London, GB: Routledge & Kegan Paul, Limited, **1969**. illus: b&w: drawings, graphs, photos, tables; front: b&w photo; [i], ix, 416p; app, bibl, gloss, index; 26 cm.

OCLC attributes authorship to The Cave Research Group of Great Britain.

A careful overview of all known areas of caving, using the best British cavers of the time to write each chapter. Contains 21 chapters, 12 of which have bibliographies. Only b&w photo is frontpiece that shows camp I at 1600 feet in Berger.

[777] **Culver, David C.**, 1944– *Adaptation and natural selection in caves: the evolution of* Gammarus minus. Kane, Thomas C.; Fong, Daniel W. Fong, Daniel W. Cambridge, MA, US: Harvard University Press, **1995**. illus: b&w: charts, drawings, graphs, maps, photomicrographs, tables; 223p; bibl, gloss, index; 25 cm.

Illus.

A detailed study of one cave organism, the book advances *Gammarus minus* as a paradigm for cave colonization and adaptation and is a general case study of the role of natural selection and adaptation in evolution.

[778] **Culver, David C.**, 1944– *Biogeography of subterranean crustaceans: the effects of different scales: a symposium held at the summer meeting of the Crustacean Society in Charleston, South Carolina, USA, in June 1992*. Holsinger, John Robert, 1934–. Dordrecht, Boston, MA, NL, US: Kluwer Academic Publishers, **1994**. Hydrobiologia v 287 no1. illus: b&w: maps; 156p; bibl; 26 cm.

[779] **Culver, David C.**, 1944– *Cave life: evolution and ecology*. Cambridge, MA, US: Harvard University Press, **1982**. illus: b&w: drawings, graphs, maps, photos, tables; viii, 189p; bibl, index; 25 cm.

Treatment of several topics in the population biology of caves, including adaptation, regressive evolution, genetic structure, and population ecology.

[780] **Curcic, Bozidar P. M.** *Cave-dwelling pseudoscorpions of the Dinaric Karst = Jamski pascipalci dinarskega krasa*. Ljubljana, SI: Slovenska akademija znanosti in umetnosti, **1988**. Opera - Academia Scientiarum et Artium Slovenica, Classis IV, Historia Naturalis. 191p; bibl; 24 cm.

Illus. Summaries also in Serbo-Croatian and Slovenian.

[781] **Currens, James C.**, 1952– *Bibliography of karst geology in Kentucky*. McGrain, Preston. Lexington, KY, US: Kentucky Geological Survey, **1979**. Kentucky Geological Survey series XI Special publication 1. [iii], 59p; app, bibl, index; 28 cm.

[782] **Currens, James C.**, 1952– *Flooding of the Sinking Creek Karst area in Jessamine and Woodford Counties, Kentucky*. Graham, C.; Douglas R. Lexington, KY, US: Kentucky Geological Survey, University of Kentucky, **1993**. Report of Investigations (Kentucky Geological Survey). Series XI, 7. illus: b&w: maps; 33p; bibl; 28 cm.

Illus.

[783] **Curro, Joseph R.** *Cavity detection and delineation research; Prepared for Office, Chief of Engineers. Report 2: Seismic methodology: Medford Cave Site, Florida*. United States Army Corps of Engineers Geotechnical Laboratory. Vicksburg, MS, US: Geotechnical Laboratory, U.S. Army Engineer Waterways Experiment Station, Jun, **1963**. Technical report U.S. Army Engineer Waterways Experiment Station GL 83 1. illus: b&w: drawings, graphs, maps; [ii], 35p, [64]; bibl; 27 cm.

Title from cover.

[784] **Curry, David A.** *Joint Mitnor Cave*. South Devon, GB: William Pengelly Cave Studies Trust, May, **1982**. Occasional publication no 1.5. illus: b&w: drawings, maps; 11p, [1]; bibl; 20.5 x 14.5 cm.

[785] **Curt Teich & Co, Inc** *Carlsbad Caverns National Park, New Mexico*. Carlsbad, NM, US: Cavern Supply Company, **1963**. illus: b&w: maps; col: photos; [24]; 28 x 21.5 cm.

[786] **Cushman, Robert Vittum** *Hydrologic problems as related to wilderness management and water supplies for recreational areas at Mammoth Cave National Park, Kentucky*. npop: United States Geological Survey, **1967**. USGS Open File Report. 16p.

The effects of wilderness designation and water supply problems/needs at Mammoth Cave National Park.

[787] **Cushman, Robert Vittum** *Present and future water supply for Mammoth Cave National Park, Kentucky*. Krieger, Robert Albert, 1918–; McCabe, John A. Washington, DC, US: np, **1965**. Geological Survey water-supply paper 1475 Q Hydrology of the Public Domain. illus: b&w: maps; 47p; bibl; 24 cm.

Prepared in cooperation with the National Park Service. pp 601-647. Illus.

[788] **Cuvay, Roxane** *Cave painting*. Shenfield, Margaret (trans). New York, NY, US: Crown Publishers, **1963**. Movements in World Art. illus: b&w: charts, drawings, maps; col: photos; 64p; bibl; 19 cm.

[789] **Cuvay, Roxane** *Prehistoric cave painting*. London, GB: Metheun, **1963**. illus: col: photos; 64p; bibl; 18 cm.

Translation of Hohlemalerei. American ed has title: Cave painting. Contains 24 colored plates of cave painting with a one page description on each. 11 page introduction.

[790] **d'Amboise, Valery (auth, illus, photo)** *Eternal caves*. Chard, Andrew (trans). Chambery Cedex, FR: Editions des Alpes, **1986**. illus: partly col: maps; col: photos; 165p, [2]; index; 19 x 25 cm.

More than half of the pages are color photos.

[791] **D'Arcy Galleries (New York, N.Y.)** *The Caves of Karawari*. [New York, NY, US]: D'Arcy Galleries, **1968**. illus: b&w: maps; xix, 105p; 34 x 14 cm.

Title from cover. German, French, and English.

[792] **Dake, Charles Laurence**, 1883– *Subterranean stream piracy in the Ozarks*. Bridge, Josiah. Rolla, MO, US: [School of Mines and Metallurgy, University of Missouri], **1924**. School of mines and metallurgy University of Missouri. Bulletin... Technical series v 7 no 1. illus:

b&w: drawings, maps; partly folded diagrams; 26p; 23 cm.

At head of title: School of Mines and Metallurgy, University of Missouri. Contains plates.

[793] Daldy, Isbister & Co. *Half hours underground: volcanoes, mines, and caves.* London, GB: Daldy, Isbister & Co, Virtue and Co, **1878**. Half Hour Library of Travel Nature and Science for Young Readers. xii, 369p, [3]; 18 cm.

Includes publisher's catalog. Juvenile.

[794] Dalquest, Walter W. *The mammal fauna of Schulze Cave, Edwards County, Texas.* Roth, Edward; Judd, Frank. Gainesville, FL, US: University of Florida, Jun, **1969**. Bulletin of the Florida State Museum vol 13 no 4. illus: b&w: drawings, maps, tables; text on lining pages; 72p; bibl.

[795] Dalton, Richard F. *Caves of New Jersey.* Sullivan, Gerardus Nicholas, 1927–[Brother Nicholas] (cont); Eckler, A. Ross (cont). Trenton, NJ, US: New Jersey Department of Environmental Protection Bureau of Geology & Topography, **1976**. New Jersey Geological Survey Bulletin 70. illus: b&w: drawings, maps, photos, tables; 7 maps folded in rear pocket; iv, 51p; 28 cm.

Has cave biology by Brother Nicholas and History and Legends of caves by Ross Eckler. 7 plates, 38 figs and 2 tables plus 7 maps in back pocket.

Catalog of caves and cave features. About 150 named things, but many are shelters, sinks, springs, and the like. Twenty-three cave maps. Miscellaneous chapters include biology, history, and legends.

[796] Dalton, Richard F. *New Jersey caves in brief.* Trenton, NJ, US: The Geological Society of New Jersey, **1970**. illus: b&w: drawings, maps, photos; partly col: drawings; [iii], [22]; bibl; 29 cm.

Includes 8 full page maps, 20 references, 3 colored block diagrams, 3 b&w photos, and very short descriptions of 8 caves. This is the first survey of N.J. caves.

[797] Damon, Paul Herbert, 1934– *Caving in America: the story of the National Speleological Society, 1941-1991: commemorating 50 years of history and growth: including a special illustrated history of cave exploration in the society entitled—the last frontier for the pioneer.* Huntsville, AL, US: The National Speleological Society, **1991**. illus: b&w: drawings, maps, photos, repros; 2 maps, 3 repros folded in rear pocket; 445p; app, index; 29 cm.

A history of the NSS published to commemorate the first 50 years of the society. Contains some omissions and inaccuracies, but nevertheless a starting point for a record of the society's work.

[798] Damon, Paul Herbert, 1934– *The history of Laurel Caverns of Fayette County, Pennsylvania, known throughout the years as Delaney's Cave.* Altoona, PA, US: Speece Productions, Jun, **1976**. Spelean history series no 2. illus: b&w: maps, photos, repros; 2 maps, 3 repros folded in rear pocket; viii, 58p, 2p pls; bibl; 28 cm.

Presented at the history session, National Speleological Society convention, Morgantown, West Virginia, June 30, 1976. History, area, and people associated with Laurel Caverns.

[799] Damon, Paul Herbert, 1934– *Thirty years with the Pittsburgh Grotto, National Speleological Society.* National Speleological Society. Pittsburgh Grotto. Pittsburgh, PA, US: Pittsburgh Grotto Press, Oct, **1977**. illus: b&w: repros; [v], 28p, 4p pls; 28 cm.

Cover title: "The thirtieth anniversary, October 1977."

[800] Daniel, Glyn Edmund *Lascaux and Carnac.* London, GB: Lutterworth Press, **1955**. illus: b&w: charts, drawings, maps, photos; map on lining pages; front: b&w photo; 127p, 14p pls; index.

Excellent, but uncommon book. This British archaeologist tells about the painted and engraved caves of the Dordogne: Lascaux, Les Eyzies, La Mouthe, Font de Gaume and Les Combarelles.

[801] Daniel, Thomas W. *Exploring Alabama caves.* Coe, William D. University, AL, US: Geological Survey of Alabama Division of Energy Resources Research, **1973**. Geological Survey of Alabama Bulletin 102. illus: b&w: drawings, photos; partly col: drawings, maps; 2 partly col maps folded in text; [viii], 99p; bibl.

89 figures.

Educational booklet on cave geology and caving techniques.

[802] Daoxian Yuan, 1933– *A brief Introduction to China's research in Karst.* Guilin, Guangxi, CN: The Institute of Karst Geology, Ministry of Geology, People's Republic of China, **1981**. illus: b&w: charts, drawings, graphs, maps, photos, tables; Photos on back lining page; [ii], 34p; bibl; 26 x 18.5 cm.

OCLC spells the authors name as: Yuan, Tao-hsien. Bibliography in Chinese. Title from cover.

[803] Daoxian Yuan, 1933– *Karst of China.* Dehao Zhu; Jintao Weng; Xuewen Zhu; Xingrui Han; Xunyi Wang; Guihong Cai; Yuanfeng Zhu; Guangzhong Cui; Ziqiang Deng. 1st. Beijing, CN: Geological Publishing House, **1991**. illus: col: charts, drawings, graphs, maps, photos, tables; v, 224p, 8p pls; bibl.

This serves to peak our curiosity about the rest of the work being done there.

[804] Darton, Nelson Horatio, 1865-1948 *Western Texas and Carlsbad Caverns.* King, Philip Burke, 1903–; Haigh, B. R.; United States Geological Survey. Washington, DC, US: United States Govt Print Office, **1932**. illus: b&w: maps, tables; folded map and folded table; iii, 38p; bibl; 22 cm.

At head of title: International Geological Congress, XVI session, United States, 1933. Guidebook 13: Excursion C-1. Published under the auspices of the United

States Geological Survey.

[805] Dasher, George R., 1952– *The caves and karst of the Buckeye Creek Basin, Greenbrier County, West Virginia.* Balfour, William M. (ed). Maxwelton, WV, US: West Virginia Association for Cave Studies, Aug, **1994**. Bulletin - West Virginia Speleological Survey no12. illus: b&w: charts, graphs, maps, photos, tables; col: photos; [8]p folded plates; 238p, 4p pls; app, bibl, index; 28 cm.

Col illus.

[806] Dasher, George R., 1952– *On station: a complete handbook for surveying and mapping caves.* Huntsville, AL, US: National Speleological Society, **1994**. illus: b&w: drawings, maps, photos; b&w drawings on lining pages; x, 242p; app, bibl, gloss, index.

188 figs.

A strong attempt to produce the penultimate book on cave surveying, but it comes up a little short.

[807] Dashwood, Francis *West Wycombe caves: [a brief history and description].* [West Wycombe, GB]: West Wycombe Caves Ltd., **1975**. illus: b&w: maps, ports, repros; 18p; 22 cm.

Title from cover. The caves were built between 1748 and 1752 for Sir Francis Dashwood.

[808] Dasque, Jean (prod) *Garden of shadows.* npop: Les Films de l'adagio; Davidson Films, **1972**. The World through the mask [Motion picture].

26 min, sd, col, 16 mm.

Examines and traces the origins and manifestations of above-water caves, and contrasts them with the submarine grottos which contain living things. Shows how life originated with microorganisms carried by the water.

[809] Daugherty, Richard D., 1922– *A data recovery study of Layser Cave (45-LE-223) in Lewis County, Washington.* Flenniken, J. Jeffrey; Welch, Jeanne M. Portland, OR, US: United States Forest Service. Pacific Northwest Region, **1987**. Studies in Cultural Resource Management no 7. illus: b&w: maps; vii, 164p; bibl; 28 cm.

Illus.

[810] Davaras, Costis *The Cave of Psychro.* Athens, GR: Ministry of Culture Archaeological Receipts Fund, **1989**. illus: b&w: ads; col: maps, photos; 15p, [1].

[811] Davenport, J. Walker *Archaeological exploration of Eagle Cave, Langtry, Texas.* Southwest Texas Archaeological Society. San Antonio, TX, US: Witte Memorial Museum, **1933**. Big Bend basket maker papers no 4; Bulletin Southwest Texas Archaeological Society no 3a. 32p; 27 cm.

Title from cover. Xerographic copy. Ann Arbor, Mich.: Xerox University Microfilms, 1974. Illus.

[812] Davey, Adrian G. *Management of Victorian caves and karst: a report to the Caves Classification Committee, Department of Conservation, Forests and Lands.* White, Susan; Victoria Department of Conservation, Forests, and Lands, Caves Classification Committee. Canberra, ACT, AU: Applied Natural Resource Management, **1986**. illus: b&w: maps; x, 74p; bibl; 30 cm.

Running title: Victorian caves & karst management. Discussion of karst problems in Victoria and recommendations for solutions.

[813] Davey, Adrian G. *Nullarbor karst, a bibliography.* Lewis, I. D. Canberra, ACT, AU: Nungarigo Resources Management, **1978**. 32p; bibl; 30 cm.

[814] Davey, Adrian G. (co-ord) *Resource management of the Nullarbor Region, W.A..* Cundy, J. G. (tm mem); Dunkley, J. R. (tm mem); Hamilton-Smith, Elery (tm mem); Lance, K. A. (tm mem); Lewis, I. D. (tm mem); Mott, K. R. (tm mem); Shoosmith, R. W. (tm mem); White, N. J. (tm mem); Williamson, K. A. (tm mem). Broadway, NSW, AU: Australian Speleological Federation, **1978**. illus: b&w: graphs, maps, photos, tables; folded maps in text; viii, 115p, [39]; app, bibl, gloss.

A report prepared for the Australian Environmental Protection Authority.

[815] Davey, Adrian G. *Victorian caves and karst: strategies for management and catalogue: a report to the Caves Classification Committee, Department of Conservation, Forests and Lands /.* White, Susan; Victoria Department of Conservation, Forests, and Lands, Caves Classification Committee. East Melbourne, VIC, AU: The Department, **1986**. illus: b&w: maps; x, 315p; index; 29 cm.

[816] Davey, Adrian G. *World Heritage significance of karst and other landforms in the Nullarbor Region.* Gray, M. R.; Grimes, K. G.; Hamilton-Smith, Elery; James, Julia M.; Spate, Andrew P. Belconnen, ACT, AU: Applied Ecology Research Group, Faculty of Applied Science, University of Canberra, **1992**. illus: b&w: charts, maps, tables; viii, 202p; app, bibl, gloss.

A report to the Commonwealth Department of The Arts, Sport, The Environment and Territories. Extensive bibliography.

[817] Davey, Adrian G. *Yallingup Cave Park - a management plan.* Hamilton-Smith, Elery; Pierce, Miles; Williamson, Kerry; Loveday, Barry (mapper); Loveday, Frank (mapper); Department of Conservation and Environment [Australia]. [AU]: Australian Speleological Federation, **1978**. illus: b&w: drawings, maps, photos, tables; 2 maps folded in back; ix, 120p, [34], 2p pls; bibl.

[818] Davidge, G. W. *Narracoorte [sic]: caves and town 20: book of views.* Naracoorte, SA, AU: G. W. Davidge, **nd**. illus: b&w: photos; text on lining pages; 20p.

Title from cover. Extremely limited text. Spelled as Naracoorte on other works.

[819] Davidson, Joseph Killworth, 1938– *CRF personnel manual.* Washington, DC, US: Cave Research Foundation, May, **1967**. illus: b&w: drawings, maps; Various pagination, ix; bibl; 28 cm.

3rd ed: 1981: xxi, 155p, annot bibl, index.

A very useful guide to expedition caving (the CRF way). Full of good ideas. Updated and reprinted several times.

[820] Davidson, Joseph Killworth, 1938– *Wilderness resources in Mammoth Cave National Park: a regional approach*. Bishop, William P. Columbus, OH, US: Cave Research Foundation, **1971**. illus: b&w: drawings, maps; partly col: maps; 34p; bibl; 29 cm.

[821] Davidson, Robert, 1808-1876 *An excursion to the Mammoth Cave, and the barrens of Kentucky: with some notices of the early settlement of the state*. Lexington, KY, US: A.T. Skillman & Son, **1840**. [iv], viii, 134p; 15 cm.

Title on spine: Mammoth Cave of Kentucky. One OCLC records lists Thomas, Cowperthwait, and Co., (Philadelphia) as the publisher. Microfilm version: Microfiche. Louisville, Ky.: Lost Cause Press, 1967. 4 microfiche. Some printings contain bibliographical references.

A very rare book because it is highly sought by historians as well as by cavers. Contains only about 50 pages on the cave.

[822] Davies, Melvyn *Cave sump index; South Wales*. Pontypool, Monmouthshire, GB, Nov, **1966**. 17p; 30 x 21 cm.

No illus. OCLC lists Leeds, England as the place of publication.

[823] Davies, Melvyn *The caves of the South Gower coast: an archaeological assessment*. npop: M. Davies, **1991**. 44p; bibl; 30 x 21 cm.

[824] Davies, Paul *A pictorial history of Swildon's Hole*. Frost, F. W. (photo). [Pangbourne, GB]: Wessex Cave Club, **1975**. illus: b&w: drawings, maps, photos, ports, repros; [ii], 94p; 22 cm.

A limited edition of 300 copies. More than 100 photographs showing this well known cave from 1901 to present.

An account of exploration, using fully annotated illustrations, of the cave from 1901 to the 1970s. Sections cover surveying and equipment used, these offering visual reminders of the way British caving has been conducted through the years.

[825] Davies, Phillip *An index of submerged cave passages; sumps*. [GB]: Cave Diving Group, Apr, **1957**. [24]; 26 x 20 cm.

No illus. Section 1 - Somerset; Section 2 - Devonshire.

[826] Davies, William Edward, 1917– *Bibliography of North American speleology*. npop: [William E. Davies], **nd**. 53p; bibl.

[827] Davies, William Edward, 1917– *Caverns of West Virginia*. 1958 ed. Morgantown, WV, US: West Virginia. Geological and Economic Survey, Jul, **1958**. West Virginia Geological and Economic Survey Reports vol 19(A) i4. illus: b&w: drawings, maps, photos; front: b&w photo; vi, 330p; index; 23 cm.

57 pls, 82 figs.

1965 reprint ed. [Morgantown, WV, US]: [West Virginia Geological and Economic Survey], **1965**. illus: b&w: drawings, maps, photos; b&w ; [v], vi, 330 + 72p; index; 23 cm.

Issued with Caverns of West Virginia (Supplement) also by Davies, 72 p., 1965. Main work has 52 pls and 82 figs; supp has 13 figs and corrections.

1st ed. Morgantown, WV, US: West Virginia Geological and Economic Survey, **1949**. West Virginia Geological and Economic Survey Reports vol 19. illus: b&w: maps, photos; b&w front; x, 353p; app, errat, index; 24 cm.

The 1949 printing is the first edition. 53 pls, 68 figs. errata sheet is loose.

Geological introduction, and catalog of over four hundred caves, with text descriptions and locations. Occasional photos and maps. The 1958 edition is actually an update to Davies 1949. The 1965 reprint is bound together with a 72-page supplement of additions and corrections. This is perhaps one of the only guides to caves of the area by a geologist.

[828] Davies, William Edward, 1917– *Caverns of West Virginia (supplement)*. [Morgantown, WV, US]: [West Virginia Geological and Economic Survey], **1965**. West Virginia Geological and Economic Survey Reports v 19A. illus: b&w: maps; [v], 72p; 23 cm.

[829] Davies, William Edward, 1917– *The caves of Maryland*. 1961 reprint ed. Baltimore, MD, US, **1961**. Maryland Dept. of Geology, Mines, and Water Resources; Bulletin 7. illus: b&w: drawings, maps; 76p; app, bibl, index; 23 cm.

1st ed. Baltimore, MD, US: Department of Geology, Mines and Water Resources, **1950**. Maryland Dept of Geology Mines and Water Resources Bulletin 7. illus: b&w: maps, photos; 8 folded maps in pocket; x, 70p; bibl, index.

Maryland cave survey with appendix in 1952 and 1961. 11 pls, 16 figs.

Catalog of 43 known caves, with descriptions and locations. Fifteen cave maps. Reprinted with 1952 supplement with corrections and seven additional caves.

[830] Davies, William Edward, 1917– *Geology of caves*. Morgan, I. M. [Arlington, VA, US]: United States Dept of the Interior, United States Geological Survey, **1977**. illus: b&w: charts, drawings, maps, photos; partly col: drawings; 19p; bibl; 23 x 11 cm.

A brief, readable overview of caves and their origins. It is of particular use to those largely unfamiliar with caves. A plus for outing clubs, land owners, etc.

[831] Davies, William Edward, 1917– *Report on sediments in Mammoth Cave, Kentucky*. Chao, E. C. T. [Washington, DC, US]: U.S. Geological Survey, Aug, **1959**. illus: b&w: drawings, graphs, maps, photos, tables; 1 map folded in text; vi, 117p; app, bibl; 27 cm.

43 figs.

[832] **Davis, Bertell D.** *Use and abandonment of habitation caves in the prehistoric settlement of southeastern Oahu: a proposed research design for the 1980 University of Hawaii Archaeological Field Program.* Kaschko, Michael W. npop: np, **1980**. illus: b&w: maps; 16p; bibl; 28 cm.

Caption title.

[833] **Davis, Harold (ed)** *Caves of Schoharie County, New York.* Haure, Pete; Hartline, Daniel; Souse, Margaret (eds). npop: np, **1966**. illus: b&w: maps, repros; 124p; bibl, index.

[834] **Davis, Henry Harrison** *An excursion from Lancaster, up the vale of Lune, and from Kirkby Lonsdale, to the caves of Yorkshire.* "New edition". London, Kirkby Lonsdale, GB: J. Foster/Simpkin Marshall & Co, **1851**. [i], 46p; 18 x 10.5 cm.

First ed very rare; included in other publications; 1849 pagination different: 1+44. Other title: An excursion to the caves of Yorkshire. OCLC lists the date as 1849.

[835] **Davis, M. J.** *Nobody here but us bats!: through Carlsbad Caverns with cartoons and comments.* [Albuquerque, NM, US]: Pony-X Press, **1954**. illus: b&w: maps; [16]; 19 cm.

Illus.

[836] **Davis, R. V.** *Limestone caves: a concise explanation.* Clapham, North Yorkshire, GB: Dalesman Books, **1978**. illus: b&w: maps, photos, tables; 32p; bibl, gloss; 22 x 14 cm.

[837] **Davis, Wilbur A.** *Archaeological surveys of Crater Lake National Park and Oregon Caves National Monument, Oregon.* npop: np, **1964**. illus: b&w: maps; 31p; [4]; 28 cm.

Illus. Report on an archeological project carried out under terms of a memorandum of agreement, Crater Lake National Park - FY 1961 (contract no. 14- 0-0434-900), between the University of Oregon and the US National Park Service "Eugene, Oregon, January, 1964." Typescript (Photocopy).

[838] **Davison, W. Donald (auth, photo)** *Majlis Al Jinn Cave, Sultanate of Oman.* Jones, Cheryl (photo). Ruwi, OM: np, Oct, **1985**. Report PAWR (Oman. Hayah al-Ammah li-Mawarid al-Miyah) no 85 20. illus: b&w: maps; col: photos, ports; front: col photo of cave interior; [viii], 16p; bibl; 30 cm.

Abstract also in Arabic. 16 pls, 4 figs.

[839] **Dawkins, William Boyd**, 1838-1929 *The British Pleistocene Mammalia: part II.* Sanford, W. Ashford. London, GB: Printed for the Palaeontographical Society, **1868**. illus: b&w: charts; 96p, [2], 14p pls; 28 x 22 cm.

VI-XIX pls; pagination odd (29-124).

[840] **Dawkins, William Boyd**, 1838-1929 *Cave hunting, researches on the evidence of caves respecting the early inhabitants of Europe.* Mellor, d. C. (fore). 1973 reprint ed. Wakefield, GB: E. P. Publishing, **1973**. Speleologia. b&w front; app, bibl, index.

Copy examined had Zephyrus Press, Inc. pasted over publisher's statement.); new foreword by D.C. Mellor. Microfilm edition: Master microform held by: UMI. Microfilm. Ann Arbor, Mich., University Microfilms International, 19—. "The evidence which is offered by the animals as to the geography and climate of Europe, which I have published from time to time in the works of the Palaeontographical Society, the Geological Journal, and in the Popular science, British quarterly, and Edinburgh reviews, is collected together in this work..."—Pref.

1st ed. London, GB: Macmillan and Company, **1874**. illus: b&w: drawings, maps, tables; col front; xxiv, 455p; app, bibl, index; 23 cm.

129 figs.

When published in 1874, this book represented the first attempt since 1823 to gather within one book the history of the use and exploration of caves since ancient times. Principally concerned with archeology, an introductory chapter details the history of cave exploration and archeological work in Europe up to the 1870s. Further details, such as the date of discovery, appear in sections concerning specific caves. See also Bishop, M.J.: The Cave Hunters (Derbyshire Museum Service, Buxton, 1982, 48pp) which is a biographical sketch of Dawkins and Dr J. Wilfred Jackson, who was also concerned with cave excavation.

[841] **Dawkins, William Boyd**, 1838-1929 *Further discoveries in the Cresswell caves.* npop: np, Nov, **1879**. 12p.

Reprinted from the Quarterly Journal of the Geological Society, pp 724-735.

[842] **Dawkins, William Boyd**, 1838-1929 *On the discovery of an ossiferous cavern of Pliocene Age at Doveholes, Buxton (Derbyshire).* Manchester, GB: J.E. Cornish, **1903**. Manchester Museum Notes 16. 24p, 5p pls; 22 cm.

[843] **Dayton, Gordon O. (comp)** *The caves of Centre County, PA.* White, William Blaine, 1934–. [State College, PA, US]: Mid-Appalachian Region of the National Speleological Society, Feb, **1979**. Bulletin 11 Mid-Appalachian Region of the National Speleological Society. illus: b&w: charts, graphs, maps, tables; 12 maps folded in text; 126p; bibl; 28 cm.

One folded map inserted.

[844] **Dayton, Gordon O. (comp)** *The caves of Mifflin County, Pennsylvania.* White, William Blaine, 1934– (comp); White, Elizabeth L. (comp). State College, PA, US: Mid-Appalachian Region of the National Speleological Society, Jul, **1981**. MAR Bulletin no.12. illus: b&w: charts, maps, photos, tables; 6 foldout maps; 5 loose folded maps; xii, 72p.

[845] **Deakin, P. R.** *British caves and potholes.* Gill, David W. Truro, GB: D. Bradford Barton Limited, **1975**. illus: b&w: photos; [96]; 23 cm.

A beautifully done photo collection about caves. All photographs are black & white and, overall, they are very good. Has great shots of cavers in waterfalls or water passages. Most vertical shots show ladders instead of SRT, and metric measurements are used throughout. Minimal text.

[846] Dean, Anabel *Going underground: all about caves and caving*. Minneapolis, MN, US: Dillon Press Inc, **1984**. Doing & Learning Books. illus: b&w: drawings, photos; 159p, [1]; app, bibl, gloss, index; 24 cm.

Juvenile.

"Discusses the various types of caves, the uses people have found for caves, and the rules, equipment, and techniques involved in exploring caves, or spelunking."

[847] Dean, Pauline, 1931– *Manitouwadge: cave of the Great Spirit*. Manitouwadge, ONT, CA: Great Spirit Writers, **1989**. vi, 249p; bibl; 29 cm.

Illus.

[848] Deeleman-Reinhold, Christa L. *Revision of the cave-dwelling and related spiders of the genus* Troglohyphantes *Joseph (Linyphiidae) with special reference to the Yugoslav species*. Ljubjana, SI: Slovenska Akademija Znanosti in Umetnosti, Academia Scientiarum et Artium Slovenica, **1978**. illus: b&w: drawings, maps, photos, tables; 218p, [5], 2p pls; bibl, index.

[849] Dehejia, Vidya *The Namakkal caves*. [Madras, IN]: State Dept. of Archaeology, Govt. of Tamilnadu, **1969**. Tamil Nadu India Dept of Archaeology T.N.D.A. publication no 6. 38p; 18 cm.

Illus.

[850] de Joly, Robert *Memoirs of a speleologist: the adventurous life of a famous French cave explorer*. Kurz, Peter (trans); Boulanger, Pierre (ed). Teaneck, NJ, US: Zephyrus Press, **1975**. Speleologia. illus: b&w: maps, photos; xxiii, 162p, 8p pls; index; 21 cm.

Translation of *Ma vie aventureuse d'explorateur d'abimes*. 66 short and lively sketches from comedy to tragedy. 21 b&w photos.

The English language translation of the account of this early French speleologist's career in the years following 1926, in the form of 66 short sketches of his adventures. In a career begun at the age of 40, he systematically explored over 800 caves and invented the flexible lightweight cable ladder for easier access to deep caves.

[851] Delderfield, Eric Raymond *The story of Brixham Cavern*. Brixham, GB: J & W.S. Lack, **c1959**. illus: b&w: maps, photos, tables; 16p; 18 x 12 cm.

Later editions were published. Cover title: The Brixham Cavern: a guide to a fantastic underworld.

Description and history of this show cave of archaeological importance.

[852] DeLoach, Ned (ed) *Ned DeLoach's diving guide to underwater Florida*. Jacksonville, FL, US: New World Publications, **1978**. illus: b&w: ads, maps, photos; 159p, [1].

[853] de Longchène, M. *The underground world of geological marvels*. Breakspear, Marjorie (trans). Box, Wiltshire, GB: Mike Breakspear, **1993**. illus: b&w: maps; [iv], 80p; app.

Translated from French. Original published at Tours in 1846. Only illustration is repro of original cover page.

[854] Demek, Jaromír *Czech Speleological Society 1986-1989: published on occasion of 10th International Speleological Congress Hungary 1989*. Herzlík, Bořivoj (trans). Praha, CZ: Czech Speleological Society, **1989**. illus: b&w: drawings, maps, photos; col: photos; col photo on rear lining page; front: col photo; [77]; bibl; 24.5 x 17 cm.

[855] den Hertog, Sonja (ed) *Walking the valley: an oral record of caving and bushwalking in the Burragorang and beyond, during the 1930s*. Broadway, NSW, AU: Sydney Speleological Society, **1994**. Sydney Speleological Society Occasional Paper no 11. illus: b&w: maps, photos; x, 62p; app, bibl, index.

Cover title: Walking the valley. Contains an epilogue.

[856] Dennewald, Fearless Fiona (comp) *Outdoor recreation II: caving: journey to the centre of the earth: 19-20 October 1985*. Prentice, Active Andy (comp); Muir, Jovial Jane (comp); Brown, Spontaneous Sandy (comp); Woods, Mysterious Marianne (comp). npop: np, **1985**. illus: b&w: ads, drawings, maps, photos; 34p; bibl.

Contains cross word puzzle.

[857] De Paepe, Duane *Gunpowder from Mammoth Cave: the saga of saltpetre mining before and during the War of 1812*. Hays, KS, US: Cave Pearl Press, **1985**. illus: b&w: drawings, maps, photos, repros; col: photos; [ii], 38p; bibl; 28 cm.

[858] [Derbyshire Caving Association] *Derbyshire caving 1987: the handbook of Derbyshire caving*. Derbyshire, GB: Derbyshire Caving Association, **1987**. illus: b&w: maps; [ii], 49p; errat; 21 x 15 cm.

[859] de Saussure, Raymond *Report of the California-Nevada Speleological Survey June 13-September 7, 1952*. Mowat, George; Lang, Arthur. npop: Western Speleological Institute, Mar, **1953**. illus: b&w: drawings, graphs, maps, photos; 6 folded maps in back; iv, 198p, [1].

[860] [Descent Magazine] *Descent handbook for cavers 1974*. Wells, GB: Descent Magazine, **[1973]**. 44p.

[861] Despain, Joel *Crystal cave: a guidebook to the underground world of Sequoia National Park*. Sequoia Natural History Association. Three Rivers, CA, US: Sequoia Natural History Association, **1995**. [vi], 49p; 28 cm.

Illus. Cover title: A guidebook to the underground world of Crystal Cave.

[862] Devil's Hole Pupfish Recovery Team (prep) *Devil's Hole Pupfish Recovery Plan*. Portland, OR, US: United States Fish and Wildlife Service, Endangered Species Program, Region 1, Jul, **1980**. illus: b&w: charts, graphs, maps, photos, tables; [iv], 46p, 30p pls; bibl.

Comment letters at the end.

[863] De Watteville, Alastair *The island of Staffa: home of the world-renowned Fingal's cave.* Romsey: Romsey Fine Art, **1993**. 42p.

[864] Dewey, Jennifer *The creatures underneath.* 1st ed. Santa Fe, NM, US: Red Crane Books, **1994**. 70p; bibl; 27 cm.

Illus. Juvenile.

[865] Dibble, David S. *Bonfire shelter: a stratified bison kill site, Val Verde County, Texas.* Lorrain, Dessamae; Frank, Ruben (app). Austin, TX, US: Texas Memorial Museum, University of Texas at Austin, Jan, **1968**. Texas Memorial Museum Miscellaneous Papers no 1. illus: b&w: charts, graphs, maps, photos, tables; text on rear lining page; 138p; bibl.

[866] Dick, Herbert W. *Bat Cave.* Santa Fe, NM, US: The School of American Research, **1965**. The School of American Research Monograph no 27. illus: b&w: charts, drawings, maps, photos, tables; xiv, 114p; bibl.

61 figs.

Important archaeological study revealing the first evidence of domesticated corn by prehistoric people in the United States.

[867] Dicken, Samuel Newton, 1901– *Soil erosion in the karst lands of Kentucky.* Brown, Harry Bates, 1914– . Washington, DC, US: US Department of Agriculture, **1938**. Circular United States Department of Agriculture no 490. illus: b&w: drawings, maps; 62p, 1p pls; 23 cm.

Contribution from Soil Conservation Service.

[868] Dickey, F. *Report of the Caver Proliferation Committee.* Huntsville, AL, US: National Speleological Society, **1974**. 16p.

A report on the increasing number of cavers in the U.S. Study design has statistical problems.

[869] Dickinson, Leo *Anything is possible.* London, GB: Jonathan Cape Ltd, **1989**. illus: b&w: photos, ports, repros; col: photos; 223p, [1]; app; 26 cm.

[870] Dickson, Don R. *The Albertson site: a deeply and clearly stratified Ozark bluff shelter.* Fayetteville, AR, US: Arkansas Archeological Survey, **1991**. Arkansas Archeological Survey research series no 41. illus: b&w: maps; xii, 315p; bibl; 28 cm.

Illus.

[871] Dilamarter, Ronald R. (ed) *Hydrologic problems in karst regions.* Csallany, Sandor C. (ed). Bowling Green, KY, US: Western Kentucky University, **1971**. illus: b&w: charts, drawings, graphs, maps, photos, photomicrographs, tables; viii, 481p; bibl, index.

A very good proceedings volume. Precursor to the NGWA Karst Conferences. Applied and theoretical.

[872] Dillaye, Stephen D. *A glance at Mammoth Cave, Kentucky.* Syracuse, NY, US: James Kinney, Printer, **1848**. 17p, [1]; 25 cm.

[873] Dimitrijevic, Vesna M. *Quaternary mammals of the Smolucka Cave in southwest Serbia.* Zagreb: Jugoslavenska akademija znanosti i umjetnosti, **1991**. Palaeontologia Jugoslavica sv 41. 88p, 17p pls; 29 cm.

Translation of Kvartarni sisari Smolucke pecine (JZ Srbija). Summary in Croatian. Illus.

[874] Dinger, James S. *Ordinance for the control of urban development in sinkhole areas in the blue grass karst region, Lexington, Kentucky.* Rebmann, James R. Lexington, KY, US: Kentucky Geological Survey, **1991**. illus: b&w: maps; 14p; bibl; 28 cm.

Illus.

[875] District of Columbia Grotto of the National Speleological Society *Members manual.* [Washington, DC, US]: The Grotto, May, **1991**. illus: b&w: maps; 22p; bibl; 22 cm.

Title from cover. Illus.

[876] Dixon, E. James *1980 progress report of archeological reconnaissance and testing of Pleistocene cave and alluvial deposits, Porcupine River, Alaska.* Thorson, Robert M.; Plaskett, David C.; National Geographic Society (U.S.). Committee for Research and Exploration. [Fairbanks, AK, US]: [University of Alaska Museum], Apr, **1981**. illus: b&w: maps; iv, 56p; 28 cm.

[877] Dixon, E. James *Report of 1979 archeological and geological reconnaissance and testing of cave deposits, Porcupine River, Alaska.* Plaskett, David C.; Thorson, Robert M.; National Geographic Society (U.S.). npop: np, Jan, **1980**. 47p; 28 cm.

Title from cover.

[878] Dixon, Richard L. *Speleological investigation of Nakimu Caves, British Columbia.* Rohrer, Thomas, A. npop: np, **1965**. illus: b&w: photos; partly col: maps; map in pocket on rear lining; iii, 53p, [2]; bibl.

b&w photos pasted in.

[879] Dobson, Phil (comp) *The Cave Diving Group annotated ten year cumulative index to the cave dives for the years 1969-1978.* Bristol, GB: Cave Diving Group, Aug, **1981**. 50p; 30 x 21 cm.

[880] Doe, Michael F. *Geology in the vicinity of Montezuma's Cave, southern Huachuca Mountains, Cochise County, Arizona: final [report].* npop: np, Aug, **1985**. illus: b&w: maps; 24p; bibl; 28 cm.

Illus."For Coronado National Memorial."

[881] Dogwood City Grotto of the National Speleological Society *Pigeon Mountain and Ellison's Cave system: a report to the Georgia Department of Natural Resources by the Dogwood City Grotto of the National Speleological Society, Post Office Box 12072, Atlanta, Georgia 30305.* npop: np, Jul, **1974**. illus: b&w: drawings, maps; map folded in rear pocket; [i], 19p.

[882] Donehoo, George Patterson, 1862-1934 *Pennsylvania's Historic Indian Cave at Franklinville, PA, Huntington County....* npop: np, **1930**. 23p; 21 cm.

OCLC lists both 1930 and 1939 as publication date.

[883] Dopp, Katherine Elizabeth, 1863– *The early cave-men*. Chicago, IL, US: Rand McNally & Company, **1904**. Industrial and Social History Series Book II. illus: b&w: drawings; 183p, [1]; 20 cm.

85 drawings. Reprinted in 1932 by W. J. Gage & Co, Ltd, Toronto. Juvenile.

[884] Dopp, Katherine Elizabeth, 1863– *The later cave men*. Chicago, IL, US: Rand McNally & Company, **1906**. Industrial and Social History Series Book III. illus: b&w: ads, drawings; front: b&w drawing; 197p, [1]; 20 cm.

114 drawings. Reprinted in 1926 and 1934 by Rand McNally and issued in London by Harrap (180 p). Juvenile.

[885] Dorrell, Margaret *Caves through the ages*. James, Julia M.; Hangay, George (illus). npop: np, **1974**. illus: b&w: drawings; 22 x 34 cm.

Coloring book containing 14 cartoons that can be colored in.

[886] Dortch, C. E. *Devils Lair: a search for ancient man in Western Australia*. Perth, WA, AU: Western Australia Museum Press, **1976**. illus: b&w: drawings, maps, photos, repros; 24p; bibl.

[887] Dortch, C. E. *Devil's Lair, a study in prehistory*. Perth, WA, AU: Western Australia Museum, **1984**. illus: b&w: charts, drawings, graphs, maps, photos, tables; col: drawings, photos; 94p, 8p pls; bibl, gloss.

[888] Dougherty, Percy H. (ed) *Caves and karst of Kentucky*. Lexington, KY, US: Kentucky Geological Survey, University of Kentucky, **1985**. Special Publication Kentucky Geological Survey ser 11 no 12. illus: b&w: charts, drawings, graphs, maps, photos, tables; [iii], 196p; bibl; 28 cm.

Published in cooperation with the National Speleological Society.

[889] Dougherty, Percy H. (ed) *Environmental karst*. Cincinnati, OH, US: Geo Speleo Publications, **1993**. illus: b&w: drawings, graphs, maps, photos, tables; viii, 167p; app, bibl; 21 cm. cm.

Photographs appear photocopied. Result of the karst symposium held at the Association of American Geographers meeting in April 1980 in Louisville, KY. 16 articles on various karst environmental problems ("valley tides", sinkhole flooding, pollution, etc).

[890] Douglas, Henry Hulbert, 1906-1989 *Caves of Virginia*. National Speleological Society. Falls Church, VA, US: Virginia Cave Survey, **1964**. illus: b&w: maps, photos; iii, 761p; bibl, index; 24 cm.

Facsimile reproductions of quadrangles. Virginia Region of the National Speleological Society included on the title page.

Catalog of 1790 Virginia caves, with locations. Brief text descriptions of half; some maps. Introductory text on geology, biology, and saltpeter mining history.

[891] Douglas, John Scott *Caves of mystery: the story of cave exploration*. New York, NY, US: Dodd, Mead, & Company, **1956**. illus: b&w: photos; xiv, 273p, [1], 16p pls; index; 21 cm.

Fourth printing, 1958, 8th printing in 1966.

A collection of miscellaneous cave lore. Dozens of short stories and anecdotes about caves. For teenagers.

[892] Dowie, H. G. *The history of Kent's Cavern; Torquay: with illustrations*. Kent's Cavern, Torquay, GB: W.F. and L.W. Powe, **1932**. illus: b&w: charts, maps, photos; [32]; 22 x 14 cm.

[893] Dowling, Richard, 1846-1898 *Catmur's cave*. New York, NY, US: Street & Smith, **1900**. Medal library no 86. 264p; 18 cm.

[894] Downing, R.A. (ed) *The hydrogeology of the chalk of north-west Europe*. Price, M.; Jones, G. P. (ed). Oxford, GB: Oxford University Press, **1993**. illus: b&w: charts, drawings, graphs, maps, photos, photomicrographs, tables; vii, 300p; bibl, index.

[895] Downs, Theodore, 1919– *Quaternary animals from Schuiling Cave in the Mojave Desert, California*. Los Angeles, CA, US: Los Angeles County Museum, **1959**. Contributions in Science no 29. illus: b&w: maps; 21p.

Illus.

[896] Dowswell, Peter N. F. *Limestone caves of Scotland, vol IV: the caves of Schichallion*. Dunfermline, GB: The Grampian Speleological Group, Nov, **1980**. illus: b&w: maps; [i], 55p; bibl, index.

23 figs.

[897] Draper, John A. *Archaeology of Chokecherry Cave (35GR500) in Grant County, southeastern Oregon*. Reid, Kenneth C. Pullman, WA, US: Center for Northwest Anthropology, Washington State University, **1989**. Project Report, Center for Northwest Anthropology Washington State University no 9. illus: b&w: maps; viii, 136p; bibl; 28 cm.

Illus.

[898] Drew, David Phillip, 1943– *Aillwee Cave and the caves of the Burren*. Dublin, IE: Eason, **1984**. The Irish Heritage Series 43. illus: b&w: drawings, maps, photos, ports; col: photos; folded map on front lining page; [28]; 25 cm.

[899] Drew, David Phillip, 1943– *Dunmore Cave, County Kilkenny: a reassessment*. Huddart, David. Dublin, IE: Royal Irish Academy, **1980**. Proceedings of the Royal Irish Academy Sect B Biological Geological and Chemical Science vol 380 no 1. 23p; 25 cm.

Title from cover. Illus.

[900] Drew, David Phillip, 1943– *Dunmore Cave: a short guide*. Huddart, David; Economic Institute for Research and Education (U.S.).; Office of Public Works; National Monuments Branch. Dublin, IE: National Parks and Monuments Service, **1978**. 20p.

1980 and 1989 editions may exist. Illus.

[901] Drew, David Phillip, 1943– *Karst processes and landforms*. Hampshire and London, GB: Macmillian Education Ltd, **1985**. Aspects of geography. illus: b&w: charts, drawings, graphs, maps, photos; [i], v, 63p; bibl; 21.5 x 14 cm.

[902] Drew, David Phillip, 1943– *Mendip karst hydrology research project: phase 3*. Newson, Malcolm David, 1945–; Smith, D. I.; Hanwell, J. D. (ed). Pangbourne, Berks, UK: Wessex Cave Club, Dec, **1968**. Series 2 no 2 Occasional Publication. illus: b&w: drawings, graphs, photos, tables; [iv], 28p; 25.5 x 20 cm.

[903] Drew, David Phillip, 1943– *Techniques for the tracing of subterranean drainage*. Smith, David I. Norwich, Norfolk, GB: Geo Abstracts for the British Geomorphological Research Group, **1969**. British Geomorphological Research Group Technical Bulletin no 2. illus: b&w: drawings, graphs, maps, tables; text on lining pages; 36p; bibl.

Some new techniques have been added since the publication of this brief guide, but the basics remain.

[904] Dreybrodt, Wolfgang *Processes in karst systems*. New York, NY, US: Springer Verlag, **1988**. illus: b&w: drawings, graphs, photos, tables; xii, 288p; 25 cm.

Limited in scope to shallow phreatic, CO_2-related karst, only through maturity. This work is strictly theoretical, with a dense text that could have benefited from additional figures and photos.

[905] Driscoll, Ian *The Binoomea Cut, Jenolan Caves: a paper read to the Jenolan Caves Historical and Preservation Society on Saturday 9th, February 1974*. Jenolan Caves, NSW, AU: Jenolan Caves Historical and Preservation Society, Apr, **1977**. Jenolan Caves Historical & Preservation Society Occasional paper number 4. illus: b&w: maps, photos, repros; 16p; 21 cm.

[906] Drnevich, Vincent P. *Location of solution channels and sinkholes at dam sites and backwater areas by seismic methods*. Lexington, KY, US: University of Kentucky, Water Resources Institute, Aug, **1972**. University of Kentucky Soil Mechanics series no15. vi, 30p; bibl; 28 cm.

Illus.

[907] Drumm, E. C. *Subsidence of residual soils in a karst terrain*. Et al. Oak Ridge, TN, US: Oak Ridge National Laboratory, Jun, **1990**. ORNL/TM-11525. illus: b&w: drawings, maps; xi, 92p; bibl, index; 28 cm.

Issued on microfiche by NTIS, Springfield, VA. Prepared by the Oak Ridge National Laboratory operated by Martin Marietta Energy Systems, Inc. for the U.S. Department of Energy under contract no. DE-AC05-84OR21400.

[908] Duddington, C. L. *Caves*. Gilham, Alan (illus). London, New York, NY, US, GB: McGraw-Hill Publishing Company Limited, **1969**. Natural History Series. illus: b&w: drawings; partly col: maps; col: drawings; 55p; bibl; 23 cm.

Juvenile.

Describes the different kinds of caves, where they are found, how they are formed, the plants and animals found in them, and their use by early man.

[909] Duff, B. M. *Ground penetrating electromagnetic tests at Medford Cave, Florida and Waterways Experiment Station, Vicksburg, Mississippi*. Suhler, Sidney A. San Antonio, TX, US: Southwest Research Institute, **1980**. iii, 35p; 28 cm.

Illus. June 18, 1980. Prepared for Waterways Experiment Station, Vicksburg, Mississippi.

[910] Dulluc, Brigitte *Discovering Lascaux*. Delluc, Gilles; Delvert, Ray (photo); Moyon, Angela (trans). Luçon, FR: Editions Sud-ouest, **1990**. illus: b&w: charts, drawings, maps, photos; 64p; bibl; 26 x 19 cm.

[911] Dumont, Kevin Allen *Karst hydrology and geomorphology of the Barrack Zourie Cave System, Schoharie County, New York*. Schoharie, NY, US: Speleobooks, **1995**. New York Cave Survey Bulletin 5. illus: b&w: charts, graphs, maps, tables; col: photos; 1 loose folded map plate; ix, 70p, 1p pls; app, bibl.

[912] Dunham, Wayland A., 1898– *Enchanted corridors*. Portland, OR, US: W.A. Dunham, **1939**. illus: b&w: photos; 64p; 19 cm.

Very rare little book about Oregon Caves. Includes 4 tipped in b & w photos; printed on unusual paper (birch bark?); soft cover with dust jacket.

[913] Dunkley, John R. (comp) *A bibliography of the Jenolan Caves: part 1: speleological literature*. Broadway, NSW, AU: Speleological Research Council Ltd, **nd**. [iii], 52p; 30 cm.

No illus. Includes abstracts.

[914] Dunkley, John R. (comp) *A bibliography of the Jenolan Caves: part 2: literature*. Broadway, NSW, AU: Speleological Research Council Ltd, **1988**. illus: b&w: repros; [vi], 45p; 30 cm.

[915] Dunkley, John R. *The caves of Jenolan: 1, the exploration and speleogeography of Mammoth Cave, Jenolan*. Anderson, Edward G.; Winglee, P. J. 2nd ed. Sydney, AU: Geological Research Council Ltd. for Sydney University Speleological Society, **1978**. illus: b&w: maps; x, 62p; 28 cm.

2nd ed, with additional material by P. J. Winglee, of The exploration and speleogeography of Mammoth Cave, Jenolan.

A well-illustrated book using nineteenth-century wood engravings. Sections deal with different aspects of the cave, using extensive quotations and background information on the caves and the people concerned with their exploration. A useful resource.

[916] Dunkley, John R. *The caves of Thailand*. Broadway, NSW, AU: Speleological Research Council Ltd, Jan, **1995**. illus: b&w: maps, photos, repros; 124p; bibl, gloss.

[917] Dunkley, John R. *Caves of the Nullarbor; a review* *of speleological investigations in the Nullarbor Plain, Southern Australia*. Wigley, T. M. L.; Dury, G. H. (forw). 1978 reprint ed. Broadway, NSW, AU: The Speleological Research Council, Ltd, **1978**. illus: b&w: drawings, maps, photos, tables; 2 folded maps facing pp 14,22.; b&w photo; viii, 64p; bibl; 27 cm.

Reprint edition of the 1967 edition with the addition of supplementary references. Produced by Sydney University Speleological Society and the Cave Exploration Group (South Australia); 32 figs.

1st ed. Wigley, T. M. L. Sydney, NSW, AU: The Speleological Research Council Ltd, **1967**. illus: b&w: drawings, maps; 2 folded maps facing pp 14,22; partly folded diagrams; vii, 61p, 2p pls; bibl; 27 cm.

[918] Dunkley, John R. *The exploration and speleogeography of Mammoth Cave, Jenolan*. Anderson, Edward G. (asst). 1st ed. Kingsford, NSW, AU: Speleological Research Council Limited and Sydney University Speleological Society, **1971**. illus: b&w: maps, photos; partly col: maps; one map folded in text; x, 53p; app, bibl; 28 cm.

2nd ed. Anderson, Edward G. (asst). Broadway, NSW, AU, **1978**. x, 62p; app, bibl.

Excellent item with many maps and photos.

[919] Dunkley, John R. *The exploration and speleography of Mammoth Cave, Jenolan*. Anderson, Edward G. Kingsford, NSW, AU: Speleological Research Council Limited for Sydney University Speleological Society, **1971**. illus: b&w: photos; partly col: maps; 1 partly col map folded in text; x, 53p; app, bibl; 28 cm.

[920] Dunkley, John R. *Jenolan caves as they were in the nineteenth century*. Broadway, NWS, AU: Speleological Research Council Ltd for Jenolan Caves Historical & Preservation Society, **1986**. illus: b&w: drawings, maps, repros; 59p; app, bibl; 21 cm.

[921] Dunkley, John R. *Preliminary checklist of references to Jenolan Caves*. npop: np, Nov, **1985**. 12p.

Issued for private circulation only. An extract from the Bibliography of the Jenolan Caves Part 2. No illus.

[922] Dunkley, John R. *Thailand caves catalogue*. Broadway, NSW, AU: Speleological Research Council, **1994**. illus: b&w: maps, photos, repros, tables; 44p.

[923] Dunkley, John R. *Caves of north-west Thailand: report of the Australian speleological expeditions, 1983-1986*. Brush, John B. Sydney, AU: Speleological Research Council, Ltd., **1986**. illus: b&w: maps, photos; 2 folded maps in text; 62p; app, gloss; 30 cm.

[924] Dunlop, Bruce Thomas *Jenolan Caves*. 1969 ed. illus: b&w: maps, photos, tables; maps on lining pages; 88p; bibl.

1972 ed.

1975 ed. illus: b&w: maps, photos, tables; maps on lining pages; 88p; bibl; 22 cm.

1977 ed. illus: b&w: maps, photos, tables; maps on lining pages; 88p; bibl.

1979 ed.

1st ed. [Sydney, NSW, AU]: New South Wales Dept of Tourism, **1950**. illus: b&w: ads, photos, tables; 94p, [1]; bibl.

Cover title: Jenolan Caves, New South Wales. Over 30 excellent photos.

2nd ed. [New South Wales Department of Tourist Activities and Immigration], **1952**. illus: b&w: ads, photos, tables; 94p, [1]; bibl.

3rd ed. 1960. illus: b&w: maps, photos, tables; b&w map folded in text; no text on lining pages; 96p, 2p pls; bibl.

4th ed. [New South Wales Government Tourist Bureau], **1961**. illus: b&w: maps, photos, tables; map folded in text; 96p, 2p pls; bibl.

Ads deleted, and cover is different.

5th ed. 1964. illus: b&w: maps, photos, tables; map folded in text; 96p, 2p pls; bibl.

6th ed. 1967. illus: b&w: maps, photos, tables; map folded in text; 94p, [2], 2p pls; bibl.

[925] Durden, C. C. *The Carlsbad Cavern of New Mexico: it's [sic] history and geology*. npop: [Durden?], **1928**. [46]; 26 cm.

Illus.

Of considerable historic interest, of more interest to bibliophiles than to geologists

[926] Durrant, Gillian A. *Himalaya underground - 1976 speleological expedition*. Smart, C. M.; Turner, J. npop: np, **1979**. illus: b&w: maps, photos; 88p, [1]; bibl; 29.5 x 21 cm.

Caving in Nepal, North Pakistan, and Northern India. 54 photos. 2nd corrected ed 1981.

[927] Dusar, Michiel *Geological and speleological reconnaissance of the East Yunnan Karst (P.R. China): preliminary results of a field trip between 17.12.1990 and 3.01.1991*. Shouyue Zhang. [Bruxelles], BE: Service Geologique de Belgique, **1991**. Professional paper (Service Geologique de Belgique) 1991/4 no 248. illus: b&w: maps; 40p; bibl; 29 cm.

At head of title: "East Yunnan 1991." Illus.

[928] Duxbury, J. *The joint Anglo-Soviet speleological expedition to central Asia 1990 "Kugitang 90"*. London, GB: The Chelsea Speleological Society, Jun, **1991**. illus: b&w: maps, repros, tables; col: photos; iii, 15p, [6]; app; 30 x 21 cm.

[929] Dyer, C. F. *Geology and occurrence of ground water at Jewel Cave National Monument, South Dakota*. Washington, DC, US: United States Government Printing Office, **1961**. Geological Survey Water Supply Paper 1475 D. illus: b&w: maps, tables; 19p; bibl; 24 cm.

[930] Dyer, William Henry *A brief story of St Clement's caves*. Hastings, Sussex, GB: Hastings Tourism and Recreation Department, **1979**. illus: b&w: drawings, maps, photos; [24]; 17 cm.

[931] Dyson, H. Jane (ed) Cover title: St Clement's caves, Hastings, Sussex. Illus.

[931] Dyson, H. Jane (ed) *Wombeyan Caves.* Ellis, Ross Andrew, 1940–(ed); James, Julia M. (ed). Broadway, NSW, AU: The Sydney Speleological Society, **1982**. Sydney Speleological Society Occasional Paper no 8. illus: b&w: drawings, maps, photos, tables; col: photos; viii, 224p.

A well-produced volume on the Wombeyan Caves, Australia, with an introductory section on early visits and the area's history; fully referenced.

[932] Eastmead, William *Historia rievallensis: containing the history of Kirkby Moorside ... to which is prefixed a dissertation on the animal remains, and other curious phenomena, in the recently discovered cave at Kirkdale.* London, Thirsk, GB: np, **1824**. xiv, 486p; 23 cm.

Printed at the office of R. Peat. Illus.

[933] Easton, T. S. *The secret of the Wallowa Cave.* Portland, OR, US: Metropolitan Press, **1934**. 127p; 20 cm.

Fiction? Illus.

[934] Eaton, Joli (ed) *GYPKAP 1987 annual report.* npop: Southwestern Region. National Speleological Society, **[1988]**. illus: b&w: charts, drawings, maps, photos, tables; 3 loose folded maps; text on front lining page (p 2); text, photo on rear lining page (p 35); 36p.

Pagination includes covers.

[935] Eatough, J. A. *Mendip's vanishing grottoes: a photographic record of Balch Cave by J. A. Eatough and Shatter Cave by A. E. Mc. R. Pearce.* Pearce, J. A.; Mc. R. Pearce, A. E. Somerset, GB: The Bristol Exploration Club, **[1972]**. Caving Report no 16. illus: b&w: maps, photos; 25.5 x 20 cm.

[936] Eavis, A. J. (comp) *Caves of Mulu '80: the limestone caves of the Gunong Mulu National Park, Sarawak.* London, GB: The Royal Geographical Society, **1981**. illus: b&w: drawings, graphs, maps, photos, tables; col: photos; map folded in center; b&w front; 52p, 4p pls; 30 cm.

Continuing Caves of Mulu (1978). Cathay Pacific Airways Mulu '80 Expedition, British-Malaysian Speleological Expedition to Sarawak.

[937] Eavis, A. J. (comp) *Caves of Mulu '84: the limestone caves of the Gunong Mulu National Park, Sarawak.* Bridgwater, GB: British Cave Research Association, **1985**. illus: b&w: maps, photos; some maps folded in text; 56p; 30 cm.

Continuing Caves of Mulu (1978) and Caves of Mulu ('80).

[938] Eberhard, Rolan *The Junee River karst system, Tasmania: a report to the Forestry Commission.* TAS, AU: Forestry Commission, **1994**. 3 maps in pocket; 125p.

An inventory and management report on a significant karst system in Tasmania. Land use problems are thoroughly covered. Well supported with maps, tables, and diagrams.

[939] Eberhard, Stefan *The cave fauna at Ida Bay and the effect of quarry operation.* Hobart, TAS, AU: Department of Parks, Wildlife, and Heritage, **nd**. Ida Bay Karst Study Report to the Department of Parks, Wildlife, and Heritage. 24p.

[940] Eberhard, Stefan *The invertebrate cave fauna of Tasmania.* Richardson, Alastair Mackenzie Martyn; Swain, Roy. Hobart, TAS, AU: Zoology Dept, University of Tasmania, **1991**. illus: b&w: maps; 174p; bibl; 29 cm.

[941] Eberhard, Stefan *Report of the Tasmanian Caverneering Club 1986 speleological reconnaissance expedition to Precipitous Bluff.* Hume, N. [Hobart, TAS, AU]: Tasmanian Caverneering Club, **1986**. 19p; bibl; 30 cm.

Title from cover. Running title: Speles spiel.

[942] Edelbrock, Gay *Oregon Caves National Monument resource database: a user's guide.* Wright, R. Gerald. Moscow, ID, US: University of Idaho in cooperation with Cooperative Parks Studies Unit, National Park Service, Pacific Northwest Region, **1993**. Technical Report NPS/MRUI/NRTR; 93/14. [19]; 28 cm.

Includes 1 computer disk (5 - 1/4") in back envelope. "Subagreement no. 8 to Cooperative Agreement No. CA-9000-8-0005; Managing resource data in the Pacific Northwest Region."

[943] Ediger, Donald *The well of sacrifice.* Garden City, NY, US: Doubleday, **1971**. illus: b&w: photos, ports; col: photos; 288p, 16p pls; index; 24 cm.

[944] Edmonds, Francis Samuel *The dream of Kent's Cavern and ballads of the Bay.* npop: Downe's Library Torquay, **1928**. 20p; 21.5 cm.

Contains 6 ballads including the Kent's Cavern one. 3 half-tone photo illus.

[945] Educational Enrichment Materials, Inc *Cave science.* Bedford Hills, NY, US: Educational Enrichment Materials, **1980**.

Title from data sheet. Sound accompaniment compatible for manual and automatic operation. 2 filmstrips (131 fr), col, 35 mm + 2 sound cassettes (24 min, 1 7/8 ips) + 1 teacher's guide.

Intended audience: Junior and senior high school students. 1. The biological cave—2. The physical cave. Examines how caves are formed and suggests safety precautions to be used during cave explorations.

[946] Edwards, Ira *Underground Geology at the Endless Caverns, New Market, VA.* Milwaukee, WI, US: Public Museum, **1925**.

Year book, 82-104, 11 photos.

[947] Ege, John R. *Formation of solution-subsidence sinkholes above salt beds.* [Reston, VA, US]: United States Dept of the Interior, Geological Survey, **1984**. Geological Survey Circular no 897. illus: b&w: charts, maps, photos; iii, 11p; bibl; 26 cm.

Illus.

[948] Eggleston, Peter *Underground Video Techniques.*

2nd ed. Shropshire, GB: Shropshire Caving and Mining Club & I. A. Recordings, **1994**. Number 19 - Shropshire Caving and Mining Club. illus: b&w: ads, drawings; 45p; app, gloss; 29.5 x 21 cm.

[949] Ehr, Bob (prod) *Guide to the caves of Wisconsin.* Madison, WI, US: Wisconsin Speleological Society, Inc, **1974**. illus: b&w: maps; Two pages of maps, folded loose in text.; [ii], 22p, [19]; bibl.

Folded maps may have been handed out at meeting.

[950] Eigenmann, Carl H., 1863-1927 *Cave vertebrates of America: a study in degenerative evolution.* Washington, DC, US: Carnegie Institution of Washington, **1909**. Carnegie Institution of Washington Publication no 104. illus: b&w: drawings, photos, tables; col: photos; b&w photos; ix, 241p, 29p pls; 30 cm.

Microfilm. Washington, D.C.: Library of Congress, 1957. 1 microfilm reel: negative ; 35 mm.

The classic study of eye anatomy of cave vertebrates, especially of the cave fish in the family Amblyopsidae.

[951] Elliot, Dave, 1946– *Caves of Northern Derbyshire: part 1: the Eldon Hill area.* Cheadle, GB: Dave Eliott, **1975**. illus: b&w: ads, maps, photos; 8 b&w photos at rear; [ii], 27p, [1], 3p pls; bibl; 25.5 x 20 cm.

20 photos; 9 maps.

[952] Elliot, Dave, 1946– *Caves of Northern Derbyshire: part 2: Giants-Oxlow System.* npop: np, Mar, **1975**. 21p.

12 photos.

[953] Elliot, Dave, 1946– *Caves of Northern Derbyshire: part 3: Perryfoot/Coal Pit Hole.* Cheshire, GB: Dave Elliot, May, **1975**. illus: b&w: maps; [iii], 26p, [2]; 25.5 x 20 cm.

[954] Elliot, Dave, 1946– *Caves of Northern Derbyshire: part 4: Rushup Edge swallets.* Cheshire, GB: Dave Elliot, Feb, **1976**. illus: b&w: maps, photos; ad on rear lining page; [iv], 28p, [1], 2p pls; 25.5 x 20 cm.

[955] Elliot, Dave, 1946– *Caves of Northern Derbyshire: part 5: Treak Cliff Hill.* Cheshire, GB: Dave Elliot, **nd**. illus: b&w: maps; [iii], 31p, [1]; 25.5 x 20 cm.

[956] Elliot, Dave, 1946– *Single rope technique rigging guide.* Lawson, Dick. Farfield Mill, GB: Lizard Speleo-Systems, **1987**. illus: b&w: ads, maps, photos; 128p; bibl.

Another copy lists publisher as Troll Safety Equipment, Ltd., Spring Mill, England.

A guide to rigging specific caves in Britain.

[957] Elliot, Dave, 1946– *Single rope technique: a training manual.* Spring Mill, GB: Troll Safety Equipment, Ltd, **1986**. illus: b&w: drawings, maps, photos; text on lining pages; 95p; app; 20 cm.

Copyright Lizard Speleo-systems; cover title: SRT.

An extremely popular, small book of techniques.

[958] Elliot, Ian *The discovery and exploration of the Yanchep Caves.* npop: np, **1977**. 35p, [3].

Unpublished manuscript: a paper read before the Royal W. A. Historical Society on February 25th, 1977. No illus.

[959] Elliott, D. M. (ed) *Tasmanian Caverneering Club Handbook.* Hobart, TAS, AU: [Tasmanian Caverneering Club], Jul, **1953**. 71p.

No illus.

[960] Elliott, Larry P. *Study of potential gas accumulation in caves in Bowling Green, including relationship to water quality.* Bowling Green, KY, US: Ogdon College of Science and Technology, Western Kentucky University, **1974**. illus: b&w: drawings, graphs, maps, tables; [ii], 111p; app, bibl.

Cover title: Final project report funded by city of Bowling Green resolution No. 73-52.

[961] Elliott, William Rawleigh, 1946– *Draft recovery plan for endangered karst invertebrates in Travis and Williamson Counties, Texas.* O'Donnell, Lisa. Austin, TX, US: United States Fish and Wildlife Service, Jun, **1993**. illus: b&w: maps, tables; 133p; app, bibl, gloss.

[962] Elliott, William Rawleigh, 1946– *A field guide to the caves of Kendall County.* [Austin, TX, US]: Texas Speleological Survey, Dec, **1985**. illus: b&w: maps; partly col: maps; 1 partly col map folded in text; [iv], 143p.

Limited edition of 71; second printing 1986.

[963] Elliott, William Rawleigh, 1946– *New cavernicolous Rhagidiidae from Idaho, Washington, and Utah (Prostigmata, Acari, Arachnida).* [Lubbock, TX, US]: Museum, Texas Tech University, **1976**. Occasional papers Texas Tech University Museum no 43. illus: b&w: drawings; 15p; bibl; 23 cm.

Illus.

[964] Elliott, William Rawleigh, 1946– *The new Melones Cave harvestman transplant.* San Bonito, TX, US: [Texas Tech Univ School of Medicine. Epidemiological Studies Program (pesticides)], **1978**. illus: b&w: graphs, maps, tables; 62p; bibl.

Contract #DACW05-78-C-007, US Army Corps of Engineers, Sacramento District.

[965] Ellis, Arthur Charles *An historical survey of Torquay from the earliest times, as illustrated by finds in Kent's Cavern, down to the present time.* 2nd ed. Torquay, GB: The "Torquay Directory", **1930**. illus: b&w: ports; viii, 506p; 30 cm.

50% cave-related. Illus.

[966] Ellis, Bryan M., 1934– *An introduction to cave surveying: a handbook of techniques for the preparation and interpretation of conventional cave surveys.* London, GB: British Cave Research Association, **1988**. Cave Studies Series number 2. illus: b&w: drawings, tables; 40p; bibl.

[967] Ellis, Bryan M., 1934– *The manufacture of lightweight caving equipment.* rev ed. Bridgwater, Somerset, GB: Bristol Exploration Club, Oct, **1962**. Caving Report no 3a. illus: b&w: drawings, tables; 23p; 26 x 20 cm.

This is a revised, rewritten edition.

[968] **Ellis, Bryan M.**, 1934– *Mendip cave registry: handbook for members of the Executive Committee.* [Somerset, GB], **[1963]**. 16p; 21 x 17 cm.

How to set up cave registers for that area. No illus.

[969] **Ellis, Bryan M.**, 1934– *A survey of headgear and lighting available for caving.* Bridgwater, Somerset, GB: Bristol Exploration Club, Oct, **1958**. Caving report no 5. illus: b&w: photos, tables; [ii], 72p; app, bibl; 26 x 20 cm.

Revised by Bull, G. 1967.

[970] **Ellis, Bryan M.**, 1934– *Surveying caves.* Somerset, GB: The British Cave Research Association, **1976**. Caving Series. illus: b&w: ads, drawings, graphs, maps, photos, tables; xv, 88p; bibl, index; 21 cm.

Last 5 pages are ads.

Introduction to principles of cave surveying. A major step in improving surveying techniques and bringing some consistency to the field.

[971] **Ellis, Edward Sylvester**, 1840-1916 *Bear cavern.* London, GB: Cassell, **1900**. vi, 151p, 4p pls; 19 cm.

Illus. 50% cave-related. Reprinted in 1982.

[972] **Ellis, Martin** *The caves of Tenerife.* [GB]: np, **1992**. illus: b&w: maps, tables; 46p; bibl.

Published as: Shepton Mallet Caving Club Journal, ser 9, no 2, spring 1992.

[973] **Ellis, Robert D. (scrn wrt)** *Carlsbad Caverns National Park.* Lindsay, Charles (narr). Whittier, CA, US: Finley-Holiday Film Corporation, **1993**. National Park and Monument Series.

1 videocassette (20 min), sd, col, 1/2 in; VHS format.

A ranger-guided tour of the enchanting Carlsbad Caverns.

[974] **Ellis, Ross Andrew**, 1940– *Australian caves and caving.* Sydney Speleological Society. Marrickville, NSW, AU: Western Colour Print Pty Ltd, **1980**. illus: b&w: photos; col: photos; text on lining pages; [94]; 28 cm.

Cover title: Australian caves and caving. Title on verso of cover: Caves and caving in Australia. "A photographic record of cave exploration in Australia with the Sydney Speleological Society." Chiefly illustrations.

[975] **Ellis, Ross Andrew**, 1940– *Bungonia Caves.* Hawkins, Lynsey (ed); Hawkins, Robert (ed); James, Julia M. (ed); Middleton, James (ed); Nurse, Benjamin (ed); Gleniss Wellings (ed). Broadway, NSW, AU: The Sydney Speleological Society, **1972**. Sydney Speleological Society Occasional Paper no 4. illus: b&w: drawings, maps, photos, ports, tables; col: photos; 1 map folded in pocket, 1 map folded in center; xii, 230p; bibl, index; 27 cm.

14 numbered figs, xxiv plates on numbered pages, 16 tables.

An excellent summary of history and exploration to the Bungonia Caves, Australia, is included, plus details of early visitors and settlers in the area.

[976] **Ellis, Ross Andrew**, 1940– *A history of the journal of the Sydney Speleological Society.* npop: np, **nd**. [30].

No illus.

[977] **Ellul, Joseph S.** *Malta's prediluvian culture at the stone age temples with special reference to Hagar Qim, Ghar Dalam cart ruts, Il-Misqa, Il-Maqluba and Creation.* MT: Printwell, **1988**. illus: b&w: charts, drawings, maps, photos; iii, 94p, 29p pls; 22 cm. cm.

[978] **Elsasser, Albert B.** *The archaeology of Bowers Cave, Los Angeles County, California.* Heizer, Robert Fleming, 1915–. Berkeley, CA, US: University of California Archaeological Survey, **1963**. California University Berkeley Archaeological Survey Reports no 59. 83p; bibl; 28 cm.

Illus.

[979] **Ely, Glen Sample. (photo, ed)** *A history of the Guadalupe Mountains and Carlsbad Caverns, 1840–1940.* Viator, Mark (original music soundtrack). Austin, TX, US: Forest Glen TV Productions, **1990**. Texas History Companion v 1.

1 videocassette (60 min, 18 sec), sc, col, 1/2 in, VHS format.

[980] **Encyclopaedia Britannica Educational Corporation** *Caves: the dark wilderness.* Herman, Matthew N. (prod). Chicago, IL, US: Encyclopaedia Britannica Educational Corporation, **1976**.

1 reel, 22 min, sd, col, 16 mm & teacher's guide.

Explores the ecology of caves, showing how animals have adapted to life in a cave. Explains the creation of internal cave formations from the traces of limestone in water drops and shows examples of caves formed by volcanoes, water, earthquakes, and desert winds.

[981] **Endless Caverns, Inc** *Endless caverns, New Market, Virginia, in the heart of the Shenandoah Valley...3 minute drive off U.S. 11 (Lee highway) over new boulevard.* npop: np, **1930**. [12].

Title from cover. "The gateway to endless thrills." OCLC lists both a 1930 and 1939 publication date.

[982] **Endless Caverns, Inc** *Endless Caverns, Virginia. "Considered the most beautiful of the Shenandoah Valley caverns"....* New York, NY, US: Nomad Publishing Co, **1939**. [20].

Title from cover. OCLC lists both 1930 and 1939 as publication date.

[983] **Endless Caverns, Inc** *Endless Caverns, wonderful and spectacular.* 1921 ed. New Market, VA, US, **1921**. 29p.

1924 ed. illus: b&w: ads; 30p, [2].

Cover illus "The snowdrift," in color; others in blue and white; illus. on p. [14], "The Lodge and Tea House." Other minor variations; [2] p. at end have different advertisements.

[984] **Endless Caverns, Inc** *Endless Caverns, wonderful and spectacular, New Market, Virginia, in the heart of the*

Shenandoah Valley. Henkel, David S. [New Market, VA, US], **nd**. illus: b&w: maps; 30p; 13 x 18 cm.

A description of some of the records and passages, arranged from a description written by David Henkel: p. 17-30. Illus.

[985] Endless Caverns, Inc *Endless Caverns: on U.S. 11, 3 miles south of New Market, Virginia: illuminated by indirect electric flood lighting*. New Market, VA, US: The Caverns, **1987**. illus: b&w: maps; 1 folded sheet; 12p; 23 cm.

[986] Endless Caverns, Inc *Exploring the Endless Caverns of New Market, Virginia in the heart of the Shenandoah Valley: two authentic and thrilling accounts of the adventures of members of the American Museum of Natural History and of the Explorers Club, who vainly sought the end to the Endless Caverns*. Ashton, Horace; Barnitz, Wirt Whitcomb, 1887–; Larimore, Betty. New York, NY, US: Nomad Pub Co, **1926**. 47p; 22 cm.

Illus. Inside the earth, by H. Ashton.—Endless Caverns of Virginia among earth's wonders, by W. W. Barnitz.—The first expedition of members of the Explorers Club and American Museum of Natural History to the Endless Caverns, by B. Larimore.

[987] Endless Caverns, Inc *Exploring the Endless Caverns of New Market, Virginia. Authentic and thrilling accounts of the adventures of members of the Explorers Club and of the American Museum of Natural History*. Larimore, Betty; Ashton, Horace; Cowling, Herford Tynes, 1890–. New Market, VA, US, **1947**. illus: b&w: maps, ports; 48p; 22 cm.

Illus. On cover: Mountain climbing underground in the Endless Caverns of New Market, Virginia. The first expedition, described by Betty Larimore.—Inside the earth, a story of the second expedition, by Horace Ashton.—Mountain climbing underground, or the third expedition, by Herford Tynes Cowling.

[988] Endless Caverns, Inc *Nature underground; the Endless Caverns in the heart of the historic Shenandoah Valley*. New York, NY, US: Eastern Printing Corporation, **1946**. 29p.

[989] Engel, Thomas Daniel, 1954– *The almost complete eclectic caver*. Siron, Mary Ann (illus). Schoharie, NY, US: Speleobooks, **1992**. illus: b&w: drawings; vii, 152p.

An absolutely wonderful collection of essays and stories, epitomizing humorous situations that have happened to all cavers. No, not electric!

[990] Engel, Thomas Daniel, 1954– *A chronicle of selected northeastern caves: a history guide for the 1979 NSS Convention, Pittsfield, Massachusetts, August 5-12, 1979*. [Albany, NY, US]: [Thomas D. Engel], Jul, **1979**. illus: b&w: maps, photos, repros; 49p; bibl; 28 cm.

Tour and history of 11 of the northeast's caverns, USA. Illus.

[991] Engelbrecht, N. P. *Caving in the Transvaal*. npop: [N. P. Engelbrecht], Aug, **1972**. illus: b&w: drawings, maps; map on rear lining page; [i], 71p, [3].

South African caves, equipment, techniques.

[992] Enger, Walter D. *Archaeology of Black Rock 3 Cave, Utah*. 1942 ed. npop: np, **1942**. Archaeology and Ethnology Papers, Museum of Anthropology, University of Utah, no 7. iv, 27p, [1]; bibl; 30 cm.

Illus.

1950 edition. Salt Lake City, UT, US: University of Utah Press, Nov, **1950**. Anthropological papers University of Utah Dept of Anthropology no 7. 20p; bibl; 28 cm. pp 74-94.

[993] England, A. W. *Caves, an anthology*. Edinburgh, GB: Oliver & Boyd, **1973**. illus: b&w: drawings; 138p; bibl.

Illus. Juvenile.

Collection of cave writings from Lovelock to Hardy; writings imaginative & factual, calculated to appeal to children 11 to 13.

[994] English, Jackie (comp, photo) *Mitchelstown Cave: its discovery and history*. Kilkenny, IE: np, **c1980**. illus: col: photos; 15p; 21 x 15 cm.

[995] Ensminger, Scott A. *Canadian caves of the Niagara Gorge*. npop: np, **1983**. illus: b&w: maps; 16p; bibl; 28 cm.

[996] Ensminger, Scott A. *The caves of Niagara County, New York*. Lockport, NY, US: Niagara County Historical Society, **1987**. Occasional Contributions of the Niagara County Historical Society no 27. illus: b&w: maps; iii, 72p, 5p pls; index; 28 cm.

[997] Environmental Consulting Engineers *East Chestnut ridge hydroelectric characterization: a geophysical study of two karst features*. Knoxville, KY, US: Environmental Consulting Engineers. illus: b&w: charts, drawings, photos, tables.

[998] Epstein, Jeremiah F. *Centipede and Damp caves: excavations in Val Verde County, Texas, 1958*. United States National Park Service; Texas Archeological Salvage Project. Austin, TX, US: np, **1960**. ix, 171p; 28 cm.

Illus.

[999] Epstein, Sam, 1909– *All about prehistoric cave men*. Epstein, Beryl Williams, 1910–; Huntington, Will (illus). New York, NY, US: Random House, **1959**. All About Books. illus: b&w: drawings, maps; partly col: drawings; partly colored drawings on lining pages; [vii], 137p, 4p pls; index; 24 cm.

Copy examined was a second printing. Juvenile.

[1000] Epton, Nina Consuelo *The Valley of Pyrene*. London, GB: Cassell and Co, **1955**. illus: b&w: maps, photos; xii, 212p.

This is an account of visiting prehistoric caves with Comte Begouen and Norbert Casteret. Has 16 pages of b&w photos and a map.

[1001] **Eraso, Adolfo (ed)** *1st international symposium of glacier caves and karst in polar regions. Proceedings.* Madrid, ES: Instituto Tecnologico GeoMinero de Espana, **1991**. illus: b&w: drawings, maps; 236p; 30 cm.

Illus. Symposium held 1 to 5 October 1990 in Madrid, as result of decision to create International Working Group of Glacier Caves and Karst in Polar Regions, by the General Assembly of 10th International Congress of Speleology, held in Budapest in August 1989—1990 Glacier Caves and Karst in Polar Regions, 1st International Symposium.

[1002] **Eraso, Adolfo (ed)** *The Stor-Glomfjord project: karst research - the possibilities of leakage and the selection of damsite alternative.* Lund, Cecilie (ed). NO: Planning Division, General Planning Office, State Power Board, Mar, **1988**. illus: b&w: graphs, photos, tables; 57p; bibl.

[1003] **Erchul, Ronald A.** *Geotechnical applications of the self potential (SP) method. Report 1: the use of self potential in the detection of subsurface flowpatterns in and around sinkholes.* [Lexington, VA, US]: [Department of civil engineering, Virginia Military Institute], Mar, **1988**. Report 1 of a series. illus: b&w: drawings, graphs, maps, tables; [ii], 24p; bibl.

Prepared for Department of the US Army Corps of Engineers, Washington, DC 20314-1000; monitored by Geotechnical laboratory, US Army Engineer waterways experiment station PO Box 631, Vicksburg, Mississippi 39180-0631 under work unit 32315.

[1004] **Erchul, Ronald A.** *Geotechnical applications of the self potential (SP) method. Report 2: the use of self potential to detect ground-water flow in karst.* Slifor, Dennis W. [Lexington, VA, US]: [Department of Civil Engineering, Virginia Military Institute], May, **1989**. Report 2 of a series. illus: b&w: drawings, graphs, maps, tables; [ii], 41p, [78]; bibl; 28 cm cm.

Prepared for Department of the US Army Corps of Engineers, Washington, DC 20314-1000; monitored by Geotechnical laboratory, US Army Engineer waterways experiment station PO Box 631, Vicksburg, Mississippi 39180-0631 under work unit 32315.

[1005] **Erickson, Jon**, 1948– *Craters, caverns, and canyons: delving beneath the earth's surface.* New York, NY, US: Facts on File, **1993**. The Changing Earth Series. illus: b&w: maps; viii, 200p; bibl, index; 25 cm.

Illus.

[1006] **Estes, James H.** *Deep Cave, Texas. A preliminary report of the 1965 Project "Deep".* npop: np, Mar, **1967**. 25p.

"This offset printed report includes the history, logistics, and accomplishments of this official TSA Project."

[1007] **Eszterhás, István** *The caves on the Tés Plateau and in the Bakony Mountains.* Tádal, Ódón (trans); Hazslinszky, Tamás (ed). Budapest, HU: 10th International Congress of Speleology, **1989**. Field Trip Guide E5. illus: b&w: maps; 23p.

[1008] **Euler, Robert C.** *Archaeological and paleobiological studies at Stanton's Cave, Grand Canyon National Park, Arizona; a report of progress.* Washington, DC, US: National Geographic Society, **1978**. 22p; bibl; 23 cm.

Reprinted from National Geographic Society Research Reports, 1969 Projects. Illus. pp 141-162.

[1009] **Euler, Robert C. (ed)** *The Archaeology, geology, and paleobiology of Stanton's Cave: Grand Canyon National Park, Arizona.* Elston, Donald Parker, 1926–. [Grand Canyon, AZ, US]: Grand Canyon Natural History Association, **1984**. Monograph Grand Canyon Natural History Association no 6. illus: b&w: charts, graphs, maps, photos, photomicrographs, tables; viii, 141p; bibl; 28 cm.

Illus.

[1010] **European Regional Conference of Speleology (1980: Sofia, Bulgaria)** *Abstracts of papers.* Sofia, BG: Published for European Regional Conference of Speleology, **1980**. 120p; bibl; 23 cm.

English, French, German, and Russian.

[1011] **Evans, Glen Louis**, 1911– *The Friesenhahn Cave.* Meade, Grayson E. Austin, TX, US: Texas Memorial Museum, **1961**. Bulletin of the Texas Memorial Museum no 2 September 1961. illus: b&w: drawings, maps, photos, tables; 60p; bibl; 30 cm.

[1012] **Evans, Gregory T.** *Oregon Caves National Monument: a manual for cave guides.* Lescalleet, David G. rev1975. [Seattle, WA, US]: [U.S. Dept. of the Interior], National Park Service, Oregon Caves National Monument, **1976**. [i], 64p; app, bibl, gloss; 23 cm.

Item 648-A. Illus.

[1013] **Evans, M. J.** *67 expedition to Crete: BUSS Report.* Et al. [GB]: Birmingham University Speleological Society, **[1967]**. illus: b&w: graphs, maps; 34p; 26 x 20 cm.

[1014] **Eversole, Jack (ed)** *The Mammoth Cave area: a planning proposal.* Hegen, E. E. (proj coord). Bowling Green, KY, US: Barren River Area Development District; Western Kentucky University, **1973**. illus: b&w: maps; partly col: maps; [ii], 47p, 1p pls; 28 cm.

[1015] **Exeter University Spelaeological Society** *Expedition to Morocco 1969.* [Exeter, GB], **[1970]**. illus: b&w: maps, photos; 4 pull-out maps; [ii], 20p, 2p pls; 20.5 x 16 cm.

Cover title: Expedition report (including cave diving) to caves in Chikker Basin, Morocco.

[1016] **Exley, Irby Sheck**, 1944-1994 *Basic cave diving: a blueprint for survival.* 1st ed. Jacksonville, FL, US: National Speleological Society, Cave Diving Section, **1979**. illus: b&w: drawings; 46p; 23 cm.

A classic, basic cave diving text; has been reprinted in five different editions and is still the most important first text for anyone interested in cave diving. Examples on what can go wrong on a cave dive. Simple, easy to read, blueprint for survival.

2nd ed rev. Branford, FL, US: Cave Diving Section of the National Speleological Society, **1980**.

4th ed. Branford, FL, Flagstaff, AZ, US: Cave Diving Section of the National Speleological Society; Best Pub Co [distributor], **1981**. 22 cm.

Best Pub Co (dist).

5th ed. Branford, FL, Flagstaff, AZ, US: Cave Diving Section of the National Speleological Society, **1986**. 22 cm.

Best Pub Co [dist].

[1017] Exley, Irby Sheck, 1944-1994 *Caverns measureless to man*. St. Louis, MO, US: Cave Books, **1994**. illus: b&w: maps, photos, tables; col: photos; xii, 326p; app, index.

Autobiography published just after the author's death while attempting to set a new depth record. Explains a great deal about the man and his motivation. An autobiographical account of cave diving around the world between the mid-1960s and 1989, but also covering earlier work at the sites visited. Appendices include progressive world cave diving records and a brief history of organized cave diving in the USA.

[1018] Exley, Irby Sheck, 1944-1994 *Mapping underwater caves*. Friedman, Bob. 2nd ed. Gainesville, FL, US: National Association for Cave Diving, **1972**. illus: b&w: charts, maps, tables; 22p; bibl; 29 cm.

1973 ed. Friedman, Bob. High Springs, FL, US: National Association for Cave Diving, **1973**. illus: b&w: maps; 16p; 28 cm.

Imprint stamped on verso of t.p. Bibliography: p. 15-16. 3rd printing from 1974 also exists.

[1019] Exley, Irby Sheck, 1944-1994 *N. S. S. cave diving manual*. Young, India F. (ed). Jacksonville, FL, US: National Speleological Society, Cave Diving Section, **1982**. illus: b&w: drawings, photos; viii, 291p; app, bibl, index.

[1020] Eyre, Jim *The cave explorers*. Calgary, Alberta, CA: The Stalactite Press, **1981**. illus: b&w: cartoons, drawings, photos; viii, 264p; 22 cm.

An autobiographical account of more than 35 years of caving in Yorkshire, France, Spain, Yugoslavia, Greece Bulgaria and Mexico.

[1021] Eyre, Jim *The Ease Gill system: forty years of exploration*. npop: British Cave Rescue Association, **1986**. Speleo-History Series no 1. illus: b&w: cartoons, maps, photos; partly col: maps; 48p; app, bibl.

Cover title: The Ease Gill system.

[1022] Eyre, Jim *Lancaster Hole and the Ease Gill Caverns, Casterton Fell, Westmorland*. Ashmead, P.; Aspin, J. (cont); Bliss, R. A. (photo). GB: Cave Research Group of Great Britain, Mar, **1967**. Transactions of the Cave Research Group of Great Britain vol 9 no 2. illus: b&w: charts, maps, photos, tables; 1 fold-out map, 1 folded map in pocket; 66p, 8p pls; bibl, index.

[1023] Eyre, Jim *Race against time: a history of the Cave Rescue Organisation*. Frankland, John; Harding, Mike (forw); Hartnup, Dave (drawing, maps); Eyre, Jim (cartoons). Dent Sebergh, Cubria, GB: Lyon Equipment (Books), **1988**. illus: b&w: cartoons, charts, drawings, maps, photos, ports, repros; maps on lining pages; 208p; app, gloss; 24 cm.

Reproductions of newspaper photos and stories; illustrations include x-rays.

Tells the sad (and humorous) side of Britain's most famous cave rescues. Includes Jim Eyre's raucous cartoons. History of the Cave Rescue Organisation, which serves the Yorkshire Dales, England and covers the organisation and its background from its formation. Appendices deal with all callouts with basic details of the outcome.

[1024] Fabre, Guilhem (comp) *Speleological conventional signs*. Audetat, Maurice (comp). Montpellier, FR: Centre d'Etudes et de Recherches Géologiques et Hydrogéologiques, Universite des Sciences et Techniques de Languedoc, **1978**. C.E.R.G.H. Mémoire no.14. illus: b&w: charts; 44p; bibl.

Trilingual: French, German, English

[1025] Fabun, Don *Shelter: the cave re-examined*. Beverly Hills, CA, US: Glencoe Press; Collier-Macmillan, **1970**. Dimensions of Change 2. illus: col: maps; 32p, [8]; 28 cm.

Illus (some col). Also pub in Great Britain.

[1026] Far West Cave Management Symposium (1979: Redding, CA) *Far West cave management symposium proceedings: Redding, California October 23-26, 1979*. Sims, Mike (prod); Sims, Lynne (prod). Oregon City, OR, US: Pygmy Dwarf Press, Jan, **1980**. illus: b&w: drawings, maps, repros, tables; [iv], 109p; bibl; 29 cm.

A collection of 31 papers. Sponsors: Bureau of Land Management and other agencies.

[1027] Far West Cave Management Symposium (1981: Portland, OR) *Far west cave management symposium proceedings: Portland Oregon April 14-16, 1981*. Sims, Mike (prod); Sims, Lynne (prod). Oregon City, OR, US: Pygmy Dwarf Press, Jul, **1980**. illus: b&w: drawings, graphs, repros, tables; brochure in rear pocket entitled "A self-guided tour of Diamond Craters: Oregon's geologic gem"; [iii], 109p; app, bibl.

[1028] Farr, Martyn, 1950– *The darkness beckons: the history and development of cave diving*. 1st ed. London, GB: Diadem Books Limited, **1980**. illus: b&w: maps, photos, ports; col: photos; b&w; 207p, 8p pls; app, bibl, gloss, index; 21 cm.

83 b&w photos, 15 col photos. History of the development of cave diving.

2nd ed. Stone, William Curtis, 1952–(forw). Diadem Books, **1991**. illus: b&w: drawings; col: photos; b&w;

279p, [1], 32p pls; gloss; 25 cm.

Hodder & Stoughton also issued the 2nd ed in 1991 with 256 pp, as did Cave Books (St. Louis, MO).

A fully updated 2nd edition of the work which first appeared in 1980. It is strikingly readable and well researched history. Includes worldwide information on most major advances in cave diving from square one to 1991. Highly important in any study of the subject.

[1029] Farr, Martyn, 1950– *The great caving adventure.* Sparkford, Yeovil, Somerset, GB: Oxford Illustrated Press Limited, **1984**. Great Adventure Series no 2. illus: b&w: maps; col: photos; v, 229p, 8p pls; app; 23 cm.

Cover subtitle: The story of his major caving and diving expeditions. Col illus.

An autobiographical account of caving in the UK, the USA, Mexico, the Bahamas and Iran, covering the period from 1968 to 1984. A number of major discoveries, and the events leading up to them, are included.

[1030] Farr, Martyn, 1950– *Wookey: the caves beyond.* Bristol, GB: Redcliff Press, Ltd, **1985**. illus: b&w: maps, photos; 38p, [2].

About two-thirds of the booklet is concerned with early exploration and diving, in particular with the development of the Cave Diving Group of Great Britain; more information appears in The Darkness Beckons.

[1031] Farrer, James William *Further explorations in the Dowkerbottom Caves, in Craven.* Denny, Henry (remarks). [US]: Peter M. Hauer, **1970**. illus: b&w: maps; fold-out b&w map; 12p; 21 x 14 cm.

Reprinted from Proceedings, York Geological Society 1865. Offprint printed in 1866. Cover says only 30 reproductions made. Author is a Member of Parliament and of Ingleborough House.

[1032] Farrer, Reginald John, 1880-1920 *On the caves of the world.* London, GB: E. Arnold, **1926**. illus: b&w: ports; folded map; fronts; Various pagination; 23 cm.

Illus.

[1033] Farwell, Robin E. *The Pictured Cliffs project: petroglyphs and talus shelters in San Juan County, New Mexico.* Wening, Karen; Cully, Anne C.; Research Section. New Mexico. State Highway Dept. Santa Fe, NM, US: Laboratory of Anthropology, Museum of New Mexico, **1985**. Laboratory of Anthropology note no 299. illus: b&w: photos; ix, 148p; bibl; 28 cm.

Submitted in fulfillment of contract no. DO1622, between the Museum of New Mexico and the New Mexico State Highway Department; Museum of New Mexico project no. 41.308, New Mexico State Highway Department project no. F-032-1(21)—T.p. verso.

[1034] Faulkner, Charles H. (ed) *The prehistoric native American art of Mud Glyph Cave.* Knoxville, TN, US: The University of Tennessee Press, **1986**. illus: b&w: drawings, maps, photos, tables; 124p, [2]; bibl, index.

29 plates, 5 figs, 4 tables.

Collection of papers describing and analyzing prehistoric (ca., 1000-1350) mudglyphs found in an eastern Tennessee cave. Considered a hallmark study.

[1035] Faulkner, Trevor, 1942– *Report of the SWETC caving club expedition to Norway 1974.* St Pierre, Shirley; South West Essex Technical College Caving Club. London, GB: The club, **1977**. Occasional Publication no 4. illus: b&w: maps; vi, 64p, [8]; 30 cm.

Illus.

[1036] Faulkner, Trevor, 1942– *Sump index Norway.* Cave Diving Group. Bristol, GB: Cave Diving Group, **1979**. illus: b&w: maps, photos; 45p; 30 cm.

[1037] Faust, Burton Sherwood, 1898-1967 *Saltpetre caves and Virginia history.* Falls Church, VA, US: Econo Print, **1964**. illus: b&w: maps, photos; 56p; bibl; 23 cm.

"Reprinted with permission from Caves of Virginia, by Henry H. Douglas."

[1038] Faust, Burton Sherwood, 1898-1967 *Saltpetre mining in Mammoth Cave, Ky.* npop: The Filson Club, Incorporated, **1967**. illus: b&w: maps; front: port; [x], 93p; bibl; 25 cm.

[1039] Fellows, L. D. *Guidebook to Ozark carbonate terrane, Rolla - Devils Elbow area, Missouri.* Williams, James Hadley, 1929–; Vineyard, Jerry Daniel, 1935–. npop: [Association of American State Geologists], **1970**. illus: b&w: charts, maps; partly col: maps; col: maps; 17p; bibl.

Cover title: Association of American State Geologists - 1970 annual meeting: Ozark carbonate terrane field trip, April 29, 1970.

[1040] Fells, Richard *A visitor's guide to underground Britain: caves, caverns, mines, tunnels, grottoes.* 1st ed. GB: Webb & Bower, **1989**. 144p.

[2nd ed]. London, GB: Bloomsbury Books, **1993**. 144p; bibl, index; 25 cm.

"First published in Great Britain 1989 by Webb & Bower"—verso t.p. Illus.

[1041] Felter, Brian *The Virginia cave locator series: book 1: Frederick Co. and Clark Co.* npop: Felter, **1979**. Virginia Cave Locator Series book 1. illus: b&w: maps; ii, 23p, [4]; 21 cm.

Limited edition. Information sources: The caves of Virginia (Douglas) and Descriptions of Virginia caves (Holsinger).

[1042] Felter, Brian *The Virginia cave locator series: book 2: Warren Co.* npop: Felter, **1979**. Virginia Cave Locator Series book 2. illus: b&w: maps; ii, 31p, [1]; 21 cm.

Limited edition. Information sources: The caves of Virginia (Douglas) and Descriptions of Virginia caves (Holsinger).

[1043] Felter, Brian *The Virginia cave locator series: book 3: Shenandoah Co.* npop: Felter, **1980**. Virginia Cave Locator Series book 3. illus: b&w: maps; iv, 53p,

[2]; 21 cm.

Limited edition. Information sources: The caves of Virginia (Douglas) and Descriptions of Virginia caves (Holsinger).

[1044] Felter, Brian *The Virginia cave locator series: book 4: Page Co.* npop: Felter, **1980**. Virginia Cave Locator Series book 4. illus: b&w: maps; iv, 35p, [1]; 21 cm.

Limited edition. Information sources: The caves of Virginia (Douglas) and Descriptions of Virginia caves (Holsinger).

[1045] Felter, Brian *The Virginia cave locator series: book 5: Rockingham Co.* npop: Felter, **1982**. Virginia Cave Locator Series book 5. illus: b&w: maps; iii, 127p, [2]; 21 cm.

Limited edition. Information sources: The caves of Virginia (Douglas) and Descriptions of Virginia caves (Holsinger). pp 63 + 64.

[1046] Felter, Brian *The Virginia cave locator series: book 6: Augusta Co.* npop: Virginia Region. National Speleological Society, **1982**. Virginia Cave Locator Series book 6. illus: b&w: maps; iv, 51p, [1]; 21 cm.

Limited edition. Information sources: The caves of Virginia (Douglas) and Descriptions of Virginia caves (Holsinger).

[1047] Ferdon, Edwin N. *An excavation of Hermit's cave, New Mexico.* Museum of New Mexico. Santa Fe, NM, US: School of American Research, **1946**. Monographs of the School of American Research no 10. 29p; bibl; 28 cm.

Microfilm. New York: New York Public Library, 197-1 microfilm reel; 35 mm.

[1048] Ferguson, George E. *Springs of Florida.* Lingham, G. E.; Love, C. W.; Love, S. K. Tallahassee, FL, US: Florida Geological Survey, **1947**. Florida. Geological Survey Geological Bulletin no 31. illus: b&w: maps; 1 map in pocket; xii, 196p, 42p pls; 24 cm.

[1049] Ferguson, Samuel, 1810-1886 *Account of Ogham inscriptions in the cave at Rathcroghan, Co. Roscommon.* Dublin, IE: McGlashan & Gill, **1865**. 11p; bibl; 22 cm.

Microopaque. New York: Readex Microprint, 1978. 1 microopaque; 23 x 15 cm. (Landmarks II) Illus; pp 99-109.

[1050] Fernandez, Carlos A. *The Tasaday: cave-dwelling food gatherers of South Cotabato, Mindanao: preliminary report submitted to the Panamin Foundation, Inc. on June 1, 1972.* Lynch, Frank, 1921–. Makati, Rizal, PH: Panamin Foundation, Inc, **1972**. illus: b&w: maps; ii, 53p; bibl; 28 cm.

Illus. Title from cover.

[1051] Fetvedt, John E. (ed) *Mystery Cave area: MSS Corn Feed August 1984.* Spring Valley, MN, US: Minnesota Speleological Survey, Aug, **1984**. MSS Special Publication number three. illus: b&w: charts, drawings, maps; one map folded loose in text; one map folded in text; [i], vi, 87p.

This was published as a guidebook for a regional gathering, but it is a comprehensive geologic and caving guide to the main cave area in the state of Minnesota. Includes map of 19 km Mystery Cave.

[1052] Fetvedt, John E. (ed) *North Country Region: Spring 1984, LaCrosse, Wisconsin.* Spring Valley, MN, US: Minnesota Speleological Survey, Apr, **1984**. MSS special publication no. 2. illus: b&w: maps, tables; [iii], 34p, [16]; 28 cm.

[1053] Fewkes, Jesse Walter, 1850-1930 *Preliminary account of an expedition to the cliff villages of the red rock country, and the Tusayan ruins of Sityatki and Awatobi, Arizona, in 1895.* [Washington], DC, US: United States Government Printing Office, **1896**. Annual report of the Board of Regents of the Smithsonian Institution ... to July, 1895. 31p, 34p pls; 24 cm.

[1054] Fewkes, Jesse Walter, 1850-1930 *The cave dwellings of the Old and New Worlds.* Washington, DC, US: Smithsonian Institution, **1911**. 22p, 11p pls; 24 cm.

Offprint: Smithsonian report (1910). Reprinted...from the American Anthropologist, vol. 12, no 3, July-Sept., 1910, pp 613-634. Illus.

[1055] Field, Henry, 1902– *Caves and rockshelters in northern Iraq..* [Coconut Grove, FL, US], **1955**. 20p; 28 cm.

Title from cover. Multigraphed.

A tabulated list of over 500 caves and rock shelters. It gives a number, name, nearest village, location etc. This is a rare item. No geology, per se, but useful.

[1056] Fieseler, Ronald G. (ed) *The caves of Brewster and western Pecos Counties.* Kunath, Carl E. (ed). Austin, TX, US: Texas Speleological Survey, Apr, **1975**. Texas Speleological Survey vol iv, no. 1. illus: b&w: maps; viii, 55p, [1]; bibl; 28 cm.

[1057] Film Associates of California *Cavern formation.* Kennedy, George C. (supervisor). Los Angeles, CA, US: Film Associates of California, **1965**. Earth science series [motion picture]; no. 11. illus: partly col: photos.

1 film loop (4 min), si, col, 8 mm, with study guide.

Shows views of a cavern, including closeups of the dripping water that causes stalactites and stalagmites. Points out many unusual formations and identifies them by name.

[1058] Film Images *Caves of Pierre Saint-Martain.* Paris, FR: Les Actualitis Francraises, **1957**.

25 min, sd, bcw, 16 mm. Released in the U.S. by Film Images.

Documentary of the 1953 speleological expedition to the world's deepest known caves in the Pyrenees Mountains in France. Scientists are shown exploring the caverns and underground rivers seeking a way of converting the caves into a reservoir for three mountain villages.

[1059] Fincham, A. *Speleological surveys in Co. Sligo*

and Co. Leitrim, Eire. Leeds, [GB]: Printed for the Leeds Philosophical and Literary Society by Chorley & Pickersgill, **1962**. Proceedings of the Leeds Philosophical and Literary Society Scientific Section vol 8, pt 10. illus: b&w: maps; 1 folded leaf of pls; 18p, 1p pls; bibl; 24 cm.

Caption title.

[1060] Fincham, Alan Goldworth, 1933– *Jamaica underground: a register of data regarding the caves, sinkholes and underground rivers of the island*. Wadge, Geoff; Draper, Gren. Kingston, JM: Geological Society of Jamaica, **1977**. illus: b&w: maps; [iv], 247p; bibl; 28 cm.

Limited edition of 300. 30p of text, intro, biblio, etc. and then a listing of caves and karst features followed by a section of 140 cave maps.

[1061] Fincham, Alan Goldworth, 1933– *The University of Leeds hydrological survey expedition to Jamaica 1963*. Ashton, K. (comp). GB: Cave Research Group of Great Britain, Jan, **1967**. Transactions of the Cave Research Group of Great Britain vol 9 no 1. illus: b&w: graphs, maps, photos, tables; front: b&w photos; 60p, 2p pls; app, bibl.

[1062] Finley, Claudette *Hand signals for diving*. Stone, Jamie; Vilece, Carol; National Association for Cave Diving, Standardization Committee. [Gainesville, FL, US]: National Association for Cave Diving, **1980**. Informational circular #2. illus: b&w: drawings; [ii], 30p; 22 cm.

Originally published in 1977 under title: Hand signals for cave diving. Third printing, 1983; fourth printing, 1986.

[1063] Finley-Holiday Film Corporation *Carlsbad Caverns National Park & Guadalupe Mountains National Park*. Whittier, CA, US: Finley-Holiday Film Corp, **1980**. National Park and Monument Series; The Holiday Video Library.

1 videocassette (VHS) (53 min), sd, col, 1/2 in. Title from cassette label. "#V27." OCLC lists both 1980 and 1988 as publication dates.

[1064] Finley-Holiday Film Corporation *Luray Caverns in Virginia's Shenandoah Valley*. Finley, Russ (photo); Armenta, Sandra Richard (ed); McCurdy, Richard (music scorer); Workman, Lindsay. Whittier, CA, US: Finley Holiday Films, **1988**. The Holiday Video Library American Adventure Series.

1 videocassette (30 min), sd, col, 1/2 in, VHS format. Container title: Luray Caverns, Virginia.

[1065] Finley-Holiday Film Corporation *Mammoth Caves*. [CA, US]: Finley-Holiday Film Corp, **1990**.

OCLC lists both 1990 and 1993 as date of publication. 10 slides, col.

[1066] Fisher, John L. (ed) *Caves of the Watertown Area, New York state*. npop: np, **1958**. Northeastern Regional Organization of the National Speleological Society publi-

cation no 5. illus: b&w: charts, drawings, maps; 52p; bibl, index.

Described from 1979 convention reprint.

[1067] Fleming, John *Blanchard Springs Caverns story*. Joe Mosby Creative Printing (design, prod); Gould, Gary W. (cover illus). [dedication ed]. Little Rock, AK, US: Gallinule Society, **1973**. 56p; 24 cm.

Illus.

[1068] Fletcher, George *Notes on caving and potholing for beginners (and for those needing a boost in safety)*. Chelsea Speleological Society. Illford, Essex, GB: Chelsea Speleological Society, **1968**. 40p; bibl; 20 cm.

17 Illus.

[1069] Flindall, Roger *The caverns and mines of Matlock Bath: 1 Nestus Mines: Rutland and Masson caverns*. Hayes, Andrew. Hartington, Buxton, Derbyshire, GB: Moorland Publishing Company, **1976**. illus: b&w: drawings, graphs, maps, photos, tables; b&w front drawing; 72p; gloss, index; 21 cm.

First accurate account of mines-cum-show caves in Masson Hill. 24 photos, 16 figs

[1070] Flood, Josephine *Archaeology of the dreamtime*. Sydney, NSW, AU: William Collins Pty Ltd, **1983**. illus: b&w: drawings, maps, photos, tables; partly col: photos; 288p, 8p pls; app, bibl, gloss, index, note.

Cave and rock shelter references scattered throughout the text. Cover title: Archaeology of the Dreamtime: The story of prehistoric Australia and her people. Epilogue.

This book provides a general prehistory of Australia for the lay person, identifying key issues and sites in an accessible manner.

[1071] Flood, Josephine *The riches of ancient Australia: a journey into prehistory*. St. Lucia, QLD, AU: University of Queensland Press, **1990**. illus: b&w: drawings, maps, photos; xii, 373p; bibl, gloss, index, note.

A lay person's guide to prehistoric sites, Australia-wide.

[1072] Florida Cave Survey *Cave locations listing*. Gainesville, FL, US: Florida Cave Survey, **1962**. illus: b&w: maps; 8p; 30 cm.

Title from cover.

[1073] Floyd, E. F. *The New Calaveras Cave, Murphys, Calaveras County, California: W.J. Mercer, proprietor: description and guide*. Mercer, W. J. [CA, US], **1887**. 26p; 23 cm.

Copyrighted 1887, by E.F. Floyd. Microfilm. Berkeley: University of California, Library Photographic Service, 1985. 1 microfilm reel; 35 mm., 10 ft.

[1074] Fogg, P. (ed) *China Caves Project 1987—1988: the Anglo—Chinese project in caves of South China*. Fogg, Tim (ed). npop: China Caves Project, **nd**. illus: partly col: maps, photos; col: maps, photos; 32p.

The product of co-operation and friendship between the Institute of Karst Geology, Guilin Guizhou Normal University [and the] British Cave Research Association.

[1075] Fogle, A. R. (ed staff) *Basic cave rescue orientation course study guide*. Bannerman, Jackie; Hempel, John Charles, 1949–(ed staff). Culloden, WV, US: Eastern Region. National Cave Rescue Commission, **1990**. illus: b&w: charts, drawings, photos, tables; Various pagination, 200p; app.

Cover title: Basic cave rescue orientation course study guide: 1990 edition: coordination, communication, training.

[1076] Folsom, Franklin Brewster, 1907– *Exploring American caves*. 1962 rev ed. New York, NY, US: Collier Books, **1962**. illus: b&w: maps, photos; 319p, 16p pls; bibl, gloss, index; 18 cm.

Cover title: Exploring American caves, their history, geology, lore and location; a spelunker's guide.

A reprise of the exploration of caves in the United States prior to 1956.

1970 rev ed. New York, NY, US: Collier Books, **1970**. Collier books; AS353. 319p; bibl; 18 cm.

Illus.

[1077] Folsom, Franklin Brewster, 1907– *Exploring American caves, their history, geology, lore, and location: a spelunker's guide*. 1st ed. New York, NY, US: Crown Publishers, Inc, **1956**. illus: b&w: drawings, photos; x, 280p, 64p pls; bibl, gloss, index; 24 cm.

100 photos.

One of the first attempts to present an all round guide for the quickly growing sport of caving. Gives a good view as to what were approved and advanced techniques of the day. A glossary of logical terms plus a listing of caves also appears.

[1078] Ford, Derek Clifford, 1935– *The 1965/66 karst hydrology expedition to Jamaica: full report*. Boon, J. M. (org); Brown, M. C. (sci off); Livesey, M. P. (sec); Morris, T. L. F.; Stoyles, T. R. npop: np, Jul, **1966**. illus: b&w: charts, maps, photos, tables; 6 folded maps in separate envelope; vi, 63p, [3], 4p pls; app, bibl; 25 x 20 cm.

It is surprising that a beautiful tropical island with as rich and diverse karst resources should have attracted as little attention from speleologists. This and Aley's work are two of the few.

[1079] Ford, Derek Clifford, 1935– *Castleguard Cave and karst, Columbia Icefield Area, Rocky Mountains of Canada: a symposium*. Boulder, CO, US: Institute of Arctic and Alpine Research, University of Colorado, **1983**. Arctic and Alpine Research, v. 15, no 4. illus: b&w: charts, graphs, maps, photos, tables; text; photo; [ii], 136p; bibl; 28 cm.

pp 425-560, illus.

This contains some material of certain interest to alpine karst students

[1080] Ford, Derek Clifford, 1935– *Castleguard Cave: challenge under the glacier*. Perou, Syd (camera, direct); Haverand Productions (prod); Strange, Cy (narr); Hurson, Tim (script); Gradidge, Havelock (ed). [CA]: National Film Board of Canada, **1974**.

1 film reel (51 min), sd, col, 16 mm.

Film of an expedition into the cave under the Columbia Icefield in Banff National Park, an exhausting crawl through more than 5 km of tunnel, the longest in Canada.

[1081] Ford, Derek Clifford, 1935– *Environmental change in karst areas*. Berlin, DE: Springer Verlag, Jun, **1993**. Environmental Geology vol 21 no 3 pp 107-182.

Editorial and 10 articles on human effects on karst by internationally known karst scientists.

[1082] Ford, Derek Clifford, 1935– *Guidebook to symposium SA6 "Karst geomorphology of the Canadian Rockies" 22nd international geographical congress Canada 1972*. Brown, M. C.; Quinlan, James Francis, 1936-1995. npop: [International Geographical Congress], Jul, **1972**. illus: b&w: charts, graphs, maps, photos, tables; 1 map folded in back; iii, 84p, 12p pls; bibl.

This guidebook is of sure interest to students of Canadian Geology and Alpine Karst and Geomorphology

[1083] Ford, Derek Clifford, 1935– *Karst geomorphology and hydrology*. Williams, Paul W. London, GB: Chapman & Hall, **1989**. illus: b&w: drawings, graphs, maps, photos, photomicrographs; xv, 601p; bibl, index; 24cm cm.

Publisher is listed as Unwin Hyman in some printings.

Driven from a systems perspective, this work is methodically laid out and executed. This is a useful technical overview, well written and readable.

[1084] Ford, Derek Clifford, 1935– *Theme and resource inventory study of the karst regions of Canada: final report upon project A*. Quinlan, James Francis, 1936-1995. [Hamilton, Ontario, CA]: [Department of Geography, McMaster University], Mar, **1973**. illus: b&w: drawings, maps; iii, 112p.

Contract 72-32 of the National and Historic Parks Branch, National Parks File No. 61/7-Pl.

This affords an excellent overview of Canadian karst and karst studies.

[1085] Ford, Trevor David, 1925– *International seminar on karst denudation*. GB: Cave Research Group of Great Britain, Nov, **1972**. Transactions of the Cave Research Group of Great Britain vol 14 no 2. illus: b&w: charts, graphs, maps, photomicrographs, tables; text on lining pages; [ii], 137p; bibl.

Cover title: Karst denudation.

[1086] Ford, Trevor David, 1925– *Caves of Derbyshire*. Allsop, David G. (assist); Travis, R. J. A. (assist); Derbyshire Caving Association. 1st ed. Clapham, via Lancaster, Yorkshire, GB: Dalesman Publishing Company Ltd, **1964**. illus: b&w: maps; 105p; bibl; 19 cm.

A listing with brief descriptions of about 200 caves in the Derbyshire area of central England. Some geologic and hydrologic observations make this of interest to serious students of karst as well as to cavers.

2nd ed. **1967**. Leicester University Dept of Geology Publication no 141. 135p, [9]; 14 cm.

Illus.

3rd ed. Derbyshire Caving Association. Clapham, GB: Dalesman, **1974**. Dalesman caving books. illus: b&w: ads, maps, photos; Ads on front and rear.; 132p, 10p pls; gloss; 19 cm.

Illus.

4th ed. Gill, David W. (comp); Derbyshire Caving Association. **1979**. A Dalesman Caving Book. illus: b&w: maps; 168p; 19 cm.

5th ed. Gill, David W. (comp); Derbyshire Caving Association. **1984**. illus: b&w: maps; 159p; 19 cm.

Illus.

[1087] Ford, Trevor David, 1925– *Caves—origins, development and formations.* Ford, Derek Clifford, 1935–; Diana Wyllie, Ltd; Carman Educational Associates. npop: np, **1973**.

38 fr, col, 35 mm, with teacher's guide. Double-frame version also issued. OCLC lists 1972 and 1973 as publication dates.

Describes caves formed in limestone.

[1088] Ford, Trevor David, 1925– *Caving and potholing techniques.* Diana Wyllie, Ltd; Carman Educational Associates. npop, **1973**.

Double-frame version also issued. 38 fr, col, 35 mm, with teacher's guide. OCLC lists 1972 and 1973 publication dates.

Deals with the equipment, techniques, and potential dangers involved in caving and potholing.

[1089] Ford, Trevor David, 1925– *Derbyshire sump index.* 1st ed. Leeds, GB: Cave Diving Group, **1968**. illus: b&w: maps; [ii], 13p, [1], 3p pls; 27 cm.

3rd ed. Murland, Jerry D. (rev); Cave Diving Group. Bristol, GB: Cave Diving Group, **1980**. illus: b&w: maps; 41p; 30 cm.

Previous ed: 1974.

[1090] Ford, Trevor David, 1925– *Ingleborough Cavern.* rev ed. Clapham via Lancaster, GB: Dalesman, **1971**. illus: b&w: maps; 30p; bibl; 22 cm.

Illus. Later editions are titled Ingleborough Cavern and Gaping Gill.

[1091] Ford, Trevor David, 1925– *Ingleborough Cavern and Gaping Gill.* 3rd enl ed. Clapham, North Yorkshire, GB: Dalesman Books, **1975**. illus: b&w: ads, drawings, maps, photos, tables; 2 b&w photos on rear lining page; 1 map on front lining page; 35p, [1].

Previous edition published as 'Ingleborough Cavern' 1967 and 1971.

[1092] Ford, Trevor David, 1925– *Ingleborough Cavern including: notes on Gaping Gill and the geology of Ingleborough.* rev ed. Clapham via Lancaster, GB: Dalesman, **1967**. illus: b&w: ads, drawings, maps, photos, tables; ads on lining pages; 32p; bibl; 22 cm.

OCLC lists a 1971 edition also. Later editions are titled Ingleborough Cavern and Gaping Gill.

[1093] Ford, Trevor David, 1925– *Limestone and caves of the Peak District.* Norwich, GB: Geo Abstracts Ltd, **1977**. The Limestones and Caves of Britain. illus: b&w: drawings, graphs, maps, photos, tables; xix, 469p; bibl, index.

106 photos, 24 tables, 99 figs.

An outline of early research; The Carboniferous limestone; Carboniferous volcanic activity; The Millstone Grit deltas; sands & solution in the Tertiary Era; The Pleistocene Ice Age; geological structure; minerals & mines; natural resources; geochronology; hydrology; biospeleology; archeology & paleontology; & 9 chapters on the caves of various regions, illustrated with many cave plans.

[1094] Ford, Trevor David, 1925– *Limestones and caves of Wales.* British Cave Research Association. Cambridge, GB: Cambridge University Press, **1989**. illus: b&w: drawings, maps, photos, tables; [v], 257p; bibl, index; 31 cm.

A well written and illustrated collection of papers by various authors on all aspects of caves. Good references for this classic area.

[1095] Ford, Trevor David, 1925– *The science of speleology.* Cullingford, Cecil Howard Dunstan, 1904-1990. London, New York, NY, GB, US: Academic Press, **1976**. illus: b&w: maps, ports; folded plate; Various pagination, xiv, 593p; bibl, index; 24 cm.

Very detailed book with 22 contributors. Contains 14 chapters with very extensive bibliographies at the end of each. Has both a subject index and a separate cave and fissure index.

[1096] Ford, Trevor David, 1925– *The story of Poole's Cavern.* Allsop, David G. Green Lane, Buxton, Derbyshire, GB: Poole's Cavern, Jul, **1977**. illus: b&w: drawings, maps, photos, tables; col: photos; col photos on lining pages; 31p, [1]; bibl.

Good descriptive account of the cave, from 1580 AD. Cover title: Guide to Poole's Cavern Buxton Country Park. Softbound pamphlet with 6 color photos on front and back inside and outside of covers plus 9 B & W old scenes from the cave.

[1097] Ford, Trevor David, 1925– *The story of the Speedwell Cavern.* Castleton, Derbyshire, GB: R.J. & D. Harrison, **1962**. illus: b&w: drawings, maps, photos; col: photos; 27p; bibl; 21 cm.

Contains 8 b&w photos, 1 map, and 5 color photos.

[1098] Ford, Trevor David, 1925– *The story of Treak Cliff Cavern.* Castleton, Derbyshire, GB: H. Harrison, **1955**. illus: b&w: drawings, maps, photos; partly col: drawings, maps, photos; 23p, 4p pls; bibl; 21 cm.

Cover title: The story of Treak Cliff Cavern: Castleton

- Derbyshire.

[1099] Ford, Trevor David, 1925– *Symposium of the origin and development of caves.* Ledbury, Herefordshire, GB: Cave Research Group of Great Britain, Jun, **1971**. Transactions of the Cave Research Group of Great Britain vol 13 no 2. illus: b&w: charts, graphs, maps, photos, photomicrographs, tables; text on rear lining page; ii, 120p; bibl.

Proceedings of a symposium held at Vaughan College Leicester, 6 Mar 1971.

[1100] Forder, John (auth, photo) *Secrets of the moors and dales.* Forder, Eliza (co-auth, photo). Cumbria, GB: Frank Peters, **1992**. illus: col: photos; 158p; 31 x 25 cm.

Half of the photos are cave-related in this picture book.

[1101] Forster, Raymond R., 1922– *Harvestmen of the sub-order Laniatores from New Zealand caves.* Dunedin, NZ: Otago Museum Trust Board, **1965**. Records of the Otago Museum Zoology no. 2. 18p; 25 cm.

Illus.

[1102] Forti, Paolo (ed) *International symposium on evaporite karst: preprints: Bologna October 22-25, 1985.* Federazione Speleologica Regionale Dell'Emilia-Romagna. Bologna, IT: Instituto Italiano di Speleologia, **1985**. 52p.

Text in Italian, English, or French. No illus.

This is one of the few meetings devoted to evaporite karst. It presents a diversity of papers not generally available in English.

[1103] Forwood, William Stump, 1830-1892 *An historical and descriptive narrative of the Mammoth Cave of Kentucky: including explanations of the causes concerned in its formation, its atmospheric conditions, its chemistry, geology, zoology, etc., with full scientific details of the eyeless fishes.* 1st ed. Philadelphia, PA, US: J.B. Lippincott & Co, **1870**. illus: b&w: drawings; front: b&w drawing; xi, 225p, [1], 12p pls; app; 20 cm.

Spine title: Mammoth Cave. Half t.p.: The Mammoth Cave. Microfilm. Madison, Wis.: State Historical Society of Wisconsin, 1984. 1 microfilm reel: negative ; 35 mm. Master microform held by: LCP. Microopaque. Louisville, Ky.: Lost Cause Press.

This book is one of the first really authoritative books on Mammoth Cave to be published. Originally published with either a green or red cover.

4th ed. J.B. Lippincott & Co, **1875**. illus: b&w: drawings; 1 map folded; front: b&w drawing; xi, 241p, [1], 12p pls; app; 19 cm.

This edition, which was only printed in wrappers, has a very significant addition compared to the first edition: a letter from Franklin Gorin is included that adds significantly to the early history of Mammoth.

[1104] Fossen, Joyce *Caves of Paha Sapa, and other findings of Dr. Knows-It.* Rapid City, SD, US: Dr. Knows-It Publishing, **1985**. 23p; 22 x 28 cm.

Illus. Juvenile.

[1105] Foster, David G. *Guidebook: 1991 National Cave Management Symposium, Bowling Green, KY Oct. 23-26.* npop: American Cave Conservation Association, **1991**. 23p.

Program abstracts and guide book to field trips.

[1106] Foster, David G. (ed) *Learning to live with caves and karst: a cave and karst curriculum and resource guide.* Foster, Debra (ed); Nimo, Peggy (ed); Forbis, Shelly (ed); Clausen, Will (ed);. npop: American Cave Conservation Association, **1994**. iv, 72p; gloss.

Useful guide for K-12. List of resource materials, organizations, and glossary. Juvenile.

[1107] Foster, J. J. *The Jenolan Caves.* New South Wales. Dept. of Mines and Agriculture. Sydney, NSW, AU: Charles Potter, Government Printer, **1890**. illus: b&w: maps; viii, 96p; 22 cm.

Microfilm edition: United States Library of Congress.

A rare book on Australian caves. Contains 32 short chapters but no illus.

[1108] Fountain, Lewis S. *Earth resistivity and hole-to-hole electromagnetic transmission tests at Medford Cave, Florida.* Herzig, Francis X. U.S. Army Engineer Waterways Experiment Station. Southwest Research Institute. San Antonio, TX, US: Southwest Research Institute, Jul, **1980**. some lvs folded; iv, 41p; 28 cm.

Illus. July 31, 1980. Prepared for: Department of the Army, Waterways Experiment Station, Corps of Engineers, Vicksburg, Mississippi.

[1109] Fowke, Gerard, 1855-1933 *Archeological investigations.* Smithsonian Institution. Bureau of American Ethnology. Washington, DC, US: United States Government Printing Office, **1922**. Smithsonian Institution Bureau of American Ethnology Bulletin 76. illus: b&w: charts, drawings, maps, photos, tables; folded map; 204p, 45p pls; index; 24 cm.

Illus. The second part is published in the Forty-fourth Annual report of the Bureau of American Ethnology. 1. Cave explorations in the Ozark region of central Missouri—2. Cave explorations in other states—3. Explorations along the Missouri River bluffs in Kansas and Nebraska—4. Aboriginal house mounds—5. Archeological work in Hawaii. Microfiche. Washington, D.C.: Microcard Editions, 1966. 3 microfiche ; 11 x 15 cm.

One of the first archeological studies to investigate the prehistoric bluff dwellers of the Ozark Region.

[1110] Fowler, Don D., 1936– *The archeology of Newark Cave, White Pine County, Nevada.* Reno, NV, US: Western Studies Center, Desert Research Institute, Mar, **1968**. Desert Research Institute: University of Nevada Technical Report series S-H. Social Sciences and Humanities publication no 3. illus: b&w: drawings, maps, photos, tables; 39p; bibl; 28 cm.

Illus.

[1111] **Fox, Robert B.**, 1918– *The Tabon Caves: archaeological excavations on Palawan Island, Philippines (1962—64)*. Manila, PH: National Science Development Board, **1964**. 165p, [5]; bibl; 31 cm.

Mounted illus.

[1112] **Fox, Robert B.**, 1918– *The Tabon Caves; archaeological explorations and excavations on Palawan Island, Philippines*. Manila, PH: [National Museum], **1970**. Monograph of the National Museum no 1 Philippines. illus: b&w: drawings, maps, photos, tables; col: drawings; 2 maps folded in text; xiii, 197p, [1], 17p pls; app, bibl; 23 cm.

23 pls.

[1113] **Fox, W. Storrs** *Notes on the excavation of Harborough Cave, near Brassington, Derbyshire*. npop: np, **1908**. illus: b&w: maps; 17p; 22 cm.

Illus. From the Proceedings of the Society of Antiquaries of London, 2nd ser., XXII, Feb. 27, 1908.

[1114] **Fraile, D. Daniel Barettino** *El Karst / Karst*. Eraso, D. Adolfo (work team); Fillol, J. Boquera (work team); Plédel, D. Bruno Martinez (collab). Madrid, ES: Instituto Technológico Geominero de España, **1994**. Series Geoenvironmental Engineering.

[1115] **Francis, Charles M. (comp)** *The management of edible bird's nest caves in Sabah*. Sabah, MY: Wildlife section, Sabah Forest Department, **1989**. illus: b&w: graphs, maps, photos, tables; xii, 217p; app, bibl.

[1116] **Francis, Timothy** *Mendip Caving Group: Belize '94*. Flavell, J.; Hesketh, J.; Hollings, P. Bristol, GB: The Mendip Caving Group, **1995**. Mendip Caving Group Occasional Publications no 1. illus: b&w: maps; 27p, [7]; 30 x 21 cm.

[1117] **Frank, Cathleen** *Legislative history for Oregon Caves National Monument, 59th Congress through 96th Congress*. Roether, Joyce; Hori, Nancy; Mentzer, Annette. Seattle, WA, US: United States National Park Service, Pacific Northwest Regional Library, Jun, **1986**. 260p; 29 cm.

[1118] **Frank, Helmut** *Dunkle Portale*. npop: Laichingen, **1966**. illus: b&w: photos; 207p.

English language? Nice small hardback about caves in Schwabischen Alb. 29 photos (b&w) and 50 figures. Printed on high quality paper.

[1119] **Frank, Ruben Milton**, 1935– *The vertebrate paleontology of Texas caves*. Austin, TX, US: Texas Speleological Association, Sep, **1964**. Texas Speleological Survey v 2 no 3. illus: b&w: maps; 43p; bibl; 28 cm.

[1120] **Franke, Herbert Werner**, 1927– *Wilderness under the earth*. Saville, Mervyn (trans). London, GB: The Scientific Book Club; English translation Lutterworth Press, **1958**. illus: b&w: drawings, maps, photos, tables; 1 map folded, loose in text; front: b&w photo; 204p, 24p pls; index; 22 cm.

Translation of: Wildnis unter der Erde. A general book on European cave regions. Contains nine cave maps. Two editions available.

[1121] **Franklin, A.G.** *Foundation considerations in siting of nuclear facilities in karst terrains and other areas susceptible to ground collapse*. Patrick, D. M.; Butler, D. K.; Strohm, W. E., Jr.; Hayers-Griffin, M. E. Washington, DC, US: Nuclear regulatory Commission, **1981**. NUREG/CR-2062. illus: b&w: maps; xiii, 229p; bibl; 28 cm.

Illus. Geotechnical Laboratory, U.S. Army Engineer Waterways Experiment Station.

[1122] **Frantz, A. P.** *Cave rescue operations*. Williams, K. M. San Jose, CA, US: Private Printing, **1980**. 153p.

[1123] **Franz, Richard** *Caves of Maryland*. Slifer, Dennis. [Baltimore, MD, US]: Maryland Geological Survey, **1971**. Educational series (Baltimore, Md.) no 3. illus: b&w: drawings, graphs, maps, photos, tables; vii, 120p; bibl, index; 28 cm.

95 figs. Reprinted 1976.

Catalog of 148 known caves in the state, with text descriptions and locations. About fifty cave maps. Introductory chapters on geology and biology.

[1124] **Fraser, L.A.** *Weyer Cave. Cave of the Fountains. Madison Cave: a descriptive sketch of the grottoes of the Shenandoah*. Shendun, VA, US: The Grottoes Company, Jul, **1892**. illus: b&w: photos; ad on front lining page; testimonials on back lining pages and back cover; xii, 2p pls; 22 cm.

[1125] **Frederikson, Rosalie** *The New Zealand glowworm*. Waitomo Caves, NZ: Waitomo Caves Museum Society, Inc., **1983**. illus: b&w: charts, drawings, maps, tables; ii, 32p; bibl, gloss.

Reprinted 1984.

[1126] **Freeman, John P. (ed)** *CRF Personnel Manual: Central Kentucky karst area*. 2nd ed. Columbus, OH, US: Cave Research Foundation, **1975**. illus: b&w: drawings, maps, photos; note on back lining page.; xiii, 109p; note.

Cover title: Cave Research Foundation personnel manual.

[1127] **Friedberg, Lionel** *Mysteries underground*. WETA-TV (Television station: Washington, D.C.). npop: National Geographic Society (U.S.), **1992**. National Geographic Special.

1 videocassette (VHS) (56 min), sd, col, 1/2 in. Recorded, with permission, from satellite broadcast November 18, 1992. Richard Kiley.

Chronicles the exploration of Lechuguilla Cave, located five miles from Carlsbad Cavern, N.M. Lechuguilla, discovered in 1986, is America's deepest limestone cave.

[1128] **Fullerton, Iona** *Cave minerals and speleothems: a review of the literature with special reference to Waitomo*. Waitomo Caves, NZ: Waitomo Caves Museum Society Inc, **1983**. illus: b&w: drawings; iii, 31p; bibl, gloss.

Cover title: Cave minerals.

[1129] Fulton, William Shirley *A ceremonial cave in the Winchester Mountains, Arizona.* Dragoon, AZ, US: The Amerind Foundation, Inc, **1941**. The Amerind Foundation, Inc Publications no 2. illus: b&w: drawings, maps, photos, tables; 35p; 27 cm.

XIII pls.

[1130] Funk, Robert E. *Archaeological and paleoenvironmental investigations in the Dutchess Quarry Caves, Orange County, New York.* Steadman, David W.; Kopper John S. Buffalo, NY, US: Persimmon Press, **1994**. Persimmon Press Monographs in Archaeology. illus: b&w: charts, maps, photos, tables; text; v, 125p; app, bibl, index; 26 cm.

Illus.

[1131] Funkhouser, William Delbert, 1881– *The so-called "ash caves" in Lee County, Kentucky.* Webb, William Snyder, 1882–. [Lexington, KY, US]: [University of Kentucky], **1929**. Publications of the Department of Anthropology and Archaeology, University of Kentucky v 1 no 2. ii, 76p; 26 cm.

Illus. At head of title: The University of Kentucky. Reports in archaeology and anthropology, number 2.

One of the first studies conducted in Kentucky discussing the unique preservation and archeology of rockshelter sites in eastern Kentucky.

[1132] Furlong, Eustace Leopold, 1874– *The exploration of Samwel Cave.* [New Haven, US]: np, **1906**. 13p; 24 cm.

Illus. Cover-title. "From the American journal of science, vol. XXII, September, 1906."

[1133] G. P. Slide Co *Luray Caverns.* Houston, TX, US: G.P. Slide Co, **1970**.

40 slides, col. + 1 sound cassette (1 7/8 ips). Sound accompaniment compatible for manual operation only.

In addition to a tour of the cavern formations, you will see and hear the great stalactite organ, the only one of its kind in the world, the carillon tower, the antique car museum, and hear the sound of the engine as you see each car.

[1134] Gabriel, Diaconu *Closani Cave: mineralogical and genetic study of carbonates and clays.* Bucuresti, RO: Institutul de Speologie "Emil Racovita", **1990**. Miscellanea Speologica Romanica 2. illus: b&w: maps; 1 folded plate; 135p, 61p pls; bibl; 25 cm.

Illus.

[1135] Gádoros, Miklós *Speleotherapic and speleclimatological centres.* Salamon, Gábor; Horváth, Tibor; Roda, István; Tardy, János; Táda;. Ödön (trans); Hazslinszky, Tamás (ed). Budapest, HU: 10th International Congress of Speleology, **1989**. Field Trip Guide D2. illus: b&w: maps, tables; 24p, 24p pls.

[1136] GAF Corporation *Carlsbad Caverns National Park.* Thomas, Lowell (ed). New York, NY, US: GAF Corporation, **1950**. National Park series United States travel. 32p.

6 stereograph reels (Viewmaster) (7 double fr each), col. + 2 booklets (16 p each), in 2 containers. Captions on reels. To be used on a View-master viewer. Booklets edited by Lowell Thomas.

[1137] GAF Corporation *Luray Caverns, Va.* New York, NY, US: GAF Corporation, **1970**.

20 slides, col, 5 x 5 cm. Sound accompaniment compatible for manual operation only. A tour of the cavern formations.

[1138] Gagné, Wayne C. *The cavernicolous fauna of Hawaiian lava tubes, part VI: Mesoveliidae or water treaders (Heteroptera).* Howarth, Francis Gard, 1940–. Honolulu, HI, US, **1974**. Technical Report U.S. International Biological Program Island Ecosystems IRP no 40. ii, 23p; bibl; 28 cm.

Illus.

[1139] Gagné, Wayne C. *The cavernicolous fauna of Hawaiian lava tubes, part VII. Emesinae or thread-legged bugs (Heteroptera: Redvuiidae).* Howarth, Francis Gard, 1940–. Honolulu, HI, US, Jul, **1974**. Technical Report (U.S. International Biological Program Island Ecosystems IRP) no 43. illus: b&w: drawings, photos; iii, 18p, [3]; bibl; 28 cm.

[1140] Gams, Ivan (ed, trans) *Man's impact in Dinaric Karst: guide-book.* Habič, Peter (ed); Kranjc, Maja (trans); Kranjc, Andrej (trans); Šušteršič, France (trans). Ljubljana, SI: Department of Geography, Faculty of Letters, University "E. Kardelj" Ljubljana, and Institute for Karst Research ZRC SAZU in Postojna, **1987**. illus: b&w: charts, graphs, maps, repros, tables; 205p; bibl.

Guide to the field excursions of the meeting of the study group (Sept 11-14 1987 in Postojna).

[1141] Gams, Ivan *Slovenska Kraska Terminologija = Slovene karst terminology.* Ljubljana, SI: Univerza, Katedra za fizično geografijo Oddelka za geografijo FF, **1973**. [i], xix, 76p; bibl; 24 cm.

A 7 language list of karst terms with definitions. Essays on the origins of the words "karst" and "dolina" are printed in full in English and Slovene, with an English summary of a third essay.

[1142] Gans, Roma, 1894– *Caves.* Maestro, Giulio (illus). New York, NY, US: Thomas Y. Crowell Company, **1976**. Let's Read and Find Out Science Books. illus: partly col: drawings; [vi], 32p, [2]; 21 x 23 cm.

Juvenile.

A simple introduction to caves, their formation, past uses, and distinguishing features.

[1143] Ganter, John Hamilton, 1962– *A systematic guide to making your first cave map.* npop: [Ganter], May, **1985**. illus: b&w: drawings, maps, photos; i, 27p; app; bibl; 28 cm.

Minor revisions, January 1986.

[1144] Ganter, John Hamilton, 1962– *The Xilitla region: an annotated bibliography.* npop: John Ganter,

1989. 12p.

[1145] García Guinea, Miguel Angel *Altamira and prehistoric art in the caves of Santander.* Madrid, ES: Raycar, **1975**. illus: b&w: maps; 1 plate folded; 71p, 50p pls; bibl; 17 cm.

Col illus.

[1146] García Guinea, Miguel Angel *Altamira, the beginning of art.* Ramírez, Eleonor Donínguez (photo). Madrid, ES: np, Mar, **1969**. illus: b&w: drawings, maps, photos, repros; col: photos; 47p, 48p pls; bibl.

[1147] García Guinea, Miguel Angel *Altamira: the origin of art.* Madrid, ES: Silex, **1971**. 47p, 24p pls; bibl; 17 cm.

Col illus. OCLC lists both 1971 and 1973 as dates of publication.

[1148] García Guinea, Miguel Angel *Santillana and Altamira.* Leon, ES: Editorial Everest, **1984**. illus: b&w: maps, ports; partly col: maps; 64p; 25 cm.

Translation of: Santillana y Altamira.

[1149] Garder, George D. *Tracing of subsurface flow in karst regions using artificially colored spores.* Gray, Richard E. Monroe, PA, US: GAI Consultants, Inc, **c1978**. illus: b&w: drawings, photos; iii, 26p, 4p pls; bibl.

[1150] Gardner, Erle Stanley *The hidden heart of Baja [Mexico].* New York, NY, US: William Morrow & Co, **1962**. illus: b&w: photos, ports; col: photos; 256p, 4p pls; 25 cm.

Dust jacket: An account of a most important and romantic archaeological discovery - the Indian caves of Baja California. Published simultaneously in Toronto, CA: George J. McLeod Limited and in London, GB: Jarrolds Publishers Ltd. 148 illus.

[1151] Gardner, James *An inventory and evaluation of Missouri state parks cave resources.* npop: Missouri Department of Natural Resources, **1982**. v, 68p.

Inventory and description of 93 caves in 13 Missouri state parks.

[1152] Gardner, James Eugene, 1953– *Cave resources of Ozark National Scenic Riverways: an inventory and evaluation phase II.* Taft, John B. [Jefferson City, MO, US]: Missouri Conservation Department, **1983**. illus: b&w: drawings, maps, tables; iii, 93p; app, bibl, gloss; 28 cm.

11 tables, 5 figs. A preliminary copy of a final report submitted to Ozark National Scenic Riverways, National Park Service in compliance with contract CA-6640-3-8001. Publishers from logos on cover. Republished in 1984 by National Park Service (Washington, DC).

[1153] Gardner, James Eugene, 1953– *An introduction to the inventory and evaluation of biological cave resources.* npop: [Missouri Department of Conservation], Oct, **1984**. 4p; bibl.

Prepared for 1984 National Cave Management Symposium, Rolla, Missouri.

[1154] Gardner, James Eugene, 1953– *An inventory and evaluation of cave resources of Mark Twain National Forest: a final report submitted to the Mark Twain National Forest, United States Department of Agriculture in compliance with a cooperative cave inventory agreement.* Gardner, Treva L. Missouri. Dept. of Conservation. [Jefferson City, MO, US]: Missouri Department of Conservation, Sep, **1982**. illus: col: cartoons; iii, 113p; bibl; 28 cm.

[1155] Gardner, James Eugene, 1953– *Invertebrate fauna from Missouri caves and springs.* Auckley, Jim (ed). [Jefferson City, MO, US]: Missouri Dept of Conservation, **1986**. Natural History Series Missouri Department of Conservation no 3. illus: b&w: maps, photos; vi, 72p; bibl; 28 cm.

[1156] Gardner, James Eugene, 1953– *Missouri Department of Conservation Cooperative cave inventory project: a final report submitted to the Missouri Department of Conservation as part of the cooperative cave inventory project: a final report.* [Jefferson City, MO, US]: Missouri Department of Conservation, Jan, **1984**. illus: b&w: drawings, tables; ix, 125p; app, bibl; 28 cm.

2 b&w drawings, 17 tables. Inventory and classification system used by state agencies.

[1157] Gargett, Robert H. *Cave bears and modern human origins: the spatial taphonomy of Pod Hradem Cave, Czech Republic.* Lanham, MD, US: University Press of America, **1996**. illus: b&w: maps; xx, 265p; bibl, index; 23 cm.

Illus.

[1158] Garlake, Peter S. *The painted caves: an introduction to the prehistoric art of Zimbabwe.* Harare, ZM: Modus Publications, **1987**. iv, 100p, 8p pls; bibl; 22 x 24 cm.

Illus (some col).

[1159] Garrard, Ian M. *Caving.* Burton, John, 1948–. Sydney, NSW, AU: Crown Lands Office of N.S.W., **1983**. An Activity Profile no 11. 17p, 6p pls; 30 cm.

Illus.

[1160] Garrett, Willie Ann *The Carlsbad Cavern National Park.* [Nacogdoches, TX, US], **1949**. Research paper: Stephen F. Austin State College 1949. illus: b&w: maps; 33p; bibl; 28 cm.

Illus.

[1161] Garrigan, George A. *Skyline Caverns and its geologic relationship to the Shenandoah Valley and paleoindian cultures.* Woodbridge, VA, US: Northern Virginia Community College, Woodbridge Campus, **1994**. 37p.

[1162] Garrod, Dorothy Anne Elizabeth, 1892– *The paleolithic of southern Kurdistan: excavations in the caves of Zarzi and Hazar Merd.* American School of Prehistoric Research. [New York, NY, US]: Kraus Reprint, **1968**. Bulletin of the American School of Prehistoric Research no 6. some illus partly folded; 43p; 23 cm.

Illus. Reprint. Originally published: New Haven,

American School of Prehistoric Research, 1930.

[1163] Garrod, Dorothy Anne Elizabeth, 1892– *The transition from Lower to Middle Palaeolithic and the origin of modern man: international symposium to commemorate the 50th anniversary of excavations in the Mount Carmel Caves by D.A.E. Garrod, University of Haifa, 6-14 October 1980*. Ronen, Avraham; Applied Scientific Research Co; Universitat Hefah. Oxford, GB: B.A.R., **1982**. BAR International Series 151. xvi, 329p; bibl; 29 cm.

Illus. Applied Scientific Research Co, University of Haifa Ltd. English and German.

[1164] Garton, Emmel Ray, 1950– *Caves of North Central West Virginia*. Garton, Mary Ellen; Carpenter, Alan. Barrackville, WV, US: West Virginia Speleological Survey Bulletin, Jun, **1976**. Bulletin 5 West Virginia Speleological Survey. illus: b&w: drawings, maps, photos; 3 maps folded in text, 2 maps folded in back pocket; viii, 108p; errat, index; 28 cm.

[1165] Garton, Emmel Ray, 1950– *The vertebrate fauna of West Virginia caves*. Grady, Frederick; Carey, Steven D. [Charlottesville, VA, US]: West Virginia Speleological Survey, **1993**. West Virginia Speleological Survey bulletin 11. 107p; bibl; 28 cm.

Illus (some col).

[1166] Gascone, William (ed) *Caves of the North Outcrop*. Gwent, GB: Cambrian Cave Register, **1982**. [11]; 29.5 x 21 cm.

In Welsh & English. Printed one side only.

[1167] Gatacre, E. V. *Wookey Hole [Including articles on] (paper making, Lady Bangor's fairground collection [and] Madame Tussaud's store)*. Crowther, Robert (design, draw); Griffiths, Ken; Parker, Chris; Webb, John; Wilson, April (photo). Hayes, Middlesex, GB: Wookey Hole Caves Ltd, Mar, **1975**. illus: b&w: drawings, photos, tables; partly col: maps; col: photos; 32p; 23 x 15 cm.

Contains 14 colored illustrations and several black or duotones also. Includes the history of the paper mill there and [4] special watermarked pages.

[1168] Gebauer, Herbert Daniel *Caves of India & Nepal: including the results of Speläologische Südasien Expedition 1981/82*. Schwabisch Gmund, DE: Herbert Daniel Gebauer, **1983**. illus: b&w: drawings, maps; vi, 165p; app, bibl; 21 cm.

Illus. English and German.

Catalog of known caves and rock shelters, with text descriptions, rough locations, and detailed maps of a few of the larger caves. Introductory text and longer descriptions are given in both English and German. This is the only work of its kind to deal with this fascinating area.

[1169] Gebauer, Herbert Daniel *Kurnool 1984: report of the speleological expedition to the district of Kurnool, Andhra Pradesh, India*. Munich, DE: Verband der deutschen Höhlen- und Karstforscher e V, **1985**. Abhandlungen zur Karst- und Höhlenkunde Reihe A Speläologie heft 21. illus: b&w: drawings, maps, photos; [iii], 77p; bibl.

"A supplement to 'Caves of India & Nepal', including the full expedition report, additional cave surveys by M. Narayan Reddy, a list of Indian rockshelters and an extensive bibliography." Bilingual: German and English.

This presents limited geomorphic and site descriptions but adequate maps. Not much geology, but a useful recon with scattered geochemical observations and references.

[1170] Gebauer, Herbert Daniel *Speleological bibliography of South Asia including the Himalayan regions*. Mansfield, Raymond Walter; Kusch, Heinrich; Chabert, Claude. Kathmandu, NP: [Armchair Adventure Press], Nov, **1995**. illus: b&w: drawings, maps, repros; [iv], 226p; bibl.

Cover title: *Speleological bibliography of South Asia: annotated bibliographical details of papers published throughout the world on speleological aspects of South Asia including the Himalaya*.

Bhutan, India, Nepal, Pakistan, Sri Lanka, and the Himalayan part of Tibet. Over 3300 references to natural caves, and about 500 to cave temples and cave art. Some annotations, introductory essays.

[1171] Geikie, James, 1839-1915 *The antiquity of man in Europe, being the Munro lectures, 1913*. Edinburgh, GB: Oliver and Boyd, **1914**. illus: b&w: drawings, maps; 4 folded maps; xx, 328p; 24 cm.

21 pls, 4 maps. Consists of 10 lectures. Lectures I & II are called "The Testimony of the Caves" and cover pp 35-100. This is mostly on materials and paintings found in caves. Text has only a few photos.

[1172] Gem State Grotto of the National Speleological Society *Caves of the gem state*. npop: The Gem State Grotto, **1968**. Idaho Speleological Survey no 1; GSG special publication no 1. illus: b&w: maps, photos; 4 maps folded in pocket or loose; [i], 30p, [1]; index; 28 cm.

32 figs.

Catalog of known caves in the state of Idaho. Most are shelters or volcanic pseudokarst. Brief descriptions with locations. Twenty cave maps.

[1173] Gemmell, Arthur, 1915– *Underground adventure*. Myers, Jack O. 1952 ed. Clapham, Yorkshire, GB: Dalesman Publishing Company, **1952**. illus: b&w: maps, photos; 141p, 9p pls; gloss; 21 cm.

Accounts of exploration of a number of Yorkshire caves.

1990 ed. Myers, Jack O. Castle Cary, GB: Mendip, **1990**. illus: b&w: maps; 140p, 8p pls; gloss; 20 cm.

Illus.

[1174] Geographical Field Group *Coast and karst in Istria Rossa; a report of geographical field work carried out during August, 1960*. [Nottingham, GB]: Geographical Field Group, **1960**. Geographical Field Group. Regional studies no 5. illus: b&w: maps; 120p; 26 cm.

Illus

[1175] George, Angelo Isham, 1944– *Bibliography of Wyandotte Cave*. Louisville, KY, US: George Publishing Co, **1991**. illus: b&w: drawings, maps, repros; b&w front (map and drawing); xiv, 74p; bibl, index; 28 cm.

[1176] George, Angelo Isham, 1944– *Mummies of Short Cave, Kentucky, and the Great Catacomb Mystery*. Louisville, KY, US: George Publishing Company, **1985**. illus: b&w: maps, repros; [iv], 72p; bibl; 21 cm.

Discusses the historical significance of Fawn Hoof, an desiccated prehistoric Native American found near Mammoth Cave, Kentucky.

[1177] George, Angelo Isham, 1944– *Mummies, catacombs, and Mammoth Cave*. Louisville, KY, US: George Publishing Company, **1994**. illus: b&w: maps, photos, repros; [i], ix, 153p; bibl.

A general summary of the prehistoric Native American remains found in the Mammoth Cave area of Kentucky

[1178] George, Angelo Isham, 1944– *The New Madrid earthquake at Mammoth Cave (1811-1812)*. O'Dell, Gary A. Louisville, KY, US: George Publishing Company, **1992**. illus: b&w: drawings, maps, photos; text on lining pages; 27p; bibl.

[1179] George, Angelo Isham, 1944– *Prehistoric mummies from the Mammoth Cave area: foundations and concepts*. Louisville, KY, US: George Pub Company, **1990**. illus: b&w: maps, repros; xi, 117p; bibl; 21 cm.

Illus.

A summary of the prehistoric Native American remains found in the Mammoth Cave area of Kentucky.

[1180] George, Angelo Isham, 1944– *Saltpeter and gunpower manufacturing in Kentucky*. Louisville, KY, US: George Publishing Company, Apr, **1986**. The Filson Club History Quarterly vol 60 no 2. illus: b&w: maps, repros, tables; text on lining pages.; [i], 29p; bibl.

Reprint of pp 189-217.

[1181] George, Angelo Isham, 1944– *Wyandotte Cave down through the centuries*. Louisville, KY, US: George Publishing Company, **1991**. illus: b&w: drawings, maps, ports, repros; iii, 68p; bibl; 22 cm.

[1182] Georgia Speleological Survey *Georgia Speleological Survey mapbook*. 2nd ed. npop, GA, US: Georgia Speleological Survey, Oct, **1991**. illus: b&w: maps, tables; folded maps in text; [400].

First edition 1989 listed 280 caves; 2nd edition contains 310 caves. Pagination: a wild guess.

One of a series of books containing all available maps of caves in the state of Georgia. The first was published in 1989 and contained 280 maps. This one contains 310. New editions appear roughly every other year. No text descriptions or locations.

[1183] Gerboth, David W. *A guidebook to the caves and sinks of the Spring Valley Caverns area, Fillmore County, Minnesota: 1989 Cornfeed Minnesota Speleological Survey*. Kramar, Matt (photo); Nordquist, Gerda (ed, layout). Minneapolis, MN, US: Minnesota Speleological Survey, Inc, Aug, **1989**. MSS Special Publication no. 4. illus: b&w: maps, photos, repros, tables; two folded maps in text; front; b&w photo; [ii], ii, 71p.

[1184] Gerrard, John *Kent's Cavern*. Sparks, J. B.; Lake, P. M. B. Torquay [Devon], GB: W.F. & L.W. Powe, Buckley & Sons, **1960**. illus: b&w: charts, maps, photos; b&w map on fold-out front; fold-out evolution chart; [20], 2p pls; 23 x 14 cm.

Published in many editions (1934 earliest?). Cover title: The story of Kents cavern, Wellswood, Torquay. "Reprinted by kind permission of Chambers's journal, time chart by kind permission of J.B. Sparks, Esq." "Kent's cavern from the survey made in 1934 by P.M.B. Lake"—Title of 1st folded leaf. Copy examined had no date. Has 11 b&w photos and Lake's 1934 map.

[1185] Gertsch, Willis J. *The spider family Nesticidae (Araneae) in North America, Central America, and the West Indies*. Austin, TX, US: Texas Memorial Museum, **1984**. Bulletin 31. illus: b&w: drawings, maps, tables; viii, 91p; bibl.

[1186] Ghosh, Leila *Ajanta and Ellora*. Roy, Dalia. Bombay, IN: India Book House, **1986**. 28 cm.

Chiefly col illus; unpaged.

[1187] Gibb, George Duncan, 1821-1876 *On Canadian caverns (read before the British Association for the Advancement of Science, at Aberdeen, 16th Sept. 1859)*. 1st ed. London, GB: "Geologist" Office, **1861**. illus: b&w: drawings, maps, tables; Engraving; [iii], 29p, 8p pls; 22 cm.

reprint ed. British Association for the Advancement of Science; Brison, David N. (intro, bibl). [Altoona, PA, US]: Speece Reprints, **1969**. front; bibl.

Reprinted by Jack H. Speece, Speece Reprint #5, Dec 1969 from the 1861 ed published in London by the "Geologist" Office.

[1188] Gibb, Hugh *The Tabon caves*. Fox, Robert B., 1918–. Honolulu, HI, US: Academics Hawaii, **1981**. The Philippine Story Videorecording 1.

1 videocassette (25 min), sd, col, 3/4 in. Title from data sheet.

Dr. Robert B. Fox, noted American archaeologist, and Dr. Alfredo Evangelista of the Philippine National Museum explain finds made during excavations in the Tabon Caves, Philippines. These caves, located in limestone cliffs high above the South China Sea, were inhabited by early Filipino man at least 50,000 years ago. The archaeological finds include decorated funereal jars, stone tools and skeletal remains which raise questions of possible links with the aborigines of Australia.

[1189] Gibbens, Ashley *Church Creek Caves*. npop: np, **1975**. 71p; 21 x 35.5 cm.

Appears to be mimeographed. This is a report on

the history, access, caving, bushwalking, geology and caves of this area within the Kanagra-Boyd National park, N.S.W.

[1190] Gibbons, Gail *Caves and Caverns.* 1st ed. San Diego, CA, US: Harcourt Brace & Company, **1993**. illus: col: drawings; [32]; 23 x 28 cm.

Juvenile.

Text and labeled illustrations describe the formation and physical features of various kinds of caves, with a brief section on spelunking.

[1191] Gidley, James William, 1866-1931 *The Pleistocene vertebrate fauna from Cumberland Cave, Maryland.* Gazin, Charles Lewis, 1904–(co-auth). Washington, DC, US: U.S. Government Printing Office, **1938**. Smithsonian Institution, United States National Museum: Bulletin no. 171. illus: b&w: drawings, maps, tables; front: map; vi, 99p; bibl, index; 24 cm.

Master microform held by: Ann Arbor, Mich: University Microfilms International. 1 reel. 35 mm.

[1192] Gidley, James William, 1866-1931 *Preliminary report on a recently discovered Pleistocene cave deposit near Cumberland, Maryland.* Washington, DC, US: Smithsonian Institution Press, **1914**. Proceedings of the United States National Museum v 46. 10p; 24 cm.

pp 93-102. Illus.

[1193] Gidley, James Williams, 1866-1931 *A pleistocene cave deposit of Western Maryland.* Washington, DC, US: Smithsonian Institution, **1918**. Annual report Smithsonian Institution 1918. 7p, 6p pls; 28 cm.

Illus. Caption title.

[1194] Gidley, James Williams, 1866-1931 *Pleistocene peccaries from the Cumberland cave deposit.* Washington, DC, US: Smithsonian Institution Press, **1921**. United States National Museum. Proceedings of the United States National Museum.1921, v 57 pp 651-678. 28p, 1p pls; 24 cm.

Illus, pl 54-55 on 1 l.

[1195] Gifford Pinchot National Forest *Layser Cave: silent voices, vital clues.* Janet Healy Video Production. [US]: United States. Department of Agriculture. Forest Service, **1991**.

1 videocassette (VHS)(20 min), sd, col, 1/2 in.

Discusses the discovery of artifacts in Layser Cave in Lewis County, Washington and the efforts of the Forest Service to unearth, decipher, preserve, and share these finds.

[1196] Gifford, James C., 1973– *Archaeological explorations in caves of the Point of Pines region, Arizona.* Tucson, AZ, US: University of Arizona Press, **1980**. Anthropological Papers of the University of Arizona no 36; Contributions to Point of Pines archaeology no 27. illus: b&w: charts, drawings, maps, photos, tables; xii, 218p; bibl, index; 28 cm.

[1197] Gilbert Group *Caves, a deeper look at our Earth.* Gilbert, Bill (prod); Aley, Thomas John, 1938–(advs). Overland Park, KS, US: Gilbert Group, **1978**.

1 reel, 15 min, sd, col, 16 mm.

Examines how caves are formed and some of the unusual animal life that lives in caves.

[1198] Gilbertson, D. D. *In the shadow of extinction: a Quaternary archaeology and paleoecology of the lake, fissures, and smaller caves at Creswell Crags SSSI.* Jenkinson, R. D. S. Sheffield, [GB]: J.R. Collis, **1984**. vii, 129p; bibl; 21 x 30 cm.

Illus.

[1199] Gilbreath, Alice Marie , 1921– *Nature's underground palaces: caves and caverns.* Eagle, Michael (Illus). New York, NY, US: David McKay Company, Inc, **1978**. illus: b&w: drawings; front; [31]; 26 cm.

Juvenile.

Discusses physical phenomena discovered in caves, the people who once inhabited them, animals which are true cave dwellers, and man's uses for 'underground palaces'.

[1200] Giles, Cathy *Cave entrance plants including Lampenflora.* Waitomo Caves, NZ: Waitomo Caves Museum Society, **1984**. illus: b&w: charts, drawings; text on lining pages; [i], 13p.

[1201] Giles, Cathy *Cave fauna of Waitomo.* Waitomo Caves, [NZ]: Waitomo Caves Museum Society, **1983**. illus: b&w: charts, drawings; [iv], 34p; bibl, gloss.

[1202] Gill, David W. *The untamed river expedition: Nakanai Mountains, East New Britain, Papua New Guinea.* Stockport, GB: The Untamed River Expedition, **1988**. illus: b&w: ads, drawings, maps, photos, ports, tables; [ii], 69p; app; 31 cm.

33 figs.

[1203] Gill, David W. (comp) *Caves of the Peak District.* Beck, John S. (comp); Derbyshire Caving Association. new ed. Clapham, GB: Dalesman, **1991**. illus: b&w: ads, maps; col: photos; 257p; bibl, gloss, index; 17 cm.

Illus.

Latest of a series of cavers' guides for part of England. Earlier editions were titled 'Caves of Derbyshire.' Format and content similar to D. Brook, *et al.* 1988, except that references are listed.

[1204] Gill, George *Stump Cross Caverns.* Greenhow Hill, Pateley Bridge, GB: G. Gill, Stump Cross Caverns, **nd**. illus: b&w: drawings, maps, photos; col: photos; text on rear lining page; 1 map on rear lining page; 26p; gloss.

Cover title: Stump Cross Caverns: the underground wonderland: official guide. 7 small (1 on cover) photos and 13 b&w maps

[1205] Gill, Thomas B. *Guide to the home of the rainbow: Cox's Cave, Cheddar, Somerset.* Cheddar, Somerset, GB: Cox's Cave, **[1936]**. illus: b&w: photos; photos on both; [28]; 19 x 12.5 cm.

Only 4 pages of text. 24 brown and white photos.

[1206] Gill, Thomas B. *Guide to the world famous*

Gough's Caves, Cheddar, Somerset. Cheddar, Somerset, GB: Gough's Caves and Cave Man Restaurant, [**1936**]. illus: b&w: drawings, photos; photos both ends; [28]; 19 x 12.5 cm.

Similar in nature to several others by same author.

[1207] Gillett, J. E. (ed) *Crewe Climbing and Potholing Club: Gouffre Berger Expedition.* [Gawsworth, Near Macclesfield, Cheshire, GB]: Crewe Climbing and Potholing Club, **c1983**. illus: b&w: cartoons, drawings, maps, photos; front: b&w photo; [ii], 23p, [9]; app.

[1208] Gillieson, D. S. *Lelet: report of the 1976 New Ireland Speleological Expedition.* Canty, R.; Landsberg, J; Page, R.; Wilson, S. Port Moresby, Papua, PG: Papua New Guinea Cave Exploration Group, Dec, **1977**. Niugini Caver vol 5 no 3. illus: b&w: graphs, maps, photos; two folded maps in text; 42p.

pp 61-101.

[1209] Gillieson, David S. (ed) *Geology, climate, hydrology and karst formation: field symposium in Australia 4 to 18 December 1992: Buchan - Mt. Gambier / Naracoorte - Nullarbor Plain humid temperate impounded karst, sub-humid temperate syngenetic karst, arid temperate karst: guidebook.* Canberra, ACT, AU: Department of Geography and Oceanography, University College, The University of New South Wales, Australian Defence Force Academy, **1992**. Project 299 of I.G.C.P.; Special Publication no 4 Department of Geography and Oceanography University College Australian Defense Force Academy Canberra. illus: b&w: drawings, graphs, maps, photos, tables; text on lining pages; viii, 115p; app, bibl.

Supported by the United Nations Educational, Scientific, and Cultural Organization (UNESCO); International Union of Geological Sciences (IUGS); University of New South Wales; University of Melbourne; Australian Geological Survey Organisation.

A wealth of data on a diverse and fascinating area, not otherwise available in book format. Valuable references to articles not in the main-line literature

[1210] Gillieson, David S. (ed) *Resource management in limestone landscapes: international perspectives proceedings of the International Geographical Union study group man's impact on karst Sydney 15-21 August 1988.* Smight, David Ingle (ed); Cochrane, Anne (prod); Ballard, Paul (prod). Canberra, ACT, AU: Department of Geography and Oceanography, University College, Australian Defence Force Academy, Canberra, **1989**. Special publication No 2 Department of Geography and Oceanography, University College Australian Defence Force Academy Canberra. illus: b&w: charts, drawings, graphs, maps, photos, tables; iii, 260p, [2]; bibl.

An introduction to 19 articles from papers given at the International Geographical Union Conference - study group on man's impact on karst.

[1211] Gillieson, David S. *Caves: processes, development, and management.* Cambridge, MA, US: Blackwell Publishers, **1996**. The Natural Environment. illus: b&w: charts, maps, photos, tables; ix, 324p; bibl, gloss, index.

[1212] Gilmore, Elizabeth (prod, ed, auth) *Silent Splendor.* Denver Museum of Natural History. Denver, CO, US: Denver Museum of Natural History, **1986**.

1 videocassette (26 min), sd, col, 1/2 in.

Presents the story of Cave of the Winds near Colorado Springs, Colorado, and the nearby, newly-discovered cave named Silent Splendor.

[1213] Gimenez Reyna, Simeon *The cave of "La Pileta": Benaojan, Malaga, Spain.* Huntingford, George Wynn Brereton. [Malaga, ES]: Delegacion Provincial de Excavaciones Arqueologicas de Malaga, **1965**. illus: b&w: maps; 58p, 16p pls; 25 cm.

Illus.

[1214] Gines, Joaquin *Proceedings of the 1986 meeting of the International Geophysical Union study group on man's impact on karst.* Palma de Mallorca, ES: Publicacio d'Espeleogia Federacio Balear d'Espeleolgia, May, **1987**. Endins vol 13.

12 papers in English from international karst researchers on impact of human activity on karst. Also contains 4 introductory papers on Mallorca caves; only 1 is in English.

[1215] Giusti, Ennio V. *Hydrogeology of the karst of Puerto Rico.* Washington, DC, US: USGS, **1978**. Geological Survey Professional Paper 1012. illus: b&w: drawings, graphs, maps, photos, tables; b&w drawing and map folded in rear pocket; vii, 68p.

59 figs, 2pls, 9 tables.

Important basic publication on this area and a good overview of karst geology, hydrology, geochemistry and geologic evolution in Puerto Rico. It focuses on the importance of early porosity and permeability, lithic nature and tectonic history in the evolution of the various caves.

[1216] Giusti, Ennio V. *Water resources of the North Coast Limestone area, Puerto Rico.* Bennett, Gordon D. Ft. Buchanan, PR: U.S. Geological Survey, Water Resources Division, **1976**. Water-resources investigation 42-75. illus: b&w: charts, drawings, graphs, maps, photos, tables; iv, 42p; bibl, errat; 28 cm.

[1217] Glanvill, Peter *First aid for cavers.* Coath, Tim (illus); Moss, Phil (Illus). Gloucester, GB: Ambit Publications Ltd, **1969**. illus: b&w: drawings; 30p.

Based on a series of articles published in Descent 85-88.
Includes basic "what to do" information.

[1218] Gleave, Joseph James *Yorkshire Caves: Victoria cave, Settle; Ingleborough Cave, Clapham; Yordas Cave, Ingleton.* Manchester, GB, US: Thomas Sowler & Sons, **c1900**. illus: b&w: drawings; front: line drawing of Victoria Cave, the entrance from the inside; 19p; 18.5 x 12.5 cm.

Cave descriptions.

[1219] Glennie, E. A. *Cave Fauna.* npop: Cave Research Group, **1945**. 36p.

Part 1 (1945) has 16 pages; part 2 (1946) has 20 pages.

[1220] Godden, Elaine *Rock paintings of Aboriginal Australia.* Malnic, Jutta (photo). 1988 reprint ed. Frenchs Forest, NSW, AU: Reed Books Pty Ltd, **1988**. illus: b&w: maps, photos, ports, repros; col: photos; bibl, index; 29 cm.

1st ed. Malnic, Jutta (photo). Frenchs Forest, NSW, AU: Reed Books Pty Ltd, **1981**. illus: col: photos; 128p, [2]; bibl, index; 26 x 37 cm.

Illus (some col). OCLC lists 1981 and 1982 as date of publication.

[1221] Goldberg, Paul *Formation processes in archaeological context.* Nash, David T.; Petraglia, Michael D. Madison, WI, US: Prehistory Press, **1993**. Monographs in World Archaeology no 17. x, 188p; bibl; 28 cm.

Illus.

[1222] Goodall, Elizabeth *Prehistoric rock art of the Federation of Rhodesia & Nyasaland.* Salisbury, ZW: National Publications Trust, Rhodesia and Nyasaland, **1959**. SOLINET/ASERL Cooperative Microfilming Project (NEH PS-20317); SOLINET/ASERL Cooperative Microfilming Project (NEH PS-20317); SOL MN03077.08 FUG. illus: b&w: maps; xix, 267p, 1p pls; bibl, index; 32 cm.

Illus (partly col). Cover title: Rock art of Central Africa. Microfilm. Atlanta, Ga.: SOLINET, 1993. 1 microfilm reel ; 35 mm.

[1223] Goode, Cecil E. *World wonder saved: how Mammoth Cave became a national park.* Mammoth Cave, KY, US: Mammoth Cave National Park Association, **1986**. 92p; bibl; 24 cm.

Illus.

[1224] Goode, Glenn T. *Archaeological testing of the cave at site 41BX22, Bexar County, Texas.* [Austin, TX, US]: Texas State Dept of Highways and Public Transportation, Highway Design Division, **1985**. illus: b&w: maps; 23p; bibl; 28 cm.

Illus.

[1225] Goodwin, Astley John Hilary, 1900-1959 *Archaeology of the Cape St. Blaize cave and raised beach, Mossel Bay.* npop: Printed for the Trustees of the South African Museum, Cape Town, by Neill and Co, Ltd, **1935**. Annals of the South African Museum v 24 pt 3 art 4 6. 1 folded illus; 30p, 1p pls; 26 cm.

Title from cover. With: Some native snuff-boxes in the South African Museum, by M. Shaw.

[1226] Goodwin, J. West *Mammoth Cave: America's great natural wonders.* Sedalia: Steam Printer. 10p.

This is an 19th century brochure that has 10 pages and two illustrations on the two covers. Pictures Mammoth Cave Hotel and picture of cave entrance.

[1227] Gordon, George Byron, 1870-1927 *Caverns of Copan, Honduras: report on explorations by the museum, 1896-97.* 1st ed. Cambridge, MA, US: The Museum, **1898**. Memoirs of the Peabody Museum of American Archaeology and Ethnology, Harvard University v1 no. 5. illus: b&w: maps; 12p, 1p pls; 36 cm.

Illus. Also paged continuously with the other numbers of the volume. Issued under one cover with v.1, no.4.

reprint ed. New York, NY, US: Kraus, **1970**. Memoirs of the Peabody Museum of American archaeology and ethnology, Harvard University, v. 1, no. 5. 12p; 36 cm.

Illus.

[1228] Gordon, Isabella *On* Spelaeogriphus, *a new cavernicolous crustacean from South Africa.* [GB]: British Museum (Natural History), **1957**. British Museum (Natural History) Bulletins Zoology Series vol 5 no 2. 47p; 25 cm.

Illus.

[1229] Gospodarič, R. (ed) *Underground water tracing: investigation in Slovenia 1972-1975.* Habič, Peter (ed). Postojna, YU: Institute for Karst Research SAZU [Slovene Academy of Sciences and Arts], **1976**. illus: b&w: charts, graphs, maps, tables; partly col: maps; col: photos; 2 folded maps in pocket, 1 of them is partly col; 309p, 12p pls; bibl.

Summaries at end in Slovene, Serbo-Croat, Macedonian.

[1230] Gough, A. G. H. *Guide to Gough's Caves, Cheddar; how they were found, the finest in the world (electrically illuminated) eighteen beautiful views of the cave prehistoric man: - how he was found.* Cheddar, Somerset, GB: A.G.H. Gough, **1912**. illus: b&w: photos, ports; ad for cheese on rear lining page; 36p; 10.5 x 14 cm.

Working from photocopy.

[1231] Gough, A. G. H. *Pictorial guide to the caves Cheddar.* Bristol, GB: E.W. Savory Ltd (Printer), **nd**. illus: b&w: photos; 32p.

Cover title: Pictorial guide the caves Cheddar opposite the Lion Rock

[1232] Gough, William *The cliffs & caves of Cheddar: a series of beautiful views in real photogravure published by William Gough, son of the discoverer of the famous cave.* Bristol, GB: printed by E. W. Savory Ltd, **nd**. illus: b&w: photos; [15].

[1233] Government of Australia *Nomination of Australian fossil sites (a serial nomination of sites at Murgon, Riversleigh and Naracoorte): The origin and evolution of Australia's mammals.* Canberra, ACT, AU: Department of the Environment, Sport and Tourism, **1993**. illus: b&w: maps, photos; Various pagination; app, bibl.

By the Government of Australia for inscription on the World Heritage List. Extensive Bibliography.

[1234] Graham, Andrew W. (ed) *Aspects of land use planning at Fanning River. With particular reference to the Fanning River caves.* Annerley, QLD, AU: np, May,

1972. illus: b&w: maps; [16]; bibl, gloss.

[1235] Graham, David Crockett *The ancient caves of Szechwan Province, China.* Washington, DC, US: Smithsonian Institution, United States National Museum, **1932**. Proceedings of the United States National Museum vol 80 art 16. 13p, 8p pls; 25 cm.

No. 2916. Title from cover.

[1236] Graham, Joseph J. *The Florida Caverns at Marianna, Florida.* Evanston, IL, US: np, **1939**. 23p; 28 x 24 cm.

Carbon copy of original.

[1237] Graham, Richard *Exploring the underground.* North Hollywood, CA, US: Center for Cassette Studies, **1970**. Geology. bibl.

1 sound cassette (31 min), 1 7/8 ips, mono. Bibliography and annotation laid in. Both 1970 and 1979 listed as date of publication in OCLC.

[1238] Grampian Speleological Group. *Library Grampian Speleological Group library catalogue.* Jeffreys, Alan Lawrence, 1940–(comp). Edinburgh: [Grampian Speleological Group], **1991**. 73p; 30 cm.

[1239] Grange Cavern Military Museum *Souvenir guide to the world's largest underground military museum: the Grange Cavern.* Holywell, Clwyd, GB: The Grange Cavern Military Museum, **1980**. [18]; 15 x 21 cm.

Title from cover. Illus. 1980 and 1988 are listed as date of publication by OCLC.

[1240] Grant, Blanche Chloe, 1874-1948 *Carlsbad Caverns.* Carlsbad ed. npop: np, **1927**. 34p; 20 cm.

On verso of t.p.: Cavern guide book. Blank pages for "Notes" (5 p. at end).

[1241] Grant, Blanche Chloe, 1874-1948 *Cavern guide book, Carlsbad Caverns, Carlsbad, N. M.* Topeka, KS, US: Crane & Co., **1928**. 43p; 20 cm.

Blank pages for "Notes" (5 at end). Illus.

[1242] Grant, James, 1831-1920 *Superficial geology of the valley of the Ottawa and the Wakefield cave.* npop: np, **1868**. CIHM/ICMH Microfiche series no 06531. 18p.

pp [17]-34. Caption title. Transactions of the Ottawa Natural History Society, Paper 2. "Read before the Society, November 25th, 1868." Microfiche. Ottawa: Canadian Institute for Historical Microreproductions, 1980. 1 microfiche (14 fr.); 11 x 15 cm.

[1243] Grattan-Smith, Thomas Edward *The cave of a thousand columns.* Uden, E. Boye (illus). London, GB: Hutchinson & Co, **1938**.

Containing five half-tone plates by E. Boye Uden.

[1244] Graves, John (ed) *The River Styx-Salt Spring Cave system.* Walsh, John M. (ed). The Texas Cave Report Series. Texas Cave Report Series no 1. illus: b&w: drawings, graphs, photos; partly col: maps; 2 folded in text.; [iii], ii, 38p, [1]; bibl; 28 cm.

[1245] Gray, Eric (comp) *Devonshire sump index.* Devon, GB: Eric Gray, **1985**. p 35 on rear lining page; [vi], 29p, [5]; index; 21 x 15 cm.

No illus.

[1246] Grayson, Donald K. *Danger Cave, Last Supper Cave, and Hanging Rock Shelter: the faunas.* New York, NY, US: [American Museum of Natural History], May, **1988**. Anthropological papers of the American Museum of Natural History, v 66, pt 1 0065-9452. illus: b&w: maps; 130p; bibl; 26 cm.

Issued May 6, 1988. Illus.

[1247] Great American Film Factory *Underground wilderness.* Williams, Bob (prod). Sacramento, CA, US: Great American Film Factory, **1977**.

110 fr, col, 35 mm & cassette (20 min). Juvenile.

Depicts a group of young spelunkers as they explore an uncharted cave. Includes comments by cave explorers and general information on various types of rock formations.

[1248] Greeley, Ronald *Geology and morphology of selected lava tubes in the vicinity of Bend, Oregon.* npop: np. illus: b&w: drawings, maps, photos; 3 pls maps (folded in back); 51p, [3].

24 figs.

[1249] Greeley, Ronald *Geology of selected lava tubes in the Bend Area, Oregon.* Portland, OR, US: Department of Geology and Mineral Industries, **1972**. State of Oregon Dept of Geology and Mineral Industries Bulletin 71. vi, 47p; 28 cm.

Illus.

[1250] Greeley, Ronald *Lava tubes of the Cave Basalt, Mount St. Helens, Washington.* Hyde, J. H. npop: United States. NASA, **1971**. Technical Memorandum NASA TM X-62 022. 33p.

[1251] Greeley, Ronald *Observations of actively forming lava tubes and associated structures, Hawaii.* npop: np, **1971**. illus: b&w: maps; bibl.

2 pts in 1 vol, illus.

[1252] Green, H. Stephen *Pontnewydd Cave: a lower palaeolithic hominid site in Wales: the first report.* National Museum of Wales. Cardiff, GB: Amgueddfa Genedlaethol Cymru, **1984**. National Museum of Wales Quaternary Studies Monographs no 1. illus: b&w: maps, repros; xvii, 227p; bibl; 31 cm.

Illus.

[1253] Green, John *The Story of Cave Hill: Grottos, VA.* npop: np, **nd**. illus: b&w: drawings, maps, photos; [16]; app, bibl.

[1254] Greenberg, Judith E. *Caves.* Carey, Helen H.; Miyake, Yoshi. Milwaukee, WI, US: Raintree Publishers, **1990**. Raintree Science Adventures. 32p; 24 cm.

Col illus. Juvenile.

Members of an expedition into a cave learn how different kinds of caves and cave formations are created in nature and what caves and cave animals are like.

[1255] Greer, John W. *Report on preliminary archeological explorations at Carlsbad Caverns National Park,*

New Mexico. [Austin, TX, US]: Department of Anthropology, University of Texas, **1966**. illus: b&w: maps; 49p, [37]; bibl; 29 cm.

Illus.

[1256] **The Gregory Publishing Co Pty Ltd** *Jenolan Caves (N. S. W.) 'Nature's Masterpiece': a complete description of the geology, discovery, development, and features of the Jenolan limestone system beautifully illustrated, with high-class maps*. 1st ed. Sydney, NSW, AU: N. S. W. Government Tourist Bureau, **nd**. illus: b&w: ads, maps, photos; b&w photos (many folded in text); 60p.

Cover title: Jenolan Caves "nature's masterpiece".

[1257] **Gregory, John Walter**, 1864-1932 *Some caves and a rock shelter at Loch Ryan and Portpatrick, Galloway*. Ritchie, James, 1882-1958; Kennedy, W. Q.; Leitch, Duncan. npop: np, **1930**. Papers from the Geological Department Glasgow University v 13; University of Glasgow Publications XX. 18p; 24 cm.

Illus. Title from cover. Reprinted from the Proceedings of the Society of Antiquaries of Scotland, Vol LXIV. (Vol. IV., Sixth series), Session 1929-30 (pages 247 to 264).

[1258] **Gregory, W. L.** *The Walkberg Cave System*. Gamble, F. M. npop: np, **1980**. 58p.

Report highlighting the cave in South Africa. Another record lists spelling as Wolkberg. 12 maps, 12 pls.

[1259] **Grenfell, Harold** *The caves of Gower*. Morris, Bernard. Swansea, GB: The Gower Society, **1971**. Publications, Gower Society. illus: b&w: maps; 16p; 21 cm.

Illus

[1260] **Gries, John Paul**, 1911- *Continued investigation of water losses to sinkholes in the Pahasapa Limestone and their relation to resurgent springs, Black Hills, South Dakota*. [Rapid City, SD, US]: [South Dakota School of Mines and Technology], Jun, **1969**. illus: b&w: maps; 15p, [21]; bibl; 28 cm.

Title from cover.

[1261] **Griffin, John Wallace** *Investigations in Russell Cave: Russell Cave National Monument, Alabama*. Huntsville, AL, US: Huntsville Chapter of the Alabama Archaeological Society, **1978**. Publications in Archeology, United States National Park Service 13. illus: b&w: drawings, graphs, maps, photos, tables; xiv, 127p; bibl, index; 26 cm.

Illus. Reprint. Originally published: Washington: National Park Service, 1974. 61 illus, 32 tables.

[1262] **Griffin, K. A.** *Confidence in the cave. A guide for a safe speleological adventure*. Houston, TX, US: Spelaean Group, **1968**. 12p.

This is a brief, well-written, manual for the beginning caver.

[1263] **Griffiths, Julian (comp)** *Cave Diving Group: northern sump index*. Bristol, GB: Cave Diving Group, **1981**. illus: b&w: maps; 111p; index; 30 x 21 cm.

Title from caption.

[1264] **Griffiths, Paul** *Horne Lake Wonder Caves: a guide to Horne Lake Caves Park*. Goth, Kathy (typist). npop: Maquinna Publishing, **1975**. illus: b&w: drawings, maps; fold-out map on rear lining page; 12p; gloss.

[1265] **Grigson, Geoffrey**, 1905- *Painted caves*. London, GB: Phoenix House Ltd, **1957**. illus: b&w: drawings, maps, photos; front: b&w photo; 223p, 16p pls; index; 23 cm.

24 pls, 7 drawings, 7 maps; introduction to painted caves in Britain and abroad.

[1266] **Grimes, K. G.** *South East Karst Province of South Australia: Australian Caves & Karst Management Association October 1995*. Hamilton-Smith, Elery; Spate, Andrew P. np: npop, Oct, **1995**. illus: b&w: drawings, graphs, maps, photos, tables; iv, 110p; app, bibl; 30 x 21 cm.

[1267] **Grimes, Ken** *Field Guide to karst features in southeast South Australia and Western Victoria: for the Karst Studies Seminar, Naracoorte, February 1996*. White, Susan. Hamilton, VIC, AU: Regolith Mapping, February, **1996**. illus: b&w: drawings, graphs, maps, photos; [vi], 34p; bibl; 30 x 21 cm.

[1268] **Grisafe, David A.** *Stabilization of Dakota Sandstone surface of the Faris Cave petroglyphs, Kanopolis Lake Project, Kansas*. Vicksburg, Springfield, MS, VA, US: U.S. Army Engineer Waterways Experiment Station, Aug, **1992**. Contract report EL-92-2 Contract report U.S. Army Engineer Waterways Experiment Station EL-92-2. 24p; 28 cm.

Illus. Title from cover. At head of title: "Environmental Impact Research Program." Pagination: 13, 11 pp.

[1269] **Grobet, André-H.** *The underground lake of St. Léonard: Valais*. Sion, CH: np, May, **1966**. illus: b&w: maps, photos, tables; 12p; 17 cm.

Title on cover: The subterranean lake of St-Leonard, Wallis.

[1270] **Groenhout, Ron (ed)** *Guide for prospective cavers*. Sims, Brailey (ed). 2nd ed. Tighes Hill, NSW, AU: Newcastle University Speleological Society, Sep, **1968**. illus: b&w: drawings; 12p; gloss.

Title from cover.

[1271] **Groom, Gillian Elisabeth** *The geomorphology and speleogenesis of the Dachstein caves*. Coleman, Alice. [Ledbury, GB]: Cave Research Group of Great Britain, **1958**. Occasional Publications The Cave Research Group of Great Britain no 2. v, 23p; bibl; 24 cm.

Illus.

[1272] **Grosscup, Gordon L.** *The culture history of Lovelock Cave, Nevada*. Berkeley, CA, US: University of California, **1960**. Reports of the University of California Archaeological Survey no 52. illus: b&w: maps; ii, 72p; bibl; 28 cm.

Illus.

A treatise about the archaeology of the Desert Archaic

and the unique organic prehistoric cultural artifacts at this Intermontane site.

[1273] Grove, David C. *The Olmec paintings of Oxtotitlan Cave Guerrero, Mexico.* Dávalos, Felipe (renderings). Washington, DC, US: Dumbarton Oaks; Trustees for Harvard University, **1970**. Studies in Pre-Columbian Art and Archeology number six. illus: b&w: drawings, maps, photos; col front (rendering); 36p; bibl.

[1274] Gruhn, Ruth1907– *The archaeology of Wilson Butte Cave south-central Idaho.* Pocatello, ID, US: Idaho State College Museum, **1961**. Occasional Papers of the Idaho State College Museum no 6. illus: b&w: charts, drawings, maps, photos, tables; folded map inserted and cover title.; iii, 202p, 41p pls; add, app, bibl; 28 cm.

39 pls.

[1275] Guernsey Press *Illustrated "Press" guide to the caves and cliffs of Sark: with map and plans of roads..* Guernsey, GB: Guernsey Press, **1931**. illus: b&w: maps; [32]; 25 x 12 cm.

Illus. Title from cover. Caption title: The Island of Sark: special guide. Illustrative matter in pocket.

[1276] Guernsey, Daniel Riley *Lost in the Mammoth Cave.* New York, NY, US: Broadway Publishing Company, **1905**. front; vii, 315p; 20 cm.

Master microform held by: Ky U. Microfilm. Lexington, Ky., University of Kentucky, 19—. 1 reel. 35 mm.

[1277] Guernsey, Samuel James, 1868-1936 *Basket-Maker caves of northeastern Arizona; report on the explorations, 1916-17.* Kidder, Alfred Vincent, 1885-1968. 1921 ed. Cambridge, MA, US: The Museum, **1921**. Papers of the Peabody Museum of American Archaeology and Ethnology, Harvard University vol VIII no 2. vii, 121p; 25 cm.

44 pls (1 col).

1967 reprint ed. New York, NY, US: Kraus Reprint Corp, **1967**. bibl.

[1278] Guernsey, Samuel James, 1868-1936 *Explorations in northeastern Arizona; report on the archaeological fieldwork of 1920-1923.* 1931 ed. Cambridge, MA, US: The Museum, **1931**. Papers of the Peabody Museum of American Archaeology and Ethnology, Harvard University vol XII no 1. illus: b&w: maps; xi, 123p; bibl; 25 cm.

66 pl (incl col front) on 34 lvs.

1967 reprint ed. New York, NY, US: Kraus Reprint Co, **1967**. bibl.

[1279] Guilday, John E. *The Baker Bluff cave deposit, Tennessee, and the late Pleistocene faunal gradient.* Hamilton, Harold W.; Anderson, Elaine; Parmalee, Paul Woodburn. Pittsburgh, PA, US: Carnegie Museum of Natural History, **1978**. Bulletin of Carnegie Museum of Natural History no 11. illus: b&w: charts, graphs, maps, photos, tables; Text on rear; 67p; bibl; 28 cm.

Illus.

[1280] Guilday, John E. *The Clark's Cave bone deposit and the late Pleistocene paleoecology of the central Appalachian Mountains of Virginia.* Parmalee, Paul Woodburn; Hamilton, Harold W. Pittsburgh, PA, US: Carnegie Museum of Natural History, **1977**. Bulletin of Carnegie Museum of Natural History no 2. illus: b&w: drawings, graphs, maps, photos, tables; 87p; bibl; 28 cm.

Illus.

[1281] Guilday, John E. *Jaguar (*Panthera onca*) remains from Big Bone Cave, Tennessee and east central North America.* McGinnis, Helen J. npop: np, **1972**. [29]; bibl; 23 cm.

Illus. Reprinted from National Speleological Society Bulletin, v.34, no.1, January 1972.

[1282] Guilday, John E. *The late Pleistocene small mammals of Eagle Cave, Pendleton County, West Virginia.* Hamilton, Harold W. Pittsburgh, PA, US: [Carnegie Museum], **1973**. Annals of the Carnegie Museum v 44 article 5. 14p; bibl; 23 cm.

Caption title. Illus.

[1283] Guilday, John E. *The Welch Cave Peccaries (*Platygonus*) and associated fauna, Kentucky Pleistocene.* Hamilton, Harold W.; McCrady, Allen D. Pittsburgh, PA, US: Carnegie Museum, **1971**. Annals of Carnegie Museum v 43 article 9. illus: b&w: charts, drawings, graphs, maps, photos, tables; 72p; bibl; 23 cm.

pp 249-320.

[1284] Guilday, John E., 1925-1982 *New Paris no. 4: a late Pleistocene cave deposit in Bedford County, Pennsylvania.* Martin, Paul S.; McCrady, Allen D. reprint. Arlington, VA, US: National Speleological Society, **1964**. National Speleological Society Bulletin v 26 no 4. 74p, 2p pls; bibl; 25 cm.

pp 121-194. Illus (some col).

[1285] Guilday, John E., 1925-1982 *A new* Peromyscus *(Rodentia: Cricetidae) from the Pleistocene of Maryland.* Pittsburgh, PA, US: Carnegie Museum, **1967**. Annals of the Carnegie Museum v 39 article 6. 12p; 23 cm.

Illus. Fossils found in Cumberland Cave

[1286] Gulden, Robert *Caves of the eastern panhandle of West Virginia.* Johnsson, Mark J. Barrackville, WV, US: West Virginia Speleological Survey, Dec, **1984**. Bulletin West Virginia Speleological Survey 8. illus: b&w: drawings, maps, tables; 3 b&w maps folded in text; [iv], 135p, [1]; index; 28 cm.

Copyright 1985.

[1287] Günay, Gültekin (ed) *Karst water resources: proceedings of a symposium held at Ankara, July 1985, and sponsored by the United Nations Development Program, the United Nations Educational, Scientific and Cultural Organization, the International Association of Hydrological Sciences, many government organizations of Turkey, and Hacettepe University of Ankara, Turkey.* Johnson, A. Ivan (ed). Wallingford, Oxon, GB: International Association of Hydrological Sciences. IAHS Publication no 161. illus: b&w: charts, drawings, graphs, maps, photos, tables; text on lining pages; xi, 642p, [6]; bibl.

International symposium on karst water resources 7-19 July 1985, Ankara-Antalya/Turkey.

An important proceedings volume.

[1288] **Gunn, John (ed)** *An introduction to British limestone karst environments.* Bridgwater, Somerset, GB: British Cave Research Association, **1994**. Cave studies series no. 5. illus: b&w: charts, graphs, maps, tables; Text on front; 40p; bibl; 21 cm.

[1289] **Gunzi, Christiane** *Cave life.* Greenaway, Frank (illus). New York, NY, US: Distributed by Houghton Mifflin Company, **1993**. Look Closer. illus: col: drawings, photos; col photos on lining pages; 29p; gloss, index.

This is the first American ed. Published also by Pymble, NSW: Collins/Angus & Robertson; and London, GB: Dorling Kindersley. Juvenile.

Discusses the plants and animals that live in caves. Includes the peacock butterfly, cave cricket, and maidenhair fern.

[1290] **Gurnee, Jeanne Marie**, 1926– *The Ensueño Cave Study: Ensueño Cave, Hatillo, Puerto Rico; expedition period: February 18-25, 1987.* Closter, NJ, US: National Speleological Foundation, **1988**. illus: b&w: drawings, graphs, maps, photos; 64p.

Survey of cave's geology, biology and plan for commercial development of the cave.

[1291] **Gurnee, Jeanne Marie**, 1926– *National Speleological Society field trip to Aguas Buenas Caves, Puerto Rico: February 1968.* npop: np, Aug, **1968**. illus: b&w: drawings, maps, photos; fold-out map; i, 35p.

This is another special NSS report of very limited publication. This is based on a Feb 1968 expedition. Included are 6 B&W photos and one large fold-out map.

[1292] **Gurnee, Jeanne Marie**, 1926– *A study of Fountain National Park and Fountain Cavern: Anguilla, British West Indies.* Closter, NJ, US: National Speleological Foundation, **1969**. illus: b&w: drawings, graphs, maps; 48p; bibl; 28 cm.

Published in the U.S.A. with a grant from the Foundation. National Park Development Committee, Government of Anguilla, British West Indies and the resources of The National Speleological Foundation.

[1293] **Gurnee, Jeanne Marie**, 1926– *A study of Harrison's Cave, Barbados, West Indies.* Closter, NJ, US: National Speleological Foundation, Jul, **1978**. illus: b&w: drawings, maps, photos, tables; 5 maps folded in text; 32p, [1].

Published in Barbados with a grant from the Ministry of Housing, Lands and the Environment, Caves Authority, Government of Barbados, West Indies and the resources of the National Speleological Foundation. Plan for the commercialization of the cave.

[1294] **Gurnee, Russell Hampton**, 1922-1995 *Cave clippings of the nineteenth century.* Closter, NJ, US: R.H. Gurnee, Inc, **1983**. illus: b&w: maps, photos, repros; viii, 197p.

Limited edition of 200.

A useful resource of early newspaper cuttings with a brief historical perspective and index.

[1295] **Gurnee, Russell Hampton**, 1922-1995 *Discovery at the Rio Camuy.* Gurnee, Jeanne Marie, 1926–. New York, NY, US: Crown Publishers, Incorporated, **1974**. illus: b&w: drawings, maps, photos; front: b&w map; viii, 183p; gloss, index; 24 cm.

Seven chapters with 53 black & white photos.

The exploration of the Rio Camuy system from 1958 through 1970. An exciting account of the exploration of a challenging underground river system.

[1296] **Gurnee, Russell Hampton**, 1922-1995 *Discovery at the Rio Camuy, Puerto Rico.* Sullivan, Gerardus Nicholas, 1927–[Brother Nicholas]; Thrailkill, John V. Washington, DC, US: National Geographic Society, **1968**. 12p; bibl; 23 cm.

Illus. Reprinted from: National Geographic Society Research Reports, 1963 projects pp 115-126.

[1297] **Gurnee, Russell Hampton**, 1922-1995 *Discovery of Luray Caverns, Virginia.* Closter, NJ, US: R. H. Gurnee, Inc. illus: b&w: maps, photos, ports, repros; [iv], 107p; index; 26 cm.

Published in both soft and hardback editions Softbound copies were only available at the cave.

The history of the search for and development of Luray Caverns since 1879, narrated in a very readable style with numerous period illustrations.

[1298] **Gurnee, Russell Hampton**, 1922-1995 *Gurnee guide to American caves: a comprehensive guide to the caves in the United States open to the public.* Gurnee, Jeanne Marie, 1926–. 1980 ed. Teaneck, NJ, US: Zephyrus Press ; distributed by Caroline House, Ossining, NY, US, **1980**. illus: b&w: maps, photos; viii, 252p; gloss, index; 23 cm.

Guide to the commercial caves of the United States. 20 cave descriptions. An update on Visiting American Caves; completely redone.

1990 ed. Gurnee, Jeanne Marie, 1926–. Closter, NJ, US: R. H. Gurnee, Inc, **1990**. illus: b&w: repros; 288p; gloss, index; 23 cm.

[1299] **Gwin, J. N.** *A visit to the Mammoth Cave.* Indianapolis, IN, US: John G. Doughty, **1875**. 16p; 23 cm.

Title from cover. Extremely rare item. Listed in Hovey & Call bibliography but not listed by Wilkes's Mammoth Cave bibliography.

[1300] **Haast, J.** *Researches and excavations carried on in and near Moa Bone Point Cave, Summer Road, in the year 1872.* Christchurch, NZ: np, **1874**. 22p.

[1301] **Habe, France** *Jame Škocjan = Grotte Skocjan*. [Portorož]: TOP Portorož TOZD Gostintvo Sežana, **1988**. illus: b&w: drawings, photos, repros; col: photos; [62]; 24 cm.

[1302] **Habe, France** *Postojna*. Beograd, YU: Jugoslavija, **1969**. illus: b&w: photos, repros; col: maps; pull-out captions to pls (unnum) & loose folded printed supp; maps on lining pages; 21p, [4], 64p pls; 17 x 12 cm.

Description of caves near Postojna, some of them show caves.

[1303] **Habe, France** *Postojna Cave*. npop: np, **1976**.

Beautiful coffee-table-type book with 93 color pls.

[1304] **Habe, France** *Postojna Caves*. 1st ed. npop: np, **1973**. 100p.

Beautiful collection of color photographs from the commercial cave located in Yugoslavia.

2nd ed. Postojna, YU: Postojnska jama, **1979**. illus: b&w: photos; partly col: maps; col: photos; [114].

94 pls. Minimal text.

3rd ed. Lenarčič, Simon (trans); Birelli (photo); Habe, France (photo); Kadrnka, Oldrich (photo). Postojna, SI: Postojnska jama, **1982**. illus: b&w: drawings, repros; partly col: maps; col: photos; front: color photo; [90].

94 numbered illustrations.

4th ed. **1989**. illus: b&w: photos; partly col: maps; col: photos; [71].

60 illus. Minimal text. 5000 copies.

[1305] **Habe, France** *The Postojna Caves and other tourist caves in Slovenia*. 1st ed. Postojna, YU: Postojnska jama, **1977**. illus: b&w: maps, repros; partly col: maps; col: photos; 1 color map folded in back; 82p, [1], 16p pls; bibl; 19 cm.

Translation of Postojnska jama.1979 version may exist.

3rd ed. Torkar, Rado (trans); Torkar, Vivienne (trans). Postojna, SI: "Postojnska jama" Tourist and Hotel Association, **1985**. illus: b&w: maps, repros; col: photos; partly col fold-out map at end; 82p, 24p pls; bibl.

[1306] **Habe, France** *The Postojna Grottes with the Planina and Predjama caves*. Golias, Jamko (trans). Postonja, YU: [Postojnska jama], **1972**. illus: b&w: maps, photos, repros; col: maps; fold-out map at rear; colored map on rear lining page; 69p, [2], 16p pls; bibl; 18.5 x 12 cm.

Cover title: The Postojna Grottos.

[1307] **Habe, France** *The Postojna Grottoes*. Lenarčič, Simon (trans). Postojna, YU: Cave of Postojna, **1973**. illus: b&w: photos, repros; partly col: maps; 1 map folded and loose; [113].

Nice little pocket size softbound booklet; quite a few excellent black and white photos. Minimal text. 30.000 copies.

[1308] **Habe, France** *Predjama: the castle and the cave*. Torkar, Rado (trans); Torkar, Vivienne (trans). Postojna, YU: Postojnska jama, Tourist and Hotel Organization, **1980**. illus: b&w: repros; col: photos; 48p, [33]; bibl.

[1309] **Habe, France** *Show caves in Slovenia - Guide*. Šaja, S.; Šlenc, S. [Ljubljana, YU]: Karst Association for Environment Preservation at Postojna, **1978**. illus: partly col: maps; col: photos; 31p; bibl, gloss.

[1310] **Hack, John Tilton**, 1913– *Geology of Luray Caverns Virginia*. Durlo, Leslie H. 1962 ed. Charlottesville, VA, US: Virginia Division of Mineral Resources, **1962**. Virginia Division of Mineral Resources Report of Investigations 3. illus: b&w: charts, drawings, maps, photos; 1 map in pocket on rear lining; viii, 43p; bibl; 23 cm.

1977 ed. bibl.

[1311] **Hacker, Bradley (ed)** *Caves of Gunung Buda: report of the joint Sarawak Forest Department and USA caving expedition to Sarawak, Malaysia 1995*. Bunnell, David Edward, 1952–(layout, prepress). npop: Buda Caves Project, [**1996**]. illus: b&w: maps, photos, tables; 4 fold-out maps, 1 loose folded map; 48p.

Cover title: Caves of Gunung Buda, Sarawak, Malaysia 1995.

[1312] **Hadingham, Evan** *Secrets of the Ice Age: the world of the cave artists*. New York, NY, US: Walker and Company, **1979**. illus: b&w: charts, drawings, graphs, maps, photos; viii, 342p; bibl, gloss, index; 24 cm.

Illus.

[1313] **Halaburda, Sue** *Mammoth Cave*. Corrie, George B.; Halaburda, Sue (illus). npop: np, **1984**. illus: b&w: drawings; 24p.

Cover title: Mammoth Cave: children's guide and coloring book. Printed by Master Printers, Inc. Juvenile.

[1314] **Halbert, Erik** *The history of the Sydney Speleological Society 1954-1994*. Bonwick, John; Members of the Sydney Speleological Society (photo). Broadway, NSW, AU: Sydney Speleological Society, Nov, **1994**. Sydney Speleological Society Occasional Paper No 12. illus: b&w: photos, ports, repros; xii, 129p; app, bibl.

[1315] **Haley, Boyd Raymond**, 1922– *Mineral resources of the Belle Starr Caves wilderness study area, Sebastian and Scott Counties, Arkansas*. Earhart, Robert L.; Stroud, Raymond B. [Denver, CO, US]: U.S. Geological Survey, **1980**. Open-File Report United States Department of the Interior Geological Survey 80-356. illus: b&w: maps; 1 map on folded leaf in pocket; ii, 19p; bibl; 28 cm.

[1316] **Halladay, Orlynn J.** *The Lehman Caves story*. Peacock, Var Lynn. Baker, NV, US: Lehman Caves Natural History Association, **1972**. illus: b&w: drawings, maps, photos; [iv], 27p; 18 x 17 cm.

Published in cooperation with the National Park Service. Thorough report, history.

[1317] **Hallberg, George R.** *Sinkholes, hydrogeology, and ground-water quality in northeast Iowa*. Hoyer, Bernard E. Iowa City, IA, US: Iowa Geological Survey, May, **1982**. Open-file report Iowa Geological Survey no 82-3. illus: b&w: maps; xi, 120p; bibl; 28 cm.

Illus.

The physical setting in both the Karst and Shallow-Bedrock regions present potential hazards for groundwater contamination. Any management strategies developed for protection of these water resources must consider both of these settings, which in total constitute about 6,800 square miles of land overlying important bedrock aquifers.

[1318] Halliday, William R., 1926– *Adventure is underground: the story of the great caves of the West and the men who explore them.* New York, NY, US: Harper & Brothers, Publishers, **1959**. illus: b&w: drawings, maps, photos; xviii, 206p, 32p pls; gloss, index; 22 cm.

The story of caves in the western half of the United States including several legendary "lost caves." Includes chapters on Neff Canyon Cave, Utah (the deepest in USA at time of publication); Carlsbad Caverns, N.M.; Calif caves; and lava tubes and ice caves. A list of U.S. commercial caves is included.

[1319] Halliday, William R., 1926– *American caves and caving: techniques, pleasures, and safeguards of modern cave exploration.* 1974 ed. New York, NY, US: Harper & Row Publishers, **1974**. illus: b&w: drawings, photos, tables; xvii, 348p; bibl, gloss, index; 24 cm.

Includes 74 b&w photographs.

Halliday was one of the few cavers to be able to convince major publishers that cave books could sell. His writing technique is clear and fits the popular press style appropriate to the publisher. This book was probably used by more beginners than any other until the publication of Caving Basics. Contains 10 chapters on the techniques, pleasures and safeguards of modern cave exploration.

1982 rev ed. 1982. bibl, gloss, index.

Barnes & Noble Books edition.

[1320] Halliday, William R., 1926– *Ape Cave and the Mount St. Helens apes.* Larson, Charles Victor, 1928– (photo). Vancouver, WA, US: ABC Printing & Publishing, **1983**. illus: b&w: maps, photos; map on front lining page; bibl on rear lining page; 24p, [1]; bibl; 22 cm.

[1321] Halliday, William R., 1926– *The basic geology of Neff Canyon Cave, Utah ; The speleogenesis of Neff Canyon Cave, Utah.* [UT, US]: Salt Lake Grotto, National Speleological Society, Jan, **1954**. Technical note National Speleological Society Salt Lake Grotto #4-#5. 8p, [1]; bibl; 28 cm.

Title from cover. Illus.

[1322] Halliday, William R., 1926– *Caves and associated features of open vertical volcanic conduits of the Kaupulehu lava flows xenolith nodule beds, Haulalai Volcano, Hawaii: basic speleological considerations.* Werner, Marlin Spike; National Speleological Society. [Hawaii, HI, US]: Hawaii Speleological Survey of the National Speleological Society, Dec, **1992**. Report (Hawaii Speleological Survey) no 92-05. 14p; bibl; 28 cm.

Title from cover. Illus.

[1323] Halliday, William R., 1926– *Caves and associated features of open vertical volcanic conduits of the Kaupulehu lava flows xenolith nodules beds, Hualalai Volcano, Hawaii: preliminary report.* Werner, Marlin Spike. [Hawaii, HI, US]: Hawaii Speleological Survey of the National Speleological Society, Apr, **1992**. Report Hawaii Speleological Survey of the National Speleological Society #92-01. 10p, [3]; bibl; 28 cm.

Title from cover. Illus.

[1324] Halliday, William R., 1926– *Caves of California: a special report of the Western Speleological Survey in cooperation with the National Speleological Society, June 1962.* Western Speleological Survey National Speleological Society. Seattle, WA, US: Western Speleological Survey, **1962**. illus: b&w: maps, photos; [vii], 194p, [51], 16p pls; add, bibl, errat, index; 28 cm.

Limited edition of 500. Microfilm. Berkeley: University of California, Berkeley, Library Photographic Service, [196-?]. 1 microfilm reel: negative; 35 mm.

Geological introduction, then catalog of approximately six hundred known caves in the state. Descriptions, locations to the section (square mile), references. About fifty cave maps. Publication coincided in time and space with birth of cave-secrecy movement, hence controversial. An extremely high demand, difficult to find book.

[1325] Halliday, William R., 1926– *Caves of Washington.* Olympia, WA, US: State Printing Plant, **1963**. Washington (State) Department of Conservation. Division of Mines and Geology. Information Circular, no. 40. illus: b&w: drawings, maps, photos; partly col: maps; 7 maps folded in pocket in back; xiv, 132p; bibl, gloss, index; 29 cm.

Survey of 110 Washington caves. 92 figs, 9 pls.

Catalog of known caves in the state, including limestone caves and (mostly) lava tubes; shelters, talus caves, etc. explicitly excluded. About 110 caves described and located to the section (square mile). Forty cave maps.

[1326] Halliday, William R., 1926– *A consideration of the effect of surface hot-cold cycles on caves.* US: Western Speleological Survey, Oct, **1958**. Western Speleological Survey Miscellaneous Series Bulletin no 4; W.S.S. serial no12. 10p; bibl; 28 cm.

Title from cover. Illus.

[1327] Halliday, William R., 1926– *Considerations of Lower Uilani Cave for inclusion in the Hawaii Natural Area Reserve System.* Hilo, HI, US: Hawaii Speleological Survey of the National Speleological Society, Apr, **1995**. Report Hawaii Speleological Survey of the National Speleological Society no 95-03. illus: b&w: maps; [1] folded leaf of plates; 13p, 1p pls; bibl; 28 cm.

Title from cover.

[1328] Halliday, William R., 1926– *Depths of the earth: caves and cavers of the United States.* New York, NY, US: Harper & Row, **1966**. illus: b&w: drawings, maps, photos; map on lining pages; b&w photo; xiii, 398p; bibl, gloss, index; 22 cm.

A very readable and interesting account of cave exploration at specific sites and/or states or regions throughout the USA, covering work to the mid-1960s. It touches on many of caving's most famous caves and stories.

[1329] Halliday, William R., 1926– *Depths of the earth: caves and cavers of the United States.* rev and enl ed. **1976**. illus: b&w: maps, photos; col: maps; xv, 432p; bibl, gloss, index; 24 cm.

The second edition has several different chapters from the first, including one on caves in the Northeastern United States. Some people joke that Halliday added the chapters so collectors would have to own both editions.

[1330] Halliday, William R., 1926– *History and publications of the Western Speleological Survey.* Western Speleological Survey, National Speleological Society. Seattle, WA, US: The Association, Feb, **1978**. Western Speleological Survey serial no 57. illus: b&w: repros; 16p, [4]; bibl; 28 cm.

Title from cover.

[1331] Halliday, William R., 1926– *Initial inventory of named caves and related features and cave-related place names in Hawaii.* [Honolulu?, HI, US]: Hawaii Speleological Survey of the National Speleological Society, Oct, **1995**. Report: Hawaii Speleological Survey of the National Speleological Society no 95-05. 33p; 28 cm.

Title from cover.

[1332] Halliday, William R., 1926– *Introduction to Hawaiian caves: field guide for the 6th international symposium on volcano-speleology: Hilo, HI.* npop: np, Aug, **1991**. illus: b&w: maps; iii, 46p.

[1333] Halliday, William R., 1926– *Littoral caves of ancient Lake Bonneville.* [UT, US]: Salt Lake Grotto, National Speleological Society, Sep, **1954**. Technical note (National Speleological Society. Salt Lake Grotto) #19. i, 17p; bibl; 28 cm.

Title from cover.

[1334] Halliday, William R., 1926– *Outline of major biological conclusions from 1980-81 Mount St. Helens eruptions.* Crawford, Ronald L., 1947–. npop: np, May, **1981**. Western Speleological Survey serial no 66; Washington Speleological Survey bulletin 19. 10p; bibl; 28 cm.

Title from cover. Presented in part at the annual meeting of the Association of American Geographers, Los Angeles, April 20, 1981.

[1335] Halliday, William R., 1926– *The Paradise Ice Caves, Mount Rainier National Park, Washington.* Anderson, Charles H. npop: W. R. Halliday & C. H. Anderson, **1972**. illus: b&w: maps, photos; map on rear lining page; 26p; bibl; 22 cm.

[1336] Halliday, William R., 1926– *A preliminary report on the caves of eight northern California counties.* [Seattle, WA, US]: California Speleological Survey, **1961**. Western Speleological Survey Serial no 20; Bulletin California Speleological Survey, Mother Lode Section no 2. 23p; 28 cm.

Preliminary version of part of The Caves of California. Title from cover.

[1337] Halliwell, Ric (ed) *Gouffre Berger: August 1994.* npop: Craven Pothole Club, c**1995**. illus: b&w: drawings, maps, photos; 80p, [4]; 29 x 21 cm.

[1338] Hamer, John Lewis *The falls and caves of Ingleton.* Clapham, Lancaster, GB: Dalesman, **1946**. Dalesman Pocket Book #2. illus: b&w: drawings, maps, tables; fold-out map; 28p; gloss.

1951 edition may exist.

[1339] Hamer, John Lewis *The moors and fells of Ingleton.* Lancaster, GB: Dalesman Publishing Co, **1951**. Daleman Pocket Book no. 9. illus: b&w: maps; 20p; 22 cm.

Guide to Ingleton cave surroundings. Contains maps and pls.

[1340] Hamilton, Elizabeth, 1900– *The first book of caves.* Davis, Bette J. (illus). 1st ed. New York, NY, US: Franklin Watts, Inc, **1956**. First Books no 54. illus: b&w: drawings, maps; partly col: drawings; drawing in lining page partly col; 63p; index; 23 cm.

Both 1955 and 1956 listed as publication dates in OCLC. Published by Bailey & Swinfen, 1957. Juvenile. Helpful information for the beginner. For the young caver, all about caves.

2nd ed. npop, 1964: E. Ward. illus: b&w: photos; 65p.

Juvenile.

[1341] Hamilton, Holman *The Cave of the Winds' and the compromise of 1850.* npop: np, **1957**. 23p; 24 cm.

Caption title. Reprint from The Journal of Southern History, Volume XXIII, August 1957, Number 3, pp 331-353.

[1342] Hamilton-Smith, Elery (ed) *Abstracts of papers: karst studies seminar Naracoorte.* Hamilton, Victoria, AU: Regolith Mapping, February, **1996**. illus: b&w: drawings; [vi], 37p, [1]; 30 x 21 cm.

Program listing abstracts.

[1343] Hamilton-Smith, Elery *Australian cave fauna: notes on collecting.* Montmorency, VIC, AU: Elery Hamilton-Smith, Oct, **1962**. illus: b&w: drawings; [i], 19p; bibl.

[1344] Hamilton-Smith, Elery *Batu Caves, Kuala Lumpur - towards a plan of management.* Kuala Lumpur, MY: Malayan Nature Society, Jun, **1989**. illus: partly col: maps; [ii]; bibl.

[1345] Hamilton-Smith, Elery *Cave Reserves of the Katherine Area.* Champion, C. Randell; Robinson, Lloyd, N. Broadway, NSW, AU: Australian Speleological Federation, Sep, **1974**. illus: b&w: drawings, maps, overlays, photos; 1 map folded in rear pocket; some maps folded in text; [vii], 68p, [8]; bibl.

A report prepared for the Northern Territory Reserves

Board. Copy examined was a photocopy.

[1346] Hamilton-Smith, Elery (ed) *Caving in Australia.* Broadway, NSW, AU: Australian Speleological Federation; University of Technology Students' Union, **1958**. 23p; bibl.

Editor not listed on title page. No illus.

[1347] Hamilton-Smith, Elery (prep) *Cutta Cutta Caves Nature Park: draft plan of management.* Holland, Ernst (prep); Mott, Kevin (prep); Spate, Andrew P. (prep). npop: Australasian Cave Management Association, Jan, **1989**. illus: b&w: maps, photos, tables; 7 maps folded in text; vi, 102p; app, bibl, gloss.

Prepared for the Conservation Commission of the Northern Territory. Copy examined was a photocopy.

[1348] [Hamilton-Smith, Elery] *Determining an environmental and social carrying capacity for Jenolan Caves Reserve: applying a visitor impact management system.* Jenolan Caves Reserve Trust. Surry Hills, NSW, AU: Manidis Roberts Consultants, Sep, **1994**. illus: b&w: charts, drawings, maps, tables; 1 chart folded in text; [v], 23p, [17]; app.

Author not listed on publication.

[1349] [Hamilton-Smith, Elery] *Draft management plan: Naracoorte Caves Conservation Park: South East South Australia.* Programmes Branch, National Parks and Wildlife Service (comps); National Parks & Wildlife Service. Adelaide, SA, AU: Department of Environment and Planning, **1966**. illus: b&w: charts, drawings, graphs, maps, photos, ports, tables; xv, 116p; app, bibl.

Author not listed on publication.

[1350] Hamilton-Smith, Elery *Mt. Etna & the caves: a plan for action.* Champion, Randell; Queensland Conservation Council. St Lucia, QLD, AU: University of Queensland Speleological Society, **1976**. illus: b&w: drawings, maps; xix, 147p; 21 x 30 cm.

A report for the Queensland Conservation Council. Cover title: Mt. Etna & the Caves; caption page title: Mt. Etna & the caves. This is a conservation book written to try and save the Mt. Etna area, Queensland, from development.

[1351] Hamilton-Smith, Elery (tm ldr) *Proposed developments at Waitomo Caves.* Melbourne, VIC, AU: Australasian Cave Management Association, Nov, **1988**. illus: b&w: charts, maps; [iii], 30p, [1]; app.

[1352] Hamilton-Smith, Elery *The Regional Caves Interpretive Centre: a planning review report for the Augusta Margaret River Tourist Bureau.* [Melbourne, VIC, AU]: Rethink Consulting P/L, Oct, **1994**. illus: b&w: tables; [i], 12p.

[1353] Hammatt, Hallett H. *Archaeological assessment and sensitivity map for the Pohakuloa training area (PTA), Hawaii Island, State of Hawaii.* Shideler, David W. npop: Cultural Surveys Hawaii, Apr, **1990**. Archeology Hawaii Lava Tubes. illus: b&w: maps, tables; maps in rear pocket; iv, 60p.

Prepared for Richard Sato and Associates. Maps in back pocket not in copy examined.

[1354] Hanihara, Kazuro, 1927- *Paleolithic site of Douara Cave and paleogeography of Palmyra Basin in Syria.* Sakaguchi, Yutaka, 1929–. [Tokyo, JP]: University of Tokyo Press, **1978**. Bulletin: University Museum University of Tokyo no 14 Bulletin Tokyo Daigaku. Sogo Kenkyu Shiryokan no 14 etc.

Includes bibliographies.

[1355] Hannan, Thomas *The beautiful Isle of Mull with Iona and the Isle of Saints.* Edinburgh, GB: R. Grant and Son, **1926**. maps on lining pages; front; 211p; bibl; 20 cm.

Covers the caves of the island.

[1356] Hanwell, James, 1935– *The great storms and floods of July 1968 in Mendip.* Newson, Malcolm David, 1945–. Priddy, Wells, Somerset, GB: The Wessex Cave Club, Oct, **1970**. Wessex Cave Club Occasional Publication Series 1 number 2. illus: b&w: drawings, graphs, maps, photos, tables; viii, 72p; app; 21 cm.

Second impression April 1984. 12 pls, 18 figs. OCLC lists place of pub as Pangbourne. Copy examined listed author as Hanwell, J. D.

[1357] Hardcastle, Raymond A. *Caves and shelters in Bonanza King Canyon.* npop: Mojave Division of the California Speleological Survey, Apr, **1977**. Western Speleological Survey Serial no 54. 14p.

Pagination includes covers.

[1358] Hardingham, B. G. *Living in caves.* npop: np, **1943**. 126p.

Descriptions of cave dwellings. Illus.

[1359] Hargrave, Lyndon Lane, 1896– *Feathers from Sand Dune Cave: a Basketmaker cave near Navajo Mountain, Utah.* Museum of Northern Arizona. 1967 ed. npop: np, **1967**. illus: b&w: maps, ports; ix, 99p, 21p pls; bibl, gloss, index; 30 cm.

Illus. "A research project conducted for the Museum of Northern Arizona." "Completed under contract no. 14-10-3:931-2 with the National Park Service."

1970 ed. Museum of Northern Arizona. Flagstaff, AZ, US: Northern Arizona Society of Science and Art, Inc, **1970**. Technical Series no 9. illus: b&w: drawings, maps, photos, tables; ix, 52p; app, bibl, gloss, index; 30 cm.

"A research project conducted for the Museum of Northern Arizona." "Completed under contract no. 14-10-3:931-2 with the National Park Service." 45 figs; 3 tables.

[1360] Harlan, Richard, 1796-1843 *Description of the fossil bones of the Megalonyx, discovered in White Cave, Kentucky.* [Philadelphia, PA, US]: np, **1831**. 27p.

Reprinted from Trans Acad Nat Sci Philad, vol vi. pls (some fldg).

[1361] Harpers Ferry Center. Division of Museum Services *Mammoth Cave National Park: collection manage-*

ment plan. Harpers Ferry, WV, US: The Division, **1976**. illus: b&w: drawings; 150p; 27 cm.

Title from cover.

Preservation and management plan for artifacts recovered from the large caves and surrounding areas in Mammoth Cave National Park.

[1362] Harrington, Mark Raymond, 1882-1971 *Gypsum cave, Nevada, report of the second sessions expedition.* Los Angeles, CA, US: Southwest Museum, Apr, **1933**. Southwest Museum Papers number 8. illus: b&w: charts, drawings, maps, photos; col: drawings; map and drawings folded in text; front: col drawing; ix, 197p, 19p pls; 27 cm.

1963 reprint may exist. 19 pls, 77 figs.

Archaeological report about the late Pleistocene archaeological remains, including ground sloth, found at Gypsum Cave.

[1363] Harris, Shane R. *UEA speleological expedition northern Turkey: August / September 1992.* [GB]: University of East Anglia Fell and Cave Club, **c1992**. illus: b&w: maps; col: photos; three folded maps in text; 54p.

[1364] Harris, Stephen *Caves of New Caledonia: report of the 1975 Australian Expedition.* Harris, S. [Hobart, TAS, AU]: Harris, Gillieson, Gleeson, Landsberg, **1976**. illus: b&w: drawings, maps, photos; partly col: maps; 1 map folded in text; [iv], 68p; bibl; 30 cm.

6 weeks on the island.

[1365] Harris, Yvonne Dorothy, 1935– *The caves of Chlingitani.* Whitehorse, Yukon, CA: Harris Pub, **1990**. 127p; 22 cm.

Illus.

[1366] Harrison, David Lakin *The world of American caves.* Chicago, IL, US: Reilly & Lee Books, **1970**. illus: b&w: photos; [viii], 152p, 8p pls; gloss, index; 22 cm.

10 chapters and 8 b&w photos.

[1367] Harrisonburg-Massanutten Corp (Harrisonburg, Va) *About Massanutten caverns; the jeweled chamber that hides the secret of Nakwisi-Gatusi.* [Harrisonburg, VA, US]: np, **1928**. [xxxii].

Title from cover.

[1368] Harrisonburg-Massanutten Corp (Harrisonburg, Va) *Massanutten Caverns, Harrisonburg, Va. The most beautiful and unusual caverns yet discovered....* [Staunton, VA, US]: The McClure Co, Inc, **1925**. [xvi].

Title from cover.

[1369] Harrisonburg-Massanutten Corp (Harrisonburg, Va) *Massanutten Caverns, Harrisonburg, Va.: the most beautiful and unusual caverns yet discovered, in the heart of the Shenandoah Valley of Virginia.* npop: Garrison Press, **1940**. illus: b&w: maps; [12]; 24 cm.

Title from cover. Bears ill (col) cover. Text on covers. OCLC lists both 1940 and 1949 as dates of publication. Illus.

[1370] Harrisonburg-Massanutten Corp (Harrisonburg, Va) *The wonderful Massanutten Caverns in the Shenandoah Valley, Harrisonburg, Va.* House, L.B. [Brooklyn, NY, US]: Albertype Co, **nd**. 1p, 12p pls.

[1371] Hart, William J. *A wilderness plan for Mammoth Cave National Park and the surrounding region.* Boardman, Walter S. Washington, DC, US: National Parks Association, Sep, **1967**. illus: b&w: maps; 13p, 3p pls.

"Prepared in anticipation of public hearings to be held by the U. S. National Park Service on the matter of establishing wilderness areas within Mammoth Cave National Park."

[1372] Harter, J. W. *Classification of lava tubes.* Harter, Russell G. npop: np, May, **1970**. illus: b&w: photos; 11p.

[1373] Harter, Russell George, 1947– *Bibliography on lava tube caves.* npop: [Western Speleological Survey], Jun, **1971**. Western Speleological Survey Bulletin no 44; Bulletin #14 Miscellaneous Series. 52p; errat.

Limited edition. Loose errata sheet dated Jul 1975.

[1374] Harter, Russell George, 1947– *Bibliography on lava tube caves, supplement.* npop: np, **1972**. 14p.

Title from cover.

[1375] Hartley, Howard W. *Tragedy of Sand Cave.* Newman, George R. (intro). 2nd ed. Louisville, KY, US: The Standard Printing Company Incorporated, **1925**. illus: b&w: ads, drawings, photos, ports; 145p, [4]; 20 cm.

First edition is rare. Microfilm edition: Atlanta, Ga.: SOLINET, 1993. 1 microfilm reel ; 35 mm.

An account of the attempted Floyd Collins rescue. It gives a contemporary view of the accident and attempts of rescue. It contains 72 photographs of newspaper type quality. The book is printed on very fragile paper. Another way people made money off of Floyd even after his death.

[1376] Hartwig, Georg, 1813-1880 *Marvels under our feet: from "the subterranean world".* London, GB: Longmans, Green, and Co, **1888**. 144p.

22 engravings.

[1377] Harvey, Edward Joseph, 1916– *Application of thermal imagery and aerial photography to hydrologic studies of karst terrane in Missouri.* Williams, James Hadley, 1929–; Dinkle, T. R. Rolla, MO, US: US Geological Survey, **1977**. Water Resources Investigations; 77 - 16. illus: b&w: drawings, maps, photos; vi, 58p; bibl; 26 cm.

Prepared in cooperation with the Missouri Department of Natural Resources.

[1378] Harvey, Edward Joseph, 1916– *Hydrology of carbonate terrane, Niangua, Ossage Fork, and Grandglaize basins, Missouri.* Skelton, John; Miller, Don E. Rolla, MO, US: Missouri Department of Natural Resources, Division of Geology and Land Survey, **1983**. Water Resources Report (Missouri Division of Geologic Survey and Water Resources); 35. illus: b&w: charts, drawings, maps, photos; col: photos; vi, 132p; 28 cm. cm.

[1379] Hassemer, Jerry Herman, 1934– *Cave development in Marble Canyon, Grand Canyon National Park, Arizona*. Lakewood, CO, US: Hassemer, Mar, **1980**. illus: b&w: drawings, maps, photos; [i], 36p, [6]; app, bibl.

Photocopy examined.

[1380] Hassemer, Jerry Herman, 1934– *Caving basics*. National Speleological Society Caver Training Committee. Huntsville, AL, US: The National Speleological Society, **1982**. illus: b&w: drawings, maps, photos, tables; viii, 125p; app, bibl, gloss, index; 28 cm.

The NSS's first attempt at instructing members in cave safety. It turned out to be a major step in the direction of the Society's publishing efforts.

[1381] Hassemer, Jerry Herman, 1934– *Caving basics: a comprehensive manual for beginning cavers*. National Speleological Society Caver Training Committee; Rea, George Thomas, 1934–(ed). 2nd ed?. Huntsville, AL, US: National Speleological Society, **1987**. illus: b&w: drawings, maps, photos; viii, 128p; app, bibl, gloss, index; 28 cm.

[1382] Hassemer, Jerry Herman, 1934– *Colorado caves and karst 1974*. Lakewood, CO, US: Jerry H. Hassener, Feb, **1975**. illus: b&w: charts, graphs, maps, tables; partly col: maps; 1 fold-out map in text, 1 folded map in rear pocket; [iv], 43p; bibl.

[1383] Hassemer, Jerry Herman, 1934– *Colorado caves and karst 1975-1976*. Lakewood, CO, US: Jerry H. Hassener, Apr, **1977**. illus: b&w: maps, photos, tables; map on rear lining page; [iv], 31p; bibl.

[1384] Hastings, Valerie *Jo and Coney's Cavern*. npop: Parrish, **1959**. 172p; 20 cm.

For adolescents. Juvenile.

[1385] Hatherley, Paul (comp) *Symposium on surveying caves*. Bristol, GB: British Cave Research Association, **1987**. Cave Science Transactions of the British Cave Research Association v 14 no 2 August 1987. illus: b&w: charts, drawings, graphs, maps, photos, tables; text on front lining page; 42p; bibl; 30 cm.

Title from cover. Illus. Symposium on Surveying Caves (1986: University of Sheffield, England).

[1386] Hatt, Robert Torrens, 1902– *Faunal and archeological researches in Yucatan caves*. Fisher, Harvey I.; Langebartel, Dave A.; Brainerd, George W. Bloomfield Hills, MI, US: Cranbrook Institute of Science, **1953**. Cranbrook Institute of Science Bulletin no 33. illus: b&w: drawings, maps, photos, tables; [iii], 119p, 23p pls; bibl; 23 cm.

12 pls.

[1387] Hauer, Peter Marshall, 1945-1975 *Caves of Massachusetts*. Altoona, PA, US: Speece Productions, **1969**. illus: b&w: drawings, maps; [i], 62p; index; 30 cm.

"Geospeleology of Massachusetts [by] Arthur N. Palmer"; leaves 4-6. Toc: "A guide to the known caves of Massachusetts with descriptions, locations, history, and legends."

Catalog of known caves, with descriptions and locations. 124 named caves, but only about 30 are solutions caves. Some maps.

[1388] Havard, Ward L. *The Orient Cave, Jenolan Caves*. NSW, AU: Government Tourist Bureau, **nd**. illus: b&w: maps, photos; 31p, [1].

[1389] Havard, Ward L. *The Romance of Jenolan Caves*. Sydney, NSW, AU: Reproduced by NSW Department of Leisure, Sport and Tourism by Osprey Enterprises, **nd**. illus: b&w: maps, photos, ports; map on lining page; 48p.

Read before the Royal Australian Historical Society, November 28, 1933. Different reprint editions exist.

[1390] Hawkins, R. J. (comp) *Australasian speleo map index no. 1*. Hawkins, L. J. (comp). Broadway, NSW, AU: Sydney Speleological Society, **1972**. vi, 33p; bibl.

No illus.

[1391] Hawkins, Thomas *The caves of Honduras: a first list*. McKenzie, Andrew. npop: British Tea Cavers, **1993**. illus: b&w: drawings; xiii, 59p; bibl, index.

..."provides references to 154 caves, and a total of 370 cave related records."

Database of known caves, with very terse descriptions, locations, and references to bibliography. No maps. Lists 154 caves; admittedly extremely preliminary list.

[1392] Hay, Deborah *Kahaluu data recovery project: excavations at site 50-10-37-7702, Kahaluu Habitation Cave, land of Kahaluu, North Kona, Island of Hawaii*. Haun, Alan E.; Rosendahl, Paul Harmer; Severance, Craig J.; Hawaii Island Office of Housing and Community Development. [Hilo, HI, US]: np, Oct, **1986**. illus: b&w: maps, photos; Various pagination; 28 cm.

Prepared for County of Hawaii, Office of Housing and Community Development. Contract No. 82-146. October 1986 Report 61-022084.

[1393] Hay, William Perry, 1872– *Observations on the crustacean fauna of Nickajack cave, Tennessee, and vicinity*. npop: [United States. National Museum], **1903**. Proceedings of the United States National Museum v 25. 23p; 24 cm.

Issued September 23, 1902, pp 417-439.

[1394] Hay, William Perry, 1872– *Observations on the crustacean fauna of the region about Mammoth Cave, Kentucky*. npop: United States National Museum, **1903**. Proceedings of the United States National Museum v 25. 14p; 24 cm.

Issued September 12, 1902, pp 223-236.

[1395] Hayami, Itaru, 1933– *Submarine cave bivalvia from the Ryukyu Islands: systematics and evolutionary significance*. Kase, Tomoki. Tokyo, JP: University Museum, University of Tokyo, **1993**. Bulletin University

Museum University of Tokyo no 35 0910- 481; Bulletin Tokyo Daigaku Sogo Kenkyu Shiryokan no 35. illus: b&w: maps; vi, 133p, 1p pls; bibl; 26 cm.

Illus (some col).

[1396] Hayllar, Bruce (proj offr) *The development of leisure and educational resources at Jenolan Caves Resort.* Brown, Peter (proj offr). npop: The Kuring-gai College of Advanced Education Centre for Leisure and Tourism Studies, Mar, **1984**. illus: b&w: charts, drawings, photos, repros; [xlvi]; app.

A report for the Department of Leisure Sport and Tourism. Copy examined was a photocopy.

[1397] Hayllar, Bruce (proj offr) *Jenolan Caves Resort teachers kit.* Brown, Peter (proj offr). [NSW, AU]: Kuring-gai College of Advanced Education Centre for Leisure & Tourism Studies, Jan, **1985**. illus: b&w: charts, drawings, maps, photos, tables; ii, 34p, [40]; bibl.

Prepared on behalf of the New South Wales Department of Leisure, Sport and Tourism. Accompanied by a set of student activity sheets.

[1398] Haynes, Caleb Vance, 1928– *Geochronology of Sandia Cave.* Agogino, George. Washington, WA, US: Smithsonian Institution Press, **1986**. Smithsonian Contributions to Anthropology no 32. viii, 32p; bibl; 28 cm.

1 microfiche ; 11 x 15 cm. Also reprinted in 1991 on microfilm by UMI. Illus.

[1399] Hazelton, Mary (ed) *British hypogean fauna and biological records of the Cave Research Group part IX (1963).* [GB]: Cave Research Group of Great Britain, **1965**. illus: b&w: photos, tables; 40p, 1p pls; bibl.

Cover says Number 1, evidently intended to be the first of a new series. Subsequent issues are part of Transactions of the Cave Research Group of Great Britain.

[1400] Hazelton, Mary *Fauna collected from caves as recorded in the C.R.G. fauna records: part III (1947).* GB: Cave Research Group, Oct, **1958**. Cave Research of Great Britain, Biological Supplement. illus: b&w: tables; 20p.

[1401] Hazelton, Mary *Fauna collected from caves, mines and wells as recorded in the C.R.G. fauna records: part IV (1948-1949).* GB: Cave Research Group, Feb, **1959**. Cave Research of Great Britain, Biological Supplement. illus: b&w: tables; 17p; errat.

Contains errata to part III.

[1402] Hazelton, Mary *Fauna collected from caves, mines and wells as recorded in the C.R.G. fauna records: part V (1950-1953).* GB: Cave Research Group, Feb, **1960**. Cave Research of Great Britain, Biological Supplement. illus: b&w: tables; 21p; errat.

Contains errata to part IV.

[1403] Hazelton, Mary *Fauna collected from caves, mines and wells as recorded in the C.R.G. fauna records: part VI (1954-1955-1956).* GB: Cave Research Group, Sep, **1960**. Cave Research of Great Britain, Biological Supplement. illus: b&w: tables; 31p; errat.

Contains errata to part V.

[1404] Hazelton, Mary *Fauna collected from caves, mines and wells as recorded in the C.R.G. fauna records: part VII (1957-1959).* GB: Cave Research Group, Nov, **1961**. Cave Research of Great Britain, Biological Supplement. illus: b&w: tables; 64p; errat.

Contains errata to part VI. Title on cover includes "C.R.G. recent invertebrate fauna records".

[1405] Hazelton, Mary *Fauna collected from caves, mines and wells as recorded in the C.R.G. fauna records: part VIII (1960-1962).* GB: Cave Research Group, Sep, **1963**. Cave Research of Great Britain, Biological Supplement. illus: b&w: tables; 41p; errat.

Contains errata to part VII.

[1406] [Hazelton, Mary (ed)] *Hypogean fauna and biological records 1964-1966.* GB: Cave Research Group of Great Britain, Sep, **1967**. Transactions of the Cave Research Group of Great Britain vol 9 no 3. illus: b&w: maps, photos, tables; [ii], 128p, 2p pls; bibl.

[1407] [Hazelton, Mary (ed)] *Hypogean fauna, biological records 1967.* GB: Cave Research Group of Great Britain, Dec, **1968**. Transactions of the Cave Research Group of Great Britain v 10 no 3. illus: b&w: drawings, maps, photos, photomicrographs, tables; text on rear lining page; 72p; bibl, errat.

[1408] [Hazelton, Mary (ed)] *Hypogean fauna, biological records 1968.* GB: Cave Research Group of Great Britain, Mar, **1970**. Transactions of the Cave Research Group of Great Britain v 12 no 1. illus: b&w: maps, photos, tables; text on rear lining page; [i], 91p, 4p pls; bibl.

Title from cover.

[1409] Hazelton, Mary (ed) *Irish hypogean fauna and Irish biological records, 1856-1971.* Hazelton, Mary (ed). [Combwich], GB: British Cave Research Association for the Cave Research Group of Great Britain, **1974**. Transactions of the Cave Research Group of Great Britain v 15 no 4. illus: b&w: maps, photos, photomicrographs, tables; text on front; [ii], 64p; bibl; 30 cm.

Title from cover. pp. 191-254. Illus.

[1410] Hazslinszky, Tamás (ed) *International conference Baradla 150: field-trip guide: Budapest-Aggtelek, 26-29.08.1975.* [Budapest, HU]: Hungarian Speleological Society, **1975**. illus: b&w: graphs, maps, tables; 145p, 11p pls.

[1411] Hazslinszky, Tamás (auth, ed, trans) *Show caves of Hungary.* Budapest, HU: 10th International Congress of Speleology, **1989**. Field Trip Guide D6. illus: b&w: maps; 36p.

[1412] Hazslinszky, Tamás *Colourful world of caves.* Rácz, Beatrix (trans); Rádai, Ödön (trans). Budapest, HU: Technológia, **1989**. illus: col: photos; [80].

Trilingual: Hungarian, German, English. 124 color photos.

[1413] Headworth, H. G. *The hydrogeology of arte-*

sian boreholes at some watercress farms in Hampshire. Worthing, Sussex, GB: Directorate of Resource Planning, Southern Water Authority, Nov, **1975**. illus: b&w: graphs, maps, photos, tables; 21p; bibl.

[1414] Heap, David *Potholing: beneath the northern Pennines.* London, GB: Routledge and Kegan Paul, Ltd, **1964**. illus: b&w: maps, photos, ports; partly col: photos; col: photos; 1 map folded after p 52; front: col photo; xvii, 206p; app, bibl, gloss, index; 23 cm.

Written by a caver, this book tells of the explorations of Yorkshire potholes. Potholes is a British term for a cave with a vertical entrance or a deep cave, usually one with many vertical drops in it.

[1415] Heap, David (comp) *Report of the British speleological expedition to arctic Norway, 1969, incorporating the work of the 1968 Hulme Schools expedition.* GB: Kendal Caving Club, **1969-1970**. illus: b&w: maps, photos, tables; 3 fold-out maps; 1 loose folded map; front: b&w photo; 37p, 5p pls.

[1416] Heath, Thomas *An abstract description and history of the Bone Caves of Creswell Crags.* Derby, GB: Wilkins and Ellis, Aug, **1879**. illus: b&w: drawings, maps, tables; 17p, 1p pls; 20 x 13 cm.

[1417] Heath, Thomas *Creswell Caves vs Professor Boyd Dawkins.* Derby, GB: Edward Clulow, Jun, **1880**. illus: b&w: repros; 37p, [3]; errat; 21.5 x 14 cm.

Controversy over geological theory.

[1418] Hedges, James *The Ice Cave at Decorah, Iowa.* Knudson, George Ellert, 1915-. [Des Monies, IA, US]: np, **1975**. Annals of Iowa 3rd ser v 43 no 2. 19p; 24 cm.

Illus, pp 113-131.

[1419] Hedinger, J. M. *A short description of Castleton in Derbyshire: its natural curiosities and mineral productions.* 16th ed. Stockport, GB, **1820**. 2 fronts: illus of Peak Cavern in 1828; 37p; 17 x 10.5 cm.

Cover title: A guide through Peak's Hole; with a description of the curiosities of Castleton. First ed c1800; 1818 =14th ed; 1820 = 16th ed; 1828 = 21st ed; other eds in [1801], [1804], [1810], and 1839.

[1420] Heizer, Robert Fleming, 1915– *Archaeological investigations in Lovelock Cave, Nevada.* Napton, Lewis K.; University of California Archaeological Research Facility. Reprint ed. Berkeley, CA, US: University of California, Dept. of Anthropology, **1970**. Contributions of the University of California Archaeological Research Facility no 10. illus: b&w: maps; 86p; bibl; 28 cm.

Illus.

An archaeological treatise about the excavations at Lovelock Cave. Data recovered helped to define the prehistoric Desert Archaic way of life.

[1421] Heizer, Robert Fleming, 1915– *The archaeology of Humboldt Cave, Churchill County, Nevada.* Krieger, Alex Dony, 1911–. 1956 ed. Berkeley, CA, US: University of California Press, **1956**. University of California Publications in American Archaeology and Ethnology v 47 no 1. 189p; bibl; 26 cm.

Illus.

An archaeological treatise about the excavations of Humbolt Cave. Data recovered helped to define the prehistoric Desert Archaic way of life.

1971 reprint ed. New York, NY, US: Kraus Reprint Co, **1971**. 2 folded maps; 23 cm.

With this is issued Kroeber, A. L. Ethnographic interpretations, 1-11. New York, 1971.

[1422] Heizer, Robert Fleming, 1915– *The archaeology of two sites at Eastgate, Churchill County, Nevada.* Baumhoff, Martin A.; Elsasser, Albert B.; Prince, E. R. 1961 ed. Berkeley, CA, US: University of California Press, **1961**. University of California publications Anthropological records v 20 no 4. 31p; bibl; 28 cm.

Illus. I. Wagon Jack shelter, by Robert F. Heizer and M. A. Baumhoff.—II. Eastgate Cave, by Albert B. Elsasser and E. R. Prince.

1976 reprint ed. Millwood, NY, US: Kraus Reprint, **1976**.

[1423] Heizer, Robert Fleming, 1915– *Massacre Lake Cave, Tule Lake Cave and shore sites.* [Washington, DC, US]: np, **1942**. illus: b&w: drawings, maps, tables; 14p; bibl; 29 cm.

Caption title. "Reprinted from Carnegie Institution of Washington Publication 538." pp 121-134

[1424] Hellenic Speleological Society *International meeting on the show caves and their problems.* Petrochilou, Anna (ed); Papadapoulis, G.R. (ed). Athens, GR: Hellenic Speleological Society, **1983**. illus: b&w: drawings, graphs, maps, photos; 319p; bibl.

Partly in Greek, French, and German. 21 papers in English. Field trip guide.

[1425] Helm, Tex *Carlsbad Caverns; National Park, New Mexico.* Carlsbad, NM, US: Caverns Supply Company, Inc, **1969**. illus: partly col: photos; col: maps, photos; [26]; 22 x 28 cm.

Copyright: Curt Teich & Co, Inc: Chicago, IL, US. Colorful booklet with color front and back covers plus 23 full page color photos. Minimal text.

[1426] Helmer, William *"There we was!....".* Austin, TX, US: University of Texas Grotto, National Speleological Society, **1961**. illus: b&w: cartoons; 21p; bibl.

2nd printing 1961. No date known for 1st printing. "This is a collection of cartoons, telling the story of the typical caving trip."

[1427] Hempel, John Charles, 1949– *Caves of Monroe County, W. Va.* Haarr, Doris (ed); Medville, Douglas Michael, 1941–(ed); Werner, Eberhard Wolfgang, 1942– (ed). Pittsburgh, PA, US: West Virginia Speleological Survey Press, **1975**. Bulletin - West Virginia Speleological Survey no 4. illus: b&w: drawings, maps, photos; partly col: photos; 7 maps folded in text, 2 folded in pocket and loose; 1 drawing folded in text; xxi, 149p, [3]; bibl, index; 28 cm.

Cover title: Caves and karst of Monroe County, W. Va. Extra sheet inserted.

One of the most cave-rich counties in the state. One of a series of catalogs by county or other portion of the state. Text descriptions with locations of about two hundred caves, with 77 cave maps.

[1428] Hempel, John Charles, 1949– *Come crawl into the caves of western Pennsylvania.* Pittsburgh, PA, US: Pittsburgh Grotto, **1969**. Netherworld News. illus: b&w: maps, tables; Various pagination, [89]p.

Published in two issues of Netherworld News: Aug 1969 and Oct 1969.

[1429] Hempel, John Charles, 1949– *Report of the 1982 expedition to Barra Honda National Park.* Schomer, Barb (ed); Randall, Bru (ed); Werner, Eberhard Wolfgang, 1942–(ed). Huntsville, AL, US: [NSS], **1989**. National Speleological Society Costa Rica Project Bulletin 1. illus: b&w: charts, drawings, maps, photos, tables; iii, 52p.

[1430] Henderson, Junius, 1865-1937 *Caverns, ice caves, sinkholes, and natural bridges, part I & II.* Boulder, CO, US: University of Colorado, Oct, **1932**. University of Colorado Studies vol 19 no 4 vol 20 nos 2-3. [i], 91p; bibl.

Part 1 - Caverns; Part 2: 16 caves and related phenomena; reprinted without change of pagination, pp 115-158, 359—405.

[1431] Henderson, Kent *The Wombeyan and Abercrombie caves.* Victoria, NSW, AU: Kent Henderson Publications, **1992**. illus: partly col: maps; col: photos; Partly colored maps on front and rear.; 24p.

Title on cover: The Wombeyan and Abercrombie caves of N.S.W.

[1432] Henderson, Kent, 1954– *The Buchan experience: a guide to the Buchan and Murrindal Caves - East Gippsland, Victoria.* Belmont, VIC, AU: Kent Henderson, **1985**. illus: b&w: ads, drawings, repros; partly col: drawings, maps; col: photos; maps on lining pages; 64p.

Extensively illustrated. Contains ads.

[1433] Henderson, Kent, 1954– *Jenolan: a guide to Australia's famous caves.* Gniel, Penny (illus). Belmont, VIC, AU: Kent Henderson Publications, **1990**. illus: b&w: maps; partly col: maps; col: photos; col photos and map on lining page.

[1434] Henderson, Kent, 1954– *The Naracoorte and Tantanoola Caves: a guide to the famous caves of Southeast South Australia.* Meldrum, David (photo). Belmont, VIC, AU: Kent Henderson Publications, **1986**. illus: partly col: maps; col: photos; maps on lining pages; 32p.

[1435] Henderson, Kent, 1954– *Princess Margaret Rose Caves - Western Victoria.* Meldrum, David. Belmont, VIC, AU: Kent Henderson Publications, **1987**. illus: b&w: maps, ports; [16]; 28 cm.

Title from cover. Illus (col).

[1436] Henderson, Kent, 1954– *The Wellington and Abercrombie Caves.* Belmont, VIC, AU: Kent Henderson Publications, **1988**. illus: col: maps, photos; photos on lining pages; 24p, [1].

Cover title: The Abercrombie and Wellington Caves New South Wales. Also published with cover title: The Wellington and Abercrombie Caves New South Wales.

[1437] Henderson, Kent, 1954– *The Wombeyan experience: a guide to the Wombeyan Caves New South Wales.* Newtown, Geelong, VIC, AU: Neptune Press Pty, Ltd in conjunction with Kent Henderson, **1985**. illus: b&w: drawings, maps; partly col: maps; col: photos; b&w maps on lining pages; 56p.

Limited text.

[1438] Hendricks, Albert C. *Quality of groundwater in a carbonate terrain in western Virginia.* Blacksburg, VA, US: Center for Environmental Studies, Department of Biology, Virginia Polytechnic Institute and State University, **1974**. illus: b&w: charts, drawings, maps, photos; 69p; 28 cm.

Prepared for the New River Valley Comprehensive Health Planning Council of the New River Valley.

[1439] Hendrix, Charles E. *The cave book.* Revere, MA, US: The Earth Science Publishing Company, **1950**. Earth Science Institute Revere Mass Special Publication no 1. illus: b&w: drawings, photos; front: b&w photo; 67p; bibl; 23 cm.

General book on caves that tells a little about each aspect involved in caving. 6 chapters and 34 figs.

[1440] Hendy, Phil *Muddy oxbows: a tourist trip into the seldom-trodden inner recesses of the human brain.* [Wells, Somerset, GB]: [Mendip Publishing], **1978**. text on front lining page; illus on rear lining page; 64p; 21 cm.

Illus.

[1441] Henkel, David S. *A description of the Endless Caverns of New Market.* New York, NY, US: Eastern Printing Co, **1919**. 35p; 16 cm.

Reproduced from 1880 ed. by Henkel & Co, Printers, New Market, Va.

[1442] Henkel, David S. *A description of the New Market Endless Caverns.* New Market, VA, US: Henkel & Co, Printers, **1880**. 36p; 17 cm.

Original edition has about 8 pp of ads in addition to 36 pp of text. Ads are not in reprints.

[1443] Hennessey, David, 1847– *The Caves of Shend.* London, GB: Hodder and Stoughton, **1900**. 304p.

[1444] Hennings, Kevin (ed) *1971 guidebook: Upper Mississippi conference UMAC Sept. 17, 18, 19 at Eagle Cave.* UMAC (Upper Mississippi Area Conference) (1971: Eagle Cave, WI). [Madison, WI, US]: Wisconsin Speleological Society, **c1971**. illus: b&w: maps; partly col: maps; one folded map in text; 14p, [22]; bibl.

[1445] Henry, Thomas *A guide to Maquoketa Caves State Park*. North Mankato, MN, US: Henry, Thomas, **1993**. illus: b&w: drawings, maps; [i], 71p, [2]; bibl, index.

[1446] Herak, Milan (ed) *Karst: important karst regions of the northern hemisphere*. Stringfield, Victor Timothy 1902–(ed). Amsterdam, NL: Elsevier Publishing Company, May, **1972**. illus: b&w: drawings, graphs, maps, photos, tables; 1 folded in text; xiv, 551p; bibl, index; 25 cm.

This classic takes a first step in presenting a broad range of karst products and processes in the Northern Hemisphere, and in presenting historic reviews of morphologic and hydrologic concepts. A wide variety of papers are presented, both in scope and readability. For those areas covered it provides good bibliography of early work.

[1447] Herman, Janet S. (ed) *Travertine-marl: stream deposits in Virginia*. Hubbard, David A. (ed). Charlottesville, VA, US: Commonwealth of Virginia, Dept. of Mines, Minerals, and Energy, Division of Mineral Resources, **1990**. Virginia Division of Mineral Resources publication 101. illus: b&w: charts, graphs, maps, photos, repros, tables; [vi], 184p; bibl, gloss; 28 cm.

[1448] Herzlík, Bořivoj (trans) *Czech Speleological Society 1982-1986: published on occasion of 9th International Speleological Congress Spain 1986*. Praha, CZ: Czech Speleological Society, **1986**. illus: b&w: maps; front; 44p.

Author's name also spelled "Herclik".

[1449] Hesler, Donald J. *A hydrogeologic study of the Knox-Skull Cave System, Albany County, New York*. NY, US: np, **1990**. New York Cave Survey Bulletin 4. illus: b&w: charts, graphs, maps, photos, tables; vi, 55p; app, bibl.

[1450] Hess, John Warren, 1947– *Hydrograph analysis of carbonate aquifers*. White, William Blaine, 1934–. University Park, Pa, US: Institute for Research on Land and Water Resources, Pennsylvania State University, Jun, **1974**. Research Publication Institute for Research on Land and Water Resources, Pennsylvania State University no 83. illus: b&w: drawings, graphs, maps, photos, tables; iv, 63p; bibl, index; 28 cm.

This deals principally with shallow Appalachian-Plateau type settings (Kentucky). It is a useful field example of how these analyses may be carried out and interpreted.

[1451] Hevrah la-hakirat Erets-Yisrael ve-atikoteha *The Expedition to the Judean Desert*. [Jerusalem, IL]: Israel Exploration Society, **1960, 1962**. illus: b&w: maps, repros; 25 cm.

On spine: Judean Desert caves/survey and excavations. Vol. 1 (1960 expedition) is revised edition of Hebrew publication in the Bulletin of the Israel Exploration Society, 25, 1961; v.2 (1962) revised edition of Hebrew publication in the Bulletin of the Israel Exploration Society, 26, 1962.

[1452] Hibben, Frank Cummings, 1910– *Evidences of early occupation in Sandia Cave, New Mexico, and other sites in the Sandia-Manzano Region*. Bryan, Kirk, 1888– (app). Washington, DC, US: The Smithsonian Institution, **1941**. Smithsonian Miscellaneous Collections v 99 no 23 Publication 3636. illus: b&w: maps, photos, tables; 1 map partly folded; vi, 64p, 8p pls; app, bibl; 25 cm.

Microfilm. Ann Arbor, Mich., University Microfilms International, 19—. 1 reel. 35 mm. OCLC lists a 1983 edition also (microfilm?).

[1453] Higgins, C.G. (ed) *Groundwater geomorphology*. Coates, D.R. (ed). npop: Geological Society of America, **1990**. Special Paper 252. illus: b&w: photos; ix, 368p.

Illus.

[1454] Hill, A. L. (ed) *Mullamullang Cave expeditions 1966*. [Valley View, SA, AU]: Cave Exploration Group, **1966**. Occasional papers: Cave Exploration Group South Australia no 4. illus: b&w: drawings, graphs, maps, photos, tables; 5 folded maps at end of text + 1 key; 47p, 5p pls; bibl; 26 cm.

[1455] Hill, Arthur H. *A brief glossary of Welsh topographic names for walkers and cavers*. npop: Cave Research Group of Great Britain, **1952**. Cave Research Group of Great Britain Publications series no 4. 19p; 10 x 16 cm.

[1456] Hill, Carol Ann, 1940– *Cave minerals*. National Speleological Society. Huntsville, AL, US: National Speleological Society, **1976**. illus: b&w: photos; col: photos; front: col photo; xiii, 137p, 4p pls; bibl, gloss, index; 28 cm.

7 col pls; 108 figs.

Very useful for the time; it is now overshadowed by her later contributions (with Paolo Forti).

[1457] Hill, Carol Ann, 1940– *Cave minerals of the world*. Forti, Paolo; Shaw, Trevor R. (hist intro). Huntsville, AL, US: National Speleological Society, **1986**. illus: b&w: drawings, photos, repros, tables; col: photos; front lining page with b&w drawing; front: col photo; x, 238p, 16p pls; bibl, gloss, index; 28 cm.

135 figs; 17 tables; 33 col pls.

A beautifully illustrated and informative work, which comprehensively presents and discusses the occurrence of cave minerals and of the speleothems they form.

[1458] Hill, Carol Ann, 1940– *The geology of Carlsbad Cavern*. Albuquerque, NM, US: Division of Continuing Education, University of New Mexico, **1988**. The Story of New Mexico Lecture Series Spring 1988 The New Mexico woman in history and today.

1 videocassette (120 min), sd, col, 1/2 in, VHS format.

Ms. Hill, cave geologist and author of Cave minerals of the world and Geology of Carlsbad Caverns, relates how the rock surrounding Carlsbad Cavern formed, how and why the cave formed in the rock and how the minerals formed in the cave.

[1459] Hill, Carol Ann, 1940– *Geology of Carlsbad Cavern and other caves in the Guadalupe Mountains, New Mexico and Texas.* Socorro, NM, US: New Mexico Bureau of Mines & Mineral Resources, **1987**. Bulletin: New Mexico Bureau of Mines & Mineral Resources 117. illus: b&w: drawings, graphs, maps, photos, tables; col: photos; 9 sheets of drawings and maps folded in rear pocket.; viii, 150p, 9p pls; bibl, index; 28 cm.

Spine title: Caves of Guadalupe Mountains. 131 figs; 31 pls (num 1-16); 31 tables.

This ambitious and generally well illustrated book presents a broad overview of the speleology of this area. It concentrates on cave mineralogy and speleogenetic processes associated with the "sulfuric acid mechanism". This work brings together a diversity of data and novel hypotheses to explain them, This work has inspired and facilitated a new wave of Guadalupian studies.

[1460] Hill, Carol Ann, 1940– *Geology of the Delaware Basin, Guadalupe, Apache, and Glass Mountains, West Texas and New Mexico.* Midland, TX, US: Permian Basin Section. Society for Sedimentary Geology, **1996**. 478p.

Col cover, 16 col pls, 213 figs.

[1461] Hill, Chris, 1947– *Caves of Wyoming.* Sutherland, Wayne; Tierney, Lee. Laramie, WY, US: University of Wyoming, **1976**. Bulletin Geological Survey of Wyoming 59. illus: b&w: drawings, maps, photos, tables; 4 map sheets folded, loose; b&w photo; 230p, 4p pls; app, bibl, gloss; 23 cm.

37 pls; 21 figs.

Catalog of 271 caves and cave-related features in the state. Geological introduction, text descriptions, locations stated only to topographic quadrangle, but shown to fraction of a township on location maps. Eighty-two cave maps.

[1462] Hill, John R. *The geologic story of Wyandotte Cave.* IN, US: State of Indiana Department of Natural Resources Geological Survey, **nd**. illus: b&w: maps, photos, tables; 10p; bibl.

Published 1983 or later.

[1463] Hills Speleology Club Ltd (prep) *A guide to Mount Fairy Caves: a speleological field guide to the limestone and caves at Mount Fairy.* Castle Hill, NSW, AU: Hills Speleology Club Ltd, **1992**. Overkarst number 3. illus: b&w: drawings, maps, tables; some maps folded in text; 72p.

Cover title: Mount Fairy Caves: Goulburn District N.S.W.

[1464] Hindle, Brian Paul *Cave formation in northern England.* Dent, Cumbria, GB: B. P. Hindle, **1960**. illus: b&w: drawings, maps; b&w maps on lining pages; [ii], 37p; bibl.

Printed by Eric Moore & Co, Ltd. About the karst in this region: structure, solution, waters, climate, formations, etc. 21 figs.

[1465] Hobbs, Horton Holcombe, 1914– *A new crayfish from Alabama caves with notes of the origin of the genera* Orcnectes *and* Cambaras *(Decapoda: Astracidae).* Washington, DC, US: Smithsonian Press, **1967**. Proceedings of the United States National Museum vol 123 no 3621. illus: b&w: drawings, maps, tables; 17p; bibl.

[1466] Hobbs, Horton Holcombe, 1914– *Origins and affinities of the troglobitic crayfishes of North America (Decapoda: Astacidae), II, genus em Orconectes.* Barr, Thomas Calhoun, 1931–. Washington, DC, US: Smithsonian Institution Press, **1972**. Smithsonian Contributions to Zoology number 105. illus: b&w: drawings, maps, tables; iii, 84p; app, bibl.

[1467] Hobbs, Horton Holcombe, 1914– *A review of the troglobitic decapod crustaceans of the Americas.* Hobbs, Horton Holcombe, 1944–; Daniel, Margaret A. Washington, DC, US: Smithsonian Institution Press, **1977**. Smithsonian Contributions to Zoology no 244. illus: b&w: drawings, maps, photos, tables; v, 183p; app, bibl, gloss, index; 26 cm.

70 figs.

A thorough review of all of the crayfish, shrimp, and crabs found in North American caves, with a detailed taxonomic key.

[1468] Hobbs, Horton Holcombe, 1944– *Assessment of the ecological resources of the caves of Russell Cave National Monument, Jackson County, Alabama and of selected caves at the Lookout Mountain unit of Chickamanga-Chattanooga National Military Park, Dade County, Georgia and Hamilton County, Tennessee.* Springfield, US: [W. Henberg University] Dept of Biology, Jan, **1994**. illus: b&w: drawings, maps, photos, tables; 1 map folded in text; xiii, 199p, [1]; bibl.

122 figs.

[1469] Hobbs, Horton Holcombe, 1944– *On the troglobitic shrimps of the Yucatan peninsula, Mexico (Decapoda: Atyidae and Palaemonidae).* Hobbs, Horton Holcombe, 1914–. Washington, DC, US: Smithsonian Institution Press, **1976**. Smithsonian Contributions to Zoology number 240. illus: b&w: drawings, maps; iii, 23p; app, bibl.

[1470] Hobbs, Horton Holcombe, 1944– *A study of environmental factors in Harrison's Cave, Barbados, West Indies.* npop: National Speleological Foundation. illus: b&w: charts, graphs, maps, photos, tables; col: photos; 3 foldout maps. 4 foldout technical drawings.; vi, 106p, 4p pls; bibl.

[1471] Hockensmith, Brenda Louise *Investigation of sinkhole occurrences at Goretown, near Loris, South Carolina.* Pelletier, Armand Michel. [Columbia, SC, US]: [South Carolina Water Resources Commission], Feb, **1987**. South Carolina Water Resources Commission open-file report no OF-11. illus: b&w: maps; 21p, [6]; bibl; 28 cm.

Illus.

[1472] Hodgkinson, Gerald William *Wookey Hole: the cave of mystery and history.* 2nd ed. Bath, GB: np, **c1964**. illus: b&w: drawings, maps, photos, repros; col: photos; 2 photos on front lining page; map on rear lining page; [12]; 17 x 12 cm.

Cover title: The great cave of Wookey Hole.

3rd ed. Bristol, GB, **c1947**. illus: b&w: drawings, maps, photos; photos on front lining page; 1 map on rear lining page; [10].

[1473] Hodgkinson, Olive *The story of Wookey Hole.* npop: Photo Precision Ltd, **1970**. illus: b&w: photos; col: photos; 32p.

Softbound colorful booklet. Includes complete short history of the cave along with a total of 25 color & b&w photos with 6 of them full page.

[1474] Hoebel, Edward Adamson, 1906– *The archaeology of Bone cave, Miller county, Missouri.* New York, NY, US: [American Museum of Natural History], **1946**. Anthropological Papers of the American Museum of Natural History v 40 pt 2. illus: b&w: drawings, maps, photos, tables; [iii], 19p, 5p pls; bibl; 27 cm.

5 pls, pp 139-157.

[1475] Hoffman, Ray J. *Chronology of diving activities and underground surveys in Devils Hole and Devils Hole Cave, Nye County, Nevada, 1950-86.* United States Geological Survey. Carson City, NV, US: United States Geological Survey, **1988**. Open-File Report U.S. Geological Survey 88-93. iv, 12p; bibl; 28 cm.

Illus. Also issued in microfiche.

[1476] Hoffman, Richard L. *New genera and species of cavernicolous diplopods from Alabama.* University, AL, US: np, **1956**. Geological Survey Museum Paper 35, Alabama Museum of Natural History. 11p; bibl; 23 cm.

Illus.

[1477] Hoffmann, Paul *The caves of north central Morgan County, Alabama: a guide for the 1974 SERA Winter Business Meeting of the National Speleological Society at Decatur, Alabama, February 1974.* Williams, Steve; National Speleological Society Decatur, Alabama Grotto. Decatur, AL, US: Decatur Alabama Grotto of the National Speleological Society, **1974**. illus: b&w: maps; some folded maps; [17]; 28 cm.

Chiefly maps. On cover: SERA '74 guidebook to caves of Morgan County, Alabama.

[1478] Hoffmann, Paul R. *Newsome sinks - our national landmark.* Decatur, AL, US: Decatur Alabama Grotto of the National Speleological Society, Mar, **1975**. illus: b&w: charts, maps, photos; [i], 34p, [1].

[1479] Hogberg, Rudolph K. *Guide to the caves of Minnesota.* Bayer, T. N. Minneapolis, MN, US: University of Minnesota, Minnesota Geological Survey, **1967**. Minnesota Geological Survey Educational series 4. illus: b&w: charts, drawings, maps, photos; ii, 61p; app, bibl; 23 cm.

17 b&w photographs and two cave maps.

An educational booklet. Not a catalog of caves; describes only show caves.

[1480] Hogg, Garry *Deep down: great achievements in cave exploration.* 1962 rev ed. New York, NY, US: Criterion Books, **1962**. illus: b&w: photos; 160p; bibl, index; 22 cm.

Revised version of a book published in Great Britain by Hutchinson & Co, Ltd of the same title. Accounts of caving throughout the world. Seven stories of cave discoveries and 2 of cave accidents. One chapter is now Carlsbad Caverns, NM and the rest of the book is about caves in Europe.

1st ed. London, GB: Hutchinson & Co (Publishers) Ltd, **1961**. illus: b&w: photos; 152p, 4p pls; bibl, index.

[1481] Holiday Film Corporation *Carlsbad Caverns, New Mexico.* Whittier, CA, US: Holiday Film Corporation, **1970 1980**.

40 slides, col + 1 sound cassette: analog. Sound accompaniment for manual operation only. OCLC lists both 1970 and 1980 as dates of publication.

[1482] Holiday Film Corporation *Luray Caverns.* Whittier, CA, US: Holiday Film Corp, **1984**.

40 slides, col + 1 sound cassette (1 7/8 ips). Sound accompaniment compatible for manual operation only.

In addition to a tour of the cavern formations, you will see and hear the great stalactite organ, the only one of its kind in the world, the carillon tower, the antique car museum, and hear the sound of the engine as you see each car.

[1483] Holiday Film Corporation *Mammoth Cave National Park.* Whittier, CA, US: Holiday Film Corp, **1970**.

40 slides, col + 1 sound cassette. Title on cassette: Mammoth Caves, Kentucky. OCLC lists both 1970 and 1979 as dates of publication.

[1484] Holiday Film Corporation *Meramec Caverns, Stanton, Missouri.* Whittier, CA, US: Holiday Film Corp, **1991**.

1 sound cassette (ca 30 min), analog, mono + 40 slides, col, 2x2 in + 1 description sheet.

[1485] Holland, Eric George *A guide to Batu caves.* Singapore, SG: D. Moore, **1955**. [viii], 16p, 4p pls; bibl; 19 cm.

Illus.

[1486] Holland, Eric George (comp) *Underground in Furness: a guide to the geology, mines, potholes and caves.* 2nd. **1967**. illus: b&w: maps; 110p; index.

Cover title: Underground in Furness: geology, mines, caves and potholes.

[1487] Holland, Eric George *Underground in Furness: a guide to the geology, mines, potholes and caves.* Clapham, Yorkshire, GB: Dalesman Publishing Co, **1960**.

illus: b&w: maps, photos; 6 folded map pages; 71p; gloss; 19 cm.

[1488] Holler, Cato Oliver, 1944– *Hollow Hills of Sunnalee: The Linville Caverns story*. Holler, Susan G.; Holler, Oliver (illus). Old Fort, NC, US: Hollow Hills Publishing Company, **1989**. illus: b&w: drawings, maps, photos; front: b&w drawing; xv, 68p; bibl; 22 cm.

Cover title: Hollow Hills of Sunnalee: The Linville Caverns story / 50th anniversary ed. 14 pls.

[1489] Holler, Cato Oliver, 1944– *North Carolina cave survey vol. 1 no. 1*. npop: North Carolina Cave Survey, **1975**. illus: b&w: maps, tables; Text on front; [ii], 80p.

[1490] Holler, Cato Oliver, 1944– *North Carolina cave survey vol. 2*. npop: North Carolina Cave Survey, **1979**. illus: b&w: maps, tables; Text on front and rear; vii, 70p.

[1491] Holman, J. Alan, 1931- *The pleistocene (Kansan) herpetofauna of Trout Cave, West Virginia*. Pittsburgh, PA, US: Carnegie Museum of Natural History, **1982**. Annals of Carnegie Museum v 51 article 20 0097-4463. bibl.
Caption title.

[1492] Holman, J. Alan, 1931– *A Mid-Pleistocene (Irvingtonian) herpetofauna from a cave in southcentral Texas*. Winkler, Alisa J. Austin, TX, US: [Texas Memorial Museum], Mar, **1987**. Pearce-Sellards series 44. bibl.

[1493] Holman, J. Alan, 1931– *The Pleistocene (Kansan) herpetofauna of Cumberland Cave, Maryland*. Pittsburgh, PA, US: Carnegie Museum of Natural History, **1977**. Annals of Carnegie Museum v 46 article 11 0097-4463. 16p; bibl; 23 cm.
Caption title. Illus; pp 157-172.

[1494] Holmes, Frank *Monkey Merry and St. Crida Cave*. npop: Victory Press, **1951**. 85p; 19 cm.
Rev ed: 1966. Juvenile.

[1495] Holsinger, John Robert, 1934– *Descriptions of Virginia caves*. Charlottesville, VA, US: Commonwealth of Virginia, Dept. of Conservation and Economic Development Division of Mineral Resources, **1975**. Bulletin 85 Commonwealth of Virginia Department of Conservation and Economic Development, Division of Mineral Resources. illus: b&w: maps, photos, tables; 7 maps folded in pocket in back; [xiv], 450p; app, bibl, index; 23 cm.
Collector's edition of 50 copies exists. 155 figs; 7 pls; 2 tables.
Updates Douglas 1964, with 2319 caves, including 680 new caves and 501 updated descriptions. Approximately 150 cave maps.

[1496] Holsinger, John Robert, 1934– *Ecological analysis of the Kentucky cave shrimp*, Palaemonias ganteri *Hay, at Mammoth Cave National Park (phase V)*. Leitheuser, Arthur T.; Olson, Rick; Pace, Norman R.; Whitman, Richard L.; Whitmore, Thomas; Old Dominion University Department of Biological Sciences. Norfolk, VA, US: Old Dominion University Research Foundation, Jan, **1985**. illus: b&w: maps; vii, 102p; 28 cm.

Progress report for the period October 1, 1983 to September 30, 1984. Prepared for the United States Department of the Interior, under Contract Number CX-5000-1-1037.

[1497] Holsinger, John Robert, 1934– *Ecological analysis of the Kentucky cave shrimp*, Palaemonias ganteri *Hay, at Mammoth Cave National Park (phase VI): preliminary observations on stream interstitial meiofauna communities and related abiotic factors*. Leitheuser, Arthur T.; Whitman, Richard L.; Gochee, Angel V.; Old Dominion University Department of Biological Sciences. Norfolk, VA, US: Old Dominion University Research Foundation, **1986**. illus: b&w: maps; xiv, 115p; bibl; 28 cm.
September 1986. Final report for the period October 1, 1984 to May 15, 1985. Prepared for the United States Department of the Interior, National Park Service, under Contract Number CX-5000-1-1037. Illus.

[1498] Holsinger, John Robert, 1934– *Ecological analysis of the Kentucky cave shrimp*, Palaemonias ganteri *Hay, Mammoth Cave National Park*. Leitheuser, Arthur T.; United States National Park Service Southeast Regional Office; Old Dominion University Department of Biological Sciences. Norfolk, VA, US: Old Dominion University Research Foundation, Oct, **1983**. illus: b&w: maps; vi, 70p; bibl; 28 cm.
Final report. At head of title: Department of Biological Sciences, School of Sciences and Health Professions, Old Dominion University. Contract no. CX-5000-1-1037. Illus.

[1499] Holsinger, John Robert, 1934– *Ecological analysis of the Kentucky cave shrimp*, Palaemonias ganteri *Hay, Mammoth Cave National Park (phase I)*. Leitheuser, Arthur T. Norfolk, VA, US: [Old Dominion University Department of Biological Sciences], **1982**. illus: b&w: maps; iii, 34p; bibl; 28 cm.
Prepared for the United States Department of the Interior...under Contract no. CX-5000-1-1037. Progress report, for the period September 25, 1981-March 31, 1982.

[1500] Holsinger, John Robert, 1934– *Ecological analysis of the Kentucky cave shrimp*, Palaemonias ganteri *Hay, Mammoth Cave National Park (phase II)*. Leitheuser, Arthur T. Norfolk, VA, US: Old Dominion University. Dept of Biological Sciences, Oct, **1983**. illus: b&w: graphs, maps, tables; v, 64p, 16p pls; bibl; 28 cm.
Prepared for the United States Department of the Interior, National Park Service...under contract no. CX-5000-1-1037. Final report, for the period September 25, 1981 - September 30, 1982. 16 figs, 14 tables.

[1501] Holsinger, John Robert, 1934– *Ecological analysis of the Kentucky cave shrimp*, Palaemonias ganteri *Hay, Mammoth Cave National Park (phase III)*. Leitheuser, Arthur T.; Old Dominion University Department of Biological Sciences. Norfolk, VA, US: Old Dominion University Research Foundation, Apr, **1983**. illus: b&w:

maps, tables; iii, 31p; bibl; 28 cm.

Progress report for the period September 30, 1982 to March 31, 1983. Prepared for the United States Department of the Interior, under Contract No. CX-5000-1-1037. 9 figs.

[1502] Holsinger, John Robert, 1934– *The freshwater amphipod crustaceans Gammaridae of North America*. [Washington, DC, US]: [U.S. Environmental Protection Agency], Apr, **1972**. Biota of Freshwater Ecosystems Identification Manual no 5; Water Pollution Control Research Series. illus: b&w: drawings, maps; viii, 89p, [2]; bibl, gloss, index; 29 cm.

Prepared by the Oceanography and Limnology Program, Smithsonian Institution, for the Environmental Protection Agency project #18050eld contract #14-12-894.

[1503] Holsinger, John Robert, 1934– *The invertebrate cave fauna of Virginia and a part of eastern Tennessee*. Culver, David Claire, 1944–. Raleigh, NC, US: North Carolina State Museum of Natural Sciences, Jun, **1988**. Brimleyana no 14. illus: b&w: charts, graphs, maps, photos, tables; text on lining pages; 162p; bibl.

[1504] Holsinger, John Robert, 1934– *The invertebrate cave fauna of West Virginia*. Baroody, Roger A.; Culver, David Claire, 1944–. npop, **1976**. Bulletin WV Speleological Survey 7. illus: b&w: maps; 3 folded lvs.; iv, 82p; bibl; 28 cm.

[1505] Holsinger, John Robert, 1934– *Morphological variation in* Gammarus minus *Say (Amphipoda, Gammaridae) with emphasis on subterranean forms*. Culver, David Claire, 1944–. New Haven, CT, US: Peabody Museum of Natural History, Yale University, **1970**. Postilla no 146. 24p.

Illus.

[1506] Holsinger, John Robert, 1934– *The origin and geographic distribution of troglobites*. [Norfolk, VA, US]: Old Dominion University, Department of Biological Sciences, Nov, **1985**. 24p, 4p pls; bibl; 28 cm.

Title from cover.

[1507] Holsinger, John Robert, 1934– *Speciation in cave fauna*. Barr, Thomas Calhoun, 1931–. Norfolk, VA, US: [Old Dominion University Dept. of Biological Sciences], Feb, **1985**. 43p, [2]; bibl; 28 cm.

Running title: Cave speciation.

[1508] Holsinger, John Robert, 1934– *The subterranean amphipod crustacean fauna of an artesian well in Texas*. Longley, Glenn, 1942–. Washington, DC, US: Smithsonian Institution Press, **1980**. Smithsonian Contributions to Zoology no 308. illus: b&w: drawings, maps, photos, tables; iii, 62p; bibl; 26 cm.

27 figs, 3 tables.

[1509] Holsinger, John Robert, 1934– *Systematics of the subterranean amphipod genus* Stygobromus *(Gammaridae): part I. species of the western United States*. Washington, DC, US: Smithsonian Institution Press, **1974**. Smithsonian Contributions to Zoology number 160. illus: b&w: drawings, maps; iii, 63p; bibl.

37 figs.

A discussion of 29 species in one of the most common genera found in caves with detailed information on evolution and biogeography. This is the first in a series of monographs on this genus by Holsinger.

[1510] Holsinger, John Robert, 1934– *Systematics of the subterranean Amphipod genus* Stygobromus *(Rangonyctidae), part II: species of the eastern United States*. Washington, DC, US: Smithsonian Institution Press, **1978**. Smithsonian Contributions to Zoology Number 266. illus: b&w: drawings, maps, photos, tables; iv, 144p; bibl.

77 figs, 4 tables.

A detailed discussion of one of the preeminent subterranean genera in North America with a description or redescription of 48 species.

[1511] Holsinger, John Robert, 1934– *Systematics, speciation, and distribution of the subterranean amphipod genus* Stygonectes *(Gammaridae)*. Washington, DC, US: Smithsonian Institution Press, **1967**. United States National Museum Bulletin 259. illus: b&w: drawings, graphs, maps, photos, tables; vi, 176p; bibl.

36 figs, 6 tables.

[1512] Holt, Robert (comp) *Howe Caverns; Howes Cave N.Y.; 60th Anniversary, 1929-1989*. Cobleskill, NY, US: Howe Caverns Corp, **1989**. illus: b&w: ads, drawings, photos, ports; 84p; 21 x 14 cm.

[1513] Home, Everard, 1756-1832 *Account of some remarkable caves in the principality of Bayreuth: likewise observations on the fossil bones by the late John Hunter*. [London, GB]: np, **1794**. 18p.

[1514] Hooks, William H. *Maria's Cave*. Juhasz, Victor (illus). New York, NY, US: Coward, McCann & Geoghegan, Inc, **1977**. illus: b&w: charts, drawings; partly col: charts; col lining pages; 60p, [4]; bibl.

Published simultaneously in Canada by Longman Canada Limited, Toronto. Juvenile.

A young Spainish girl discovers the first prehistoric painting ever seen modern times in the cave at Altamira, Spain.

[1515] Hooper, John *Caves in Buckfastleigh Quarries*. Joint, Wilfrid. 2nd ed. South Devon, GB: William Pengelly Cave Studies Trust, Apr, **1985**. Occasional Publication no 1.2. illus: b&w: drawings, maps; [iv], 25p; bibl; 21.5 x 14.5 cm.

[1516] Horn, Edward Courtright, 1871- *Mazes and marvels of Wind cave*. Cross, W. R. (photos). Hot Springs, SD, US: Press of the Hot Springs Star, **1901**. illus: b&w: ads, photos, ports; front; 58p; 20 cm.

Contains contemporary advertisements. This is a very rare item with an unusual creased paper cover. It contains 23 illus but only 6 are underground scenes. All photos where taken by W.R. Cross who also had ad in the back of the book.

[1517] Horne, Julia *Jenolan Caves: when the tourists came.* Crows Nest, NSW, AU: Kingsclear Books, **1994**. illus: b&w: photos, repros; ii, 72p; bibl, index.

[1518] Horne, Peter *Barnoolut Estate sinkholes environmental assessment project, 27 December 1988.* Melbourne, VIC, AU: Cave Divers Association of Australia, Research Group, **1988**. Report Cave Divers Association of Australia Research Group no 5. illus: b&w: maps; 20p; bibl; 21 cm.

Illus.

[1519] Horne, Peter *Fossil cave, 5L81 underwater palaeontological and surveying project, 1987-1988.* South Australian Underwater Speleological Society. Elisabeth Downs, SA, AU: South Australian Underwater Speleological Society, **1988**. Project Report South Australian Underwater Speleological no 1. illus: b&w: maps; 29p; bibl; 30 cm.

Title from cover. Illus.

[1520] Horne, Peter *Gouldens Hole - 5L8 - mapping project, 1987 - 1988.* Elizabeth Downs, SA, AU: South Australian Speleological Society, **1988**. Project Report / South Australian Underwater Speleological Society no 2. illus: b&w: maps; 20p, 1p pls; bibl.

Illus.

[1521] Horne, Peter (comp) *Lower South East cave reference book: an illustrated catalogue of the registered caves, sinkholes and associated karst features of the Lower South East Region of South Australia.* Adelaide, SA, AU: Peter Horne with the support of the Cave Exploration Group of South Australia (Inc), Aug, **1993**. illus: b&w: maps, photos; app, bibl.

Produced from an electronic database upon request. Cover title: Lower South East Cave reference book. Unnumbered.

[1522] Horne, Peter *Piccaninnie Ponds mapping project, 1984/85.* North Adelaide, SA, AU: Cave Divers Association of Australia, Research Group, **1985**. Report,Cave Divers Association of Australia Research Group no 3. illus: b&w: maps; 21p, 2p pls; bibl; 21 cm.

Illus.

[1523] Horne, Peter *A report on South Australian diving fatalities.* rev 2nd ed. Hove, SA, AU: P. Horne, **1980**. illus: b&w: maps; 30p; 29 cm.

Title from cover. Illus.

[1524] Horne, Peter *Research handbook for cave divers: an introduction to the realm and techniques of the underwater speleologist.* Adelaide, SA, AU: P. Horne, **1990**. illus: b&w: cartoons, charts, drawings, photos; col: photos; 46p; bibl; 30 cm.

Illus.

[1525] Horne, Peter *South Australian diving fatalities 1950-1985.* rev 2nd ed. Hove, SA, AU: P. Horne, **1987**. illus: b&w: maps; v, 74p; 21 cm.

Previous ed.: Hove, S. Aust.: P. Horne, 1980. Cover subtitle: A report compiled in the interests of diving safety. Illus.

[1526] Horsley, Dave *Oxford University Cave Club Juracao expedition 28th June-17th August 1989.* npop: The Club, **1989**.

[1527] Hosford, Jessie *Listen to the rain; a story of the Carlsbad Caverns.* Lindborg, E. (illus). New York, NY, US: Exposition Press, **1953**. illus: b&w: drawings; 45p; 21 cm.

[1528] Hoskins, R. Taylor *Caverns of enchantment.* [Mammoth Cave, KY, US]: Mammoth Cave Operating Committee, **1940**. illus: b&w: maps, photos; 96p; 20 cm.

Copy examined was a photocopy. This booklet about Mammoth Cave contains 96 pages of about 60 b&w photos, descriptions of the routes, shrubs, forms, birds, trees & animals found in the park, etc. Excellent.

[1529] Hosley, Robert J. *Cave surveying and mapping.* Indianapolis, IN, US: Crown Press (the author), **1971**. illus: b&w: drawings, maps, photos; partly col: drawings, maps; vii, 136p; app, bibl, index; 23 cm.

Published under the auspices as the Natural Sciences Resources Studies Group.

The first United States attempt to standardize cave surveying techniques.

[1530] Hötzl, H. (ed) *Tracer hydrology: proceedings of the 6th international symposium on water tracing Karlsruhe, Germany 21-26 September 1992.* Werner, A. (ed). Rotterdam, NL: A.A. Balkema, **1992**. illus: b&w: drawings, graphs, maps, photos, tables; xiii, 464p; bibl, index.

Several papers on karst.

[1531] Houshold, Ian (prep) *Cave & karst guidebook and documentation: Australasian Cave and Karst Management Association eleventh Australasian Conference on cave and karst management Tasmania 29 April to 7 May 1995 hosted by The Parks & Wildlife Service - Tasmania.* Kiernan, Kevin (co-prep); Hamilton-Smith, Elery (co-prep); Henderson, Kent (co-prep). [Mole Creek Cave, TAS, AU]: Tasmania Parks and Wildlife Service, **1995**. illus: b&w: graphs, maps; [ii], 109p; bibl.

Cover title: Guidebook: Australasian Cave and Karst Management Association Eleventh Australasian Conference on cave and karst management Tasmania 29 April to 7 May 1995

[1532] Houshold, Ian *Karst Landforms of the Lemonthyme and Southern Forests, Tasmania: report to the Commonwealth Commission of Inquiry into the Lemonthyme and Southern Forests.* Davey, Adrian. Canberra, ACT, AU: Applied Natural Resource Management, October, **1987**. illus: b&w: drawings, maps, tables; iv, 154p, [6]; app, bibl, index; 30 x 21 cm.

6 maps.

[1533] Housman, John *A descriptive tour, and guide to the lakes, caves, mountains, and other natural curiosities,* in Cumberland, Westermoreland, Lancashire, and a part of the West Riding of Yorkshire. 1st. Carlisle, GB: F. Jollie, **1800**. [vii], 226p.

2nd. 1802. [vii], 226p, 2p pls.

3rd ed. Carlisle, GB: Printed by and for F. Jollie, **1808**. illus: b&w: maps; folded pls, maps; [iv], iv, 323p; 20 cm.

6th. 1814.

7th. 1816.

8th ed. London, GB: Longman, Hurst & Co, **1817**. illus: b&w: drawings, maps; 11 fold-outs.

At least 15 caves are mentioned.

[1534] **Hovey, Edmund Otis**, 1801-1877 *Great American caverns. The underground beauties and wonders of Mammoth, Wyandotte, Luray, and other famous caves ....* New York, NY, US: np, **1902**. 13p; 25 cm.

From The Junior Munsey and The Puritan, vol. XI, no. 5, February 1902.

[1535] **Hovey, Horace Carter**, 1833-1914 *Celebrated American caverns: especially Mammoth, Wyandot, and Luray: together with historical, scientific, and descriptive notices of caves and grottoes in other lands.* Halliday, William Ross, 1926–(intro). reprint ed. New York, NY, US: Johnson Reprint Corporation, **1970**. Classics in Speleology. illus: b&w: drawings, maps, ports; 3 partly folded maps; xxxviii, 228p; app, index.

A reprint of Havey's scarce book with a new introduction by Halliday. Based on the 1896 printing.

[1536] **Hovey, Horace Carter**, 1833-1914 *Celebrated American caverns: especially Mammoth, Wyandot, and Luray: together with historical, scientific, and descriptive notices of caves and grottoes in other lands.* Cincinnati, OH, US: Robert Clarke & Company, **1882**. illus: b&w: ads, drawings, maps, ports; partly folded maps; front: b&w drawing; xii, 228p, 8, 28p pls; app, index; 24 cm.

28 pls.

An important account of exploration, also covering Howe's Cave and (in less detail) other tourist caves, including Cacahuamilpa in Mexico.

[1537] **Hovey, Horace Carter**, 1833-1914 *Guide book to the Mammoth Cave of Kentucky: historical, scientific, and descriptive.* 13th ed; rev and enl. **1891**. front: folded map; 75p, 6p pls; app; 24 cm.

6 pls.

14th ed; rev and enl. 1891. front: folded map; 75p, 6p pls; app; 24 cm.

6 pls. Cover title: Mammoth Cave.

15th ed; rev and enl. 1892. front: folded map; 75p, 6p pls; app; 23 cm.

6 pls. Cover title: Mammoth Cave.

16th ed; rev and enl. 1895. folded map; 75p, 6p pls; app; 24 cm.

6 pls. Cover title: Mammoth Cave.

1st ed. Cincinnati, OH, US: Robert Clarke & Company, **1882**. illus: b&w: maps; folded map in front; 70p, 5p pls; app; 24 cm.

At least 18 editions are known with the 18th ed dated 1897.

5th ed. 1887. app.

6th ed. [1887]. app.

9th ed. 1889. app.

[1538] **Hovey, Horace Carter**, 1833-1914 *Hovey's handbook of the Mammoth Cave of Kentucky: a practical guide to the regulation routes with maps and illustrations.* Louisville, KY, US: John P. Morton & Company, **1909**. Travels in the New South. illus: b&w: ads, maps, photos; ad on back lining page; 63p, [1]; 19 cm.

Microfilm edition: 1965; Master microform held by: LCP. Microopaque. Louisville, Ky., Lost Cause Press, 1965. 2 cards. 8 x 13 cm. (Travels in the New South). Microfilm edition: 1983 Washington, D.C.: Library of Congress Photoduplication Service, 1987. 1 microfilm reel; 35 mm.

Excellent booklet with 17 photographs and 5 maps.

[1539] **Hovey, Horace Carter**, 1833-1914 *Mammoth Cave of Kentucky: with an account of Colossal Cavern.* Rev ed. Louisville, KY, US: John P. Morton & Company Incorporated, **1912**. illus: b&w: drawings, maps, photos; 1 map folded in front; v, 131p, 46p pls.

Revised edition of Hovey & Calls's Original 1897 book on Mammoth cave: *Mammoth Cave of Kentucky; an illustrated manual*.

[1540] **Hovey, Horace Carter**, 1833-1914 *Mammoth Cave of Kentucky; an illustrated manual.* Call, Richard Ellsworth, 1856–. 1st. Louisville, KY, US: J. P. Morton and Company, **1897**. Travels in the New South. illus: b&w: maps, photos; folded map in back; front: b&w photo; v, 107p, [6], 23p pls; 22 cm.

"With historical notes, scenic accounts and descriptive and scientific matters of interest to visitors, based upon new and original explorations." Title vignette. This edition has more photos and more updated maps. Also there is a breakdown of the four tours that are currently (1912) available. Some versions have different covers. Microfilm Edition: Ann Arbor, University of Michigan, University Library, Preservation Office Microfilming Unit, 1985. 1 reel. 35 mm. Microopaque. Louisville, Ky., Lost Cause Press, 19—. 2 cards. 8 x 13 cm. (Kentucky culture series, 101)

Although largely an account of the cave, a historical section is included and many details of exploration are found throughout the volume.

3rd ed. 1899.

4th ed. 1902.

5th ed. 1904.

6th ed. 1906.

[1541] **Hovey, Horace Carter**, 1833-1914 *One hundred*

miles in Mammoth Cave - in 1880: an early exploration of America's most famous cavern. Golden, CO, US: Outbooks, **1982**. 24p; 23 cm.

Cover title. Originally published in Scribner's Monthly, 1880 ; illustrations added from other sources.

[1542] Howard, John Eliot *The caves of South Devon and their teachings*. London, GB: Hardwicke & Bogue, **1879**. illus: b&w: ads, drawings; [ii], 40p; bibl; 21 x 13.5 cm.

OCLC says 48 p. On the bones of mankind and animals found in caves, and their implications for contemporeity. Offprint of paper in J Trans Victoria Inst, 13 1886:163-210.

[1543] Howarth, Francis Gard, 1940– *Ecological studies on Hawaiian lava tubes*. Honolulu, HI, US: Dept of Botany, University of Hawaii, **1972**. Technical Report Island Ecosystems IRP U.S. International Biological Program no 16. ii, 20p; bibl.

Photo reproduction. Illus.

[1544] Howe Caverns, Inc *Marvelous Howe Caverns*. Cobleskill, NY, US: Howe Caverns, Inc, **nd**. illus: b&w: photos; [13].

Cover title: The marvelous Howe Caverns, Cobleskill, New York. Limited text

[1545] Howe Caverns, Inc *The story of Howe caverns....* Cobleskill, NY, US: Howe Caverns, Inc, **1930**. illus: b&w: drawings, photos; partly col: maps; col: photos; 15p; 22 cm.

Title from cover.

[1546] Howe, Bruce, 1912– *A stone age cave site in Tangier: preliminary report on the excavations at the Mugharet el 'Aliya, or High Cave, in Tangier*. Movius, Hallam Leonard, 1907–. reprint ed. New York, NY, US: Kraus Reprint Co, **1974**. Peabody Museum of Archaeology and Ethnology Papers v 28 no 1. 32p, 4p pls; bibl; 27 cm.

Reprint of the 1947 ed published by Peabody Museum of American Archaeology and Ethnology, Cambridge, Mass., which was issued as v. 28, no. 1 of Papers of the Peabody Museum, Harvard University.

[1547] Howe, Warren *The Howe Family From Massachusetts to New York: A story of Lester Howe, his ancestors, and the Howe Caverns*. Cold Springs, NY, US: [Warren Howe], **1988**. illus: b&w: drawings, maps, photos; iii, 23p.

Also has a copyright date of 1986.

[1548] Howell, C. *A Burrington cave atlas*. Irwin, David J.; Stuckey, D. [Somerset, GB]: Bristol Exploration Club, Jul, **1973**. Caving Report no 17. illus: b&w: maps, photos; 35p, 2p pls; bibl.

[1549] Howes, Chris, 1951– *Cave photography: a practical guide*. Buxton, GB: Caving Supplies, **1987**. illus: b&w: drawings, photos; col: photos; 67p, [1]; index; 21 cm.

[1550] Howes, Chris, 1951– *Cave references in Scientific American*. Crymych (Rhychydwr), GB: Anne Oldham, **1987**. 41p; bibl, index; 22 cm.

[1551] Howes, Chris, 1951– *CMW Bran CC Journal index vol. 1-20; 1967-1994 including newsletters, newssheets & the Silver Jubilee Journal*. [Cardiff, GB]: [Wildplaces Publishing], **1995**. 30p; 21 x 14.5 cm.

[1552] [Howes, Chris, 1951– *The Descent caver's handbook*. Gloucester, GB: Ambit Publications, **1994**. 113p; 17 x 10 cm.

Lists of information for British cavers. No illus. Numerous annual editions exist.

[1553] Howes, Chris, 1951– *The Descent index: issues 1-100.*. npop: Ambit Company, **1991**. [56].

[1554] Howes, Chris, 1951– *ISCA CC journal index volumes 1-17; 1984-1994*. [Cardiff, GB]: [Wildplaces Publishers], **1995**. 24p; 21 x 14.5 cm.

[1555] Howes, Chris, 1951– *The Red Dragon Journal of the Cambrian Caving Council index issues (1)-(20) 1974-1993*. Vstradgynlais, GB: Cambrian Caving Council, **1994**. 21p; 29.5 x 21 cm.

Sponsorship notice in back.

[1556] Howes, Chris, 1951– *To photograph darkness: the history of underground and flash photography*. Gloucester, GB: Alan Sutton, **1989**. illus: b&w: drawings, maps, photos; xxi, 330p; app, bibl, gloss, note.

Simultaneously published by Southern Illinois University Press, Carbondale, IL, US.

A detailed study of the development of equipment and techniques involved in underground photography, including biographical information on the photographers concerned. It covers the period from 1865 to 1989, but concentrates on work prior to the Second World War (including cine photography). A chronology of events, and full references and notes, appear in the appendices. For research into early cave photographs this volume should be used as a starting-point.

[1557] Howes, Chris, 1951– *Wide world magazine Index volumes 1-50 1898-1923*. [Cardiff, GB]: [Wildplaces Publishing], **1994**. 12p; 20.5 x 14 cm.

[1558] Howison, Allan Moore *The Grottoes of the Shenandoah, consisting of the Weyer and the Fountain Caves, the two most wonderful caverns in the world, at Grottoes Station of Shenandoah Valley Railway, Augusta County, Virginia. A. M. Howison, Secretary.*. Strother, David Hunter, 1816-1888; Townsend, Mary Ashley (Van Voorhis) 1832-1901. Staunton, VA, US: Valley Virginian Print, **1889**. illus: b&w: maps; [24]; 23 cm.

Title from cover. "The Weyer Cave:—By Porte Crayon": p. [9]-[21] "Weyer Cave:—By Xariffa": p. [21]-[24].

[1559] Hrdlicka, Ales, 1869-1943 *Exploration of mummy caves in the Aleutian Islands ....* [Lancaster, PA, US]: The Science Press, **1941**. illus: b&w: maps, ports;

25 cm.

Caption title. Detached from the Scientific Monthly, v. 52, no. 1-2, Jan.-Feb. 1941. 2 pts in 1 vol. Illus.

[1560] **Hubbard Scientific Company** *Cavern formation*. 1972 ed. npop, **1972**. Water—soil Motion picture.

4 min, si, col, super 8 nan, loop film in cartridge. With study guide.

Shows how ground water dissolves underground limestone deposits and caries them away in solution, leaving underground caves, caverns, and sink holes.

1987 ed. Northbrook, IL, US: Hubbard Scientific Co, **1987**. Hydrogeology Series Videorecording Single Concept Science Videorecording.

1 videocassette (VHS) (9 min), sd, col, 1/2 in, Hubbard Scientific: no. 9023-VHS.

Shows how ground water dissolves underground limestone deposits and carries them away in solution, leaving underground caves, caverns, and sink holes.

[1561] **Hubbell, Theodore Huntington**, 1897-1989 *The systematics and biology of the cave-crickets of the North American tribe Hademoecini (Orthoptera Saltatoria, Ensifera, Rhaphidophoridae, Dolichopodinae)*. Norton, Russell M. Ann Arbor, MI, US: Museum of Zoology, Univ of Michigan, Aug, **1978**. Miscellaneous Publications Museum of Zoology Univ of Michigan no 156. illus: b&w: charts, drawings, graphs, photomicrographs, tables; vii, 124p, 9p pls; bibl.

25 figs, 12 tables.

A thorough discussion of both taxonomy and ecology of "cave crickets," important cave inhabitants and important sources of food for many terrestrial invertebrates.

[1562] **Hubbs, Carl Leavitt**, 1894-1979 *Fishes from the caves of Yucatan*. npop: [Carnegie Institution of Washington], **1938**. 35p; bibl; 30 cm.

Reprinted from the Carnegie Institution of Washington Publication no 491, pp 261-295. Title from cover.

[1563] **Huber, Gary** *Sinkholes: landowner perceptions of a unique source of groundwater contamination*. npop: Iowa Natural Heritage Foundation, Oct, **1989**. 20p; bibl; 30 cm.

Illus.

[1564] **Hudgens, Bruce R.** *The archeology of Exhausted Cave: a study of prehistoric cultural ecology on the Coconino National Forest, Arizona*. [Albuquerque, NM, US]: Dept of Agriculture, Forest Service, Southwestern Region, **1975**. Archeological Report no 8, United States Forest Service. illus: b&w: maps; ix, 95p; bibl; 27 cm.

[1565] **Hudson, George Henry**, 1855-1934 *Joint caves of Valcour Island: their age and their origin*. Albany, NY, US: University of the State of New York, **1911**. New York State Museum Museum Bulletin 140 and Education Department Bulletin no 473. 36p, 22p pls; 23 cm.

Illus; pp 161-196.

[1566] **Hudson, Stephen Edward**, 1950– *Manual of U.S. cave rescue techniques*. Hempel, John Charles, 1949–; Lawrence, Judi (illus); Writing Team. 2nd ed. Huntsville, AL, US: National Cave Rescue Commission, National Speleological Society, **1988**. illus: b&w: drawings, graphs, maps, photos; 260p; app, bibl, index; 28 cm.

Spine title: Cave rescue. First ed in 1981; same title, by Williams, Toni.

Although this covers most major areas of cave rescue well, it is difficult to teach or learn from in a week long class. It is an excellent reference.

[1567] **Hudson, Travis** *Guide to Painted Cave*. Santa Barbara, CA, US: McNally & Loftin Publishers, **1982**. illus: b&w: maps, photos; [iii], 27p, [1]; bibl.

[1568] **Hughes, Thomas McKenny** *On caves*. npop: np, **1887**. 30p; 21 cm.

"Being a Paper Read before the Victoria Institute or Philosophical Society of Great Britain."

[1569] **Humphrey, Patricia (ed)** *A revision of the caver's code and a collection of articles on cave conservation*. npop: The Central Indiana Grotto of the National Speleological Society, **nd**. [ii], 26p, [1].

[1570] **Humphrey, Philip Strong**, 1926– *Avifauna of three Holocene cave deposits in southern Chile*. Pefaur, Jaime E.; Rasmussen, Pamela C. Lawrence, KS, US: Museum of Natural History, the University of Kansas, Feb, **1993**. Occasional Papers of the Museum of Natural History, the University of Kansas no 154. 37p; bibl; 23 cm.

Title from cover. Illus.

[1571] **Humphrey, Richard V.** *Mystery Hill: myth and mythology in the land of academe*. Nashua, NH, US: Gamesmasters Publishers Assn, **1989**. America's Stonehenge at Mystery Hill Monograph series no 3. 16p; 22 cm.

[1572] **Humphreys, A. J. B.** *The re-excavation of Powerhouse Cave and an assessment of Dr. Frank Peabody's work on Holocene deposits in the Taung area*. Grahamstown, ZA: Published jointly by the Cape Provincial Museums at the Albany Museum, **1978**. Annals of the Cape Provincial Museums Natural History v 11 pt 12 Cape of Good Hope. illus: b&w: maps; 28p; 27 cm.

Caption title. Illus; pp 217-244.

[1573] **Humphreys, William F. (ed)** *The biogeography of Cape Range Western Australia: being the proceedings of a symposium held under the auspices of the Western Australian Museum in Perth on 21 November 1992 at the Art Gallery of Western Australia*. Perth, WA, AU: Western Australian Museum, Dec, **1993**. Records of the Western Australian Museum Supplement no 45 1993. illus: b&w: charts, drawings, graphs, maps, tables; col: maps; front: col map; x, 248p; bibl.

[1574] **Humphreys, William F.** *Survey of caves in Cape Range North West Cape Peninsula Western Australia*. Perth, WA, AU: [Western Australian Museum Dept of Biogeography and Ecology], Apr, **1991**. illus: b&w: maps; 39, 193p; bibl; 30 cm.

Illus (some col).

[1575] Hunt, Geoffrey, 1952– *Cave gating: a handbook*. Stitt, Robert R. 1975 ed. Huntsville, AL, US: National Speleological Society. Committee on Conservation, **1975**. illus: b&w: drawings, photos; v, 43p; bibl; 28 cm.

Booklet has grown out of the Cave Gating Workshop held during the conservation short course at the 1972 NSS Convention. 20 figs.

rev 2nd ed. 1981. illus: b&w: drawings, graphs, photos; v, 60p; bibl; 28 cm.

Was an excellent attempt to present safe (to both cavers and to cave life) gates. It is now badly out-of-date and in severe need of revision. Many current problems with bats and vandals are not addressed.

[1576] Huntsville Grotto of the National Speleological Society *30 favorite Alabama caves*. [Huntsville], AL, US: [The Huntsville Grotto of the National Speleological Society, Inc], **1967**. illus: b&w: maps, tables; [50]; index.

Pagination approximate.

[1577] Huntsville Grotto of the National Speleological Society *Cave preservation and residential development: a compatible approach*. Huntsville, AL, US: Huntsville Grotto of the National Speleological Society, May, **1979**. illus: b&w: graphs, maps; 12p.

Presented to: Slope Study Committee of the Huntsville Planning Commission.

[1578] Huppert, George Nixon, 1944– *Cave and karst-related theses in United States and Canadian Universities, 1899-1988*. State College, PA, US: [National Speleological Society. Cave Geology and Geography Section], Jun, **1988**. Cave Geology v 2, no 1. 82p; index; 28 cm.

[1579] Huppert, George Nixon, 1944– *Selected bibliography of cave conservation and management*. La Crosse, WI, US: University of Wisconsin-La Crosse, **1985**. Occasional Paper University of Wisconsin-La Crosse, Department of Geography and Earth Science no 1. 41p; 28 cm.

Prepared for the Cave Management Seminars.

[1580] Huppert, George Nixon, 1944– *Speleography of Papoose Cave, Idaho County, Idaho*. Moscow, ID, US: Idaho Department of Lands, Idaho Bureau of Mines and Geology, **1981**. Idaho Bureau of Mines and Geology, Idaho Department of Lands Information Circular 35. illus: b&w: maps, photos, tables; 2 folded maps in pocket inside back cover; v, 25p; bibl; 28 cm.

Abstract; 17 figs; 2 pls; 5 tables.

[1581] Hurley, Frank, 1890-1962 *The Blue Mountains and Jenolan Caves: a camera study*. Sydney, NSW, AU: Angus and Robertson, **1952**. illus: b&w: maps, photos, ports; col: photos; maps on lining pages; front: col photo; 116p; 26 cm.

3 col, 29 b&w photos.

[1582] Hurley, Frank, 1890-1962 *Gems of Jenolan*. Sydney, NSW, AU: New South Wales Government Tourist Bureau, **nd**. illus: b&w: photos; [40].

Limited text.

[1583] Hurtt, Howard A. *Rescue system outline: Lilburn Cave Project*. [Berkeley, CA, US]: Cave Research Foundation, **1977**. [i], 32p, 3p pls; app.

No illus.

[1584] Husmann, Siegfried *1st international symposium on groundwater ecology, Schlitz, 1975*. Amsterdam, NL: Swets & Zeitlinger, **1976**. International Journal of Speleology v 8 no 1/2. 228p; bibl; 24 cm.

English, German, or French. Illus.

[1585] Hutchison, Victor Hobbs, 1931– *Notes on the plethodontid salamanders*, Eurycea lucifuga *(Rafinesque) and* Eurycea longicauda longicauda *(Green)*. [Huntsville, AL, US]: National Speleological Society, Nov, **1956**. National Speleological Society Occasional Papers no 3. illus: b&w: drawings, graphs, maps, tables; 24p; bibl; 23 cm.

Illus.

[1586] Hutton, John, 1740-1806 *A tour to the caves in the environs of Ingleborough and settle in the West Riding of Yorkshire*. Mellor, D. C. (forw). reprint ed. Wakefield, Yorkshire, GB: S. R. Publishers Limited, **1970**. 104p; gloss.

This is a reprint of the 2nd ed (1781). It is a nicely bound facsimile. Book concerns the caves and potholes in the Settle and Ingleton area of West Riding of Yorkshire. No illus. "One of the earliest comprehensive works on the caves accessible at the time..." (dust jacket).

[1587] Hutton, John, 1740-1806 *A tour to the caves in the environs of Ingleborough and Settle in the West Riding of Yorkshire: with some philosophical conjectures on the deluge, remarks on the origin of fountains, and observations on the ascent and descent of vapours, occasioned by facts peculiar to the places visited. Also a large glossary of old and original words made use of in common conversation in the north of England. In a letter to a friend*. 1st. London, GB: Richardson and Urquhart, **1780**. iv, 49p, [1]; 18 cm cm.

No illus.

[1588] Hutton, John, 1740-1806 *A tour to the caves in the environs of Ingleborough and Settle in the West—Riding of Yorkshire: with some philosophical conjectures on the deluge, remarks on the origin of fountains, and observations on the ascent and descent of vapours, occasioned by facts peculiar to the places visited. Also a large glossary of old and original words made use of in common conversation in the north of England. In a letter to a friend*. 2nd ed, with large additions. London, GB: Richardson and Urquhart, **1781**. iv, 100p; gloss; 19 cm.

Dedication signed: J.H. [i.e. John Hutton]. No illus.

[1589] Hydrological and Engineering Geological Team of the Geological Bureau of the Kwangsi Chuan Autonomous Region *On the underground river system of the Tisu karst area, Tu-an County, Kwangsi, China*. Peking, CN: np, Aug, **1976**. illus: b&w: drawings, graphs, maps, photos, tables; 20p; 26 cm.

Title from cover.

[1590] Hyman, Libbie Henrietta, 1888-1969 *North American triclad Turbellaria, XIII: three new cave planarians.* Washington, DC, US: Smithsonian Institution Press, **1956**. Proceedings of the United States National Museum ; v. 103. 10p; 24 cm.

Illus; pp 563-572.

[1591] Hyslop, R. C. *Augusta Jewel Caves and other points of interest.* Spackman, C.J. (collab). Osborne Park, WA, AU: Lamb Publications Pty Ltd, **1967**. illus: b&w: maps, photos; 31p, [1]; index.

Softbound guidebook, well illustrated.

[1592] Ibberson, Dale (ed) *The caves of Perry County, Pennsylvania.* State College, PA, US: Mid-Appalachian Region of the National Speleological Society, Jan, **1983**. MAR Bulletin no13; also published as York Grotto Newsletter vol18 nos 3-4. illus: b&w: charts, maps, photos, repros, tables; 1 foldout map.; 34p.

[1593] Imperial Film Company *Mammoth Cave, Kentucky.* Lakeland, FL, US: Imperial Film Company, Inc, **1965**. Earth Science, Series I Our National Parks.

Filmstrip: 33frs, captioned color 35mm.These materials have been reviewed by the Specialized Office for the Deaf and judged appropriate for use with the hearing impaired at the specified level(s). This filmstrip, 1 in a series of 4, provides a tour of Mammoth Cave National Park in Kentucky. It surveys over 150 miles of underground lakes, rivers, and passages, noting stalactites, stalagmites, limestone, gypsum, and various other unique stone formations.

[1594] India. Public Works Department *The caves of Elephanta.* [Bombay, IN]: Government Central Press, **nd**. 20p; 22 cm.

Illus.

[1595] Indiana Cave Capers. Guidebook (1971) *1971—18th Cave Cavers: location Bedford.* Wilson, Bill (ed); Stamper, Scott (ed). [IN, US]: [Central Indiana Grotto of the National Speleological Society], **1971**. CIG Newsletter, August 1971, pp 117-162. illus: b&w: maps; text on rear lining page; map, text on front cover; text on rear cover; 46p.

Title from cover. Editorship information gleaned from 1978 guidebook bibliography. Main body pagination includes rear cover.

[1596] Indiana Cave Capers. Guidebook (1974) *1974 Cave Capers guidebook: for Harrison & Crawford counties.* Morthland, Dave (ed). Indianapolis, IN, US: SSG Grotto of the National Speleological Society, **1974**. illus: b&w: maps; repro of map on front cover; 32p; bibl.

Title from cover.

[1597] Indiana Cave Capers. Guidebook (1975) *Guidebook: Cave Capers 1975.* Ash, Paul (layout, design). Bloomington, IN, US: Indiana Cave Capers, **1975**. illus: b&w: cartoons, charts, maps; photo on front cover; drawing on rear cover; iii, 28p, [1]; bibl.

Title from cover.

[1598] Indiana Cave Capers. Guidebook (1976) *Guidebook: Cave Capers 1976.* [IN, US]: Indiana Cave Capers, **1976**. illus: b&w: ads, maps; photo on front cover; drawing on rear cover; 21p, [1]; bibl.

Title from cover.

[1599] Indiana Cave Capers. Guidebook (1977) *24th annual CIG Cave Capers.* Sims, Mike (ed); Tozer, Bill (ed). [IN, US]: Central Indiana Grotto of the National Speleological Society, **1977**. illus: b&w: ads, drawings, maps; drawing on front cover; ad on rear cover; ii, 9p, [6].

Title from cover.

[1600] Indiana Cave Capers. Guidebook (1978) *Cave Capers 1978.* Tozer, Bill (ed); Gumm, Mike [prod asst]. [IN, US]: [Central Indiana Grotto of the National Speleological Society], **1978**. illus: b&w: ads, maps; photo on front cover; ad on rear cover; 24p, [2].

Title from cover.

[1601] Indiana Cave Capers. Guidebook (1979) *Cave Capers 1979 guidebook: Milltown, Indiana.* Tozer, Bill (ed); Lindamood, Fishman. [IN, US]: [Central Indiana Grotto of the National Speleological Society], **1979**. illus: b&w: ads, drawings, maps; photo on front cover; ad on rear cover; 26p, [2]; bibl.

Title from cover.

[1602] Indiana Cave Capers. Guidebook (1980) *27th annual Cave Capers: 1980 guidebook.* Tozer, Bill (ed); Lindamood, Fishman. [IN, US]: [Central Indiana Grotto of the National Speleological Society], **1980**. illus: b&w: ads, charts, maps; drawing on front cover; ad on rear cover; 24p; bibl.

Title from cover.

[1603] Indiana Cave Capers. Guidebook (1983) *Cave Capers, 1983.* [IN, US], **1983**. illus: b&w: maps; 48p; 28 cm.

Typescript (photocopy). Title from cover.

[1604] Indiana Cave Capers. Guidebook (1987) *Wander Indiana underground: 1987 Cave Capers, Spring Valley.* [IN, US]: [Central Indiana Grotto of the National Speleological Society], **1987**. illus: b&w: maps; 89p; bibl; 28 cm.

Title from cover. Illus.

[1605] Indiana Cave Capers. Guidebook (1989) *Indiana Cave Capers guidebook: Hickory Hill Campground, Owen County, Indiana July 14-16, 1989.* Indianapolis, IN, US: Central Indiana Grotto of the National Speleological Society, **1989**. 36th Indiana Cave Capers. illus: b&w: ads, drawings, maps; [i], 89p; 28 cm.

[1606] Indiana. Dept of Natural Resources. Engineering Division *Master plan: Wyandotte Caves, Harrison-Crawford, Blue River Recreation Complex.* Schellie Associates. [Indianapolis, IN, US]: The Division, **1968**. illus:

col: maps; some col maps folded; 146p; 28 cm.
Illus.

[1607] Indiana. Division of Geology *Guide[s] to the Indiana caverns.* [Indianapolis, IN, US]: Indiana Department of Conservation; Division of Geology, **1939**. iii, 16p, [1]; 28 cm.

Title from cover. Mimeographed. One collector lists Esarey, Ralph as the author.

[1608] Indiana. Division of State Parks *Exploring the cave world: cave packet for teachers, intermediate level.* rev ed. Mitchell, IN, US: The Park, **1992**. 19p; bibl; 28 cm.

Cover title.

[1609] Innes-Smith, Robert *Castleton and its caves.* 1976 ed. Derby, GB: Derbyshire Countryside Ltd, **1976**. illus: b&w: maps, photos; col: photos; [ii], 16p, [1]; 24 cm.

Booklet about the caves in this section of England. Caves described include: Peak, Treak Cliff, Blue John, Speed-well, Eldon Hole and others. Contains 5 col, 25 b&w photos not all of caves.

1979 ed. 16p, [4].
Coat of arms, 2 maps.

1986 ed. 2 maps on lining pages; 16p.
Cover title. 1 col coat of arms.

1994 ed. npop. 16p.

[1610] Inside Earth Publications *The Wilderness below.* Binney, Frank H. (prod); Arnold, Jay (direct, writer). Bloomington, IN, US: Indiana University Audio-Visual Center, **1977, 1973**.

1 reel, 12 min, sd, col, 16 mm.

An impressionistic view of caving. Follows one spelunker as he explores a cavern and shows the beauty, mystery, and solitude to be enjoyed in caving. Without narration.

[1611] Inskeep, R. R. *Nelson Bay Cave, Cape Province, South Africa: the Holocene levels.* Avery, Graham. Oxford, GB: B.A.R., **1987**. BAR International Series 357 (i-ii). xvii, 485p, 31p pls; bibl; 30 cm.
Illus.

[1612] Institute of Hydrogeology and Engineering Geology. Chinese Academy of Geological Sciences *Karst in China.* npop: Shanghai People's Publications Publishing House, **1976**. illus: b&w: photos, photomicrographs, tables; col: photos; [149].

Attributed to Lu Yaoru.

[1613] International Association of Engineering Geology *Symposium: sink - holes and subsidence engineering - geological problems related to soluble rocks.* Kronprinzenstrabe: Deutsche Gesellschaft für Erd - und Grundbau together with Dr. R. Wolters, **1973**. illus: b&w: drawings, graphs, maps, photos; [iii]; bibl, errat.

[1614] International Association of Hydrogeologists *International Association of Hydrogeologists 12th international congress: karst hydrogeology: abstracts and program 21-27 September 1975, Huntsville, Alabama, 28 September-3 October 1975, Gulf Shores, Alabama.* University, AL, US: Geological Survey of Alabama, **1975**. [iii], 95p; bibl.

[1615] International Association of Hydrogeologists *Karst hydrogeology: 12th international congress: International Association of Hydrogeologists 21-27 September 1975, Huntsville, Alabama U.S.A..* Tolson, Janyth S. (ed). University, AL, US: Geological Survey of Alabama, **1975**. illus: b&w: drawings, graphs, maps; [ii], vi, 84p; bibl.
23 figs.

[1616] International Association of Hydrogeologists *Proceedings of the IAH 21st congress.* China National Committee for IAH; Geological Society of China. Beijing, CN: Geological Publishing House, **1988**. International Association of Hydrological Sciences Publication no 176; Proceedings of the IAH 21st Congress, vol xxi, parts 1 and 2. illus: b&w: charts, drawings, graphs, maps, photos, tables; 1261p; bibl.

Part 1: [vi], xxvi, 552, [1]pp; part 2: [vi], xv, 555—1261 + [2]p. Cover title: Karst hydrogeology and karst environment protection: proceedings of the IAH 21st congress, Guilin China 10-15 October 1988

These volumes contain a treasure trove of work dealing with karst ground water, hydrothermal water, pollution, remote sensing, modeling and more. Much of the work is by authors whose work is not generally available in the west, on areas of which we hardly ever hear. Some papers are outstanding.

[1617] International Association of Hydrogeologists *Proceedings of the twelfth international congress: karst hydrogeology: Huntsville, Alabama.* Tolson, Janyth S. (ed); Doyle, F.L. (ed); Geological Survey of Alabama. Huntsville, AL, US: The University of Alabama in Huntsville, **1977**. International Association of Hydrogeologists Memoirs Volume XII. illus: b&w: charts, drawings, graphs, maps, photos, tables; xiii, 578p, [2]; bibl, index.

A useful proceedings volume. Some in French with English abstracts.

[1618] International Conference (2nd: 1989: Blaubeuren, Germany) *Proceedings of the karst-symposium-Blaubeuren.* Chardon, Michel; Sweeting, Marjorie Mary, 1920–; Pfeffer, Karl-Heinz. Tubingen, DE: Im Selbsrverlag des Geographischen Instituts der Universitat Tubingen, **1992**. Tubingen Geographische Studien heft 109. illus: b&w: maps; vii, 130p; bibl; 24 cm.

English and French; illus.

A broad range of authors and areas are represented. The emphasis is on planation, pedogenesis, base-level changes and other surface-related processes and products.

[1619] International Conference on Geomorphology (2nd: 1989 Frankfurt am Main, Germany) *Karst.*

Nicod, Jean; Pfeffer, Karl-Heinz; Sweeting, Marjorie Mary, 1920–. Berlin, DE: Gebruder Borntraeger, **1992**. Zeitschrift fur Geomorphologie = Annals of geomorphology = Annales de geomorphologie. Supplementband 85; Geomorphology and geoecology vol 7. illus: b&w: maps; vi, 144p; bibl; 25 cm.

Illus. Papers presented in the karst sessions at the 2nd International Geomorphological Congress in Frankfurt in September 1989. Articles in English and French ; summaries in English, French and German.

[1620] **International Congress of Speleology (2nd: 1958: Bari - Lecce - Salerno, Italy)** *Actes du deuxième congrès international de spèlèologie Bari - Lecce - Alerno 5-12 Octobre 1958*. Anelli, Franco (ed); Consiglio Nazionale delle Ricerche; Cassa per il Mezzogiorno. Gastellana - Grotte (Bari), IT: Instituto Italiano de Speleologia, **1963**. illus: b&w: charts, drawings, graphs, maps, photos, tables; 2 maps folded in text; Various pagination; bibl, index.

Largely in French and Italian. vol 1: Organisation développement du congrés; I: Hydrologie et morphologie karstique; II: Chimie, météorologie souterraine et géophysique; vol 2: III: Biologie (faune et flore); IV: Palontologie et habitat humain; V: Documentation et technique; Appendice: Excursions dans les Murges, le Salente, l'Alburno et sur la côte de Salerno. vol 1: lii, 3—509pp; vol 2: 373; appendice: 68, [4]pp.

[1621] **International Congress of Speleology (4th: 1965: Postojina, Yugoslavia)** *Proceedings of the 4th International Congress of Speleology in Yugoslavia, Postojina - Ljubljana - Dubrovnik 12-26 ix 1965*. Ljubljana, YU: Speleological Society of Yugoslavia, **1968**. illus: b&w: drawings, graphs, maps, photos, tables; some maps folded in text; Various pagination, xi, [1]; bibl.

Papers in Russian, French, English, German, Italian. Published in six vols: vols 1-2, Lectures of the Plenary Session (149 p); vol 3, Physical Speleology (624 p); vols 4-5, Biospeleology, Prehistoric Speleology (407 p); vol 6, Technics of Cave Research Work, Tourism in Caves (163 p).

[1622] **International Congress of Speleology (5th: 1969: Stuttgart, Germany)** *5 Internationaler Kongress für Speläologie Stuttgart 1969*. München, DE: Verband der Deutschen Höhlen- und Karstforscher e.V, **1969**. illus: b&w: drawings, graphs, maps, photos; some maps folded in text; Various pagination; bibl.

vol 1: Morphologie des Karstes; vol 2: Speläogenese I; vol 3: Speläogenese II / Höhlenbesiedelung; vol 4: Biospeläologie; vol 5: Hydrologie des Karstes; vol 6: Dokumentation / Höhlentouristik

[1623] **International Congress of Speleology (6th: 1973: Olomouc, Czechoslovakia)** *Pantheon of Czech speleologists*. Burkhardt, Rudolf; Král, Ing; Zdeněk, Trávníček; Hruška, Jiří; Lenochová, Alena; Peprník, Jaroslav (trans). Olomouc, CZ: Publishing House of the Palackÿ University Olomouc University Press, **1973**. v, 17p, [1].

No illus. "Edited by the Organizing and Scientific Committees for the 6th International Congress of Speleology 1973 and for the Commemoration of the 400th Anniversary of the University in Olomou."

[1624] **International Congress of Speleology (6th: 1973: Olomouc, Czechoslovakia)** *Proceedings of the 6th International Congress of speleology = Actes du 6e Congrès international de spéléologie*. Panoš, Vladimír (ed); International Union of Speleology. Praha, CZ: Academia (Czechoslovak Academy of Sciences), **1975-1977**. illus: b&w: charts, drawings, graphs, maps, photos, tables; vol 2 has [2]pp maps folded in text; vol 4 has maps, graph folded, loose; Various pagination; bibl, index; 25 cm.

At head of title: International Union of Speleology. International Speleology, 1973. Contributions in various languages (including English) with summaries in English. 8 vols: 1-3 Reports on the Congress and on the General Assembly; Lists of participants and organizers; Papers of the section Geology of Karst.—4. Papers of the Section Karst Hydrology and Climatology.—5. Papers of the Section Karst Biology and Paleontology.—6. Papers of the Section Speleo-Archeology.—7-8. Papers of the Section Applied Speleology. Pagination: vol 1: 575, [1]pp; vol 2: 474, [3]; vol 3: 334, [2]pp; vol 4: 331 + [5]pp pls; vol 5: 306, [2]pp; vol 6: 154, [2]pp; vol 7: 344, [2]pp; vol 8: 260, [2]pp.

[1625] **International Congress of Speleology (7th: 1977: Sheffield, England)** *Proceedings of the 7th International Speleological Congress, Sheffield, 1977*. Ford, Trevor David, 1925–. Bridgwater, Somerset, GB: British Cave Research Association, **1977**. illus: b&w: drawings, graphs, maps, photos, tables; [xv], 444p; bibl; 30 cm.

English, French, German, Italian, or Spanish. The Second Circular for this Congress contains [41]p, text on front lining page; and is half English, half French.

[1626] **International Congress of Speleology (8th: 1981: Bowling Green, Kentucky, USA)** *Excursion E11 supplemental materials: karst of the central Appalachians: field excursion on the occasion of the eighth International Congress of Speleology*. Werner, Eberhard Wolfgang, 1942–(trip lead); Heller, Sara (trip lead); International Union of Speleology. Huntsville, AL, US: National Speleological Society, **1981**. illus: b&w: maps; 2 folded maps in text; [22].

[1627] **International Congress of Speleology (8th: 1981: Bowling Green, Kentucky, USA)** *Proceedings of the eighth international congress of speleology: a meeting of the International Union of Speleology*. Beck, Barry Frederic, 1944–(ed); International Union of Speleology. Huntsville, AL, US: National Speleological Society, **1981**.

illus: b&w: charts, drawings, graphs, maps, photos, photomicrographs, tables; xxiv, 820p; bibl; 28 cm.

In English, French or German. Summaries in English, French, German, Chinese or Spanish. 2 vols.

[1628] International Congress of Speleology (8th: 1981: Bowling Green, Kentucky, USA) *Program*. Kastning, Ernst H., 1944–(ed); International Union of Speleology. Huntsville, AL, US: National Speleological Society, **1981**. illus: b&w: ads, charts, maps; 72p.

[1629] International Congress of Speleology (8th: 1981: Bowling Green, Kentucky, USA) *Second circular*. International Union of Speleology. Huntsville, AL, US: National Speleological Society, **1981**. [ii], 13p, [8].

[1630] International Congress of Speleology (8th: 1981: Bowling Green, Kentucky, USA) *Third circular*. International Union of Speleology. Huntsville, AL, US: National Speleological Society, **1981**. 20p.

[1631] International Congress of Speleology (9th:1977: Sheffield, England) *Caves and karst of Ireland: guidebook for the International Congress of Speleology at Sheffield, 1977*. Drew, David Phillip, 1943–; Jones, Gareth Llwyd; O'Reilly, P. M. Bridgwater, Somerset, GB: 7th I.S.C. Committee [Distributed by] British Cave Research Association], **1977**. illus: b&w: charts, maps, photos; 27p; bibl; 22 cm.

Illus. Cover: "Guidebook for the International Speleological Congress 1977."

[1632] International Congress of Speleology (9th:1977: Sheffield, England) *Caves and karst of the Yorkshire Dales: guidebook for the International Congress of Speleology at Sheffield 1977*. Glover, R. R.; Pitty, Alistair F.; Waltham, Anthony Clive, 1942–[Tony]. Bridgwater, Somerset, GB: 7th I.S.C. Committee ; [Distributed by] British Cave Research Association, **1977**. illus: b&w: drawings, maps; 1 map on rear lining page; 37p; 21 cm.

Cover title: "Caves and karst of the Yorkshire Dales: guidebook for the international speleological congress 1977."

[1633] International Congress of Speleology (9th: 1986: Barcelona, Spain) *Communications 1-7 August 1986*. Barcelona, ES, **1986**. illus: b&w: charts, drawings, graphs, maps, tables; Various pagination.

Vol 1: 309pp; vol 2: 330pp.

[1634] International Congress of Speleology (10th: 1989: Budapest, Hungary) *Proceedings: 13-20 August 1989*. Budapest, HU, **1989**. illus: b&w: drawings, graphs, maps, tables; v, 707p, [2]; bibl.

2 volumes.

[1635] International Congress of Speleology (11th: 1993: Beijing, China) *Proceedings of the XI international congress of speleology, August 2 to 8, 1993, Beijing China*. Beijing, CN: General Secretary of the Organizing Committee of the XI International Union of Speleology, **1993**. illus: b&w: charts, drawings, maps, photos, photomicrographs, tables; viii, 250p; bibl.

These proceedings, and those of the previous ten conferences contain a broad selection of papers on all aspects of speleology. They often give a hint of the direction in further studies as well as the names and addresses of some of the principle actors. Previous conferences whose proceedings were not examined include: 1st - 1953, Paris, FR; 3rd - 1963, Vienna, AT.

[1636] International Film Bureau *Lascaux, cradle of man's art*. Chicago, IL, US: The Bureau, **1976**.

1 videocassette (VHS) (17 min), sd, col, 1/2 in.

A documentary about the Lascaux Cave paintings uncovered in the Dordogne region in France in 1940. Recreates the circumstances of the discovery. Two small boys chasing a rabbit with their dog fall through the ground into caverns and find that the walls are covered with the forms of animals. Shows photographs of the paintings and explains that they are one of the most significant finds of prehistoric remains.

[1637] International Geographical Union *International symposium on karren landforms: Mallorca 1995*. Commission on Environmental Changes and Conservation in Karst Areas. Palma de Mallorca, ES: Federació Balear d'Espeleolgia, Sep, **1995**. Endins Publicació d'Espeleologia no. 20; Monografies de la Societat d'Història Natural de les Balears no 3.

[1638] International Geographical Union *Symposium on karst - morphogenesis: papers: European regional conference*. Jakucs, L'aszl'o (ed). Szeged, HU: Faculty of Science at "Attila Józsof" University, **1973**. illus: b&w: drawings, graphs, maps, tables; [i], 304p; bibl.

Some papers in French or German.

This deals mostly with surface-related processes/products. About half the papers are presented in English, although most of the authors are Eastern European. It provides a glimpse of the nature and direction of interests in this area, before the exchange of ideas and scholars became easier.

[1639] International Geographical Union *Symposium on karst - morphogenesis: papers: European regional conference (minus papers): a detailed exposition of the study tours*. Jakucs, László (ed). Szeged, HU: Faculty of Science at "Attila J ózsof" University, **1977**. [i], 21p, [1].

Half English, half Russian. No illus.

[1640] International Geographical Union *Symposium on karst-morphogenesis: European regional conference: Budapest - Aggtelek 5-9 August 1971*. Szeged, HU: Faculty of Science at "Attila József" University, **1971**. 13p.

No illus.

[1641] International Geographical Union Study Group on Man's Impact in Karst *Karst and man: proceedings of the international symposium on human influence in karst, 11-14th September 1987, Postojna, Yugoslavia*.

Kunaver, Jurij (ed). Ljubljana, YU: Department of Geography, Philosophical Faculty, University of E Kardalj of Ljubljana, **1987**. illus: b&w: drawings, graphs, maps, photos, tables; 265p; bibl.

Limited edition of 250 copies. 21 articles, 20 in English, on human impact on karst terrain.

[1642] International Meeting on Karstic Sedimentology (1987: Han-sur- Lesse, Belgium) *Colloque international de sedimentologie karstique = international meeting on karstic sedimentology, Han-sur-Lesse, Belgique, 18-22 Mai 1987.* Bastin, B. Liege, BE: [Centre Belge d'Etudes Karstologiques; Societe Geologique de Belgique], Sep, **1988**. Annales de la Societe Geologique de Belgique, t 111, fasc 1 0037-9395. iv, 199p; bibl; 30 cm.

English and French ; summaries in English and French. Illus.

Some interesting and new work on an underrepresented field.

[1643] International Speleological Conference (1964- : Brno) *Problems of the speleological research: proceedings of the international speleological conference held in Brno June 29-July 4, 1964.* Štelcl, Otakar (ed); Ložek, Vojen (ed); Košáková, Jana (ed); Odehnal, Ludvík (ed). Prague, CZ: Academia, Publishing House of the Czechoslovak Academy of Sciences, **1965**. illus: b&w: drawings, graphs, maps, photos, tables; [3]pp folded graphs, maps ; Various pagination; bibl, errat; 25 cm.

Vol 1: 220 + [18]pp pls; vol 2: 319 + [1] errata + [20]pp pls.

[1644] International Speleologist *Caves of Okinawa.* Halliday, William Ross, 1926–; Sprague, S. S. New York, NY, US: International Speleologist, Jun, **1962**. International Speleologist vol 1 no 2. 13p; 30 cm.

Speleo-archaeology and the West Indies.

[1645] International Symposium of Underground Water Tracing (3rd: 1976: Ljubljana, Yugoslavia) *Undergound water tracing: investigations in Slovenia 1972—1975.* Gospodarič, Rado (ed); Habič, Peter (ed). Postojna, YU: Institute for Karst Research, **1976**. Yugoslav Committee for International Hydrological Program Third International Symposium of Underground Water Tracing (3. SUWT) Ljubljana, Bled 1976 Yugoslavia. illus: b&w: drawings, graphs, maps; col: drawings, photos; 309p, 23p pls; bibl.

11 pls, 85 figs. Hardbound ed with color cover and maps in back pocket. 32 authors have papers presented in the volume. In English with 3 other languages (Slovene, Serbocroation, and Macedonian) summarized in the back. Contains 57 color photos. Results of 4 years research in water economics and supply.

Much practical and interesting material.

[1646] International Symposium on Cave Biology and Cave Paleontology (1975: Oudtshoorn, South Africa) *Proceedings, U.I.S. international symposium on cave biology and cave paleontology, held in Oudtshoorn, South Africa from the 3rd- 6th August, 1975.* [Cape Town, ZA]: South African Speleological Association, **1975**. illus: b&w: charts, drawings, graphs, maps, photos, tables; [ii], 72p; bibl; 29 cm.

Summaries in English, French, and Afrikaans. Cover title: U.I.S. International Symposium on Cave Biology and Cave Paleontology.

[1647] International Symposium on Changing Karst Environments (1994: Oxford, England and Huddersfield, England) *Conference abstracts: [of papers presented at Changing karst environments, hydrogeology, geomorphology and conservation, an international symposium held at the Universities of Oxford and Huddersfield, September 1994].* Lowe, David J. (ed); Gunn, John. London, GB: British Cave Research Association, Aug, **1994**. Cave and Karst Science v 21 no 1. 22p; 30 cm.

Title from cover. 9 articles and an editorial on karst environments; 5 of which deal with human impact. International group of authors.

[1648] International Symposium on Changing Karst Environments (1994: Oxford, England and Huddersfield, England) *Papers presented at the international symposium on changing karst environments, Oxford and Huddersfield, September 1994.* London, GB: British Cave Research Association, Mar, **1995**. Cave and Karst Science v 21 no 2. illus: b&w: charts, graphs, maps, photos, tables; text; 46p; bibl; 30 cm.

Title from cover. Illus.

[1649] International Symposium on Human Influence in Karst *Karst and man: proceedings of the international symposium on human influence in karst, 11 - 14 September, 1987, Postojna, Yugoslavia.* Kunaver, Jurij (ed in ch). Ljubljana, YU: Dept. of Geography, Philosophical Faculty, University E. Kardelj of Ljubljana, **1987**. 265p; 24 cm.

At head of title: International Geographical Union. Study Group on Man's Impact in Karst. English or French; summaries also in German and Italian. Illus.

[1650] International Symposium on Karst Hydrogeology (1st: 1979: Oymapinar) *International seminar on karst hydrogeology: Oymapinar, October 1979: proceedings.* Gŭnay, G. (ed). Ankara, TK: np, **1980**. DSI-UNDP Project Technical Report no 27. illus: b&w: drawings, graphs, maps, photos, tables; xxiv, 385p; app, bibl.

State Hydraulic Works (DSI) United Nations Development Programme (UNDP), DSI-UNDP Project, TUR 77/015.

[1651] International Symposium on Karst of Inner Plate Region with Monsoon Climate (1991: Guilin, China) *World karst correlation: proceedings of the international symposium on karst of the Inner Plate Region with monsoon climate, July 1991, Guilin, China.* Daoxian

Yuan, 1933–(chief ed); Tan-hsien Yuan; National Science Foundation of China. Guilin, CN: Institute of Karst Geology, **1992**. Supplementary issue of Carsologica Sinica no 16; Chung-kuo yen jung Supplement no 16. illus: b&w: drawings, graphs, maps, photos, tables; 230p; bibl; 26 cm.

Supported by NSFC; IGCP Project 299; CN45-1157/P. Title from cover.

[1652] International Symposium on Karsthydrology (1978: Budapest, Hungary) *[Proceedings] nemzetkozi karszthidrologiai szimpozium = international symposium on karsthydrology = Mezhdunarodnyi simpozium po gidrologii karsta*. Bŏcker, Tivadar (ed); Hazslinszky, Tamás (ed); Hungarian Geological Society; Hungarian Meteorological Society. Budapest, HU: Hungarian Speleological Society, **1978**. illus: b&w: drawings, graphs, maps, photos, tables; Various pagination; bibl; 29 cm.

In English, Russian or Hungarian; with summaries in English, Russian and Hungarian. Includes abstracts only of some articles. v. 1. Karst water regime.—v. 2. Karst water use and protection. Vol 1: 288, [1]pp; vol 2: 210, [1]pp.

[1653] International Symposium on Underground Water Tracing (5th: 1986: Athens) *Abstracts*. Athens, GR: The Institute of Geology and Mineral Exploration, **1986**. illus: b&w: maps; 133p, [4]; index.

[1654] International Symposium on Underground Water Tracing (5th: 1986: Athens) *Proceedings of the 5th International Symposium on Underground Water Tracing: Athens 1986*. Morfis, A.; Paraskevopoulou, P. Athens, GR: The Institute of Geology and Mineral Exploration, **1986**. illus: b&w: maps; 473p; bibl; 25 cm.

Introductory matter in Greek. Cover title: 5th international symposium on underground water tracing. At head of cover title: Institute of Geology and Mineral Exploration (Greece). International Working Group on Tracer Methods in Hydrology. Illus.

[1655] International Symposium on Underground Water Tracing (5th: 1986: Athens) *Program*. Athens, GR: The Institute of Geology and Mineral Exploration, **1986**. [19].

[1656] International Symposium on Vulcanospeleology (3rd: 1982: Bend, Oregon, USA) *Proceedings of the third international symposium on volcanospeleology: a special session of the 39th annual convention of the National Speleological Society Bend, Oregon, USA: July 30-August 1, 1982: with related biovulcanospeleological papers also presented at the 39th annual meeting of the National Speleological Society*. Halliday, William R. Seattle, WA, US: International Speleological Foundation, ABC Publishing, **1993**. illus: b&w: drawings, maps, photos, tables; 132p; bibl.

[1657] International Symposium on Vulcanospeleology (5th: 1988: Izunagaoka Hot Springs, Japan) *Excursion guide book: 5th international symposium on vulcanospeleology: [pre-activity: November 4-9, 1988: post-activity: November 13-20, 1988]*. Sameshima, Teruhiko; Ogawa, Takanori; Kashima, Naruhiko; Japan Volcanspeleological Society. Tokyo, JP: Secretariat: 5° International Symposium on Vulcanospeleology, **1988**. illus: b&w: drawings, maps, photos, tables; 3 maps folded laid in and 2 folded in text; [ii], 80p; bibl; 30 cm.

[1658] International Symposium on Vulcanospeleology (5th: 1988: Izunagaoka Hot Springs, Japan) *5th International symposium on vulcanospeleology: program Izunagaoka: November 9-11, 1988*. Tokyo, JP: Secretariat: 5° International Symposium on Vulcanospeleology, **1988**. 21p.

In English and Japanese.

[1659] International Symposium on Vulcanospeleology (6th: 1991: Hilo, Hawaii) *6th international symposium on vulcanospeleology, Hilo, Hawaii August 1991*. Rea, George Thomas,1934–(ed); Halliday, William Ross, 1926–(chair). Huntsville, AL, US: National Speleological Society, **1992**. illus: b&w: drawings, maps, photos, tables; iv, 286p; bibl.

[1660] International Symposium on Vulcanospeleology (7th: 1994: Canary Islands) *VII International Symposium on Volcanospeleology, La Palma - El Hierro - Tenoride Canary Islands, 4-11 November, 1994, Program and Abstracts*. Grupo de Espeleología "junonia" Federación Territorial Canaria de Espeleología Universidad de La Laguna. [Canary Islands]: Hawaii Speleological Survey of the National Speleological Society, **1994**. [51].

Some pages blank; illus.

[1661] International Symposium on Vulcanospeleology and Its Extraterrestrial Applications (1972: White Salmon, Washington) *Proceedings of the international symposium on vulcanospeleology and its extraterrestrial applications: a special session of the 29th Annual Convention of the National Speleological Society, White Salmon, Washington, 16 August 1972*. Halliday, William Ross, 1926–. Seattle, WA, US: Western Speleological Survey, in cooperation with the National Speleological Society, **1977**. illus: b&w: drawings, graphs, maps, photos, tables; 4 maps folded in text; [vi], 85p; bibl; 29 cm.

Twenty five years later and they are still talking about caves on Mars: never-the-less, a few good pieces on pseudokarst right here on old Terra.

[1662] [International Union of Speleology] *Caves in fine art exhibition*. Budapest, HU: Organizing Committee of the 10th International Congress of Speleology, **1989**. 20p; 21 x 15 cm.

No illus.

[1663] International Union of Speleology. Commission for Cave Protection and Cave Tourism *Cave tourism: proceedings of international symposium at 170- anniversary of Postojnska jama, Postojna, Yugoslavia, Nov. 10-12, 1988*. Institut za Raziskovanje Krasa (Slovenska akademija znanosti in umetnosti); Postojnska jama Tourist and Hotel Organization. Postojna, YU: Postojnska jama, Tourist and Hotel Organization, **1989**. 204p; bibl; 24 cm.

Illus. At head of title: International Union of Speleology, Commission or Cave Protection and Cave Tourism.

[1664] **International Union of Speleology. Commission for Physical, Chemical and Hydrological Research of Karst** *Communications = mitteilungen = soobshcheniia: international symposium on physical, chemical and hydrological research of karst: Kosice, Czechoslovakia, May 10-15, 1988*. Liptovsky Mikuláš, CZ: International Union for Speleology; Slovenska Speleologicka Společnost, **1989**. illus: b&w: charts, drawings, graphs, maps, photos; 216p; bibl; 24 x 17 cm.

At head of title: International Union for Speleology, Commission for Physical, Chemical and Hydrological Research of Karst, Slovak Society for Speleology, Commission for Physical, Chemical and Hydrological Research of Karst, Centre of the State Protection of Nature in Liptovsky Mikulas. Title from cover. In English, Slovak, German and Russian.

[1665] **Iorio, Ralph** *The history of Timpanogos Cave National Monument: American Fork Canyon, Utah*. [American Fork, UT, US]: Timpanogos Cave National Monument, **1968**. 53p; 27 cm.

Title from cover. Loose-leaf typed report on this cave.

[1666] **Iowa. Department of Agriculture and Land Stewardship** *Agricultural drainage well research and demonstration project. Sinkhole/karst area demonstration project*. [Des Moines, IA, US]: [Iowa. Department of Agriculture and Land Stewardship], Jul, **1989**. 27p; 29 cm.

Illus. Title from cover.

[1667] **Ireton, Frank** *Tee-Maze cave system, Lincoln County, Idaho, U.S.A*. [US]: Western Speleological Survey, Jul, **1973**. Miscellaneous Series, Bulletin of the Western Speleological Survey no 15, W.S.S Serial no 46. illus: b&w: maps; 21p; bibl; 28 cm.

Title from cover. Illus.

[1668] **Irving, David** *A guide to the caves in the Oak Ridge-Knoxville area*. npop: np, **nd**. Speleotype vol 5, no. 1. illus: b&w: maps; 52p.

[1669] **Irwin, David J.** *Bristol Exploration Club caving report no 13 [St. Cuthbert's report]: Part H: Rabbit Warren extension*. Turner, D. B. (photo, co-auth). Somerset, GB: Bristol Exploration Club, Aug, **1970**. illus: b&w: maps, photos, tables; 1 fold-out map; 9p, [1], 2p pls; 25 x 20 cm.

[1670] **Irwin, David J.** *Bristol Exploration Club caving report no 13: St. Cuthbert's report: Part E: Rabbit Warren*. Somerset, GB: Bristol Exploration Report, Jun, **1970**. illus: b&w: maps, photos, tables; 4 fold-out maps; 20p, 5p pls; 25 x 20 cm.

[1671] **Irwin, David J. (comp)** *A catalogue of ephemera relating to Wookey Hole*. [GB]: [David J. Irwin], Aug, **1995**. illus: b&w: repros; [32]; 30 x 21 cm.

Printed on one side only. Limited publication. No illus.

[1672] **Irwin, David J. (comp)** *A catalogue of postcards of caves in Yorkshire & Derbyshire (including some misc sites in the Midlands)*. Rhychydwr, Crymych, Dyfed, GB: Anne Oldham, Mar, **1983**. illus: b&w: repros; 246p; app, bibl.

Copy examined was a stapled photocopy.

[1673] **Irwin, David J.** *A catalogue of the postcards of Gough's Cave, Cox's Cave & Wookey Hole, Somerset 1900-1980*. Crymych, GB: Anne Oldham, **1981**. illus: b&w: repros; [iii], xx, 132p; 26 cm.

Written December 1980; copyright 1981.

A limited history of the caves is covered as background to the cards themselves.

[1674] **Irwin, David J. (ed)** *Cave notes 74*. Somerset, GB: Bristol Exploration Club, Oct, **1974**. Caving Report no 18. illus: b&w: maps, tables; 27p, 1p pls; 30 x 21 cm.

[1675] **Irwin, David J.** *Cheddar Caves bibliography*. [GB]: David J. Irwin, Apr, **1986**. [1]; 30 x 21 cm.

Printed one side only. No illus. Revised July 86, March 87, April 87.

[1676] **Irwin, David J.** *The discovery and exploration of St. Cuthbert's Swallet*. Stenner, R. D.; Tilly, G. D. Bridgwater, Somerset, GB: Bristol Exploration Club, Oct, **1968**. Caving Report no 13 part A. illus: b&w: ads, drawings, maps, photos; 36p, [1], 3p pls; bibl; 20 x 26 cm.

[1677] **Irwin, David J.** *An index to the publications of The Bristol Exploration Club 1947-1987*. [GB]: Bristol Exploration Club, Dec, **1987**. BEC Caving Report no 22. 37p; 30 x 21 cm.

No illus.

Published to celebrate forty years of the Belfry Bulletin.

[1678] **Irwin, David J.** *Mendip underground: a caver's guide*. Knibbs, Anthony J. Wells, Somerset, GB: Mendip Publishing, **1977**. illus: b&w: ads, maps, photos; 176p; app, bibl; 19 cm.

Very detailed descriptions; 70 caves, 35 surveys.

3rd ed. Jarrett, Anthony. Mendip Publishing, **1993**. illus: col: photos; Front and rear lining pages have ads; 240p; bibl.

The third and latest edition of cavers' guide to the Mendip Hills in southeast England. Descriptive text with locations (grid coordinates and locations maps) and references for 85 caves. Small-scale maps of some.

[1679] **J., J. M.** *How the grottoes of an ancient church were discovered in the convent of the "Dames de Nazareth" at Nazareth in Galilee*. Beyrouth, [LB]: Imprimerie Catholique, **1941**. illus: b&w: maps, photos; 23p; 18 cm.

[1680] **Jack, James L.** *Moore's Cave: its discovery and significance*. npop: np, **nd**. illus: b&w: maps, photos; folded map; [ii], 14p; bibl.

[15] partially numbered pls.

[1681] Jackson, Donald Dale, 1935– *Underground worlds*. The Editors of Time-Life Books. Alexandria, VA, US: Time-Life Books, **1982**. Planet Earth 6. illus: b&w: photos, repros; partly col: drawings, photos; col: drawings, maps, photos; 176p; bibl, index; 29 cm.

A readable, well-illustrated account of the history of speleology with a worldwide overview. Specific sections deal with Pierre St Martin (France), Floyd Collins, Mammoth Cave (USA), and others. Produced by the hundred thousands of copies for the general public. It included good photography and fairly good information. It was even printed in French and German. The popular press binding had a tendency to fall apart, unlike the library binding.

[1682] Jackson, George Frederick, 1906-1981 *The history and exploration of Wyandotte Cave*. npop, **1972**. illus: b&w: ads, photos; [16]; bibl.

Brief, but interesting booklet on Wyandotte.

[1683] Jackson, George Frederick, 1906-1981 *The story of Wyandotte Cave*. Albuquerque, NM, US: Speleobooks, **1975**. illus: b&w: maps, photos, repros; front: b&w map repro; vi, 96p, [1]; app, bibl.

[1684] Jackson, George Frederick, 1906-1981 *Wyandotte Cave*. Narberth, PA, US: Livingston Publishing Company, **1953**. illus: b&w: maps, photos; map on lining pages; front: b&w photo; 66p; gloss.

[1685] Jackson, Harold *Sudwala*. Sullivan, Gerardus Nicholas, 1927–[Brother Nicholas]. Durban, ZA, **c1969**. illus: b&w: drawings, maps, photos, repros; 23p, [1]; 21 x 16.5 cm.

Description of Sudwala show caves in the Transvaal, South Africa.

[1686] Jackson, John Wilfrid, 1880-1978 *[Fist [sic] to third] reports on the exploration of "Dog Holes" Cave, Warton Crag, near Carnforth, Lancashire*. Manchester, GB: Richard Gill, **1910-1913**. illus: b&w: maps; Various pagination; 22 cm.

Reprinted from the Transactions of the Lancashire and Cheshire Antiquarian Society, vol. 27, 28 and 30. 3 vols; pls.

[1687] Jackson, John Wilfrid, 1880-1978 *Report on the animal remains found in the Kilgreany Cave, Co. Waterford*. Bristol, GB: University of Bristol Spelaeological Society, **1928**. Proceedings of the University of Bristol Spelaeological Society, vol 3, no 3, 1928, pp 137-153. 17p.

[1688] Jacobsen, T. W. *Excavations at Franchthi Cave, Greece*. Bloomington, IN, US: Indiana University Press, **1987**. bibl; 28 cm.

Illus.

[1689] Jacobsen, T. W. *Franchthi Cave and Paralia: maps, plans, and sections*. Farrand, William R., 1931– . Bloomington, IN, US: Indiana University Press, **1987**. Excavations at Franchthi Cave Greece fasc 1. some lvs of pls folded; xiii, 33p, 71p pls; bibl, errat; 30 cm.

Erratum sheet laid in. Illus (some col).

[1690] Jacobson, Donald *Caving: an introductory guide to spelunking*. Stral, Lee Philip. Boyne City, MI, US: Harbor House Publishers, **1986**. illus: b&w: drawings, photos; [xvi], 143p; app, gloss, index; 22 cm.

Cover title: Caves and caving; a handbook and guide to American caves - from simply enjoying them to professional spelunking. Title on half t.p.: Meet the enchanting world of caves and caverns.

The second worst caving book ever written. NSS Board members instructed the Society to remove the book from the shelves. Included a section on photography with no information on how to take pictures underground and several illustrations printed upside down.

[1691] Jagnow, David Henry *Cavern development in the Guadalupe Mountains*. Columbus, OH, US: Cave Research Foundation, Jan, **1979**. illus: b&w: charts, drawings, maps, photos, repros, tables; 3 fold-out maps in body; 3 folded map pls in pocket; ix, 55p; bibl; 28 cm.

Originally submitted as an unpublished master's thesis at the University of New Mexico, 1977, under the title: Geologic factors influencing speleogenesis in the Captain reef complex, New Mexico and Texas. Printed by Adobe Press, Albuquerque, NM.

Much useful data, although the main genetic model (major cave development by sulfuric acid from pyrite oxidation) has largely been abandoned in this area.

[1692] Jagnow, David Henry *Stories from stone: the geology of the Guadalupe Mountains*. Jagnow, Rebecca Rohwe. Carlsbad, NM, US: Carlsbad Caverns/Guadalupe Mountains Natural History Association, **1992**. illus: b&w: maps, photos; partly col: drawings, maps, photos; col: maps, photos; 40p; 28 cm.

The geologic story of Carlsbad Cavern and the Guadalupe Mountains.

A useful introductory book geared for young readers or those with little geologic background.

[1693] Jakeman, Anthony John, 1951– *Time series models for the prediction and characterisation of stream flow from rainfall in the Cave Creek system*. Greenaway, Mark Alwyn, 1953–; Jennings, Joseph Newell, 1916-1984. Canberra, ACT, AU: Centre for Resource and Environmental Studies, Australian National University, **1983**. CRES Working Paper 1983/10 0313-7414. 11p, [4]; bibl; 30 cm.

ISSN misprinted in book. Illus.

[1694] Jakucs, László *Morphogenetics of karst regions: variants of karst evolution*. Balkay, Balint (trans). Rev and enl. New York, NY, US: John Wiley, **1977**. illus: b&w: charts, drawings, graphs, maps, photos, tables; 283p, [1]; bibl, index; 25 cm.

Translation and revision of: em A karsztok morfogenetik'aja. Budapest: Akad'emiai Kiad'o, 1971. Pease: this hardback is written to be a type of textbook in its field. Pub. in Budapest, it describes the variants of karst evolution and includes a 24 p. reference list in the back. The book is worth buying if only for its extensive bibliography. Not intended for the sport caving public.

Highly technical, well illustrated, systematic, with an excellent bibliography (especially rich in Central European literature). A very useful text, though dated in certain aspects.

[1695] James, C. H., (photo) *Electric light views in the Caverns of Luray at Luray Station (Page Co., Virginia), Shenandoah Valley Railroad.* Philadelphia, PA, US: C. H. James, **1882**.

Title from label. Views in Luray Caverns, Virginia: walkways and stairways in the caverns, stalactites and other rock formations, wires for the electric lights are visible in some of the views. 40 stereographs (photoprints), albumen, b&w, 107 x 177 mm.

[1696] James, Ian (ed) *Army caving expedition to South East Java: July and August 1986: exercise Phreatic Diamond.* Et al. npop: Royal Corps of Transport Armoured Division Transport Regiment; British Army of the Rhone; Army Caving Association, **[1986]**. illus: b&w: ads, charts, drawings, maps, photos; 31p, [36]; app; 30 x 21 cm.

Cover title: 1986 Java Expedition. Variously paginated, some blank pages.

[1697] James, Julia M. *Caves and karst of the Muller Range: report of the 1978 speleological expedition to the Atea Kananda, Southern Highlands, Papua New Guinea.* Dyson, H. Jane. Newtown, NSW, AU: Atea 78 in conjunction with the Speleological Research Council, **1980**. illus: b&w: charts, drawings, graphs, maps, photos, tables; 2 maps folded in pocket; viii, 150p; app, bibl, gloss, index; 30 cm.

Cover title: Exploration in Papua New Guinea. 38 b&w maps, 29 figs, 71 pls, 25 tables.

[1698] James, Julia M. *Caves and karst of the Muller Range: report of the 1978 Speleological Expedition to the Atea Kananda, Southern Highlands, Papua New Guinea.* Dyson, H. Jane. Newtown, NSW, AU: Atea 78 in conjunction with the Speleological Research Council, **1980**. two maps on folded leaves in pocket; viii, 150p; bibl; 30 cm.

Cover subtitle: Exploration in Papua New Guinea. Illus.

[1699] James, Julia M. *Papua New Guinea Speleological Expedition NSRE 1973: the report of the 1973 Niugini Speleological Research Expedition to the Muller Range.* Dorrell, Margaret (lit ed); Montgomery, Neil Robert, 1953– (tech ed); Pavey, Andrew J. (photo ed); Dyson, H. Jane (assist ed). Kingsford, NSW, AU: Speleological Research Council Ltd, **1974**. illus: b&w: ads, drawings, maps, photos, ports, tables; 1 b&w map folded in text; ads on lining pages; v, 70p; bibl; 30 cm.

[1700] James, Julia M. (ed) *The report of the 1973 Niugini Speleological research expedition to the Muller Range.* Papua New Guinea Speleological Expedition, 1973. Kingsford, NSW, US: Speleological Research Council, **1974**. illus: b&w: maps; v, 69p; bibl; 30 cm.

Illus.

[1701] James, Julia M. *Timor Caves.* Middleton, Gregory J.; Montgomery, Neil Robert, 1953–; Parker, Fred; Rolls, Diane; Scott, Peter; Wellings, Gleniss. Broadway, NSW, AU: Sydney Speleological Society, **1976**. Sydney Speleological Society Occasional Paper no 6. illus: b&w: maps, photos, tables; vi, 50p; bibl.

This is a report on proposed development of Timor Caves, neat Murrurundi, N.S.W. Prepared for Murrurundi Shire Council. Contains 14 figs and 8 pls. There are 27 b&w photos on the 8 pls.

[1702] James, Noel P. *Paleokarst.* Choquette, Philip W. New York, NY, US: Springer-Verlag, **1988**. illus: b&w: charts, drawings, graphs, maps, photos, photomicrographs, tables; xi, 416p; bibl, index; 28 cm.

Based on a symposium convened at the 1985 Mid-year Meeting of the Society of Economic Paleontologists and Mineralogists at Colorado School of Mines.

A useful explanation of paleokarst covering (1) development, preservation, modification and recognition, and (2) case studies, for which Pt I does not always prepare the reader. This is a good introduction to paleokarst.

[1703] Jameson, Roy Alan, 1951– *The waters of Mystery Cave, Forestville State Park, Minnesota.* Alexander, E. Calvin; Minnesota Department of Natural Resources. Minneapolis, MN, US: Dept of Geology and Geophysics, University of Minnesota, **1994—1995**. illus: b&w: maps; Various pagination; bibl; 28 cm.

Illus. "Mystery Cave resources evaluation (groundwater)." 3 vols, illus.

[1704] Jammal & Associates, Inc *The Winter Park sinkhole: a report to the city of Winter Park, Florida.* Winter Park, FL, US: Jammal & Associates, Inc, Mar, **1982**. illus: b&w: drawings, graphs, maps, photos, tables; col: maps; some illus folded in text; front: col; Various pagination, xxi; app, bibl.

Cover title: The Winter Park sinkhole: a report of the investigation, findings, evaluation and recommendations for: The City Commission City of Winter Park, Florida.

[1705] Janetski, Joel C. *An archaeological and geological assessment of Antelope Cave (NA 5507), Mohave County, Northwestern Arizona.* Hall, Michael J. St. George, UT, US: Bureau Of Land Management, Arizona Strip District, **1983**. Brigham Young University Department of Anthropology Technical Series no 83-73. illus: b&w: maps; vi, 95p; bibl; 28 cm.

Illus.

[1706] Jankowski, Laurence (exe prod) *Let's explore*

a cave. Covey, Lynn (videography); Whitley, Steven (videography); Tiell, Brian (videography); Hewlett, Brian (music); Corrigan, Vincent (music). Maumee, OH, US: Instruction[al] Video, **1985**. Earth Sciences Series.

1 videocassette (20 min, 45 sec), sd, col, 1/2 in, VHS. "Earth sciences series"—Cover of container. Juvenile.

An introduction to caves and caverns as well as their formations such as stalagmites and stalactites. Explores the chemical and geological processes to form a cave.

[1707] Jantschke, Herbert *Tunel de la Atlantida, Haria, Lanzarote, Canary Islands: the hydrodynamic, the chemistry and the minerals of the lava tube, the population density of* Munidopsis polymorpha. Nohlen, Christine; Schafheutle, Markus. npop: np, **1994**. 38p.

A documentation of the GHS expedition 1994.

[1708] Jarman, R. A. *Ingleborough Cavern and Gaping Gill.* Jarman, S. A. [Lancaster, GB]: Designed and printed by ETW Dennis & Sons Ltd Scarborough, **1991**. illus: b&w: maps; col: photos; maps on lining pages; [16]; 21 x 15 cm.

[1709] Jasek, James F. *Inner Space Cavern.* Georgetown Corporation. Georgetown, TX, US: Georgetown Corp, **1982**. illus: b&w: maps; 18p; 28 cm.

Caption title. Some of the text and all of the photography by James Jasek. Illus.

[1710] Jasinski, Marc *Caves and caving: a guide to the exploration, geology and biology of caves.* Maxwell, Bill (eng adap). Feltham, Middlesex, GB: Paul Hamlyn Publishing Group Ltd, **1967**. A Little Guide in Colour. illus: b&w: charts, tables; col: drawings, photos; 160p; app, bibl, gloss, index; 16 cm.

1969 edition may exist.

[1711] Jean, David *Underground worlds.* Loze, Keith; Youens, Paula; Regional Resources Centre (Exeter). Exeter, GB: Regional Resources Centre, **1976**. illus: b&w: maps, repros; 34p; index; 24 cm.

A "how-to" book for young people.

[1712] Jeffery, R. G. *Caves of Plymouth and District.* Devonport, Plymouth, GB: Plymouth Caving Group, **nd**. Plymouth Caving Group Special Publication no 4. illus: b&w: drawings, maps; 1 b&w map folded in text; 77p; [5]; bibl.

[1713] Jeffreys, Alan Lawrence, 1940– *A bibliography of technical articles referring to practical caving subjects.* npop: Grampian Speleological Group, Mar, **1970**. Occasional Publication no 1 Grampian Speleological Group. [61].

From documents held in the G.S.G. Library. No illus.

[1714] Jeffreys, Alan Lawrence, 1940– *The Caves of Assynt.* npop: Grampian Speleological Group, Apr, **1972**. Grampian Speleological Group Occasional Publication #2. illus: b&w: maps; viii, 59p.

36 caves described.

[1715] Jeffreys, Alan Lawrence, 1940– *Scotland underground: a caver's guide to the limestone caves of Scotland.* Crymych, Rhychydŵr, GB: Anne Oldham, Jan, **1984**. illus: b&w: maps; xvi, 52p, [1]; app, bibl, index; 30 cm.

[1716] Jenkins, D. W. (ed) *South Wales Caving Club tenth anniversary publication.* npop: South Wales Caving Club, **1957**. illus: b&w: charts, drawings; [iv], 54p; 25.5 x 20 cm.

Cover title: South Wales Caving Club 1946-1956.

[1717] Jenkins, David William *Caves in Wales and the Marches.* Williams, Ann Mason [Edington, M. Ann]. [2nd ed]. Clapham via Lancaster, Yorkshire, GB: Dalesman Publishing Company Ltd, **1963**. illus: b&w: maps; 80p; bibl, gloss; 19 cm.

[3rd ed]. **1967**. 128p.

[1718] Jennings, Jesse David, 1909– *Cowboy Cave.* Holmer, Richard N. Salt Lake City, UT, US: University of Utah Press, **1980**. Anthropological Papers Salt Lake City Utah no 104. xiv, 224p; bibl; 28 cm.

Archaeological investigations led to the definition of the prehistoric Desert Archaic culture.

[1719] Jennings, Jesse David, 1909– *Danger Cave.* 1957 ed. Salt Lake City, UT, US: University of Utah Press, Oct, **1957**. Memoirs of the Society for American Archaeology no 14; Supplement to American Antiquity, volume XXIII, number 2, part 2. illus: b&w: maps; xii, 328p; bibl, errat; 27 cm.

A hallmark study discussing the archaeology of Danger Cave and the Desert Archaic culture.

1970 [reprint] ed. **1970**. University of Utah Dept of Anthropology Anthropological Papers no 27. 28 cm.

1974 reprint ed. Millwood, NY, US: Kraus Reprint, **1974**.

[1720] Jennings, Jesse David, 1909– *Excavation of Cowboy Caves, June 3-July 26, 1975: preliminary report.* United States Bureau of Land Management. Salt Lake City, UT, US: Dept of Anthropology, University of Utah, Sep, **1975**. illus: b&w: maps; 15p, 8p pls; 28 cm.

Submitted in fulfillment of the provisions of permit no. 74-UT011 issued by Bureau of Land Management and U.S. Dept of the Interior. "September 10, 1975." Illus (8 lvs of pls).

[1721] Jennings, Jesse David, 1909– *Mogollone material, Tularosa Cave, New Mexico.* Halstead, Whitney (photo). npop: [University of Wisconsin. Bureau of Audio-Visual Instruction], **1951**.

39 slide, col, 2 x 2 in, with notes.

Shows Mogollon materials from the Tularosa Cave in New Mexico, including views of blankets, sandals, sherds, tools, wooden spoons, and pottery.

[1722] Jennings, Jesse David, 1909– *Shaw Cave, Wyoming.* reprint. npop: University of Utah, Oct, **1957**. University of Utah Anthropological Papers no 27. [15]; 28 cm.

Special reprint ... for distribution by the National Park

Service. Illus.

[1723] Jennings, Joseph Newell, 1916-1984 *Caves*. Brooks, Ronald (illus). London, GB: Oxford University Press, **1970**. Life in Australia. illus: b&w: drawings; 32p.

[1724] Jennings, Joseph Newell, 1916-1984 *CEGSA 1965-6 Nullarbor expedition*. [Adelaide, SA, AU]: Cave Exploration Group, Aug, **1961**. Occasional Paper Cave Exploration Group South Australia no 2. illus: b&w: maps; 45p, 5p pls; bibl; 30 cm.

Illus (5 lvs of pls).

[1725] Jennings, Joseph Newell, 1916-1984 *Karst*. Canberra, ACT, AU: Australian National University Press, **1971**. An Introduction to Systematic Geomorphology v 7. illus: b&w: drawings, graphs, maps, photos, tables; xviii, 252p; bibl, index; 22 cm.

The successor is entitled Karst geomorphology. Second printing 1973. 40 pls, 7 tables, 69 figs. also published by M.I.T. Press, Cambridge, Massachusetts. Pease: An excellent brand new textbook on Karst and its ship to caves.

A classic work in introductory karst processes and products as they were view then: thorough and wide-ranging.

[1726] Jennings, Joseph Newell, 1916-1984 *Karst geomorphology*. Oxford; New York, NY, GB, US: Basil Blackwell Ltd, **1985**. illus: b&w: drawings, graphs, maps, photos, tables; x, 293p; bibl, index; 24 cm.

91 figs, 45 pls, 6 tables.

A highly readable introductory text, enlarging on the 1971 classic *Karst*. This book provides a review of the subject, particularly as it relates to water resources. The bibliography focuses principally on the British and American literature.

[1727] Jennings, Joseph Newell, 1916-1984 *The limestone ranges of the Fitzroy Basin, Western Australia; a tropical semi-arid karst*. Sweeting, Marjorie Mary, 1920– . Bonn, DE: Ferd. Dümmlers Verlag, **1963**. Bonner Geographische Abhandlungen heft 32. illus: b&w: maps, photos, tables; partly col: maps; 2 b&w, 1 partly col maps folded in rear pocket; 60p, [3], 12p pls; app, bibl; 23 cm.

Summary in German. 18 b&w photos, 10 figs.

[1728] Jennings, Joseph Newell, 1916-1984 *A preliminary report of the karst morphology of the Nullarbor Plains*. npop: Cave Exploration Group (South Australia), Aug, **1961**. illus: b&w: drawings, maps, photos, tables; some b&w maps folded in text; [iii], 45p; app, bibl.

Some very interesting observations and relationships are presented here, which set the stage for subsequent work on the area and elsewhere.

[1729] Jennings, Joseph Newell, 1916-1984 *Wee Jasper Caves*. James, Julia M. (ed); Martin, David John 1957– (ed); Welch, Bruce R. (ed). rev 2nd ed. Broadway, NSW, AU: Speleological Research Council Limited, **1987**. illus: b&w: drawings, maps, photos, tables; [iii], 48p; bibl; 30 cm.

Incorporating research papers originally published in Helictite, The Journal of Australasian Cave Research. First published: Broadway, N.S.W., Speleological Research Council, 1985.

[1730] Jensen, Douglas N. *Discover Timpanogos Cave National Monument*. JTV Productions. [UT, US]: JTV Productions, **1988**.

1 videocassette (22 min), sd, col, 1/2 in, VHS format.

Beautiful photography, computer animation of cave evolution, and a reenactment of the cave's discovery in 1887 have all combined to give the viewer a deeper understanding and appreciation of one of the most popular attractions in the state of Utah.

[1731] Jessey, Gary D. *Letcher County's Pine Mountain caves: an in-depth pictorial essay of Letcher County's Pine Mountain caves and caverns*. Cromona, KY, US: Superior Printing and Publishing Co, **1973**. illus: b&w: maps, photos; 128p; errat; 22 cm.

Over 250 pictures and over 30 maps.

[1732] Jewish Museum (New York, N.Y.) *Masada and the finds from the Bar-Kokhba caves: struggle for freedom*. Finkelstein, Louis, 1895–; Field Museum of Natural History. [New York, NY, US]: np, **1967**. illus: b&w: maps, repros; 48p; 25 cm.

Exhibition sponsored by the Jewish Theological Seminary of America and shown at the Jewish Museum Oct. 12, 1967-Feb. 18, 1968; participating institutions: Field Museum of Natural History, Chicago, and others. Illus (some partly col).

[1733] Jillson, Willard Rouse, 1890– *A bibliography of Mammoth Cave, 1798-1949*. [1st ed.]. Frankfort, KY, US: Roberts Printing Co, **1953**. 81p; 23 cm.

"Of this volume one hundred copies have been printed and the type destroyed April 6, 1953."

[1734] Jillson, Willard Rouse, 1890– *Geology of Crystal Cave in southern Pulaski County, Kentucky*. Frankfort, KY, US: Roberts Print Co, **1954**. 39p; 23 cm.

Illus.

[1735] JLM Visuals *Caves and other ground water features*. Grafton, WI, US: JLM Visuals, **1979**. Natural Science Slide Program 146.

20 slide, col + descriptive notes Photos of sinkholes, stalactites, stalagmites, flowstone, crystal formations, and cave bats.

[1736] JLM Visuals *Caves and other ground water features II*. Grafton, WI, US: JLM Visuals, **1979**. Natural Science Slide Program 147.

20 slides, col. + descriptive notes. Photos of sinkholes, vertical joints, stalactites, stalagmites, and flowstone.

[1737] Johnson, Cuthbert, 1946– *The history of Mendip caving*. Newton Abbot, Devonshire, GB: David & Charles, **1967**. illus: b&w: maps, photos, ports; 196p, [1]; app, bibl, index; 22.5 cm.

Author also listed on title page as Peter Johnson. Cuthbert from OCLC. 16 pls.

Mendip complete from Roman days to the 1960's. Detailed account of cave exploration. A useful beginning to any study of Mendip caving history. However, where detail is important some further checks will be necessary. General chapters are followed by detailed histories of specific caves.

[1738] Johnson, Elden *The Lee Mill Cave.* Taylor, Philip S. Saint Paul, MN, US: Science Museum of the Saint Paul Institute, **1956**. Spring Lake Archeology Science Bulletin no 3 pt 2; Saint Paul Institute Science Museum Science Bulletin no 3 pt.2. i, 31p, 7p pls; 28 cm.

Illus.

[1739] Johnson, Rogor *Rogor Johnson's cave scrapbook.* npop, **[1979]**. illus: b&w: repros.

Only two copies of this exist. Date R. J. donated copy to NSS library - photocopy of scrapbook of newspaper articles, correspondence. 185 reproductions.

[1740] Johnston, David Alexander, 1949-1980 *Guide to some volcanic terranes in Washington, Idaho, Oregon and Northern California.* Donnelly-Nolan, Julie M. [Reston, VA, US]: U.S. Geological Survey, **1981**. U.S. Geological Survey Circular 838. x, 189p; bibl; 26 cm.

[1741] Johnston, Frances Benjamin, 1864-1952 *Mammoth Cave by flash-light.* Washington, DC, US: Gibson Brothers, **1893**. 59p, 1p pls; 21 x 26 cm.

Microfilm: Atlanta, Ga.: SOLINET, 1994. 1 microfilm reel ; 35 mm. Illus.

[1742] Joiner, Thomas J. *Hydrology of limestone terranes: geophysical investigations.* Scarbrough, W. Leon. University, AL, US: Geological Survey of Alabama, Division of Water Resources, **1969**. Bulletin 94 part D. illus: b&w: charts, graphs, maps, photos, tables; partly col: maps; 8 folded maps in pocket, 1 chart in pocket; [ii], 43p; app, bibl.

[1743] Jones, Gareth Llwyd *Cave Diving Group: Irish sump index.* npop: Speleological Union of Ireland, **1988**. SUI Journal vol 4 no 2 special edition. 76p; 30 x 21 cm.

Cover title: Cave diving in Ireland. No illus.

[1744] Jones, Gareth Llwyd *The caves of Fermanagh and Cavan.* Enniskillen, GB: The Watergate Press, **1974**. illus: b&w: maps, photos; x, 117p, 8p pls; bibl, gloss, index.

[1745] Jones, Gareth Llwyd *Irish sump index.* npop: Cave Diving Group in conjunction with the Spelaeological Union of Ireland, **1987**. illus: b&w: maps; 76p; 30 x 21 cm.

[1746] Jones, H. (auth, ed) *Honduras recce 1994: a short report.* Barrington, Piers; Iles, Pete. npop: np, **c1994**. illus: b&w: maps, tables; [27]; app, bibl.

[1747] Jones, Ivor Wynne *Llechwedd.* [2nd?]. Blaenau Ffestiniog, GB: Llechwedd Slate Caverns, **1986**. illus: b&w: ports, repros; col: ports, repros; text and illus on front lining page; [24]; 21 cm.

Title from cover. Disused slate mines rather than natural caves.

[1748] Jones, Ivor Wynne *Llechwedd Slate Caverns.* Blaenau Ffestiniog, GB: Quarry Tours Ltd, **1976**. illus: col: ports, repros; 25p; 19 cm.

Illus (chiefly col). Disused slate mines rather than natural caves.

[1749] Jones, Keith *Caves and mines of the Sychrhyd Gorge.* Rhychydwr, Crymych, Dyfed, GB: Anne Oldham, **1992**. Limestones and Caves of South Wales part 9. illus: b&w: maps; 1 folded map; 43p; 22 cm.

Illus.

[1750] Jones, R. *Wilderness caves of the Gordon - Franklin River System.* TAS, AU: Board of Environmental Studies; University of Tasmania, **1974**. illus: b&w: maps, photos; 4 folded maps in text; [viii], 107p, [4]; bibl; 30 x 21 cm.

[1751] Jones, Thomas Rupert, 1819-1911 *Lecture on the antiquity of man: illustrated by the contents of caves and relics of cave-folk.* rev and aug ed. London, GB: John Van Voorst, **1877**. illus: b&w: charts, maps; 1 folded pl; 52p; index; 21 cm.

Delivered before the members of the Croydon Microscopical Club, April 26th, 1876. Revised and augmented. Title from cover. Illus.

[1752] Jones, W. F. *Some caves of Cyprus.* Rigby, J. (prod and ed). npop: W. F. Jones, **1962**. illus: b&w: maps; cover on reprint folds out to give information on reprint; 18p; 17 x 17.5 cm.

Reprint photocopy of original. Printed one side of sheets only.

[1753] Jones, Walter Bryan, 1895-1977 *Caves of Madison County, Alabama.* Varnedoe, William Whitfield, 1923–. University, AL, US: Geological Survey of Alabama, Division of Economic Geology, **1968**. Geological Survey of Alabama Circular 52. illus: b&w: drawings, maps, photos, tables; col: drawings; 1 map folded in pocket in back; [viii], 177p; app, bibl.

Catalog of the 140 known caves in that county, with locations, text descriptions, and about 100 cave maps.

[1754] Jones, Walter Bryan, 1895-1977 *Caves of Morgan County, Alabama.* Varnedoe, William Whitfield, 1923–. University, AL, US: Geological Survey of Alabama, Mineral Resources Division, **1960**. Bulletin Geological Survey of Alabama 112. illus: b&w: maps, photos; 5 maps folded in pocket in back; [xii], 205p; app, bibl; 23 cm.

178 figs, 5 pls.

Catalog of 208 caves in the county, with locations. Maps of three quarters of them, but written descriptions of only one quarter.

[1755] Jones, William Basil, 1822-1897 *Wonderful curiosity, or, a correct narrative of the celebrated Mammoth Cave of Kentucky with incidents and anecdotes.* Russellville, KY, US: Smith & Rhea, **1844**. Kentucky Culture

Series no 23. 67p, [1]; 24 cm.

Microopaque: Louisville, Ky., Lost Cause Press, 19—. 2 cards.

[1756] Jones, William K., 1945– *Hydrology of limestone karst in Greenbriar County, West Virginia*. [Charleston, WV, US]: U. S. Geological Survey, **1973**. West Virginia Geological and Economic Survey Bulletin 36. illus: b&w: maps, photos, tables; col: photos; 2 maps in back pocket; vi, 49p; bibl; 23 cm.

Produced in cooperation with the West Virginia Geological and Economic Survey. 23 figs, 5 tables, 3 col photos, 2 maps.

[1757] Jones, William K., 1945– *Karst hydrogeology: A summary of the principals of ground-water flow through soluble limestone aquifers prepared for the Smithsonian Institution short course on speleology*. Smithsonian Institution. [Charleston, WV, US]: West Virginia Geological and Economic Survey, Mar, **1975**. illus: b&w: charts, drawings, graphs, maps, photos; [iv], [30]; bibl, gloss; 28 cm.

[1758] Jordan, George *The great cave of Dry Fork of Cheat River, Virginia: one of the greatest wonders of the world examined and explored Giving a complete description of this subterraneous cavern, of its rooms and passages, which are of enormous size, together with a great stream of water, a lake, and springs, minerals, fossils, and many curiosities, that were found in singular abundance, &c.*. 1972 ed. npop: American Spelean History Association, **1972**.

Reprint of 1855 ed., published by Hanzsche & Co., Baltimore. Cover notes printing limited to 26 copies.

1974 ed. Morgantown, WV: Copycat, **1974**.

Reprint of 1855 ed.

1st ed. Baltimore, MD, US: Hanzsche & Co, **1855**. 48p; 22 cm.

"... Situated ... in the county of Randolph ... about 40 miles from Beverly and 76 from Romney" [West Virginia]—Pref. Printed by Hanzsche. Ed Z. notes, "Report on one of the greatest wonders of the world".

[1759] Jordan, J. Murray *The Caverns of Luray and vicinity*. Philadelphia, PA, US: J. Murray Jordan, Publisher, **1897**. International Souvenir Series. [24]; 20 x 25 cm.

Title from cover. Publisher's name, copyright date, and series title from wrapper.

[1760] Jordon, Donald George, 1926– *Water and copper-mine tailings in karst terrane of Rio Tanama Basin, Puerto Rico*. Washington, DC, US: US Geological Survey, **1970**. illus: b&w: maps; [1] folded leaf of pls; 24p, [1]; 28 cm.

Prepared in cooperation with Commonwealth of Puerto Rico. Illus.

[1761] Juberthie, Christian (ed) *Encyclopaedia biospeologica tome 1*. Decu, Vasile (ed). Moulis, FR: Société de Biospéologie, **1994**. illus: b&w: drawings, maps, photos, tables; col: photos; xii, 834p; bibl.

Several chapters in French.

A multi-authored volume that reviews the many groups of subterranean invertebrates, and history of biospeleological research in Europe and the Americas.

[1762] Judson, David *Caving and potholing*. Champion, Arthur. London, GB: Granada Publishing Limited, **1961**. A Mayflower Book. illus: b&w: maps; 192p, 16p pls; bibl, gloss, index; 18 cm.

[1763] Judson, David (ed) *Caving practice and equipment*. 1st ed. Newton Abbot, [GB]: David & Charles (Publishers) Limited, **1984**. illus: b&w: drawings, graphs, maps, photos; 238p; app, bibl, index; 26 cm.

Continuing the long-term British Research Cave Association attempt to provide up-to-date general techniques books for the British caving community. Well written and detailed.

1991 rev ed. British Cave Research Association. Leicester, GB: Published jointly by Cordee [and] British Cave Research Association, **1991**. illus: b&w: charts, drawings, photos, tables; col: photos; iv, 296p, 8p pls; bibl, index; 24 cm.

An updated and improved edition.

1995 rev ed. Leicester; Birmingham, AL, GB; US: Published jointly by Cordee [and] British Cave Research Association; Menasha Ridge Press, **1995**. 295p, 4p pls; bibl, index; 24 cm.

[1764] Judson, David M. *Cave surveying for expeditions*. [London, GB]: [Royal Geographical Society], **1974**. Geographical Journal vol 140 pt 2. 11p; bibl; 24 cm.

Illus.

[1765] Judson, David M. *Ghar Parau*. London, GB: Cassell, **1973**. illus: b&w: drawings, maps, photos, ports; col: photos; [viii], 216p, 32p pls; app, index; 25 cm.

52 illus, 8 apps. Also issued by Macmillan Publishing Co. (NY) in 1973.

Account of the well-organized 1972 British expedition to explore Ghar Parau in Iran. Several appendices are included covering expedition medical needs, water tracing and analysis, photography, food, and equipment.

[1766] Kaczmarak, Michael B. *Basic guidelines for spelunking*. npop: Shining Mountains Grotto, Nov, **1973**. 10p.

Written for the novice cavers. No illus.

[1767] Kahrau, Wolfgang *Australian caves and caving*. Melbourne, VIC, AU: Lansdowne Press Pty Ltd, **1972**. Periwinkle Colour Guides. illus: b&w: drawings, maps, photos, tables; col: photos; front: b&w photo; 111p; bibl, gloss; 19 cm.

Contains 9 chapters, 33 figs: 28 col, 11 b&w photos.

[1768] Karst Hydrogeology Symposium (1977: Oymapinar) *Karst hydrogeology symposium: Oymap-*

inar, October 1977: proceedings*. Günay, G.; Karanjac, J. (co-eds). Ankara, TR, **1978**. DSI-UNDP Project Technical report no 27. illus: b&w: drawings, graphs, maps, photos, tables; partly col: drawings, graphs, maps; [xxiii], xxiii, 293p, [293]; bibl, index.

Contains reprints. DSI-UNDP Project, TUR 77/015/ Strengthening DSI ground water investigative capabilities Phase II. Not found in OCLC.

[1769] Kastning, Ernst H., 1944– *Appalachian Karst: proceedings of the Appalachian karst symposium: Radford, Virginia, March 23-26, 1991*. Kastning, Karen M. (ed). Huntsville, KY, US: The National Speleological Society, **1991**. illus: b&w: charts, drawings, graphs, maps, photos, tables; viii, 239p; bibl; 28 cm.

Sponsored by the National Speleological Society in celebration of its 50th anniversary; hosted by the Department of Geology, Radford University.

A valuable summary of this diverse area of geological, speleological and historic importance.

[1770] Kastning, Ernst H., 1944– *Cavern development in the Helderberg Plateau, East-central New York*. npop: [New York Cave Survey], **1975**. New York Cave Survey Bulletin 1. illus: b&w: charts, graphs, maps, photos, tables; 1 folded loose map plate; front: photo; xviii, 194p, 7p pls; bibl; 27 cm.

[1771] Kastning, Ernst H., 1944– *Caverns of the Shawangunk (Shon-gum) and its environs, Southeastern New York*. Cohen, Steven M. (co-auth). Radford, VA, US: National Speleological Society, Jun, **1988**. illus: b&w: maps; 20p, [1]; add, bibl, errat.

Cover title: Caverns of the Shawangunk (Shon-gum) and its environs, Southeastern New York: a field guide to the caves and karst of Orange, Rockland, Sullivan, and Ulster Counties, New York. Errata sheet dated 24 May, 1989.

[1772] Kastning, Ernst H., 1944– *Caves and karst of the New River Valley, Virginia: guidebook for a geologic fieldtrip, eighth annual New River symposium, Radford, Virginia*. Lenhart, Stephen W. Radford, VA, US: Department of Geology, Radford University, Apr, **1989**. illus: b&w: charts, maps; 1 fold-out newspaper reproduction glued to back lining; [25]; bibl.

[1773] Kastning, Ernst H., 1944– *Caves, karst, and environmental impact in the new river drainage basin of Virginia and West Virginia: guidebook for a geologic fieldtrip, Appalachian karst symposium, Radford, Virginia, 23 March 1991*. Jameson, Roy A. (cont); Koerschner III, William F. (cont); Lenhart, Stephen W. (cont); Ortiz, R. Keith (cont); Saunders, Joseph W. (cont). VA, US: Department of Geology, Radford University, Mar, **1991**. illus: b&w: charts, graphs, maps; 19p, [17]; bibl.

[1774] Kastning, Ernst H., 1944– *Geochemistry of karst waters: a basic bibliography*. npop: [Ernst H. Kasting], May, **1977**. iv, 57p.

[1775] Kastning, Ernst H., 1944– *Karst geomorphology and hydrogeology*. 2nd ed. npop: np, Jul, **1989**. 10p; bibl.

Nice, but essentially a pamphlet.

[1776] Kauffman, D. L. *The beautiful Luray Caverns, Luray, Virginia*. Luray, VA, US: Published and sold by D.L. Kauffman, **1913**. [28]; 20 x 26 cm.

Title from cover. Minimal text.

[1777] Kawiti, Walter Brown *Waiomio's limestone caves; linked with Maori tribal legend, rich in natural history*. Kaikohe, NZ: Kaikohe News Print, **1969**. illus: b&w: ports; 48p; 21 cm.

Cover-title. Illus.

[1778] Keck, Ken *Abercrombie Caves: cave chronicles: a new look at an old wonder: including an account of the Bathurst convict rebellion of 1830*. Cubitt, Barry. Trunkey Creel, NSW, AU: Abercrombie Caves Historical Research Group, **1991**. illus: b&w: maps, ports; 58p, 1p pls; bibl, index; 30 cm.

[1779] Keck, Ken E. *An introduction to Abercrombie Caves*. North Paramatta, NSW, AU: Metropolitan Speleological Society, Nov, **1985**. illus: b&w: drawings, maps; 1 b&w map folded on back cover; text on front lining pages; 17p.

Limited edition of 300 copies. Prepared for Abercrombie Caves Resort.

[1780] Kehret, Roger Alan, 1942– *Crystal Lake Cave Iowa*. Chatfield, MN, US: R. Kehret, **1976**. illus: b&w: charts, drawings, maps, photos; 20p; 21 cm.

Other copy examined entitled Crystal Lake Cave. Cover title: Your guide to Crystal Lake Cave. One copy lists 24 pages and notes, "Tour guide to an Iowa Commercial cave."

[1781] Kehret, Roger Alan, 1942– *Minnesota caves of history and legend: a collection of unique cave stories*. Chatfield, MN, US: R. Kehret, **1974**.

Ed Z. notes, "Collection of unique cave stories" 19 black & white photos

[1782] Kehret, Roger Alan, 1942– *Minnesota karst country tours*. Chatfield, MN, US: Roger Kehret, **1988**. illus: b&w: maps, photos, tables; Text on front lining page; b&w photos on rear lining page; [24].

Title from cover.

[1783] Kehret, Roger Alan, 1942– *Your guide to Mystery Cave*. Chatfield, MN, US: R. Kehret, **1974**. illus: b&w: drawings, maps, photos; 28p.

Ed Z. notes, "Commercial guide with maps and history." Cover title: Your guide to Mystery Cave: with maps, photos, and descriptions of your tour.

[1784] Kehret, Roger Alan, 1942– *Your guide to Spook Cave*. 1976 ed. Chatfield, MN, US: R. Kehret, **1976**. illus: b&w: drawings, maps, photos; text and photo on front lining page, photo on rear lining page; 24p.

Cover title: Your guide to Spook Cave: with photos, maps and descriptions of your tour. Tour guide to a Northern Iowa commercial cave.

1985 ed. npop, **1985**. illus: b&w: drawings, maps, photos; col: photos; text and photo on front lining page, drawing on rear lining page.

[1785] Keith, Arthur, 1866-1955 *History from caves: a new theory of the origin of modern races of mankind, by Arthur Keith, being the presidential address given at Buxton, Derbyshire to the first speleological conference.* npop: British Speleological Association, **1936**. 18p; 17.5 x 11 cm.

Cover title: British Speleological Association presidential address.

[1786] Kell, Neil *The Winston Churchill Memorial Trust of Australia: project: to study recent developments in the design and installation of lighting to enhance the public education and enjoyment of show caves.* [AU]: The Winston Churchill Memorial Trust of Australia, **[1993]**. 32p; bibl.

No illus.

[1787] Keller, Charles M. *Montagu Cave in prehistory: a descriptive analysis.* Berkeley, CA, US: University of California Press, **1973**. Anthropological Records v 28. illus: b&w: drawings, maps, tables; [ix], 98p, 139p pls; app, bibl; 28 cm.

53 figs, 53 pls.

[1788] Keller, David R. (ed) *Paleokarst, karst-related diagenesis, reservoir development, and exploration concepts: examples from the Paleozoic section of the southern mid-continent: 1993 annual fieldtrip guidebook Permian Basin Section - SEPM Arbuckle Mountains, Oklahoma.* Reed, Christy L. (co-ed). Midland, TX, US: Society of Economic Paleontologists, Permian Basin Section, **1993**. PSB-SEPM Publication no 93-34. illus: b&w: charts, drawings, graphs, maps, photos, tables; vii, 109p, [3]; bibl.

[1789] Kemmerly, Philip R. *Subhold collapse in Montgomery County, Tennessee: an overview for the planning process.* npop: Division of Geology. Dept of Conservation Tennessee, **1980**. State of Tennessee Division of Geology Environmental Geology series no 6. illus: b&w: drawings, graphs, maps, photos; iv, 42p; app, bibl, gloss.

[1790] Kempe, David Ronald Charles, 1927– *Living underground: a history of cave and cliff dwelling.* London, GB: The Herbert Press Ltd, **1988**. illus: b&w: charts, maps, photos, repros; 256p; bibl, index; 26 cm.

[1791] Kenk, Roman *Freshwater Triclads (Turbellaria) of North America, IX, the genus* Sphalloplana. Washington, DC, US: Smithsonian Institution Press, **1977**. Smithsonian Contributions to Zoology Number 246. illus: b&w: drawings, photos, photomicrographs, tables; iii, 38p; bibl.

62 figs, 1 table.

[1792] Kennard, Don (ed) *Devil's Sinkhole area: headwaters of the Nueces River.* Smith, Griffin (cont); Johnston, Marshall (cont); Story, Dee Ann (cont); Tunnell, Curtis (cont). Austin, TX, US: Dept. of Natural Resources and Environment, University of Texas at Austin, **1975**. Natural Area Survey part 8. illus: b&w: maps, photos, tables; col: photos; front: col photo; [ix], 117p; bibl.

[1793] Kenney, C. Howard, 1927-1980 *Caving log 1942-1950.* Hendy, P. G. (ed). Somerset, GB, US: The Wessex Cave Club, **1985**. Wessex Cave Club Occasional Publication Series 1 no 3. illus: b&w: photos, repros; Various pagination, 216p, [4]; 28 x 21 cm.

Copy of handwritten log book from 1942-1950.

[1794] Kent, Paul *Tour caves of the Ozarks: Missouri Arkansas: the secret beneath the surface.* Cassville, MO, US: Tour Caves, **1983**. illus: b&w: maps, tables; 24p.

[1795] Kentucky Geological Survey *Memoirs of the Geological Survey of Kentucky.* Shaler, Nathaniel Southgate, 1841-1906; Allen, Joel Asaph, 1838–1921. Cambridge, KY, US: University Press, **1876**. Kentucky Culture Series. 246p; 32 cm.

Pt. I. On the antiquity of the caverns and cavern life of the Ohio Valley, by N.S. Shaler.—Pt. II. The American bisons, living and extinct, by J.A. Allen. Microfiche. Louisville, Ky., Lost Cause Press, 19—. 4 sheets. 11 x 15 cm. (Kentucky culture series)

[1796] Kentucky Speleofest. Guidebook (1972) *Guidebook to the Kentucky Speleofest, Meade County, Kentucky.* George, Angelo Isham, 1944–(ed). Louisville, KY, US: Louisville Grotto of the National Speleological Society, **1972**. illus: b&w: cartoons, graphs, maps, photos, tables; photo on front lining page; map on rear lining page; photo, cartoon on rear cover; photo on front cover; vi, 51p, [1]; bibl.

Cover title: Guidebook: Kentucky Speleofest 1972.

[1797] Kentucky Speleofest. Guidebook (1973) *Guidebook to the Kentucky Speleofest, Warren County, Kentucky.* George, Angelo Isham, 1944–(ed). Louisville, KY, US: Louisville Grotto of the National Speleological Society, **1973**. illus: b&w: charts, drawings, graphs, maps, photos, tables; text on front lining page; photo on front cover; [iii], 70p, [1]; bibl.

Cover title: Guidebook: Kentucky Speleofest 1973. Describes 21 caves and includes maps to many of them. Total of 46 figures of which only a few are actual photos.

[1798] Kentucky Speleofest. Guidebook (1974) *Kentucky Speleofest guidebook: selected caves for the annual summer field conference, the Louisville Grotto, NSS: Squire Boon Caverns, Indiana.* George, Angelo Isham, 1944–(ed); McCarty, L. R. (ed). Louisville, KY, US: Louisville Grotto of the National Speleological Society, **1974**. illus: b&w: ads, cartoons, maps, photos; photo, text on front lining page; photo on rear lining page; photo on front cover; ad, drawing on rear cover; [iv], ii, 28p, [1].

Cover title: Guidebook: Kentucky Speleofest 1974. Rear cover included in numbered main body pages.

[1799] Kentucky Speleofest. Guidebook (1975) *Ken-*

tucky Speleofest guidebook: annual summer field conference, the Louisville Grotto NSS. George, Angelo Isham, 1944–(ed). Louisville, KY, US: Louisville Grotto of the National Speleological Society, **1975**. illus: b&w: ads, cartoons, charts, drawings, maps, photos, tables; photo on front lining page; text on rear lining page; ad, drawing on rear cover; photo on front cover; [iv], 70p, [1]; bibl.

Cover title: Guidebook: Kentucky Speleofest 1975, Larue County, Kentucky. Main body pagination includes rear cover.

[1800] Kentucky Speleofest. Guidebook (1976) *Guidebook to the 5th annual Kentucky Speleofest. Annual summer field conference of the Louisville Grotto NSS: Camp Carlson, Meade County, Kentucky.* Gilkey, John L. (ed); Price, Wayne (assist). Louisville, KY, US: Louisville Grotto of the National Speleological Society, **1976**. illus: b&w: ads, cartoons, charts, graphs, maps, photos, tables; ads on front lining page; text on rear cover; photo on front cover; [ii], 92p, [2]; bibl.

Cover title: Guidebook: Kentucky Speleofest '76.

[1801] Kentucky Speleofest. Guidebook (1977) *Guidebook to the 1977 Kentucky Speleofest, Cedar Hill Camp, Edmonson Co., Ky.* George, Angelo Isham, 1944–(ed). Louisville, KY, US: Louisville Grotto of the National Speleological Society, **1977**. illus: b&w: ads, cartoons, charts, drawings, maps, photos, repros, tables; photos on lining pages; ad on rear cover; photo on front cover; Various pagination, 89p, [2]; bibl.

Cover title: Guidebook: Kentucky Speleofest '77: 6th annual summer field conference of the Louisville Grotto NSS.

[1802] Kentucky Speleofest. Guidebook (1978) *Guidebook to the 1978 Kentucky Speleofest: Camp Carlson, Meade County, Kentucky.* George, Angelo Isham, 1944–(ed). Louisville, KY, US: Louisville Grotto of the National Speleological Society, **1978**. illus: b&w: ads, cartoons, graphs, maps, tables; ad on front lining page; cartoon on rear lining page; ad on rear cover; photo on front cover; [iv], 49p, [1]; bibl.

Cover title: Guidebook to the 1978 Kentucky Speleofest: Camp Carlson, Meade Co. Ky.

[1803] Kentucky Speleofest. Guidebook (1979) *Guidebook: Kentucky Speleofest '79.* Gilkey, John L. (ed). Louisville, KY, US: Louisville Grotto of the National Speleological Society, **1979**. illus: b&w: ads, cartoons, charts, drawings, maps, photos; drawing on front lining page; photo on rear lining page; ad on rear cover; drawing on front cover; Various pagination; bibl.

Title from cover.

[1804] Kentucky Speleofest. Guidebook (1980) *Guidebook to the 1980 Kentucky Speleo-fest: annual summer field conference, Pulaski Co. Park, Ky.* George, Angelo Isham, 1944–(ed). Louisville, KY, US: Louisville Grotto of the National Speleological Society, **1980**. illus: b&w: ads, cartoons, drawings, maps, photos, tables; 4 loose folded map pls; drawing, text on front lining page; ad on rear cover; drawing on front cover; [ii], ii, 42p, [2].

Cover title: Speleofest 80.

[1805] Kentucky Speleofest. Guidebook (1987) *Guidebook for the 1987 Kentucky Speleofest: Renfro Valley KOA, Kentucky, May 22-25, 1987.* O'Dell, Gary A. (ed). npop: Blue Grass Grotto of the National Speleological Society, **1987**. illus: b&w: maps; 51p; 28 cm.

Illus.Title from cover.

[1806] Kentucky Speleofest. Guidebook (1989) *1989 Speleofest guidebook. May 26-29, 1989, Sandhill Campground, Whitley City, Kentucky.* Stecko, Doug (ed); Kessel, Jay (ed). [KY, US]: Miami Valley Grotto of the National Speleological Society, **1989**. v 18 Speleofest Guidebook. illus: b&w: ads, charts, drawings, maps, photos, tables; ads on lining pages and rear cover; drawing on front cover; 54p, [2]; bibl; 28 cm.

Cover title: Speleofest '89, Whitley City, Kentucky.

[1807] Kentucky Speleofest. Guidebook (1997) *1997 Speleofest: Cadiz, Ky, Trigg County Recreational Complex, Cadiz, Kentucky.* npop: Golden Pond Grotto and Louisville Grotto of the National Speleological Society, **1997**. 60p.

[1808] Kerans, C., 1954– *Depositional systems and karst geology of the Ellenburger Group (Lower Ordovician), subsurface west Texas.* Austin, TX, US: Bureau of Economic Geology, University of Texas at Austin, **1990**. Rept. of Investigations University of Texas at Austin Bureau of Economic Geology no 193. illus: b&w: maps; Six folded leaves of plates in pocket; v, 63p, 6p pls; bibl; 28 cm.

[1809] Kerans, C., 1954– *Karst-controlled reservoir heterogeneity and an example from the Ellenburger Group (Lower Ordovician) of west Texas.* Austin, TX, US: Bureau of Economic Geology, University of Texas at Austin, **1989**. Report of Investigations (University of Texas at Austin Bureau of Economic Geology) no186. iv, 40p; bibl; 28 cm.

[1810] Kerbo, Ronal Carrel, 1944– *Cave resources management, planning, restoration, and redevelopment, Cathedral Caverns State Park, Grant, Alabama, 10-14 January 1994.* Santa Fe, NM, US, Mar, **1994**. [66].

Attachment 1: Carlsbad Caverns National Park, Cave management plan, pp 32-54. Attachment 2: Lint in caves pp 55-60. Attachment 3: Survey standards for Carlsbad Caverns National Park, pp 61-66.

[1811] Kerbo, Ronal Carrel, 1944– *Caves.* Chicago, IL, US: Childrens Press, **1981**. A Radlauer Geo Book. An Elk Grove Book. illus: col: photos; 48p; gloss, index; 25 cm.

Juvenile.

"Discusses how caves are formed and what a spelunker might find while exploring a cavern."

[1812] Kerry, Charles (photo) *Right Imperial: Carlotta*

Arch: Devil's Coach House. npop: np, **nd**. illus: b&w: photos; [32].

Photo captions are the only text. Date is approximately 1905 (Ellis, per comm).

[1813] Kessler, Hubert, 1907-1994 *Aggtelek*. Helvey, Elek (trans); Scott, Len (trans re); Csiby, Mihály (design); Várhegyi, Sándor (map-design). Budapest, HU: Panoráma, **1971**. illus: b&w: drawings, photos, repros; partly col: maps; col: photos; 29p, 19p pls.

[1814] Keyes, Elizabeth *Titania's cave; a fairy story of the Endless Caverns of Virginia*. New York, NY, US: Nomad, **1930**. 20p; 21 cm.

Illus.

[1815] Kidder, Daniel Parish, 1815-1891 *The caves of the earth: their natural history, features, and incidents*. New York, NY, US: Published by Lane & Tippett, for the Sunday School Union of the Methodist Episcopal Church, **1848**. Youth's Library 418. illus: b&w: ads; front; 187p, [4]; 15 cm.

Some versions list the Religious Tract Society as the only visible author. Copy viewed had Carlton and Porter as the publisher and no date. Includes publisher's advertisements: [4] p. at end. On spine: 418. Juvenile.

[1816] Kiernan, Kevin *An atlas of Tasmanian karst*. [TAS, AU]: Tasmanian Forest Research Council Inc, **1995**. Research Report no 10. illus: b&w: maps; partly col: charts; 255+351p.

2 vols, 2 maps.

[1817] Kiernan, Kevin *Bibliography of Tasmanian karst*. Ellis, Ross Andrew, 1940–(asst). Sydney, AU: Sydney Speleological Society in association with the Australian Speleological Federation Commission on Bibliography, Sep, **1989**. Australian Speleological Federation Commission on Bibliography Regional Bibliography Series 1. illus: b&w: ads, maps; viii, 18p; add, bibl; 30 x 21 cm.

[1818] Kiernan, Kevin *Karst, caves and management at Mole Creek, Tasmania*. Tasmania Forestry Commission, National Parks and Wildlife Service, Tasmania. Hobart, TAS, AU: Department of Parks, Wildlife and Heritage, Nov, **1989**. Department of Parks Wildlife and Heritage Occasional paper no 22. illus: b&w: charts, maps, photos, repros, tables; ix, 130p; app, bibl; 30 cm.

A report to the Forestry Commission and National Parks & Wildlife Service, Tasmania, 1984.

[1819] Kiernan, Kevin *The management of soluble rock landscapes: an Australian perspective*. Sydney, NSW, AU: The Speleological Research Council Ltd, **1988**. illus: b&w: tables; vi, 61p; bibl.

Extremely useful compilation of karst geomorphology and associated management problems. The extensive bibliography is very complete.

[1820] Kimball, Herbert Harvey, 1862– *Ice caves and frozen wells as meteorological phenomena ....* [Washington, US]: np, **1901**. 15p, 7p pls.

Reprinted from the monthly Weather review for August, 1901. Caption title.

[1821] King, Mary Elizabeth *The Salts Cave textiles: a preliminary account*. New York, NY, US: Academic Press, **1974**. 10p; 24 cm.

"Reprinted from Archeology of the Mammoth Cave area: New York, San Francisco, London: Academic Press, c1974." illus.

A chapter contained within Patty Jo Watson's book, *Archaeology of the Mammoth Cave Area*, and one of the first and most thorough discussions about prehistoric textiles in the U.S. Eastern Woodlands.

[1822] Kirk, Keith G. *Handbook of geophysical cavity-locating techniques: with emphasis on electrical resistivity*. Werner, Eberhard Wolfgang, 1942–. Washington, DC, US: U.S. Department of Transportation, Federal Highway Administration, Apr, **1981**. FHWA-IP-81-3. illus: b&w: maps; viii, 175p; bibl; 28 cm.

"Implementation package."

[1823] Kirk, Ruth *The oldest man in America: an adventure in archeology*. New York, NY, US: Harcourt Brace Jovanovich, Inc, **1970**. illus: b&w: drawings, maps, photos; 95p, [1]; 22 cm.

".. exploration of the Marmes rock shelter on the Palouse River in Southeastern Washington."

[1824] Kiver, Eugene P. *Summit firn cave study, 1970-1973, Mount Rainier, Washington*. npop: np, **1973**. 51p; bibl; 29 cm.

Illus.

[1825] Klamath Mountains Conservation Task Force *A preliminary report on the caves of the Marble Mountain Wilderness*. Knutson, Richard Stephen; Sims, Lynne; Sims, Mike. npop: np, **1977**. KMCTF Technical Report No. 2. illus: b&w: maps; [58]; gloss.

[1826] Klippel, Walter E. *The paleontology of Cheek Bend Cave, Maury County, Tennessee: phase II: report to the Tennessee Valley Authority*. Parmalee, Paul Woodburn. [Chattanooga, TN, US]: Tennessee Valley Authority, **nd**. TVA/ONRED/L&ER/86/54. illus: b&w: maps; xiii, 249p; bibl; 28 cm.

"Investigations conducted in accordance with Tennessee Valley Authority contract (TVA TV-49244A—-fieldwork; TVATV-53013A—-laboratory analyses)." illus.

[1827] Klippel, Walter E. *The unmodified vertebrate fauna from Granite Quarry Cave (23CT36), Carter County, Missouri*. Snyder, Lynn M.; Parmalee, Paul Woodburn; Midwest Archeological Center (U.S.) University of Tennessee (Knoxville campus). Knoxville, TN, US: The University, Aug, **1986**. ii, 59p; bibl; 28 cm.

Title from cover. Cover title. Purchase order no. PX-6115-6-0087. illus.

[1828] Kmecl, Matjaž *Postojna caves: enter traveller into this immensity!*. Habič, Peter (ed); Šajn, Srečko (ed);

Čeh, Anne (trans); Golob, Franc (photo);et al. (photo). Postojna, YU: Postojna jama, Turizem, **1990**. illus: b&w: photos; partly col: maps; col: drawings, photos; Repros front & rear; 140p; 23 cm.

English ed; other eds published in Slovene, Serbo-Croatian, German, French, Italian, and Spanish.

[1829] Knapp, Brian J. *Cave.* 1992 ed. South Melbourne, VIC, AU: Macmillan Education Australia, **1992**. Landshapes 8. illus: col: maps; 37p; index; 29 cm.

Col illus. Also issued by Atlantic Europe in 1992. Juvenile.

1993 ed. Danbury, CT, US: Grolier Educational Corp, **1993**. Landshapes 8. illus: col: maps; illus on lining papers; 37p; 29 cm.

Col illus. Juvenile.

Describes the underground world of limestone caves that are forever hidden from the light of day.

[1830] Knauth, Otto *Cold Water Cave.* Kelly, Herb. [Des Moines, IA, US]: [Public Library of Des Moines], **1975**. Iowa Oral History Project.

Includes typewritten summary. Knauth, a reporter for the Des Moines Register, discusses the discovery of the Cold Water Cave discovered in northeast Iowa in 1967. 1 sound cassette: 1 7/8 ips, mono.

[1831] Knight, E. Leslie *Caves of Mississippi.* Irby, Bobby N.; Carey, Steven D. Hattiesburg, MS, US: University of Southern Mississippi, in cooperation with Southern Mississippi Grotto of the National Speleological Society, **1974**. illus: b&w: drawings, maps, photos; [vi], 93p; bibl; 22 cm.

Catalog of 43 known caves, all small, in this cave-poor state, with maps of 30 of them.

[1832] Knight, J. C. *The Lapiaz Superieure du Pla Segoune.* [Ledbury, GB]: Cave Research Group of Great Britain, **1962**. Occasional Publications - Cave Research Group of Great Britain no 6. illus: b&w: maps; iii, 21p; 24 cm.

[1833] Knisel, Walter G. *Response of karst aquifers to recharge.* For t Collins, CO, US: Colorado State University, **1972**. Hydrology Papers (Colorado State University) no 60. illus: b&w: maps; vi, 50p; 29 cm.

Illus.

[1834] Knoepfel, William H. *An account of Knoepfel's Schoharie Cave, Schoharie County, New York: with the history of its discovery, subterranean lake, minerals and natural curiosities.* New York, NY, US: W.E. & J. Sibell, **1853**. illus: b&w: drawings, maps; 16p, 2p pls; 24 cm.

Cover title: Knoepfel's subterranean cave and lake, at Schoharie. Copyrighted by W.H. Knoepfel. Illustrated with engravings.

[1835] Knott, David L. *Current foundation engineering practice for structures in karst areas.* Rojas-Gonzalez, Luis F.; Newman, F. Barry. Harrisburg, PA, US: Pennsylvania Dept. of Transportation, Office of Research and Special Studies, **1993**. illus: b&w: maps; Various pagination, vii, 442p; 28 cm.

Illus.

[1836] Knott, David L. *Guide for foundation engineering in Pennsylvania karst.* Rojas-Gonzalez, Luis F.; Newman, F. Barry. Harrisburg, PA, US: Pennsylvania Dept. of Transportation, Office of Research and Special Studies, **1992**. illus: b&w: maps; Various pagination, 191p; 28 cm.

Illus.

[1837] Knox, Orion *Natural Bridge Caverns.* Berkeley, CA, US: Mike Roberts Color Productions, **1976**. illus: col: photos; 28p.

[1838] Knutson, Richard Stephen (comp) *Bibliography of Oregon speleology: initial compilation.* Halliday, William Ross, 1926–(comp). [Seattle, WA, US]: [Western Speleological Survey; Oregon Speleological Survey], **1966**. Western Speleological Survey Serial 37; Oregon Speleological Survey Bulletin 3. 12p; 28 cm.

[1839] Knutson, Richard Stephen *The caves of Deschutes County, Oregon.* npop: [Western Speleological Survey], nd. Western Speleological Survey Serial 35; Oregon Speleological Survey Bulletin 2.

[1840] Knutson, Richard Stephen *The speleological potential of the Big Bar Limestone deposit, Hell's Canyon, Oregon-Idaho.* npop: [Western Speleological Survey], **nd**. Western Speleological Survey Serial 33; Oregon Speleological Survey Bulletin 1.

[1841] Kobert, Bill *Caverns of the Ozarks.* Berkeley, CA, US: Mike Roberts Color Production, **nd**. illus: col: photos; 24p; 14 x 18 cm.

Contains 31 col photos.

[1842] Koch, Donald L. *Report on Cold Water Cave: a summary of research results with inclusion of information on related to potential development of a new recreational facility by the state of Iowa.* Case, James D. npop: np, Dec, **1974**. illus: b&w: drawings, graphs, maps, tables; 1 map folded in pocket; [i], iii, 80p.

16 figs, 9 tables.

[1843] Koehler, W. H. *Lure of the labyrinth.* Costa Mesa, CA, US: Professional Association of Diving Instructors, **1971**. 35p; 22 cm.

Illus.

[1844] Kohary, Jana *Archaeology and ecology of Rose Cottage Cave, Orange Free State.* npop: np, **1988**. vii, 362p; bibl; 30 cm.

Illus.

[1845] Kohler, S. D. F. *The Crystal Cave at Virginsville.* reprint. npop: American Speleal History Association, **1971**. 10p, [1].

Reprint of a 1874 publication. Cover says printing limited to 35 reproductions.

[1846] Kollár, Attila K. (comp) *Subaquatic caves in Hungary for divers.* Budapest, HU: 10th International

Congress of Speleology, **1989**. Field Trip Guide D5. illus: b&w: maps; [ii], 36p; bibl.

[1847] Komatsu, Toshihiro *Cave spiders of Japan; their taxonomy, chorology, and ecology.* Toa Kumo Gakkai. Nagano City, JP: Arachnological Society of East Asia, **1961**. illus: b&w: tables; 91p; 27 cm.

Title from cover. Illus.

[1848] Kornicker, Louis S., 1919– *New Ostracoda (Halocyprida: Thaumatocyprididae and Halocyprididae) from anchialine caves in the Bahamas, Palau, and Mexico.* Iliffe, Thomas M. Washington, DC, US: Smithsonian Institution Press, **1989**. Smithsonian Contributions to Zoology no 470. illus: b&w: charts, drawings, maps, photos, photomicrographs, tables; text front & rear; iii, 47p; bibl; 28 cm.

Illus.

[1849] Kornicker, Louis S., 1919– *Ostracoda (Halocyprididae) from anchialine caves in the Bahamas.* Yager, Jill; Williams, Dennis. Washington, DC, US: Smithsonian Institution Press, **1990**. Smithsonian Contributions to Zoology no 495. iii, 51p; bibl; 28 cm.

Microfiche. [Washington, D.C.?]: SUPTDOCS/GPO, [1991].11 x 15 cm. Illus.

[1850] Kornicker, Louis S., 1919– *Ostracoda (Halocypridina, Cladocopina) from anchialine caves in Jamaica, West Indies.* Iliffe, Thomas M. Washington, DC, US: Smithsonian Institution Press, **1992**. Smithsonian Contributions to Zoology no 530. illus: b&w: maps; iii, 22p; bibl; 28 cm.

Illus.

[1851] Kornicker, Louis S., 1919– *The troglobitic halocyprid Ostracoda of anchialine caves in Cuba.* Yager, Jill. Washington, DC, US: Smithsonian Institution Press, **1996**. Smithsonian Contributions to Zoology no 580. illus: b&w: maps; iii, 16p; bibl; 28 cm.

Illus.

[1852] Kornicker, Louis S., 1919– *Troglobitic Ostracoda (Myodocopa: Cypridinidae, Thaumatocyprididae) from anchialine pools on Santa Cruz Island, Galapagos Islands.* Iliffe, Thomas M. Washington, DC, US: Smithsonian Institution Press, **1989**. Smithsonian Contributions to Zoology no 483. iii, 38p; bibl; 28 cm.

Illus.

[1853] Kosa, Attila (ed) *Bir Al Ghanam karst study project: final report 1981.* Tripoli, LY: Socialist People's Libyan Arab Jamahiriya, **1981**. illus: b&w: maps, photos, tables; col: photos; many fold-out maps in text; fold-out maps in rear pocket; 79p, 80p pls; bibl; 29 x 20.5 cm.

[1854] Kósa, Attila *The caver's living dictionary.* Chabert, Claude (cont); Hazslinszky, Tamás (cont); Holzmann, Heinz (cont); Marek-Limagne, Edit (cont); Oldham, Anthony Clive, 1939–(cont). Budapest, HU: Hungarian Speleological Society, **1995**. test on front lining page; 50p; bibl.

Limited ed of approx 100 copies. No illus.

[1855] Koul, Samsar Chand *The mysterious cave of Amar Nath, Kashmir India.* Koul, Lokesh (rev). 4th ed. Srinagar, IN: np, **c1980**. illus: b&w: ads; [xiii], 26p, [1]; 18 x 13.5 cm.

Printed by India Packaging Printers. Photocopy examined.

[1856] Kozlowski, Janusz Krzysztof *Excavation in the Bacho Kiro Cave (Bulgaria): final report.* Ginter, Boleslaw. Warszawa, PL: Panstwowe Wydawnictwo Naukowe, **1982**. 172p; bibl, errat; 28 cm.

Illus.

[1857] Kozlowski, Janusz Krzysztof *Meso and Neolithic sequence from the Odmut Cave (Montenegro).* Kozlowski, Stefan Karol; Radovanovic, Ivana. Warszawa, PL: Wydawnictwa Uniwersytetu Warszawskiego, **1994**. 70p; bibl; 24 cm.

Illus.

[1858] Kozlowski, Janusz Krzysztof *Temnata cave: excavations in Karlukovo karst area Bulgaria.* Laville, Henri; Ginter, Boleslaw. Krakow, PL: Jagellonian University Press, **1992**. Jagellonian University Press Varia tome 299. illus: b&w: maps; 11 folded lvs of pls; 501p, 11p pls; bibl; 24 cm.

Illus.

[1859] Kramer, Stephen P. *Caves.* Day, Kenrick L. (photo). Minneapolis, MN, US: Carolrhoda Books, Inc, **1995**. Nature in Action. illus: b&w: repros, tables; partly col: drawings, maps; col: photos; color photo; 48p; gloss, index; 22 x 25 cm.

Juvenile.

[1860] Kranjc, Andrej (ed) *Proceedings of international symposium at 170th anniversary of Postojnska jama, Postojna (Yugoslavia) Nov. 10-12 1988.* Centre of Scientific Research of SAZU, The Institute of Karst (ed). Postojna, YU: Postojaska Jama, Tourist and Hotel Organization, **1989**. illus: b&w: graphs, maps, photos, repros, tables; 204p; 24 x 17 cm.

Limited ed of 200 copies. Cover title: Cave tourism. Contains papers given at symposium. 42 papers - 29 in English. Well illustrated articles on show caves.

[1861] Kranjc, Andrej, 1943– *The contribution to the history of the speleological explorations of the West Indies at 500—anniversary of the discovery of America.* [Ljubljana, SI]: Slovene Academy of Sciences and Arts, **1993**. Acta Carsologica volume 22. illus: b&w: drawings, graphs, maps, photos, ports, repros; 209p, [2]; bibl, errat; 24 cm.

An extremely important work on the caving history of the West Indies and Bermuda, separated into sections covering different islands or regions. This should serve as the initial reference for any work on the caves of this area.

[1862] Kranjc, Andrej, 1943– *Proceedings of international symposium "man on karst," Postojana, Septem-*

ber 23-25, 1993. Ljubljana, SI: Karst Research Institute, Slovenian Academy of Science and Art, **1995**. Acta Carsologica vol 24. illus: b&w: ads, charts, drawings, graphs, maps, photos, repros, tables; 592p; bibl.

English with Slovene summaries.

[1863] Kranjc, Andrej, 1943– *Recent fluvial cave sediments, their origin and role in speleogenesis = Recentni fluvialni jamski sedimenti, njihovo nastajanje in vloga v speleogenezi*. Ljubljana, SI: Slovenska Akademija Znanosti in Umetnosti, **1989**. (Slovenska Akadenija in Umetnosti. Razred za Naravoslovne Vede) ; Dela 27. 167p; bibl; 24 cm.

Summary in Slovenian. Illus.

[1864] Kranzel, Rich *Caves of Bucks County, Pennsylvania*. State College, PA, US: National Speleological Society, Mid-Appalachian Region, **1983**. MAR Bulletin no 14. illus: b&w: maps, photos, repros, tables; b&w photo on rear lining page; 42p; bibl; 28 cm.

[1865] Kranzel, Rich *Caves of Lehigh County, Pennsylvania*. State College, PA, US: National Speleological Society, Mid-Appalachian Region, Apr, **1985**. MAR Bulletin no 15. illus: b&w: maps, photos, tables; 1 foldout map; 34p, 1p pls; bibl; 28 cm.

[1866] Kristofferson, N. R. *Kaver komix*. Carta Valley, TX, US: Carta Valley Press, **1972**. illus: b&w: drawings; col: drawings; 20p.

One of the only x rated cave books ever published.

[1867] Kühn, Herbert, 1895– *The rock pictures of Europe*. Brodrick, Alan Houghton (trans). [1st] ed. London, GB: Sidgwick and Jackson, **1956**. illus: b&w: drawings, maps, photos; xxix, 230p, 80p pls; index; 22 cm.

91 pls; 144 figs; Translation of Die Felsbilder Europas. Also published in the U.S. by Essential Books, Inc., Fairlawn, NJ (US 1st ed.)

[2nd] ed. 1966.

[3rd] ed. New York, NY, US: October House, Inc, **1967**.

[1868] Kühn, Herbert, 1895– *Symposium on cave sites and explorations in Upper Rhine regions*. Kahrs, Ernst; Zepp, Peter; Loeser, Rudolf. Dusseldorf, DE: L. Schwann, **1932**. Nachrichtenblatt fur Rheinische HeimAtpflege ... 4. jahrg hft7/8. illus: b&w: maps; Various pagination, [40]; 24 cm.

Arbitrary title. Illus. Die rheinischen hohlen und die vorgeschichte Von H. Kuhn p. 306- 309.—Die Von E. Kahrs. p. 321-326.—Hohlen der Eifel. Von P. Zepp. p.339-351.—Hohlen des Hunsrucks. Von P. Zepp. p. 356-364.—Hohlen des Saargebiets. Von R. Loeser. p. 365-371.

[1869] Kuhns, Roger J. *Speleology of Lewis and Clark Caverns, Montana*. Butte, MT, US: Montana Bureau of Mines and Geology, **1982**. Open-file Report Montana Bureau of Mines and Geology 88. illus: b&w: maps; Seven folded maps in pocket; 30p; bibl; 28 cm.

Illus. Photocopy.

[1870] Kunath, Carl Edwin, 1940– *The caves of Carta Valley, Edwards and Val Verde Counties, Texas*. Austin, TX, US: Texas Speleological Survey, Oct, **1995**. illus: b&w: charts, graphs, photos, tables; col: maps, photos; 1 foldout col map; 2 foldout b&w maps; 109p; bibl, errat; 28 cm.

Illus.

[1871] Kunath, Carl Edwin, 1940– *The caves of McKittrick Hill, Eddy County, New Mexico: history and results of field work 1965-1976*. Austin, TX, US: Texas Speleological Survey, **1978**. illus: b&w: drawings, maps, photos; 8 maps folded and tipped into text, 4 maps folded and loose in back; front; 87p; bibl; 28 cm.

619 numbered copies.

A survey, with descriptions and maps, of these mazy caves. Includes chapters on biology, geology and paleontology.

[1872] Kunath, Carl Edwin, 1940– *The caves of the Stockton Plateau, Texas*. Smith, Richard A. Austin, TX, US: Texas Speleological Association, Apr, **1968**. Texas Speleological Survey v 3 no 2. illus: b&w: maps; partly col: maps; 111p; bibl; 28 cm.

Illus.

[1873] Kunath, Carl Edwin, 1940– *Ye olde history*. TX, US: np, Mar, **1965**. illus: b&w: tables; [iii], 60p; bibl.

This booklet tells the history of the Texas Speleological Association from its beginning.

[1874] Kunsky, Josef, 1903-1977 *Homes of primeval man: wandering in the caves of Czechoslovakia*. Prague, CZ: Artia, **1954**. illus: b&w: photos; [200]; 29 cm.

Coffee table b&w photo book on Bohemia, Moravia/Silesia and Slovakia. Also published in other languages.

[1875] Kurtén, Björn *The Carnivora of the Palestine caves*. Helsinki, FI: np, **1965**. Acta Zoologica Fennica 107. 74p; bibl; 26 cm.

Illus.

[1876] Kurtén, Björn *The cave bear story: life and death of a vanished animal*. NY, US: Columbia University Press, **1976**. illus: b&w: charts, drawings, graphs, maps, photos, tables; xi, 163p; bibl, index.

[1877] Kurtén, Björn, 1924-1988 *Life and death of the Pleistocene cave bear: a study in paleoecology*. Helsinki, FI: Geological Institute of Helsingfors University, **1958**. Acta Zoologica Fennica 95. illus: b&w: drawings, graphs, tables; 59p; bibl; 26 cm.

Illus.

[1878] L'Association Francaise de Karstologie, Museum d'Histoire Naturelle *Actes du symposium international sur l'erosion karstique = proceedings of the international symposium on karstic erosion: Aix—en-Provence-Marseille-Nimes 10-14 Septembre 1979*. Union Internationale de Spéléologie, Commission de l'Erosion du Karst; Association Francaise de Karstologie, Commission des Phénomènes Karstiques. Nîmes, FR: np, **1979**. Mémoire no 1 de l'A.F.K. illus: b&w: charts, graphs, maps, photos, photomicrographs, tables; iii, 290p, 1p pls; bibl.

Le Campion-Alsumard, Th'er'ese, le biokarst marin: role des organismes perforants, pp133-140. Partly in English.

[1879] **La Rock, Edward J.** *The caves of Snyder County, Pennsylvania.* Philadelphia, PA, US: np, Dec, **1976**. Mar Bulletin no 10, Philadelphia Grotto Digest vol 15. illus: b&w: maps, photos; photo on rear lining page; i, 30p; bibl.

[1880] **LaFleur, R.G. (ed)** *Groundwater as a geomorphic agent.* Boston, MA, US: Allen and Unwin, **1984**. The "Binghamton" symposia in geomorphology, no 13. illus: b&w: photos; xvi, 390p; bibl, index; 25 cm.

Illus.

[1881] **Laine, Edwin F.** *Detection of water-filled and air-filled underground cavities.* United States Army Engineer Waterways Experiment Station; United States Bureau of Mines. Livermore; Springfield, CA; VA, US: Lawrence Livermore Laboratory, University of California, **1980**. 15p; 27 cm.

Manuscript date: December 1, 1980. Work performed for the United States Dept. of Interior, Bureau of Mines. Illus.

[1882] **Lamb, Susan** *Lava Beds National Monument.* Tulelake, CA, US: Lava Beds National Monument, **1991**. illus: b&w: drawings; partly col: maps; col: photos; 48p; 28 x 23.5 cm.

[1883] **Lambert, Marjorie F.** *A survey and excavation of caves in Hidalgo County, New Mexico.* Ambler, J. Richard. Santa Fe, NM, US: The School of American Research, **1961**. Monograph no 25 The School of American Research. illus: b&w: drawings, maps, photos, tables; xvi, 107p; bibl; 28 cm.

Illus.

[1884] **Laming-Emperaire, Annette**, 1917-1977 *The cave of Lascaux.* Roussel, Monique. Paris, FR: Caisse National des Monuments Historiques, **1950**. 30p; 18 cm.

Translation of La Grotte de Lascaux. Illus.

[1885] **Laming-Emperaire, Annette**, 1917-1977 *Lascaux: paintings and engravings.* Armstrong, Frances Eleanore (trans). Hammondsworth, Middlesex, GB: Penguin Books Ltd, **1959**. illus: b&w: maps, photos, repros, tables; 208p, 48p pls; bibl, index; 18 cm.

[1886] **LaMoreaux, Philip Elmer**, 1920– *Annotated bibliography of karst terranes: volume five: with three review articles.* Assaad, Fakhry A (ed); McCarley, Ann (ed); International Association of Hydrogeologists. Hannover, DE: Verlag Heinz Heise GmbH & Co KG, **1993**. International Contributions to Hydrogeology v 14. illus: b&w: drawings, graphs; [v], xvi, 425p; bibl.

See also LaMoreaux, et al. (1970, 1975, 1984, 1986, 1989). This, and the other works by LaMoreaux, represents a tremendous tool: lots of good material from odd places.

[1887] **LaMoreaux, Philip Elmer**, 1920– *Guide to the hydrology of carbonate rocks.* Wilson, Betty Morere; Memon, Beshir A. (co-eds). Paris, FR: United Nations Educational, Scientific and Cultural Organization, **1984**. Studies and Reports in Hydrology 41. illus: b&w: ads, drawings, graphs, tables; 345p, [2]; bibl, index.

Updating and editorial responsibility for the final version provided by Phillip E. LaMoreaux.

One part of a very comprehensive, indexed bibliography of limestone hydrology. Has many obscure references not found elsewhere. See other listings for P. LaMoreaux, et al. (1970, 1975, 1986, 1989, 1993). Also has listing of important karst meetings and symposia.

[1888] **LaMoreaux, Philip Elmer**, 1920– *Hydrology of limestone terranes: annotated bibliography of carbonate rocks.* Raymond, Dorothy; Joiner, Thomas J. University, AL, US: Geological Survey of Alabama, **1970**. Geological Survey of Alabama Bulletin 94 Part A State publications program no EN000-057. illus: b&w: photos, tables; col: photos; col and b&w photo on front lining page; viii, 242p; bibl, index; 23 cm.

A U.S. contribution to the international hydrological decade. In cooperation with Department of Geology and Geography, University of Alabama; Office of Water Resources Research of the Department of the Interior; Water Resources Research Institute of Auburn University; and the U.S. National Committee for the International Hydrologic Decade of the U.S. National Academy of Sciences-National Research Council. Microfiche. Englewood, Colo.: Information Handling Services, 1977. 3 microfiches: negative; 11 x 15 cm. (State publications program; EN000;057).

See also LaMoreaux, et al. (1975, 1984, 1986, 1989, 1993).

[1889] **LaMoreaux, Philip Elmer**, 1920– *Hydrology of limestone terranes: annotated bibliography of carbonate rocks, volume 4.* Prohic, Esad; Zoetl, Josef; Tanner, J. Mark; Roche, Bernadette N. Hannover, DE: Vorlan Heinz Heise GmbH & Co, **1989**. International Contribution to Hydrogeology volume 10. illus: b&w: drawings, maps, photos, repros; [iii], xvi, 267p; bibl, index.

See also LaMoreaux, et al (1970, 1975, 1984, 1989, 1993). Also has listing of important karst meetings and symposia.

[1890] **LaMoreaux, Philip Elmer**, 1920– *Hydrology of limestone terranes: annotated bibliography of carbonate rocks, volume three.* Tanner, J. Mark (ed); ShoreDavis, P. (ed); International Association of Hydrogeologists, Karst Commission. Hannover, DE: Verlag Heinz Heise GmbH, **1986**. International Contributions to Hydrogeology v 2. [iii], xix, 341p; bibl, index.

Contribution to UNESCO IHP Project A.1.13.

See also LaMoreaux, *et al.* (1970, 1975, 1984, 1989, 1993).

[1891] LaMoreaux, Philip Elmer, 1920– *Hydrology of limestone terranes: progress of knowledge about hydrology of carbonate terranes with an annotated bibliography of carbonate rocks.* LeGrand, Harry Elwood, 1917– ; Stringfield, Victor Timothy, 1902–; Tolson, Janyth S.; Warren, William M.; Moore, James D. (ann bib). University, AL, US: Geological Survey of Alabama, Jan, **1975**. Geological Survey of Alabama Bulletin 94 Part E. illus: b&w: tables; vii, 168p; bibl, index; 23 cm.

In cooperation with U.S. National Committee for the International Hydrologic Decade of the U.S. National Academy of Sciences-National Research Council.

A handy book for ferreting out some of the earlier studies: obviously dated. See also LaMoreaux, *et al.* (1970, 1984, 1986, 1989, 1993).

[1892] Lan-po Chia *The cave home of Peking Man.* Peking, CN: Foreign Languages Press, **1975**. illus: b&w: charts, maps, photos, repros, tables; partly col: photos; col: photos; [viii], 52p, 18p pls; bibl; 19 cm.

Archeology of Peking Man Cave.

[1893] Lan-po Chia *The story of Peking Man, from archaeology to mystery.* Wei-wen Huang. Beijing; HongKong; New York, CN; HK; US: Foreign Languages Press; Oxford University Press, **1990**. vi, 270p; 25 cm.

Translation from Chinese of Chou-kou-tien fa chueh chi. Illus.

[1894] Lane, Edward, 1935– *Environmental geology and hydrogeology of the Ocala area, Florida.* Hoenstine, Ronald. Tallahassee, FL, US: State of Florida, Dept. of Natural Resources, Division of Resource Management, Florida Geological Survey, **1991**. Special Publications (Florida Geological Survey) no. 31. illus: b&w: maps; col: maps; vii, 71p; 28 cm.

Some col illus and maps.

Lane has done a lot in Florida- this just scratches the surface of his work.

[1895] Lane, Edward, 1935– *Karst in Florida.* Florida Geological Survey. Tallahassee, FL, US: Published for the Florida Geological Survey [by] State of Florida, Dept. of Natural Resources, Division of Resource Management, Bureau of Geology, **1986**. Special publication Florida Bureau of Geology no 29. illus: b&w: graphs, maps, photos, tables; partly col: drawings; viii, 100p; bibl; 23 cm.

[1896] Lange, Arthur L. *Stream piracy and cave development along Baker Creek, Nevada.* Santa Barbara, CA, US: Western Speleological Institute, Jul, **1958**. Western Speleological Institute Bulletin no 1. illus: b&w: maps, photos; foldout at end w/map on front, b&w photos on rear; 20p, 4p pls; bibl; 25 cm.

[1897] Lange, Arthur L. *Studies on the origin of Montezuma Well and cave, Arizona.* San Francisco, CA, US: Cave Research Associates, **1957**. Cave Studies no 9. illus: b&w: maps; 1 folded map; 14p; 28 cm.

Project of the Western Speleological Institute.

[1898] Langford, R.L. (ed) *Karst geology in Hong Kong: proceedings of a conference on "Karst geology in Hong Kong" held at the University of Hong Kong on 6 January 1990.* Hansen, A. (ed); Shaw, R. (ed). Hong Kong, HK: Geological Society of Hong Kong, **1990**. Bulletin - Geological Society of Hong Kong no 4, 1990. illus: b&w: charts, drawings, graphs, maps, photos, photomicrographs, tables; text on lining pages; 243p; bibl; 30 cm.

[1899] Langford, R. L. (ed) *Karst geology in Hong Kong: programme and abstracts: a conference organized by the Geological Society of Hong Kong and Department of Geography and Geology, University of Hong Kong: Programme and Abstracts: University of Hong Kong: 4-6 January 1990.* Hansen, A. (ed); Shaw, R. (ed); University of Hong Kong Department of Geography and Geology. Hong Kong, HK: Geological Society of Hong Kong, **1990**. Abstracts - Geological Society of Hong Kong no 6. illus: b&w: charts, drawings, graphs, maps, tables; 58p; 30 cm.

[1900] Larimore, Betty *Exploring the Endless Caverns of New Market Virginia on US 11 in the heart of the Shenandoah Valley.* Ashton, Horace; Cowling, Herford Tynes, 1890–. New Market, VA, US: Endless Caverns, Inc, **1947**. illus: b&w: maps, photos, repros; 40p.

Cover title: Mountain climbing underground in the Endless Caverns of New Market, Virginia. Compilation of stories of the original exploration of the cave.

[1901] Larsen, Helge *Trail Creek: final report on the excavation of two caves on Seward Peninsula, Alaska.* Kobenhavn, [AK, US]: Ejnar Munksgaard, **1968**. Acta Arctica fasc 15. 78p, 10p pls; bibl; 29 cm.

Illus.

[1902] Larson, Charles Victor, 1928– *Bibliography of Oregon speleology, 1977.* Vancouver, WA, US: Oregon Speleological Survey, **1977**. Bulletin Oregon Speleological Survey 6; Western Speleological Survey serial no 55. illus: b&w: maps, photos; x, 95p; bibl; 28 cm.

[1903] Larson, Charles Victor, 1928– *A bibliography of Sea Lion Cave.* npop, [US]: np, **nd**. Bulletin Oregon Speleological Survey no 5; Western Speleological Survey Serial no 52.

[1904] Larson, Charles Victor, 1928– *Caves of Oregon.* Larson, Jo. Vancouver, WA, US: np, **1975**. Western Speleological Society Serial Bulletin no 49. illus: b&w: drawings, maps; 43p; 28 cm.

Originally prepared by the Audio-Visual Aids Committee of the National Speleological Society for use with set of 103 accompanying slides.

[1905] Larson, Charles Victor, 1928– *Central Oregon caves.* Larson, Jo. Vancouver, WA, US: ABC Printing &

Publishing, **1987**. illus: b&w: drawings, maps, photos; 44p; bibl, gloss; 24 cm.

[1906] **Larson, Charles Victor**, 1928– *An illustrated glossary of lava tube features.* Vancouver, WA, US: [Western Speleological Survey], **1993**. Western Speleological Survey Bulletin no 87. illus: b&w: photos; 56p; bibl, gloss, index.

174 references in glossary, 121 photos, alphabetical list of over 1100 citations.

[1907] **Larson, Charles Victor**, 1928– *Lava Beds caves.* Larson, Jo. Vancouver, WA, US: ABC Publishing, **1989**. illus: b&w: drawings, maps, photos; partly col: maps; 56p; bibl, gloss; 23.5 x 15.5 cm.

Reprinted 1990.

[1908] **Larson, Charles Victor**, 1928– *Lava River Cave.* Larson, Jo. Vancouver, WA, US: ABC Print.ing & Publishing, **1987**. illus: b&w: maps, photos; 24p; bibl; 22 cm.

[1909] **Larson, Lane**, 1953– *Caving: the Sierra Club guide to spelunking.* Larson, Peggy, 1931–. San Francisco, CA, US: Sierra Club Books, **1982**. illus: b&w: drawings, graphs, photos, tables; vii, 311p; bibl, index; 20 cm.

[1910] **Latrobe, G.** *Guide to the Coast, Caves, and Bays of Sark.* Latrobe, L. 3rd ed. npop: np, **1964**. illus: b&w: maps, photos; 75p.

12 b&w photos. 14 maps of the island. 1st ed was July 1941.

[1911] **Latrobe, G.** *A revised and enlarged guide to Sark containing a detailed description of the coast, caves and bays.* Latrobe, L. 4th ed. npop: The Guernsey Press Co Ltd, **1968**. illus: b&w: maps, photos; 1 map folded in front; 72p, 12p pls; index.

Cover title: Guide to the Coast, Caves, and Bays of Sark containing 14 maps of the island and numerous illustrations. 4th edition, revised and enlarged, 1968.

5th ed, rev. Latrobe, L. [St Peter Port]: Guernsey Press, **1975**. illus: b&w: maps, photos; 1 map folded in back; vi, 75p; index; 18 cm.

Cover title: Guide to the Coast, Caves and Bays of Sark.

6th ed. Latrobe, L. Stanford-le-Hope, GB: Trace-Lejins, **1994**. illus: b&w: maps; x, 58p; 19 cm.

[1912] **Lavaur, Guy de**, 1908-1986 *Caves and cave diving.* Mason, Edmund J (trans). New York, NY, US: Crown Publishers, Inc, **1956**. illus: b&w: drawings, maps, photos; 175p, 8p pls; app, bibl, gloss, index; 22 cm.

12 illus. Alt record says R. Hale is publisher. Reprinted 1958.

[1913] **Laville, Henri** *Rock shelters of the Perigord: geological stratigraphy and archaeological succession.* Rigaud, Jean Philippe; Sackett, James. New York, NY, US: Academic Press, **1980**. Studies in Archaeology. illus: b&w: charts, drawings, graphs, maps, tables; xii, 371p; bibl, index; 25 cm.

This work is one of the few on the sedimentologic and stratigraphic interpretation of caves and rock shelters. These shelters provide the best paleolithic sites in Europe. Very well done, professional, well illustrated.

[1914] **Lawrence, Joseph**, 1924– *The caves beyond: the story of the Floyd Collins' Crystal Cave exploration.* Brucker, Roger Warren, 1929–. New York, NY, US: Funk & Wagnalls, **1955**. illus: b&w: charts, drawings, maps, photos, tables; ii, 283p; app, bibl; 25 cm.

For The National Speleology Society. Floyd Collins' Crystal Cave Exploration, 1954.

The official account of the 1954 NSS expedition, the first large-scale exploratory expedition in the USA, as related first hand by participants. Appendices include expedition organization, medical, supply, communications, geological, biological and meteorological reports.

[1915] **Lawrence, Joseph**, 1924– *The caves beyond: the story of the Floyd Collins' Crystal Cave exploration.* Brucker, Roger Warren, 1929–; National Speleological Society. Teaneck, NJ, US: Zephyrus Press, **1975**. Speleologia Series. illus: b&w: charts, maps, photos; xxviii, 290p; bibl, index; 34 cm.

Reprint of the 1955 ed published by Funk & Wagnalls, New York. Contains new intro by Roger Warren Brucker.

[1916] **Lawson, Andrew J.** *Cave art.* Princes Risborough, Buckinghamshire, GB: Shire Publications. Shire Archaeology series. illus: b&w: charts, drawings, graphs, maps, photos, tables; 64p; bibl, index.

[1917] **Lawson, T. J.** *Caves of Assynt.* Jeffreys, Alan Lawrence, 1940–. Edinburgh, GB: Grampian Speleological Group, **1988**. Occasional Publication, Grampian Speleological Group no 6. illus: b&w: maps, photos; [ii], 90p; bibl, index; 29 x 20.5 cm.

Earlier edition by A. L. Jeffreys, 1977.

Limestone Caves of Scotland, part 2.

[1918] **Laycock, George** *Caves.* Burt, DeVere E. (illus). New York, NY, US: Four Winds Press, **1976**. illus: b&w: drawings, photos; viii, 101p; app, bibl, index; 24 cm.

Juvenile.

George Laycock is an extremely popular juvenile writer who has written about both caves and bats quite well. Discusses aspects of caves: their formation and structure.

[1919] **Leakey, Richard E.** *A new era.* British Broadcasting Corporation; Television Service Time-Life Films. New York, NY, US: Time Life Video, **1981**. The Making of Mankind 5.

1 videocassette (55 min), sd, col, 3/4 in. Also issued as 2 film reels, 16 mm, sd, col.

Richard Leakey details the emergence of our species, *Homo sapiens*, and explores some of the astonishing art our ancestors left behind, including the beautiful cave of Lascaux.

[1920] **Leakey, Robert Dove** *Potholer's songs.* Giggleswick, Yorkshire, GB: R.D. Leakey, **[1947]**. [34]; 30 x 21 cm.

Photo copy examined.

[1921] Leck Studio *The wonderful Carlsbad Caverns: near Carlsbad, New Mexico.* Davis, Ray V. (photo). Brooklyn, NY, US: Albertype Co, **1920**. illus: b&w: photos; [1], 12p pls; 21 x 26 cm.

Title from cover. Bound with green cord. OCLC lists both 1920 and 1929 as date of publication.

[1922] Leclerc, Guy *Derivation of hydrologic frequency caves.* Schaake, John C. Cambridge, MA, US: Ralph M. Parsons Laboratory for Water Resources and Hydrodynamics, Massachusetts Institute of Technology, **1972**. Report no 142 DSR 72943 R72-6. 151p; bibl, errat; 28 cm.

Prepared under Grant No. 14-31-0001-3403 Office of Water Resources Research, U.S. Dept.of the Interior.

[1923] Lee, G. *The rock art of Petroglyph Point and Fern Cave, Lava Beds National Monument: final report.* Hyder, William D.; Benson, Arlene; Edwards, E. Los Osos, CA, US: Associated Rock Art Consultants, **1988**. iv, 162p; bibl; 28 cm.

Illus. Contract no. PX8400-7-0694, PX8410-8-0185. "Original report, color slides, black and white photographs and negatives and a set of black line copies of the drawings are on file at the Rock Art Archives at the University of California, Los Angeles. A second original report, color slides ... are curated at Lava Beds National Monument"— Letter of transmittal.

[1924] Lee, Joli (Eaton) (ed) *GYPKAP report volume 3: January 1992—April 1996.* npop: Southwestern Region of the National Speleological Society, **1996**. illus: b&w: charts, drawings, maps, photos, tables; 7 loose folded maps; text on lining pages; 68p; bibl.

Cover lacks word volume.

[1925] Lee, Thomas A. *San Pablo Cave and El Cayo on the Usumacinta River, Chiapas, Mexico.* Hayden, Brian. Provo, UT, US: New World Archaeological Foundation, Brigham Young University, **1988**. Papers of the New World Archaeological Foundation no 53. illus: b&w: maps; ix, 79p; bibl; 27 cm.

Illus.

[1926] Lee, Willis Thomas *A visit to Carlsbad Cavern: recent explorations of a limestone cave in the Guadalupe Mountains of New Mexico reveal a natural wonder of the first magnitude.* National Geographic Magazine. Washington, DC, US: National Geographic Society, **1924**. illus: b&w: photos; 32p; 26 cm.

From National Geographic Magazine, Vol.XLV, no.1, January, 1924.

[1927] Legault, Myrna Diaz (comp) *Bibliography of Puerto Rican caves, karst and limestone geology; Rio Camuy Cave System.* St. Pierre, David (comp); Veve, Thalia (comp). High Bentham, North Yorkshire, GB: [David St. Pierre], Oct, **1990**. [20]; 30 x 21 cm.

Printed one side only. No illus.

[1928] Lehman Caves National Monument (Nev.) *Proposed natural resources management plan and environmental assessment, Lehman Caves National Monument, Nevada.* Rev. Feb. 1980. [Baker, NV, US]: Lehman Caves National Monument, **1980**. illus: b&w: maps; one leaf folded; i, 26p; 28 cm.

[1929] Lehman Caves National Monument *Proposed natural resources management plan, Lehman Caves National Monument, Nevada.* United States National Park Service. [San Francisco, CA, US]: The Monument, **1974**. illus: b&w: graphs, maps; i, 50p; 27 cm.

At head of title: Environmental assessment. Cover title: Natural resources management plan, Lehman Caves National Monument, Nevada.

[1930] Leitch, David Edwin *Ogof Agen Allwedd in relation to the Mynydd Llangattwg.* npop, [GB]: Cave Research Group of Great Britain, **1960**. Cave Research Group of Great Britain Publications no 10. 56p; 26 cm.

Illus.

[1931] Lénárt, László *A Bukki barlangok kutatasanak, vedelmenek es hasznositasanak legujabb eredmenyei = Newest results in research, protection and use of caves of Bukk Mountains: Miskolci Egyetem, 1993. November 11-13.* Miskolc, HU: Miskolci Egyetem, **1993**. Various pagination; bibl; 24 cm.

Illus. Proceedings from a conference at the Miskolc University in Miskolc, Hungary.

[1932] Lénárt, László *Caves in the Bükk Mountains.* Szenthe, István; Varga, Csaba; Szilvássy, Zoltán (trans); Hazslinszky, Tamás (ed). Budapest, HU: 10th International Congress of Speleology, **1989**. Field Trip Guide E2. illus: b&w: maps; 32p; bibl.

[1933] Leontiadis, J. *The use of radioisotopes in tracing karst groundwater in Greece.* Dimitroulas, Ch. Athens, GR: Greek Atomic Energy Commission, Nuclear Research Center "Democritus", Jan, **1972**. [23]; bibl; 28 cm.

Illus.

[1934] Leroi-Gourhan, André, 1911-1986 *Treasures of prehistoric art.* Guterman, Norbert, (trans). New York; Paris, NY, US; FR: H. N. Abrams Inc, **1967**. illus: b&w: charts, drawings, maps, photos; col: photos; 543p; bibl, index; 32 cm.

739 photos, 187 drawings, 56 maps and charts. Translation of Prehistoire de l'art occidental.

[1935] LeRoux, J. P. *The report of the 1992 caving expedition to the Chimanimani Mountains in the eastern highlands of Zimbabwe.* Koliasnikoff, A.; Truluck, T. E. Capetown, ZA: South African Speleological Association, Jan, **1993**. illus: b&w: charts, maps, photos, repros; foldout map in text; front: b&w photo: "The four intrepid cavers under 'arrest' at Chimanmani Hotel"; [ii], iv, 78p; app; 21 x 29.5 cm.

[1936] Letcher, Montgomery E. *Wonderful discovery: being an account of a recent exploration of the celebrated*

Mammoth Cave, in Edmonson County, Kentucky, by Dr. Rowan, Professor Simmons and others, of Louisville, to its termination in an inhabited region, in the interior of the earth. New York, NY, US: R.H. Elton, **1839**. 24p; 20 cm.

[1937] **Lewis, Charles C. (ed)** *Chronicles of the Reading Grotto, in which we go to the California convention, June 3rd to September 6th, 1966.* Austin, TX, US: C. C. Lewis, **1993**. illus: b&w: drawings; v, 146p; gloss; 21.6 x 28 cm.

[1938] **Lewis, Ian D.** *Cave diving in Australia.* Stace, Peter. 1982 rev ed. Adelaide, SA, AU, **1982**. illus: b&w: cartoons, charts, graphs, photos, tables; text on rear lining page; front: map; 174p; app, index.

1st ed. Stace, Peter. Adelaide, SA, AU: Ian Lewis, **1980**. illus: b&w: drawings, maps; front: map; i, 162p; app; 21 cm.

First book on cave diving in Australia.

[1939] **Lewis, Ian D.** *Discover Naracoorte caves.* Hocknell, P. (ed). Edwardstown, SA, AU: Subterranean Foundation (Australia), **1977**. illus: b&w: drawings, maps, photos, repros; vi, 74p; 14 x 21 cm.

[1940] **Lewis, Ian D.** *South Australian cave reference book.* Valley View , SA, AU: Cave Exploration Group of South Australia Inc, **1976**. Occasional paper number 5 Cave Exploration Group (South Australia) Inc. illus: b&w: maps, tables; xi, 140p; app, bibl; 30 cm.

[1941] **Libra, Bob** *1990 Friends of karst field trip: the Big Spring Basin, northeast Iowa.* Littke, John; Rowden, Bob; Bounk, Micheal [sic]; Friends of Karst. [IA, US]: Geological Survey Bureau Iowa Department of Natural Resources, **[1990]**. illus: b&w: charts, graphs, maps; [17].

Title from cover.

[1942] **Lillis, Brian** *Poul Na Gollor Cave at Inch Ennis, Co Clare.* [IE]: Limerick Caving Club, nd. illus: b&w: drawings, maps, photos; fold-out map in back; index on front lining page; 12p; bibl; 21.5 x 15 cm.

[1943] **Limbert, Howard** *Mexico 85/86.* Others. [GB], **[1986]**. illus: col: photos; 6 folded maps; poster in back; [ii], 122p, 9p pls; 30 x 21 cm.

[1944] **Limbert, Howard** *The 1994 British/Vietnamese speleological expedition report March/April 1994.* Others. npop, [GB]: British/Vietnamese Speleological Expedition, **[1995]**. illus: b&w: maps, photos, tables; col: photos; 3 fold-out maps, printed both sides; [48]; 30 x 21 cm.

Cover title: Vietnam '94.

[1945] **Lindsley, R. P. (ed)** *Survey and assessment of cave resources at Buffalo National River, Arkansas.* Welbourn, W. Calvin (ed). Cedar Falls, IA, US: Cave Research Foundation, **1977**. 106p.

[1946] **Link, Louis W.** *Lewis and Clark Cavern, Montana.* 1961. [Cardwell, MT, US]: [L. W. Link]. 96p.

For the 1961 edition, OCLC lists L. W. Link as the publisher and [Cardwell, MT?] as the place of publication. Copies examined were the 1964 edition and the 1956 revised in 1967 edition.

1964. npop, [MT, US]: np. illus: b&w: drawings, maps, photos; 112p.

"A commemorative issue honoring the State of Montana, celebrating centennial as a territory, 1864-1964, diamond jubilee as a State,1889-1964."

1967. npop, [MT, US]: np. illus: b&w: drawings, maps, photos; 112p.

1st. Oakland, CA, US: Fontes Printing Company, **1956**. illus: b&w: charts, drawings, maps, photos, ports, repros; 72p; 23 cm.

[1947] **Lino, Clayton F.** *Cavernas, o fascinante Brasil subterraneo = Caves, the fascination of underground Brazil.* Sao Paulo, BR: Editora Rios, **1989**. illus: b&w: charts, drawings, maps, photos, repros, tables; col: drawings, photos; loose folded table of caves; photo on lining page; 279p; bibl; 36 cm.

Text in English and Portuguese.

[1948] **Little, William Henry**, 1919-1992 *Coming to South Wales caves?* South Wales Cave Rescue Organisation. Swansea Valley, GB: South Wales Cave Rescue Organisation, **1970**. illus: b&w: maps; 1 folded plate; v, 13p; bibl; 21 cm.

[1949] **Lively, Richard** *Radon concentrations, radon decay product activity, meteorological conditions and ventilation in Mystery Cave.* Krafthefer, Brian. [MN, US]: R. Lively and B. Krafthefer, **1993**. illus: b&w: maps; Various pagination; 28 cm.

Illus.Title from cover.

[1950] **Livesay, Elizabeth Ann** *Geology of the Mammoth Cave National Park area.* McGrain, Preston, 1917–(rev). 1962 rev ed. Kentucky Geological Survey; College of Arts and Sciences, University of Kentucky, **1962**. illus: b&w: charts, drawings, maps, photos; bibl.

1st ed. Lexington, KY, US: Kentucky Geological Survey, **1953**. Kentucky Geological Survey Series X Special Publication 7. illus: b&w: drawings, maps, photos; 40p; bibl; 22 cm.

Master microform held by: LCP. Microfiche. Louisville, Ky.: Lost Cause Press, 19—. Two copies inspected were different sizes, but with identical contents.

[1951] **Lloyd, Oliver Cromwell**, 1911-1985 *A cave diver's training manual.* Bristol, GB: The Cave Diving Group, Jul, **1975**. The Cave Diving Group Technical Review no 2 July 1975. illus: b&w: drawings; text on front lining page; i, 88p, [4], 11p pls; index; 30 cm.

16 figs. OCLC also lists 1983.

[1952] **Lloyd, Oliver Cromwell**, 1911-1985 *Doolin-St. Catherine's Cave: Co. Clare, Eire.* Bristol, GB: University of Bristol Speleological Society, Nov, **1964**. illus: b&w: photos; 4 folded maps in text; 30p, 3p pls; bibl; 26 x 20 cm.

[1953] **Lobeck, Armin Kohl**, 1886-1958 *The geology and physiography of the Mammoth Cave national park.*

Frankfort, KY, US: The Kentucky Geological Survey, **1928**. Kentucky Geological Survey Ser VI Pamphlet XXI. illus: b&w: drawings, maps, photos; [xv], 69p; 23 cm. 37 figs.

[1954] Lockett, H. Claiborne *Woodchuck Cave: A basketmaker II site in Tsegi Canyon, Arizona.* Hargrave, Lyndon L.; Colton, Harold S.; Euler, Robert C. (eds). Flagstaff, AZ, US: Northern Arizona Society of Science and Art, Inc, **1953**. Museum of Northern Arizona Bulletin 26. illus: b&w: maps, photos; v, 33p; bibl, index.

Forward says that this edition is limited to 700 copies. 18 figs.

[1955] Logan, Wilfred David, 1923– *Graham Cave, an archaic site in Montgomery County, Missouri.* Chapman, Carl Haley (app). Columbia, MI, US: Missouri Archaeological Society, **1952**. Memoir of the Missouri Archaeological Society no 2. illus: b&w: drawings, maps, photos, tables; 101p; app, bibl; 23 cm.

Microfiche, Greeley, University of Northern Colorado, 1978. 24 pls.

Reports on the archaeological excavations at a stratified Archaic (8000-1000 B.C.) prehistoric site located between St. Louis and Columbia, Missouri.

[1956] Logan, William Steve *Self-belaying cable ladder - a rope climbing aid.* Kell, H. S.; Harris, B. J. npop: np, **1971**. illus: b&w: drawings, photos, tables; ii, 29p.

Prepared for ME 484, winter quarter 1971.

[1957] Long, Abijah *The big cave: early history and authentic facts concerning the history and discovery of the world famous Carlsbad Caverns of New Mexico.* Long, Joe N. Long Beach, CA, US: Cushman Publications, **1956**. illus: b&w: drawings, maps, photos, repros; 126p, [1]; bibl; 20 cm.

OCLC lists: 1958: 2nd ed, 1961: 3rd ed, 1964: 5th ed, 1965: 6th ed, 1966: 7th ed, 1968: 8th ed, 1969: 9th ed, 1971:10th ed (title listed as The big cave; size is 21 cm). 7th edition: 1966. 9th edition: 1969. 10th edition: 1971 Early history and facts on Carlsbad Caverns.

Long's account of the discovery of Carlsbad Caverns, New Mexico (compare this with other accounts, notably by White, Jim).

[1958] Long, Gary F. (auth, des, photo) *Tuckaleechee Caverns: 31 photos in natural color: Townsend, Tennessee: greatest sight under the Smokies.* Gatlinburg, TN, US: W. M. Cline Company, **nd**. illus: partly col: maps; col: photos; text on front lining page; partly col map on rear lining page; [16].

Title from cover. Published 1966 or after. Minimal text.

[1959] Longley, Glenn *Preliminary report of biological investigation Valdina Farms sink hole - Medina Co., Texas.* San Marcos, TX, US: Longley, Nov, **1977**. illus: b&w: drawings, maps, tables; [ii], 16p, [20]; app, bibl.

For Edwards Underground Water District, San Antonio, Texas.

[1960] Longley, Glenn *Status of Trogloglanis pattersoni Eigenmann, the toothless blindcat, and status of Satan eurystomus Hubbs and Bailey, the widemouth blindcat.* Karnei, Henry. [San Marcus, TX, US]: [Longley], Jun, **1978**. illus: b&w: charts, drawings, graphs, maps, tables; vi, 48+54p; app, bibl.

Prepared for the Fish and Wildlife Service, Albuquerque, New Mexico, Contract # 14-16-0002-77-035. Two parts, separately numbered. 16 figs.

[1961] Longley, Glenn, 1942– *A list of the arthropoda in the limestone caves in Kantô-Mountainland, with the description of a new genus and three species.* Nomura, Sizumu. Tokyo, JP: The Japanese Society for the Study of Insects, Jan, **1952**. "The Chōhō" no 6. illus: b&w: drawings, photos, tables; 1 fold-out table; 8p, 2p pls.

[1962] Longley, Glenn, 1942– *Status of Typhlomolge (=Eurycea) rathbuni.* Albuquerque, NM, US: United States Fish and Wildlife Service, Feb, **1977**. Endangered Species Report 2. illus: b&w: charts, drawings, graphs, maps, photos, photomicrographs, tables; vi, 45p; app, bibl.

Cover title: Status of the Texas blind salamander.

[1963] Longsworth, Polly *Exploring caves.* Schrotter, Gustar (illus). New York, NY, US: Thomas Y. Crowell Company, **1959**. illus: b&w: maps, photos; 176p; bibl, gloss, index; 21 cm.

Juvenile.

[1964] Longul, Wally *The last horizon.* TVOntario. [Ontario, CA]: TVOntario, **1987**.

VHS format. Available from CIN Services. Cassettes 1-2 include 2 30-minute programs each; cassette 3 includes 1 30-minute program. [1]. Heat life ; Desert creatures—[2]. Caves down under ; Rare animals of the Arc—[3]. Fueled by fire. NHK producer, Yasuji Hamagami ; music, Keiko Ota. Explores Australia's varied and strange animals in their natural surroundings; its underground and underwater phenomena; its rare lakes and vast desert; and the unique people who live there. 3 videocassettes, sd, col ; 1/2 in. Juvenile.

[1965] Lothson, Gordon A., 1939– *The archaeology of the Fallen Arches Cave Site, 45SA41, Borigo timber sale, Gifford Pinchot National Forest: phase I survey and reconnaissance.* Wilkinson, Robert. Pullman, WA, US: Washington Archaeological Research Center, Washington State University, **1980**. Project Report Washington Archaeological Research Center no 94. illus: b&w: drawings, maps, photos, tables; xii, 125p; app, bibl, gloss; 28 cm.

[1966] Loubser, J. H. N. *A guide to the rock paintings of Tandjesberg.* Bloemfontein, ZA : Nasionale Museum, Oct, **1993**. Navorsinge van die Nasionale Museum v 9 pt 11. 38p; bibl; 26 cm.

[1967] Loud, Llewellyn Lemont, 1879-1946 *Lovelock Cave.* 1965 reprint ed. New York, NY, US: Kraus Reprint

Corp, **1965**.

Title page includes original imprint: Berkeley, University of California press, 1931.

1st ed. Harrington, Mark Raymond, 1882-1971. Berkeley, CA, US: University of California Press, **1929**. University of California Publications in American Archaeology and Ethnology v 25 no1. illus: b&w: drawings, maps, tables; some maps partly folded; viii, 183p, 35p pls; bibl; 28 cm.

Title from cover. 68 pls.

[1968] Loud, Llewellyn Lemont, 1879-1946 *Lovelock Cave: The republication of the rare 1929 Nevada classic.* Harrington, Mark Raymond, 1882-1971. Sparks, NV, US: Falcon Hill Press, Feb, **1991**. illus: b&w: drawings, maps, photos; some folded maps; viii, 183p, 68p pls; app, bibl; 27 cm.

Reprint. Originally published: University of California Publications in American archaeology and ethnology v 25 no1.

[1969] Love, Douglas L. *The spelunker's guide to the caves of the Garrison Chapel Valley.* Bloomington, IN, US: Cave Information Service, **1972**. Cave Information Service Guidebook IN-1. illus: b&w: maps; 24p; 28 cm.

3rd (rev) printing, October 1975; 4th (rev) printing, October, 1979 (Worthington, IN); 6th (rev) printing, March, 1990 (Greenbelt, MD).

[1970] Love, Douglas P. *The spelunker's guide to the caves north of Campbellsburg, Indiana.* rev 2nd print. npop: np, **1977**. 48p; 29 cm.

Illus.Sixth (rev) printing, March, 1990.

[1971] Love, Douglas P. *The spelunker's guide to the caves North of Campbellsburg, Washington County, Indiana.* Love, Douglas P. Bloomington, IN, US: Cave Information Service, **1973**. Cave Information Service Guidebook IN-2. illus: b&w: maps, photos; 48p; 28 cm.

[1972] Lovelock, James *Caving.* London, GB: B. T. Batsford, Limited, **1969**. illus: b&w: drawings, photos; 144p, 16p pls; app, bibl, gloss, index; 23 cm.

Listing of caving clubs. 37 pls.

[1973] Lovelock, James *A caving manual.* London, GB: B.T. Batsford, LTD, **1981**. illus: b&w: photos; vi, 144p; app, bibl, gloss, index; 26 cm.

96 pls.

[1974] Lovelock, James *Life and Death Underground.* London, GB: G. Bell and Sons, Ltd, **1963**. illus: b&w: drawings, photos; 157p; index.

21 b&w photos and drawings.

A gruesome, but well-written collection of some of the most famous caving fatalities. Readable accounts of the attempted rescue of Neil Moss (Peak Cavern, England,1959), Floyd Collins (Sand Cave, KY, 1925), Marcel Loubens (Pierre St Martin, France, 1952) are given, plus details of successful rescues and cave diving.

[1975] Lowe, David (comp) *A dictionary of karst and caves.* Waltham, Anthony Clive, 1942–[Tony]. [GB]: British Cave Research Association, **1995**. Cave Studies Series number 6. illus: b&w: charts, photos, tables; text on front lining page; table on rear lining page; 40p; gloss.

[1976] Lowe, David John *Tonga '86, Tonga '87 expedition report.* Keyworth, [GB]: British Geological Survey, **1988**. illus: b&w: drawings, maps, photos, ports; Various pagination; 30 cm.

Title from cover.

[1977] Lowry, D. C. *Geology of the Western Australian part of the Eucla Basin.* [AU]: Geological Survey of Western Australia, **1970**. Geological Survey of Western Australia Bulletin 122. illus: b&w: charts, graphs, maps, photos, tables; col: maps; col map folded in rear; three pp tables and chart folded in text; 200p, [3]; app, bibl.

Geology of the Nullarbor plain.

[1978] Lu, Yaoru *Engineering and geological conditions of karst.* Et al. Peking, CN: np, **1972**. illus: b&w: maps; folded map; 19p; 26 cm.

Illus.

[1979] Lübke, Anton, 1890– *The world of caves.* Bullock, Michael (trans). New York, NY, US: Coward-McCann Inc, **1958**. illus: b&w: photos; 295p, 24p pls; index; 23 cm.

[1st American edition]. British edition published by Weidenfeld and Nicolson, London, GB. Translation of Geheimnisse des unterirdischen. 24 pp b&w photos.

[1980] Luce, A. A. *Berkeley's description of the cave of Dunmore.* npop: np, **1931**. 13p, 1p pls; 22 cm.

Reprinted from Hermathena, v. 2, 1931, pp. 149-161. The essay discussed originally appeared in Berkeley's Commonplace book.

[1981] Luce, Pat (comp) *Let's see Marvel Cave: your personal guide to America's largest privately owned cave.* Lynch, William Henry. npop: np, **1940**. 31p; 24 cm.

Illus.

[1982] Lumley, Mark *Mexico: the black holes expedition: the 1988 British expedition to explore the caves of the Sierra de Zongolica.* Bradshaw, Dany; Miner, Steve. npop: Bristol Exploration Club and Northern Caving Club, **[1989]**. illus: b&w: drawings, maps, photos; col: photos; rear lining is p 37; 37p; 30 x 21 cm.

Limbert, Howard, trip lead.

[1983] Lundelius, Ernest L., 1927– *The mammalian fauna of Madura Cave, Western Australia.* Turnbull, William D. 1982 ed. Chicago, IL, US: Field Museum of Natural History, **1982**. Fieldiana Geology New Series, no 11. vii, 32p; bibl; 24 cm.

Caption title. Illus.

1989 ed. Turnbull, William D.; Field Museum of Natural History. Chicago, IL, US: Field Museum of Natural History, Mar, **1989**. Publication Field Museum of Natural History no 1399; Fieldiana Geology, New Series no17. iv, 71p; bibl; 24 cm.

Accepted March 11, 1987. Illus.

[1984] Lundelius, Ernest L., 1927– *The mammalian fauna of Madura Cave, Western Australia, part VI, Macropodidae: Potoroinae.* Turnbull, William D. Chicago, IL, US: Field Museum of Natural History, **1984**. Fieldiana Geology New Series no 14; Field Museum of Natural History Publication 1354. ix, 63p; bibl; 24 cm.

Caption title. Illus.

[1985] Lundelius, Ernest L., 1927– *The mammalian fauna of Madura Cave, Western Australia: part I.* Turnbull, William D. [Chicago, IL, US]: Field Museum of Natural History, Feb, **1973**. illus: b&w: drawings, graphs, maps, tables; 35p; bibl.

Three tables, 13 figs.

[1986] Lundelius, Ernest L., 1927– *The mammalian fauna of Madura Cave, Western Australia: part II.* Turnbull, William D. [Chicago, IL, US]: Field Museum of Natural History, Aug, **1975**. illus: b&w: drawings, graphs; 117p; app, bibl.

15 tables, 21 figs, pp 37-117.

[1987] Lundelius, Ernest L., 1927– *The mammalian fauna of Madura Cave, Western Australia: part III.* Turnbull, William D. Chicago, IL, US: Field Museum of Natural History, Dec, **1978**. Fieldiana Geology vol 38. illus: b&w: drawings, graphs, photos, tables; [x], 120p; bibl.

29 tables.

[1988] Lundelius, Ernest L., 1927– *The mammalian fauna of Madura Cave, Western Australia: Part IV.* Turnbull, William D. Chicago, IL, US: Field Museum of Natural History, Feb, **1981**. Fieldiana Geology, New Series no 6; Field Museum of Natural History Publication 1315. illus: b&w: drawings, graphs, maps, tables; vii, 72p; bibl; 24 cm.

[1989] Lundelius, Ernest L., 1927– *Natural history of Texas caves.* Slaughter, Bob H. (ed). Dallas, TX, US: Gulf Natural History, **1971**. illus: b&w: maps; 171p; 28 cm.

On spine: Texas caves.

A collection of papers on the fauna of Texas caves from a variety of perspectives.

[1990] Luray Caverns Corporation *The beautiful caverns of Luray, Luray, Virginia, in the Shenandoah Valley.* npop: np, **1953**. illus: partly col: maps; [24]; 13 x 19 cm. Illus.

[1991] Luray Caverns Corporation *The beautiful caverns of Luray, Luray, Virginia. In the Shenandoah Valley, three miles of subterranean splendor, brilliantly lighted by electricity.* Luray, VA, US: np, **1906**. 15p, [1]; 14 x 20 cm.

Title from cover.

[1992] Luray Caverns Corporation *The beautiful caverns of Luray: Luray, Virginia.* [Luray, VA, US]: The Corporation, Marken & Bielfeld, **1949**. illus: b&w: maps, photos; 1 folded sheet; 12p; 23 cm.

Illus.

[1993] Luray Caverns Corporation *The beautiful caverns of Luray: Luray, Virginia, in the Shenandoah Valley: miles of subterranean splendor, brilliantly lighted by electricity.* Luray, VA, US: The Corporation, Marken & Bielfeld, **1930**. illus: b&w: maps; [22]; 14 x 20 cm.

Illus, music.

[1994] Luray Caverns Corporation *Caverns of Luray.* npop: np, **nd**. [17].

One OCLC record list [20]p and a possible date of 1900. Illus.

[1995] Luray Caverns Corporation *Electric light views in the Caverns of Luray. The Caverns of Luray (at Luray, Page County, Virginia, a station on the Shenandoah Valley Railroad) as a resort for tourists ... are unexcelled, in their wonderful attractiveness, by any other creation of nature.* [Philadelphia, PA, US]: np, **1882**. 30 cm.

Cover title: Caverns of Luray. One vol, unpaged, mounted pls.

[1996] Luray Caverns Corporation *Luray Caverns.* [Luray, VA, US]: np, **1900**. illus: b&w: ads; [16]; 14 x 20 cm.

Advertisement of Norfolk and Western Railway on p. [4] of cover. Illus.

[1997] Luray Caverns Corporation *Luray Caverns of the beautiful Shenandoah Valley on the Norfolk and Western Railway, Luray, Virginia. Luray, Virginia Mansion Inn.* [Roanoke, VA, US]: The Stone Printing & Manufacturing Company, **1900, 1909**. 20p.

[1998] Luvless, Dungass F. (pseud) *The speleobopper's guide to the caves of the Gruesome Chapel Valley.* [IN, US]: Cave Desecration Service, **1972**. Cave Desecration Service Guidebook IND-1. illus: b&w: drawings, photos, repros; i, 27p, [1].

Reproductions of newspaper articles. A parody of Douglas Love's The spelunker's guide to the caves of Garrison Chapel Valley.

[1999] Lynch, Thomas F., 1938– *Guitarrero Cave: early man in the Andes.* New York, NY, US: Academic Press, **1980**. Studies in Archaeology. illus: b&w: charts, drawings, maps, photos, tables; xviii, 328p; bibl, index; 25 cm.

[2000] Lynott, Mark J. *Archeological investigations at Limekiln Cave, 23SH109, Ozark National Scenic Riverways, southeast Missouri.* Monk, Susan M.; Midwest Archeological Center (U.S.). Lincoln, NE, US: Midwest Archeological Center, National Park Service, **1986**. illus: b&w: maps; ii, 26p; bibl; 28 cm.

Illus.

[2001] Lyon, Ben *Venturing underground: the new speleo's guide.* Yorkshire, GB: EP Publishing Limited, **1983**. illus: b&w: drawings, maps, photos; partly col: photos; 160p; bibl, gloss; 25 cm.

Some col illus.

[2002] Lyons, David *Floyd Collins, greatest cave explorer ever known.* Medley, Joy. [Bowling Green, KY,

US]: The Cockrel Corp, **1981**. Kentucky Culture Series. illus: b&w: maps; 39p; 29 cm.

Microfiche. Louisville, Ky: Lost Cause Press, 1981. 2 microfiches ; 10 x 15 cm. (Kentucky culture series). Illus.

[2003] Lyons, Joy Medley *Mammoth cave*. Van Camp, Mary L. Las Vegas, NV, US: KC Publications, **1991**. The Story Behind the Scenery. illus: b&w: maps; col: maps; 48p; 31 cm.

Col illus.

[2004] Lyons, R. G. *Rock to stalactite*. [Waitomo, NZ]: Waitomo Caves Museum Society, Inc, **1985**. illus: b&w: drawings, photos; text on lining pages; 21p.

[2005] MacCulloch, Donald B. *The island of Staffa*. Edinburgh, GB: Moray Press, **1951**. illus: b&w: photos; Photo "Fingal's Cave Staffa. Lona on the horizon"; 60p, [2]; 18.5 x 12.5 cm.

Describes a cave in basalt. Nine photos.

[2006] MacCulloch, Donald B. *The island of Staffa: with 12 photographs, two plans and one map, by the author*. Glasgow, GB: Alwx. Maclaren and Sons, **1927**. illus: b&w: maps, photos; front: b&w photo; 64p, 14p pls.

12 b&w photos.

[2007] MacCurdy, George Grant, 1863– *La Combe, a paleolithic cave in the Dordogne*. Lancaster, GB: The New Era Printing Co, **1914**. 27p.

Reprinted from the American Anthropologist (N.S.), pp 157-184. Illus.

[2008] MacDougall, J. H. *Western Australia's wonderland*. npop, **[1922]**. illus: b&w: drawings, photos; tourist info on "coupon tours" on lining pages; [18]; 14 x 22.5 cm.

Working from photocopy. Additional title page: Western Australia's wonderland and south west caves.

[2009] MacEnery, John M., 1796-1841 *Cavern researches, or, discoveries of organic remains, and of British and Roman reliques, in the caves of Kent's Hole, Anstis Cove, Chudleigh, and Berry Head*. Vivian, E. (ed). Strand, Torquay, GB: Simpkin, Marshall and Company, **1859**. illus: b&w: drawings; vi, 78p, 18p pls; 23 cm.

Photocopy examined. Larger sized edition (38.8 cm) also printed in 1859.

[2010] MacGregor, John R. *Ecology of a limestone cave*. Belinky, Charles R. Lyons Fall, NY, US: Educational Images, **1978**.

Offers a look at the various organisms found in a limestone cave. 20 col slides, 2x2 in & script.

[2011] Mackie, Charles, 1864– *Historical description of the chapel and castle of Roslin, and the caverns of Hawthornden*. St. Clair family. London, GB: J. Leslie, **1830**. 32p; 20 cm.

Also has engraved t.-p. The castle was the home of the St. Clair family. OCLC notes some pls.

[2012] Mackie, Evan W. *The caves at East Wemyss, Fife: an interim report on new investigations in 1980*. Glaister, Jane M. Glasgow, GB: Hunterian Museum, University of Glasgow; Kirkcaldy Museum and Art Gallery, **1981**. illus: b&w: charts, drawings, maps; 17p, 6p pls; errat; 30 x 21 cm.

Cover title: The Wemyss Caves, Fife. Loose errata sheet. OCLC lists Evan and Euan. 7 figs.

[2013] MacLean, John Patterson, 1848-1939 *An historical, archaeological and geological examination of Fingal's Cave in the island of Staffa*. Subscriber's ed. Cincinnati, OH, US: Robert Clarke & Company, **1890**. illus: b&w: drawings, maps, photos; front: b&w photo; 49p; 23 cm.

Rewritten and enlarged from the original report made to the Smithsonian Institution, in the year 1887.

[2014] Macleay, Kenneth, 1819– *Description of the spar cave, lately discovered in the Isle of Skye: with some geological remarks relative to that island*. Leyden, John, 1775-1811. Edinburgh, GB: Printed by James Clarke ... for Thomas Bryce & Co ..., **1811**. 88p; 22 cm.

On title page: "... to which is joined the mermaid, a poem" by John Leyden.

[2015] MacMaster University Caving and Climbing Club *A description of the Sotano del Rio Iglesia*. [Hamilton, Ontario, CA]: np, **[1968]**. MUCCC Publication No. 1. illus: b&w: maps, tables; large loose folded map; [11], 2p pls.

[2016] Madacheen, L. Michael *New Melones cave inventory and evaluation study: preliminary report: archeological caves*. Grady, Mark A. npop: Archeology Research Program, Dept of Anthropology, Southern Methodist University, **1977**. illus: b&w: maps, tables; vii, 90p; bibl.

[2017] Maddox, Gary L. *Karst features of Florida Caverns State Park and Falling Waters State Recreation Area, Jackson and Washington Counties, Florida*. Tallahassee, FL, US: Southeastern Geological Society, Nov, **1993**. Guidebook, Southeastern Geological Society no 34. 38p; bibl; 28 cm.

Illus.

[2018] Maempel, George Zammit *Ghar Dalam: cave and deposits*. Malta, MT: George Zammit Maempel, **1989**. 74p; bibl, index; 21 cm.

Illus.

[2019] Maempel, George Zammit *Pioneers of Maltese geology*. Malta, MT: George Zammit Maempel, **1989**. illus: b&w: drawings, photos, ports, repros, tables; partly col: maps; col: photos; 302p; bibl, index; 22 x 15 cm.

In covering the early workers in the field of geology, Maempel frequently notes their involvement in cave studies, both with respect to archaeology and exploration as well as giving a historical overview. An important work in English for the island's cave history.

[2020] Maih, Lindsay (ed) *The New Zealand Cave atlas: volume 2: South Island*. Waitomo Caves, NZ: New

Zealand Speleological Society Inc, **1993**. Occasional Publications number 8. illus: b&w: maps; 13 folded plates,10 printed one side only; [iv], 20p, [202], 10p pls; 20 x 22 cm.

List of caves paginated, all else not.

[2021] Maire, Richard (coord, surv, art) *Report of the French Speleological expeditions to PNG in collaboration with the Committee of French Speleological Expedition and the Scientific Committee of the French Federation of Speleology.* Pernette, Jean-François (coord); Rigalde, Christian (surv, art); Gratté, Dummy Lucien (surv, art). Paris, FR: Fédératon Française de Spéléologie, Jul, **1981**. Spelunca Supplement no 3. illus: b&w: charts, drawings, graphs, maps, photos, tables; three b&w maps folded in center; 47p; bibl.

Cover title: Papua New Guinea

[2022] Mais, Stuart Peter Brodie *12 natural colour photographs of Cheddar and Cox's Cave "The home of the rainbow".* npop, [GB], **c1934**. illus: col: photos; [3], 12p pls; 23.5 x 14.5 cm.

Three page intro and 12 color pictures are mounted on thick gray paper.

[2023] Malatesta, Alberto *Dwarf deer and other late Pleistocene fauna of the Simonelli Cave in Crete.* Rome, IT: Academia Nazionale dei Lincei, **1980**. Problemi Attuali di Scienza e di Cultura Quaderno n 249; Problemi Attuali di Scienza e di Cultura Sezione missioni ed Esplorazioni 4. illus: b&w: maps; 126p, 30p pls; bibl; 27 cm.

Illus.

[2024] Malayan Nature Society *Malaysian caves.* Kuala Lumpur, MY: Malayan Nature Society, May, **1965**. The Malyan Nature Journal vol 19 no 1. illus: b&w: drawings, maps, photos, tables; four folded maps in text; [ii], 112p; bibl, errat.

Special issue devoted to Malaysian caves.

[2025] Malott, Clyde Arnett, 1887-1950 *A geologic profile of Sloans Valley, Pulaski County, Kentucky.* McGrain, Preston. Lexington, KY, US: Kentucky Geological Survey, **1977**. Series X Report of Investigations 20. illus: b&w: charts, maps, photos; one map folded in pocket on rear lining; [iv], 11p; bibl.

[2026] Malott, Clyde Arnett, 1887-1950 *Lost River at Wesley Chapel Gulf, Orange County, Indiana.* reprint. Indianapolis, IN, US: GIG Publications, **1970**. illus: b&w: maps; 1 folded map in text; text on lining pages; [ii], 32p; 23 cm.

Originally published in: Proceedings of the Indiana Academy of Science, vol. 41, 1932, pp 285-316.

[2027] Malott, Clyde Arnett, 1887-1950 *Significant features of the Indiana karst.* reprint. Indianapolis, IN, US: GIG Publications, **1974**. illus: b&w: maps, photos; text on lining pages; [ii], 17p; bibl; 23 cm.

Originally published in: Proceedings of the Indiana Academy of Science, vol. 54, 1945, pp 8-24.

[2028] Malott, Clyde Arnett, 1887-1950 *The swallow-holes of Lost River, Orange County, Indiana.* reprint. Indianapolis, IN, US: GIG Publications, **1970**. illus: b&w: charts, maps, photos; text on lining pages; [ii], 45p; bibl; 23 cm.

Originally published in: Proceedings of the Indiana Academy of Science, vol. 61, 1952, pp 187-231. Cover title: The swallow-holes of Lost River.

[2029] Mammoth Cave National Park Association *Mammoth Cave National Park Association campaign, 1927-1928.* npop: np, **1928**. [350].

OCLC also lists 1927 date of publication. This record book contains copies of correspondence, lists of contributors, and other materials used by the Mammoth Cave National Park Association in its campaign to have the cave declared a national park. The materials date from 1927 to 1928. The correspondence concerns the organization of local offices across the state of Kentucky, and contributions to the Association. The book also contains samples of forms and broadsides used in the campaign.

[2030] Mammoth Cave National Park Association *A national park in Kentucky.* npop: np, **1985**. illus: b&w: maps; [16], 8p pls; 27 cm.

Reprint of 1900 publication? Microfiche. Louisville, [Ky.]: Lost Cause Press, 1969. 2 microfiches ; 11 x 15 cm. Illus.

[2031] Mammoth Cave National Park Association *Now how! A national park in Kentucky.* [Louisville, KY, US], **1927**. illus: b&w: maps; 29p; 24 cm.

Title from cover. Caption title: A national park in Kentucky.

[2032] Mammoth Cave National Park Association *A privilege—a duty—an opportunity for Kentucky.* [Louisville, KY, US]: np, **nd**. 16p; 27 cm.

Cover-title. Master microform held by: Louisville, Ky., Lost Cause Press, 19—. 1 card. Illus.

[2033] Mammoth Cave Operating Committee, Mammoth Cave, Ky *Caverns of enchantment.* Mammoth Cave, KY, US, **1940**. illus: b&w: maps; 96p; 20 cm.

Text on p. [2] and [3] of cover. Illus.

[2034] Mammoth Cave Operating Committee, Mammoth Cave, Ky *Mammoth cave.* npop, **1859**. 116p; 20 cm.

Illus. OCLC shows 1850 edition as well.

[2035] [Manchester University Speleological Society] *Report of the 1974 British speleological expedition to Matienzo area of Santander Province of northern Spain.* [GB]: Manchester University Speleological Society, **[1975]**. illus: b&w: maps, photos, repros; 6 folded maps, some loose and some in text; [iii], iii, 32p, 9p pls; 30 x 21 cm.

Cover title: Matienzo North Spain: the 1974 British expedition report.

[2036] Maness, Lindsey Vance *North Carolina coastal*

plain caves and their impact on mining and quarrying. Denver Mining Club. Golden, CO, US: Maness, Lindsey Vance, **1994**. 52p; bibl; 28 cm.

Caption title. "Presented at the Denver Mining Club, February 3rd, 1994".

[2037] Mann, Ian (ed) *Speleotec '87 guidebook: a guide to the 16th biennial conference of the Australian Federation Inc.* Broadway, NSW, AU: Speleotec '87 Organizing Committee, **[1987]**. illus: b&w: cartoons, charts, maps; [i], 24p; bibl.

Published on behalf of Australian Speleological Federation Inc. Held at Macquarie University, Sydney January 4-11, 1987.

[2038] Mann, Paul (ed) *Oxford University Cave Club Cabeza Julagua expedition final report 28 June - 20 August 1993.* Lowe, Gavin (ed). npop: Oxford University Cave Club, **1994**. illus: b&w: maps; ii, 38p; 30 x 21 cm.

[2039] Mansfield, Kay Patricia Russell (comp) *Caving periodicals in the Central Collection of Caving Publications.* Mansfield, Raymond Walter (comp). npop, [GB]: The Mendip Cave Registry, Mar, **1968**. 14p; 26 x 20.5 cm.

No illus.

"A catalogue of the caving periodicals held by Bristol Central Reference Library and a selected short title catalog of books."

[2040] Mansfield, Raymond Walter *Mendip cave bibliography and survey catalogue, January, 1901—December, 1963.* Reynolds, T. E. (comp); Standing, I. J. (comp). Ledbury, Herefordshire, GB: The Cave Research Group of Great Britain, **1965**. Cave Research Group of Great Britain Publication no 13. [ii], 164p; index; 26 cm.

Cover title: Mendip Cave bibliography and survey catalogue1901-1963.

[2041] Mansfield, Raymond Walter *Somerset sump index.* Bristol, GB: Cave Diving Group, **1978**. illus: b&w: maps; [ii], 43p, [1], 2p pls; 26 x 21 cm.

1964: 2nd ed; 1977: rev ed.

[2042] Mansfield, Raymond Walter (ed) *Speleological abstracts: key to Britain's speleological literature.* Settle, Yorkshire, GB: British Speleological Association, Jul, **1970**. Vol 1 no 6. x, 159p; 26 x 20 cm.

No illus. Cover title: The literature of 1967.

[2043] Mansfield, Raymond Walter (ed) *Speleological abstracts: the literature of 1968.* Settle, Yorkshire, GB: British Speleological Association, Dec, **1971**. Journal of the British Speleological Association vol 6 no 47. i, 114p; 26 x 20 cm.

No illus.

[2044] Mansfield, Raymond Walter *Sump index: section 1 (Somerset).* GB: Caving Diving Group, Apr, **1964**. [xliv]; app, bibl; 25 x 20.5 cm.

Revised again in 1977. No illus.

[2045] Marcus, Rebecca Brian *Prehistoric cave paintings.* 1968 ed. New York, NY, US: F. Watts, **1968**. illus: b&w: maps; 88p; 23 cm.

Illus.Microfiche: New York, N.Y.: Watts, 1968, 10 x 15 cm. Juvenile.

1970 ed. London, GB: Franklin Watts Ltd, **1970**. illus: b&w: drawings, maps, photos; [vi], 89p; gloss, index; 23 cm.

Previous ed: New York: Watts, 1968. Illus. Juvenile.

[2046] Marker, Margaret E. *Echo cave: a tentative Quaternary chronology for the Eastern Transvaal.* Brook, G. A. Johannesburg, ZA: University of the Witwatersrand, Dept. of Geography and Environmental Studies, **1970**. University of the Witwatersrand. Dept of Geography and Environmental Studies, Occasional Paper no 3. illus: b&w: drawings, maps, photos, tables; 38p; 24 cm.

Illus.

A short book on geomorphology, cave geology, paleoclimatology, and cave stratigraphy. One of the few in English. Some similarities to Appalachian-style speleogenesis. Somewhat awkwardly written.

[2047] Marker, Margaret E. *Lower southeast of South Australia: a karst province.* Johannesburg, ZA: Dept of Geography and Environmental Studies, University of the Witwatersrand, **1975**. University of the Witwatersrand. Dept. of Geography and Environmental Studies, Occasional Paper no 13. illus: b&w: charts, graphs, maps, photos, tables; Various pagination, 66p; app, bibl, index; 23 cm.

18 figs, 9 tables.

A description of middle- to upper-Cenozoic carbonate rocks in a high rainfall area. Karst is related to emergence history, water table depth, regional geology, and surface topography. It is a descriptive rather than an interpretive work.

[2048] Marochov, Nick (ed) *Below Belize: Queen Mary College speleological expedition to Belize, 1988 and the British speleological expedition to Belize, 1989.* Williams, Nick (ed). npop: [The Wessex Cave Club?], **1992**. illus: b&w: maps, photos; col: photos; ii, 58p, 8p pls; bibl.

[2049] Marochov, Nick (ed) *Below Belize: Queen Mary College Speleological Expedition to Belize 1988 and the British Speleological Expedition to Belize 1989.* Williams, Nick (ed). Priddy, nr Wells, Somerset, GB: The Wessex Cave Club, **[1992]**. illus: b&w: maps, photos; col: photos; ii, 58p; 29 x 21 cm.

18 col photos.

[2050] Marsh, Dorothy *Life at Russell Cave.* New York, NY, US: Eastern Acorn Press, **1980**. illus: b&w: charts; 31p; bibl.

[2051] Marsh, John, 1942– *A history of Glacier House and Nakimu Caves, Glacier National Park, British Columbia.* Peterborough, Ontario, CA: Canadian Recreation Services, **1979**. [30]; bibl; 22 cm.

Illus.

[2052] Marsh, John, 1942– *Nakimu Caves*. Golden & District Historical Society. [Golden, British Columbia, CA]: Golden & District Historical Society, **1973**. illus: b&w: maps; 15p; bibl; 22 cm.

Issued with Woods, M. Redgrave, sheriff. Illus.

[2053] Marshall, Des *Mallorca Caves; an interim guide*. Derbyshire, GB: Des Marshall, **1994**. illus: b&w: maps; [18]; 29.5 x 21 cm.

Pages printed on one side only.

[2054] Marshall, Des *Vercors caves*. Deakin, P. R. (photo). Leicester, GB: Cordee, **1993**. Classic French caving v 1. illus: b&w: charts, maps, photos, tables; col: photos; b&w photo; 100p; bibl; 20 cm.

Presumably because of extensive publication in French, a widely understood language, there is little on French caves in English. This cavers' guide is supposed to be the first of a series. Location and permission information, descriptions and rigging guides, and small-scale maps of twenty caves in the area, including Gouffre Berger.

[2055] Marshall, Wayne (ed) *Sixth annual Florida caver's re-union (Cave Cavort): April 7-9, 1989, Hog Island Campground, Withlacoochee State Forest*. Tampa, FL, US: Tampa Bay Area Grotto of the National Speleological Society, **1989**. illus: b&w: drawings, maps; Various pagination, [37].

[2056] Martel, Edouard Alfred, 1859-1938 *Cavern of the dragon*. npop: np, **c1950**. illus: b&w: photos; 32p; 17 x 12 cm.

Title from cover. Description of Cueva del Drach (Mallorca) by the 1896 explorer. Translated extracts from annuaire du Club Alpin Francais 23 for 1896 (pub 1897) [368]-413. Also 2 later editions [1960]s with text reset but unchanged except that one of them has a map of the island on the back cover. More editions may exist.

[2057] Martel, Edouard Alfred, 1859-1938 *The caverns of Derbyshire: being an extract from Irlande at Cavernes Anglaises*. Winder, Francis A. (trans); Phillips, Stanley (trans). Sheffield, GB: W. Hartley Seed, **1914**. illus: b&w: maps, photos; 2 fold-out maps; front: photo of Peak Cavern Gorge, Castleton; 46p; bibl; 21.5 x 13.5 cm.

[2058] Martel, Edouard Alfred, 1859-1938 *The Grottoes of the Drach near Manacor (Majorca)*. Palma, Majorca, ES: Printed by Empress Soler, **1923**. illus: b&w: photos; partly col: maps; Partly col fold-out map at back; 48p; 17.5 x 12 cm.

Record originally published in the year book of the French Alpine Club, vol xxiii, 1896.

[2059] Martel, Edouard Alfred, 1859-1938 *Padirac; its history and a short description*. "Les Editions Quercynoises". Saint-Cave, FR: Tardy-Quercy (S.A.) Cahors , **1980**. illus: b&w: photos; partly col: maps; 30p, [2]; 21 x 13 cm.

Originally pub in 1925.

[2060] Martin, Dave (ed) *Orpheus Caving Club songbook*. Leicester, GB: Orpheus Caving Club, Feb, **1968**. [i], v, 74p; 30 x 21 cm.

No illus.

[2061] Martin, George Castor, 1885– *Archaeological exploration of the Shumla caves; report of the George C. Martin expedition... June, July and August, 1933*. Southwest Texas Archaeological Society. San Antonio, TX, US: Witte Memorial Museum, **1933**. Big Bend Basket Maker Papers no 3; Bulletin Southwest Texas Archaeological Society 3. 90p; 22 cm.

Caption title. Illus. Xerographic copy. Ann Arbor, Mich.: Xerox University Microfilms, 1974.

[2062] Martin, George V. *The Timpanogos Cave story: the romance of its exploration*. Salt Lake City, UT, US: Hawkes Publications, **1973**. illus: b&w: photos; 64p.

"Guide and history of this Utah cave. Author's account of early exploration."

[2063] Martin, Horace *Pictorial guide to the Mammoth Cave, Kentucky*. Wallen, S.; Andrew, J. W.; Orr, N. (illus). New York, NY, US: Stringer & Townsend, **1851**. illus: b&w: drawings; front: drawing; 114p, 10p pls; 20 cm.

Cover title: Mammoth Cave. Added tp, engraved. Microfilm edition: 1993: Printing Master B92-140. Atlanta, Ga.: SOLINET, 1993. 1 microfilm reel ; 35 mm.Microfilm edition: 1958: Microfiche. Louisville: Lost Cause Press, 1958 4 microfiche: 11 x 15 cm.

[2064] Martin, Paul Sidney, 1899– *Caves of the Reserve area*. Rinaldo, John Beach, 1912–; Bluhm, Elaine A. [Chicago, IL, US]: Chicago Natural History Museum, **1954**. Publication 731 Fieldiana: anthropology, v 42 Chicago Natural History Museum. illus: b&w: maps; 227p; bibl; 24 cm.

Illus.

[2065] Martin, Paul Sidney, 1899– *Mogollon cultural continuity and change; the stratigraphic analysis of Tularosa and Cordova caves*. [Chicago, IL, US]: Chicago Natural History Museum, **1952**. Fieldiana Anthropology v 40; Chicago Natural History Museum Publication 699. 528p; 24 cm.

Illus.

[2066] Martin, Ronald L. *Cave development in the Bull Creek drainage basin of southwest Missouri*. Springfield, MO, US: Heart of the Ozarks Grotto, **1972**. Ozark Caver volume 4 number 1. illus: b&w: charts, maps, photos; [iv], 27p, [1]; bibl.

Deals with morphology of the caves. 44 figs.

[2067] Martin, Ronald L. *Marvel Cave: Silver Dollar City, Branson, Missouri*. Springfield, MO, US: Ozark Mountain Publishers, **1990**. illus: b&w: drawings, photos, repros, tables; partly col: maps; col: charts, photos; 64p.

"Discovery and exploration of Missouri's deepest cave. 1st edition published 1974"—t.p. verso.

[2068] Martin, Ronald L. *Official guide to Marvel Cave, Silver Dollar City, Missouri.* Springfield, MO, US: Ozark Mountain Publishers, **1974**. illus: b&w: drawings, maps, photos, repros, tables; 55p; 22 cm.

[2069] Martinez del Rio, Pablo *A preliminary report on the mortuary cave of Candelaria, Coahuila, Mexico.* [Austin, TX, US]: [Texas Archeological Society], **1953**. 41p; bibl; 23 cm.

Reprinted from: Bulletin of Texas Archeological Society, vol. 24 October, 1953, pp 209-252. Illus.

[2070] Maryland Governor's Advisory Committee on Promoting Paleontology *Report.* Eshelman, Ralph E.; Weishampel, David B. (co-chairs). npop: The Committee, Oct, **1993**. illus: b&w: maps; Various pagination; 28 cm.

[2071] Mason, Edmund John, 1911-1993 *Bone Cave (Ogof-yr-Esgyrn) Dan-yr-Ogof: the history of the Bone Cave (Ogof yr Esgyrn): a guide and background to the exhibits in Bone Cave.* Edington, M. Ann. Abercrave, GB: Dan-yr-Ogof Showcave, **1979**. illus: b&w: charts; col: maps, photos; col map rear lining page; 28p, 8p pls; 21 cm.

Cover title: Ogof-yr-Esgyrn Bone Cave: Archaeological Tour. A guide and background to the exhibits in Bone Cave.

[2072] Mason, Edmund John, 1911-1993 *Caves and caving in Britain.* London, GB: Robert Hale, Ltd, **1977**. illus: b&w: photos; 208p, 12p pls; bibl, gloss, index; 23 cm.

Label mounted on tp: Transatlantic Arts, Levittown,N.Y., sole distributor for the U.S.A. Good general introduction to the caves of Britain; 38 pls, 6 maps.

[2073] Mason, Edmund John, 1911-1993 *The story of Wookey Hole.* Bath, GB: G.W. Hodgkinson, Jul, **1963**. illus: b&w: drawings, maps, photos, repros; 48p; 23 x 14.5 cm.

Cover title: The story of Wookey Hole: fully illustrated. 1966: 2nd ed; 1970: 3rd ed.

[2074] Mason, Gregory, 1889– *Pottery and other artifacts from caves in British Honduras and Guatemala.* New York, NY, US: Museum of the American Indian, Heye Foundation, **1928**. Indian Notes and Monographs Miscellaneous no 47. 45p; 17 cm.

Illus.

[2075] Mason, Otis Tufton, 1838-1908 *The caverns of Luray.* 1884. npop: [Shenandoah Valley Railroad Co]. illus: b&w: ads, maps; 15p; 23 cm.

Cover title: *The caverns of Luray, located on line of the Shenandoah Valley Railroad.* The original report, written by O. T. Mason and printed in the Report of the Smithsonian Institution for 1880, has been slightly abridged in reprinting. Illus.

1886. Shenandoah Valley Railroad County. npop: [Shenandoah Valley Railroad Co]. illus: b&w: ports; 12p; 24 cm.

Caption title. Map on p. [4] of cover. On cover: The Caverns of Luray at Luray, Virginia, on the line of the Shenandoah Valley R. R. The original report, written by O. T. Mason and printed in the Report of the Smithsonian Institution for 1880, has been slightly abridged in reprinting. Illus.

[2076] Mason, Otis Tufton, 1838-1908 *The caverns of Luray.* Shenandoah Valley Railroad County. npop: [Shenandoah Valley Railroad Co], **1882**. illus: b&w: maps; 13p, [1]; 23 cm.

Cover title. The original report, written by O. T. Mason and printed in the Report of the Smithsonian Institution for 1880, has been slightly abridged in reprinting. Double map. Illus.

[2077] Mason, Otis Tufton, 1838-1908 *Report of a visit to the Luray Cavern, in Page County, Virginia, under the auspices of the Smithsonian institution, July 13 and 14, 1880.* Washington, DC, US: Smithsonian Institution, **1881**. Smithsonian Institution. Annual Report, 1880. 12p; 24 cm.

Illus, pp 449-460.

[2078] Mason, Revil J. *Cave of Hearths, Makapansgat, Transvaal.* Brain, Charles Kimberlin. Johannesburg, ZA, **nd**. University of the Witwatersrand Archaeological Research Unit Occasional papers 21. illus: b&w: maps; vi, 713p; bibl; 31 cm.

Reports on Early, Middle and Late Stone Age, Iron Age and historic data from the Cave of Hearths and adjacent historic cave. Illus.

[2079] Mason, Revil J. *Kruger Cave, late Stone Age, Magaliesberg.* Steel, R. H. Johannesburg, ZA, Sep, **1988**. Occasional paper of the Archaeological Research Unit University of the Witwatersrand Johannesburg no 17. illus: b&w: maps; ix, 374p, [4]; bibl; 31 cm.

Illus.

[2080] Mason-Williams, Ann *A preliminary investigation into the bacterial & botanical flora of caves in South Wales.* Benson-Evans, Kathryn. Berkhamsted, GB: Cave Research Group of Great Britain, Sep, **1958**. Cave Research Group of Great Britain Publication no 8. illus: b&w: drawings, maps, photos, tables; [iv], v, 70p; bibl, note; 26 x 20 cm.

[2081] Mass, Nuri *The wizard of Jenolan.* Mass, Celeste (illus). 1946 ed. Sydney, NSW, AU: Angus and Robertson Ltd, **1946**. illus: b&w: drawings; col: drawings; front: col drawing; 93p, 3p pls.

Juvenile.

1993 rev ed. Kubbos, Vivien (illus). Chatswood, NSW, AU: Just Solutions Pty Ltd, **1993**. illus: col: drawings; col drawing on lining pages; front: col drawing; [i], 60p, 3p pls.

Juvenile.

[2082] Masschelein, Jan (ed) *Teng Long Dong, the longest cave in China: report of the first Belgian-Chinese* *speleological expedition in 1988.* Shouyue Zhang (ed). npop: Belgian-Chinese Karst and Caves Association, **nd**. illus: b&w: charts, graphs, maps, tables; col: maps, photos; 1 loose folded map; front: drawing; 48p.

[2083] Massey, William C. *A burial cave in Baja California, the Palmer Collection 1887.* Osborne, Carolyn M. reprint. Berkeley, CA, US: University of California Press, Kraus Reprint, **1976**. Anthropological Records v 16 no 8, pp 339-363. illus: b&w: drawings, maps, photos; 25p; bibl; 28 cm.

Reprint of 1961 ed.

[2084] Mathpal, Yashodhar *Prehistoric rock paintings of Bhimbetka, Central India.* New Delhi, IN: Abhinav Publications, **1984**. illus: b&w: maps, photos; xviii, 236p, 32p pls; bibl, index; 28 cm.

Some col illus.

[2085] Mattheij, Johannes Adrianus Maria, 1940– *The functional cell types in the pars distalis analogue of the pituitary gland in the blind mexican cave fish,* Anoptichthys jordani. [Utrecht, NL]: np, **1970**. Various pagination; bibl.

Summary in Dutch. Proefschrift, Rijksuniversiteit te Utrecht. Illus.

[2086] Matthews, Geoff (ed) *Report of the Nottingham University Students Union Spelaeological Expedition 1970, Picos de Europa North-West Spain.* Nottingham, GB: Nottingham University Caving Club in conjunction with expeditions Co-ordinating Committee, May, **1971**. illus: b&w: drawings, maps, photos; 3 fold-out maps; [iv], 64p, 7p pls; add, app, errat; 24 x 17 cm.

Cover title: Exploration '70. Addendum is a loose-identification of specimens found in Cueva Dobros.

[2087] Matthews, Larry Edwin, 1946– *Bibliography of Tennessee speleology.* Tennessee Cave Survey. [1st]. npop: Tennessee Cave Survey, **1974**. [21]; bibl; 28 cm.

Cover title. Covers more than 100 years of Tennessee speleology.

[2nd ed]. Knoxville, TN, US, Aug, **1975**. text on rear lining page; iii, 65p; bibl, index.

Title from cover. No illus.

[3rd ed]. Tennessee Cave Survey. Apr, **1980**. iv, 288p; index; 29 cm.

At head of title: Tennessee Cave Survey, Bulletin 1. Pages 130-131, 262-264, 287-288 blank. Covers more than 100 years of Tennessee speleology. 1482 entries.

[4th ed]. Knoxville, TN, US, Nov, **1994**. [xi], 520p; bibl, index.

No illus.

[2088] Matthews, Larry Edwin, 1946– *Cumberland Caverns.* Huntsville, AL, US: National Speleological Society, **1989**. illus: b&w: maps, photos, repros; 317p; app, bibl, gloss, index; 21 cm.

An account of the cave's exploration, including a good historical introduction and an extensive chronology of events.

[2089] Matthews, Larry Edwin, 1946– *Descriptions of Tennessee caves.* Nashville, TN, US: State of Tennessee, Department of Conservation, Division of Geology, **1971**. Tennessee Division of Geology Bulletin 69. illus: b&w: maps, photos; 3 folded facing on page 60; folded maps in back pocket; 150p; app; 29 cm.

Update to Barr 1961. Over three hundred additional caves; forty cave maps. Eighteen pages of photographs in introductory chapter on saltpeter mining.

[2090] Matthews, Peter Gahan, 1938– *Australian karst index 1985.* Melbourne, VIC, AU: Australian Speleological Federation, Inc, **1985**. illus: b&w: charts, graphs, maps, tables; Various pagination, vi, 495p; app, bibl, gloss; 30 cm.

Limited distribution.

Catalog of 6639 caves (mostly) and other karst features in Australia, with brief text descriptions and rough locations. List of known maps, but none printed. Thousand-item bibliography.

[2091] Matthews, Peter Gahan, 1938– *Australian karst index data preparation manual.* Broadway, NSW, AU: Australian Speleological Federation, Inc, Jul, **1983**. illus: b&w: maps, photos, tables; Various pagination; app, bibl.

Five sections, only three of which exist in this book; other two to come later.

[2092] Matthews, Peter Gahan, 1938– *Check-list of Australian caves and karst, 1979.* Broadway, NSW, AU: Australian Speleological Federation, **1979**. 73p; 30 cm.

No illus.

[2093] Matthews, Peter Gahan, 1938– *Speleo handbook.* Broadway, NSW, AU: Australian Speleological Federation, Jan, **1968**. illus: b&w: drawings, maps, tables; v, 322p; app, bibl, gloss; 26 x 20.5 cm.

Includes much introductory material about caves and caving, but the majority of the book is a catalog of caves in Australia, including Papua and New Guinea. Brief text descriptions and references.

[2094] Matthews, William Henry, 1919– *The geologic story of Longhorn Cavern.* Austin, TX, US: Bureau of Economic Geology, University of Texas, Feb, **1963**. Bureau of Economic Geology University of Texas Guidebook 4. illus: b&w: charts, drawings, maps, photos; [iv], 46p, [4]; bibl, index; 26 cm.

Excellent booklet with many photos.

[2095] Maturango Museum of Indian Wells Valley *Excavation of two sites in the Coso Mountains of Inyo County, California.* Panlaqui, Carol (The Ray Cave site); Berry, Kristin H. (The Baird site); Hillebrand, Timothy Shaw (Floras of the Cave area). China Lake, CA, US: The Museum, **1974**. Monograph / Maturango Museum of Indian Wells Valley no 1. illus: b&w: drawings; 86p; 28 cm.

Illus.

[2096] Mazonowicz, Douglas *Cave art of France and Spain.* Shorewood Fine Art Reproductions, Inc.; Shorewood Educational Programs. Sandy Hook, CT, US: Shorewood Fine Art Reproductions, **1984**. Shorewood Art Programs for Education. illus: b&w: maps, repros; 6+8p; 58+23 x 73 cm.

Title from accompanying guide. In storage portfolio, 71 x 75 x 4 cm. On portfolio: Shorewood educational programs. Accompanying guide has bibliography (p. [8]). Standing bison (Altamira, Spain)—Engraved deer (Les Combarelles, France)—Two reindeer (Font-de-Gaume, France)—Yellow horse (Lascaux, France)—Group of deer (Covalanas, Spain)—Head of a bull (Lascaux, France).

[2097] Mazonowicz, Douglas *In search of cave art*. Rohnert Park, CA, US: Gallery of Prehistoric Paintings, **1973**. illus: b&w: charts, drawings, maps, photos; [112]; bibl.

Most illustrations are photographs of serigraphs by author of cave and rock art.

[2098] Mazonowicz, Douglas *On the rocks: prehistoric art of France and Spain*. Villiers, Berna; Gallery of Prehistoric Art [New York, N.Y.]. New York, NY, US: Gallery of Prehistoric Art, **1989**.

VHS format. Producer, Berna Villiers ; narrator, Douglas Mazonowicz ; Music, Lou Harrison. Focuses on the cave art of Southwestern Europe with particular attention to Lascaux and Altamira. Using photographs and artwork, the film celebrates the remarkable artistic ability of prehistoric man. 1 videocassette (25 min.): sd., col. ; 1/2 in.

[2099] Mazonowicz, Douglas *The painted caves of France & Spain*. [New York, NY, US]: [Gallery of Prehistoric Art, distributor], **1979**.

Title on guide: The Genius of mankind's earliest artists: the painted caves of France & Spain. 40 slides: col. + 1 sound cassette (1 7/8 ips, 2 track, mono.) + 1 instructor's guide.

[2100] Mazonowicz, Douglas *Prehistoric paintings of France and Spain: a description of 34 actual size copies hand screenprinted by Douglas Mazonowicz*. npop: Douglas Mazonowicz, **1970**. illus: b&w: maps, photos; 38p pls.

Illustrations mostly photos of author's serigraphs of cave art. Book is really his catalog.

[2101] Mazonowicz, Douglas *Prehistoric paintings: a catalog of actual size copies in silkscreen*. Valencia: Impr. Nacher, **1966**. illus: b&w: maps; [23]; 17 x 23 cm.

Cover title. Illus.

[2102] Mazonowicz, Douglas *The prehistoric rock paintings of Tassili n'Ajjer: a description of 15 actual size copies, hand screenpainted*. Petaluma, CA, US: Douglas Mazonowicz, **1970**. Booklet no 2. illus: b&w: maps; [39]; 24 cm.

Contents chiefly illus.

[2103] Mazonowicz, Douglas *Voices from the stone age: a search for cave and canyon art*. London, GB: G. Allen and Unwin, **1974**. illus: b&w: drawings, maps, photos; partly col: photos; col: photos; viii, 211p, 8p pls; bibl, index.

Published simultaneously in Canada by Fitzhenry & Whiteside, Toronto and in the United States by Thomas Y. Crowell Company in NY. 200 b&w photos, 24 pp col photos. Second printing lacks the [8]p of color prints totally and the partly colored photos are b&w.

[2104] McAlpine, Donald F. *Status and conservation of solution caves in New Brunswick*. [Saint John, New Brunswick, CA]: The New Brunswick Museum, **1983**. Publications in Natural Science (New Brunswick Museum) no 1. 28p; bibl; 28 cm.

Summary also in French. Illus.

[2105] McCann, Gerald *In my torchlight: a guide's guide to the Cango Caves*. npop: np, **[1992]**. illus: b&w: drawings, maps; 16p; 21.5 x 15 cm.

Description and history of this South African show cave.

[2106] McCarthy, Frederick D., 1905– *Australian aboriginal rock art*. 3rd ed. **1967**. 72p; bibl.

[2107] McCarthy, Frederick D., 1905– *Australian aboriginal rock art*. Sydney, AU: Trustees of the Australian Museum, **1958**. 68p; bibl; 24 cm.

Illus.

[2108] McClurg, David Robert, 1929– *Adventure of caving: a practical guide for advanced and beginning cavers*. McClurg, David Robert, 1929–(photo); McClurg, Janet (photo). Carlsbad, NM, US: D&J Press, **1986**. illus: b&w: drawings, photos, tables; vi, 332p; app, bibl, gloss, index; 22 cm.

[2109] McClurg, David Robert, 1929– *Adventure of caving: new updated edition*. McClurg, Janet (photo). updated ed. Carlsbad, NM, US: D&J Press, **1996**. illus: b&w: drawings, maps, photos, tables; 251p.

Update edition of *Adventure of caving: a practical guide for advanced and beginning cavers*.

A good basic text.

[2110] McClurg, David Robert, 1929– *The amateur's guide to caves & caving; skill-building ways to finding and exploring the underground wilderness*. Harrisburg, PA, US: Stackpole Books, **1973**. illus: b&w: drawings, maps, photos, tables; 191p; bibl, gloss; 22 cm.

Endorsed by the National Speleological Society.

Second, revised edition with this title corrects some of the more egregious errors.

[2111] McClurg, David Robert, 1929– *Caving short course: 1982 NSS Convention*. npop: np, **1982**. illus: b&w: charts, drawings, graphs, maps, photos, tables; Various pagination, 60p; bibl.

[2112] McClurg, David Robert, 1929– *Exploring caves: a guide to the underground wilderness*. Harrisburg, PA, US: Stackpole Books, **1980**. illus: b&w: drawings, photos; v, 287p; app, bibl, gloss, index; 21 cm.

Vastly improved version of "Amateurs Guide to Caves and Caving". Getting underground, finding caves, clothes, equipment, SRT, etc.

[2113] McClurg, James E. *Caves and their mysteries.* Kenyon, Norman (illus). Racine, WI, US: Whitman Publish Company, **1962**. A Whitman Learn About Book. illus: partly col: drawings, maps; 58p; 22 cm.

Juvenile.

[2114] McCracken, Harold, 1894– *The Mummy Cave Project in northwestern Wyoming.* Wedel, Waldo; Edgar, Robert; Moss, John H.; Wright, H. E.; Husted, Wilfred H.; Mulloy, Willliam. Cody, WY, US: Buffalo Bill Historical Center, **1978**. illus: b&w: charts, drawings, maps, photos; b&w photo; 160p; bibl; 28 cm.

65 numbered pls.

[2115] McCrady, Edward, 1906– *New finds of Pleistocene jaguar skeletons from Tennessee caves.* Washington, WA, US, **1954**. illus: b&w: drawings; 15p; 24 cm.

OCLC notes some pls. pp 497-511.

[2116] McDonald, Alvin F., 1873-1893 *Private account of A. F. McDonald, permanent guide of Wind Cave.* [Hot Springs, SD, US]: [Wind Cave National Park], **nd**. 64p.

..."an early explorer and guide of Wind Cave ... the son of J.D. McDonald, one of the original explorators of Wind Cave ... the pages that follow were typed from Alvin's diary. The original diary is stored in the archives at Wind Cave National Park, Hot Springs, South Dakota." No illus.

[2117] McDonald, Donald L. *Official guide to Marvel Cave: the only complete authentic history of Marvel Cave ever published.* Springfield, MO, US: Ozark Mountain Publishers, **1974**. illus: b&w: drawings, maps, photos, repros; 54p, [1].

Cover title: Official guide to Marvel Cave: Silver Dollar City, Missouri: Its discovery and exploration: the only complete and authentic history of Marvel Cave ever published.

[2118] McDonald, M. C. (prep) *The Somerset sump index.* updated ed. [GB]: Somerset Section of Cave Diving Group, Dec, **1993**. illus: b&w: maps; 56p; 30 x 21 cm.

1991: 3rd ed.

[2119] McDonald, Mike *The journal of the Joint Bristol Exploration Club (United Kingdom) and the National Mountaineering Federation of the Philippines caving expedition to the Philippines, January to April 1992.* Bristol, GB: Bristol Exploration Club, **c1993**. illus: b&w: maps; col: photos; col photocopy of photos (4); [i], 46p, [1], 7p pls; 30 x 21 cm.

Cover title: Speleo Philippines 1992.

[2120] McEachern, J. Michael *An inventory and evaluation of the cave resources to be impacted by the New Melones Reservoir Project, Calaveras and Tuolumne Counties, California.* Grady, Mark A.; United States Army Corps of Engineers Sacramento District. [Dallas, TX, US]: Archaeology Research Program, Southern Methodist University, **1978**. Archaeology Research Program research report 109. illus: b&w: maps; maps folded in pocket; Various pagination.

Contract DACW05-77-C00038. Illus, 2 vols.

[2121] McEldowney, Holly *Survey of lava tubes in the former Puna Forest Reserve and on adjacent State of Hawaii lands.* Stone, F. D. ; Hawaii Division of Water Resource Management. npop: np, Oct, **1991**. illus: b&w: maps; 53p; bibl; 28 cm.

Survey extends into Campbell Estate property on which geothermal resource development is being planned. Running title: Puna lava tube survey. Prepared for State Historic Preservation Division, Division of Water Resource Management, Dept. of Land and Natural Resources, State of Hawaii.

[2122] McEwan, Graham *Crypts, caves and catacombes: subterrenea of Derbyshire & Nottingham.* Cheshire, GB: Sigma Leisure, **1994**. illus: b&w: drawings, maps, photos, repros; vi, 146p; 21 x 15 cm.

[2123] McFarlane, Donald A. *A preliminary catalog of the caves on Anguilla, British West Indies.* npop: [Donald A. McFarlane], Oct, **1989**. illus: b&w: maps; 18p; bibl.

[2124] McGill, Greg *Alabama caves.* Domnanovich, Joe. 1987 ed. Birmingham, AL, US: Alabama Cave Survey, **1987**. illus: b&w: maps; Various pagination; 29 cm.

Photocopy, 3 vols. "This is the second complete survey published by the Alabama Cave Survey. Updates have been made available to members each year since the original 1980 publication."—preface to the 1987 ed. "This book contains 1615 cave maps."—"Statistics."

[2125] McGill, William Mahone, 1897-1962 *Caverns of Virginia.* Richmond, VA, US: State Commission on Conservation and Development, **1933**. Commonwealth of Virginia State Commission on Conservation and Development; Bulletin 35 Educational Series no 1. illus: b&w: drawings, maps, photos, tables; 187p, 42p pls; bibl, index; 28 cm.

First survey of Virginia caves. Also, the first mention of a type of formations known as shields or palettes.

Chapters on geology of caves and the Appalachian valleys. Surveys mainly the developed show caves of the state, with maps and photographs.

[2126] McGrain, Preston (comp) *Itinerary: geology of the Mammoth Cave Region, Barren, Edmonson, and Hart Counties, Kentucky.* Walker, Frank H. (comp). Lexington, KY, US: Kentucky Geological Survey, Apr, **1954**. illus: b&w: charts, drawings, graphs, maps, tables; 32p; bibl.

Cover title: Geological Society of Kentucky Field Trip: Itinerary. 14 figs.

[2127] McGrain, Preston, 1917– *The geologic story of Diamond Caverns.* Lexington, KY, US: College of Arts and Sciences, University of Kentucky, **1961**. Kentucky Geological Survey Series X Special Publication 6. illus:

b&w: charts, drawings, maps, photos; 24p; bibl; 23 cm.

[2128] McGrain, Preston, 1917– *Geology of the Carter and Cascade Caves area*. Lexington, KY, US: [Kentucky Geological Survey, University of Kentucky], **1954**. Kentucky Geological Survey Series IX. Special Publication no. 5. illus: b&w: charts, drawings, maps, photos; 32p; bibl.

Microopaque or microfiche Louisville, Ky., Lost Cause Press, 19—. Reprinted 1966. Copy examined was 1966 edition.

[2129] McGregor, Duncan J. *Some features of karst topography in Indiana*. Rarick, R. Dee. npop: Indiana Geological Survey, **1962**. illus: b&w: drawings, maps, photos; 20p; bibl.

From page 2: "This guidebook was prepared for the annual spring geologic field trip of the Indiana Academy of Science. It does not constitute an official publication of the Indiana Geological Survey and, therefore, should not be cited as a reference."

[2130] McKenzie, David (ed) *The caves of Bell and Coryell Counties*. Reddell, James Russell, 1938–. Austin, TX, US: Texas Speleological Association, **1964**. Texas Speleological Survey v 2 no 4. illus: b&w: maps; 3 maps folded in pocket; 63p; bibl; 28 cm.

Illus.

[2131] McKinnon, Judith *Under the ground*. Tripp, Tim; Williamson, Fraser. Auckland; Crystal Lake, IL, NZ; US: Shortland Publications; Distributed in the United States by Rigby, **1991**. A Read-about. 24p; 23 cm.

Col illus. Juvenile.

[2132] McLane, Alvin Ray *A bibliography of Nevada caves*. University of Nevada Center for Water Resources Research. Reno, NV, US: Center for Water Resources Research, Desert Research Institute, University of Nevada System, **1974**. illus: b&w: maps, photos; partly col: maps; partly col folded map in rear pocket; vi, 99p; index.

Cave index: p. 97-99. Excellent reference with about 950 entries and about 300 caves.

[2133] McLean, John Scott, 1942– *Factors altering the microclimate in Carlsbad Caverns, New Mexico*. Office of Natural Sciences, Southwest Region, National Park Service. Albuquerque, NM, US: United States Geological Survey, Feb, **1976**. Open-file report (United States Geological Survey) 76-171. illus: b&w: graphs, maps, tables; b&w graphs folded text; ii, 56p; bibl; 28 cm.

NPS Research Project CACA-N-la.

[2134] McLean, John Scott, 1942– *The microclimate in Carlsbad Caverns, New Mexico*. Albuquerque, NM, US: United States. National Park Service. Geological Survey (U.S.), May, **1971**. U.S. Geological Survey Open-file Report. illus: b&w: drawings, graphs, maps; 67p; bibl; 30 cm.

19 figs. Prepared by the U.S. Geological Survey for the National Park Service under NPS research project CACA-N-la.

[2135] McLoyd & Graham *Catalogue and description of a very large collection of prehistoric relics: obtained in the cliff houses and caves of Southeastern Utah*. Durango, CO, US: McLoyd & Graham, **1894**. [38] ; 22 cm.

Title from cover. Original printed red wrappers, side-stapled. No first names given for authors.

[2136] McMurtrie, James *The Mendip Caverns*. npop: Office of Bolton and Partners, Aug, **1882**. fold-out map; 29p.

Reprinted from The Times and the Proceedings of the Somersetshire Archeological and Natural History Society 1880 Part II. Copy examined was a photocopy. Part I is Anon; part II by McMurtrie. No illus.

[2137] Mead, Jim I. *The late Wisconsinan vertebrate fauna from Deadman Cave, southern Arizona*. [San Diego, CA, US]: San Diego Society of Natural History, Nov, **1984**. Transactions of the San Diego Society of Natural History v 20 no 14 pp [247]-276. 30p; bibl; 26 cm.

Caption title: "20 November 1984." Illus.

[2138] Medville, Douglas Michael, 1941– *Caves and karst hydrology in northern Pocahontas County*. Medville, Hazel E. Cliffside Park, NJ, US: West Virginia Speleological Survey Bulletin, Mar, **1976**. Bulletin West Virginia Speleological Survey 6. illus: b&w: maps, photos, tables; 2 folded maps in back pocket; ii, 174p; bibl, index; 28 cm.

Very good regional descriptive guide. Includes many cave maps. The southern part of this West Virginia county is covered in Storrick 1992.

[2139] Medville, Douglas Michael, 1941– *Caves and karst of Randolph County*. Medville, Hazel E. [2nd ed]. Barrackville, WV, US: West Virginia Speleological Survey, **1995**. West Virginia Speleological Survey Bulletin 13. illus: b&w: maps, photos, repros; maps folded to 26 x 20 cm; [ii], 250p; bibl, index; 28 cm.

One of the most cave-rich counties in state of West Virginia. Descriptions and locations of 520 caves; maps of 110 of them. Updates Medville and Medville 1971.

[2140] Medville, Douglas Michael, 1941– *Caves of Randolph County*. Medville, Hazel E. Charlottesville, VA, US: Pittsburgh Press, **1971**. Bulletin (West Virginia Speleological Survey). illus: b&w: maps, photos, repros; 218p; bibl, index; 28 cm.

[2141] Mehl, Maurice Goldsmith, 1887-1966 *Missouri's Ice Age animals*. Rolla, MO, US: Missouri Geological Survey and Water Resources, Dec, **1962**. Educational Series no 1. illus: b&w: charts, drawings, maps, photos, tables; partly col: drawings; i, 104p; app, bibl.

[2142] Mehl, Maurice Goldsmith, 1887-1966 *Notes on Missouri Pleistocene peccaries*. npop: np, **1969**. 21p; 28 cm.

Title from cover. "Reprinted from Missouri speleology, vol. 8, no. 2, April 1966", pp54-74. Illus.

[2143] Meleen, Elmer E. *A preliminary report on rock shelters in Fall River County, South Dakota*. Pruitt, James; JWH Over Museum, South Dakota State University; United States Work Projects Administration. [Vermillion, SD, US]: Museum, University of South Dakota, Apr, **1941**. Archaeological Studies Circular no 1. illus: b&w: maps; 17p; bibl; 28 cm.

OCLC also lists 1938 as date of pub. Prepared at the museum as a report on Work Projects Administration O.P. no. 665-74-3-136; W.P. no. 3740. Assistance in the preparation of these materials furnished by the personnel of the Archaeological Research Project. Illus.

[2144] Mellors, P. T. *Legal aspects of access underground: a guide to the legal rights and obligations of people who explore cave, potholes and disused mine, and of those people who control access to them*. National Caving Association; Lane, C. (illus). npop: British Cave Research Association, **1989**. illus: b&w: cartoons, drawings, tables; 32p.

[2145] Meloy, Harold Raymond, 1913-1985 *Mummies of Mammoth Cave: an account of the Indian mummies discovered in Short Cave, Salts Cave, and Mammoth Cave, Kentucky*. 1971 ed. Shelbyville, IN, US: Micron Publishing Company. illus: b&w: drawings, maps, photos, repros; text on rear lining page; map on front lining page; 40p, [1]; bibl; 22 cm.

"Published as a part of a Historical Research Project on Mammoth Cave, Kentucky." Cover title: Mummies of Mammoth Cave: facts about the Indian mummies found in Short Cave, Salts Cave, and the Mammoth Cave, Ky. A new facet of Mammoth Cave history. Illus.

A history of the prehistoric human remains found in the Mammoth Cave area during the 19th and early 20th centuries. All of Meloy's publications are historically accurate but written for a popular audience.

1973 ed (4th). illus: b&w: drawings, maps, photos, repros; Map on front; text on rear; 40p; bibl.

1977 ed.

[2146] Meloy, Harold Raymond, 1913-1985 *Mummies of Mammoth Cave: Fawn Hoof, Little Alice, Lost John and others. An account of the Indian mummies discovered in Short Cave, Salts Cave, and Mammoth Cave, Kentucky*. Shelbyville, IN, US: Micron Publishing Co, **1968**. illus: b&w: drawings, photos, repros; text on rear lining page; 40p, [1]; bibl.

Little known facts on Mammoth Cave mummies.

[2147] Mera, H. P. *Reconnaissance and excavation in southeastern New Mexico*. Menasha, WI, US: American Anthropological Association, **1938**. Memoirs of the American Anthropological Association Number 51; Supplement to American Anthropologist, vol 40 no 4 part 2. Contributions from the Laboratory of Anthropology 3. illus: b&w: drawings, maps, photos, tables; 70p, 24p pls; bibl.

24 pls of b&w photos, 5 tables, 9 figs.

[2148] Meramec Valley Conservation Task Force. National Speleological Society *An endangered heritage: a story of the Meramec Valley, its caves, and their possible destruction by the U. S. Army Corps of Engineers*. [MO, US]: Missouri Speleological Survey, **[1973]**. illus: b&w: maps, tables; [23]; bibl.

[2149] Mercer, Henry Chapman, 1856-1930 *The bone cave at Port Kennedy, Pennsylvania, and its partial excavation in 1894, 1895 and 1896*. Philadelphia, PA, US: P.C. Stockhausen, **1899**. illus: b&w: maps; 1 folded pl; 16p, 5p pls; 35 cm.

Title from cover. Reprinted from: Journal of the Academy of Natural Sciences of Philadelphia ; v. 11, pt. 2, pp 270-285, [5]lvs pls.

[2150] Mercer, Henry Chapman, 1856-1930 *Cave hunting in Yucatan: a lecture delivered before the Society of Arts of the Massachusetts Institute of Technology, on December 10,1896*. Boston, MA, US, **1897**. illus: b&w: maps, ports; front; 19p, 6p pls; 24 cm.

Cover title: Cave hunting in Yucatan. "Reprinted from the Technology Quarterly for December, 1897, Vol. X, No. 4." Caption title. No. 4 in a vol. bound with: The history of the beginnings of the science of prehistoric anthropology. [Washington, D.C.?: U.S. National Museum, 1899?].

[2151] Mercer, Henry Chapman, 1856-1930 *An exploration of Durham Cave in 1893*. Boston, MA, US: Ginn & Company. The Athenaeum Press, **1897**. illus: b&w: maps; 29p; 22 cm.

Title from cover. Caption title: An exploration of Durham Cave, Bucks County, Pennsylvania, in 1893."Reprinted from Publications of the University of Pennsylvania, vol. VI." pp 149-178

[2152] Mercer, Henry Chapman, 1856-1930 *The finding of the remains of the fossil sloth at Big Bone Cave, Tennessee, in 1896*. American Philosophical Society. Philadelphia, PA, US: Press of MacCalla and Co, Inc, **nd**. illus: b&w: drawings, maps, photos; 39p.

Reprinted from vol. 36, no. 154, Proceedings American Philosophical Society. 26 numbered figs.

[2153] Mercer, Henry Chapman, 1856-1930 *The hill-caves of Yucatan, a search for evidence of man's antiquity in the caverns of Central America. Being an account of the Corwith expedition of the Department of Archaeology and Palaeontology of the University of Pennsylvania*. Pennsylvania University Department of Archaeology. 1st ed. Philadelphia, PA, US: J.B. Lippincott Company, **1896**. illus: b&w: drawings, maps, photos; front: map; 183p; index; 24 cm.

74 illus.

[2154] Mercer, Henry Chapman, 1856-1930 *The hill-caves of Yucatan, a search for evidence of man's antiquity in the caverns of Central America. Being an account of the*

Corwith expedition of the Department of Archaeology and Paleontology of the University of Pennsylvania. Thompson, Eric S. (intro). 1975 reprint ed. Teaneck, NJ, Norman, OK, US: Zephyrus Press; University of Oklahoma Press, **1975**. Speleologia. front: map; xliv, 183p, 10p pls; errat, index; 22 cm.

New 38pp introduction by Sir Eric S. Thompson. 74 illus. Master microform held by: DLC. Microfilm. Washington, D.C., United States Library of Congress, 19—. 1 reel. 35 mm.

[2155] Mercer, W. J. *The New Calaveras Cave, Murphys, Calaveras County, California: W.J. Mercer, proprietor: description and guide.* Floyd, E. F. [CA, US]: np, **1887**. 26p; 23 cm.

Copyrighted 1887, by E.F. Floyd. Bacon & Company, Printers.

[2156] Meredith, Michael Edward, 1943– *Giant caves of Borneo.* Wooldridge, Jerry; Lyon, Ben. Kuala Lumpur, MY: Tropical Press, **1992**. illus: col: maps, photos; 142p; index; 26 cm.

[2157] Meredith, Michael Edward, 1943– *Java caves 1983: report of a visit to Indonesia by Australian and British cavers.* Salzburg, AT: [Ernst Koschier], **[1985]**. illus: b&w: maps, photos; 3 maps folded in text; 26p, 2p pls; 30 x 21 cm.

Photocopy examined.

[2158] Meredith, Michael Edward, 1943– *Vertical caving.* 1st ed. npop: np, **c1980**. illus: b&w: charts, drawings, photos; 63p; bibl.

2nd (rev and enlg) ed. Martinez, Dan. Cumbria, GB: np, **1986**. illus: b&w: charts, drawings, photos; 63p; bibl; 21 cm.

[2159] Merk, Conrad *Excavations at the Kesserloch near Kesserloch near Thayngen Switzerland: a cave of the Reindeer Period.* Lee, John Edward, 1808-1887 (trans). London, GB: Longmans, Green, and Co, **1876**. illus: b&w: drawings, tables; front; b&w drawing; viii, 68p, 15p pls; app.

Description of the cave.

[2160] Merriam, John Campbell, 1869-1945 *Recent cave explorations in California.* Putnam, Frederic Ward, 1839-1915. Lancaster, PA, US: New Era Print Co, **1906**. 15p.

Title from cover. "Reprinted from the American Anthropologist (N.S.) vol. 8, no. 2, April-June, 1906." Master microform held by: DLC. Microfilm. Washington, D.C., United States Library of Congress, 19—. 1 reel. 35 mm. pp 221-235.

[2161] Meshram, Pradip Shaligram, 1954– *Early caves of Maharashtra: a cultural study.* Delhi, IN: Sundeep Prakashan, **1991**. vi, 172p, 19p pls; bibl, index; 25 cm.

Architecture, costumes, footwear, ornaments, and vessels as depicted in the cave drawings of Maharashtra, India; covers the period 2nd cent. B.C.-3rd cent. A.D. Illus.

[2162] Messling, Gordon *Caverns of Sonora.* Messling, Brenda. Sonora, TX, US: Gordon and Brenda Messling, **1978**. illus: col: photos; 44p; 22 cm.

Title taken from cover.

[2163] Meulen, A. J. van der *Microtus and Pitymys (Arvicolidae) from Cumberland Cave, Maryland, with a comparison of some new and old world species.* Pittsburgh, PA, US: Carnegie Museum of Natural History, **1978**. Annals of the Carnegie Museum v 47 article 6. 44p; bibl; 23 cm.

Illus.

[2164] Meyer-Rochow, Victor Benno *The New Zealand glowworm.* Waitomo Caves, NZ: Waitomo Caves Museum Society Inc, **1990**. illus: b&w: drawings, graphs, maps, photos, photomicrographs, tables; 60p; bibl, gloss, index.

[2165] Miccolis, Maria *The Castellana Grottoes.* Milian, IT: Commune of Castellana-Grotte, **1980**. illus: b&w: maps; col: photos; front; col photo; 38p.

[2166] Middleton, Gregory J. *Oliver Trickett: doyen of Australia's cave surveyors 1847-1934.* Broadway, NSW, AU: Sydney Speleological Society, **1991**. Sydney Speleological Society Occasional Paper no 10. illus: b&w: maps, photos, repros; partly col: maps; map in rear pocket, 2 folded maps tipped in; port on lining page; front: b&w; viii, 156p; app, bibl, index; 29.5 x 20.5 cm.

Published in association with the Jenolan Caves Historical & Preservation Society. 104 figs.

An excellent, detailed examination of Trickett's work and influence, which encompassed much of Australia's cave surveying of the time. Effectively included in the study is much of the background to Australian cave exploration; an essential reference.

[2167] Middleton, Gregory J. *Speleology.* npop: np, **1965—1966**. illus: b&w: charts, maps; partly col: maps; some maps folded in text; Various pagination; app, bibl, gloss.

Unnumbered. A project submitted for the Rover Scout Badge to the Lane Cove District Rover Crew. Approximately 3 cm thick. Receipts and correspondence included.

[2168] Middleton, Gregory J. *Wilderness caves: a preliminary survey of the caves of the Gordon-Franklin River System, South-West Tasmania.* Hobart, Tasmania, AU: Board of Environmental Studies, University of Tasmania, **1979**. Environmental Studies Occasional Paper 11, Centre for Environmental Studies University of Tasmania. illus: b&w: maps; col: maps; folded maps; 107p; bibl; 30 cm.

Cover title: Wilderness caves of the Gordon-Franklin River System. Illus.

[2169] Middleton, Gregory J. *Yarrangobilly Caves, Kosciusko National Park.* [Sydney, NSW, AU]: National Parks and Wildlife Service of NSW, **1970**. illus: b&w: maps, photos; partly col: ads; col: photos; map on front lining page; 16p.

[2170] **Middleton, Gregory. J. (comp)** *Index to references to caves in three New South Wales government publications 1870-1919*. Broadway, NSW, AU: Sydney Speleological Society, Sep, **1988**. Sydney Speleological Society Occasional Paper no 9. 22p.

No illus.

[2171] **Middleton, John**, 1942– *The underground atlas: a gazetteer of the world's cave regions*. Waltham, Anthony Clive, 1942–[Tony]. 1986 reprint ed. Leicester, GB: Promotional Reprint Co. 24p pls; app, bibl, gloss; 25 cm.

1992 reprint ed. Waltham, Anthony Clive, 1942–[Tony]. London, GB, **1992**. 24p pls; app, bibl, gloss; 25 cm.

Publisher is an unknown reprint co.

1st U.S. ed. Waltham, Anthony Clive, 1942–[Tony]. New York, NY, US: St. Martin's Press, **1986**. illus: b&w: maps, photos; 239p, 24p pls; app, bibl, gloss; 24 cm.

Illus. Also published by R. Hale, London, GB.

A country-by-country survey of the cave areas of the world. A useful summary source, but unfortunately no references. Fifty very small-scale cave maps.

[2172] **Middleton, John**, 1942– *The underground atlas: a gazetteer of the world's cave regions*. Waltham, Anthony Clive, 1942–[Tony]. London, GB: R. Hale, **nd**. illus: b&w: maps, photos; 239p, 24p pls; app, bibl, gloss; 25 cm.

[2173] **Middleton, T. C.** *Cave development and formations*. Elmira, NY, US: Educational Images, **1979**.

T. C. Middleton discusses cave formation and various shapes created by deposits in caves. 40 slides col and b & w ; 5 x 5 cm. + 1 script.

[2174] **Mijatovió, B.F.** *Hydrology of the Dinaric karst*. Heise, Hanover, DE: International Association of Hydrogeologists, **1984**. illus: b&w: photos; 254p.

Illus.

[2175] **Mikulec, Stjepan** *Karst hydrology and water resources: proceedings of the U. S.—Yugoslavian symposium, Dubrovnik, June 2-7, 1975*. Sarić, Avdo (ed); Šunjić, Jakov (ed); Trumić, Aleksander (ed); Yevjevich, Vujica (ed English ed). Fort Collins, CO, US: Water Resources Publications, **1976**. illus: b&w: charts, drawings, graphs, maps, photos, ports, tables; Various pagination, xiv, 873p; bibl, index.

Vol 1: Karst hydrology, xiv, pp 1-439; vol 2: Karst water resources, viii, pp 443-873. Title of Serbocroation language edition: *Hidrologija and Vodno Bogatstvo Krša*, published by the Institute of Water Resources Engineering of the University of Sarajevo, Sarajevo, Yugoslavia.

An excellent source of all sorts of interesting data. It is somewhat patchy in terms of readability, illustration and clarity.

[2176] **Milanović, Petar T.** *Karst hydrogeology*. Buhac, J. J. (trans). Littleton, CO, US: Water Resources Publications, **1981**. illus: b&w: charts, drawings, graphs, maps, photos, tables; x, 434p, [2]; bibl, errat, index; 24 cm.

Translation of: *Hidrogeologija karsta i metode i straživanja*. 199 figs, 6 tables.

A text translated from Serbo-Croatian. Contains many applied techniques including tracing and boreholes. Although dense and technical, this is a good translation of the original Hidrogeologija Karsta. It is a practical look at the classic Serbo-Croatian karst. It emphasizes water resource evaluation, monitoring, tracing, management, water budgets, etc.

[2177] **Miles, Sibella Elizabeth** *The Grotto of Neptune ("Antro Di Nettuno"), Sardinia; a poem illustrative of three views of this interesting cavern, taken in July 1824 by the late Commander Alfred Miles and dedicated to his memory by his widow, Sibella Elizabeth Miles*. London, GB: Day and Son, **1864**. illus: b&w: maps, repros; list of subscribers on back lining page; 14p, [2], 4p pls; add; 37.5 x 27 cm.

Tinted lithographs.

[2178] **Mill, Lloyd (ed)** *Victorian caves and karst: a guidebook to the 13th A.S.F. conference: Melbourne 1980*. White, Susan (ed); Mackey, Phil (ed); Victorian Speleological Association; Australian Speleological Federation. [Melbourne, VIC, AU]: Cave Convict Ad Hoc Committee, **1980**. illus: b&w: maps, photos; vi, 86p, 4p pls; bibl, errat.

[2179] **Millar, Ian R.** *General policy and guidelines for cave and karst management in areas managed by the Department of Conservation*. Wilde, Kevan A. Wellington, NZ: Department of Conservation, **1989**. illus: b&w: charts, drawings, maps, tables; [iv], iii, 72p; app, bibl, gloss; 30 x 21 cm.

Unbound.

[2180] **Miller, Carter H.** *Preliminary seismic-velocity and magnetic studies of a carbonate rock-sinkhole area in Shelby County, Alabama*. [Denver, CO, US]: U.S. Geological Survey, **1984**. Open-file Report United States Department of the Interior Geological Survey no 84-409. illus: b&w: maps; 22p; bibl; 28 cm.

[2181] **Miller, Edward Sherman**, 1866– *Joint thesis upon Maidstone Cave*. Redenbaugh, William Alfred, 1871–. [Hanover, NH, US], **1893**. illus: b&w: maps; 12p; 21 x 34 cm.

Typescript. Illus.

[2182] **Miller, Gerrit Smith**, 1869-1923 *A second collection of mammals from caves near St. Michel, Haiti*. Washington, DC, US: The Smithsonian Institution, Mar, **1929**. Smithsonian Miscellaneous Collections v 81 no 9. illus: b&w: photos; [ii], 30p, 10p pls; 24.5 cm.

"With 10 plates". Publication 3012. Master microform held by: Ann Arbor, Mich, University Microfilms International, 19—. 1 reel. 35 mm.

[2183] **Miller, Loye Holmes**, 1874-1970 *The Pleistocene*

birds of San Josecito cavern, Mexico. Berkeley, CA, US: University of California Press, **1980**. University of California Publications in Zoology v 47 no 5, pp 143-167. 25p; bibl.

Photocopy. Originally published in 1943. Ann Arbor, Mich.: University Microfilms International, 1980.

[2184] Miller, Tom *The karst development and associated archeology of the Chiquibul, Belize: (preliminary report of the results of grant #2742-83)*. [WA, US]: Tom Miller; Eastern Washington University, **[1984]**. illus: b&w: charts, drawings, maps, photos, tables; [46]; app, bibl.

Published in photocopy form.

[2185] Mills, M. T. *The subterranean wonders of Sardinia*. npop: np, Jul, **1990**. illus: b&w: maps; partly col: drawings; 22p; bibl.

[2186] Mills, Martin Taylor *The subterranean wonders of Hawaii*. GB: M.T. Mills, **1974**. illus: b&w: maps; index on rear lining page; 24p; bibl, index; 30 x 21 cm.

One copy examined was published December 1979.

[2187] Mills, Martin Taylor *The subterranean wonders of Iceland*. GB: M.T. Mills, Dec, **1981**. illus: b&w: maps; 19p; index; 30 x 21 cm.

[2188] Mills, Martin Taylor *The subterranean wonders of Kenya*. [Shepton Mallet, GB]: [Shepton Mallet Caving Club], **1977**. Shepton Mallet Caving Club SMCC Occasional Publication no 8. illus: b&w: maps; Text on rear lining page; 19p; bibl, index.

Summary of caves and cave features, mainly volcanic pseudokarst, in this East African country, mainly from bibliographic search. Includes 88-item bibliography.

[2189] Milner, A. J. *Caves of the Alum Pot area*. [GB]: University of Leeds Speleological Association, **1972**.

[2190] Mindick, Robert R. *Commercial caves in Southern Indiana*. Hanover, IN, US: Geology Dept, Hanover College, **1977**. Hanover College Geology Dept Field Guide 2. illus: b&w: maps; 39p; bibl; 29 cm.

Illus. Title from cover.

[2191] Mindick, Robert R. *Four non-commercial caves in Southern Indiana*. Hanover, IN, US: Geology Dept, Hanover College, **1977**. Hanover College Hanover Indiana Geology Department Field Guide 1. illus: b&w: maps; 27p; 29 cm.

Illus.Title from cover.

[2192] Miotke, F. M. *Carbon dioxide and the soil atmosphere*. Blaubeuren, DE: In Kommission bed F. Mangold, **1974**. Abhandlungen zur Karst-und Hohlenkunde Reihe A heft 9. illus: b&w: charts, graphs, tables; 49p; bibl, index; 21 cm.

A thorough, technical examination of CO_2 in soils, its production, distribution, measurement, temporal variation, exchange, etc. A useful, scholarly work.

[2193] Miotke, Franz-Dieter, 1934– *Genetic relationship between caves and landforms in the Mammoth Cave National Park area; a preliminary report*. Palmer, Arthur Nicholas, 1940–. Hannover, DE: Geographischen Institut der Technischen Universität Hannover, **1972**. illus: b&w: drawings, graphs, maps, photos, tables; partly col: maps; map folded in pocket; [x], 69p; bibl; 21 cm.

58 figs.

This is one of the first works to consider this important area in what has come to be considered the "standard model", and sets the stage for more neat work by diverse authors.

[2194] Miss, Christian J. *Archaeological evaluations of the Riparia (45WT1) and Ash Cave (45WW61) sites on the Lower Snake River*. Cochran, Bruce D. Pullman, WA: Laboratory of Archaeology and History, Washington State University, **nd**. Project report Washington State University Laboratory of Archaeology and History no 14. viii, 156p; 28 cm.

Undertaken for the U.S. Army Corps of Engineers, Walla Walla District, in fulfillment of contract number DACW-81-C-0108—-T.p. verso. Illus.

[2195] Missouri Civil Defense Agency *Missouri underground shelter space: a fallout shelter survey of mines and caves in Missouri*. npop, **1962**. illus: b&w: maps; two folded maps; 153p; 28 cm.

Illus.

[2196] Missouri Division of Geological Survey and Water Resources *Partial catalog of caves in Missouri*. [Rolla], MO, US: Missouri Division of Geological Survey and Water Resources, Mar, **1952**. illus: b&w: tables; [ii], 44p; 28 cm.

[2197] Missouri Geological Survey and Water Resources *Guidebook to the geology of the Rolla area emphasizing solution phenomena*. Rolla, MO, US: Missouri Geological Survey and Water Resources, Dec, **1960**. illus: b&w: charts, maps, tables; 35p; bibl.

"Prepared for the fifth annual midwest groundwater conference field trip." Title from cover.

[2198] Missouri Speleological Survey *Archaeological investigations: cave explorations in the Ozark region of central Missouri*. Fowke, Gerard, 1855-1933 (ed); Weaver, Herman Dwight, 1938–(new intro). reprint ed. MO, US: Missouri Speleological Survey, Dec, **1977**. Missouri Speleology vol 17(3-4). illus: b&w: drawings, maps, photos; text on rear lining page; [viii], 122p; app, bibl.

Reprint of 1922 edition.

[2199] Missouri Speleological Survey *The art of cave mapping*. Thomson, Kenneth C. (ed); Taylor, Robert L. (ed). St Louis, MO, US: Missouri Speleological Survey, January-December, **1991**. Missouri Speleology vol 31(1-4). illus: b&w: maps; ii, 182p; bibl, index; 29 cm.

Illus.

A reprint with revisions of, *An introduction to cave mapping*.

[2200] **Missouri Speleological Survey** *A biological study of Cathedral Cave, Crawford County, Missouri.* Schwartz, John S. (ed). MO, US: Missouri Speleological Survey, **1976**. Missouri Speleology vol 6(4). illus: b&w: charts, maps, photos, tables; text on lining pages; [ii], 20p, 8p pls; bibl.

[2201] **Missouri Speleological Survey** *Caves of Christian County, Missouri: part 1 - eastern half.* Thomson, Kenneth C. (ed); Martin, Ronald L. (ed). MO, US: Missouri Speleological Survey, Jul, **1975**. Missouri Speleology vol 15(3). illus: b&w: charts, maps, photos; text on lining pages; [ii], 36p; bibl.

[2202] **Missouri Speleological Survey** *Caves of Christian County, Missouri: part 2- western half.* Thomson, Kenneth C. (ed); Martin, Ronald L. (ed). MO, US: Missouri Speleological Survey, Oct, **1975**. Missouri Speleology vol 15(4). illus: b&w: maps, photos; text on lining pages; [ii], 51p.

[2203] **Missouri Speleological Survey** *Caves of Crawford County.* Eddleman, William R. (ed). MO, US: Missouri Speleological Survey, Jul, **1981**. Missouri Speleology vol 21(3-4). illus: b&w: maps, photos; 6 foldout maps; text; 96p; bibl.

[2204] **Missouri Speleological Survey** *Caves of the Current River Valley.* [Lohman, MO, US]: Missouri Speleological Society, July-December, **1982**. Missouri Speleology vol 22(3-4). illus: b&w: maps, photos; 3 fold-out maps; text on lining pages; 119p; 28 cm.

[2205] **Missouri Speleological Survey** *Caves of the Jacks Fork Valley.* [Eddleman, William R. (ed)]. Lohman, MO, US: Missouri Speleological Survey, January-June, **1982**. Missouri Speleology vol 22(1-2). illus: b&w: maps, photos; 4 foldout maps in body; text; [iv], 63p; 28 cm.

Title from cover.

[2206] **Missouri Speleological Survey** *A checklist of invertebrate species recorded from Missouri subterranean habitats.* Craig, John L. (ed). Missouri Speleological Survey, Apr, **1975**. Missouri Speleology vol 15(2). illus: b&w: photos, tables; text on lining pages; [ii], 10p; bibl.

[2207] **Missouri Speleological Survey** *The commercial caves of Camden County, Missouri.* Weaver, Herman Dwight, 1938–(ed). Rolla, MO, US: Missouri Speleological Survey, January-December, **1987**. Missouri Speleology vol 27(1-4). illus: b&w: maps, photos, repros; 2 foldout maps in body; text on lining pages; iv, 136p; 28 cm.

Title from cover.

[2208] **Missouri Speleological Survey** *A history of the caves of Camden County, Missouri.* Weaver, Herman Dwight, 1938–(ed). MO, US: Missouri Speleological Survey, **1970**. Missouri Speleology vol 11(1-2). illus: b&w: maps, photos, repros, tables; text on lining pages; [ii], 51p; bibl.

[2209] **Missouri Speleological Survey** *Hydrogeologic controls on solution of carbonate rocks in Christian County, Missouri.* Dreiss, Shirley Jean (ed). MO, US: Missouri Speleological Survey, Jan, **1976**. Missouri Speleology vol 16(1-2). illus: b&w: charts, graphs, maps, photos, tables; text on lining pages; 46p; bibl.

[2210] **Missouri Speleological Survey** *An introduction to cave mapping.* Thomson, Kenneth C. (ed); Taylor, Robert L. (ed). 2nd ed. [Columbia, MO, US]: Missouri Speleological Survey, **1981**. Missouri Speleology vol 21(1-2). illus: b&w: charts, drawings, graphs, maps, photos, tables; text on lining pages; iii, 127p; bibl; 28 cm.

Replaces the issue done by Lang Brod in 1962 called "Cave Mapping: A Systematic Approach", Volume IV, Numbers 1-3, January to April, 1962.

An excellent special issue of Missouri Speleology. It has been reprinted and updated several times. It was the most comprehensive text on surveying until the publication of On Station.

[2211] **Missouri Speleological Survey** *An introduction to caving: a guide for beginners.* Thomson, Kenneth C. (ed); Martin, Ronald L. (ed). MO, US: Missouri Speleological Survey, Jan, **1980**. Missouri Speleology vol 20(1-2). illus: b&w: charts, drawings, tables; text on lining pages; ii, 43p; bibl.

[2212] **Missouri Speleological Survey** *The invertebrate fauna of Mystery Cave, Perry County, Missouri.* Lewis, Jerry (ed). MO, US: Missouri Speleological Survey, Oct, **1974**. Missouri Speleology vol 14(4). illus: b&w: maps, photos, tables; text on lining pages; [ii], 19p; bibl.

[2213] **Missouri Speleological Survey** *An investigation into ground water pollution in caves.* Thenhaus, Paul (ed). MO, US: Missouri Speleological Survey, Apr, **1973**. Missouri Speleology vol 13(2). illus: b&w: charts, drawings, graphs, maps, tables; text on lining pages; 14p, 4p pls; bibl.

[2214] **Missouri Speleological Survey** *J Harlen Bretz: a geologist's encounters with Missouri caves.* Bretz, J Harlen, 1882-1981 (ed); Vineyard, Jerry Daniel, 1935– (ed). Lohman, MO, US: [Missouri Speleological Survey], October-December, **1979**. Missouri Speleology vol 19(3-4). illus: b&w: drawings, maps, photos; text on rear lining page; ii, 69p; bibl; 28 cm.

Caption title.

Memoirs of J Harlen Bretz.

[2215] **Missouri Speleological Survey** *Jefferson County caves.* [Webster, David (ed)]. MO, US: Missouri Speleological Survey, Jan, **1971**. Missouri Speleology vol 12(1). illus: b&w: maps, photos; 2 fold-out maps; text on lining pages; [ii], 33p; bibl.

[2216] **Missouri Speleological Survey** *Laclede County caves.* [Webster, David (ed)]. MO, US: Missouri Speleological Survey, Jan, **1973**. Missouri Speleology vol 13(1). illus: b&w: drawings, maps; text on lining pages; [iv], 40p; bibl.

[2217] Missouri Speleological Survey *Miller County caves.* [Vineyard, Jerry Daniel, 1935–(ed)]. MO, US: Missouri Speleological Survey, Apr, **1961**. Missouri Speleology vol 3(2). illus: b&w: maps; 2 fold-out maps; [iv], 25p; bibl.

[2218] Missouri Speleological Survey *Mystery Cave.* Walsh, Joseph E. (ed). St. Louis, MO, US: Missouri Speleological Survey, **1992**. Missouri Speleology vol 28(1-4). illus: b&w: maps; col: photos; iv, 142p; app, bibl; 28 cm.

"January-December 1988"—p. i. "Journal of the Missouri Speleological Survey" Illus.

[2219] Missouri Speleological Survey *Origin and development of Cave Spring, Shannon County, Missouri.* Vineyard, Jerry Daniel, 1935–(ed). [Rolla, MO, US]: Missouri Speleological Society, July-December, **1986**. Missouri Speleology vol 26(3-4). illus: b&w: charts, maps, photos, tables; text on lining pages; v, 51p; bibl; 28 cm.

Title from cover.

[2220] Missouri Speleological Survey *Phelps County caves.* [Weaver, Herman Dwight, 1938–(ed)]. MO, US: Missouri Speleological Survey, Jan, **1963**. Missouri Speleology vol 5(1). illus: b&w: maps, photos; 5 fold-out maps; text on front lining page; [ii], 28p; bibl.

[2221] Missouri Speleological Survey *Pike County caves.* [Weaver, Herman Dwight, 1938–(ed)]. MO, US: Missouri Speleological Survey, **1963**. Missouri Speleology vol 5(2-3). illus: b&w: drawings, maps; 4 fold-out maps; text on front lining; [ii], 29p; bibl.

[2222] Missouri Speleological Survey *Pit caves of Jefferson County.* [Webster, David (ed)]. MO, US: Missouri Speleological Survey, Apr, **1971**. Missouri Speleology vol 12(2). illus: b&w: maps; 3 fold-out maps; [ii], 34p.

[2223] Missouri Speleological Survey *A pleistocene fauna from Zoo Cave, Taney County, Missouri.* Hood, Clark H. (ed); Hawksley, Oscar (ed). MO, US: Missouri Speleological Survey, Jan, **1975**. Missouri Speleology vol 15(1). illus: b&w: charts, maps, photos, tables; text; [ii], 42p; bibl.

[2224] Missouri Speleological Survey *Pulaski County caves.* [Webster, David (ed)]. MO, US: Missouri Speleological Survey, Apr, **1964**. Missouri Speleology vol 6(2). illus: b&w: maps; 4 fold-out maps; text on front lining page; [ii], 36p; bibl.

[2225] Missouri Speleological Survey *Ralls County caves.* [Webster, David (ed)]. MO, US: Missouri Speleological Survey, Jan, **1965**. Missouri Speleology vol 7(1). illus: b&w: maps, tables; [ii], 22p; bibl.

[2226] Missouri Speleological Survey *Remains of Quaternary vertebrates from Ozark caves and other miscellaneous sites.* Hawksley, Oscar (ed). Rolla, MO, US: Missouri Speleological Survey, January-June, **1986**. Missouri Speleology vol 26(1-2). illus: b&w: charts, maps, photos, tables; text on lining pages; v, 67p; bibl; 28 cm.

[2227] Missouri Speleological Survey *Report of the Devil's Icebox - Rockbridge Park conservation task force of the National Speleological Society.* Hargrove, Gene (ed). MO, US: Missouri Speleological Survey, Jul, **1973**. Missouri Speleology vol 13(3). illus: b&w: drawings, maps, photos, tables; 1 foldout map; text on lining pages; vi, 51p; bibl.

[2228] Missouri Speleological Survey *Rimstone River Cave.* Walsh, Joseph E. (ed). St. Louis, MO, US: Missouri Speleological Survey, **1992**. Missouri Speleology vol 29(1-4). illus: b&w: drawings, graphs, maps, tables; text on lining pages; v, 158p; app, bibl; 29 cm.

"January-December 1989"—p.i.

[2229] Missouri Speleological Survey *Sandstone caves of Jefferson County.* [Webster, David (ed)]. MO, US: Missouri Speleological Survey, Jul, **1971**. Missouri Speleology vol 12(3). illus: b&w: charts, maps, photos; [ii], 30p; bibl.

[2230] Missouri Speleological Survey *Shannon County caves.* [Webster, David (ed)]. MO, US: Missouri Speleological Survey, [Dec, **1966**]. Missouri Speleology vol 8(4). illus: b&w: charts, maps, photos; 2 fold-out maps; text on lining pages; [ii], 40p; bibl, index.

[2231] Missouri Speleological Survey *St. Charles County caves.* [Webster, David (ed)]. MO, US: Missouri Speleological Survey, Apr, **1965**. Missouri Speleology vol 7(2). illus: b&w: maps; 4 fold-out maps; text on lining pages; [ii], 34p.

[2232] Missouri Speleological Survey *St. Genevieve County caves.* [Webster, David (ed)]. MO, US: Missouri Speleological Survey, Jul, **1965**. Missouri Speleology vol 7(3). 1 fold-out map; text on lining pages; [ii], 21p; bibl.

[2233] Missouri Speleological Survey *St. Louis County caves.* [Webster, David (ed)]. MO, US: Missouri Speleological Survey, Jan, **1966**. Missouri Speleology vol 8(1). illus: b&w: maps; 5 fold-out maps; text on lining pages; [ii], 50p; bibl.

[2234] Missouri Speleological Survey *Washington County caves.* [Webster, David (ed)]. MO, US: Missouri Speleological Survey, Oct, **1967**. Missouri Speleology vol 9(4). illus: b&w: drawings, maps; 6 fold-out maps in text; text on lining pages; [ii], 61p; bibl, index.

[2235] Missouri Speleological Survey *The wild caves of Camden County, Missouri.* Weaver, Herman Dwight, 1938–(ed). Lohman, MO, US: Missouri Speleological Survey, January-December, **1983**. Missouri Speleology vol 23(1-4). illus: b&w: maps, photos; text on front lining page; ix, 141p; 28 cm.

Title from cover.

[2236] Mitchell, Albert *Yorkshire caves and potholes: 1, North Ribblesdale.* 1st ed. [**1937**]. [iii], xii, 78p; 18 cm.

2nd ed. Skipton, GB: Albert Mitchell, **1948**. illus: b&w: maps, photos; front: b&w photo "In Alum Pot"; xi, 66p, [2]; index; 17.5 x 12 cm.

[2237] **Mitchell, Albert** *Yorkshire caves and potholes: 2, under Ingleborough.* Skipton, GB: Albert Mitchell, **1950**. illus: b&w: maps, photos, tables; xiii, 138p, [2]; index; 17.5 x 12 cm.

[2238] **Mitchell, Jack** *Jack Mitchell, caveman.* Mitchell, Ida; Bailey, Alberta; Timberlake, Cynthia. Torrance, CA, US: Lewellen Press, **1964**. illus: b&w: photos, ports; 164p; 22 cm.

Illus.

[2239] **Mitchell, Robert Wetsel**, 1933– *Mexican eyeless Characin fishes, genus* Astyanax: *environment, distribution, and evolution.* Russell, William Hart, 1937–; Elliott, William Rawleigh, 1946–. Lubbock, TX, US: Texas Tech Press, Feb, **1977**. Special Publications The Museum Texas Tech University no12. illus: b&w: drawings, graphs, maps, photos, tables; folded map in text; 89p, 1p pls; bibl; 26 cm.

21 figs.

A thorough discussion of the ecology and biogeography of the Mexican cave Charazin. Probably the most studied cave organism. This is a critical reference for any study.

[2240] **Mitchell, Robert Wetsel**, 1933– *A new family, genus, and species of cave-adapted planarian from Mexico (Turbellaria, Tricladida, Maricola).* Kawakatsu, Masaharu. Lubbock, TX, US, **1972**. Texas Tech University Museum Occasional Papers no 8. 16p; bibl; 23 cm.

Caption title.

[2241] **Mitchell, Robert Wetsel**, 1933– *Studies on the cavernicole fauna of Mexico.* Reddell, James Russell, 1938–(ed). Austin, TX, US: [Association for Mexican Cave Studies], **1971**. Association for Mexican Cave Studies Bulletin 4. 239p; bibl; 29 cm.

Illus.

[2242] **Mitchell, Robert Wetsel**, 1933– *Studies on the cavernicole fauna of Mexico and adjacent regions.* Reddell, James Russell, 1938–(ed). Austin, TX, US: [Association for Mexican Cave Studies], Jul, **1973**. Association for Mexican Cave Studies Bulletin 5. illus: b&w: drawings, maps, photos, photomicrographs, tables; col: photos; front: col photo; [viii], 201p, 1p pls; bibl; 29 cm.

[2243] **Mitchell, William Reginald**, 1928– *The hollow mountains: the story of man's conquest of the caves and potholes of Northwest Yorkshire throughout 10,000 years.* npop: William Reginald Mitchell, **1961**. illus: b&w: photos; [33]; 14.5 x 21 cm.

24 illus.

[2244] **Mitchell, William Reginald**, 1928– *Yorkshire's hollow mountains.* Settle, Yorkshire, GB: W. R. Mitchell, **1989**. A Castleberg Publication. illus: b&w: drawings, maps, photos, ports, repros; 48p; 24 cm.

Previous ed: The hollow mountains, 1961.

[2245] **Mockford, D. P.** *Caves of the Bristol region.* Male, A. J. Bristol, GB: The Bristol Cave Register, Oct, **1974**. 74p; 21 x 15 cm.

Alphabetical list of caves at end.

[2246] **Mohapatra, R. P.**, 1939– *Udayagiri and Shandagiri caves.* Delhi, IN: D. K. Publications, **1981**. illus: b&w: charts, drawings, maps, photos; xxii, 270+20p, [4]; bibl, index; 25 cm.

Based on the author's doctoral thesis.

Few works on this region exist.

[2247] **Mohler, John Leonard**, 1840-1937 *A description of Weyer's Cave in Augusta County, Virginia.* Weyer, Bernard. npop, **1852**. 11p; 22 cm.

The cave was discovered in 1804, by Bernard Weyer.

[2248] **Mohr, Charles E.**, 1907– *Cave life.* 1960 ed. Garden City, NY, US: N. Doubleday. National Audubon Society Nature Program. illus: b&w: photos; col: photos; 64p; 21 cm.

Col stickers.

1965 ed. Gurnee, Russell Hampton, 1922-1995. New York, NY, US: N. Doubleday. National Audubon Society Nature program. illus: partly col: maps; 64p; 21 cm.

Illus.

1st ed. Garden City, NY, US: N. Doubleday, **1956**. National Audubon Society Nature Program. illus: col: photos; 48p; 21 cm.

This is the first edition of this book. It is softbound and contains no black and white photos but has place for 39 color stamps to be put in. Stamps are on one sheet.

[2249] **Mohr, Charles E.**, 1907– *Celebrated American caves.* Sloane, Howard Norman, 1909-1972. New Brunswick, NJ, US: Rutgers University Press, **1955**. illus: b&w: drawings, maps, photos, ports; xii, 339p; index; 23 cm.

94 b&w photos, 6 maps, 9 drawings, ports.

One of the most heavily distributed general press books on caves ever published in the United States. Reprinted at least five times over a period of at least 25 years.

[2250] **Mohr, Charles E.**, 1907– *The life of the cave.* Poulson, Thomas Layman, 1934–. New York, NY, US: McGraw-Hill Book Company, **1966**. Our Living World of Nature. illus: b&w: drawings, photos; partly col: drawings, maps; col: photos; color; 232p; app, bibl, gloss, index.

Excellent introduction to cave faunas and cave ecology written for the general reader. It also contains many excellent photographs. Examines living things that inhabit caves, their environment, adaptations for survival, food supply, and habits. Includes descriptions of cave formations, and a list of caves to be visited and explored.

[2251] **Moller, Jacob**, 1935– *Coastal caves, marine limits, and ice retreat in Lofotenvesteralen, North Norway.* McMillan, Nora Fisher. Tromso, NO: Universitetet i Tromso, Institutt for Museumsvirksomhet, **1982**. Tromura Tromso Museums Rapportserie Naturvitenskap. 34 +17p; 30 cm.

Abstracts in Norwegian. Illus.

[2252] **Molyneux, Arthur John Charles** *The Sinoia caves, Lomagundi district.* Salisbury, GB: Geological Survey Office, **1935**. Southern Rhodesia Geological Survey Short Report no 8. one plate folded; 11p, 4p pls; 25 cm.

Issued 30th July, 1920. Reprinted 1935.

[2253] **Monico, Paul (comp)** *CDG northern sump index 1995.* [GB]: Cave Diving Group of Great Britain, **1995**. illus: b&w: charts, maps; 9 fold-out maps in body; vi, 286p, 9p pls; bibl, index.

[2254] **Monico, Paul (auth, comp)** *ULSA explorations: journal II.* Leeds, GB: University of Leeds Speleological Association, **1969**. illus: b&w: maps, photos; 3 folded maps in text; [viii], 130p, [1], 24p pls; 30 x 21 cm.

[2255] **Monroe, Watson Hiner**, 1907– *A glossary of karst terminology.* Washington, DC, US: United States Government Printing Office, **1970**. USGS Water Supply Paper 1899-K. 26p; bibl, gloss; 24 cm.

A well done and very handy glossary of early terminology, in some need of a new edition.

[2256] **Monroe, Watson Hiner**, 1907– *The karst landforms of Puerto Rico: a discussion of a solution landscape formed in a tropical climate of moderately high rainfall.* Arlington, VA, US: United States Geological Survey, **1976**. Geological Survey Professional Paper 899. illus: b&w: drawings, maps, photos, tables; partly col: maps; one folded map, in pocket; iv, 69p; bibl, gloss, index; 29 cm.

Cover title: The Karst Landforms of Puerto Rico; Prepared in cooperation with the Department of Natural Resources of the Commonwealth of Puerto Rico.

[2257] **Montgomery, Neil R.** *Single rope techniques: a guide for vertical cavers.* Mroczkowski, Donna (asst). Broadway, NSW, AU: The Sydney Speleological Society, **1977**. Sydney Speleological Society Occasional Paper no 7. illus: b&w: drawings, photos; x, 122p; add, app, bibl, index; 27 cm.

A dramatic advance in the distribution of SRT information worldwide. A bible of SRT for many years.

[2258] **Montz, Gary R.** *The aquatic invertebrates of Mystery Cave, Forestville State Park, Minnesota.* [Minneapolis, MN, US]: [Minnesota Dept of Natural Resources], Jun, **1993**. illus: b&w: maps; 17p; bibl; 28 cm.

[2259] **Moody, Larry D.** *Warrens Cave Reserve stewardship plan: Gainesville, Alachua County, Florida.* Gainesville, FL, US: Florida Speleological Society, Nov, **1977**. illus: b&w: maps, photos, tables; [xx], 25p, [47]; app, bibl.

Some pages blank.

[2260] **Moore, George William**, 1928– *The origin of helictites.* Trenton, NJ, US: National Speleological Society, **1954**. National Speleological Society Occasional Papers no 1. illus: b&w: charts, photos, photomicrographs; 16p; bibl; 23 cm.

Illus.

[2261] **Moore, George William**, 1928– *Origin of limestone caves; a symposium with discussion.* Alexandria, VA, US: National Speleological Society, Jan, **1960**. Bulletin of the National Speleological Society v 22 pt 1. 84p; bibl; 25 cm.

Title from cover. Illus.

[2262] **Moore, George William**, 1928– *Speleology: caves and the cave environment.* Sullivan, Gerardus Nicholas,1927–; Schoenherr, John C. (Illus). 3rd rev ed. St Louis, MO, US: Cave Books, **1997**. illus: b&w: charts, drawings, graphs, maps, photomicrographs, tables; xiv, 176p; app, bibl, index; 21 x 24 cm.

Prepared in cooperation with the National Speleological Society. Third edition of Speleology: the study of caves.

[2263] **Moore, George William**, 1928– *Speleology: the study of caves.* Sullivan, Gerardus Nicholas, 1927– [Brother Nicholas]; Schoenherr, John C. (Illus). 1st ed. Boston, MA, US: D. C. Heath and Co, **1964**. Science Resource Series. illus: b&w: charts, drawings, maps; vii, 120p; app, bibl, index; 21 cm.

2nd rev ed. Sullivan, Gerardus Nicholas,1927–; Schoenherr, John C. (Illus). St Louis, MO, US: Cave Books, **1978**. Science Resource Series. illus: b&w: drawings, graphs, maps, photomicrographs; xiii, 150p; app, bibl, index; 22 cm.

Prepared in cooperation with the National Speleological Society. Simple foundation in biology, geology, physics, and more. A highly readable introduction to cave science in many of its guises. Although necessarily brief and in a few aspects, dated, it provides an excellent overview of the diverse subject. It remains the preeminent laypersons introduction to the basics of speleology.

[2264] **Moore, Gerald K.** *Limestone hydrology in the Upper Stones River Basin, central Tennessee.* Burchett, Charles R.; Bingham, Roy H. Nashville, TN, US: State of Tennessee Department of Conservation Division of Water Resources, **1969**. illus: b&w: charts, graphs, maps, photos, tables; folded graph in text; [iv], 58p; bibl.

Prepared in a cooperative program between the Tennessee Department of Conservation, Division of Water Resources and U. S. Geological Survey.

[2265] **Moore, James D.** *Hydrology of limestone terranes: quantitative studies.* Moravec, George F.; LaMoreaux, Philip Elmer, 1920–. University, AL, US: Geological Survey of Alabama, **1975**. Bulletin 94-F. illus: b&w: charts, graphs, maps, photomicrographs, tables; partly col: charts, graphs, maps; 16 b&w or partly col maps, graphs in pockets in separate cover; 99p, [8]; bibl.

[2266] **Moore, Michael C.** *Southern Indiana Karst field trip guide, March 11 & 12, 1972.* [IL, US]: [University of Illinois], **1972**. illus: b&w: maps; 20p; bibl; 28 cm.

At head of title: University of Illinois.

[2267] **Morana, Martin** *The prehistoric cave of Gher*

Dalam. Fgura, MT, **1987**. illus: b&w: charts, drawings, maps, photos; 23p; 21.5 x 15 cm.

A bone cave. B&W plans.

[2268] Morfis, A. (ed) *Karst hydrogeology of the central and eastern Peloponnesus (Greece).* Zojer, H. (ed). Wien, AT: Springer-Verlag, **1986**. Steirische Beiträge zur Hydrogeologie 37/38. illus: b&w: drawings, graphs, photos, tables; nine pls folded in rear pocket; 301p, 9p pls; bibl.

5th international symposium on underground water tracing, Athens 1986.

[2269] Morfis, A. (ed) *Proceedings of the 5th international symposium on underground water tracing, Athens 1986.* Paraskevopoulou, P. (ed); Institute of Geology and Mineral Exploration (Greece); International Working Group on Tracer Methods in Hydrology. Athens, GR: Institute of Geology and Mineral Exploration, **1986**. illus: b&w: charts, drawings, graphs, maps, photos, tables; 473p, [5]; bibl, index.

[2270] Morgan, Llewellyn E. (auth, photo) *Guide to Dan-yr-Ogof Caves: Swansea Valley Caves.* Davies, T. J. (photo). Swansea Valley, GB: Dan-yr-Ogof Caves Limited, **nd**. illus: b&w: maps, photos; text on lining pages; 28p; 18.5 x 12 cm.

Cover title on some versions: The Dan-yr-Ogof Caves Official Guide. Cover title on others: The Dan-yr-Ogof Caves: situated between Brecon and Swansea on the A. 4067 Road (750 yards North of Craig-y-Nos Castle): Swansea Valley Caves. Many editions printed, lining pages and cover vary, photos used vary, some editions 32 pages, others 28 pages, most editions have no dates, but those examined were c1930-1970.

[2271] Morgan, Robert *Caves in world history.* npop: [National Speleological Society], **nd**. National Speleological Society Bulletin 5. 16p.

Caption-title: "Reprint from National speleological society. Bulletin number five."

[2272] Morgan, Robert *Through Mammoth Cave, giving a graphic description of all the routes, avenues, passes, passages, pits, domes, streams, rivers, seas, lakes, and other wonders to be found in that greatest of all natural curiosities in the whole realm of this great world.* 5th ed. Louisville, KY, US: [Courier-Journal Job Rooms], **nd**. illus: b&w: ads; ads on rear lining page and outside back cover.; 45p.

13 new pp of ads.

[2273] Morgan, Robert *Through Mammoth Cave, giving a graphic description of all the routes, avenues, passes, passages, pits, domes, seas, lakes, rivers, streams, and other wonders of that greatest natural curiosity in the world.* Louisville, KY, US: [Courier-Journal Job Rooms], **1881**. illus: b&w: ads; ads on rear lining page and outside back cover.; 32p.

"Presented by H. McKenna, distiller of Nelson Co. Pure Old Line Sour-Mash Whisky [sic]." The Morgan guides are among the rarest of Mammoth Cave guides.

[2274] Morland, Timothy Edward *A glossary of French speleological terms.* npop: G. Platten, **1959**. 13p; 26 cm.

Typewriter script.

[2275] Morris, David (comp) *Cave Diving Group: Welsh sump index.* Adams, John (comp). 2nd ed. North Yorkshire, GB: Cave Diving Group, Dec, **1985**. illus: b&w: maps; 53p; 30 x 21 cm.

[3rd]. Knaresborough, **1986**.

Extent of changes between 1985 and 1986 printings is unknown.

[2276] Morris, Deborah, 1956– *Trapped in a cave!: a true story.* Nashville, TN, US: Broadman & Holman Publishers, **1993**. illus: b&w: photos; 128p; 21 cm.

Juvenile.

[2277] Morris, Linda *The Hayes Cave site, South Maitland, Nova Scotia.* Nova Scotia Dept of Education, Nova Scotia Museum. Halifax, NS: Nova Scotia Museum, **1985**. Curatorial report Nova Scotia Museum no 50. 128p; bibl; 28 cm.

At head of title: Nova Scotia, Dept. of Education, Nova Scotia Museum Complex. Illus.

[2278] Morris, Neil *Caves.* New York, NY, US: Crabtree Publishing Co, **1996**. The Wonders of the World. illus: b&w: repros; partly col: maps; col: charts, drawings, photos; 32p; gloss, index.

Juvenile.

[2279] Morrison, Geo. D. *New entrance to Mammoth Cave: a history of the cave together with descriptive guide to routes and illustrations of points of interest.* npop, **c1920**. illus: b&w: maps, photos; 63p; 23 cm.

The microfilm version lists date "? 1920 1929". No date given on original piece. Souvenir postcards front and back. One copy examined had a separate [16]p booklet with some pictures from original and no author, some text.

[2280] Moseley, John Judy *History, geology of Carlsbad Caverns: what' a hole.* Dunagan, W. A. npop: np, **1948**. [16]; 19 cm.

Title from cover. In envelope. Booklet intended for kids. Written in a Q&A format. Illus.

[2281] Moseley, John Judy *What' a hole: condensed Carlsbad Cavern [i.e. Caverns] information.* Dunagan, W. A. Carlsbad, NM, US: Myers Print Co, **1941**. illus: b&w: drawings, maps; [16]; 19 cm.

Title from cover. In envelope.

[2282] Mostardi, Michael (comp) *Caves of Berks County, Pennsylvania.* Durant, Joe (comp). State College, PA, US: Mid-Appalachian Region. National Speleological Society, Dec, **1991**. MAR bulletin no 18. illus: b&w: charts, maps, photos, tables; 4 fold-out maps in body; photo on front lining page; ii, 84p; bibl; 28 cm.

[2283] Motaş, C. *Emil Racoviţă 1868-1947.* Ghica, C. Bucharest, HU: Meridiare Publishing House, **1968**. illus: b&w: photos; 65p; add; 20 x 13 cm.

[2284] Mott, Kevin (ed) *Nullarbor caving atlas.* Pilkington, Graham (ed); Cave Exploration Group (South Australia). 1st ed. Rostrevor, SA, AU: Subterranean Foundation (Australia), **1982**. Occasional Paper Cave Exploration Group (South Australia) 7. illus: b&w: maps; Various pagination; 18 x 27 cm.

Exists as looseleaf (updated once) and bound editions. Edited by Kevin Mott and Graham Pilkington. Atlas (2 vol).

[2285] Moulin, Raoul-Jean *Prehistoric painting.* 1966 ed. London, GB: Heron Books, **1966**. History of painting. illus: b&w: maps; 207p; 28 cm.

Partly col illus. Originally published as *Sources de la peinture*. Lausanne, Editions Pencoutre, 1965.

1967 British ed. London, GB: Heron. illus: b&w: maps, tables; 208p; bibl; 27 cm.

1967 Swiss ed. Geneva, CH: Edito-Service, **1967**. History of painting. illus: b&w: maps, tables; 207p; bibl; 27 cm.

Some col illus. Translation of Sources de la peinture.

1969 ed. Rhodes, Anthony (trans). New York, NY, US: Funk & Wagnalls. illus: b&w: charts, maps, photos, tables; col: photos; 207p; bibl, gloss.

[2286] Mount, Tom *Cave diving manual.* Miami, FL, US: National Association for Cave Diving, **1973**. illus: b&w: drawings, maps, photos; ii, 35p; app, bibl.

[2287] Mount, Tom *Safe cave diving.* High Springs, FL, US: The National Association for Cave Diving, **1973**. illus: b&w: ads, charts, drawings, graphs, photos, tables; viii, 196p, [1], 7p pls; app, bibl.

16 essays by the nation's foremost cave divers.

[2288] Mountain Rescue Committee (Great Britain) *Mountain and cave rescue, with lists of official rescue teams and posts: the handbook of the Mountain Rescue Committee.* 1975 ed. Buxton, GB: Mountain Rescue Committee, **1975**. illus: b&w: maps; col: maps; 95p; bibl; 18 cm.

Title from cover.

1986 ed. New Mills, Stockport, Cheshire, GB: Mountain Rescue Committee, **1984**. 40p; 21 cm.

Title from cover. Contains a directory of the Mountain Rescue Committee bound in the middle of the text. Illus.

[2289] Mountain Rescue Committee (Great Britain) *Mountain rescue & cave rescue.* 1964 ed. [Manchester, GB], **1964**. 60p; 19 cm.

Title from cover. Illus.

1968 ed. GB: Mountain Rescue Committee, **1968**. illus: b&w: drawings; partly col: maps; text on front lining page; 68p; 19 cm.

Title from cover. Illus.

1970 ed. Manchester, GB: Mountain Rescue Committee, **1970**. illus: b&w: maps; col: maps; 72p; bibl; 19 cm.

Title from cover. Illus.

[2290] Mountain Rescue Committee (Great Britain) *Mountain rescue, cave rescue.* npop: Mountain Rescue Committee, **1963**. illus: b&w: drawings; partly col: maps; 56p; 19 cm.

Illus.

[2291] Mountain Rescue Council *Mountain & cave rescue: the handbook of the Mountain Rescue Council.* Ulverston, [GB], **1994**. illus: b&w: maps; 52p, 44p pls; bibl; 21 cm.

Illus.

[2292] Movius, Hallam Leonard, 1907– *The Mousterian cave of Teshik-Tash, Southeastern Uzbekistan, Central Asia.* Okladnikov, Aleksei Pavlovich, 1908–. [Cambridge, MA, US]: np, **1953**. The whole of: American School of Prehistoric Research Bulletin 17 1953, pp 11-71. illus: b&w: maps; 61p; bibl; 28 cm.

Caption title. "Based on translation of Chester S. Chard of the Russian excavation report by A. P. Okladnikov and others." Illus.

[2293] Mroczkowski, Donna Marie, 1950– *Safety and techniques.* [Hollister, MO, US]: Safety and Technique Committee. National Speleological Society, **1975**. illus: b&w: cartoons, drawings, tables; ii, 116p.

"...assembled as an aid for the sessions being offered by the Safety and Technique Committee at the 1975 NSS convention..."

[2294] Mturi, A. A. *A guide to Tongoni ruins: with notes on other antiquities in Tanga and Pangani districts including the Amboni Caves: pamoja na maelezo kwa Kiswahili.* [Dares Salaam, TZ]: Division of Antiquities, Ministry of National Culture and Youth, **1975**. 29p; bibl; 21 cm.

Title from cover. Illus.

[2295] Muchmore, William B. *North American cave pseudoscorpions of the genus* Kleptochthonius, *subgenus* Chamberlinochthonius *(Chelonethida, Chthoniidae).* New York, NY, US: American Museum of Natural History, **1965**. American Museum Novitiates no 2234. 27p; 24 cm.

Illus.

[2296] Muir, Robert Dalton *Castleguard.* Ford, Derek Clifford, 1935–. Ottawa, CA: Minister of Supply and Services Canada, **1985**. illus: col: maps, photos; 267p, [267]; 23 x 29 cm.

Limited text. 133 col photos.

[2297] Mull, D. S. *Application of dye-tracing techniques for determining solute-transport characteristics of ground water in karst terranes.* Liebermann, T. D.; Smoot, J. L.; Woosley, L. H. Atlanta, GA, US: U.S. Environmental Protection Agency. Ground-Water Protection Branch Region IV, Oct, **1988**. illus: b&w: charts, drawings, graphs, maps, tables; x, 103p; app, bibl.

EPA904/6-88-001. 18 illus.

One of the earliest EPA publications to cover applied karst hydrology. A very good reference.

[2298] Müller, H. O. (comp, ed) *Dolomite caves of the*

eastern Transvaal (South Africa): a free caver monograph 1985. [Zurfontein, ZA]: Free Caver, **1985**. illus: b&w: maps; iv, 71p; 30 cm.

Cover title: Dolomite caves of the eastern Transvaal (South Africa).

[2299] Munich, Frederick J. (auth, prod) *Fumaroles, ice caves and time: the variety and scope of research on Mt. Baker.* [Cheney, WA, US]: Instructional Media Center, Eastern Washington State College, **1977**.

Describes research and monitoring of Mt. Baker's volcanic, hydrothermal, seismic, glacial, and avalanche activities by members of the Eastern Washington State College Dept. of Geology and the University of Washington Geophysics Department. 1 videocassette (44 min), sd, col,; 3/4 in.

[2300] Munson, Patrick J. *The prehistoric and early historic archaeology of Wyandotte Cave and other caves in southern Indiana.* Munson, Cheryl Ann. Indianapolis, IN, US: Indiana Historical Society, **1990**. Prehistory Research Series v 7 no1. illus: b&w: charts, drawings, maps, photos, photomicrographs, tables; vii, 101p; app, bibl; 26 cm.

Illus.

Description of archeological study conducted by Patrick and Cheryl Munson and Ken Tankerslkey at Indiana's Wyandotte Cave following earlier studies by P.J.Watson at Mammoth Cave

[2301] Murdoch, Judy *Light on dark: caves of the south east of South Australia.* Mount Gambier, SA, AU: South East Book Promotion Group, Apr, **1994**. illus: b&w: drawings; [iii], 38p; bibl.

[2302] Murie, Adolph, 1899– *Field investigation report: Lehman Caves - Wheeler Peak: portion of southern section of Snake Range, White Pine County, Nevada.* Schumacher, Paul J. F.; Cole, James E. npop: U. S. National Park Service, Region Four, **1959**. three pockets of folded maps; 97p; 28 cm.

Illus.

[2303] Murie, Adolph, 1899– *Second field investigation report, Lehman Caves - Wheeler Peak: October 13 to 17, 1958, October 29 to November 13, 1958: portion of southern section of Snake Range, White Pine County, Nevada.* Schumacher, Paul J. F.; Cole, James E. [San Francisco, CA, US]: U. S. National Park Service, Region Four, Feb, **1959**. illus: b&w: maps; col: maps; Various pagination; bibl; 27 cm.

Title from cover. Some col illus.

[2304] Murray, Peter *Pleistocene vertebrate remains from a cave near Montagu, N.W. Tasmania.* Goede, Albert. [Launceston, TAS, AU]: Queen Victoria Museum, **1977**. Records of the Queen Victoria Museum no 60. illus: b&w: maps; 30p; bibl; 30 cm.

Title from cover. Illus.

[2305] Murray, Robert K. *Trapped!: the story of the struggle to rescue Floyd Collins from a Kentucky cave in 1925, an ordeal which became one of the most sensational events of modern times.* Brucker, Roger Warren, 1929–. 1st ed. New York, NY, US: G. P. Putnam's Sons, **1979**. illus: b&w: maps, photos, ports; 335p, 8p pls; app, index; 23 cm.

The sensational story of the entrapment, attempted rescue, and death of Floyd Collins. Murray's analysis of Collins' sexual life seems oddly out of place in this account.

[2306] Murray, Robert K. *Trapped!: the story of the struggle to rescue Floyd Collins from a Kentucky cave in 1925, an ordeal which became one of the worst sensational events of modern times.* 1982 reprint ed. Lexington, KY, US: University Press of Kentucky, **1982**.

[2307] Mussaeus, Thomas Allen, 1887– *The lure of cave lore.* Strasburg, VA, US: np, **1939**. illus: b&w: drawings, maps, photos, tables; viii, 65p, 16p pls; app; 24 cm.

Illus. Printed by Shenandoah Pub House, Inc.

[2308] Mussaeus, Thomas Allen, 1887– *The lure of cave lore; being a random narrative upon caverns in general and in particular the properties of the Skyline Caverns, Front Royal, Virginia.* rev ed. Limeton, VA, US: np, **1950**. illus: b&w: maps, photos, tables; map folded in text; [x], 57p, 8p pls; app; 24 cm.

Printed by Shenandoah Publishing House, Inc.

[2309] MVOR (Mississippi Valley Ozark Regional). Guidebook (1968) *Fall 1968 MVOR guidebook: Stone County, Arkansas.* St. Louis, MO, US: St. Louis University Grotto, **1968**. illus: b&w: charts, maps, photos; 2 fold-out maps; ii, 23p, [2].

Title from cover.

[2310] MVOR (Mississippi Valley Ozark Regional). Guidebook (1969) *MVOR Guidebook Spring 1969.* npop: Chouteau Grotto. National Speleological Society, **1969**. Foresight vol 12 no 2. illus: b&w: cartoons, maps, photos, tables; 2 fold-out maps; text on front lining page; 26p, [2].

[2311] MVOR (Mississippi Valley Ozark Regional). Guidebook (1970) *1970 guidebook UMAC-MVOR: Upper Mississippi area conference-Mississippi Valley Ozarks Regional Sept. 25, 26, & 27 at Eagle Cave.* Kugler, Ralph (ed). [Madison, WI, US]: Wisconsin Speleological Society, **nd**. illus: b&w: maps; partly col: maps; [ii], 14p, [14]; errat.

[2312] MVOR (Mississippi Valley Ozark Regional). Guidebook (1970) *MVOR Spring "70" Guidebook.* [Carbondale, IL, US]: Little Egypt Student Grotto, **1970**. illus: b&w: maps; [10].

[2313] MVOR (Mississippi Valley Ozark Regional). Guidebook (1971) *Fall 1971 MVOR: Glover's Cave.* Evansville, IN, US: Evansville Metropolitan Grotto. National Speleological Society, **1971**. illus: b&w: drawings, maps; [10].

[2314] MVOR (Mississippi Valley Ozark Regional).

Guidebook (1971) *MVOR Spring 1971, April 30, May 1 & 2: Pulaski County.* Buss, James C.; Christensen, Conway B.; Vetz, Edwin B.; Vetz, Michael A. (guidebook staff). St. Louis, MO, US: Hondo Underground Rescue Team, **1971**. illus: b&w: charts, maps, photos; text on front lining page; map on rear lining page; 44p.

Title from cover.

[2315] MVOR (Mississippi Valley Ozark Regional). Guidebook (1972) *Guidebook: Fall 1972 MVOR, Perry County, Missouri.* St. Louis, MO, US: Middle Mississippi Valley Grotto, **1972**. illus: b&w: charts, maps, photos, tables; 32p.

Cover title: MVOR Fall 1972 Guidebook.

[2316] MVOR (Mississippi Valley Ozark Regional). Guidebook (1973) *Guidebook: MVOR Spring '73.* Neff, David; O'Brien, John; Walley, Bill; Warshauer, Susan (eds). Pt. Lookout, MO, US: The School of the Ozarks Troglophiles, **1973**. The Underground Leader vol 3 no 1. illus: b&w: ads, maps, photos; text on front lining page, ad on rear lining page; 30p, [4].

[2317] MVOR (Mississippi Valley Ozark Regional). Guidebook (1974) *Lake of the Ozarks MVOR, Fall 1974: Camden County, Miller County, Morgan County.* Weaver, Herman Dwight, 1938–; Scott, Curtis E. (eds). Eldon, MO, US: Lake of the Ozarks Grotto. National Speleological Society, Sep, **1974**. Ozark Speleograph vol 4 no 3. text on front lining page; 46p.

Title from cover.

[2318] MVOR (Mississippi Valley Ozark Regional). Guidebook (1975) *MVOR: Spring 1975 in the Shawnee Hills.* Carbondale, IL, US: Little Egypt Student Grotto, **1975**. Crawlway Courier vol no 8 no 2. illus: b&w: ads, charts, maps, photos, tables; one map folded loose in text; text on front lining page; [ii], 25p, [2].

Title from cover.

[2319] MVOR (Mississippi Valley Ozark Regional). Guidebook (1976) *Fall 1976 MVOR guidebook, Perry Co., Mo.* Eddleman, William R. (ed). Columbia, MO, US: Southeast Missouri Grotto. National Speleological Society, Sep, **1976**. Southeast Caver vol 2 no 3. illus: b&w: ads, maps, photos, tables; [40].

Cover title: Fall '76 MVOR Guidebook.

[2320] MVOR (Mississippi Valley Ozark Regional). Guidebook (1977) *Spring 1977 MVOR Guidebook: Middle Mississippi Valley Grotto, Shannon Co., Missouri.* Eddleman, William R. (ed). St. Louis, MO, US: Middle Mississippi Valley Grotto. National Speleological Society, **1977**. illus: b&w: ads; [30]; bibl.

Title from cover.

[2321] MVOR (Mississippi Valley Ozark Regional). Guidebook (1979) *Perry County: spring M.V.O.R. '79.* Quamen, Al (vol ed); Hopper, Pat (vol ed). [US]: Little Egypt Student Grotto. National Speleological Society, **1979**. Crawlway Courier vol 13 no 1. illus: b&w: ads, maps, photos; 34p.

Title from cover.

[2322] MVOR (Mississippi Valley Ozark Regional). Guidebook (1980) *Guidebook to spring 1980 MVOR, Shannon County, Missouri.* House, Scott (ed). [US]: Southeast Missouri Grotto, **1980**. The Southeast Caver vol 6 no 3. illus: b&w: maps; [26].

Title from cover.

[2323] MVOR (Mississippi Valley–Ozark Region). Guidebook (1984) *MVOR Guidebook: 1984 Fall.* Mylroie, John Eglinton, 1949–(ed). Evansville, IL, US: Evansville Metropolitan Grotto, Mississippi Valley—Ozark Region, **1984**. illus: b&w: drawings, maps, photos; 2 maps folded in back; iv, 23p.

Held at Gover's Cave, Christian County, Kentucky, September 28,29,30 1984.

[2324] MVOR (Mississippi Valley–Ozark Region). Guidebook (1989) *MVOR Guidebook: 1989 Spring.* Springston, Bob (ed); Addison, Aaron (ed). [IL, US]: Little Egypt Student Grotto, Mississippi Valley—Ozark Region, **1989**. illus: b&w: ads, charts, drawings, maps; iv, 23p.

[2325] Myers, Arthur John, 1918– *Guide to Alabaster Cavern and Woodward County, Oklahoma.* 1982 rev ed. University of Oklahoma, **1982**. bibl.

[2326] Myers, Arthur John, 1918– *Guide to Alabaster Cavern and Woodward County, Oklahoma.* Norman, OK, US: University of Oklahoma, Oklahoma Geological Survey, **1969**. Oklahoma Geological Survey Guidebook 15. illus: b&w: charts, drawings, maps, photos, ports, repros, tables; partly col: maps; 38p; bibl; 28 cm.

Cover title: Oklahoma: Alabaster Cavern and Woodward County. 41 figs. 2nd printing 1972.

[2327] Mylroie, John Eglinton, 1949– *Field guide to the karst geology of San Salvador Island, Bahamas: prepared for the 10th Friends of Karst meeting, February 11-15, 1988, College Center of the Finger Lakes, Bahamian Field Station, San Salvador Island, Bahamas.* MS, US: Department of Geology & Geography, Mississippi State University, in cooperation with College Center of the Finger Lakes Bahamian Field Station San Salvador Island, Bahamas, **1988**. illus: b&w: charts, drawings, maps, photos; 2 maps folded in back; [iii], 108p; bibl; 28 cm.

67 figures, plus 2 on back cover.

Some good data and observations are presented, and a number of field locales described for this Caribbean paradise. Comparison to other islands would have been good.

[2328] Mylroie, John Eglinton, 1949– *First international cave management symposium: proceedings [held at] College of Environmental Sciences, Murray State University, July 15-18, 1981.* Murray, KY, US: Murray State University, Department of Geoscience, **1983**. illus: b&w: cartoons, charts, drawings, graphs, maps, photos; [iii], 171p.

Eighth International Congress of Speleology Bio/article conservation.

[2329] Mylroie, John Eglinton, 1949– *Speleogenesis and karst geomorphology of the Helderberg Plateau, Schoharie County, New York*. [Austin, TX, US]: New York Cave Survey, **1977**. New York Cave Survey Bulletin 2. illus: b&w: charts, graphs, maps, tables; xx, 336p; add, bibl; 28 cm.

Reprint of the author's thesis, Rensselaer Polytechnic Institute, with addendum. Limited edition of 500 copies.

A capable review of the field relationships, regional geology.

[2330] Mylroie, John Eglinton, 1949– *Western Kentucky Speleological Survey: annual report 1978*. Murray, KY, US: College of Environmental Sciences, Murray State University , **1979**. illus: b&w: maps; 55p; bibl; 29 cm.

[2331] Mylroie, John Eglinton, 1949– *Western Kentucky Speleological Survey: annual report 1979*. npop: np, **1979**. illus: b&w: maps; 84p; 29 cm.

Includes 18 surveys.

[2332] Myrick, Donal Richard, 1938– *Fern cave: the history of the discovery, exploration and mapping of the Fern Cave system*. Huntsville, AL, US: Donal Richard Myrick, **1972**. illus: b&w: drawings, maps, photos; iii, 106p; app, bibl; 23 cm.

[2333] Myron, Robert *Prehistoric art*. Hopkins, John (illus). New York, NY, US: Pitman Publishing Corporation, **1964**. illus: b&w: drawings, photos; front: b&w photo; 92p, [3].

42 illus.

[2334] Naden, C. J. *I can read about caves*. McWilliams, Virginia (Illus). Mahwah, NJ, US: Troll Associates, **1979**. Troll Read-alongs Kit. illus: b&w: drawings; [44].

Copy examined was just the book. Book also published separately. Juvenile.

[2335] Naegele, Thomas A. *Caving: a child's adventure*. Perkins, James (illus). [Calumet, MI, US]: Thomas A. Naegele, **1987**. illus: b&w: drawings; [i], 19p.

Juvenile.

[2336] Nash, D. A. *A Peak Cavern bibliography*. Beck, John S. Derbyshire, GB: British Cave Research Association, Sep, **1989**. [iv], 79p, [12]; 30 x 21 cm.

No illus.

[2337] National Cave Management Symposium (1975) *National cave management symposium proceedings: Albuquerque, New Mexico, October 6-10, 1975*. Cave Research Foundation. Albuquerque, NM, US: Speleobooks, **1976**. illus: b&w: drawings, graphs, photos, tables; [ii], iv, 146p; bibl; 28 cm.

[2338] National Cave Management Symposium (1976) *Proceedings: national cave management symposium, Mountain View, Arkansas, October 26 - 29, 1976*. Aley, Thomas John, 1938–(ed); Rhodes, Douglas Warren, 1944–(ed). Albuquerque, NM, US: Speleobooks, **1977**. illus: b&w: charts, drawings, graphs, maps, photos, tables; [viii], 106p; bibl; 28 cm.

[2339] National Cave Management Symposium (1977) *National cave management symposium proceedings: Big Sky, Montana, October 3-7, 1977*. Zuber, Ron (ed); Chester, James (ed); Gilzbert, Stephanie (ed); Rhodes, Douglas Warren, 1944–(ed). Albuquerque, NM, US: Adobe Press, **1978**. illus: b&w: drawings, graphs, maps, photos, repros, tables; [iii], 140p; app, bibl; 28 cm.

[2340] National Cave Management Symposium (1978+80) *National cave management symposia: proceedings*. Wilson, Ronald C. (ed); Lewis, Julian J. (ed). Oregon City, OR, US: Pygmy Dwarf Press, **1982**. illus: b&w: drawings, graphs, maps, photos, repros, tables; [vii], 234p; bibl; 28 cm.

This volume contains papers from both the Fourth National Cave Management Symposium held in Carlsbad, New Mexico on October 16- 20, 1978 and the Fifth National Cave Management Symposium held in Mammoth Cave National Park, Kentucky on October 14-17, 1980. Cover title: National cave management symposia proceedings: Carlsbad, New Mexico 1978: Mammoth Cave, Kentucky 1980.

[2341] National Cave Management Symposium (1980) *Fifth national cave management symposium: Mammoth Cave National Park, October 14-17, 1980*. npop: np, **1980**. illus: b&w: drawings; 4p, [27]; bibl; 28 cm.

Title from cover. No illus. Program; see also proceedings.

[2342] National Cave Management Symposium (1982) *Cave management symposium: proceedings: Harrisonburg, Virginia November 5-7, 1982*. Thornton, Helen (ed); Thornton, Jer (ed); National Speleological Society. Richmond, VA, US: American Cave Conservation Association, **1985**. illus: b&w: drawings, graphs, maps, photos, repros, tables; ii, 119p; app, bibl; 28 cm.

Cover title: National cave management proceedings. Illus.

[2343] National Cave Management Symposium (1984) *Proceedings of the 1984 national cave management symposium, Rolla, MO*. Rolla, MO, US: Missouri Speleological Survey, Jan, **1985**. Missouri Speleology v 25(1-4). illus: b&w: graphs, maps, repros, tables; text on lining pages; vi, 236p; bibl; 28 cm.

Title from cover.

[2344] National Cave Management Symposium (1987) *1987 cave management symposium: Rapid City, South Dakota, October 1987*. Huppert, George Nixon, 1944– (prog chair). Huntsville, AL, US: National Speleological Society, **1989**. illus: b&w: drawings, maps, repros, tables; 160p; app, bibl; 28 cm.

[2345] National Cave Management Symposium (1989) *Proceedings of the 1989 national cave management symposium: New Braunfels, Texas, U.S.A..* Jorden, Jay R.

(ed); Obele, Robert K. (ed); Walsh, John M.(ed). New Braunfels, Austin, TX, US: Texas Cave Management Association and Texas Parks and Wildlife Department, **1993**. illus: b&w: drawings, graphs, maps, photos, tables; 157p; bibl; 28 cm.

Cover title: 1989 National Cave Management Symposium proceedings. Running title: 1989 cave management proceedings. (9/93)—T.p. verso. 1992 printing not widely distributed because of errors.

[2346] National Cave Management Symposium (1991) *1991 national cave management symposium proceedings: Bowling Green, Kentucky, October 23-26, 1991.* Foster, Debora L. (ed); Foster, David G. (ed); Snow, Mary M. (ed); Snow, Richard K. (ed). Horse Cave, KY, US: American Cave Conservation Association, Inc, **1993**. illus: b&w: charts, drawings, graphs, maps, photos, tables; [ii], 405p; bibl, errat; 28 cm.

Cover title: Proceedings of the national cave management symposium October 23-26 Bowling Green, Kentucky

[2347] National Cave Management Symposium (1993) *Proceedings of the 1993 national cave management symposium: held in Carlsbad, New Mexico, October 27-30, 1993.* Pate, Dale L. (ed). [Carlsbad, NM, US]: National Cave Management Symposium Steering Committee, **1995**. illus: b&w: drawings, graphs, maps, photos, photomicrographs, tables; [ii], vi, 357p, [1]; bibl.

[2348] National Cave Management Symposium (1995) *Proceedings of the 1995 national cave management symposium: Spring Mill State Park: Mitchell, Indiana: October 25-28, 1995.* Rea, George Thomas, 1934–(ed); Bowman, Bruce. Indianapolis, IN, US: Indiana Karst Conservancy, Inc, **1995**. illus: b&w: charts, drawings, graphs, maps, photos, photomicrographs, repros, tables; [ii], 318p; bibl.

[2349] National Caving Association *Leadership and instructor qualifications in caving.* [GB]: National Caving Association, **[1988]**. [i], 23p; 30 x 21 cm.

No illus.

[2350] National Caving Association *What do cavers think and do?* Bridgwater, GB: National Caving Association, Jul, **1990**. illus: b&w: charts, maps; 36p; app; 30 x 21 cm.

Results of the questionnaires issued by the NCA's structure working party 1987-1990.

[2351] National Conservation Commission of Barbados *Harrison's Cave: Barbados.* BB: National Conservation Commission of Barbados, **[1986]**. illus: b&w: maps, photos; col: photos, 19p, 1p pls; 22 x 14 cm.

[2352] National Parks and Public Land Division, Department of Conservation & Environment *Draft strategy for the management of caves and karst in Victoria.* East Melbourne, VIC, AU: National Parks and Public Land Division, Department of Conservation & Environment, Mar, **1991**. illus: b&w: charts, tables; text on lining pages; [iii], 21p, [10]; app.

[2353] National Research Council (U.S.). Transportation Research Board *Subsidence over mines and caverns, moisture and frost actions, and classification.* Washington, DC, US: National Academy of Sciences, **1976**. Transportation Research Record 612. iii, 83p; bibl, index; 28 cm.

Twelve reports prepared for the 55th annual meeting of the Transportation Research Board. Illus.

[2354] National Speleological Society *Caving information series.* Washington, DC, US: National Speleological Society, **1960—1978**. Various pagination; bibl.

Largely reprints and single sheet informational papers; 29 nos in one vol. Updated periodically. Illus.

[2355] National Speleological Society (comp) *America's wonderland of caves: palaces under the earth: a directory of commercially operated caves.* Washington, DC, US: National Speleological Society, **1951**. 19p; bibl.

Cover title: Visit Palaces under the Earth: the wonderland of caves.

[2356] National Speleological Society *Black Hills symposium.* Huntsville, AL, US: National Speleological Society, Inc. NSS bulletin v 51 no 2 pp 71-142. 72p; bibl, index; 28 cm.

Cover title. "Special issue". Illus.

[2357] National Speleological Society *The caves of Texas.* Washington, DC, US: The National Speleological Society, Apr, **1948**. Bulletin 10. illus: b&w: cartoons, drawings, maps, photos, repros; 1 map folded in text; [ii], 136p, [1]; bibl, index; 28 cm.

[2358] National Speleological Society *Cumulative index for the National Speleological Society Bulletin, volumes 1 through 45, and Occasional papers of the N.S.S., numbers 1 through 4.* Sasowsky, Ira Daniel, 1959–(ed). Huntsville, AL, US: National Speleological Society, **1986**. [iii], 200p; 28 cm.

Spine title: National Speleological Society Bulletin index.

[2359] National Speleological Society *A scrap book of articles about the C3 expedition into Floyd Collins Crystal Cave, Kentucky; from Sunday February 14 to February 20, 1954.* npop: [National Speleological Society], Mar, **1977**. illus: b&w: repros; [111].

One of a kind.

[2360] National Speleological Society *Wind Cave: the world below.* Huntsville, AL; Lead, SD, US: National Speleological Society; Historical Footprints, **1990**.

1 videocassette (VHS) (33 min), sd, col, 1/2 in.

[2361] National Speleological Society. American Caving Accidents (1967) *American caving accidents 1967.* Curl, Rane L. (ed). npop: National Speleological Society, Safety and Techniques Committee, Aug, **1969**. illus: b&w: tables; 16p.

Pagination includes covers.

[2362] National Speleological Society. American Caving Accidents (1967-1970) *American caving accidents 1967-1970.* Curl, Rane L. (ed). Albuquerque, NM: Speleobooks, **1974**. 44p.

[2363] National Speleological Society. American Caving Accidents (1968) *American caving accidents 1968.* Curl, Rane L. (ed). npop: National Speleological Society, Safety and Techniques Committee, Apr, **1970**. illus: b&w: tables; 15p.

Pagination includes covers.

[2364] National Speleological Society. American Caving Accidents (1969) *American caving accidents 1969.* Curl, Rane L. (ed). npop: National Speleological Society, Safety and Techniques Committee, **nd**. illus: b&w: tables; 16p.

[2365] National Speleological Society. American Caving Accidents (1970) *American caving accidents 1970.* Curl, Rane L. (ed). npop: National Speleological Society, Safety and Techniques Committee, Nov, **1971**. illus: b&w: tables; 14p.

[2366] National Speleological Society. American Caving Accidents (1971) *American caving accidents 1971.* Anderson, Jennifer Ann, 1942–(ed). npop: National Speleological Society, Aug, **1974**. illus: b&w: tables; 14p.

[2367] National Speleological Society. American Caving Accidents (1972) *American caving accidents 1972: a report of the National Speleological Society.* Breisch, Richard Lewis, 1943–(ed). Albuquerque, NM, US: Speleobooks, Aug, **1975**. illus: b&w: graphs, maps, tables; folded detachable report at end; [iv], 24p.

[2368] National Speleological Society. American Caving Accidents (1973) *American caving accidents 1973: a report of the National Speleological Society.* Breisch, Richard Lewis, 1943–(ed). Albuquerque, NM, US: Speleobooks, Aug, **1975**. illus: b&w: maps, tables; [iii], 22p.

[2369] National Speleological Society. American Caving Accidents (1974) *American caving accidents 1974: a report of the National Speleological Society.* Breisch, Richard Lewis, 1943–(ed). Albuquerque, NM, US: Speleobooks, Aug, **1976**. illus: b&w: maps, tables; [iii], 24p; app, bibl.

[2370] National Speleological Society. American Caving Accidents (1975) *American caving accidents 1975: a report of the National Speleological Society.* Breisch, Richard Lewis, 1943–(ed). Albuquerque, NM, US: Speleobooks, Aug, **1977**. illus: b&w: graphs, maps, tables; [iv], 44p; app.

[2371] National Speleological Society. American Caving Accidents (1976-1979) *American caving accidents 1976 through 1979.* Knutson, Richard Stephen (ed). 1976—1979. Huntsville, AL, US: National Speleological Society, May, **1981**. NSS News vol 39 no. 5 pt 2. ii, 82p.

ACA 1980-81 appears as part of *NSS News*, vol 43(3), March 1983 and ACA 1982 appears as part of *NSS News*, vol 43(10), October 1983.

[2372] National Speleological Society. American Caving Accidents (1983) *1983 American caving accidents.* Knutson, Richard Stephen (ed). [Huntsville, AL, US]: National Speleological Society, Nov, **1984**. NSS News vol 42 no 11 part 2. illus: b&w: charts, tables; 12p.

Title from cover.

[2373] National Speleological Society. American Caving Accidents (1984) *1984 American caving accidents.* Knutson, Richard Stephen (ed). [Huntsville, Al, US]: National Speleological Society, Nov, **1985**. NSS News vol 43 no 11 part 2. illus: b&w: tables; 13p.

[2374] National Speleological Society. American Caving Accidents (1985) *1985 American caving accidents.* Knutson, Richard Stephen (ed). [Huntsville, Al, US]: National Speleological Society, Nov, **1986**. NSS News vol 44 no 11 part 2. illus: b&w: ads, charts, tables; 12p.

Title from cover.

[2375] National Speleological Society. American Caving Accidents (1986) *American caving accidents [for 1986].* Knutson, Richard Stephen (ed). [Huntsville, Al, US]: National Speleological Society, Nov, **1987**. NSS News vol 44 no 11 part 2. illus: b&w: ads, charts, tables; 14p.

Incorrectly labeled vol 44 really vol 45.

[2376] National Speleological Society. American Caving Accidents (1987) *American caving accidents [for 1987].* Knutson, Richard Stephen (ed). [Huntsville, Al, US]: National Speleological Society, Dec, **1988**. NSS News vol 46 no 12 part 2. illus: b&w: ads, charts, tables; 17p.

[2377] National Speleological Society. American Caving Accidents (1988) *American caving accidents [for 1988].* Knutson, Richard Stephen (ed). [Huntsville, Al, US]: National Speleological Society, Dec, **1989**. NSS News vol 47 no 12 part 2. illus: b&w: ads, charts, tables; 22p.

[2378] National Speleological Society. American Caving Accidents (1989) *American caving accidents 1989.* Knutson, Richard Stephen (ed). [Huntsville, Al, US]: National Speleological Society, Dec, **1990**. NSS News vol 48 no 12 part 2. illus: b&w: ads, charts, tables; 20p.

Title from cover. Title page incorrectly says 1990 American caving accidents.

[2379] National Speleological Society. American Caving Accidents (1990) *American caving accidents 1990.* Knutson, Richard Stephen (ed). [Huntsville, Al, US]: National Speleological Society, Dec, **1991**. NSS News vol 48 no 12 part 2 [sic]. illus: b&w: ads, charts, tables; 28p.

This issue of the NSS news was incorrectly labeled as Vol 48, 1990, it is really vol 49, 1991.

[2380] National Speleological Society. American Caving Accidents (1991) *American caving accidents 1991.* Knutson, Richard Stephen (ed). [Huntsville, Al, US]: National Speleological Society, Dec, **1992**. NSS News vol 50 no 12 part 2. illus: b&w: ads, charts, tables; 26p.

[2381] National Speleological Society. American Caving Accidents (1992) *American caving accidents.* Knutson, Richard Stephen (ed). [Huntsville, Al, US]: National Speleological Society, Dec, **1992**. NSS News vol 51 no 12 part 2. illus: b&w: ads, tables; 26p.

[2382] National Speleological Society. American Caving Accidents (1993) *American caving accidents 1993.* Knutson, Richard Stephen (ed). [Huntsville, Al, US]: National Speleological Society, Dec, **1994**. NSS News vol 52 no 12 part 2. illus: b&w: ads, charts, tables; 34p.

[2383] National Speleological Society. Annual Convention. Guidebook (1960: NM) *A guide book to Carlsbad Caverns National Park.* Spangle, Paul F. Washington, DC, US: National Speleological Society, **1960**. National Speleological Society Guide Book Series no 1. illus: b&w: charts, drawings, maps, photos, tables; 43p; bibl; 24 cm.

Pagination includes covers.

Guidebook for the 1960 NSS annual convention in New Mexico. First of the annual guidebooks series.

[2384] National Speleological Society. Annual Convention. Guidebook (1962: [Black Hills], SD) *A guide to the caves of the Black Hills, S. D.* Wilbur, Robert. [Custer, SD, US]: National Speleological Society, **1962**. illus: b&w: maps; iv, 21p; bibl; 25 cm.

Prepared for the pre-convention field trip of the 19th Annual Convention of the NSS.

[2385] National Speleological Society. Annual Convention. Guidebook (1963: Mountain Lake, VA) *Guide book to the major caves in the vicinity of Mountain Lake, Virginia.* Holsinger, John Robert, 1934–. Falls Church, VA, US: EconoPrint for National Speleological Society, **1963**. illus: b&w: maps, photos; 23p; 28 cm.

Title from cover. Prepared for the 1963 annual convention of the National Speleological Society.

[2386] National Speleological Society. Annual Convention. Guidebook (1964: New Braunfels, TX) *A guide to the caves of Texas and 1964 N.S.S. convention (New Braunfels, TX) field trips.* Reddell, James Russell, 1938– (ed). US: [National Speleological Society], May, **1964**. illus: b&w: maps; 63p; bibl; 25 cm.

Cover title: Prepared for the 1964 National Speleological Society Convention, New Braunfels, Texas, June 14-20.

[2387] National Speleological Society. Annual Convention. Guidebook (1965: Bloomington, IN) *National Speleological Society convention 1965 guidebook, Indiana University Bloomington, Indiana June 12-20, 1965.* Powell, Richard L. (ed, illus); Rea, G. Thomas (ed). US: [National Speleological Society], **[1965]**. illus: b&w: graphs, maps, tables; [ii], 38p.

Cover title: 1965 convention Bloomington Indiana National Speleological Society.

[2388] National Speleological Society. Annual Convention. Guidebook (1966) *Caves of the Sequoia Region California, issued in conjunction with the 25th anniversary convention of the National Speleological Society, David R McClurg, convention chairman.* McClurg, David Robert, 1929– (conv chair); Reardon, Bob. Washington, DC, US: National Speleological Society, **1966**. National Speleological Society Guidebook number 7. illus: b&w: maps, photos; [i], iv, 38p.

Guidebook for the 1966 NSS convention.

[2389] National Speleological Society. Annual Convention. Guidebook (1967: Birmingham and Huntsville, AL) *The caves of Alabama: a guide for the 1967 convention of the National Speleological Society at Birmingham and Huntsville, Alabama, June 1967.* Veitch, John David, 1931–; Cooper, John Edward, 1929– (cont). npop: [National Speleological Society], **1967**. illus: b&w: maps; [iv], 51p; 23 cm.

Cover title: The caves of Alabama: a guide for the 1967 convention of the National Speleological Society.

[2390] National Speleological Society. Annual Convention. Guidebook (1968: Southwestern MO) *Guidebook to selected caves in southwestern Missouri: prepared for the 25th annual convention of the National Speleological Society.* Vineyard, Jerry Daniel, 1935– (ed). MO, US: Missouri Speleological Survey, Jul, **1968**. Missouri Speleology vol 10(3). illus: b&w: charts, maps, photos; partly col: maps; col: maps; 4 fold-out maps in text; text on lining pages; [ii], 45p; bibl.

[2391] National Speleological Society. Annual Convention. Guidebook (1969: Lovell, WY) *Caves of the Big Horn - Pryor Mountains of Montana and Wyoming. Issued in conjunction with the 28th anniversary convention of the National Speleological Society at Lovell, Wyoming, Harley A Leach, convention chairman.* Schultz, Robert L. (ed). Washington, DC, US: National Speleological Society, **1969**. National Speleological Society Guidebook number 10. illus: b&w: charts, maps, photos; [i], iv, 28p; bibl.

[2392] National Speleological Society. Annual Convention. Guidebook (1970: Pittsburgh, PA) *NSS 1970 guidebook: Mid-Appalachian Region.* Schleicher, Donald P. (ed). Pittsburgh, PA, US: The National Speleological Society, **1970**. illus: b&w: charts, maps, photos, tables; b&w photo on lining page; 95p; 28 cm.

1970 Field trip guidebook. Illus.

Annual convention was in states of West Virginia and Pennsylvania.

[2393] National Speleological Society. Annual Convention. Guidebook (1971) *NSS 71 guidebook.* Whittemore,

R.E. (ed). [US]: npop, **1971**. illus: b&w: charts, maps, photos, tables; 3 foldout map sheets; 69p; bibl, index.

Published as The Region Record v1 no4.

[2394] National Speleological Society. Annual Convention. Guidebook (1972: White Salmon, WA) *Selected caves of the Pacific Northwest with particular reference to the vulcanospeleology of the state of Washington.* Larson, Charles Victor, 1928–(chair); Nieland, James Raymond, 1949–(cart); Halliday, William Ross, 1926–(ed). Huntsville, AL, US: National Speleological Society, **1972**. Guidebook for the Convention National Speleological Society 1972 no13. illus: b&w: maps, photos; 2 maps folded in text; ii, 75p; bibl, index; 28 cm.

Illus.

[2395] National Speleological Society. Annual Convention. Guidebook (1973: Bloomington, IN) *NSS 73 convention guidebook; National Speleological Society convention, Bloomington, Indiana, June 16-24, 1973.* [Huntsville, AL, US]: National Speleological Society, **1973**. illus: b&w: drawings, maps, photos, repros, tables; three folded map plates in rear pocket; four folded maps in text; [ii], 80p, [1]; bibl; 28 cm.

Cover title: NSS 73 convention guidebook Bloomington, Indiana June 16-24, 1973. Illus.

[2396] National Speleological Society. Annual Convention. Guidebook (1974: Decorah, IO) *National Speleological Society 1974 convention Luther College, Decorah, Iowa 10-18 August, 1974.* Hedges, James (comp, ed). [Huntsville, AL, US]: The National Speleological Society, Inc, **1974**. Field trip guidebook: 1974 National Speleological Society. Guidebook: [no.15]. illus: b&w: charts, maps, photos, tables; four folded maps in text; b&w photos on lining pages; 139p; bibl; 28 cm.

Cover title: NSS 1974 convention Decorah, Iowa: Upper Mississippi Valley cave region. Issued at the 1974 convention held Aug. 10-18, 1974 at Luther College, Decorah, Iowa.

[2397] National Speleological Society. Annual Convention. Guidebook (1975: Frogtown, CA) *National Speleological Society 1975 Convention, Frogtown, Ca.* Rogers, Bruce W. (ed). npop: Albany Press, **1975**. National Speleological Society Guidebook no 16. illus: b&w: maps; 244p; 28 cm.

Cover title. Spine title: NSS 1975 convention guidebook. Illus.

[2398] National Speleological Society. Annual Convention. Guidebook (1976: Morgantown, WV) *1976 annual convention guidebook, Morgantown, W. Va.* Garton, Emmel Ray, 1950–(ed). Huntsville, AL, US: The National Speleological Society, **1976**. illus: b&w: ads, charts, drawings, graphs, maps, photos, tables; 1 loose folded map; 6 map foldouts; vi, 117p; errat; 28 cm.

Provides information on caves; geology; hydrology; and scenic and historical attractions within a two hour driving radius of Morgantown. Included are ...directions and maps of 26 selected caves in West Virginia, Pennsylvania, and Maryland.

[2399] National Speleological Society. Annual Convention. Guidebook (1977: Alpena, MI) *Official 1977 guidebook, Alpena, Michigan: '77, the Lakeshore convention: Alpena, Michigan, NSS.* [Fritz, Carol (ed)]. Alpena, MI, US: National Speleological Society, **1977**. illus: b&w: ads, drawings, maps, photos; col: maps; 1 map folded; b&w photos on lining pages; 68p; 29 cm.

Includes description and tours to Ontario caves. Illus.

Annual convention guidebook. Includes geology field-trip log for the significant karst in the Alpena area.

[2400] National Speleological Society. Annual Convention. Guidebook (1978: New Braunfels, TX) *An introduction to the caves of Texas: prepared for the 1978 National Speleological Society Convention New Braunfels, Texas.* Fieseler, Ronald Glenn, 1948–(ed); Jasek, James (ed); Jasek, Mimi (ed). Austin, TX, US: Texas Speleological Society, **1978**. NSS Convention Guidebook number 19. illus: b&w: cartoons, maps, photos; 7 maps folded; front: b&w photo; 115p, [2]; 18 cm.

[2401] National Speleological Society. Annual Convention. Guidebook (1979: Pittsfield, MA) *An introduction to caves of the northeast: guidebook for the 1979 National Speleological Convention, Pittsfield, Massachusetts.* Evans, John (ed); Quick, Peter (ed); Sloane, Bruce Charles, 1935–(ed). [Pittsfield, MA, US]: National Speleological Society, **1979**. NSS Convention Guidebook number 20. illus: b&w: maps, photos, repros, tables; 76p; 28 cm.

Accompanied by a small comic "Wowee Caverns", written and drawn by Doug Kirby, inserted in pocket.

Guidebook for annual NSS convention in Massachusetts.

[2402] National Speleological Society. Annual Convention. Guidebook (1980: White Bear Lake, MN) *An introduction to caves of Minnesota, Iowa, and Wisconsin: guidebook for the 1980 National Speleological Society Convention, Lakewood Community College, White Bear Lake, Minnesota.* Alexander, E. Calvin (ed). Huntsville, AL, US: National Speleological Society, Jul, **1980**. NSS Convention Guidebook number 21. illus: b&w: charts, drawings, graphs, maps, photos, tables; five folded maps in rear pocket; vi, 190p; bibl; 28 cm.

[2403] National Speleological Society. Annual Convention. Guidebook (1982: Bend, OR) *An introduction to caves of the Bend area: guidebook of the 1982 NSS Convention.* Larson, Charles Victor, 1928–(ed); Nieland, James Raymond, 1949–(cart). Huntsville, AL, US: National Speleological Society, **1982**. illus: b&w: drawings, maps, photos; partly col: maps; 4 maps folded in text; v, 74p; index; 28 cm.

Guidebook for the National Speleological Society Annual Convention, Bend, Oregon, June 27 - July 3, 1982.

[2404] National Speleological Society. Annual Convention. Guidebook (1983: Elkins, WV) *An introduction to the caves of east-central West Virginia: guidebook for the 1983 National Speleological Society Convention, Elkins, West Virginia.* Medville, Douglas Michael, 1941–(ed); Werner, Eberhard Wolfgang, 1942–(ed); Dasher, George R., 1952–(ed). npop: West Virginia Speleological Survey, **1983**. illus: b&w: ads, charts, drawings, graphs, maps, photos, tables; 1 loose folded map; 7 foldout maps; photo on rear lining page; 146p; bibl; 28 cm.

[2405] National Speleological Society. Annual Convention. Guidebook (1984: Sheridan, WY) *Guidebook of the 1984 NSS convention.* Flurkey, Andrew J. (ed); Gilbert, B. Miles (cont). Sheridan, WY, US: [National Speleological Society], **1984**. illus: b&w: drawings, maps, photos; 2 maps folded in text; vi, 76p; bibl; 28 cm.

Illus.

[2406] National Speleological Society. Annual Convention. Guidebook (1985: Frankfort, KY) *The caves of south eastern Kentucky: guidebook for the 1985 National Speleological Society Convention, Frankfort, Kentucky.* Hacker, Christopher (ed); Vore, Duane (ed); Vore, Debbie (ed); Ohio Valley Region. [Huntsville, AL, US]: [National Speleological Society], **1985**. NSS Convention Guidebook no 25. illus: b&w: ads, cartoons, charts, drawings, maps, photos, tables; folded map loose in text; 76p; bibl; 28 cm.

[2407] National Speleological Society. Annual Convention. Guidebook (1986: Tularosa, NM) *1986 NSS convention guidebook, Tularosa, New Mexico.* Belski, David Stanley, 1937–(ed); Belski, Carol (ed); Hardy, Fritzi (ed). Albuquerque, NM, US: National Speleological Society, **1986**. NSS Convention Guidebook number 25. illus: b&w: ads, drawings, graphs, maps, photos, tables; vii, 66p.

[2408] National Speleological Society. Annual Convention. Guidebook (1987: Sault Sainte Marie, MI) *National Speleological Society official 1987 guidebook, August 3-7, 1987, Sault Sainte Marie, Michigan.* [Rea, George Thomas, 1934–(ed)]. [Huntsville, AL, US]: National Speleological Society, **1987**. illus: b&w: drawings, maps, photos, tables; text on rear lining page; [iv], 40p; bibl.

[2409] National Speleological Society. Annual Convention. Guidebook (1988: Hot Springs, SD) *1988 NSS convention guidebook.* Schilberg, Gary (ed); Springhetti, David (ed). Huntsville, AL, US: National Speleological Society, **1988**. illus: b&w: graphs, maps, photos, repros, tables; col: photos; 7 foldout maps (1 has graphs on reverse); b&w photos on lining pages; iv, 156p; bibl; 28 cm.

The convention was in South Dakota. Bears a second title: Caves and Associated Features of the Black Hills Area, South Dakota and Wyoming.

[2410] National Speleological Society. Annual Convention. Guidebook (1989: Sewanee, TN) *Caves and caving in TAG: a guidebook for the 1989 Convention of the National Speleological Society ; Sewanee, Tennessee.* Putnam, William O. (ed). [Sewanee, TN, US]: [National Speleological Society], **1989**. illus: b&w: maps; 240p; 28 cm.

Illus.

Annual NSS convention guidebook. Locations and descriptions of selected caves in Tennessee, Alabama, and Georgia in the south eastern United States. Numerous maps. Introductory material on geology and archeology.

[2411] National Speleological Society. Annual Convention. Guidebook (1990: Yreka, CA) *NSS 1990 convention guidebook: Yreka, California: "where the lava meets the limestone".* Johnson, Victoria S. (ed). [Huntsville, AL, US]: National Speleological Society, **1990**. illus: b&w: ads, drawings, maps, photos, tables; 6 map plates in separate envelopes; b&w photos on lining pages; [iv], 203p, [1].

Cover title: NSS 1990 convention guidebook.

[2412] National Speleological Society. Annual Convention. Guidebook (1991: Cobleskill, NY) *Guide to the caves and karst of the Northeast.* Nardacci, Michael (ed in chief); Palmer, Arthur Nicholas, 1940–(layout); Palmer, Margaret V. (layout). Huntsville, AL, US: National Speleological Society, **1991**. illus: b&w: charts, drawings, maps, photos, tables; col: photos; 2 loose folded maps; [ii], 167p; bibl, errat.

Cover title: 50th Anniversary NSS Convention, Cobleskill, New York, 1991.

[2413] National Speleological Society. Annual Convention. Guidebook (1992: Salem, IN) *Caving in the heartland: a guidebook for the 1992 Convention of the National Speleological Society, Salem, Indiana.* Rea, George Thomas, 1934–(ed, design, layout). Huntsville, AL, US: The National Speleological Society, **1992**. illus: b&w: charts, drawings, maps, photos, tables; four maps folded in text; vi, 255p; app; 28 cm.

Cover title: Caving in the heartland: 1992 NSS Convention Guidebook. Illus.

[2414] National Speleological Society. Annual Convention. Guidebook (1993: Pendleton, OR) *The 1993 NSS Convention guidebook.* Kline, Thomas (ed). Huntsville, AL, US: National Speleological Society, **1993**. illus: b&w: drawings, maps, photos; 2 maps folded in rear pocket; [v], 149p; app, index.

[2415] National Speleological Society. Annual Convention. Guidebook (1994: Brackettville, TX) *The caves and karst of Texas: a guidebook for the 1994 Convention of the National Speleological Society with emphasis on the Southwestern Edwards Plateau: Brackettville, Texas, June 19-24, 1994.* Elliott, William Rawleigh, 1946–(ed);

Veni, George, 1957–(ed). Huntsville, AL, US: National Speleological Society, **1994**. illus: b&w: maps; viii, 342p, 7p pls; bibl, errat, index; 28 cm.

Illus.

[2416] National Speleological Society. Annual Convention. Guidebook (1995: Blacksburg, VA) *Underground in the Appalachians: a guidebook for the 1995 Convention of the National Speleological Society, Blacksburg, Virginia, July 17-21, 1995.* Zokaites, Carol (ed); Hansen, Kimberly S. (des); Vermeulen, Susan E. (photo ed). Huntsville, AL, US: The National Speleological Society, **1995**. illus: b&w: maps; ii, 217p; bibl, index; 28 cm.

Accompanying maps in separate envelope. Illus.

[2417] National Speleological Society. Annual Convention. Guidebook (1996: Salida, CO) *The caves and karst of Colorado: a guidebook for the 1996 convention of the National Speleological Society.* Kolstad, Rob (ed); Rhinehart, Richard (asst ed); Hose, Louise (asst ed). Huntsville, AL, US: National Speleological Society, **1996**. illus: b&w: cartoons, charts, drawings, graphs, maps, photos, repros; iii, 214p; bibl.

[2418] National Speleological Society. Annual Convention. Guidebook (1997: Sullivan, MO) *Exploring Missouri caves. A guidebook for the 1997 convention of the National Speleological Society Sullivan, Missouri June 23-27, 1997.* Taylor, Robert L. (ed, layout & des); Beard, Jonathan B. (assoc ed). Huntsville, AL, US: National Speleological Society, **1997**. illus: b&w: drawings, graphs, maps, photos, ports; col: maps, photos; [ii], x, 278p; bibl, index.

A nice convention guidebook, with material on a hundred caves and many of the big karst springs for which Missouri is famous. Besides cave descriptions and maps, there is good mix of material on geology, biology, and history.

[2419] National Speleological Society. Annual Convention. Guidebook. Geology and Biology Field Trip (1982: Bend, OR) *Caves and other volcanic landforms of Central Oregon.* Sims, Lynne (ed); Benedict, Ellen M. (ed). Huntsville, AL, US: National Speleological Society, **1982**. illus: b&w: charts, drawings, graphs, maps, photos; 47p; bibl; 28 cm.

Guidebook of the Geology and Biology Field Trip, National Speleological Society Annual Convention, Bend, Oregon, June 27 - July 3, 1982.

There is surprisingly little in book form on lava tubes, etc. This has relatively little geology, per se, but much to read between the lines. So too, does it provide a number of subsequent avenues of study.

[2420] National Speleological Society. Annual Convention. Guidebook. Geology Field Excursion (1978: New Braunfels, TX) *Caves and karst hydrogeology of the southeastern Edwards Plateau, Texas: guidebook, geology field excursion, National Speleological Society annual convention, New Braunfels, Tex., June 18-23, 1978.* Kastning, Ernst H., 1944–; Knox, Jan.; Byrd, Thomas M. (excur co-lead). Austin, TX, US: Department of Geological Sciences and Texas Natural Areas Survey, The University of Texas at Austin, **1978**. Guidebook NSS Geology Field Excursion 1978. illus: b&w: charts, drawings, graphs, maps, photos; 2 map pls folded in back; front: b&w photo; iv, 46p; bibl; 28 cm.

18 figs, 2 pls.

[2421] National Speleological Society. Annual Convention. Guidebook. Geology Field Trip (1979: Pittsfield, MA) *Karst hydrogeology and geomorphology of eastern New York: a guidebook to the geology field trip, National Speleological Annual Convention, Pittsfield, Massachusetts August 5-12, 1979.* Cullen, James J.; Mylroie, John Eglinton, 1949–; Palmer, Arthur Nicholas, 1940–. Pittsfield, MA, US: [National Speleological Society], Aug, **1979**. illus: b&w: charts, drawings, maps, photos, tables; 83p; bibl.

36 figs.

[2422] National Speleological Society. Annual Convention. Guidebook. Geology Field Trip (1986: Tularosa, NM) *Geology field trip guidebook, 45th National Speleological Society (1986) Tularosa, New Mexico, convention.* Jagnow, David Henry; DuChene, H. R.; Anderson, R. Y.; Sares, S. W.; Wells, Stephen G. Albuquerque, NM, US: Adobe Press, **1986**. 102p.

[2423] National Speleological Society. Annual Convention. Guidebook. Geology Field Trip (1990: Yreka, CA) *Geologic features of Mount Shasta and Medicine Lake Volcanoes.* Tinsley, John C. (comp). npop: [National Speleological Society], **1990**. illus: b&w: drawings, maps, photos; [iii], 82p; bibl.

Cover title: NSS 1990 Convention Geologic Field Guide. 38 figs.

[2424] National Speleological National Speleological Society. Annual Convention. Guidebook. Geology Field Trip (1995: Blacksburg, VA) *Origin of caves and karst in the Shenandoah Valley, Rockingham and Augusta Counties, Virginia: guidebook for a geologic fieldtrip, National Speleological Society Annual Convention, Blacksburg, VA, 16 July 1995.* Kastning, Kass; Hubbard, David A.; Kastning, Ernst H., 1944–; Kastning, Karen M. [Blacksburg, VA, US]: [National Speleological Society], **1995**. illus: b&w: charts, graphs, maps, photos, repros, tables; 1 loose folded map; 3 foldout maps; ii, 50p, 3p pls; bibl.

[2425] National Speleological Society. Annual Convention. Guidebook. Preconvention Field Trip (1973: Bloomington, IN) *C. H. U. G. Crawford Harrison Underground Gang presents twenty-two favorite Harrison Crawford caves: 1973 pre-convention activities.* Powell, Richard L.; Spalding, G.; Steele, William. npop:

Crawford Harrison Underground Gang, **1973**. illus: b&w: charts, repros, tables; [i], 24p.

Cover title: Crawford-Harrison Underground Gang preconvention guidebook.

[2426] National Speleological Society. Annual Convention. Postconvention Guidebook (1992: Salem, IN) *1992 post convention guidebook: enter the karst of Indiana, you may never want to come out.* Cook, Holly (ed); Black, David (ed). npop: Harrison Crawford Grotto, National Speleological Society, **1992**. illus: b&w: drawings, maps, photos; 50p, [1].

Supplement to the 1992 NSS convention guidebook.

[2427] National Speleological Society. Annual Convention. Proceedings (1976: Morgantown, WV) *Proceedings of the 1976 NSS annual convention at Morgantown, West Virginia 24 June - 3 July 1976.* Werner, Eberhard Wolfgang, 1942–(ed). npop: West Virginia Speleological Survey, **1977**. illus: b&w: drawings, graphs, maps, photos, tables; iii, 60p; bibl.

[2428] National Speleological Society. Annual Convention. Program (1976: Morgantown, WV) *1976 National Speleological Society 35th Annual Convention Guidebook* [program] *with Abstracts.* Cullinan, M. Huntsville, AL, US: National Speleological Society, **1976**. 55p.

[2429] National Speleological Society. Annual Convention. Program (1977: Alpena, MI) *Program 1977 Annual Meeting of the National Speleological Society, August 1-5, Alpena, Michigan].* Huntsville, AL, US: National Speleological Society, **1977**. illus: b&w: ads, charts, maps; 33p, [1].

[2430] National Speleological Society. Annual Convention. Program (1978: New Braunfels, TX) *Deep in the karst of Texas: general information program: abstracts NSS-New Braunfels 1978.* Kastning, Ernst H., 1944–(ed); Kastning, Karen M. (ed). npop: National Speleological Society, **1978**. 74p; 28 cm.

Title from cover. Illus.

[2431] National Speleological Society. Annual Convention. Program (1979: Pittsfield, MA) *The program of the 1979 N.S.S. Convention, August 5-12, 1979, Pittsfield, Mass.* Cullen, Jim (ed); Engel, Thom (asst ed); Hauser (asst ed). Huntsville, AL, US: National Speleological Society, **1979**. illus: b&w: ads, charts, drawings, maps; 100p, [1].

Title from cover.

[2432] National Speleological Society. Annual Convention. Program (1980: White Bear Lake, MN) *1980 NSS Convention Program, Caving in the City, July 27 - August 1, Lakewood College, White Bear Lake, Minnesota.* Huntsville, AL, US: National Speleological Society, **1980**. illus: b&w: ads, drawings, maps; 47p, [1].

Title from cover. 1981 convention was held in conjunction with the International Congress of Speleology and did not have a separate program.

[2433] National Speleological Society. Annual Convention. Program (1982: Bend, OR) *The program of the annual convention of the National Speleological Society, June 27-July 3, Bend, Oregon.* Benedict, Ellen (ed). Huntsville, AL, US: National Speleological Society, **1982**. illus: b&w: ads, charts, drawings, maps, photos; 41p.

Title from cover.

[2434] National Speleological Society. Annual Convention. Program (1983: Elkins, WV) *Program, 1983 NSS Annual Convention.* Werner, Eberhard Wolfgang, 1942– (ed); Medville, Hazel E. (ed). Huntsville, AL, US: National Speleological Society, **1983**. illus: b&w: ads, charts, maps, tables; text on front lining page; map on rear lining page; 98p.

Title from cover.

[2435] National Speleological Society. Annual Convention. Program (1984: Sheridan, Wyoming) *The program of the 1984 NSS Convention, June 25-29, 1984, Sheridan, Wyoming.* Pygmy Dwarf Press [Mike Sims] (ed). Huntsville, AL, US: National Speleological Society, **1984**. illus: b&w: charts, maps, tables; map on rear lining page; 32p.

Title from cover.

[2436] National Speleological Society. Annual Convention. Program (1985: Frankfort, KY) *Program for the 1985 NSS Convention, Frankfort, Kentucky, June 23-29.* Dougherty, P. H. (ed). Huntsville, AL, US: National Speleological Society, **1985**. illus: b&w: maps; 51p; 28 cm.

Title from cover.

[2437] National Speleological Society. Annual Convention. Program (1986: Tularosa, NM) *Program, 1986 NSS Convention, June 22-28, 1986, Tularosa, New Mexico.* Peerman, Kathy (prog chair); Peerman, Steve (prog chair). Huntsville, AL, US: National Speleological Society, **1986**. illus: b&w: charts, maps, tables; chart on rear lining page; ii, 53p.

Title from cover.

[2438] National Speleological Society. Annual Convention. Program (1987: Sault Sainte Marie, MI) *Program 1987 NSS Convention, August 3-7, 1987, Sault Sainte Marie, Michigan.* Huntsville, AL, US: National Speleological Society, **1987**. illus: b&w: ads, charts, maps, tables; text on front lining page; ad on rear lining page; map on rear; ii, 53p.

Title from cover.

[2439] National Speleological Society. Annual Convention. Program (1988: Hot Springs, SD) *The program of the 1988 NSS convention: June 27-July 1, 1988: Hot Springs, South Dakota.* Schilberg, Gary (ed); Springhetti, David (ed). [Hot Springs, SD, US]: [National Speleological Society], **1988**. illus: b&w: charts, maps; 48p.

Title from cover.

[2440] **National Speleological Society. Annual Convention. Program (1989: Swanee, TN)** *1989 NSS convention program: July 31 - August 4, 1989.* Eck, Greg (ed); Olsen, Laurie (ed). Plainfield, IN, US: [National Speleological Society], **1992**. illus: b&w: charts, maps; 76p.

Title from cover.

[2441] **National Speleological Society. Annual Convention. Program with Abstracts (1990: Yreka, CA)** *Program of the National Speleological Society convention July 8 through July 14, 1990: Yreka, California.* Frantz, W. S. (ed). [Los Gatos, CA, US]: National Speleological Society, **1990**. illus: b&w: ads, charts, drawings, maps; text on lining pages; [iii], 50p, [1].

Cover title: Program NSS 1990: where the lava meets the limestone.

[2442] **National Speleological Society. Annual Convention. Program (1991: Cobleskill, NY)** *Program: 1991 National Speleological Society convention: July 1-5: Cobleskill, NY.* Nardacci, Mike (pub chair). [Albany, NY, US]: [National Speleological Society], **1991**. illus: b&w: ads, charts, drawings, maps, repros, tables; map on rear, chart on front lining pages; 104p.

[2443] **National Speleological Society. Annual Convention. Program (1992: Salem, IN)** *Program: 1992 NSS convention: the heartland of the karstland: Salem, Indiana.* Rea, George Thomas, 1934–(ed). Plainfield, IN, US: [National Speleological Society], **1992**. illus: b&w: ads, charts, maps; 64p.

Title from cover.

[2444] **National Speleological Society. Annual Convention. Program (1993: Pendleton, OR)** *[The program of the annual convention of the National Speleological Society, 1-6 August, Pendleton, Oregon].* [Pendleton, OR, US]: [National Speleological Society], **1993**. illus: b&w: ads, charts, maps; 55p, [1].

On cover: "The Wagon Trail to Caver Hospitality"

[2445] **National Speleological Society. Annual Convention. Program (1994: Brackettville, TX)** *National Speleological Society, program: 1994 NSS Convention: Brackettville, Texas: June 18-25.* Mixon, William Weston, 1940—(ed). [TX, US]: [National Speleological Society], **1994**. illus: b&w: ads, charts, drawings, maps; map on rear, chart on front lining pages; 60p.

Title from cover.

[2446] **National Speleological Society. Annual Convention. Program with Abstracts (1986: Tularosa, NM)** *National Speleological Society convention program with abstracts.* Belski, David Stanley, 1943–(ed); Eaton, J. (ed); Rhodes, Douglas Warren, 1944–(ed). Albuquerque, NM, US: Adobe Press, **1986**. 54p.

[2447] **National Speleological Society. Annual Convention. Supplemental Guidebook (1970: Pittsburgh, PA)** *Supplemental guidebook: 1970 NSS convention.* Thrun, Robert, 1940–(ed). npop: np, **1970**. illus: b&w: maps; [23].

Unofficial guidebook. Title from cover.

[2448] **National Speleological Society. Annual Convention. Supplemental Guidebook (1989: Sewanee, TN)** *A guide to East Tennessee Caves: prepared for the 1989 National Speleological Society Convention, Sewanee, Tennessee.* East Tennessee Grotto; Bowers, Jeff (ed). [Huntsville, AL, US]: [National Speleological Society], **1989**. illus: b&w: charts, maps, photos, repros, tables; 58p, [2]; errat.

[2449] **National Speleological Society. Speleo Digest** *Speleo Digest 25 - year cumulative index, 1956 through 1980.* Sasowsky, Ira Daniel, 1959–; Wheeland, Keith D. Huntsville, AL, US: National Speleological Society, **1991**. ii, 565p; index.

Limited edition of 200. Keyword and author indexes.

[2450] **National Speleological Society. Speleo Digest (1956)** *Speleo digest, 1956.* Dunn, John R. (ed); White, William Blaine, 1934–(ed); Howard, S. Russell (ed). [Pittsburgh, PA, US]: National Speleological Society, Pittsburgh Grotto Press, Mar, **1957**. illus: b&w: drawings, maps, photos; [iii]; bibl, index.

A collection of speleological writing taken from the publications of chapters of the National Speleological Society. Illus. Incl cartes. Was reprinted.

[2451] **National Speleological Society. Speleo Digest (1957)** *Speleo digest, 1957.* Dunn, John R. (ed); White, William Blaine, 1934–(ed). 1st ed. [Pittsburgh, PA, US]: National Speleological Society, Pittsburgh Grotto Press, Apr, **1958**. illus: b&w: cartoons, drawings, graphs, maps, photos, tables; Various pagination; bibl, index; 28 cm.

reprint ed. nd. illus: b&w: maps.

"... reprint of the 1957 Speleo Digest."

[2452] **National Speleological Society. Speleo Digest (1958)** *Speleo digest, 1958.* Dunn, John R. (ed); McCrady, Allen D. (ed). [Pittsburgh, PA, US]: National Speleological Society, Pittsburgh Grotto, May, **1959**. illus: b&w: maps, tables; approximately 14 maps folded in text; Various pagination; bibl, index; 28 cm.

[2453] **National Speleological Society. Speleo Digest (1959)** *Speleo digest, 1959.* Dunn, John R. (ed); McCrady, Allen D. (ed). [Pittsburgh, PA, US]: National Speleological Society, Pittsburgh Grotto, Apr, **1961**. illus: b&w: drawings, maps, tables; partly col: maps; approx 13 b&w and partly col maps folded in text; Various pagination; bibl, index.

[2454] **National Speleological Society. Speleo Digest (1960)** *Speleo digest, 1960.* Black, Herbert L. (ed); Haarr, Allan P. (ed); McGrew, Wesley C. (ed). [Pittsburgh, PA, US]: National Speleological Society, Pittsburgh Grotto, Apr, **1962**. illus: b&w: drawings, maps, photos, tables; approx 11 b&w maps folded in text; Various pagination; bibl, index.

[2455] **National Speleological Society. Speleo Digest**

(1961) *Speleo digest, 1961*. Black, Herbert L. (ed); Haarr, Allan P. (ed). [Pittsburgh, PA, US]: National Speleological Society, Pittsburgh Grotto, Apr, **1963**. illus: b&w: cartoons, maps, photos, tables; 9 b&w maps folded in text; Various pagination; bibl, index.

[2456] **National Speleological Society. Speleo Digest (1962)** *Speleo digest, 1962*. Haarr, Allan P. (ed); Plummer, William T. (ed). [Pittsburgh, PA, US]: National Speleological Society, Pittsburgh Grotto, Apr, **1964**. illus: b&w: cartoons, drawings, graphs, maps, tables; 16 maps folded in text; Various pagination; bibl, index.

[2457] **National Speleological Society. Speleo Digest (1963)** *Speleo digest, 1963*. Haarr, Allan P. (ed); McGrew, Wesley C. (ed). [Pittsburgh, PA, US]: National Speleological Society, Pittsburgh Grotto, Apr, **1965**. illus: b&w: cartoons, drawings, graphs, maps, photos, tables; 2 maps folded in text; Various pagination; bibl, index.

[2458] **National Speleological Society. Speleo Digest (1964)** *Speleo digest, 1964*. McGrew, Wesley C. (ed); Haarr, Allan P. (ed). Vienna, VA, US: National Speleological Society, The SpeleoDigest, Apr, **1966**. illus: b&w: cartoons, drawings, graphs, maps, photos, tables; 1 map folded in text; Various pagination; bibl, index.

[2459] **National Speleological Society. Speleo Digest (1965)** *Speleo digest, 1965*. Black, Herbert L. (ed); Haarr, Allan P. (ed). Vienna, VA, US: National Speleological Society, The SpeleoDigest, Aug, **1967**. illus: b&w: cartoons, drawings, graphs, maps, tables; Various pagination; bibl, index.

[2460] **National Speleological Society. Speleo Digest (1966)** *Speleo digest, 1966*. Plummer, William T. (ed). Vienna, VA, US: National Speleological Society, The SpeleoDigest, Apr, **1969**. illus: b&w: cartoons, drawings, graphs, maps, photos; Various pagination; bibl, index.

[2461] **National Speleological Society. Speleo Digest (1967)** *Speleo digest, 1967*. McCutchen, Gary D. (ed). Huntsville, AL, US: National Speleological Society, The SpeleoDigest, Jun, **1974**. illus: b&w: cartoons, drawings, graphs, maps, photos, tables; Various pagination; bibl, index.

[2462] **National Speleological Society. Speleo Digest (1968)** *Speleo digest, 1968*. Davis, John O. (ed); Smith, A. Richard (ed); Sherborne, William B. (ed); Power, Louise (ed). Huntsville, AL, US: National Speleological Society, Jul, **1974**. illus: b&w: cartoons, drawings, graphs, maps, tables; Various pagination; bibl, index.

[2463] **National Speleological Society. Speleo Digest (1969)** *Speleo digest, 1969*. Burns, Eleanor W. (ed). Huntsville, AL, US: National Speleological Society, **1976**. illus: b&w: cartoons, drawings, graphs, maps, tables; Various pagination; bibl, index.

[2464] **National Speleological Society. Speleo Digest (1970)** *Speleo digest, 1970*. Kramer, Jim (ed). Huntsville, AL, US: National Speleological Society, **1970**. illus: b&w: cartoons, drawings, graphs, maps, tables; ix, 338p; bibl, index.

[2465] **National Speleological Society. Speleo Digest (1971)** *Speleo digest, 1971*. Cooper, John Edward, 1929– (ed). Huntsville, AL, US: National Speleological Society, **1978**. illus: b&w: cartoons, drawings, graphs, maps, tables; xi, 415p; bibl, index.

[2466] **National Speleological Society. Speleo Digest (1972)** *Speleo digest, 1972*. Moody, Jill (ed). Huntsville, AL, US: National Speleological Society, **1980**. illus: b&w: cartoons, drawings, graphs, maps, tables; ix, 336p; bibl, index.

[2467] **National Speleological Society. Speleo Digest (1973)** *Speleo digest, 1973*. Baz-Dresch, John (ed); Mixon, William Weston, 1940–(ed). Huntsville, AL, US: National Speleological Society, **1980**. illus: b&w: cartoons, drawings, graphs, maps, photos, tables; xv, 380p; bibl, index.

[2468] **National Speleological Society. Speleo Digest (1974)** *Speleo digest, 1974*. Mixon, William Weston, 1940–(ed). Huntsville, AL, US: National Speleological Society, **1981**. illus: b&w: cartoons, drawings, graphs, maps, tables; ix, 305p; bibl, index.

[2469] **National Speleological Society. Speleo Digest (1975)** *Speleo digest, 1975*. Fawley, J. Philip (ed); Long, Kenneth M. (ed). Huntsville, AL, US: National Speleological Society, **1982**. illus: b&w: cartoons, drawings, graphs, maps, tables; vii, 389p; bibl, index.

[2470] **National Speleological Society. Speleo Digest (1976)** *Speleo digest, 1976*. Ehr, Bob (ed). Huntsville, AL, US: National Speleological Society, **1983**. illus: b&w: cartoons, drawings, graphs, maps, photos, tables; xiii, 462p.

[2471] **National Speleological Society. Speleo Digest (1977)** *Speleo digest, 1977*. Mixon, William Weston, 1940–(ed). Huntsville, AL, US: National Speleological Society, **1983**. illus: b&w: cartoons, drawings, graphs, maps, tables; xiii, 430p; bibl, index.

[2472] **National Speleological Society. Speleo Digest (1978)** *Speleo digest, 1978*. The Mountain State Grotto (ed); The West Virginia Speleological Society (ed). Huntsville, AL, US: National Speleological Society, **1980**. illus: b&w: cartoons, drawings, graphs, maps, photos, tables; 4 maps folded in text; vi, 301p, [3] ; bibl, index.

[2473] **National Speleological Society. Speleo Digest (1979)** *Speleo digest, 1979*. Garton, Emmel Ray, 1950– (ed); Garton, Mary Ellen (ed); Deem, Kelley (ed); Deem, Dixie (ed). Huntsville, AL, US: National Speleological Society, **1981**. illus: b&w: cartoons, drawings, graphs, maps, photos, tables; 6 maps folded in text; xiv, 334p; bibl, index.

[2474] **National Speleological Society. Speleo Digest (1980)** *Speleo digest 1980*. The SpeleoDigest Committee (ed). Huntsville, AL, US: National Speleological Society,

1985. illus: b&w: cartoons, drawings, graphs, maps, photos; 6 maps folded in text; [xi], 253p; bibl, index.

[2475] **National Speleological Society. Speleo Digest (1981)** *Speleo digest, 1981.* Balfour, William (ed). Huntsville, AL, US: National Speleological Society, **1984**. illus: b&w: cartoons, drawings, graphs, maps, photos, tables; 1 map folded in text; [xi], 302p; bibl, index.

[2476] **National Speleological Society. Speleo Digest (1982)** *Speleo digest, 1982.* Deem, Kelley (ed); Deem, Dixie (ed); Garton, Emmel Ray, 1950–(ed); Garton, Mary Ellen (ed); Sowers, John (ed). Huntsville, AL, US: National Speleological Society, **1986**. illus: b&w: cartoons, drawings, maps, photos; 5 maps folded in text; [xiv], 382p; bibl, index.

[2477] **National Speleological Society. Speleo Digest (1983)** *Speleo digest, 1983.* Stitt, Robert R. (ed). Huntsville, AL, US: National Speleological Society, **1991**. illus: b&w: cartoons, drawings, graphs, maps, tables; 2 maps folded in text; x, 458p; bibl, index.

[2478] **National Speleological Society. Speleo Digest (1985)** *Speleo digest, 1985.* Kambesis, Pat (ed). Huntsville, AL, US: National Speleological Society, **1987**. illus: b&w: cartoons, drawings, graphs, maps, photos, tables; v, 374p; bibl, index.

[2479] **National Speleological Society. Speleo Digest (1986)** *Speleo digest, 1986.* Bozeman, Sue (ed). Huntsville, AL, US: National Speleological Society, **1990**. illus: b&w: cartoons, drawings, graphs, maps, photos, tables; viii, 426p; bibl, index.

[2480] **National Speleological Society. Speleo Digest (1987)** *Speleo digest, 1987.* Fee, Scott (ed). Huntsville, AL, US: National Speleological Society, **1991**. illus: b&w: cartoons, drawings, graphs, maps, photos, tables; v, 468p; bibl, index.

[2481] **National Speleological Society. Speleo Digest (1989)** *Speleo digest, 1989.* Fee, Scott (ed). Huntsville, AL, US: National Speleological Society, **1990**. illus: b&w: cartoons, drawings, graphs, maps, photos, tables; [v], 462p; bibl, index.

[2482] **National Speleological Society. Speleo Digest (1993)** *Speleo digest, 1993.* Fee, Scott (ed). Huntsville, AL, US: National Speleological Society, **1994**. illus: b&w: cartoons, drawings, graphs, maps, photos, tables; vi, 550p; bibl, index.

[2483] **National Water Well Association** *Proceedings of environmental problems in karst terranes and their solutions conference, October 28-30, 1986, Bowling Green, Kentucky.* Dublin, OH, US: National Water Well Association, **1987**. illus: b&w: charts, drawings, maps, photos, tables; ii, 525p; bibl; 29 cm.

Sponsors: National Water Well Association, Eastern Kentucky University, Friends of the Karst, Kentucky Division of Water, Western Kentucky University, National Parks Service.

Worthwhile proceedings from one of the important applied karst conferences. This volume contains a variety of papers - theoretical, case studies, legal aspects, modeling, mitigation, etc. Although unedited, this is still a valuable contribution to the available literature.

[2484] **National Water Well Association** *Proceedings of the second conference on environmental problems in karst terranes and their solutions conference, November 16-18, 1988, Maxwell House Hotel, Nashville, Tennessee.* Dublin, OH, US: National Water Well Association, **1988**. illus: b&w: charts, drawings, graphs, maps, tables; 441p; bibl, index; 29 cm.

Presented by the Association of Ground Water Scientists and Engineers, a division of the National Water Well Association.

Some interesting material, but so hard to find.

[2485] **Natural Resources Seminar (3rd: 1976: Santa Fe)** *Radon radiation situation in NPS caves: January 19, 1976.* United States National Park Service Southwest Region. Santa Fe, NM, US: National Park Service, Southwest Region, **1976**. vii, 16p, 22p pls; 27 cm.

Illus.

Considering the variety of origins and settings of the caves in the National Park System, it seems reasonable to conclude that the data presented probably reflects values found in similar caves elsewhere.

[2486] **Nazarieff, de Serge**, 1935– *Clair de Roche.* Strinati, Pierre (photo); Letu, Bernard (ed). Geneve, CH: B. Letu, **1981**. illus: col: photos; 39p; 27 cm.

27 col photos of nudes in caves and on mountains.

[2487] **Neighbor, Frank** *Major limestone caverns of the Black Hills, South Dakota.* [UT, US]: Salt Lake Grotto, National Speleological Society, Nov, **1954**. Technical Note National Speleological Society Salt Lake Grotto no 26. 15p, 7p pls; bibl; 28 cm.

Title from cover. Illus.

[2488] **Nelson, Douglas R.** *Introductory cave surveying.* [Minneapolis, MN, US]: [Minnesota Speleological Survey], **1974**. Minnesota Speleology Monthly Supplement. 24p, 3p pls; 28 cm.

Illus.

[2489] **Nelson, Nels Christian**, 1875– *Contributions to the archaeology of Mammoth Cave and vicinity, Kentucky.* New York, NY, US: American Museum of Natural History, **1917**. Anthropological Papers of the American Museum of Natural History. 73p; bibl; 25 cm.

Microfilm. Atlanta, Ga.: SOLINET, 1992. 1 microfilm reel ; 35 mm. Illus.

Earliest professional archeological study of the Mammoth Cave area. Work served as a hallmark study for future research at Mammoth Cave and surrounding area.

[2490] **Nelson, Raymond L.** *Mammoth Cave tour leader manual.* npop: np, Nov, **1959**. 95p.

[2491] **Netherton, Shaaron** *Cave management plan for*

the Ely District. Wood, Edward E.; Colcord, Diane. 1985 ed. Ely, NV, US: U.S. Bureau of Land Management, Ely District, Aug, **1985**. illus: b&w: maps; 39p; 28 cm.

Illus. Incl forms.

1986 ed. Mar, **1986**. 41p.

[2492] **New Mexico Geological Society** *Guidebook of southeastern New Mexico: fifth field conference, October 21-22-23 & 24, 1954*. [Socorro, NM, US]: The Society, **1954**. Field conference, New Mexico Geological Society; 5th. illus: b&w: maps; 209p; 28 cm.

Microfiche: Socorro, N.M. New Mexico Geological Society, 1982. 4 microfiche; 11 x 15 cm. Illus.

[2493] **New Mexico Resource Management and Development Division** *Bandera Crater/Ice Cave: a state park feasibility study*. Santa Fe, NM, US: New Mexico Natural Resources Dept, Resource Management and Development Division, **1986**. illus: b&w: maps; 46p; 28 cm.

Illus.

[2494] **New South Wales Department of Tourism** *Jenolan Caves, New South Wales*. 1973 ed. Sydney, AU: Department of Tourism, **1973**. [11]; 21 cm.

Cover title. Chiefly col illus.

1977 ed.

[2495] **New South Wales Immigration and Tourist Bureau** *The systems of limestone caves: Jenolan, Yarrangobilly, Wombeyan, Abercrombie and Wellington*. Sydney, NSW, AU: New South Wales Immigration and Tourist Bureau, **1915**. illus: b&w: photos; 23p, [1]; 21 x 13 cm.

Cover title: The limestone caves of New South Wales, Australia. Another copy examined listed 1917 as the date of publication.

[2496] **New South Wales National Parks and Wildlife Service** *Cooleman Plain karst area management plan: a supplementary plan to the Kosciusko National Park plan of management*. Sydney, NSW, AU: New South Wales National Parks and Wildlife Service, Jul, **1987**. illus: b&w: maps, tables; vi, 110p; app, bibl.

[2497] **New South Wales Teaching Resources Centre** *A visit to the Jenolan Caves: reference notes for the teacher*. [Sydney, NSW, AU]: [Govt. Pr.], **1971**. illus: b&w: drawings, maps; 13p; 21 cm.

Title from cover. To be used in conjunction with film strip.

[2498] **New South Wales. Department of Tourism** *Limestone caves of New South Wales*. Sydney, AU: New South Wales Department of Tourism, **1982**. [18]; 21 x 9 cm.

Title from cover. Col illus.

[2499] **New Zealand Speleological Society** *The little red cavers book*. Worthy, Trevor H. (ed); Worthy, Cathy (ed); Pugsley, Chris (ed). 2nd ed. Waitomo Caves, NZ: New Zealand Speleological Society (Inc), Feb, **1989**. New Zealand Speleological Society Occasional Publication no 4. text on lining pages; [i], 21p; app.

Title from cover. Limited edition of 500 copies. No illus.

[2500] **New Zealand Speleological Society** *The little red cavers book: NZ Speleological Society information manual*. Waitomo Caves, NZ: New Zealand Speleological Society, **1985**. text on lining pages; [i], 21p; app.

Title from cover. No illus.

[2501] **Newbould, Ronald L.** *The discovery of Barralong Cave, Jenolan Caves, N.S.W.: based on the diary of Ronald L. Newbould, guide, Jenolan Caves, co-discoverer with John P. Culley, senior guide, Jenolan Ceves [sic] of the "Barralong Cave" on the 7th June, 1964*. Culley, John P. (co-auth, photo); Ralston, B. (photo); Rawlinson, N. (photo). Jenolan Caves, NSW, AU: Jenolan Caves Historical and Preservation Society, Mar, **1975**. Jenolan Caves Historical and Preservation Society Occasional Paper no 3. illus: b&w: maps, photos; [16]; 28 cm.

Read before the society on the 11th May, 1974.

[2502] **Newbould, Ronald L.** *Steam cleaning of Orient Cave*. 2nd ed. Jenolan Caves, NSW, AU: Jenolan Caves Historical and Preservation Society, **1976**. Occasional Paper Jenolan Caves Historical and Preservation Society no 1. 15p; bibl; 27 cm.

Title from cover. "Based on a paper read before the Society on the 10th February, 1973"—Verso t.p. Illus.

[2503] **Newbould, Ronald L.** *Steam cleaning of Orient Cave, Jenolan Caves, N.S.W.* Ellis, Ross Andrew, 1940– (ed); Williamson, L. (typist); Wawman, D. (photo); Bramston, D. (photo). 1974 ed. Jenolan Caves, NSW, AU: Jenolan Caves Historical and Preservation Society, May, **1974**. Jenolan Caves Historical and Preservation Society Occasional Paper no 1. illus: b&w: photos; iv, 17p; bibl; 30 cm.

Limited edition of 200 copies. Based on a paper read before the Society on the 10th February, 1973.

1978 ed. Ellis, R. (ed); Williamson, L. (typist); Wawman, D. (photo); Bramston, D. (photo). npop: np, Jan, **1978**. illus: b&w: photos; 24p; bibl.

Limited edition of 500 copies. Info about other printings/editions: 2nd printing (of 1st edition) (500 copies) Nov 1974.

[2504] **Newell, Amy L.** *The caves of Put-in-Bay*. Put-in-Bay, OH, US: Lake Erie Originals, **1995**. xi, 109p; app; 21 cm.

Illus.

[2505] **Newton, John G.**, 1929– *Development of sinkholes resulting from man's activities in the eastern United States*. [Reston, VA, US]: U.S. Geological Survey, **1987**. U.S. Geological Survey Circular no 968. illus: b&w: drawings, maps, photos, tables; iv, 54p; bibl; 26 cm.

Regardless of the title, this focuses on the central- and southeastern US. It is still a useful summary of recent sinkhole development. Some "state-line" geological phenomena make one question the rigor with which the various data were integrated.

[2506] Newton, John G., 1929– *Sinkhole problem along proposed route of interstate highway 459 near Greenwood, Alabama.* Copeland, Charles W.; Scarbrough, L. W. University, AL, US: Geological Survey of Alabama, Division of Paleontology and Stratigraphy, **1973**. Geological Survey of Alabama Circular no 83. illus: b&w: ads; 2 folded illus in pocket; 63p; bibl; 23 cm.

Illus.

[2507] Newton, John G., 1929– *Sinkhole problem in and near Roberts Industrial subdivision, Birmingham, Alabama: a reconnaissance.* Hyde, L. W. University, AL, US: US Geological Survey, Geological Survey of Alabama, **1971**. Geological Survey of Alabama Circular 68. illus: b&w: graphs, maps, photos, tables; partly col: graphs, maps; [v], xlii; bibl.

[2508] Nguyen Tai Duong (comp) *Vietnam cavern tourism.* Nguyan Quang My; Vu Van Phai. npop: Trinh Minh Son, **[1994]**. illus: partly col: maps; col: photos; 44p; 27 x 19 cm.

English translation of text in back. No illus.

[2509] Nicholson, Frank Ernest *The exploration of Carlsbad Cavern.* 1st and 2nd printings. Wichita Falls, TX, US: Railey Printing Company, Inc, Oct, **1930**. illus: b&w: photos, ports; partly col: photos; part col photo on back lining page; front: part col; 48p, [1]; 19 cm.

Nicholson, leader of the New York Times - Carlsbad Expedition.

An account of Nicholson's involvement; to be treated as suspect in some instances.

[2510] Nicol, Allen Hankins, 1907– *Application of geophysical methods to location of cavities in residual soil and caverns in coralline limestone.* United States Far East Command; United States Geological Survey Military Geology Branch. npop: Prepared by Military Geology Branch, U.S. Geological Survey for Intelligence Division, Office of the Engineer, General Headquarters, Far East Command, **1951**. some plates folded in pocket; 25p, [25]; 27 cm.

Illus.

[2511] Nicoll, Robert S. *Guidebook to the caves of southeastern New South Wales and eastern Victoria and caves around Canberra.* Brush, John B.; Jennings, Joseph Newell, 1916-1984. npop: np, Dec, **1976**. Australian Speleological Federation Guidebook 1. illus: b&w: maps; [iii], 85+23p; bibl.

Prepared for Cavconact-76. The eleventh biennial convention of the Australian Speleological Federation held at the Australian National University, Canberra. Two vols together; 54 figs.

[2512] Nieland, James Raymond, 1949– *Little known caves of Oregon.* Canby, OR, US: Jim Nieland, **1969**. illus: b&w: maps; [ii], 19p, 10p pls.

This mimeo lists 53 caves in 12 counties and is the first cave list from Oregon.

[2513] Noah, Samuel *A description of Weast's Cave.* Weast, Edward. npop: np, **1835**. 16p; 21 cm.

Caption title. The cave was discovered in 1835 by Edward Weast.

[2514] Noble, Bruce J. *Cultural resource management in Mammoth Cave National Park: a National Park Service— Kentucky Heritage Council cooperative project.* Kentucky Heritage Council. Washington, DC, US: U.S. National Park Service, Interagency Resources Division, Preservation Planning Branch, **1991**. illus: b&w: maps, photos; partly col front; i, 119p; 28 cm.

[2515] Nodine-Zeller, Doris E. *Karst-derived early Pennsylvanian conglomerate in Ness County, Kansas: subsurface Mississippian-Pennsylvanian boundary delineated in well core.* Lawrence, KS, US: Kansas Geological Survey, Aug, **1991**. Kansas Geological Survey Bulletin 222. illus: b&w: charts, maps, photos, tables; 1 map in rear pocket ; text on front lining page; [vi], 30p; bibl.

[2516] North Country Region *North Country Region MECCA Summer 1979.* [Chicago, IL, US]: [Windy City Grotto], **1979**. illus: b&w: ads, maps, photos, repros, tables; text on front lining page; 22p; errat.

Title from cover.

[2517] Northcott, T. C. *How nature made Luray Caverns, Virginia.* Luray, VA, US: Luray Caverns Corp, **c1995**. illus: b&w: photos; col: photos; text on rear lining page; [16]; 23 x 15 cm.

Cover title: How nature made Luray Caverns Virginia: illustrated in full color. Material for this publication was gathered from "How Nature Makes a cave".

[2518] Northcott, T. C. *How nature makes a cave.* Luray, VA, US: Educational Department, Luray Caverns Corporation, **1934**. illus: b&w: photos; 15p.

Several editions published, of which this is probably the earliest.

[2519] Northeast Regional Organization [NRO]. National Speleological Society *New Hampshire Caves.* preliminary ed. npop: [NRO], Jul, **1979**. NRO Publication 19. illus: b&w: drawings; [29].

[2520] Northeastern Regional Organization [NRO]. National Speleological Society (comp) *Gage Caverns.* [US]: [Northeastern Regional Organization], **nd**. illus: b&w: maps; two large fold-outs; 8p, 2p pls.

Compiled by NSS members and NRO.

[2521] Northup, Diana Eleanor, 1948– *Lechuguilla Cave: biological inventory.* Carr, Deborah L.; Crocker, Tad M.; Hawkins, Lauraine K.; Leonard, Patricia; Welbourn, W. Calvin. Albuquerque, NM, US: Diana Eleanor Northup, **1992**. illus: b&w: drawings, graphs, maps, photomicrographs, tables; col: photos; viii, 161p; app, bibl.

15 pp col and b&w pls.

[2522] Northwest Cave Research Institute *A management plan for Bighorn Caverns, Montana.* Bighorn Research Project. npop: np, Jan, **1987**. illus: b&w: graphs, tables; one table folded loose in rear; ix, 197p, 13p pls; app.

14 figs, 14 tables.

[2523] Northwest Caving Association. National Speleological Society (1987) *Northwest Caving Association regional meeting guidebook: caves of the Peterson Prairie area.* Crawford, Rodney Lee, 1951–. npop: Cascade Grotto, National Speleological Society, **1987**. illus: b&w: maps; [i], 40p.

2nd edition, 1991, 73p.

[2524] Northwest Caving Association. National Speleological Society (1992) *1992 Northwest Caving Association regional meeting cave guide [22-24 may 1992].* Wendell, ID, US: Gem State Research, **1992**. illus: b&w: maps; [51].

Several pages blank.

[2525] Northwest Regional Association Symposium on Cave Science and Technology *Northwest Regional Association symposium on cave science and technology: 1980: Feb 16-18, 1980, Seattle, WA, 1981: Feb 14-16, 1981, Seattle, WA.1982: Feb 13-15, 1982, Boise, Idaho.* National Speleological Society. npop: np, **1980—1982**. 20+12+10p.

Abstracts of papers.

[2526] Norton, John S. *The Cave of the Winds.* Carey, Grant. Manitou Springs, CO, US: [Cave of the Winds], **1990**. illus: b&w: maps, repros; col: drawings, photos; text on lining pages; 16p; gloss; 21.5 x 14 cm.

Cover title (?): Cave of the Winds: God's underground gift to Colorado: the history and geology of Cave of the Winds.

[2527] Nota, Dirk Johannes Gregorius *A hydrogeological study in the basin of the Gulp Creek: a reconnaissance in a small catchment area: 1, Groundwater flow characteristics.* Weerd, B. van de. Wageningen, NL: H. Veenman & Zonen B.V., **1978**. Mededelingen Landbouwhogeschool Geologische Instituut Mededelingen 78-20. bibl.

Summary in French. No format in OCLC.

[2528] Nott, David *Into the lost world - a descent into prehistoric time.* Englewood Cliffs, NJ, US: Prentice-Hall, Inc, **1975**. illus: b&w: photos; 186p; app.

An exciting adventure in the jungles of Venezuela. How these guys got out alive is more than I can tell. Description of the descent of The Pit on the Sarisarinama Plateau, Venezuela.

[2529] Nougier, Louis René, 1912– *The Cave of Rouffignac.* Romain, Robert; Scott, David (trans). London, GB: George Newnes Limited, **1958**. illus: b&w: photos, repros, tables; [x], 230p, 24p pls; app.

39 figs. Translation, from the French: Rouffignac, ou la guerre des mammouths.

As well as detailing the prehistoric art at Rouffignac, France, there is an extensive, well-researched section dealing with the cave's history and the furor in accepting the authenticity of the site.

[2530] Nougier, Louis-René, 1912– *Art treasures of prehistoric man in the caves of France and Spain.* Robert, Romain (photo). npop: Cultural History Research [distributor], **1961**.

Presents a collection of 25 important, decorated caves of France and Spain. 335 slides: col. + 1 list + 335 guide cards, (2 portfolios). OCLC also lists 1962 as date of pub.

[2531] The NSW Cave Rescue Squad Inc *The NSW Cave Rescue Squad Inc. Annual Report July 1988-June 1989.* Bankstown, NSW, AU: The NSW Cave Rescue Squad Inc, **nd**. illus: b&w: photos, ports, tables; b&w photos (photocopied) on lining pages; 51p; app.

Cover title: Cave rescue annual report July 1988-June 1989.

[2532] The NSW Cave Rescue Squad Inc. *The NSW Cave Rescue Squad Inc.* Bankstown, NSW, AU: The NSW Cave Rescue Squad Inc, **nd**. illus: b&w: charts, photos, tables; [ii], 5p, [11].

[2533] Nucolorvue Productions *Mt. Gambier and district: Port MacDonnell and Tantanoola caves.* [Mulgrave, VIC, AU]: Nucolorvue Productions, **1974**. map on lining page; [24]; 13 x 16 cm.

Cover from title. Chiefly col illus.

[2534] Nunez Jimenez, Antonio *Karstological investigations in Cuba.* Washington, DC, US: np, **1966**. United States Joint Publications Research Service Translations on Cuba no 518. 57p; 27 cm.

Translation of a book by Antonio Nunez Jimenez, Vladimir Panos & Otakar Stelcl, entitled Investigaciones Carsologicas en Cuba

[2535] Nusbaum, Jesse Logan *The 1926 re-excavation of Step House Cave, Mesa Verde National Park.* Mesa Verde National Park, CO, US: Mesa Verde Museum Association, **1981**. Mesa Verde Research Series Paper no 1 1981. illus: partly col: photos; xvi, 33p; 29 cm.

[2536] Nusbaum, Jesse Logan *A basket-maker cave in Kane County, Utah.* Kidder, Alfred Vincent, 1885-1968; Guernsey, Samuel James, 1868–1936. reprint? of 1922 ed. New York, NY, US: Museum of the American Indian, Heye Foundation, **1985**. Indian Notes and Monographs Miscellaneous. illus: b&w: maps; 153p; bibl; 17 cm.

Most of the plates printed on both sides. Photocopy. [Berkeley, Calif: University of California, Library. Photographic Service, 1985.] 24 cm. Illus.

[2537] Nymeyer, Robert, 1910-1983 *Carlsbad Cavern, the early years: a photographic history of the cave and its people.* Halliday, William Ross, 1926–. Carlsbad, NM, US: Carlsbad Caverns, Guadalupe Mountains Association, **1991**. illus: b&w: maps, photos; 155p; 26 cm.

A well-illustrated account of the exploration of Carlsbad Caverns, New Mexico, and the inauguration of the Na-

[2538] **Nymeyer, Robert**, 1910-1983 *Carlsbad, caves, and a camera*. Teaneck, NJ, US: Zephyrus Press, **1978**. Speleologia. illus: b&w: photos; [vi], 318p; index; 26 cm.

The exploration of the Carlsbad region of the USA during the 1930s. Contains numerous, well annotated early photographs. The personal account of exploration of the caves of the Guadalupe Mountains during the 1930s by the photographer, Robert Nymeyer. Contains a photographic essay of 185 large format photos of Guadalupe caves, which are poignant in light of the extensive vandalism in the caves discussed since his time.

[2539] **Oakley, Carey B.** *Archaeological testing at site 1Ms357, Cathedral Caverns, Marshall County, Alabama*. Atchison, Robert B. Moundville, AL, US: Alabama Archaeological Society, **1990**. Journal of Alabama Archaeology vol 36 no 1. illus: b&w: charts, maps, photos, tables; Text on front; v, 62p; bibl; 25 cm.

[2540] **Oberhansley, Frank R.** *Crystal Cave in Sequoia National Park*. 1946. Visalia, CA, US: Sequoia Natural History Association, May. illus: b&w: maps, photos; [32]; bibl.

One of the series of publications on the human and natural history of Sequoia-Kings National Park, California, v.1, no.1, May 1946. History and geology for the commercial tour.

   1961. Three Rivers, CA, US. 26p; 23 cm.
   1978. Sequoia Natural History Association, c**1978**. illus: b&w: maps, photos; Text on front; map on rear; [42]; bibl; 28 cm.

Copy examined says 14th printing since original publication in 1946. Revisions have been made.

   1982. Chapa, Lynette N. (photo). Sequoia Natural History Association, **1982**. illus: b&w: maps, photos; col: photos; Color photo front; text rear; [28]; bibl; 28 cm.

Title from cover. Published in cooperation with National Park Service.

   1987. [28]; bibl; 23 cm.
   Col illus.

[2541] **Obermaier, Hugo**, 1877-1946 *Bushman art; rock paintings of south-west Africa, based on the photographic material collected by Reinhard Maack*. Kuhn, Herbert, 1895–; Maack, Reinhard. London, GB: Oxford University Press, **1930**. illus: partly col: photos; [i], xi, 69p, [1]; bibl; 35 cm.

Col illus.

[2542] **Obermaier, Hugo**, 1877-1946 *The caves of Altamira*. Madrid, ES: Blass, **1928**. Publications of the Patronato Nacional de Turismo no 1. illus: b&w: maps; partly col: maps; maps on lining papers; 24p; 22 cm.

Illus.

[2543] **Ockenden, Allen C.** *Caves of the Mellte Valley*. 2nd ed. Cymych, Dyfed, GB: Anne Oldham, **1991**. Limestones and Caves of South Wales 7. illus: b&w: maps; [vi], 48p; bibl; 22 cm.

1st ed appeared serially in the Croydon Caving Club Journal Pelobates', nos. 51-56 (1988-91).

[2544] **O'Connor, Donal** *Crag Cave: Castle Island: Co. Kerry: Ireland's most exciting show cave*. IE: [Crag Cave Ltd], **1991**. illus: b&w: drawings, photos; col: photos; location map on rear lining page; 16p; 21 x 15 cm.

[2545] **O'Donnell, Lisa** *Recovery plan for endangered karst invertebrates in Travis and Williamson Counties, Texas*. Elliott, William Rawleigh, 1946–; Stanford, Ruth A.; Shull, Alisa (ed). Albuquerque, NM, US: U.S. Fish and Wildlife Service, Aug, **1994**. illus: b&w: maps, tables; vi, 154p; app, bibl, gloss.

Prepared for U.S. Fish and Wildlife Service Region 2, Albuquerque, NM. Cover title: Endangered karst invertebrates (Travis and Williamson Counties, Texas) Recovery Plan.

[2546] **Ohio Caverns, Inc.** *Ohio Caverns: where nature carved a fairyland*. Chicago, IL, US: Curt Teich & Co, **1971**. [16].

Col illus.

[2547] **Oklahoma Department of Wildlife Conservation** *Hydrogeology of Ozark cavefish caves*. [1st ed]. [Oklahoma City, OK, US]: Oklahoma Dept of Wildlife Conservation, **1990**. illus: b&w: maps; 73p; 28 cm.

Federal aid project E-6-1. Contract date: June 1, 1989 - May 31, 1990. Microfiche. Oklahoma City: Oklahoma Dept. of Libraries, Micrographics Unit, 1990. 1 microfiche ; 11 x 15 cm.

   [2nd ed]. **1991**. 49p, 23p pls.

Federal aid project E-6. Period covered: June 1, 1989 - June 30, 1991. Microfiche. Oklahoma City: Oklahoma Dept. of Libraries, Micrographics Unit, 1991. 1 microfiche ; 11 x 15 cm. Illus.

[2548] **Oklahoma Department of Wildlife Conservation** *Survey and species determination of cave crayfish in Oklahoma*. [Oklahoma City, OK, US]: Oklahoma Dept of Wildlife Conservation, **1990**. illus: b&w: maps; 17+11p, [38]; bibl; 28 cm.

Cover title: "Federal aid project E-5, June 1, 1989-May 31, 1990." Appendix II: A biochemical genetic analysis of troglobitic crayfish (*Cambarus* spp) in Missouri, Oklahoma and Arkansas / Jeffrey B. Koppelman.Microfiche. Oklahoma City: Oklahoma Dept. of Libraries, Micrographics Unit, 1990. 1 microfiche ; 11 x 15 cm.

[2549] **Oldham, Anthony Clive [Tony]**, 1939– *The caves of Carmarthen*. Eastville, Bristol, GB: Anne Oldham, **1975**. Limestones and Caves of South Wales no 3. illus: b&w: maps; 1 map folded in text; xii, 49p; bibl, index; 26 cm.

Over 60 sites of speleological interest. Welsh introduction.

[2550] **Oldham, Anthony Clive [Tony]**, 1939– *The*

caves of Clydach. Crymych, Wales, GB: Anne Oldham, **1981**. Limestones and Caves of South Wales no 4. illus: b&w: ads, maps; vii, 22p; bibl, index; 26 cm.

Information on the caves of Clydach Gorge and Blaenavon areas.

[2551] **Oldham, Anthony Clive [Tony]**, 1939– *The caves of Co. Cork*. Crymych, Wales, GB: Anne Oldham, **1981**. Caves of Ireland pt 1. illus: b&w: maps, photos, ports; xvi, 66p, 2p pls; bibl, index; 26 cm.

Includes a chapter in Irish.

[2552] **Oldham, Anthony Clive [Tony]**, 1939– *The caves of Gower*. 1st ed. Crymych, Wales, GB: Anne D. Oldham, **1978**. Limestones and Caves of South Wales no 2. illus: b&w: maps; xii, 68p, 1p pls; bibl, index; 26 cm.

Introduction also in Welsh. History of exploration and descriptions of over 100 caves. 1 map loose. OCLC also lists 1977 as date of pub.

2nd ed. **1982**. Limestones and Caves of South Wales no 2. illus: b&w: maps, ports, repros; xvi, 101p, 2p pls; bibl.

[2553] **Oldham, Anthony Clive [Tony]**, 1939– *The caves of north Wales*. 1st ed. Crymych, Wales, GB: Anne Oldham, **1977**. illus: b&w: ads, maps; x, 63p, [3]; 26 cm.

Introduction in English and Welsh. Over 120 caves described.

[2nd ed]. **1981**. illus: b&w: ads, maps; xii, 69p, [1]; bibl, index; 26 cm.

[2554] **Oldham, Anthony Clive [Tony]**, 1939– *The caves of Scotland (except Assynt)*. Bristol, GB: Tony Oldham, **1975**. illus: b&w: drawings, maps, photos; 1 map folded in text; b&w front; xvi, 174p, 19p pls; bibl, gloss, index; 25 cm.

Multiple editions; minor variations common. Cover title: Caves of Scotland. Some maps printed on col paper.

Catalog of over 400 caves from the literature. Locations, brief descriptions, extensive references. A few maps.

[2555] **Oldham, Anthony Clive [Tony]**, 1939– *Caves of the central northern outcrop*. 1st ed. Crymych, Wales, GB: Anne Oldham, **1986**. Limestones and Caves of South Wales no 6. illus: b&w: ads, maps, photos; viii, 58p, [2]; bibl, index; 22 cm.

2nd ed rev with addenda. Anne Oldham, **1990**. add.

[2556] **Oldham, Anthony Clive [Tony]**, 1939– *The caves of the Little Neath Valley*. Crymych, Wales, GB: Anne Oldham, **1993**. Limestones and Caves of South Wales no 10. illus: b&w: maps; one folded leaf; vi, 30p, 1p pls; bibl, index; 30 cm.

[2557] **Oldham, Anthony Clive [Tony]**, 1939– *The caves of the south eastern outcrop*. Jones, Keith. Crymych, Wales, GB: Anne Oldham, **1991**. Limestones and Caves of South Wales no 8. illus: b&w: maps; vi, 40p; bibl; 22 cm.

Illus.

[2558] **Oldham, Anthony Clive [Tony]**, 1939– *The caves of the south eastern outcrop and caves and mines in the Forest of Dean*. Jones, Keith. 2nd ed. Crymych, Wales, GB: Anne Oldham, **1992**. Limestones and Caves of South Wales no 8. illus: b&w: maps; xii, 74p; index; 30 cm.

Originally published: 1991. Previously entitled: The caves of the South Eastern outcrop.

[2559] **Oldham, Anthony Clive [Tony]**, 1939– *The caves of the southern outcrop*. 1st ed. Crymych, Wales, GB: Anne Oldham, Jan, **1985**. Limestones and Caves of South Wales no 5. illus: b&w: maps; vi, 36p, [1]; index; 26 cm.

2nd rev ed. **1992**. illus: b&w: drawings, maps; vi, 90p; bibl, index; 22 cm.

[2560] **Oldham, Anthony Clive [Tony]**, 1939– *The caves of west Wales*. Crymych, Wales, GB: Anne Oldham, **1979**. illus: b&w: ads, maps, photos, repros; xii, 75p, [1]; bibl, index; 29 cm.

Describes over 150 caves, with locations.

[2561] **Oldham, Anthony Clive [Tony]**, 1939– *The complete caves of Devon*. npop: T. Oldham, **1989**. [12].

[2562] **Oldham, Anthony Clive [Tony]**, 1939– *The concise caves of north Wales*. 1990 rev ed. Crymych, Wales, GB: Anne Oldham, **1990**. illus: b&w: maps; x, 80p; bibl; 22 cm.

Originally published 1989. Illus.

1991 rev ed. **1991**.

1st ed. Crymych, GB: Anne Oldham, **1989**. illus: b&w: maps, repros; x, 77p; bibl, index.

[2563] **Oldham, Anthony Clive [Tony]**, 1939– *Discovering caves*. Oldham, Joyce Elizabeth Anne. Aylesbury, Bucks, GB: Shire Publications Ltd, **1972**. illus: b&w: drawings, maps, photos; 51p, [1], 8p pls; bibl, index; 18 cm.

Cover title: Discovering caves: a guide to the show caves of Britain.

[2564] **Oldham, Joyce Elizabeth Anne** *The caves of Devon*. Smart, James. npop: np, Nov, **1972**. illus: b&w: maps, photos; front: b&w photo; [xiv], 134p, 10p pls; bibl.

Reference on over 270 caves, with information on each.

[2565] **Oldham, Joyce Elizabeth Anne** *The concise caves of Devon*. Smart, James. Crymych, Wales, GB: Anne Oldham, Apr, **1985**. illus: b&w: maps; ii, 68p, 3p pls.

First published November 1972. French and German summaries. 200+ caves in the county of Devonshire.

[2566] **Oldham, Joyce Elizabeth Anne** *The limestones and caves of Devon*. Smart, James. [Crymmych, Wales, GB]: [Anne Oldham], **1978**. illus: b&w: maps; xviii, 143 + 6p, 2p pls; bibl, index; 26 cm.

Originally published: 1972.

[2567] **Oliva, Deane** *A visitor's guide to Mammoth Cave National Park.* Genter, Karen. Bowling Green, KY, US: Cogenisys, **1991**. illus: b&w: maps; 72p; bibl, index; 22 cm.

Illus; 23 maps.

[2568] **Oliver, Douglas L.** *A boy scout in the grand cavern.* npop, **1930**. illus: b&w: ports; front; x, 148p.

[2569] **Ollerenshaw, Arthur Edward** *Blue John Caverns and Blue John Mine, Castleton, Sheffield.* [GB]: A.E. Ollerenshaw, **1964**. 23p; 21 cm.

[2570] **Ollerenshaw, Arthur Edward** *The history of Blue John Stone: methods of mining and working ancient and modern.* Harrison, R. J.; Harrison D. Castleton, GB: A.E. Ollerenshaw and R.J. & D Harrison, **c1963**. illus: b&w: maps, photos; col: photos; col photo on rear lining page; 23p, [1]; 22 cm.

This ornamental stone is mined from the natural Blue John Caverns in limestone.

[2571] **Ollier, Clifford David** *Speleochronology.* reprint. NSW, AU: Helictite, Oct, **1966**. Helictite vol 5 no 1. 10p; bibl; 26 cm.

Stamped on title page: Department of Geology, University of Melbourne, Victoria, Australia. Title from cover. Illus.

[2572] **Ong, Johnney** *Mysterious caves of Langkawi, Malaysia.* Petaling Jaya, Selangor, MY: Dept of Irrigation and Drainage, Ministry of Agriculture, Malaysia; and Design Dimension Sdn Bhd, **1994**. illus: col: photos, ports; [ii], 119p, [3]; 28 x 21 cm.

[2573] **Orchard, William C.** *Sandals and other fabrics from Kentucky caves.* New York, NY, US: Museum of the American Indian, Heye Foundation, **1920**. Indian Notes and Monographs; Miscellaneous no 4 Kentucky Culture Series. 20p; 17 cm.

Illus. Reprinted: Microfiche. Louisville, Ky.: Lost Cause Press, 1980. 2 microfiche: 10 x I5 cm. (Kentucky culture series).

One of the first paleoethnobotanical studies about the unique organic cultural artifacts recovered from the Mammoth Cave and eastern Kentucky sites.

[2574] **O'Reilly, P. M.** *Ogof Ffynnon Ddu: Penwyllt, Breconshire.* O'Reilly, S. E.; Fairbairn, C. M.; South Wales Caving Club Members. Penwyllt, Swansea, GB: South Wales Caving Club, **1969**. illus: b&w: drawings, graphs, maps, photos, tables; 1 folded map tipped in after page 32; front: photo; xii, 52p; app, bibl; 25.5 x 20 cm.

18 pls, 9 figs.

[2575] **Orghidan, Traian**, 1917-1985 *Proceedings of the first symposium on theoretical and applied karstology held in Bucharest, Romania, 22-24 April 1983.* npop: Institute of Biological SciencesBucharest "Emil Racovita" Speleological Institute, and Entpreprise (sic) for geological and geophysical prospectingBucharest, **1984**. Theoretical and applied karstology v 1. illus: b&w: drawings, graphs, maps, tables; seven maps folded in text; 254p; bibl; 24 cm.

About half of text in French.

[2576] **Orr, Phil C.** *Excavations in Moaning Cave.* Santa Barbara, CA, US: Museum of Natural History, **1952**. Bulletin - Dept. of Anthropology Santa Barbara Museum of Natural History no 1. illus: b&w: drawings, maps, photos; [i], 19p; bibl; 24 cm.

Title from cover.

[2577] **Orr, Phil C.** *Pleistocene man in Fishbone Cave, Pershing County, Nevada.* Nevada State Museum. Dept. of Archeology. Carson City, NV, US: Nevada State Museum, Dept. of Archeology, **1956**. Bulletin - Nevada State Museum Dept of Archeology no 2. illus: b&w: maps; 20p; bibl; 23 cm.

Title from cover.

[2578] **Orr, Phil C.** *Preliminary excavations of Pershing County caves.* Nevada State Museum. Dept. of Archeology. Carson City, NV, US: Nevada State Museum, Dept of Archeology, **1952**. Bulletin - Nevada State Museum Dept of Archeology no 1. 21p; 22 cm.

Title from cover. Illus.

[2579] **Orrock, Clive** *Imperial College Caving Club Peru '84 Expedition (19th July - 19th Sept 1984).* Lane, Steve. London, GB: Imperial College Caving Club, **c1985**. illus: b&w: charts, maps, photos; 3 fold-out maps in text; [iv], 54p, [1]; 30 x 21 cm.

[2580] **Osterreichische Fremden Verkehrswerbung (ed)** *Caves in Austria.* 1970. npop: Austrian Federal Government. illus: b&w: photos; partly col: maps; 30p; gloss; 18 x 12 cm.

1973. Vienna, AT: Austrian Tourist Board. illus: b&w: maps, photos; text on lining pages; pagination starts on front lining page; 30p.

No apparent differences between the1973 edition and the 1970.

[2581] **Osterreichische Fremden Verkehrswerbung (ed)** *Caves in Austria.* npop: Austrian Federal Government, **nd**. illus: b&w: photos; list of Austrian show caves on front lining page; 31p; gloss; 18 x 12 cm.

Cover title: Show Caves in Austria.

[2582] **Owen, Daniel (comp)** *Proyek Kelelawar: final report of the Oxford University expedition to the Togian Islands, Sulawesi, Indonesia summer 1987.* Bilton, David (comp); Lonsdale, Kate (comp); Strathdee, Stuart (comp). npop: The compilers, **1987**. illus: b&w: maps, photos, tables; one map folded in text; 47p, [3]; app, bibl.

[2583] **Owen, Luella Agnes** *Cave regions of the Ozarks.* Vineyard, Jerry Daniel, 1935–(new intro). 1968 reprint ed. MO, US: Missouri Speleological Survey, Apr, **1968**. Missouri Speleology vol 10 no 2. illus: b&w: charts, drawings, photos; 1 fold-out map; text on lining pages; [ii], 65p.

Partial reprint of the 1898 edition with some new illus and intro.

[2584] **Owen, Luella Agnes** *Cave regions of the Ozarks and Black Hills*. Vineyard, Jerry D. (intro). 1970 reprint ed. New York, NY, US: Johnson Reprint Corporation, **1970**. Classics in Speleology. [xvi], xlii, 228p; bibl.

1st ed. Cincinnati, OH, US: The Editor Publishing Co, **1898**. illus: b&w: maps, photos; 2 maps folded in text; b&w front; [xvi], 228p; bibl; 20 cm.

[2585] **Oxford University Cave Club** *Oxford University Cave Club: Huerta del Rey expedition final report*. [Oxford, GB, US]: Oxford University Cave Club, **1993**. illus: b&w: maps, tables; [i], 34p, [1]; 30 x 21 cm.

[2586] **Oxford University Cave Club** *Oxford University Cave Club: La Verdelluenga 1994 final report*. [Oxford, GB]: Oxford University Cave Club, **c1994**. illus: b&w: photos; loose maps (1 double page printed both sides); [32]; 30 x 21 cm.

[2587] **Oxford University Expedition to Northern Spain (1961)** *Oxford University expedition to northern Spain, 1961*. Ledbury, Herefordshire, GB: Cave Research Group of Great Britain, Nov, **1965**. Cave Research Group of Great Britain Publication no 14. illus: b&w: maps, photos, tables; [iv], 42p, 3p pls; bibl; 25 cm.

8 pls, 9 figs.

[2588] **Ozark National Forest (AR)** *Blanchard Springs Caverns: the amazing world below*. Finley-Holiday Film Corporation. Whittier, CA, US: Finley-Holiday Film Corp, **1980**. The Holiday Video Library National Park & Forest Series.

Cassette label and container carry subtitle first. "A Finley-Holiday video presentation"—Cassette label. "#V108"—Cassette label. Explores the ecological world of Blanchard Springs Caverns in the Ozark hill country of Arkansas. VHS. 1 videocassette (ca 20 min), sd, col, 1/2 in. OCLC also lists 1990 as date of pub.

[2589] **P. Tarrant Ltd** *Waitomo Caves New Zealand: a souvenir booklet of your Waitomo visit with story of life cycle of New Zealand glow-worm*. Waitomo, NZ: P. Tarrant Ltd, **nd**. illus: b&w: drawings; partly col: maps; col: maps, photos; text on front lining page, text and map on rear lining page; [16].

Contains isometric map dated 1973. Another version exists with [12pp] that contains many of the same photos but is missing isometric map.

[2590] **Packard, Alpheus Spring**, 1839-1905 *The cave fauna of North America, with remarks on the anatomy of the brain and origin of the blind species*. 1977 reprint ed. New York, NY, US: Arno Press, **1977**. History of Ecology. [v], 156p, [3], 27p pls.

1st ed. Washington, DC, US: National Academy of Sciences, **1888**. National Academy of Sciences (U.S.) Memoirs v 4 First Memoir pt 1 1888. illus: b&w: drawings, maps; partly col: drawings; 1 folded map in text; 156p, 27p pls; bibl; 24 cm.

The classic study of the North American cave fauna, written during the heyday of North American Neolamarckians and contains a species list from the few caves studies and a detailed discussion of Packard's views of evolution of cave organisms.

[2591] **Packard, Alpheus Spring**, 1839-1905 *The Mammoth Cave and its inhabitants, or, descriptions of the fishes, insects and crustaceans found in the cave with figures of the various species and an account of allied forms, comprising notes upon their structure, development and habits, with remarks upon subterranean life in general*. Putnam, Frederic Ward, 1839-1915. Salem, MA, US: Naturalists' Agency, **1872**. 62p; 24 cm.

Reprint Edition: 1958. Master microform held by: LCP. Microopaque. Louisville, Ky.: Lost Cause Press, 19—. Illus.

[2592] **Packard, Alpheus Spring**, 1839-1905 *On a new cave fauna in Utah: and on new phyllopod Crustacea from the West*. author's ed. Washington, WA, US: U.S. Geological and Geographical Survey, **1877**. 23p; 23 cm.

Title from cover. "Extracted from the Bulletin of the survey, v. 3, no. 1" pp157-179. Illus.

[2593] **Padgett, Allen** *On rope: North American vertical rope techniques for caving, search and rescue, mountaineering*. Smith, Bruce, 1948–; Williams, Pandra (illus). Huntsville, AL, US: National Speleological Society, Vertical Section, **1987**. illus: b&w: cartoons, charts, drawings, graphs, photos, ports; [ii], 341p; bibl, gloss, index; 29 cm.

2nd printing, 1988, containing minor corrections; 3rd printing, 1989.

The NSS's first foray into mass distribution of a publication. An excellent presentation of United States vertical techniques; weak in European methods. Brought a great deal of acclaim to the Society.

[2594] **Page, Lawrence M.** *Status of the cypres darter,* Etheostoma proeliare, *and comments on the spring cavefish,* Chologaster agassizi, *in Max Creek, Johnson County, Illinois*. Burr, Brooks M. Urbana, IL, US: Illinois Natural History Survey, **1977**. 25p; bibl; 28 cm.

Illus.

[2595] **Palacio, Joseph O.** *Excavation at Hokeb Ha, Belize*. Belize City, BZ: Belize Institute for Social Research and Action, **1977**. Occasional Publications—Belize Institute for Social Research and Action no 3. 48p; bibl; 21 cm.

Title from cover. Illus.

[2596] **Palmer, Arthur Nicholas**, 1940– *A geological guide to Mammoth Cave National Park*. Teaneck, NJ, US: Zephyrus Press, **1981**. Speleologia. illus: b&w: drawings, maps, photos; xiv, 196p; bibl, index; 23 cm.

103 figs.

This guide is written for those with a greater background and interest in speleology than most of the National Park guides. Other parks should encourage such works to supplement the laymans publications.

[2597] **Palmer, Arthur Nicholas**, 1940– *Geology and origin of Mystery Cave, Forestville State Park, Minnesota*. Palmer, Margaret V. npop: A.N. and M.V. Palmer, **1993**. illus: b&w: maps; Various pagination; 28 cm.

Illus. Title from cover. "LCMR Mystery Cave resources evaluation." Accompanied by "Extended profile of Mystery Cave" (20 sheets). Report performed for Department of Natural Resources. 3 vols: Interpretive report—Technical report—Management report.

[2598] **Palmer, Arthur Nicholas**, 1940– *The geology of Wind Cave*. Wind Cave National Park, SD, US: Wind Cave Natural History Association, **1981**. illus: b&w: drawings, maps, photos; [ii], 44p; bibl, index; 28 cm.

With his equally well presented geologic studies of Jewel and Mammoth Caves, Palmer has earned the Triple Crown of National Park Speleology!

[2599] **Palmer, Arthur Nicholas**, 1940– *A guide to the historic section of Mammoth Cave*. Palmer, Margaret V.; White, William Blaine, 1934–. Huntsville, AL, US: National Speleological Society, **1982**. illus: b&w: charts, drawings, graphs, maps, photos, tables; [ii], 59p; bibl; 22 cm.

In English, French, and German. Cover title: Guidebook to the historic section of Mammoth Cave. Prepared for the Eighth International Congress of Speleology, Bowling Green, Kentucky, U.S.A. July 18 to 24, 1981.

[2600] **Palmer, Arthur Nicholas**, 1940– *Jewel Cave: a gift from the past*. 1st ed. Hot Springs, SD, US: Wind Cave/Jewel Cave Natural History Association, **1984**. illus: b&w: charts, maps, photos; partly col: maps; col: photos; text on lining pages; 40p; bibl.

Well written and illustrated introduction to the fascinating geology and exploration of this great cave. This and his Wind Cave publication are examples of informative yet simply written national park books as they should be.

[2nd ed]. Black Hills Parks and Forests Association, [**1995**]. illus: b&w: charts, photos; partly col: charts, maps; col: photos; col photos on lining pages; 56p; bibl.

[2601] **Palmer, Arthur Nicholas**, 1940– *Knox Cave: Albany County, NY*. Northeastern Regional Organization, National Speleological Society (comp). Pittsfield, MA, US: NRO NSS, Jun, **1963**. illus: b&w: drawings, maps, photos; one map folded in text; figs on rear lining page; 22p; bibl.

[2602] **Palmer, Arthur Nicholas**, 1940– *Wind Cave: an ancient world beneath the hills*. 1995 ed. Black Hills Parks and Forests Association. illus: b&w: photos; col: drawings; b&w photos, text on lining pages; 64p.

1st ed. Hot Springs, SD, US: Wind Cave/Jewel Cave Natural History Association, **1988**. illus: b&w: charts, maps; partly col: charts, maps; col: photos; text on lining pages; 49p; bibl; 21 x 25 cm.

Title from caption.

[2603] **Palmer, Robert John**, 1951-1997 *The blue holes of the Bahamas*. London, GB: Jonathan Cape, **1985**. illus: b&w: maps; col: photos; 183p, [1], 40p pls; 25 cm.

82 pls, 7 maps.

[2604] **Palmer, Robert John**, 1951-1997 *Deep into blue holes: the story of the Andros Project*. London, GB: Unwin Hyman Limited, **1989**. illus: b&w: maps, photos, ports; col: ports; xii, 164p, 16p pls; app, index; 26 cm.

Col illus.

[2605] **Palmer, Robert John**, 1951-1997 *The report of the 1981 and 1982 British cave diving expeditions to Andros Island, Bahamas*. Other members of the expedition (comp). Bridgwater, Somerset, GB: British Cave Research Association, Mar, **1984**. Cave Science: Transactions of the British Cave Research Association vol 11 no 1. illus: b&w: charts, graphs, maps, photos, tables; text on front lining page; [ii], 54p; bibl.

[2606] **Palmer, Robert John**, 1951-1997 *Report of the 1987 international blueholes research project*. Bristol, GB: The Andros project, **1988**. illus: b&w: maps, photos; col: photos; center fold and covers; [ii], 28p, [2]; 30 x 21 cm.

Cover title: The Andros Project.

[2607] **Palmer, William O.** *Springs, caves and their related features*. [US]: np, **1970 1981**.

[United States s. n., between 1970 and 1981]. 81 slides: sd, col + 1 cassette (2-track: mono) and script (8 p).

[2608] **Palmer, William Thomas**, 1877– *The complete hill walker, rock climber and cave explorer*. London, GB: Sir I. Pitman & Sons, Ltd, **1934**. front; xi, 219p; 22 cm.

Illus. 50

[2609] **Panayiotakis, Yioryos I.** *The Dictaean Cave*. Provatakis, Theocharis M. English ed. Lassithi, GR: Y. Panayiotakis, **1988**. 118p; 21 cm.

Col illus.

[2610] **Panton, James Hoyes**, 1847-1898 *The Mammoth Cave of Kentucky*. Louisville, KY, US: C.G. Darnell, Commercial Photographer, **1890**. [67]; bibl; 14 cm.

Several pages blank. Reprint edition: 1969. Date of publication derived from Wilkes.

[2611] **Papadakis, Peggy** *The story of Crystal Ice Caves: within the Great Rift National Landmark, American Falls, Idaho*. Papadakis, Jim (photo). American Falls, ID, US: Power County Press, May, **1969**. illus: b&w: maps, photos; col: photos; [16].

[2612] **Parizek, R. R.** *Hydrogeology and geochemistry of folded and faulted carbonate rocks of the central Appalachian type and related land use problems*. White, William Blaine, 1934–; Langmuir, Donald. University Park, PA, US: Pennsylvania State University, Nov, **1971**. Pennsylvania State University College of Earth and Mineral Sciences Circular 82. illus: b&w: drawings, graphs, maps, photos, tables; col: maps, photos; 1 map folded in text; Various pagination, xxvi, 182+30p; app, bibl.

Prepared in cooperation with the Pennsylvania Geological Survey, Mineral Conservation Section, and Institute for Research on Land and Water Resources, The Pennsylvania State University for the Annual Meeting of the Geological Society of America and associated societies.

Definitive work on this much studied area. Includes sections on fracture trace analysis, well yield, and field trips. Difficult to obtain.

[2613] Parmalee, Paul Woodburn *Pleistocene and recent faunas from the Brynjulfson Caves, Missouri.* Oesch, Ronald D. Springfield, IL, US: Illinois State Museum, **1972**. Illinois State Museum Reports of Investigations, No. 25. illus: b&w: maps, photos, tables; vii, 52p; bibl.

16 figs, 8 tables.

[2614] Parmalee, Paul Woodburn *Pleistocene and recent vertebrate faunas from Crankshaft Cave, Missouri.* Oesch, Ronald D.; Guilday, John E. Springfield, IL, US: Illinois State Museum, **1969**. Illinois State Museum Reports of Investigations no 14. illus: b&w: maps, photos, tables; iv, 37p; bibl; 27 cm.

15 figs, 5 tables.

[2615] Parris, Lloyd E. *Caves of Colorado.* Boulder, CO, US: Pruett Publishing Company, **1973**. illus: b&w: drawings, maps, photos; map on lining page; xiv, 247p; gloss, index; 29 cm.

Paperback edition was issued in 1981; differs only in the absence of the map on the lining pages.

Although this is a caver's guide to the 265 known caves in the state, extensive mention is made of the history of discovery of individual sites. Rough locations. Forty-seven cave maps. More photographs than usual in such a book. Only state-wide cave guide published by a commercial firm, and highly controversial for that reason when it appeared.

[2616] Paterson, J. *Teana-au caves information.* Wakelin, D.; Lewis, J. R. (illus). Te Anau, NZ: Fiordland National Park, **1983**. illus: col: drawings, maps; map on front lining page; 16p; gloss; 21 x 15 cm.

[2617] Paterson, Keith *New directions in karst: proceedings of the Anglo-French Karst Symposium, September 1983.* Sweeting, Marjorie Mary, 1920–. Norwich, GB: Geo Books, **1986**. illus: b&w: maps; xxii, 613p; bibl; 22 cm.

Text in English and French.

[2618] Patton, Thomas Hudson, 1934– *Fossil vertebrates from Miller's Cave, Llano County, Texas.* Austin, TX, US: Texas Memorial Museum, University of Texas, Sep, **1963**. Bulletin of the Texas Memorial Museum no 7. illus: b&w: drawings, graphs, maps, photos, tables; Text on rear; 41p; bibl; 30 cm.

Illus.

[2619] Pavey, Andrew (comp) *An index to cave maps in N.S.W.* Australian Speleological Federation. NSW, AU: The N. S. W. Liaison Council of the Australian Speleological Federation, Nov, **1972**. illus: b&w: drawings; [i], 52p.

[2620] Payen, Louis A. *Preliminary report on the archeological investigation of Pinnacle Point Cave, Toulumne [sic] County, California.* [Sacramento, CA, US]: np, **1963**. 8p, [17]; 29 cm.

Caption title. Illus.

[2621] Payne, R. M. *Wonder cave of Tennessee.* Monteagle, TN, US: R.M. Payne & Son, **nd**. 13 x 19 cm.

Illus, unpaged.

[2622] Peabody, Charles, 1867-1939 *The exploration of Jacobs Cavern, McDonald County, Missouri.* Moorehead, Warren King, 1866-1939. Andover, MA, US: Phillips Academy, **1904**. Phillips Academy Department of Archaeology Bulletin 1. illus: b&w: drawings, maps, photos, tables; 1 map folded in text; iv, 29p, 11p pls; 25 cm.

Printed for the Academy by Norwood Press.

[2623] Peabody, Charles, 1867-1939 *Pt. I: the exploration of Bushey Cavern near Cavetown, Maryland.* Moorehead, Warren King, 1866-1939. Andover, MA, US: Phillips Academy, **1908**. Phillips Academy Department of Archaeology Bulletin IV. illus: b&w: drawings, maps, photos, tables; map folded in front; 166p; bibl; 25 cm.

[2624] Peacock, Norma Dee *Studies in the Rio Corredor Basin: 1988 to 1991.* Hempel, John Charles, 1949–. Huntsville, AL, US: The Costa Rica Project, **1994**. National Speleological Society Costa Rica Project Bulletin 2. some maps folded; Various pagination; 28 cm.

Special issue of: The NSS Bulletin, v 55, no 1-2 (June/December 1993). Illus.

[2625] Pearman, Harry *Caves and tunnels in Kent.* London, GB: Chelsea Spelaeological Society, **1968**. Records of the Chelsea Spelaeological Society v 6. illus: b&w: maps; 106p; 26 cm.

Illus.

[2626] Pearman, Harry *Caves and tunnels in south-east England.* London, GB: Chelsea Speleological Society, **1976**. Records of the Chelsea Speleological Society v 7-8, 11, 13-15. bibl, index; 26 cm.

[2627] Pearman, Harry *Secret tunnels in Surrey.* Chelsea Speleological Society. London, GB: Chelsea Speleological Society, **1963**. Records of the Chelsea Speleological Society v 3. illus: b&w: maps; 61p; 25 cm.

Illus.

[2628] Pearman, Harry *Underground gazetteer of south-east England.* London, GB: Chelsea Speleological Society, **1983**. Records of the Chelsea Speleological Society v 12. 44p; index; 30 cm.

Illus.

[2629] Pearse, Arthur Sperry, 1877-1956 *The cenotes of Yucatan.* Washington, DC, US: Carnegie Institution of Washington, **1936**. illus: b&w: drawings, graphs, photos, tables; 28p, 2p pls; bibl.

Reprinted from Carnegie Institution of Washington

Publication no 457. pp.1-28, February 5, 1936; 2 pls, 11 figs.

[2630] Pearse, Arthur Sperry, 1877-1956 *The cenotes of Yucatan: a zoological and hydrological study*. Creaser, Edwin P.; Hall, F.G. Washington, DC, US: Carnegie Institution of Washington, Feb, **1936**. Carnegie Institution of Washington Publication no 457, pp.1-28, February 5, 1936. illus: b&w: drawings, graphs, maps, photos, tables; [iv], 304p, 19p pls; bibl; 29 x 22 cm.

[2631] Pearse, Arthur Sperry, 1877-1956 *Fauna of the caves of Yucatan*. Banks, Nathan, B. 1868–(collab); et al. [Washington, DC, US]: Carnegie Institution of Washington, Jun, **1938**. Carnegie Institution of Washington Publication no 491. illus: b&w: drawings, graphs, maps, photos, photomicrographs, tables; iii, 304p, 8p pls; bibl; 29 x 23 cm.

Microfilm version exists. Illus.

[2632] Pearson, Les *Chillagoe souvenir guide*. Carins, QLD, AU: L. M. Pearson, **1983**. illus: b&w: maps, photos, ports; col: photos; col photos on lining pages; 32p; gloss.

Limited edition of 3000 copies.

[2633] Peck, Dallas L. *Karst hydrogeology in the United States of America: 21st congress of the International Association of Hydrogeologists: karst hydrogeology and karst environment protection October 1988 Guilin, China*. Troester, Joseph W.; Moore, John E.; United States Geological Survey; International Association of Hydrogeologists. Denver, CO, US: U.S. Geological Survey, Oct, **1988**. U.S. Geological Survey Open-File Report 88-476. illus: b&w: drawings, maps; iii, 19p; bibl.

Presented at the 1988 Congress of the International Association of Hydrogeologists, Guilin, China.

This booklet gives a brief overview of karst hydrologic studies in the US, as of 1988. This is suitable for non-hydrologists or those not familiar with related work in the US.

[2634] Peck, Steward Blaine, 1942– *A systematic revision and evolutionary biology of the* Ptomaphagus *(Adelops) beetles of North America (Coleoptera; Leiodidae; Catopinae), with emphasis on cave-inhabiting species*. Cambridge, MA, US: Harvard University, May, **1973**. Bulletin of the Museum of Comparative Zoology v 145 no 2 pp 29-162. illus: b&w: charts, drawings, maps; text on front lining page; 134p; bibl, index.

[2635] Pelech, Johann E. *The valley of Stracena and the Dobschau ice-cavern (Hungary)*. Klein, Samuel; Lowe, Walter Bezant. London, GB: Trubner & Co, **1879**. illus: b&w: drawings; 31p; 24 cm.

[2636] Pemberton, Clive *Kents Caverns, home of prehistoric man and animals: the origin, story, and descriptive tour of the caves*. Torquay, Devon, GB: L. W. Powe, **1972**. illus: b&w: charts, drawings, graphs, maps, photos; partly col: drawings; col: drawings, photos, repros; "Plan of Kents Cavern by DMB Lake" folded in center; evolution chart folded in rear; blue & black print map of area on rear lining page; [32]; 21.5 x 13.5 cm.

Cover title: Kents Cavern, its origin and story. Copyright 1950 - New revised ed 1964; readapted 1972. Similar to other editions.

[2637] Pemberton, Clive *The origin and story of Kents Cavern*. Sparks, J. B. ("Time chart by kind permission of ..."). 1947 ed. Torquay, Devon, GB: L. W. Powe and W. F. Powe, **1947**. illus: b&w: charts, photos; time chart in rear: "A new plan of Kent's Cavern" by P. M. B. Lake with text on back folded in text; [20]; 22 x 14 cm.

Cover title: The story of Kents Cavern. Looks the same as 1934 ed, but different author.Many editions. Note: One copy dated 1950 had blank last page; one had map or area in black blue & white on last page; same in all else.

1950 ed. L. W. Powe, **1950**. illus: col: repros; [28]; 22 x 14.5 cm.

[2638] Pemberton, Clive *The origin, story and descriptive tour of the caves*. 1964 ed. Torquay, Devon, GB: L. W. Powe, **1964**. illus: b&w: photos; partly col: drawings; col: repros; "Plan of Kents Cavern by DMB Lake" folded in front; evolution chart folded in rear; blue & black print map of area on rear lining page; [26], 2p pls; 21.5 x 13.5 cm.

Cover title: Kents Cavern, its origin and story. Copyright 1950 - New revised ed 1964. Similar to other editions.

1981 rev ed. npop, **1981**. illus: b&w: charts, maps, photos; partly col: drawings; col: photos; evolution chart folded in rear; map folded in center; [30].

Copyright 1950 a new and revised edition 1981. Photos are different than in the1950 edition. Major layout changes exist.

[2639] Pemble, Edna R. *Limestones caves and cavers*. Caringbah, NSW, AU: [Edna R. Pemble], Nov, **1972**. [ix], 355p; bibl, gloss.

Hand typed manuscript; appears intended for publication. No illus.

[2640] Pendergast, David M. *Excavations at Eduardo Quiroz Cave, British Honduras (Belize)*. Savage, Howard G. (faunal analysis). Toronto, CA: Royal Ontario Museum, **1971**. Royal Ontario Museum Art and Archaeology Occasional Paper 21. illus: b&w: drawings, maps, photos, tables; [iv], 123p; app, bibl.

10 pls, 17 figs.

[2641] Pendergast, David M. *A. H. Anderson's excavations at Rio Frio Cave, British Honduras (Belize)*. Beard, Forial (ed). [Toronto, Ontario, CA]: Art and Archaeology, Royal Ontario Museum, **1970**. Royal Ontario Museum Art and Archaeology Occasional Paper 20. illus: b&w: drawings, maps, photos; [iv], 59p; bibl; 25 cm.

Illus,10 pls, 12 figs.

[2642] Pendergast, David M. *Excavations of Actum Pol-*

bilche, Belize. Luther, Elizabeth. Toronto, Ontario, CA: Royal Ontario Museum, Feb, **1974**. Royal Ontario Museum Archaeology Monograph 1. illus: b&w: drawings, maps, photos, tables; [v], 103p; add, bibl.

15 pls, 11 figs, 2 tables.

[2643] **Pengelly, Hester (ed)** *A memoir of William Pengelly, of Torquay, F.R.S., Geologist, with a selection from his correspondence*. Bonney, Rev. London, GB: John Murray, **1897**. illus: b&w: photos, ports; front: b&w port of William Pengelly; x, 341p, 12p pls; bibl, index; 23 x 15 cm.

[2644] **Pengelly, William** *Kent's Cavern: a lecture*. London, GB: np, [**1872**]. Science Lectures For The People. 17p, [3]; 18.5 x 12.5 cm.

Bound with other lectures. No illus. pp 276-292.

[2645] **Pengelly, William** *Kent's Cavern and its wonders: a lecture delivered in the public hall Warrington on December 22nd 1873*. Warrington, GB: np, **1874**. Science Lectures For The People. 17p, [3]; 18 x 11.5 cm.

Printed by Guardian SteamPrinting Works. No illus.

[2646] **Pengelly, William** *Kent's Cavern: its testimony to the antiquity of man: a lecture delivered in the City Hall, Glasgow on Wednesday, 22nd December, 1875*. Glasgow Science Lectures Association. London; Glasgow, GB: W. Collins, Sons & Co, **1876**. 32p.

Illus.

[2647] **Pengelly, William** *The literature of Kent's Cavern prior to 1859*. npop: np, [**1870**]. 54+310p; 22 x 14 cm.

Part I and part II bound together. Reprinted from the Transactions of the Devonshire Association for the Advancement of Science Literature and Art 1868: Part II - 1869. Part I: 54pp; Part II: 310 pp. No illus.

[2648] **Pennick, Nigel** *The subterranean kingdom: a survey of man-made structures beneath the earth*. Wellingborough, Northamptonshire, GB: Turnstone Press, **1981**. 160p; index; 22 cm.

Illus.

[2649] **Pennington, Rooke** *Notes on the barrows and bone-caves of Derbyshire with an account of a descent into Elden Hole*. London, GB: Macmillan and Co, **1877**. [vii], 124p; app, errat; 23 cm.

[36]p of Macmillan and Co catalog (dated Dec 1876 but published 1877) at back of two copies of issue A examined. TS: "There is only one edition of this book but two distinct issues are known. The prelims and text of both are identical and the differences lie in the cover, advertisements and errata slips. In the absence of contemporary inscriptions the date of the second issue remains unknown although it remains as 1877 on the title page."

[2650] **Pennsylvania Railroad** *A summer series of personally-conducted pleasure tours to the Luray Caverns or Gettysburg via Pennsylvania Railroad*. npop: The Company, **1885**. 16p; 17 cm.

Illus.

[2651] **The Pennsylvania State University** *Field trip guidebook for the central Appalachian carbonate hydrology workshop, The Pennsylvania State University, University Park, Pennsylvania, May 21, 22, 1970*. University Park, PA, US: The Pennsylvania State University, **1970**. illus: b&w: charts, drawings, graphs, maps; text on front lining page; [19].

[2652] **Pereda de la Reguera, Manuel** *Guide Santillane on the sea and Altamira*. English ed. Santander, [ES]: Cantabria, **1966**. illus: b&w: maps, photos; six folded leaves; 46p, 40p pls; 17 cm.

Illus. 50

[2653] **Pericot Garcia, Luis**, 1899– *Prehistoric art of the western Mediterranean and the Sahara*. Ripoll Perello, Eduardo (ed). Chicago, IL, US: Aldine, **1964**. Viking Fund Publications in Anthropology no 39. xiv, 262p; bibl; 26 cm.

Various languages. Illus. Wartenstein Symposium on Rock Art of Western Mediterranean and Sahara (1960: Wartenstein Castle)

[2654] **Perkal, Malissa Faye** *Uranium decay series studies on archaeological materials: application to thermoluminescence dating and dating of cave deposits*. New Haven, CT, US: Yale University Department of Geology and Geophysics, **1978**. Senior paper series - Department of Geology and Geophysics Yale University 1978:5. illus: b&w: drawings, graphs, maps; [54]; bibl; 28 cm.

Title from cover.Illus.

[2655] **Perry, Clair Willard [Clay]**, 1887-1961 *New England's buried treasure*. New York, NY, US: Stephen Daye Press, **1946**. The American Caves Series 1. illus: b&w: maps, photos, ports; map on lining pages of some copies; front: b&w port; 348p, 45p pls; app, index; 24 cm.

55 photos.

An early attempt to survey an area for all of its major caves. Perry was a newspaper man and writes in a dramatic and readable style.

[2656] **Perry, Clair Willard [Clay]**, 1887-1961 *Underground empire: wonders and tales of New York caves*. New York, NY, US: Stephen Daye Press, **1948**. The American Cave Series 2. illus: b&w: drawings, maps, photos, ports; maps on lining pages; xvii, 221p, 23p pls; app, index; 24 cm.

Author's name listed as Clay Perry on title page. Reissued in 1966 by Ira J. Friedman, Inc. by arrangement with Frederick Ungar Publishing Company, Empire State historical publications, series 41.

The history and exploration of the caves of New York state, USA, with appendices covering speleogenesis, minor sites, and caves known in 1948.

[2657] **Perry, Clair Willard [Clay]**, 1887-1961 *Underground New England*. Brattleboro, VT, US: Stephen Daye Press, **1939**. illus: b&w: maps, photos; front: b&w map and photo; 247p, 14p pls; app; 22 cm.

Author's name on title page: Clay Perry.

The history and exploration of the caves of New England, USA, with appendices covering geology, wildlife, and caves known in 1939.

[2658] Perry, Eugene S. *Morrison Cave, Lewis and Clark Cavern State Park, Montana.* Butte, MN, US: Tom Greenfield Inc, **1946**. illus: b&w: maps; some folded maps; 18p; 25 cm.

Cover title reads: Lewis and Clark Cavern State Park, Montana. Folded map tipped in after p 18. I: Cavern growth—II: Discovery—III: Trip through the cavern—IV: Geology. Illus.

[2659] Petersen, David *Carlsbad Caverns National Park.* Chicago, IL, US: Childrens Press, **1994**. A New True Book. 45p; index; 23 cm.

Some col illus. Juvenile.

[2660] Petrocheilou, Anna *The Diros Caves of Mani: Alepotrypa and Glyphada.* Athens, GR: Petrochilos, **1970**. illus: b&w: maps, photos; col: photos; 32p, 16p pls; 17 cm.

Illustrations have captions in English, German and Greek. 3rd ed publication of supplement, 1978.

[2661] Petrocheilou, Anna *The Greek caves.* Karpodini-Dimitriad, E. (ed); Turner, Louise (trans); Doomas, A. (cap trans). Athens, GR: Ekdotike Athenon, **1984**. illus: b&w: maps; col: photos; 160p; 24 cm.

Translation of: Spelaia tes Helladas.135 illus.

Mainly a show-cave guide to Greece, with numerous color photos, but includes a number of undeveloped caves. By no means a complete catalog of caves.

[2662] Petrocheilou, Anna *The Perama Caves of Loannina.* McCallum, Mary (trans). Athens, GR: np, Dec, **1972**. illus: b&w: maps, photos; col: photos; 26p, 38p pls.

Illustrations have captions in English, German and Greek.

[2663] Peyrony, E. *Les Eyzies and the Vezere Valley; an illustrated guide for scholars and tourists.* Smith P. (trans). Montignac, [FR]: Imprimerie de la Vezere, **1959**. illus: b&w: drawings, maps, photos; 60p; 18.5 x 13.5 cm.

Show cave booklet; date of publication and number of pages may vary.

[2664] Pfeffer, Karl-Heinz (ed) *International atlas of karst phenomena: sheets 8 - 12.* International Union of Speleology. Berlin, DE: Gebruder Borntraeger, **1990**. Zeitschrift fur Geomorphologie = Annals of Geomorphology Supplementband 77. illus: b&w: drawings, maps, photos; col: maps; five col maps folded in pocket; 105p; bibl; 25 cm.

Of the various publications of the Atlas of Karst Phenomena, this is the only one principally in English. Works by diverse authors on diverse and often obscure areas.

[2665] Pfeffer, Karl-Heinz (ed) *Karst.* 1st. Berlin, DE: Gebruder Borntraeger, **1989**. Zeitschrift fur Geomorphologie Supplementband, Annals of Geomorphology 75. illus: b&w: maps; 135p; bibl; 25 cm.

Illus.

Offerings in French and English cover a broad range of karst-topics: much here, much left out.

[2666] Phillips, Harry *Jenolan Caves, (the underground wonderland) New South Wales, Australia.* Katoomba, NSW, AU: H. Phillips, **nd**. illus: b&w: photos; some photos folded in text.

Minimal text, unnumb.

[2667] Phillips, Harry *Nature's masterpiece: Jenolan Caves N.S.W.* Willoughby, NSW, AU: H. Phillips, **nd**. illus: b&w: maps, photos; [26].

Title from cover. Order of pagination varies in different versions. Illus.

[2668] Phillips, Richard, 1767-1840 *A view of the earth, containing an account of its internal structure; its caves and subterranean passages; its mountains, its rivers and cataracts. Together with a brief view of the universe, to which are added, problems on the globes, directions for drawing maps and tables of latitudes and longitudes.* Philadelphia, PA, US: Johnson and Warner, **1810**. illus: b&w: drawings; 51p; 18 cm.

Some pls. 50% cave related.

[2669] Photochrom Co., Tunbridge Wells *Cheddar Gorge and caves.* Tunbridge Wells, [GB]: Photochrom Co, Ltd, **1900**. 16p; 23 cm.

23 pls. Reprinted in 1987?

[2670] Picknett, R.G. *Symposium on cave hydrology and water-tracing.* GB: Cave Research Group of Great Britain, May, **1968**. Transactions of the Cave Research Group of Great Britain v 10 no 2. illus: b&w: charts, graphs, maps, tables; 1 folded map; text on rear lining page; 79p; bibl.

Proceedings of symposium held at Vaughan College, Leicester, 3 Feb 1968.

[2671] Picknett, R.G. (fore) *Symposium on cave surveying.* GB, Jul, **1970**. Transactions of the Cave Research Group of Great Britain vol 12 no 3. 113p.

3rd annual symposium proceedings held on 7 Mar 1970 in Leicester, UK.

[2672] Pictet, Marc-Auguste, 1752-1825 *On the ice-caves or natural ice-houses found in some of the caverns of the Jura and the Alps.* Edinburgh, GB: Printed for A. Constable, **1823**. 17p; 22 cm.

Microopaque. New York: Readex Microprint, 1979. 23 x 15 cm. (Landmarks II).

[2673] Pierpont, E. de *The wonders of the Grotto of Han (Belgium): guidebook, and description with engravings and plan and a notice upon the very interesting Grotto of Rochefort.* Bruxelles, BE: Societé Anomyme de Rotogravure d' Art, **1947**. illus: b&w: drawings, maps, photos; partly col: maps; photo on rear lining page; 47p, 8p pls; 20.5 x 13 cm.

Printed in a green black ink (photos & cover).

[2674] Pilkington, Graham (ed) *Nullarbor caving atlas.* Mott, Kevin (ed); Cave Exploration Group South Australia Incorporated (CEGSA). 2nd edition. Adelaide, SA, AU: Subterranean Foundation (Australia) Incorporated, **1986**. Occasional Paper Cave Exploration Group South Australia 7. illus: b&w: maps; col: ads; Various pagination.

Library and field editions exist. First edition edited by Kevin Mott and Graham Pilkington.

[2675] Pilkington, Graham *Speleovision field notes.* Adelaide, SA, AU: Cave Exploration Group of South Australia Inc, **1982**. Occasional Paper Cave Exploration Group of S.A. Inc no 6. illus: b&w: maps; x, 76p; 29 cm.

Illus.

[2676] Pinkham, Mary R. *From the cradle to the cave: the life story of "Dad" Truitt, "Cave Man of the Ozarks".* Noel, MO, US: McDonald County Press, **1954**. 91p; 19 cm.

[2677] Pinney, Roy *The complete book of cave exploration: an authoritative guide to the wonders, mysteries and excitement of caves and caving.* Sanderson, Ivan T. (fore). New York, NY, US: Coward-McCann, Inc, **1962**. illus: b&w: photos, repros; 256p, 16p pls; app, bibl, gloss; 22 cm.

[2678] Pinnock, Rick (ed) *A guide to Glenrock caves: a speleological field guide to the limestone caves on Glenrock Station.* Baulkham Hills, NSW, AU: Hills Speleological Club Ltd, **1987**. Overkarst no 1. illus: b&w: cartoons, maps, photos, ports, tables; [vi], 43p, [[100]].

Cover title: Glenrock Caves: Upper Hunter Valley N.S.W.

[2679] Pipkin, Turk (ed) *Barton Springs eternal: the soul of a city.* Frech, Marshall (ed). Austin, TX, US: Softshoe Publishing, The Hill Country Foundation, **1993**. illus: b&w: cartoons, drawings, maps, photos, ports, repros; viii, 136p.

[2680] Pistole, Nancy (ed) *Proyecto Cheve 1986-1993.* [Atlanta, GA, US]: Cheve Project, **1995**. illus: b&w: maps; one loose map pl not included in page count; 45p.

This nice report summarizes the exploration of Sistema Cheve in Oaxaca, Mexico, which was, at 1386 meters, for a few years the deepest cave in the Western Hemisphere. It also covers a number of smaller caves in the area. There are short sections on archaeology and geology.

[2681] Pitty, Alistair F. *An approach to the study of karst water: illustrated by results from Poole's Cavern, Buxton.* Willerby, Hull, Yorkshire, GB: University of Hull, **1966**. Hull University. Dept of Geography Occasional Papers in Geography no 5. illus: b&w: graphs, maps, tables; viii, 70p; bibl; 22 cm.

[2682] Plante, Alan R. (comp) *Berkshire Cave Guide.* Northeast Regional Organization, National Speleological Society. [Gorham, NH, US]: Berkshire Cave Survey, **1990**. illus: b&w: maps, photos; three maps folded in text; ii, 61p.

Prepared for the NRO, NSS.

[2683] Plew, Mark G. *The archaeology of Nahas Cave: material culture and chronology.* Boise, ID, US: Boise State University, **1986**. Archaeological reports / Boise State University ; no. 13. illus: b&w: maps; vi, 109p; bibl; 28 cm.

Illus.

[2684] Plimpton, George (host) *The Los Tayos challenge.* Visual Productions 80 Ltd SelectVideo (Firm). Englewood, CO, US: SelectVideo, **1987**. Challenge (Englewood, Colo).

Cataloged from contributor's data. A foreign videorecording (Canada) VHS. 1 videocassette (30 min), sd, col, 1/2 in.

Tells how the story that the Los Tayos caves in Ecuador were a subterranean base for extraterrestrial visitors sparked a small army of soldiers and scientists to mount an exploratory expedition deep into the tropical jungle of South America to examine these caves.

[2685] Pohl, Erwin Robert, 1904– *Itinerary, geologic features of the Mississippian Plateaus in the Mammoth Cave and Elizabethtown areas.* [Lexington, KY, US]: [Geological Society of Kentucky], **1964**. illus: b&w: maps; 32p; 28 cm.

[2686] Pohl, Erwin Robert, 1904– *Vertical shafts in limestone caves.* [Trenton, NJ, US]: National Speleological Society, Apr, **1955**. Occasional Papers National Speleological Society no 2. illus: b&w: charts, maps, photos; 24p; bibl; 23 cm.

Title from cover. Illus.

[2687] Pollack, John C., 1949– *National Speleological Society (USA) 1973 field trip to Greece.* [Huntsville, AL, US]: [National Speleological Society], **1977**. [ii], 33p, [2]; 28 cm.

[2688] Pollard, Anthony James, 1941– *Smoo Cave.* Glasgow, Scotland, GB: Glasgow University, Department of Archaeology, **1992**. GUARD 60. illus: b&w: maps; 19p; 30 cm.

Illus.

[2689] Pond, Alonzo William, 1894– *Caverns of the world.* New York, NY, US: W.W. Norton and Company Inc, **1969**. illus: b&w: photos; xiv, 178p; index; 24 cm.

Describes the formation of different types of caves and the plant and animal life which exists in them. Also discusses exploration and the ways in which man has used caverns since ancient times.

[2690] Pond, Alonzo William, 1894– *Guide book of Cave of the Mounds ... Blue Mound, Wis.* 1941 ed. Milton Junction, WI, US: Frantz Print Co, **1941**. illus: b&w: maps; folded map; 34p; 22 cm.

Cover title: Illustrated guide book: newly discovered Cave of the Mounds, Blue Mounds, Wisconsin. Illus.

1945 ed. 1945. bibl.

Cover title: Illustrated guide book, Cave of the Mounds, Blue Mounds, Wisconsin. Illus.

1951 ed. Printed by Straus Print Co.

[2691] Poole, Gary A. (ed) *Bexar County speleology vol 1*. Passmore, C. G. (ed); Association for Bexar Cave Studies. San Antonio, TX, US: Ream and a Prayer Press, **1978**. illus: b&w: maps; [iii], 54p, [1]; errat, gloss.

[2692] Poole, Lynn *Deep in caves and caverns*. Poole, Gray Johnson. New York, NY, US: Dodd, Mead & Company, **1962**. illus: b&w: drawings, photos; 158p; index; 21 cm.

16 illus.

[2693] Pope, Joyce *Life in the dark*. Stilwell, Stella; Ward, Helen, 1962–. 1991 Australian ed. South Melbourne, VIC, AU: Macmillan, **1991**. Curious Creatures. 46p; index; 27 cm.

Col illus. Describes the habitats and activities of creatures that live in darkness. Juvenile.

1992 American ed. Austin, TX, US: Steck-Vaughn Co, **1992**. bibl, index.

Juvenile.

[2694] Popham, William Lee, 1885-1953 *Mammoth Cave romance*. Whidby, Jim (new intro). 1993 reprint ed. Maryville, TN, US: Bryon's Graphic Arts, **1993**. 110p, [1]; 20 cm.

1st ed. Louisville, KY, US: The World Supply Company, Mayes Printing Company, **1911**. illus: b&w: photos, ports; 110p; 20 cm.

Copyright date from verso of t.p. Photographic illustrations on p. [1] and [2] of paginated sequence. "Books by the same author": p. [8]."..it was one of a series of seven books in which he [Popham] referred to as his 'seven Wonders of the World (American)."

[2695] Porteous, Crichton *Caves and caverns of Peakland*. Derby, GB: The "Come-to-Derbyshire" Association, **1950**. illus: b&w: photos; 104p; index; 19 cm.

Cover title: Caves & caverns of Peakland.

[2696] PostCard Videos, Inc *Howe Caverns and New York Central Leatherstocking Region*. St. Armands, FL, US: PostCard Videos, **nd**.

Title on container: Howe Caverns. One videocassette (VHS) (17 min), sd, col, 1/2 in.

Presents Howe Caverns and the Central Leatherstocking Region in all their natural beauty. Enjoy the majesty of the Caverns and learn how they were discovered and the difficulties that were overcome to make them accessible to the public. View ancient rock formations and take a boat ride 200 feet below the ground. See the Bridal Alter, site of hundreds of subterranean weddings.

[2697] Potton, Craig *Images from a limestone landscape: a journey into the Punakaiki-Paparosa Region*. Dennis, Andy. Nelson, NZ: Craig Potton, **1987**. illus: partly col: maps; col: maps, photos; 118p.

Minimal text.

[2698] Poulianos, Aris N. *The cave of the Petralonian archanthropinae: a guide to the science behind the excavations*. Athens, GR: Library of the Anthropological Association of Greece, **1982**. 80p; 21 cm.

Col illus.

[2699] Poulson, Thomas Layman, 1934– *Symposium on life histories of cave beetles: symposium held at the 1973 annual convention of the National Speleological Society, Bloomington, Indiana*. National Speleological Society. reprint. Amsterdam, NE: Swets & Zeitlinger, **1975**. International journal of speleology v 7, no 1-2. illus: b&w: graphs, maps, photos, tables; 78p; bibl; 24 cm.

A collection of papers that describe in detail population ecology of North American cave beetles. Particular emphasis paid to the interaction between beetles and cave crickets.

[2700] Pound, Louise, 1872-1958 *Nebraska cave lore*. [Lincoln, NE, US]: Nebraska State Historical Society, **1948**. 25p; 23 cm.

Reprinted from Nebraska history, v 29, no 4, Dec 1948 pp 299-323.

[2701] Powell, Hazel Rowena *Adventures underground in the caves of Missouri*. [1st ed]. New York, NY, US: Pageant Press, **1953**. 63p; 21 cm.

Juvenile.

[2702] Powell, Richard Lewis, 1936– *Caves of Indiana*. Bloomington, IN, US: Indiana Geological Survey, Indiana Department of Conservation, Oct, **1961**. Geological Survey Circular no 8. illus: b&w: drawings, maps, photos, tables; 1 map folded in pocket; 127p; bibl, index; 23 cm.

58 figs.

Catalog of 398 known caves in the state. Text descriptions of somewhat fewer than half of them; approximately 60 cave maps.

[2703] Powell, Richard Lewis, 1936– *A guide to the selection of limestone caverns and springs in the United States as national landmarks: for the National Park Service*. United States National Park Service. Bloomington, IN, US: Indiana Geological Survey, Indiana Department of Conservation, Dec, **1970**. illus: b&w: drawings, maps, tables; map folded in back pocket, 12 maps folded in text; [xii], 292p; bibl, index; 28 cm.

106 figs, 1 pl. A rare item - only 20 copies produced. Has a thorough literature survey plus a listing of numerous caves as possible landmarks. Many maps and cave descriptions.

[2704] Powers, Richard M. *The cave dwellers in the old stone age*. New York, NY, US: Coward-McCann, **1963**. Life Long Ago. illus: b&w: charts, drawings, maps; partly col: drawings; partly col drawing on lining pages; 62p; bibl, index.

Cover title: The cave dwellers: a re-creation through pictures and text of life in the old stone age. Juvenile.

[2705] **Pratchett, Nick** *International expedition to Gouffre Berger, 1956: the exploits of the British members, Nick Pratchett & Bob Powell*. Powell, Bob. Berkhamstead, Herts, GB: The Cave Research Group of Great Britain, **1956**. Occasional publications, the Cave Research Group of Great Britain, no 1. illus: b&w: drawings, maps; Loose folded map.; [ii], 25p; 24 cm.

Illus.

[2706] **Preble, John Wesley** *Sinks of Gandy Creek*. Parson, WV, US: McClain Printing Company, **1969**. illus: b&w: maps, photos; col: photos; 26p; bibl; 22 cm.

Delightful history of this well known cave.

[2707] **Prentice, Guy** *Mammoth Cave archeological inventory project interim report - 1987 investigations*. Tallahassee, FL, US: National Park Service, Southeast Archeological Center, Jan, **1988**. National Park Service, Southeast Archeological Center. illus: b&w: maps; xvi, 193p; bibl; 28 cm.

Chiefly maps.

Interim archaeological study of the prehistoric and historic cultural resources of Mammoth Cave National Park Kentucky.

[2708] **Prentice, Guy** *Mammoth Cave National Park archeological inventory project interim report-1988 investigations*. Tallahassee, FL, US: National Park Service, Southeast Archeological Center, **1989**. illus: b&w: maps; bibl; 28 cm.

[2709] **Prestwich, Joseph** *Report on the exploration of Brixham Cave, conducted by a committee of the geological society, and under the superintendence of Wm. Pengelly, etc.* [reprint]. npop: np, **1872**. 101p; 29 cm.

pp 471-572.

[2710] **Přibyl, Jan** *Karst utilization in practice*. Praha, CZ: Academia, Publishing House of the Czechoslovak Academy of Sciences, **1986**. Acta Scientiarum Naturalium Academiae Scientiarum Bohemoslovacae Brno v 20 New Series no 11. illus: b&w: drawings, maps, photos, tables; [ii], 50p, 4p pls.

[2711] **Přibyl, Jan (ed)** *Largest cave system of the Czech Socialist Republic in the Moravský kras (Moravian karst)*. Brno, CZ: Institute of Geography Czechoslovak Akademy of Sciences, **1973**. Studia Geographica 35. illus: b&w: drawings, tables; 1 map folded, loose; text on lining pages; 83p, [4], 23p pls; bibl; 25 cm.

Summaries in Czech and Russian. Presented to 6th International Congress of Speleology participants by the Institute of Geography Czechoslovak Akademy of Sciences. Some col illus.

[2712] **Přibyl, Jan** *Paleohydrography of the caves in the Moravsky Kras (Moravian Karst)*. Brno, CZ: Czechoslovak Academy of Sciences, Institute of Geography, **1973**. Studia Geographica 28. illus: b&w: maps; 64p; 25 cm.

Summary in Czech. Illus.

[2713] **Přibyl, Jan** *Regularities of karst processes (karst landscape)*. Praha, CZ: Academia, Publishing House of the Czechoslovak Academy of Sciences, **1986**. Acta Scientiarum Naturalium Academiae Scientiarum Bohemoslovacae Brno v 20, New Series no 12. illus: b&w: drawings, maps, photos; two maps folded, loose; 43p, [2], 2p pls.

[2714] **Přibyl, Jan** *Věkåum budoucím*. Keprt, Jiři. Brno, CZ: Geografický ústav ČSAV, **nd**. illus: b&w: photos; col: photos; [110].

Picture book on Moravian Karst; trilingual: Czech, Russian (?), English.

[2715] **Price, Elizabeth Ann** *Club history 1956—1981 25th anniversary publication: Cerberus Spelaeological Society*. Bath, Avon, GB: The Cerberus Spelaeological Society, Sep, **1981**. Occasional Publication 3. illus: b&w: charts, graphs; 28p; app; 29.5 x 21 cm.

Includes member lists and list of pubs as appendix.

[2716] **Price, Elizabeth Ann [Liz]** *Bath freestone workings*. Radstock, GB: The Resurgence Press, **1984**. illus: b&w: maps; 74p; bibl; 30 cm.

Illus.

[2717] **Price, Graham (ed)** *Fairy Cave Quarry: a study of the caves*. Radstock, Bath, GB: The Cerberus Spelaeological Society, Aug, **1977**. The Cerberus Spelaeological Society Occasional Publication no 1. illus: b&w: charts, drawings, maps, photos, tables; 72p; app, bibl; 30 x 21 cm.

62 b&w photos, 12 figs.

[2718] **Price, Graham** *The Holloch: a general account and trip report*. [GB]: Cerberus Spelaeological Society, Sep, **1981**. Occasional Publication no 4. illus: b&w: charts, graphs, maps, photos; 1 fold-out map at end; 34p, 1p pls.

Originally in CSS Journal vol 11 no 4, July/Aug 1981, with additions.

[2719] **Prinz, W.** *The crystalline structures of the caves of Belgium*. Melmore, Sidney (trans). State College, PA, US: Published in cooperation with the Section of Cave Geology & Geography, National Speleological Society, **1980**. Cave Geology v 1, no 7. 68p; bibl; 28 cm.

Translation of: Les cristallisations des grottes de Belgique. Caption title. "Translation first published in Cave Science ... Vol. II. No. 9 (July 1949); No. 10 (Oct. 1949); No. 11 (Jan. 1950); and No.13 (July 1950)." "Translation No. 12." "July 1980." Reprint of these 4 issues; appears to be retyped (205p + [9] p plates). Illus. pp191-258

[2720] **Proctor, Christopher J. (comp)** *Atlas of the Berry Head Caves*. Hampton, Middlesex, GB: C. J. Proctor, Jul, **1987**. illus: b&w: maps; 5 fold out maps in text; illus of Sweet Water Pot on front; [iii], 42p; 30 x 21 cm.

[2721] **Proctor, Christopher J.** *The caves of Chudleigh and Kingsteignton*. Devon, GB: C.J. Proctor, **1987**. Devon Caves vol 2. illus: b&w: maps; [iv], 36p; 21 x 15 cm.

[2722] **Proctor, Christopher J.** *Caves of East Devon*.

Devon, GB: C. J. Proctor, **1987**. Devon Caves vol 1. illus: b&w: maps; [iv], 39p; bibl, index; 21 x 15 cm.

No illus.

[2723] Proctor, Christopher J. *Caves of North Tor Bay*. Hampton, Middlesex, GB: Christopher J. Proctor, **1987**. Devon Caves vol 3. illus: b&w: maps; [iv], 45p; 21 x 15 cm.

[2724] Prosser, J. Joseph *Cave diving communications*. Grey, H. V. (auth, illus); McKinnon, Wayne (illus). 1st ed. Branford, FL, US: Cave Diving Section of the National Speleological Society, Inc, **1990**. illus: b&w: cartoons, drawings; viii, 59p; bibl, index; 22 cm.

[2725] Prosser, J. Joseph *NSS cave diving manual: an overview*. Grey, H. V. 1st ed. Branford, FL, US: Cave Diving Section of the National Speleological Society, **1992**. xxix, 377p; index; 21 cm.

Illus.

[2726] Prosser, J. Joseph (ed) *The NSS instructor's training manual*. Branford, FL, US: Cave Diving Section of the National Speleological Society, Inc, **1986**. illus: b&w: drawings, tables; Various pagination.

Three-ring binder. 2nd edition: October 1, 1989.

[2727] Prosser, J. Joseph *NSS student cave diver workbook: designed specifically for use by the student cave diver participating in the NSS full cave diver course*. Leonard, Mark D.; Hires, H. Lamar (comp). Branford, FL, US: Cave Diving Section of the National Speleological Society, Inc, Oct, **1989**. illus: b&w: ads, charts, drawings, graphs, maps, tables; one map folded in text; Various pagination, iii, 160p.

Cover title: NSS student cave diver workbook: designed specifically for use by the student cave diver participating in the NSS Full Cave Diver Course.

[2728] Prufer, Olaf H. *Krill Cave: a stratified rockshelter in Summit County, Ohio*. Long, Dana A.; Metzger, Donald J. Kent, OH, US: Department of Sociology and Anthropology, Kent State University, **1989**. Kent State Research Papers in Archaeology no 8. illus: b&w: drawings, maps, photos, tables; [vi], 109p; bibl.

24 figs, 12 tables.

[2729] Publications Team (ed) *Cave Notes 75*. [GB]: Bristol Exploration Club, May, **1977**. Caving Report no 21. illus: b&w: charts, maps; 41p, [1]; 30 x 21 cm.

[2730] Pulina, Marian (ed) *2nd international symposium of glacier caves and karst in polar regions: proceedings*. Eraso, Adolfo (ed). Sosnowiec, PL: Department of Karst Geomorphology University of Silesia, **1992**. illus: b&w: charts, graphs, maps, photos, tables; 127p; bibl.

Sponsored by: University of Silesia (Poland), Ministry of National Education (Poland), Politechnic University, E.T.S.I. de Minas, Madrid (Spain).

[2731] Pulina, Marian (ed) *The dynamic of the contemporary karstic processes in the tropical area of Cuba: preliminary report of the field investigations performed by the expedition Guajaibon '84 in the winter season 1984*. Bierwiaczonek, Boguslaw (trans); Fagundo, J Reynerio (co-ed). Sosnowiec, PL: Uniwersytet Slaski, Katedra Geomorfologii Krasu, **1984**. illus: b&w: graphs, maps, tables; maps, graphs, folded in rear pocket; one table folded in text; [iv], 42p; bibl; 24 cm.

The Cuban-Polish expedition Guajaibon '84 was conducted between January 21 and February 7, 1984. 5 figs, 20 tables.

[2732] Purcell, David *Guide to the lava tube caves of central Oregon*. Astiasuain, Linda (ed); Brogan, Phil F., 1896–. 1st ed. npop: David Purcell, **1977**. illus: b&w: cartoons, drawings, maps, photos; 53p; gloss; 23 cm.

"Most of the information in this book was obtained from newspaper articles written by Phil Brogan."—tp verso. Illus.

2nd ed. Astiasuain, Linda (ed); Brogan, Phil F., 1896–. Corvallis, OR, US: High Mountain Press, **1978**. 55p; 23 cm.

[2733] Purcell, David *Spelunking (the exploration of caves) guide to central Oregon*. Astiasuain, Linda (ed). 1st ed. Washougal, WA, US: High Mountain Press, **1977**. illus: b&w: drawings, maps, photos; 44p; gloss; 22 cm.

2nd ed, 1978; 3rd ed, 1980. Illus. 3rd ed title may be rearranged to read: Spelunking guide to Central Oregon: the exploration of caves.

[2734] Putnam, Frederic Ward, 1839-1915 *Archaeological researches in Kentucky and Indiana, 1874*. Boston Society of Natural History. npop: np, **1875**. Kentucky Thousand. 18p; 25 cm.

"From the Proceedings of the Boston Society of Natural History, Vol. XVII, 1875." Microfiche. Louisville, Ky.: Lost Cause Press, 1961. 1 microfiche. (Kentucky Thousand).

Early base-level study of the prehistoric and historic cultural resources of Kentucky and Indiana by one of the "fathers" of North American archaeology.

[2735] Queensland Conservation Council (prep) *The case against the Pike Creek Dam*. University of Queensland Speleological Society (prep). [QLD, AU]: np, Mar, **1973**. illus: b&w: maps, tables; map folded in rear; [ii], 41p; bibl.

Report on the impact of the proposed flooding of the Texas caves. With supplements.

[2736] Queensland Conservation Council (prep) *Pike Creek Dam: a preliminary criticism of the Queensland Irrigation and Water Supply's Commission's environmental impact study*. University of Queensland Speleological Society (prep). [QLD, AU]: np, Jun, **1973**. illus: b&w: repros, tables; Various pagination, 26p; app; 27 cm.

Title from cover. Report on the impact of the proposed flooding of the Texas caves.

[2737] Quick, Dell (ed) *SCG history: summer 1981*. South Pasadena, CA, US: Southern California Grotto of the National Speleological Society, **1981**. 51p; 28 cm.

Cover is reproduction of Grotto Charter. No illus.

[2738] Quick, Peter Gunder *Vermont caves: a geologic and historical guide*. Ypsilanti, MI, US: Paleoflow Press, **1994**. illus: b&w: charts, maps, photos; col: photos; 5 fold-out maps; [iv], 74p, 5p pls; 28 cm.

Catalog of known solution caves, with text descriptions and locations. Maps of most of them. About one hundred entries, some of which are groups of small, adjacent caves. Geological introduction.

[2739] Quinlan, James Francis, 1936-1995 *Ground water flow in the Mammoth Cave Area, Kentucky with emphasis on principles, contaminant dispersal, instrumentation for monitoring water quality, and other methods of study*. Ewers, Ralph O. npop: National Water Well Association, **1986**. illus: b&w: charts, drawings, graphs, maps, photos; ii, 94p; bibl.

Field trip A: environmental problems in karst terranes and their solutions October 28-30, 1986 Western Kentucky University Bowling Green, Kentucky. Sponsored by the National Water Well Association along with Eastern Kentucky University, Friends of the Karst, Kentucky Division of Water, National Park Service, and Western Kentucky University. An updated, partially edited reprint of a guidebook first published in 1983 (Field Trips in Midwestern Geology: Bloomington, Indiana, Geological Society of America and Indiana Geological Survey, edited by Shaver, R.H. and Sunderman, J.A., Volume 1 pp 1-85.)

Good overview of work in the region. Difficult to obtain.

[2740] Quinlan, James Francis, 1936-1995 *Ground water monitoring in karst terranes: recommended protocols and implicit assumptions*. Koglin, Eric N. (proj of). Mammoth Cave, KY, US: [U. S. National Park Service], Feb, **1989**. illus: b&w: drawings, graphs, maps; [i], ix, 79p; bibl.

Study done in cooperation with the National Park Service. U.S. Environmental Protection Agency IAG No. DW 14932604-01-0, EPA 600/X-89/050.

This is a proof draft of an EPA guidance document. Very useful, never yet issued as final.

[2741] Quinlan, James Francis, 1936-1995 *Groundwater hydrology and geomorphology of the Mammoth Cave Region, Kentucky, and of the Mitchell Plain, Indiana*. npop: np, **1980**. 84p; bibl; 28 cm.

Field trip 7. Illus. OCLC lists both 1980 and 1991 as date of pub.

[2742] Quinlan, James Francis, 1936-1995 *Hydrogeology and geomorphology of the Mammoth Cave area, Kentucky: Southeastern Friends of the Pleistocene 1990 field excursion, November 17, 1990*. Ewers, Ralph O.; Palmer, Arthur Nicholas, 1940–. Nashville, TN, US: Southeastern Friends of the Pleistocene, **1990**. illus: b&w: drawings, graphs, maps, photos; b&w map folded in rear pocket; ii, 102p; bibl.

Cover title: A magnificent panorama of the Mammoth Cave in Kentucky. A slightly modified version of Quinlan and Ewer's Ground water flow in the Mammoth Cave Area, Kentucky with emphasis on principles, contaminant dispersal, instrumentation for monitoring water quality, and other methods of study.

[2743] Quinlan, James Francis, 1936-1995 *Hydrology and water quality in the central Kentucky karst: phase 1*. Rowe, Donald R. Mammoth Cave, KY, US: United States National Park Service, Southeast Regional, Uplands Field Research Laboratory, **1977**. Management Report United States National Park Service no 12; Research Report University of Kentucky Water Resources Research Institute no 101. illus: b&w: charts, graphs, maps, tables; col: charts, graphs, maps; viii, 91p, [2]; bibl; 27 cm.

22 figs, 12 tables. Reprinted from Kentucky Water Resources Research Institute, Research report no. 101: 1977. project no.: A-062-KY (Complete report), agreement numbers: 14-31-0001-5017 (FY 1975) 14-34-0001-6018 (FY 1976). University of Kentucky, Water Resources Research Institute.

[2744] Quinlan, James Francis, 1936-1995 *Hydrology of the Turnhole Spring Groundwater Basin and vicinity, Kentucky: an area that includes part of the Mammoth Cave National Park*. [Mammoth Cave, KY, US]: [National Park Service, Southeast Regional, Uplands Field Research Laboratory], **1976**. Management Report no 11 United States Southeast Regional Uplands Field Research Laboratory, Great Smoky Mountains National Park. Management Reports, Uplands Field Research Laboratory. illus: b&w: charts, maps; partly col: maps; one map folded in rear pocket; [i], 22p; bibl; 22 cm.

Prepared for a Field Excursion conducted on April 27, 1976 as part of the International Symposium on Hydrologic Problems in Karst Regions. "This is a reprint of the field excursion guidebook that was made out-of-date within a week after the conclusion of the symposium. It has been partially up-dated and numerous pages from have been slightly revised and corrected. This is an interim report."

[2745] Quinlan, James Francis, 1936-1995 *Practical karst hydrogeology, with emphasis on groundwater monitoring, February 8-11, 1994, Holiday Inn West, Gainesville, Florida*. Ewers, Ralph O. (lect); Aley, Thomas John, 1938–(lect); Huntoon, Peter W. (lect); Wilson, William L. (lect); Environmental Education Enterprises Institute (spon); The Association of Engineering Geologists (spon). [Columbus, OH, US]: [Environmental Education Enterprises Institute], **1994**. illus: b&w: charts, drawings, maps, photos; Various pagination; bibl.

Several sections are photocopies of journal articles, etc.

Mainly a collection of reprints. Useful, but virtually impossible to obtain.

[2746] **Radcliffe, F.G.** *The wonder caves of Maoriland: Waitomo, Ruakuri & Aranui.* Radcliffe, F. G. (photos). npop: Frank Duncan, **1925**. Tourist Series no 31. [24]; 19 x 23 cm.

Illus.

[2747] **Radlauer, Ruth**, 1926– *Carlsbad Caverns National Park.* Radlauer, Ed (photo); Zillmer, Rolf (design and map). Chicago, IL, US: Childrens Press, **1981**. Parks for People An Elk Grove Book. illus: b&w: drawings, maps; partly col: maps; col: photos; 48p, [3]; gloss; 24 cm.

Juvenile.

"Describes the special physical features and the plant and animal life of this national park located in the New Mexican desert."

[2748] **Radlauer, Ruth**, 1926– *Mammoth Cave National Park.* 1986 rev ed. **1986**.

Juvenile.

1st ed. Radlauer, Ed (photo); Zillmer, Rolf (design, map). Chicago, IL, US: Childrens Press, **1978**. Parks for People An Elk Grove Book. illus: partly col: maps; col: photos; 48p, [3]; 24 cm.

"Discusses the rock formations, animal life, vegetation, and history of the world's longest cave." Juvenile.

[2749] **Railton, Courtenaye Lewis**, 1907-1971 *The Ogof Ffynnon Ddu system: its discovery and exploration (1927-53): and a theory of its development.* npop: Cave Research Group of Great Britain, **1953**. Cave Research Group of Great Britain Publications Series no 6. 49p; 26 cm.

Illus.

[2750] **Railton, Courtenaye Lewis**, 1907-1971 *The survey of Tunnel Cave: S. Wales.* npop: Cave Research Group of Great Britain, **1958**. Cave Research Group of Great Britain Publications no 7. 35p; 25 cm.

Typewriter script. Illus.

[2751] **Raines, Terry W. (comp)** *Caves of Mexico.* Austin, TX, US: Association for Mexican Cave Studies, **1989**. illus: b&w: photos, tables; col: photos; [xviii], 259p; index.

Less information than title and page count suggest. Computer-generated from the beginning of a database. Best on vicinities of Valles and Xilitla.

[2752] **Raines, Terry W.** *First reports: 1984-1988.* Austin, TX, US: Association for Mexican Cave Studies, Sep, **1988**. illus: b&w: photos, repros; col: photos; front: col photo; Various pagination, 60p.

"Unedited and assembled by Terry Raines: mainly a facsimile repro of caver logbooks from restaurants in Mexico."

[2753] **Raines, Terry W.** *Sotanito de Ahuacatlán: Sierra Madre Oriental; Jalpan; Ahuacatlán.* Austin, TX, US: Association for Mexican Caves Studies, **1972**. Association for Mexican Cave Studies Cave Report Series no 1. illus: b&w: photos; partly col: maps; col: photos; 2 maps folded in back pocket; [iii], 20p; bibl; 28 cm.

[2754] **Raines, Terry W. (comp)** *Sotano de las Golondrinas: description.* Austin, TX, US: Association for Mexican Cave Studies, Aug, **1968**. Association for Mexican Cave Studies Bulletin 2. illus: partly col: maps; col: photos; 1 map folded in back pocket; [v], 20p; bibl; 29 cm.

[2755] **Rains, Geo. W.** *Notes on making saltpetre from the earth of the caves.* Augusta, GA, US: SteamPower Press Chronicle & Sentinel, **1861**. [10].

Reprinted 1973 by Huntsville Grotto, with one preface page added. No illus.

[2756] **Raistrick, Arthur, 1897?–1991** *Grassington and Upper Wharfedale.* npop: Dalesman, **1948**. 28p.

Guide to this popular Yorkshire karst area.

[2757] **Ralston, Basil** *Jenolan: the golden ages of caving.* 1989 ed. Winmalee, NSW, AU: Three Sisters Productions, **1989**. illus: b&w: drawings, maps, photos, ports; 63p, 4p pls; app, bibl, index.

A partly biographical account of early exploration at Jenolan, Australia.

1990 ed. 1990. illus: b&w: maps, ports; app, bibl, index; 24 cm.

[2758] **Randall, Robert L.** *Along the way to the caves: Timpanogos Cave National Monument, Utah.* United States National Park Service. Globe, AZ, US: Southwest Parks and Monuments Association, **1970**. [24]; 13 x 18 cm.

Title from cover. Published in cooperation with the National Park Service. Illus.

[2759] **Randolph, Helen Fitz** *Mammoth Cave and the cave region of Kentucky, illustrated: with bibliography of Mammoth Cave (Willard Rouse Jillson); first accurate underground survey (H. Bruce Hoffman); introduction (H. C. Nelson).* Louisville, KY, US: The Standard Printing Company, Incorporated, **1924**. illus: b&w: ads, maps, photos; 2 maps folded in text; front: b&w photos, 1 map folded; 153p, [6]; bibl, index; 19 cm.

Cover title: Mammoth Cave and the cave region of Kentucky, illustrated. Microfilm edition: Lost Cause Press.

[2760] **Rankin, Frank** *Guide to the Wemyss caves.* Save the Wemysss Ancient Caves Society. [Markinch, GB]: [Wemyss Environmental Education Centre], **1989**. illus: b&w: maps; text on front lining page; 40p; 21 cm.

Copyright: Frank Rankin. Title from cover.

[2761] **Raphael, Max**, 1889-1952 *Prehistoric cave paintings.* Guterman, Norbert, 1900–(trans). New York, NY, US: Pantheon Books, **1945**. The Bollingen Series IV. illus: b&w: drawings; [x], 100p; 31 x 24 cm.

Photocopy. Ann Arbor, Mich.: University Microfilms, 1981, 28 cm, 92 pp.

[2762] **Rasquin, Priscilla** *The influence of light and darkness on thyroid and pituitary activity of the characin*

Astyanax mexicanus *and its cave derivatives*. New York, NY, US: American Museum of Natural History, **1949**. Bulletin of the American Museum of Natural History v 93 article 7. 31p, 3p pls; bibl; 27 cm.

pp 501-531. Illus.

[2763] Rauch, Henry W. *Lithologic controls on the development of solution porosity in carbonate aquifers.* White, William Blaine, 1934–. University Park, PA, US: Institute for Research on Land and Water Resources, Pennsylvania State University, **nd**. Reprint Series no 13. illus: b&w: charts, graphs, maps, tables; 18p; bibl.

Reprint from Water Resources Research vol 6 no 4 Aug 1970.

[2764] Rauch, Henry William (ed) *Proceedings of the fourth conference on karst geology and hydrology.* Werner, Eberhard Wolfgang, 1942–(ed);The Department of Geology and Geography of West Virginia University (sponsors). [Morgantown, WV, US]: West Virginia Geological and Economic Survey, **1974**. illus: b&w: charts, drawings, graphs, maps, photos, tables; 1 map folded in text; iv, 187p, [1]; bibl.

The conference was held on May 3-5, 1974.

Contains many interesting papers and extended abstracts from well-known researchers.

[2765] Rautjoki, Harri *North-West Nelson State Forest Park Honeycomb Hill Cave System: concept plan: West Coast Conservancy.* Millar, Ian; Wilde, Kevan A. (rev); Worthy, Trevor H. (rev). [NZ]: [New Zealand Department of Conservation], **nd**. illus: b&w: maps; [iii], 32p, [3]; app, bibl.

1991: rev ed

[2766] Ray, Michael Allen *Underground worlds: tour guide training manual.* New York, NY, US: American Museum of Natural History, **nd**. illus: b&w: drawings, tables; [i], 22p, [41].

[2767] Rea, G. Thomas (ed) *Caving basics: a comprehensive guide for beginning cavers.* National Speleological Society. Special Publications Committee. 3rd rev ed. Huntsville, AL, US: [National Speleological Society], **1992**. viii, 187p; bibl, index; 28 cm.

Produced by the NSS Special Publications Committee. Illus.

With each revision this book gets better and better.

[2768] Rebmann, James R. *Caves of Rockcastle County, Kentucky.* O'Dell, Gary A. [Lexington, KY, US]: Blue Grass Grotto? Geology Department, University of Kentucky, **1972**. ii, 101p, [2]; 28 cm.

Title from cover. Col illus.

[2769] Rechtien, Richard D. *The detection and mapping of subterranean water bearing channels: completion report.* Gardner, Larry W.; United States Office of Water Resources Research; University of Missouri Rolla Water Resources Research Center. Rolla, MO, US: Missouri Water Resources Research Center, University of Missouri-Rolla, **1972**. 13p, [2]; bibl; 28 cm.

Title from cover. Project no. A-051, OWRR agreement no 14-31-0001-3525.

[2770] Reddell, James Russell, 1938– *A bibliographic guide to Texas speleology.* [Austin, TX, US]: Texas Speleological Association, Oct, **1968**. Texas Speleological Survey v 3 no 3. [i], 173p; bibl, index; 28 cm.

[2771] Reddell, James Russell, 1938– *Catalogue, bibliography, and generic revision of the order Schizomida (Arachnida).* Cokendolpher, James C. Austin, TX, US: Texas Memorial Museum, College of Natural Sciences, University of Texas at Austin, **1995**. Speleological Monographs no 4. illus: b&w: drawings, maps, photos, tables; 170p; bibl, index.

[2772] Reddell, James Russell, 1938– *The caves of Langtry.* Russell, William Hart, 1937–(ed). Austin, TX, US: The Texas Region of the National Speleological Society, **nd**. Texas Speleological Survey v1 no 2. illus: b&w: charts, maps, tables; 2 fold-out maps, 1 fold-out chart; 29p; bibl.

[2773] Reddell, James Russell, 1938– *The caves of Bexar County.* Knox, Orion (ed). Austin, TX, US: Texas Speleological Association, Mar, **1962**. Texas Speleological Survey v 1 no 4. illus: b&w: maps, tables; 3 fold-out maps; 38p; bibl.

[2774] Reddell, James Russell, 1938– *The caves of Comal county.* Austin, TX, US: Texas Speleological Association, Jun, **1964**. Texas Speleological Survey v 2 no 2. illus: b&w: maps; partly col: maps; 6 maps folded in text; 60p; bibl; 28 cm.

Illus.

[2775] Reddell, James Russell, 1938– *The caves of Edwards County.* Austin, TX, US: Texas Speleological Association, Nov, **1965**. Texas Speleological Survey v 2 no 5-6. illus: b&w: maps; 3 maps folded in text; 70p; bibl.

[2776] Reddell, James Russell, 1938– *The caves of far west Texas.* Fieseler, Ronald Glenn, 1948–(ed). Austin, TX, US: Texas Speleological Survey, Apr, **1977**. Texas Speleological Survey v 4 no 2. illus: b&w: maps, tables; x, 103p; bibl.

Spelled Far-West on cover.

[2777] Reddell, James Russell, 1938– *The caves of Lubbock County.* Austin, TX, US: Texas Speleological Association, Mar, **1970**. Texas Speleological Survey v 3 no 4. illus: b&w: maps; partly col: maps; [i], 16p, [4]; 28 cm.

[2778] Reddell, James Russell, 1938– *The caves of Medina county.* Austin, TX, US: Texas Speleological Association, Apr, **1967**. Texas Speleological Survey v 3 no 1. illus: b&w: maps; col: maps; 6 maps folded in text; 58p; bibl; 28 cm.

2nd printing, August 1970.

[2779] Reddell, James Russell, 1938– *The caves of northwest Texas.* Russell, William Hart, 1937–(ed). Austin, TX, US: Texas Speleological Association, Jun,

1963. Texas Speleological Survey v 1 no 8. illus: b&w: maps, tables; 11 fold-out maps; 56p; bibl.

[2780] **Reddell, James Russell**, 1938– *The caves of San Saba County.* 2nd ed. Austin, TX, US: Association for Mexican Cave Studies, Feb, **1973**. Texas Speleological Survey v 3 no 7-8. illus: b&w: maps, photos; partly col: maps; 6 maps folded in text; [ii], 127p, [9]; bibl, index; 28 cm.

[2781] **Reddell, James Russell**, 1938– *The caves of San Saba County part I.* Estes, James H. (ed). Austin, TX, US: Texas Speleological Association, Jul, **1962**. Texas Speleological Survey v 1 no 6. illus: b&w: maps, tables; 3 fold-out maps; 42p.

[2782] **Reddell, James Russell**, 1938– *The caves of Travis County.* Russell, William Hart, 1937–(ed). Austin, TX, US: The Texas Region of the National Speleological Society, **nd**. Texas Speleological Survey v 1 no 1. illus: b&w: maps, tables; 2 fold-out maps; 31p; bibl.

[2783] **Reddell, James Russell**, 1938– *The caves of Val Verde County.* Austin, TX, US: Texas Speleological Association, Feb, **1963**. Texas Speleological Survey v 1 no 7. illus: b&w: maps, tables; 4 fold-out maps; 53p; bibl.

[2784] **Reddell, James Russell**, 1938– *The caves of Willamson County.* Finch, Richard (ed). Austin, TX, US: Texas Speleological Association, Oct, **1963**. Texas Speleological Survey v 2 no 1. illus: b&w: maps; partly col: maps; col: maps; 1 map folded in text; 61p; bibl; 28 cm.

Reprinted August 1970.

[2785] **Reddell, James Russell**, 1938– *A checklist and annotated bibliography of the subterranean aquatic fauna of Texas.* Mitchell, Robert Wetsel, 1933–; Water Resources Center. Lubbock, TX, US: Department of Biology, Texas Tech University, Aug, **1969**. Special Report Number 24 International Center for Arid and Semi-Arid Land Studies. [i], 48p.

[2786] **Reddell, James Russell**, 1938– *A checklist of the cave fauna of Texas.* [Austin, TX, US]: np, **1965**. bibl; 23 cm.

Title from cover. Vols. 1-2 reprinted from The Texas journal of science, vol. XVII, no. 2, June 1965 and vol. XVIII, no. 1, May 1966. v. 1. The invertebrata (exclusive of insecta)—v. 2. Insecta.

[2787] **Reddell, James Russell**, 1938– *A checklist of the caves of Texas.* Russell, William Hart, 1937–(ed). Austin, TX, US: Texas Speleological Association, Apr, **1962**. Texas Speleological Survey v1 no 5. illus: b&w: tables; 27p.

[2788] **Reddell, James Russell**, 1938– *A field guide to caves of Blanco, Gillespie, and Llano counties, Texas.* Elliott, William Rawleigh, 1946–(ed); Smith, A. Richard (ed). [Austin, TX, US]: Texas Speleological Survey, Apr, **1989**. illus: b&w: drawings, graphs, maps, tables; 4 fold-out maps; [iv], 101p, [35]; bibl.

[2789] **Reddell, James Russell**, 1938– *Further studies on the cavernicole fauna of Mexico and adjacent regions.* Austin, TX, US: [Association for Mexican Cave Studies, Texas Memorial Museum; The University of Texas at Austin], Mar, **1982**. Bulletin 8 Association for Mexican Cave Studies; Bulletin 28 Texas Memorial Museum. illus: b&w: maps, photos, tables; [v], 288p; bibl, index; 29 cm.

Partly in Spainish.

[2790] **Reddell, James Russell**, 1938– *A preliminary bibliography of Mexican cave biology with a checklist of published records.* Austin, TX, US: Association for Mexican Cave Studies, **1971**. Association for Mexican Cave Studies Bulletin 3. illus: col: photos; col front; [iii], 184p; bibl, index.

Arranged by taxonomic categories. Includes cave index.

[2791] **Reddell, James Russell**, 1938– *A preliminary survey of the caves of the Yucatan Peninsula.* reprint. Austin, TX, US: Association for Mexican Cave Studies, **1977**. Association for Mexican Cave Studies Bulletin 6, pp 215-296. illus: b&w: maps; 81p; bibl.

Illus.

Catalog of about ninety known caves in the area. Descriptions and location maps. No cave maps. Introductory text covers especially archaeology and biology. Reprinted from Reddell, *Studies on the caves and cave fauna of the Yucatan Peninsula* (AMCS Bulletin 6) 1977.

[2792] **Reddell, James Russell**, 1938– *A review of the cavernicole fauna of Mexico, Guatemala, and Belize.* Austin, TX, US: Texas Memorial Museum, The University of Texas at Austin, Jul, **1981**. Bulletin 27 Texas Memorial Museum University of Texas at Austin. illus: b&w: maps, tables; [iii], 327p; app, bibl; 29 cm.

87 figs, 31 tables, supp.

A list of and distribution data for cavernicolous fauna of three countries, including nearly three hundred troglobitic species. The standard reference for the region.

[2793] **Reddell, James Russell**, 1938– *A revised checklist of Texas caves.* Smith, A. Richard. Austin, TX, US: Texas Speleological Association, Jul, **1966**. Texas Speleological Survey v 2 no 8. Various pagination, i, 37p; index.

No illus.

[2794] **Reddell, James Russell**, 1938– *Studies on the cave and endogean fauna of North America.* Austin, TX, US: Texas Memorial Museum, The University of Texas at Austin, Dec, **1986**. Texas Memorial Museum Speleological Monographs 1. illus: b&w: drawings, maps, photos, tables; [v], 167p; bibl, index; 29 cm.

[2795] **Reddell, James Russell**, 1938– *Studies on the cave and endogean fauna of North America II.* Austin, TX, US: Texas Memorial Museum, Dec, **1992**. Texas Memorial Museum Speleological Monographs 3. illus: b&w: charts, drawings, maps, photomicrographs, tables; vii, 257p; bibl, index; 29 cm.

[2796] Reddell, James Russell, 1938– *Studies on the cavernicole fauna of Mexico.* Mitchell, Robert Wetsel, 1933–. Austin, TX, US: Association for Mexican Cave Studies, Sep, **1971**. Association for Mexican Cave Studies Bulletin 4. illus: b&w: drawings, photos, tables; front: b&w photo; [vii], 239p, [3]; bibl, errat.

[2797] Reddell, James Russell, 1938– *Studies on the caves and cave fauna of the Yucatan Peninsula.* Austin, TX, US: Association for Mexican Cave Studies, **1977**. Association for Mexican Cave Studies Bulletin 6. illus: b&w: drawings, maps, photos, tables; col: photos; front: col photo; [viii], 296p; bibl; 29 cm.

[2798] Reed, Alan D. *A stabilization assessment of Mantel's Cave, site 5MF1, Dinosaur National Monument, Colorado.* Alpine Archaeological Consultants. Montrose, CO, US: Alpine Archaeological Consultants, Inc, **1988**. illus: b&w: photos; [67+6]; 29 cm.

Prepared for Dinosaur Nature Association, and submitted to [the] National Park Service, Dinosaur National Monument.

[2799] Reeds, Chester Albert, 1882– *The Endless Caverns of the Shenandoah Valley, an account of the wonderful work of water: with special reference as to how the caverns and the Shenandoah Valley were formed.* New York, NY, US: Eastern Printing Corporation, **1925**. illus: b&w: maps, photos; front: b&w photo; 48p; 25 cm.

One copy examined lists the publisher as The Evans-Brown Company, Inc. Microfilm. Atlanta, Ga.: SOLINET, 1992. 1 microfilm reel ; 35 mm. Cover title: The Endless Caverns of the Shenandoah Valley.

[2800] Reeds, Chester Albert, 1882– *Rivers that glow underground.* New York, NY, US: The American Museum of Natural History, **1928**. Reprinted from Natural history vol XXVIII no 2 1928 pp 131-146. 16p; 25 cm.

Illus.

[2801] Reeve, Terry *Caves & swallets in chalk.* London, GB: Chelsea Speleological Society, **1979**. Records of the Chelsea Speleological Society v 9. 69p; 21 cm.

No illus.

[2802] Reich, J. R. (comp) *Caves of Pennsylvania's Piedmont region.* Marietta, PA, US: Speleo-Research Associates, **1969**. illus: b&w: photos, tables; partly col: maps; ii, 67p.

[2803] Reich, J. R. (comp) *Caves of southeastern Pennsylvania.* Harrisburg, PA, US: Commonwealth of Pennsylvania, Department of Environmental Resources, Bureau of Topographic and Geologic Survey, **1974**. General geology report 65. illus: b&w: drawings, maps, photos, ports, tables; folded maps in pocket; [i], xi, 120p; bibl, errat, gloss; 23 cm.

Pennsylvania Geological Survey, fourth series. 47 figs, 17 pls, 2 ports.

[2804] Reid-Cowan Productions *Caves: the dark wilderness.* Herman, Matthew N. (prod). Chicago, IL, US: Encyclopaedia Britannica Educational Corp, **1976**.

Explores the ecology of caves, showing how animals have adapted to life in a cave. Explains the creation of internal cave formations from the traces of limestone in water drops and shows examples of caves formed by volcanoes, water, earthquakes, and desert winds. 1 reel, 22 min, sd, col, 16 mm & guide and as 1 videocassette (VHS) (24 min), sd, col, 1/2 in.

[2805] Religious Tract Society *The caves of the earth: their natural history, features, and incidents.* London, GB: np, **1800**. front; [i], vi, 192p; 15.5 x 10 cm.

Date of 1800 was supplied from the OCLC record. May be the same at the book with the same title, written by Daniel Parish Kidder. Also reissued in [1852], together with a second part "Mines and minin", under the title "The caves and mines of the earth." Another edition [nd] was published by the American Sunday School Union.

[2806] Rentz, David C. *A new genus and species of camel cricket from the Farallon Islands of California (Orthoptera: Gryllacrididae).* San Francisco, CA, US: The Academy, **1972**. Occasional Papers of the California Academy of Sciences no 93. 13p; add, bibl; 26 cm.

Title from cover. Illus.

[2807] Repenning, Charles Albert, 1922– *The microtine rodents of the Cheetah Room fauna, Hamilton Cave, West Virginia, and the spontaneous origin of* Synaptomys. Grady, Frederick. Washington, Denver, WA, CO, US: U.S. Government Printing Office , **1989**. U.S. Geological Survey Bulletin 1853. iv, 32p; bibl, index; 28 cm.

Illus.

[2808] Reynolds, Sidney Hugh *A monograph of the British Pleistocene Mammalia vol II part II: the bears.* London, GB: For the Palaeontographical Society, **1906**. illus: b&w: charts, drawings; 35p, [2], 14p pls; 28 x 22 cm.

Bound with other monograph in series.

[2809] Reynolds, Sidney Hugh *A monograph of the Pleistocene mammalia vol II, part I: the cave hyaena.* London, GB: printed for the Palaeontographical Society, **1902**. illus: b&w: charts, drawings; 25p, [2], 14p pls; bibl; 28 x 22 cm.

Pagination is odd.

[2810] Rheams, Karen F. *Hydrogeologic and biologic factors related to the occurrence of the Alabama cave shrimp* (Palaemonias alabamae), *Madison County, Alabama.* Moser, Paul H.; McGregor, Stuart W. Tuscaloosa, AL, US: Geological Survey of Alabama, Environmental Geology Division, **1994**. Bulletin Geological Survey of Alabama 161. illus: b&w: maps; 147p, 4p pls; bibl; 28 cm.

Illus.

[2811] Rhodes, Andrew Jackson, 1829-1907 *Presenting the wonders of Lost River.* Armstrong, Robert R., 1934–. reprint. Indianapolis, IN, US: GIG Publications, **1973**. illus: b&w: repros; text and copyright on front lining page; text on rear lining page; front: repro drawing; [iv], 21p; app; 18 cm.

This is a reprint, with a new intro of *The wonders of Lost River: 40 miles of underground river with its caves and hidden water falls, blind fish and other wonders* by A.J. Rhodes in 1905 in Indiana.

[2812] **Rhodes, Andrew Jackson**, 1829-1907 *The wonders of Lost River: 40 miles of underground river with its caves and hidden waterfalls, blind fish and other wonders.* [IN, US]: np, **1905**. 21p; 15 cm.

[2813] **Richard, Colette** *Climbing blind.* Dale, Norman (trans); Herzog, Maurice (pref); Casteret, Norbert, 1897-1987 (intro). 1966 ed. London, GB, US: Hodder & Stoughton. illus: b&w: photos, ports; 159p; 21 cm.

Originally published as Des cimes aux cavernes, Mulhouse, Editions Salvador, 1965. Epilog.

1967 ed. New York, NY, US: E.P. Dutton & Co, Inc. illus: b&w: photos, ports.

[2814] **Richards, James Harray** *Waitomo caves.* 1953 ed. Wellington, NZ: A. H. & A. W. Reed, Dec. illus: b&w: drawings, maps, photos; text on front lining page, text and map on rear lining page; 48p; 25 cm.

Cover title: Waitomo Caves: New Zealand's Underground Wonderland.

1954 ed. illus: b&w: charts, maps, photos; col: photos.

First edition 1953. Reprinted 1955, 1956, 1958, 1961.

1958 ed. illus: b&w: drawings, maps, photos, ports; col: photos; text on front lining page; maps on rear lining page.

1961 ed.

[2815] **Richardson, Douglas Turnbull (comp)** *White Rose Pothole Club library catalog November 1962.* Skipton, Yorkshire, GB: [White Rose Pothole Club], Nov, **1962**. vi, 46p; 25 x 20 cm.

No illus.

[2816] **Rickels, Curtis Eddie** *The three ring circus: how Rickwood Caverns became a state park.* Warrior, AL, US: C.E. Rickels, **1992**. ii, 289p; 21 cm.

Title from cover; illus (some col).

[2817] **Riddell, Francis A.** *The archaeology of two Kern County sites.* Heizer, Robert Fleming, 1915–. Berkeley, CA, US: University of California, Department of Anthropology, **1951**. Reports of the University of California Archaeological Survey no 10. ii, 36p; bibl; 28 cm.

Contents: The archaeology of site Ker-74. / F.A. Riddell—A cave burial from Kern County / R.F. Heizer. Photocopy. Salinas, Calif.: Coyote Press, 1985? Illus.

[2818] **Riddell, William Hatton** *Altamira: a note upon the Palaeolithic paintings in the Cave of Altamira near Santillane del Mar in the Spanish province of Santander.* Edinburgh, London, GB: Oliver and Boyd, **1938**. illus: b&w: drawings; col: repros; folded col repro in pocket in rear; 50p, [1]; 18.5 x 12.5 cm.

[2819] **Riely, Samuel L.** *Story of Wyandotte Cave.* Wyandotte, IN, US: Samuel L. Riely Co, **1945**. 32p; 23 cm.

Illus.

[2820] **Rigby, Susan** *Caves.* Burns, Robert (illus); et al. Mahwah, NJ, US: Troll Associates, **1992**. Our Planet. illus: col: drawings, maps, photos; 32p; index; 23 cm.

"Examines the formation, ecology, and folklore of caves." Juvenile.

[2821] **Ringer, Árpád** *Prehistoric remains in Hungary.* Tádal, Ödön (trans); Hazslinszky, Tamás (ed). Budapest, HU: 10th International Congress of Speleology, **1989**. Field Trip Guide D3. illus: b&w: charts, maps; 20p, 20p pls.

[2822] **Ripoll Perello, Eduardo** *The Cave of Las Monedas in Puente Viesgo (Santander).* Jacques, Richard; Matthews, John. Barcelona, ES: Instituto de Prehistoria y Arqueologia, Diputacion Provincial de Barcelona ; Wenner-Gren Foundation for Anthropological Research, **1980**. Monographs on Cave Art Paleolithic Art I. 67p; 31 cm.

Illus. Also published in New York, NY, US.

[2823] **Ripoll Perello, Eduardo** *The painted shelters of La Gasulla (Castellon).* Barcelona, ES: Instituto de Prehistoria y Arqueologia de la Diputacion Provincial de Barcelona ; Wenner-Gren Foundation for Anthropological Research, **1968**. Monographs on Cave Art Levantine Art no 2. seven folded lvs of pls; 59p, 35p pls; bibl; 32 cm.

Translation of Pinturas rupestres de la Gasulla, Castellon. Col illus. Also published in New York, NY, US.

[2824] **Roberson, Gary** *ISS report on Binkley's Cave system.* New Albany, IN, US: Indiana Speleological Survey, **c1970**. illus: b&w: drawings, maps, photos; [40].

Pagination approximate.

[2825] **Roberson, Gary T.** *Official souvenir book of Marengo Cave: U.S. National Landmark: Marengo, Indiana, the heart of cave country.* Marengo, IN, US: Marengo Cave Park, **c1989**. illus: b&w: charts, maps, photos; text on lining pages; [i], 32p, [1]; 22 cm.

Title from cover.

[2826] **Roberts, Allan** *Underground life.* Chicago, IL, US: Childrens Press, **1983**. I Want to Know About ... 26 New True Book. 45p; index; 22 cm.

Describes the characteristics and habits of a variety of insects and animals that live in the soil or in caves. Col illus. Also published (reprinted?) in 1985. Juvenile.

[2827] **Roberts, Jon** *Mendip Caving Group: 1982 library list.* Bristol, GB: Mendip Caving Group, Nov, **1982**. [ii], 60p; 30 x 21 cm.

MCG Recorder (was by author name). No illus.

[2828] **Robinson, Donald** *Caving and potholing.* Greenbank, Anthony; Marchant, L. F. (illus). London, GB: Constable & Company, Ltd, **1964**. illus: b&w: maps, photos; 171p; app, bibl, gloss, index.

15 b&w photos, maps.

[2829] Robinson, Donald *Potholing and caving*. Regional Councils of Caving Clubs in Great Britain. 1967 ed. Wakefield, England, GB: E.P. Publishing Co Ltd for Educational Productions Limited, London. Know the Game Series. illus: b&w: drawings; partly col: drawings, maps, photos; ads on lining pages; 40p; bibl; 14 x 21 cm.

Cover title: Potholing & caving. "Published in collaboration with the Regional Councils of Caving Clubs in Great Britain." 40 figs.

1974 ed. 40p.

[2830] Robinson, E. Russell, 1932-1981 *The story of the Shoshone Indian ice caves*. [ID, US]: [E. Russell Robinson], **1978**. illus: b&w: drawings, maps, photos, ports; partly col: maps; in memoriam on front lining page; [30]; 23 cm.

Printed by Ace Printing July 1964 per letter to William Halliday from the author. Illus.

[2831] Robison, Mabel Otis *The hole in the mountain*. New York, NY, US: Dodd, Mead & Company, **1966**. illus: b&w: maps, photos, ports; front: b&w photo; 127p, [1]; 24 cm.

Cover title: The hole in the mountain: the story of the Carlsbad Caverns. Illus. Juvenile.

[2832] Robison, Mabel Otis *Mystic wonderlands*. Minneapolis, MN, US: T. S. Denison & Company, Inc, **1959**. illus: b&w: photos; 95p, [1]; 29 cm.

[2833] Robson, M. *The decomposition of limestone breccia from the cave of Les Eyzies, Dordogne, France*. London, GB: Institute of Archaeology, University of London, **1975**. xiv, 98p; bibl.

Microfiche. London: Institute of Archaeology, University of London, [1984?] 2 microfiches: negative ; 11 x 15 cm. Illus.

[2834] Rocky Mountain Region. Guidebook (1991) *Top secret, for cavers' eyes only: 1991 Rocky Mountain Regional guidebook, Black Hills, South Dakota*. Hanson, Mike (ed). [Custer, SD, US]: Paha Sapa Grotto, **1991**. illus: b&w: maps, tables; 24p.

Title from cover.

[2835] Rocky Mountain Region. Guidebook (1994) *Caves of the eastern White River Plateau: guidebook to the 1994 Rocky Mountain regional, Willow Peak, Colorado*. Rhinehart, Richard (ed). npop: Front Range Grotto, **nd**. illus: b&w: maps; 1 loose folded map; maps on lining pages; 16p.

Title form cover.

[2836] Rodda, Jan *Mulka's Cave site management project: emphasising visitor survey, April-June 1988, with management evaluation and further recommendations for management; with conclusions of an archaeological test excavation*. [Perth, WA, US]: Western Australian Heritage Committee, Western Australian Museum, Department of Aboriginal Sites, **1989**. illus: b&w: maps; viii, 88p; 30 cm.

Col illus.

[2837] Rodriguez, Ortega Eduardo (auth, photo) *The cave of Nerja*. [Malaga, ES]: np, **1970**. illus: b&w: drawings, maps, photos; 181p; 25 cm.

Illus.

[2838] Rodway School Speleological Group *Caves of the Avon Gorge Bristol, part 2: eastern or Clifton side*. Mangotsfield, Bristol, GB: Rodway School, **1974**. illus: b&w: drawings, maps, photos; 1 folded map loose; front: b&w photo; 76p; bibl, index.

7 pls, 5 figs, 14 maps.

[2839] Rodway School Speleological Group *Caves of the Avon Gorge Bristol: part 1: western or Leigh Woods side*. Bristol, GB: Rodway School, **1973**. illus: b&w: cartoons, drawings, maps, photos; 1 pull out map mounted inside back cover; v, 64p; bibl, index; 21 x 14.5 cm.

8 pls, 17 figs, 16 surveys.

[2840] Roemer, Ferdinand Carl, 1818-1891 *The bone caves of Ojcow in Poland*. Lee, John Edward, 1808-1887. London, GB: Longmans, Green, and Co, **1884**. illus: b&w: maps; [iii], viii, 41p, [23]; 31 cm.

Each plate (12 pls) accompanied by leaf with descriptive letterpress.

[2841] Rogers, Bruce W. *Karst features of Pohn Pei, Chuuk, and Waqab: a survey of caves and karst features in the States of Pohn Pei, Chuuk, and Waqab, Federated States of Micronesia*. Legge, Charmaine J. San Francisco, CA, US: Ring of Fire Press, Jan, **1992**. Pacific Basin Speleological Survey Bulletin 1. 10p; bibl.

Submitted to Division of Cultural Affairs, Federated States of Micronesia. Bulletins 1-5 bound together.

[2842] Rogers, Bruce W. *Karst features of Rota (Luta) Island, CNMI, A survey of caves and karst features on Rota (Luta) Island, Commonwealth of the Northern Mariana Islands*. Legge, Charmaine J. San Francisco, CA, US: Ring of Fire Press, May, **1992**. Pacific Basin Speleological Survey Bulletin 4. 20p; bibl.

Submitted to Cultural and Historical Preservation Department, Rota Island, Commonwealth of the Northern Mariana Islands. Bulletins 1-5 bound together.

[2843] Rogers, Bruce W. *Karst features of Saipan, CNMI: a survey of caves and karst features on Saipan Island, Commonwealth of the Northern Mariana Islands*. Legge, Charmaine J. San Francisco, CA, US: Ring of Fire Press, May, **1992**. Pacific Basin Speleological Survey Bulletin 2. 26p; bibl.

Submitted to Division of Cultural Affairs, Saipan Island, Commonwealth of the Northern Mariana Islands. Bulletins 1-5 bound together.

[2844] Rogers, Bruce W. *Karst features of the Palau Islands, a survey of caves and karst features in the Palau Archipelago*. Legge, Charmaine J. San Francisco, CA, US: Ring of Fire Press, May, **1992**. Pacific Basin Speleological Survey Bulletin 3. 131p, [2].

Submitted to Division of Cultural Affairs of Belau. Bulletins 1-5 bound together.

[2845] Rogers, Bruce W. *Karst features of the Territory of Guam: a survey of caves and karst features on Guam Island, Territory.* Legge, Charmaine J. San Francisco, CA, US: Ring of Fire Press, May, **1992**. Pacific Basin Speleological Survey Bulletin 5. 33p; bibl.

Submitted to Department of Parks and Recreation, Territory of Guam. Bulletins 1-5 bound together.

[2846] Rogers, Bruce W. *A preliminary report on Vanished River Cave, Santa Cruz.* CA, US: [private printing], **1970**. 36p.

[2847] Rogers, Edmund B. *Mammoth Cave National Park, Kentucky: history of legislation through the 82nd Congress.* United States National Park Service. [Washington, DC, US]: U.S. National Park Service, **1958**. History of Legislation Relating to the National Park System through the 82d Congress v 56. Various pagination; 23 cm.

No illus.

[2848] Rogers, Edmund B. *Oregon Caves National Monument, Oregon: history of legislation through the 82nd Congress.* [Washington, DC, US]: U.S. National Park Service, **1958**. History of Legislation Relating to the National Park System Through the 82d Congress v 72. Various pagination; 23 cm.

First in the vol with binder's title: Oregon Caves/Pea Ridge/Perry's Victory. No illus.

[2849] Rogers, Edmund B. *Wind Cave National Park, South Dakota: history of legislation through the 82nd Congress.* [Washington, DC, US]: U.S. National Park Service, **1958**. History of Legislation Relating to the National Park System Through the 82d Congress v 93. Various pagination; 23 cm.

Second in the vol. with binder's title: Washington - Gettysburg Boulevard / Wind Cave. No illus.

[2850] Rohrer, Thomas A. *Caves and cave surveying.* npop: np, **1961**. 6 folded maps; 17p; 22 cm.

[2851] Rolsh Photographies *Cave wonderlands of Western Australia; Jewel Caves, Lake Cave, Mammoth Cave, Yallingup Cave.* npop: np, **nd**. illus: col: photos; text on front lining page; map on rear lining page; 26p; 28 x 21 cm.

List of "other West Australian caves".

[2852] Romberg, Frederick *The detection of subsurface voids by gravimetry.* Dallas, TX, US: Texas Instruments, **1961**. Texas Instruments Inc. illus: b&w: maps; 5 maps in pocket; Various pagination, 58p; bibl; 30 cm.

AFCRL no. 1014. Final report, contract no. AF19(604)-8348. Illus.

[2853] Ronquillo, Wilfredo P. *The technological and functional analyses of lithic flake tools from Rabel Cave, northern Luzon, Philippines.* Manila, PH: np, Dec, **1981**. Anthropological Papers National Museum Philippines no 13. 41p; bibl; 26 cm.

Illus.

[2854] Rose, David, 1959– *Beneath the mountains: exploring the deep caves of Asturias.* Gregson, Richard, 1956–. 1987 ed. London, GB: Hodder & Stoughton. illus: b&w: maps, ports; col: photos; 414p, 8p pls; gloss; 23 cm.

The story of the Oxford University Cave Club's discoveries in the caves of the Picos de Europa in Northern Spain from 1980 through 1986.

1988 ed. Leicester, GB: Ulverscroft Large Print. Ulverscroft Large Print. illus: b&w: maps; col: photos, ports; gloss.

Large print. Illus (some col).

[2855] Rosen, Donn Eric, 1929– *Comments on the relationships of the North American cave fishes of the family Amblyopsidae.* New York, NY, US: American Museum of Natural History, Oct, **1962**. American Museum Novitiates no 2109. illus: b&w: drawings, photos, photomicrographs, tables; Text front & rear; 35p; bibl; 23 cm.

Illus.

[2856] Ross, Richard E. *The archeology of Eagle Cave.* Parsons, Mark L.; Ross, Richard E. (apps). Austin, TX, US: Texas Archeological Salvage Project, The University of Texas, Oct, **1965**. illus: b&w: drawings, graphs, photos, tables; [ii], vii, 163p; app, bibl.

Report submitted to the National Park Service, Memorandum of agreement 14-10-0333-1121. 34 figs, 17 tables.

[2857] Ross, Sylvia H. *Introduction to Idaho caves and caving.* Huppert, George Nixon, 1944–(comp add reading). Moscow, ID, US: Idaho Bureau of Mines and Geology, **1969**. Earth Science Series no 2. illus: b&w: drawings, maps, photos; 1 map folded in back pocket; [iv], 54p; bibl; 28 cm.

33 figs.

[2858] Roswell Geological Society *Slaughter Canyon, New Cave and Capitan Reef exposures, Carlsbad Caverns National Park: field trip no. 10, April 13, 1957.* Stipp, T. F. [Roswell, NM, US]: The Society, **1957**. Roswell Geological Society Field Trip Guidebooks no 10. 19p; 29 cm.

Mimeographed.

[2859] Rother, Hubert *Lost caves of St. Louis.* Rother, Charlotte. St. Louis, MO, US: Virginia Pub Co, **1996**. illus: b&w: maps; 143p; bibl, index; 22 cm.

Cover title: History of the City's forgotten caves.

[2860] Rothrock, H. W. *Wyandotte Cave: where it is and what it is.* npop: np, **1889**. illus: b&w: ads; 12p, [2]; 15 cm.

Caption title. Cover title: Twenty-three miles underground: Wyandotte cave, what it is, where it is, and how to reach it. Typesigned at end: H.A. Rothrock. Possible dates based on ads for U.S. Mail Steamers on p. [4] of wrappers. Advertisements: [1] p. at end and p. [2]-[4] of wrappers.

[2861] Rouse, Irving, 1913– *Excavations at Maria de la Cruz Cave and Hacienda Grande Village Site, Loiza, Puerto Rico*. Alegria, Ricardo E.; Wing, Elizabeth S. New Haven, CN, US: np, **1990**. Yale University Publications in Anthropology no 80. illus: b&w: maps; viii, 133p; app, bibl; 25 cm.

[2862] Royse, John *Ancient Castleton Caves: their place in time and history, also what people want to know about Castleton*. Sheffield, GB: Chas Swinburn & Co Ltd, **nd**. illus: b&w: drawings, maps, tables; 76p.

[2863] Royse, W. *Descriptive account of the Blue John Mines and Caverns at Castleton, Derbyshire*. Sheffield, Derbyshire, GB: R. Taylor & Son, **1891**. 16p; 20 x 12 cm.
No illus.

[2864] Rozarie, Charles *The Archaeology at Lehman Caves National Monument: Nevada State Museum report*. United States National Park Service, Lehman Caves National Monument. [Carson City, NV, US]: [Nevada State Museum, Department of Archaeology], **1964**. Archives E78 U5n L52 no 4 NPS. illus: b&w: maps, photos; ii, 68p, 42p pls; app; 29 cm.

43 original photos. Nine plates folded in back. Appendices (leaves 17-62): A. An evaluation of the human skeletal remains from Lehman Caves, Nevada.—B. Animal bones from Lehman Caves National Monument.

[2865] Ruhe, R. V. *Geohydrology of karst terraine, Lost River watershed southern Indiana*. Bloomington, IN, US: Indiana University Water Resources Research Center, Dec, **1975**. Indiana University Water Resources Research Center Report of Investigation no 7. illus: b&w: charts, drawings, graphs, maps, tables; vii, 91p; bibl.

Title II Grant No. 14-31-0001-3689, Project No. C-3122 Office of Water Research and Technology U. S. Department of the Interior Completion Report.

[2866] Rusling, James Fowler *A trip to the Mammoth Cave, Ky*. Nashville, TN, US: np, **1864**. 40p; 21 cm.

[2867] Ruspoli, Mario *The cave of Lascaux: the final photographs*. Coppens, Yves (pref); Wormell, Sebastian (trans). NY, US: Harry N. Abrams, **1987**. illus: b&w: drawings, photos, tables; partly col: charts, maps; col: photos, ports; 1 fold-out map; 208p; app, bibl, index.

Also published in London by Thames and Hudson, perhaps under a different title.

[2868] Russell, D. J. *Role of the Sylvania Formation in sinkhole development, Essex County*. [Sudbury, Ontario, CA]: Ontario Geological Survey, Ministry of Northern Development and Mines, **1993**. Open File Report Ontario Geological Survey 5861. xi, 122p; bibl; 28 cm.
Illus.

[2869] Russell, Donald R. *An essay on the caves of the Cherokee Nation: a conservation plea*. [Broken Arrow, OK, US]: [Donald R. Russell], **1971**. illus: b&w: maps, photos; photo on rear lining page; [i], 50p; 28 cm.

Cover title: An essay on the caves of the Cherokee Nation. Illus.

[2870] Russell, William Hart, 1937– *The Buttercup Creek karst: Travis and Williamson Counties, Texas: geology, biology, and land development*. Austin, TX, US: University Speleological Society, Jul, **1993**. illus: b&w: charts, photos, tables; 1 fold-out map; [iv], 76p, 1p pls; bibl.

The University Speleological Society is the University of Texas Student Grotto of the NSS. Title from cover.

[2871] Russell, William Hart, 1937– *Caves of the Inter-American Highway: Nuevo Laredo, Tamulipas to Tamazunchale, San Luis Potosi*. Raines, Terry W. Austin, TX, US: [Association for Mexican Cave Studies], **1967**. Bulletin of the Association for Mexican Cave Studies 1. illus: b&w: maps; some maps folded; 126p; bibl, index; 28 cm.
Illus.

Ground-breaking survey of the caves of northeastern Mexico. Road logs, area maps, geology and biology notes. Descriptions of perhaps sixty caves, maps of 23. Still sought after as the best general guide to the area.

[2872] Rust, Pauline Carrington *Legend of the Luray Caverns, founded upon the discovery of a skeleton in one of its chasms*. Philadelphia, PA, US: Loughead & Co, **1887**. illus: b&w: drawings, repros; 15p; 16 cm.

[2873] Rutter, John, 1796-1851 *Delineations of the north western division of the county of Somerset, and of its antediluvian bone caverns, with a geological sketch of the district*. Shaftesbury, London, GB: Printed and published by the author; Longman, Rees, and Co [etc], **1829**. illus: b&w: maps, ports; col: maps; folded map; fronts; xxiv, 349p; app; 24 cm.

Half-title: Delineations of ... Somerset,and of the Mendip caverns. 13 pls. Contains 12 chapters and 6 apps. 30 beautiful plates are included but the list of them in the front is not correct. There are beautiful hand colored maps with hand-lettered descriptions.

[2874] Ryder, Peter F. *Caves of Skye*. Jeffreys, Alan Lawrence, 1940–. Edinburgh, GB: Grampian Speleological Group, **1995**. The Limestone Caves of Scotland Part 5; Occasional Publication (Grampian Speleological Group) no 7. illus: b&w: maps, ports; vii, 73p; bibl, index; 23 cm.
Illus.

[2875] St. Pierre, David *The caves of Graatadalen, northern Norway: report of the Southwest Essex Technical College Caving Club expeditions 1963-5*. Et al. GB: Cave Research Group of Great Britain, **1966**. Transactions of the Cave Research Group of Great Britain v 8 no 1. illus: b&w: charts, maps, photos, tables; 2 fold-out maps; [iii], 64p, 2p pls.

[2876] St. Pierre, David *The Caves of Rana, Nordland, Norway*. St. Pierre, Shirley. GB: Cave Research Group of Great Britain, Mar, **1969**. Transactions of the Cave Research Group of Great Britain v 11 no 1. illus: b&w: maps, photos, tables; text on rear lining page; [vi], 71p; app, bibl.

[2877] **Salamon, Gábor** *Aslóhegy shafts*. Kosa, Attila (trans); Rádai, Ödön (trans); Hazslinszky, Tamás (ed). Budapest, HU: 10th International Congress of Speleology, **1989**. illus: b&w: charts, maps; 39p, 39p pls.

[2878] **Salamon, Gábor** *The Stalactite Caves of Jósvafő and Aggtelek*. Hazslinszky, Tamás (ed); Tádal, Ödön (trans). Budapest, HU: 10th International Congress of Speleology, **1989**. Field Trip Guide E4. illus: b&w: charts, maps; 30p.

[2879] **Saltsman, Dayton** *The art of safe cave diving*. Gainesville, FL, US: National Association for Cave Diving , **1995** . xxi, 221p; bibl; 28 cm.

Updated ed of: Safe cave diving. 1973. Illus.

[2880] **Salvona, J.** *Scottish cave guides; the southern highlands*. Young, I. W. Lothian, GB: Ivan Young, **1988**. illus: b&w: maps; ii, 34p; index; 21 x 15 cm.

[2881] **Salvona, J.** *The Southern highlands*. Young, Ivan. reprint with rev. Kirkliston, [GB]: I. Young, **1989**. Scottish Cave Guides. illus: b&w: maps; ii, 34p; index; 21 cm.

First published 1988.

[2882] **Sameshima, Teruhiko** *5th international symposium on vulcanospeleology excursion guide book*. Ogawa, Takanori; Kashima, Naruhiko. Tokyo, JP: Secreteriete 5 International Symposium on Vulcanospeleology, **1988**. illus: b&w: maps, photos; 3 loose large maps and 2 foldouts; [ii], 80p; 30 x 21 cm.

[2883] **Sandak, Inc.** *Cave paintings—Lascaux and Altamira*. npop: np, **1973**.

With teacher's guide. Presents examples of cave paintings and reliefs found in the Spanish caves of Altamira and Castillo and in the French cave of Lascaux. 30 slides, color, 2 x 2 in.

[2884] **Santamarta Cuenca, Pedro** *The grottos of Majorca*. Leon, ES: Editorial Everest, **1978**. Coleccion Iberica. illus: b&w: maps, repros; partly col: maps; col: charts, maps, photos; front: col map; 64p; bibl; 25 cm.

Descriptions of Majorca caves. 55 col photos. Published in Spanish as "Las cuevas de Mallorca: (Hams, Drach, y Arta)".

[2885] **Sarawak Museum** *Summary of archaeological work in Sarawak: with special reference to Niah Caves*. Kuching, [MY]: Borneo Literature Bureau, **1973**. 13p; bibl; 19 cm.

Illus.

[2886] **Sartor, James Doyne** *Meteorological-geological investigations of the Wupatki blowhole system*. Lamar, D. L. Santa Monica, CA, US: The Rand Corporation, **1962**. Research Memorandum Rand Corporation RM-3139-RC. 41p.

Illus.

[2887] **Sasowsky, Ira Daniel**, 1959– *Breakthroughs in karst geomicrobiology and redox geochemistry: abstracts and field trip guide for the symposium held February 16 - 19, 1994, Colorado Springs, Colorado*. Palmer, Margaret V. Charles Town, WV, US: Karst Waters Institute, **1994**. Special publications of the Karst Waters Institute no 1. illus: b&w: charts, drawings, graphs, maps, photos, photomicrographs, tables; xiv, 111p; bibl, index; 28 cm.

An exciting collection of short papers addressing a broad range of karst-related topics., including the mixing of natural ground waters, redox geochemistry, the role, distribution and nature of microbes in caves, and more. Much of this material may not appear in the "open" literature for some years, although these studies reflect the sense of present interests and research directions.

[2888] **Sasowsky, Ira Daniel**, 1959– *Index to volumes 46 through 50 of the National Speleological Society bulletin*. Huntsville, AL, US: National Speleological Society, **1991**. text on front; 64p.

[2889] **Sattler, Robert A.** *Quantitative analysis of a late-Quaternary cave deposit, Porcupine River, Alaska*. [Fairbanks, AK, US]: Geist Fund Committee, University of Alaska Museum, Apr, **1989**. illus: b&w: maps; 14p; bibl; 28 cm.

Cover title. "Final Report to the Geist Committee, University of Alaska Museum, 1 April 1989."

[2890] **Sauro, U. (ed)** *Proceedings of the international conference on environmental changes in karst areas I.C.E.C.K.A. Italy, September 15-27th, 1991*. Bondesan, A (ed); Meneghel, M. (ed); International Geographical Union Study Group; International Speleological Union Commission for Physics-Chemistry and Hydrology of Karst. Padova, IT: Dipartimento di Geografia, Universitìdegli Studi di Padova, **1991**. illus: b&w: charts, drawings, graphs, maps, photos, tables; col: maps; one col map folded, loose; xvi, 414p; bibl, index.

A collection of 43 articles and 9 abstracts on environmental karst - all in English. A preliminary edition was issued with 23 papers, 214p.

[2891] **Sauter, Martin** *Quantification and forecasting of regional groundwater flow and transport in a karst aquifer (Gallusquelle, Maim, southwestern Germany)*. Tübingen, DE: Tübingen Geolwissenschaftliche Arbeiten, University of Tübomgem, **1992**. Series C no13. viii, 151p.

b&w figs.

[2892] **Savidge, Bob** *Cwmbran Caving Club New Mexico expedition report; caving in the Guadalupe Mountains and Carlsbad National Park April 1994*. Et al. npop: [Cwmbran Caving Club], [**1995**]. illus: b&w: maps; 38p, [1]; 29.5 x 21 cm.

Front cover included in pagination.

[2893] **Savory, James Henry**, 1889-1962 *A man deep in Mendip: the caving diaries of Harry Savory, 1910—1921*.

Savory, John, 1943– (ed). Gloucester, GB: Alan Sutton, **1989**. illus: b&w: drawings, maps, photos; xviii, 150p; index; 24 x 26 cm.

American edition: 1990: Carbondale, IL: Southern Illinois University Press.

The illustrated, edited diaries of professional photographer Harry Savory, arguably the best of the early British cave photographers. Diary entries cover how important discoveries were made on Mendip, photographed as they occurred.

[2894] Sawtell, Ruth Otis *Primitive hearths in the Pyrenees; the story of a summer's exploration in the haunts of prehistoric man.* Treat, Ida; Vaillant-Couturier, Paul (illus). New York, NY, US: D. Appleton, **1927**. illus: b&w: drawings, maps, photos; b&w drawing; xiv, 306p, 17p pls; 21 cm.

Visits to French caves by 2 American ladies.

[2895] Say, L. W. (ed) *Cheddar Caves Museum.* npop: np, **1976**. illus: col: photos; printer's info & museum info on rear lining page; 24p.

[2896] Sbordoni, Valerio (ed) *Symposium on speciation and adaptation to cave life: gradual vs. rectangular evolution: a symposium organized by the Unione Zoologica Italiana and the Société de Biospéologie, Roma, October 6-11, 1986: part I.* [IT]: Societá Speleologica Italiana, **1987**. International Journal of Speleology v 16 no 1-2. illus: b&w: charts, drawings, graphs, maps, tables; text on lining pages; iv, 66p.

[2897] Sbordoni, Valerio (ed) *Symposium on speciation and adaptation to cave life: gradual vs. rectangular evolution: a symposium organized by the Unione Zoologica Italiana and the Société de Biospéologie, Roma, October 6-11, 1986: part II.* [IT]: Societá Speleologica Italiana, **1988**. International Journal of Speleology v 17 no 1-4. illus: b&w: charts, drawings, graphs, maps, tables; [ii], 80p; errat.

[2898] Schaik, David Cornelis van *The old town of Maastricht and the caves of mount St. Peter.* Hear near Maastricht, NL: The author, **1945**. maps on pp [2] and [3] of cover; 48p; 22 cm.

On cover: Souvenir-edition in honour of our American liberators. Illus.

[2899] Scheele, William E. *The cave hunters.* Cleveland, OH, US: The World Publishing Company, **1959**. illus: b&w: drawings; 63p, [1].

Juvenile.

[2900] Scheltens, John (auth, comp) *Wind Cave 1972.* Stock, Mark (auth, comp); Flurkey, Andy (cave map prep). [Chicago, IL, US]: [Windy City Grotto], **[1973]**. illus: b&w: maps, photos, tables; 1 loose folded map; iii, 26p.

Title from cover.

[2901] [Scheltens, John(ed)] *Wind Cave expedition 1971.* [Chicago, IL, US]: [Windy City Grotto], **[1971]**. illus: b&w: maps, photos; i, 18p.

Title from cover.

[2902] [Scheltens, John (ed)] *Wind Cave expedition August 1970: a report to the National Park Service by the Windy City Grotto of the National Speleological Society on the results of the August 1970 Wind Cave expedition.* [Chicago, IL, US]: [Windy City Grotto], **[1970]**. illus: b&w: maps, photos; 17p.

Title from cover.

[2903] Schenck, Bill *Proceedings of the 6th annual seminar, Lindenwood College, St. Charles, Missouri, June 16-17, 1973; Proceedings of the 7th annual seminar, Ramada Inn West, Jacksonville, Florida, June 15- 16, 1974 ; Research papers by N.A.C.D. instructor candidates.* Gainesville, FL, US: National Association for Cave Diving, **1975**. 174p; bibl; 24 cm.

Cover title: Proceedings of the sixth and seventh annual N.A.C.D. seminars and research papers by N.A.C.D. instructor candidates. Illus.

[2904] Schiffer, Donna M. *Hydrology of the Wolf Branch sinkhole basin, Lake County, east-central Florida.* Tallahassee, FL, US: U.S. Geological Survey Information Services, **1996**. U.S. Geological Survey Open-File Report 96-143. illus: b&w: maps; iv, 29p; bibl; 28 cm.

Illus.

[2905] Schindel, Geary Michael, 1957– *Guidebook: environmental hydrogeology of karst terranes in the vicinity of Nashville Tennessee.* Hannah, E. D.; Crawford, N. C.; Hoffelt, J. F.; Quinlan, James Francis, 1936-1995. Nashville, TN, US: Friends of the Karst, Dec, **1991**. Friends of the Karst Occasional Publication no 3. illus: b&w: charts, maps; Various pagination, ii; bibl.

Guidebook for field trip: December 5, 1991: hydrogeology, ecology, monitoring, and management of ground water in karst terranes: Third conference: Nashville, Tennessee: December 4-6, 1991.

[2906] Schmidt, Jeremy *Lehman Caves.* [Baker, NV, US]: Great Basin Natural History Association, **1987**. illus: b&w: maps, photos, repros; partly col: charts, maps; col: photos; text on front lining page; map and text on rear lining page; 32p; 20 x 23 cm.

[2907] Schoon, Theo *Maori rock drawing: the Theo Schoon interpretations.* Christchurch, NZ: Robert McDougall Art Gallery, **1985**. [11]; bibl; 30 cm.

Illus. Prepared to accompany an exhibit of the artist's copies of rock drawings and his works based on their images.

[2908] Schram, Frederick R., 1943– *Structural composition and dental variations in the murids of the Broom Cave fauna, late Pleistocene, Wombeyan Caves area, N.S.W., Australia.* Turnbull, William D. Sydney, AU: Australian Museum, **1970**. illus: b&w: drawings, graphs, tables; 24p, [3]; 24 cm.

Title from cover. Illus.

[2909] **Schreiber, Richard** *Ellison's Cave: Georgia's finest.* McGuffin, Della. Atlanta, GA, US: Dogwood City Grotto of the National Speleological Society, Sep, **1969**. Georgia Underground v 6 no 3, pp 55-108. illus: b&w: maps, photos, tables; folded map in rear pocket; b&w photos on front lining page; text on rear lining page; 54p.

[2910] **Schröder, Karl-Heinz** *Guide through the Kubacher Crystal Cave.* Giessen, DE: M.G. - Schmitz-Verlag, 10, **1980**. Glossy Paper w/Photo. illus: b&w: maps; text on lining pages; [16]; 21 x 10 cm.
 Cover includes in pagination.

[2911] **Schroeder, Albert H.** *Pratt Cave studies: Guadalupe National Park, Texas.* Bronaugh, Mitchell. 1978 ed. Santa Fe, NM, US: U. S. National Park Service, Southwest Region, **1978**. illus: b&w: maps; 348p; 30 cm.
 Illus.
 1983 ed. El Paso, TX, US: El Paso Archaeological Society, **1983**. Artifact v 21 no 1-4. xi, 241p; bibl; 28 cm.
 Illus.

[2912] **Schuetz, Mardith K.** *An analysis of Val Verde County cave material.* Texas Archaeological Society. reprint. TX, US: Witte Memorial Museum, **1956**. Bulletin of the Texas Archaeological Society 1956 pp129-160. 32p; 24 cm.
 Illus.

[2913] **Schultz, Charles Bertrand**, 1908– *The fauna of Burnet Cave, Guadalupe Mountains, New Mexico.* Howard, Edgar B., 1887–. offprint. Philadelphia, PA, US: np, **1935**. Proceedings of the Academy of Natural Sciences of Philadelphia, v 87, 1935, pp 273-298. 26p, 6p pls; bibl; 28 cm.
 Illus.

[2914] **Schultz, Ronald**, 1951– *Looking inside caves and caverns.* Gadbois, Nick; Aschwanden, Peter (illus). Santa Fe, NM, US: John Muir Publications, **1993**. X-ray Vision. illus: col: drawings, photos; [ii], 46p; gloss, index; 28 cm.
 Juvenile.
 An excellent children's book which explores the world of caves and describes the work of speleologists and cavers.

[2915] **Schuurmans, Theo J. C.** *A caver's view of the Clydach River: an introduction to the underground waters of Pwll Ddu, Llangynid and Llangattwg mountains.* 2nd ed. Cwmbran, GB: Cwmbran Caving Club, **1987**. illus: b&w: maps, ports; 150p; bibl; 21 cm.
 Previous ed.: 1986. Illus.

[2916] **Schuurmans, Theo J. C.** *Postcards of the Caves of Han printed by Nels of Brussels between 1905 and 1940.* 1st ed. Cwmbran, Gwen, GB: Th. J. C. Schumrmans, **1987**. illus: b&w: repros; [iii], 83p; app; 21 cm.
 From the collections of B. Mooy, D. de Swart and T. Schuurmans.

[2917] **Schuurmans, Theo J. C.** *Postcards of the Caves of Han printed by Nels of Brussels between 1905-1940.* 2nd ed. Rotterdam, NL: Th. J. C. Schuurmans, **1989**. illus: b&w: repros; [iv], 85p; 30 x 21 cm.

[2918] **Schwartz, Douglas W.** *Prehistoric man in Mammoth Cave.* Woodson, Jack (illus). npop: Eastern National Park & Monument Association, **1965**. Interpretative Series no 2. illus: b&w: drawings, maps, photos, tables; text on front lining page; drawing and text on rear lining page; [iv], 26p, [2].
 Published in cooperation with Mammoth Cave National Park Juvenile.

[2919] **Schwartz, Jeffrey H.** *A diverse hominoid fauna from the Late Middle Pleistocene breccia cave of Tham Khuyen, Socialist Republic of Vietnam.* New York, NY, US: np, Jan, **1994**. Anthropological Papers of the American Museum of Natural History no 73. 11p; bibl; 26 cm.
 Issued January 14, 1994. Illus.

[2920] **Schwatka, Frederick** *In the land of cave and cliff dwellers.* New York, NY, US: The Cassell Publishing Company, **1893**. x, 385p.
 Illus.

[2921] **Schweiker, Roioli** *Caves of Albany County, New York.* Anderson, Richard R.; Van Note, Peter; Jurgens, Robert. 1969. npop: np. Northeastern Regional Organization of the National Speleological Society Publication no 8. illus: b&w: maps; 26p; index.
 Revised edition 1969, described from 1976 reprint.

[2922] **Schweiker, Roioli** *Field guide to New York caves: Schoharie County.* Anderson, Richard; van Note, Peter. NY, US: np, **1958**. NRO Publication no 6. illus: b&w: charts, drawings, maps; 86p; 22 x 14 cm.
 Limited ed; mimeograph printed.

[2923] **Scott, Fred**, 1942– *Preliminary investigations at Hayes Cave, Hants County, Nova Scotia, in 1978.* Halifax, CA: Nova Scotia Museum, **1979**. Curatorial Report no 38. illus: b&w: maps; 14p; add, bibl; 28 cm.
 Illus.

[2924] **Scott, John**, 1929– *Caves in Vermont, a spelunker's guide to their location and lore.* Hancock, VT, US: Killooleet, **1959**. 45p; 23 cm.
 Illus.

[2925] **Scott, Thea** *Fingal's cave.* Leicester, GB: Toni Savage of the Pandora Press, **1961**. [51]; 27 cm.
 This edition is limited to 150 copies. Col, b&w illus.

[2926] **Scott, W. Ray** *A pictorial study of Mammoth Cave and Mammoth Cave National Park.* 1950 ed. Mammoth Cave, KY, US: Mammoth Cave National Park National Park Concessions. illus: b&w: maps; 38p; 22 cm.
 Cover title: Mammoth Cave National Park, Kentucky. Illus. OCLC also lists 1959 as date of pub.
 1966 rev ed. Mammoth Cave, KY, US: National Park Concessions, Inc. illus: b&w: maps; 44p.
 Illus.

[2927] **Seger, Joe D.** *Gezer V: the field I caves.* Lance,

Hubert Darrell, 1935–. Jerusalem, IL: Hebrew Union College Nelson Glueck School of Biblical Archaeology, **1988**. Annual of the Nelson Glueck School of Biblical Archaeology. four folded plans in front pocket; xv, 243p; bibl; 29 cm.

49 pp of pls. Illus.

[2928] Selby, Paul (ed) *Sieben Hengste 74: the karst and caves of the southwestern zone of the Sieben Hengste Ridge: a report by the members of the Croydon Caving Club on an expedition to the Seefeld area, north of Interlaken, Switzerland, summer 1974.* [GB]: np, Feb, **1975**. illus: b&w: charts, drawings, maps; folded map, plate glued inside back cover; [iii], 41p; app, bibl, gloss.

[2929] Self, Charles A. *Caves of County Clare.* 1978 ed. [Bristol, GB]: University of Bristol Spelaeological Society. illus: b&w: maps; [18] pp of pls (some folded); 225p; index; 25 cm.

Catalog of 251 caves in this Irish county, with descriptions and locations. About 25 cave maps. Includes County Clare, Gort Lowlands of County Galway, and the Aran Islands.

1981 ed. illus: b&w: maps; col: maps; index.

[2930] Self, Charles A. *Not the UBSS song book.* Saxton, Claire. Bristol, GB: University of Bristol Speleological Society, Dec, **1979**. 71p, [1]; 30 x 21 cm.

Photocopy examined. No illus.

[2931] Selfinger, Edward John *Experimental geology applied to speleogenesis.* npop: np, **1964**. iii, 22p; bibl; 29 cm.

Includes abstract. (Research Paper)—South Dakota School of Mines and Technology, Rapid City, 1964. Illus.

[2932] Sellers, O. R. *A Roman-Byzantine burial cave in Northern Palestine (the joint excavation of the American School of Oriental Research in Jerusalem and McCormick Theological Seminary at Silet edhDhahr.* Baramki, D. C. New Haven, CN, US: American Schools of Oriental Research, **1953**. Bulletin of the American Schools of Oriental Research Supplementary Studies nos 15-16. illus: b&w: drawings, maps, photos; 55p.

62 figs.

[2933] Semken, Holmes A. *Mammalian remains from Rattlesnake Cave, Kinney County, Texas.* Austin, TX, US: Texas Memorial Museum, **1967**. The Pearce-Sellards Series no 7. illus: b&w: graphs, tables; Text on rear; 11p; bibl.

Title from cover. Illus.

[2934] Senior, Kevin J. (ed) *The Yangtze Gorges expedition: China caves project.* Bridgwater, Somerset, GB: British Cave Research Association, Oct, **1995**. Cave Science: Transactions of the British Cave Research Association v 22 no 2. illus: b&w: charts, maps, photos, tables; text on lining pages; 42p; app, bibl.

[2935] Seppala, Matti *Glacier cave observations on Llewellyn Glacier, British Columbia.* 1973 reprint ed. Vammala, FI: Vammalan Kirjapaino, **1973**. Turun Yliopiston Maantieteen Laitoksen Julkaisuja no 60. 15p; 25 cm.

Reprint from Acta Geographica 27 (1972). Illus.

1st ed. Helsinki, FI: Societas Geographica Fenniae, **1972**. Acta Geographica Helsinki, Finland 27. 15p; 25 cm.

Illus.

[2936] Šerko, Alfred, 1910-1948 *The Cave of Postojna and other curiosities of the karst.* Ivan, Michler; Klemenčič, Slavo (trans). 1953 ed. [Postojna, SI]: [Kraške jame Slovenije], **1953**. illus: b&w: maps; six fold-out maps in rear pocket; 196p; 18 cm.

Translation of Postojnska jama in druge zanimivosti Krasa.

2nd rev ed. 1958. illus: b&w: maps; five folded maps in rear pocket; 187p; 17 cm.

[2937] Šerko, Alfred, 1910-1948 *The Postojna grottoes and the other marvels of the karst.* Michler, Ivan; Pahor, Božidar (trans). Postojna, SI: [Postojnska jama], **1967**. illus: b&w: drawings, photos, repros, tables; 58p, 28p pls; bibl; 19 cm.

[2938] Seward, William *A tour to Yordes Cave.* Kirkby Lonsdale, GB: np, **1801**. iv, 30p; 17 x 10.5 cm.

Printed by A. Foster for the author. Written in verse. No illus.

[2939] Sexton, Robert T. *Cave surveying in South Australia.* [SA, AU]: Cave Exploration Group (South Australia), Dec, **1958**. Occasional Paper/Cave Exploration Group South Australia no 1. illus: b&w: drawings, maps, photos, tables; maps, drawings folded in text; iii, 21p, 5p pls; bibl; 29 cm.

[2940] Sexton, Robert T. *Caves of the coastal areas of South Australia.* Broadway, NSW, AU: Journal of Australasian Cave Research, **1965**. Occasional Papers Cave Exploration Group (South Australia) no 3, pp 45-59. illus: b&w: maps; 15p; bibl; 29 cm.

OCLC also lists 1977 as date of pub.

[2941] Shafer, Harry J. *Archeological and botanical studies at Hinds Cave, Val Verde County, Texas.* Bryant, Vaughn M. [College Station, TX, US]: [Texas A&M University], **1977**. Texas A&M University Anthropology Research Laboratory Special series no 1. illus: b&w: tables; vi, 137p; bibl; 28 cm.

Annual report of research under the auspices of the National Science Foundation (Grant no. BN576-10293), submitted by the Texas A&M Research Foundation. Illus.

[2942] Shafer, Harry J. *A preliminary report of Hinds Cave, Val Verde County, Texas.* Shafer, Harry J. [College Station, TX, US]: Anthropology Research Laboratory, Texas A&M University, **1975**. Technical Report Anthropology Laboratory Texas A&M University no 8. illus: b&w: maps; v, 56p; 28 cm.

Illus.

[2943] **Shahan, Myrtle** *Creation of Carlsbad Caverns.* Carlsbad, NM, US: Shahan, **1949**. [20]; 23 cm.
Illus.

[2944] **Shaler, Nathaniel Southgate**, 1841-1906 *On the antiquity of the caverns and cavern life of the Ohio valley.* Cambridge, KY, US: University Press, **1876**. Memoirs of the Geological Society of Kentucky v 1 pt 1. 13p, 1p pls; 31 cm.
Illus.

[2945] **Shannon, Robert Terry** *About caves.* Payzant, Charles (illus). Chicago, IL, US: Melmont Publishers, **1960**. Look Read Learn. 46p; 22 cm.
Juvenile.

[2946] **Shannon, Robert Terry** *About caves.* London, GB: Muller, **1966**. Junior Look Read and Learn Books Second Series no 8. i, 46p; 24 cm.
Illus. Juvenile.

[2947] **Shapcott, L. E.** *The story of Yanchep: the western wonderland.* [Perth, WA, AU]: The State Gardens Board of Western Australia, **[1934]**. illus: b&w: maps, photos; col: maps; 30p, [2], 2p pls.
Cover title: Yanchep.

[2948] **Sharp, Mabel** *White Scar Cave: a famous underground system which you can explore in your best clothes.* Clapham via Lancaster, GB: Dalesman, **1968**. illus: b&w: maps, ports; 32p; 22 cm.
Illus.

[2949] **Sharp, Mabel** *White Scar Cave: a picture guide.* Clapham, Yorkshire, GB: Dalesman, **1959**. illus: b&w: ads, charts, maps, photos; 31p.

[2950] **Sharp, Mabel** *White Scar Cave: a picture guide to a famous underground system you can explore in your best clothes.* Clapham via Lancaster, GB: Dalesman Publishing Company Ltd, **1965**. illus: b&w: ads, drawings, maps, photos, ports; ads on lining pages; 31p, [1]; 22 cm.
first published 1962 (from title page).

[2951] **Sharpe, Grant William**, 1925– *Evaluation of Mount St. Helens lava casts and caves, Cowlitz and Skamania Counties, Washington, for eligibility for registered natural landmark designation.* Seattle, WA, US: College of Forest Resources, University of Washington, **1973**. illus: col: maps; one col map folded in pocket; 20p; bibl.
Illus.

[2952] **Shaw, Trevor Royle**, 1928– *Cave illustrations before 1900: a catalogue of non-photographic illustrations of caves.* Settle, Yorkshire, GB: British Speleological Association, **1967**. British Speleological Association Monograph. illus: b&w: drawings, repros; b&w front; iii, 184p, 2p pls; index; 25.3 cm.
In 2 vols.
A catalogue listing of cave illustrations, useful to the historian in helping trace the first publication of a specific view or in comparisons of depicted scenes.

[2953] **Shaw, Trevor Royle**, 1928– *The deepest caves in the world and caves which have held the world depth record.* [Ledbury, GB]: Cave Research Group of Great Britain, **1961**. Occasional Publications - Cave Research Group of Great Britain no 5. illus: b&w: tables; 16p; bibl; 24 cm.

[2954] **Shaw, Trevor Royle**, 1928– *History of cave science: the exploration and study of limestone caves, to 1990.* 2nd ed. npop: Sydney Speleological Society, **1992**. illus: b&w: charts, drawings, graphs, maps, photos, repros, tables; partly col: maps; 2 fold-out charts in text; xiv, 338p; bibl, index.
Probably the single most important piece of historical research into the science of speleology to be published in book format. First published in 1979, but updated and with a better index in this 2nd edition. Detailing the development of scientific theories, the history of cave exploration since the 16th century is also covered; fully referenced, this volume should form a major part of any serious examination of cave history. An essential reference for all aspects of cave history.

[2955] **Shaw, Trevor Royle**, 1928– *History of cave science: the scientific investigation of limestone caves, to 1900.* Oldham, Joyce Elizabeth Anne. Cyrmych, Dyfied, Wales, GB: Anne Oldham, **1979**. illus: b&w: charts, maps, photos; partly col: maps, photos; text folded; xvi, 490p, 71p pls; bibl, index; 30 cm.
In 2 vols. 2nd ed:1992 Sydney Speleological Society.

[2956] **Shaw, Trevor Royle**, 1928– *Mendip Cave bibliography: part II: books pamphlets, manuscripts and maps; 3rd Century to December 1968.* Herefordshire, GB: Cave Research Group of Great Britain, Jul, **1972**. v 14 no 3 The Transactions of the Cave Research Group of Great Britain. list of pubs available on rear lining page; viii, 226p; 30 x 21 cm.
No illus.

[2957] **Shear, William A.** *North American cave millipeds II, an unusual new species (Dorypetalidae) from Southern California, and new records of* Speodesmus tuganbius *(Trichopolydesmidae) from New Mexico.* San Francisco, CA, US: California Academy of Sciences, **1974**. Occasional Papers of the California Academy of Sciences no 112. 9p; bibl; 26 cm.
Cover title. Illus.

[2958] **Shear, William A.** *A synopsis of the cave millipeds of the United States: with an illustrated key to genera.* Cambridge, MA, US: Cambridge Entomological Club, **1969**. Psyche v 76 no 2 pp 126-143. 28p; bibl; 24 cm.
Caption title. Illus. Described in pt 2 as being pt. 1 of the author's series, North American cave millipeds.

[2959] **Shelley, Maryann** *Karst and caves in the Caucasus.* npop: Maryann Shelley, Dec, **1956**. Study Field Research Projects no 3. illus: b&w: charts, drawings, maps, photos; two maps folded in pocket in rear; vi, 74p; bibl;

28 cm.

One of the only books in English on this remote area. It touches on the geomorphology, hydrology and climatology, reviews the division of the area into karst provinces, and provided limited cave descriptions, locations and maps. Additionally, it reviews the literature to 1956, almost all in Russian but with the titles translated.

[2960] Shenandoah Caverns Corporation *Legends of Shenandoah Caverns: in the Valley of Virginia.* Shenandoah Caverns, VA, US: Shenandoah Caverns Corp, **1922**. [24]; 14 x 20 cm.

Title from cover. Caption title: Tales and legends of the Shenandoah Caverns. Illus.

[2961] Shenandoah Caverns Corporation *Shenandoah Caverns of Virginia.* Shenandoah Caverns, VA, US: The Corporation, Roanoke Printing Co, **1932**. pages folded to 23x10 cm; 14p; 23 x 20 cm.

[2962] Shenandoah Caverns Corporation *Shenandoah Caverns: Valley of Virginia.* Shenandoah Caverns, VA, US: The Corporation, Marken & Bielfeld, **1928**. pages folded to 23 x 10 cm; 15p; 23 x 20 cm.

Running title: A symphony in stone: Shenandoah Caverns: the Valley of Virginia. Illus.

[2963] Shenandoah Caverns Corporation, Shenandoah Caverns, Va *Shenandoah Caverns, Valley of Virginia. A symphony in stone.* Frederick, MA, US: Marken & Bielfeld, Inc, **1929**. 15p.

Title from cover. Running title: A symphony in stone. Caption title: Shenandoah Caverns, a symphony in stone.

[2964] Shenandoah Caverns Corporation, Shenandoah Caverns, Va *Shenandoah Caverns, Valley of Virginia. The grotto of the gods.* [Frederick, MD, US]: Marken & Bielfeld, Inc, **1926**. [16].

Title from cover. Caption title: "A symphony in stone" Shenandoah Caverns. No illus.

[2965] Shenandoah Valley Railroad *Through the Shenandoah Valley: Caverns of Luray, Natural Bridge, grottoes of the Shenandoah and the chronicle of a leisurely journey through the uplands of Virginia.* npop: Giles, Litho. & Liberty Print. Co, **1889**. illus: b&w: maps; map in pocket; 72p; 20 cm.

Cover title: The Shenandoah Valley of Virginia. Illus.

[2966] Shepard, Dave *Songs poems and curses.* Nuttall, Kev; Perr, Dave; Penny, Jim (cartoonist). Burnley, GB: The Burnley Caving Club, Mar, **1971**. illus: b&w: drawings; 22p pls; 30 x 21 cm.

Photocopy examined.

[2967] Shepard, Edward M. *Underground waters of Missouri; their geology and utilization.* Washington, DC, US: U.S. Government Printing Office, **1907**. Water-supply and Irrigation Paper no 195. illus: b&w: maps, photos; 224p; 24 cm.

[2968] Sherman, Geraldine *Caverns: a world of mystery and beauty.* New York, NY, US: Julian Messner, **1980**. illus: b&w: maps, photos; 63p; index; 22 cm.

"Discusses the plants, animals, rock formations, and other features an explorer might encounter inside a cavern." Juvenile.

[2969] Shimek, S. (ed) *Bower Cave and Cave of Orpheus file.* npop: Private printing, **1971**. 132p.

[2970] Shippee, J. M. *The archaeology of Arnold Research Cave, Callaway County, Missouri.* Columbia, MO, US: The Missouri Archaeologist, Dec, **1966**. illus: b&w: maps; 107p; bibl; 23 cm.

Illus.

[2971] Shoemaker, Henry Wharton, 1880-1958 *The legends of the caverns of Centre County, Pennsylvania.* Reading, PA, US: Reading Eagle Press, **1929**. 23p; 23 cm.

[2972] Shoemaker, Henry Wharton, 1880-1958 *Penn's grandest cavern: the history, legends and description of Penn's Cave in Centre County, Pennsylvania.* 1900 ed. Altoona, PA, US: Altoona Tribune Co. 96p; 20 cm.

Master microform held by: NN. Microfilm. New York, N.Y.: New York Public Library, 19—. 1 microfilm reel ; 35 mm. Illus.

1914 ed. Altoona Tribune Press. illus: b&w: drawings, photos; photo in front cover; 78p, [16], 13p pls; 20 cm.

1916 rev ed. Altoona Tribune Press. Pennsylvania History on Microfilm. 94p, 14p pls; 20 cm.

Filmed with: Gen. Joseph Warren / by Charles W. Stone. [S.l.: s.n., 1910?] Microfilm. University Park, Pa.: Pennsylvania State University, 1993. 1 microfilm reel ; 35 mm. Illus.

1919 ed. Altoona Tribune Co. 96p; 20 cm.
Illus.

1921 ed. Altoona Tribune Co. 96p; 19 cm.
Illus.

1923 rev ed. Altoona Tribune Co. 96p; 20 cm.
Illus incl pls.

1930 ed. The Times Tribune Co. 101p; 21 cm.

Cover title: Penn's cave; Pennsylvania's grandest cavern and beautiful Lake Karoondinha: the history, legends and description of Penn's cave and its magnificent lake in Centre County, Pennsylvania.

[2973] Shoemaker, Henry Wharton, 1880-1958 *Penn's grandest cavern: the history, legends photographs and description of Penn's Cave in Centre County, Pennsylvania.* 1950 ed. Harrisburg, PA, US: The Telegraph Press. illus: b&w: photos; 109p; 22 cm.

Cover title: Penn's cave: Pennsylvania's grandest cavern and beautiful Lake Nitanee; "the history, legends, photos, and description of Penn's cave in the heart of central Pennsylvania." Illus.

1971 ed. Centre Hall, PA, US: Penn's Cave, Inc. illus: b&w: photos; 109p; 22 cm.

Cover title: Penn's cave: Pennsylvania's grandest cavern and beautiful Lake Nitanee; "the history, legends, photos, and description of Penn's cave in the heart of central Pennsylvania." Illus.

[2974] **Shutler, Mary Elizabeth (ed)** *Deer Creek cave, Elko County, Nevada*. Shutler, Richard (ed). Carson City, NV, US: Nevada State Museum, Oct, **1963**. Nevada State Museum Anthropological Papers no 11. illus: b&w: charts, graphs, photos, tables; 1 fold-out table in book; 2 fold-out charts in pocket; text on front lining page; 70p; bibl; 27 cm.

Illus.

[2975] **Siddiqui, Safiuddin** *A pictorial guide to Aurangabad, Daulatabad, Ellora, and Ajanta*. Dongerkery, Shri S.R. Aurangabad, IN: A. H. S. Mohamedbhoy, **1971**. iv, 43p; bibl; 21 cm.

Illus.

[2976] **Sieveking, Ann** *The cave artists*. London, GB: Thomes and Hudson Ltd, **1979**. Ancient Peoples and Places. illus: b&w: drawings, maps, photos, tables; partly col: photos; vi, 221p; bibl, index; 25 cm.

Col, b&w illus.

[2977] **Sieveking, Ann** *The caves of France and northern Spain*. Sieveking, Gale de Giberne. London, GB: Vista Books, **1962**. illus: b&w: charts, drawings, maps; vii, 269p, [3], 16p pls; app, bibl, gloss; 23 cm.

Illus.

A brief description of a large number of caves. More archeology/anthropology than geology, it is still of interest to earth scientists. It presents very basic maps of caves, and good black lines of the cave art.

[2978] **Sieveking, Ann** *The caves of France and northern Spain, a guide*. Sieveking, Gale de Giberne. Philadelphia, PA, US: Dufour, **1966**. illus: b&w: maps; 269p; 23 cm.

Illus.

[2979] **Siffre, Michel**, 1939– *Beyond time*. Briffault, Herma (trans, ed). New York; London, NY, US; GB: McGraw-Hill Book Company; Chatto & Windus Ltd, London, **1965**. illus: b&w: drawings, graphs, maps, photos; [iii], 228p.

The Hero: adventure of a scientist's 63 days spent in darkness and solitude in a cave 375 feet undergound.

[2980] **Siffre, Michel**, 1939– *Preliminary results of an experiment on human chronobiology and neurobiology in a subterranean environment. 1. Life on a bicircadian rhythm (P. Englender). 2. Life in continuous light (J. Chabert). Longitudinal analysis and computer correlation of neurologic, psychological and physiologic data collected in beyond-time cave experiments from 1968-1969*. Washington, DC, US: U. S. National Aeronautics and Space Administration, Jun, **1976**. N76-27838; NASA Technical Translation NASA TT F-15 499. illus: b&w: drawings, graphs, maps, photos, tables; [iii], iii, 327p; app, bibl.

Photocopy from microform. Translation of "Resultats préliminaires d'une expérience de chronobiologie et de neurobiologie humaines en milieu souterrain. 1. Vie en rythme bi-circadien: 48 heures: P. Englender. 2. Vie en lumière continue: Jacques Chabert. Analyse longitudinale et mise en correlaton par ordinateur des paramètres neurologiques, psychologiques et physiologiques recuillis au cours des exprinces hors du temps en caverne 1968-1969," Convention de Recherche de la Delegation Générale à la Recherche Scientifique et Technique, DGRST Report No. 70/02/183, 1976 (?), pp. 1-346.

[2981] **Silver, Constance S.** *The pictographs of Fern Cave, Lava Beds National Monument: agents of deterioration and prospects for conservation*. San Francisco, CA, US: National Park Service, Western Region Office, Mar, **1982**. illus: b&w: maps; [29]; bibl; 28 cm.

Illus.

[2982] **Silver, Donald M.**, 1947– *Cave*. Wynne, Patricia J. (illus). New York, NY, US: Scientific American Books for Young Readers W.H. Freedom and Company, **1993**. illus: partly col: drawings; [ii], 48p; bibl, index; 24 cm.

Col, b&w illus. Juvenile.

[2983] **Silverman, Sharon Hernes** *Going underground: your guide to caves in the Mid-Atlantic*. Philadelphia, PA, US: Camino Books, **1991**. illus: b&w: maps, photos; col: photos; iv, 119p; gloss; 22 cm.

[2984] **Simpson, George Gaylord**, 1902– *A fossil deposit in a cave in St. Louis*. New York, NY, US: American Museum of Natural History, Feb, **1949**. American Museum Novitiates no 1408, Feb 4, 1949. 46p; 24 cm.

[2985] **Simpson, James Young**, 1811-1870 *Account of some ancient sculptures on the walls of caves in Fife*. Edinburgh, GB: [Neill], **1867**. 22p; 24 cm.

Illus.

[2986] **Simpson, Lou** *Sex, lies, and survey tape*. npop: Lou Simpson, **1995**. illus: b&w: cartoons, drawings, maps, photos; iv, 99p; bibl, index.

[2987] **Sinclair, William C.** *Types, features, and occurrence of sinkholes in the karst of west-central Florida*. Stewart, J. W.; Knutilla, R. L.; Gilboy, A. E.; Miller, R. L. Tallahassee, FL, US: U. S. Geological Survey, **1985**. Water-Resources Investigations Report 85-4126. illus: b&w: charts, drawings, graphs, maps, photos, tables; vi, 81p; bibl.

Prepared in cooperation with the Southwest Florida Water Management District.

[2988] **Sinclair, William Campbell**, 1928– *Sinkhole development resulting from ground-water withdrawal in the Tampa area, Florida*. Tallahassee, FL, US: U. S. Geological Survey Water Resources Division, **1982**. Water-Resources Investigations 81-50. illus: b&w: maps; iv, 19p; bibl; 28 cm.

Illus.

[2989] **Sinclair, William John**, 1877-1935 *Euceratherium: a new ungulate from the Quaternary caves* *of California*. Furlong, Eustace Leopold, 1874–. 1st ed. Berkeley, CA, US: University of California Press, **1904**. University of California, Berkeley University of California Publications on Geological Sciences, v 3 no 20. 8p, 2p pls; 27 cm.

Title from cover. Illus.

reprint ed. New York, NY, US: Johnson Reprint, **1966**. 24 cm.

[2990] **Sinclair, William John**, 1877-1935 *New mammalia from the Quaternary caves of California*. Berkeley, CA, US: University of California Press, **1905**. University of California, Berkeley University of California Publications on Geological Sciences v 4 no 7. 17p, 5p pls; bibl; 27 cm.

Title from cover. Illus.

[2991] **Single, Michael** *Castles of the underworld*. Wild South (Firm) New Dimension Media; Donovan, Paul (camera); Penniket, Andrew (underwater camera); Animation Research Unit (graphics); Danks, Libby (ed); Copland, Neville (music); Fraser, Eugene (music). Eugene, OR, US: Distributed by New Dimension Media, **1992**.

A spectacular look at the strange and wonderful world of caves and the scientific information that comes from them. VHS, 1 videocassette (26 min), sd, col, 1/2 in.

[2992] **Sinise, Jerry** *Texas show caves*. Sinise, Dorothy. Austin, TX, US: Eakin Press, **1963**. illus: b&w: photos; 85p; gloss; 22 cm.

Cover title: Texas show caves: Natural Bridge Caverns, Inner Space, Longhorn Cavern, Cascade Caverns, Cave Without a Name, Wonder Cave, Caverns of Sonona

[2993] **Skelton, Harold E.** *Weyer's Cave's first century, 1874-1974*. [Harrisonburg, VA, US]: Park View Press, **1974**. vi, 82p; 24 cm.

Illus.

[2994] **Skinner, A. D.** *The Mole Creek Caves*. Skinner, R. K. npop: np, Jul, **1978**. illus: b&w: maps, photos; 36p.

[2995] **Skinner, Morris F.** *The fauna of Papago Springs Cave, Arizona, and a study of Stockoceros; with three new Antilocaprines from Nebraska and Arizona*. New York, NY, US: American Museum of Natural History, **1942**. Bulletin of the American Museum of Natural History v 80 article 6 pp 143-220. illus: b&w: graphs; 78p; 24 cm.

Cover title. Illus.

[2996] **Skinner, R. K.** *Hastings Caves State Reserve Tasmania: a visitors' guide*. Skinner, A. D. npop: [Huon News Print], **nd**. illus: b&w: photos; 16p.

Title from cover.

[2997] **Skjolsvold, Arne** *Excavations of Habitation Cave*. Bellwood, Peter S. Honolulu, HI, US: Department of Anthropology, Bernice P. Bishop Museum, Feb, **1972**. Pacific Anthropological Records no 16. illus: b&w: drawings, maps, photos, tables; [iii], ii, 51p; app, bibl.

31 figs, 4 tables.

[2998] **Skřivánek, František** *Caves in Czechoslovakia*. Rubín, Josef; Zárubová, H. (trans). Prague, CZ: Academia: Publishing House of the Czechoslovak Academy of Sciences; Organizing Committee of the Sixth International Congress of Speleology in Olomouc, **1973**. illus: b&w: maps, photos; [1]p folded in text; 133p, 32p pls; add, bibl, index; 24 cm.

21 b&w maps, 59 b&w photos.

[2999] **Slaven, John F.** *The speleoguide*. Fresno, CA, US: John F. Slaven (private printing), **1971**. illus: b&w: charts, drawings, graphs, maps, tables; 92p, [4]; app, bibl.

Filled a gap in beginner information which was much needed at the time. Compact and useful introduction to caving for novices

[3000] **Slaven, John F.** *The world within*. Crawford, Paul C.; Osiris Productions. npop: np, **nd**.

An adventure in cave exploration—-Cover. Don Messick. A group of explorers discuss the appeals and challenges of spelunking as they explore a cave. 1 videocassette (VHS) (28 min), sd, col, 1/2 in.

[3001] **Slifer, Dennis William**, 1947– *Sinkhole dumps and the risk to groundwater in Virginia's karst areas: final report to the Virginia Environment Endowment*. Erchul, Ronald A. npop: np, Apr, **1989**. illus: b&w: drawings, maps; 14p, [60]; 28 cm.

Illus.

[3002] **Sloan, Tacoma G.** *Archaeological survey of Mammoth Cave National Park*. Schwartz, Douglas Wright, 1929–. Richmond, VA, US: US National Park Service, Southeast Region, **1960**. illus: b&w: maps; 38p; bibl; 28 cm.

Overview of the archaeological resources of Mammoth Cave National Park.

[3003] **Sloane, Bruce**, 1935– *Cavers, caves, and caving*. New Brunswick, NJ, US: Rutgers University Press, **1977**. illus: b&w: drawings, photos; xiii, 409p; bibl, gloss, index; 22 cm.

Cave exploration in the U.S.; also articles on conservation, pollution, research, etc. Anthology of American caving exploits, discoveries, problems, and people.

[3004] **Sloane, Howard N. (d. 1972)** *Visiting American caves*. Gurnee, Russell Hamtpon, 1922-1995. New York, NY, US: Bonanza Books, **1966**. illus: b&w: maps, photos; maps on lining pages; [ii], x, 246p; bibl, index; 24 cm.

Endorsed by National Speleological Society. Cover subtitle: A comprehensive guide to all the American caves open to the public.

[3005] **Smart, James** *A Philippine cave index and bibliography*. Crymych, GB: Anne Oldham, **1994**. illus: b&w: maps; [iv], 27p; bibl.

Catalog of caves in the Philippines from literature search. Very brief descriptions with references.

[3006] **Smart, Peter L. (comp)** *Cathay Pacific Airways*

Mulu '80 expedition: British Malaysian speleological expedition to Sarawak. Willis, R. G. (comp). Bridgwater, Somerset, GB: British Cave Research Association, Jun, **1982**. Cave Science: Transactions of the British Cave Research Association v 9 no 2. illus: b&w: charts, graphs, maps, photos, tables; text on lining pages; 112p; bibl.

[3007] **Smeltzer, Bernard L.** *Caves of the southern Cumberland valley.* State College, PA, US: Mid-Appalachian Region of the National Speleological Society, Oct, **1964**. MAR Bulletin no 6. illus: b&w: maps, photos, tables; 5 loose folded maps; 126p.

[3008] **Smith Daniel Irving**, 1946– *Handbook of cave rescue operations.* Petaluma, CA, US: National Cave Rescue Commission, National Speleological Society, **1978**. illus: b&w: ads, charts, drawings, graphs, maps, tables; [iii], 138p; bibl.

[3009] **Smith, A. Richard (ed)** *The caves of Kimble county.* Austin, TX, US: Texas Speleological Association, **1971**. Texas Speleological Survey v 3 no 6. illus: b&w: maps; partly col: maps; [ii], i, 47p; bibl; 28 cm.

Illus.

[3010] **Smith, A. Richard (ed)** *The caves of Kinney county.* Reddell, James Russell, 1938–. Austin, TX, US: Texas Speleological Association, **1965**. Texas Speleological Survey v 2 no 7. illus: b&w: maps; partly col: maps; 34p; 28 cm.

Illus.

[3011] **Smith, Allyn Goodwin** *Snails from California caves.* San Francisco, CA, US: The Academy, **1957**. Proceedings of the California Academy of Sciences 4th ser v 29 no 2. 26p; 26 cm.

Title from cover.

[3012] **Smith, Anthony**, 1926– *Blind white fish in Persia.* 1966 ed. George Allen and Unwin. 207p; 18 cm.

Reprint without the photo pls.

1990 ed. Penguin Books. illus: b&w: drawings, maps, photos; 243p; 20 cm.

Many of the photos different from those in 1953 ed. New postscript tells of later actual finding of blind white fish in Persia. 16pp pls.

[3013] **Smith, Anthony**, 1926– *Blind white fish in Persia.* London, GB: George Allen and Unwin, **1953**. illus: b&w: photos; 231p; 21 cm.

Adventure-travel classic on seeking cave fish in caves and qanats of Persia. They didn't find any. Records in OCLC for this edition differ on whether there are 231 or 256 pp. Also published in 1953 by E. P. Dutton and Co, NY, NY, US: 256pp, 16pp pls, illus: b&w photos, maps, drawings.

[3014] **Smith, Burrell F.** *The cave and her men: Oregon Caves National Monument: its history and formation.* npop: Burrell F. Smith, Burrell a D & Z Production, **1988**. illus: b&w: photos, ports; b&w port; 28p, [2].

[3015] **Smith, Christopher** *Mid Argyll cave and rock shelter survey.* Newcastle, GB: Department of Archaeology, University of Newcastle upon Tyne, **1989**. illus: b&w: maps; Various pagination; bibl; 30 cm.

Title from cover. Reports written by: Christopher Smith. Contents: Report no. 1, 1985—Report no. 2, 1986—Report no. 3, 1987—Report no. 4, 1988.

[3016] **Smith, David Ingle (comp, ed)** *Limestones and caves of the Mendip Hills.* Drew, David Phillip, 1943– (asst). Newton Abbot; North Pomfret, GB: David & Charles for The British Cave Research Association, **1975**. The Limestones and Caves of Britain. illus: b&w: maps; 424p; bibl, index; 23 cm.

Illus.

[3017] **Smith, Elmer Richard**, 1909– *The archaeology of Deadman Cave, Utah: a revision.* Salt Lake City, UT, US: Department of Anthropology, University of Utah, **1952**. Utah University Dept of Anthropology Anthropological Papers no 10. illus: b&w: maps; 41p; bibl; 28 cm.

A revision of an article which appeared in the University of Utah bulletin, v 32, no 4. Specifically the new material embraces p1, part of 2,6,7, part of 8, 37-40. Foreward.

[3018] **Smith, Gerald Arthur**, 1915– *The archaeology of Newberry Cave, San Bernardino County, Newberry, California.* San Bernardino, CA, US: San Bernardino County Museum Association, **1957**. Scientific Series San Bernardino County Museum Association no 1. 26p, 20+13p pls.

Cover title: Newberry Cave, California. Illus.

[3019] **Smith, Gerald Arthur**, 1915– *Preliminary report of the Schuiling Cave, Newberry, California.* Downs, Theodore, 1919–; Howard, Hildegarde, 1901–. [Bloomington, CA, US]: np, **1955**. Quarterly San Bernardino County Museum Association v 3 no 2. 29p; 29 cm.

Illus.

[3020] **Smith, H.** *Kent's Cavern.* npop: np, **1882**. illus: b&w: photos; front: photo; 9p; 11.5 x 18 cm.

Copy examined was a photocopy.

[3021] **Smith, J. P.** *Paper on Dunald Mill Hole: read by J. P. Smith before the Barrow Naturalists Field Club.* Barrows-in-Furress, GB: np, **1890**. illus: b&w: charts, maps; 28p, 3p pls; add; 21 x 14 cm.

[3022] **Smith, James** *The early days of Olsen's Caves as seen through a geologist's eyes.* npop: np, Oct, **1964**. illus: b&w: photos, ports; 53p.

Title from cover.

[3023] **Smith, Jim** *From Katoomba to Jenolan Caves: the Six Foot Track, 1884-1984.* Katoomba, NSW, AU: np, [**1984**]. A Megalong Book. illus: b&w: maps, ports, repros; b&w maps on lining pages; front: b&w repro; 166p, [1]; app, bibl; 24 cm.

Concerned principally with the route taken to Jenolan, Australia, a large number of obscure accounts of the journey are reproduced and include references to underground visits. Any social study of the caves should refer to this volume. A wonderful assembly of stuff.

[3024] Smith, John *Monograph of the stalactites and stalagmites of the Cleaves Cove, near Dalry, Ayrshire.* [London, GB]: Elliot Stock, London, **1894**. v, 34p, 36p pls.

1st on the subject.

[3025] Smith, John Moyr *The Hades of Ardenne: a visit to the Caves of Han described and illustrated.* The T. T. Club. London, GB: Sampson Low, Marston, Searle, & Rivington, **1883**. illus: b&w: drawings, maps; folded map; 160p; 19 cm.

OCLC record lists author as Smith, John Moyr. Pseudonyms in text for authors are: Daubiton, M. A.; Drypoynter, B.E.; & Hinkityne, W. Excellent little hardback with a large fold-out map of the caves in the back plus 8 full page plates and numerous sketches. Not a common book.

[3026] Smith, Lee *Down under Texas.* Garcia, Cleo Zoe (prod, dir, video); Garrison, Bill (field prod, dir, video); McKann, Belva (script consul); Andrle, Fred (script consul); Roberts, Richard (exec prod); Benning, Anne E. (exec prod); Smith, Lee (exec prod). [Austin, TX, US]: Texas Parks and Wildlife Department, **nd**. Made in Texas 41.

Non broadcast version.—Container label. VHS format, 28 min, sd, col, 1/2 in. Guided by members of the Texas Speleological Association, some of the more than 3,000 caves in Texas are explored; their processes of formation, fragile ecosystems, beauty and awesomeness are shown and discussed.

[3027] Smith, Marion Otis, 1942– *The exploration & survey of Ellison's Cave, Georgia.* Birmingham, AL, US: [Marian O. Smith], **1977**. illus: b&w: maps, photos; ii, 140p; app, bibl, index; 28 cm.

[3028] Smith, Marion Otis, 1942– *Letters from TAG 1966-1969.* Anthony, Patricia J. (comp). npop: np, **1992**. illus: b&w: photos; col: photos; 295p; index.

Back and front covers col photos.

[3029] Smith, Marion Otis, 1942– *Saltpeter mining in East Tennessee.* Knoxville, TN, US: Marion O. Smith, **1990**. illus: b&w: repros, tables; title page on front lining page; 32p; bibl.

repro on cover.

[3030] Smith, Marion Otis, 1942– *TAG pits.* [Atlanta, GA, US]: Marion O. Smith, **c.1973**. [ii], 146p; index.

Locations and brief descriptions (no maps) of 409 vertical caves in the states of Tennessee, Alabama, and Georgia. Duplicated and comb bound in relatively limited edition by one of the most fanatical vertical cavers in the area.

[3031] Smith, Peter B. *The P8 Cave: Castleton.* Derbyshire, GB: British Speleological Association, **1973**. Cave Science, Journal of British Speleological Association no 50. illus: b&w: maps, photos; 2 fold-out maps; 28p, 6p pls; 25 x 21.5 cm.

b&w photo on front cover.

[3032] Smith, Philip M. *Speleological research in the Mammoth Cave Region, Kentucky: elements of an integrated program.* Yellow Springs, OH, US: Cave Research Foundation, **1960**. vi, 18p; bibl.

[3033] Smith, Philip Wayne, 1921– *A summary of the life history and distribution of the spring cavefish, Chologaster agassizi Putnam, with population estimates for the species in southern Illinois.* Welch, Norbert M. Urbana, IL, US: Illinois Natural History Survey, **1978**. Illinois. Natural History Survey Division Biological Notes no 104. illus: b&w: maps; 8p; bibl; 28 cm.

Cover title. Illus.

[3034] Smith, Rodney *Waitomo Caves.* Wellington, NZ: New Zealand Pictorial Books Ltd, **1981**. illus: b&w: maps, ports, repros; partly col: maps; col: photos; text on front lining page, maps on back lining page; [16].

Cover title: Waitomo Caves: colour souvenir.

[3035] Smith, Rodney *Waitomo: Glow-worm Cave, Aranui Cave, Ruakuri Cave and surrounding district.* Wellington, NZ: A. H. & A. W. Reed Ltd, **1980**. A Baldwin Book. illus: b&w: maps, photos, ports, repros; partly col: maps; col: photos; text on lining pages and col photo on front lining page; [32].

Title from cover.

[3036] Smith, William Hovey *Geology of the Tennille Lime Sinks, Washington County, Georgia ; with an introduction to local geology.* Sandersville, GA, US: Whitehall Press—Budget Publications, **1983**. Washington County Geology Series no 1. illus: b&w: maps; 34p; bibl, gloss; 21 cm.

Illus.

[3037] Smithsonian Institution *The caverns of Luray, located on line of the Shenandoah Valley R.R.* [Washington, WA, US]: The Smithsonian, **1880**. illus: b&w: ads, maps; 15p; 22 cm.

Illus.

[3038] Snider, George Washington *How I found the Cave of the winds ; being the true story as set forth by the discoverer, George W. Snider.* Denver, CO, US: Press of Carson-Harper, **1916**. illus: b&w: ports; 88p; 24 cm.

Part 1 written by Jewel Fouke (cf. Foreword, dated September 1915) "Part 2. How I lost the Cave of the Winds": p.29-88. (Printed in Los Angeles?)

[3039] Snider, George Washington *Manitou Grand Caverns, Manitou Springs, Colorado.* npop: np, **1888**. [12]; 20 cm.

Cover-title. Text on p [2-3] of cover. From the Manitou Springs Daily Journal, September 1, 1888. Signed: George W. Snider, proprietor.

[3040] Snider, George Washington *The tourists' gem: the Manitou Grand Caverns, the largest and most wonderful subterranean in the Rocky Mountains, and other attractions for tourists.* [Manitou Springs, CO, US]: np,

1885. 16p; 14 cm.
Illus.

[3041] Snodgrasse, Richard Montgomery *The human skeletal remains from Pictograph and Ghost caves, Montana*. Mulloy, W. T. Laramie, WY, US: Graduate School, University of Wyoming, **1958**. University of Wyoming Publications v 22 no 2. 29p; 26 cm.

A preliminary historical outline for the northwestern plains. Caption title. Cover title: The skeletal remains from Pictograph and Ghost caves, Montana.

[3042] Snyder, Dean H. (comp, ed) *The caves of Northampton County, Pennsylvania*. State College, PA, US: np, Feb, **1989**. MAR Bulletin 16. illus: b&w: maps, photos, repros, tables; 1 map folded in text, 1 map folded loose; 46p; bibl; 28 cm.

Illus.

[3043] Snyder, R. P., 1932– *Evaluation of breccia pipes in southeastern New Mexico and their relation to the Waste Isolation Pilot Plant (WIPP) site*. Gard, L. M.; Mercer, J. A. Reston, VA, US: U.S. Geological Survey, **1982**. Open-file Report United States Geological Survey 82-968. illus: b&w: charts, drawings, graphs, maps, photos; iii, 73p; bibl, index; 28 cm.

A well done review of the geology of the eastern Delaware Basin - stratigraphy, hydrogeology, paleokarst, and present karst as it effects the WIPP site. It also discusses breccia pipes in other North American locales.

[3044] Society for Visual Education *Carlsbad Caverns National Park*. Chicago, IL, US: Society for Visual Education, **1945**. National Parks Filmstrip.

Locates the park on the map; shows the area around the caverns as well as the interior; emphasizes stalactite and stalagmite formations. 1 filmstrip (32 fr), b&w, 35 mm.

[3045] Soepadmo, E. *A guide to Batu Caves*. Ho, Thian Hua. Kuala Lumpur, MY: Malayan Nature Society, **1971**. iii, 29p; 21 cm.

Title from cover. Illus.

[3046] Solecki, Ralph S., 1917– *Shanidar: the first Flower people*. Knoph, A (ed). London, GB: Allen Lane, **1972**. illus: b&w: charts, drawings, maps, photos; xiv, 222p, [16]; bibl, index, note; 23 cm.

Originally published, New York: Alfred A. Knopf, 1971.

Archaeological and physical anthropological study of the oldest planned human burials in the world. A personal narration of one of the most important and exciting archeological discoveries of recent years.

[3047] Sollas, William Johnson, 1849-1936 *Paviland Cave: an Aurignacian station in Wales (The Huxley memorial lecture for 1913)*. London, GB: The Royal Anthropological Institute of Great Britain and Ireland, **1913**. Huxley Memorial Lecture 1913; Journal of the Royal Anthropological Institute v XLIII. 50p.

Title from cover. Illus.

[3048] Somers, Lee H. *Cave diving: equipment and procedures*. [Ann Arbor, MI, US]: University of Michigan, **1971**. Multidisciplinary Research in the Great Lakes Basin Technical Report Underwater Operations Project no 2. illus: b&w: photos, tables; ii, 28p, 6p pls; bibl.

Supported by University of Michigan Sea Grant Program, and National Science Foundation. Cooperative Research Program sponsored by University of Michigan and National Oceanic Atmospheric Administration. Illus.

[3049] Sonderegger, John L. *Hydrology of limestone terranes: geologic investigations*. Kelley, James C. University, AL, US: Geological Survey of Alabama, Division of Water Resources, **1970**. Bulletin 94 part B. illus: b&w: charts, graphs, maps, photos, tables; partly col: maps; 6 folded partly col maps in pocket; [vi], 146p; app, bibl.

[3050] Sonderegger, John L. *Hydrology of limestone terranes: photogeologic investigations*. University, AL, US: Geological Survey of Alabama, Division of Water Resources, **1970**. Bulletin 94 part C. illus: b&w: charts, graphs, maps, tables; partly col: maps; 5 folded part col maps in pocket; [v], 27p; app, bibl.

[3051] Sotnak, Lewann *Carlsbad Caverns*. Mankato, MN, US: Crestwood House, **1988**. National Parks Mankato, Minn. illus: partly col: maps, photos; iv, 47p; bibl, gloss, index; 28 cm.

Describes the formations and plant and animal life of Carlsbad Caverns and examines the history of this national park.

[3052] South African Spelaelogical Association Members *Cango; the story of the Cango Caves of South Africa*. 3rd ed. **1968**. [viii], 83p, 4p pls; index.

[3053] South African Spelaeological Association *Cango; the story of the Cango Caves of South Africa*. Walker, A. S. (ed). 4th ed. **1970**. illus: b&w: charts, photos; col: photos.

No visible changes from 3rd ed.

[3054] South African Spelaeological Association Members *Cango; the story of the Cango Caves of South Africa*. 1st ed. Capetown, ZA: Maskew Miller Ltd, **1958**. illus: b&w: charts, photos; col: photos; maps of cave on lining page; front: col photo; [viii], 77p, 4p pls; index; 19 x 12.5 cm.

2nd ed. 1960. illus: col: photos.

No visible changes from first edition, but it says revised ed.

[3055] South Australia National Parks and Wildlife Service *Draft management plan, Tantanoola Caves Conservation Park, lower South East, South Australia*. Adelaide, SA, AU: Department of Environment and Planning, Mar, **1983**. illus: b&w: maps, ports; ix, 71p; bibl; 30 cm.

Illus.

[3056] South Australia National Parks and Wildlife Service *Naracoorte Caves Conservation Park management plan, South East, South Australia*. [Adelaide, SA,

AU]: National Parks and Wildlife Service, Department of Environment and Planning, **1990**. illus: b&w: maps; 22p; bibl; 30 cm.

Illus.

[3057] South Australia National Parks and Wildlife Service *Naracoorte Caves Conservation Park, South East, South Australia: draft management plan.* South Australia Department of Environment and Planning. Adelaide , SA, AU: np, Feb, **1986**. illus: b&w: maps; xv, 116p, 1p pls; bibl; 30 cm.

Illus.

[3058] South Australia National Parks and Wildlife Service *Tantanoola Caves Conservation Park management plan, South East, South Australia.* [Adelaide, SA, AU]: National Parks and Wildlife Service, Department of Environment and Planning, **1990**. illus: b&w: maps; 13p; bibl; 30 cm.

[3059] South Wales Caving Club *Members handbook: South Wales Caving Club.* [GB]: South Wales Caving Club, Feb, **1993**. illus: b&w: maps; 25p; 21 x 15 cm.

Cover title: South Wales Caving Club handbook. Members list in center.

[3060] South Wales Caving Club *Some technical aids for cave exploration.* Church Stretton, Salop, GB: Cave Research Group of Great Britain, Nov, **1962**. Publication no 11 Cave Research Group of Great Britain. illus: b&w: drawings, graphs, photos; 126p, [2]; bibl; 25 cm.

[3061] South Wales Caving Club *South Wales Caving Club twenty first anniversary publication.* [GB]: South Wales Caving Club, **c1968**. illus: b&w: maps, photos, tables; 3 large maps folded, loose in back; 260p; 26.5 x 21 cm.

[3062] South Wales Caving Club *South Wales Caving Club, 1946—1956.* npop: South Wales Caving Club, **1957**. 54p; 26 cm.

Illus.

[3063] Southeastern Regional Association (S.E.R.A.) Cave Carnival (1968: TN) *1968 SERA cave carnival.* [TN], US: [Tennessee Tech Speleophiles], [**1968**]. illus: b&w: maps; [25].

[3064] Southeastern Regional Association (S.E.R.A.) Cave Carnival (1975: Monterey, TN) *Guidebook to the caves of the Monterey, Tennessee Area.* Smyre, John; Yarbrough, Ed. Nashville, TN, US: Nashville Grotto, **1975**. illus: b&w: charts, maps, photos; index to caves loose in text; ix, 27p; index.

Cover title: 1975 SERA cave carnival.

[3065] Southeastern Regional Association (S.E.R.A.) Cave Carnival (1976: Guntersville, AL) *SERA 1976 guidebook.* Birmingham, AL, US: Birmingham Grotto, National Speleological Society, **1976**. illus: b&w: ads, drawings, photos, tables; ads on lining pages; index.

Title from cover.

[3066] Southeastern Regional Association (S.E.R.A.) Cave Carnival (1978: TN) *Guidebook to the caves of the 1978 summer SERA cave carnival.* Kerr, Chris (ed). [Oak Ridge, TN, US]: Smoky Mountain Grotto, **1978**. illus: b&w: ads, charts, maps, photos; 40p, [6].

Cover title: '78 SERA Cave Carnival.

[3067] Southeastern Regional Association (S.E.R.A.) Cave Carnival (1979: Ketner's Mill, TN) *1979 Southeastern Regional Association (S.E.R.A.) cave carnival at Ketner's Mill, TN.* Wolinsky, Mark (exe ed); Buice, Rick (consult ed); Moni, Gerald (consult ed); Durham, Dave (print, layout). Chattanooga, TN, US: Chattanooga Grotto, **1979**. illus: b&w: maps, photos, tables; 3 folded maps in text; text on front lining page, map on rear lining page; 54p, [4]; bibl; 28 cm.

Cover title: 1979 SERA cave carnival.

[3068] Southeastern Regional Association (S.E.R.A.) Cave Carnival (1983: AL) *1983 SERA cave carnival: 5-7 August 1983.* Torode, William Wallace, 1943–(ed). [Huntsville, AL, US]: [Huntsville Grotto, National Speleological Society], Aug, **1983**. illus: b&w: ads, drawings, maps, photos, tables; table of contents on front lining page; b&w maps on rear lining page; 68p.

[3069] Southeastern Regional Association (S.E.R.A.) Cave Carnival (1989: [TN]) *SERA 89.* Cannon, Mark (ed); Paris, Sandy (ed); Park, Ray (ed); Shrewsbury, William (ed). Chattanooga, TN, US: Chattanooga Grotto, National Speleological Society, [**1989**]. illus: b&w: ads, drawings, maps, photos; [vi], 94p; 28 cm.

Title from cover. Col photos on covers. Illus.

[3070] Southeastern Regional Association (S.E.R.A.) Cave Carnival (1990: Goose Pond Campground, Scottsboro, AL) *SERA cave carnival, 1990.* Torode, William Wallace, 1943–(ed); Skipworth, Joe (ed); Youree, Roger (ed); Morgan, Angela (ed); Moss, Tom (ed); White, Tim (ed). [Huntsville, AL, US]: [Huntsville Grotto of the National Speleological Soceity], **1990**. illus: b&w: ads, drawings, maps, photos; ads on lining pages; 75p; index; 28 cm.

Title from cover. "Reference points" ([1] leaf) inserted. Illus.

[3071] Southeastern Regional Association (S.E.R.A.) Cave Carnival (1992: AL) *SERA cave carnival 1992.* Pope, Tim (ed). [AL, US]: Northeast Alabama Grotto, National Speleological Society, [**1992**]. illus: b&w: ads, drawings, maps, photos; 56p.

Title from cover.

[3072] Southeastern Regional Association (S.E.R.A.) Cave Carnival (1994: Short Mountain, TN) *Guidebook to the 1994 SERA cave carnival: June 10-12, 1994 CCWHA Campground at Short Mountain Cannon County, Tennessee.* Lance, Don (ed). npop: Tennessee Central Basin Grotto, National Speleological Society, **1994**. 43rd Annual SERA Guidebook. illus: b&w: ads, cartoons, charts, drawings, maps, photos; map on front lining page;

ad on back lining page; 70p.

Cover title: SERA 1994: June 10-12, 1994 CCWHA Campground Short Mountain, Tennessee.

[3073] Southeastern Regional Association (S.E.R.A.) Cave Carnival (1995: Lafayette, GA) *1995 SERA summer cave carnival: the party on the farm: Smokey and Tina Cadwell's farm: Lafayette, Georgia May 19-21, 1995.* Wilkes, Chris (ed). [Athens, GA, US]: Athens Speleological Society, National Speleological Society, [**1995**]. illus: b&w: ads, maps, photos, repros; 40p.

Title from cover.

[3074] Southeastern Regional Association (S.E.R.A.) Cave Carnival (1996: Camp Jackson, Jackson Co. AL) *45th Annual SERA Summer cave carnival: July 19-21, 1996, Camp Jackson, Jackson County, Alabama.* Fee, Scott (ed). npop: Birmingham Grotto, National Speleological Society, **1996**. illus: b&w: ads, charts, maps, photos, tables; partly col: maps; col: photos; ads on lining pages; 72p.

Pagination includes covers.

[3075] Southhampton University Exploration Society *Southhampton University Exploration Society: Peru expedition.* Surrey, GB: Southhampton University Exploration Society Peru Expedition, **1982**. illus: b&w: drawings, graphs, maps, photos, tables; Various pagination, [90]; index; 30 x 21 cm.

Cover title: Peru '82.

[3076] Southwest Georgia Planning & Development Commission *Glory Hole Caverns.* Woodham, Robert Earl. [Camilla, GA, US]: np, **1968**. 74p, [1]; 22 cm.

[3077] Southwest Missouri State University Dept. of Geography and Geology *Speleology workshop.* Springfield, MO, US: Ozark Highlands Grotto, National Speleological Society, **1979**. illus: b&w: cartoons, drawings, graphs, maps, photos; 168p; bibl; 30 cm.

Title from cover. Illus. 1979 Speleological workshop (2nd: April 6-8, 1979: Springfield, Mo.).

[3078] Sowers, George F. *Building on sinkholes: design and construction of foundations in karst terrain.* New York, NY, US: American Society of Civil Engineers, **1996**. 202p; bibl, index; 23 cm.

[3079] Sowers, J. M. *Cave management plan and environmental assessment, Lava Beds National Monument, California.* Sydoriak, C. San Francisco, CA, US: U.S. National Park Service, **1990**. 101p.

[3080] Sparrow, Andy *A Mendip caver's ropework guide.* GB: [Andy Sparrow], **1991**. illus: b&w: ads, drawings, maps, photos; col: photos; ad on rear lining page; 86p; 20.5 x 15 cm.

[3081] Spate, Andy *A brief introduction to speleology.* Dovey, Liz (cont). Queanbeyan, NSW, AU: [New South Wales National Parks and Wildlife Service], **1988**. illus: b&w: charts, drawings, graphs, maps, tables; 40p; bibl.

Prepared for presentation to the Metropolitan Speleological Society at Abercrombie Caves 18-19 June 1988.

[3082] Spate, Andy *Kubla Khan Cave State Reserve, Mole Creek, Tasmania: pilot management study.* Houshold, Ian (cont); Eberhard, Stefan (cont). Hobart, TAS, AU: Department of Parks, Wildlife and Heritage, **1991**. illus: b&w: maps, tables; iv, 64p; app, bibl, gloss; 30 cm.

Prepared for The Department of Parks, Wildlife and Heritage.

[3083] Speare, Eva A. *The story of Polar Caves.* Littleton, NH, US: Resort Enterprises Inc, **nd**. illus: b&w: maps, photos; 64p; 15 x 10 cm.

Cover title: Polar Caves; one of New Hampshire's natural wonders.

[3084] Speece, Jack H. *The cave of Delaware.* Forney, Gerald Glenn. Altoona, PA, US: Speece Productions, **1977**. Spelean History Series no 3.

[3085] Speece, Jack H. (comp, ed) *The caves of Blair County, Pennsylvania.* Cullinan, Mike (comp, ed). State College, PA, US: Mid-Appalachian Region of the National Speleological Society, Jul, **1972**. MAR Bulletin no 8. illus: b&w: drawings, maps, tables; [iii], 90p; bibl.

[3086] Speece, Jack H. (ed) *Fulton County, Pennsylvania.* State College, PA, US: Mid-Appalachian Region of the National Speleological Society, Feb, **1969**. MAR Bulletin no.7. illus: b&w: maps; partly col: maps; 1 loose folded map; [16]; bibl.

[3087] Speece, Jack H. *George Washington Cave.* npop: np, **1978**. 16p.

A small cave in Jefferson County, West Virginia

[3088] Speece, Jack Howard, 1947– *Alexander Caverns: the Carlsbad of Pennsylvania.* npop: np, **1973**. illus: b&w: photos; 24p.

[3089] Speece, Jack Howard, 1947– *Special convention issue: Symposium on the History of American Caving, Pittsfield, Mass., August 5-12, 1979.* [Seattle, WA, US]: American Spelean History Association, **1979**. The Journal of Spelean History v 13 no 1-2. illus: b&w: maps; 55p; bibl; 28 cm.

Illus.

[3090] Speleological Conference (1987) *Proceedings - 16th biennial conference Australian Speleological Federation Inc.* Broadway, NSW, AU: Speleological Research Council Ltd, **1987**. Helictite volume 25 no 2. illus: b&w: drawings, photos, ports, tables; text on lining pages; 56p; bibl.

[3091] Speleological Conference (1991: Margaret River, WA, AU) *Proceedings of the 18th biennial speleological conference: 30 December 1990 to 5 January 1991 at Margaret River, W.A.* Brooks, Steven; Western Australian Speleological Group; Speleological Research Group of Western Australia. [AU]: Australian Speleological Federation Inc, **1991**. ii, 113p; bibl; 30 cm.

Cover title: Cave Leeuwin 1991. Illus.

[3092] Speleology Workshop (1979: 2nd: Southwest

Missouri State University) *1979 Speleology workshop manual.* Southwest Missouri State University. [Springfield, MO, US]: Ozark Highlands Grotto, National Speleological Society, **1979**. 168p; bibl; 30 cm.

Illus.

[3093] **Spencer, Edgar Winston** *Guidebook to the Natural Bridge and Natural Bridge Caverns.* Lexington, KY, US: Poorhouse Mountain Studios, **1985**. 48p; 22 cm.

Cover title: Natural Bridge and Natural Bridge Caverns. Illus.

[3094] **Spigner, B. C.** *Review of sinkhole-collapse problems in a carbonate terrane.* United States Soil Conservation Service. [Columbia, SC, US]: The Commission, Jul, **1978**. Open-file Report South Carolina Water Resources Commission no 78-2. illus: b&w: maps; 16p; bibl; 28 cm.

U.S. Department of Agriculture Soil Conservation Service Combined Engineering and Sedimentation Geology Workshop (1978: Charleston, S.C.).

[3095] **Spink, Walter M.** *Ajanta: a brief history and guide.* [Ann Arbor, MI, US]: Asian Art Archives of the University of Michigan, **1990**. 43p; 23 cm.

Col illus.

[3096] **Sprague, Stuart** *Caves of Herkimer County.* npop: NRO, **c1960**. NRO report. illus: b&w: maps; 12p.

[3097] **Sprent, J. K., (ed)** *Mount Etna caves: a collection of papers covering several aspects of the Mt. Etna and Limestone Ridge caves area of central Queensland.* St Lucia, QLD, AU: University of Queensland Speleological Society, **1970**. illus: b&w: drawings, graphs, maps, photos; iii, 116p; app, bibl; 25 cm.

[3098] **Squire, Ralph E.** *Report of study by National Speleological Society, Cave Conservation Task Force: New Melones Project.* Cave Conservation Task Force of the National Speleological Society. npop: np, Nov, **1971**. illus: b&w: maps; 18p; 28 cm.

Title from cover.

[3099] **Squire, Ralph E.** *Stanislaus cave country: report of study.* New Melones Task Force. npop: National Speleological Society, **1972**. illus: b&w: maps, photos; [iv], 25p; bibl; 28 cm.

Study of Melones Dam on a major cavernous belt. Standing, em et al. Little Neath River Cave. UBSS.

[3100] **Standing, Ian James** *Council of Southern Caving Clubs handbook.* [Nailsworth, Glos, GB]: Council of Southern Caving Clubs, Sep, **1966**. 21p.

No illus.

[3101] **Standing, Ian James (ed)** *Speleological abstracts: the publications of 1963.* Settle, Yorkshire, GB: British Speleological Association, **1963**. v 1 no 2. ix, 87p; 26 x 20 cm.

No illus.

[3102] **Standing, Ian James (ed)** *Speleological abstracts: the publications of 1964.* npop: np, **1964**. v 1 no 3. xiii, 143p; 26 x 20 cm.

No illus.

[3103] **Standing, Ian James (ed)** *Speleological abstracts: the publications of 1965.* npop: np, **1969**. v 1 no 4. ix, 112p; 26 x 20 cm.

No illus.

[3104] **Standing, Ian James (ed)** *Speleological abstracts: the publications of 1966.* npop: np, Sep, **1969**. v 1 no 5. ix, 118p; 26 x 20 cm.

No illus.

[3105] **Standing, Peter** *Medical aspects of speleology.* Bridgwater, Somerset, GB: British Cave Research Association, Aug, **1975**. Transactions of the British Cave Research Association v 2 no2. illus: b&w: charts, drawings, graphs, photos, tables; text on front lining page; 61p.

[3106] **Stanton, William Iredale** *Cheddar Caves: illustrated official guide.* Grinsell, L. V. St. Ives, Cambria, GB: Produced in Great Britain by Photo Precision Ltd, **1980**. illus: col: photos; [32]; 24 x 17.5 cm.

Earlier versions (editions?) may exist. No illus.

[3107] **Stanton, William Iredale** *Pioneer under the Mendips: Herbert Ernest Balch of Wells, a short biography.* Pangbourne, Berk, GB: The Wessex Caving Club, Oct, **1969**. Wessex Cave Club Occasional Publication Series 1 no 1. illus: b&w: drawings, maps, photos, repros; b&w front (port drawing); [vi], vi, 123p; app, bibl, index. 34 pls, 2 figs.

[3108] **Stanton, William Iredale** *Timeless Cheddar Caves.* Jacobi, R. M. npop: np, **c1987**. illus: b&w: repros; col: drawings, maps, photos, ports; text & photos on lining pages; [20].

Title from cover.

[3109] **Starr, Joan**, 1927– *The Wellington caves: treasure trove of fossils.* McMillan, Doug, 1935–. Dubbo, NSW, AU: Macquarie Publications, **1985**. illus: b&w: drawings, maps, photos; col: photos; b&w maps on lining pages; 52p, [1]; 24 cm.

Cover title: The Wellington Caves.

[3110] **Steel, William Gladstone** *Oregon, no 32, vol 1.* npop: np, **1931**. illus: b&w: maps, ports; 36 cm.

OCLC also lists 1907 as date of pub. Title from spine. Includes inscription of compiler: "Oregon State Library, compliments of William Gladstone Steel, Medford, Oregon, October 12, 1931", and bookplate: "Steel Points Library. Property of William Gladstone Steel." A scrapbook of clippings, primarily from the Oregonian, on a variety of Oregon subjects, especially the Oregon Caves. (The source of each clipping is handwritten in the margin.) Also includes typescript and holograph correspondence between the compiler and John J. Cuvriet regarding the disposition of the hull of the ship Bengal, and a copy of Oregon, the land of opportunity, published by the Portland Chamber of Commerce. Illus.

[3111] **Steele, C. William** *Yochib, the river cave: an account of the exploration of the Sumidero Yochib of Mexico,* *a dangerous and difficult cave.* St Louis, MO, US: Cave Books, **1985**. ix, 164p; 27 cm.

Illus.

[3112] Steen, Charlie R. *Ruins stabilization records for Canyon de Chelly National Monument 1942: Antelope House and Mummy Cave Ruins.* United States National Park Service. [Coolidge, AZ, US]: Southwestern National Monuments, **1942**. Ruins Stabilization: Records 1921-1983. two lvs folded; 14p; 28 cm.

Illus. 12 orig photos.

[3113] Steidtmann, Edward, 1881-1948 *Humidity and waters of a limestone cavern near Lexington, Virginia.* Charlottesville, VA, US: University of Virginia, **1936**. Virginia Geological Survey Bulletin 46-E. 9p; 23 cm.

Read before the Virginia Academy of Science, Section of Geology, Lexington meeting, May 1, 1936.

[3114] Stein, Aurel, 1862-1943 *Serindia: detailed report of explorations in Central Asia and westernmost China.* Delhi, CN: Motilal Banarsidass, **1980**. xxxix, 1580+59p, 175p pls; bibl.

Tun-huang Caves. 5 vols, 94 maps. Reprint. Originally published: Oxford: Clarendon Press, 1921. Bibliography: v. 1, p. xxv-xxviii. v. 1-3. Text—v. 4. Plates—v. 5. Maps.

[3115] Stekelis, Moshe, 1898-1967 *The Abu Usba Cave (Mount Carmel).* Haas, G. Jerusalem, IL: Israel Exploration Society, **1952**. illus: b&w: maps; 33p; 25 cm.

Offprint from: Israel Exploration Journal, v 2 no 1 (1952). Illus.

[3116] Stelle, James Parish *The Wyandotte Cave of Crawford County, Ind.* Cincinnati, OH, US: Moore, Wilstach & Baldwin, Printers, **1884**. iv, 85p, [1]; 19 cm.

First published in a series of articles in the Waverley Magazine Boston. Also as Speece Reprint #2, February 1969; limited edition of 55.

[3117] Stellmack, John Arnold, 1926– *Sort of an annual report from the standing committees and the secretary-treasurers of the NSS, including a summary annual report of the National Speleological Foundation, 1969-70.* npop: National Speleological Society, Jul, **1970**. 19p.

Only one of these ever published.

[3118] Stenner, Roger D. *Some smaller Mendip caves: vol 1.* Et al. Bridgwater, Somerset, GB: Bristol Exploration Club, Oct, **1961**. Caving Report no 6. illus: b&w: maps; 25p, [1]; 26 x 20 cm.

[3119] Stenuit, Robert *Caves and the marvelous world beneath us.* Jasinski, Marc ; Pearman, Harry (trans). British ed. London, GB: Nicholas Vane Ltd, **1966**. illus: b&w: drawings, maps, photos, photomicrographs, repros, tables; partly col: maps; col: photos; col front photo; 93p, [3]; gloss; 24 cm.

Translation of Merveilleux monde souterrain. One source lists a 1969 ed.

US edition. Jasinski, Marc. South Brunswick, [NJ], US: A.S. Barnes.

[3120] Sterling, Dorothy, 1913– *The story of caves.* Lubell, Winifred (illus). 1956 ed. Garden City, NY, US: Doubleday & Company, Inc. illus: partly col: maps; partly col drawings on lining pages; 121p; index.

Reprinted several times. Juvenile.

1956 reprint ed. Glenview, IL, US: Scott Foresman. Scott Foresman Invitations to Personal Reading Program A. illus: partly col: photos; 121p; index; 24 cm.

Reprint of the 1956 ed published by Doubleday, Garden City, New York. "Caves to visit": pp 113-116. Juvenile.

1966 ed. New York, NY, US: Scholastic Book Services. 121p; index; 20 cm.

Illus. Juvenile.

[3121] Stern, Philip Van Doren, 1900– *The beginnings of art.* New York, NY, US: Four Winds Press, **1973**. 239p; 26 cm.

Discusses prehistoric cave paintings of Europe, their discovery, meanings, and the methods used to produce them. Juvenile.

[3122] Stevens, John (ed) *An exploration journal of Llangattwg Mountain.* GB: Chelsea Spelaeological Society, Jan, **1992**. The Chelsea Spelaeological Society Records v 19. illus: b&w: maps; v, 133p, 2p pls; app; 30 x 21 cm.

Cover title: A detailed description of the cave of Llangattwg explored by CSS since 1980. Publications list tipped in back.

[3123] Stevens, Martin V. B. *The history, guide and description of Wyandotte Cave.* Stevens, Georgiana Lord Richardson. Harrisburg, IL, US: Patriot Publishing Co., **1876**. 24p; 22 cm.

Cover title: A history and description of Wyandotte Cave ..., attributes joint authorship to Mrs. G.L.R. Stevens. pp14-24 devoted primarily to phrenology. Phrenological ads on pp [2-4] of wrappers.

[3124] Stevens, Paul James, 1944– *Caves of the Organ Cave Plateau: Greenbrier County, West Virginia.* West Virginia Speleological Survey. npop: District of Columbia Grotto, National Speleological Society, **1988**. West Virginia Speleological Survey Bulletin 9. ix, 200p; bibl; 28 cm.

"Organ Save System Project of the National Speleological Society; a joint project of the District of Columbia Grotto and the West Virginia Association for Cave Studies"—Cover. Illus.

[3125] Steward, Julian Haynes, 1902-1972 *Ancient caves of the Great Salt Lake region.* Washington, DC, US: U.S. Government Printing Office; Smithsonian Institution Bureau of American Ethnology, **1937**. Smithsonian Institution Bureau of American Ethnology Bulletin 116. folded plan; xiv, 131p, 5p pls; bibl; 24 cm.

Englewood, Colo.: Microcard edition, 1966. 2 microfiche 11 x 15 cm.

[3126] **Stewart, John**, 1920– *Secret of the bats; the exploration of Carlsbad Caverns*. Philadelphia, PA, US: Westminster Press, **1972**. 124p, [1]; bibl; 21 cm.

Describes the role of New Mexican cowboy Jim White in exploring and publicizing the vast Carlsbad Caverns. Illus.

[3127] **Stewart, Thomas Dale**, 1901– *The Neanderthal skeletal remains from Shanidar Cave, Iraq: a summary of findings to date*. [Philadelphia, PA, US]: The American Philosophical Society, Apr, **1977**. Proceedings of the American Philosophical Society v 121 no 2 pp 121-165. 45p; bibl; 28 cm.

[3128] **Stirling, James** *On the caves perforating marble deposits, Limestone Creek (read 12th April 1883)*. Melbourne, VIC, AU: Royal Society of Victoria, **1884**. illus: b&w: drawings, maps; 3 fold-out maps with vignettes; 11p, 3p pls.

Copy examined was a photocopy. Also published in Trans & Proc of the Royal Society of Victoria, vol 20, 1884, pp[7]-17.

[3129] **Stock, Chester**, 1892-1950 *Quaternary antelope remains from a second cave deposit in the Organ Mountains and further study of the quaternary antelopes of Shelter Cave, New Mexico*. Los Angeles, CA, US: Los Angeles Museum, **1930, 1932**. Los Angeles County Museum Los Angeles Paleontology Publications, Science Series Publication no 2 and 3.

Illus. 2 vols in 1.

[3130] **Stoddard, Seneca Ray**, 1844-1917 *Howe's Cave*. Troy, NY, US: Nims & Knight, **1889**. 16 x 22 cm.

15 pls. Pagination unclear.

[3131] **Stoddard, Sheena** *An introduction to cave photography*. London, GB: British Cave Research Association, **1994**. Cave Studies Series no 4. illus: b&w: charts, photos; col: photos; text on lining pages; 32p; app, bibl; 21 cm.

[3132] **Stohr, C. J.** *Delineation of sinkholes using thermal infrared imagery*. West Lafayette, IN, US: Laboratory for Applications of Remote Sensing, Purdue University, **1974**. LARS Information Note no 032574. illus: b&w: maps; 13p; bibl; 28 cm.

Illus. Title from cover.

[3133] **Stone, Andrea Joyce** *Images from the underworld: Naj Tunich and the tradition of Maya cave painting*. Austin, TX, US: University of Texas Press, **1995**. illus: b&w: cartoons, charts, drawings, maps, photos, tables; col: photos; x, 284p, 8p pls; bibl, index; 29 cm.

[3134] **Stone, John M.** *The underground passages, caverns, etc., of Greenwich and Blackheath*. London, GB: The Blackheath Press, **1914**. 18p; 22 cm.

Lecture delivered before the Greenwich Antiquarian Society.

[3135] **Stone, Ralph Walter**, 1876-1964 *Caves of Pennsylvania*. National Speleological Society. [Trenton, PA, US]: National Speleological Society, **1953**. The American Caver December 1953. 143p.

'Bulletin fifteen of the National Speleological Society.' Title from cover. Illus.

[3136] **Stone, Ralph Walter**, 1876-1964 *Descriptions of Pennsylvania's undeveloped caves*. Washington, DC, US: National Speleological Society, **1953**. Bulletin of the National Speleological Society, pp 51-139. illus: b&w: maps; 89p.

Originally an article in Bulletin of the National Speleological Society (no. 15, December 1953). Microfilm. Tucson, Ariz.: University of Arizona Library, Photographic Services, 1972. 1 microfilm reel; 35 mm.

[3137] **Stone, Ralph Walter**, 1876-1964 *Pennsylvania caves*. 1st ed. Harrisburg, PA, US: Department of Internal Affairs, Topographic and Geologic Survey, **1930**. Pennsylvania Geological Survey 4th 1919 Bulletin G3. 63p; bibl; 23 cm.

Illus.

2nd ed. Barnsley, Edward Roberts, 1906–; Hickok, WIlliam Orville, 1905–; Mohr, Charles Edward, 1907– (Pennsylvania cave fauna). **1932**. Pennsylvania Geological Survey (4th 1919-) Bulletin G-3. illus: b&w: drawings, maps; iv, 143p; 24 cm.

Illus.

Revised and considerably expanded from Stone 1930. Catalog of 65 known caves. Descriptions and locations; maps of about half. Chapters on geology and biology, the latter by Charles Mohr.

[3138] **Stone, William Curtis**, 1952– *Caves of the San Juan Plateau*. Jameson, Roy. Austin, TX, US: AMCS, **1977**. Bulletin Association for Mexican Cave Studies 7. illus: b&w: maps; seven folded lvs of maps in pocket; 59p; bibl; 28 cm.

b&w, col illus.

[3139] **Stone, William Curtis**, 1952– *Inner space: the last terrestrial frontier*. Gaithersburg, MD, US: U.S. National Institute of Standards and Technology, **1992**.

VHS format. At a NIST Colloquium held February 14, 1992, at the National Institute of Standards and Technology, Dr. Stone, leader of the U.S. Deep Caving Team, discusses recent high-tech speleological expeditions. He also discusses sensory deprivation, circadian cycle dilation, team selection and training, biohazards, and closed cycle life support systems used by the team to explore deep into inner space.1 videocassette (60 min), sd, col, 1/2 in.

[3140] **Stone, William Curtis**, 1952– *A report to the government of Mexico on the 1994 San Augustin expedition to Huautla de Jimenez, Oaxaca*. am Ende, Barbara Anne. Gaithersburg, MD, US: United States Deep Caving Team, Aug, **1994**. illus: b&w: charts, drawings, graphs, tables; col: photos; [x], 127p, 10p pls; app, bibl.

[3141] **Stone, William Curtis**, 1952– *The Wakulla Springs project*. Derwood, MD, US: U.S. Deep Caving

Team, **1989**. illus: b&w: ads, graphs, photos, ports, tables; partly col: photos; col: maps; 1 col photo, 3 maps folded in text; vii, [212]; app; 29 cm.

An excellent project report which was printed in limited quantities.

[3142] Stoneman, John (narr) *The cave divers*. Foundation for Ocean Research. Morris Plains, NJ, US: Lucerne Media, **1990**. The Last Frontier.

VHS. Hosted by John Stoneman and members of the Foundation for Ocean Research. In an underwater Florida cave, the team investigates a find of ancient bones and fossils believed to be the remains of a Mastodon. 1 videocassette (ca. 25 min), sd, col, 1/2 in.

[3143] Storey, James Welborn, 1935-1992 *Advanced cave diving*. npop: James W. Storey, Jun, **1971**. illus: b&w: cartoons, charts, graphs, maps, photos, tables; Various pagination, 39p.

Title from cover. Published for the 1971 NSS convention.

[3144] Storey, James Welborn, 1935-1992 *American caving, illustrated; caving, climbing, camping*. Atlanta, GA, US: np, **1965**. illus: b&w: ports; xviii, 302p; bibl; 22 cm.

Illus.

[3145] Storey, James Welborn, 1935-1992 *Cave diving notes: equipment - methods - danger*. npop: James W. Storey, [**1970**]. illus: b&w: cartoons, charts, drawings, maps, photos, tables; Various pagination, 20p.

Published for the 1970 NSS convention. Title from cover. Pagination includes covers.

[3146] Storrick, Gary D. *The caves and karst hydrology of southern Pocahontas County and the upper Spring Creek Valley*. West Virginia Speleological Survey. Reston, VA, US, May, **1992**. Bulletin West Virginia Speleological Survey 10. illus: b&w: maps; vi, 214p; index; 28 cm.

Illus.

The northern part of this West Virginia county is covered in Medville and Medville 1976.

[3147] Story, Isabelle Florence, 1888-1970 *Carlsbad Caverns National Park*. United States National Park Service. Washington, DC, US: U.S. Government Printing Office, **1935**. illus: b&w: maps; 27p, [1]; 24 cm.

At head of title: United States Dept. of the Interior, National Park Service. Illus.

[3148] Stow, Marcellus Henry, 1902– *Description of points of scenic, historic, and geologic interest between Washington, D. C., and Luray, Virginia*. McGill, William Mahone, 1896–(Luray Caverns in Virginia. Luray Caverns in Virginia). Washington, DC, US: np, [**1939**]. illus: col: maps; col map in pocket; 20p; bibl; 25 cm.

American Geophysical Union Excursion to Luray, Virginia, Sunday, "American Geophysical Union Excursion to Luray, Virginia, Sunday, September 10, 1939."

[3149] Stratford, Tim *Caves of South Wales*. 1st ed. Leicester, GB: Cordee, **1978**. A Caver's Guide Book. illus: b&w: maps; 92p, 4p pls; index; 19 cm.

Illus.

2nd ed. 1982. 96p.

Illus. New ed, fully updated.

3rd ed. 1986. illus: b&w: maps; 117p.

Illus.

4th ed. Baker, Tony (photo). **1995**. illus: b&w: maps, tables; col: photos; x, 149p, 8p pls; gloss.

[3150] Stratford, Tim (comp) *Toros 89-92; the Camlik Project: a report on the speleological investigations carried out by the Swindon Speleological Society in the Camlik Region of Toros Mountain, Konya Turkey from 1989-1992*. Swindon, Wiltshire, GB: Swindon Speleological Society, **1992**. Swindon Speleological Society Occasional Publication. illus: b&w: drawings, maps, photos; ii, 58p; bibl; 30 x 21 cm.

[3151] Stringfield, Victor Timothy, 1902– *Hydrology of limestone terranes in the coastal plain of the southeastern states*. LeGrand, Harry Elwood, 1917–. 1964 ed. New York, NY, US: Geological Society of America, **1964**. illus: b&w: charts, maps; ii, 54p; bibl, index; 27 cm.

Preprinted for Annual Meeting of Geological Society of America; November 1964.

1966 ed. Special Paper 93. 46p.

[3152] Stringfield, Victor Timothy, 1902– *Hydrology of limestone terranes in the coastal plain of the southeastern United States*. LeGrand, Harry Elwood, 1917–. New York, NY, US: [Geological Society of America], **1966**. Regional Studies Special Papers 93. illus: b&w: charts, drawings, maps, photos; 46p; bibl, index; 23 cm.

On cover: Regional studies.

This is an early review of the karst geohydrology in a Tertiary-to-Recent Coastal Plain setting. Although the processes are comparable with those in the Mammoth Cave area, karst evolution has been significantly different, having been effected by sea-level fluctuations.

[3153] Stringfield, Victor Timothy, 1902– *Hydrology of limestone terranes: development of karst and is effects on the permeability and circulation of water in carbonate rocks, with special reference to the southeastern states*. La Moreaux, Philip Elmer, 1920–; LeGrand, Harry Elwood, 1917–. University, AL, US: Geological Survey of Alabama, **1977**. Geological Survey of Alabama Bulletin 94 Part G. illus: b&w: drawings, maps, photos; partly col: maps; [vi], 68p, [1]; bibl; 23 cm.

[3154] Stringfield, Victor Timothy, 1902– *Karst and paleohydrology of carbonate rock terranes in semiarid and arid regions with a comparison to humid karst of Alabama*. La Moreaux, Philip Elmer, 1920–; LeGrand, Harry Elwood, 1917–; Tolson, Jan (ed); Rayfield, C. Raymond (Graphics). University, AL, US: Geological Survey of Alabama, **1974**. Geological Survey of Alabama Bulletin 105. illus: b&w: drawings, maps, photos, tables; partly col: drawings, maps, photos; col: photos; b&w maps incl 3 folded in text; [ix], 106p; bibl.

Unique comparison of the Alabama karst to that of Arizona, Yucatan, Australia, and Egypt.

[3155] **Studley, Cordelia A.** *Notes upon the human remains from the caves of Coahuila, Mexico.* Salem, MA, US: Printed at the Salem Pr, **1884**. XVI Report of the Peabody Museum of American Archaeology and Ethnology; Cambridge, Mass, 1883 pp 233-59. illus: b&w: tables; 27p; 23 cm.

[3156] **Styles, Frank Showell**, 1908– *How underground Britain is explored.* London, GB: Routledge and Kegan Paul, **1959**. The How Series. xii, 147p, 8p pls; app, bibl, gloss.

Illus.

[3157] **Sullivan, Gerardus Nicholas**, 1927– *Appendix: checklist of macroscopic troglobitic organisms of the United States.* reprint ed. Notre Dame, IN, US: University of Notre Dame Press, Jul, **1960**. The American Midland Naturalist, vol 64, no 1 pp 123-160. 38p; bibl; 23 cm.

[3158] **Sullivan, Gerardus Nicholas**, 1927– *Checklist of troglobitic organisms of Middle America.* reprint ed. Notre Dame, ID, US: University of Notre Dame Press, **1962**. Reprinted from The American Midland Naturalist, vol 68, no 1 (July 1962) pp165-188. 24p; bibl; 23 cm.

Illus.

[3159] **Sumia Konda** *The report of lava flow and lava caves on Mauna Loa.* The Whole Earth Club (Japan). JP: np, Jul, **1993**. illus: b&w: drawings, maps, photos; [6]p folded maps; [ii], 32p; app, bibl.

English section followed by Japanese section.

[3160] **Surin, Phukhachon** *Preliminary report of excavations at Moh-Khiew Cave, Krabi Province, Sakai Cave, Trang Province, and ethnoarchaeological research of hunter-gatherer group, socall[ed] Sakai or Semang at Trang Province: the Hoabinnian Research Project in Thailand.* Bangkok, TH: Faculty of Archaeology, Silpakorn University, **1991**. 31 cm.

Cover title: Raingan buangton kankhutkhon thi Tham Mo Khiao, Cho. Krabi, Tham Sakai, Cho. Trang lae kansuksa chattiphanwitthaya thang borannakhadi chon klum noi phao Sakai, Cho. Trang. 1 vol. Illus.

[3161] **Sutcliffe, Antony John** *Joint Mitnor Caves: Buckfastleigh.* npop: np, **c1942**. Reprinted from the Transactions of Torquay National History Society XIII part 1. illus: b&w: photos; front: b&w photos-both sides; 28p, [4]; 22 x 14 cm.

Cover subtitle: A report on the excavations carried out during 1939-1941 by the late A.H. Ogilvie.

[3162] **Sutton, L. J.** *Meteorological conditions in caves and ancient tombs in Egypt.* Cairo, EG: Government Press, **1945**. Egypt Public Works Dept Physical Dept Paper no 50. illus: b&w: drawings, maps, tables; 81p; 27 cm.

[3163] **Suzuki, Hisashi**, 1912– *The Palaeolithic site at Douara Cave in Syria: report of the fourth season of the Tokyo University Scientific Expedition to Western Asia.* Takai, Fuyuji, 1911–(ed). [Tokyo, JP]: University of Tokyo Press, **1973**. bibl; 27 cm.

Illus, 2 vols, pls.

[3164] **Sweatman, Cheyenne (ed)** *8th Annual TAG fall cave-in.* npop: [TAG Region], Oct, **1985**. illus: b&w: ads, drawings, maps; 52p; bibl.

[3165] **Sweeting, Marjorie Mary**, 1920-1994 *Karst geomorphology.* Stroudsburg, PA, US: Hutchinson Ross Publishing Company, **1981**. Benchmark Papers in Geology 59. illus: b&w: charts, graphs, maps, photos, tables; xv, 427p; bibl, index; 27 cm.

An outstanding collection of classic papers on karst processes and products. The originals were published in a wide variety of journals, many largely non-English, so that the present work reflects a terrific advance in the availability and readability of these classics. A highly technical but exciting book.

[3166] **Sweeting, Marjorie Mary**, 1920-1994 *Karst in China: its geomorphology and environment.* Pfeffer, Karl-Heinz. Berlin; New York, NY, DE; US: Springer Verlag, **1995**. Springer Series in the Physical Environment 15. illus: b&w: drawings, photos; xi, 265p; bibl, index; 25 cm.

An excellent overview of the subject, combining the author's own research with much work previously only reported in Chinese.

[3167] **Sweeting, Marjorie Mary**, 1920-1994 *Karst landforms.* London, GB: The Macmillan Press Ltd, **1972**. illus: b&w: charts, drawings, graphs, maps, photos, photomicrographs, tables; xvi, 362p; bibl, gloss, index; 24.5 x 18.5 cm.

American edition published 1973 by Columbia University Press, New York, US. 57 photos, 127 figs, 16 tables.

The classic text on the subject. Although they skip around a bit, the eighteen chapters touch on a broad range of karst-related topics. The title does not do justice to the breadth and depth of this work. Although largely limited to shallow phreatic and vadose, CO_2-driven dissolution of limestone terranes, this well written and illustrated book is an early classic.

[3168] **Sweeting, Marjorie Mary**, 1920-1994 *Karst processes.* Pfeffer, Karl-Heinz. Berlin, DE: Gebr Borntraege, **1976**. Zeitschrift fur Geomorphologie Supplement band n F 26. 210p; bibl; 25 cm.

English, German, or French. Illus.

Papers on karst geomorphology since 1960, in various languages. Abstracts in English, French, and German. References and well done black and white photos and drawings make even non-English works useful. English papers deal largely with tropical karst.

[3169] **Sydney Speleological Society** *Prospective's handbook*. Broadway, NSW, AU: Sydney Speleological Society, **nd**. illus: b&w: drawings, tables; Various pagination.

[3170] **Sydney Speleological Society** *Timor Caves*. npop: np, **1976**. illus: b&w: maps; 50p.

Proposed development near Nurrurundi. Illus. Incl surveys.

[3171] **Sydney University Speleological Society** *Nibicon log book reports December 1972 - February 1973*. Sydney, NSW, AU: [Sydney University Speleological Society], Nov, **1975**. SUSS Bulletin v 15 no 7. illus: b&w: maps, photos; [iii], 33p; bibl.

Logbook of fieldtrips of the Ninth Biennial Conference of the Australian Speleological Federation, held Christmas 1972-New Year 1973.

[3172] **Sykes, Les** *C. N. C. C. eco-resin rigging system no 1*. [GB]: The Council of Northern Caving Clubs, [**1994**]. illus: b&w: drawings; [ii], 35p; 21 x 15 cm.

[3173] **Symonds, William Samuel**, 1818-1887 *The Seven Straits; or, notes on glacial drifts, bone caverns, and old glaciers: some within reach of the Malvern Hills*. Tewkesbury [Gloucestershire]; London, GB: William North; Simpkin, Marshall, **1883**. 65p; bibl; 22 cm.

[3174] **Symposium on Detection of Subsurface Cavities (1977: Vicksburg, MS)** *Symposium on Detection of Subsurface Cavities, 12-15 July, 1977*. Vicksburg, MI, US: Department of Defense, Department of the Army, Corps of Engineers, Waterways Experiment Station, Oct, **1977**. Various pagination, 187p; 27 cm.

Illus.

[3175] **Szentes, Georg** *Sandstone caves in Nigeria*. Frankfurt am Main, DE: Höhlenforschergruppe Rhein-Main, Jul, **1989**. illus: b&w: maps, photos, tables; text on lining pages; b&w photo on rear lining page; 35p; bibl.

Bilingual: German and English. Sandsteinhöhlen in Nigeria. Title page and some other material in German. Poor quality photos.

[3176] **Szentes, Georg** *Tropical karst and caves in the central Cordillera, Colombia*. Frankfurt am Main, DE: Höhlenforschergruppe Rhein-Main, **1992**. illus: b&w: charts, maps, photos, tables; text on lining page; 67p; bibl.

Bilingual: German and English. Höhlen und tropischer karst in der mittleren Kordillere, Kolumbien.

[3177] **Taketaro Shinkai**, 1868-1927 *Rock-carvings from the Yun-kang caves*. Tadayori Nakagawa, 1873-1928. Tokyo, JP: Bunkyudo, **1921**. illus: b&w: maps; folded pls and plan; 16p; 30 cm.

Japanese and English. Some of the 200 plates accompanied by guard sheets with outline drawings. Issued in portfolio.

[3178] **Talent, J. A.** *The Buchan Caves*. VIC, AU: Victoria Mines Department, **nd**. illus: partly col: drawings, maps, photos; map on front lining page; photo and text on back lining page; [12].

Title from cover.

[3179] **Talley, John H.** *Sinkholes, Hockessein area, Delaware*. Newark, DE, US: Delaware Geological Survey, Mar, **1981**. Open File Report Delaware Geological Survey no 14. 16p; 28 cm.

Illus.

[3180] **Tankersley, Ken** *The cavernous karst land forms of Jackson County, Kentucky*. [Cincinnati, OH, US]: np, **1975**. illus: b&w: maps, repros; [89]; 29 cm.

Includes excerpt from the University of Kentucky Geological Survey's Report of the geology of parts of Jackson County. Illus, facsims.

[3181] **Taralon, Jean** *The grotto of Lascaux*. Paris, FR: Caisse Nationale des Monuments Historiques, Service Commercial, **1962**. [37], 18p pls; bibl; 18 cm.

Cover title: Lascaux. b&w, col illus.

[3182] **Tardy, Janos** *Geological nature conservation cave-protection*. Budapest, HU: Institute for Speleology, Ministry of Environment and Water Management, **1989**. illus: b&w: maps; col: maps; 53p; 29 cm.

One col map on 1 sheet laid in. English and Hungarian. Table of contents also in Hungarian. Illus.

[3183] **Tarkington, Terry Warren**, 1925– *Alabama caves*. Varnedoe, William Whitfield, 1923–; Veitch, John David, 1931–. Huntsville, AL, US: Huntsville Grotto, National Speleological Society, Jun, **1965**. illus: b&w: maps, tables; Various pagination; 28 cm.

Mimeographed. Illus.

First of a series of catalogs of all known caves in the state of Alabama. Each contains a printout of computer data-base giving cave names, locations, and very limited additional information, plus all available maps. No text descriptions. This 1965 version lists 617 caves and includes roughly 350 maps. The 1973 version (Vernedoe 1973) has 1386 caves, with 1034 maps. The 1980 version (Varnedoe 1980) has 1969 caves, with 1337 maps, one per page. Generally the maps are crude, but adequate to identify or explore the caves. There were also occasional intervening supplements published, and there may be later issues.

[3184] **Tarkington, Terry Warren**, 1925– *A manual for beginners*. McElroy, Jim. npop: Terrapin Trail Club, University of Maryland, May, **1965**. illus: b&w: drawings, photos; text on front lining page; [i], 29p; bibl.

Pagination includes back cover.

[3185] **Tarkington, Terry Warren**, 1925– *So you want to go caving! A manual for beginners*. Varnedoe, William Whitfield,1923–; Veitch, John David, 1931–. [Huntsville, AL, US]: Huntsville Grotto, National Speleological Society, **1965**. illus: b&w: drawings; 11p; bibl.

[3186] **Tásler, Radko** *Owen 90: New Zealand*. Trutnov, CZ: Česka Speleologická Společnost, **1991**. illus: b&w: drawings, maps, photos, repros; col: photos; 1 folded map stapled at center; 3 loose maps unfolded; 2p, [58]; bibl; 29.5 x 21 cm.

Expedition report to caves at Mount Owen.

[3187] Tasmanian Caverneering Club *Caverneering handbook of the Tasmanian Caverneering Club.* [Tasmania, AU]: [Tasmanian Caverneering Club], **1963**. illus: b&w: drawings; 49p; gloss.

Part I.

[3188] Taylor, Alisa Johanna *The mammalian fauna from the mid-Irvingtonian Fyllan Cave local fauna, Travis County, Texas.* npop: np, **1982**. ix, 106p; bibl; 28 cm.

Illus.

[3189] Taylor, Kirk R. (comp) *Caves of Bedford County, Pennsylvania.* Ganter, John H. (comp); White, William Blaine, 1934–(comp); White, Elizabeth L. (comp). State College, PA, US: National Speleological Society, Mid-Appalachian Region, Jan, **1993**. MAR Bulletin no 19. illus: b&w: maps, tables; 5 fold-out maps; vii, 67p; bibl; 27 cm.

Illus.

[3190] Taylor, Maurice Clague *Three below Gower: the story of cave exploration in Gower by The Taylors'.* Crymych, GB: Anne Oldham, **1991**. illus: b&w: maps; 89p, 11p pls; 22 cm.

Illus. Intro by M.C. Taylor: "my sisters and I (we are the three' of the title)."

[3191] Taylor, Michael Ray, 1959– *Cave passages: roaming the underground wilderness.* New York, NY, US: Scribner, **1996**. illus: b&w: photos; 285p, 8p pls; gloss, index; 23 cm.

Also issued by Vintage Books, New York,1997; slightly revised.

[3192] Taylor, Michael Ray, 1959– *Lechuguilla: jewel of the underground.* Widmer, Urs (design). Basel, SZ: Speleo Projects, Caving Publications International, **1991**. illus: b&w: drawings, maps, photos; col: drawings, maps, photos; 144p; bibl; 31 cm.

[3193] Tazieff, Haroun, 1914– *Caves of adventure.* Hodge, Alan (trans). 1953 American ed. New York, NY, US: Harper, **1953**. 222p.

Illus.

1953 British ed. London, GB: Hamish Hamilton, **1953**. 139p; 22 cm.

Illus.

An account of the 1951 and 1952 expeditions to Pierre Saint-Martin by the geologist and expedition member Haroun Tazieff.

1960 ed. New York, NY, US: Viking Press, **1960**. 222p; 20 cm.

Explorer books edition. Illus.

[3194] Technical Aids in Caving Symposium (1972: Buxton High Peak College) *Technical aids in caving symposium held on Sunday 5th March, 1972 at Buxton High Peak College, Harpur Hill, Buxton, Derbyshire.* Sports Council (East Midlands Sports Council); Cave Research Group of Great Britain. Buxton, Derbyshire, Gb: Buxton High Peak College, **1972**. 23p; 30 cm.

Illus.

[3195] Tell, Leander *Erosionsforloppet, med sarskild hansyn till Lummelundagrottorna. The rate of erosion, with special reference to the caves of Lummelunda.* Norrkoping, SE: Centrocommerce, **1961**. Arkiv fur Svensk Grottforskning 14. illus: b&w: maps; 46p; index; 21 cm.

Illus.

Brief descriptions of geologic setting and geomorphology. Few location maps, no cave maps, and a short but useful text.

[3196] Tell, Leander, 1895– *Fifty typical Swedish caves.* Norrkoping, SE: Centrocommerce, **1976**. Archives of Swedish Speleology 14 Norrkoping. illus: b&w: maps, photos; 41p; 21 cm.

Illus.

[3197] Tennessee Academy of Science *Journal of the Tennessee Academy of Science. Cave number.* Nashville, TN, US: Tennessee Academy of Science, Jul, **1930**. v 5 no 3 pp 81-136. illus: b&w: charts, maps, photos, tables; 56p.

The nature and formation of caves, by Berlen C. Moneymaker.—Underground in Tennessee and Kentucky, by Erwin R. Pohl.—Cave animals of Tennessee, by J.D. Ives.—The cave man in Tennessee, by P.E. Cox.—Tennessee caves in historical times, by John L. Morris.—Caves in mythology, by Charles Little.

[3198] Thayer, Charles W. *Mud stalagmites and the conulite, a new speleothem.* [Hanover, NH, US]: np, **1966**. 9p; bibl; 29 cm.

Reproduced from typescript. Award winning paper for Upham Geology Prize contest. Illus.

[3199] Thomas, Alan, 1931– *The last adventure.* Wells, Somerset, GB: INA Books, **1989**. illus: b&w: maps, photos, ports; partly col: photos; col: ports; vi, 75p; 23.5 x 15.5 cm.

8 different authors on cave diving exploration.

[3200] Thomas, David Hurst *The archaeology of Hidden Cave, Nevada.* New York, NY, US: np, Jun, **1985**. Anthropological Papers of the American Museum of Natural History v 61 pt 1. illus: b&w: maps; 430p; bibl, index; 26 cm.

Issued June 20, 1985. Illus.

[3201] Thomas, Harold S. *The geology of Kankee Caverns [sic]: a geological report on the underground storage of LP-gas on the Phillips pipe line terminal, Kankakee-Illinois.* [Bartlesville, OK, US]: Phillips Petroleum Company, **1953**. illus: b&w: maps; some folded lvs of pls; 62p, 23p pls; 30 cm.

Cover title: Geology of Kankakee Caverns. b&w, col illus.

[3202] Thompson, Edward Herbert, 1860-1935 *Cave*

of Loltun, Yucatan: report of explorations by the Museum, 1888-89 and 1890-91. 1897 ed. Cambridge, MA, US: The Museum, **1897**. Memoirs of the Peabody Museum of American Archaeology and Ethnology, Harvard University v 1 no 2. illus: b&w: maps; 24p, 8p pls; 36 cm.

Also paged continuously with the other numbers of the volume. Illus.

1970 reprint ed. New York, NY, US: Kraus Reprint, **1970**.

[3203] Thompson, James B. *The geology of Jewel Cave.* Custer, SD, US: Wind Cave National Park and Jewel Cave National Monument Natural History Associations, **1958**. illus: b&w: charts, photos; 18p; 23 cm.

Tourist-oriented outline of geology and speleothems. Illus. Rev ed in 1978; no apparent changes from OCLC record.

[3204] Thompson, John, 1874– *Mammoth Cave, Kentucky; an historical sketch containing a brief description of some of the principal places of interest in the Mammoth Cave. Also a short description of Colossal Cavern.* 4th ed. Smiths Grove, KY, US: Times Publishing Co, **1909**. illus: b&w: drawings, maps, photos; map folded in back; front; [iii], 48p, [1], 17p pls; 23 cm.

Another OCLC record indicates a 1909 copy with 26 pages. Pages 44-47 blank for "Signatures of our party", Louisville, Courier-journal Job Print. Microfilm edition: Printing Master B92-128. Atlanta, Ga.: SOLINET, 1993. 1 microfilm reel ; 35 mm. One copy examined was 48pp, printed by Courier-Journal Job Print, Louisville and had C.R.Blackall Map

[3205] Thompson, Peter, 1943– *Cave exploration in Canada: a special issue of the Canadian Caver magazine.* The Canadian caver magazine. Edmonton, Alberta, CA: Canadian Caver Magazine, Dept of Geography, University of Alberta, **1976**. illus: b&w: maps, photos, tables; 7 maps folded; viii, 183p; bibl, index; 29 cm.

Includes some text in French. "A special issue of the Canadian caver magazine."

A detailed summary of cave exploration in all the important regions of Canada between 1965 and 1975.

[3206] Thompson, Ralph Seymour *The sucker's visit to Mammoth Cave.* Bridge, John F. (intro). New York, NY, US: Johnson Reprint Corp, **1970**. Classics in Speleology. illus: b&w: maps; xix, 128p; 19 cm.

"Originally published in 1879." The 1879 ed was a revision of a work published in 1871 under title: Western Kentucky above ground and below.

[3207] Thompson, Ralph Seymour *The sucker's visit to the Mammoth Cave including a history of the experience and adventures of a party who undertook to see the cave and have some fun going there. A full and accurate description of the cave and the science of its formation with an account of the living inhabitants.* Springfield, OH, US: Live Patron Publishing Office, **1879**. Travels in the New South. illus: b&w: maps; [iii], xix, 128p; 19 cm.

The 1879 ed was a revision of a work published in 1871 under title: Western Kentucky above ground and below. Microopaque version, Louisville, Ky, Lost Cause Press, 1965. 2 cards.

Warm and humorous account of a cave expedition.

2nd ed. 1881. The Economist Library v 1 no 4 December 1886. 64p; 17 cm.

5th ed. New Era Company, **1886**. The Economist Library v 1 no 4 December 1886. 64p; 19 cm.

[3208] Thompson, Robert S. *Paleoenvironmental investigations at Seed Cave (Windust Cave H- 45FR46), Franklin County, Washington.* United States Army Corps of Engineers Walla Walla District. Cheney, WA, US: Eastern Washington University, Archaeological and Historical Services, **1985**. Eastern Washington University Reports in Archaeology and History no 100-41. illus: b&w: maps; v, 52p; bibl; 28 cm.

Report prepared for the U.S. Army Corps of Engineers, Walla Walla District. Illus.

[3209] Thornber, Norman *Britain underground.* Stride, A.H.; Stride, R. D.; Myers, Jack O.; Gemmell, Arthur, 1915–(map, illus). Clapham, Lancaster, GB: Dalesman Publishing Company; Blandford Press, **1953**. illus: b&w: maps; ix, 246p, 20p pls; 14 cm.

[3210] Thornber, Norman *Pennine underground.* Gemmell, Arthur, 1915–(map illus). 1959 ed. Clapham, Lancaster, GB: Dalesman Publishing Company, **1959**. illus: b&w: ads, maps; b&w ads on lining pages; 208p; 14 cm.

One source lists a 1947 edition.

1965 ed. 224p.

[3211] Thornycroft, L. B. *The story of Wookey Hole in fact, fiction and photo.* Taunton, [GB]: Barnicotts, Limited, **1948**. illus: b&w: maps; 147p; app.

1956 printing: [xii], 147pp, folded map attached to inside rear; biblio; illus: b&w photos, drawings, maps, table, repro. Notes: first edition, reprinted 1951, 1954, 1956.

[3212] Thrailkill, John Vernon, 1930– *Hydrogeology and environmental geology of the inner Bluegrass Karst Region Kentucky: field guide for the annual meeting of the southeastern and north-central sections Geological Society of America: Lexington, Kentucky April 4-6 1984.* npop: [Geology Society of America Southeastern and North-Central Sections], **[1984]**. illus: b&w: charts, maps, tables; [i], 31p; bibl.

[3213] Thrailkill, John Vernon, 1930– *Introduction to caving.* 1st ed. Ward, CO, US: Gerry Mountaineering Equipment Co, **1954**. 28p; 22 cm.

Illus.

2nd ed. [Boulder, CO, US]: Highlander Publishing Co, **1962**. illus: b&w: charts, drawings; 30p; bibl, gloss.

Illus.

3rd ed. [Boulder, CO, US]: Colorado Outdoor Sports

Corp, **1962**. illus: b&w: maps; 30p; bibl.

[3214] Thrailkill, John Vernon, 1930– *Solution geochemistry of the water of limestone terrains*. Lexington, KY, US: University of Kentucky Water Resources Institute, **1970**. Research Report no 19. illus: b&w: charts, graphs, maps, tables; ix, 125p; app, bibl.

[3215] Thrailkill, John Vernon, 1930– *Studies in dye-tracing techniques and karst hydrogeology*. Byrd, Phillip E.; Sullivan, Stephen B.; Spangler, Lawrence E.; Taylor, Charles J.; Nelson, Greta K.; Pogue, Kevin R. Lexington, KY, US: University of Kentucky Water Resources Research Institute, **1983**. University of Kentucky Water Resources Research Institute Research Report No 140. illus: b&w: drawings, graphs, tables; [i], viii, 89p; bibl.

[3216] Thrun, Robert, 1940– *Prusiking*. Huntsville, AL, US: National Speleological Society, **1973**. illus: b&w: drawings, graphs, maps, photos; 75p, [3]; app, bibl, index; 28 cm.

One of the first US publications to attempt to give a basic overview to rope-climbing techniques; contains nothing about rappelling. It did a very good job for the time.

[3217] Thurgate, Mia E. *Sinkholes, caves and spring lakes: an introduction to the unusual aquatic ecosystems of the lower south east of South Australia: South Australian Underwater Speleological Society occasional paper number 1*. npop: np, **1995**. South Australian Underwater Speleological Society Occasional Paper number 1. illus: b&w: charts, drawings, graphs, maps, tables; col: photos, photomicros; front: partly colored photo of The Chasm, Piccaninnie Ponds; also the title page; iii, 44p, [4]; app, bibl; 30 x 21 cm.

[3218] Thynne, Daphne Winifred Louise (Marchioness of Bath) *Cheddar caves*. 1956 ed. Longleat, GB: Longleat Estate, **1956**. illus: b&w: drawings, photos, repros; col: repros; 32p; 18 x 12 cm.

1960 ed. illus: b&w: photos; 35p, [1].

[3219] Tintilozov, Zurab *Akhali Atoni's cave*. Tbilisi, GE: Sabchota Sakartvelo, **1978**. illus: b&w: maps; col: photos; col photo on front lining page, text on rear lining page; 19p; 16 cm.

[3220] Toll, Henry Wolcott *An analysis of variability and condition of cavate structures in Bandelier National Monument*. McKenna, Peter J.; Crowder, June. [Santa Fe, NM, US]: Anthropology Program, U.S. National Park Service, **1995**. Intermountain Cultural Resources Center Professional Paper no 53; Bandelier Archeological Project Contribution no 3. illus: b&w: maps; xvi, 300p; bibl, index; 28 cm.

[3221] Toops, Bonnie *Let's explore: caves and caverns: a young explorer series*. Trembay, Dan (illus). Conway, AR, US: Explorers Press, **1990**. illus: b&w: drawings, maps, photos; 32p.

Juvenile.

[3222] Torode, William Wallace, 1943– *Ellison's Cave: Georgia*. [Huntsville, AL, US]: [Torode], nd. illus: b&w: repros; map folded in rear pocket; [v], 84p, [78].

[3223] Torres González, Arturo *Geohydrology of the Rio Camuy Cave system Puerto Rico*. Aguilar, E.; Pannela, G.; United States Office of Water Policy. Mayaguez, PT: Water Resources Research Institute, School of Engineering, University of Puerto Rico, Mayaguez Campus, Mar, **1983**. illus: b&w: drawings, graphs, maps, tables; iii, 57p; bibl; 28 cm.

Final technical report to U.S. Dept. of the Interior. Partially funded by the U.S. Dept. of the Interior under the Water Research and Development Act of 1978. Project A-053-PR, grant agreement no. 14-34-0001-8041.

[3224] Towler, George (photo) *Views of Chapman Cave*. [GB]: George Towler, **1890**. illus: b&w: photos; [12]; 40 x 30 cm.

Title from cover. No text. May be unique; second example seen with some different photos. One copy has letter tipped in front from photographer, January 1890.

[3225] Traister, Robert J. *Cave exploring*. Blue Ridge Summit, PA, US: TAB Books, **1983**. illus: b&w: drawings, maps, photos; viii, 184p; gloss, index; 25 cm.

Without any doubt the worst caving book ever written. Loaded with dangerous information including the recommendation of the use of 1 inch manilla rope for vertical work. One section of the NSS Members manual was used without copyright permission. Many of the books were destroyed by the publisher to avoid a lawsuit. Bears mention only as a book to be assiduously avoided as a serious introduction to caving. Replete with dangerous safety information, inaccuracies regarding caves, and egregious cave conservation practices, the NSS was forced to take action to prevent its further dissemination. Only of interest as a collector's piece.

[3226] Tratman, Edgar Kingsley, 1899-1978 *The Caves of North West Clare, Ireland*. University of Bristol Spelaeological Society. New Abbot, GB: David and Charles, **1969**. [xxviii], iv, 256p, 51p pls.

Really a scientific monograph on this part of a county, where most of the significant caves lie. But the last half consists of cave descriptions, with locations, rigging information, and some maps. This book is basic and predates much modern karst hydrologic and geomorphologic theory. The geologic accounts provide an interesting picture of the evolution of these wet, active, largely post-Glacial caves.

[3227] Tratman, Edgar Kingsley, 1899-1978 *Reports on the investigations of Pen Park Hole, Bristol*. Church Stretton, GB: The Cave Research Group of Great Britain, Feb, **1963**. Cave Research Group of Great Britain Publication no 12. illus: b&w: drawings, graphs, maps, photos, repros; 3 maps folded, loose, in rear; [iv], 54p; bibl; 25 x 20 cm.

[3228] Treganza, Adan Eduardo, 1916-1968 *An ethno-*

archaeological examination of Samwel Cave. Castro Valley, CA, US: Cave Research Associates, **1964**. Cave Studies no 12. illus: b&w: drawings, maps, photos; color photo; v, 29p; bibl; 23 cm.

[3229] Trexler, Keith A. *Lehman Caves...its human story: from the beginning through 1965.* rev ed. Baker, NV, US: Lehman Caves National Monument, **1975**. 121p; 29 cm.

Updated through 1975 by NPS staff. 121 loose leaves in notebook.

[3230] Trezise, P. J. *Rock art of South-east Cape York.* Canberra, ACT, AU: Australian Institute of Aboriginal Studies, **1971**. Australian Aboriginal Studies no 24; Prehistory and Material Culture Series no 4. illus: b&w: drawings, maps, photos; col: photos; b&w photos folded in text; col photo front; 132p; bibl.

[3231] Trickett, Oliver *The Abercrombie Caves.* Sydney, NSW, AU: The Government Tourist Bureau, C. D. Paterson, Tourist Agent, **1906**. illus: b&w: ads, maps, photos; advertisement for cave on front lining page; map on rear lining page; 12p; 16.5 x 10 cm.

Limited edition of 10,000, but today is extremely rare. Cover title: The Abercrombie Caves, near Newbridge, New South Wales: from a description prepared by O. Trickett L.S., M.S., for the Dept of Mines & Agriculture, N. S. W. Copy examined was phcpy.

[3232] Trickett, Oliver *Guide to the Jenolan Caves New South Wales.* 1st ed. Sydney, NSW, AU: Geological Survey of New South Wales, **1899**. illus: b&w: maps, photos; partly col: maps; text on front lining page; ad on rear cover; b&w front; 63p, [1]; 18 x 24.5 cm.

Cover title: Album of picturesque Jenolan Caves in the state of New South Wales Australia. This edition also includes a description of the Tuglow Cave which was deleted from later editions.

2nd ed. Jan, **1905**. illus: b&w: ads, maps, photos; partly col: maps; 76p, [6].

Cover title: Jenolan Caves New South Wales Australia. 2nd ed varies in cover color and material; several printings exist with slightly different photos.

3rd ed. np, Jan, **1915**. illus: b&w: ads, maps, photos; partly col: maps; col: photos; 82p, [7].

Covers vary.

4th ed. **1922**. illus: b&w: ads, maps, photos; col: ads; front: b&w ads (1 partly col); 86p, [1].

[3233] Trickett, Oliver *Guide to the limestone caverns of New South Wales: Jenolan, Wombeyan, and Yarrangobilly.* New South Wales. Geological Survey. Sydney, NSW, AU: Australia Immigration and Tourist Bureau, **1906**. illus: b&w: photos; partly col: maps; col: drawings, maps; 76+34+28p.

Originally published as three separate titles by Oliver Trickett. At head of separate title pages: Geological survey of New South Wales. Guide to the Jenolan caves, New South Wales / O. Trickett. 2d ed. (1905).—Guide to the Wombeyan caves, New South Wales / O. Trickett (1906).—Guide to the Yarrangobilly caves, New South Wales / O. Trickett (1906).

[3234] Trickett, Oliver *Guide to Wombeyan Caves NSW.* npop: np, **1906**. illus: b&w: maps, photos; 34p, 2p pls; 18 x 25 cm.

Photocopy examined.

[3235] Trickett, Oliver *Guide to Yarrangobilly Caves New South Wales.* 2nd ed. [Sydney, NSW, AU]: [New South Wales Government Tourist Bureau], **1917**. illus: b&w: maps, photos; partly col: ads, maps; 41p, [3], 2p pls.

Originally published in 1906.

[3236] Trickett, Oliver *Notes on the limestone caves of New South Wales with plans.* Sydney, NSW, US: New South Wales Department of Mines and Agriculture. Geological Survey, **1898**. illus: b&w: maps; [i], 12p; bibl.

Very rare.

[3237] Trickett, Oliver *The Wellington Caves.* Sydney, NSW, AU: The Government Tourist Bureau, C. D. Paterson, Tourist Agent, **1906**. illus: b&w: ads, maps, photos; 12p.

Limited edition of 10,000, but today is extremely rare. Cover title: The Wellington Caves, near Wellington, New South Wales: from a description prepared by O. Trickett L.S., M.S., for the Dept of Mines & Agriculture, N. S. W. Copy examined was phcpy.

[3238] Trimble, Stephen, 1950– *Window into the earth: Timpanogos Cave.* [Globe, AZ, US]: Southwest Parks and Monuments Association, **1983**. [48]; bibl; 21 x 22 cm.

Cover title: Timpanogos Cave. b&w, col illus.

[3239] Trinkaus, Erik *The Shanidar Neandertals.* Weston, CN, US: Pictures of Record, **1985**.

Title from data sheet. The Neandertals are one of the best known of archaic human ancestors, appearing across Europe and the Near East about 100,000 years ago. They were replaced by modern humans between 30,000 and 40,000 years ago. Their skeletal remains provide a knowledge of their anatomy, insights into their behavior, and a background against which to view the emergence of people like ourselves. Nine partial Neanderthal skeletons were excavated by Ralph Solecki in Shanidar Cave in northeastern Iraq. Set includes views of skulls, dentitions, vertebrae, pelvis and hand bones, plus views of the withered arm of Shanidar 1 and the injured rib of Shanidar 3. 20 slides, col, 2 x 2 in + discussion guide (9 p).

[3240] Trinkaus, Erik *The Shanidar Neandertals.* New York, NY, US: Academic Press, **1983**. xxiv, 502p; bibl, index; 24 cm.

[3241] Trommer, J. T. *Potential for pollution of the Upper Floridan aquifer from five sinkholes and an internally drained basin in west-central Florida.* Tallahassee, FL, US: U.S. Geological Survey, **1987**. Water-Resources In-

vestigations Report no 87-4013. illus: b&w: maps; vii, 103p; bibl; 28 cm.
Illus.

[3242] Trower, Harold Edward *Capri: the story of the blue grotto*. Naples, IT: np, **c1928**. 24p; 21.5 x 14 cm.

Printed by R. Stabilimento Tipografico Francesco Gianni & Figii.

[3243] Trudgill, Stephen Thomas, 1947– *A bibliography of British karst, 1960-1977*. Brack, E. V. Norwich, GB: GeoAbstracts, LTD, **1977**. illus: b&w: maps; 33p; index; 21 cm.

Compiled on the occasion of the 7th International Speleological Congress in Sheffield, 1977.

An excellent reference work, in need of updating. It describes material much of which is in relatively obscure sources, not often available except by special order.

[3244] Trudgill, Stephen Thomas, 1947– *Limestone geomorphology*. Clayton, Keith M. (ed). London; New York, NY, GB; US: Longman Group Limited, **1985**. Geomorphology Texts 8. illus: b&w: charts, drawings, graphs, maps, photos, photomicrographs, tables; x, 196p; bibl, index; 25 cm.

This is a well presented, informative text dealing essentially with surface karst. Technical but not overwhelming, this work presents karst geomorphology from both carbonate and British slants. It emphasizes the relation of processes and descriptive forms in terrestrial and marginal marine environments, with a needed discussion of marine biokarst. Non-British examples provide some breadth. Generally well illustrated. Extensive bibliography.

[3245] Truluck, T. F. *SASA (Cape) expedition to Chinhoyi, Zimbabwe August 1992*. Capetown, ZA: South African Spelaeological Association (Cape Section), Oct, **1992**. illus: b&w: maps, tables; 38p; app, bibl; 21 x 29.5 cm.

Lists of longest and deepest caves of Zimbabwe in back.

[3246] Tsen, Darrell Nyuk Choi *The show caves of Mulu Sarawak*. 2nd ed. **1993**. illus: col: maps, photos.

[3247] Tsen, Darrell Nyuk Choi *The show caves of Mulu Sarawak*. Kuala Lumpur, Sarawak, MY: Pro Art & Design, **1991**. illus: col: maps, photos; [80]; 30 x 21 cm.

Very rare; only seen once outside of Malaysia.

[3248] Tucker, J. H. *Some smaller Mendip caves: vol 2*. Bridgwater, Somerset, GB: Bristol Exploration Club, Aug, **1962**. Caving Report no 9. illus: b&w: maps, photos; 24p; 26 x 20 cm.

[3249] Tulis, J. (comp) *Slovakia: the karst and speleology*. Mikuláš, SK: Slovak Speleological Society, **1993**. illus: b&w: maps; col photos on lining pages; 11p, [1]; 22.5 x 16.5 cm.

Contains lists of longest/deepest Slovakian caves.

[3250] Tullis, Edward L. *Black Hills caves*. Gries, John Paul, 1911–. Rapid City, SD, US: South Dakota State School of Mines, **1940**. The Black Hills Engineer v XXIV no 4 pp 233-271. 39p; 24 cm.
Illus.

[3251] Turnbull, William D. *Broom County Cercartetus, with observations on pygmy possum dental morphology, variation, and taxonomy*. Schram, Frederick R. Sydney, New South Wales, AU: Australian Museum, **1973**. Australian Museum Records v 28 no 19 pp 437-464. illus: b&w: drawings, maps, photos, tables; 28p; 24 cm.
Illus.

[3252] Turner, Brian B. *Natural stone bridge and caves; fascinating geology, history & legends*. Beckler, Edward B. (comp). Burlington, VT, US: Printed by G. Little Press, **1973**. illus: b&w: drawings, maps, photos, tables; b&w photos on lining pages; b&w front photos; v, 31p, [1]; app; 23 cm.

Cover title: Natural stone bridge and caves: The only cave attraction in the Adirondacks: fascinating geology, history, & legends

[3253] Turner, Cleon *Discovery of the Onyx Cave and my biography*. Park City, KY, US: Cleon Turner, **1968**. illus: b&w: photos; 123p.

[3254] Turner, James William, 1848– *Wonders of the great Mammoth Cave of Kentucky containing thorough and accurate historical and descriptive sketches of this marvelous underground world with a chapter on the geology of cave formation*. Carrier Mills, IL, US: Turner Publishing Company, **1912**. 116p, 6p pls; 16 cm.

Master microform held by: DLC. Microfilm. Washington, D.C.: Library of Congress Photoduplication Service, 1987. 1 microfilm reel 35 mm.

[3255] Turner, Percy (auth, comp) *History of Manitou's caves, the Grand Caverns and Cave of the Winds: and points of interest in and about Manitou*. Denver, CO, US: Merchants Pub. Co., **1895**. illus: b&w: ads, maps, photos; 108p; 20 cm.

Illus. "Written and compiled from notes in the discoverer's diary, by Percy Turner." Very rare.

[3256] Turpin, Sloveig A. (comp) *Seminole Sink (41VV620): excavation of a vertical shaft tomb, Val Verde County, Texas*. Austin, TX, US: Texas Archeological Survey, **1985**. illus: b&w: charts, graphs, maps, photos, photomicrographs, tables; xii, 216p; app, bibl.

University of Texas at Austin Research Report 93

[3257] [Tyers, John] *A guide to Wind Cave and Wind Cave National Park. 19th annual convention of the National Speleological Society self guiding tour of the commercial areas of Wind Cave*. npop: National Speleological Society, **1962**. illus: b&w: maps; 12p.

[3258] Tyler, Larry H. *History of Rockhouse Cave*. [Cassville, MO, US]: OMi Museum, **1980**. illus: b&w: maps, ports; 168p; 28 cm.
Illus.

[3259] Tyler, Ronald M. *Caving handbook: a guide to the underworld*. Villa Park, IL, US: Willowbrook High

School, Dec, **1971**. illus: b&w: charts, maps, photos, ports, tables; text on lining pages; [v], 104p; bibl; 21 cm.
Illus.

[3260] United States Army Corps of Engineers *Geophysical phase IV monitoring of potential sinkhole collapse at military ocean terminal. Sunny Point Access Railroad, Boiling Springs, N.C..* Wilmington District; Technos (Firm) Consultants in Applied Earth Sciences. Miami, FL, US: np, Dec, **1987**. Various pagination; 28 cm.
Illus.

[3261] U.S. Army, Engineer Intelligence Division *Natural caves of West Germany Baden - Wuerttemberg.* npop: np, Aug, **1962**. Tactical Study of the Weather and Terrain of Central Europe, Part 2 Underground Facilities, Pits, and Quarries. illus: b&w: maps; some maps folded in text; iv, 67p, [1].

[3262] U.S. Bureau of Land Management *Final Dark Canyon environmental impact statement.* Santa Fe, NM, US: U.S. Bureau of Land Management, New Mexico State Office, Dec, **1993**. illus: b&w: charts, drawings, graphs, maps, tables; 1 folded chart in text; 3 folded tables in text; 22 folded maps in text; 2 folded maps in pocket; Various pagination, xv, 500p; app, bibl, gloss, index.
Photo on cover.

[3263] U.S. Bureau of Land Management, Barstow Resource Area *Shoshone Cave (Whip-scorpion Habitat) wildlife habitat management plan.* California Department of Fish and Game. npop: np, **1982**. illus: b&w: maps; 19p; bibl; 28 cm.
Cover title. "A Sikes Act Project." "CA-06-WHMA-15."

[3264] U.S. Bureau of Land Management, Roswell District Office *Bureau of Land Management interim guide for oil & gas drilling operations in cave and karst areas.* [Roswell, NM, US]: Bureau of Land Management, Roswell District, Feb, **1993**. illus: b&w: charts, graphs, maps, tables; 2 folded maps in pocket; [iii], 20p; bibl; 28 cm.
Title from cover. Illus. [Washington, D.C.]: Supt. of Docs. US GPO: [1995] 3 microfiches: negative.

[3265] U.S. Bureau of Land Management, Roswell District Office *Cave inventory and classification systems.* npop: U.S. Bureau of Land Management, Roswell Office, **[1982]**. 24p; app.
This is the BLM's comprehensive data processing system dealing with cave management.

[3266] U.S. Bureau of Land Management, Shoshone District *T-maze cave management plan.* npop: U.S. Bureau of Land Management, Shoshone District, **1987**. iii, 15p.

[3267] U.S. Congress, House Committee on Interior and Insular Affairs *Protecting cave resources on federal lands, and for other purposes: report (to accompany H.R. 1975) (including the cost estimate of the Congressional Budget Office).* Washington, DC, US: US Government Printing Office, Mar, **1988**. Report 100th Congress, 2d session, House of Representatives 100-534. 11p; 24 cm.
Distributed to some depository libraries in microfiche.

[3268] U.S. Congress, House Committee on Natural Resources *Lechuguilla Cave Protection Act of 1993: report (to accompany H.R. 698) (including cost estimate of the Congressional Budget Office).* Washington, DC, US: US Government Printing Office, May, **1993**. Report 103d Congress 1st session House of Representatives 103-86. 7p; 23 cm.
Caption title. Distributed to some depository libraries in microfiche.

[3269] U.S. Congress, House Committee on Public Lands *Shenandoah, Great Smoky Mountains, and Mammoth Cave National Parks: hearings before the Committee on the Public Lands, House of Representatives, Sixty-ninth Congress, first session, on H.R. 11287 [and] H.R. 12020, May 11, 1926.* Washington, DC, US: U.S. Government Printing Office, **1926**. 33p; 23 cm.
Running title: Providing for establishment of national parks.

[3270] U.S. Congress, Senate Committee on Energy and Natural Resources *Federal Cave Resources Protection Act of 1988: report (to accompany H.R. 1975).* Washington, DC, US: U.S. Government Printing Office, Sep, **1988**. Report 100th Congress 2d session Senate 100-559. 12p; 24 cm.
Caption title. Distributed to some depository libraries in microfiche.

[3271] U.S. Congress, Senate Committee on Energy and Natural Resources *Lechuguilla Cave Protection Act of 1993: report (to accompany H.R. 698).* Washington, DC, US: U.S. Government Printing Office, Dec, **1993**. Report 103d Congress 1st session Senate 103-213. 9p; 24 cm.
Caption title.

[3272] U.S. Congress, Senate Committee on Energy and Natural Resources, Subcommittee on Public Lands, National Parks and Forests *Federal Cave Resources Protection Act and restriction of dams in parks and monuments hearing before the Subcommittee on Public Lands, National Parks, and Forests of the Committee on Energy and Natural Resources, United States Senate, One Hundredth Congress, second session, on S. 927/H.R. 1975 ... H.R. 1173 ... June16,1988.* Washington, DC, US: US Government Printing Office, **1988**. United States Congress Senate Hearing 100-863. iii, 126p; 24 cm.
2 microfiches, negative 11 x 15 cm.

[3273] U.S. Environmental Protection Agency, Region 4 *Environmental impact statement, Mammoth Cave area, Kentucky: wastewater facilities.* Atlanta, GA, US: np, Aug, **1981**. U.S. Environmental Protection Agency Region 4 EPA/904/9-81-076. illus: b&w: charts, maps, tables; 2 fold-out maps in body; [ii], xii, 157p; 28 cm.

PB82-143793 NTIS 2 microfiche. Illus. Executive summary. Cover title: Mammoth Cave area wastewater facilities, Kentucky.

[3274] **U.S. Environmental Protection Agency, Region 4** *Environmental impact statement; Mammoth Cave area, Kentucky, wastewater facilities.* Atlanta, GA, US: United States. Environmental Protection Agency, Region 4, Apr, **1981**. illus: b&w: charts, drawings, maps, tables; 1 folded map in pocket; 13 fold-out maps; [i], xvi, 200p.

[3275] **U.S. Environmental Protection Agency, Region 4** *Mammoth Cave area, Kentucky, 201 facilities plan environmental impact statement.* Atlanta, GA, US: U.S. Environmental Protection Agency. Region IV, Apr, **1981**. Various pagination; 28 cm.

Illus. Title from cover. Vol. 1. Environmental inventory, technical reference document. Vol. 3. Alternatives evaluation technical reference document.

[3276] **U.S. Forest Service** *Operation of Blanchard Springs Caverns, Ozark-St. Francis National Forest, Arkansas: final environmental statement.* npop: U.S. Forest Service, **1973**. WO 73-62. illus: b&w: maps; iii, 83p, 2p pls; bibl; 27 cm.

Title from cover.

[3277] **U.S. Forest Service** *The Oregon Caves, Siskiyou National Forest.* 1924 ed. Washington, DC, US: U.S. Forest Service, **1924**. 15p; bibl; 23 cm.

Title from cover. Illus.

1926 ed. US Government Printing Office. illus: b&w: maps; 16p.

At head of title: U.S. Department of Agriculture. Forest Service. Illus.

[3278] **U.S. Forest Service** *The Oregon Caves: Siskiyou National Forest.* 1900 ed. Washington, DC, US: U.S. Government Printing Office, **1900**. illus: b&w: maps; 16p; bibl; 23 cm.

At head of title: U.S. Department of Agriculture. Forest Service. "Bibliography of the Oregon caves": p. 15. Microfilm. Washington, D.C., United States Library of Congress, 1983. 1 reel. 35 mm. Illus.

[3279] **U.S. Forest Service** *A preliminary plan for the development of Blanchard Springs Cavern: Sylamore District, Ozark National Forest.* Washington, DC, US: US Forest Service, **1964**. illus: b&w: maps; partly col: maps; some folded maps; 26p, [19]; 38 cm.

Col, b&w Illus.

[3280] **U.S. Forest Service, Southern Region** *Construction of phases II and III of the Blanchard Springs Caverns Project final environmental statement.* npop: U.S. Department of Agriculture, Forest Service, **1975**. USDA-FS-R8 FES (ADM) 75-6. illus: b&w: charts, maps, photos, tables; 2 fold-out maps; front: b&w photo; Various pagination, vi, 87p, [6]; app; 27 cm.

[3281] **U.S. Forest Service, Southern Region** *Construction of phases II and III of the Blanchard Springs Caverns Project: draft environmental statement.* Atlanta, GA, US: The Region, **1974**. illus: b&w: maps; Various pagination, 138p; bibl; 27 cm.

Includes: Operation of Blanchard Springs Caverns. Illus.

[3282] **U.S. Forest Service, Southern Region** *Welcome to Blanchard Springs Caverns.* [Atlanta, GA, US]: US Forest Service, Southern Region, **1973**. illus: partly col: drawings, maps; col: drawings, photos; text on lining pages; [16]; 20.5 x 22 cm.

Title from cover. Col illus.

[3283] **U.S. Geological Survey** *Bibliography of the Ground Water Resources of New York through 1967, with subject and location cross-reference to the annotated bibliography.* MacNish, Robert D. Albany, NY, US: State of New York; Water Resources Commission, **1967**. illus: b&w: maps; iii, 186p; 28 cm.

Has an annotated bibliography and both subject and location indexes. "Prepared by United States Department of the Interior, Geological Survey in cooperation with New York Water Resources Commission."

[3284] **U.S. Geological Survey** *Geology of caves.* Davies, William Edward, 1917-1990. 1977 ed. [Washington, US]: U.S. Geological Survey, **1977**. 19p; 23 cm.

Prepared from materials provided by W. E. Davies and I. M. Morgan. Illus.

1986 ed. Washington, Alexandria, Denver, DC, VA, CO, US: The Survey; Eastern Distribution Branch, U.S. Geological Survey; Western Distribution Branch, U.S. Geological Survey], **1986**. illus: b&w: maps; bibl; 23 x 11 cm.

[3285] **U.S. Geological Survey** *Indiana cave areas: topographic quadrangle maps.* Chicago, IL, US: Windy City Grotto, National Speleological Society, **1980**. illus: b&w: maps; [67].

Reduced b&w reprints of 67 USGS topo maps for cave areas of Indiana. Limited edition.

[3286] **U.S. National Park Service** *Carlsbad Caverns National Park, New Mexico: wilderness reevaluation study: preliminary - subject to change.* Denver, CO, US: [U.S. National Park Service, Denver Service Center], Sep, **1980**. U.S. Dept of the Interior National Park Service Denver Service Center NPS 1657. illus: b&w: maps; 22p; bibl; 28 cm.

September 1980.

[3287] **U.S. National Park Service** *Carlsbad Caverns National Park: final interpretive prospectus.* npop: The Service, **1976**. United States National Park Service NPS 1084. illus: b&w: maps; 51p; bibl; 27 cm.

Illus.

[3288] **U.S. National Park Service** *Carlsbad Caverns: official map & guide.* [Washington, DC, US]: np, **1986**.

illus: b&w: maps; folded to 10x21 cm; 51 x 42 cm.

Carlsbad Caverns National Park, New Mexico. Col, b&w illus.

[3289] U.S. National Park Service *Caves and their conservation.* Rothrock, H. E. Washington, DC, US: np, Mar, **1937**. United States National Park Service Geology Memorandum no 2. 22p; 27 cm.

No illus. Gives methods for preparing a cave for commercialization. Out of date, but perhaps the first cave conservation book in the U.S.

[3290] U.S. National Park Service *Development concept: Slaughter Canyon, Carlsbad Caverns National Park, New Mexico.* npop: U.S. National Park Service, Aug, **1975**. 1 fold-out map in body; text on front lining page; Various pagination, 25p.

Includes attached environmental assessment.

[3291] U.S. National Park Service *Draft general management plan, environmental impact statement: Carlsbad Caverns National Park, New Mexico.* [Denver, CO, US]: U.S. National Park Service. Denver Service Center, Nov, **1995**. illus: b&w: charts, drawings, graphs, maps, photos, repros, tables; 9 fold-out maps in body; text on rear lining page; ix, 216p; app, bibl, index.

[3292] U.S. National Park Service *Draft, environmental impact statement, general management plan, development concept plan Timpanogos Cave National Monument.* Denver, CO, US: U.S. National Park Service, Denver Service Center, Feb, **1993**. illus: b&w: maps; xiii, 142p; bibl, index; 28 cm.

Title from cover. Distributed to depository libraries in microfiche.Microfiche. [Washington, D.C.?]: Supt. of Docs., U.S. G.P.O., [1993] 7 microfiche: negative.

[3293] U.S. National Park Service *Draft, general management plan, environmental impact statement Jewel Cave National Monument, South Dakota.* Denver, CO, US: U.S. National Park Service, Denver Service Center, Jun, **1993**. illus: b&w: maps; ix, 67p; bibl; 28 cm.

Title from cover. Distributed to depository libraries in microfiche. Illus.

[3294] U.S. National Park Service *Final environmental statement: Mammoth Cave National Park master plan and wilderness suitability study, Kentucky.* Washington, DC, US: U.S. National Park Service Denver Service Center, **1976**. illus: b&w: drawings, graphs, maps, photos, tables; 2 fold-out maps; text on rear lining page; [ii], xvii, 275p; app, bibl; 27 cm.

At head of title: Department of the Interior. Stamped "FES 76-27." Illus.

[3295] U.S. National Park Service *Final environmental statement: proposed wilderness, Carlsbad Caverns National Park, New Mexico.* npop: U.S. National Park Service, **1973**. illus: b&w: maps; some maps folded; 55p; 27 cm.

Title from cover. "FES 73-23."

[3296] U.S. National Park Service *Final master plan: Mammoth Cave National Park.* [Denver, CO, US]: U.S. National Park Service [Denver Service Center], Apr, **1976**. illus: b&w: drawings, graphs, maps, photos, repros, tables; 1 fold-out map in body; text on rear lining page; [ii], 80p; bibl; 27 cm.

Illus.

[3297] U.S. National Park Service *Final, general management plan, environmental impact statement Jewel Cave National Monument, South Dakota.* [Denver, CO, US]: U.S. National Park Service, Denver Service Center, Jun, **1994**. illus: b&w: maps; ix, 111p; bibl; 28 cm.

Title from cover. U.S. G.P.O., [1996] 2 microfiches. Illus.

[3298] U.S. National Park Service *General management plan, wilderness suitability study: El Malpais National Monument, New Mexico.* Eury, D. [Denver, CO, US]: U.S. National Park Service, **1990**. illus: b&w: maps; vii, 177p, [1]; bibl; 28 cm.

Illus (some col).

[3299] U.S. National Park Service *The Great Onyx Job Corps Civilian Conservation Center alternative relocation sites, Mammoth Cave National Park, Kentucky.* Mammoth Cave, KY, US: U.S. National Park Service, Mammoth Cave National Park, Apr, **1979**. illus: b&w: maps, tables; 4 fold-out maps; text on rear lining page; [vi], 44p, [6]; app.

Cover title: Environmental assessment alternative relocation sites, Mammoth Cave, The Great Onyx.

[3300] U.S. National Park Service *Lehman Caves National Monument.* 1971 ver. Washington, DC, US: U.S. National Park Service, **1971**. illus: partly col: maps, photos.

Colorful underground photography focuses on the formation of the Lehman limestone caves to help viewers understand geological formation of these and other caves. Distributed by National Audiovisual Center. 1 reel, 12 min, sd, col, 16 mm.

1979 ver. Washington, WA, US: Division of Audiovisual Arts; distributed by National Audiovisual Center, **1979**. illus: partly col: photos.

Issued in 1971 as 16 mm motion picture. Focuses on the formation of limestone caves by highlighting the Lehman Caves in Lehman Caves National Monument in Nevada. 1 cassette, 12 min, sd, col, 3/4 in.

[3301] U.S. National Park Service *Mammoth Cave National Park.* Finley-Holiday Film Corporation. Whittier, CA, US: Finley-Holiday Film Corporation, **1987**. National Park and Monument series. illus: partly col: photos.

VHS format. Title on cassette: Mammoth Cave National Park, Kentucky. C. Lindsay Workman. Tour of Mammoth Cave National Park in Kentucky, which is called the most beautiful of underground caves. 1 videocassette (30 min), sd, col,1/2 in.

[3302] **U.S. National Park Service** *Mammoth Cave National Park, Kentucky.* Denver, CO, US: U.S. National Park Service, Denver Service Center, **1974**. NPS 805. three folded maps; 13+7p, [28]; 27 cm.

At head of title: Wilderness recommendation.

[3303] **U.S. National Park Service** *Mammoth Cave National Park: collection management plan.* Harpers Ferry, WV, US: US National Park Service, Division of Museum Services, **1978**. 17p, [106]; 27 cm.

Illus.

A summary and management plan of the cultural resources (prehistoric and historic curated at Mammoth Cave National Park, Kentucky).

[3304] **U.S. National Park Service** *Master plan: Carlsbad Caverns National Park, New Mexico.* npop: [U.S. National Park Service], **[1973]**. illus: partly col: maps; 2 folded partly col maps in body; [ii], 49p; app, bibl.

[3305] **U.S. National Park Service** *National Cave and Karst Research Institute Study: a draft report to Congress as required by Public Law 101-578 of November 15, 1990.* [Washington, DC, US]: U.S. National Park Service, Southwest Region, Office of the Associate Regional Director, Planning and Professional Services, Dec, **1994**. illus: b&w: maps; 37p, [32]; 28 cm.

"Consideration of the feasibility of establishing a National Cave and Karst Institute in association with the National Park Service's Cave Research Program."

[3306] **U.S. National Park Service** *Natural resources management program: an addendum to the natural resources management plan for Lehman Caves National Monument, Nevada.* rev ed. Baker, NV, US: Lehman Caves National Monument, Feb, **1980**. Archives QH 75 USn L52 no. 4 NPS. i, 23p; 28 cm.

No illus.

[3307] **U.S. National Park Service** *Oregon Caves.* npop: np, **1976**. 14p; 9 cm.

Illus.

[3308] **U.S. National Park Service** *Oregon Caves National Monument, master plan.* Denver, CO, US: U.S. National Park Service, Denver Service Center, **1975**. NPS 506. illus: b&w: maps; col: maps; col, b&w maps folded; 31p; 27 cm.

[3309] **U.S. National Park Service** *Oregon Caves National Monument, Oregon: final interpretive prospectus.* Denver, CO, US: The Service, **1976**. United States National Park Service NPS 1095 cn. illus: b&w: maps; one map folded; 40p; 27 cm.

Illus.

[3310] **U.S. National Park Service** *Oregon Caves National Monument, Oregon: official map and guide.* Washington, DC, US: US National Park Service, **1992**. illus: b&w: maps; folded to 21x10 cm; [12]; 21 x 59 cm.

Illus.

[3311] **U.S. National Park Service** *Paleontology.* Washington, DC, US: U.S. National Park Service, **1992**. Various pagination, 87p; 28 cm.

No collective t.p. Forest ecology—Geology of Wind Cave National Park—Cave ecology—History of Wind Cave National Park—History of Jewel Cave National Monument—Cave management at Wind Cave—Prairie ecology—Wildlife management. Illus.

[3312] **U.S. National Park Service** *Pollution abatement project, Carlsbad Caverns National Park, New Mexico.* npop: np, **1973**. illus: b&w: maps; 33p.

FES 73-27. Title from cover. At head of title: Department of the Interior.

[3313] **U.S. National Park Service** *Proposed Mammoth Cave National Park master plan and wilderness study, Kentucky: draft environmental impact statement.* Denver, CO, US: U.S. National Park Service Denver Service Center, **nd**. illus: b&w: maps; col: maps; one folded leaf of plates; xx, 194p; bibl; 27 cm.

DES 74-43.

[3314] **U.S. National Park Service** *Proposed wilderness, Carlsbad Caverns National Park, New Mexico.* npop: [U.S. National Park Service], **1973**. illus: b&w: photos; maps partly folded; 55p.

Title from cover. "Final environmental statement." "FES 72-73. At head of title: Department of the Interior.

[3315] **U.S. National Park Service** *Statement for management: Jewel Cave National Monument.* Jewel Cave National Monument. Denver, CO, US: [U. S. National Park Service], **1987**. illus: b&w: maps; 41p; 28 cm.

Cover title. Cover title. Shipping list no.: 86-495-P. "March 1987." "NPS D-6C"—P. [3] of cover. Illus.

[3316] **U.S. National Park Service** *Statement for management: Wind Cave National Park.* Washington, DC, US: U.S. National Park Service, Jul, **1986**. NPS D-8c. illus: b&w: maps; 48p; 28 cm.

Title from cover. Illus.

[3317] **U.S. National Park Service** *Timpanogos Cave National Monument, Utah.* [Denver, CO, US]: The Service, **1976**. United States National Park Service NPS 1031. illus: b&w: maps; [96], 11p pls; 27 cm.

Cover title. At head of title: Environmental assessment, rockfall protection. Illus.

[3318] **U.S. National Park Service** *Timpanogos Cave National Monument: statement for management.* [Denver, CO, US]: The Service, Aug, **1986**. illus: b&w: maps; 19p; 28 cm.

Title from cover. "NPS D-1d"—P. [3] of cover. Illus.

[3319] **U.S. National Park Service** *Voices of the cave.* Warties, Burleigh (photo); Eger, Mixican (ed); McBridge, Robert (narr); Bell, Jonathan (music). Washington, DC, US: Division of Audiovisual Arts, Earthrise Entertainment (Firm) National Audiovisual Center, **1984**.

Title from data sheet. Intended audience: Visitors to

Mammoth Cave National Park. Issued also as videorecording. Tells the story of man's continuous involvement with Mammoth Cave in Kentucky from prehistory to the present. Viewed from within the cave, offers glimpses of the cave's many uses and long history which range from Indian burials to mining, medical, religious, and tourist activities. Distributed by National Audiovisual Center. 1 film reel (12 min), sd, col, 16 mm.

[3320] **U.S. National Park Service** *Wilderness recommendation, Carlsbad Caverns National Park, New Mexico.* npop: [U.S. National Park Service], Aug, **1972**. illus: b&w: maps, tables; partly col: maps; 4 partly col fold-out maps; [i], 21p, [15], 4p pls; 27 cm.

Cover title: Carlsbad Caverns National Park, New Mexico: wilderness recommendation.

Recommendation that 30, 210 acres with the park be designated a wilderness area.

[3321] **U.S. National Park Service** *Wind Cave National Park, South Dakota.* Washington, DC, US: U.S. National Park Service, **1937**. Handbook United States National Park Service Division of Publications 104. illus: b&w: charts, maps, photos, tables; [iv], 16p; gloss; 23 cm.

Title from cover.

[3322] **U.S. National Park Service** *Wind Cave: National Park Service, South Dakota.* Washington, DC, US: The Division, **1979**. illus: b&w: drawings, ports, repros; partly col: maps; col: photos; text on lining pages; [iii], ii, 144p; bibl, index; 22 cm.

[3323] **U.S. National Park Service** *Development concept plan, Slaughter Canyon, Carlsbad Caverns National Park, New Mexico: environmental assessment.* Denver, CO, US: U.S. National Park Service, Denver Service Center, Sep, **1974**. illus: b&w: maps, photos, tables; some maps folded; iii, 127p, [16]; bibl; 27 cm.

Cover title: Carlsbad Caverns National Park, New Mexico, Slaughter Canyon: development concept, environmental assessment.

[3324] **U.S. Office of Strategic Services, Research and Analysis Branch** *The caves of greater Germany.* npop: U.S. Office of Strategic Services, Research and Analysis Branch, Apr, **1945**. R & A no 2936. illus: b&w: maps; partly col: maps; 32p, 3p pls; 27 cm.

Title from cover. Illus.

[3325] **U.S. Office of Water Resources Research** *The detection and mapping of subterranean water bearing channels: phase 2: final report.* Rolla, MI, US: Missouri Water Resources Research Center, University of Missouri, **1975**. 20p; 29 cm.

Project no. B-087, agreement no. 14-31-0001, funded partly by U.S. Office of Water Resources Research. Illus. Pagination listed as: 22, 249-252, 20 lvs.

[3326] **Ucko, Peter J.** *Palaeolithic cave art.* Rosenfeld, Andrée. London, GB: Weidenfeld & Nicolson, **1967**. World University Library. illus: b&w: drawings, maps, photos; partly col: charts, drawings, maps, photos; front: b&w photo; 256p; bibl, index, note; 19 cm. 106 figs.

[3327] **Umbreit, Tom** *Wisconsin Speleological Society present: Hodag Hunt 1976, Flying J Campgrounds, Gotham, Wisconsin.* npop: The Wisconsin Speleological Society, **1976**. illus: b&w: maps; 20p; 28 cm.

[3328] **United Air Lines, Inc.** *Luray Caverns in Virginia's Shenandoah Valley.* Maupintour (Firm) Modern Video Programs; Schecter, Irving (prod). St Petersburg, FL, US: Modern Video Programs, **1985**.

VHS format. "T002" Videocassette release of the films by Luray Caverns (1975); Maupintour, Inc. (1984) and United Airlines (1980?). Luray Caverns in Virginia's Shenandoah Valley. Unforgettable Maui: executive producer, John Grember ; producer-director, Robert Bagley ; writer, Peter Altschiller. Visit scenes from Luray Caverns in Virginia, tour New England during the autumn season and see the island of Maui, part of the Hawaiian Islands. 1 videocassette (60 min), sd, col, 1/2 in.

[3329] **Universitah ha-petuhah** *Israeli scenery Soreq Cavern.* Tel Aviv; New York, NY, IL; US: Everyman's University, **1988**.

VHS format. Title from cassette label. Distributed by Sisu Home Entertainment. 1 videocassette (25 min), sd, col, 1/2 in.

[3330] **University of Bristol** *Karst hydrology expedition to Jamaica: preliminary report.* GB: The University Bristol Department of Geography, Dec, **1967**. illus: b&w: maps, photos; [ii], 17p, 3p pls; 19 x 25 cm.

Copy examined was a photocopy.

[3331] **University of California Archaeological Research Facility, Department of Anthropology** *Papers on Great Basin prehistory.* Heizer, Robert F.; Baumhoff, M.A.; Clewlow, C.W.; Roust, N.L.; Hoover, R.L. Berkeley, CA, US: University of California Archaeological Research Facility, Department of Anthropology, **1968**. University of California Archaeological Survey Reports no 71. 119p; bibl.

Illus. I. Heizer, R.F. Baumhoff, M.A. and Clewlow, C.W. Archaeology of South Fork shelter (NV-EL-11), Elko County, Nevada.—II. Heizer, R.F. and Clewlow, C.W. Projectile points from site NV-CH-15, Churchill County, Nevada.—III. Clewlow, C.W. Projectile points from Lovelock Cave, Nevada.—IV. Roust, N.L. Projectile points from Hidden Cave (NV-CH-16), Churchill County, Nevada.—V. Hoover, R.L. An unusual Chumahs pictograph. Photocopy. Ann Arbor, Mich.: Xerox University Microfilms, 1975. 24 cm.

[3332] **University of California Archaeological Survey** *Papers in California archaeology: 17-18.* Heizer, Robert F. Berkeley, CA, US: University of California, **1952**. University of California Berkeley California Archaeological Survey. Reports, no 15. 39p; 28 cm.

17. A survey of cave archaeology in California / Robert F. Heizer.—18. Excavation of Sis-13, a rock-shelter in Siskiyou County, California. Illus.

[3333] University of California Archaeological Survey *Papers on California archaeology: 30-31*. Meighan, Clement W.; Gonsalves, William C. Berkeley, CA, US: University of California, Department of Anthropology, **1955**. Reports of the University of California Archaeological Survey no 29. 45p, [3]; bibl; 28 cm.

30. Excavation of Isabella Meadows Cave, Monterey County, California / Clement W. Meighan—31. Winslow Cave, a mortuary site in Calaveras County, California / William C. Gonsalves. Photocopy. Salinas, Calif.: Coyote Press, 1985? Illus.

[3334] University of California Archaeological Survey *Papers on California archaeology: 32-33*. Meighan, Clement W.; Baumhoff, M.A. Berkeley, CA, US: University of California, Department of Anthropology, **1955**. Reports of the University of California Archaeological Survey no 30. 73p, 11p pls; bibl; 28 cm.

Archaeology of the North Coast Ranges, California / Clement W. Meighan—33. Excavation of site Teh-1 (Kingsley Cave) / M.A. Baumhoff. Photocopy. Salinas, Calif.: Coyote Press, [1985?]. Illus.

[3335] University of California Santa Barbara, Art Galleries *Prehistoric rock art of the Santa Barbara region: [exhibition] Art Gallery, University of California, Santa Barbara, October 12-Nov. 7, 1965*. Grant, Campbell, 1909–. Santa Barbara, CA, US: The Gallery, **1965**. illus: b&w: maps; [21]; bibl; 25 cm.

Title from cover. "Planned under the general direction of Campbell Grant and the staff of the Art Gallery, University of California, Santa Barbara." Illus.

[3336] University of Nevada System, Laboratory of Desert Biology *Final reports on the Lehman Caves studies*. [Reno, NV, US]: The Laboratory, Dec, **1968**. illus: b&w: maps; Various pagination; bibl; 28 cm.

Cover title: Final reports: Lehman Caves studies. Typescript. Illus.

[3337] [University of Nottingham] *University of Nottingham Spelaeological Expedition Exploration '64*. Nottingham, GB: np, **[1965]**. illus: b&w: drawings, maps, photos; 39p; 21.5 x 14 cm.

Title from cover. Exploration of Picos de Europa NW Spain. Summary of British exploration on p 29. Printed by Wm J. Butler & Co

[3338] [University of Nottingham Union] *University of Nottingham Union report of the expedition's co-ordinating committee for 1968*. Derby, GB: np, **1969**. illus: b&w: ads, maps, photos; 5 folded, 21 x 29 map sheets; 52p; 22 x 14 cm.

Cover title: Exploration 1968.

[3339] University of Wisconsin, Recreation Resources Center *Seen a good cave lately?: Cave of the Mounds, Blue Mounds, Wis.: "a visitor study", 1977*. Gray, Jack; Bierly, Pierce. [Madison, WI, US]: Recreation Resources Center, University of Wisconsin-Extension, **1977**. illus: b&w: maps; [ii], 25p; 28 cm.

Title from cover. Demographic profile of visitors to the cave.

[3340] Upper Valley Regional Park Authority (VA) *The story of Cave Hill*. Grottoes, VA, US: Upper Valley Regional Park Authority, **1975**. [16]; 22 cm.

Cover title. "This booklet was prepared by the Upper Valley Regional Park Authority." Describes the caverns located in the Cave Hill area in Augusta County, Virginia. Grand Caverns, Fountain Cave and Madison's Cave are described in detail. Illus.

[3341] Vachell, Eustace Tanfield *Kents Cavern: its origin and history*. npop: Torquay Natural History Society, **1959**. 23p; 22 cm.

No illus.

[3342] Vale, W. P. *Aspex '90: Anglo -Soviet Pamirs Expedition 1990: Bajsuntai Khrebetuzbekistan SR, USSR*. Cheshire, GB: Aspex '90, **[1990]**. illus: b&w: maps; [12]; 30 x 21 cm.

Printed one side only.

[3343] Valentine, Joseph Manson, 1902– *New genera of Anophthalmid beetles from Cumberland Caves*. University, AL, US: University of Alabama, Nov, **1952**. Alabama Geological Survey Museum Paper 34. illus: b&w: drawings; 41p; bibl.

[3344] Valli, Eric *The nest gatherers of Tiger Cave*. Summers, Diane. London, GB: Thames and Hudson, **1990**. illus: col: photos; double fold-out title page 108 x 38 cm; [65]; 39 cm.

Oversized coffee table book. One OCLC record listed pages as [108].

[3345] Valsero, Juan José Durán *Cueva de Nerja: English edition*. Cantos, Francisco Carrasco; Turton, Mary (trans); Turton, Basil (trans). Málaga, ES: Patronato de la Cueva de Nerja, **nd**.

Illus with photos.

[3346] Van Couvering, John A. *Characteristics of large springs in Kentucky*. [Lexington, KY, US]: [University of Kentucky], **1962**. Kentucky Geological Survey Series X; College of Arts and Sciences, University of Kentucky Information Circular 8. illus: b&w: drawings, graphs, maps; 37p; bibl; 22 x 28 cm.

Prepared by the U. S. Geological Survey and the Kentucky Geological Survey with the cooperation of the Dept. of Economic Development. Illus.

[3347] Van Everdingen, R. O. *Morphology, hydrology and hydrochemistry of Karst in permafrost terrain near Great Bear Lake, Northwest Territories*. National Hydrology Research Institute (Canada). Calgary, CA: Environment Canada, Inland Waters Directorate, **1981**. NHRI paper no 11 0713-2816; IWD Scientific Series (Canada. Inland Waters Directorate) no 114. illus: b&w: maps; partly col: maps; two maps on 2 folded sheets in pocket; ix, 53p; bibl; 28 cm.

Includes abstract in French. Issued by National Hydrology Research Institute. Illus.

[3348] Van Everdingen, R. O. *Thermal and mineral springs in the Southern Rocky Mountains of Canada.* Ottawa, CA: Department of the Environment, Water Management Service, **1972**. illus: b&w: drawings, maps, photos; viii, 151p; bibl, index; 25 cm.

Interesting occurrences are described, but interpretation is limited and dated. This data is not easily available elsewhere.

[3349] Van Voris, Arthur H. *The lesser caverns of Schoharie County.* Troy, NY, US: Mohawk-Hudson Grotto, **1970**. 32p; 27 cm.

[3350] Vandel, Albert, 1894-1980 *Biospeleology: the biology of cavernicolous animals.* Freeman, B. E. (trans). Oxford, GB: Pergamon Press, **1965**. International Series of Monographs in Pure and Applied Biology Division Zoology v 22. illus: b&w: drawings, maps, photos, tables; xxiv, 524p; bibl, index; 24 cm.

Translation of: Biospeologie: la biologie des animaux cavernicoles. 1964. 80 figs, 11 pls.

A landmark study of biospeleology, from a European perspective. Despite outdated and wrong ideas about evolution, it contains a huge amount of useful information of about the biogeography, systematics, ecology, and physiology of cave organisms.

[3351] Vandike, James E. *Using digital watershed modeling to estimate sinkhole-flooding potential at Perryville, Missouri.* Rolla, MO, US: Missouri Dept. of Natural Resources, Division of Geology and Land Survey, Oct, **1988**. Open File Report Series Missouri Department of Natural Resources, Geology and Land Survey Division OFR-88-73-WR. illus: b&w: maps; 117p; bibl; 28 cm.

[3352] Varnedoe, William Whitfield, 1923– *Alabama caves and caverns.* Huntsville, AL, US: [Alabama Cave Survey], **1973**. illus: b&w: graphs, maps; bibl; 29 cm.

"This book lists all caves numbers 1 thru [sic] 1,421, as logged by the Alabama Cave Survey." cf. Introd. Sequel to Alabama Caves. 1,000+ unnumbered pages (and we were not going to count them). Loose-leaf.

See Tarkington, *et al.* 1965.

[3353] Varnedoe, William Whitfield, 1923– *Alabama caves, 1980.* Birmingham, AL, US: Alabama Cave Survey, Mar, **1981**. illus: b&w: maps, tables.

Lots (1,500?) of unnumbered pages. Lists 2,000+ caves. Highest cave number is 2016.

See Tarkington, *et al.* 1965.

[3354] Varnedoe, William Whitfield, 1923– *Interim Alabama cave survey: report number 1.* Huntsville, AL, US: Huntsville Grotto of the National Speleological Society, Jan, **1972**. illus: b&w: tables; [60]; 28 cm.

Unnumbered main body pages are approximate.

[3355] Varnedoe, William Whitfield, 1923– *Interim report number 2: Alabama cave survey.* Huntsville, AL, US: Huntsville Grotto, National Speleological Society, **1975**. illus: b&w: graphs, maps, tables; 82p, [205]; 28 cm.

Unnumbered main body pages are approximate.

[3356] Vaughan, Jennifer *Caves.* Appleton, Marion (illus). London, GB: Macdonald Educational, **1973**. Macdonald Starters 51. illus: partly col: drawings; col: drawings; [ii], 25p; gloss, index; 21 cm.

Chiefly col illus. 2nd ed: 1974. Juvenile.

[3357] Veni, George, 1957– *The caves of Bexar County.* Texas Memorial Museum. 2nd ed. Austin, TX: Texas Memorial Museum, University of Texas at Austin, **1988**. Texas Memorial Museum Speleological Monographs 2. illus: b&w: drawings, maps, photos, tables; 10 pp folded maps; xx, 300p, 10p pls; bibl, gloss; 29 cm.

184 maps, 42 b&w photos, 6 figs.

Geological and biological introductions. Descriptions of 208 small caves in that Texas county, with maps to almost all of them. It is not clear what publication was the 'first edition.'

[3358] Verneau, Rene, 1852-1938 *The men of the Barma-grande (Baousse-Rousse) An account of the objects collected in the Museum praehistoricum, founded by Commendatore Th. Hanbury near Mentone.* 2nd ed. Baousse-Rousse, IT: F. Abbo, **1908**. 192p; 19 cm.

On t.-p. of 2d edition, 1908: Translated ... by O. C. & B. C. Illus. Grimaldi Caves.

[3359] Vernon, Robert Orion, 1912– *Florida Caverns: a nature-made wonderland.* Tallahassee, FL, US: Florida Park Service, **1950**. illus: b&w: maps; [28]; 20 cm.

On cover: Florida Caverns State Park, Marianna, Florida, a nature-made underground wonderland. Illus.

[3360] Vessely, Carol *Proyecto Papalo expedition report, 1986-1989: the exploration of Sistema Cuicateca - second deepest cave in the western hemisphere.* npop: np, **[1989]**. illus: b&w: drawings, maps, photos; text on rear lining page; 14p, 3p pls.

[3361] Vickery, Margaret Ray *Ozark stories of the Upper Current River.* Salem, MO, US: The Salem News, **1960**. illus: b&w: photos; partly col: maps; folded map partly col; [iv], xii, 95p, 1p pls; bibl; 26 cm.

Describes many caves along the river. OCLC also lists 1969 as date of pub.

[3362] Vietzen, Raymond Charles, 1907– *The saga of Glover's Cave.* Vietzen, Ruth G.; Me too. [Wahoo, NE, US]: np. illus: b&w: maps, ports; 306p; 31 cm.

Illus.

[3363] Vineyard, Jerry Daniel, 1935– *Catalogue of the caves of Missouri.* Stevens, D'Jeanne (comp); Wills, Ronnie L. (comp); Koenig, John W. (comp). 1957 ed. Rolla, MO, US: Division of Geological Survey and Water Resources, State of Missouri. illus: b&w: tables; [ii], 50p.

1964 ed. [Jefferson City, MO, US]: Missouri Speleological Survey. illus: b&w: maps; 99p; bibl; 28 cm.

Cover title.

1968 ed. Brod, Langford G. (comp). MO, US: Missouri Speleological Survey, Dec. illus: b&w: maps, tables; 139p; bibl.

[3364] **Vineyard, Jerry Daniel**, 1935– *Guidebook to the geology between Springfield and Branson, Missouri, emphasizing stratigraphy and cavern development*. Fellows, Larry Dean, 1934–. Rolla, MO, US: Dept. of Business and Administration, Division of Geological Survey and Water Resources, **1967**. Missouri Division of Geological Survey and Water Resources. Report of Investigations no 37. illus: b&w: maps; 39p; bibl; 28 cm.

Illus.

[3365] **Vineyard, Jerry Daniel**, 1935– *Springs of Missouri*. Feder, Gerald L.; Pflieger, William L.; Lipscomb, Robert G. 1974 ed. Rolla, MO, US: Missouri Geologic Survey and Water Resources, **1974**. Water Resources Report (Missouri. Division of Geological Survey and Water Resources) no 29. illus: b&w: drawings, photos, tables; partly col: charts, drawings, graphs, maps; col: photos; text on front lining page; text (last p of index) on rear lining page; vi, 266p; bibl, index; 28 cm.

Illus. Reference book on karst hydrology. "A cooperative effort of Missouri Geological Survey & Water Resources, U.S. Geological Survey, Missouri Department of Conservation." With sections on fauna and flora by William L. Pflieger and Robert G. Lipscomb.

1982 ed. Missouri Department of Natural Resources, **1982**. 212p; 2 cm.

[3366] **Vineyard, Jerry Daniel**, 1935– *Supplement to the catalogue of the caves of Missouri*. [Rolla, MO, US]: Missouri Speleological Survey, **1960**. illus: b&w: tables; 31p; bibl; 28 cm.

Title from cover. No illus.

See Vineyard, *et al.* 1957.

[3367] **Virginia Commission on the Conservation of Caves** *Report of the Virginia Commission on the Conservation of Caves to the Governor and the General Assembly of Virginia*. Richmond, VA, US: Commonwealth of Virginia, Division of Purchases and Supply, **1979**. House document Commonwealth of Virginia General Assembly House of Delegates no 5. illus: b&w: maps, photos, tables; 44p; app, bibl, errat; 28 cm.

Illus.

[3368] **Virginia Region. National Speleological Society** *A history of the Virginia Region*. Johnson City, TN, US: Virginia Region of the National Speleological Society. The Region Record Summer 1979 v 3 no 4 issue 12 pp 131-295. 165p; index; 28 cm.

Special issue. Illus.

[3369] **Vishoek (South Africa)** *The Peers' cave, tunnel cave and rock shelters at Skildergat, Fish Hoek, the home of pre-historic man*. Jager, H. S. (ed). 4th ed. Fish Hoek, ZA: np, **1949**. illus: b&w: maps; 20p.

Illus.

[3370] **Wagenaar Hummelinck, P.** *De Grotten Van de Nederlande Antillen - Caves of the Netherland Antilles*. Zaltbommel, NL: np Natuurwetenschappelijke Studiekrung voor Suraname en de Nederlandse Antillen (NSSNA), **1979**. Publications of the Foundation for Scientific Research in Surinam and the Netherlands Antilles no 97; Natural history series no 1. illus: b&w: maps, photos, tables; front: photo of Cave of Hato; 176p; bibl, index; 24 x 16 cm.

In Dutch & English; most text translated.d

English and Dutch. Descriptive, with little geology. Never-the-less it is one of the few for this area.

[3371] **Wagner, Georg**, 1883-1912 *The bears cave of Erpfingen*. Erpfingen: np, **1966**. illus: b&w: drawings, maps, photos; 36p, 16p pls; bibl; 24 x 16 cm.

[3372] **Wagoner, John J.** *Mammoth Cave*. Cutliff, Lewis D.; Clark, Chip (photo). Flagstaff, AZ: Interpretive Publications, **1985**. illus: b&w: drawings, repros; partly col: drawings, maps; col: maps, photos; col photos on lining pages; 48p; 30 cm.

Caption title. "Produced for Eastern National Park & Monument Association."

[3373] **Wainwright, Alfred** *Wainwright in the Limestone Dales*. npop: Michael Joseph, **1991**. 192p.

No illus.

[3374] **Waitomo District Council** *District plan discussion paper: tourism issues*. NZ: np, March, **1994**. illus: b&w: tables; [i], 75p; app, bibl; 30 x 21 cm.

[3375] **Walker, Kevin** *Caving in the Crickhowell area*. Buttling, Helene (illus). Powys, GB: Kevin Walker, **1983**. Heritage Guides. illus: b&w: drawings; [ii], 23p, [2]; 21 x 15 cm.

Cover title: Local Caving. Revised edition: 1985; second revised edition: 1985.

[3376] **Walker, Kevin** *The Llangattock escarpment*. 1st ed. Powys, GB: Kevin Walker, **1983**. Heritage Guides. illus: b&w: drawings, maps; map loose in back; [ii], 16p, [1]; 21 x 15 cm.

Revised ed: 1985.

[3377] **Wallace, Malcolm (ed)** *Back in time for tea and medals*. GB: np, [**1995**]. illus: b&w: ports; col: ports; [108]; 21 x 15 cm.

Stories, tributes and expedition tales about Sgt Ian Michael Rolland BEM FRGS RAF: one of the finest cave explorers of modern times.

[3378] **Wallis, G. R.** *Glass sand occurrences, Kurnell Peninsula: preliminary investigations ; Guano deposits in Willi Willi Caves, Kempsey ; Geological report on Wallent's Somersby Clay Pit*. npop: Department of Mines, New South Wales: V.C.N. Blight, Government Printer,

1965. Geological Survey Report (Geological Survey of New South Wales: 1965) no 27. illus: b&w: maps; 19p; bibl; 25 cm.

Issued under the authority of the Hon. T.L. Lewis, M.L.A., Minister for Mines.One diagram and one plan in pocket. Illus.

[3379] **Walsh, Frank K.** *Discovery and exploration of the Oregon caves: Oregon Caves National Monument.* Halliday, William Ross, 1926–. [Grants Pass, OR, US]: Te-Cum-Tom Enterprises, **1971**. illus: b&w: maps, ports; iv, 27p; 22 cm.

Illus.

[3380] **Walsh, Frank K.** *Oregon Caves discovery & exploration: Oregon Caves National Monument.* 3rd ed. Coos Bay, OR, US: Te-Cum-Tom Publications, **1982**. illus: b&w: maps, ports; 27p; 22 cm.

With today's tour map. Illus.

[3381] **Walsh, Frank K.** *Oregon Caves: discovery and exploration.* Halliday, William Ross, 1926–. bicentennial ed. Grants Pass, OR, US: Te-Cum-Tom Publications, **1976**. illus: b&w: maps, photos, ports, tables; iv, 27p; 22 cm.

A complete account.

[3382] **Walsh, John Michael**, 1947– *Mexican caving, 1966-1971.* San Marcos, TX, US: Southwest Texas Grotto, National Speleological Society, **1972**. four folded maps; 146p; 28 cm.

Cover title: Mexican caving of the Southwest Texas Grotto, 1966—1971. Illus.

Similar in style to Russell and Raines 1967, which it nicely supplements. About ninety caves in northeastern Mexico, with eighteen cave maps.

[3383] **Walsh, John Michael**, 1947– *A proposed cave access and cave management plan for the Devil's Sinkhole State Natural Area & Kickapoo Caverns State Park.* New Braunfels, TX, US: Texas Cave Management Association, **1991**. illus: b&w: drawings, maps; [29].

Title from cover.

[3384] **Walsh, John Michael**, 1947– *Texas cave humor, the first twenty-five years.* Spence, J. (ed); Tolar, D. J. (ed). Austin, TX, US: Texas Speleological Association, **1982**. 80p.

No illus.

[3385] **Walter, Erin** *A guide to Austin's most visited caves, with maps and cross-sections of area caves.* Austin, Texas, US: Parks and Recreation Department, Natural Resources Division, City of Austin, Jul, **1996**. illus: b&w: maps, photos, tables; text on front lining page, map and text on rear lining page; 16p; bibl.

Covers are really part of body.

[3386] **Walters, R. (comp)** *The Crocodile Caves of Ankarana: 1986: an expedition to study and explore the limestone massif of Ankarana in northern Madagascar.* [GB]: np, **1986**. illus: b&w: drawings, graphs, maps, photos; b&w photos on lining pages; 90p, 4p pls; bibl; 30 cm.

[3387] **Waltham, Anthony Clive**, 1942– *British karst research expedition to the Himalaya 1970: full report.* Nottingham, GB: A. C. Waltham, Jun, **1971**. illus: b&w: ads, charts, drawings, graphs, photos, tables; 99p; bibl; 24 cm.

Cover title: British karst research expedition to the Himalaya 1970 report.

[3388] **Waltham, Anthony Clive**, 1942– *Caves and karst of the Yorkshire Dales: an excursion guidebook to the karst landforms and some accessible caves within Yorkshire Dales.* Davies, Martin. Bridgwater, Yorkshire, GB: British Cave Research Association, **1987**. illus: b&w: maps, photos; 32p; 20 x 15 cm.

Note: This is a major revision of the1977 guide by Glover, Pitty and Waltham used for the International Congress.

A fine guide book to one of the classic karst areas of Europe; it demonstrates that even classic areas benefit from reexamination and representation.

[3389] **Waltham, Anthony Clive**, 1942– *Caves, crags, and gorges: a guide to the limestone country of England and Wales.* London, GB: Constable and Company Ltd, **1984**. illus: b&w: maps, photos; 335p; 18 cm.

Minimal geology for non-geologists, but contains some interesting geologic ruminations, maps of locations, black-lines showing setting, topography, drainage, outcrops, and basic cave maps.

[3390] **Waltham, Anthony Clive**, 1942– *China caves '85: the first Anglo-Chinese project in the caves of south China.* London, GB: Royal Geographical Society, **1986**. illus: b&w: maps, photos; col: photos; 3 folded maps incl in page ct; 60p; 30 cm.

The product of friendship and co-operation between: The Institute of Karst Geology, Guilin, Guizhou Normal University, British Cave Research Association. Col, b&w illus.

[3391] **Waltham, Anthony Clive**, 1942– *Karst and caves in the Yorkshire Dales National Park.* Grassington, North Yorkshire, GB: Yorkshire Dales National Park Committee in conjunction with the British Cave Research Association, **1987**. Yorkshire Caves National Parks Topic Series. illus: b&w: drawings, photos; partly col: drawings, maps; title page info on rear lining page; 32p; bibl; 20 x 21 cm.

Photo on cover. Cover title: Karsts and caves.

[3392] **Waltham, Anthony Clive**, 1942– *Limestone and caves of Northwest England.* Sweeting, Marjorie Mary, 1920–. Newton Abbot, GB: David & Charles for the British Cave Research Association, **1974**. Limestones and Caves of Britain. illus: b&w: drawings, graphs, maps, photos, tables; 477p; bibl, index; 22 x 14.5 cm.

54 illus, 86 figs, 14 tables.

A good general introduction to the caves and geology of this area, readable and well illustrated.

[3393] **Waltham, Anthony Clive**, 1942– *The world of*

caves. London; New York, GB; US: Orbis Publishing Limited; G.P. Putnam's Sons, **1976**. illus: partly col: maps, photos; col: photos; vi, 128p; bibl, gloss, index; 31 cm.

Speleology on a global scale.

[3394] **Waltham, Anthony Clive**, 1942– *Xingwen: China caves project 1989-1992*. Willis, R. G. (ed). Bridgwater, GB: British Cave Research Association, **1993**. illus: b&w: charts, maps, photos; col: photos; 48p; bibl.

[3395] **Waltham, Anthony Clive**, 1942– *Yorkshire Dales: limestone country*. rev ed. London, GB: Constable, **1987**. illus: b&w: maps; 186p; index; 18 cm.

Previous ed.: published as Caves, crags and gorges, 1984. Illus.

[3396] **Ward, John A**. *The people of Burrows Cave: who they were, where they came from and when*. Vincennes, IN, US: Burrows Cave Research Center, **1990**. 27p; bibl; 28 cm.

Illus.

[3397] **Ward, William Cruse**, 1933– *Geology and hydrology of the Yucatan and Quaternary geology of Northeast Yucatan Peninsula with a part on the History of Northern Quintana Roo*. Weidie, A. E., 1931–; Black, William, 1925–; Andrews, Anthony P. New Orleans, LA, US: New Orleans Geological Society. illus: b&w: maps; v, 160p; bibl, errat; 28 cm.

Illus.

[3398] **Warden, D. E.** *Cave surveying*. npop: np, **nd**. illus: b&w: drawings, maps, photos, tables; 13p; bibl.

Copy examined was a photocopy. Cover title: Surveying caves for the beginner.

[3399] **Warild, Alan** *Vertical: a technical manual for cavers*. 1st ed. Sydney, NSW, AU: Speleological Research Council, **1988**. illus: b&w: charts, graphs, maps, photos, tables; viii, 152p; bibl, index; 30 cm.

An attempt to compile vertical techniques from all over the world and compare them. Not as detailed as On Rope but more global in nature. An excellent text that was updated several times. The comparative charts are especially useful.

2nd ed. Speleological Research Council, **1990**. viii, 128p; 30 cm.

Illus.

3rd ed. 1994. illus: b&w: drawings; viii, 128p.

[3400] **Warnell, Norman** *Mammoth Cave: forgotten stories of it's [sic] people*. [Brownsville, KY, US]: N. Warnell, **1997**. illus: b&w: maps, ports; 230p; 23 cm.

[3401] **Warren, William M.**, 1945– *Sinkhole occurrence in western Shelby County, Alabama*. University, AL, US: Geological Survey of Alabama, Water Resources Division, **1976**. Circular Geological Survey of Alabama no 101. 4 folded in pocket; 45p; bibl; 23 cm.

Illus.

[3402] **Warton, Mike** *A preliminary report of findings for the karst terrains feature known as Rugh Cavern, located along northern Seco Creek, northwestern Medina County, Texas*. Cedar Park, TX, US: Mike Warton & Associates, **1993**. illus: b&w: maps; 6p, [5]; 28 cm.

Caption title: The gating and exploration of Rugh Cavern, Medina County, Texas. "By Mike Warton"—Leaf 1 (1st group). "August 23, 1993."

[3403] **Waterhouse, J. D.** *Pollution of underground water*. Parkside, SA, AU: South Australia Department of Mines and Energy, **nd**. Mineral Information Series. illus: b&w: drawings, photos; partly col: drawings; text on lining pages and back cover; 12p.

Title from cover.

[3404] **Waters, Aaron Clement**, 1905– *Selected caves and lava tube systems in and near Lava Beds National Monument, California*. Donnelly-Nolan, Julie M.; Rogers, Bruce W. Washington; Denver, DC;CO, US: U.S. Geological Survey, **1990**. U.S. Geological Survey Bulletin 1673. illus: b&w: charts, drawings, maps, photos, tables; partly col: maps; col: photos; twenty maps on 6 folded leaves; text on front lining page; ix, 103p; bibl, index; 27 cm.

Detailed visitors' guide to about thirty lava tubes and other pseudokarst features, with large, detailed maps. Notable as probably the only real cave survey published by the United States Geological Survey. Well written, beautifully illustrated.

[3405] **Watkinson, P.** *Expedition 67 to the Gou*. Parker, B; Wicks, A; Lord, H; Kidd, H. npop: The Pegasus C, **1968**. illus: b&w: ads, drawings, maps, photos; map folded in back; iv, 30p, 24p pls.

[3406] **Watson, J. R. (ed)** *Cave tourism in Western Australia: proceedings of a seminar held at Busselton November 10-11, 1978*. WA, AU: Western Australia Department of Tourism, Feb, **1979**. illus: b&w: tables; [i], 28p.

Incl forms.

[3407] **Watson, Patty Jo**, 1932– *Archeology of the Mammoth Cave Area*. New York, NY, US: Academic Press, **1974**. Studies in Archeology. illus: b&w: graphs, maps, photos, tables; [vi], xxi, 255p; bibl, index; 25 cm.

Account of investigations in Mammoth Cave National park since 1969. Continues the editor's The prehistory of Salts Cave, Kentucky.

A detailed edited description of archeological research carried out by Watson and her colleagues between 1969 and 1973 in Mammoth Cave National Park, but especially within Mammoth, Salts, Lee, and Bluff caves.

[3408] **Watson, Patty Jo**, 1932– *The prehistory of Salts Cave: Kentucky*. Yarnell, Richard Asa. Springfield, IL, US: [Illinois State Museum], **1969**. Illinois State Museum Reports of Investigations no 16. illus: b&w: drawings, maps, photos, tables; 1 map folded in back pocket; [xiii], ii, 86p, 16p pls; app, bibl; 28 cm.

Intensive archeological study of Salts Cave (in MCNP,

Kentucky), including a discussion of excavations and surface studies within and around Salts Cave.

[3409] Watson, Richard Allan, 1931– *The Cave Research Foundation 1969-1973*. St. Louis, MO, US: Cave Books, **1984**. illus: b&w: charts, graphs, maps, photos, tables; iv, 264p; bibl; 28 cm.

Reprint of CRF Annual Reports for those years.

[3410] Watson, Richard Allan, 1931– *The Cave Research Foundation 1974-1978*. St. Louis, MO, US: Cave Books, **1984**. illus: b&w: charts, graphs, maps, photos, tables; iv, 341p; bibl; 28 cm.

Reprint of CRF Annual Reports for those years.

[3411] Watson, Richard Allan, 1931– *The Cave Research Foundation: origins and the first twelve years [1957-1968]*. Mammoth Cave, KY, US: Cave Research Foundation, **1981**. illus: b&w: charts, drawings, graphs, maps, photos, tables; 494p; bibl; 28 cm.

Mostly reprinted documents, including CRF Annual Reports 1959-1968.

[3412] Watson, Richard Allan, 1931– *Caving*. St. Louis, MO, US: Cave Books, **1994**. 17p; 15 cm.

Also published as Dust from the ego trip, no. 18—1st prelim. No illus.

[3413] Watson, Richard Allan, 1931– *The Mammoth Cave National Park research center*. Smith, Phillip M. Yellow Springs, OH, US: Cave Research Foundation, **1963**. 50p.

[3414] Watson, Richard Allan, 1931– *The Mammoth Cave National Park Research Center: a Cave Research Foundation study*. Smith, Philip M. Yellow Springs, OH, US: Cave Research Foundation, **1963**. illus: b&w: drawings, maps; 50p; 23 cm.

[3415] Watson, Richard Allan, 1931– *The preservation of wilderness karst in central Kentucky, U.S.A*. Washington, DC, US: The Cave Research Foundation, May, **1967**. illus: b&w: maps; 12p; bibl; 28 cm.

"Statement of Dr. Richard A. Watson, president of the Cave Research Foundation at the Symposium on the application of the Wilderness Act as a means of preserving the surface and the underground features of Mammoth Cave National Park, Kentucky, 22-26 May 1967."

[3416] Watson, Sally, 1954– *Secret underground Bristol*. Bristol Junior Chamber. Bristol, GB: Bristol Junior Chamber, **1991**,. illus: b&w: maps, ports, repros; col: maps; 120p; 21 x 22 cm.

[3417] Watters, David Robert *Final report on the archaeology of Fountain Cavern, Anguilla, West Indies: a report*. [Anguilla]: Anguilla Archaeological and Historical Society, Dec, **1987**. 31p, [15], 6p pls; bibl; 28 cm.

Illus.

[3418] Wayland, John Walter, 1872-1962 *The master sculptor; a brief treatise on erosion, and its wondrous effects in the Shenandoah Valley*. New York, NY, US: Nomad Publishing Co, **1930**. illus: b&w: maps; 19p; 18 cm.

This booklet is presented with the compliments of Endless Caverns, Inc., Endless Caverns, Va.—p. 19. Illus.

[3419] Weaver, Herman Dwight, 1938– *Adventures at Mark Twain Cave*. Johnson, Paul A. (illus). Jefferson City, MO, US: Discovery Enterprises, **1972**. Dark Pathways Series. illus: b&w: drawings, maps, photos, ports, repros; viii, 64p; bibl; 23 cm.

1st printing May 1972, 2nd printing May 1973, 3rd May 1975, 4th, May 1977. Illus.

[3420] Weaver, Herman Dwight, 1938– *Great American show caves*. [McMinnville, TN, US]: National Caves Association, **1982**. illus: b&w: maps, photos; col: photos; text and photos on lining pages; 27p; 28 cm.

Title from cover. Illus.

[3421] Weaver, Herman Dwight, 1938– *Meramec Caverns: legendary hideout of Jesse James*. Johnson, Paul A. (illus). Jefferson City, MO, US: Discovery Enterprises, Aug, **1977**. illus: b&w: drawings, maps, photos, ports, repros; partly col: maps; col: photos; front: black & white photo; 126p, 2p pls; bibl; 28 cm.

[3422] Weaver, Herman Dwight, 1938– *Missouri: the cave state*. Johnson, Paul A. (illus). Jefferson City, MO, US: Discovery Enterprises, Apr, **1980**. Dark Pathways Series book 4. illus: b&w: drawings, maps, photos, ports, repros, tables; col: photos; 336p; app, bibl, index; 28 cm.

Includes history, show caves, longest, deepest.

[3423] Weaver, Herman Dwight, 1938– *Onondaga, the mammoth cave of Missouri*. Jefferson City, MO, US: Discovery Enterprises. Dark Pathways Series no 2. 94p; bibl; 23 cm.

[3424] Weaver, Herman Dwight, 1938– *The wilderness underground: caves of the Ozark plateau*. Huckins, James N. (photo ed); Walk, Rickard [sic] L. (photo ed). Columbia, MO, US: University of Missouri Press, **1992**. illus: b&w: maps; col: photos; front: col photo; xiv, 113p; bibl, gloss, index; 30 cm.

A beautifully illustrated coffee-table book on Ozark caves. Covers geology, biology, history, hydrology, and human impact.

[3425] Webb, Donald W. *The biological resources of Illinois caves and other subterranean environments: determination of the diversity, distribution, and status of the subterranean faunas of Illinois caves and how these faunas are related to groundwater quality*. Taylor, Steven J.; Krejca, Jean K. Springfield, IL, US: The Office, May, **1994**. Technical Report Illinois Natural History Survey Center for Biodiversity no 8. ix, 157p; bibl; 28 cm.

This is perhaps the most thorough account available of the subterranean fauna for any state. It is also noteworthy in its attention to non-cave subterranean habitats.

[3426] Webb, Donald W. *Status report on the cave Amphipod* Gammarus acherondytes *Hubricht and Mackin (Crustacea: Amphipoda) in Southern Illinois*. Champaign, IL, US: Center for Biodiversity, Illinois Natural

History Survey , **1995**. Technical Report 1995 (22) Illinois Natural History Survey, Center for Biodiversity. illus: b&w: tables; 22p; bibl; 28 cm.

[3427] **Webb, William Snyder**, 1882– *The occurrence of the fossil remains of Pleistocene vertebrates in the caves of Barren County, Kentucky*. Funkhouser, William Delbert, 1881-1948. [Lexington, KY, US]: [University of Kentucky]. The University of Kentucky Reports in Archaeology and Anthropology v 3 no 2; Publications of the Department of Anthropology and Archaeology, University of Kentucky v3 no 2 pp [39]—64. 26p, [1]; 26 cm.
Illus.

[3428] **Wefer, Fred L.** *An annotated bibliography of cave meteorology*. State College, PA, US: Cave Geology, published in cooperation with the Section of Cave Geology and Geography, National Speleological Society, Jun, **1991**. Cave Geology v 2 no 2 pp 83—119. 37p; bibl; 28 cm.
Unfortunately, most of the material on cave meteorology postdates this, but this is a good source.

[3429] **Wegrzyn, Mikolaj** *Rainstorm related terrain failures in Puerto Rico: final technical report*. Perez, Juan A.; Andrews, Alejandro Soto. [Mayaguez, PR]: Water Resources Research Institute, School of Engineering, University of Puerto Rico, Mayaguez Campus, Jun, **1984**. illus: b&w: maps; iv, 126p, 35p pls; bibl; 28 cm.
Illus.

[3430] **Weigh, Cliff** *The South Street caves, Dorking*. Dorking and Leith Hill District Preservation Society, Local History Group. npop: Local History Group of the Dorking and Leith Hill Preservation Society, **1988**. 28p.
No illus.

[3431] **Weinel, John** *Pittsburgh Grotto Picnic Fall M.A.R. '79: Guidebook, Sept. 14-15, 1979*. Held, Fred (ed); Long, Pam (ed). [Pittsburgh, PA, US]: Pittsburgh Grotto, **1979**. illus: b&w: maps; one folded map; iii, 53p.
Illus.

[3432] **Welch, Bruce R. (ed)** *The northern limestone*. Sydney, NSW, AU: Sydney University Speleological Society in conjunction with the Speleological Research Council Ltd, **1976**. The caves of Jenolan ; 2. illus: b&w: drawings, maps, photos, tables; ix, 131p; app; 28 x 22 cm.

[3433] **Weller, James Marvin**, 1899– *The geology of Edmonson County: a detailed presentation of the physical, stratigraphic, structural, and economic geology of this district*. Frankfort, KY, US: The Kentucky Geological Survey, **1927**. Kentucky Geological survey ser 6 Geologic Reports v 28. illus: b&w: drawings, maps; one map, diagram folded; front; 246p; 24 cm.
'First edition, 2000 copies'. Some discussion of cave development. Illus.

[3434] **Wells, Oliver C. (ed)** *A history of the exploration of Swildon's Hole*. [Pittsburgh, PA, US]: Pittsburgh Grotto, NSA [sic for NSS], **1960**. illus: b&w: drawings, maps; Various pagination, [80]; bibl; 28 x 21 cm.
Numbered copies (#3 examined). Pagination by chapter, some pages not numbered.

[3435] **Wells, Patrick H.** *Responses to light by cave crayfishes*. Huntsville, AL, US: National Speleological Society, Apr, **1959**. National Speleological Society Occasional Paper no 4. illus: b&w: graphs, photos; 15p; bibl; 23 cm.
Illus.

[3436] **Wells, R. T.** *World famous Fossil Cave Naracoorte*. Naracoorte, SA, AU: Norman Hansen Printing House, **1975**. illus: b&w: photos; partly col: photos; text on front lining pages, text and photos on rear lining page; [12]; bibl.
Title from cover.

[3437] **Wen-chung Pei**, 1904-1982 *The upper cave fauna of Choukoutien*. Chungking, CN: Geological Survey of China, **1940**. China Geological Survey New Ser C Palaeontologia Sinica no10, Whole Ser no 125. illus: b&w: tables; folded diagram; iv, 100p; bibl; 30 cm.
On cover: Peiping (Peking), October, 1940. Illus, 8 pls.

[3438] **Wendy, Herbert** *In search of Adam: the story of mans quest for the truth about his earliest ancestors*. Cleugh, James (trans). npop: np, **1956**. illus: b&w: charts, drawings, photos, repros; xv, 540p, 47p pls; index.

[3439] **Wermund, E. G.** *Regional distribution of fractures in the southern Edwards Plateau and their relationship to tectonics and caves*. Cepeda, Joseph C.; Luttrell, P. E. [Austin, TX, US]: Bureau of Economic Geology, University of Texas at Austin, **1978**. University of Texas Bureau of Economic Geology Geological Circular 78-2. illus: b&w: maps; 14p; bibl; 24 x 28 cm.

[3440] **Werner, Eberhard Wolfgand**, 1942– *Index of the literature pertaining to West Virginia caves and karst*. [WV, US]: [West Virginia Speleological Survey], **1974**. Bulletin West Virginia Speleological Survey Bulletin 3. xv, 140p; index; 28 cm.
No illus. Printed by The Speleo Press, Austin, TX.

[3441] **Werner, Eberhard Wolfgang** 1942– *Development of solution features, Cloverlick Valley, Pocahontas County*. npop: [West Virginia Speleological Survey], **1972**. Bulletin 2 (West Virginia Speleological Survey). illus: b&w: drawings, maps, photos; partly col: maps; viii, 53p; app, bibl; 28 cm.
Results of field work done in the period before 1965 and 1971. Illus. Printed by the Pittsburgh Grotto Press.

[3442] **Werner, Eberhard Wolfgang**, 1942– *Guidebook to the karst of the Central Appalachians; prepared for the Eighth International Congress of Speleology, Bowling Green, Kentucky, U.S.A., July 18 to 24, 1981*. Beck, Barry Frederic, 1944–(ed); Lipp, Astrid (trans); Williams, Carolyn G. (trans). [Huntsville, AL, US]: National Speleological Society, Jul, **1981**. illus: b&w: charts, graphs, maps, photos; 51p; bibl; 22 cm.
In English, French and German. Cover title: Guide-

book to karst of the central Appalachians. One record lists 1982 as date of pub.

[3443] Wernerus, Mathias *The grottos at Dickeyville.* [Dickeyville, WI, US]: np, **1929**. [20]; 23 cm.

Title from cover. OCLC also lists 1920 as date of pub. Minimal text.

[3444] West Texas Geological Society *Delaware Basin exploration: Guadalupe Mts., Hueco Mts., Franklin Mts., geology of the Carlsbad Caverns ; Nov. 6th, 7th and 8th, 1969.* Midland, TX, US: The Society, **1969**. Supplement West Texas Geological Society Field Trip Guidebook 1969. 24p; 29 cm.

There has been a partial revision of mileage and stops of the 1968 field trip.—-cf. p. 3. Illus.

[3445] West Texas Geological Society *Delaware Basin exploration: Guadalupe Mts., Hueco Mts., Franklin Mts., geology of the Carlsbad Caverns, Nov. 6th, 7th and 8th, 1969.* Stewart, W. J. rev ed. Midland, TX, US: The Society, **1969**. Publication West Texas Geological Society no 68-55a; 1969 Supplement Publication West Texas Geological Society no 68-55a. illus: b&w: charts, maps, overlays; one map, one chart on 2 folded leaves in pocket; 170+24p; bibl; 29 cm.

Combined revised publication of original plus supplement. Illus. "There has been a partial revision of mileage and stops of the 1968 Field Trip"-p. 11. Four overlaps. Originally published without supplement in 1968.

[3446] West Texas Geological Society *Guadalupe Mts., Hueco Mts., Franklin Mts., geology of the Carlsbad Caverns, Delaware Basin exploration, Oct. 31st, Nov. 1st and 2nd, 1968.* Midland, TX, US: np, **1968**. West Texas Geological Society Publication no 68-55. illus: b&w: maps; maps folded in pocket; 170p; 29 cm.

Illus.

[3447] West Texas Historical and Scientific Society *Publication no. 4.* Alpine, TX, US: Sul Ross State Teachers College, **1932**. Bulletin Sul Ross State University no 44. 56p; 23 cm.

Conkling Cavern: the discoveries in the Bone Cave at Bishop's Gap, New Mexico / Roscoe P. Conkling—Perishable artifacts of the Hueco Caves / Eileen E. Alves—Early border elections / E.E. Townsend—Old Fort Lancaster / Henry T. Fletcher—Survey of the Jeff Davis - Presidio County line / Barry Scobee. Illus.

[3448] Westall, William *Views of the caves near Ingleton, Gordale Scar and Malham Cove in Yorkshire.* 1st ed. London, GB: John Murray, **1818**. illus: b&w: drawings; 8p, 12p pls; 37.5 x 27 cm.

Drawn and engraved by William Westall. 12 engravings.

reprint ed. Shaw, Trevor Royle, 1928–(intro). Anne Oldham, **1983**. illus: b&w: ports; xii, 8p, 12p pls; bibl; 29.5 x 20.5 cm.

A facsimile with a new, historical introduction by Trevor Shaw.

[3449] Westcott, Richard L. *A new subfamily of blind beetle from Idaho ice caves, with notes on its bionomics and evolution (Coleoptera: Leiodidae).* Los Angeles, CA, US: Los Angeles County Museum of Natural History, Jun, **1968**. Contributions in Science no 141. illus: b&w: drawings, photos, tables; text on lining pages; 14p; bibl; 24 cm.

Illus.

[3450] Western Australia Department of Conservation and Land Management *Leeuwin-Naturaliste National Park: cave permits draft issue plan.* Como, WA, AU: Department of Conservation and Land Management, **1990**. 20p; 30 cm.

Title from cover. Illus. Three forms.

[3451] Western Land Surveys, Utah Division *Timpanogos Cave environmental impact report.* npop: Western Land Surveys - Utah Division, **1977**. illus: b&w: maps; 98p; 29 cm.

[3452] Wetterhall, W. S. *Reconnaissance of springs and sinks in west-central Florida.* Tallahassee, FL, US: np, **1965**. Florida Geological Survey Report of Investigations no 39. illus: b&w: maps; vi, 42p; bibl; 23 cm.

Illus.

[3453] Weyer, Bernard *A description of Weyer's Cave in Augusta County, Va.* Mohler, John Leonard, 1840-1937. Staunton, VA, US: Valley Virginian Motor Print, **1881**. 15p; 22 cm.

Cover title: A description of the famed Weyer's Cave. Illus.

[3454] Weyer, Bernard *A description of Weyer's Cave, in Augusta County, Virginia.* Winchester, VA, US: Printed at the Republican Office, **1849**. 11p; 22 cm.

Microfilm. Atlanta, Ga.: SOLINET, 1993. 1 microfilm reel ; 35 mm.

[3455] Whealdon, Everett *The legend of Parker's Cave.* Richford, VT, US: Samisdat, **1983**. Samisdat Series v 34 no 2. 22p; 22 cm.

No illus.

[3456] Wheeler, Joy *Oberon-Jenolan district historical notebook.* Garland, Blue. Barraba, NSW, AU: Wendy Rene, **1969**. Australian Historical Series. illus: b&w: maps; 44p; 19 cm.

Illus.

[3457] Whelchel, Sandy *A day at the cave.* Brandt, Bill. npop: np, **1985**. illus: b&w: drawings; 26p; 28.5 x 22 cm.

Cover title: Coloring book; Cave of the Winds; a day at the cave.

[3458] Whidbey, Joseph *On some fossil bones discovered in caverns in the lime-stone quarries of Oreston.* Clift, William, 1775-1849. reprint ed. London, GB: W. Nicol, **1823**. Philosophical Transactions. 12p, 7p pls.

[3459] White, George W. *The limestone caves and caverns of Ohio.* reprint ed. Columbus, OH, US: np, **1926**. Ohio Journal of Science vol XXVI no 2 1925 pp 73-116.

44p.

[3460] White, Jim, 1882-1946 *The discovery and history of Carlsbad Caverns, New Mexico*. Nicholson, Frank Ernest (comp); Kennicott, H.C. (photo). [Carlsbad, NM, US]: [Carlsbad Caverns National Park], **1932**. illus: b&w: drawings, photos; [32]; 25 cm.

On cover: Jim White's own story; the discovery and history of Carlsbad Caverns, by James Larkin White, the Cave's original discoverer and explorer as told to Frank Ernest Nicholson. First edition cover has "Twin Domes and giant stalagmites in Big Room" as center picture; later editions have "Rock of Ages, Carlsbad Caverns." Some records indicate that Curt Teich, printer in Chicago, IL is the publisher. Records for later versions include: 1936,1938, 1940, 1944,1946, 1947, 1948 (all with some pagination and size). In 1951 the format changed. Col photo and drawing on front cover. 1938 editions contains a table of visitation up to 1937. Illus.

[3461] Chisholm, T.O. *Grand Avenue Cave. A description in detail of one of America's greatest natural wonders*. [Franklin, KY, US]: [by the Author], **1892**. illus: b&w: ads; 43p, [1].

Printed by Brandon Printing Company, Nashville, TN. Cover title: Grand Avenue Cave: on line of Mammoth Cave Railroad, Edmonson County, KY. Very rare.

[3462] White, Jim, 1919– *Carlsbad Caverns National Park, New Mexico: its early explorations as told by Jim White*. White, Charlie L. [Chicago, IL, US]: [C.T. Photo-Platin], **1951**. illus: b&w: drawings, maps, photos, ports; col: photos; [32]; 24 cm.

Jim White's story of Carlsbad Caverns. Copyright Charlie L. White and Jim White, Jr. Publisher is from OCLC, but it is probable that this was privately published by Jim White or his estate. Illus.

[3463] White, John *Nature preserve potential of the Burton Cave area, Adams County, Illinois*. [Rockford, IL, US]: Illinois Nature Preserves Commission, **1973**. illus: b&w: maps; 13p, [1]; 28 cm.

Description of the cave and surrounding area as a possible state nature preserve.

[3464] White, John *Nature preserve potential of Twin Culvert Cave, Pike County, Illinois*. [Rockford, IL, US]: Illinois Nature Preserves Commission, **1973**. illus: b&w: maps; 10p, [1]; bibl; 28 cm.

Caption title. Caption title: "INPC 53, item 26". Illus.

[3465] White, John *Preservation of caves in Illinois*. [Rockford, IL, US]: [Illinois Nature Preserves Commission], **1973**. 11p; 28 cm.

Caption title. "49th INPC, item 11."

[3466] White, John *Preservation of Fogelpole Cave, Monroe County, Illinois*. [Rockford, IL, US]: Illinois Nature Preserves Commission, **1973**. illus: b&w: maps; 20p; 28 cm.

Caption title. "49th INPC, item 21."

[3467] White, M. *Western Region speleo-education seminar program with abstracts*. Benedict, Ellen M. Vancouver, WA, US: Private printing, **1978**. 13p.

[3468] White, Susan *A bibliography of Victorian caves & karst*. Parkville, VIC, US: np, Aug, **1986**. 35p.

No illus.

[3469] White, William Blaine, 1934– *The Appalachian valleys of central Pennsylvania*. White, Elizabeth L. State College, PA, US: Published in cooperation with the Section of Cave Geology and Geography, National Speleological Society, May, **1992**. Guidebooks to Selected Karst Regions no 1; Cave Geology v 2 no 3 pp 121-191. 71p; bibl; 27 cm.

[3470] White, William Blaine, 1934– *Cave and karst-related papers in the mainstream scientific literature: a bibliography*. White, Elizabeth L. State College, PA, US: Published in cooperation with the Section of Cave Geology & Geography, National Speleological Society, **1984**. Cave Geology v 1 no 9. 102p; bibl, index; 28 cm.

Before the days of high-speed on-line searches of every data base in the literary world, this was quite a contribution.

[3471] White, William Blaine, 1934– *Caves of Pennsylvania*. State College, PA, US: National Speleological Society, Mid-Appalachian Region, **1964**. MAR Bulletin nos 5-6. illus: b&w: maps; 28 cm.

Binders title. Illus. 2 vols in 1. OCLC also lists 1960 as date of pub.

[3472] White, William Blaine, 1934– *Caves of Western Pennsylvania compiled by members of the Mid-Appalachian Region of the National Speleological Society*. Harrisburg, PA, US: Commonwealth of Pennsylvania, Department of Environmental Resources, Bureau of Topographic and Geologic Survey, **1976**. Fourth Series General Geology Report 67. illus: b&w: charts, drawings, maps, photos, tables; seven folded plates in envelope in book; viii, 97p; bibl, index; 23 cm.

41 figs, 7 maps, 12 tables.

This presents some good background materials, not easily gathered otherwise.

[3473] White, William Blaine, 1934– *Geology and biology of Pennsylvania caves*. Harrisburg, PA, US: Commonwealth of Pennsylvania, Department of Environmental Resources, Bureau of Topographic and Geologic Survey, **1976**. Pennsylvania Geological Survey Fourth Series General Geology Report 66. illus: b&w: drawings, maps, photos, tables; viii, 103p; bibl; 23 cm.

26 figs, 1 table, bibl after each section. White, W. B. The geology of caves—Holsinger, J. R. The cave fauna of Pennsylvania—Guilday, J. E. Appalachian bone caves. Microfiche. Englewood, Colo.: Information Handling Services, 1977. 2 microfiche: negative, ill.; 11 x 15 cm.

[3474] White, William Blaine, 1934– *Geomorphology and hydrology of karst terrains*. New York, NY, US: Oxford University Press, **1988**. ix, 464p; bibl, index; 25 cm. Illus.

The first college-level North American textbook. Highly comprehensive, written in an engaging tone. Very strong chemistry section.

[3475] White, William Blaine, 1934– *Karst hydrology: concepts from the Mammoth Cave area.* White, Elizabeth L. (ed). New York, NY, US: Van Nostrand Reinhold, **1989**. illus: b&w: charts, drawings, graphs, maps, photos, tables; partly col: maps; Two partly col maps on folded lvs in pocket; xiii, 346p; bibl, index; 24 cm.

Illus.

The definitive publication on the region which many consider to be the most significant karst in the world. Twenty plus years of work. Fold-out dye trace map. Our basic understanding of shallow phreatic solutional systems and products and their relationship to surface karst and hydrology is largely based on studies of this area. These well written and illustrated papers are by those who know the area best. It covers as much hydrology as geology.

[3476] White, William Blaine, 1934– *Reconnaissance geology of Timpanogos Cave: Wasatch County, Utah.* Van Gundy, James J. reprint. npop: National Speleological Society, Jan, **1974**. NSS Bulletin vol 36 no 1 January 1974 pp 5-17. 13p; 28 cm.

Illus.

[3477] Whitehouse, Frederic Cope *Is Fingal's Cave artificial?* New York, NY, US: np, **1882**. 12p; 27 cm.

Address American Association for the Advancement of Sciences, Montreal. Aug. 30. Address Academy of Science. N.Y. Oct. 9, 1882. Reprint Popular Science Monthly. Dec. 1882.

[3478] Whitfield, Henry C. *King Caverns report.* Dunn, Walter; Halliday, William Ross, 1926–. [Mapa, CA, US]: [Whitfield], Jul, **1963**. illus: b&w: maps; col: photos; 8p, [10].

[3479] Whitfield, Roderick *Dunhuang: caves of the singing sands: Buddhist art from the Silk Road.* Otsuka, Seigo. London, GB: Textile & Art Publications, **1995**. illus: b&w: maps; Various pagination; bibl, index; 37 cm.

Col illus; plans. Translated from the Japanese.

[3480] Whittemore, Anne *A history of the Virginia Region.* Johnson City, PA, US: Virginia Region of the National Speleological Society, **1979**. The Region Record v 3 n 4 Summer 1979 Special Issue pp 131-293. illus: b&w: drawings, photos; 163p.

Covers approximately 1950-1970.

[3481] Wickersham, David L. *A survey of the caves of Roanoke County, Virginia.* Roanoke, VA, US: Blue Ridge Grotto (Va.), National Speleological Society, **1988**. illus: b&w: maps; 51p; bibl; 22 cm.

Illus.

[3482] Wilde, Kevan A. *The cave and karst resource of New Zealand.* npop: np, **1981**. illus: b&w: charts, drawings, maps; 51p, [11]; bibl, gloss; 30 x 21 cm.

Extract from a dissertation, required by Lincoln College in part fulfillment of Diploma Requirement for Diploma in Parks and Recreation, National Parks Option.

[3483] Wilde, Kevan A. *Environmental monitoring of karst and caves.* Williams, Paul. npop: np, **nd**. illus: b&w: maps; [11]; bibl; 30 x 21 cm.

No illus.

[3484] Wilde, Kevan A. *West coast cave and karst management strategy and operational guidelines.* Worthy, Trevor H. [NZ]: [New Zealand Department of Conservation], **1992**. illus: b&w: maps, tables; [iv], 86p; app, bibl, gloss.

[3485] Wilford, Gerald Edward *The geology of Sarawak and Sabah caves.* [Kuching, MY]: Geological Survey, Borneo Region, Malaysia, **1964**. Malaysia Geological Survey Borneo Region Bulletin 6. illus: b&w: charts, drawings, maps, photos, tables; [iii], viii, 181p, 21p pls; bibl, index; 26 cm.

On spine: Sarawak and Sabah caves. Illus.

Geological introduction, then catalog of known caves in these Malaysian states on island of Borneo. Text descriptions of several hundred caves, with roughly a hundred cave maps. Location maps.

[3486] Wilkes, Frank G. *Bibliography of Mammoth Cave National Park, Mammoth Cave, Kentucky.* Louisville, KY, US: Potamological Institute, University of Kentucky, **1962**. 63p; 28 cm.

No illus.

[3487] Wilkins, Bob *Cuba Contact '88: an investigative visit by members of the Westminster Speleological Group to the karst of the Sierra de los Organos 18/9/88 - 7/10/88.* Westminster, GB: Westminster Speleological Group, **1989**. vol 5 no 5 1989 Journal of the Westminster Speleological Group. illus: b&w: maps, photos; 46p; 30 x 21 cm.

[3488] Wilkins, Frances *Caves.* Oxford, GB: Blackwell, **1977**. Blackwell's Learning Library no 98. illus: b&w: maps; 59p, 4p pls; index; 21 cm.

Col, b&w illus. Juvenile.

[3489] Wilkinson, Charles Smith *Photographs of the Jenolan Caves (interior views photographed by means of the electric and magnisium [sic] lights.* Sydney, New South Wales, AU: Charles Potter, **1887**. illus: b&w: photos; [46]; 44 x 56 cm.

Note: sequence of photos in different order than other variation. Dyer was the photographer for at least some of the photos. Pagination includes title page, 1 page of text, and 44 leaves of photos. Mitchell Library version has 51 photos.

[3490] Wilkinson, Tony J. *Franchthi Paralia—the sediments, stratigraphy, and offshore investigations.* Duhon, Susan T. Bloomington, IN, US: Indiana University Press,

1990. Excavations at Franchthi Cave Greece fasc 6. xvi, 207p, 8p pls; bibl; 28 cm.

1 folded leaf. Spine title: Franchthi. Illus.

[3491] William Pengelly Cave Studies Trust *Caves in Buckfastleigh quarries*. Buckfastleigh, South Devon, GB: William Pengelly Cave Studies Trust, Apr, **1985**. Occasional Publication no 1 2. illus: b&w: drawings, maps, ports, repros; [iv], 25p; bibl.

[3492] William Pengelly Cave Studies Trust *Joint Mitnor Cave*. Buckfastleigh, South Devon, GB: William Pengelly Cave Studies Trust, May, **1982**. Occasional Publication no 1 5. illus: b&w: drawings, maps; 11p, [1]; bibl.

[3493] Williams, Dave *Waitomo Caves: Glowworm Cave, Ruakuri Cave, Aranui Cave*. Omaru, NZ: Colourview Publications Ltd, **1987**. illus: b&w: maps, ports, repros; col: photos; col photo on front lining page; [32].

Title from cover. May be related to *Waitomo: Glowworm Cave, Aranui Cave, Ruakuri Cave and surrounding district* by Rodney Smith (1980).

[3494] Williams, David *Some account of the fissures and caverns hitherto discovered in the western district of the Mendip range of hills: comprised in a letter from Rev. D. Williams to the Rev. Patterson*. Shaftsbury, GB: np, **1829**. 16p; 19.5 x 11 cm.

Printed by John Rutter.

[3495] Williams, James Henry, 1843– *Legend of the Cave of the Winds*. Colorado Springs, CO, US: np, **1914**. illus: b&w: ports; 40p; 18 cm.

[3496] Williams, Nick (auth, ed) *Below Belize 1991: the report of the 1991 expedition to the Little Quartz Ridge Southern Belize February to April 1991*. Frew, Duncan; Hardy, Ern; Buskens, Olaf; Arveschoug, Dave; Boomer, Douglas; Sims, Mark. Priddy, Somerset, GB: Below Belize 1992, **1992**. illus: b&w: maps; text on front lining page; [i], 22p; bibl.

Copy examined was a photocopy lacking the 10 pls and maps.

[3497] Williams, Nick *Bibliography of Belizean caving*. Somerset, GB: Nick Williams, Jan, **1994**. 19p; 30 x 22 cm.

No illus.

[3498] Williams, Nick (comp) *British Cave Research Association Cave Radio and Electronics Group bibliography of underground communications*. Bridgwater, Somerset, GB: British Cave Research Association, Mar, **1992**. 31p; 21 x 15 cm.

No illus.

[3499] Williams, Nick (ed) *Queen Mary College: below Belize 1988; the logistics report of the 1988 Speleological Expedition to Belize*. Clark, Tina (ed). London, GB: Queen Mary Speleological Expedition to Belize, **1990**. illus: b&w: drawings, graphs, maps, photos, tables; 213p; app, bibl, index; 30 x 21 cm.

Photos photocopied.

[3500] Williams, Paul W. *Contributions to the study of Karst*. Jennings, Joseph Newell, 1916-1984. Canberra, ACT, AU: Australian National University Canberra, Research School of Pacific Studies, **1968**. Department of Geography Publication G 5 1968. illus: b&w: drawings, graphs, maps, photos, tables; x, 110p; bibl; 25 x 20 cm.

2nd printing 1969.

[3501] Williams, Toni Lewis (ed) *Manual of U.S. cave rescue techniques*. Huntsville, AL, US: National Cave Rescue Commission of the National Speleological Society, **1981**. illus: b&w: charts, drawings, graphs, tables; iv, 106p; app, bibl, index; 28 cm.

Spine title: Cave rescue. "Supplement #1R": [7] p. Illus.

[3502] Williamson, George Charles, 1858-1942 *The Guildford Caverns*. Guilford, England, GB: The Corporation of Guildford, **1930**. folded maps; 19p; 19 cm.

Illus.

[3503] Willis, Dick *Caving expeditions*. London, GB: Expedition Advisory Centre in association with the British Cave Research Association, **1986**. illus: b&w: ads, charts, drawings, photos; 132p, 2p pls; app, bibl.

[3504] Willis, L. D. *A recovery plan for the Ozark cavefish* (Amblyopsis rosae). Atlanta, GA, US: U.S. Fish and Wildlife Service Region IV, Dec, **1986**. illus: b&w: maps; 42p; bibl; 28 cm.

Cover title: Ozark cavefish recovery plan. "Dec 17 1986."

[3505] Wilmut, John (ed) *Caves and conservation*. Birmingham, [GB]: The National Caving Association, **1972**. illus: b&w: charts, graphs, maps, photos, tables; 38p; app, bibl; 30 x 21 cm.

A review of the status of cave conservation problems and efforts in the UK. Gives information on cave classification; types of damage; use levels; and protection efforts.

[3506] Wilson, B. E. (comp) *The Bradford Pothole Club library list*. Leeds, GB: Bradford Pothole Club, Nov, **1965**. [30]; 33 x 21 cm.

No illus.

[3507] Wilson, George Herbert, 1874?–1958 *Cave hunting holidays in Peakland*. npop: np, **nd**. illus: b&w: photos, repros; 93p.

Some adventures and discoveries.

[3508] Wilson, George Herbert, 1874?–1958 *Some caves and crags of Peakland*. Chesterfield, GB: Wilfred Edmunds Limited, **c1926**. illus: b&w: maps, photos, repros; front: illus of Thors Cave on title page; 60p; 18 x 12.5 cm.

"The story of adventures and discoveries in the Peakland underworld with 30 illustrations and special maps by Bartholomew and Sons."

[3509] Wilson, Jane *Lemurs of the lost world: exploring the forests and Crocodile Caves of Madagascar*. 1st ed. London, GB: Impact Books, **1990**. Travellers' Tales.

illus: b&w: maps; col: photos; 216p, 8p pls; app, bibl, gloss; 20 cm.

2nd ed. 1995. app, bibl, gloss.

[3510] Wilson, John, 1800-1849 *A visit to the Mammoth Cave of Kentucky*. Edinburgh, GB: np, **1849**. 19p, [1]; 18 cm.

"Scotia's dirge, being verses on the death of John Wilson, Esq., the Scottish vocalist, who died at Quebec on the 9th July 1849": p. 20. Signed at end: W.J., Gordon Schoolhouse.

[3511] Wilson, Paul (prep) *Managing the limestone caves of Chillagoe and Mungana*. [QLD, AU]: National Parks and Wildlife Service of Queensland, **1977**. illus: b&w: maps, photos, tables; 43+19p, [8]; app, bibl.

One of the prefatory pages is a note stating that Paul Wilson has written permission to reproduce this report and that it is not the official policy of the NP&WLS. Copy examined was a photocopy. Col slides stapled to front cover.

[3512] Wilson, William L. *Investigation of karst-related subsidence near the southwest landfill, Alachua County, Florida, using ground penetrating radar: a report*. Beck, Barry Frederic, 1944–. Orlando, FL, US: Florida Sinkhole Research Institute, College of Engineering, University of Central Florida, Jul, **1987**. Reports Florida Sinkhole Research Institute no 87-88-1. illus: b&w: drawings, maps, photos; ii, 52p, [6]; bibl; 28 cm.

Illus.

[3513] Wilson, William L. (ed) *Karst of the Orlando area: a guidebook prepared for the engineering and geology of karst terranes short course, Orlando, Florida, February 1-5, 1988*. The Florida Sinkhole Research Institute. Orlando, FL, US: College of Extended Studies, University of Central Florida, **1988**. Report No. 87-88-5. illus: b&w: drawings, graphs, maps, photos, tables; vi, 75p; bibl.

37 figs, 9 tables.

[3514] Windels, Fernand *The Lascaux Cave paintings*. Breuil, Abbé Henri, 1877-1961 (pers note); Hawkes, C.F.C. (pref); Leroi-Gourham, A. (intro); Laming, Annette (text in collab). British ed. London, GB: Faber & Faber, **1949**. [iv], 139p.

US ed. New York, NY, US: The Viking Press, **1950**. illus: b&w: drawings, maps, tables; partly col: photos; [iv], 139p; bibl.

Describes various aspects of Lascaux Cave in SW France. Printed in England

[3515] Winder, Francis A. *Unconventional guide to the Caverns of Castleton and Disaster*. npop: np, **1938**. illus: b&w: maps, photos; 81p, 17p pls.

Illus.

[3516] Wintner, Radar H. *Lake Shasta Caverns*. 3rd ed. Obrian, CA, US: Lake Shasta Properties, **1977**. illus: b&w: maps; partly col: photos; 26p.

Account of an unusual show cave.

[3517] Wisconsin Speleological Society *1978 hodag hunt guidebook*. [Madison, WI, US]: [Wisconsin Speleological Society], **1978**. illus: b&w: maps, tables; 33p; bibl.

Title from cover.

[3518] Witcombe, Richard *Who Was Aveline Anyway? Mendip's cave names explained*. Castle Cary, Somerset, GB: Mendip Publishing, **1992**. [Wessex Cave Club Occasional Publication Series 2 no 1]. illus: b&w: photos; [iii], 95p; 21 cm.

18 b&w photos. A5 format, softback.

[3519] Wolfe, James E. *Map location and dimensional definition of subsurface caverns*. npop: np, **1975**. Hawaii University Honolulu Dean Prize for Undergraduate Research Papers 1976. illus: b&w: maps; Various pagination, 98p; bibl.

Science category: 1st prize (tie) Photocopy of typescript. Illus.

[3520] Wood, Andrew *The mammoth cavern*. npop: Parrish, **1957**. Max Parrish Readers. 160p; 19 cm.

Illus.

[3521] Wood, Frances Elizabeth *Rocky Mountain, Mesa Verde, Carlsbad Caverns*. Chicago, IL, US: Follett Pub Co, **1963**. Our National Parks. 32p; 26 cm.

Illus.

[3522] Wood, Jenny *Caves*. 1990 [British] ed. npop: Franklin Watts, **1990**. Jump Nature Book. [32].

Juvenile.

British ed. npop: Two-Can, **1990**. Jump Nature Book. 27p, [5].

No illus. Juvenile.

North Am ed. Milwaukee, WI, US: Gareth Stevens Children's Books, **1991**. Wonderworks of Nature. 27p, [5]; index; 28 cm.

Subtitle: an underground wonderland. Text and pictures present the formation, types, inhabitants of, and exploration of caves. Col illus. Juvenile.

US ed. New York, NY, US: Puffin Books, **1990**. A Puffin Original. illus: partly col: drawings; col: drawings, photos; [32]; gloss, index; 27 cm.

Cover title: Caves: Facts Stories Projects. A brief look at caves, including how they are formed and the animals that live in them. Col illus. Records for the American editions differ in terms of pages (27 vs 32) and publisher (Penguin vs Puffin). Juvenile.

[3523] Wood, Jenny *Caves*. Australian ed. Sydney, NSW, AU: Doubleday, **1990**. Jump Nature Books no 2. index.

For children. Reprinted by Ashton Scholastic in 1992. Other reprints appear to exist. Juvenile.

[3524] Woodall, Brian *Peak Cavern: a guide to this famous show-cave; its formation, history & folklore*. Buxton, GB: Brian Woodall, **1976**. illus: b&w: maps, photos, repros; partly col: photos; repros on front & back lining

pages; 12p; 21 x 15 cm.
Photo in black & yellow.

[3525] Woods, Julian Edmund *Geological observations in South Australia: principally in the district south-east of Adelaide*. London, GB: Longman, Green, Longman, Roberts and Green, **1862**. illus: b&w: drawings; b&w front drawing; xviii, 404p, 5p pls; app, index.

Shortly after the publication of this book, he began calling himself Tenison-Woods. One of the seminal works on speleogenesis of caves in soft limestones whose worth first began to be recognized in the 1960s.

[3526] Workman, C. Lindsay (narr) *Carlsbad Caverns National Park*. MMI Corporation [dis]. Baltimore, MD, US: Holiday Video Library [prod], **1991**. National Park and Monument Series.

Divided into 3 parts: part 1 explores Carlsbad Caverns, part 2, New Cave, part 3, Guadalupe Mountains, Texas. 1 videocassette (20 min) sd col.

[3527] World Travelogues Corporation *Lewis and Clark Caverns*. Whittier, CA, US: Holiday Film Corp, **1984**.

40 slides, col. + 1 sd cassette (1 7/8 ips). Sound accompaniment compatible for manual operation only.

"The caverns are very unusual, large, and beautiful. The narration is filled with humor and is told just like the ranger guides tell it during your tour."

[3528] Worthy, Trevor *Fossils of Honeycomb Hill*. Wellington, NZ: Museum of New Zealand Te Papa Tongarewa, **1993**. illus: b&w: maps; xxx, 56p; bibl.
Illus.

[3529] Worthy, Trevor *Inventory of New Zealand caves and karst of international, national and regional importance*. Joint Earth Science Societies Working Group on Geopreservation. 1st ed. Lower Hutt, NZ: Geological Society of New Zealand, **1989**. Geological Society of New Zealand Miscellaneous Publication no 45. 42p; bibl, index; 30 cm.

At head of title: Joint Earth Science Societies Working Group on Geopreservation. "Produced with the financial assistance of the New Zealand Lottery Board, Lottery Science and the New Zealand Geological Survey". Illus.

2nd ed. Geological Society of New Zealand; New Zealand Geological Survey; Joint Earth Science Societies Working Group on Geopreservation. Lower Hutt, NZ: Geological Society of New Zealand, **1989**. Geological Society of New Zealand miscellaneous publication ; no. 47. 42p; bibl, index; 30 cm.

[3530] Wreschner, E. *The Geula Caves. Mount Carmel*. Avnimelech, M.; Schmid, E.; Haas, G.; Dart, R. A. Roma, IT: np, **1967**. 72p; 24 cm.

Extrait de: Quaternaria, Roma, 1967, 9. Illus. pp 69-140.

[3531] Wright, Charles W. *A guide manual to the Mammoth Cave of Kentucky*. 1876 ed. npop: C.T. Dearing. 30p; 23 cm.

Cover title: Complete guide to Mammoth Cave and the leading business houses and hotels of Louisville.

[3532] Wright, Charles W. *A guide manual to the Mammoth Cave of Kentucky*. Louisville, KY, US: Bradley & Gilbert (Printers), **1860**. Kentucky Culture Series no 56. vi, 63p; 16 cm.

Contains chemistry, geology, and zoology of Mammoth Cave, for the layperson, and a description of major features of the cave. One copy examined did not have pages 61-63 on Indian cave. The 1867 version has extra title page with 1867, followed by the 1860 title page; contains only 60p. Reprint , 1979, University Microfilms International, Ann Arbor, Michigan. Microopaque. Louisville, Ky., Lost Cause Press, 1957, 2 cards 8 x 13 cm.

[3533] Wright, Charles W. *The Mammoth Cave of Kentucky*. Vincennes, IN, US: Harvey, Mason & Co, **1858**. 68p; 16 cm.

Reprinted by Jack Speece (limited edition of 55), July 1970. There is some confusion concerning the date of publications of the original and other facsimiles.

[3534] Wright, David P. (ed) *1st Claygate "Selachii" Venture Scout Unit expedition to Iceland: 21st July - 9th August 1987*. np: J. Dutton, **[1988]**. illus: b&w: ads, cartoons, maps, photos, tables; ad on front lining page; [ii], 56p; app.

[3535] Wright, R. V. S. *Archaeology of the Gallus Site, Koonalda Cave*. Canberra, ACT, AU: Australian Institute of Aboriginal Studies, **1971**. Australian Aborigines Studies no 26 Prehistory Series no 5. illus: b&w: charts, drawings, graphs, maps, photos, photomicrographs, tables; ix, 133p; bibl.

Presents a detailed site report and material culture inventory of excavations carried out between 1957-1968, as part of the Nullarbor project of the Cave Exploration Group in South Australia. Also includes some significant ethnographic accounts of the Yirela Meening tribe, who occupied the region until the late 19th century

[3536] Wright, V. Paul, 1953– *Palaeokarsts and palaeokarstic reservoirs*. Estaban, Mateu (ed); Smart, P. L. (ed). Reading, GB: Postgraduate Research Institute for Sedimentology, University of Reading, **1991**. PRIS Contribution no 152. illus: b&w: charts, drawings, graphs, maps, tables; (incl 1 folded loose in text); [iv], 158p; bibl; 30 cm.

A course book prepared for a University of Reading Short Course, "A course book prepared for a University of Reading Short Course, May 2 1991."

[3537] Wyllie, Diana *Caves—origins, development and formations*. Carman Educational Associates; Ford, Trevor David, 1925–; Ford, Derek Clifford, 1935–. npop: np, **1973**.

Double-frame version also issued. With teacher's guide. Describes caves formed in limestone. 38 fr col 35 mm.

[3538] Wynne, Annette *The trip thru fairyland; Great Onyx cave.* Edwards, L. P. Mammoth Cave, KY, US: np, **1924**. illus: b&w: photos, ports; [i], vi, 27p; 19 cm.

[3539] Yale Speleological Society *The caves of Connecticut.* 2nd ed. New Haven, CT, US: The Society, **1963**. illus: b&w: maps, tables; 26p; index; 31 cm.

Originally published in the May 1961 issue of the "Journal of the Yale Speleological Society."

[3540] Yauru Lu, 1931– *Karst in China - landscapes, types, rules.* Beijing, CN: Institute of Hydrogeology & Engineering Geology, Chinese Academy of Geological Sciences, **1986**. illus: b&w: charts, drawings, maps, photos, repros, tables; partly col: maps; col: maps, photos; col photos on lining pages; 288p; 24 x 34 cm.

In Chinese, with 172 page English translation bound separately. Over 400 col photos.

[3541] Yeadon, Geoff (auth, illus) *The Cave Diving Group technical review no.3: line laying and following.* Bristol, GB: Cave Diving Group, **1981**. illus: b&w: drawings; 10p, 9p pls; 30 x 21 cm.

[3542] Yevjevich, Vujica *Karst water research needs.* Littleton, CO, US: Water Resources Publications, **1981**. illus: b&w: maps, tables; vii, 266p; bibl, index.

[3543] Yorkshire Subterranean Society (Great Britain) *Caves and caving, 1.* npop: np, **1976**. Journal Yorkshire Subterranean Society no 1. illus: b&w: maps; 59p; 30 cm.

Title composed. Illus.

[3544] Yorkshire Subterranean Society (Great Britain) *Caves and caving, 2.* npop: np, **1979**. Journal Yorkshire Subterranean Society Great Britain no 2. illus: b&w: maps; 69p; 30 cm.

Illus.

[3545] Young, Ivan (ed) *Appin cave guide.* Edinburgh, GB: Grampian Speleological Group, **1978**. Special Publications Grampian Speleological Group no 1. illus: b&w: maps; ii, 30p; index; 21 cm.

Map on back cover.

[3546] Young, James Jay, 1948– *Caves in Kansas.* Beard, Jonathan. Lawrence, KS, US: Kansas Geological Survey, **1993**. Educational Series 9. illus: b&w: drawings, graphs, maps, photos; col: photos; text on front lining page; iv, 48p, 4p pls; gloss; 28 cm.

[3547] Youngsteadt, Norman W. *A survey of some cave invertebrates from northern Arkansas.* Youngsteadt, Jean O. Searcy, AR, US: Association for Arkansas Cave Studies, Oct, **1978**. Arkansas Cave Studies number 1. illus: b&w: maps; text on lining pages; 13p; bibl.

Pagination includes covers.

[3548] Yuan, Daoxian, 1933– *Problems of environmental protection of karst area.* npop: np, **1983**. illus: b&w: drawings, maps, photos; b&w photos on lining pages; [ii], 14p; bibl.

Paper presented to the Annual Meeting of AAAS, Degradation and Rehabilitation of Fragile Environments: From title page: "Karst areas and desert margins (Part I) Detroit, Michigan, USA, 28th May, 1983 Arranged by William Back and Yuan Daoxian."

[3549] Zadnikar, Marjan *The castle of Predjama.* Postojna, SI: [Postojnska Jama], **1960**. illus: b&w: photos, repros; 25p, 8p pls; bibl; 19 x 12 cm.

Description of this medieval castle in a cave and of the cave itself.

[3550] Zámbó, László (ed) *Conference on the karst and cave research activities of education and research institutions in Hungary: papers: Jósvafő 17-19 May 1991.* Veress, Márton (ed). [Jósvafő, HU]: [Berzsenyi College, Jósvafő, Hungary], **1993**. illus: b&w: charts, drawings, graphs, maps, tables; 155p; bibl.

[3551] Zannes, Tom (prod, dir) *Spirit of exploration: discovering the wonders of Carlsbad Caverns National Park.* Carlsbad, NM, US: Carlsbad Caverns-Guadalupe Mountains Association, **1993**.

Produced and Directed by Tom Zannes, Electronic Films. Richard Allen In "Spirit of Exploration", videographer Tom Zannes takes us on an adventure where we discover rarely visited areas of the 46,000 acre National Park. Included are scenes of fascinating plants and animals of the Chihuahuan Desert, steep walled canyons, breathtaking scenic vistas and the mass flight of over a million Mexican free-tailed bats. Included is the recently discovered Lechuguilla Cave. VHS format. 1 videocassette (ca 51 min), sd, col, 1/2 in.

[3552] Zawislak, Ronald L. *The Tennessee pit survey.* Nashville, TN, US, **1967**. illus: b&w: maps, photos; B&W photos; 74p; 28 cm.

Cover title. 2 vols.

[3553] Zeoderborg, Harry *The phantoms of stork fontein [sic].* npop: Printed by Damax Printers, West Krugersdorp, nd. illus: b&w: drawings, maps, photos, ports; 19p.

[3554] Zhaoyang Wang *Underground worlds: Guizhou.* Deming Jin (photo). Beijing, CN: Guizhou People's Publishing House ; distributed by China International Book Trading Corp (Guoji Shudian), **1983**. Underground Worlds 1. illus: col: maps, photos; [vi], 128p; 31 cm.

In English and Chinese. 124 photos, 1 map. Minimal text.

A beautiful coffee-table book of Chinese caves and karst.

[3555] Zhu Xuewen, 1932- *Guilin karst.* Ruoyon Shang (trans). Shanghai, CN: Shanghai Scientific & Technical Publishers, **1988**. illus: b&w: drawings, graphs, maps, photos, tables; partly col: graphs; col: maps, photos; b&w photos on lining pages; [iv], 188p; 27 cm.

Translation of: Kuei-lin yen jung. 360+ col photos, 35 figures. OCLC lists author as: Chu, Hsueh-wen, 1932—.

Largely a photodocumentation of the spectacular mature humid tropical karst. The text is good, well written and translated. The color in the photos is confusing, as is some terminology. Both surface and subsurface karst are well documented.

[3556] Zim, Herbert Spencer, 1909– *Caves and life*. Cuffari, Richard, 1925–(illus). New York, NY, US: William Morrow and Company, **1978**. illus: b&w: charts, drawings, maps; 62p, [2]; index; 22 cm.

Discusses the formation and structure of caves, the animals that dwell in them, and man's use of caves from ancient times to today. Juvenile.

[3557] Zingg, Robert Mowry, 1900-1957 *Report on archaeology of southern Chihuahua*. Weltfish, Gene, 1902- Archaeology of southern Chihuahua. Cave-dweller twill-plaited basketry. [Denver, CO, US]: np, **1940**. Center of Latin American Studies Publication I. vi, 95p; bibl; 24 cm.

At head of title: The University of Denver. Contributions of the University of Denver. Seven full-page illustrations not included in paging. Reproduced from typewritten copy. "A summary and a resume ... of field work done in ... 1931."—pref. Weltfish, Gene: Archaeology of southern Chihuahua. Cave-dweller twill-plaited basketry.

[3558] Zumrick, John L. *NSS cavern diving manual*. Prosser, J. Joseph; Grey, H. V. (illus); McKinnon, Wayne (illus). Branford, FL, US: Cave Diving Section of the National Speleological Society, **1988**. illus: b&w: charts, drawings; vi, 121p; app, bibl, index; 22 cm.

Illus.

16 Index of Secondary Authors, Editors, Illustrators, Etc.

Adams, John: 2275.
Addison, Aaron: 2324.
Adovasio, J. M.: 76.
African Spelaeological Association. Cape Section: 681.
Agogino, George: 1398.
Aguilar, E.: 3223.
Alegria, Ricardo E.: 2861.
Alexander, Cecil Hume: 464.
Alexander, E. Calvin: 1703, 2402.
Alexander, Scott C.: 26.
Aley, Catherine: 32, 35, 42, 43, 44.
Aley, Thomas John, 1938–: 1197, 2338, 2745.
Allen, Joel Asaph, 1838–1921: 1795.
Allsop, David G.: 1086, 1096.
Alon, David: 318.
Alpine Archaeological Consultants: 2798.
am Ende, Barbara Anne: 3140.
Ambler, J. Richard: 1883.
American Philosophical Society: 2152.
American School of Prehistoric Research: 1162.
Anderson, Charles H.: 1335.
Anderson, Edward G.: 915, 918, 919.
Anderson, Elaine: 1279.
Anderson, Jennifer Ann, 1942–: 2366.
Anderson, R. Y.: 2422.
Anderson, Richard: 2922.
Anderson, Richard R.: 2921.
Andree, John, 1699?–1785: 373.
Andrew, J. W.: 2063.
Andrews, Alejandro Soto: 3429.
Andrews, Anthony P.: 3397.
Andrle, Fred: 3026.
Anelli, Franco: 1620.
Animation Research Unit: 2991.
Anthony, Patricia J.: 3028.
Appleton, Marion: 3356.
Applied Scientific Research Co: 1163.
Arias de la Canal, Fredo: 391.
Arkansas Game and Fish Commission: 44.
Arkansas Game and Fish Commission, Federal Aid in Fish and Wildlife Restoration: 516.
Armenta, Sandra Richard: 1064.
Armstrong, Frances Eleanore: 1885.
Armstrong, Robert R., 1934–: 2811.
Arnold, Jay: 1610.
Arveschoug, Dave: 3496.
Asarco Inc: 682.
Aschwanden, Peter: 2914.
Ash, Paul: 1597.
Ashmead, P.: 1022.
Ashton, Horace: 986, 987, 1900.
Ashton, K.: 1061.
Ashworth, H. W. W.: 300.
Aspin, J.: 1022.
Assaad, Fakhry A: 1886.
Association for Bexar Cave Studies: 2691.
Association Francaise de Karstologie, Commission des Phénomènes Karstiques: 1878.
Association of Ground Water Scientists and Engineers (U.S.): 689.
Astiasuain, Linda: 2732, 2733.
Atchison, Robert B.: 2539.
Atkinson, Vernon: 243.
Auckley, Jim: 1155.
Audetat, Maurice: 1024.
Audy, Igor: 248.
Australian ed: 3523.
Australian Speleological Federation: 255, 2178, 2619.
Australian Speleological Foundation: 265.
Avery, Graham: 1611.
Avnimelech, M.: 3530.

Bacon, Mary R.: 31.
Bahn, Paul G.: 642, 643.
Bailey, Alberta: 2238.
Bailey, Florence Merriam, 1863–: 280.
Baker, Bruce: 453.
Baker, Linda: 70.
Balch, Herbert Ernest: 283.
Balfour, William: 2475.
Balfour, William M.: 805.
Balkay, Balint: 1694.
Ballard, Paul: 1210.
Banks, Nathan, B. 1868–: 2631.
Bannerman, Jackie: 1075.
Baramki, D. C.: 2932.
Barfus, Brian: 356.
Barnes, John Wykeham, 1921–: 737.
Barnitz, Wirt Whitcomb, 1887–: 986.
Baroody, Roger A.: 1504.
Barr, Thomas Calhoun, 1931–: 740, 1466, 1507.
Barrington, Piers: 1746.
Basin Research Associates: 320.
Bassett, Terry: 433.
Bastin, B.: 1642.
Baumhoff, M.A.: 3331, 3334.
Baumhoff, Martin A.: 1422.
Bayer, T. N.: 1479.
Baz-Dresch, John: 2467.
Beard, Jonathan B.: 2418.
Beard, Forial: 2641.

Beard, Jonathan: 3546.
Beck, Barry Frederic, 1944–: 1627, 3442, 3512.
Beck, John S.: 1203, 2336.
Beckler, Edward B.: 3252.
Bedford, Bruco L.: 301.
Belinky, Charles R.: 2010.
Bell, Harold: 455.
Bell, Jonathan: 3319.
Bellwood, Peter S.: 2997.
Belski, Carol: 2407.
Belski, David Stanley, 1937–: 2407.
Belski, David Stanley, 1943–: 2446.
Benedict, Ellen: 2433.
Benedict, Ellen M.: 2419, 3467.
Bennett, Gordon D.: 1216.
Benning, Anne E.: 3026.
Benson, Arlene: 1923.
Benson-Evans, Kathryn: 2080.
Berdal Stromme Consulting Engineers: 395.
Berry, Kristin H.: 2095.
Bertolino, Daniel: 592.
Bierly, Pierce: 3339.
Bierwiaczonek, Boguslaw: 2731.
Bighorn Research Project: 2522.
Bilton, David: 2582.
Binet, Michel: 247.
Bingham, Gaby L.: 65.
Bingham, Roy H.: 2264.
Binney, Frank H.: 1610.
Bishop, William P: 820.
Bitman, Roman: 595.
Black, David: 697, 2426.
Black, Herbert L.: 2454.
Black, Herbert L.: 2455, 2459.
Black, William, 1925–: 3397.
Bliss, R. A.: 1022.
Bluhm, Elaine A.: 2064.
Board of County Commissioners of Sem: 71.
Boardman, Walter S.: 1371.
Böcker, Tivadar: 1652.
Bondesan, A: 2890.
Bonney, Rev: 2643.
Bonwick, John: 1314.
Boomer, Douglas: 3496.
Boon, J. M.: 1078.
Bosted, Peter, 1954–: 725.
Boston Society of Natural History: 2734.
Boulanger, Pierre: 850.
Bounk, Micheal [sic]: 1941.
Bourke, R. M.: 515.
Bowers, Jeff: 2448.
Bowman, Bruce: 2348.
Boyer, Paul: 216.
Boyle, Mary Elizabeth: 470.

Bozeman, Sue: 2479.
Brack, E. V.: 3243.
Bradshaw, Dany: 1982.
Brain, Charles Kimberlin: 2078.
Brainerd, George W: 1386.
Bramston, D.: 2503.
Brandt, Bill: 3457.
Bray, L. G.: 672.
Breakspear, Marjorie: 853.
Breisch, Richard Lewis, 1943–: 2367, 2368, 2369, 2370.
Bretz, J Harlen, 1882-1981: 2214.
Breuil, Abbé Henri, 1877-1961: 3514.
Bricker, Harvey M.: 707.
Bridge, John F.: 3206.
Bridge, Josiah: 792.
Briffault, Herma: 2979.
Bristol Exploration Club: 513.
Bristol Junior Chamber: 3416.
British Broadcasting Corporation: 1919.
British Broadcasting Corporation. Television Service: 483.
British Cave Research Association: 1094.
Brodrick, Alan Houghton: 1867.
Brogan, Phil F., 1896–: 2732.
Bronaugh, Mitchell: 2911.
Brook, David, 1944–: 499, 500, 501.
Brook, G. A.: 2046.
Brooks, Ronald: 1723.
Brooks, Steven: 3091.
Brown, Harry Bates, 1914–: 867.
Brown, M. C.: 1078, 1082.
Brown, Peter: 1396, 1397.
Brown, Spontaneous Sandy: 856.
Brucker, Roger Warren, 1929–: 765, 1914, 1915, 2305.
Brush, John B.: 923, 2511.
Bryan, Kirk, 1888–: 1452.
Bryant, Vaughn M.: 2941.
Buchan Caves Advisory Committee: 255.
Bufton, Bill: 304.
Buhac, J. J.: 2176.
Buice, Rick: 3067.
Bullock, Michael: 1979.
Bunnell, David Edward, 1952–: 1311.
Burchett, Charles R.: 2264.
Bureau of Geology, Division of Resource Management, Florida Department of Natural Resources: 71.
Burkhardt, Rudolf: 1623.
Burns, Conor: 153.
Burns, Denver P.: 535.
Burns, Eleanor W.: 2463.
Burns, Robert: 2820.
Burr, Brooks M.: 2594.
Burt, DeVere E.: 1918.
Burton, John, 1948–: 1159.

Busby, Colin I.: 319, 320.
Buskens, Olaf: 3496.
Buss, James C.: 2314.
Bussleton Tourist Bureau: 254.
Butler, D. K.: 1121.
Buttling, Helene: 3375.
Byrd, Phillip E.: 3215.
Byrd, Thomas M.: 2420.

Caisse National des Monuments Historiques et des Sites: 249.
California Department of Fish and Game: 3263.
Call, Richard Ellsworth, 1856–: 1540.
Canada. Dept of Indian Affairs and Northern Development: 594.
Caniff, Milton Arthur, 1907–: 493.
Cannon, Mark: 3069.
Cantos, Francisco Carrasco: 3345.
Canty, R.: 1208.
Carey, Grant: 2526.
Carey, Helen H.: 1254.
Carey, Steven D.: 1165, 1831.
Carman Educational Associates: 1087, 1088, 3537.
Carman, S.: 235.
Carpenter, Alan: 1164.
Carr, Deborah L.: 2521.
Case, James D.: 1842.
Cassa per il Mezzogiorno: 1620.
Casteret, Norbert, 1897-1987: 2813.
Causey, Nell B.: 329.
Cave Conservation Task Force of the National Speleological Society: 3098.
Cave Diving Group: 301, 1036.
Cave Exploration Group: 2284.
Cave Exploration Group South Australia Incorporated: 2674.
Cave Research Foundation: 2337.
Cave Research Group of Great Britain: 3194.
Cavin, Scott: 356.
Čeh, Anne: 1828.
Central Institute of Art and Design: 497.
Centre of Scientific Research of SAZU, The Institute of Karst: 1860.
Cepeda, Joseph C.: 3439.
Chabert, Claud, 1939–: 725.
Chabert, Claude: 1170, 1854.
Chamberlin, Joseph C.: 329.
Champion, Arthur: 1762.
Champion, C. Randell: 1345.
Champion, Randell: 1350.
Chao, E. C. T.: 831.
Chapman, Carl Haley: 1955.
Chard, Andrew: 790.
Chardon, Michel: 1618.

Charleston, B. M.: 424.
Checkley, Dave: 438.
Chelsea Speleological Society: 1068, 2627.
Chester, James: 2339.
China National Committee for IAH: 1616.
Choquette, Philip W.: 1702.
Christensen, Conway B.: 2314.
Christiansen, Kenneth: 329.
Clark, Chip: 23, 3372.
Clark, Michael R.: 408.
Clark, Tina: 3499.
Clausen, Will ;: 1106.
Clay, R. C. C.: 294.
Clayton, Keith M.: 3244.
Cleugh, James: 3438.
Clewlow, C.W.: 3331.
Clift, William, 1775-1849: 3458.
Clottes, Jean: 642, 643.
Coates, D.R.: 1453.
Coath, Tim: 1217.
Cobb, David: 775.
Cochran, Bruce D.: 2194.
Cochrane, Anne: 1210.
Cockerill, R.: 262, 263.
Coe, R. G.: 509.
Coe, William D.: 801.
Cohen, Steven M.: 1771.
Cokendolpher, James C.: 2771.
Colcord, Diane: 2491.
Cole, James E.: 2302, 2303.
Coleman, Alice: 1271.
Collard, Derek: 437.
Collett, John, 1828-1899: 730.
Colton, Harold S.: 1954.
Commission on Environmental Changes and Conservation in Karst Areas: 1637.
Comstock, V. N.: 31.
Consiglio Nazionale delle Ricerche: 1620.
Cook Laboratories: 698.
Cook, Emory, 1915–: 698.
Cook, Holly: 2426.
Cook, Jill: 75.
Cooper, John Edward, 1929–: 2389, 2465.
Cooper, N. C.: 294.
Copeland, Charles W.: 2506.
Copland, Neville: 2991.
Coppens, Yves: 2867.
Cordingley, J. N.: 603.
Cordingley, John N.: 301.
Corrie, George B.: 1313.
Corrigan, Vincent: 1706.
Courtin, Jean: 667.
Covey, Lynn: 1706.
Cowling, Herford Tynes, 1890–: 987, 1900.

Crabtree, H.: 504.
Craig, John L.: 2206.
Crawford, N. C.: 2905.
Crawford, Paul C.: 3000.
Crawford, Rodney Lee, 1951–: 2523.
Crawford, Ronald L., 1947–: 1334.
Creaser, Edwin P.: 2630.
Creswell Crags Visitor Centre: 512.
Crocker, Tad M.: 2521.
Cross, W. R.: 1516.
Crowder, June: 3220.
Crowther, Robert: 1167.
Csallany, Sandor C.: 871.
Csiby, Mihály: 1813.
Cubitt, Barry: 1778.
Cuffari, Richard, 1925–: 3556.
Cullen, James J.: 2421.
Cullen, Jim: 2431.
Cullen, Vera: 770.
Culley, John P.: 2501.
Cullinan, M.: 2428.
Cullinan, Mike: 3085.
Cullingford, Cecil Howard Dunstan, 1904-1990: 1095.
Cully, Anne C.: 1033.
Culver, David Claire, 1944–: 1503, 1504, 1505.
Cundy, J. G.: 814.
Curl, Rane L.: 2361, 2362, 2363, 2364, 2365.
Cutliff, Lewis D.: 3372.

Dale, Norman: 2813.
Damon, Paul Herbert, 1934–: 646.
Daniel, Glyn Edmund: 471.
Daniel, Margaret A.: 1467.
Danks, Libby: 2991.
Daoxian Yuan, 1933–: 1651.
Dart, R. A.: 3530.
Dasher, George R., 1952–: 2404.
Dávalos, Felipe: 1273.
Davey, Adrian: 1532.
Davies, C. M.: 509.
Davies, G. M.: 500, 501, 505, 506, 507, 508, 511.
Davies, Martin: 3388.
Davies, T. J.: 2270.
Davies, William Edward, 1917-1990: 3284.
Davis, Bette J.: 1340.
Davis, John O.: 2462.
Davis, Marsha: 25, 26, 28.
Davis, Nevin W.: 565.
Davis, Ray V.: 1921.
Dawkins, William Boyd, 1838–: 299, 756.
Day, Kenrick L.: 1859.
de Courval, M.: 628.
de Saussure, R.: 616.
De Vivo, Antonio: 393.

Deakin, P. R.: 2054.
Dearne, M. J.: 463.
Decu, Vasile: 1761.
Deem, Dixie: 2473, 2476.
Deem, Kelley: 2473, 2476.
Dehao Zhu: 803.
Dell, David: 618.
Delluc, Gilles: 910.
Delvert, Ray: 910.
Deming Jin: 3554.
Dénes, György: 8.
Dennis, Andy: 2697.
Denny, Henry: 1031.
Denver Mining Club: 2036.
Denver Museum of Natural History: 353, 1212.
Denver Museum of Natural History Audio/Video Department: 352.
Department of Conservation and Environment [Australia]: 817.
Department of Conservation, Buller District, West Coast Region: 250.
Derbyshire County Council: 512.
Derbyshire Caving Association: 1086, 1203.
Desautels, David A.: 477.
Deschamps, Eliette Brunel: 642, 643.
Diana Wyllie, Ltd: 671, 1087, 1088.
Dimitroulas, Ch.: 1933.
Dinkle, T. R.: 1377.
Dinnage, Rosemary: 609.
Dirkmaat, Dennis C.: 76.
Domnanovich, Joe: 2124.
Donaldson, D. R.: 213.
Donati, Anne: 235.
Dongerkery, Shri S.R.: 2975.
Donnelly-Nolan, Julie M.: 1740, 3404.
Donovan, Paul: 2991.
Doomas, A.: 2661.
Dorking and Leith Hill District Preservation Society, Local History Group: 3430.
Dorrell, Margaret: 1699.
Dossey Domes Cavern and Dossey Cliff Walk: 111.
Dougherty, P. H.: 2436.
Douglas R.: 782.
Douty, Bill: 62.
Dovey, Liz: 3081.
Downs, Theodore, 1919–: 3019.
Doyle, F.L.: 1617.
Draper, Gren: 1060.
Dreiss, Shirley Jean: 2209.
Drew, David Phillip, 1943–: 245, 1631, 3016.
Drummond, Ian M.: 726.
Dubertret, L.: 561, 562.
DuChene, H. R.: 2422.
Duhon, Susan T.: 3490.

Dunagan, W. A.: 2280, 2281.
Duncan, Dan: 682.
Dunkley, J. R.: 261, 814.
Dunn, John R.: 2450, 2451, 2452, 2453.
Dunn, Walter: 3478.
Durant, Joe: 2282.
Durham, Dave: 3067.
Durlo, Leslie H.: 1310.
Dury, G. H.: 917.
Dyson, H. Jane: 1697, 1698, 1699.

Eagle, Michael: 1199.
Earhart, Robert L.: 1315.
East Tennessee Grotto: 2448.
Eaton, J.: 2446.
Eberhard, Rolan: 557.
Eberhard, Stefan: 3082.
Eck, Greg: 2440.
Eckler, A. Ross: 795.
Economic Institute for Research and Education (U.S.).: 900.
Eddleman, William R.: 2203, 2319, 2320.
[Eddleman, William R.]: 2205.
Edgar, Robert: 2114.
Edington, A.: 672.
Edington, M. Ann: 2071.
Edwards, E.: 1923.
Edwards, L. P.: 3538.
Eger, Mixican: 3319.
Ehr, Bob: 2470.
Eller, P. Gary: 765.
Elliott, William Rawleigh, 1946–: 2239, 2415, 2545, 2788.
Ellis, Ross Andrew, 1940–: 931, 1817, 2503.
Elsasser, Albert B.: 1422.
Elston, Donald Parker, 1926–: 1009.
Engel, Thom: 2431.
Engineering and Design Associates: 35, 42, 43.
Environmental Education Enterprises Institute: 2745.
Epstein, Beryl Williams, 1910–: 999.
Eraso, Adolfo: 2730.
Eraso, D. Adolfo: 1114.
Erchul, Ronald A.: 3001.
Ericksen, Annette G.: 658.
Eshelman, Ralph E.: 2070.
Estaban, Mateu: 3536.
Estes, James H.: 2781.
Et al: 907.
Et al.: 47, 236, 281, 1013, 1696, 1978, 2631, 2820, 2875, 2892, 3118.
Et al. (20 other authors): 309.
Euler, Robert C.: 1954.
Eury, D.: 3298.
Evans, John: 2401.

Ewer, R. F.: 460.
Ewers, Ralph O.: 2739, 2742, 2745.
Eyre, Jim: 492, 1023.

Fagundo, J Reynerio: 2731.
Fairbairn, C. M.: 2574.
Fakui Yao: 133.
Farr, M.: 672.
Farrand, William R., 1931–: 1689.
Fawley, J. Philip: 2469.
Feder, Gerald L.: 3365.
Federazione Speleologica Regionale Dell'Emilia-Romagna: 1102.
Fee, Scott: 3074.
Fee, Scott: 2480, 2481, 2482.
Fellows, Larry Dean, 1934–: 3364.
Field Museum of Natural History: 1732.
Fieseler, Ronald Glenn, 1948–: 2400, 2776.
Fillol, J. Boquera: 1114.
Finch, Richard: 2784.
Finkelstein, Louis, 1895–: 1732.
Finley, Russ: 1064.
Finley-Holiday Film Corporation: 2588, 3301.
Fisher, Harvey I.: 1386.
Flanagan, J. R.: 527.
Flavell, J.: 1116.
Flenniken, J. Jeffrey: 809.
Floquet, Francois: 592.
Florida Geological Survey: 1895.
Florida sinkhole Research Institute: 356.
Florida Sinkhole Research Institute, College of Engineering, Univ of Central Florida: 363.
Floyd, E. F.: 2155.
Flurkey, Andrew J.: 2405.
Flurkey, Andy: 2900.
Fogg, Tim: 1074.
Fong, Daniel W. Fong, Daniel W.: 777.
Fontaine, J. P.: 415, 416, 417.
Forbis, Shelly: 1106.
Ford, Derek Clifford, 1935–: 441, 594, 1087, 2296, 3537.
Ford, Trevor David, 1925–: 1625, 3537.
Forder, Eliza: 1100.
Forney, Gerald Glenn: 3084.
Forti, Paolo: 1457.
Foster, David G.: 2346.
Foster, Debora L.: 2346.
Foster, Debra: 1106.
Foster, S.: 645.
Foundation for Ocean Research: 3142.
Fowke, Gerard, 1855-1933: 2198.
Fox, Robert B., 1918–: 1188.
Frank, Ruben: 865.
Franke, Herbert W., 1927–: 424.
Frankland, John: 1023.

Frantz, W. S.: 2441.
Fraser, Eugene: 2991.
Frech, Marshall: 2679.
Freeman, B. E.: 3350.
Fregeau, Annette: 313.
Frew, Duncan: 3496.
Friedman, Bob: 1018.
Friends of Karst: 1941.
[Fritz, Carol]: 2399.
Frost, F. W.: 824.
Frost, Frank: 304.
Funkhouser, William Delbert, 1881-1948: 3427.
Furlong, Eustace Leopold, 1874–: 2989.

Günay, G.: 1768.
Gadbois, Nick: 2914.
Gallery of Prehistoric Art [New York, N.Y.]: 2098.
Gamble, F. M.: 1258.
Ganter, John H.: 3189.
Garcia, Cleo Zoe: 3026.
Gard, L. M.: 3043.
Gardner, Larry W.: 2769.
Gardner, Treva L. Missouri. Dept. of Conservation: 1154.
Garland, Blue: 3456.
Garner, Marilyn: 667.
Garrison, Bill: 3026.
Garton, Emmel Ray, 1950–: 2398, 2473, 2476.
Garton, Mary Ellen: 1164, 2473, 2476.
Gazin, Charles Lewis, 1904–: 1191.
Gemmell, Arthur, 1915–: 237, 3209, 3210.
Genter, Karen: 2567.
Geological Society of China: 1616.
Geological Society of Iowa: 446.
Geological Survey of Alabama: 1617.
Geological Survey of Georgia: 637.
George, Angelo Isham, 1944–: 1796, 1797, 1798, 1799, 1801, 1802, 1804.
Georgetown Corporation: 1709.
Ghica, C.: 2283.
Gibbs, Philip: 217.
Gilbert, B. Miles: 2405.
Gilbert, Bill: 1197.
Gilboy, A. E.: 2987.
Gilham, Alan: 908.
Gilkey, John L.: 1800, 1803.
Gill, David W.: 438, 845.
Gillieson, Dave S.: 264.
Gilmore, Elizabeth: 353.
Gilzbert, Stephanie: 2339.
Ginter, Boleslaw: 1856, 1858.
Giovannoli, Leonard: 280.
Glaccum, Robert A.: 388.
Glaister, Jane M: 2012.
Glasgow Science Lectures Association: 2646.

Glazek, Jerzy: 441.
Gleniss Wellings: 975.
Glover, R. R.: 1632.
Gniel, Penny: 1433.
Gochee, Angel V.: 1497.
Goede, Albert: 262, 263, 2304.
Golden & District Historical Society: 2052.
Goldstein, Arthur G.: 348.
Golias, Jamko: 1306.
Golob, Franc ;et al. (photo): 1828.
Gonsalves, William C.: 3333.
Goodnight, Clarence J.: 329.
Goodnight, Marie L: 329.
Gospodariç, Rado: 1645.
Goth, Kathy: 1264.
Gould, Gary W.: 1067.
Gradidge, Havelock: 594, 1080.
Grady, Frederick: 1165, 2807.
Grady, Mark A.: 2016, 2120.
Graham, C.: 782.
Graham, R.: 616.
Grant, Campbell, 1909–: 3335.
Gratté, Dummy Lucien: 2021.
Gravatt, T.: 213.
Gray, Jack: 3339.
Gray, M. R.: 816.
Gray, Richard E.: 1149.
Greater Allentown Grotto of the National Speleological Society: 233.
Green, John W.: 339.
Greenaway, Frank: 1289.
Greenaway, Mark Alwyn, 1953–: 1693.
Greenbank, Anthony: 2828.
Gregory, Adrian: 438.
Gregson, Richard, 1956–: 2854.
Grey, H. V.: 2724, 2725, 3558.
Gries, John Paul, 1911–: 3250.
Griffin, H.: 424.
Griffiths, J.: 499, 510.
Griffiths, Ken: 1167.
Grimes, K. G.: 816.
Grimes, Ken: 271.
Grinsell, L. V.: 3106.
Groba, E.: 605.
Groves, Christopher G.: 752.
Grupo de Espeleología "junonia" Federación Territorial Canaria de Espeleología Universidad de La Laguna: 1660.
Guangzhong Cui: 803.
Guernsey, Samuel James, 1868–1936: 2536.
Guiging Zhou: 133.
Guihong Cai: 803.
Guilday, John E.: 2614.
Gumm, Mike [prod asst]: 1600.
Günay, G.: 1650.

Gunn, John: 1647.
Gurnee, Jeanne Marie, 1926–: 1295, 1298.
Gurnee, Russell Hamtpon, 1922-1995: 3004.
Guterman, Norbert,: 1934.
Guterman, Norbert, 1900–: 2761.

Haarr, Allan P.: 2454, 2455, 2456, 2457, 2458, 2459.
Haarr, Doris: 1427.
Haas, G.: 3115, 3530.
Habič, Peter: 1140, 1229, 1645, 1828.
Hacker, Christopher: 2406.
Hagen, Adrianne: 356.
Haigh, B. R.: 804.
Halaburda, Sue: 1313.
Hall, F.G.: 2630.
Hall, Michael J.: 1705.
Halliday, William R.: 1656.
Halliday, William Ross, 1926–: 291, 1535, 1644, 1659, 1661, 1838, 2394, 2537, 3379, 3381, 3478.
Halstead, Whitney: 1721.
Hamilton, Harold W.: 1279, 1280, 1282, 1283.
Hamilton-Smith, Elery: 251, 255, 814, 816, 817, 1266, 1531.
Hangay, George: 885.
Hannah, E. D.: 2905.
Hansen, A.: 1898, 1899.
Hansen, Kimberly S.: 2416.
Hanson, Mike: 2834.
Hanwell, J. D.: 902.
Harding, Mike: 1023.
Hardy, Ern: 3496.
Hardy, Fritzi: 2407.
Hargrave, Lyndon L.: 1954.
Hargrove, Gene: 2227.
Harrington, Mark Raymond, 1882–1971: 341.
Harrington, Mark Raymond, 1882–1971: 1968.
Harris, B. J.: 1956.
Harris, "Digger": 304.
Harris, S.: 257, 1364.
Harris, Seymour Edwin, 1897–: 468.
Harrison D.: 2570.
Harrison, R. J.: 2570.
Harter, Russell G.: 1372.
Hartline, Daniel: 833.
Hartnup, Dave: 1023.
Hašek, Zdeněk: 660.
Hassall, John: 220, 299.
Hatt, E. M.: 647.
Hauakukaua, J. P.: 709.
Haun, Alan E.: 1392.
Haure, Pete: 833.
Hauser: 2431.
Havel, Hugo: 564.
Haverand Productions: 1080.

Hawaii Division of Water Resource Management: 2121.
Hawaii Island Office of Housing and Community Development: 1392.
Hawkes, C.F.C.: 3514.
Hawkins, L. J.: 1390.
Hawkins, Lauraine K.: 2521.
Hawkins, Lynsey: 975.
Hawkins, Robert: 975.
Hawksley, Oscar: 2223, 2226.
Hayden, Brian: 1925.
Hayers-Griffin, M. E.: 1121.
Hayes, Andrew: 1069.
Hazelton, Mary: 1409.
Hazslinszky, Tamás: 288, 1007, 1135, 1652, 1854, 1932, 2821, 2877, 2878.
Hedges, James: 2396.
Hegen, E. E.: 1014.
Heizer, Robert F.: 3331, 3332.
Heizer, Robert Fleming, 1915–: 978, 2817.
Held, Fred: 3431.
Heller, Sara: 1626.
Helvey, Elek: 1813.
Hempel, John Charles, 1949–: 313, 1075, 1566, 2624.
Henderson, Kent: 1531.
Hendy, P. G.: 1793.
Henkel, David S.: 984.
Herman, Janet S.: 270.
Herman, Matthew N.: 980, 2804.
Heron, Michael: 392.
Herzig, Francis X. U.S. Army Engineer Waterways Experiment Station. Southwest Research Institute: 1108.
Herzlík, Bořivoj: 854.
Herzog, Maurice: 2813.
Hesketh, J.: 1116.
Hewlett, Brian: 1706.
Hibben, Frank C.: 542.
High, Colin: 245.
Hillaire, Christian: 642, 643, 643.
Hillebrand, Timothy Shaw: 2095.
Hinds, Henry: 391.
Hires, H. Lamar: 2727.
Ho, Thian Hua: 3045.
Hoadley, Ann D.: 348.
Hobbs, Horton Holcombe, 1914–: 329, 1469.
Hobbs, Horton Holcombe, 1944–: 1467.
Hocknell, P.: 1939.
Hodge, Alan: 3193.
Hoenstine, Ronald: 1894.
Hoffelt, J. F.: 2905.
Holland, Ernst: 1347.
Holler, Oliver: 1488.
Holler, Susan G.: 1488.
Hollings, P.: 1116.
Holmer, Richard N.: 1718.

Holsinger, John Robert, 1934–: 550, 778, 2385.
Holzmann, Heinz: 1854.
Hong Kong. Geotechnical Control Office: 395.
Hood, Clark H.: 2223.
Hoover, R.L.: 3331.
Hopkins, John: 2333.
Hopkins, Mark S.: 350.
Hopper, Pat: 2321.
Horáček, Ivan: 441.
Hori, Nancy: 1117.
Horváth, Tibor: 1135.
Hose, Louise: 2417.
House, L.B.: 1370.
House, Scott: 2322.
Houshold, Ian: 3082.
Howard, Edgar B., 1887–: 2913.
Howard, Hildegarde, 1901–: 3019.
Howard, S. Russell: 2450.
Howarth, Francis Gard, 1940–: 1138, 1139.
Hoyer, Bernard E.: 1317.
Hruška, Jiří: 1623.
Hubbard, David A.: 1447, 2424.
Huckins, James N.: 3424.
Huddart, David: 899, 900.
Hughes, G. H.: 71.
Hume, N.: 941.
Hungarian Geological Society: 1652.
Hungarian Meteorological Society: 1652.
Huntingford, George Wynn Brereton: 1213.
Huntington, Will: 999.
Huntoon, Peter W.: 2745.
Huppert, George Nixon, 1944–: 2344, 2857.
Hurson, Tim: 594, 1080.
Hurst, B. P.: 763.
Husted, Wilfred H.: 2114.
Hyde, J. H.: 1250.
Hyde, L. W.: 2507.
Hyder, William D.: 1923.
Hyman, Libbie H.: 329.

IAH Commission for Hydrogeology of Karst: 562.
Iles, Pete: 1746.
Iliffe, Thomas M.: 1848, 1850, 1852.
Institut National de L'Audiovisuel: 249.
Institut za Raziskovanje Krasa: 1663.
Institute of Geology and Mineral Exploration: 2269.
International Association of Hydrogeologists: 1886, 2633.
International Association of Hydrogeologists, Karst Commission: 1890.
International Geographical Union Study Group: 2890.
International Speleological Union Commission for Physics-Chemistry and Hydrology of Karst: 2890.
International Union of Speleology: 1624, 1626, 1627, 1628, 1629, 1630, 2664.
International Working Group on Tracer Methods in Hydrology: 2269.
Irby, Bobby N.: 1831.
Irving, Robert Lock Graham, 1877–: 606, 610.
Irwin, David J.: 385, 386, 1548.
Ivan, Michler: 2936.

Jackson, J. Wilfred: 294.
Jacobi, R. M.: 3108.
Jacques, Richard: 2822.
Jager, H. S.: 3369.
Jagnow, Becky: 453.
Jagnow, David Henry: 2422.
Jagnow, Rebecca Rohwe: 1692.
Jakucs, László: 1639.
Jakucs, L'aszl'o: 1638.
James, Julia M.: 816, 885, 931, 975, 1729.
Jameson, Roy: 3138.
Jameson, Roy A.: 1773.
Janet Healy Video Production: 1195.
Japan Volcanspeleological Society: 1657.
Jarman, S. A.: 1708.
Jasek, James: 2400.
Jasek, Mimi: 2400.
Jasinski, Marc: 3119.
Jeannin, Pierre-Yves: 405.
Jeffreys, Alan Lawrence, 1940–: 1238, 1917, 2874.
Jenkinson, R. D. S.: 1198.
Jennings, Jesse D.: 20.
Jennings, Joseph Newell, 1916-1984: 1693, 2511, 3500.
Jenolan Caves Reserve Trust: 1348.
Jewel Cave National Monument: 3315.
Jewett, A.: 237.
Jiemin Xie: 133.
Jintao Weng: 803.
Joe Mosby Creative Printing: 1067.
Johnson, A. Ivan: 1287.
Johnson, Frederick: 707.
Johnson, Kenneth S.: 453.
Johnson, Paul A.: 3419, 3421, 3422.
Johnson, Victoria S.: 2411.
Johnsson, Mark J: 1286.
Johnston, Marshall: 1792.
Joiner, Thomas J.: 1888.
Joint Earth Science Societies Working Group on Geopreservation: 3529.
Joint, Wilfrid: 1515.
"JOK" [pseud for Jock Orr]: 45, 46.
Jones, Ceris: 722.
Jones, Cheryl: 838.
Jones, G. P.: 894.
Jones, Gareth Llwyd: 1631.

Jones, H.: 645.
Jones, John Clive, 1934–: 30.
Jones, Keith: 2557, 2558.
Jones, Steve: 438.
Jorden, Jay R.: 2345.
JTV Productions: 1730.
Judd, Frank: 794.
Judson, David: 672.
Juhasz, Victor: 1514.
Jurgens, Robert: 2921.
JWH Over Museum, South Dakota State University: 2143.

Kácha, Stenislav: 660.
Kahrs, Ernst: 1868.
Kambesis, Pat: 2478.
Kandó, Michael: 8.
Kane, Thomas C.: 777.
Karanjac, J.: 1768.
Karnei, Henry: 1960.
Karpát, József: 9.
Karpodini-Dimitriad, E.: 2661.
Kaschko, Michael W.: 242, 832.
Kase, Tomoki: 1395.
Kashima, Naruhiko: 1657, 2882.
Kastning, Ernst H., 1944–: 1628, 2420, 2424, 2430.
Kastning, Karen M.: 1769, 2424, 2430.
Kastning, Kass: 2424.
Kawakatsu, Masaharu: 2240.
Kell, H. S.: 1956.
Kelley, James C.: 3049.
Kelly, Herb: 1830.
Kennedy, George C.: 1057.
Kennedy, Jim "Crash": 646.
Kennedy, W. Q.: 1257.
Kenney, C. Howard: 544.
Kennicott, H.C.: 3460.
Kentucky Heritage Council: 2514.
Kenyon, Norman: 2113.
Keprt, Jiří: 2714.
Kerr, Chris: 3066.
Kessel, Jay: 1806.
Kidd, H.: 3405.
Kidder, Alfred Vincent, 1885-1968: 1277, 2536.
Kiernan, Kevin: 1531.
King, Philip Burke, 1903–: 804.
King, R. S.: 386.
Kiss, Attila: 9.
Kleespie, Tom: 682.
Klein, Richard G.: 701.
Klein, Samuel: 2635.
Klemenčič, Slavo: 2936.
Kline, Thomas: 2414.
Knibbs, Anthony J.: 1678.

Knoph, A: 3046.
Knox, Jan.: 2420.
Knox, Orion: 2773.
Knudson, George Ellert, 1915–: 1418.
Knutilla, R. L.: 2987.
Knutson, Richard Stephen: 1825, 2371, 2372, 2373, 2374, 2375, 2376, 2377, 2378, 2379, 2380, 2381, 2382.
Kobori, Larry S.: 319, 320.
Koenig, John W.: 3363.
Koep, Ann E.: 312.
Koerschner III, William F.: 1773.
Koglin, Eric N.: 2740.
Koliasnikoff, A.: 1935.
Kolstad, Rob: 2417.
Kopper John S.: 1130.
Košáková, Jana: 1643.
Kosa, Attila: 2877.
Koul, Lokesh: 1855.
Kowalski, Gary: 453.
Kozlowski, Stefan Karol: 1857.
Krafthefer, Brian: 1949.
Král, Ing: 1623.
Kramar, Matt: 1183.
Kramer, Jim: 2464.
Kranjc, Andrej: 1140.
Kranjc, Maja: 1140.
Krejca, Jean K.: 3425.
Krekeler, Carl H., 1920-: 325.
Krieger, Alex Dony, 1911–: 1421.
Krieger, Robert Albert, 1918–: 787.
Kugler, Ralph: 2311.
Kuhlen, Barbara: 451.
Kuhn, Herbert, 1895–: 2541.
Kunath, Carl E.: 1056.
Kunaver, Jurij: 1641, 1649.
Kurz, Peter: 850.
Kusch, Heinrich: 1170.

La Moreaux, Philip Elmer, 1920–: 3153, 3154.
Lake, P. M. B.: 1184.
Lamar, D. L.: 2886.
Lamberg-Karlovsky, C. C.: 707.
Laming, Annette: 3514.
LaMoreaux, Philip Elmer, 1920–: 2265.
Lancaster University Speleological Society. Universidad Politecnica de Madrid. Seccion Espeleologia Ingenieros Industriales: 645.
Lance, Don: 3072.
Lance, Hubert Darrell, 1935–: 2927.
Lance, K. A.: 814.
Landis, Carolyn: 713.
Landis, Charles: 713.
Landsberg, J: 1208.
Lane, C.: 2144.

Lane, Steve: 2579.
Lang, Arthur: 859.
Lange, A. R.: 616.
Langebartel, Dave A.: 1386.
Langmuir, Donald: 2612.
Larimore, Betty: 986, 987.
Larson, Charles Victor, 1928–: 1320, 2394, 2403.
Larson, Jo: 1904, 1905, 1907, 1908.
Larson, Peggy, 1931–: 1909.
Latrobe, L.: 1910, 1911.
Laville, Henri: 1858.
Lawrence, Judi: 1566.
Lawson, Dick: 956.
Lee, Alexander Y.: 57, 58.
Lee, John Edward, 1808-1887: 2159, 2840.
Legge, Charmaine J.: 2841, 2842, 2843, 2844, 2845.
LeGrand, Harry Elwood, 1917–: 1891, 3151, 3152, 3153, 3154.
Leitch, Duncan: 1257.
Leitheuser, Arthur T.: 1496, 1497, 1498, 1499, 1500, 1501.
Lél-Óssy, Sándor: 9.
Lél-Össy, Szabolcs: 8.
Lénárt, László: 9.
Lenarčič, Simon: 1307.
Lenhart, Stephen W.: 1772, 1773.
Lenochová, Alena: 1623.
Leonard, Mark D.: 2727.
Leonard, Patricia: 2521.
Leroi-Gourham, A.: 3514.
Lescalleet, David G: 1012.
Letu, Bernard: 2486.
Levette, Gilbert M.: 730.
Lewis, I. D.: 813, 814.
Lewis, J. R.: 2616.
Lewis, Jerry: 2212.
Lewis, Julian J.: 2340.
Leyden, John, 1775-1811: 2014.
Liebermann, T. D.: 2297.
Lindamood, Fishman: 1601, 1602.
Lindborg, E.: 1527.
Lindsay, Charles: 973.
Lindsley, Karen Bradley, 1948–: 474, 725.
Lingham, G. E.: 1048.
Lipp, Astrid: 3442.
Lipscomb, Robert G.: 3365.
Littke, John: 1941.
Livesey, M. P.: 1078.
Lóczy, Dénes: 289.
Loeser, Rudolf: 1868.
Long, Dana A.: 2728.
Long, Joe N.: 1957.
Long, Kenneth M.: 2469.

Long, M. H.: 499, 500, 501, 505, 506, 507, 508, 509, 510, 511.
Long, Pam: 3431.
Longley, Glenn, 1942–: 1508.
Longshaw, Rose,: 231.
Lonsdale, Kate: 2582.
Lord, H: 3405.
Lorrain, Dessamae: 865.
Loud, Llewellyn Lemont, 1879-1946: 341.
Love, C. W.: 1048.
Love, Douglas P.: 1971.
Love, S. K.: 1048.
Loveday, Barry: 817.
Loveday, Frank: 817.
Lowe, David J.: 1647.
Lowe, Gavin: 2038.
Lowe, Walter Bezant: 2635.
Loze, Keith: 1711.
Ložek, Vojen: 1643.
Lubell, Winifred: 3120.
Lund, Cecilie: 1002.
Luther, Elizabeth: 2642.
Luttrell, P. E.: 3439.
Lynch, Frank, 1921–: 1050.
Lynch, William Henry: 1981.
Lyon, Ben: 2156.

Maack, Reinhard: 2541.
Mackey, Phil: 2178.
MacNish, Robert D.: 3283.
Maestro, Giulio: 1142.
Malcolm, David R.: 329.
Male, A. J.: 2245.
Malnic, Jutta: 1220.
Mansfield, Kay: 480.
Mansfield, Raymond Walter: 480, 1170, 2039.
Mansker, William L.: 394.
Marchant, L. F.: 2828.
Marek-Limagne, Edit: 1854.
Markoe, Glenn, 1951–: 320.
Marshall, Bette Yvonne: 691.
Martin, David John 1957–: 1729.
Martin, Paul S.: 1284.
Martin, Ronald L.: 2201, 2202, 2211.
Mason, E.J.: 672.
Mason, Edmund J: 1912.
Mass, Celeste: 2081.
Massello, James W.: 37.
Massingill, Renda C.: 285.
Matthews, John: 2822.
Maucha, László: 309.
Maupintour Modern Video Programs: 3328.
Maxwell, Bill: 1710.
Mayer, Stanislaw: 564.

Mc. R. Pearce, A. E.: 935.
McAtee, Waldo Lee, 1883-1962: 315.
McBridge, Robert: 3319.
McCabe, John A.: 787.
McCallum, Mary: 2662.
McCallum, Tom: 595.
McCarley, Ann: 1886.
McCarty, L. R.: 1798.
McClurg, David Robert, 1929–: 2108, 2388.
McClurg, Janet: 2108, 2109.
McCrady, Allen D.: 1283, 1284, 2452, 2453.
McCulloch, Stuart J.: 641.
McCurdy, Richard: 1064.
McCutchen, Gary D.: 2461.
McElroy, Jim: 3184.
McGill, William Mahone, 1896–(Luray Caverns in Virginia. Luray Caverns in Virginia): 3148.
McGinnis, Helen J: 1281.
McGrain, Preston: 781, 2025.
McGrain, Preston, 1917–: 1950.
McGregor, Stuart W.: 2810.
McGrew, Wesley C.: 2454, 2457, 2458.
McGuffin, Della: 2909.
McKann, Belva: 3026.
McKenna, Peter J.: 3220.
McKenzie, Andrew: 1391.
McKinnon, Wayne: 2724, 3558.
McMillan, Doug, 1935–: 3109.
McMillan, Nora Fisher: 2251.
McWilliams, Virginia: 2334.
Me too: 3362.
Meade, Grayson E: 1011.
Medley, Joy: 2002.
Medville, Douglas Michael, 1941–: 1427, 2404.
Medville, Hazel E.: 2138, 2139, 2140, 2434.
Meighan, Clement W.: 3333, 3334.
Meldrum, David: 1434, 1435.
Mellor, D. C.: 284, 840, 1586.
Melmore, Sidney: 2719.
Meloy, Harold: 552.
Members of the Sydney Speleological Society: 1314.
Memon, Beshir A.: 1887.
Meneghel, M.: 430, 2890.
Mentzer, Annette: 1117.
Mercer, J. A.: 3043.
Mercer, W. J.: 1073.
Merkle, Virginia: 356.
Messling, Brenda: 2162.
Metzger, Donald J.: 2728.
Meyers, Karlin: 405.
Michler, Ivan: 2937.
Middleton, Gregory J.: 252, 1701.
Middleton, James: 975.
Midwest Archeological Center (U.S.): 2000.

Midwest Archeological Center (U.S.) University of Tennessee: 1827.
Millar, Ian: 2765.
Miller, Don E.: 1378.
Miller, R. L.: 2987.
Miner, Steve: 1982.
Minnesota Department of Natural Resources: 1703.
Mitchell, Ida: 2238.
Mitchell, Robert Wetsel, 1933–: 2785, 2796.
Mixon, William Weston, 1940–: 2467, 2468, 2471.
Miyake, Yoshi: 1254.
MMI Corporation [dis]: 3526.
Mohler, John Leonard, 1840-1937: 3453.
Molosky, Ann: 676.
Moni, Gerald: 3067.
Monk, Susan M.: 2000.
Montagu, Adelinae: 421.
Montana Bureau of Mines and Geology: 590.
Montgomery, Neil Robert, 1953–: 1699, 1701.
Moody, Jill: 2466.
Moore, James D.: 636, 1891.
Moore, John E.: 2633.
Moorehead, Warren King, 1866-1939: 2622, 2623.
Moravec, George F.: 2265.
Morfis, A.: 1654.
Morgan, Angela: 3070.
Morgan, Eric L.: 442.
Morgan, I. M: 830.
Morris, Bernard: 1259.
Morris, T. L. F.: 1078.
Morthland, Dave: 1596.
Moser, Paul H.: 2810.
Moss, John H.: 2114.
Moss, Phil: 1217.
Moss, Tom: 3070.
Mott, K. R.: 814.
Mott, Kevin: 271, 1347, 2674.
Movius, Hallam Leonard, 1907–: 1546.
Mowat, George: 859.
Moyon, Angela: 247, 910.
Mroczkowski, Donna: 2257.
Muir, Jovial Jane: 856.
Mulloy, W. T.: 3041.
Mulloy, William: 2114.
Munson, Cheryl Ann: 2300.
Museum of New Mexico: 1047.
Museum of Northern Arizona: 1359.
Musil, František: 564.
Muzeon Yisrael: 317.
Myers, Jack O.: 1173, 3209.
Mylroie, John Eglinton, 1949–: 769, 2323, 2421.

Napton, Lewis K.: 341, 1420.
Nardacci, Michael: 2412.

Nash, David T.: 1221.
Nass, John P.: 658.
National Association for Cave Diving, Standardization Committee: 1062.
National Caving Association: 2144.
National Geographic Magazine: 1926.
National Geographic Society (U.S.): 877.
National Geographic Society (U.S.). Committee for Research and Exploration: 876.
National Groundwater Association [U. S.]: 358.
National Hydrology Research Institute: 3347.
National Monuments Branch: 900.
National Museum of Wales: 1252.
National Parks & Wildlife Service: 1349.
National Science Foundation of China: 1651.
National Speleological Society: 890, 1322, 1456, 1915, 2342, 2525, 2699, 3135.
National Speleological Society Caver Training Committee: 1380, 1381.
National Speleological Society Decatur, Alabama Grotto: 1477.
National Speleological Society. Pittsburgh Grotto: 799.
National Speleological Society. Special Publications Committee: 2767.
Neace, Bob: 422.
Neff, David: 2316.
Nelson, Greta K.: 3215.
Nevada State Museum. Dept. of Archeology: 2577, 2578.
New Melones Task Force: 3099.
New South Wales. Dept. of Mines and Agriculture: 1107.
New South Wales. Geological Survey: 3233.
Newman, F. Barry: 1835, 1836.
Newman, Gaj: 213.
Newman, Gavin: 438.
Newman, George R.: 1375.
Newson, Malcolm David, 1945–: 902, 1356.
Nguyan Quang My: 2508.
Nicholson, Frank Ernest: 3460.
Nicod, Jean: 1619.
Nieland, James Raymond, 1949–: 2394, 2403.
Nimo, Peggy: 1106.
Nohlen, Christine: 1707.
Nomura, Sizumu: 1961.
Nordquist, Gerda: 1183.
Nordstrom, Darrell Kirk: 305.
Northeast Regional Organization, National Speleological Society: 2682.
Northeastern Regional Organization, National Speleological Society: 2601.
Norton, Russell M.: 1561.
Nottinghamshire County Council: 512.
Nova Scotia Dept of Education, Nova Scotia Museum: 2277.
Nurse, Benjamin: 975.

Nuttall, Kev: 2966.

Obele, Robert K.: 2345.
O'Brien, John: 2316.
Odehnal, Ludvík: 1643.
O'Dell, Gary A.: 1178, 1805, 2768.
O'Donnell, Lisa: 961.
Oesch, Ronald D.: 2613, 2614.
Office of Natural Sciences, Southwest Region, National Park Service: 2133.
Office of Public Works: 900.
Ogawa, Takanori: 1657, 2882.
Ohio Dept of Natural Resources: 658.
Ohio Valley Region: 2406.
Okladnikov, Aleksei Pavlovich, 1908–: 2292.
Old Dominion University Department of Biological Sciences: 1496, 1497, 1498, 1501.
Oldham, Anthony Clive, 1939–: 1854.
Oldham, Joyce Elizabeth Anne: 2563, 2955.
Olsen, Laurie: 2440.
Olson, Rick: 1496.
Oosthuizen, Hans: 553.
O'Reilly, P. M.: 1631.
O'Reilly, S. E.: 2574.
Orpheus Caving Club: 513.
Orr, N.: 2063.
Ortiz, R. Keith: 1773.
Osborne, Carolyn M.: 2083.
Osiris Productions: 3000.
Other members of the expedition: 2605.
Others: 1943, 1944.
Otsuka, Seigo: 3479.
Ozark Underground Laboratory: 41.

P & L Systems, Inc: 658.
P. E. LaMoreaux & Associates: 358.
Pace, Norman R.: 1496.
Page, R.: 1208.
Pahor, Božidar: 2937.
Palmer, Arthur Nicholas, 1940–: 769, 2193, 2412, 2421, 2742.
Palmer, Margaret V.: 2412, 2597, 2599, 2887.
Palmer, R. J.: 301.
Paloc, Henri: 270.
Panlaqui, Carol: 2095.
Pannela, G.: 3223.
Panoš, Vladimír: 564, 1624.
Papadakis, Jim: 2611.
Papadapoulis, G.R.: 1424.
Papakis, Nicholas: 558, 559.
Papua New Guinea Speleological Expedition, 1973: 1700.
Paraskevopoulou, P.: 1654, 2269.
Paris, Sandy: 3069.
Park, Orlando: 329.

Park, Ray: 3069.
Parker, B: 3405.
Parker, Chris: 1167.
Parker, Fred: 1701.
Parker, Marion L.: 6.
Parmalee, Paul Woodburn: 1279, 1280, 1826, 1827.
Parsons, James J.: 40.
Parsons, Mark L.: 2856.
Passmore, C. G.: 2691.
Pate, Dale L.: 2347.
Patrick, D. M.: 1121.
Pavey, Andrew J.: 1699.
Payzant, Charles: 2945.
Peacock, Var Lynn: 1316.
Pearce, J. A.: 935.
Pearman, Harry: 3119.
Pearson, Felicity M.: 358.
Peerman, Kathy: 2437.
Peerman, Steve: 2437.
Pefaur, Jaime E.: 1570.
Pelletier, Armand Michel: 1471.
Pennell, Joseph, 1857-1926: 57, 58.
Penniket, Andrew: 2991.
Pennsylvania University Department of Archaeology: 2153.
Penny, Jim: 2966.
Peprník, Jaroslav: 1623.
Perez, Juan A.: 3429.
Perkins, James: 2335.
Pernette, Jean-François: 2021.
Perou, Syd: 594, 1080.
Perr, Dave: 2966.
Petraglia, Michael D.: 1221.
Petrochilou, Anna: 1424.
Pfeffer, Karl-Heinz: 1618, 1619, 3166, 3168.
Pflieger, William L.: 3365.
Phillips, Stanley: 2057.
Pierce, Miles: 817.
Pilkington, Graham: 2284.
Pitty, Alistair F.: 1632.
Plante, Alan: 444.
Plaskett, David C.: 876, 877.
Plédel, D. Bruno Martinez: 1114.
Plummer, William T.: 2456, 2460.
Pogue, Kevin R.: 3215.
Pololy, Judit: 289.
Pommerantz, Inna: 316.
Pond, Alonzo W.: 718.
Poole, Gray Johnson: 2692.
Pope, Tim: 3071.
Postojnska jama Tourist and Hotel Organization: 1663.
Potash, Laura L.: 17.
Poulson, Thomas Layman, 1934–: 2250.
Powell, Bob: 2705.
Powell, "Mossy": 304.
Powell, Richard L.: 2387, 2425.
Power, Louise: 2462.
Prentice, Active Andy: 856.
Price, M.: 894.
Price, Wayne: 1800.
Prince, E. R.: 1422.
Programmes Branch, National Parks and Wildlife Service: 260, 1349.
Prohic, Esad: 1889.
Prosser, J. Joseph: 3558.
Provatakis, Theocharis M.: 2609.
Pruitt, James: 2143.
Pugsley, Chris: 2499.
Putnam, Frederic Ward, 1839-1915: 2160, 2591.
Putnam, William O.: 2410.
Pygmy Dwarf Press [Mike Sims]: 2435.

Quamen, Al: 2321.
Queensland Conservation Council: 1350.
Queensland National Parks and Wildlife Service: 649.
Quick, Peter: 2401.
Quinlan, James Francis, 1936-1995: 27, 1082, 1084, 2905.

Rácz, Beatrix: 1412.
Rádai, Ödön: 9, 1412, 2877.
Ráday, Ödön: 289.
Radcliffe, F. G.: 2746.
Radlauer, Ed: 2747.
Radovanovic, Ivana: 1857.
Railton, C. L.: 568.
Raines, Terry W.: 2871.
Ralston, B.: 2501.
Ramírez, Eleonor Donínguez: 1146.
Randall, Bru: 1429.
Rarick, R. Dee: 2129.
Rasmussen, Pamela C.: 1570.
Rasquin, Priscilla: 465.
Rawlinson, N.: 2501.
Rayfield, C. Raymond: 3154.
Raymond, Dorothy: 1888.
Rea, G. Thomas: 2387.
Rea, George Thomas, 1934–: 1381, 2348, 2408, 2413, 2443.
Rea, George Thomas,1934–: 1659.
Reardon, Bob: 2388.
Rebmann, James R.: 874.
Reddell, James Russell, 1938–: 2130, 2241, 2242, 2386, 3010.
Redenbaugh, William Alfred, 1871–: 2181.
Reed, Christy L.: 1788.
Regional Councils of Caving Clubs in Great Britain: 2829.

Regional Resources Centre: 1711.
Reid, Kenneth C.: 897.
Reiter, Paul: 29.
Research Section. New Mexico. State Highway Dept: 1033.
Reynolds, T. E.: 2040.
Rhinehart, Richard: 2417, 2835.
Rhodes, Douglas Warren, 1944–: 2338, 2339, 2446.
Rice, John A.: 519.
Richardson, Alastair Mackenzie Martyn: 940.
Rickind, David H.: 31.
Rigalde, Christian: 2021.
Rigaud, Jean Philippe: 1913.
Rigby, J.: 1752.
Riley, Patty: 453.
Rinaldo, John Beach, 1912–: 2064.
Ripoll Perello, Eduardo: 2653.
Ritchie, James, 1882-1958: 1257.
Robert, Romain: 2530.
Roberts, Richard: 3026.
Robinson, A. C.: 253.
Robinson, Lloyd, N.: 1345.
Robson, Eric: 661.
Roche, Bernadette N.: 1889.
Roda, István: 1135.
Roether, Joyce: 1117.
Rogers, Bruce W.: 2397, 3404.
Rohrer, Thomas, A.: 878.
Rojas-Gonzalez, Luis F.: 1835, 1836.
Rolls, Diane: 1701.
Romain, Robert: 2529.
Romejn, E.: 605.
Ronen, Avraham: 1163.
Rosendahl, Paul Harmer: 1392.
Rosenfeld, Andrée: 3326.
Ross, Richard E.: 2856.
Roth, Edward: 794.
Rother, Charlotte: 2859.
Rothrock, H. E.: 3289.
Rothrock, H. W.: 730.
Rouiller, Phillipe: 405.
Roussel, Monique: 1884.
Roust, N.L.: 3331.
Rowden, Bob: 1941.
Rowe, Donald R.: 2743.
Roy, Dalia: 1186.
Royal Geographical Society - Sarawak Government Expedition 1977-78: 503.
Rubín, Josef: 2998.
Ruhir, Ed: 216.
Ruoyon Shang: 3555.
Ruspoli, Mario: 249.
Russell, William Hart, 1937–: 2239, 2772, 2779, 2782, 2787.
Rutter, J. G.: 48, 49.
Ryan, Joseph Edward: 65, 66, 67.
Ryder, P. F.: 507, 510, 511.
Ryšav, Přemysl: 564.

Sackett, James: 1913.
Šaja, S.: 1309.
Šajn, Srečko: 1828.
Sakaguchi, Yutaka, 1929–: 1354.
Salamon, Gábor: 1135.
Sameshima, Teruhiko: 1657.
Sampson, Clavil Garth, 1941–: 589.
Sanderson, Ivan T.: 2677.
Sanford, W. Ashford: 839.
Sares, S. W.: 2422.
Sarić, Avdo: 2175.
Sasowsky, Ira Daniel, 1959–: 2358, 2449.
Saunders, Joseph W.: 1773.
Sauro, U.: 430.
Savage, Howard G.: 2640.
Save the Wemysss Ancient Caves Society: 2760.
Saville, Mervyn: 1120.
Savory, J. H.: 299.
Savory, John, 1943–: 2893.
Saxton, Claire: 2930.
Sayed, Sayed: 361.
Scarbrough, L. W.: 2506.
Scarbrough, W. Leon: 1742.
Schaake, John C.: 1922.
Schafheutle, Markus: 1707.
Schecter, Irving: 3328.
Schellie Associates: 1606.
Schick, Tamar: 318.
Schilberg, Gary: 2409, 2439.
Schleicher, Donald P.: 2392.
Schmid, E.: 3530.
Schmid, June C.: 423.
Schneider, Rex: 462.
Schoenherr, John C.: 2262, 2263.
Schomer, Barb: 1429.
Schram, Frederick R.: 3251.
Schrotter, Gustar: 1963.
Schultz, Robert L.: 2391.
Schumacher, Paul J. F.: 2302, 2303.
Schwartz, Douglas Wright, 1929–: 3002.
Schwartz, John S.: 2200.
Scott, Curtis E.: 2317.
Scott, David: 2529.
Scott, Len: 1813.
Scott, Peter: 1701.
Senior, Kev: 438.
Sequoia Natural History Association: 861.
Severance, Craig J.: 1392.
Shafer, Harry J.: 2942.

Shaler, Nathaniel Southgate, 1841-1906: 50, 1795.
Shane, Orrin C.: 678.
Shannon, C. H. C.: 515.
Shaw, R.: 1898, 1899.
Shaw, Trevor R.: 1457.
Shenandoah Valley Railroad: 57, 58.
Shenandoah Valley Railroad County: 2076.
Shenfield, Margaret: 788.
Sherborne, William B.: 2462.
Shideler, David W.: 1353.
Shoosmith, R. W.: 814.
ShoreDavis, P.: 1890.
Shorewood Educational Programs: 2096.
Shorewood Fine Art Reproductions, Inc.: 2096.
Shouyue Zhang: 927, 2082.
Shrewsbury, William: 3069.
Shull, Alisa: 2545.
Shutler, Richard: 2974.
Sieveking, Gale de Giberne: 2977, 2978.
Simičková, Hedvika: 248.
Sims, Brailey: 1270.
Sims, Lynne: 1026, 1027, 1825, 2419.
Sims, Mark: 3496.
Sims, Mike: 1026, 1027, 1599, 1825.
Sinclair, William Campbell, 1928–: 362.
Sinise, Dorothy: 2992.
Siron, Mary Ann: 989.
Sizumu Nomura: 21.
Skelton, John: 1378.
Skinner, A. D.: 2996.
Skinner, R. K.: 2994.
Skinner, Shaune M.: 658.
Skipworth, Joe: 3070.
Slaughter, Bob H.: 1989.
Šlenc, S.: 1309.
Slifer, Dennis: 1123.
Slifor, Dennis W.: 1004.
Sloane, Bruce Charles, 1935–: 2401.
Sloane, Howard Norman, 1909-1972: 2249.
Smart, C. M.: 513, 926.
Smart, James: 2564, 2565, 2566.
Smart, P. L.: 244, 3536.
Smight, David Ingle: 1210.
Smith P.: 2663.
Smith, D. I.: 902.
Smith, A. Richard: 2462, 2788, 2793.
Smith, Bruce, 1948–: 2593.
Smith, David I.: 903.
Smith, Dick: 243.
Smith, Fred L.: 571.
Smith, Griffin: 1792.
Smith, James H.: 751.
Smith, Lee: 3026.
Smith, Marian O.: 586.
Smith, Philip M.: 3414.
Smith, Phillip M.: 3413.
Smith, Richard A: 1872.
Smithsonian Institution: 1757.
Smithsonian Institution. Bureau of American Ethnology: 1109.
Smoot, J. L.: 2297.
Smyre, John: 3064.
Snow, Mary M.: 2346.
Snow, Richard K.: 2346.
Snyder, Lynn M.: 1827.
Souse, Margaret: 833.
South Australia Department of Environment and Planning: 3057.
South Australian Underwater Speleological Society: 1519.
South Dakota School of Mines and Technology: 24.
South Dakota State Historical Society: 426.
South Wales Cave Rescue Organisation: 1948.
South Wales Caving Club Members: 2574.
South West Essex Technical College Caving Club: 1035.
Southwest Missouri State University: 3092.
Southwest Texas Archaeological Society: 811, 2061.
Sowers, John: 2476.
Spackman, C.J.: 1591.
Spalding, G.: 2425.
Spangle, Paul F.: 2383.
Spangler, Lawrence E.: 3215.
Sparks, J. B.: 1184.
Sparks, J. B. ("Time chart by kind permission of ..."): 2637.
Spate, Andrew P.: 816, 1266, 1347.
Speece, Jack Howard, 1947–: 772.
Speleological Research Group of Western Australia: 3091.
Spence, J.: 3384.
Spencer, Lee: 319.
Sports Council: 3194.
Sprague, S. S.: 1644.
Springhetti, David: 2409, 2439.
Springston, Bob: 2324.
St Pierre, Shirley: 1035.
St. Clair family: 2011.
St. Pierre, David: 1927.
St. Pierre, Shirley: 2876.
Stace, Peter: 1938.
Stahl, Ben F.: 498.
Stamper, Scott: 1595.
Standing, I. J.: 2040.
Stanford, Ruth A.: 2545.
Stanton, William: 333, 334.
Steadman, David W.: 1130.
Stecko, Doug: 1806.
Steel, R. H.: 2079.

Steele, William: 2425.
Štelcl, Otakar: 1643.
Stenner, R. D.: 386, 1676.
Stevens, D'Jeanne: 3363.
Stevens, Georgiana Lord Richardson: 3123.
Stevenson, R. A.: 301.
Stewart, J. W.: 2987.
Stewart, W. J.: 3445.
Stilwell, Stella: 2693.
Stipp, T. F.: 2858.
Stitt, Robert R.: 1575, 2477.
Stock, Mark: 2900.
Stone, F. D.: 2121.
Stone, Jamie: 1062.
Story, Dee Ann: 1792.
Stoyles, T. R.: 1078.
Stral, Lee Philip: 1690.
Strange, Cy: 594, 1080.
Strathdee, Stuart: 2582.
Stride, A.H.: 3209.
Stride, R. D.: 3209.
Strinati, Pierre: 2486.
Stringfield, Victor Timothy 1902–: 1446.
Stringfield, Victor Timothy, 1902–: 1891.
Strohm, W. E., Jr.: 1121.
Strother, David Hunter, 1816-1888: 1558.
Stroud, Raymond B.: 1315.
Stuckey, D.: 1548.
Šušteršič, France: 1140.
Suhler, Sidney A.: 909.
Sullivan, Gerardus Nicholas, 1927–[Brother Nicholas]: 329, 795, 1296, 1685, 2263.
Sullivan, Gerardus Nicholas,1927–: 2262.
Sullivan, Michele: 313.
Sullivan, Stephen B.: 3215.
Summers, Diane: 3344.
Šunjić, Jakov: 2175.
Sutcliffe, J. R: 508.
Sutcliffe, J. R.: 506.
Sutherland, Wayne: 1461.
Swain, Roy: 940.
Swan, Robert McNair Wilson, 1858-1904: 389.
Sweeting, Marjorie Mary, 1920–: 1618, 1619, 1727, 2617, 3392.
Sydney Speleological Society: 974.
Sydney Speleological Society. Palawan Expedition, stage I.: 534.
Sydoriak, C.: 3079.
Székely, Kinga: 289, 427.
Szenthe, István: 9, 1932.
Szilvássy, Zoltán: 1932.

Táda;. Ödön: 1135.
Tádal, Ödön: 288, 309, 1007, 2821, 2878.

Tadayori Nakagawa, 1873-1928: 3177.
Taft, John B.: 1152.
Takacs-Solner, Katalin: 9.
Takai, Fuyuji, 1911–: 3163.
Tan-hsien Yuan: 1651.
Tanner, J. Mark: 1889, 1890.
Tardy, János: 1135.
Tasmania Forestry Commission, National Parks and Wildlife Service, Tasmania: 1818.
Taylor, Audrey: 676.
Taylor, Charles J.: 350, 3215.
Taylor, Philip S.: 1738.
Taylor, Robert L.: 2199, 2210.
Taylor, Robert L. (ed, layout & des): 2418.
Taylor, Steven J.: 3425.
Technos Consultants in Applied Earth Sciences: 3260.
Technos [Firm]: 388.
Television Service Time-Life Films: 1919.
Tennessee Cave Survey: 2087.
Tennessee Technological University. Agency for Instructional Television: 734.
Texas Archaeological Society: 2912.
Texas Archeological Salvage Project: 998.
Texas Memorial Museum: 3357.
Thaon, Maurice: 496.
The Association of Engineering Geologists: 2745.
The Canadian caver magazine: 3205.
The Cave Research Group of Great Britain: 776.
The Editors of Time-Life Books: 1681.
The Florida Sinkhole Research Institute: 3513.
The Mountain State Grotto: 2472.
The SpeleoDigest Committee: 2474.
The T. T. Club: 3025.
The West Virginia Speleological Society: 2472.
The Whole Earth Club: 3159.
Thenhaus, Paul: 2213.
Thomas, Lowell: 1136.
Thompson, A. G.: 48.
Thompson, Eric S.: 2154.
Thomson, Kenneth C.: 38, 2199, 2201, 2202, 2210, 2211.
Thornton, Helen: 2342.
Thornton, Jer: 2342.
Thorson, Robert M.: 876, 877.
Thrailkill, John V.: 1296.
Thrun, Robert, 1940–: 2447.
Ti chih yen chiu so: 179.
TI-IN Network (Television Station: San Antonio, Tex.): 383.
Tiell, Brian: 1706.
Tierney, Lee: 1461.
Tilly, G. D.: 1676.
Timberlake, Cynthia: 2238.
Tinsley, John C.: 2423.
Toa Kumo Gakkai: 1847.

Tolar, D. J.: 3384.
Tolson, Jan: 3154.
Tolson, Janyth S.: 1615, 1617, 1891.
Torkar, Rado: 1308.
Torkar, Vivienne: 1308.
Torode, William Wallace, 1943–: 3068, 3070.
Tourism Commission of New South Wales. New South Wales. Dept of Mines: 587.
Townsend, Mary Ashley 1832-1901: 1558.
Tozer, Bill: 1599, 1600, 1601, 1602.
Travis, R. J. A.: 1086.
Treat, Ida: 2894.
Trembay, Dan: 3221.
Tripp, Tim: 2131.
Troester, Joseph W.: 2633.
Tropicon in Australia's bicentenary year: 265.
Truluck, T. E.: 1935.
Trumić, Aleksander: 2175.
Tucker, J. H.: 287.
Tucknott, Bill: 304.
Tunnell, Curtis: 1792.
Tuohy, Donald R.: 341.
Turnbull, William D.: 1983, 1984, 1985, 1986, 1987, 1988, 2908.
Turner, D. B.: 1669.
Turner, J.: 926.
Turner, Louise: 2661.
Turner, T. Hal: 554.
Turton, Basil: 3345.
Turton, Mary: 3345.
TVOntario: 1964.

Uden, E. Boye: 1243.
UMAC (1971: Eagle Cave, WI): 1444.
Union Internationale de Spéléologie, Commission de l'Erosion du Karst: 1878.
United States Army Corps of Engineers Geotechnical Laboratory: 308, 570, 571, 710, 783.
United States Army Corps of Engineers Nashville District: 16.
United States Army Corps of Engineers Sacramento District: 2120.
United States Army Corps of Engineers Walla Walla District: 3208.
United States Army Engineer Waterways Experiment Station: 388, 1881.
United States Bureau of Land Management: 1720.
United States Bureau of Mines: 1881.
United States Department of Energy: 269.
United States Far East Command: 2510.
United States Fish and Wildlife Service: 44.
United States Geological Survey: 804, 1475, 2633.
United States Geological Survey Military Geology Branch: 2510.
United States Interagency Archeological Services [Atlanta]: 16.
United States National Park Service: 33, 34, 35, 42, 43, 56, 327, 618, 998, 1929, 2703, 2758, 2847, 3112, 3147.
United States National Park Service Southeast Regional Office: 1498.
United States National Park Service Southwest Region: 600, 2485.
United States National Park Service, Lehman Caves National Monument: 2864.
United States Office of Water Policy: 3223.
United States Office of Water Resources Research: 2769.
United States Soil Conservation Service: 3094.
United States Work Projects Administration: 2143.
Universitat Hefah: 1163.
University of Bristol Spelaeological Society: 3226.
University of California Archaeological Research Facility: 1420.
University of Hong Kong Department of Geography and Geology: 1899.
University of Missouri Rolla Water Resources Research Center: 2769.
University of Nevada Center for Water Resources Research: 2132.
University of New Mexico Biology Department: 577.
University of Queensland Speleological Society: 2735, 2736.

Vaillant-Couturier, Paul: 2894.
Valley Land and Improvement Company [Luray, Va.]: 57, 58.
Van Camp, Mary L.: 2003.
Van Gundy, James J.: 3476.
Van Note, Peter: 2921, 2922.
Varga, Csaba: 9, 1932.
Várhegyi, Sándor: 1813.
Varnedoe, William Whitfield, 1923–: 1753, 1754, 3183.
Varnedoe, William Whitfield,1923–: 3185.
Veitch, John David, 1931–: 2389, 3183, 3185.
Veni, George, 1957–: 2415.
Vér, Zsolt: 9.
Veress, Márton: 3550.
Vermeulen, Susan E.: 2416.
Vertut, Jean: 274.
Vetz, Edwin B.: 2314.
Vetz, Michael A.: 2314.
Veve, Thalia: 1927.
Viator, Mark: 979.
Victoria Department of Conservation, Forests, and Lands, Caves Classification Committee: 812, 815.
Victorian Speleological Association: 2178.
Vidović-Čulić, Zjena: 429.
Vietzen, Ruth G.: 3362.
Vilece, Carol: 1062.

Villiers, Berna: 2098.
Vineyard, Jerry D.: 2584.
Vineyard, Jerry Daniel, 1935–: 738, 1039, 2214, 2217, 2219, 2390, 2583.
Visual Productions 80 Ltd SelectVideo: 2684.
Vivian, E.: 2009.
Vore, Debbie: 2406.
Vore, Duane: 2406.
Vu Van Phai: 2508.

Wadge, Geoff: 1060.
Wainhouse, Austryn: 342.
Wakelin, D.: 2616.
Walk, Rickard [sic] L.: 3424.
Walker, A. S.: 3053.
Walker, Frank H.: 2126.
Wallace, Bob: 595.
Wallen, S.: 2063.
Walley, Bill: 2316.
Walsh, John M.: 1244, 2345.
Walsh, Joseph E.: 2218, 2228.
Waltham, Anthony Clive, 1942–[Tony]: 503, 1632, 1975, 2171, 2172.
Ward, Helen, 1962–: 2693.
Warren, William M.: 1891.
Warrington, John: 608.
Warshauer, Susan: 2316.
Warties, Burleigh: 3319.
Water Resources Center: 2785.
Waters, Aaron C.: 235.
Watershed Management Coordinating Committee: 549.
Watson, Patty Jo, 1932–: 602.
Watson, R. J.: 254.
Watson, Richard Allan, 1931–: 765.
Watson, Tony: 271.
Wawman, D.: 2503.
Weast, Edward: 2513.
Weaver, Herman Dwight, 1938–: 2198, 2207, 2208, 2220, 2221, 2235, 2317.
Webb, John: 1167.
Webb, William Snyder, 1882–: 1131.
Webber, T. R.: 49.
[Webster, David]: 2215, 2216, 2222, 2224, 2225, 2229, 2230, 2231, 2232, 2233, 2234.
Webster, M.: 385.
Wedel, Waldo: 2114.
Weerd, B. van de: 2527.
Wei-wen Huang: 1893.
Weidie, A. E., 1931–: 3397.
Weishampel, David B.: 2070.
Welbourn, W. Calvin: 1945, 2521.
Welch, Bruce R.: 1729.
Welch, Jeanne M.: 809.
Welch, Norbert M.: 3033.

Wellings, Gleniss: 1701.
Wells, Stephen G.: 765, 2422.
Weltfish, Gene, 1902- Archaeology of southern Chihuahua. Cave-dweller twill-plaited basketry.: 3557.
Wening, Karen: 1033.
Werner, A.: 1530.
Werner, Eberhard Wolfgang, 1942–: 1427, 1429, 1626, 1822, 2404, 2427, 2434.
Werner, Eberhard Wolfgang, 1942–;The Department of Geology and Geography of West Virginia University (sponsors): 2764.
Werner, Marlin Spike: 1322, 1323.
West Virginia Speleological Survey: 3124, 3146.
West, R. G.: 763.
Western Australian Speleological Group: 3091.
Western Kentucky University. Center for Local Government.: 751.
Western Speleological Institute: 616.
Western Speleological Survey National Speleological Society.: 1324.
Western Speleological Survey, National Speleological Society: 1330.
WETA-TV (Television station: Washington, D.C.): 1127.
Weyer, Bernard: 2247.
Wheeland, Keith D.: 2449.
Whidby, Jim: 2694.
White, Charlie L.: 3462.
White, Elizabeth L.: 844, 3189, 3469, 3470, 3475.
White, Jim, 1919–: 576.
White, N. J.: 814.
White, Susan: 271, 812, 815, 1267, 2178.
White, Tim: 3070.
White, William Blaine, 1934–: 565, 843, 844, 1450, 2450, 2451, 2599, 2612, 2763, 3189.
Whitehead, William: 595.
Whitley, Steven: 1706.
Whitman, Richard L.: 1496, 1497.
Whitmore, Thomas: 1496.
Whittemore, R.E.: 2393.
Whitten, Charlie B.: 571.
Wicks, A: 3405.
Widmer, Urs: 3192.
Wigley, T. M. L.: 917.
Wilbur, Robert: 2384.
Wilcox John P.: 765.
Wild South New Dimension Media: 2991.
Wilde, Kevan A.: 250, 2179, 2765.
Wilkes, Chris: 3073.
Wilkinson, Robert: 1965.
Willemont, Jacques: 249.
Williams, Ann Mason [Edington, M. Ann]: 1717.
Williams, Bob: 1247.
Williams, Carolyn G.: 3442.
Williams, Dennis: 1849.

Williams, James Hadley, 1929–: 37, 1039, 1377.
Williams, K. M.: 1122.
Williams, Nick: 2048, 2049.
Williams, Pandra: 2593.
Williams, Paul: 3483.
Williams, Paul W.: 1083.
Williams, Steve: 1477.
Williamson, Fraser: 2131.
Williamson, K. A.: 814.
Williamson, Kerry: 817.
Williamson, L.: 2503.
Willis, Lawrence D.: 516.
Willis, R. G.: 3006, 3394.
Willis, Simon: 438.
Wills, Ronnie L.: 3363.
Wilmington District: 3260.
Wilson, April: 1167.
Wilson, Betty Morere: 1887.
Wilson, Bill: 1595.
Wilson, Ronald C.: 2340.
Wilson, S.: 1208.
Wilson, William L.: 360, 2745.
Winder, Francis A.: 2057.
Wing, Elizabeth S.: 2861.
Winglee, P. J.: 915.
Winkler, Alisa J.: 1492.
Wolinsky, Mark: 3067.
Wood, Edward E.: 2491.
Woodham, Robert Earl: 3076.
Woods, Mysterious Marianne: 856.
Woodson, Jack: 2918.
Woodward, Arthur Smith, 1864-1944: 703.
Wooldridge, Jerry: 2156.
Woolridge, Gerry: 228.
Woosley, L. H.: 2297.
Workman, Lindsay: 1064.
World Travelogues Corporation.: 207.
Wormell, Sebastian: 2867.
Worthy, Cathy: 2499.

Worthy, Trevor H.: 2499, 2765, 3484.
Wright, H. E.: 2114.
Wright, R. Gerald: 942.
Writing Team: 1566.
Wroth, James Stewart, 1885–: 65, 66, 67.
WSJK (Television station: Knoxville, Tenn.): 734.
Wyeth, John: 438.
Wynne, Patricia J.: 2982.

Xingrui Han: 803.
Xuewen Zhu: 803.
Xunyi Wang: 803.

Yager, Jill: 1849, 1851.
Yarbrough, Ed: 3064.
Yarnell, Richard Asa: 3408.
Yen jung yen chiu tsu: 179.
Yevjevich, Vujica: 2175.
Youens, Paula: 1711.
Young, I.: 2880.
Young, India F.: 1019.
Young, Ivan: 2881.
Youngsteadt, Jean O.: 3547.
Youree, Roger: 3070.
Yuanfeng Zhu: 803.
Yule, Lauray: 682.

Zachman, Dieter W.: 305.
Zárubová, H.: 2998.
Zdeněk, Trávníček: 1623.
Zepp, Peter: 1868.
Zi Meng: 133.
Zillmer, Rolf: 2747.
Ziqiang Deng: 803.
Zoetl, Josef: 1889.
Zojer, H.: 2268.
Zokaites, Carol: 2416.
Zollweg, James: 766.
Zuber, Ron: 2339.

17 Title Index

12 natural colour photographs of Cheddar and Cox's Cave "The home of the rainbow": 2022

The 1926 re-excavation of Step House Cave, Mesa Verde National Park: 2535

The 1965/66 karst hydrology expedition to Jamaica: full report: 1078

1967 expedition to the Gouffre Berger: 489

1968 SERA cave carnival: 3063

1970 guidebook UMAC-MVOR: Upper Mississippi area conference-Mississippi Valley Ozarks Regional Sept. 25, 26, & 27 at Eagle Cave: 2311

1971 guidebook: Upper Mississippi conference UMAC Sept. 17, 18, 19 at Eagle Cave: 1444

1971—18th Cave Cavers: location Bedford: 1595

1974 Cave Capers guidebook: for Harrison & Crawford counties: 1596

1976 annual convention guidebook, Morgantown, W. Va: 2398

1976 National Speleological Society 35th Annual Convention Guidebook *[program]* with Abstracts: 2428

1978 hodag hunt guidebook: 3517

1979 Southeastern Regional Association (S.E.R.A.) cave carnival at Ketner's Mill, TN: 3067

1979 Speleology workshop manual: 3092

1980 NSS Convention Program, Caving in the City, July 27 - August 1, Lakewood College, White Bear Lake, Minnesota: 2432

1980 progress report of archeological reconnaissance and testing of Pleistocene cave and alluvial deposits, Porcupine River, Alaska: 876

1983 American caving accidents: 2372

1983 SERA cave carnival: 5-7 August 1983: 3068

1984 American caving accidents: 2373

The 1984 Army Caving Association expedition to the Gouffre Berger: 227

1985 American caving accidents: 2374

1986 NSS convention guidebook, Tularosa, New Mexico: 2407

1987 cave management symposium: Rapid City, South Dakota, October 1987: 2344

1988 NSS convention guidebook: 2409

1989 NSS convention program: July 31 - August 4, 1989: 2440

1989 Speleofest guidebook. May 26-29, 1989, Sandhill Campground, Whitley City, Kentucky: 1806

1990 Friends of karst field trip: the Big Spring Basin, northeast Iowa: 1941

1991 national cave management symposium proceedings: Bowling Green, Kentucky, October 23-26, 1991: 2346

1992 Northwest Caving Association regional meeting cave guide [22-24 may 1992]: 2524

1992 post convention guidebook: enter the karst of Indiana, you may never want to come out: 2426

1993 Jamaica expedition: 624

The 1993 NSS Convention guidebook: 2414

The 1994 British/Vietnamese speleological expedition report March/April 1994: 1944

1995 SERA summer cave carnival: the party on the farm: Smokey and Tina Cadwell's farm: Lafayette, Georgia May 19-21, 1995: 3073

1997 Speleofest: Cadiz, Ky, Trigg County Recreational Complex, Cadiz, Kentucky: 1807

1st Claygate "Selachii" Venture Scout Unit expedition to Iceland: 21st July - 9th August 1987: 3534

1st international symposium of glacier caves and karst in polar regions. Proceedings: 1001

1st international symposium on groundwater ecology, Schlitz,1975: 1584

2,000-year-old duck decoys from Lovelock Cave, Nevada: 341

24th annual CIG Cave Capers: 1599

27th annual Cave Capers: 1980 guidebook: 1602

2nd international symposium of glacier caves and karst in polar regions: proceedings: 2730

30 favorite Alabama caves: 1576

37 views of Naracoorte caves and Naracoorte South Australia: 78

45th Annual SERA Summer cave carnival: July 19-21, 1996, Camp Jackson, Jackson County, Alabama: 3074

5 Internationaler Kongress für Speläologie Stuttgart 1969: 1622

5th annual Hog-Fest: Lickford Valley Campgrounds, Corydon, Indiana, May 1-5, 1992: 697

5th international symposium on vulcanospeleology excursion guide book: 2882

5th International symposium on vulcanospeleology: program Izunagaoka: November 9-11, 1988: 1658

67 expedition to Crete: BUSS: 402

67 expedition to Crete: BUSS Report: 1013

6th international symposium on vulcanospeleology, Hilo, Hawaii August 1991: 1659

8th Annual TAG fall cave-in: 3164

The 9th ACKMA conference proceedings; Margaret River, WA, Australia: 377

A. H. Anderson's excavations at Rio Frio Cave, British Honduras (Belize): 2641

Abbé Breuil: prehistorian: a biography: 494

Aberbrothok illustrated: 7

The Abercrombie Caves: 3231
The Abercrombie Caves, New South Wales: a guide to these remarkable caves near Bathurst, N.S.W. and a description of the surrounding gold bearing country of Tuena and Trunkey Creek: 344
Abercrombie Caves: cave chronicles: a new look at an old wonder: including an account of the Bathurst convict rebellion of 1830: 1778
Aboriginal art galleries of western New South Wales: 410
About caves: 2945, 2946
About Massanutten caverns; the jeweled chamber that hides the secret of Nakwisi-Gatusi: 1367
An abstract description and history of the Bone Caves of Creswell Crags: 1416
Abstracts: 1653
Abstracts of papers: 1010
Abstracts of papers: karst studies seminar Naracoorte: 1342
Abstracts of the fourth conference on karst geology and hydrology: West Virginia University, Mountainlair Theater of Mountainlair Building: May 3, 4, and 5, 1974: 690
The Abu Usba Cave (Mount Carmel): 3115
An account of Knoepfel's Schoharie Cave, Schoharie County, New York: with the history of its discovery, subterranean lake, minerals and natural curiosities: 1834
Account of Ogham inscriptions in the cave at Rathcroghan, Co. Roscommon: 1049
Account of some ancient sculptures on the walls of caves in Fife: 2985
Account of some remarkable caves in the principality of Bayreuth: likewise observations on the fossil bones by the late John Hunter: 1513
An account of the caves of Ballybunian, county of Kerry: with some mineralogical details: 20
Actes du deuxième congrès international de spèlèologie Bari - Lecce - Alerno 5-12 Octobre 1958: 1620
Actes du symposium international sur l'erosion karstique = proceedings of the international symposium on karstic erosion: Aix—en-Provence-Marseille-Nimes 10-14 Septembre 1979: 1878
Acute toxicity of cadmium, zinc, and total residual chlorine to epigean and hypogean isopods (Asellidae): 442
Adaptation and natural selection in caves: the evolution of *Gammarus minus*: 777
Administrative handbook: 258
Advanced cave diving: 3143
Adventure is underground: the story of the great caves of the West and the men who explore them: 1318
Adventure of caving: a practical guide for advanced and beginning cavers: 2108
Adventure of caving: new updated edition: 2109
Adventure underground: 331
Adventures at Mark Twain Cave: 3419

Adventures in underground fairylands: a fantasy tour of the Carlsbad Caverns: 767
Adventures under ground [sic]: 475
Adventures underground in the caves of Missouri: 2701
Aggtelek: 1813
Agricultural drainage well research and demonstration project. Sinkhole/karst area demonstration project: 1666
Aillwee Cave and the caves of the Burren: 898
Aillwee Cave: Ballyvaughan, Burren Co., Clare: 693
Aillwee Cave: Ireland's premier show cave: 80
Ajanta and Ellora: 1186
Ajanta: a brief history and guide: 3095
Akhali Atoni's cave: 3219
Alabama caves: 2124, 3183
Alabama caves and caverns: 3352
Alabama caves, 1980: 3353
The Albertson site: a deeply and clearly stratified Ozark bluff shelter: 870
ALCADI '92 international conference on speleo history field trip guide: 427
Alexander Caverns: the Carlsbad of Pennsylvania: 3088
All about Poole's Cavern, Buxton. An official guide to this unique natural curiosity, together with an inventory of the museum: 756
All about prehistoric cave men: 999
The almost complete eclectic caver: 989
Along the way to the caves: Timpanogos Cave National Monument, Utah: 2758
Altamira and prehistoric art in the caves of Santander: 1145
Altamira, the beginning of art: 1146
Altamira: a note upon the Palaeolithic paintings in the Cave of Altamira near Santillane del Mar in the Spanish province of Santander: 2818
Altamira: the origin of art: 1147
The amateur's guide to caves & caving; skill-building ways to finding and exploring the underground wilderness: 2110
American caves and caving: techniques, pleasures, and safeguards of modern cave exploration: 1319
American caving accidents: 2381
American caving accidents 1967: 2361
American caving accidents 1967-1970: 2362
American caving accidents 1968: 2363
American caving accidents 1969: 2364
American caving accidents 1970: 2365
American caving accidents 1971: 2366
American caving accidents 1972: a report of the National Speleological Society: 2367
American caving accidents 1973: a report of the National Speleological Society: 2368
American caving accidents 1974: a report of the National Speleological Society: 2369

American caving accidents 1975: a report of the National Speleological Society: 2370
American caving accidents 1976 through 1979: 2371
American caving accidents 1989: 2378
American caving accidents 1990: 2379
American caving accidents 1991: 2380
American caving accidents 1993: 2382
American caving accidents [for 1986]: 2375
American caving accidents [for 1987]: 2376
American caving accidents [for 1988]: 2377
American caving, illustrated; caving, climbing, camping: 3144
America's wonderland of caves: palaces under the earth: a directory of commercially operated caves: 2355
An analysis of Val Verde County cave material: 2912
An analysis of variability and condition of cavate structures in Bandelier National Monument: 3220
Ancient Castleton Caves: their place in time and history, also what people want to know about Castleton: 2862
The ancient caves of Szechwan Province, China: 1235
Ancient caves of the Great Salt Lake region: 3125
The Anglo-Canadian Rocky Mountains Speleological Expeditions 1983 and 1984: a report on recent discoveries made by two caving expeditions to the Rocky Mountains of Canada by combined British and Canadian teams: 77
Animal life of the Carlsbad Cavern: 279
An annotated bibliography of cave meteorology: 3428
Annotated bibliography of karst terranes: volume five: with three review articles: 1886
Annotated bibliography of North Carolina speleology: 11
An annotated bibliography of Pacific Northwest speleobiology: 753
The antiquity of man in Europe, being the Munro lectures, 1913: 1171
The antiquity of the deposits in Jacob's Cavern: 53
Anything is possible: 869
Ape Cave and the Mount St. Helens apes: 1320
Appalachian Karst: proceedings of the Appalachian karst symposium: Radford, Virginia, March 23-26, 1991: 1769
The Appalachian valleys of central Pennsylvania: 3469
Appendix: checklist of macroscopic troglobitic organisms of the United States: 3157
Appin cave guide: 3545
Application of dye-tracing techniques for determining solute-transport characteristics of ground water in karst terranes: 2297
Application of geophysical methods to location of cavities in residual soil and caverns in coralline limestone: 2510
Application of radar and seismic reflection techniques to cavity detection, Medford Cave site, Florida: 388
Application of thermal imagery and aerial photography to hydrologic studies of karst terrane in Missouri: 1377
Applied karst geology: proceedings of the fourth Multidisciplinary Conference on Sinkholes and Engineering and Environmental Impacts of Karst, Panama City, Florida 25-27 January 1993: 354
An approach to the study of karst water: illustrated by results from Poole's Cavern, Buxton: 2681
The aquatic invertebrates of Mystery Cave, Forestville State Park, Minnesota: 2258
The archaeological and faunal material from Williams Cave, Guadalupe Mountains, Texas: 268
An archaeological and geological assessment of Antelope Cave (NA 5507), Mohave County, Northwestern Arizona: 1705
Archaeological and paleobiological studies at Stanton's Cave, Grand Canyon National Park, Arizona; a report of progress: 1008
Archaeological and paleoenvironmental investigations in the Dutchess Quarry Caves, Orange County, New York: 1130
Archaeological assessment and sensitivity map for the Pohakuloa training area (PTA), Hawaii Island, State of Hawaii: 1353
An archaeological assessment of Carlsbad Caverns National Park: 447
Archaeological evaluations of the Riparia (45WT1) and Ash Cave (45WW61) sites on the Lower Snake River: 2194
Archaeological excavation at North Face Cave, Little Ormes Head, Gwynedd 1962-1976: 419
The archaeological excavation of two rock shelters near Cave Lake, Ely, Nevada: 554
Archaeological exploration of Eagle Cave, Langtry, Texas: 811
Archaeological exploration of the Shumla caves; report of the George C. Martin expedition... June, July and August, 1933: 2061
Archaeological explorations in caves of the Point of Pines region, Arizona: 1196
Archaeological investigations in Lovelock Cave, Nevada: 1420
Archaeological investigations in Turkey Cave (NA2520) Navajo National Monument, 1963: 467
Archaeological investigations: cave explorations in the Ozark region of central Missouri: 2198
Archaeological researches in Kentucky and Indiana, 1874: 2734
Archaeological survey of Mammoth Cave National Park: 3002
Archaeological surveys of Crater Lake National Park and Oregon Caves National Monument, Oregon: 837
Archaeological testing at site 1Ms357, Cathedral Caverns, Marshall County, Alabama: 2539
Archaeological testing at two rockshelters in the Tombigbee River Multi-Resource District, Alabama and Mississippi: an interim report: 16

Archaeological testing of the cave at site 41BX22, Bexar County, Texas: 1224

Archaeological work in Sarawak: with special reference to Niah Caves: 654

Archaeology and ecology of Rose Cottage Cave, Orange Free State: 1844

The Archaeology at Lehman Caves National Monument: Nevada State Museum report: 2864

The archaeology of Arnold Research Cave, Callaway County, Missouri: 2970

Archaeology of Black Rock 3 Cave, Utah: 992

The archaeology of Bone cave, Miller county, Missouri: 1474

The archaeology of Bowers Cave, Los Angeles County, California: 978

Archaeology of Chokecherry Cave (35GR500) in Grant County, southeastern Oregon: 897

The archaeology of Deadman Cave, Utah: a revision: 3017

The archaeology of Hidden Cave, Nevada: 3200

The archaeology of Humboldt Cave, Churchill County, Nevada: 1421

The archaeology of Nahas Cave: material culture and chronology: 2683

The archaeology of Newberry Cave, San Bernardino County, Newberry, California: 3018

The archaeology of the Beaver Creek shelter (39CU779), Wind Cave National Park, South Dakota: a preliminary statement: 24

Archaeology of the Cape St. Blaize cave and raised beach, Mossel Bay: 1225

Archaeology of the dreamtime: 1070

The archaeology of the Fallen Arches Cave Site, 45SA41, Borigo timber sale, Gifford Pinchot National Forest: phase I survey and reconnaissance: 1965

Archaeology of the Gallus Site, Koonalda Cave: 3535

The archaeology of two Kern County sites: 2817

The archaeology of two sites at Eastgate, Churchill County, Nevada: 1422

The archaeology of Wilson Butte Cave south-central Idaho: 1274

The Archaeology, geology, and paleobiology of Stanton's Cave: Grand Canyon National Park, Arizona: 1009

Archeological and botanical studies at Hinds Cave, Val Verde County, Texas: 2941

Archeological investigations: 1109

Archeological investigations at Limekiln Cave, 23SH109, Ozark National Scenic Riverways, southeast Missouri: 2000

Archeological investigations at Parida Cave, Val Verde County, Texas: 31

The archeology of Eagle Cave: 2856

The archeology of Exhausted Cave: a study of prehistoric cultural ecology on the Coconino National Forest, Arizona: 1564

The archeology of Newark Cave, White Pine County, Nevada: 1110

Archeology of the Mammoth Cave Area: 3407

Army Caving Association: Peru 1987: 228

Army caving expedition to South East Java: July and August 1986: exercise Phreatic Diamond: 1696

The art of cave mapping: 2199

The art of safe cave diving: 2879

Art of the cave dweller—a study of the earliest artistic activities of man: 523

The art of the cave dweller: a study of the earliest artistic activities of man: 524

The art of the stone age: forty thousand years of rock art: 312

Art treasures of prehistoric man in the caves of France and Spain: 2530

Ash Hollow Cave: a study of stratigraphic sequence in the central great plains: 633

Aslóhegy shafts: 2877

Aspects of land use planning at Fanning River. With particular reference to the Fanning River caves: 1234

Aspex '90: Anglo-Soviet Pamirs Expedition 1990: Bajsuntai Khrebetuzbekistan SR, USSR: 3342

Assessment of the ecological resources of the caves of Russell Cave National Monument, Jackson County, Alabama and of selected caves at the Lookout Mountain unit of Chickamanga-Chattanooga National Military Park, Dade County, Georgia and Hamilton County, Tennessee: 1468

Atea: in search of the world's deepest cave: 483

An atlas of Tasmanian karst: 1816

Atlas of the Berry Head Caves: 2720

Atlas of the great caves of the world: 725

Augusta Jewel Caves and other points of interest: 1591

Australasian speleo map index no. 1: 1390

Australian aboriginal rock art: 2106, 2107

Australian cave fauna: notes on collecting: 1343

The Australian cavers diary: 514

Australian caves and caving: 974, 1767

Australian caves, cliffs, and waterfalls: 330

Australian karst index 1985: 2090

Australian karst index data preparation manual: 2091

Australian Ranger Bulletin: feature: cave management: 256

Australian Speleological Handbook: 81

Australian speleology 1972: 264

Avifauna of three Holocene cave deposits in southern Chile: 1570

The B. E. C. method of caving ladder construction: 674

Back in time for tea and medals: 3377

The Baker Bluff cave deposit, Tennessee, and the late Pleistocene faunal gradient: 1279
The Baker extension to the Banwell Bone Cave: 287
Balague '70: 387
Balankanche, throne of the tiger priest: 74
Bandera Crater/Ice Cave: a state park feasibility study: 2493
Barma Grande; the great cave and its inhabitants: 452
Barnoolut Estate sinkholes environmental assessment project, 27 December 1988: 1518
Barton Springs eternal: the soul of a city: 2679
Basic cave diving: a blueprint for survival: 1016
Basic cave rescue orientation course study guide: 1075
The basic geology of Neff Canyon Cave, Utah ; The speleogenesis of Neff Canyon Cave, Utah: 1321
Basic guidelines for spelunking: 1766
Basic underwater cave surveying: 560
A basket-maker cave in Kane County, Utah: 2536
Basket-Maker caves of northeastern Arizona; report on the explorations, 1916-17: 1277
Bat Cave: 866
Bath freestone workings: 2716
Batu Caves, Kuala Lumpur - towards a plan of management: 1344
BCRA library catalogue: 484
Bear cavern: 971
The bears cave of Erpfingen: 3371
The beautiful caverns of Luray, Luray, Virginia, in the Shenandoah Valley: 1990
The beautiful caverns of Luray, Luray, Virginia. In the Shenandoah Valley, three miles of subterranean splendor, brilliantly lighted by electricity: 1991
The beautiful caverns of Luray: Luray, Virginia: 1992
The beautiful caverns of Luray: Luray, Virginia, in the Shenandoah Valley: miles of subterranean splendor, brilliantly lighted by electricity: 1993
The beautiful Isle of Mull with Iona and the Isle of Saints: 1355
Beautiful Kweilin: 82
The beautiful Luray Caverns, Luray, Virginia: 1776
The beginnings of art: 3121
Below Belize 1991: the report of the 1991 expedition to the Little Quartz Ridge Southern Belize February to April 1991: 3496
Below Belize: Queen Mary College speleological expedition to Belize, 1988 and the British speleological expedition to Belize, 1989: 2048
Below Belize; Queen Mary College Speleological Expedition to Belize 1988 and the British Speleological Expedition to Belize 1989: 2049
Beneath the mountains: exploring the deep caves of Asturias: 2854
Berkeley's description of the cave of Dunmore: 1980
Berkshire Cave Guide: 2682
Bexar County speleology vol 1: 2691
Beyond the bounds of history: scenes from the old stone age: 470
Beyond time: 2979
A bibliographic guide to Texas speleology: 2770
Bibliographie Speleologique Belge: 415
Bibliographie Speleologique Belge. Editions speleologiques belges. 1907-1964. Avec en appendice la liste des ouvrages a caractere speleologique edites en Belgique: 417
Bibliographie Speleologique Belge: editions speleologiques belges 1975-1979: 416
Bibliography of Belizean caving: 3497
A bibliography of British karst, 1960-1977: 3243
Bibliography of karst geology in Kentucky: 781
Bibliography of Mammoth Cave National Park, Mammoth Cave, Kentucky: 3486
A bibliography of Mammoth Cave, 1798-1949: 1733
A bibliography of Nevada caves: 2132
Bibliography of North American speleology: 826
Bibliography of Oregon speleology, 1977: 1902
Bibliography of Oregon speleology: initial compilation: 1838
Bibliography of Puerto Rican caves, karst and limestone geology; Rio Camuy Cave System: 1927
A bibliography of Sea Lion Cave: 1903
Bibliography of Tasmanian karst: 1817
A bibliography of technical articles referring to practical caving subjects: 1713
Bibliography of Tennessee speleology: 2087
Bibliography of the Ground Water Resources of New York through 1967, with subject and location cross-reference to the annotated bibliography: 3283
A bibliography of the Jenolan Caves: part 1: speleological literature: 913
A bibliography of the Jenolan Caves: part 2: literature: 914
A bibliography of Victorian caves & karst: 3468
Bibliography of Wyandotte Cave: 1175
Bibliography on lava tube caves: 1373
Bibliography on lava tube caves, supplement: 1374
The big cave: early history and authentic facts concerning the history and discovery of the world famous Carlsbad Caverns of New Mexico: 1957
The Binoomea Cut, Jenolan Caves: a paper read to the Jenolan Caves Historical and Preservation Society on Saturday 9th, February 1974: 905
The biogeography of Cape Range Western Australia: being the proceedings of a symposium held under the auspices of the Western Australian Museum in Perth on 21 November 1992 at the Art Gallery of Western Australia: 1573

Biogeography of subterranean crustaceans: the effects of different scales: a symposium held at the summer meeting of the Crustacean Society in Charleston, South Carolina, USA, in June 1992: 778

A biological inventory of the Warsaw Caves area of natural and scientific interest, Peterborough County, Ontario: 538

The biological resources of Illinois caves and other subterranean environments: determination of the diversity, distribution, and status of the subterranean faunas of Illinois caves and how these faunas are related to groundwater quality: 3425

A biological study of Cathedral Cave, Crawford County, Missouri: 2200

Biospeleology: the biology of cavernicolous animals: 3350

Bir Al Ghanam karst study project: final report 1981: 1853

Black Hills caves: 3250

Black Hills symposium: 2356

Blanchard Springs Caverns story: 1067

Blanchard Springs Caverns: the amazing world below: 2588

Blind white fish in Persia: 3012, 3013

The Blue holes of Andros: 595

The blue holes of the Bahamas: 2603

Blue John Caverns and Blue John Mine, Castleton, Sheffield: 2569

The Blue Mountains and Jenolan Caves: a camera study: 1581

Bone Cave (Ogof-yr-Esgyrn) Dan-yr-Ogof: the history of the Bone Cave (Ogof yr Esgyrn): a guide and background to the exhibits in Bone Cave: 2071

The bone cave at Port Kennedy, Pennsylvania, and its partial excavation in 1894, 1895 and 1896: 2149

The bone caves of Ojcow in Poland: 2840

Bonfire shelter: a stratified bison kill site, Val Verde County, Texas: 865

A book of songs and poems for the Hunters Lodge and similar places: 83

Bower Cave and Cave of Orpheus file: 2969

A boy scout in the grand cavern: 2568

The Bradford Pothole Club library list: 3506

Bramabiau's underground river: 73

Breakthroughs in karst geomicrobiology and redox geochemistry: abstracts and field trip guide for the symposium held February 16 - 19, 1994, Colorado Springs, Colorado: 2887

A brief glossary of Welsh topographic names for walkers and cavers: 1455

A brief history of the Dunhuang Caves: a millennium of Chinese Buddhist art: 286

A brief Introduction to China's research in Karst: 802

A brief introduction to speleology: 3081

A brief story of St Clement's caves: 930

Bristol Exploration Club caving report no 13 [St. Cuthbert's report]: Part H: Rabbit Warren extension: 1669

Bristol Exploration Club caving report no 13: St. Cuthbert's report: Part E: Rabbit Warren: 1670

Bristol Exploration Club caving report no.13 [St. Cuthbert's Report]: part F: Gour Hall area: 385

Bristol exploration club: caving report no.19: 1975 expedition to the Pierre Saint-Martin: 481

Britain underground: 3209

British Cave Research Association Cave Radio and Electronics Group bibliography of underground communications: 3498

British Cave Research Association, library catalogue: 487

British caves and potholes: 845

British caving: an introduction to speleology: 773

British hypogean fauna and biological records of the Cave Research Group part IX (1963): 1399

British karst research expedition to the Himalaya 1970: full report: 3387

The British New Guinea speleological expedition: 502

The British Pleistocene Mammalia: part II: 839

British Speleological Expedition to the Cantabrian Mountains, northwest Spain 1965: 492

Broken River karst: a speleological field guide North Queensland Australia: 649

Broom County *Cercartetus*, with observations on pygmy possum dental morphology, variation, and taxonomy: 3251

The Buchan Caves: 3178

The Buchan experience: a guide to the Buchan and Murrindal Caves - East Gippsland, Victoria: 1432

Building on sinkholes: design and construction of foundations in karst terrain: 3078

A Bukki barlangok kutatasanak, vedelmenek es hasznositasanak legujabb eredmenyei = Newest results in research, protection and use of caves of Bukk Mountains: Miskolci Egyetem, 1993. November 11-13: 1931

Bungonia Caves: 975

Bureau of Land Management interim guide for oil & gas drilling operations in cave and karst areas: 3264

A burial cave in Baja California, the Palmer Collection 1887: 2083

Burnsville Cove Symposium: 565

A Burrington cave atlas: 1548

Bushman art; rock paintings of south-west Africa, based on the photographic material collected by Reinhard Maack: 2541

The Buttercup Creek karst: Travis and Williamson Counties, Texas: geology, biology, and land development: 2870

C. H. U. G. Crawford Harrison Underground Gang presents twenty-two favorite Harrison Crawford caves: 1973 pre-convention activities: 2425

C. N. C. C. eco-resin rigging system no 1: 3172
Cambrian Caving Council handbook 1969: 657
Cambrian Caving Council handbook 1993: 273
Canadian caves of the Niagara Gorge: 995
Cango Cave, Oudtshoorn district of the Cape Province, South Africa: an assessment of its development and management 1780-1992: 735
Cango; the story of the Cango Caves of South Africa: 3052, 3053, 3054
Capital area cavers bulletin number 1: 70
Capri: the story of the blue grotto: 3242
Carbon dioxide and the soil atmosphere: 2192
Carbonate sediments and their diagenesis: 345
The Carlsbad Cavern National Park: 1160
Carlsbad Cavern National Park. Guadalupe Mountains National Park: 84
The Carlsbad Cavern of New Mexico: its history and geology: 65
The Carlsbad Cavern of New Mexico: its history and geology, formations of the cavern, discovery and exploration, bats of the cavern, cavern geology, administration: the land nobody knows: 66
The Carlsbad Cavern of New Mexico: its history and geology; formations of the cavern, discovery and exploration, bats of the cavern, cavern geology, administration; the land nobody knows: 67
The Carlsbad Cavern of New Mexico: it's [sic] history and geology: 925
Carlsbad Cavern, the early years: a photographic history of the cave and its people: 2537
Carlsbad Caverns: 412, 1240, 3051
Carlsbad Caverns National Park: 85, 240, 323, 973, 1136, 2659, 2747, 3044, 3147, 3526
Carlsbad Caverns National Park & Guadalupe Mountains National Park: 1063
Carlsbad Caverns National Park, Carlsbad, New Mexico: silent chambers, timeless beauty: 324
Carlsbad Caverns National Park, New Mexico: 86, 785
Carlsbad Caverns National Park, New Mexico: its early explorations as told by Jim White: 3462
Carlsbad Caverns National Park, New Mexico: wilderness reevaluation study: preliminary - subject to change: 3286
Carlsbad Caverns National Park: final interpretive prospectus: 3287
Carlsbad Caverns, New Mexico: 87, 241, 599, 1481
Carlsbad Caverns: official map & guide: 3288
Carlsbad Caverns; National Park, New Mexico: 1425
Carlsbad, caves, and a camera: 2538
The Carnivora of the Palestine caves: 1875
The case against the Pike Creek Dam: 2735
The Castellana Grottoes: 2165
Castle Guard Cave, challenge under the glacier: 594
The castle of Predjama: 3549

Castleguard: 2296
Castleguard Cave and karst, Columbia Icefield Area, Rocky Mountains of Canada: a symposium: 1079
Castleguard Cave: challenge under the glacier: 1080
Castles of the underworld: 2991
Castleton and its caves: 1609
Catalogue and description of a very large collection of prehistoric relics: obtained in the cliff houses and caves of Southeastern Utah: 2135
A catalogue of ephemera relating to Wookey Hole: 1671
Catalogue of literature on Chinese karst caves: 88
A catalogue of postcards of caves in Yorkshire & Derbyshire (including some misc sites in the Midlands): 1672
Catalogue of the caves of Missouri: 3363
A catalogue of the postcards of Gough's Cave, Cox's Cave & Wookey Hole, Somerset 1900-1980: 1673
Catalogue, bibliography, and generic revision of the order Schizomida (Arachnida): 2771
Cathay Pacific Airways Mulu '80 expedition: British Malaysian speleological expedition to Sarawak: 3006
Cathedral Cave situated between Brecon and Swansea on the A. 4067 road (750 yards North of Craig-y-Nos Castle) Dan-yr-Ogof Swansea Valley Caves: 670
Catmur's cave: 893
Cave: 381, 1829, 2982
Cave & karst guidebook and documentation: Australasian Cave and Karst Management Association eleventh Australasian Conference on cave and karst management Tasmania 29 April to 7 May 1995 hosted by The Parks & Wildlife Service - Tasmania: 1531
The cave and her men: Oregon Caves National Monument: its history and formation: 3014
Cave and karst hydrology assessment project for Horsethief Cave, Wyoming. Report to the Worland District, BLM: 32
The cave and karst resource of New Zealand: 3482
Cave and karst-related papers in the mainstream scientific literature: a bibliography: 3470
Cave and karst-related theses in United States and Canadian Universities, 1899-1988: 1578
Cave art: 1916
Cave art of France and Spain: 2096
The cave artists: 2976
The cave at Vari: excavations by American School of Classical Studies at Athens in February, 1901: 89
The cave bear story: life and death of a vanished animal: 1876
Cave bears and modern human origins: the spatial taphonomy of Pod Hradem Cave, Czech Republic: 1157
The cave beneath the sea: paleolithic images at Cosquer: 667
The cave book: 1439
Cave Capers 1978: 1600
Cave Capers 1979 guidebook: Milltown, Indiana: 1601

Cave Capers, 1983: 1603
Cave clippings of the nineteenth century: 1294
The cave community: 325
Cave development and formations: 2173
Cave development in Marble Canyon, Grand Canyon National Park, Arizona: 1379
Cave development in the Bull Creek drainage basin of southwest Missouri: 2066
The cave divers: 3142
A cave diver's training manual: 1951
The cave divers: illustrated with photographs and drawings: 563
Cave Diving Association of Australia, Inc: conference 1977 held at Mount Gambier, South Australia: 613
Cave diving communications: 2724
The Cave Diving Group annotated ten year cumulative index to the cave dives for the years 1969-1978: 879
The Cave Diving Group technical review no.3: line laying and following: 3541
Cave Diving Group: Derbyshire sump index 1987: 338
Cave Diving Group: Irish sump index: 1743
Cave Diving Group: northern sump index: 1263
Cave Diving Group: Peak District sump index 1994: 603
Cave Diving Group: Welsh sump index: 2275
Cave diving in Australia: 1938
Cave diving manual: 2286
Cave diving notes: equipment - methods - danger: 3145
Cave diving on air: 434
Cave diving: equipment and procedures: 3048
Cave diving: the Cave Diving Group manual: 301
Cave drawings: an exhibition of drawings by the Abbé Breuil of palaeolithic paintings and engravings: 471
The cave dwellers in the old stone age: 2704
The cave dwellers of Southern Tunisia: recollections of a sojourn with the Khalifa of Matmata: 539
The cave dwellings of the Old and New Worlds: 1054
Cave ecology: 717
Cave ecology: a course book for the cave ecology class at Malheur Environmental Field Station: 382
Cave entrance plants including Lampenflora: 1200
Cave exploration in Canada: a special issue of the Canadian Caver magazine: 3205
Cave explorations in Iran, 1949: 705
The cave explorers: 1020
Cave Explorers in Co. Fermanagh 1907: 281
Cave exploring: 69, 3225
Cave Fauna: 1219
The cave fauna at Ida Bay and the effect of quarry operation: 939
The cave fauna of North America, with remarks on the anatomy of the brain and origin of the blind species: 2590
Cave fauna of Waitomo: 1201
Cave formation in northern England: 1464
Cave gating: a handbook: 1575

The cave home of Peking Man: 1892
The cave hunters: 2899
The cave hunters: biographical sketches of the lives of Sir William Boyd Dawkins (1837-1929) and Dr. J. Wilfrid Jackson (1880-1978): 403
Cave hunting holidays in Peakland: 3507
Cave hunting in Yucatan: a lecture delivered before the Society of Arts of the Massachusetts Institute of Technology, on December 10,1896: 2150
Cave hunting, researches on the evidence of caves respecting the early inhabitants of Europe: 840
Cave illustrations before 1900: a catalogue of non-photographic illustrations of caves: 2952
A cave in the desert, Nahal Hemar: 9,000-year-old finds: [exhibition, the Israel Museum, Jerusalem, spring 1985]: 317
Cave inventory and classification systems: 3265
Cave life: 1289, 2248
Cave life of Carter Caves State Park: 691
Cave life of Kentucky: mainly in the Mammoth cave region: 280
The cave life of Oklahoma: a preliminary study (excluding chiroptera): 409
Cave life: evolution and ecology: 779
Cave light: 584
Cave locations listing: 1072
Cave Management in Australasia VIII: proceedings of the eighth Australian conference on cave tourism and management, Paparoa National Part, Punakaiki, New Zealand, April 1989: 250
Cave Management in Australia II: proceedings of the second Australian Conference on cave tourism and management, Hobart, Tasmania, 3rd-5th May, 1977: 252
Cave Management in Australia III: proceedings of the third Australian Conference on cave tourism and management, Mount Gambier, South Australia, 30th April - 4th May, 1979: 253
Cave Management in Australia IV: proceedings of the fourth Australian Conference on cave tourism and management, Yallingup, Western Australia Australia, September 1981: 254
Cave Management in Australia V: proceedings of the Fifth Australian Conference on cave tourism and management, Lakes Entrance, Victoria, April 1983: 255
Cave Management in Australia: proceedings of the first Australian Conference on cave tourism, Jenolan Caves, N.S.W. 10th-13th July, 1973: 251
Cave management investigations on the Ozark National Scenic Riverways, Missouri: 33, 34
Cave management plan and environmental assessment, Lava Beds National Monument, California: 3079
Cave management plan for the Ely District: 2491
Cave management symposium: proceedings: Harrisonburg, Virginia November 5-7, 1982: 2342

Cave men new and old: 606
Cave minerals: 1456
Cave minerals and speleothems: a review of the literature with special reference to Waitomo: 1128
Cave minerals of the world: 1457
Cave notes 74: 1674
Cave Notes 75: 2729
The cave of a thousand columns: 1243
The cave of Altamira and other caves with paintings in the province of Santander: 598
The cave of Delaware: 3084
Cave of Hearths, Makapansgat, Transvaal: 2078
The cave of "La Pileta": Benaojan, Malaga, Spain: 1213
The Cave of Las Monedas in Puente Viesgo (Santander): 2822
The cave of Lascaux: 1884
The cave of Lascaux: the final photographs: 2867
Cave of Loltun, Yucatan: report of explorations by the Museum, 1888-89 and 1890-91: 3202
The cave of Manaccora, Monte Gargano: 349
The cave of Nerja: 2837
The Cave of Postojna and other curiosities of the karst: 2936
The Cave of Psychro: 810
The Cave of Rouffignac: 2529
The cave of the Petralonian archanthropinae: a guide to the science behind the excavations: 2698
The cave of the treasure: the finds from the caves in Nabal Mishmar: 316
The Cave of the Winds: 2526
The Cave of the Winds' and the compromise of 1850: 1341
Cave painting: 788
Cave paintings of Baja: 758
The cave paintings of Baja California: 759
The cave paintings of Baja California: the great murals of an unknown people: 760
Cave paintings of the Chumash Indians: 90
Cave paintings—Lascaux and Altamira: 2883
Cave passages: roaming the underground wilderness: 3191
Cave photography: a practical guide: 1549
Cave preservation and residential development: a compatible approach: 1577
Cave references in Scientific American: 1550
Cave regions of the Ozarks: 2583
Cave regions of the Ozarks and Black Hills: 2584
Cave rescue operations: 1122
The Cave Research Foundation 1969-1973: 3409
The Cave Research Foundation 1974-1978: 3410
The Cave Research Foundation: origins and the first twelve years [1957-1968]: 3411
Cave Reserves of the Katherine Area: 1345

Cave resources in British Columbia: a discussion paper: 488
Cave resources management, planning, restoration, and redevelopment, Cathedral Caverns State Park, Grant, Alabama, 10-14 January 1994: 1810
Cave resources of Ozark National Scenic Riverways: an inventory and evaluation phase II: 1152
Cave science: 945
Cave spiders of Japan; their taxonomy, chorology, and ecology: 1847
Cave Studies No. 1-11: 616
Cave sump index; South Wales: 822
Cave survey: 569
Cave surveying: 568, 3398
Cave surveying and mapping: 1529
Cave surveying for expeditions: 1764
Cave surveying in South Australia: 2939
Cave surveying techniques: 339
Cave tourism in Western Australia: proceedings of a seminar held at Busselton November 10-11, 1978: 3406
Cave tourism: proceedings of international symposium at 170- anniversary of Postojnska jama, Postojna, Yugoslavia, Nov. 10-12, 1988: 1663
Cave vertebrates of America: a study in degenerative evolution: 950
Cave wonderlands of Western Australia; Jewel Caves, Lake Cave, Mammoth Cave, Yallingup Cave: 2851
The cave-deposit at Sha Kuo Tun in Fengtien: 72
Cave-dwelling pseudoscorpions of the Dinaric Karst = Jamski pascipalci dinarskega krasa: 780
The cave: what lives there: 498
Cavecraft: an introduction to caving and potholing: 694
Cavefish (*Amblyopsis rosae*) in Arkansas: populations, incidence, habitat requirements, and mortality factors: a final report: 516
Cavern development in the Guadalupe Mountains: 1691
Cavern development in the Helderberg Plateau, East-central New York: 1770
Cavern formation: 367, 1057, 1560
Cavern guide book, Carlsbad Caverns, Carlsbad, N. M: 1241
Cavern of the dragon: 2056
Cavern researches, or, discoveries of organic remains, and of British and Roman reliques, in the caves of Kent's Hole, Anstis Cove, Chudleigh, and Berry Head: 2009
Cavern: a pictorial souvenir of Carlsbad Caverns: 91
Cavernas, o fascinante Brasil subterraneo = Caves, the fascination of underground Brazil: 1947
Caverneering handbook of the Tasmanian Caverneering Club: 3187
The cavernicolous fauna of Hawaiian lava tubes, part VI: Mesoveliidae or water treaders (Heteroptera): 1138

The cavernicolous fauna of Hawaiian lava tubes, part VII. Emesinae or thread-legged bugs (Heteroptera: Redvuiidae): 1139
The cavernous karst land forms of Jackson County, Kentucky: 3180
Caverns and geysers: 396
The caverns and mines of Matlock Bath: 1 Nestus Mines: Rutland and Masson caverns: 1069
Caverns measureless to man: 1017
Caverns of Copan, Honduras: report on explorations by the museum, 1896-97: 1227
The caverns of Derbyshire: being an extract from Irlande at Cavernes Anglaises: 2057
Caverns of enchantment: 1528, 2033
The Caverns of Luray: 92, 1994, 2075, 2076
The Caverns of Luray and vicinity: 1759
The Caverns of Luray located on line of the Shenandoah Valley R.R.: 93
The caverns of Luray, located on line of the Shenandoah Valley R.R: 3037
The caverns of Luray, the property of the Valley Land and Improvement Co., Luray, Va: the manner of their formation, their peculiar growths, their geology, chemistry, etc: an illustrated guide book: 57
The caverns of Luray. The property of the Valley Land and Improvement Co., Luray, Va. The manner of their formation, their peculiar growths, their geology, chemistry, etc. An illustrated guide book: 58
The caverns of Luray: an illustrated guide-book to the caverns, explaining the manner of their formation, their peculiar growths, their geology, chemistry, etc: 59
Caverns of Sonora: 2162
Caverns of Sonora: an awe-inspiring adventure you will remember forever: 518
Caverns of the Ozarks: 1841
Caverns of the Shawangunk (Shon-gum) and its environs, Southeastern New York: 1771
Caverns of the world: 2689
The caverns of Upper Ease Gill: 237
Caverns of Virginia: 2125
Caverns of West Virginia: 827
Caverns of West Virginia (supplement): 828
Caverns, ice caves, sinkholes, and natural bridges, part I & II: 1430
The Caverns, rocks and ruins of America's Southwest: 94
Caverns: a world of mystery and beauty: 2968
The caver's living dictionary: 1854
A caver's view of the Clydach River: an introduction to the underground waters of Pwll Ddu, Llangynid and Llangattwg mountains: 2915
Cavers, caves, and caving: 3003
Caves: 1, 95, 96, 462, 908, 1142, 1254, 1723, 1811, 1859, 1918, 2278, 2820, 3356, 3488, 3522, 3523
Caves of Mississippi: 1831

Caves & swallets in chalk: 2801
Caves and associated features of open vertical volcanic conduits of the Kaupulehu lava flows xenolith nodule beds, Haulalai Volcano, Hawaii: basic speleological considerations: 1322
Caves and associated features of open vertical volcanic conduits of the Kaupulehu lava flows xenolith nodules beds, Hualalai Volcano, Hawaii: preliminary report: 1323
Caves and cave diving: 1912
Caves and cave life: 638
Caves and cave surveying: 2850
Caves and Caverns: 1190
Caves and caverns of Peakland: 2695
Caves and Caving City Museum Queens Road Bristol 8: 97
Caves and caving in Britain: 2072
Caves and caving in TAG: a guidebook for the 1989 Convention of the National Speleological Society ; Sewanee, Tennessee: 2410
Caves and caving, 1: 3543
Caves and caving, 2: 3544
Caves and caving: a guide to the exploration, geology and biology of caves: 1710
Caves and conservation: 3505
Caves and karst hydrogeology of the southeastern Edwards Plateau, Texas: guidebook, geology field excursion, National Speleological Society annual convention, New Braunfels, Tex., June 18-23, 1978: 2420
Caves and karst hydrology in northern Pocahontas County: 2138
The caves and karst hydrology of southern Pocahontas County and the upper Spring Creek Valley: 3146
The caves and karst of Colorado: a guidebook for the 1996 convention of the National Speleological Society: 2417
Caves and karst of Ireland: guidebook for the International Congress of Speleology at Sheffield, 1977: 1631
Caves and karst of Kentucky: 888
Caves and karst of Randolph County: 2139
Caves and karst of southern England and South Wales: guidebook for the International Congress of Speleology at Sheffield, 1977: 244
The caves and karst of Texas: a guidebook for the 1994 Convention of the National Speleological Society with emphasis on the Southwestern Edwards Plateau: Brackettville, Texas, June 19-24, 1994: 2415
The caves and karst of the Buckeye Creek Basin, Greenbrier County, West Virginia: 805
Caves and karst of the Muller Range: report of the 1978 speleological expedition to the Atea Kananda, Southern Highlands, Papua New Guinea: 1697, 1698
Caves and karst of the New River Valley, Virginia: guidebook for a geologic fieldtrip, eighth annual New River symposium, Radford, Virginia: 1772

Caves and karst of the Yorkshire Dales: an excursion guidebook to the karst landforms and some accessible caves within Yorkshire Dales: 3388
Caves and karst of the Yorkshire Dales: guidebook for the International Congress of Speleology at Sheffield 1977: 1632
Caves and life: 3556
Caves and mines of the Sychrhyd Gorge: 1749
Caves and other ground water features: 1735
Caves and other ground water features II: 1736
Caves and other volcanic landforms of Central Oregon: 2419
Caves and rockshelters in northern Iraq.: 1055
Caves and shelters in Bonanza King Canyon: 1357
Caves and stories of central Asia: 393
Caves and the marvelous world beneath us: 3119
Caves and their conservation: 3289
Caves and their mysteries: 2113
Caves and tunnels in Kent: 2625
Caves and tunnels in south-east England: 2626
The caves at East Wemyss, Fife: an interim report on new investigations in 1980: 2012
The caves beyond: the story of the Floyd Collins' Crystal Cave exploration: 1914, 1915
Caves for kids: in historic New York: illustrated with maps and photographs: 669
Caves in Austria: 2580, 2581
Caves in Buckfastleigh Quarries: 1515, 3491
Caves in Czechoslovakia: 2998
The caves in Derbyshire: 379
Caves in fine art exhibition: 1662
Caves in Kansas: 3546
Caves in the Bükk Mountains: 1932
Caves in the Richlands area of Greenbrier County, West Virginia: 232
Caves in Vermont, a spelunker's guide to their location and lore: 2924
Caves in Wales and the Marches: 1717
Caves in world history: 2271
The caves of Langtry: 2772
Caves of adventure: 3193
The caves of Alabama: a guide for the 1967 convention of the National Speleological Society at Birmingham and Huntsville, Alabama, June 1967: 2389
Caves of Albany County, New York: 2921
The caves of Altamira: 2542
The caves of Arta situated on the sea coast of the municipal confines of Capdepera: 98
The Caves of Assynt: 1714, 1917
Caves of Bedford County, Pennsylvania: 3189
The caves of Bell and Coryell Counties: 2130
Caves of Berks County, Pennsylvania: 2282
The caves of Bexar County: 2773, 3357
The caves of Blair County, Pennsylvania: 3085

The caves of Brewster and western Pecos Counties: 1056
Caves of Bucks County, Pennsylvania: 1864
The caves of Buda: 8
The caves of Budapest: 9
Caves of California: a special report of the Western Speleological Survey in cooperation with the National Speleological Society, June 1962: 1324
The caves of Capistrano: 572
The caves of Carmarthen: 2549
The caves of Carta Valley, Edwards and Val Verde Counties, Texas: 1870
The caves of Centre County, PA: 843
The caves of Chinoyi: a Rhodesian story: 231
The caves of Chlingitani: 1365
Caves of Christian County, Missouri: part 1 - eastern half: 2201
Caves of Christian County, Missouri: part 2- western half: 2202
The caves of Chudleigh and Kingsteignton: 2721
The caves of Clydach: 2550
The caves of Co. Cork: 2551
Caves of Colorado: 2615
The caves of Comal county: 2774
The caves of Connecticut: 3539
Caves of County Clare: 2929
Caves of Crawford County: 2203
Caves of Derbyshire: 1086
The caves of Deschutes County, Oregon: 1839
The caves of Devon: 2564
Caves of East Devon: 2722
The caves of Edwards County: 2775
The caves of Elephanta: 1594
The caves of far west Texas: 2776
The caves of Fermanagh and Cavan: 1744
The caves of France and northern Spain: 2977
The caves of France and northern Spain, a guide: 2978
The caves of Georgia: 637
The caves of Gower: 1259, 2552
The caves of Graatadalen, northern Norway: report of the Southwest Essex Technical College Caving Club expeditions 1963-5: 2875
The caves of greater Germany: 3324
Caves of Gunung Buda: report of the joint Sarawak Forest Department and USA caving expedition to Sarawak, Malaysia 1995: 1311
Caves of Herkimer County: 3096
The caves of Honduras: a first list: 1391
The caves of Huntingdon County, Pennsylvania: 772
Caves of Illinois: 468
Caves of India & Nepal: including the results of Speläologische Südasien Expedition 1981/82: 1168
Caves of Indiana: 2702
The caves of Ireland: 679

The caves of Jenolan: 1, the exploration and speleogeography of Mammoth Cave, Jenolan: 915
The Caves of Karawari: 791
The caves of Kentucky: Mammoth Cave, Colossal Cavern, White's Cave: 99
The caves of Kimble county: 3009
The caves of Kinney county: 3010
Caves of Lehigh County, Pennsylvania: 1865
The caves of Lubbock County: 2777
Caves of Madison County, Alabama: 1753
The caves of Maryland: 829, 1123
Caves of Massachusetts: 1387
The caves of McKittrick Hill, Eddy County, New Mexico: history and results of field work 1965-1976: 1871
The caves of Medina county: 2778
The caves of Mendip: 293, 332
Caves of Mexico: 2751
The caves of Mifflin County, Pennsylvania: 844
Caves of Missouri: 469
Caves of Monroe County, W. Va: 1427
Caves of Montana: 590
Caves of Morgan County, Alabama: 1754
Caves of Mulu '80: the limestone caves of the Gunong Mulu National Park, Sarawak: 936
Caves of Mulu '84: the limestone caves of the Gunong Mulu National Park, Sarawak: 937
Caves of Mulu: the limestone caves of the Gunong Mulu National Park, Sarawak: 503
Caves of mystery: the story of cave exploration: 891
Caves of New Caledonia: report of the 1975 Australian Expedition: 257, 1364
Caves of New Jersey: 795
The caves of Niagara County, New York: 996
The caves of north central Morgan County, Alabama: a guide for the 1974 SERA Winter Business Meeting of the National Speleological Society at Decatur, Alabama, February 1974: 1477
Caves of North Central West Virginia: 1164
Caves of North Tor Bay: 2723
The caves of north Wales: 2553
The Caves of North West Clare, Ireland: 3226
Caves of north-west Thailand: report of the Australian speleological expeditions, 1983-1986: 923
The caves of Northampton County, Pennsylvania: 3042
Caves of Northern Derbyshire: part 1: the Eldon Hill area: 951
Caves of Northern Derbyshire: part 2: Giants-Oxlow System: 952
Caves of Northern Derbyshire: part 3: Perryfoot/Coal Pit Hole: 953
Caves of Northern Derbyshire: part 4: Rushup Edge swallets: 954
Caves of Northern Derbyshire: part 5: Treak Cliff Hill: 955
The Caves of Northumberland County, Pennsylvania: 233
The caves of northwest Texas: 2779
Caves of Okinawa: 1644
Caves of Oregon: 1904
Caves of Paha Sapa, and other findings of Dr. Knows-It: 1104
Caves of Pennsylvania: 3135, 3471
Caves of Pennsylvania's Piedmont region: 2802
The caves of Perry County, Pennsylvania: 1592
Caves of Pierre Saint-Martain: 1058
Caves of Plymouth and District: 1712
The caves of Put-in-Bay: 2504
The Caves of Rana, Nordland, Norway: 2876
Caves of Randolph County: 2140
Caves of Rockcastle County, Kentucky: 2768
The caves of San Saba County: 2780
The caves of San Saba County part I: 2781
Caves of Schoharie County, New York: 833
The caves of Scotland (except Assynt): 2554
The Caves of Shend: 1443
Caves of Skye: 2874
The caves of Snyder County, Pennsylvania: 1879
The Caves of Sof Omar: 661
The caves of South Devon and their teachings: 1542
The caves of south eastern Kentucky: guidebook for the 1985 National Speleological Society Convention, Frankfort, Kentucky: 2406
Caves of South Wales: 3149
Caves of southeastern Pennsylvania: 2803
The caves of Table Mountain: a report on their present condition, potential, and conservation requirements: 681
The caves of Tenerife: 972
Caves of Tennessee: 326
The caves of Texas: 2357
The caves of Thailand: 916
Caves of the Alum Pot area: 2189
Caves of the Avon Gorge Bristol, part 2: eastern or Clifton side: 2838
Caves of the Avon Gorge Bristol: part 1: western or Leigh Woods side: 2839
Caves of the Big Horn - Pryor Mountains of Montana and Wyoming. Issued in conjunction with the 28th anniversary convention of the National Speleological Society at Lovell, Wyoming, Harley A Leach, convention chairman: 2391
Caves of the Bristol region: 2245
Caves of the Catabrian Mountains, north-west Spain: 731
Caves of the central northern outcrop: 2555
Caves of the coastal areas of South Australia: 2940
Caves of the Current River Valley: 2204
The caves of the earth: their natural history, features, and incidents: 1815, 2805
Caves of the eastern panhandle of West Virginia: 1286

Caves of the eastern White River Plateau: guidebook to the 1994 Rocky Mountain regional, Willow Peak, Colorado: 2835
Caves of the gem state: 1172
The caves of the great hunters: 347
Caves of the Inter-American Highway: Nuevo Laredo, Tamulipas to Tamazunchale, San Luis Potosi: 2871
Caves of the Jacks Fork Valley: 2205
The caves of the Little Neath Valley: 2556
Caves of the Mellte Valley: 2543
Caves of the North Outcrop: 1166
Caves of the Nullarbor; a review of speleological investigations in the Nullarbor Plain, Southern Australia: 917
Caves of the Organ Cave Plateau: Greenbrier County, West Virginia: 3124
Caves of the Peak District: 1203
Caves of the Reserve area: 2064
Caves of the San Juan Plateau: 3138
Caves of the Sequoia Region California, issued in conjunction with the 25th anniversary convention of the National Speleological Society, David R McClurg, convention chairman: 2388
The caves of the south eastern outcrop: 2557
The caves of the south eastern outcrop and caves and mines in the Forest of Dean: 2558
The caves of the South Gower coast: an archaeological assessment: 823
Caves of the southern Cumberland valley: 3007
The caves of the southern outcrop: 2559
The caves of the Stockton Plateau, Texas: 1872
The caves of the Tennessee State Natural Areas system: 677
Caves of the Upper Gila and Hueco areas in New Mexico and Texas: 720
Caves of the Watertown Area, New York state: 1066
Caves of Thunder expedition report: an expedition to the world's largest underground river, Irian Jaya, Indonesia 1992: 438
The caves of Travis County: 2782
The caves of Val Verde County: 2783
Caves of Virginia: 890
Caves of Washington: 1325
The caves of west Wales: 2560
The caves of Western Australia: 100
Caves of Western Pennsylvania compiled by members of the Mid-Appalachian Region of the National Speleological Society: 3472
The caves of Willamson County: 2784
Caves of Wyoming: 1461
The caves on the Tés Plateau and in the Bakony Mountains: 1007
Caves through the ages: 885
Caves, a deeper look at our Earth: 1197
Caves, an anthology: 993
Caves, crags, and gorges: a guide to the limestone country of England and Wales: 3389
Caves, ground water and karst features of the Cretaceous in South central Texas: fourteenth annual S.A.S.G.S. field conference, April 6-8, 1973: 408
Caves, karst, and environmental impact in the new river drainage basin of Virginia and West Virginia: guidebook for a geologic fieldtrip, Appalachian karst symposium, Radford, Virginia, 23 March 1991: 1773
Caves—origins, development and formations: 1087, 3537
Caves: processes, development, and management: 1211
Caves: the dark wilderness: 980, 2804
Caving: 449, 774, 1159, 1972, 3412
Caving and potholing: 1762, 2828
Caving and potholing techniques: 671, 1088
Caving basics: 1380
Caving basics: a comprehensive manual for beginning cavers: 1381
Caving basics: a comprehensive guide for beginning cavers: 2767
Caving expeditions: 3503
Caving for beginners: 723
Caving handbook: a guide to the underworld: 3259
Caving in America: the story of the National Speleological Society, 1941-1991: commemorating 50 years of history and growth: including a special illustrated history of cave exploration in the society entitled—the last frontier for the pioneer: 797
Caving in Australia: 259, 1346
Caving in Italy: 659
Caving in the abode of the clouds: the caves and karst of Meghalaya, north east India: 513
Caving in the Crickhowell area: 3375
Caving in the heartland: a guidebook for the 1992 Convention of the National Speleological Society, Salem, Indiana: 2413
Caving in the Transvaal: 991
Caving in West Virginia: 101
Caving information series: 2354
Caving log 1942-1950: 1793
A caving manual: 1973
Caving periodicals in the Central Collection of Caving Publications: 2039
Caving periodicals in the central collection of caving publications: a catalogue of the caving periodicals held by Bristol Central Reference Library and a selected short title catalogue of books: 480
Caving practice and equipment: 1763
Caving safety code: first aid cave rescue procedure: 615
Caving short course: 1982 NSS Convention: 2111
Caving, episodes of underground exploration: 284
Caving: a child's adventure: 2335
Caving: an introductory guide to spelunking: 1690
Caving: the Sierra Club guide to spelunking: 1909

Cavity detection and delineation research: Report 1: Microgravimetric and magnetic surveys: Medford Cave Site, Florida: 570

Cavity detection and delineation research: Report 3: Acoustic resonance and self-potential applications: Medford Cave and Manatee Springs sites, Florida: 710

Cavity detection and delineation research: Report 4: Microgravimetric survey: Manatee Springs site, Florida: 571

Cavity detection and delineation research: report 5: electromagnetic (radar) techniques applied to cavity detection: 308

Cavity detection and delineation research; Prepared for Office, Chief of Engineers. Report 2: Seismic methodology: Medford Cave Site, Florida: 783

CDG northern sump index 1995: 2253

CEGSA 1965-6 Nullarbor expedition: 1724

Celebrated American caverns: especially Mammoth, Wyandot, and Luray: together with historical, scientific, and descriptive notices of caves and grottoes in other lands: 1535, 1536

Celebrated American caves: 2249

Cenote of sacrifice: Maya treasures from the sacred well at Chichen Itza: 678

The cenotes of Yucatan: 2629

The cenotes of Yucatan: a zoological and hydrological study: 2630

Centipede and Damp caves: excavations in Val Verde County, Texas, 1958: 998

Central Oregon caves: 1905

Cerberus series: Maypole series: 386

A ceremonial cave in the Winchester Mountains, Arizona: 1129

Challenge underground: 369

Characteristics of large springs in Kentucky: 3346

Chauvet Cave: the discovery of the world's oldest cave paintings: 642

Check-list of Australian caves and karst, 1979: 2092

A checklist and annotated bibliography of the subterranean aquatic fauna of Texas: 2785

A checklist of invertebrate species recorded from Missouri subterranean habitats: 2206

A checklist of the cave fauna of Texas: 2786

A checklist of the caves of Texas: 2787

Checklist of troglobitic organisms of Middle America: 3158

Cheddar: 102

Cheddar caves: 3218

Cheddar Caves bibliography: 1675

Cheddar caves in the Cheddar Gorge: an illustrated survey & guide: 675

Cheddar Caves Museum: 2895

Cheddar Caves: illustrated official guide: 3106

Cheddar Gorge and caves: 2669

Chiang Dao Cave: 103

Chillagoe souvenir guide: 2632

China caves '85: the first Anglo-Chinese project in the caves of south China: 3390

China Caves Project 1987—1988: the Anglo—Chinese project in caves of South China: 1074

A chronicle of selected northeastern caves: a history guide for the 1979 NSS Convention, Pittsfield, Massachusetts, August 5-12, 1979: 990

Chronicles of the Reading Grotto, in which we go to the California convention, June 3rd to September 6th, 1966: 1937

Chronology of diving activities and underground surveys in Devils Hole and Devils Hole Cave, Nye County, Nevada, 1950-86: 1475

Church Creek Caves: 1189

Clair de Roche: 2486

The Clapham Cave with illustrations from photographs by the author: 520

The Clark's Cave bone deposit and the late Pleistocene paleoecology of the central Appalachian Mountains of Virginia: 1280

Classification of lava tubes: 1372

Cliff castles and cave dwellings of Europe: 322

The cliffs & caves of Cheddar: a series of beautiful views in real photogravure published by William Gough, son of the discoverer of the famous cave: 1232

Climbing blind: 2813

Closani Cave: mineralogical and genetic study of carbonates and clays: 1134

Club history 1956—1981 25th anniversary publication: Cerberus Spelaeological Society: 2715

CMW Bran CC Journal index vol. 1-20; 1967-1994 including newsletters, newssheets & the Silver Jubilee Journal: 1551

Coast and karst in Istria Rossa; a report of geographical field work carried out during August, 1960: 1174

Coastal caves, marine limits, and ice retreat in Lofoten-vesteralen, North Norway: 2251

Cold Water Cave: 1830

Collected caving songs: volume 1: the songs of Mendip: 397

Colloque international de sedimentologie karstique = international meeting on karstic sedimentology, Han-sur-Lesse, Belgique, 18-22 Mai 1987: 1642

The Colong story: 686

Colorado caves and karst 1974: 1382

Colorado caves and karst 1975-1976: 1383

Colourful world of caves: 1412

Come crawl into the caves of western Pennsylvania: 1428

Coming to South Wales caves?: 1948

Comments on the relationships of the North American cave fishes of the family Amblyopsidae: 2855

Commercial caves in Southern Indiana: 2190

The commercial caves of Camden County, Missouri: 2207

Communications 1-7 August 1986: 1633

Communications = mitteilungen = soobshcheniia: international symposium on physical, chemical and hydrological research of karst: Kosice, Czechoslovakia, May 10-15, 1988: 1664

Comparative optic development in *Astyanax mexicanus* and in two of its blind cave derivatives: 575

Comparative studies in the light sensitivity of blind characins from a series of Mexican caves: 465

A comparative study of stylistic and extrapictorial elements in the murals of Cave I, Ajanta and of the Brancacci Chapel: 411

The complete book of cave exploration: an authoritative guide to the wonders, mysteries and excitement of caves and caving: 2677

The complete caves of Devon: 2561

The complete caves of Mendip: 333

The complete hill walker, rock climber and cave explorer: 2608

A computer based inventory of recorded recent sinkholes in Florida: 355

The Conard fissure, a Pleistocene bone deposit in northern Arkansas: with descriptions of two new genera and twenty new species of mammals: 517

The concise caves of Devon: 2565

The concise caves of north Wales: 2562

Conference abstracts: [of papers presented at Changing karst environments, hydrogeology, geomorphology and conservation, an international symposium held at the Universities of Oxford and Huddersfield, September 1994]: 1647

Conference on the karst and cave research activities of education and research institutions in Hungary: papers: Jósvafő 17-19 May 1991: 3550

Confidence in the cave. A guide for a safe speleological adventure: 1262

The conquest (the riddle of an abyss and its sinking river): 4

The conquest of the caves and underground rivers of Czechoslovakia's Macocha Abyss: a historical and technical study of their exploration: 5

A consideration of the effect of surface hot-cold cycles on caves: 1326

Considerations of Lower Uilani Cave for inclusion in the Hawaii Natural Area Reserve System: 1327

Construction and testing of an underground radio: a report submitted to the Research and Educational Branch of Workers' Health, Safety and Compensation, Alberta Government: 726

Construction of phases II and III of the Blanchard Springs Caverns Project final environmental statement: 3280

Construction of phases II and III of the Blanchard Springs Caverns Project: draft environmental statement: 3281

Contemporaneity of man with the extinct Mammalia, as taught by recent cavern exploration, and its bearing upon the question of man's antiquity: 583

Continued investigation of water losses to sinkholes in the Pahasapa Limestone and their relation to resurgent springs, Black Hills, South Dakota: 1260

The contribution to the history of the speleological explorations of the West Indies at 500—anniversary of the discovery of America: 1861

Contributions to the archaeology of Mammoth Cave and vicinity, Kentucky: 2489

Contributions to the study of Karst: 3500

Control of exotic plants in Oregon Caves, Oregon Caves National Monument: 35

Cooleman Plain karst area management plan: a supplementary plan to the Kosciusko National Park plan of management: 2496

Correlation of the deposits of Sandia Cave, New Mexico, with the glacial chronology: 542

Cougar Mountain Cave in south central Oregon: 727

Council of Southern Caving Clubs handbook: 3100

Cowboy Cave: 1718

Cox's Stalactite Cavern, Cheddar: 104

Crag Cave: Castle Island: Co. Kerry: Ireland's most exciting show cave: 2544

Craters, caverns, and canyons: delving beneath the earth's surface: 1005

Creation of Carlsbad Caverns: 2943

The creatures underneath: 864

Creswell Caves vs Professor Boyd Dawkins: 1417

Crewe Climbing and Pot Holing Club; Peak rigging guide: 105

Crewe Climbing and Potholing Club: Gouffre Berger Expedition: 1207

CRF personnel manual: 617, 819

CRF Personnel Manual: Central Kentucky karst area: 1126

The Crocodile Caves of Ankarana: 1986: an expedition to study and explore the limestone massif of Ankarana in northern Madagascar: 3386

Crypts, caves and catacombes: subterrenea of Derbyshire & Nottingham: 2122

The Crystal Cave at Virginsville: 1845

Crystal Cave in Sequoia National Park: 2540

Crystal cave: a guidebook to the underground world of Sequoia National Park: 861

Crystal Lake Cave Iowa: 1780

The crystalline structures of the caves of Belgium: 2719

Cuba Contact '88: an investigative visit by members of the Westminster Speleological Group to the karst of the Sierra de los Organos 18/9/88 - 7/10/88: 3487

Cueva de Nerja: English edition: 3345

Cultural resource management in Mammoth Cave National Park: a National Park Service—Kentucky Heritage Council cooperative project: 2514

The culture history of Lovelock Cave, Nevada: 1272

Cumberland Caverns: 2088

Cumulative index for the National Speleological Society Bulletin, volumes 1 through 45, and Occasional papers of the N.S.S., numbers 1 through 4: 2358

Current foundation engineering practice for structures in karst areas: 1835

Cutta Cutta Caves Nature Park: draft plan of management: 1347

Cwmbran Caving Club New Mexico expedition report; caving in the Guadalupe Mountains and Carlsbad National Park April 1994: 2892

Czech Speleological Society 1982-1986: published on occasion of 9th International Speleological Congress Spain 1986: 1448

Czech Speleological Society 1986-1989: published on occasion of 10th International Speleological Congress Hungary 1989: 854

Czechoslovak speleological expedition to Nepal Himalaya 85: 660

The D. C. Jester cave system: 453

Dan yr Ogof and its associated caves: 672

Dan-yr-ogof show caves: 106

Danger Cave: 1719

Danger Cave, Last Supper Cave, and Hanging Rock Shelter: the faunas: 1246

The darkness beckons: the history and development of cave diving: 1028

The darkness under the earth: 607

A data recovery study of Layser Cave (45-LE-223) in Lewis County, Washington: 809

Daugherty's Cave: a stratified site in Russell County, Virginia: 390

Dawn of art: the Chauvet Cave: the oldest known paintings in the world: 643

A day at the cave: 3457

De Grotten Van de Nederlande Antillen - Caves of the Netherland Antilles: 3370

The decomposition of limestone breccia from the cave of Les Eyzies, Dordogne, France: 2833

Deep Cave, Texas. A preliminary report of the 1965 Project "Deep": 1006

Deep down: great achievements in cave exploration: 1480

Deep in caves and caverns: 2692

Deep in the karst of Texas: general information program: abstracts NSS-New Braunfels 1978: 2430

Deep into blue holes: the story of the Andros Project: 2604

The deepest caves in the world and caves which have held the world depth record: 2953

Deer Creek cave, Elko County, Nevada: 2974

Delaware Basin exploration: Guadalupe Mts., Hueco Mts., Franklin Mts., geology of the Carlsbad Caverns ; Nov. 6th, 7th and 8th, 1969: 3444

Delaware Basin exploration: Guadalupe Mts., Hueco Mts., Franklin Mts., geology of the Carlsbad Caverns, Nov. 6th, 7th and 8th, 1969: 3445

Delineation and hydrogeologic study of the key cave aquifer Lauder Dale County, Alabama: 36

Delineation of sinkholes using thermal infrared imagery: 3132

Delineations of the north western division of the county of Somerset, and of its antediluvian bone caverns, with a geological sketch of the district: 2873

Depositional systems and karst geology of the Ellenburger Group (Lower Ordovician), subsurface west Texas: 1808

Depths of the earth: caves and cavers of the United States: 1328

Depths of the earth: caves and cavers of the United States: 1329

Derbyshire caving 1987: the handbook of Derbyshire caving: 858

Derbyshire sump index: 1089

Derivation of hydrologic frequency caves: 1922

The Descent caver's handbook: 1552

Descent handbook for cavers: 368

Descent handbook for cavers 1971/1972: 370

Descent handbook for cavers 1974: 860

The Descent index: issues 1-100.: 1553

The Descent magazine's caving yearbook: 371

The descent of Pierre Saint-Martin: 608

A description of Howe's Cave: with a popular treatise on the formation of caves in lime rock, from the size of a quill to a mammoth: 107

Description of points of scenic, historic, and geologic interest between Washington, D. C., and Luray, Virginia: 3148

A description of the Endless Caverns of New Market: 1441

A description of the fluor spar or Blue John Mine at Castleton: 540

Description of the fossil bones of the Megalonyx, discovered in White Cave, Kentucky: 1360

A description of the grand Devonshire cavern at Matlock: 541

A description of the Mammoth Cave of Kentucky, the Niagara River and Falls, and the Falls in summer and winter; the prairies, or life in the West; the Fairmont Water Works and scenes on the Schuylkill, xc. xc.: to illustrate Brewer's Panorama: 108

A description of the Mammoth Cave of Kentucky, the Niagara river and falls, and the falls in summer and winter; the prairies, or life in the West; the Fairmount Water

Works and scenes on the Schuykill, etc. etc.: to illustrate Brewer's panorama.: 109
A description of the New Market Endless Caverns: 1442
A description of the Sotano del Rio Iglesia: 2015
Description of the spar cave, lately discovered in the Isle of Skye: with some geological remarks relative to that island: 2014
A description of Weast's Cave: 2513
A description of Weyer's Cave: 704
A description of Weyer's Cave in Augusta County, Va: 3453
A description of Weyer's Cave in Augusta County, Virginia: 2247
A description of Weyer's Cave, in Augusta County, Virginia: 3454
Descriptions of Pennsylvania's undeveloped caves: 3136
Descriptions of Tennessee caves: 2089
Descriptions of Virginia caves: 1495
A descriptive account of a descent made into Penpark-hole, in the parish of Westbury-upon-Trim, in the county of Gloucester, in the year 1775: 612
Descriptive account of the Blue John Mines and Caverns at Castleton, Derbyshire: 2863
A descriptive tour, and guide to the lakes, caves, mountains, and other natural curiosities, in Cumberland, Westermoreland, Lancashire, and a part of the West Riding of Yorkshire: 1533
The detection and mapping of subterranean water bearing channels: completion report: 2769
The detection and mapping of subterranean water bearing channels: phase 2: final report: 3325
Detection of subsurface cavities: 343
The detection of subsurface voids by gravimetry: 2852
Detection of water-filled and air-filled underground cavities: 1881
Determining an environmental and social carrying capacity for Jenolan Caves Reserve: applying a visitor impact management system: 1348
Development concept plan, Slaughter Canyon, Carlsbad Caverns National Park, New Mexico: environmental assessment: 3323
Development concept: Slaughter Canyon, Carlsbad Caverns National Park, New Mexico: 3290
Development of caves: 110
The development of leisure and educational resources at Jenolan Caves Resort: 1396
Development of sinkholes resulting from man's activities in the eastern United States: 2505
Development of solution features, Cloverlick Valley, Pocahontas County: 3441
Devil's Hole Pupfish Recovery Plan: 862
Devil's Lair, a study in prehistory: 887
Devils Lair: a search for ancient man in Western Australia: 886
Devil's Sinkhole area: headwaters of the Nueces River: 1792
Devonshire sump index: 1245
The Dictaean Cave: 2609
A dictionary of karst and caves: 1975
Directions of ground-water flow and locations of ground-water divides in the Lost River watershed near Orleans, Indiana: 350
The Diros Caves of Mani: Alepotrypa and Glyphada: 2660
Discover Naracoorte caves: 1939
Discover Timpanogos Cave National Monument: 1730
Discovering caves: 2563
Discovering Lascaux: 910
The discovery and exploration of St. Cuthbert's Swallet: 1676
Discovery and exploration of the Oregon caves: Oregon Caves National Monument: 3379
The discovery and exploration of the Yanchep Caves: 958
The discovery and history of Carlsbad Caverns, New Mexico: 3460
Discovery at the Rio Camuy: 1295
Discovery at the Rio Camuy, Puerto Rico: 1296
The discovery of Barralong Cave, Jenolan Caves, N.S.W.: based on the diary of Ronald L. Newbould, guide, Jenolan Caves, co-discoverer with John P. Culley, senior guide, Jenolan Ceves [sic] of the "Barralong Cave" on the 7th June, 1964: 2501
Discovery of Luray Caverns, Virginia: 1297
Discovery of the Onyx Cave and my biography: 3253
District plan discussion paper: tourism issues: 3374
A diverse hominoid fauna from the Late Middle Pleistocene breccia cave of Tham Khuyen, Socialist Republic of Vietnam: 2919
Dolomite caves of the eastern Transvaal (South Africa): a free caver monograph 1985: 2298
Doolin-St. Catherine's Cave: Co. Clare, Eire: 1952
Dossey Domes: most beautiful of all the caverns at Green River Landing (One-half mile from Mammoth Cave, Ky.); Cliff Walk along Green River: finest scenery in the state: 111
Down to the sunless sea: 435
Down under Texas: 3026
Draft management plan: Tantanoola Caves Conservation Park: Lower South East, South Australia: 260
Draft general management plan, environmental impact statement: Carlsbad Caverns National Park, New Mexico: 3291
Draft management plan, Tantanoola Caves Conservation Park, lower South East, South Australia: 3055
Draft management plan: Naracoorte Caves Conservation Park: South East South Australia: 1349
Draft recovery plan for endangered karst invertebrates in Travis and Williamson Counties, Texas: 961

Draft strategy for the management of caves and karst in Victoria: 2352

Draft, environmental impact statement, general management plan, development concept plan Timpanogos Cave National Monument: 3292

Draft, general management plan, environmental impact statement Jewel Cave National Monument, South Dakota: 3293

The dream of Kent's Cavern and ballads of the Bay: 944

Dunhuang: caves of the singing sands: Buddhist art from the Silk Road: 3479

Dunkle Portale: 1118

Dunmore Cave: 10

Dunmore Cave, County Kilkenny: a reassessment: 899

Dunmore Cave: a short guide: 900

Dwarf deer and other late Pleistocene fauna of the Simonelli Cave in Crete: 2023

The dynamic of the contemporary karstic processes in the tropical area of Cuba: preliminary report of the field investigations performed by the expedition Guajaibon '84 in the winter season 1984: 2731

E.A. Martel 1859 - 1939 bibliographie: 628

The early cave-men: 883

Early caves of Maharashtra: a cultural study: 2161

The early days of Olsen's Caves as seen through a geologist's eyes: 3022

Earth resistivity and hole-to-hole electromagnetic transmission tests at Medford Cave, Florida: 1108

The Ease Gill system: forty years of exploration: 1021

East Chestnut ridge hydroelectric characterization: a geophysical study of two karst features: 997

The easy guide to Chichen Itza, Balankanchen and Izamal: 418

Echo cave: a tentative Quaternary chronology for the Eastern Transvaal: 2046

Ecological analysis of the Kentucky cave shrimp, *Palaemonias ganteri* Hay, at Mammoth Cave National Park (phase V): 1496

Ecological analysis of the Kentucky cave shrimp, *Palaemonias ganteri* Hay, at Mammoth Cave National Park (phase VI): preliminary observations on stream interstitial meiofauna communities and related abiotic factors: 1497

Ecological analysis of the Kentucky cave shrimp, *Palaemonias ganteri* Hay, Mammoth Cave National Park: 1498

Ecological analysis of the Kentucky cave shrimp, *Palaemonias ganteri* Hay, Mammoth Cave National Park (phase I): 1499

Ecological analysis of the Kentucky cave shrimp, *Palaemonias ganteri* Hay, Mammoth Cave National Park (phase II): 1500

Ecological analysis of the Kentucky cave shrimp, *Palaemonias ganteri* Hay, Mammoth Cave National Park (phase III): 1501

An ecological and taxonomic account of the algae of a semi-marine cavern, Paradise Cave, Queensland: 755

Ecological effects of water pollutants in Mammoth Cave: final technical report to the National Park Service: 327

Ecological studies on Hawaiian lava tubes: 1543

Ecology of a limestone cave: 2010

The el cheapo book of home brew diving equipment: 700

El Karst / Karst: 1114

Electric light views in the Caverns of Luray at Luray Station (Page Co., Virginia), Shenandoah Valley Railroad: 1695

Electric light views in the Caverns of Luray. The Caverns of Luray (at Luray, Page County, Virginia, a station on the Shenandoah Valley Railroad) as a resort for tourists ... are unexcelled, in their wonderful attractiveness, by any other creation of nature: 1995

Ellison's Cave: Georgia: 3222

Ellison's Cave: Georgia's finest: 2909

Emil Racoviță 1868-1947: 2283

Enchanted corridors: 912

Encyclopaedia biospeologica tome 1: 1761

An endangered heritage: a story of the Meramec Valley, its caves, and their possible destruction by the U. S. Army Corps of Engineers: 2148

The Endless Caverns of the Shenandoah Valley, an account of the wonderful work of water: with special reference as to how the caverns and the Shenandoah Valley were formed: 2799

Endless caverns, New Market, Virginia, in the heart of the Shenandoah Valley...3 minute drive off U.S. 11 (Lee highway) over new boulevard: 981

Endless Caverns, Virginia. "Considered the most beautiful of the Shenandoah Valley caverns" ...: 982

Endless Caverns, wonderful and spectacular: 983

Endless Caverns, wonderful and spectacular, New Market, Virginia, in the heart of the Shenandoah Valley: 984

Endless Caverns: on U.S. 11, 3 miles south of New Market, Virginia: illuminated by indirect electric flood lighting: 985

Engineering and environmental impacts of sinkholes and karst: proceedings of the third Multidisciplinary Conference on Sinkholes and the Engineering and Environmental Impacts of Karst, St. Petersburg Beach, Florida, 2-4 October 1989: 356

Engineering and geological conditions of karst: 1978

The Ensueño Cave Study: Ensueño Cave, Hatillo, Puerto Rico; expedition period: February 18-25, 1987: 1290

The entombment of Floyd Collins in Sand Cave, Kentucky: 464

Environmental change in karst areas: 1081

Environmental Education Enterprises, Inc. short course showcase January 26-27-28, 1994, Grosvenor Resort Orlando, Fl: Winter Park sinkhole guidebook: 112

Environmental geology and hydrogeology of the Ocala area, Florida: 1894

Environmental geology and water sciences: special issue on sinkholes: 113

Environmental impact statement, Mammoth Cave area, Kentucky: wastewater facilities: 3273

Environmental impact statement; Mammoth Cave area, Kentucky, wastewater facilities: 3274

Environmental karst: 889

Environmental monitoring of karst and caves: 3483

Erosionsforloppet, med sarskild hansyn till Lummelundagrottorna. The rate of erosion, with special reference to the caves of Lummelunda: 3195

An essay on the caves of the Cherokee Nation: a conservation plea: 2869

Eternal caves: 790

An ethno-archaeological examination of Samwel Cave: 3228

Euceratherium: a new ungulate from the Quaternary caves of California: 2989

Evaluation of breccia pipes in southeastern New Mexico and their relation to the Waste Isolation Pilot Plant (WIPP) site: 3043

Evaluation of Mount St. Helens lava casts and caves, Cowlitz and Skamania Counties, Washington, for eligibility for registered natural landmark designation: 2951

Evidences of early occupation in Sandia Cave, New Mexico, and other sites in the Sandia-Manzano Region: 1452

Excavation at Hokeb Ha, Belize: 2595

Excavation in the Bacho Kiro Cave (Bulgaria): final report: 1856

Excavation of Cowboy Caves, June 3-July 26, 1975: preliminary report: 1720

An excavation of Hermit's cave, New Mexico: 1047

The Excavation of Squirt Cave, 45WW25: 687

Excavation of two sites in the Coso Mountains of Inyo County, California: 2095

Excavation techniques in Pin Hole Cave, Creswell Crags S.S.S.I., Derbyshire: 512

Excavations at Chelm's Combe: Cheddar: conducted under the Excavations Committee of the Somerset Archaeological and Natural History Society 1925-26: 294

Excavations at Eduardo Quiroz Cave, British Honduras (Belize): 2640

Excavations at Franchthi Cave, Greece: 1688

Excavations at Maria de la Cruz Cave and Hacienda Grande Village Site, Loiza, Puerto Rico: 2861

Excavations at the Kesserloch near Kesserloch near Thayngen Switzerland: a cave of the Reindeer Period: 2159

Excavations in Moaning Cave: 2576

Excavations of Actum Polbilche, Belize: 2642

Excavations of Habitation Cave: 2997

Excursion E11 supplemental materials: karst of the central Appalachians: field excursion on the occasion of the eighth International Congress of Speleology: 1626

An excursion from Lancaster, up the vale of Lune, and from Kirkby Lonsdale, to the caves of Yorkshire: 834

Excursion guide book: 5th international symposium on vulcanospeleology: [pre-activity: November 4-9, 1988: post-activity: November 13-20, 1988]: 1657

An excursion to the Mammoth Cave, and the barrens of Kentucky: with some notices of the early settlement of the state: 821

Expedition 67 to the Gou: 3405

Expedition to Morocco 1969: 1015

The Expedition to the Judean Desert: 1451

Experimental geology applied to speleogenesis: 2931

Exploration '66; University of Nottingham; biospeleological research expedition to Ireland; Riverview work project in Portugal; British speleological expeditions to Turkey: 114

The exploration & survey of Ellison's Cave, Georgia: 3027

The exploration and speleogeography of Mammoth Cave, Jenolan: 918

The exploration and speleography of Mammoth Cave, Jenolan: 919

The exploration and survey of McBridge Cave, Jackson County, Alabama: 586

Exploration in Wind Cave: 531

An exploration journal of Llangattwg Mountain: 3122

The exploration of Carlsbad Cavern: 2509

An exploration of Durham Cave in 1893: 2151

The exploration of Jacobs Cavern, McDonald County, Missouri: 2622

Exploration of mummy caves in the Aleutian Islands ...: 1559

The exploration of Samwel Cave: 1132

Exploration of the caves and rivers of New South Wales (minutes, reports, correspondence, accounts): 115

Explorations in northeastern Arizona; report on the archaeological fieldwork of 1920-1923: 1278

The explorations journal of The University of Leeds Speleological Association: 504

Exploring Alabama caves: 801

Exploring American caves: 1076

Exploring American caves, their history, geology, lore, and location: a spelunker's guide: 1077

Exploring caves: 775, 1963

Exploring caves: a guide to the underground wilderness: 2112

Exploring Missouri caves. A guidebook for the 1997 convention of the National Speleological Society Sullivan, Missouri June 23-27, 1997: 2418

Exploring the cave world: cave packet for teachers, intermediate level: 1608

Exploring the Endless Caverns of New Market Virginia on US 11 in the heart of the Shenandoah Valley: 1900
Exploring the Endless Caverns of New Market, Virginia in the heart of the Shenandoah Valley: two authentic and thrilling accounts of the adventures of members of the American Museum of Natural History and of the Explorers Club, who vainly sought the end to the Endless Caverns: 986
Exploring the Endless Caverns of New Market, Virginia. Authentic and thrilling accounts of the adventures of members of the Explorers Club and of the American Museum of Natural History: 987
Exploring the underground: 1237
Ezra's Retreat: a rockshelter/cave occupation site in the north central Great Basin: 319

Factors affecting sinkhole formation: 634
Factors altering the microclimate in Carlsbad Caverns, New Mexico: 2133
Fairy Cave: 116
Fairy Cave Quarry: a study of the caves: 2717
Fall 1968 MVOR guidebook: Stone County, Arkansas: 2309
Fall 1971 MVOR: Glover's Cave: 2313
Fall 1976 MVOR guidebook, Perry Co., Mo: 2319
The falls and caves of Ingleton: 1338
Famous caverns and grottoes: described and illustrated: 13
Famous caves and catacombs: described and illustrated: 14
Far west cave management symposium proceedings: Portland Oregon April 14-16, 1981: 1027
Far West cave management symposium proceedings: Redding, California October 23-26, 1979: 1026
Father of prehistory: the Abbé Henri Breuil: his life and times: 495
Fauna collected from caves as recorded in the C.R.G. fauna records: part I (1938-39): 619
Fauna collected from caves as recorded in the C.R.G. fauna records: part II (war years 1940-46 and 1945-46): 620
Fauna collected from caves as recorded in the C.R.G. fauna records: part III (1947): 1400
Fauna collected from caves, mines and wells as recorded in the C.R.G. fauna records: part IV (1948-1949): 1401
Fauna collected from caves, mines and wells as recorded in the C.R.G. fauna records: part V (1950-1953): 1402
Fauna collected from caves, mines and wells as recorded in the C.R.G. fauna records: part VI (1954-1955-1956): 1403
Fauna collected from caves, mines and wells as recorded in the C.R.G. fauna records: part VII (1957-1959): 1404
Fauna collected from caves, mines and wells as recorded in the C.R.G. fauna records: part VIII (1960-1962): 1405
The fauna of Burnet Cave, Guadalupe Mountains, New Mexico: 2913
The fauna of Mayfield's Cave: 314
The fauna of Papago Springs Cave, Arizona, and a study of Stockoceros; with three new Antilocaprines from Nebraska and Arizona: 2995
Fauna of the caves of Yucatan: 2631
Faunal and archeological researches in Yucatan caves: 1386
Feathers from Sand Dune Cave: a Basketmaker cave near Navajo Mountain, Utah: 1359
Federal Cave Resources Protection Act and restriction of dams in parks and monuments hearing before the Subcommittee on Public Lands, National Parks, and Forests of the Committee on Energy and Natural Resources, United States Senate, One Hundredth Congress, second session, on S. 927/H.R. 1975 ... H.R. 1173 ... June16,1988: 3272
Federal Cave Resources Protection Act of 1988: report (to accompany H.R. 1975): 3270
Fern cave: the history of the discovery, exploration and mapping of the Fern Cave system: 2332
A field guide to caves of Blanco, Gillespie, and Llano counties, Texas: 2788
Field Guide to karst features in southeast South Australia and Western Victoria: for the Karst Studies Seminar, Naracoorte, February 1996: 1267
Field guide to New York caves: Schoharie County: 2922
A field guide to the caves of Kendall County: 962
Field guide to the karst geology of San Salvador Island, Bahamas: prepared for the 10th Friends of Karst meeting, February 11-15, 1988, College Center of the Finger Lakes, Bahamian Field Station, San Salvador Island, Bahamas: 2327
Field investigation report: Lehman Caves - Wheeler Peak: portion of southern section of Snake Range, White Pine County, Nevada: 2302
Field trip guidebook for the central Appalachian carbonate hydrology workshop, The Pennsylvania State University, University Park, Pennsylvania, May 21, 22, 1970: 2651
Fifth national cave management symposium: Mammoth Cave National Park, October 14-17, 1980: 2341
Fifty typical Swedish caves: 3196
Final Dark Canyon environmental impact statement: 3262
Final environmental statement: Mammoth Cave National Park master plan and wilderness suitability study, Kentucky: 3294
Final environmental statement: proposed wilderness, Carlsbad Caverns National Park, New Mexico: 3295
Final master plan: Mammoth Cave National Park: 3296
Final report on the archaeology of Fountain Cavern, Anguilla, West Indies: a report: 3417
Final report on the survey and assessment of the prehistoric and historic archaeological remains in Big Bear Cave, Van Buron County, Tennessee: 764

Final reports on the Lehman Caves studies: 3336
Final, general management plan, environmental impact statement Jewel Cave National Monument, South Dakota: 3297
The finding of the remains of the fossil sloth at Big Bone Cave, Tennessee, in 1896: 2152
Fingal's cave: 2925
First aid for cavers: 1217
The first book of caves: 1340
First international cave management symposium: proceedings [held at] College of Environmental Sciences, Murray State University, July 15-18, 1981: 2328
First reports: 1984-1988: 2752
Fishes from the caves of Yucatan: 1562
[Fist [sic] to third] reports on the exploration of "Dog Holes" Cave, Warton Crag, near Carnforth, Lancashire: 1686
The Flint Ridge Cave System: Mammoth Cave National Park, Kentucky; a Cave Research Foundation cartographic project in cooperation with the National Park Service: 535
Flooding of the Sinking Creek Karst area in Jessamine and Woodford Counties, Kentucky: 782
The Florida Caverns at Marianna, Florida: 1236
Florida Caverns: a nature-made wonderland: 3359
Floyd Collins, greatest cave explorer ever known: 2002
Focus on the Cango Caves: 117
Formation of caves: 267
Formation of solution-subsidence sinkholes above salt beds: 947
Formation of the Wink Sink: a salt dissolution and collapse feature, Winkler County, Texas: 348
Formation processes in archaeological context: 1221
Fossil cave, 5L81 underwater palaeontological and surveying project, 1987-1988: 1519
A fossil deposit in a cave in St. Louis: 2984
Fossil vertebrates from Miller's Cave, Llano County, Texas: 2618
Fossils of Honeycomb Hill: 3528
Foundation considerations in siting of nuclear facilities in karst terrains and other areas susceptible to ground collapse: 1121
Four non-commercial caves in Southern Indiana: 2191
Fourteen years at the Badger Hole: 300
Franchthi Cave and Paralia: maps, plans, and sections: 1689
Franchthi Paralia—the sediments, stratigraphy, and offshore investigations: 3490
The freshwater amphipod crustaceans Gammaridae of North America: 1502
Freshwater Triclads (Turbellaria) of North America, IX, the genus *Sphalloplana*: 1791
The Friesenhahn Cave: 1011

From Katoomba to Jenolan Caves: the Six Foot Track, 1884-1984: 3023
From the cradle to the cave: the life story of "Dad" Truitt, "Cave Man of the Ozarks": 2676
From the mouth of the dark cave: commemorative sculpture of the late classic Maya: 340
Fulton County, Pennsylvania: 3086
Fumaroles, ice caves and time: the variety and scope of research on Mt. Baker: 2299
The functional cell types in the pars distalis analogue of the pituitary gland in the blind mexican cave fish, *Anoptichthys jordani*: 2085
Further discoveries in the Cresswell caves: 841
Further excavations (1939) at the Mumbwa caves, Northern Rhodesia: 663
Further explorations in the Dowkerbottom Caves, in Craven: 1031
Further studies on the cavernicole fauna of Mexico and adjacent regions: 2789

Gage Caverns: 2520
Gaping Gill: 150 years of exploration: 366
Garden of shadows: 808
Gems of Jenolan: 1582
General management plan, wilderness suitability study: El Malpais National Monument, New Mexico: 3298
General policy and guidelines for cave and karst management in areas managed by the Department of Conservation: 2179
Genetic relationship between caves and landforms in the Mammoth Cave National Park area; a preliminary report: 2193
Geochemistry of karst waters: a basic bibliography: 1774
Geochronology of Sandia Cave: 1398
Geohydrology of karst terraine, Lost River watershed southern Indiana: 2865
Geohydrology of the Rio Camuy Cave system Puerto Rico: 3223
Geologic features of Mount Shasta and Medicine Lake Volcanoes: 2423
A geologic profile of Sloans Valley, Pulaski County, Kentucky: 2025
The geologic story of Diamond Caverns: 2127
The geologic story of Longhorn Cavern: 2094
The geologic story of Wyandotte Cave: 1462
Geological and speleological reconnaissance of the East Yunnan Karst (P.R. China): preliminary results of a field trip between 17.12.1990 and 3.01.1991: 927
A geological guide to Mammoth Cave National Park: 2596
Geological nature conservation cave-protection: 3182
Geological observations in South Australia: principally in the district south-east of Adelaide: 3525
Geology and biology of Pennsylvania caves: 3473

The geology and fossil vertebrates of Ventana Cave: 543

Geology and hydrology of the Yucatan and Quaternary geology of Northeast Yucatan Peninsula with a part on the History of Northern Quintana Roo: 3397

Geology and morphology of selected lava tubes in the vicinity of Bend, Oregon: 1248

Geology and occurrence of ground water at Jewel Cave National Monument, South Dakota: 929

Geology and origin of Mystery Cave, Forestville State Park, Minnesota: 2597

The geology and physiography of the Mammoth Cave national park: 1953

The geology and speleogenesis of Red Run Cave Tucker County, West Virginia: 428

Geology explained in the Peak District: 712

Geology field trip guidebook, 45th National Speleological Society (1986) Tularosa, New Mexico, convention: 2422

Geology in the vicinity of Montezuma's Cave, southern Huachuca Mountains, Cochise County, Arizona: final [report]: 880

The geology of Carlsbad Cavern: 1458

Geology of Carlsbad Cavern and other caves in the Guadalupe Mountains, New Mexico and Texas: 1459

The geology of caves: 118, 830, 3284

Geology of Crystal Cave in southern Pulaski County, Kentucky: 1734

The geology of Cumberland Caverns, Warren County, Tennessee: 328

The geology of Edmonson County: a detailed presentation of the physical, stratigraphic, structural, and economic geology of this district: 3433

The geology of Jewel Cave: 3203

The geology of Kankee Caverns [sic]: a geological report on the underground storage of LP-gas on the Phillips pipe line terminal, Kankakee-Illinois: 3201

Geology of Luray Caverns Virginia: 1310

The geology of Sarawak and Sabah caves: 3485

Geology of selected lava tubes in the Bend Area, Oregon: 1249

Geology of the Capitan Reef Complex of the Guadalupe Mountains, Culberson County, Texas and Eddy County, New Mexico. Field trip guidebook May 6, 7, 8, 9, 1964: 456

Geology of the Carter and Cascade Caves area: 2128

Geology of the Delaware Basin, Guadalupe, Apache, and Glass Mountains, West Texas and New Mexico: 1460

Geology of the Mammoth Cave National Park area: 1950

Geology of the Tennille Lime Sinks, Washington County, Georgia ; with an introduction to local geology: 3036

Geology of the Western Australian part of the Eucla Basin: 1977

The geology of Wind Cave: 2598

Geology, climate, hydrology and karst formation: field symposium in Australia 4 to 18 December 1992: Buchan - Mt. Gambier / Naracoorte - Nullarbor Plain humid temperate impounded karst, sub-humid temperate syngenetic karst, arid temperate karst: programme and abstracts: 119

Geology, climate, hydrology and karst formation: field symposium in Australia 4 to 18 December 1992: Buchan - Mt. Gambier / Naracoorte - Nullarbor Plain humid temperate impounded karst, sub-humid temperate syngenetic karst, arid temperate karst: guidebook: 1209

Geomorphology and hydrology of karst terrains: 3474

The geomorphology and speleogenesis of the Dachstein caves: 1271

Geophysical phase IV monitoring of potential sinkhole collapse at military ocean terminal. Sunny Point Access Railroad, Boiling Springs, N.C.: 3260

George Washington Cave: 3087

Georgia Speleological Survey mapbook: 1182

Geotechnical applications of the self potential (SP) method. Report 1: the use of self potential in the detection of subsurface flowpatterns in and around sinkholes: 1003

Geotechnical applications of the self potential (SP) method. Report 2: the use of self potential to detect ground-water flow in karst: 1004

The Geula Caves. Mount Carmel: 3530

Gezer V: the field I caves: 2927

Ghar Dalam: cave and deposits: 2018

Ghar Parau: 1765

Giant caves of Borneo: 2156

Glaciéres; or, freezing caverns: 291

Glacier cave observations on Llewellyn Glacier, British Columbia: 2935

Glacier Park cave study: 591

A glance at Mammoth Cave, Kentucky: 872

Glass sand occurrences, Kurnell Peninsula: preliminary investigations ; Guano deposits in Willi Willi Caves, Kempsey ; Geological report on Wallent's Somersby Clay Pit: 3378

Gleanings from nature: ten Indiana caves and the animals that inhabit them: 413

Glorious Kangaroo Island: its caves and beauty spots: 455

Glory Hole Caverns: 3076

A glossary of French speleological terms: 2274

A glossary of karst terminology: 2255

Going underground: all about caves and caving: 846

Going underground: your guide to caves in the Mid-Atlantic: 2983

Gouffre Berger: August 1994: 1337

Gough's Caves, Cheddar, Somerset: 120

Gouldens Hole - 5L8 - mapping project, 1987 - 1988: 1520

Gower Caves: a survey of the Gower Caves with an account of recent excavations: part I: 48

Gower Caves: a survey of the Gower Caves with an account of recent excavations: part II: 49

Graham Cave, an archaic site in Montgomery County, Missouri: 1955

The Grampian caving manual: an introduction to caving and potholing for the beginner: 588

Grampian Speleological Group library catalogue: 1238

Grand Avenue Cave. A description in detail of one of America's greatest natural wonders: 3461

Grand Caverns: wonders of the subterranean world: Grottoes, Virginia: 121

Grand excursion to Mammoth Cave: 579

The Grand Kentucky junction: memoirs: 765

Grassington and Upper Wharfedale: 2756

Great American caverns. The underground beauties and wonders of Mammoth, Wyandotte, Luray, and other famous caves ...: 1534

Great American show caves: 3420

The great cave of Dry Fork of Cheat River, Virginia: one of the greatest wonders of the world examined and explored Giving a complete description of this subterraneous cavern, of its rooms and passages, which are of enormous size, together with a great stream of water, a lake, and springs, minerals, fossils, and many curiosities, that were found in singular abundance, &c.: 1758

The great caverns of Kentucky: Diamond Cave, Mammoth Cave, Hundred Dome Cave: 275

The great caving adventure: 1029

The great Dan-yr-ogof day out: 122

The Great Onyx Job Corps Civilian Conservation Center alternative relocation sites, Mammoth Cave National Park, Kentucky: 3299

The Great San Agustín Rescue: 436

The great storms and floods of July 1968 in Mendip: 1356

The Greek caves: 2661

The grotto of Lascaux: 3181

The Grotto of Neptune ("Antro Di Nettuno"), Sardinia; a poem illustrative of three views of this interesting cavern, taken in July 1824 by the late Commander Alfred Miles and dedicated to his memory by his widow, Sibella Elizabeth Miles: 2177

The grottoes of Adelsberg and the *Proteus* of Anguinus: a few pages from the journal of a continental tour: 123

The Grottoes of the Drach near Manacor (Majorca): 2058

The Grottoes of the Shenandoah, consisting of the Weyer and the Fountain Caves, the two most wonderful caverns in the world, at Grottoes Station of Shenandoah Valley Railway, Augusta County, Virginia. A. M. Howison, Secretary.: 1558

The grottos at Dickeyville: 3443

Grottos of Bétharram: 2

The Grottos of Han and of Rochefort (Belgium): 124

The grottos of Majorca: 2884

Ground penetrating electromagnetic tests at Medford Cave, Florida and Waterways Experiment Station, Vicksburg, Mississippi: 909

Ground water flow in the Mammoth Cave Area, Kentucky with emphasis on principles, contaminant dispersal, instrumentation for monitoring water quality, and other methods of study: 2739

Ground water monitoring in karst terranes: recommended protocols and implicit assumptions: 2740

Ground-water hydrology and geomorphology of the Mammoth Cave Region, Kentucky, and of the Mitchell Plain, Indiana: 2741

Groundwater as a geomorphic agent: 1880

Groundwater contamination and sinkhole collapse induced by leaky impoundments in soluble rock terrain: 37

Groundwater geomorphology: 1453

Guadalupe Mts., Hueco Mts., Franklin Mts., geology of the Carlsbad Caverns, Delaware Basin exploration, Oct. 31st, Nov. 1st and 2nd, 1968: 3446

Guide book for the Diamond Cave, Barren County, Kentucky. Located near "Old Bell Tavern," now the Mammoth Cave and Glosgow Junctions, on the L & N railroad, two miles from the R.R. immediately on the Mammoth Cave Road: 125

Guide book of Cave of the Mounds ... Blue Mound, Wis: 2690

A guide book to Carlsbad Caverns National Park: 2383

Guide book to the major caves in the vicinity of Mountain Lake, Virginia: 2385

Guide book to the Mammoth Cave of Kentucky: historical, scientific, and descriptive: 1537

Guide for foundation engineering in Pennsylvania karst: 1836

Guide for prospective cavers: 1270

A guide manual to the Mammoth Cave of Kentucky: 126, 127, 3531, 3532

Guide Santillane on the sea and Altamira: 2652

Guide through the Kubacher Crystal Cave: 2910

Guide to Alabaster Cavern and Woodward County, Oklahoma: 2325, 2326

A guide to Austin's most visited caves, with maps and cross-sections of area caves: 3385

A guide to Batu caves: 1485, 3045

Guide to cavern engineering: 395

Guide to Dan-yr-Ogof Caves: Swansea Valley Caves: 2270

A guide to East Tennessee Caves: prepared for the 1989 National Speleological Society Convention, Sewanee, Tennessee: 2448

A guide to Glenrock caves: a speleological field guide to the limestone caves on Glenrock Station: 2678

Guide to Gough's Caves, Cheddar; how they were found, the finest in the world (electrically illuminated) eighteen beautiful views of the cave prehistoric man: - how he was found: 1230

A guide to major Wisconsin caves: 451

A guide to Maquoketa Caves State Park: 1445

A guide to Mount Fairy Caves: a speleological field guide to the limestone and caves at Mount Fairy: 1463
Guide to Painted Cave: 1567
Guide to some volcanic terranes in Washington, Idaho, Oregon and Northern California: 1740
A guide to the Cave of Altamira and the town of Santillana del Mar (Province of Santander, Spain): 128
Guide to the caves and karst of the Northeast: 2412
A guide to the caves in the Oak Ridge-Knoxville area: 1668
Guide to the caves of Minnesota: 1479
A guide to the caves of Texas and 1964 N.S.S. convention (New Braunfels, TX) field trips: 2386
A guide to the caves of the Black Hills, S. D: 2384
Guide to the caves of Wisconsin: 949
Guide to the Coast, Caves, and Bays of Sark: 1910
A guide to the Craven Dales: 604
Guide to the Grand Cavern within the mountain of Abraham's Heights, Matlock Bath: 129
A guide to the historic section of Mammoth Cave: 2599
Guide to the home of the rainbow: Cox's Cave, Cheddar, Somerset: 1205
Guide to the hydrology of carbonate rocks: 1887
Guide to the Jenolan Caves New South Wales: 3232
Guide to the Kalk Bay and Muizenberg Mountains; (walks, caves, camp sites): 130
Guide to the lava tube caves of central Oregon: 2732
Guide to the limestone caverns of New South Wales: Jenolan, Wombeyan, and Yarrangobilly: 3233
A guide to the Museum of the Torquay Natural History Society: 131
A guide to the rock paintings of Tandjesberg: 1966
A guide to the selection of limestone caverns and springs in the United States as national landmarks: for the National Park Service: 2703
Guide to the sporting caves, potholes and mines of Derbyshire: 306
Guide to the Wemyss caves: 2760
Guide to the world famous Gough's Caves, Cheddar, Somerset: 1206
A guide to Tongoni ruins: with notes on other antiquities in Tanga and Pangani districts including the Amboni Caves: pamoja na maelezo kwa Kiswahili: 2294
A guide to Wind Cave and Wind Cave National Park. 19th annual convention of the National Speleological Society self guiding tour of the commercial areas of Wind Cave: 3257
Guide to Wombeyan Caves NSW: 3234
Guide to Yarrangobilly Caves New South Wales: 3235
Guidebook for the 1987 Kentucky Speleofest: Renfro Valley KOA, Kentucky, May 22-25, 1987: 1805
Guidebook of southeastern New Mexico: fifth field conference, October 21-22-23 & 24, 1954: 2492
Guidebook of the 1984 NSS convention: 2405

Guidebook to karst and caves of Tennessee: emphasis on the Cumberland plateau escarpment region and guidebook to karst and caves of the Ozark Region of Missouri and Arkansas: prepared for the Eighth International Congress of Speleology, Bowling Green, Kentucky, U. S. A. July 18 to 24 1981: 738
Guidebook to Ozark carbonate terrane, Rolla - Devils Elbow area, Missouri: 1039
Guidebook to selected caves in southwestern Missouri: prepared for the 25th annual convention of the National Speleological Society: 2390
Guidebook to spring 1980 MVOR, Shannon County, Missouri: 2322
Guidebook to symposium SA6 "Karst geomorphology of the Canadian Rockies" 22nd international geographical congress Canada 1972: 1082
Guidebook to the 1977 Kentucky Speleofest, Cedar Hill Camp, Edmonson Co., Ky: 1801
Guidebook to the 1978 Kentucky Speleofest: Camp Carlson, Meade County, Kentucky: 1802
Guidebook to the 1980 Kentucky Speleo-fest: annual summer field conference, Pulaski Co. Park, Ky: 1804
Guidebook to the 1994 SERA cave carnival: June 10-12, 1994 CCWHA Campground at Short Mountain Cannon County, Tennessee: 3072
Guidebook to the 5th annual Kentucky Speleofest. Annual summer field conference of the Louisville Grotto NSS: Camp Carlson, Meade County, Kentucky: 1800
A guidebook to the caves and sinks of the Spring Valley Caverns area, Fillmore County, Minnesota: 1989 Cornfeed Minnesota Speleological Survey: 1183
Guidebook to the caves of southeastern New South Wales and eastern Victoria and caves around Canberra: 2511
Guidebook to the caves of the 1978 summer SERA cave carnival: 3066
Guidebook to the caves of the Monterey, Tennessee Area: 3064
Guidebook to the geology between Springfield and Branson, Missouri, emphasizing stratigraphy and cavern development: 3364
Guidebook to the geology of the Rolla area emphasizing solution phenomena: 2197
Guidebook to the karst features and stratigraphy of the Rolla area: 239
Guidebook to the karst of the Central Appalachians; prepared for the Eighth International Congress of Speleology, Bowling Green, Kentucky, U.S.A., July 18 to 24, 1981: 3442
Guidebook to the Kentucky Speleofest, Meade County, Kentucky: 1796
Guidebook to the Kentucky Speleofest, Warren County, Kentucky: 1797
Guidebook to the Natural Bridge and Natural Bridge Caverns: 3093

Guidebook: 1991 National Cave Management Symposium, Bowling Green, KY Oct. 23-26: 1105
Guidebook: Cave Capers 1975: 1597
Guidebook: Cave Capers 1976: 1598
Guidebook: environmental hydrogeology of karst terranes in the vicinity of Nashville Tennessee: 2905
Guidebook: Fall 1972 MVOR, Perry County, Missouri: 2315
Guidebook: Kentucky Speleofest '79: 1803
Guidebook: MVOR Spring '73: 2316
Guide[s] to the Indiana caverns: 1607
The Guildford Caverns: 3502
Guilin karst: 3555
Guilin tourist album: 132
Guilin: the crown of superb landscapes in China: 133
Guitarrero Cave: early man in the Andes: 1999
Gunpowder from Mammoth Cave: the saga of saltpetre mining before and during the War of 1812: 857
Gurnee guide to American caves: a comprehensive guide to the caves in the United States open to the public: 1298
GYPKAP 1987 annual report: 378, 934
GYPKAP report volume 3: January 1992—April 1996: 1924
Gypsum cave, Nevada, report of the second sessions expedition: 1362

The Hades of Ardenne: a visit to the Caves of Han described and illustrated: 3025
Hadfields Cave: a perspective on Late Woodland culture in northeastern Iowa: 384
Half hours underground: volcanoes, mines, and caves: 793
Hand signals for diving: 1062
Handbook for caves and guiding staff: 134
Handbook of cave rescue operations: 3008
Handbook of geophysical cavity-locating techniques: with emphasis on electrical resistivity: 1822
Handbook of karst hydrogeology with special reference to the carbonate aquifers of the Mediterranean Region: 558
Hang Vietnam; report of the British Speleological Expedition to the Bac-Sun Massiflang: 51
The Har Dalam Cavern, Malta and its fossiliferous contents by John H. Cooke with a report on the organic remains by Arthur Smith Woodward: 703
Harrison's Cave: Barbados: 2351
Hart County solid waste management plan: 54
Harvestmen of the sub-order Laniatores from New Zealand caves: 1101
Hastings Caves State Reserve Tasmania: a visitors' guide: 2996
Hastings Newdegate cave rehabilitation plan: 135
The Hayes Cave site, South Maitland, Nova Scotia: 2277
The hidden heart of Baja [Mexico]: 1150
The Hilina Pali petroglyph cave, Hawai'i island: a report on preliminary archaeological investigations: 665
The hill-caves of Yucatan, a search for evidence of man's antiquity in the caverns of Central America. Being an account of the Corwith expedition of the Department of Archaeology and Palaeontology of the University of Pennsylvania: 2153
The hill-caves of Yucatan, a search for evidence of man's antiquity in the caverns of Central America. Being an account of the Corwith expedition of the Department of Archaeology and Paleontology of the University of Pennsylvania: 2154
Himalaya underground - 1976 speleological expedition: 926
Historia rievallensis: containing the history of Kirkby Moorside ... to which is prefixed a dissertation on the animal remains, and other curious phenomena, in the recently discovered cave at Kirkdale: 932
The historic Blue Grottoes of Virginia; Harrisonburg VA: 136
An historical and descriptive narrative of the Mammoth Cave of Kentucky: including explanations of the causes concerned in its formation, its atmospheric conditions, its chemistry, geology, zoology, etc., with full scientific details of the eyeless fishes: 1103
Historical description of the chapel and castle of Roslin, and the caverns of Hawthornden: 2011
An historical survey of Torquay from the earliest times, as illustrated by finds in Kent's Cavern, down to the present time: 965
An historical, archaeological and geological examination of Fingal's Cave in the island of Staffa: 2013
History and description of the Luray Cave (illustrated), including explanations of the manner of its formation, its peculiar growths, its geology, chemistry, &c.; also a map. The whole so arranged as to serve as a guide.: 60
History and description of the Luray Cave (illustrated), including explanations of the manner of its formation, its peculiar growths, its geology, chemistry, etc.; also a map. The whole so arranged as to serve as a guide: 61
The history and exploration of Wyandotte Cave: 1682
History and publications of the Western Speleological Survey: 1330
History from caves: a new theory of the origin of modern races of mankind, by Arthur Keith, being the presidential address given at Buxton, Derbyshire to the first speleological conference: 1785
The history of Blue John Stone: methods of mining and working ancient and modern: 2570
History of cave science: the exploration and study of limestone caves, to 1990: 2954
History of cave science: the scientific investigation of limestone caves, to 1900: 2955
A history of Gaping Gill and Ingleborough Cave: 454

A history of Glacier House and Nakimu Caves, Glacier National Park, British Columbia: 2051
The history of Kent's Cavern; Torquay: with illustrations: 892
The history of Laurel Caverns of Fayette County, Pennsylvania, known throughout the years as Delaney's Cave: 798
History of Manitou's caves, the Grand Caverns and Cave of the Winds: and points of interest in and about Manitou: 3255
The history of Mendip caving: 1737
History of Morrell Cave: Part I Tennessee: 12
History of Rockhouse Cave: 3258
The history of Ruby Falls: 479
History of the 7 caves: 493
A history of the caves of Camden County, Missouri: 2208
A history of the exploration of Swildon's Hole: 3434
A history of the Guadalupe Mountains and Carlsbad Caverns, 1840—1940: 979
A history of the journal of the Sydney Speleological Society: 976
The history of the Lockport caves: 582
The history of the Sydney Speleological Society 1954-1994: 1314
A history of the Virginia Region: 3368, 3480
The history of Timpanogos Cave National Monument: American Fork Canyon, Utah: 1665
History of Wind Cave: 425
History, geology of Carlsbad Caverns: what' a hole: 2280
The history, guide and description of Wyandotte Cave: 3123
Hodag hunt 1980: guidebook September 12-14 1980: 431
Hogup Cave: 18
The hole in the mountain: 2831
Holiday caving in Mallorca: 664
The Holloch: a general account and trip report: 2718
Hollow Hills of Sunnalee: The Linville Caverns story: 1488
The hollow mountains: the story of man's conquest of the caves and potholes of Northwest Yorkshire throughout 10,000 years: 2243
Homes of primeval man: wandering in the caves of Czechoslovakia: 1874
Honduras recce 1994: a short report: 1746
Horne Lake Wonder Caves: a guide to Horne Lake Caves Park: 1264
Hovey's hand-book of the Mammoth Cave of Kentucky: a practical guide to the regulation routes with maps and illustrations: 1538
How I found the Cave of the winds ; being the true story as set forth by the discoverer, George W. Snider: 3038
How nature made Luray Caverns, Virginia: 2517
How nature made the beautiful caverns: 137
How nature makes a cave: 2518

How the grottoes of an ancient church were discovered in the convent of the "Dames de Nazareth" at Nazareth in Galilee: 1679
How underground Britain is explored: 3156
Howe Caverns and New York Central Leatherstocking Region: 2696
Howe Caverns; Howes Cave N.Y.; 60th Anniversary, 1929-1989: 1512
The Howe Family From Massachusetts to New York: A story of Lester Howe, his ancestors, and the Howe Caverns: 1547
Howe's Cave: 3130
The human skeletal remains from Pictograph and Ghost caves, Montana: 3041
The human skeletal remains from the Sha Kuo T'un cave deposit in comparison with those from Yang Shao Tsun and with recent North China skeletal material: 406
Humidity and waters of a limestone cavern near Lexington, Virginia: 3113
Hundred Dome Cave ("partial reproduction of the last known copy of the book, 'the great caverns of Kentucky'"): 276
The hunters or the hunted? An introduction to African cave taphonomy: 459
HWYL [songbook]: 715
Hydrogeologic and biologic factors related to the occurrence of the Alabama cave shrimp (*Palaemonias alabamae*), Madison County, Alabama: 2810
Hydrogeologic controls on solution of carbonate rocks in Christian County, Missouri: 2209
Hydrogeologic mapping of unincorporated Greene County, Missouri, to identify areas where sinkhole flooding and serious groundwater contamination could result from land development: 38
Hydrogeologic problems resulting from development upon karst terrain, Bowling Green, Kentucky. Guidebook prepared for Karst Hydrogeology Workshop August 31 - September 3, 1982 Nashville, Tennessee: 739
Hydrogeologic study of Jewel Cave/Wind Cave: final report - phase I, 3rd and 4th quarters of phase I, 1 November 1985 through 30 April 1986: 28
Hydrogeologic study of Jewel Cave/Wind Cave: final report -year 2, 4th quarter of year 2, February 1, 1987 through April 30, 1987: 25
A hydrogeologic study of the Knox-Skull Cave System, Albany County, New York: 1449
A hydrogeological study in the basin of the Gulp Creek: a reconnaissance in a small catchment area: 1, Groundwater flow characteristics: 2527
Hydrogeology and environmental geology of the inner Bluegrass Karst Region Kentucky: field guide for the annual meeting of the southeastern and north-central sections Geological Society of America: Lexington, Kentucky April 4-6 1984: 3212

Hydrogeology and geochemistry of folded and faulted carbonate rocks of the central Appalachian type and related land use problems: 2612
Hydrogeology and geomorphology of the Mammoth Cave area, Kentucky: Southeastern Friends of the Pleistocene 1990 field excursion, November 17, 1990: 2742
The hydrogeology of artesian boreholes at some watercress farms in Hampshire: 1413
Hydrogeology of karstic terrains: 561
Hydrogeology of karstic terrains: with a multilingual glossary of specific terms: 562
Hydrogeology of karstic terranes, case histories: 605
Hydrogeology of Ozark cavefish caves: 2547
Hydrogeology of selected karst regions: 270
The hydrogeology of the chalk of north-west Europe: 894
Hydrogeology of the karst of Puerto Rico: 1215
Hydrogeology of the Snail Shell Cave: overall creek drainage basin and the ecology of the Snail Shell Cave System: 740
Hydrograph analysis of carbonate aquifers: 1450
Hydrologic problems as related to wilderness management and water supplies for recreational areas at Mammoth Cave National Park, Kentucky: 786
Hydrologic problems in karst regions: 871
Hydrologic study of Jewel Cave/Wind Cave: final report: 26
Hydrology and water quality in the central Kentucky karst: phase 1: 2743
Hydrology of carbonate terrane, Niangua, Ossage Fork, and Grandglaize basins, Missouri: 1378
Hydrology of limestone karst in Greenbriar County, West Virginia: 1756
Hydrology of limestone terranes in the coastal plain of the southeastern states: 3151
Hydrology of limestone terranes in the coastal plain of the southeastern United States: 3152
Hydrology of limestone terranes: annotated bibliography of carbonate rocks: 1888
Hydrology of limestone terranes: annotated bibliography of carbonate rocks, volume 4: 1889
Hydrology of limestone terranes: annotated bibliography of carbonate rocks, volume three: 1890
Hydrology of limestone terranes: development of karst and is effects on the permeability and circulation of water in carbonate rocks, with special reference to the southeastern states: 3153
Hydrology of limestone terranes: geologic investigations: 3049
Hydrology of limestone terranes: geophysical investigations: 1742
Hydrology of limestone terranes: photogeologic investigations: 3050

Hydrology of limestone terranes: progress of knowledge about hydrology of carbonate terranes with an annotated bibliography of carbonate rocks: 1891
Hydrology of limestone terranes: quantitative studies: 2265
Hydrology of the cavernous limestones of the Mammoth Cave area, Kentucky: 530
Hydrology of the Dinaric karst: 2174
Hydrology of the Turnhole Spring Groundwater Basin and vicinity, Kentucky: an area that includes part of the Mammoth Cave National Park: 2744
Hydrology of the Wolf Branch sinkhole basin, Lake County, east-central Florida: 2904
Hydrology of three sinkhole basins in Southwestern Seminole County Florida: 71
Hypogean fauna and biological records 1964-1966: 1406
Hypogean fauna and biological records 1970, [and] fauna of Gibraltar caves: 621
Hypogean fauna and biological records 1970-71: 622
Hypogean fauna, biological records 1967: 1407
Hypogean fauna, biological records 1968: 1408

I can read about caves: 2334
I don't play golf: recollections of a rescue volunteer: 407
The Ice Cave at Decorah, Iowa: 1418
Ice caves and frozen wells as meteorological phenomena ...: 1820
Ice caves and the causes of subterranean ice: 292
Ice-caves of France and Switzerland. A narrative of subterranean exploration: 532
Ida Bay - Benders Quarry. An estimate of reserves within existing quarry limits: 351
An illustrated glossary of lava tube features: 1906
Illustrated guide book of your tour through Carlsbad Caverns: 461
Illustrated "Press" guide to the caves and cliffs of Sark: with map and plans of roads.: 1275
Illustrated souvenir grid to the caverns and glens at the foot of Ingleborough (with map): 138
Images from a limestone landscape: a journey into the Punakaiki-Paparosa Region: 2697
Images from the underworld: Naj Tunich and the tradition of Maya cave painting: 3133
Images of the ice age: 274
Imperial College Caving Club Peru '84 Expedition (19th July - 19th Sept 1984): 2579
Imperial College karst research expedition to the Peruvian Andes, 1972: 490
In my torchlight: a guide's guide to the Cango Caves: 2105
In Salzburg's Netherworld: 421
In search of Adam: the story of mans quest for the truth about his earliest ancestors: 3438
In search of cave art: 2097

In the land of cave and cliff dwellers: 2920
In the shadow of extinction: a Quaternary archaeology and paleoecology of the lake, fissures, and smaller caves at Creswell Crags SSSI: 1198
An index of submerged cave passages; sumps: 825
Index of the literature pertaining to West Virginia caves and karst: 3440
An index to cave maps in N.S.W: 2619
Index to references to caves in three New South Wales government publications 1870-1919: 2170
An index to the publications of The Bristol Exploration Club 1947-1987: 1677
Index to volumes 46 through 50 of the National Speleological Society bulletin: 2888
Indiana cave areas: topographic quadrangle maps: 3285
Indiana Cave Capers guidebook: Hickory Hill Campground, Owen County, Indiana July 14-16, 1989: 1605
[Indiana cave list]: 79
Indiana caves and their fauna: 414
Indiana caves and unique geological features: 52
The influence of light and darkness on thyroid and pituitary activity of the characin *Astyanax mexicanus* and its cave derivatives: 2762
Ingleborough Cavern: 1090
Ingleborough Cavern and Gaping Gill: 1091, 1708
Ingleborough Cavern including: notes on Gaping Gill and the geology of Ingleborough: 1092
Ingleborough Cavern: the finest show cave in Yorkshire: 139
Initial inventory of named caves and related features and cave-related place names in Hawaii: 1331
Inner Space Cavern: 1709
Inner space: the last terrestrial frontier: 3139
Inside the caves: Lava Beds National Monument: 229
Interim Alabama cave survey: report number 1: 3354
Interim report number 2: Alabama cave survey: 3355
International Association of Hydrogeologists 12th international congress: karst hydrogeology: abstracts and program 21-27 September 1975, Huntsville, Alabama, 28 September-3 October 1975, Gulf Shores, Alabama: 1614
International atlas of karst phenomena: sheets 8 - 12: 2664
International conference Baradla 150: field-trip guide: Budapest-Aggtelek, 26-29.08.1975: 1410
International expedition to Gouffre Berger, 1956: the exploits of the British members, Nick Pratchett & Bob Powell: 2705
International meeting on the show caves and their problems: 1424
International seminar on karst denudation: 1085
International seminar on karst hydrogeology: Oymapinar, October 1979: proceedings: 1650
International symposium on evaporite karst: preprints: Bologna October 22-25, 1985: 1102

International Symposium on groundwater biology: symposium held in 1978 in Blacksburg, Virginia, U.S.A.: 140
International symposium on karren landforms: Mallorca 1995: 1637
Into the lost world - a descent into prehistoric time: 2528
An introduction to Abercrombie Caves: 1779
An introduction to British limestone karst environments: 1288
An introduction to cave mapping: 2210
An introduction to cave photography: 3131
An introduction to cave surveying: a handbook of techniques for the preparation and interpretation of conventional cave surveys: 966
An introduction to caves and cave exploring in Georgia: 357
An introduction to caves of Minnesota, Iowa, and Wisconsin: guidebook for the 1980 National Speleological Society Convention, Lakewood Community College, White Bear Lake, Minnesota: 2402
An introduction to caves of the Bend area: guidebook of the 1982 NSS Convention: 2403
An introduction to caves of the northeast: guidebook for the 1979 National Speleological Convention, Pittsfield, Massachusetts: 2401
Introduction to caving: 3213
An introduction to caving and potholing for novices: 544
An introduction to caving: a guide for beginners: 2211
Introduction to Hawaiian caves: field guide for the 6th international symposium on volcano-speleology: Hilo, HI: 1332
Introduction to Idaho caves and caving: 2857
An introduction to the caves of east-central West Virginia: guidebook for the 1983 National Speleological Society Convention, Elkins, West Virginia: 2404
An introduction to the caves of Texas: prepared for the 1978 National Speleological Society Convention New Braunfels, Texas: 2400
An introduction to the inventory and evaluation of biological cave resources: 1153
An introduction to the limestone sinkholes of northeastern Michigan: 529
Introductory cave surveying: 2488
An inventory and evaluation of cave resources of Mark Twain National Forest: a final report submitted to the Mark Twain National Forest, United States Department of Agriculture in compliance with a cooperative cave inventory agreement: 1154
An inventory and evaluation of Missouri state parks cave resources: 1151
An inventory and evaluation of the cave resources to be impacted by the New Melones Reservoir Project, Calaveras and Tuolumne Counties, California: 2120
An inventory of caves, Harry S. Truman dam and reservoir: 766

Inventory of New Zealand caves and karst of international, national and regional importance: 3529
The invertebrate cave fauna of Tasmania: 940
The invertebrate cave fauna of Virginia and a part of eastern Tennessee: 1503
The invertebrate cave fauna of West Virginia: 1504
Invertebrate fauna from Missouri caves and springs: 1155
The invertebrate fauna of Mystery Cave, Perry County, Missouri: 2212
An investigation into ground water pollution in caves: 2213
Investigation of karst-related subsidence near the southwest landfill, Alachua County, Florida, using ground penetrating radar: a report: 3512
Investigation of sinkhole occurrences at Goretown, near Loris, South Carolina: 1471
Investigations in Russell Cave: Russell Cave National Monument, Alabama: 1261
Ireland 1936 the record of the party of S.J. Pick (Leicester) in County Clare, Easter 1936: 302
Ireland 1937 the record of S.J. Pick (Leicester) in County Clare May 1937: 303
Irish hypogean fauna and Irish biological records, 1856-1971: 1409
Irish sump index: 1745
Is Fingal's Cave artificial?: 3477
ISCA CC journal index volumes 1-17; 1984-1994: 1554
The island of Staffa: 2005
The island of Staffa: home of the world-renowned Fingal's cave: 863
The island of Staffa: with 12 photographs, two plans and one map, by the author: 2006
Israeli scenery Soreq Cavern: 3329
ISS report on Binkley's Cave system: 2824
Itinerary, geologic features of the Mississippian Plateaus in the Mammoth Cave and Elizabethtown areas: 2685
Itinerary: geology of the Mammoth Cave Region, Barren, Edmonson, and Hart Counties, Kentucky: 2126

J Harlen Bretz: a geologist's encounters with Missouri caves: 2214
Jack Mitchell, caveman: 2238
Jaguar (*Panthera onca*) remains from Big Bone Cave, Tennessee and east central North America: 1281
Jamaica underground: a register of data regarding the caves, sinkholes and underground rivers of the island: 1060
Jame Škocjan = Grotte Skocjan: 1301
Java caves 1983: report of a visit to Indonesia by Australian and British cavers: 2157
Jefferson County caves: 2215
Jenolan Caves: 924, 1107
Jenolan Caves (N. S. W.) 'Nature's Masterpiece': a complete description of the geology, discovery, development, and features of the Jenolan limestone system beautifully illustrated, with high-class maps: 1256
Jenolan Caves and the Blue Mountains: 225
Jenolan caves as they were in the nineteenth century: 920
Jenolan Caves Australia: 336
Jenolan Caves New South Wales Australia: 335
Jenolan Caves Reserve: plan of management.: 587
Jenolan Caves resort - some management issues: 688
Jenolan Caves Resort teachers kit: 1397
Jenolan Caves, (the underground wonderland) New South Wales, Australia: 2666
Jenolan Caves, New South Wales: 2494
Jenolan Caves, New South Wales, Australia: 662
The Jenolan caves: an excursion in Australian wonderland: 699
Jenolan Caves: Australia's underground fairyland: 141
Jenolan Caves: when the tourists came: 1517
Jenolan: a guide to Australia's famous caves: 1433
Jenolan: the golden ages of caving: 2757
The Jewel Cave adventure: fifty miles of discovery under South Dakota: 692
Jewel Cave: a gift from the past: 2600
Jo and Coney's Cavern: 1384
The joint Anglo-Soviet speleological expedition to central Asia 1990 "Kugitang 90": 928
Joint caves of Valcour Island: their age and their origin: 1565
Joint Mitnor Cave: 784, 3492
Joint Mitnor Caves: Buckfastleigh: 3161
Joint thesis upon Maidstone Cave: 2181
The journal of the Joint Bristol Exploration Club (United Kingdom) and the National Mountaineering Federation of the Philippines caving expedition to the Philippines, January to April 1992: 2119
Journal of the Tennessee Academy of Science. Cave number: 3197
The Junee River karst system, Tasmania: a report to the Forestry Commission: 938

Kahaluu data recovery project: excavations at site 50-10-37-7702, Kahaluu Habitation Cave, land of Kahaluu, North Kona, Island of Hawaii: 1392
Kango: the story of the caves: 553
Karst: 1619, 1725, 2665
Karst and cave resource significance assessment: Ketchikan Area, Tongass National Forest, Alaska: 39
Karst and caves in the Caucasus: 2959
Karst and caves in the Yorkshire Dales National Park: 3391
Karst and man: proceedings of the international symposium on human influence in karst, 11-14th September 1987, Postojna, Yugoslavia: 1641

Karst and man: proceedings of the international symposium on human influence in karst, 11 - 14 September, 1987, Postojna, Yugoslavia: 1649

Karst and paleohydrology of carbonate rock terranes in semiarid and arid regions with a comparison to humid karst of Alabama: 3154

Karst bauxites: bauxite deposits on carbonate rocks: 321

Karst curriculum resource guide: 55

The karst development and associated archeology of the Chiquibul, Belize: (preliminary report of the results of grant #2742-83): 2184

Karst features of Florida Caverns State Park and Falling Waters State Recreation Area, Jackson and Washington Counties, Florida: 2017

Karst features of Pohn Pei, Chuuk, and Waqab: a survey of caves and karst features in the States of Pohn Pei, Chuuk, and Waqab, Federated States of Micronesia: 2841

Karst features of Rota (Luta) Island, CNMI, A survey of caves and karst features on Rota (Luta) Island, Commonwealth of the Northern Mariana Islands: 2842

Karst features of Saipan, CNMI: a survey of caves and karst features on Saipan Island, Commonwealth of the Northern Mariana Islands: 2843

Karst features of the Palau Islands, a survey of caves and karst features in the Palau Archipelago: 2844

Karst features of the Territory of Guam: a survey of caves and karst features on Guam Island, Territory: 2845

Karst geohazards: engineering and environmental problems in karst terrane: proceedings of the fifth Multidisciplinary Conference on Sinkholes and the Engineering and Environmental Impact of Karst, Gatlinburg, Tennessee, 2-5 April 1995: 358

Karst geology in Hong Kong: proceedings of a conference on "Karst geology in Hong Kong" held at the University of Hong Kong on 6 January 1990: 1898

Karst geology in Hong Kong: programme and abstracts: a conference organized by the Geological Society of Hong Kong and Department of Geography and Geology, University of Hong Kong: Programme and Abstracts: University of Hong Kong: 4-6 January 1990: 1899

Karst geomorphology: 1726, 3165

Karst geomorphology and hydrogeology: 1775

Karst geomorphology and hydrology: 1083

Karst groundwater investigations: Greece: 142

The karst groundwater resources of Parnassoc-Ghiona, Greece: a report: 559

Karst hydrogeology: 2176

Karst hydrogeology and environmental problems in the Bowling Green area: 741

Karst hydrogeology and geomorphology of eastern New York: a guidebook to the geology field trip, National Speleological Annual Convention, Pittsfield, Massachusetts August 5-12, 1979: 2421

Karst hydrogeology in the United States of America: 21st congress of the International Association of Hydrogeologists: karst hydrogeology and karst environment protection October 1988 Guilin, China: 2633

Karst hydrogeology of central and northern Florida: a field trip guidebook produced in conjunction with the 1985 G.S.A. annual meeting and exposition, Orlando, FL., October, 1985: 359

Karst hydrogeology of Tennessee. Guidebook prepared for Karst Hydrogeology Workshop August 31 - September 3, 1982 Nashville, Tennessee: 742

Karst hydrogeology of the central and eastern Peloponnesus (Greece): 2268

The karst hydrogeology of the Cumberland Plateau Escarpment of Tennessee. Part IV. Erosional processes associated with subterranean stream invasion, conduit cavern development and slope retreat: 743

The karst hydrogeology of the Cumberland Plateau escarpment of Tennessee: Part I: subterranean stream invasion, conduit cavern development, and slope retreat in the Lost Creek Cove area, White County, Tennessee: 744

The karst hydrogeology of the Cumberland Plateau escarpment of Tennessee: Part II: karst valley development and the headward advance of the Sequatchie Valley in the Grassy Cove area, Cumberland County, Tennessee: 745

The karst hydrogeology of the Cumberland Plateau escarpment of Tennessee: Part III: karst valley development in the Lost Cove area, Franklin County, Tennessee: 746

The karst hydrogeology of the Cumberland Plateau escarpment of Tennessee: subterranean stream invasion, conduit cavern development, and slope retreat in the Lost Creek Cove area, White County, Tennessee: 747

Karst hydrogeology symposium: Oymapinar, October 1977: proceedings: 1768

Karst hydrogeology: engineering and environmental applications: proceedings of the second Multidisciplinary Conference on Sinkholes and the Environmental Impacts of Karst, Orlando, Florida 9-11 February 1987: 360

Karst hydrogeology: 12th international congress: International Association of Hydrogeologists 21-27 September 1975, Huntsville, Alabama U.S.A.: 1615

Karst hydrogeology: A summary of the principals of ground-water flow through soluble limestone aquifers prepared for the Smithsonian Institution short course on speleology: 1757

Karst hydrologic problems of south central Kentucky: ground water contamination, sinkhole flooding, and sinkhole collapse: field trip guide: 748

Karst hydrological and speleological features: 309

Karst hydrology and geomorphology of eastern New York: a guidebook to the geology field trip, National Speleological Society Annual Convention, Pittsfield, Massachusetts, August 5-12, 1979: 769

Karst hydrology and geomorphology of the Barrack Zourie Cave System, Schoharie County, New York: 911

Karst hydrology and physical speleology: 423

Karst hydrology and water resources: proceedings of the U. S.—Yugoslavian symposium, Dubrovnik, June 2-7, 1975: 2175

Karst hydrology expedition to Jamaica: preliminary report: 3330

Karst hydrology of the lower Maligne Basin, Jasper, Alberta: 528

Karst hydrology with special reference to the Dinaric karst: 429

Karst hydrology: concepts from the Mammoth Cave area: 3475

Karst in China: 1612

Karst in China - landscapes, types, rules: 3540

Karst in China: its geomorphology and environment: 3166

Karst in evaporites in southeastern New Mexico.: 269

Karst in Florida: 1895

Karst landforms: 3167

The karst landforms of Puerto Rico: a discussion of a solution landscape formed in a tropical climate of moderately high rainfall: 2256

Karst Landforms of the Lemonthyme and Southern Forests, Tasmania: report to the Commonwealth Commission of Inquiry into the Lemonthyme and Southern Forests: 1532

The karst landscape of Warren County: 749

Karst of China: 803

Karst of the Orlando area: a guidebook prepared for the engineering and geology of karst terranes short course, Orlando, Florida, February 1-5, 1988: 3513

Karst processes: 3168

Karst processes and land forms: 901

Karst utilization in practice: 2710

Karst water research needs: 3542

Karst water resources: proceedings of a symposium held at Ankara, July 1985, and sponsored by the United Nations Development Program, the United Nations Educational, Scientific and Cultural Organization, the International Association of Hydrological Sciences, many government organizations of Turkey, and Hacettepe University of Ankara, Turkey: 1287

Karst, caves and management at Mole Creek, Tasmania: 1818

Karst-controlled reservoir heterogeneity and an example from the Ellenburger Group (Lower Ordovician) of west Texas: 1809

Karst-derived early Pennsylvanian conglomerate in Ness County, Kansas: subsurface Mississippian-Pennsylvanian boundary delineated in well core: 2515

Karst: important karst regions of the northern hemisphere: 1446

Karstification on the Silurian Escarpment in Fayette County, northeastern Iowa: 446

Karstological investigations in Cuba: 2534

Kaver komix: 1866

Kent's Cavern: 1184, 3020

Kent's Cavern and its wonders: a lecture delivered in the public hall Warrington on December 22nd 1873: 2645

Kents Cavern Wellswood Torquay Devon, home of prehistoric man and animals: 143

Kent's Cavern: a lecture: 2644

Kents Cavern: its origin and history: 3341

Kent's Cavern: its testimony to the antiquity of man: a lecture delivered in the City Hall, Glasgow on Wednesday, 22nd December, 1875: 2646

Kents Caverns, home of prehistoric man and animals: the origin, story, and descriptive tour of the caves: 2636

Kentucky Speleofest guidebook: annual summer field conference, the Louisville Grotto NSS: 1799

Kentucky Speleofest guidebook: selected caves for the annual summer field conference, the Louisville Grotto, NSS: Squire Boon Caverns, Indiana: 1798

King Caverns report: 3478

Knox Cave: Albany County, NY: 2601

Krill Cave: a stratified rockshelter in Summit County, Ohio: 2728

Kruger Cave, late Stone Age, Magaliesberg: 2079

Kubla Khan Cave State Reserve, Mole Creek, Tasmania: pilot management study: 3082

Kurnool 1984: report of the speleological expedition to the district of Kurnool, Andhra Pradesh, India: 1169

La Combe, a paleolithic cave in the Dordogne: 2007

La Sima 56 (Picos de Europa-España): 645

Laclede County caves: 2216

Lake of the Ozarks MVOR, Fall 1974: Camden County, Miller County, Morgan County: 2317

Lake Shasta Caverns: 3516

Lancaster Hole and the Ease Gill Caverns, Casterton Fell, Westmorland: 1022

Land protection plan: Carlsbad Caverns National Park: prepared by Carlsbad Caverns National Park with assistance from the Southwest Regional Office, National Park Service.: 600

The Lapiaz Superieure du Pla Segoune: 1832

Largest cave system of the Czech Socialist Republic in the Moravský kras (Moravian karst): 2711

Lascaux and Carnac: 800

The Lascaux Cave paintings: 3514

Lascaux revisited: 249

Lascaux, cradle of man's art: 1636

Lascaux: a commentary: 496

Lascaux: paintings and engravings: 1885

The last adventure: 3199

The last horizon: 144, 1964

The late Pleistocene small mammals of Eagle Cave, Pendleton County, West Virginia: 1282
The late Wisconsinan vertebrate fauna from Deadman Cave, southern Arizona: 2137
The later cave men: 884
Lava Beds caves: 1907
Lava Beds National Monument: 1882
Lava beds underground: 235
Lava River Cave: 1908
Lava tubes of the Cave Basalt, Mount St. Helens, Washington: 1250
Layser Cave: silent voices, vital clues: 1195
Leadership and instructor qualifications in caving: 2349
Learning to live with caves and karst: a cave and karst curriculum and resource guide: 1106
Lechuguilla Cave Protection Act of 1993: report (to accompany H.R. 698): 3271
Lechuguilla Cave Protection Act of 1993: report (to accompany H.R. 698) (including cost estimate of the Congressional Budget Office): 3268
Lechuguilla Cave: biological inventory: 2521
Lechuguilla cave: the hidden giant: 353
Lechuguilla: jewel of the underground: 3192
Lecture on the antiquity of man: illustrated by the contents of caves and relics of cave-folk: 1751
The Lee Mill Cave: 1738
Leeuwin-Naturaliste National Park: cave permits draft issue plan: 3450
Legal aspects of access underground: a guide to the legal rights and obligations of people who explore cave, potholes and disused mine, and of those people who control access to them: 2144
The legend of Parker's Cave: 3455
Legend of the Cave of the Winds: 3495
Legend of the Luray Caverns, founded upon the discovery of a skeleton in one of its chasms: 2872
Legends of Shenandoah Caverns: in the Valley of Virginia: 2960
The legends of the caverns of Centre County, Pennsylvania: 2971
Legislative history for Oregon Caves National Monument, 59th Congress through 96th Congress: 1117
Lehman Caves: 2906
Lehman Caves National Monument: 3300
The Lehman Caves story: 1316
Lehman Caves...its human story: from the beginning through 1965: 3229
Lelet: report of the 1975 New Ireland Speleological Expedition: 515
Lelet: report of the 1976 New Ireland Speleological Expedition: 1208
Lemurs of the lost world: exploring the forests and Crocodile Caves of Madagascar: 3509

Les Eyzies and the Vezere Valley; an illustrated guide for scholars and tourists: 2663
The lesser caverns of Schoharie County: 3349
Letcher County's Pine Mountain caves: an in-depth pictorial essay of Letcher County's Pine Mountain caves and caverns: 1731
Let's explore a cave: 1706
Let's explore: caves and caverns: a young explorer series: 3221
Let's look at Jenolan Caves: an introduction to the features of Australia's famous limestone caverns: 573
Let's see Marvel Cave: your personal guide to America's largest privately owned cave: 1981
Letters from TAG 1966-1969: 3028
Lewis and Clark Cavern, Montana: 1946
Lewis and Clark Caverns: 3527
Life and death of the Pleistocene cave bear: a study in paleoecology: 1877
Life and Death Underground: 1974
Life at Russell Cave: 2050
The life history of the cave salamander, *Spelerpes maculicaudus* (Cope): 315
Life in the dark: 2693
The life of the cave: 2250
Life underground: caves, mines, minerals: 682
Light on dark: caves of the south east of South Australia: 2301
Limestone and caves of Northwest England: 3392
Limestone and caves of the Peak District: 1093
Limestone caverns: 718
The limestone caves and caverns of Ohio: 3459
Limestone caves of New South Wales: 2498
Limestone caves of Scotland, vol IV: the caves of Schichallion: 896
Limestone caves: a concise explanation: 836
Limestone geomorphology: 3244
Limestone geomorphology: a study in Jamaica: University of Bristol karst expedition to Jamaica 1967: 491
Limestone hydrology in the Upper Stones River Basin, central Tennessee: 2264
The limestone ranges of the Fitzroy Basin, Western Australia; a tropical semi-arid karst: 1727
The limestones and caves of Devon: 2566
Limestones and caves of the Mendip Hills: 3016
Limestones and caves of Wales: 1094
Limestones caves and cavers: 2639
Limnological survey of cavern pools: proposal to National Park Service, Carlsbad Caverns National Park: 577
Line Fork falls and caves: 223
Linville Caverns through the ages: the geological story: 525
A list of the arthropoda in the limestone caves in Kantô-Mountainland, with the description of a new genus and three species: 1961

A list of the Arthropoda in the limestone caves in Kantô-Mountainland, with the descriptions of a new genus and three species: 21
Listen to the rain; a story of the Carlsbad Caverns: 1527
The literature of Kent's Cavern prior to 1859: 2647
Lithologic controls on the development of solution porosity in carbonate aquifers: 2763
Little known caves of Oregon: 2512
The little red cavers book: 2499
The little red cavers book: NZ Speleological Society information manual: 2500
Littoral caves of ancient Lake Bonneville: 1333
Living in caves: 1358
Living underground: a history of cave and cliff dwelling: 1790
The Llangattock escarpment: 3376
Llechwedd: 1747
Llechwedd Slate Caverns: 1748
Location of solution channels and sinkholes at dam sites and backwater areas by seismic methods: 906
The Log of the Wookey Hole: exploration expedition 1935: 304
The longest cave: 536
Looking inside caves and caverns: 2914
The Los Tayos challenge: 2684
Lost caves of St. Louis: 2859
Lost in the Mammoth Cave: 1276
Lost River at Wesley Chapel Gulf, Orange County, Indiana: 2026
The lost rivers of London: a study of their effects upon London and Londoners, and the effects of London and Londoners upon them: 337
Lovelock Cave: 1967
Lovelock Cave: The republication of the rare 1929 Nevada classic: 1968
Lower South East cave reference book: an illustrated catalogue of the registered caves, sinkholes and associated karst features of the Lower South East Region of South Australia: 1521
Lower southeast of South Australia: a karst province: 2047
Lua Nunu o Kamakalepo: a cave of refuge in Kau, Hawaii: 432
Luminous darkness: the wonderful world of caves: 424
Luray Caverns: 145, 146, 1133, 1482, 1996
Luray Caverns in Virginia's Shenandoah Valley: 1064, 3328
Luray Caverns of the beautiful Shenandoah Valley on the Norfolk and Western Railway, Luray, Virginia. Luray, Virginia Mansion Inn: 1997
Luray Caverns, Va: 1137
The lure of cave lore: 2307
The lure of cave lore; being a random narrative upon caverns in general and in particular the properties of the Skyline Caverns, Front Royal, Virginia: 2308
Lure of the labyrinth: 1843
LUSS expeditions to Tresviso and the Picos de Europa in northern Spain, 1974-1977: 644

The magic of Sarawak's Mulu Caves; discover the latest wonder of the world: 147
Majlis Al Jinn Cave, Sultanate of Oman: 838
The major caves of France and their relationships with climatic factors: 713
Major limestone caverns of the Black Hills, South Dakota: 2487
Malaysian caves: 2024
Mallorca Caves; an interim guide: 2053
Malta's prediluvian culture at the stone age temples with special reference to Hagar Qim, Ghar Dalam cart ruts, Il-Misqa, Il-Maqluba and Creation: 977
The mammal fauna of Schulze Cave, Edwards County, Texas: 794
The mammal fauna of the early middle Pleistocene cavern infill site of Westbury-sub-Mendip, Somerset: 404
The mammalian fauna from the mid-Irvingtonian Fyllan Cave local fauna, Travis County, Texas: 3188
The mammalian fauna of Madura Cave, Western Australia: 1983
The mammalian fauna of Madura Cave, Western Australia, part VI, Macropodidae: Potoroinae: 1984
The mammalian fauna of Madura Cave, Western Australia: part I: 1985
The mammalian fauna of Madura Cave, Western Australia: part II: 1986
The mammalian fauna of Madura Cave, Western Australia: part III: 1987
The mammalian fauna of Madura Cave, Western Australia: Part IV: 1988
Mammalian remains from Rattlesnake Cave, Kinney County, Texas: 2933
Mammoth Cave: 1313, 2003, 2034, 3372
The Mammoth Cave and its denizens: a complete descriptive guide: 399
The Mammoth Cave and its inhabitants, or, descriptions of the fishes, insects and crustaceans found in the cave with figures of the various species and an account of allied forms, comprising notes upon their structure, development and habits, with remarks upon subterranean life in general: 2591
Mammoth Cave and the cave region of Kentucky, illustrated: with bibliography of Mammoth Cave (Willard Rouse Jillson); first accurate underground survey (H. Bruce Hoffman); introduction (H. C. Nelson): 2759
Mammoth Cave archeological inventory project interim report - 1987 investigations: 2707

Mammoth Cave area, Kentucky, 201 facilities plan environmental impact statement: 3275
The Mammoth Cave area: a planning proposal: 1014
Mammoth Cave by flash-light: 1741
Mammoth Cave in third dimension: 148
Mammoth Cave National Park: 1483, 2748, 3301
Mammoth Cave National Park archeological inventory project interim report-1988 investigations: 2708
Mammoth Cave National Park Association campaign, 1927-1928: 2029
The Mammoth Cave National Park research center: 3413
The Mammoth Cave National Park Research Center: a Cave Research Foundation study: 3414
Mammoth Cave National Park, Kentucky: 3302
Mammoth Cave National Park, Kentucky: history of legislation through the 82nd Congress: 2847
Mammoth Cave National Park: collection management plan: 1361, 3303
Mammoth Cave of Kentucky: 149, 2610, 3533
Mammoth Cave of Kentucky: with an account of Colossal Cavern: 1539
Mammoth Cave of Kentucky: the world's greatest subterranean wonder: 150
Mammoth Cave of Kentucky; an illustrated manual: 1540
Mammoth Cave romance: 2694
Mammoth Cave tour leader manual: 2490
Mammoth Cave, America's great natural wonder: 151
Mammoth Cave, Kentucky: 580, 1593
The Mammoth Cave, Kentucky; a sketch: 581
Mammoth Cave, Kentucky; an historical sketch containing a brief description of some of the principal places of interest in the Mammoth Cave. Also a short description of Colossal Cavern: 3204
Mammoth Cave... kids love it!: 666
Mammoth Cave: America's great natural wonders: 1226
Mammoth Cave: forgotten stories of it's [sic] people: 3400
Mammoth Cave: Kentucky's buried treasure: 23
The mammoth cavern: 3520
Mammoth Caves: 1065
A man deep in Mendip: the caving diaries of Harry Savory, 1910—1921: 2893
The management of edible bird's nest caves in Sabah: 1115
The management of soluble rock landscapes: an Australian perspective: 1819
Management of Victorian caves and karst: a report to the Caves Classification Committee, Department of Conservation, Forests and Lands: 812
A management plan for Bighorn Caverns, Montana: 2522
Managing the limestone caves of Chillagoe and Mungana: 3511
Manitou Grand Caverns: 311
Manitou Grand Caverns, Manitou Springs, Colorado: 3039
Manitouwadge: cave of the Great Spirit: 526, 847
Man's impact in Dinaric Karst: guide-book: 1140
Mansion Inn, Luray Caverns: 152
A manual for beginners: 3184
The manual of basic caving: 457
Manual of Caving Techniques: 776
Manual of U.S. cave rescue techniques: 1566, 3501
The manufacture of lightweight caving equipment: 967
Maori rock drawing: the Theo Schoon interpretations: 2907
Map location and dimensional definition of subsurface caverns: 3519
Mapping underwater caves: 1018
Marble Arch Caves: 153
Maria's Cave: 1514
Maribel Caves: an ideal health and summer resort.: 154
Mark Twain Cave: an adventure: 422
Marvel Cave: Silver Dollar City, Branson, Missouri: 2067
Marvel Cave: the story of America's largest privately owned cave: 155
Marvelous Howe Caverns: 1544
Marvelous Howe Caverns near Cobleskill, New York: 156
Marvels under our feet: from "the subterranean world": 1376
Masada and the finds from the Bar-Kokhba caves: struggle for freedom: 1732
Massacre Lake Cave, Tule Lake Cave and shore sites: 1423
Massanutten Caverns, Harrisonburg, Va. The most beautiful and unusual caverns yet discovered...: 1368
Massanutten Caverns, Harrisonburg, Va.: the most beautiful and unusual caverns yet discovered, in the heart of the Shenandoah Valley of Virginia: 1369
Master plan for the development of Jenolan Caves Reserve: 261
Master plan: Carlsbad Caverns National Park, New Mexico: 3304
Master plan: Wyandotte Caves, Harrison-Crawford, Blue River Recreation Complex: 1606
The master sculptor; a brief treatise on erosion, and its wondrous effects in the Shenandoah Valley: 3418
Matienzo, Spain: 485
Mazes and marvels of Wind cave: 1516
Medical aspects of speleology: 3105
Members handbook: South Wales Caving Club: 3059
Members manual: 875
A memoir of William Pengelly, of Torquay, F.R.S., Geologist, with a selection from his correspondence: 2643
Memoirs of a speleologist: the adventurous life of a famous French cave explorer: 850
Memoirs of the Geological Survey of Kentucky: 1795
Men of Pierre Saint-Martin: 246

The men of the Barma-grande (Baousse-Rousse) An account of the objects collected in the Museum praehistoricum, founded by Commendatore Th. Hanbury near Mentone: 3358

Mendip cave bibliography and survey catalogue, January, 1901—December, 1963: 2040

Mendip Cave bibliography: part II: books pamphlets, manuscripts and maps; 3rd Century to December 1968: 2956

Mendip cave registry: handbook for members of the Executive Committee: 968

The Mendip Caverns: 2136

A Mendip caver's ropework guide: 3080

The Mendip Caves: 295

Mendip Caving Group: 1982 library list: 2827

Mendip Caving Group: Belize '94: 1116

Mendip Karst Hydrology Project: phases one and two: 245

Mendip karst hydrology research project: phase 3: 902

Mendip underground: a caver's guide: 1678

Mendip, its swallet caves and rock shelters: 296

Mendip-Cheddar, its gorge and caves: 297

Mendip: the complete caves and a view of the hills: 334

Mendip: the great cave of Wookey Hole: 298

Mendip's vanishing grottoes: a photographic record of Balch Cave by J. A. Eatough and Shatter Cave by A. E. Mc. R. Pearce: 935

Meramec Caverns, Stanton, Missouri: 1484

Meramec Caverns: legendary hideout of Jesse James: 3421

Meso and Neolithic sequence from the Odmut Cave (Montenegro): 1857

Meteorological conditions in caves and ancient tombs in Egypt: 3162

Meteorological-geological investigations of the Wupatki blowhole system: 2886

Mexican caving, 1966-1971: 3382

Mexican eyeless Characin fishes, genus *Astyanax*: environment, distribution, and evolution: 2239

Mexico 85/86: 1943

Mexico: the black holes expedition: the 1988 British expedition to explore the caves of the Sierra de Zongolica: 1982

Mexico's caves and caverns: 157

The microclimate in Carlsbad Caverns, New Mexico: 2134

The microtine rodents of the Cheetah Room fauna, Hamilton Cave, West Virginia, and the spontaneous origin of *Synaptomys*: 2807

Microtus and *Pitymys* (Arvicolidae) from Cumberland Cave, Maryland, with a comparison of some new and old world species: 2163

Mid Argyll cave and rock shelter survey: 3015

A Mid-Pleistocene (Irvingtonian) herpetofauna from a cave in southcentral Texas: 1492

Miller County caves: 2217

Mineral resources of the Belle Starr Caves wilderness study area, Sebastian and Scott Counties, Arkansas: 1315

Minnesota caves of history and legend: a collection of unique cave stories: 1781

Minnesota karst country tours: 1782

Missouri Department of Conservation Cooperative cave inventory project: a final report submitted to the Missouri Department of Conservation as part of the cooperative cave inventory project: a final report: 1156

Missouri underground shelter space: a fallout shelter survey of mines and caves in Missouri: 2195

Missouri: the cave state: 3422

Missouri's Ice Age animals: 2141

Mitchell Caverns State Reserve: 578

Mitchelstown Cave: its discovery and history: 994

Mitigation of the Brady Run Rockshelter 3: a multicomponent site in Washington Township, Lawrence County, Ohio: 658

Mogollon cultural continuity and change; the stratigraphic analysis of Tularosa and Cordova caves: 2065

Mogollone material, Tularosa Cave, New Mexico: 1721

The Mole Creek Caves: 2994

Monkey Merry and St. Crida Cave: 1494

A monograph of the British Pleistocene Mammalia vol II part II: the bears: 2808

A monograph of the Pleistocene mammalia vol II, part I: the cave hyaena: 2809

Monograph of the stalactites and stalagmites of the Cleaves Cove, near Dalry, Ayrshire: 3024

Montagu Cave in prehistory: a descriptive analysis: 1787

Monteagle Wonder Cave, or, The Cave of the Cumberland: 158

Montpelier and Bear Lake County: 159

Moore's Cave: its discovery and significance: 1680

The moors and fells of Ingleton: 1339

Moors, crags & caves of the High Peak and the neighbourhood: 282

The Moravian karst: time and stone: 248

More years under the earth: 609

Morphogenetics of karst regions: variants of karst evolution: 1694

Morphological variation in *Gammarus minus* Say (Amphipoda, Gammaridae) with emphasis on subterranean forms: 1505

Morphology of extinct lava tubes and the implications for tube evolution, Chain of Craters Road, Hawaii Volcanoes National Park, Hawaii: 709

Morphology, hydrology and hydrochemistry of Karst in permafrost terrain near Great Bear Lake, Northwest Territories: 3347

Morphometric analysis of dolines: 430

Morrison Cave, Lewis and Clark Cavern State Park, Montana: 2658
The most beautiful caves in the world: 160
Mother and sun: the Cornish Fogou: 702
Mount Etna action book: time is running out: 161
Mount Etna caves: a collection of papers covering several aspects of the Mt. Etna and Limestone Ridge caves area of central Queensland: 3097
Mountain & cave rescue: the handbook of the Mountain Rescue Council: 2291
Mountain and cave rescue, with lists of official rescue teams and posts: the handbook of the Mountain Rescue Committee: 2288
Mountain rescue & cave rescue: 2289
Mountain rescue, cave rescue: 2290
The Mousterian cave of Teshik-Tash, Southeastern Uzbekistan, Central Asia: 2292
Mt. Etna & the caves: a plan for action: 1350
Mt. Gambier and district: Port MacDonnell and Tantanoola caves: 2533
Mud stalagmites and the conulite, a new speleothem: 3198
Muddy oxbows: a tourist trip into the seldom-trodden inner recesses of the human brain: 1440
Mulka's Cave site management project: emphasising visitor survey, April-June 1988, with management evaluation and further recommendations for management; with conclusions of an archaeological test excavation: 2836
Mullamullang Cave expeditions 1966: 1454
Mummies of Mammoth Cave: an account of the Indian mummies discovered in Short Cave, Salts Cave, and Mammoth Cave, Kentucky: 2145
Mummies of Mammoth Cave: Fawn Hoof, Little Alice, Lost John and others. An account of the Indian mummies discovered in Short Cave, Salts Cave, and Mammoth Cave, Kentucky: 2146
Mummies of Short Cave, Kentucky, and the Great Catacomb Mystery: 1176
Mummies, catacombs, and Mammoth Cave: 1177
The Mummy Cave Project in northwestern Wyoming: 2114
MVOR Guidebook Spring 1969: 2310
MVOR Guidebook: 1984 Fall: 2323
MVOR Guidebook: 1989 Spring: 2324
MVOR Spring 1971, April 30, May 1 & 2: Pulaski County: 2314
MVOR Spring "70" Guidebook: 2312
MVOR: Spring 1975 in the Shawnee Hills: 2318
My caves: 610
Mysteries underground: 1127
The mysterious cave of Amar Nath, Kashmir India: 1855
Mysterious caves of Langkawi, Malaysia: 2572
The mysterious world of caves: 346
Mystery Cave: 2218
Mystery Cave area: MSS Corn Feed August 1984: 1051
Mystery cave of many faces: first in a series on the saga of Burrows' Cave: 566
Mystery Hill: myth and mythology in the land of academe: 1571
Mystic wonderlands: 2832

N. S. S. cave diving manual: 1019
Nahal Hemar Cave: 318
Nahanni: 592
Nakimu Caves: 2052
The Nakimu Caves, Glacier Dominion Park, B.C.: 593
The Namakkal caves: 849
Named caves of Oregon: 433
The Naracoorte and Tantanoola Caves: a guide to the famous caves of South-east South Australia: 1434
Naracoorte Caves Conservation Park management plan, South East, South Australia: 3056
Naracoorte Caves Conservation Park, South East, South Australia: draft management plan: 3057
The Naracoorte Caves: "how to reach them": 162
Narracoorte [sic]: caves and town 20: book of views: 818
National Cave and Karst Research Institute Study: a draft report to Congress as required by Public Law 101-578 of November 15, 1990: 3305
National cave management seminar: Albuquerque, New Mexico, March 12-16, 1990: 56
National cave management symposia: proceedings: 2340
National cave management symposium proceedings: Albuquerque, New Mexico, October 6-10, 1975: 2337
National cave management symposium proceedings: Big Sky, Montana, October 3-7, 1977: 2339
A national park in Kentucky: 2030
National parks: v. 2: Carlsbad Caverns: 754
National Speleological Society (USA) 1973 field trip to Greece: 2687
National Speleological Society 1974 convention Luther College, Decorah, Iowa 10-18 August, 1974: 2396
National Speleological Society 1975 Convention, Frogtown, Ca: 2397
National Speleological Society convention 1965 guidebook, Indiana University Bloomington, Indiana June 12-20, 1965: 2387
National Speleological Society convention program with abstracts: 2446
National Speleological Society field trip to Aguas Buenas Caves, Puerto Rico: February 1968: 1291
National Speleological Society official 1987 guidebook, August 3-7, 1987, Sault Sainte Marie, Michigan: 2408
National Speleological Society, program: 1994 NSS Convention: Brackettville, Texas: June 18-25: 2445
Natural Bridge Caverns: 1837
Natural Bridge Caverns: Texas' largest caverns: One of the great show caves of the world: 163

Natural caves of West Germany Baden - Wuerttemberg: 3261

The natural history of biospeleology: 585

The natural history of Hartz-Forest, in His Majesty King George's German dominions. Being a succinct account of the caverns, lakes, springs, rivers, mountains, rocks, quarries, fossils, castles, gardens, the famous pagan idol Pustrich or spit-fire, dwarf-holes, etc....in the said forest: with several useful and entertaining physical observations: 373

Natural history of Texas caves: 1989

Natural resources management program: an addendum to the natural resources management plan for Lehman Caves National Monument, Nevada: 3306

Natural stone bridge and caves; fascinating geology, history & legends: 3252

Nature preserve potential of the Burton Cave area, Adams County, Illinois: 3463

Nature preserve potential of Twin Culvert Cave, Pike County, Illinois: 3464

Nature underground: the Endless Caverns in the heart of the historic Shenandoah Valley: 527

Nature underground; the Endless Caverns in the heart of the historic Shenandoah Valley: 988

Nature's masterpiece: Jenolan Caves N.S.W: 2667

Nature's underground palaces: caves and caverns: 1199

The Neanderthal skeletal remains from Shanidar Cave, Iraq: a summary of findings to date: 3127

Nebraska cave lore: 2700

Ned DeLoach's diving guide to underwater Florida: 852

Nelson Bay Cave, Cape Province, South Africa: the Holocene levels: 1611

The nest gatherers of Tiger Cave: 3344

The netherworld of Mendip: explorations in the great caverns of Somerset, Yorkshire, Derbyshire, and elsewhere: 283

A new analysis of Kent's Cavern, Devonshire, England: 589

New and little-known false scorpions, principally from caves, belonging to the families Chthoniidae and Neobisiidae (Arachnida, Chelonethida): 629

The New Calaveras Cave, Murphys, Calaveras County, California: W.J. Mercer, proprietor: description and guide: 1073, 2155

New cavernicolous Rhagidiidae from Idaho, Washington, and Utah (Prostigmata, Acari, Arachnida): 963

A new crayfish from Alabama caves with notes of the origin of the genera *Orcnectes* and *Cambaras* (Decapoda: Astracidae): 1465

New directions in karst: Anglo-French symposium, 1983: field excursion notes: 164

New directions in karst: proceedings of the Anglo-French Karst Symposium, September 1983: 2617

New England's buried treasure: 2655

New entrance to Mammoth Cave: a history of the cave together with descriptive guide to routes and illustrations of points of interest: 2279

A new era: 1919

A new family, genus, and species of cave-adapted planarian from Mexico (Turbellaria, Tricladida, Maricola): 2240

New finds of Pleistocene jaguar skeletons from Tennessee caves: 2115

New genera and species of cavernicolous diplopods from Alabama: 1476

New genera of Anophthalmid beetles from Cumberland Caves: 3343

A new genus and species of camel cricket from the Farallon Islands of California (Orthoptera: Gryllacrididae): 2806

A new geophiloid chilopod from Potter Creek Cave, California: 631

New Hampshire Caves: 2519

New Jersey caves in brief: 796

The New Madrid earthquake at Mammoth Cave (1811-1812): 1178

New mammalia from the Quaternary caves of California: 2990

The new Melones Cave harvestman transplant: 964

New Melones cave inventory and evaluation study: preliminary report: archeological caves: 2016

New Ostracoda (Halocyprida: Thaumatocyprididae and Halocyprididae) from anchialine caves in the Bahamas, Palau, and Mexico: 1848

New Paris no. 4: a late Pleistocene cave deposit in Bedford County, Pennsylvania: 1284

A new subfamily of blind beetle from Idaho ice caves, with notes on its bionomics and evolution (Coleoptera: Leiodidae): 3449

The New Zealand cave atlas: 763

The New Zealand cave atlas: North Island: 761

New Zealand cave atlas: South Island: 762

The New Zealand Cave atlas: volume 2: South Island: 2020

The New Zealand glowworm: 1125, 2164

A new *Peromyscus* (Rodentia: Cricetidae) from the Pleistocene of Maryland: 1285

Newsome sinks - our national landmark: 1478

Nibicon log book reports December 1972 - February 1973: 3171

Nobody here but us bats!: through Carlsbad Caverns with cartoons and comments: 835

Nomination of Australian fossil sites (a serial nomination of sites at Murgon, Riversleigh and Naracoorte): The origin and evolution of Australia's mammals: 1233

North Alabama's caves & caverns: 165

North American cave millipeds II, an unusual new species (Dorypetalidae) from Southern California, and new records of *Speodesmus tuganbius* (Trichopolydesmidae) from New Mexico: 2957

North American cave pseudoscorpions of the genus *Kleptochthonius*, subgenus *Chamberlinochthonius* (Chelonethida, Chthoniidae): 2295

North American triclad Turbellaria, XIII: three new cave planarians: 1590

North Carolina cave survey vol. 1 no. 1: 1489

North Carolina cave survey vol. 2: 1490

North Carolina coastal plain caves and their impact on mining and quarrying: 2036

North Country Region MECCA Summer 1979: 2516

North Country Region: Spring 1984, LaCrosse, Wisconsin: 1052

North-West Nelson State Forest Park Honeycomb Hill Cave System: concept plan: West Coast Conservancy: 2765

Northern cave handbook: 721

Northern caves volume 2: the three peaks: 499

Northern caves volume 4a: Scales Moor and King Scale: 505

Northern caves volume 4b: Leck and Casterton Fells: 506

Northern caves volume five: the Northern Dales: 507

Northern caves volume four: Whernside and Gragareth: 508

Northern caves volume one: Wharfedale and Nidderdale: 509

Northern caves volume three: Ingleborough: 500

Northern caves volume two: Penyghent and Malham: 501

Northern Caves: volume 3: the three counties system and the north-west: 510

Northern caves: volume1. Wharfedale and the North-East: 511

Northern caving; handbook of the Council of Northern Caving Clubs: 722

The northern limestone: 3432

Northwest Caving Association regional meeting guidebook: caves of the Peterson Prairie area: 2523

Northwest Regional Association symposium on cave science and technology: 1980: Feb 16-18, 1980, Seattle, WA,1981: Feb 14-16, 1981, Seattle, WA.1982: Feb 13-15, 1982, Boise, Idaho: 2525

Not the UBSS song book: 2930

Notes on caving and potholing for beginners (and for those needing a boost in safety): 1068

Notes on making saltpetre from the earth of the caves: 2755

Notes on Mexican cave Pseudosinella (Collembola: Entomobrydae) with the description of six new species: 656

Notes on Missouri Pleistocene peccaries: 2142

Notes on the barrows and bone-caves of Derbyshire with an account of a descent into Elden Hole: 2649

Notes on the excavation of Harborough Cave, near Brassington, Derbyshire: 1113

Notes on the limestone caves of New South Wales with plans: 3236

Notes on the plethodontid salamanders, *Eurycea lucifuga* (Rafinesque) and *Eurycea longicauda longicauda* (Green): 1585

Notes upon the human remains from the caves of Coahuila, Mexico: 3155

Notice of the occurrence of mammoth and other animal remains: discovered under limestone in a bone cave at Shandon, near Dungarvan, in the county of Waterford: 466

Now how! A national park in Kentucky: 2031

NSS 1970 guidebook: Mid-Appalachian Region: 2392

NSS 1990 convention guidebook: Yreka, California: "where the lava meets the limestone": 2411

NSS 71 guidebook: 2393

NSS 73 convention guidebook; National Speleological Society convention, Bloomington, Indiana, June 16-24, 1973: 2395

NSS cave diving manual: an overview: 2725

NSS Cave Diving Section 1986 members' manual: 614

NSS cavern diving manual: 3558

The NSS instructor's training manual: 2726

NSS student cave diver workbook: designed specifically for use by the student cave diver participating in the NSS full cave diver course: 2727

The NSW Cave Rescue Squad Inc: 2532

The NSW Cave Rescue Squad Inc. Annual Report July 1988-June 1989: 2531

Nullarbor caving atlas: 2284, 2674

Nullarbor karst, a bibliography: 813

Numerical index - area maps: 650

Oberon-Jenolan district historical notebook: 3456

Observations and significance of sinkhole development at Jefferson Island: 266

Observations of actively forming lava tubes and associated structures, Hawaii: 1251

Observations on the crustacean fauna of Nickajack cave, Tennessee, and vicinity: 1393

Observations on the crustacean fauna of the region about Mammoth Cave, Kentucky: 1394

The occurrence of the fossil remains of Pleistocene vertebrates in the caves of Barren County, Kentucky: 3427

Of caves & cavemen: 698

Of caves and shell mounds: 602

Official 1977 guidebook, Alpena, Michigan: '77, the Lakeshore convention: Alpena, Michigan, NSS: 2399

The official guide to Cheddar Somerset with map and eleven illustrations: 166

Official guide to Dan-yr-Ogof showcave: 673

Official guide to Marvel Cave, Silver Dollar City, Missouri: 2068

Official guide to Marvel Cave: the only complete authentic history of Marvel Cave ever published: 2117

Official souvenir book of Marengo Cave: U.S. National Landmark: Marengo, Indiana, the heart of cave country: 2825

Ogof Agen Allwedd in relation to the Mynydd Llangattwg: 1930

The Ogof Ffynnon Ddu system: its discovery and exploration (1927-53): and a theory of its development: 2749

Ogof Ffynnon Ddu: Penwyllt, Breconshire: 2574

Ohio Caverns: 167

The Ohio Caverns: where nature carved a fairyland: 168, 2546

The old town of Maastricht and the caves of mount St. Peter: 2898

The oldest man in America: an adventure in archeology: 1823

Oligio-Nunk, the place of caves. In the heart of Honeycomb Mountain ... Carter County, Kentucky: 169

Oliver Trickett: doyen of Australia's cave surveyors 1847-1934: 2166

The Olmec paintings of Oxtotitlan Cave Guerrero, Mexico: 1273

On a new cave fauna in Utah: and on new phyllopod Crustacea from the West: 2592

On an extinct type of dog from Ely Cave, Lee County, Virginia: 50

On Canadian caverns (read before the British Association for the Advancement of Science, at Aberdeen, 16th Sept. 1859): 1187

On caves: 1568

On centipeds and millipeds from Mexican caves: 632

On rope: North American vertical rope techniques for caving, search and rescue, mountaineering: 2593

On some fossil bones discovered in caverns in the limestone quarries of Oreston: 3458

On station: a complete handbook for surveying and mapping caves: 806

On the antiquity of the caverns and cavern life of the Ohio valley: 2944

On the caves of the world: 1032

On the caves perforating marble deposits, Limestone Creek (read 12th April 1883): 3128

On the cold caves of the Monte Testaccio at Rome: 170

On the collection of cavern insects: 3

On the contents of a bone cave in the island of Anguilla (West Indies): 711

On the discovery of an ossiferous cavern of Pliocene Age at Doveholes, Buxton (Derbyshire): 842

On the ice-caves or natural ice-houses found in some of the caverns of the Jura and the Alps: 2672

On the rocks: prehistoric art of France and Spain: 2098

On the troglobitic shrimps of the Yucatan peninsula, Mexico (Decapoda: Atyidae and Palaemonidae): 1469

On the underground river system of the Tisu karst area, Tu-an County, Kwangsi, China: 1589

On *Spelaeogriphus*, a new cavernicolous crustacean from South Africa: 1228

One hundred miles in Mammoth Cave - in 1880: an early exploration of America's most famous cavern: 1541

One man's dream; the story of Jim White, discoverer and explorer of the Carlsbad Caverns: a biography: 576

One thousand metres down: 574

Onondaga, the mammoth cave of Missouri: 3423

Operation of Blanchard Springs Caverns, Ozark-St. Francis National Forest, Arkansas: final environmental statement: 3276

Ordinance for the control of urban development in sinkhole areas in the blue grass karst region, Lexington, Kentucky: 874

Oregon Caves: 3307

Oregon Caves discovery & exploration: Oregon Caves National Monument: 3380

Oregon Caves forest and fire history: 17

Oregon Caves National Monument collections management plan: 567

Oregon Caves National Monument resource database: a user's guide: 942

Oregon Caves National Monument, master plan: 3308

Oregon Caves National Monument, Oregon: final interpretive prospectus: 3309

Oregon Caves National Monument, Oregon: history of legislation through the 82nd Congress: 2848

Oregon Caves National Monument, Oregon: official map and guide: 3310

Oregon Caves National Monument: a manual for cave guides: 1012

The Oregon Caves, Siskiyou National Forest: 3277

Oregon Caves: a pictorial souvenir guide: 171

Oregon Caves: discovery and exploration: 3381

The Oregon Caves: Siskiyou National Forest: 3278

Oregon, no 32, vol 1: 3110

The Orient Cave, Jenolan Caves: 1388

Origin and development of Cave Spring, Shannon County, Missouri: 2219

The origin and geographic distribution of troglobites: 1506

Origin and hydrology of caves in the White Limestone of north central Jamaica: report of field work carried out under ONR Contract 3656 (03) NR 388 067: 40

The origin and story of Kents Cavern: 2637

The origin and use of the Royston Cave, being the substance of a report some time since presented to the Royal Society of Antiquaries by the late Joseph Beldam ...: 374

Origin of caves and karst in the Shenandoah Valley, Rochingham and Augusta Counties, Virginia: guidebook for a geologic fieldtrip, National Speleological Society Annual Convention, Blacksburg, VA, 16 July 1995: 2424

The origin of helictites: 2260

Origin of limestone caves; a symposium with discussion: 2261

The origin, story and descriptive tour of the caves: 2638

Origins and affinities of the troglobitic crayfishes of North America (Decapoda: Astacidae), II, genus em Orconectes: 1466

Orpheus Caving Club songbook: 2060

Ostracoda (Halocyprididae) from anchialine caves in the Bahamas: 1849

Ostracoda (Halocypridina, Cladocopina) from anchialine caves in Jamaica, West Indies: 1850

Oudtshoorn and the Cango Caves: 172

Outdoor recreation II: caving: journey to the centre of the earth: 19-20 October 1985: 856

Outline of major biological conclusions from 1980-81 Mount St. Helens eruptions: 1334

Owen 90: New Zealand: 3186

Owls, caves, and fossils: predation, preservation, and accumulation of small mammal bones in caves, with an analysis of the Pleistocene cave faunas from Westbury-sub-Mendip, Somerset, UK: 75

Oxford University Cave Club Cabeza Julagua expedition final report 28 June - 20 August 1993: 2038

Oxford University Cave Club Juracao expedition 28th June-17th August 1989: 1526

Oxford University Cave Club: Huerta del Rey expedition final report: 2585

Oxford University Cave Club: La Verdelluenga 1994 final report: 2586

Oxford University expedition to northern Spain, 1961: 2587

Ozark stories of the Upper Current River: 3361

The P8 Cave: Castleton: 3031

Padirac Chasm and underground river: 22

Padirac; its history and a short description: 2059

Painted caves: 1265

The painted caves of France & Spain: 2099

The painted caves: an introduction to the prehistoric art of Zimbabwe: 1158

The painted shelters of La Gasulla (Castellon): 2823

Palaeokarsts and palaeokarstic reservoirs: 3536

Palaeolithic cave art: 3326

The Palaeolithic site at Douara Cave in Syria: report of the fourth season of the Tokyo University Scientific Expedition to Western Asia: 3163

Paleoenvironmental investigations at Seed Cave (Windust Cave H- 45FR46), Franklin County, Washington: 3208

Paleohydrography of the caves in the Moravsky Kras (Moravian Karst): 2712

Paleokarst: 1702

Paleokarst in Hungary: 288

Paleokarst, karst-related diagenesis, reservoir development, and exploration concepts: examples from the Paleozoic section of the southern mid-continent: 1993 annual fieldtrip guidebook Permian Basin Section - SEPM Arbuckle Mountains, Oklahoma: 1788

Paleokarst: a systematic and regional review: 441

The paleolithic of southern Kurdistan: excavations in the caves of Zarzi and Hazar Merd: 1162

Paleolithic site of Douara Cave and paleogeography of Palmyra Basin in Syria: 1354

Paleontology: 3311

The paleontology of Cheek Bend Cave, Maury County, Tennessee: phase II: report to the Tennessee Valley Authority: 1826

Pantheon of Czech speleologists: 1623

Paper on Dunald Mill Hole: read by J. P. Smith before the Barrow Naturalists Field Club: 3021

Papers in California archaeology: 17-18: 3332

Papers on California archaeology: 30-31: 3333

Papers on California archaeology: 32-33: 3334

Papers on Great Basin prehistory: 3331

Papers presented at the international symposium on changing karst environments, Oxford and Huddersfield, September 1994: 1648

Papua New Guinea Speleological Expedition NSRE 1973: the report of the 1973 Niugini Speleological Research Expedition to the Muller Range: 1699

The Paradise Ice Caves, Mount Rainier National Park, Washington: 1335

Partial catalog of caves in Missouri: 2196

A partial guidebook to the caves of the Berkshires: 1993 spring N.R.O. hosted by the Berkshire Hills Grotto (May 14th, 15th, 16th): 444

Particulars of a first exploration of the extensive and newly discovered cavern, at Stainton, Low Furness: 445

Paviland Cave: an Aurignacian station in Wales (The Huxley memorial lecture for 1913): 3047

Peak and Speedwell Caverns, exploration and science: including the talks presented at the Peak-Speedwell symposium, Sheffield, November 1989: 486

A Peak Cavern bibliography: 2336

The Peak Cavern system - a caver's guide: 714

Peak Cavern: a guide to this famous show-cave; its formation, history & folklore: 3524

The Peers' cave, tunnel cave and rock shelters at Skildergat, Fish Hoek, the home of pre-historic man: 3369

Pennine underground: 3210

Penn's grandest cavern: the history, legends and description of Penn's Cave in Centre County, Pennsylvania: 2972

Penn's grandest cavern: the history, legends photographs and description of Penn's Cave in Centre County, Pennsylvania: 2973

Pennsylvania caves: 3137

Pennsylvania's caves & caverns: an activity book: 676

Pennsylvania's Historic Indian Cave at Franklinville, PA, Huntington County...: 882

The people of Burrows Cave: who they were, where they came from and when: 3396

The Perama Caves of Loannina: 2662

Perishable industries from Hinds Cave, Val Verde County, Texas: 76

Perry County: spring M.V.O.R. '79: 2321

Phalangodidae from caves in the Sierra Nevada (California) with a redescription of the type genus (Opiliones, Phalangodidae): 478

The phantoms of stork fontein [sic]: 3553

Phelps County caves: 2220

A Philippine cave index and bibliography: 3005

Photographic views of some of the important points of Mammoth Cave situated in Edmondson, Co. Kentucky, USA: the wonders of this cave and its magnitude, cannot be described, and must be seen to be appreciated. It is reached only by the Louisville & Nashville Railroads. These photographs were taken by magnesium light by W.H. Sesser, St. Joseph, Mich., with a Collins camera: 173

Photographs of the Jenolan Caves (interior views photographed by means of the electric and magnisium [sic] lights: 3489

The physico-chemical evolution of moonmilk: 394

Piccaninnie Ponds mapping project, 1984/85: 1522

The pictographs of Fern Cave, Lava Beds National Monument: agents of deterioration and prospects for conservation: 2981

A pictorial guide to Aurangabad, Daulatabad, Ellora, and Ajanta: 2975

Pictorial guide to the caves Cheddar: 1231

Pictorial guide to the Mammoth Cave, Kentucky: 2063

Pictorial guide to the Mammoth Cave, Kentucky: a complete historic, descriptive and scientific account of the greatest subterranean wonder of the western world: 400

A pictorial history of Swildon's Hole: 824

A pictorial study of Mammoth Cave and Mammoth Cave National Park: 2926

The pictured cave of La Crosse Valley, near West Salem, Wisconsin: 519

The Pictured Cliffs project: petroglyphs and talus shelters in San Juan County, New Mexico: 1033

Pictures of Cheddar and Cox's cave by pen and camera: 174

Pigeon Mountain and Ellison's Cave system: a report to the Georgia Department of Natural Resources by the Dogwood City Grotto of the National Speleological Society, Post Office Box 12072, Atlanta, Georgia 30305: 881

Pike County caves: 2221

Pike Creek Dam: a preliminary criticism of the Queensland Irrigation and Water Supply's Commission's environmental impact study: 2736

Pioneer under the Mendips: Herbert Ernest Balch of Wells, a short biography: 3107

Pioneers of Maltese geology: 2019

Pit caves of Jefferson County: 2222

Pittsburgh Grotto Picnic Fall M.A.R. '79: Guidebook, Sept. 14-15, 1979: 3431

The Pleistocene (Kansan) herpetofauna of Cumberland Cave, Maryland: 1493

The pleistocene (Kansan) herpetofauna of Trout Cave, West Virginia: 1491

Pleistocene and recent faunas from the Brynjulfson Caves, Missouri: 2613

Pleistocene and recent vertebrate faunas from Crankshaft Cave, Missouri: 2614

The Pleistocene birds of San Josecito cavern, Mexico: 2183

A pleistocene cave deposit of Western Maryland: 1193

A pleistocene fauna from Zoo Cave, Taney County, Missouri: 2223

Pleistocene man in Fishbone Cave, Pershing County, Nevada: 2577

Pleistocene peccaries from the Cumberland cave deposit: 1194

The Pleistocene vertebrate fauna from Cumberland Cave, Maryland: 1191

Pleistocene vertebrate remains from a cave near Montagu, N.W. Tasmania: 2304

Policy organisation and rules: 420

Pollen analysis of a stratigraphic section from Bat Cave, New Mexico: 6

Pollution abatement project, Carlsbad Caverns National Park, New Mexico: 3312

Pollution of underground water: 3403

The Polnagollum cave, Co. Clare: 680

Pontnewydd Cave: a lower palaeolithic hominid site in Wales: the first report: 1252

Poole's Cavern: Buxton: 175

Postcards of the Caves of Han printed by Nels of Brussels between 1905 and 1940: 2916

Postcards of the Caves of Han printed by Nels of Brussels between 1905-1940: 2917

Postojna: 1302

Postojna Cave: 1303

Postojna Caves: 1304

The Postojna Caves and other tourist caves in Slovenia: 1305

Postojna caves: enter traveller into this immensity!: 1828

The Postojna Grottes with the Planina and Predjama caves: 1306

The Postojna Grottoes: 1307

The Postojna grottoes and the other marvels of the karst: 2937

Potential for pollution of the Upper Floridan aquifer from five sinkholes and an internally drained basin in west-central Florida: 3241

Potholer's songs: 1920

Potholing and caving: 2829

Potholing: beneath the northern Pennines: 1414

Pottery and other artifacts from caves in British Honduras and Guatemala: 2074

Pottery from certain caves in eastern Santo Domingo, West Indies: 439

Poul Na Gollor Cave at Inch Ennis, Co Clare: 1942

Practical karst hydrogeology, with emphasis on groundwater monitoring, February 8-11, 1994, Holiday Inn West, Gainesville, Florida: 2745

Practical tracing of groundwater, with emphasis on karst terranes: a short course manual presented on the occasion of the annual meeting of the Geological Society of America, October 24, 1992, Cincinnati, Ohio: 27

Pratt Cave studies: Guadalupe National Park, Texas: 2911

A predictive hydrologic model for evaluating the effects of land use and management on the quantity and quality of water from Ozark Springs: final report: 41

Predjama: the castle and the cave: 1308

The prehistoric and early historic archaeology of Wyandotte Cave and other caves in southern Indiana: 2300

Prehistoric art: 2333

Prehistoric art of the western Mediterranean and the Sahara: 2653

Prehistoric cave art in northern Spain, Asturias: 391

The prehistoric cave of Gher Dalam: 2267

Prehistoric cave painting: 789

Prehistoric cave paintings: 2045, 2761

Prehistoric magic: 734

Prehistoric man and his art: the caves of Ribadesella: 392

Prehistoric man in Mammoth Cave: 2918

Prehistoric mummies from the Mammoth Cave area: foundations and concepts: 1179

The prehistoric native American art of Mud Glyph Cave: 1034

Prehistoric painting: 2285

Prehistoric painting, part 2: 176

Prehistoric painting: with 56 plates in colour and monochrome and 7 line illustrations in the text: 497

Prehistoric painting: Lascaux, or, the birth of art: 342

Prehistoric paintings of France and Spain: a description of 34 actual size copies hand screenprinted by Douglas Mazonowicz: 2100

Prehistoric paintings: a catalog of actual size copies in silkscreen: 2101

Prehistoric remains in Hungary: 2821

Prehistoric rock art of the Federation of Rhodesia & Nyasaland: 1222

Prehistoric rock art of the Santa Barbara region: [exhibition] Art Gallery, University of California, Santa Barbara, October 12-Nov. 7, 1965: 3335

Prehistoric rock paintings of Bhimbetka, Central India: 2084

The prehistoric rock paintings of Tassili n'Ajjer: a description of 15 actual size copies, hand screenpainted: 2102

Prehistoric sites in Perigord: 247

Prehistoric upland bird hunters: archaeological inventory survey and testing for the MPRC project area and the Bobcat Trail Road, Pohakuloa Training Area, Island of Hawaii: 242

The prehistory of Salts Cave: Kentucky: 3408

Preliminary account of an expedition to the cliff villages of the red rock country, and the Tusayan ruins of Sityatki and Awatobi, Arizona, in 1895: 1053

A preliminary bibliography of Mexican cave biology with a checklist of published records: 2790

A preliminary catalog of the caves on Anguilla, British West Indies: 2123

Preliminary checklist of references to Jenolan Caves: 921

Preliminary excavations of Pershing County caves: 2578

A preliminary investigation into the bacterial & botanical flora of caves in South Wales: 2080

Preliminary investigations at Hayes Cave, Hants County, Nova Scotia, in 1978: 2923

A preliminary plan for the development of Blanchard Springs Cavern: Sylamore District, Ozark National Forest: 3279

Preliminary report of biological investigation Valdina Farms sink hole - Medina Co., Texas: 1959

Preliminary report of excavations at Moh-Khiew Cave, Krabi Province, Sakai Cave, Trang Province, and ethnoarchaeological research of hunter-gatherer group, socall[ed] Sakai or Semang at Trang Province: the Hoabinnian Research Project in Thailand: 3160

A preliminary report of findings for the karst terrains feature known as Rugh Cavern, located along northern Seco Creek, northwestern Medina County, Texas: 3402

A preliminary report of Hinds Cave, Val Verde County, Texas: 2942

A preliminary report of the karst morphology of the Nullarbor Plains: 1728

Preliminary report of the Schuiling Cave, Newberry, California: 3019

Preliminary report on a recently discovered Pleistocene cave deposit near Cumberland, Maryland: 1192

A preliminary report on rock shelters in Fall River County, South Dakota: 2143

Preliminary report on the archeological investigation of Pinnacle Point Cave, Toulumne [sic] County, California: 2620

A preliminary report on the caves of eight northern California counties: 1336

A preliminary report on the caves of the Marble Mountain Wilderness: 1825

A preliminary report on the mortuary cave of Candelaria, Coahuila, Mexico: 2069

A preliminary report on Vanished River Cave, Santa Cruz: 2846

Preliminary results of an experiment on human chronobiology and neurobiology in a subterranean environment. 1. Life on a bicircadian rhythm (P. Englender). 2. Life in continuous light (J. Chabert). Longitudinal analysis and computer correlation of neurologic, psychological and physiologic data collected in beyond-time cave experiments from 1968-1969: 2980

Preliminary seismic-velocity and magnetic studies of a carbonate rock-sinkhole area in Shelby County, Alabama: 2180

Preliminary survey of Palawan, Philippines, by Traditional Explorations and the Sydney Speleological Society, January/February 1980: 534

A preliminary survey of the caves of the Yucatan Peninsula: 2791

Preprints of papers for the 17th Biennial Conference of the Australian Speleological Federation Tropicon Conference, Lake Tinaroo, Far North Queensland. 27th to 31st December 1988: 265

Present and future water supply for Mammoth Cave National Park, Kentucky: 787

The presentation of cave survey data: 683

Presenting the wonders of Lost River: 2811

Preservation of caves in Illinois: 3465

Preservation of Fogelpole Cave, Monroe County, Illinois: 3466

The preservation of wilderness karst in central Kentucky, U.S.A: 3415

Primitive hearths in the Pyrenees; the story of a summer's exploration in the haunts of prehistoric man: 2894

Princess Margaret Rose Caves - Western Victoria: 1435

Private account of A. F. McDonald, permanent guide of Wind Cave: 2116

A privilege—a duty—an opportunity for Kentucky: 2032

Problems of environmental protection of karst area: 3548

Problems of the speleological research: proceedings of the international speleological conference held in Brno June 29-July 4, 1964: 1643

Proceedings - 16th biennial conference Australian Speleological Federation Inc: 3090

[Proceedings] nemzetkozi karszthidrologiai szimpozium = international symposium on karsthydrology = Mezhdunarodnyi simpozium po gidrologii karsta: 1652

Proceedings of environmental problems in karst terranes and their solutions conference, October 28-30, 1986, Bowling Green, Kentucky: 2483

Proceedings of international symposium at 170th anniversary of Postojnska jama, Postojna (Yugoslavia) Nov. 10-12 1988: 1860

Proceedings of international symposium "man on karst," Postojana, September 23-25, 1993: 1862

Proceedings of the 18th biennial speleological conference: 30 December 1990 to 5 January 1991 at Margaret River, W.A: 3091

Proceedings of the 1976 NSS annual convention at Morgantown, West Virginia 24 June - 3 July 1976: 2427

Proceedings of the 1984 national cave management symposium, Rolla, MO: 2343

Proceedings of the 1986 meeting of the International Geophysical Union study group on man's impact on karst: 1214

Proceedings of the 1989 national cave management symposium: New Braunfels, Texas, U.S.A.: 2345

Proceedings of the 1993 national cave management symposium: held in Carlsbad, New Mexico, October 27-30, 1993: 2347

Proceedings of the 1995 national cave management symposium: Spring Mill State Park: Mitchell, Indiana: October 25-28, 1995: 2348

Proceedings of the 4th International Congress of Speleology in Yugoslavia, Postojina - Ljubljana - Dubrovnik 12-26 ix 1965: 1621

Proceedings of the 5th international symposium on underground water tracing, Athens 1986: 2269

Proceedings of the 5th International Symposium on Underground Water Tracing: Athens 1986: 1654

Proceedings of the 6th annual seminar, Lindenwood College, St. Charles, Missouri, June 16-17, 1973; Proceedings of the 7th annual seminar, Ramada Inn West, Jacksonville, Florida, June 15- 16, 1974 ; Research papers by N.A.C.D. instructor candidates: 2903

Proceedings of the 6th International Congress of speleology = Actes du 6e Congrès international de spéléologie: 1624

Proceedings of the 7th International Speleological Congress, Sheffield, 1977: 1625

Proceedings of the eighth biennial conference of the Australian Speleological Federation: 262

Proceedings of the eighth international congress of speleology: a meeting of the International Union of Speleology: 1627

Proceedings of the fifth annual seminar on cave diving: 310

Proceedings of the first annual seminar on cave diving: 477

Proceedings of the first annual watershed conference, May 20-21, 1987, Springfield, Missouri: 549

Proceedings of the first symposium on theoretical and applied karstology held in Bucharest, Romania, 22-24 April 1983: 2575

Proceedings of the fourth conference on karst geology and hydrology: 2764

Proceedings of the IAH 21st congress: 1616

Proceedings of the international conference on environmental changes in karst areas I.C.E.C.K.A. Italy, September 15-27th, 1991: 2890

Proceedings of the international symposium on vulcanospeleology and its extraterrestrial applications: a special session of the 29th Annual Convention of the National Speleological Society, White Salmon, Washington, 16 August 1972: 1661

Proceedings of the karst-symposium-Blaubeuren: 1618

Proceedings of the second conference on environmental problems in karst terranes and their solutions conference, November 16-18, 1988, Maxwell House Hotel, Nashville, Tennessee: 2484

Proceedings of the tenth biennial conference of the Australian Speleological Federation: 263

Proceedings of the Third Conference on Hydrogeology, Ecology, Monitoring and Management of Ground Water in Karst Terranes: December 4 - 6, 1991, Maxwell House/Clarion, Nashville, Tennessee: 689

Proceedings of the third conference on hydrogeology, ecology, monitoring, and management of ground water in karst terranes, December 4-6, 1991, Maxwell House/Clarion, Nashville, Tennessee: 238

Proceedings of the third international symposium on volcanospeleology: a special session of the 39th annual convention of the National Speleological Society Bend, Oregon, USA: July 30-August 1, 1982: with related biovulcanospeleological papers also presented at the 39th annual meeting of the National Speleological Society: 1656

Proceedings of the twelfth international congress: karst hydrogeology: Huntsville, Alabama: 1617

Proceedings of the XI international congress of speleology, August 2 to 8, 1993, Beijing China: 1635

Proceedings, U.I.S. international symposium on cave biology and cave paleontology, held in Oudtshoorn, South Africa from the 3rd- 6th August, 1975: 1646

Proceedings: 13-20 August 1989: 1634

Proceedings: national cave management symposium, Mountain View, Arkansas, October 26 - 29, 1976: 2338

Processes in karst systems: 904

Program: 1628, 1655

Program 1977 Annual Meeting of the National Speleological Society, August 1-5, Alpena, Michigan]: 2429

Program 1987 NSS Convention, August 3-7, 1987, Sault Sainte Marie, Michigan: 2438

Program for the 1985 NSS Convention, Frankfort, Kentucky, June 23- 29: 2436

The program of the 1979 N.S.S. Convention, August 5-12, 1979, Pittsfield, Mass: 2431

The program of the 1984 NSS Convention, June 25-29, 1984, Sheridan, Wyoming: 2435

The program of the 1988 NSS convention: June 27-July 1, 1988: Hot Springs, South Dakota: 2439

The program of the annual convention of the National Speleological Society, June 27-July 3, Bend, Oregon: 2433

Program of the National Speleological Society convention July 8 through July 14, 1990: Yreka, California: 2441

Program, 1983 NSS Annual Convention: 2434

Program, 1986 NSS Convention, June 22-28, 1986, Tularosa, New Mexico: 2437

Program: 1991 National Speleological Society convention: July 1-5: Cobleskill, NY: 2442

Program: 1992 NSS convention: the heartland of the karstland: Salem, Indiana: 2443

A proposed cave access and cave management plan for the Devil's Sinkhole State Natural Area & Kickapoo Caverns State Park: 3383

Proposed developments at Waitomo Caves: 1351

Proposed Mammoth Cave National Park master plan and wilderness study, Kentucky: draft environmental impact statement: 3313

Proposed natural resources management plan and environmental assessment, Lehman Caves National Monument, Nevada: 1928

Proposed natural resources management plan, Lehman Caves National Monument, Nevada: 1929

Proposed wilderness, Carlsbad Caverns National Park, New Mexico: 3314

Prospective's handbook: 3169

Protecting cave resources on federal lands, and for other purposes: report (to accompany H.R. 1975) (including the cost estimate of the Congressional Budget Office): 3267

Proteus; the mysterious ruler of karst darkness: 47

Proyecto Cerro Rabon: 405

Proyecto Cheve 1986-1993: 2680

Proyecto Papalo expedition report, 1986-1989: the exploration of Sistema Cuicateca - second deepest cave in the western hemisphere: 3360

Proyek Kelelawar: final report of the Oxford University expedition to the Togian Islands, Sulawesi, Indonesia summer 1987: 2582

Prusiking: 3216

Pt. I: the exploration of Bushey Cavern near Cavetown, Maryland: 2623

Publication no. 4: 3447

Pulaski County caves: 2224

The quadrangle maps of Carlsbad Cavern, New Mexico: 618

Quality of groundwater in a carbonate terrain in western Virginia: 1438

Quantification and forecasting of regional groundwater flow and transport in a karst aquifer (Gallusquelle, Maim, southwestern Germany): 2891

Quantitative analysis of a late-Quaternary cave deposit, Porcupine River, Alaska: 2889
Quaternary animals from Schuiling Cave in the Mojave Desert, California: 895
Quaternary antelope remains from a second cave deposit in the Organ Mountains and further study of the quaternary antelopes of Shelter Cave, New Mexico: 3129
Quaternary mammals of the Smolucka Cave in southwest Serbia: 873
Queen Mary College: below Belize 1988; the logistics report of the 1988 Speleological Expedition to Belize: 3499

Rabies transmission by air in bat caves: 695
Race against time: a history of the Cave Rescue Organisation: 1023
Radon concentrations, radon decay product activity, meteorological conditions and ventilation in Mystery Cave: 1949
Radon radiation situation in NPS caves: January 19, 1976: 2485
Rainstorm related terrain failures in Puerto Rico: final technical report: 3429
Ralls County caves: 2225
Rambles about Ingleton, in caves by rivers and on mountains in the spring of 1865.: 601
Rambles in Mammoth Cave, during the year 1844, by a visitor: 757
Rambles in the Mammoth Cave, during the year 1844: 552
Rambles in the North Yorkshire Dales: 548
The re-excavation of Powerhouse Cave and an assessment of Dr. Frank Peabody's work on Holocene deposits in the Taung area: 1572
Recent cave explorations in California: 2160
Recent fluvial cave sediments, their origin and role in speleogenesis = Recentni fluvialni jamski sedimenti, njihovo nastajanje in vloga v speleogenezi: 1863
Reconnaissance and excavation in southeastern New Mexico: 2147
Reconnaissance geology of Timpanogos Cave: Wasatch County, Utah: 3476
Reconnaissance of springs and sinks in west-central Florida: 3452
Recovery plan for endangered karst invertebrates in Travis and Williamson Counties, Texas: 2545
Recovery plan for the Alabama cavefish, *Speoplatyrhinus poulsoni* Cooper and Kuehne 1974: 708
A recovery plan for the Ozark cavefish (*Amblyopsis rosae*): 3504
The Red Dragon Journal of the Cambrian Caving Council index issues (1)-(20) 1974-1993: 1555
Red hills: 177
The Redcliff stone age site, Rhodesia: 701
Reflections: a look at the 'spelaeodes' and other caving sagas: 45

The Regional Caves Interpretive Centre: a planning review report for the Augusta Margaret River Tourist Bureau: 1352
Regional distribution of fractures in the southern Edwards Plateau and their relationship to tectonics and caves: 3439
Regularities of karst processes (karst landscape): 2713
Reliquiae diluvianae, or observations on the organic remains contained in caves, fissures, and diluvial gravel, and on other geological phenomena attesting the action of an universal deluge /: 545
Reliquiae diluvianae: or, observations on the organic remains contained in caves, fissures, and diluvial gravel and on other geological phenomena attesting the action of a universal deluge: 546
Reliquiae diluvianae: or, observations on the organic remains contained in caves, fissures, and diluvial gravel and on other geological phenomena attesting the action of an universal deluge: 547
Remains of Quaternary vertebrates from Ozark caves and other miscellaneous sites: 2226
The remarkable Howe Caverns story: 768
Report: 2070
Report of 1979 archeological and geological reconnaissance and testing of cave deposits, Porcupine River, Alaska: 877
Report of a visit to the Luray Cavern, in Page County, Virginia, under the auspices of the Smithsonian institution, July 13 and 14, 1880: 2077
The report of lava flow and lava caves on Mauna Loa: 3159
Report of study by National Speleological Society, Cave Conservation Task Force: New Melones Project: 3098
The report of the 1973 Niugini Speleological research expedition to the Muller Range: 1700
Report of the 1974 British speleological expedition to Matienzo area of Santander Province of northern Spain: 2035
The report of the 1981 and 1982 British cave diving expeditions to Andros Island, Bahamas: 2605
Report of the 1982 expedition to Barra Honda National Park: 1429
Report of the 1987 international blueholes research project: 2606
The report of the 1992 caving expedition to the Chimanimani Mountains in the eastern highlands of Zimbabwe: 1935
Report of the British speleological expedition to arctic Norway, 1969, incorporating the work of the 1968 Hulme Schools expedition: 1415
Report of the California-Nevada Speleological Survey June 13-September 7, 1952: 859
Report of the Caver Proliferation Committee: 868
Report of the Devil's Icebox - Rockbridge Park conservation task force of the National Speleological Society: 2227

Report of the French Speleological expeditions to PNG in collaboration with the Committee of French Speleological Expedition and the Scientific Committee of the French Federation of Speleology: 2021
Report of the Nottingham University Students Union Spelaeological Expedition 1970, Picos de Europa North-West Spain: 2086
Report of the Queen Mary College Society spelaeological expedition to Yugoslavia: 641
Report of the SWETC caving club expedition to Norway 1974: 1035
Report of the Tasmanian Caverneering Club 1986 speleological reconnaissance expedition to Precipitous Bluff: 941
Report of the Virginia Commission on the Conservation of Caves to the Governor and the General Assembly of Virginia: 3367
Report on archaeology of southern Chihuahua: 3557
Report on Cold Water Cave: a summary of research results with inclusion of information on related to potential development of a new recreational facility by the state of Iowa: 1842
Report on preliminary archeological explorations at Carlsbad Caverns National Park, New Mexico: 1255
Report on sediments in Mammoth Cave, Kentucky: 831
A report on South Australian diving fatalities: 1523
Report on the animal remains found in the Kilgreany Cave, Co. Waterford: 1687
Report on the "birds' nest" caves and industry of British North Borneo: with special reference to the Gomantong caves: 640
Report on the British speleological expedition to Vietnam; March April 1990: 458
Report on the caves of the eastern highland rim and Cumberland Mountains: 278
Report on the excavation of Jemez Cave, New Mexico, Santa Fe, N.M.: 29
Report on the exploration of Brixham Cave, conducted by a committee of the geological society, and under the superintendence of Wm. Pengelly, etc: 2709
A report to the government of Mexico on the 1994 San Augustin expedition to Huautla de Jimenez, Oaxaca: 3140
Reports on the investigations of Pen Park Hole, Bristol: 3227
Rescue from remote places: cave rescue and sport diving emergencies: 482
Rescue system outline: Lilburn Cave Project: 1583
Research handbook for cave divers: an introduction to the realm and techniques of the underwater speleologist: 1524
Research of China karst: 178, 179
Researches and excavations carried on in and near Moa Bone Point Cave, Summer Road, in the year 1872: 1300
Resource management in limestone landscapes: international perspectives proceedings of the International Geographical Union study group man's impact on karst Sydney 15-21 August 1988: 1210
Resource management of the Nullarbor Region, W.A.: 814
Response of karst aquifers to recharge: 1833
Responses to light by cave crayfishes: 3435
Restoration of natural cave features, Oregon Caves National Monument, Oregon: final report: 42
Restoration of natural microclimate in Oregon Caves, Oregon Caves National Monument: final report: 43
Review of sinkhole-collapse problems in a carbonate terrane: 3094
A review of the cavernicole fauna of Mexico, Guatemala, and Belize: 2792
A review of the deposition and alteration of filled-sink deposits of east-central Missouri: 234
A review of the troglobitic decapod crustaceans of the Americas: 1467
A revised and enlarged guide to Sark containing a detailed description of the coast, caves and bays: 1911
A revised checklist of Texas caves: 2793
A revision of the cave fishes of North America: 732
Revision of the cave-dwelling and related spiders of the genus *Troglohyphantes* Joseph (Linyphiidae) with special reference to the Yugoslav species: 848
A revision of the caver's code and a collection of articles on cave conservation: 1569
Rhodesian archaeological expedition (1929): excavations in Bambata Cave and researches on prehistoric sites in Southern Rhodesia: 226
The riches of ancient Australia: a journey into prehistory: 1071
Right Imperial: Carlotta Arch: Devil's Coach House: 1812
Rimstone River Cave: 2228
The River Styx-Salt Spring Cave system: 1244
Rivers that glow underground: 2800
The rock art of Petroglyph Point and Fern Cave, Lava Beds National Monument: final report: 1923
Rock art of South-east Cape York: 3230
Rock art: the state of research in rock art - 1993: archetypes, constants and universal paradigms: 63
Rock paintings of Aboriginal Australia: 1220
The rock paintings of southern Africa: 472
The rock pictures of Europe: 1867
Rock shelters of the Perigord: geological stratigraphy and archaeological succession: 1913
Rock to stalactite: 2004
Rock-carvings from the Yun-kang caves: 3177
Rocky Mountain, Mesa Verde, Carlsbad Caverns: 3521
Rogor Johnson's cave scrapbook: 1739
Role of the Sylvania Formation in sinkhole development, Essex County: 2868

A Roman-Byzantine burial cave in Northern Palestine (the joint excavation of the American School of Oriental Research in Jerusalem and McCormick Theological Seminary at Silet edhDhahr: 2932
The Romance of Jenolan Caves: 1389
Romano-British cavemen: cave use in Roman Britain: 463
The ruined cities of Mashonaland; being a record of excavation and exploration in 1891: 389
Ruins stabilization records for Canyon de Chelly National Monument 1942: Antelope House and Mummy Cave Ruins: 3112
Ruskin and Jewel Caves: a brief history: 64
Russell Cave in northern Alabama: 533

Safe cave diving: 2287
Safety and techniques: 2293
Safford Centennial Society field trip, fall 1979: guidebook: karst hydrogeology of the Cumberland plateau escarpment in the Lost Creek Cave and Karst Cave areas of Tennessee: 750
The saga of Glover's Cave: 3362
Saint Mary's University Speleological Society: 1954-1959: 19
Saltpeter and gunpower manufacturing in Kentucky: 1180
Saltpeter mining in East Tennessee: 3029
Saltpetre caves and Virginia history: 1037
Saltpetre mining in Mammoth Cave, Ky: 1038
The Salts Cave textiles: a preliminary account: 1821
San Pablo Cave and El Cayo on the Usumacinta River, Chiapas, Mexico: 1925
Sandals and other fabrics from Kentucky caves: 2573
Sandstone caves in Nigeria: 3175
Sandstone caves of Jefferson County: 2229
Santillana and Altamira: 1148
SASA (Cape) expedition to Chinhoyi, Zimbabwe August 1992: 3245
SCG history: summer 1981: 2737
The science of speleology: 1095
Scotland underground: a caver's guide to the limestone caves of Scotland: 1715
Scottish cave guides; the southern highlands: 2880
A scrap book of articles about the C3 expedition into Floyd Collins Crystal Cave, Kentucky; from Sunday February 14 to February 20, 1954: 2359
Sea caves of Anacapa Island: 555
Sea caves of Santa Cruz Island: 556
Sea Lion Caves Oregon coast: 180
Sea Lion Caves: America's largest sea cave, on U.S. Highway 101...just North of Florence, Oregon, Oregon coast: 181
Second circular: 1629
A second collection of mammals from caves near St. Michel, Haiti: 2182

Second field investigation report, Lehman Caves - Wheeler Peak: October 13 to 17, 1958, October 29 to November 13, 1958: portion of southern section of Snake Range, White Pine County, Nevada: 2303
Secret of the bats; the exploration of Carlsbad Caverns: 3126
The secret of the Wallowa Cave: 933
Secret tunnels in Surrey: 2627
Secret underground Bristol: 3416
Secrets of the Ice Age: the world of the cave artists: 1312
Secrets of the moors and dales: 1100
Seen a good cave lately?: Cave of the Mounds, Blue Mounds, Wis.: "a visitor study", 1977: 3339
Selected bibliography of cave conservation and management: 1579
Selected caves and lava tube systems in and near Lava Beds National Monument, California: 3404
Selected caves of Orange county: 625
Selected caves of the Pacific Northwest with particular reference to the vulcanospeleology of the state of Washington: 2394
Selected views of the beautiful caverns of Luray: 182
Self-belaying cable ladder - a rope climbing aid: 1956
Seminole Sink (41VV620): excavation of a vertical shaft tomb, Val Verde County, Texas: 3256
SERA 1976 guidebook: 3065
SERA 89: 3069
SERA cave carnival 1992: 3071
SERA cave carnival, 1990: 3070
Serindia: detailed report of explorations in Central Asia and westernmost China: 3114
The seven caves: archaeological explorations in the Middle East: 706
The Seven Straits; or, notes on glacial drifts, bone caverns, and old glaciers: some within reach of the Malvern Hills: 3173
Seventy-five years at Wind Cave: a history of the national park: 426
Sex, lies, and survey tape: 2986
The Shanidar Neandertals: 3239, 3240
Shanidar: the first Flower people: 3046
Shannon County caves: 2230
Shaw Cave, Wyoming: 1722
Sheffield University Speleological Society Central Crete Expedition: Greece 1984: 236
Shelter: the cave re-examined: 1025
Shenandoah Caverns of Virginia: 2961
Shenandoah Caverns, the grotto of the gods, in the heart of the Valley of Virginia: 183
Shenandoah Caverns, Valley of Virginia. A symphony in stone: 2963
Shenandoah Caverns, Valley of Virginia. The grotto of the gods: 2964
Shenandoah Caverns: a brief history: 184

Shenandoah Caverns: Valley of Virginia: 2962
Shenandoah, Great Smoky Mountains, and Mammoth Cave National Parks: hearings before the Committee on the Public Lands, House of Representatives, Sixty-ninth Congress, first session, on H.R. 11287 [and] H.R. 12020, May 11, 1926: 3269
The shoring of swallet cave entrances: 684
A short description of Castleton in Derbyshire: its natural curiosities and mineral productions: 1419
A short history of Kents Cavern: 185
Shoshone Cave (Whip-scorpion Habitat) wildlife habitat management plan: 3263
Show caves in Slovenia - Guide: 1309
Show caves of Hungary: 1411
The show caves of Mulu Sarawak: 3246, 3247
Shri Amar Nath Ji Guide: 186
Sieben Hengste 74: the karst and caves of the southwestern zone of the Sieben Hengste Ridge: a report by the members of the Croydon Caving Club on an expedition to the Seefeld area, north of Interlaken, Switzerland, summer 1974: 2928
Significant features of the Indiana karst: 2027
Significant natural landscape features of Warren County, Kentucky: 751
Silent Splendor: 352, 1212
Single rope technique rigging guide: 956
Single rope technique: a training manual: 957
Single rope techniques: a guide for vertical cavers: 2257
Sinkhole development resulting from ground-water withdrawal in the Tampa area, Florida: 2988
Sinkhole dumps and the risk to groundwater in Virginia's karst areas: final report to the Virginia Environment Endowment: 3001
The sinkhole hazard in Pinellas County: a geological summary for planning purposes: 361
Sinkhole occurrence in western Shelby County, Alabama: 3401
Sinkhole problem along proposed route of interstate highway 459 near Greenwood, Alabama: 2506
Sinkhole problem in and near Roberts Industrial subdivision, Birmingham, Alabama: a reconnaissance: 2507
Sinkholes in Florida: an introduction: 362
Sinkholes, caves and spring lakes: an introduction to the unusual aquatic ecosystems of the lower south east of South Australia: South Australian Underwater Speleological Society occasional paper number 1: 3217
Sinkholes, Hockessein area, Delaware: 3179
Sinkholes, hydrogeology, and ground-water quality in northeast Iowa: 1317
Sinkholes: landowner perceptions of a unique source of groundwater contamination: 1563
Sinkholes: their geology, engineering and environmental impact, proceedings of the first Multidisciplinary Conference on Sinkholes, Orlando, Florida, 15-17 October 1984: 363
Sinks of Gandy Creek: 2706
The Sinoia caves, Lomagundi district: 2252
Sixth annual Florida caver's re-union (Cave Cavort): April 7-9, 1989, Hog Island Campground, Withlacoochee State Forest: 2055
Skyline Caverns: 187
Skyline Caverns and its geologic relationship to the Shenandoah Valley and paleoindian cultures: 1161
Slaughter Canyon, New Cave and Capitan Reef exposures, Carlsbad Caverns National Park: field trip no. 10, April 13, 1957: 2858
Slovakia: the karst and speleology: 3249
Slovenska Kraska Terminologija = Slovene karst terminology: 1141
Smoo Cave: 2688
Snails from California caves: 3011
So you want to go caving! A manual for beginners: 3185
The so-called "ash caves" in Lee County, Kentucky: 1131
Soil erosion in the karst lands of Kentucky: 867
Solution geochemistry of the water of limestone terrains: 3214
Some account of the fissures and caverns hitherto discovered in the western district of the Mendip range of hills: comprised in a letter from Rev. D. Williams to the Rev. Patterson: 3494
Some caves and a rock shelter at Loch Ryan and Portpatrick, Galloway: 1257
Some caves and crags of Peakland: 3508
Some caves of Cyprus: 1752
Some features of karst topography in Indiana: 2129
Some smaller Mendip caves: vol 1: 3118
Some smaller Mendip caves: vol 2: 3248
Some technical aids for cave exploration: 3060
Somerset sump index: 2041, 2118
Songs poems and curses: 2966
Sort of an annual report from the standing committees and the secretary-treasurers of the NSS, including a summary annual report of the National Speleological Foundation, 1969-70: 3117
Sotanito de Ahuacatlán: Sierra Madre Oriental; Jalpan; Ahuacatlán: 2753
Sotano de las Golondrinas: description: 2754
South Australian cave reference book: 1940
South Australian diving fatalities 1950-1985: 1525
South China caves: information on the cave and karst of South China and a report on the 1988 joint expedition between the Institute of Karst Geology, the Speleological Society of South China Normal University, and the Cave Research Foundation: 474
South East Karst Province of South Australia: Australian Caves & Karst Management Association October 1995: 1266

The South Street caves, Dorking: 3430
South Wales Caving Club Newsletter: index to numbers 1 - 100: 1946 - 1985: 376
South Wales Caving Club tenth anniversary publication: 1716
South Wales Caving Club twenty first anniversary publication: 3061
South Wales Caving Club, 1946—1956: 3062
The Southern highlands: 2881
Southern Indiana Karst field trip guide, March 11 & 12, 1972: 2266
Southhampton University Exploration Society: Peru expedition: 3075
Souvenir book of romantic Bridal Cave: 188
Souvenir book of the Oregon Caves National Monument: Marble Halls of Oregon: 189
Souvenir guide to the world's largest underground military museum: the Grange Cavern: 1239
Souvenir of Cox's Stalactite Caves: 728
Souvenir of Cox's Stalactite Caves "visited by his majesty the late King Edward VII": 729
Special convention issue: Symposium on the History of American Caving, Pittsfield, Mass., August 5-12, 1979: 3089
Special issue on the occasion of 10th international congress held in Hungary 1989: 289
Special issue on the occasion of 7th international congress held in England 1977: 290
Speciation in cave fauna: 1507
The spelaeodes: 46
Spelean stamps: a special topical collection of postage stamps of interest to cavers: 770
Speleo digest 1980: 2474
Speleo Digest 25 - year cumulative index, 1956 through 1980: 2449
Speleo digest, 1956: 2450
Speleo digest, 1957: 2451
Speleo digest, 1958: 2452
Speleo digest, 1959: 2453
Speleo digest, 1960: 2454
Speleo digest, 1961: 2455
Speleo digest, 1962: 2456
Speleo digest, 1963: 2457
Speleo digest, 1964: 2458
Speleo digest, 1965: 2459
Speleo digest, 1966: 2460
Speleo digest, 1967: 2461
Speleo digest, 1968: 2462
Speleo digest, 1969: 2463
Speleo digest, 1970: 2464
Speleo digest, 1971: 2465
Speleo digest, 1972: 2466
Speleo digest, 1973: 2467
Speleo digest, 1974: 2468
Speleo digest, 1975: 2469
Speleo digest, 1976: 2470
Speleo digest, 1977: 2471
Speleo digest, 1978: 2472
Speleo digest, 1979: 2473
Speleo digest, 1981: 2475
Speleo digest, 1982: 2476
Speleo digest, 1983: 2477
Speleo digest, 1985: 2478
Speleo digest, 1986: 2479
Speleo digest, 1987: 2480
Speleo digest, 1989: 2481
Speleo digest, 1993: 2482
Speleo handbook: 2093
The speleobopper's guide to the caves of the Gruesome Chapel Valley: 1998
Speleochronology: 2571
Speleogenesis and karst geomorphology of the Helderberg Plateau, Schoharie County, New York: 2329
Speleography of Papoose Cave, Idaho County, Idaho: 1580
The speleoguide: 2999
Speleological Abstracts: key to Britain's speleological literature: 733, 2042
Speleological abstracts: the literature of 1968: 2043
Speleological abstracts: the publications of 1963: 3101
Speleological abstracts: the publications of 1964: 3102
Speleological abstracts: the publications of 1965: 3103
Speleological abstracts: the publications of 1966: 3104
Speleological bibliography of South Asia including the Himalayan regions: 1170
The Speleological Club in Brno 1945-1973: 564
Speleological conventional signs: 1024
A speleological field guide of the towers and caves of Chillagoe - Munanga - Rookwood areas in Far North Queensland, Australia: 651
A speleological field guide of the towers and caves of Chillagoe - Munanga - Rookwood areas in Far North Queensland, Australia. Supplement for caves CH301-CH400: 652
Speleological investigation of Nakimu Caves, British Columbia: 878
The speleological potential of the Big Bar Limestone deposit, Hell's Canyon, Oregon-Idaho: 1840
Speleological research in the Mammoth Cave Region, Kentucky: elements of an integrated program: 3032
Speleological surveys in Co. Sligo and Co. Leitrim, Eire: 1059
Speleology: 2167
Speleology in France: Ressources [sic] in Limousin Quercy Perigord: 190
Speleology of Lewis and Clark Caverns, Montana: 1869
Speleology workshop: 3077
Speleology: caves and the cave environment: 2262

Speleology: the study of caves: 2263
Speleotec '87 guidebook: a guide to the 16th biennial conference of the Australian Federation Inc: 2037
Speleotherapic and speleclimatological centres: 1135
Speleovision field notes: 2675
The spelunker's guide to the caves north of Campbellsburg, Indiana: 1970
The spelunker's guide to the caves North of Campbellsburg, Washington County, Indiana: 1971
The spelunker's guide to the caves of the Garrison Chapel Valley: 1969
Spelunking: 191
Spelunking (the exploration of caves) guide to central Oregon: 2733
The spider family Nesticidae (Araneae) in North America, Central America, and the West Indies: 1185
Spirit of exploration: discovering the wonders of Carlsbad Caverns National Park: 3551
The spirit of Lost River: fact and legend about the Lost River Cave and Valley: 736
Spring 1977 MVOR Guidebook: Middle Mississippi Valley Grotto, Shannon Co., Missouri: 2320
Springs in Alabama: 636
Springs of Florida: 1048
Springs of Missouri: 3365
Springs of Texas: 537
Springs, caves and their related features: 2607
The spur book of caving: 307
St. Charles County caves: 2231
St. Genevieve County caves: 2232
St. Louis County caves: 2233
A stabilization assessment of Mantel's Cave, site 5MF1, Dinosaur National Monument, Colorado: 2798
Stabilization of Dakota Sandstone surface of the Faris Cave petroglyphs, Kanopolis Lake Project, Kansas: 1268
The Stalactite Caves of Jósvafő and Aggtelek: 2878
Stanislaus cave country: report of study: 3099
Statement for management: Jewel Cave National Monument: 3315
Statement for management: Wind Cave National Park: 3316
Status and conservation of solution caves in New Brunswick: 2104
Status of the cypres darter, *Etheostoma proeliare*, and comments on the spring cavefish, *Chologaster agassizi*, in Max Creek, Johnson County, Illinois: 2594
Status of Trogloglanis pattersoni Eigenmann, the toothless blindcat, and status of *Satan eurystomus* Hubbs and Bailey, the widemouth blindcat: 1960
Status of *Typhlomolge* (=Eurycea) *rathbuni*: 1962
Status report on the cave Amphipod *Gammarus acherondytes* Hubricht and Mackin (Crustacea: Amphipoda) in Southern Illinois: 3426
Steam cleaning of Orient Cave: 2502
Steam cleaning of Orient Cave, Jenolan Caves, N.S.W: 2503
A stone age cave site in Tangier: preliminary report on the excavations at the Mugharet el 'Aliya, or High Cave, in Tangier: 1546
The Stor-Glomfjord project: karst research - the possibilities of leakage and the selection of damsite alternative: 1002
Stories from stone: the geology of the Guadalupe Mountains: 1692
Storm water drainage wells in the karst areas of Kentucky and Tennessee: extended inventory of drainage wells in Kentucky and Tennessee: underground water source protection program Grant No. G004358-83-0: 752
The story of Brixham Cavern: 851
The story of Cave Hill: 3340
The Story of Cave Hill: Grottos, VA: 1253
The story of caves: 3120
The story of Crystal Ice Caves: within the Great Rift National Landmark, American Falls, Idaho: 2611
The story of Howe Caverns: 192, 668
The story of Howe caverns...: 1545
The story of Mammoth Cave National Park, Kentucky: a brief history: 476
The story of most interesting discoveries commencing August 1923; White Skar Cave; under Ingleborough, Yorkshire, 1 1/2 miles from Ingleton on the Hawes Road: 193
The story of newly discovered White Skar Caverns; under Ingleborough, Yorkshire, 1 1/2 miles from Ingleton on the Hawes Road: 194
The story of Peking Man, from archaeology to mystery: 1893
The story of Polar Caves: 3083
The story of Poole's Cavern: 1096
The story of the Shoshone Indian ice caves: 2830
The story of the Speedwell Cavern: 1097
The story of Treak Cliff Cavern: 1098
The story of Wookey Hole: 1473, 2073
The story of Wookey Hole in fact, fiction and photo: 3211
The story of Wyandotte Cave: 1683, 2819
The story of Yanchep: the western wonderland: 2947
A stratified rock shelter in Vermont: 277
Stream piracy and cave development along Baker Creek, Nevada: 1896
Structural composition and dental variations in the murids of the Broom Cave fauna, late Pleistocene, Wombeyan Caves area, N.S.W., Australia: 2908
Studies in dye-tracing techniques and karst hydrogeology: 3215
Studies in the Rio Corredor Basin: 1988 to 1991: 2624
Studies on Gammaridea II: proceedings of the 4th International Colloquium on Gammarus and Niphargus, Blacksburg, Virginia, U.S.A., 10-16 September 1978: 550

Studies on the cave and endogean fauna of North America: 2794

Studies on the cave and endogean fauna of North America II: 2795

Studies on the cavernicole fauna of Mexico: 2241, 2796

Studies on the cavernicole fauna of Mexico and adjacent regions: 2242

Studies on the caves and cave fauna of the Yucatan Peninsula: 2797

Studies on the origin of Montezuma Well and cave, Arizona: 1897

A study for the National Speleological Society: Knox Cave: Albany County, New York: 15

A study of environmental factors in Harrison's Cave, Barbados, West Indies: 1470

A study of Fountain National Park and Fountain Cavern: Anguilla, British West Indies: 1292

A study of Harrison's Cave, Barbados, West Indies: 1293

Study of potential gas accumulation in caves in Bowling Green, including relationship to water quality: 960

Stump Cross Caverns: 1204

Stump Cross Caverns - the underground wonderland, their development and exploration: 596

Stump Cross Caverns: the underground wonderland: 195

Stygofauna mundi: a faunistic, distributional, and ecological synthesis of the world fauna inhabiting subterranean waters (including the marine interstitial): 443

Subaquatic caves in Hungary for divers: 1846

Subhold collapse in Montgomery County, Tennessee: an overview for the planning process: 1789

Submarine cave bivalvia from the Ryukyu Islands: systematics and evolutionary significance: 1395

Subsidence of residual soils in a karst terrain: 907

Subsidence over mines and caverns, moisture and frost actions, and classification: 2353

The subterranean amphipod crustacean fauna of an artesian well in Texas: 1508

Subterranean Britain: aspects of underground archeology: 737

Subterranean climbers; twelve years in the world's deepest chasm: 647, 648

The subterranean Crustacea of New Zealand: with some general remarks on the fauna of caves and wells: 653

Subterranean fauna of Mexico: Part I: some results of the first Italian zoological mission to Mexico sponsored by the National Academy of Lincei (October 10 - December 9, 1969): 196

Subterranean fauna of Mexico: Part II: further results of the first Italian zoological mission to Mexico sponsored by the National Academy of Lincei (1969 and 1971): 197

Subterranean fauna of Mexico: Part III: further results of the first Italian zoological mission to Mexico sponsored by the National Academy of Lincei (1973 and 1975): 198

The subterranean kingdom: a survey of man-made structures beneath the earth: 2648

Subterranean stream piracy in the Ozarks: 792

The subterranean wonders of Hawaii: 2186

The subterranean wonders of Iceland: 2187

The subterranean wonders of Kenya: 2188

The subterranean wonders of Sardinia: 2185

Subterranean wonders: Mammoth Cave and Colossal Cavern, Kentucky: 199

The sucker's visit to Mammoth Cave: 3206

The sucker's visit to the Mammoth Cave including a history of the experience and adventures of a party who undertook to see the cave and have some fun going there. A full and accurate description of the cave and the science of its formation with an account of the living inhabitants: 3207

Sudwala: 1685

Summary of archaeological work in Sarawak: with special reference to Niah Caves: 655, 2885

A summary of the life history and distribution of the spring cavefish, *Chologaster agassizi* Putnam, with population estimates for the species in southern Illinois: 3033

A summer series of personally-conducted pleasure tours to the Luray Caverns or Gettysburg via Pennsylvania Railroad: 2650

Summit firn cave study, 1970-1973, Mount Rainier, Washington: 1824

Sump index Norway: 1036

Sump index: section 1 (Somerset): 2044

Superficial geology of the valley of the Ottawa and the Wakefield cave: 1242

Supplement to the catalogue of the caves of Missouri: 3366

Supplemental guidebook: 1970 NSS convention: 2447

Survey and assessment of cave resources at Buffalo National River, Arkansas: 1945

A survey and excavation of caves in Hidalgo County, New Mexico: 1883

Survey and species determination of cave crayfish in Oklahoma: 2548

A survey of headgear and lighting available for caving: 969

Survey of caves in Cape Range North West Cape Peninsula Western Australia: 1574

Survey of lava tubes in the former Puna Forest Reserve and on adjacent State of Hawaii lands: 2121

Survey of Lopinot caves: 224

The survey of Pant Mawr Pot, South Wales: 30

A survey of some cave invertebrates from northern Arkansas: 3547

A survey of the caves of Roanoke County, Virginia: 3481

A survey of the vegetation: 771

The survey of Tunnel Cave: S. Wales: 2750

Surveying caves: 970

Surveying in Red Cliffe Caves: Bristol 1953-1954: 685
The swallow-holes of Lost River, Orange County, Indiana: 2028
A symphony in limestone: the beautiful Olsen's Caves: 18 miles North of Rockhampton, Queensland: A description of Olsen's Caves: 200
Symposium of the origin and development of caves: 1099
Symposium on cave hydrology and water-tracing: 2670
Symposium on cave photography: 201
Symposium on cave sites and explorations in Upper Rhine regions: 1868
Symposium on cave surveying: 2671
Symposium on Detection of Subsurface Cavities, 12-15 July, 1977: 3174
Symposium on karst - morphogenesis: papers: European regional conference: 1638
Symposium on karst - morphogenesis: papers: European regional conference (minus papers): a detailed exposition of the study tours: 1639
Symposium on karst-morphogenesis: European regional conference: Budapest - Aggtelek 5-9 August 1971: 1640
Symposium on life histories of cave beetles: symposium held at the 1973 annual convention of the National Speleological Society, Bloomington, Indiana: 2699
Symposium on speciation and adaptation to cave life: gradual vs. rectangular evolution: a symposium organized by the Unione Zoologica Italiana and the Société de Biospéologie, Roma, October 6-11, 1986: part I: 2896
Symposium on speciation and adaptation to cave life: gradual vs. rectangular evolution: a symposium organized by the Unione Zoologica Italiana and the Société de Biospéologie, Roma, October 6-11, 1986: part II: 2897
Symposium on surveying caves: 1385
Symposium: sink - holes and subsidence engineering - geological problems related to soluble rocks: 1613
Symposium: speciation and raciation in cavernicoles: 329
A synopsis of the cave millipeds of the United States: with an illustrated key to genera: 2958
A systematic guide to making your first cave map: 1143
A systematic revision and evolutionary biology of the *Ptomaphagus* (Adelops) beetles of North America (Coleoptera; Leiodidae; Catopinae), with emphasis on cave-inhabiting species: 2634
The systematics and biology of the cave-crickets of the North American tribe Hademoecini (Orthoptera Saltatoria, Ensifera, Rhaphidophoridae, Dolichopodinae): 1561
Systematics of the subterranean amphipod genus *Stygobromus* (Gammaridae): part I. species of the western United States: 1509
Systematics of the subterranean Amphipod genus *Stygobromus* (Rangonyctidae), part II: species of the eastern United States: 1510
Systematics, speciation, and distribution of the subterranean amphipod genus *Stygonectes* (Gammaridae): 1511

The systems of limestone caves: Jenolan, Yarrangobilly, Wombeyan, Abercrombie and Wellington: 2495
T-maze cave management plan: 3266
The Tabon caves: 1188
The Tabon Caves: archaeological excavations on Palawan Island, Philippines (1962—64): 1111
The Tabon Caves; archaeological explorations and excavations on Palawan Island, Philippines: 1112
TAG pits: 3030
A tale of two caves: 440
Tantanoola Caves Conservation Park management plan, South East, South Australia: 3058
The Tasaday: cave-dwelling food gatherers of South Cotabato, Mindanao: preliminary report submitted to the Panamin Foundation, Inc. on June 1, 1972: 1050
Tasmanian Caverneering Club Handbook: 959
Teana-au caves information: 2616
Technical aids in caving symposium held on Sunday 5th March, 1972 at Buxton High Peak College, Harpur Hill, Buxton, Derbyshire: 3194
Techniques for the tracing of subterranean drainage: 903
The technological and functional analyses of lithic flake tools from Rabel Cave, northern Luzon, Philippines: 2853
Tee-Maze cave system, Lincoln County, Idaho, U.S.A: 1667
Temnata cave: excavations in Karlukovo karst area Bulgaria: 1858
Ten years under the earth: 611
Teng Long Dong, the longest cave in China: report of the first Belgian-Chinese speleological expedition in 1988: 2082
The Tennessee pit survey: 3552
A Test excavation at Mulka's Cave (Bate's Cave) near Hyden, Western Australia: a report to the Department of Aboriginal Sites, Western Australian Museum: 448
Test excavations at Painted Cave, Pershing County, Nevada: for Bureau of Land Management, Winnemucca District Office: 320
Texas cave humor, the first twenty-five years: 3384
Texas show caves: 2992
Thailand caves catalogue: 922
[The program of the annual convention of the National Speleological Society, 1-6 August, Pendleton, Oregon]: 2444
Theme and resource inventory study of the karst regions of Canada: final report upon project A: 1084
There is a cavern in the town: 202
"There we was!....": 1426
Thermal and mineral springs in the Southern Rocky Mountains of Canada: 3348
They words, they words they 'orrible words: an anthology of caving songs: 716
Third circular: 1630

Thirty years with the Pittsburgh Grotto, National Speleological Society: 799

Three below Gower: the story of cave exploration in Gower by The Taylors': 3190

The three ring circus: how Rickwood Caverns became a state park: 2816

Through Mammoth Cave, giving a graphic description of all the routes, avenues, passes, passages, pits, domes, seas, lakes, rivers, streams, and other wonders of that greatest natural curiosity in the world: 2273

Through Mammoth Cave, giving a graphic description of all the routes, avenues, passes, passages, pits, domes, streams, rivers, seas, lakes, and other wonders to be found in that greatest of all natural curiosities in the whole realm of this great world: 2272

Through the Shenandoah Valley: Caverns of Luray, Natural Bridge, grottoes of the Shenandoah and the chronicle of a leisurely journey through the uplands of Virginia: 2965

Time series models for the prediction and characterisation of stream flow from rainfall in the Cave Creek system: 1693

Timeless Cheddar Caves: 3108

Timor Caves: 1701, 3170

Timpanogos Cave environmental impact report: 3451

Timpanogos Cave National Monument, Utah: 3317

Timpanogos Cave National Monument: statement for management: 3318

Timpanogos Cave public use and impact: 203

The Timpanogos Cave story: the romance of its exploration: 2062

Titania's cave; a fairy story of the Endless Caverns of Virginia: 1814

To photograph darkness: the history of underground and flash photography: 1556

Tonga '86, Tonga '87 expedition report: 1976

Top secret, for cavers' eyes only: 1991 Rocky Mountain Regional guidebook, Black Hills, South Dakota: 2834

Toros 89-92; the Camlik Project: a report on the speleological investigations carried out by the Swindon Speleological Society in the Camlik Region of Toros Mountain, Konya Turkey from 1989-1992: 3150

Tour caves of the Ozarks: Missouri Arkansas: the secret beneath the surface: 1794

A tour to the caves in the environs of Ingleborough and settle in the West Riding of Yorkshire: 1586

A tour to the caves in the environs of Ingleborough and Settle in the West Riding of Yorkshire: with some philosophical conjectures on the deluge, remarks on the origin of fountains, and observations on the ascent and descent of vapours, occasioned by facts peculiar to the places visited. Also a large glossary of old and original words made use of in common conversation in the north of England. In a letter to a friend: 1587

A tour to the caves in the environs of Ingleborough and Settle in the West—Riding of Yorkshire: with some philosophical conjectures on the deluge, remarks on the origin of fountains, and observations on the ascent and descent of vapours, occasioned by facts peculiar to the places visited. Also a large glossary of old and original words made use of in common conversation in the north of England. In a letter to a friend: 1588

A tour to Yordes Cave: 2938

The tourists' gem: the Manitou Grand Caverns, the largest and most wonderful subterranean in the Rocky Mountains, and other attractions for tourists: 3040

Tracer hydrology: proceedings of the 6th international symposium on water tracing Karlsruhe, Germany 21-26 September 1992: 1530

Tracing of subsurface flow in karst regions using artificially colored spores: 1149

Tragedy of Sand Cave: 1375

Trail Creek: final report on the excavation of two caves on Seward Peninsula, Alaska: 1901

The transition from Lower to Middle Palaeolithic and the origin of modern man: international symposium to commemorate the 50th anniversary of excavations in the Mount Carmel Caves by D.A.E. Garrod, University of Haifa, 6-14 October 1980: 1163

The Transvaal ape-man-bearing deposits: 460

Trapped in a cave!: a true story: 2276

Trapped!: the story of the struggle to rescue Floyd Collins from a Kentucky cave in 1925, an ordeal which became one of the most sensational events of modern times: 2305

Trapped!: the story of the struggle to rescue Floyd Collins from a Kentucky cave in 1925, an ordeal which became one of the worst sensational events of modern times: 2306

Travertine-marl: stream deposits in Virginia: 1447

Treasures of prehistoric art: 1934

A trip through time: a drive-yourself guide to the landforms and geology of the Waitomo Caves area: 635

The trip thru fairyland; Great Onyx cave: 3538

A trip to the Mammoth Cave, Ky: 2866

The troglobitic halocyprid Ostracoda of anchialine caves in Cuba: 1851

Troglobitic Ostracoda (Myodocopa: Cyprinidinidae, Thaumatocyprididae) from anchialine pools on Santa Cruz Island, Galapagos Islands: 1852

Tropical karst and caves in the central Cordillera, Colombia: 3176

Tuckaleechee Caverns: 31 photos in natural color: Townsend, Tennessee: greatest sight under the Smokies: 1958

Tunel de la Atlantida, Haria, Lanzarote, Canary Islands: the hydrodynamic, the chemistry and the minerals of the lava tube, the population density of *Munidopsis polymorpha*: 1707

Twelve pictures of Cheddar and Cox's Cave: 204

Two burial caves of the Proto-Urban Period at Azor, 1971: the first season of excavations at Tell-Yarmuth, 1970: 380
Types, features, and occurrence of sinkholes in the karst of west-central Florida: 2987

Udayagiri and Shandagiri caves: 2246
UEA speleological expedition northern Turkey: August / September 1992: 1363
ULSA explorations: journal II: 2254
Unconventional guide to the Caverns of Castleton and Disaster: 3515
Undara Volcano and its lava tubes: a geological wonder of Australia in Undara Volcanic National Park, North Queensland: 243
Under the ground: 437, 2131
Undergound water tracing: investigations in Slovenia 1972—1975: 1645
Underground adventure: 1173
The underground atlas: a gazetteer of the world's cave regions: 2171, 2172
Underground Britain: a guide to the wild caves and show caves of England, Scotland and Wales: 372
Underground empire: wonders and tales of New York caves: 2656
Underground gazetteer of south-east England: 2628
Underground Geology at the Endless Caverns, New Market, VA: 946
Underground in Furness: a guide to the geology, mines, potholes and caves: 1486, 1487
Underground in the Appalachians: a guidebook for the 1995 Convention of the National Speleological Society, Blacksburg, Virginia, July 17-21, 1995: 2416
The underground lake of St. Léonard: Valais: 1269
Underground life: 2826
Underground New England: 2657
The underground passages, caverns, etc., of Greenwich and Blackheath: 3134
The underground streams in the neighbourhood of Clapham and Malham: 521
Underground Video Techniques: 948
Underground water tracing: investigation in Slovenia 1972-1975: 1229
Underground waters of Missouri; their geology and utilization: 2967
Underground wilderness: 1247
The underground world of geological marvels: 853
Underground worlds: 1681, 1711
Underground worlds: Guizhou: 3554
Underground worlds: tour guide training manual: 2766
Underneath the arches: caving in the RibbleHead District: 639
The underworld of Oregon Caves National Monument: 696

Universal study guide for use with basic orientation and basic team member courses: 313
The University of Leeds hydrological survey expedition to Jamaica 1963: 1061
University of Nottingham Spelaeological Expedition Exploration '64: 3337
University of Nottingham Union report of the expedition's co-ordinating committee for 1968: 3338
The unmodified vertebrate fauna from Granite Quarry Cave (23CT36), Carter County, Missouri: 1827
The untamed river expedition: Nakanai Mountains, East New Britain, Papua New Guinea: 1202
Unusual cave and cavern formations: 205
The upper cave fauna of Choukoutien: 3437
The Upper Wharfedale Fell Rescue Association 1948-1968: 597
Uranium decay series studies on archaeological materials: application to thermoluminescence dating and dating of cave deposits: 2654
Use and abandonment of habitation caves in the prehistoric settlement of southeastern Oahu: a proposed research design for the 1980 University of Hawaii Archaeological Field Program: 832
Use of ground penetrating radar for detecting and evaluating the sinkhole hazard in Florida: 364
The use of radioisotopes in tracing karst groundwater in Greece: 1933
User's guide for flowchart and cavemap computer programs: 62
Using digital watershed modeling to estimate sinkhole-flooding potential at Perryville, Missouri: 3351
Using trigonometric functions in cave surveying: 383

Věkåum budoucím: 2714
The Vachoniidae—a new family of false scorpions: two new species from the caves of Yucatan: 630
The Valley of Pyrene: 1000
The valley of Stracena and the Dobschau ice-cavern (Hungary): 2635
Venture caving: 450
Venturing underground: the new speleo's guide: 2001
Vercors caves: 2054
Vermont caves: a geologic and historical guide: 2738
The vertebrate fauna of West Virginia caves: 1165
The vertebrate paleontology of Texas caves: 1119
Vertical caves of Tasmania: a caver's guidebook: 557
Vertical caving: 2158
Vertical shafts in limestone caves: 2686
Vertical: a technical manual for cavers: 3399
Victoria Cave: its history and exploration: 522
Victorian caves and karst: a guidebook to the 13th A.S.F. conference: Melbourne 1980: 2178

Victorian caves and karst: strategies for management and catalogue: a report to the Caves Classification Committee, Department of Conservation, Forests and Lands /: 815

Vietnam cavern tourism: 2508

A view of the earth, containing an account of its internal structure; its caves and subterranean passages; its mountains, its rivers and cataracts. Together with a brief view of the universe, to which are added, problems on the globes, directions for drawing maps and tables of latitudes and longitudes: 2668

Views of Chapman Cave: 3224

Views of the Buchan Caves and pyramids: 206

Views of the caves near Ingleton, Gordale Scar and Malham Cove in Yorkshire: 3448

VII International Symposium on Volcanospeleology, La Palma - El Hierro - Tenoride Canary Islands, 4-11 November, 1994, Program and Abstracts: 1660

The Virginia cave locator series: book 1: Frederick Co. and Clark Co: 1041

The Virginia cave locator series: book 2: Warren Co: 1042

The Virginia cave locator series: book 3: Shenandoah Co: 1043

The Virginia cave locator series: book 4: Page Co: 1044

The Virginia cave locator series: book 5: Rockingham Co: 1045

The Virginia cave locator series: book 6: Augusta Co: 1046

Virginia Luray Caverns: 207

A visit to Carlsbad Cavern: recent explorations of a limestone cave in the Guadalupe Mountains of New Mexico reveal a natural wonder of the first magnitude: 1926

A visit to the Jenolan Caves: reference notes for the teacher: 2497

A visit to the Mammoth Cave: 1299

A visit to the Mammoth Cave of Kentucky: 3510

Visiting American caves: 3004

Visitor's guide to Laurel Caverns Park: 646

A visitor's guide to Mammoth Cave National Park: 2567

A visitor's guide to underground Britain: caves, caverns, mines, tunnels, grottoes: 1040

Voices from the stone age: a search for cave and canyon art: 2103

Voices of the cave: 3319

A volume of essays presented to Brigadier E. A. Gelnnie on the occasion of his 80th birthday, July 18th 1969: 623

Vulcon Guidebook: lava features and limestone karst of Victoria and south-eastern South Australia. Vulcon 1995 20th Biennial Conference. Australian Speleological Federation, Inc, Hamilton, Victoria 2-6 January 1995: 271

Vulcon precedings: papers submitted for presentation. Vulcon 20th Biennial Conference. Australian Speleological Federation, Inc, Hamilton, Victoria 2-6 January 1995: 272

Wainwright in the Limestone Dales: 3373

Waiomio's limestone caves; linked with Maori tribal legend, rich in natural history: 1777

Waitomo caves: 2814, 3034

Waitomo Caves management plan 1982: 208

Waitomo Caves New Zealand: a souvenir booklet of your Waitomo visit with story of life cycle of New Zealand glow-worm: 2589

Waitomo Caves Wonderland: 398

Waitomo Caves: a century of tourism: 230

Waitomo Caves: Glowworm Cave, Ruakuri Cave, Aranui Cave: 3493

Waitomo Day 1982: summary of papers: Waitomo Caves research programme: 209

Waitomo Ruakuri & Aranui New Zealand: 210

Waitomo tourist caves: 401

Waitomo: Glow-worm Cave, Aranui Cave, Ruakuri Cave and surrounding district: 3035

The Wakulla Springs project: 3141

The Walkberg Cave System: 1258

Walking the valley: an oral record of caving and bushwalking in the Burragorang and beyond, during the 1930s: 855

Wander Indiana underground: 1987 Cave Capers, Spring Valley: 1604

Warrens Cave Reserve stewardship plan: Gainesville, Alachua County, Florida: 2259

Washington County caves: 2234

WATEQ4F: a personal computer FORTRAN transformation of the geochemical model WATEQ2 with revised data base: 305

Water and copper-mine tailings in karst terrane of Rio Tanama Basin, Puerto Rico: 1760

Water features of the European Upper Paleolithic decorated cave environment: 627

Water on and under the ground: an introduction to the urban hydrogeology of the Orlando area: 365

Water quality protection studies, Logan Cave, Arkansas: final report: 44

Water resources of the North Coast Limestone area, Puerto Rico: 1216

Water resources publications of the USGS for Tennessee, 1907-1987: 285

The waters of Mystery Cave, Forestville State Park, Minnesota: 1703

Wee Jasper Caves: 1729

The weird wonders of Wookey Hole Caves: 211

The Welch Cave Peccaries (Platygonus) and associated fauna, Kentucky Pleistocene: 1283

Welcome to Blanchard Springs Caverns: 3282

Welcome to Waitomo Caves: a photographic insight into this spectacular region of New Zealand: 212

Welkome (sic) to our Slovak caves: 626

The well of sacrifice: 943
The Wellington and Abercrombie Caves: 1436
The Wellington Caves: 3237
Wellington Caves, resource and management study: 68
The Wellington caves: treasure trove of fossils: 3109
West coast cave and karst management strategy and operational guidelines: 3484
West Wycombe caves: [a brief history and description]: 807
Western Australia's wonderland: 2008
Western Kentucky Speleological Survey: annual report 1978: 2330
Western Kentucky Speleological Survey: annual report 1979: 2331
Western Region speleo-education seminar program with abstracts: 3467
Western Texas and Carlsbad Caverns: 804
Weyer Cave. Cave of the Fountains. Madison Cave: a descriptive sketch of the grottoes of the Shenandoah: 1124
Weyer's Cave's first century, 1874-1974: 2993
What do cavers think and do?: 2350
What'a hole: condensed Carlsbad Cavern [i.e. Caverns] information: 2281
White Rose Pothole Club library catalog November 1962: 2815
White Scar Cave: a famous underground system which you can explore in your best clothes: 2948
White Scar Cave: a picture guide: 2949
White Scar Cave: a picture guide to a famous underground system you can explore in your best clothes: 2950
White Scar Caves: Britain's biggest tourist cave: 213
Who Discovered Carlsbad Cavern?: 551
Who Was Aveline Anyway? Mendip's cave names explained: 3518
Wide world magazine Index volumes 1-50 1898-1923: 1557
The wild caves of Camden County, Missouri: 2235
The Wilderness below: 1610
Wilderness caves of the Gordon - Franklin River System: 1750
Wilderness caves: a preliminary survey of the caves of the Gordon-Franklin River System, South-West Tasmania: 2168
A wilderness plan for Mammoth Cave National Park and the surrounding region: 1371
Wilderness recommendation, Carlsbad Caverns National Park, New Mexico: 3320
Wilderness resources in Mammoth Cave National Park: a regional approach: 820
Wilderness under the earth: 1120
The wilderness underground: caves of the Ozark plateau: 3424
Wind Cave 1972: 2900

The Wind Cave and its territory; Garfagnana, the Apuan Alps, the Serchio Valley (Lucca, Tuscany, Italy): 214
Wind Cave expedition 1971: 2901
Wind Cave expedition August 1970: a report to the National Park Service by the Windy City Grotto of the National Speleological Society on the results of the August 1970 Wind Cave expedition: 2902
Wind Cave National Park, South Dakota: 3321
Wind Cave National Park, South Dakota: history of legislation through the 82nd Congress: 2849
Wind Cave: an ancient world beneath the hills: 2602
Wind Cave: National Park Service, South Dakota: 3322
Wind Cave: the world below: 2360
Window into the earth: Timpanogos Cave: 3238
The Winston Churchill Memorial Trust of Australia: project: to study recent developments in the design and installation of lighting to enhance the public education and enjoyment of show caves: 1786
The Winter Park sinkhole: a report to the city of Winter Park, Florida: 1704
Wisconsin Speleological Society present: Hodag Hunt 1976, Flying J Campgrounds, Gotham, Wisconsin: 3327
The witch of Wookey Hole: 375
The wizard of Jenolan: 2081
The Wombeyan and Abercrombie caves: 1431
Wombeyan Caves: 931
The Wombeyan Caves and the Bowral, Mittagong & Moss Vale tourist districts: 719
The Wombeyan experience: a guide to the Wombeyan Caves New South Wales: 1437
Wonder cave of Tennessee: 2621
The wonder caves of Maoriland: Waitomo, Ruakuri & Aranui: 2746
The wonderful Carlsbad Caverns: near Carlsbad, New Mexico: 1921
Wonderful curiosity, or, a correct narrative of the celebrated Mammoth Cave of Kentucky with incidents and anecdotes: 1755
Wonderful discovery: being an account of a recent exploration of the celebrated Mammoth Cave, in Edmonson County, Kentucky, by Dr. Rowan, Professor Simmons and others, of Louisville, to its termination in an inhabited region, in the interior of the earth: 1936
The wonderful Massanutten Caverns in the Shenandoah Valley, Harrisonburg, Va: 1370
Wonderful, famous, spectacular Crystal Cave, Kutztown, Penna [sic]: 215
The wonders of Castleton: 724
The wonders of Lost River: 40 miles of underground river with its caves and hidden waterfalls, blind fish and other wonders: 2812
Wonders of the great Mammoth Cave of Kentucky containing thorough and accurate historical and descriptive

sketches of this marvelous underground world with a chapter on the geology of cave formation: 3254
The wonders of the Grotto of Han: 216
The wonders of the Grotto of Han (Belgium): guidebook, and description with engravings and plan and a notice upon the very interesting Grotto of Rochefort: 2673
Wonders of the world: 217
Woodchuck Cave: A basketmaker II site in Tsegi Canyon, Arizona: 1954
Wookey Hole Caves: sixteen exclusive camera studies: 218
Wookey Hole Caves: Wells, Somerset: 219, 220
Wookey Hole [Including articles on] (paper making, Lady Bangor's fairground collection [and] Madame Tussaud's store): 1167
Wookey Hole: its caves and cave dwellers: 299
Wookey Hole: the cave of mystery and history: 1472
Wookey Hole; the caves and mill: 221
Wookey: the caves beyond: 1030
World famous Fossil Cave Naracoorte: 3436
World Heritage significance of karst and other landforms in the Nullarbor Region: 816
World karst correlation: proceedings of the international symposium on karst of the Inner Plate Region with monsoon climate, July 1991, Guilin, China: 1651
The world of American caves: 1366
The world of caves: 1979, 3393
The world within: 3000
World wonder saved: how Mammoth Cave became a national park: 1223
Wupatki National Monument earth cracks: 473

Wyandote Cave, pronounced by eminent scientists and noted writers of wide experience and extensive travel to be the greatest natural attraction of its kind in the world: 730
Wyandotte Cave: 1684
Wyandotte Cave down through the centuries: 1181
The Wyandotte Cave of Crawford County, Ind: 3116
Wyandotte Cave: where it is and what it is: 2860

The Xilitla region: an annotated bibliography: 1144
Xingwen: China caves project 1989-1992: 3394

Yallingup Cave Park - a management plan: 817
The Yangtze Gorges expedition: China caves project: 2934
Yarrangobilly Caves, Kosciusko National Park: 2169
Yarrangobilly Caves—Kosciusko National Park: 222
Ye olde history: 1873
Yengema Cave Report: 707
Yochib, the river cave: an account of the exploration of the Sumidero Yochib of Mexico, a dangerous and difficult cave: 3111
Yorkshire caves and potholes: 1, North Ribblesdale: 2236
Yorkshire caves and potholes: 2, under Ingleborough: 2237
Yorkshire Caves: Victoria cave, Settle; Ingleborough Cave, Clapham; Yordas Cave, Ingleton: 1218
Yorkshire Dales: limestone country: 3395
Yorkshire's hollow mountains: 2244
Your guide to Mystery Cave: 1783
Your guide to Spook Cave: 1784

ISO Country Codes and USPS State Codes

AD	Andorra
AE	United Arab Emirates
AF	Afghanistan
AG	Antigua And Barbuda
AI	Anguilla
AL	Albania
AM	Armenia
AN	Netherlands Antilles
AO	Angola
AQ	Antarctica
AR	Argentina
AS	American Samoa
AT	Austria
AU	Australia
AW	Aruba
AZ	Azerbaijan
BA	Bosnia And Herzegowina
BB	Barbados
BD	Bangladesh
BE	Belgium
BF	Burkina Faso
BG	Bulgaria
BH	Bahrain
BI	Burundi
BJ	Benin
BM	Bermuda
BN	Brunei Darussalam
BO	Bolivia
BR	Brazil
BS	Bahamas
BT	Bhutan
BV	Bouvet Island
BW	Botswana
BY	Belarus
BZ	Belize
CA	Canada
CC	Cocos (Keeling) Islands
CF	Central African Republic
CG	Congo
CH	Switzerland
CI	Cote D'ivoire
CK	Cook Islands
CL	Chile
CM	Cameroon
CN	China
CO	Colombia
CR	Costa Rica
CU	Cuba
CV	Cape Verde
CX	Christmas Island
CY	Cyprus
CZ	Czech Republic
DE	Germany
DJ	Djibouti
DK	Denmark
DM	Dominica
DO	Dominican Republic
DZ	Algeria
EC	Ecuador
EE	Estonia
EG	Egypt
EH	Western Sahara
ER	Eritrea
ES	Spain
ET	Ethiopia
FI	Finland
FJ	Fiji
FK	Falkland Islands (Malvinas)
FM	Micronesia, Federated States Of
FO	Faroe Islands
FR	France
FX	France, Metropolitan
GA	Gabon
GB	United Kingdom
GD	Grenada
GE	Georgia
GF	French Guiana
GH	Ghana
GI	Gibraltar
GL	Greenland
GM	Gambia
GN	Guinea
GP	Guadeloupe
GQ	Equatorial Guinea
GR	Greece
GS	South Georgia And The South Sandwich Islands
GT	Guatemala
GU	Guam
GW	Guinea-bissau
GY	Guyana
HK	Hong Kong
HM	Heard And Mc Donald Islands
HN	Honduras
HR	Croatia (local name: Hrvatska)
HT	Haiti
HU	Hungary
ID	Indonesia
IE	Ireland
IL	Israel
IN	India
IO	British Indian Ocean Territory
IQ	Iraq
IR	Iran (Islamic Republic Of)

IS	Iceland	NF	Norfolk Island
IT	Italy	NG	Nigeria
JM	Jamaica	NI	Nicaragua
JO	Jordan	NL	Netherlands
JP	Japan	NO	Norway
KE	Kenya	NP	Nepal
KG	Kyrgyzstan	NR	Nauru
KH	Cambodia	NU	Niue
KI	Kiribati	NZ	New Zealand
KM	Comoros	OM	Oman
KN	Saint Kitts And Nevis	PA	Panama
KP	Korea, Democratic People's Republic Of	PE	Peru
		PF	French Polynesia
KR	Korea, Republic Of	PG	Papua New Guinea
KW	Kuwait	PH	Philippines
KY	Cayman Islands	PK	Pakistan
KZ	Kazakhstan	PL	Poland
LA	Lao People's Democratic Republic	PM	St. Pierre And Miquelon
		PN	Pitcairn
LB	Lebanon	PR	Puerto Rico
LC	Saint Lucia	PT	Portugal
LI	Liechtenstein	PW	Palau
LK	Sri Lanka	PY	Paraguay
LR	Liberia	QA	Qatar
LS	Lesotho	RE	Reunion
LT	Lithuania	RO	Romania
LU	Luxembourg	RU	Russian Federation
LV	Latvia	RW	Rwanda
LY	Libyan Arab Jamahiriya	SA	Saudi Arabia
MA	Morocco	SB	Solomon Islands
MC	Monaco	SC	Seychelles
MD	Moldova, Republic Of	SD	Sudan
MG	Madagascar	SE	Sweden
MH	Marshall Islands	SG	Singapore
MK	Macedonia, The Former Yugoslav Republic Of	SH	St. Helena
		SI	Slovenia
		SJ	Svalbard And Jan Mayen Islands
ML	Mali	SK	Slovakia (Slovak Republic)
MM	Myanmar	SL	Sierra Leone
MN	Mongolia	SM	San Marino
MO	Macau	SN	Senegal
MP	Northern Mariana Islands	SO	Somalia
MQ	Martinique	SR	Suriname
MR	Mauritania	ST	Sao Tome And Principe
MS	Montserrat	SV	El Salvador
MT	Malta	SY	Syrian Arab Republic
MU	Mauritius	SZ	Swaziland
MV	Maldives	TC	Turks And Caicos Islands
MW	Malawi	TD	Chad
MX	Mexico	TF	French Southern Territories
MY	Malaysia	TG	Togo
MZ	Mozambique	TH	Thailand
NA	Namibia	TJ	Tajikistan
NC	New Caledonia	TK	Tokelau
NE	Niger		

ISO Country Codes and USPS State Codes

TM	Turkmenistan		HI	Hawaii
TN	Tunisia		IA	Iowa
TO	Tonga		ID	Idaho
TP	East Timor		IL	Illinois
TR	Turkey		IN	Indiana
TT	Trinidad And Tobago		KS	Kansas
TV	Tuvalu		KY	Kentucky
TW	Taiwan, Province Of China		LA	Louisiana
TZ	Tanzania, United Republic Of		MA	Massachusetts
UA	Ukraine		MD	Maryland
UG	Uganda		ME	Maine
UM	United States Minor Outlying Islands		MI	Michigan
US	United States		MN	Minnesota
UY	Uruguay		MO	Missouri
UZ	Uzbekistan		MS	Mississippi
VA	Vatican City State (Holy See)		MT	Montana
VC	Saint Vincent And The Grenadines		NC	North Carolina
VE	Venezuela		ND	North Dakota
VG	Virgin Islands (British)		NE	Nebraska
VI	Virgin Islands (U.S.)		NH	New Hampshire
VN	Viet Nam		NJ	New Jersey
VU	Vanuatu		NM	New Mexico
WF	Wallis And Futuna Islands		NV	Nevada
WS	Samoa		NY	New York
YE	Yemen		OH	Ohio
YT	Mayotte		OK	Oklahoma
YU	Yugoslavia		OR	Oregon
ZA	South Africa		PA	Pennsylvania
ZM	Zambia		PR	Puerto Rico
ZR	Zaire		RI	Rhode Island
ZW	Zimbabwe		SC	South Carolina
			SD	South Dakota
			TN	Tennessee
AK	Alaska		TX	Texas
AL	Alabama		US	United States
AR	Arkansas		UT	Utah
AZ	Arizona		VA	Virginia
CA	California		VI	Virgin Islands of USA
CO	Colorado		VT	Vermont
CT	Connecticut		WA	Washington
DC	District of Columbia		WI	Wisconsin
DE	Delaware		WV	West Virginia
FL	Florida		WY	Wyoming
GA	Georgia			

DISCARDED
URI LIBRARY

REF Z 6033 .C3 G85 1998

A guide to speleological
literature of the English